46th IUPAC World Polymer Congress (MACRO 2016)

Bridging Continents & Bridging Molecules

Istanbul, Turkey
17 - 21 July 2016

ISBN: 978-1-5108-3975-5

Printed from e-media with permission by:

Curran Associates, Inc.
57 Morehouse Lane
Red Hook, NY 12571

CURRAN ASSOCIATES INC.
proceedings
.com

Some format issues inherent in the e-media version may also appear in this print version.

Printed by Curran Associates, Inc. (2017)

Turkish Chemical Society (TCS)
Halaskargazi Caddesi No.: 15 D.
8 Uzay Apt. Harbiye
Istanbul / Turkey

Phone: +90 212 240 73 31
Fax: +90 212 231 70 37

tkd@turchemsoc.org

Additional copies of this publication are available from:

Curran Associates, Inc.
57 Morehouse Lane
Red Hook, NY 12571 USA
Phone: 845-758-0400
Fax: 845-758-2633
Email: curran@proceedings.com
Web: www.proceedings.com

TURKISH
CHEMICAL SOCIETY

İTÜ

IUPAC
International Union of Pure and Applied Chemistry

Bridging Continents & Bridging Molecules

MACRO2016

World Polymer Congress Istanbul 17-21 July

46th IUPAC WORLD POLYMER CONGRESS

July 17-21, 2016 / Istanbul / Turkey

Halic Congress Center

BOOK OF ABSTRACT

www.macro2016.org

Bridging Continents & Bridging Molecules

MACRO 2016

World Polymer Congress Istanbul 17-21 July

INDEX

Welcome Letter

Dear Colleagues,

On behalf of the Turkish Chemical Society, Istanbul Technical University and the Local Organizing Committee, we would like to invite th you to Istanbul, Turkey to participate in the 46th IUPAC World Polymer Congress (MACRO 2016) that will be held from 17th to 21st July 2016. It is the largest international multi-symposium conference dedicated to all aspects of polymer science and engineering and formed the latest of the biennial IUPAC Polymer Division meetings.

After the successful MACRO2014, Chiang Mai, Thailand, it is now a great pleasure to welcome you to Istanbul, one of the most attractive, historical cities in the world. Istanbul is the city that bridges Europe and Asia. The venue is in the heart of the city and there are many hotels within walking distance.

We are looking forward to welcoming polymer scientists from all over the world to Istanbul to witness an unforgettable scientific event and feel the fascinating atmosphere of this metropolis.

Prof. Yusuf Yagci
Istanbul Technical University
Chairman of the MACRO2016

Committees

Local Organising Committee

Prof. Faruk Yilmaz
Prof. Levent Demirel
Assoc. Prof. Ali Cirpan
Assoc. Prof. Ali Durmus
Assoc. Prof. Amitav Sanyal
Assoc. Prof. Baris Kiskan
Assoc. Prof. Binnur AydoganTemel
Assoc. Prof. Nurcan Tuzun
Assoc. Prof. Mustafa Sözbilir
Assoc. Prof. Remzi Becer
Assoc. Prof. Tamer Uyar
Assoc. Prof. Tuba Erdogan Bedri
Asst. Prof. Deniz Ceylan Tuncaboylu
Asst. Prof. Huseyin Esen
Asst. Prof. Muhammet Kahveci

International Advisory Committee

AFRICA
Akinola Akinlabi (Nigeria)
Bert Klumperman (South Africa)
Hatem Ben Romdhane (Tunusia)
Safaa El-din H. Etaiw (Egypt)
Suprakas Sinha Ray (South Africa)

ASIA
Abdullah M. Asiri (Saudi Arabia)
Chaobin He (Singapore)
Chan-Chin Han (Malaysia)
Chain-Shu Hsu (Taiwan)
Doo Sung Lee (Korea)
Jiasong He(China)
Masami Kamigaito (Japan)
Michael Silverstein (Israel)
Mitsuo Sawamoto (Japan)
Muhammad Siddiq (Pakistan)
Shadpour E. Mallakpour (Iran)
Supawan Tantayanon (Thailand)
Takeshi ENDO (Japan)
Voravee P. Hoven (Thailand)

AUSTRALIA
Cyrille Boyer (Australia)
Graeme Moad (Australia)
Greg Russell (New Zealand)
Martina Stenzel (Australia)
Sebastien Perrier (Australia)
Wayne Cook (Australia)

EUROPE
Roger Hiorns (France)
Asterios Pispas (Greece)
Brigitte Voit (Germany)
Burak Erman (Turkey)
Daniel Taton (France)
Natalie Stingelin (UK)
Dieter Schlüter (Switzerland)
Igor Lacik (Slovakia)
Eva M. Jonsson (Sweden)
Giancarlo Galli (Italy)
Patrick Thetao (Germany)
Igor Potemkin (Russia)
Jiri Vohlídal (Chezc Republic)
José A. Pomposo (Spain)
Michael Buback (Germany)
Natalie Stingelin (UK)
Philippe Dubois (Belgium)
Richard G. Jones (UK)
Robert Liska (Austria)
Søren Hvilsted (Denmark)
T. J. Dingemans (Netherlands)
S. Slolowski (Poland)

NORTH AMERICA
Christine Luscombe (USA)
Christopher K. Ober (USA)
Dilhan M. Kalyon (USA)
Emilio Bucio (Mexico)
Francoise M. Winnik (Canada)
Michael G. Walter (USA)
Mario Gauthier (Canada)
Rigoberto Advincula (USA)

SOUTH AMERICA
Alessandro Gandini (Brazil)
Cláudio G. dos Santos (Brazil)
Mirta I. Aranguren (Argentina)
Norma E. Marcovich (Argentina)

Scientific Sessions

1 - Recent Developments in Polymer Synthesis

a) Controlled/Living Polymerization
Co-organizer: Prof. Masami Kamigaito
Coordinator: Prof. Remzi Becer

b) Complex Macromolecular Structures
Co-organizer: Prof. Filip Du Prez
Coordinator: Assoc. Prof. Amitav Sanyal

c) Light-induced Reactions in Polymer Science
Co-organizer: Prof. Xavier Allonas
Coordinator: Assoc. Prof. Binnur Aydogan Temel

d- Condensation Polymerization and Thermoset
Co-organizer: Prof. Tsutomu Yokozawa
Coordinator: Assoc. Prof. Baris Kiskan

2 - Polymer Characterization by New and Combined Techniques
Co-organizer: Prof. Michael Hess
Coordinator: Prof. Levent Demirel

3 - Macromolecules & Nanotechnology
Co-organizer: Prof. Gyula Julius Vancso
Coordinator: Assoc. Prof. Tamer Uyar

4 - Macromolecules in Biotechnology & Medicine
Co-organizer: Prof. Martina Stenzel
Coordinator: Asst. Prof. Muhammet Kahveci

5 - Energy, Optics, & Optoelectronics
Co-organizer: Prof. Levent Toppare
Coordinator: Assoc. Prof. Ali Çırpan

6 - Polymer and Polymer-Based Membranes
Co-organizer: Prof. S. Birgul Tantekin-Ersolmaz
Coordinator: Assoc. Prof. Tuba Erdogan Bedri

7 - Smart and Functional Polymers
Co-organizer: Prof. Richard Hoogenboom
Coordinator: Prof. Faruk Yilmaz

8 - Renewable Resources and Biopolymers
Co-organizer: Prof. Henri Cramail
Coordinator: Assoc. Prof. Huseyin Esen

9 - Polymer Engineering, Processing, and Characterization
Co-organizer: Prof. Mukerrem Cakmak
Coordinator: Assoc. Prof. Ali Durmus

10 - Polymer Physics
Co-organizers: Prof. Dimitris Vlassopoulos and Kurt Kremer
Coordinator: Asst. Prof. Deniz Ceylan Tuncaboylu

11 - Polymer Education
Co-organizers: Prof. Christopher Ober and Patrick Theato
Coordinator: Assoc. Prof. Mustafa Sözbilir

12 - Industrial Polymers
Co-organizer: Dr. Johan-FGA Jansen (DSM)
Coordinator: Asst. Prof. Huseyin Esen

13 - Porous Polymer and Gels
Co-organizer: Prof. Michael S. Silverstein
Coordinator: Prof. Ali Tuncel

14 - Modeling and Simulation of Polymers
Co-organizer: Prof. Stefano Valdo Meille
Coordinator: Assoc. Prof. Nurcan Tuzun

Overall Program

July 17, 2016 - Sunday

Hermann Staudinger Hall (Sadabad)		
16.30-17.00	Opening Remarks	
17.00-18.00	**Plenary Speaker #1**	
	Krzysztof Matyjaszewski *Carnegie Mellon University, USA*	Controlled Radical Polymerization: from Mechanisms to Applications
18.00-19.00	**Plenary Speaker #2**	
	Markus Antonietti *Max Planck Institute, Germany*	Polymer Ionic Liquids: From Green Polymer Chemistry to Stabilizers to Actuators
19.00-20.00	Opening Reception	

July 18, 2016 - Monday

Hermann Staudinger Hall (Sadabad)		
10.30-12.00	**Awards Ceremony**	
10:30-11:00	***Richard Hoogenboom***	**AWARD-O-001** - Poly(2-oxazoline)s as basis for responsive polymers
11:00-11:30	***Brent Sumerlin***	**AWARD-O-002** - Streamlined and Accelerated Synthesis for Responsive and Dynamic-Covalent Polymers
11:30-12:00	***Moon Jeong Park***	**AWARD-O-003** - Design, Synthesis, and Morphology of Ion-Containing Polymers
10.30-12.30	Parallel Sessions (@ Parallel Hals)	
12.30-13.30	Lunch	
13.30-14.30	**Bob Stepto Lecturer**	
	Michael Buback *(Bob Stepto Lecturer)*	Radical Polymerization Kinetics and Mechanisms studied by Pulsed Laser Initiation in Conjunction with Highly Time-Resolved EPR Spectroscopy
13.30-16.00	Parallel Sessions (@ Parallel Hals)	
16:00-16:30	Break	
16.30-18.00	Parallel Sessions (@ Parallel Hals)	
18.00-19.30	Poster Session - 1 (@ Poster Area)	

July 19, 2016 - Tuesday

Hermann Staudinger Hall (Sadabad)		
09.00-10.00	**Plenary Speaker #3**	
	Kristi S. Anseth *University of Colorado Boulder, USA*	Hydrogels as dynamic cell niches through photo-click and photo-clip reactions
10:00-10:30	Break	
10.30-12.30	Parallel Sessions (@ Parallel Hals)	
12.30-13.30	Lunch	
13.30-14.30	**Plenary Speaker #4**	
	Nikos Hadjichristidis *KAUST, Saudi Arabia*	Novel Strategies Towards Well-Defined Polyethylene-Based Complex Macromolecular Architectures
14.30-16.00	Parallel Sessions (@ Parallel Hals)	
16:00-16:30	Break	
16.30-18.00	Parallel Sessions (@ Parallel Hals)	
18.00-19.30	Poster Session - 2 (@ Poster Area)	

July 20, 2016 - Wednesday

Hermann Staudinger Hall (Sadabad)		
09.00-10.00	**Plenary Speaker #5**	
	Niyazi Serdar Sariciftci *Johannes Kepler University, Austria*	From Organic To Bioorganic Electronic Devices
10:00-10:30	Break	
10.30-12.30	Parallel Sessions (@ Parallel Hals)	
12.30-13.30	Lunch	
13.30-14.30	**Plenary Speaker #6**	
	Takuzo Aida *University of Tokyo, Japan*	Stimuli-Responsive Smart Soft Materials
14.30-16.00	Parallel Sessions (@ Parallel Hals)	
16:00-16:30	Break	
16.30-18.00	Parallel Sessions (@ Parallel Hals)	
18:00-19.00	Poster Awards and Closing Ceremony	
19.00-22.00	Gala Dinner	

Bridging Continents & Bridging Molecules

MACRO 2016

World Polymer Congress Istanbul 17-21 July

TECHNICAL PROGRAM

Schedule for Invited and Oral Speakers

Short Codes	
RDPS-A	Recent Developments in Polymer Synthesis-A-Controlled/Living Polymerization
RDPS-B	Recent Developments in Polymer Synthesis-B-Complex Macromolecular Structures
RDPS-C	Recent Developments in Polymer Synthesis-C-Light-induced Reactions in Polymer Science
RDPS-D	Recent Developments in Polymer Synthesis-D-Condensation Polymerization and Thermoset
PCNCT	Polymer Characterization by New and Combined Techniques
MN	Macromolecules & Nanotechnology
MBM	Macromolecules in Biotechnology & Medicine
EOO	Energy, Optics, & Optoelectronics
PPBM	Polymer and Polymer-Based Membranes
SFP	Smart and Functional Polymers
RRB	Renewable Resources and Biopolymers
PEPC	Polymer Engineering, Processing, and Characterization
PP	Polymer Physics
PE	Polymer Education
IP	Industrial Polymers
PPG	Porous Polymer and Gels
MSP	Modeling and Simulation of Polymers

Recent Developments in Polymer Synthesis A-Controlled/Living Polymerization

July 18, 2016, Monday / Michael Szwarc Hall (Balat)

Session: Controlled/Living Polymerization

	Chairman: Daniel Taton	
10:30-11:00	**Filip Du Prez**	**RDPS-A-I-002** - Click/transclick reactions in combination with CRP: "TAD's the spirit"
11:00-11:20	**Metin H Acar**	**RDPS-A-O-001** - Polymer Electrolite Membrane Synthesis by ATRP and Iniferter Methods
11:20-11:40	**Metwally Ezzat**	**RDPS-A-O-002** - Anti-fog and anti-frost performance of zwitterionic polymer brushes prepared by SI-ATRP polymerization
11:40-12:00	**Julien Poly**	**RDPS-A-O-003** - Recent developments in the field of light-controlled radical polymerizations
12:00-12:20	**Tuncer Caykara**	**RDPS-A-O-013** - Ibuprofen imprinted ultrathin poly(N-(2-hydroxypropyl) methacrylamide) films
12:20-12:40	**Youhua Tao**	**RDPS-A-O-010** - Controlled Polymerization of Lysine Monomer
12.30-13.30	Lunch	
	Chairman: Stephane Carlotti	
14:30-15:00	**Daniel Taton**	**RDPS-A-I-010** - Organopolymerization by N-Heterocyclic Carbenes: Selectivity and Monomer Scope
15:00-15:20	**Christopher Michael Fellows**	**RDPS-A-O-004** - Polymerisation behaviour of 'switchable' RAFT agents
15:20-15:40	**Hossein Roghani Mamaqani**	**RDPS-A-O-005** - Reversible addition-fragmentation chain transfer polymerization of N-isopropylacrylamide and acrylic acid from nanocrystalline cellulose to prepare a dual sensitive material
15:40-16:00	**Atsushi Kajiwara**	**RDPS-A-O-006** - Step by Step Investigation of Radical Polymerization Processes by a Combination of Electron Spin Resonance and Controlled Radical Polymerization Techniques
16:00-16:30		**BREAK**
	Chairman: Junpo He	
16:30-17:00	**Tae Lim Choi**	**RDPS-A-I-011** - Cyclopolymerization of 1,6-hepatadiynes: Synthesis, Mechanism and Application
17:00-17:20	**Jan Merna**	**RDPS-A-O-011** - Hyperbranched polyethylene by living ethylene polymerization initiated by diimine and amino-imine nickel and palladium complexes
17:20-17:40	**Igor Lacik**	**RDPS-A-O-007** - Propagation rate coefficient for acrylamide radical polymerization in aqueous solution
17:40-18:00	**Dietmar Hertsen**	**RDPS-A-O-012** - The CROP reaction of 2-oxazolines *in silico*

July 19, 2016, Tuesday / Michael Szwarc Hall (Balat)

Session: Controlled/Living Polymerization

	Chairman: Stephane Carlotti	
10:30-11:00	**Takashi Ishizone**	**RDPS-A-I-006** - Anionic Polymerization of Exomethylene Hydocarbon Monomers
11:00-11:30	**Stephane Carlotti**	**RDPS-A-I-009** - Marrying Anionic Ring-Opening Polymerization with Chain-Growth Condensation Polymerization for the Synthesis of Aliphatic-Aromatic Copolyamides
11:30-12:00	**Junpo He**	**RDPS-A-I-008** - New Possibility of Living Anionic Polymerization: from Living Dendrimer to Sequence Controlled Polymer
12:00-12:20	**Rosa María Sebastián**	**RDPS-A-O-008** - Controlled living anionic polymerization of cyanoacrylates by frustrated Lewis pair based initiator
12:20-12:40	**Zhenjiang Li**	**RDPS-A-O-009** - H-bonding donor and Brønsted acid co-catalysis in ring-opening polymerization
12:30-13:30		**LUNCH**

Complex Macromolecular Structures

July 18, 2016, Monday / Wallace Carothers Hall (Kasimpasa-5)

Session: Complex Macromolecular Structures

	Chairman: Søren Hvilsted	
10:30-11:00	**Brigitte Voit**	**RDPS-B-I-001** - Highly aromatic hyperbranched polymers designed for film applications in opto-electronics
11:00-11:30	**Yasuyuki Tezuka**	**RDPS-B-I-002** - Topological Polymer Chemistry toward Macromolecular Graph Constructions
11:30-12:00	**Patrick Theato**	**RDPS-B-I-003** - From Simple Chemistries to Complex Functional Polymers
12:00-12:20	**Donatien De Rochambeau**	**RDPS-B-O-001** - Sequence-control: an access to novel polymeric supramolecular nanostructures
12:30-13:30		**LUNCH**
	Chairman: Brigitte Voit	
14:30- 15:00	**Filip Du Prez**	**RDPS-B-I-004** - Thiolactone Groups as Functional Handles: From Double End Group Modification to Macromolecular Lineups
15:00- 15:30	**Andrew P Dove**	**RDPS-B-I-005** - Thiol-yne Click Chemistry for the Synthesis of Tough Materials
15:30-15:50	**Nezha Badi**	**RDPS-B-O-003** - From star polymers to multifunctional dendrimers using thiolactone chemistry
15:50-16:10	**Hakan Durmaz**	**RDPS-B-O-002** - Main Chain Post-Functionalization Of Synthetic Polyesters Through Metal-Free Cycloaddition Reactions
16:00-16:30		**BREAK**
	Chairman: Andrew P Dove	
16:30-17:00	**Søren Hvilsted**	**RDPS-B-I-006** - Inverted grafting-to PDMS block copolymers in a PDMS matrix key to superior aqueous lubrication
17:00-17:30	**Javid Rzayev**	**RDPS-B-I-007** - Building nanostructured materials from molecular bottle brushes
17:30-17:50	**Deniz Kaya**	**RDPS-B-O-005** - Viscoelastic Properties of Peptide Nanofiber Gels via Microrheology

July 19, 2016, Tuesday / Wallace Carothers Hall (Kasimpasa-5)

Session: Complex Macromolecular Structures

	Chairman: Javid Rzayev	
10:30-11:00	**Remzi Becer**	**RDPS-B-I-008** - Precision glycopolymers and their interactions with lectins
11:00-11:30	**Amitav Sanyal**	**RDPS-B-I-009** - Thiol-based Conjugations: A Versatile Tool for Fabrication and Functionalization of Polymeric Nanomaterials
11:30-11:50	**Sananda Chanda**	**RDPS-B-O-006** - Controlling inter-lamellar spacing in periodically grafted amphiphilic copolymers
12:30-13:30		**LUNCH**

Light-induced Reactions in Polymer Science

July 19, 2016, Tuesday / Hermann Staudinger Hall (Sadabad)

Session: Light-induced Reactions in Polymer Science

	Chairman: Xavier Allonas	
10:30-11:00	**Christopher N Bowman**	**RDPS-C-I-001** - Photo-curable triazole-containing crosslinked networks from structure-property relationship to tough industrial thermosets
11:00-11:30	**Nergis Arsu**	**RDPS-C-I-002** - In-situ preparation of metal/polymer nanocomposite thin films by photopolymerization method
11:30-11:50	**Filip Du Prez**	**RDPS-C-O-001**- Remoldable Polymer Networks Using Light-responsive Anthracene Derivatives
11:50-12:10	**Gorkem Yilmaz**	**RDPS-C-O-002** - New Photochemical Strategies for Macromolecular Syntheses
12:10-12:30	**Nadia Binti Adrus**	**RDPS-C-O-003** - Comparative Study for Photopolymerization of Polyacrylamide Hydrogels using different UV light sources
12:30-13:30		**LUNCH**
	Chairman: Christopher N Bowman	
14:30-15:00	**Xavier Allonas**	**RDPS-C-I-003** - One-Pot Three-Step Polymerization System Using Double Click Michael Addition and Radical Photopolymerization
15:00- 15:20	**Binnur Aydogan Temel**	**RDPS-C-O-004** - A Coumarin Bearing Amphiphillic Block Copolymer: Synthesis, Characterization and Micellization
15:20-15:40	**Kilian N. R. Wuest**	**RDPS-C-O-005** - Light-Induced Grafting of Functional Polymers to Nanodiamonds
15:40-16:00	**David Perrot**	**RDPS-C-O-006** - Light-triggered polymerization of dopamine through a photoinduced pH jump
16:00-16:30		**BREAK**
	Chairman: Nergis Arsu	
16:30-17:00	**Jürgen Rühe**	**RDPS-C-I-004** - Surface-attached polymer layers via C,H insertion crosslinking (CHIC)
17:00-17:20	**Jamarosliza Binti Jamaluddin**	**RDPS-C-O-007** - Hydrophobic coating based on fluorinated acrylate monomer using UV-LED polymerization

Condensation Polymerization and Thermoset

July 19, 2016, Tuesday / Herman Francis Mark Hall (Kasimpasa-1)

Session: Condensation Polymerization and Thermoset

	Chairman: Mikhail Zolotukhin	
14:30-15:00	**Hatsuo Ishida**	**RDPS-D-I-001** - The Most Versatile Polymer Chemistry, Polybenzoxazines: Various Preparation Methods and New Polymerization Schemes for Advanced Material Design
15:00- 15:20	**Baris Kiskan**	**RDPS-D-O-001**- Adapting Polybenzoxazines for Smart Applications
15:20- 15:40	**Andrea Hufendiek**	**RDPS-D-O-002** - High molecular weight processable and thermally stable poly(cyclo)acetals
15:40-16:00	**Ming Li**	**RDPS-D-O-003** - Semi-crystalline Polyamide Thermosets with High Temperature Shape Memory Properties
16:00-16:30		**BREAK**
	Chairman: Hatsuo Ishida	
16:30-17:00	**Mikhail Zolotukhin**	**RDPS-D-I-002** - Metal free, superacid catalyzed step-growth polymerization: State of the art

17:00-17:20	**Hannes A. Houck**	**RDPS-D-O-004** - Novel thermoreversible and thermoadjustable blocking agents for triazolinediones: crosslinking on demand
17:20-17:40	**Xavier Fernández Francos**	**RDPS-D-O-005** - Sequential dual-curing processes based on thiol-click reactions
17:40-18:00	**William Guérin**	**RDPS-D-O-006** - New mechanism of liquid polysulfide cross-linking: a combined IR spectroscopy and rheology study

July 20, 2016, Wednesday / Herman Francis Mark Hall (Kasimpasa-1)

Session: Condensation Polymerization and Thermoset

	Chairman: Mark Soucek	
10:30-11:00	**Tsutomu Yokozawa**	**RDPS-D-I-003** - Synthesis of Conjugated Polymer by Alternating Intramolecular and Intermolecular Catalyst-Transfer Condensation Polymerization under Unstoichiometric Conditions
11:00-11:30	**Iskender Yılgör**	**RDPS-D-I-004** - Hyperbranched segmented polyurethaneureas: A versatile platform for the preparation of novel functional materials
11:30-11:50	**Guy Koeckelberghs**	**RDPS-D-O-007** - Controlled Polymerization of Conjugated Polymers: Towards All-conjugated (Co)polymers
11:50-12:10	**Tine Hardeman**	**RDPS-D-O-008** - Copolymerization of conjugated polymers: adding a new asset to their controlled synthesis
12:10-12:30	**Serkan Unal**	**RDPS-D-O-009** - Design and Synthesis of Novel Waterborne Polyurethane Dispersions for High Performance Coating and Adhesive Applications
12:30-13:30		**LUNCH**
	Chairman: Tsutomu Yokozawa	
14:30-15:00	**Mark Soucek**	**RDPS-D-I-005** - Comparison of In Situ to Ex situ Prepared Sol-gel Inorganic/Organic Hybrid Coatings
15:00- 15:20	**Wenjun Gan**	**RDPS-D-O-010** - Effects of nanocomponent on the curing reaction and reaction-induced phase separation of epoxy composites
15:20- 15:40	**Shankar Prasad Khatiwada**	**RDPS-D-O-011** - Transparent epoxy thermoset toughened by modified star block copolymer
15:40-16:00	**Rajesh Pandit**	**RDPS-D-O-012** - Investigation on Morphology and Mechanical Properties of Nanostructured Epoxy Resin Blends
16:00-16:30		**BREAK**
	Chairman: Iskender Yilgor	
16:00-16:20	**Asma Iqbal**	**RDPS-D-O-013** - Synthesis and characterization of functionalized Graphene-oxide/Polyimide nanocomposites as potential candidates for optoelectronics
16:20-16:40	**Çağla Koşak Söz**	**RDPS-D-O-014** - Mussel-inspired, tannic acid based coatings modified with l-cysteine
16:40-17:00	**Lucie Imbernon**	**RDPS-D-O-015** - Vinylogous Urethanes, a New Chemistry for High-Performance Vitrimers
17:00-17:20	**Yohana Kurnia Endah**	**RDPS-D-O-016** - Solid-state polymerization and characterization of copolyamide based on adipic acid, 1,4-butanediamine, and 2,5-furan dicarboxylic acid
17:20- 17:40	**Ali Osman Konuray**	**RDPS-D-O-017** - Room-temperature dual-curing processing of amine-acrylate-methacrylate mixtures

Polymer Characterization by New and Combined Techniques

July 18, 2016, Monday / Herman Francis Mark Hall (Kasimpasa-1)

Session: Polymer Characterization by New and Combined Techniques

Chairman: Taihyun Chang		
10:30-11:00	**Holger Schönherr**	**PCNCT-I-001** - Combined AFM - confocal fluorescence lifetime imaging microscopy (FLIM) to interrogate nanoenvironments in bacteria-sensing polymeric vesicles
11:00-11:30	**Helmut Münstedt**	**PCNCT-I-002** - Characterization of Polyethylenes by Thermorheological Methods
11:30-12:00	**Sven Henning**	**PCNCT-I-003** - Micromechanics of Polymers: Electron Microscopic Methods to Investigate Micro- and Nanoscopic Processes of Deformation and Fracture
12:00-12:20	**Marcus Stein**	**PCNCT-O-001** - About the thermo-mechanical Coupling of PET at the Yield-Point
12:20-12:40	**Roman Fuehrer**	**PCNCT-O-002** - New Applications for Science and Industry using a Universal Extensional- Fixture (UXF) with a counter rotation MCR Rheometer
12:30-13:30		**LUNCH**
Chairman: Holger Schönherr		
14:30-15:00	**Taihyun Chang**	**PCNCT-I-004** - Precise characterization of polymers prepared by anionic polymerization and controlled radical polymerization
15:00- 15:30	**Peter Kilz**	**PCNCT-I-005** - Investigation of Polymer Reactions by GPC/SEC with Mass Spectrometry
15:30-15:50	**Jale Hacaloglu**	**PCNCT-O-003** - The Use of Direct Pyrolysis Mass Spectrometry Technique to Investigate the Nanostructure of Polymer Nanoclay Composites
15:50-16:10	**Subin Damodaran**	**PCNCT-O-004** - Application of gel permeation chromatography in polymer analysis
16:00-16:30		**BREAK**
Chairman: Helmut Münstedt		
16:30-17:00	**Betty Lucy Lopez**	**PCNCT-I-006** - Study of the effect of physicochemical parameters on chitosan-oleic acid nanoparticles size for encapsulation of anti-inflammatory hydrophobic drug
17:00-17:20	**Chin Han Chan**	**PCNCT-O-008** - Dielectric and thermal properties of poly(ethylene oxide)-based polymer electrolytes
17:20-17:40	**Davide Ret**	**PCNCT-O-009** - Characterization of Hyaluronic Acid derivatives – Novel fundamental inside
17:40-18:00	**Jean Baptiste Henri Lena**	**PCNCT-O-010** - Characterization of Branching in Poly(Acrylic Acid) by 13C NMR Spectroscopy and Capillary Electrophoresis

July 19, 2016, Tuesday / Herman Francis Mark Hall (Kasimpasa-1)

Session: Polymer Characterization by New and Combined Techniques

Chairman: Sven Henning		
10:30-11:00	**Young Wan Kwon**	**PCNCT-I-007** - Magnetic Properties of π-stacked polymers
11:00-11:30	**Bernhard Bluemich**	**PCNCT-I-008** - Polymer Characterization with Compact NMR
11:30-11:50	**Esra Altuntas**	**PCNCT-O-011** - Structural characterization of synthetic polymers via mass spectrometry (MS)-based strategies
11:50-12:10	**Ezgi Eren Belgin**	**PCNCT-O-012** - A Novel Polymeric Radiation Shield reinforced with WO3: Design, Performance and Structural-Thermal Characterization
12:10-12:30	**Rached Jaafar**	**PCNCT-O-013** - Force spectroscopy with real-time modeling and analysis on soft biological and polymer materials
12:30-13:30		**LUNCH**

Macromolecules & Nanotechnology

July 18, 2016, Monday / Giulio Natta Hall (Kasimpasa-2)

Session: Macromolecules & Nanotechnology

	Chairman: Iskender Yilgor	
10:30-11:00	**Stephen Craig**	**MN-I-001** - Single Molecule Properties of Covalently Responsive Polymers
11:00-11:20	**Jens Gaitzsch**	**MN-O-001** - Bright and Patchy Polymersomes Enabled by a Symbiosis of Organic and Polymer Chemistry
11:20-11:40	**Xin Huang**	**MN-O-002** - Engineering Proteinosomes towards Biomimetic Protocells
11:40-12:00	**Gokhan Yilmaz**	**MN-O-003** - Synthesis of glyconanoparticles with different morphologies and their interactions with DC-SIGN
12:00-12:20	**Jani Markus Malho**	**MN-O-004** - Supramolecular engineering of cellulose nanocrystals for responsive hybrid materials
12:20-12:40	**Selçuk Poyraz**	**MN-O-033** - Preparation of Carbon Nanotube Decorated Fiber Reinforced Polymeric Composite Structures by Using Microwave Energy
12:30-13:30		**LUNCH**
	Chairman: Roy Shenhar	
14:30-15:00	**Iskender Yılgör**	**MN-I-002** - Simple processes for the preparation of superhydrophobic polymer surfaces
15:00- 15:20	**Cafer T Yavuz**	**MN-O-005** - Nanoporous polymers as caging supports for catalytically active nanoparticles
15:20-15:40	**Bernhard V. K. J. Schmidt**	**MN-O-006** - Novel self-assemblies via double-hydrophilic block copolymers in aqueous solution
15:40-16:00	**Yue Zhao**	**MN-O-007** - CO2-Stimulable Block Copolymer Nanostructures
16:00-16:30		**BREAK**
	Chairman: Cafer T Yavuz	
16:30-17:00	**Roy Shenhar**	**MN-I-003** - {Nano-Patterned} Polyelectrolyte Multilayers Spanning Square Centimetre Areas
17:00-17:20	**Li Zhao**	**MN-O-008** - Fabrication of biodegradable Poly(lactic acid) stereocomplex microcapsules using Layer-by-Layer technique in non-aqueous medium
17:20-17:40	**Mei Hui Tsai**	**MN-O-009** - Synthesis and characterization of Au/Zirconium phosphate nanoplatelets for catalytic reduction of 4-nitrophenol
17:40-18:00	**A Levent Demirel**	**MN-O-010** - Hydrogen-bonded layer-by-layer films of poly(2-alkyl-2-oxazoline)s

July 19, 2016, Tuesday / Giulio Natta Hall (Kasimpasa-2)

Session: Macromolecules & Nanotechnology

	Chairman: Roy Shenhar	
10:30-11:00	**Julius G Vancso**	**MN-I-004** - Charge and surface fouling at the nanoscale: Isoelectric points of proteins by Atomic Force Microscopy
11:00-11:20	**Marina Tsianou**	**MN-O-011** - Macromolecular Templates in Biomineral Formation
11:20-11:40	**Muhammad Anwaar**	**MN-O-012** - Self-assembled hydroxyapatite-chitosan nanocomposites: Promising materials for bone tissue engineering applications
11:40-12:00	**Erhan Bat**	**MN-O-013** - Direct Write Patterning of Multiple Proteins and Polymers at the Micrometer and Nanometer Scale
12:00-12:20	**Ncapayi Vuyelwa**	**MN-O-014** - Simple green synthesis of CdTe/CdSe/ZnSe core-multi shell with reduced cytotoxicity for bio imaging
12:20-12:40	**Syed Shahabuddin**	**MN-O-015** - Synthesis and characterization of polyaniline/ SrTiO3 nanocomposite and its photocatalytic activity for degradation of methylene blue dye under visible light
12:30-13:30		**LUNCH**
	Chairman: Julius G Vancso	

14:30-15:00	**Gilbert Walker**	**MN-I-005** - ear Field Infrared Microscopy of Polymer Films and Boron Nitride
15:00- 15:20	**Noor Aniza Harun**	**MN-O-016** - Simultaneous Enhanced SERS and Fluorescence Effects of Co-Encapsulated Au-Ag Alloy Nanoparticles and Silicon Quantum Dots within Polymer Nanoparticles
15:20- 15:40	**Hanna J Maria**	**MN-O-017** - Dispersion of nanoclay and its extend of compatbilization in immiscible natural rubber /nitrile rubber blends
15:40-16:00	**Tsung Yen Tsai**	**MN-O-032** - An advanced preparation and characterization of the PET/ MgAl-LDH nanocomposites
16:00-16:30		**BREAK**
	Chairman: Julius G Vancso	
16:30-17:00	**Cenk Aktas**	**MN-I-006** - Bi-phasic Hybrid Nanostructures for Functional Applications
17:00-17:20	**Ali Akbari**	**MN-O-019** - Nano hybrid organic-inorganic polyhedral oligomeric silsesquioxanes (POSS) based nanocatalyst
17:20-17:40	**Ali Demirci**	**MN-O-020** - Controlled Synthesis of Silsesquioxane Based Hybrid Copolymers for Nanostructured Spheres and Fibers
17:40-18:00	**Bin Sun**	**MN-O-021** - Sodium Alginate Hydrogel Membrane Hybrided by Silver Nanoparticle as a Highly Effective Photocatalysis Microreactor for Large Scale Dye Wastewater Treatment

July 20, 2016, Wednesday / Giulio Natta Hall (Kasimpasa-2)

Session: Macromolecules & Nanotechnology

	Chairman: Cenk Aktas	
10:30-11:00	**Ali Osman Sezer**	**MN-I-007** - Electropolymerization and characterization of polyaniline-titania nanocomposites for sensing applications
11:00-11:20	**Ilke Anac**	**MN-O-022** - Superhydrophobic Polysiloxane Nanofilament Growth on Hydrophobic and Hydrophilic Polymer Coatings
11:20-11:40	**Sevde Altuntaş**	**MN-O-023** - Detection of Alzheimer`s Protein on Polycarbonate Nanopillared Films by Using Surface Enhanced Raman Spectroscopy
11:40-12:00	**Rosnani Hasham**	**MN-O-024** - Design of Virgin Coconut Oil Based Nanostructured Lipid Carriers to co-deliver phytochemical extracts
12:00-12:20	**A. Başak Kayitmazer**	**MN-O-025** - Non-stoichiometric complex coacervates
12:20-12:40	**Zeynep Eryılmaz**	**MN-O-026** - Polymer Coated CdSe/CdS Nanocyristals: Synthesis And Optical Properties
12:30-13:30		**LUNCH**
	Chairman: Ilke Anac	
14:30-14:50	**Filip Du Prez**	**MN-O-027** - Sustainable design of nanoparticles using the triazolinedione (TAD)-ene click reaction
14:50- 15:10	**Ahmet Üner**	**MN-O-028** - Synthesis and Surface Properties of PEG Based-Amphiphilic Star-shaped Polymers with POSS (Polyhedral Oligomeric Silsesquioxane) Core
15:10- 15:30	**Ersan Eyiler**	**MN-O-029** - Magnetic Nanoparticles Surface Modified with Stimuli-Responsive Polymers: An Initial Study of Stimuli-Triggered Phase Transfer
15:30- 15:50	**Ahmed Youssef**	**MN-O-030** - Synthesis and characterization of eco-friendly bionanocomposites based on PVA and Zn/Al-LDHs
15:50- 16:10	**Tonguç Özdemir**	**MN-O-031** - Nano lead oxide and epdm rubber composite for development of lightweight radiation shielding material: gamma irradiation
16:00-16:30		**BREAK**

Macromolecules in Biotechnology & Medicine

July 18, 2016, Monday / Robert Bruce Merrifield Hall (Haskoy)

Session: Macromolecules in Biotechnology & Medicine

	Chairman: Tanja Weil	
10:30-11:00	**Jürgen Rühe**	**MBM-I-001** - Engineered biointerfaces through tailormade surface-attached polymer networks – From to new diagnostic tools to implantable materials
11:00-11:20	**Shumaila Razzaque**	**MBM-O-008** - Synthesis and development of microporous organic polymeric capsules as a drug carrier
11:20-11:40	**Rana Sanyal**	**MBM-O-002** - Micellar Drug Delivery Systems
11:40-12:00	**Guoying Zhang**	**MBM-O-003** - Acid-disintegratable polymersomes for intracellular drug delivery
12:00-12:20	**Paschalis Alexandridis**	**MBM-O-004** - Block Copolymer Micelles and Nanoparticles for Drug Delivery and Imaging
12:20-12:40	**Ghislaine Barouti**	**MBM-O-005** - Polyhydroxyalkanoate-based diblock and triblock copolymers: Chemical composition and macromolecular architecture, an easy way to tune the nanoparticles properties for drug delivery applications
12:30-13:30		**LUNCH**
	Chairman: Jürgen Rühe	
14:30- 15:00	**Tanja Weil**	**MBM-I-002** - Synthesis of Nature-inspired Biohybrid Architectures for Nanomedicine and Precision Sensing
15:00-15:20	**Pramendra Kumar**	**MBM-O-006** - Synthesis and characterization of psyllium-g-poly(acrylamide-co-acrylic acid) through microwave route for anticancer treatment
15:20-15:40	**Fatos Cigdem Kip**	**MBM-O-030** - A New Microfluidic Extraction System Equipped with a Multiple Feeding System for Human Genomic DNA (*hg*-DNA) Isolation From Whole-Blood
15:40-16:00	**Sanogo Brahima**	**MBM-O-031** - Riboflavin Release from Interpenetrated Network Poly(NIPAAm-co-AAc) Hydrogel and Release Data Modeling by using Response Surface Methodology and Artificial Neural Networks
16:00-16:30		**BREAK**
	Chairman: Fenghua Meng	
16:30-16:50	**Metin Tülü**	**MBM-O-032** - A study of Dendritic Properties on Drug Solubility: Concentration, Generation Number and Surface Functionality
16:50-17:10	**Yasemin Yuksel Durmaz**	**MBM-O-029** - Development of Nanodroplets for Histotripsy-Mediated Cell Ablation
17:10-17:30	**Tuğba Endoğan Tanır**	**MBM-O-009** - Platelet adhesion and protein adsorption on heparin immobilized graphene enhanced poly(acrylonitrile-co-ethyl methacrylate) copolymer membranes
17:30-17:50	**Maziar Mohammadi**	**MBM-O-010** - easuring the Growth Kinetics of Layer-by-Layer Assemblies of Organic Nanoparticles and Polyelectrolytes
17:50-18:10	**Marcos Lopes Dias**	**MBM-O-011** - Synthesis of fluorescent peguilated lactide copolymers by Ugi four-component condensation

July 19, 2016, Tuesday / Robert Bruce Merrifield Hall (Haskoy)

Session: Macromolecules in Biotechnology & Medicine

	Chairman: Rana Sanyal	
10:30-11:00	**Fenghua Meng**	**MBM-I-004** - Robust Polymersomal Doxorubicin for Active Tumor Targeted Chemotherapy
11:00-11:30	**Jürgen Rühe**	**MBM-I-009** - Surface-attached polymer networks for the detection of rare circulating tumor cells in whole blood

11:30-11:50	**Oswald Prucker**	**MBM-O-012** - Surface-attached coatings on biological tissues and their potential in glaucoma treatment
11:50-12:10	**Melis Kesik**	**MBM-O-013** - Multi-Functional Fluorescent Scaffold as a Multicolor Probe: Design and Application in Targeted Cell Imaging
12:10-12:30	**Jin Hu**	**MBM-O-014** - Elastin-like polypeptide fusion of Interferon-alpha with enchanced pharmacokinetics profiles and superior antitumor efficacy
12:30-13:30		**LUNCH**
	Chairman: Fenghua Meng	
14:30-15:00	**Karsten Haupt**	**MBM-I-005** - Controlled and localized photopolymerization of molecularly imprinted polymer nanocomposites as plastic antibodies for bioimaging
15:00-15:20	**Peter Charles Griffiths**	**MBM-O-016** - How mucin structures the mucus barrier
15:20-15:40	**Andreea Madalina Pandele**	**MBM-O-017** - New porous graphene oxide biomaterials with potential application in bone repair
15:40-16:00	**Erde Can**	**MBM-O-018** - Preparation and characterization of novel poly(proylene fumarate) (PPF) based scaffolds for bone tissue engineering
16:00-16:30		**BREAK**
	Chairman: Karsten Haupt	
16:30-17:00	**Mustafa O Guler**	**MBM-I-006** - Biomimetic Self-Assembled Peptide Nanostructures
17:00-17:20	**Olcay Mert**	**MBM-O-019** - Synthesis of Biodegradable PEG Based Poly(Substituted Glycolides)
17:20-17:40	**Vusala Ibrahimova**	**MBM-O-020** - Development of block copolymer based multifunctional nano-assemblies for a theranostic approach
17:40-18:00	**Kate E Belsey**	**MBM-O-021** - Intrinsically antifouling ionically cross-linked silicones

July 20, 2016, Wednesday / Robert Bruce Merrifield Hall (Haskoy)

Session: Macromolecules in Biotechnology & Medicine

	Chairman: Mustafa O Guler	
10:30-11:00	**Volga Bulmuş**	**MBM-I-007** - Well-defined Fatty Acid Polymers as Potential Membrane Destabilizing Agents
11:00-11:20	**Muhammet U. Kahveci**	**MBM-O-022** - Ultrafast tetrazine ligation for preparation bio-functional polymers
11:20-11:40	**Sevgi Balcioglu**	**MBM-O-023** - Hydrophilic and Hydrophobic Balance in Tissue Adhesives
11:40-12:00	**Hatice Betul Bingol**	**MBM-O-024** - Phosphonic Acid-Functionalized Acrylamides for Biomedical Applications
12.00-12:20	**Kouroush Salımı**	**MBM-O-025** - Ti(IV) Attached-Polydopamine–Coated Magnetic SiO2 Microspheres for Enrichment of Phosphopeptides via Immobilized Metal Affinity Chromatography
12:30-13:30		**LUNCH**
	Chairman: Volga Bulmuş	
14:30-15:00	**Devon A. Shipp**	**MBM-I-008** - Biodegradable Polyanhydrides from Radical Thiol-Ene Polymerizations: Surface Eroding Materials with Many Interesting Properties
15:00- 15:20	**Christian Hoffmann**	**MBM-O-026** - Area selective surface functionalization and enzyme immobilization in thiol-ene-epoxy microfluidic devices
15:20- 14:40	**Eda Ayse Aksoy**	**MBM-O-027** - Improving Antibacterial Efficacy of Chitosan Grafted Polyurethanes with Silver Nanoparticles
15:40-16:00	**Tarik Eren**	**MBM-O-028** - New generation of antibiotics
16:00-16:30		**BREAK**

Energy, Optics, & Optoelectronics

July 18, 2016, Monday / Karl Ziegler Hall (Kasimpasa-4)

Session: Energy, Optics, & Optoelectronics

Chairman: Martin Heeney		
10:30-11:00	**Barry C Thompson**	**EOO-I-001** - Polymer and Active Layer Design for Optimized Polymer-Fullerene Solar Cells
11:00-11:30	**Iain McCulloch**	**EOO-I-002** - Semiconducting Polymers and Small Molecules for Transistors and Solar Cells
11:30-11:50	**Murat Ates**	**EOO-O-001** - Supercapacitor device performances of graphene/conducting polymers
11:50-12:10	**Oswald Prucker**	**EOO-O-002** - Surface-attached polymer networks for planar optronic systems
12:10-12:30	**Mustafa M. Demir**	**EOO-O-003** - Integration of triboluminescent crystals into transparent polymers
12:30-13:30	**LUNCH**	
Chairman: Barry C Thompson		
14:30-15:00	**Martin Heeney**	**EOO-I-003** - Tuning the performance of conjugated polymers for transistor and solar cells applications
15:00-15:20	**Takeo Sasaki**	**EOO-O-004** - Dynamic Hologram Formation in Photorefractive Ferroelectric Liquid Crystal Blends
15:20-15:40	**Francesco Piana**	**EOO-O-005** - Cyano-functionalized polymer / barium titanate composites for organic electronics
15:40-16:00	**Yunus Karataş**	**EOO-O-006** - 4-(Hydroxymethyl)-1,3-Dioxolane-2-on and Polyethylene Oxide Containing Borate Ester Based Polymer Electrolytes
16:00-16:30	**BREAK**	
Chairman: Barry C Thompson		
16:30-17:00	**Yingping Zou**	**EOO-I-004** - New Polymers for Photovoltaics and NIR-II Imaging
17:00-17:30	**Jianxin Geng**	**EOO-I-005** - Composites of Carbon Nanomaterials and Conjugated Polymers: Preparation and Properties
17:30-17:50	**Selmiye Alkan Gürsel**	**EOO-O-007** - Graphene based Electrodes for Fuel Cells and Li-ion Batteries
17:50-18:10	**Sumanta Kumar Karan**	**EOO-O-008** - Improvement of output performance of PVDF by incorporating Fe-oxides doped reduced graphene oxide: an efficient piezoelectric nanogenerator

July 19, 2016, Tuesday / Karl Ziegler Hall (Kasimpasa-4)

Session: Energy, Optics, & Optoelectronics

Chairman: Turan Öztürk		
10:30-11:00	**Levent Toppare**	**EOO-I-006** - Conjugated polymers and their applications
11:00-11:20	**Ünsal Koldemir**	**EOO-O-010** - End Group Effects on Molecular Order and Charge Transport in Conjugated Polymers
11:20-11:40	**Serife O. Hacioglu**	**EOO-O-011** - The effect of the triphenylamine unit on electrochemical behaviors of benzotriazole bearing conjugated polymers
11:40-12:00	**Drahomir Vyprachticky**	**EOO-O-012** - Resonance Energy Transfer from Quinolinone Modified Polystyrene-*block*-Poly(styrene-*alt*-maleic anhydride) Copolymer to Terbium Metal Ions
12:00-12:20	**Mehmet Can Zeybek**	**EOO-O-024** - Conductive Polymer Nanotubes as Chemiresistive Sensors
12:30-13:30	**LUNCH**	
Chairman: Derya Baran		
14:30- 15:00	**Turan Öztürk**	**EOO-I-007** - Thienothiophenes (TT) and Dithienothiophenes (DTT) for Electronic and Optoelectronic

15:00-15:20	**Gönül Hızalan**	**EOO-O-013** - Solution Processed Orange Light Emitting Diodes Using Polyfluorene Derivative
15:20-15:40	**Jinyi Lin**	**EOO-O-014** - Polyfluorene-based Optoelectronic Device
15:40-16:00	**Cyril Brochon**	**EOO-O-015** - Design of Squaraine-Based π-Conjugated Polymers For Organic Electronic: From Cross-Coupling Reactions to Metal-Free Polycondensation
16:00-16:30		**BREAK**
	Chairman: Huseyin Bekir Yildiz	
16:30-17:00	**Derya Baran**	**EOO-I-008** - Exceeding the Limits of Fullerene Solar Cells
17:00-17:20	**Mao Li**	**EOO-O-016** - Electrochemical Coupling Layer by Layer Assembly of Multiscale Components for Fabrication of Optical Thin Films
17:20-17:40	**Amit Kumar Das**	**EOO-O-017** - High-performance ternary composite electrode material based on polyaniline (PANI), molybdenum trioxide (MoO3) and graphene nanoplatelets (GNP) for next-generation supercapacitor prepared by sono-chemical method
17:40-18:00	**Zuhal Er**	**EOO-O-018** - Comparative Study for Monthly Average of Daily Global Radiation in Istanbul, Turkey

July 20, 2016, Wednesday / Karl Ziegler Hall (Kasimpasa-4)

Session: Energy, Optics, & Optoelectronics

	Chairman: Murat Ates	
10:30-11:00	**Věra Cimrová**	**EOO-I-009** - Semiconducting Conjugated Copolymers for Optoelectronics: Effects of Side Chains and Molecular Weight
11:00-11:20	**Tzu Tien Huang**	**EOO-O-019** - Highly Transparent and Flexible Biobased Polyimide/TiO2 and ZrO2 Hybrid Films with Tunable Refractive Index, Abbe Number, and Memory Properties
11:20-11:40	**Salih Ertan**	**EOO-O-020** - A Novel Hybrid Type Conjugated Polymer Bearing Polyhedral Oligomeric Silsesquioxane Pendant Group
11:40-12:00	**Noreen Sajjad**	**EOO-O-021** - Facile Synthesis of Cyclotriphosphazene Based on Pyrole and Thiophene Monomers as Precursors for Novel Potential Conducting Polymers
12:00-12:20	**Huan Shen Liu**	**EOO-O-022** - Highly Transparent to Truly Black Electroactive Ambipolar Devices Based on Polyamides and Viologen
12:20-12:40	**Ilknur Demirtas**	**EOO-O-023** - Syntheses, Electrochemical and Spectroscopic Properties of Fused-Thiophene Based Materials Possesing Benzonytrile Functional Group
12:30-13:30		**LUNCH**

Polymer and Polymer-Based Membranes

July 18, 2016, Monday / Leo Hendrik Baekeland Hall (Marmara)

Session: Polymer and Polymer-Based Membranes

	Chairman: Volker Abetz	
10:30-11:00	**Benny Dean Freeman**	**PPBM-I-001** - Ion Solubility, Diffusion, Permeation and Conductivity in Charged Polymers
11:00-11:30	**Steven Holdcroft**	**PPBM-I-002** - Hydrocarbon, Solid Polymer Electrolytes for Acidic and Alkaline Electrochemical Devices
11:30-11:50	**Aneela Sabir**	**PPBM-O-001** - Improved permeation flux, salt rejection and chlorine resistant properties of ZnO nanoparticles infused polymeric thin film RO desalination membranes
11:50-12:10	**Md Mushfequr Rahman**	**PPBM-O-002** - Gas separation performance, sorption behavior and thermal transitions of semicrystalline PolyActiveTM

12:10-12:30	**Sadiye Halitoğlu Velioğlu**	**PPBM-O-003** - Propylene/Propane Plasticization in Unary and Binary Gas Sorption on 6FDA-Based Polyimide Membranes
12:30-13:30		**LUNCH**
	Chairman: Benny Dean Freeman	
14:30-15:00	**Volker Abetz**	**PPBM-I-003** - Integral Asymmetric Block Copolymer Membranes: Structure Formation and Preparation by Casting, Spinning, and Spraying
15:00- 15:30	**Suzana P Nunes**	**PPBM-I-004** - New polymers and blends for block copolymer membranes
15:30-15:50	**Christian Höhme**	**PPBM-O-004** - High Thermally Stable Isoporous Integral Asymmetric Block Copolymer Membranes
15:50-16:10	**Merve Cetintas**	**PPBM-O-005** - S-*b*-PNIPAM block copolymers for thermoresponsive membrane applications
16:00-16:30		**BREAK**
	Chairman: Steven Holdcroft	
16:30-17:00	**Gregory C Rutledge**	**PPBM-I-005** - Electrospun Fiber Membranes for Clean Water Applications
17:00-17:30	**Tamer Uyar**	**PPBM-I-006** - Bacteria encapsulated/immobilized electrospun polymeric nanofibrous webs for wastewater treatment
17:30-17:50	**Shahram Mehdipour Ataei**	**PPBM-O-006** - Microphase Separated Proton Exchange Membranes from Dual Electrospun Mats from Sepiolite-Based Nanocomposite Blends of Fully Sulfonated Poly(ether ketone)/ Non-sulfonated Poly(ether sulfone)

July 19, 2016, Tuesday / Leo Hendrik Baekeland Hall (Marmara)

Session: Polymer and Polymer-Based Membranes

	Chairman: Suzana P Nunes	
10:30-11:00	**Richard J Spontak**	**PPBM-I-007** - On the Morphological Control and Transport Properties of Highly Tunable Block Ionomers as Membranes
11:00-11:30	**Adriana Boschetti De Fierro**	**PPBM-I-008** - Medium Cut-off Membranes for Hemodialysis - Closer to the Natural Kidney's Removal Function
11:30-11:50	**Bruno A. Pulido**	**PPBM-O-008** - Porous polyisatinbiphenyl membranes for harsh environments
11:50-12:10	**Mian Farrukh Mehmood**	**PPBM-O-009** - The promising chemical stable membranes developed through non-solvent induced phase separation (NIPS) and UV curing of acrylic functionalized polysulfone
12:10-12:30	**Tahir Jamil**	**PPBM-O-007** - Thin film nanofiltration polyamide membranes for water purification using Na2SO4 draw solution
12:30-13:30		**LUNCH**
	Chairman: Richard J Spontak	
14:30-15:00	**Peter Martin Budd**	**PPBM-I-009** - Membrane separations with PIMs and graphene
15:00- 15:20	**Alessio Fuoco**	**PPBM-O-011** - Gas diffusion as a probe for intrinsic microporosity
15:20- 15:40	**Bekir Satilmis**	**PPBM-O-012** - Tailoring the selectivity of PIM-1 via chemical modification
15:40-16:00	**Johannes Carolus Jansen**	**PPBM-O-013** - A novel method to determine the transport parameters of individual gas mixture components in polymers of intrinsic microporosity
16:00-16:30		**LUNCH**
	Chairman: Adriana Boschetti De Fierro	
16:30-17:00	**Volkan Filiz**	**PPBM-I-010** - Ionic liquids supported by porous and isoporous membranes for CO2/N2 gas separation applications
17:00-17:20	**Marcel Balçık**	**PPBM-O-014** - CO2-Induced Plasticization in Copolyimides Containing a Sulfone Group
17:20-17:40	**Faqih Muhamad Sukma**	**PPBM-O-015** - Cellulose Membranes For Organic Solvent Nanofiltration
17:40-18:00	**Elif Nur Durmaz**	**PPBM-O-016** - Phase Inversion Kinetics of Polymer-Ionic Liquid Solutions in Relation with Membrane Morphology and Macrovoid Formation

July 20, 2016, Wednesday / Leo Hendrik Baekeland Hall (Marmara)

Session: Polymer and Polymer-Based Membranes

Chairman: Tamer Uyar		
14:40-15:00	**Ş. Birgül Tantekin Ersolmaz**	**PPBM-O-011** - Residual solvent effect on the CO2/CH4 and CO2/N2 separation properties of 6FDA-DAM membranes
15:10-15:30	**Fatma Kurşun**	**PPBM-O-017** - Permeation and Separation Characteristics of IPA/Water Mixtures Through Poly(vinyl alcohol)-g-Poly(N-Hydroxymethylacrylamide) Copolymer Membrane by Pervaporation, Evapomeation and TDEV
15:30-15:50	**Ali Delibaş**	**PPBM-O-018** - Preparation of styrene based membranes including substituted anilidic and amidic acids, characterization and protein adsorption

Smart and Functional Polymers

July 18, 2016, Monday / Otto Bayer Hall (Kasimpasa-3)

Session: Smart and Functional Polymers

Chairman: Meifang Zhu		
11:00-11:30	**Frank Wiesbrock**	**SFP-I-001** - Contact Biocides Based on Poly(2-oxazoline)s and Poly(2-oxazine)s
11:30-11:50	**Shaojian Lin**	**SFP-O-001** - CO2 induced UCST transition of amphiphilic triblock polymer and their self-assembly
11:50-12:10	**M. Khaled Arafeh**	**SFP-O-002** - A Photoresponsive Biomimetic Dry Adhesive Based on Doped PDMS Microstructures
12:10-12:30	**Syara Kassim**	**SFP-O-003** - Synthesis PMMA-Gold Core-Shell Based Metallodielectric Photonic Crystals As Substrates For Surface-Enhanced Raman Spectroscopy
12:30-13:30	**LUNCH**	
Chairman: Frank Wiesbrock		
14:30-15:00	**Meifang Zhu**	**SFP-I-002** - Multifunctional Organic/Inorganic Hybrid-gels: from Polymer Design to Fiber Preparation and Applications
15:00- 15:30	**Christopher Barner Kowollik**	**SFP-I-003** - Exploiting Photochemical Selectivity in Macromolecular Synthesis
15:30-15:50	**Demet Asil Alptekin**	**SFP-O-025** - Temperature- and Voltage-Induced Ligand Rearrangement of a Dynamic Electroluminescent Metallopolymer
15:50-16:10	**Nikhil K Singha**	**SFP-O-005** - Modified Smart Carbon Nanotube via Reversible-Deactivation Radical Polymerization
16:00-16:30	**BREAK**	
Chairman: Christopher Barner Kowollik		
16:30-17:00	**Martin Moeller**	**SFP-I-004** - From light empowered to self oscillating hydrogel objects - on the rate and directionality control of microscopic morphing
17:00-17:20	**Ilker S. Bayer**	**SFP-O-006** - Graphene-polymer biocomposites for versatile high-performance flexible electronic applications
17:20-17:40	**Dipak Khastgir**	**SFP-O-007** - Flexible Composites for Multi functional Applications
17:40-18:00	**Jordi Royes Mir**	**SFP-O-008** - Reversible stiffness photocontrollable materials

July 19, 2016, Tuesday / Otto Bayer Hall (Kasimpasa-3)

Session: Smart and Functional Polymers

Chairman: Michael J Serpe		
10:30-11:00	**Devon A. Shipp**	**SFP-I-005** - Functional polymer colloids prepared using radical thiol-ene/yne 'click' chemistry

11:00-11:20	**Nick Dibbert**	**SFP-O-009** - Dextran hydrogels - from polysaccharides to *biocompatible* and *biodegradable* scaffolds
11:20-11:40	**Kaihuan Zhang**	**SFP-O-010** - Redox-active poly(ferrocenylsilane)s as hydrogel actuators
11:40-12:00	**Seher Şenada**	**SFP-O-011** - Development of Superhydrophobic Surfaces Based on Polyurethane Chemistry For Coating Materials
12:00-12:20	**Özgül Gök**	**SFP-O-012** - Multi-arm Polymers for Conjugation and Release of Therapeutic Cargo
12:20-12:40	**RK Dey**	**SFP-O-013** - Polyamidoamine/shellac: responsive hydrogel for controlled release applications
12:30-13:30		**LUNCH**
	Chairman: Devon A. Shipp	
14:30-15:00	**Michael J Serpe**	**SFP-I-006** - Stimuli-Responsive Polymer-Based Sensors, Muscles, and Drug Delivery Platforms
15:00- 15:20	**Sergio E Moya**	**SFP-O-014** - Thermal Annealing of Polyelectrolyte Multilayers: An Effective Approach for the Enhancement of Cell Adhesion
15:20- 15:40	**Banu Iyisan**	**SFP-O-015** - Multifunctional Polymersomes as Smart Nanodevices for Biomedical Applications: Probing Responsive Behaviour and Mechanical Properties
15:40-16:00	**Zhibo Li**	**SFP-O-016** - Thermal and Oxidation Responsive Polypeptide Materials
16:00-16:30		**BREAK**
	Chairman: Michael J Serpe	
16:30-16:50	**Daniel Frank**	**SFP-O-023** - Thiolactones: Sustainable building blocks in advanced polymer design
16:50-17:10	**Göksenin Kurt Çömlekçi**	**SFP-O-017** - Encapsulation of Alkyd Resin in Urea Formaldehyde Resin for Self-Healing Coatings
17:10-17:30	**Suset Barroso Solares**	**SFP-O-018** - Magnetic polymer nanocomposite fibers for oil removal from stable water-oil emulsions
17:30-17:50	**Lawrence Mzukisi Madikizela**	**SFP-O-019** - Multi-template molecularly imprinted polymer: Synthesis, characterization and removal of selected acidic pharmaceuticals from wastewater
17:50-18:10	**Anna Isakova**	**SFP-O-022** - Design, synthesis and RAFT polymerisation of a quinoline-based monomer for use in metal-binding composite microfibers

Renewable Resources and Biopolymers

July 19, 2016, Tuesday / Wallace Carothers Hall (Kasimpasa-5)

Session: Renewable Resources and Biopolymers

	Chairman: Henri Cramail	
14:30-15:00	**Paschalis Alexandridis**	**RRB-I-001** - Cellulose Swelling and Dissolution
15:00-15:20	**Yelda Meyva**	**RRB-O-023** - Blending of Polylactide with Different Elastomeric Materials for Toughening
15:20-15:40	**Azlin Mohmad Azman**	**RRB-O-002** - Process control in the production of engineered palm-based fibreboard
15:40-16:00	**Siti Fairus Mohd Yusoff**	**RRB-O-003** - Properties and Potential Applications of Modified Liquid Natural Rubber
16:00-16:30		**BREAK**
	Chairman: Paschalis Alexandridis	
16:30-17:00	**Henri Cramail**	**RRB-I-002** - Fully bio-based Non-Isocyanate Polyurethanes (NIPU) via cyclic carbonate/amine route
17:00-17:20	**Tatsuo Kaneko**	**RRB-O-004** - Renewable super-strong plastics derived from exotic amino acids

17:20-17:40	**Neslihan Alemdar**	**RRB-O-005** - Oxygen-generating photocrosslinkable hydrogel for cardiac progenitor cells survival under hypoxic conditions
17:40-18:00	**Mona Hassan Abdel Rehim**	**RRB-O-006** - Multifunctional nanocellulose / paper composites

July 20, 2016, Wednesday / Wallace Carothers Hall (Kasimpasa-5)

Session: Renewable Resources and Biopolymers

	Chairman: Henri Cramail	
10:30-10:50	**Alessandro Pellis**	**RRB-O-022** - *Thermobifida cellulosilytica* cutinase as a powerful tool for the synthesis and functionalization of biobased polyesters
10:50-11:10	**Andrea Hufendiek**	**RRB-O-008** - ADMET and TAD chemistry: A Sustainable Alliance
11:10-11:30	**Francisco Cardona**	**RRB-O-009** - Characterization of Sustainable Phenolic Resins Synthesized with Renewable Bioresources
11:30-11:50	**Swapnil Shukla**	**RRB-O-010** - Sustainable Benzoxazine and Sulphur copolymers as Li-S Battery Cathode Materials
11:50-12:10	**Jan-Georg Rosenboom**	**RRB-O-011** - High molecular weight polyethylene furanoate (PEF) synthesis via ring-opening polymerization
12:10-12:30	**Ilknur Gönenç**	**RRB-O-012** - Evaluation of Thermal, Adsorption and Microstructural Properties of Corn Starch Based Cross-linked Composite Films
12:30-13:30		**LUNCH**
	Chairman: Paschalis Alexandridis	
14:30- 14:50	**Erde Can**	**RRB-O-013** - Novel thermosetting resins based on cardanol
14:50- 15:10	**Mohammad Asif Ali**	**RRB-O-014** - Syntheses of renewable polyamides with controlled optical activity from itaconic acid and amino acid
15:10-15:30	**Ersen Göktürk**	**RRB-O-015** - Polyglycolic Acid and Its Copolymers from Cationic Polymerization of Formaldehyde and Carbon Monoxide
15:30-15:50	**Amit Kumar Sonker**	**RRB-O-016** - Synergistic effect of nano reinforcement and cross-linking towards polyvinyl alcohol (PVA) properties
15:50-16:10	**Katherine Stephanie Encalada Flores**	**RRB-O-017** - Mechanical Behavior Approach of Biodegradable Materials Based on Polyvinyl Alcohol, Achira (*Canna edulis*) Starch and Collagen Recovered from Leather Shavings
16:00-16:30		**BREAK**
	Chairman: Tatsuo Kaneko	
16:30-16:50	**Gunasunderi Raju**	**RRB-O-026** - Preparation and Characterization of "*Hibiscus Cannabinus*" (Kenaf) Grafted Natural Rubber Based Composites via Latex Route
16.50-17.10	**Brindusa Balanuca**	**RRB-O-019** - Design of new camelina oil-based hydrophilic monomers for novel polymeric materials
17:10-17:30	**Dildare Basalp**	**RRB-O-020** - Renewable Resource based Biocomposites from Natural Luffa Cylindrica Fiber and Poly(hydroxybutyrate-co-valerate) Biopolymer
17:30-17:50	**Yeşim Müge Şahin**	**RRB-O-021** - Mechanical Properties of Marine Sourced Hydroxyapatite/ Polycaprolactone Electrospun Biocomposites
17:50-18:10	**Zafer Söyler**	**RRB-O-024** - Catalytic Transesterification of Maize Starch with Plant Oils: Sustainable Access to Termoplastic Starches

Polymer Engineering, Processing, and Characterization

July 19, 2016, Tuesday / Michael Szwarc Hall (Balat)

Session: Polymer Engineering, Processing, and Characterization

Chairman: Giuseppe Titomanlio		
14:30- 15:00	**Dietmar Auhl**	**PEPC-I-004** - Materials, process, and product design of thermoplastic polymers using hyphenated rheology
15:00-15:20	**Ewa Kicko Walczak**	**PEPC-O-011** - Flame retardands nanocomposites synergy effect of combining conventional antypirenes with carbon nanofillers
15:20-15:40	**Zitouni Safidine**	**PEPC-O-012** - Elaboration and characterization of nanostructured polyurethane rigid foams within the inclusion of O-MMT: impact and fire proof behavior
15:40-16:00	**Malik Sajjad Mehmood**	**PEPC-O-013** - Study of thermal properties of UHMWPE/MWCNTs nano composites: Effect of γ-rays modified MWCNTs incorporation
16:00-16:30		**BREAK**

Chairman: Dietmar Auhl		
16:30-16:50	**Sabu Thomas**	**PEPC-O-016** - Engineering at the Nanoscale: State of the Art, Challenges and New Opportunities
16:50-17:10	**Loïc Le Pluart**	**PEPC-O-017** - Effect of organoclays on the degradation of polyoxymethylene during melt processing
17:10-17:30	**Lívia Mesquita Dias Loiola**	**PEPC-O-018** - Composites based on amphiphilic block copolymers of PLLA, PEO and PPO blocks and nanohydroxyapatite
17:30-17:50	**Enrique Herrero Acero**	**PEPC-O-019** - Exploiting mild enzymatic hydrolysis to functionalize synthetic polymers
17:50-18:10	**Zehra Oluz**	**PEPC-O-020** - Optical Biosensing Applications on Molecularly Imprintied Poly (Methacrylic Acid-co-Cyanurate Ester) Nanorods via Thermal Polymerization

July 20, 2016, Wednesday / Michael Szwarc Hall (Balat)

Session: Polymer Engineering, Processing, and Characterization

Chairman: Mukerrem Cakmak		
10:30-11:00	**Giuseppe Titomanlio**	**PEPC-I-001** - Analysis and simulation of iPP morphology with fast temperature evolution at mold surface
11:00-11:20	**Bernabe Quevedo Sanchez**	**PEPC-O-001** - Twin-Screw Extrusion Isolates High Viscosity Polymers
11:20-11:40	**Vishwa Pratap Singh**	**PEPC-O-002** - Rheological Behavior of HDPE/Natural Needle Like Clay Nanocomposites Under Shear and Extensional Flow
11:40-12:00	**A S Mohammad Sayem Mozumder**	**PEPC-O-003** - Optimization of injection molding process conditions while manufacturing of HDPE/TiO2 nanocomposites
12:00-12:20	**Davut Demir**	**PEPC-O-021** - Simultaneously Improvement of the Mechanical Strength and Fractural Toughness Properties of Recycled Polycarbonate
12:20-12:40	**Merve Dandan Doganci**	**PEPC-O-014** - Improving Mechanical and Thermal Properties of Poly(Lactic Acid) By Using POSS Cored Star-Shaped Poly(ε-Caprolactone) Polymers
12:30-13:30		**LUNCH**
Chairman: Yakup Ulcer		
14:30-15:00	**Mukerrem Cakmak**	**PEPC-I-002** - Large scale R2R fabrication of piezoresistive films (Ni/PDMS) with enhanced through thickness electrical and thermal properties by applying magnetic field
15:00- 15:20	**Chong Min Koo**	**PEPC-O-005** - Graphene-Based Polymer Nanocomposites for Thermal Conductivity and Electromagnetic Interference Shielding Applications
15:20-15:40	**Cyrille Sollogoub**	**PEPC-O-006** - Effect of polymer on the existence of a critical thickness in coextruded multilayer films

15:40-16:00	**Jean Charles Sébileau**	**PEPC-O-007** - Consolidation by Spark Plasma Sintering of Poly-Ether-Ether-Ketone
16:00-16:30		**BREAK**
Chairman: Mukerrem Cakmak		
16:30-17:00	**Yakup Ulcer**	**PEPC-I-003** - Injection Molding of Slowly Crystallizing Polymers
17:00-17:20	**Padmanabhan Ramachandran**	**PEPC-O-008** - Structure- Property Relationship in Ethylene Octene Copolymer (EOC) - Polydimethyl Siloxane (PDMS) based TPEs for cable insulation applications
17:20-17:40	**Kinsuk Naskar**	**PEPC-O-009** - Novel High Performance TPV: a new generation thermoplastic elastomer for automotive applications
17:40-18:00	**Mohammad Taghipourfard**	**PEPC-O-010** - Determining the effect of width of waisted section on waisted tensile test specimen using HDPE non-linear elastic-plastic behaviour in FEA

Polymer Physics

July 18, 2016, Monday / Pierre-Gilles de Gennes Hall (Cibali-1)

Session: Polymer Physics

Chairman: Mehmet Sayar		
10:30-11:00	**Andreas Walther**	**PP-I-001** - Static and Dynamic Bioinspired Material Systems
11:00-11:20	**Marina Tsianou**	**PP-O-001** - Charging and Uncharging of Neutral Polymers in Solution Probed by SANS
11:20-11:40	**Guillaume Miquelard Garnier**	**PP-O-002** - Kinetics of thin polymer film rupture: model experiments for a better understanding of layer breakups in the multilayer coextrusion process
11:40-12:00	**Sara Aid**	**PP-O-003** - Study and modeling of the coalescence of two different polymer particles PVDF-PMMA
12:00-12:20	**Jaime Martín**	**PP-O-004** - Rational manipulation of the microstructure of organic semiconductors by nanoscale confinement
12:20-12:40	**Debashish Mukherji**	**PP-O-010** - Two paradoxical phenomena of (smart) polymers in solvent mixtures: Effects of preferential adsorption and depleted depletion
12:30-13:30		**LUNCH**
Chairman: Andreas Walther		
14:30-15:00	**Jiaping Lin**	**PP-I-006** - Self-Assembled Nanostructures of Polypeptide-Based Polymers
15:00-15:30	**Mehmet Sayar**	**PP-I-005** - Interface Driven Conformational Change and Aggregation in Proteins/Peptides
15:30-15:50	**Yanchun Han**	**PP-O-009** - Control over Microphase Separation, Crystallization and Molecular Orientation of All-Conjugated Diblock Copolymers
15:50-16:10	**Mostafa Ahmadi**	**PP-O-011** - Dynamics of entangled Supramolecular Networks based on Poly (n-butyl acrylate-g-Ureidopyrimidinone)

July 20, 2016, Wednesday / Pierre-Gilles de Gennes Hall (Cibali-1)

Session: Polymer Physics

Chairman: Evelyne van Ruymbeke		
10:30-11:00	**Kurt Kremer**	**PP-I-002** - Topological constraints do matter: polymer melts, elastomers, collapsed polymer globules, chromosome territories etc
11:00-11:30	**Natalie Stingelin**	**PP-I-003** - The principles of manipulating the phase transformations, solid-state order and properties of functional organic matter
11:30-11:50	**Sissi De Beer**	**PP-O-006** - Particle sorption in shaken polymer films
11:50-12:10	**Antonis Gitsas**	**PP-O-008** - Morphological effects of microscale matrix-filler interactions in thermoplastic polymers
12:10-12:30	**Theo Tervoort**	**PP-O-005** - High-performance polyethylene fibers "al dente": improved gel-spinning of ultra-high molecular weight polyethylene using vegetable oils
12:30-13:30	**LUNCH**	

Polymer Education

July 20, 2016, Wednesday / Otto Bayer Hall (Kasimpasa-3)

Session: Polymer Education

Chairman: Gabriel Jeffrey Summers		
10:30-11:00	**Patrick Theato**	**PE-I-001** - Unifying the polymer curriculum at universities: Pros and Cons
11:00-11:30	**Eric Goethals**	**PE-I-002** - The Rise of Polymer Science Education in Belgium
11:30-12:00	**Teoman Tincer**	**PE-I-003** - A Brief History of Chemistry and Polymer Science in Turkey
12:30-13:30	**BREAK**	
Chairman: Eric Goethals		
14:30-15:00	**Gabriel Jeffrey Summers**	**PE-I-004** - The Teaching and Learning of Organic Polymer Chemistry via the Open and Distance e-Learning Model at the University of South Africa
15:00-15:20	**Djafer Benachour**	**PE-O-001** - Impact of Nanotechnology on Higher Education Curricula: Cases of "Materials Science" and "Polymer Science & Engineering" Curricula
15:20-15:40	**Emine Erdem**	**PE-O-002** - Awareness on polymers among high school students

Industrial Polymers

July 19, 2016, Tuesday / Pierre-Gilles de Gennes Hall (Cibali-1)

Session: Industrial Polymers

Chairman: Kurt Dietliker		
10:30-11:00	**Andreas Taden**	**IP-I-001** - Polymer systems with latent reactivity for adhesive applications
11:00-11:30	**Norbert Moszner**	**IP-I-002** - New Components for Dental Composites and Adhesives
11:30-11:50	**Melek Bulut**	**IP-O-001** - Mechanical Properties of New Trends Thermotropic Liquid Crystalline Copolyesters with PET Blends
11:50-12:10	**M. Göktug Ahunbay**	**IP-O-002** - *In-silico* Screening of Green Plasticizers for PVC
12:10-12:30	**Varol Intasanta**	**IP-O-003** - Single-Step Multiple-Colloid Coating for Multifunctional Biodefense Textiles: A Novel Application in Mosquito-Borne Disease Prevention
12:30-13:30	**LUNCH**	

Chairman: Norbert Moszner		
14:30-15:00	**Kurt Dietliker**	**IP-I-003** - Photoinitiators – Key Components Enabling Novel Accesses to Crosslinked Industrial Polymers
15:00-15:20	**Sonja Ketin**	**IP-O-004** - Up-to Date Procedures for Polypropylene Production
15:20-15:40	**Gözde Salkıç**	**IP-O-005** - Optimization of Gas Phase Ethylene Polymerization: A Micro/ Mini Reactor Study
15:40-16:00	**Luis Alexandro Valencia Lopez**	**IP-O-006** - Isospecific polymerization of p-methylstyrene by means of a bis(phenolato)titanium based catalyst activated by MMAO: A systematic and kinetical study
16:00-16:30	**BREAK**	
Chairman: Norbert Moszner		
16:30-17:00	**Yakup Ulcer**	**IP-I-004** - Trends in Compounding
17:00-17:20	**Emel Başkent**	**IP-O-007** - Investigation of Parameters Affecting the Bitumen Modifications by Using Response Surface Method
17:20-17:40	**Gökhan Akbulut**	**IP-O-008** - Core-Shell Type Styrene-Acrylic Emulsions for Reduction of Formaldehyde Emission in Pigment Printing
17:40-18:00	**Fatih Mehmet Erguney**	**IP-O-009** - Multifunctional polycarboxylate dispersants for the optimum soil anti-redeposition performance in new generation of powder detergents

July 20, 2016, Wednesday / Pierre-Gilles de Gennes Hall (Cibali-1)

Session: Industrial Polymers

Chairman: Andreas Taden		
14:30-15:00	**Mustafa Yasin Sen**	**IP-I-005** - Novel Production of Aliphatic Polyketone in Water
15:00-15:20	**Regis M. Gauvin**	**IP-O-010** - Bridges between homogeneous and heterogeneous catalytic systems for polyolefins production: design and advanced understanding of supported aluminic activators
15:20-15:40	**Orcun Yucel**	**IP-O-011** - Nanoreinforced Rigid Polyurethane-Organoclay Composite and Characterization Thereof
15:40-16:00	**Ilhan Özen**	**IP-O-012** - Fertilizer Release Profile of Pol(vinyl alcohol)/Ammonium Sulfate Coated Nonwoven Structures

Porous Polymer and Gels

July 19, 2016, Tuesday / Paul John Flory Hall (Cibali-2)

Session: Porous Polymer and Gels

	Chairman: Michael S. Silverstein	
14:30-15:00	**Darrin Pochan**	**PPG-I-002** - Injectable solid hydrogels constructed with peptide folding and solution assembly
15:00-15:20	**Luminita Marin**	**PPG-O-005** - Self-structuring chitosan-cinnamaldehyde hydrogels
15:20-15:40	**Daniela Ailincai**	**PPG-O-006** - Dual crosslinked chitosan based hydrogels with efficient antifungal activity
15:40-16:00	**Olga Philippova**	**PPG-O-007** - Self-assembled networks composed of polymer and supramolecular chains
16:00-16:30		**BREAK**
	Chairman: Darrin Pochan	
16:30-17:00	**Oguz Okay**	**PPG-I-003** - Nanostructural evolution of self-healing hydrogels formed in micellar solutions
17:00-17:20	**Mohammad Saleem Khan**	**PPG-O-008** - Synthesis of metal entrapped stable Polymeric hydrogels for catalytic reduction of organic dyes
17:20-17:40	**Mostafa Rezaei**	**PPG-O-009** - Visual Observation on the Foaming Process of Styrene-Methyl Methacrylate Copolymer / n-Pentane system Using Temperature Increment
17:40-18:00	**Onur Buyukcakir**	**PPG-O-010** - Porous Polymers For CO2 Capture and Conversion

July 20, 2016, Wednesday / Paul John Flory Hall (Cibali-2)

Session: Porous Polymer and Gels

	Chairman: Oguz Okay	
10:30-11:00	**Michael S. Silverstein**	**PPG-I-001** - Porous polymers through emulsion templating
11:00-11:20	**Cafer T Yavuz**	**PPG-O-001** - Chemistry in the confined spaces of porous polymers
11:20-11:40	**Peter Krajnc**	**PPG-O-002** - Open porous composites and ceramic derivatives by emulsion templating
11:40-12:00	**Ahmed M. Eissa**	**PPG-O-003** - Emulsion-Templated Porous Polymer Scaffolds for 3D *In-Vitro* Cell Culture and Tissue Engineering
12:00-12:20	**Musa Şölener**	**PPG-O-004** - Synthesis and characterization of new polymeric HPLC column
12:30-13:30		**LUNCH**

	Chairman: Cafer T Yavuz	
14:30-14:50	**Tom Hasell**	**PPG-O-011** - Porous inverse vulcanised polymers for mercury capture
14:50-15:10	**Ali Coskun**	**PPG-O-012** - Microporous Graphene Frameworks For Gas and Energy Storage Applications
15:10-15:30	**Mehmet Murat Ozmen**	**PPG-O-013** - Preparation and Antimicrobial Properties of Macroporous Polypeptide-based Cryogels
15:30-15:50	**Ahmet Erdem**	**PPG-O-014** - Evaluations of molecular weight and reduction effects on Jeffamine-glutaraldehyde cryogels for potential application as scaffolds in cartilage tissue engineering
15:50-16:10	**Sadik Kaga**	**PPG-O-015** - Dendron-polymer conjugates as precursors for designing functional hydrogels
16:00-16:30		**BREAK**

Modeling and Simulation of Polymers

July 18, 2016, Monday / Paul John Flory Hall (Cibali-2)

Session: Modeling and Simulation of Polymers

	Chairman: Juan J De Pablo	
10:30-11:00	**Murugappan Muthukumar**	**MSP-I-001** - Simulation-Directed Discoveries of Concepts in Polymers: Primordial Nucleation, Translocation, and Charged Macromolecular Assemblies
11:00-11:30	**Burak Erman**	**MSP-I-002** - Rubber Elasticity and Protein Behavior
11:30-12:00	**Masao Doi**	**MSP-I-003** - Onsager Principle -A useful principle in soft matter dynamics
12:00-12:20	**Stefano Valdo Meille**	**MSP-O-007** - etermining and modelling polymer crystal structures: the perspective of a polymer crystallographer
12:30-13:30		**LUNCH**
	Chairman: Murugappan Muthukumar	
14:30-15:00	**Juan J De Pablo**	**MSP-I-004** - Directed Assembly of Polymeric Materials - Non-Equilibrium States and On-The-Fly Characterization of Assembly
15:00-15:30	**Paola Carbone**	**MSP-I-005** - Thermodynamics and dynamics of polymers at soft interfaces
15:30-15:50	**Gökhan Kaçar**	**MSP-O-003** - Hierarchical multi-scale simulation approaches for properties of polymers at different interfaces
15:50-16:10	**Chitrita Kundu**	**MSP-O-004** - Simulation Study of Diblock Copolymer Crystallization in Thin Films
16:00-16:30		**BREAK**
	Chairman: Masao Doi	
16:30-17:00	**Guido Raos**	**MSP-I-006** - Glassy and nonlinear dynamics of polymers on random and patchy surfaces
17:00-17:20	**Kostas Daoulas**	**MSP-O-005** - Thermodynamics of polymer nematics: comparing particle-based simulations with mean-field calculations
17:20-17:40	**Isa Degirmenci**	**MSP-O-006** - The Unusual Contrasting of the Same Group Element Chemical Behavior
17:40-18:00	**Ashok Kumar Dasmahapatra**	**MSP-O-009** - Crystallization of Double Crystalline Binary Polymer Blends

July 19, 2016, Tuesday / Paul John Flory Hall (Cibali-2)

Session: Modeling and Simulation of Polymers

	Chairman: Paola Carbone	
10:30-11:00	**Gregory C Rutledge**	**MSP-I-007** - Process-structure-property relationships for semicrystalline polymers: Crystallization from the melt and deformation mechanics
11:00-11:30	**Friederike Schmid**	**MSP-I-008** - Self-assembly of polymeric nanoparticles in flow
11:30-11:50	**Ping Tang**	**MSP-O-008** - Study on the physics of polymer crystallization based on the rod-coil multi-blocks chain model
12:30-13:30		**LUNCH**

Bridging Continents & Bridging Molecules

MACRO2016

World Polymer Congress Istanbul 17-21 July

POSTER PROGRAM

Poster Session 1 / July 18, 2016 - Monday	
RDPS-A	Recent Developments in Polymer Synthesis-A-Controlled/Living Polymerization
RDPS-B	Recent Developments in Polymer Synthesis-B-Complex Macromolecular Structures
RDPS-C	Recent Developments in Polymer Synthesis-C-Light-induced Reactions in Polymer Science
RDPS-D	Recent Developments in Polymer Synthesis-D-Condensation Polymerization and Thermoset
PCNCT	Polymer Characterization by New and Combined Techniques
MN	Macromolecules & Nanotechnology
MBM	Macromolecules in Biotechnology & Medicine
EOO	Energy, Optics, & Optoelectronics
Poster Session 2 / July 19, 2016 - Tuesday	
PPBM	Polymer and Polymer-Based Membranes
SFP	Smart and Functional Polymers
RRB	Renewable Resources and Biopolymers
PEPC	Polymer Engineering, Processing, and Characterization
PP	Polymer Physics
IP	Industrial Polymers
PPG	Porous Polymer and Gels
MSP	Modeling and Simulation of Polymers

Recent Developments in Polymer Synthesis
A-Controlled/Living Polymerization

RDPS-A-P-001	RAFT Polymerization of Styrene mediated by Naphthyl Functionalized RAFT Agents Gabriel J Summers, Thembinkosi S Mdletshe, Carol A Summers
RDPS-A-P-002	RAFT Polymerization of Styrene mediated by Diphenyl Functionalized RAFT Agents Gabriel Jeffrey Summers, Teboho Simon Motsoeneng, Carol Ann Summers
RDPS-A-P-003	One-step synthesis of a mid-chain functional macrophotoinitiator of a polystyrene-poly(ε-caprolactone) diblock copolymer via simultaneous ATRP and ROP using a dual-functional photoinitiator Mustafa Degirmenci, Mustafa Durgun, Zafer Uyar
RDPS-A-P-004	Synthesis of an A2B2 Type Star-shaped Copolymer by Combination of ROP, Click Chemistry and Photoinitiated Polymerization Zafer Uyar, Mustafa Değirmenci, Bahattin Abacı
RDPS-A-P-005	Structure-Acitivity Relationships in Squaramide Catalyzed Ring Opening Polymerization of Lactide: The Effect of Tertiary Amine Group Ali Rostami, Shaghayegh Ahmadi, Elahe Sadeh
RDPS-A-P-006	Simplifying the synthesis of polypeptoids Afroditi Doriti, Sarah Mae Brosnan, Helmut Schlaad
RDPS-A-P-007	Surface Properties of Silicon Disc Coated with Fluoro-Based Polymers Mehtap Evci, Tuncer Caykara
RDPS-A-P-008	Preparation of Polyhedral Oligomeric Silsesquioxane (POSS) Materials with Fluorogenic Boronic Acid Ahmet Gultek, Selda Sezer, Turgay Seckin, Fadime Nilufer Kivilcim
RDPS-A-P-009	Preparation of stereoblock olefin copolymers and their model blends with high impact polypropylene Anatolij Sokolohorskyj, Pavla Doubravová, Zdeněk Buráň, Jan Merna
RDPS-A-P-010	Hydrophilic Pentablock Copolymers Synthesized by One-Pot RAFT Polymerization of Methacrylates Povilas Radzevicius, Medeina Steponaviciute, Tatjana Krivorotova, Ricardas Makuska
RDPS-A-P-011	Star Polymers; Synthesis, Characterization and Applications Cansel Tuncer, Vural Bütün
RDPS-A-P-012	Synthesis of Highly Crosslinked Polymers and Their Usage as Dye Adsorbent Surfaces Mehtap Kılınç, Cansel Tuncer, Vural Bütün
RDPS-A-P-013	Synthesis and Solution Behavior of a Novel Polymer of Phosphorus Containing Yasemin Samav, Vural Bütün
RDPS-A-P-014	Synthesis, Characterization and Antibacterial Activity of Reduced Graphene Oxide-Poly(n-butylacrylate-block-n-vinylpyrrolidone) Nanocomposite Films Elif Vargün, Cihan Açık, Emine Sonay Elgin
RDPS-A-P-015	α-Diimine binuclear Ni-based catalysts for ethylene polymerization: effect of ortho-substituent Mostafa KHOSHSEFAT, Saeid Ahmadjo, Seyed Mohammad Mahdi Mortazavi, Gholamhossein Zohuri, Majid Zahmaty, Khosrow Valieghbal
RDPS-A-P-016	Towards "Ultimate Control": Controlled Radical Polymerization in Pores Towards Length, Sequence and Tacticity Defined Macromolecules Hui Chun Lee, Bernhard V. K. J. Schmidt, Markus Antonietti
RDPS-A-P-017	TBD catalyzed ring-opening polymerization in microreactor Ning Zhu, Xin Hu, Weiyang Feng, Kai Guo
RDPS-A-P-018	Synthesis and Characterization of Poly(n-butyl acrylate)-b-Poly(4-vinyl pyridine) Block Copolymers Hümeyra Mert, Nicolay V. Tsarevsky, Krzysztof Matyjaszewski
RDPS-A-P-019	Synthesis and characterization of antibacterial polysulfone via photoiniferter method Serhat Oran, Mehmet Atilla Tasdelen
RDPS-A-P-020	Modification of Cell Culture Dishes by Consecutive Grafting of Poly(AAm)/Poly(NIPAAm) for Cell Sheet Recovery Murat Barsbay, Olgun Güven
RDPS-A-P-021	Well-defined Proton Exchange Membranes with enhanced performance via RAFT polymerization Gökçe Çelik, Murat Barsbay, Olgun Güven
RDPS-A-P-023	Synthesis and application of magnesium and zinc complexes with imino(phenoxide) ligands in L-lactide polymerization Marcos Lopes Dias, Alana L. C. Oliveira, Leonardo C. Ferreira, Marco A. Chaer Nascimento, Rodrigo S. Bitzer, Iryna Grafova, Andriy Grafov
RDPS-A-P-024	The First Successful Cyclopolymerization of 1,6-Heptadiynes by Grubbs 1st Generation Catalyst Cheol Kang, Eun Hye Kang, Tae Lim Choi
RDPS-A-P-025	Exploration of microstructure of olefin block copolymers made by combination of chain-walking and chain-shuttling reactions in ethylene polymerization Mostafa Ahmadi, Mahsa Karimi, Saeid Ahmadjo, Majid Zahmati, Davoud Jafarifar, Mohammad Mahdi Mortazavi
RDPS-A-P-026	Effect of reversible transfer reaction on comonomer incorporation in coordinative chain transfer polymerization Mostafa Ahmadi, Ali Ebrahimi, Mahsa Karimi, Saeid Ahmadjo, Majid Zahmati, Khosrow Valieghbal, Mohammad Mahdi Mortazavi
RDPS-A-P-027	RAFT Polymerization of Styrene mediated by Diphenyl Functionalized RAFT Agents Teboho Simon Motsoeneng, Gabriel Jeffrey Summers, Carol Ann Summers
RDPS-A-P-028	Novel Poly(methylmethacrylate)-Based Two-Armed Macroinitiators via ATRP: Synthesis and Characterization Çiğdem Yağcı, Murat Mısır, Ahmet Bilgin, Nagihan Acıerik
RDPS-A-P-029	Synthesis of Linear and Star Shaped Block Copolymers with ε-caprolactone and L-lactide by using Bismuth(III)Acetate Catalyst Dildare Basalp, Funda Tıhmınlıoğlu
RDPS-A-P-030	Graft copolymers from commercial chlorinated polypropylene via Cu(0)-mediated ATRP Gokhan Acik, Mehmet Atilla Tasdelen
RDPS-A-P-031	Synthesis and Characterization of New Olefinic Block Copolymers Mahsa Karimi, Ali Ebrahimi, Mostafa Ahmadi, Saeid Ahmadjo, Majid Zahmati, Khosrow Valieghbal, Mohammad Mahdi Mortazavi

RDPS-A-P-032	Polymer Bottlebrushes with Complex Architecture as Templates for Nanoparticle Synthesis Yaoming Zhang, Joanna Pietrasik, Krzysztof Matyjaszewski
RDPS-A-P-033	Ethylene polymerization using a binary catalytic system; activity and microstructure study Mostafa Khoshsefat, Saeid Ahmadjo, Mohammad Mahdi Mortazavi, Gholamhossein Zohuri, Meisam Haghpanah
RDPS-A-P-034	One-pot synthesis of well-defined polyether/polyester block copolymers and terpolymers by a highly efficient catalyst switch approach Haleema Alamri, Nikos Hadjichristidis

Recent Developments in Polymer Synthesis
B-Complex Macromolecular Structures

RDPS-B-P-001	Theoretical investigation of organic schiff base compounds on corrosion inhibition of copper Saifi Issaadi, Tahar Douadi
RDPS-B-P-002	SYNTHESIS, SPECTROSCOPY, X-RAY CRYSTALLOGRAPHY, ELECTROCHEMESTRY AND THEORETICAL INVESTIGATION OF A TETRADENTATE COPPER SCHIFF BASE COMPLEX Cu2L2.2H2O Samra Rahmouni, Amel Djedouani, Salima Thabti, Abderrahmen Bendaas
RDPS-B-P-003	Ring-Opening Reactions of Backbone Epoxidized Polyoxanorbornene Ufuk Saim Gunay, Erhan Demirel, Gurkan Hizal, Umit Tunca, Hakan Durmaz
RDPS-B-P-004	Post-Functionalization Of Perfluorophenyl Ester-Functional Acyclic Diene Metathesis Polymer Ozgun Daglar, Hakan Durmaz, Gurkan Hizal, Umit Tunca
RDPS-B-P-005	Main Chain Post-Functionalization of Reactive Triple Bond Containing Polyester via Diels-Alder Cycloaddition Reactions Cansu Esen, Ozgun Daglar, Muge Cetin, Hakan Durmaz, Gurkan Hizal, Umit Tunca
RDPS-B-P-006	Active Ester Substitution And Nitrene Addition Reactions On To Multiwalled Carbon Nanotubes For Double Modification Pinar Sinem Omurtag, Hakan Durmaz, Gürkan Hızal, Ümit Tunca
RDPS-B-P-007	Multiarm Star Polymer as Novel Polymeric Photoinitiator for Free Radical Polymerization Neşe Çakır, Duygu S. Esen, Ümit Tunca, Nergis Arsu, Gürkan Hızal
RDPS-B-P-008	Synthesis and Self-Assembly of Fluorene-Vinylene Alternating Copolymers in "Hairy-Rod" Architecture: Side Chain – Mediated Tuning of Conformation, Microstructure and Photophysical Properties Demet Göen Colak, Ioan Cianga, Luminita Cianga, Yusuf Yagci
RDPS-B-P-009	Synthesis of Dendronized Polymers Containing Polyphenylene Dendron via Macromonomer Approach by ROMP and Their Characterization Ki Taek Bang, Inho Choi, Tae Lim Choi
RDPS-B-P-010	Synthesis of Polythioether with Clickable Side Chain Through Thiol Michael Condensation Reaciton Serter Luleburgaz, Hande Tınas, Hakan Durmaz, Umit Tunca, Gurkan Hizal
RDPS-B-P-011	Synthesis and characterization of porphyrin containing commercial polymers Aylin Bilgili, Fehmi Saltan, Hakan Akat
RDPS-B-P-012	Coacervation between biopolymers Ayse Basak Kayitmazer, Alaaddin Faruk Koksal, Elif Kilic Iyilik, Fatma Akcay Ogur

Recent Developments in Polymer Synthesis
C-Light-induced Reactions in Polymer Science

RDPS-C-P-002	Synthesis and and evaluation of novel thioxanthone-based methacrylic acids Tuğçe Nur Eren, Duygu Avci, Jacques Lalevée
RDPS-C-P-003	Extension of Linear PIBs to Higher Molecular Weight Linear and Network Products via Thiol-ene Chemistry Elif Kurnaz, Nihan Nugay, Turgut Nugay, Joseph P. Kennedy
RDPS-C-P-004	The use of 1-(2-anthryl)-1-phenylethylene moiety at the chain end to perform chain extension and block copolymerization reactions Çimen Özgüç, Yusuf Yağcı, Turgut Nugay
RDPS-C-P-005	Synthesis of Clickable Hydrogels and Linear Polymers by Type II Photoinitiation Mustafa Ciftci, Eljesa Murtezi, Yusuf Yagci, Yusuf Yagci
RDPS-C-P-006	Dibenzoyldiethylgermane as a Visible Light Photo-reducing Agent for CuAAC Click Reactions Mustafa Arslan, Gorkem Yilmaz, Yusuf Yagci
RDPS-C-P-008	Visible and LED Light-Induced Metal Free ATRP Using Electron Acceptor Dyes in Conjunction with Amines Ceren Kutahya, F. Simal Aykac, Gorkem Yilmaz, Yusuf Yagci
RDPS-C-P-009	Benzodioxinone Photochemistry for Macromolecular Synthesis Volkan Kumbaraci, Naciye Talinli, Yusuf Yagci
RDPS-C-P-011	The investigation of photophysical and photochemical properties of thioxanthone-anthracene -9-carboxylic acid Saliha Mutlu, Tolga Çeper, Nergis Arsu, Shigeru Takahara
RDPS-C-P-012	Photocoating in The Presence of Organosilica Nanoparticles from 3-Mercaptopropyltrimethoxysilane Nurcan Karaca
RDPS-C-P-013	Synthesis of a novel benzophenone derived UV absorber acrylate monomer Tuğrul Cem Biçak, Ersin Acar
RDPS-C-P-014	Photo-initiated Metal-Free Controlled Living Radical Polymerization Using Polynuclear Aromatic Compounds Andrit Allushi, Steffen Jockusch, Gorkem Yilmaz, Yusuf Yagci, Yagci Yagci
RDPS-C-P-015	Novel photoinitiators based on Irgacure 2959 and methacrylates Sesil Çınar, Duygu Avcı, Viktorya Aviyente
RDPS-C-P-016	Synthesis of Hyperbranced Polymer Using Benzodioxine Chemistry Cansu Aydogan, Mustafa Ciftci, Volkan Kumbaracı, Naciye Talınlı, Yusuf Yagcı
RDPS-C-P-017	Photoinitiated Radical Polymerization with Iodonium Salts Bearing Polymerizable Anions Annett Halbhuber, Bernd Strehmel
RDPS-C-P-018	Synthesis of 1D tubular polymers by a supramolecular polymerization approach Hao Yu, Robert Häner

RDPS-C-P-019	Synthesis of Amide Linkage Directly from Aldehyde and Amines by Visible Light Photocatalysis Naeem Iqbal
Recent Developments in Polymer Synthesis D-Condensation Polymerization and Thermoset	
RDPS-D-P-001	Synthesis, Characterization and Spectroscopic Investigation of Novel Nanoporous Pyrazinoporphyrazine Network Polymer-Supported Metal (II)-Based Catalysts Hamada H Abdel Razik, Basim H Asghar
RDPS-D-P-002	Epoxy Functionalized Hyperbranched polyurethane and its co-curing with benzoxazine: cure kinetics and thermal properties Yichao Wang, Xue Jiang, Yuhong Liu
RDPS-D-P-003	Expression of Chirality in All-conjugated Block Copolymers without any excess of chiral centers Marie-Paule Robert Van Den Eede, Anjan Bedi, Guy Koeckelberghs
RDPS-D-P-004	Blocked triazolinediones for easy handling, heat curable network formulations Lucie Imbernon, Hannes A. Houck, Hannes A. Houck, Filip E. Du Prez
RDPS-D-P-005	Synthesis and characterization of amphiphilic Polyglycerol sebacate derivatives Merve Gultekinoglu, Kezban Ulubayram
RDPS-D-P-006	Synthesis, Characterization and Investigation of Thermal Properties of POSS Containing Graphene/Polycaprolactone Composites Fehmi Saltan, Aylin Bilgili, Hakan Akat
RDPS-D-P-007	SYNTHESIS and CHARACTERIZATION of WATERBORNE, BRANCHED, FUNCTIONAL POLYURETHANES for TEXTILE COATING APPLICATIONS Ayşe Durmuş, Deniz Anıl, Yusuf Menceloğlu, Serkan Ünal
RDPS-D-P-008	Synthesis and Characterization of Novel Cyclomatrix-Type Polyphosphazene Microspheres Cross Linked With Octachlorocyclotetraphosphazene via Self-Assembly Approach Simge Metinoglu, Yasemin Süzen
RDPS-D-P-009	Preparation and Characterization of Tyramine Based Inorganic-Organic Hybride Polyphosphazene Microspheres Simge Metinoglu, Yasemin Süzen
RDPS-D-P-010	Investigating cure advancement of a dual-cure Polyurethane-Acrylate (PUA) coatings used for metal substrates Babak Kaffashi, Mohsen Sarrafi
Polymer Characterization by New and Combined Techniques	
PCNCT-P-001	Characteristics and synthesis of fluorene copolymer thin films via plasma polymerization Utku Özenir
PCNCT-P-002	Polyether Polyols as GPC Calibration Standards for Determination of Molecular Weight Distribution of Polyether Polyols Mohd Azmil Mohd Noor, Vahid Sendijarevic, Hazimah Abu Hassan, Ibrahim Sendijarevic, Tuan Noor Maznee Tuan Ismail, Seng Soi Hoong, Nurul 'ain Hanzah, Razmah Ghazali
PCNCT-P-003	Determination of linear programmed temperature gas chromatography retention indices for some plasticizers Farhi Halaimia, Imane Lakehal
PCNCT-P-004	Sorbent for concentrating lead(II) Rafiqa Alirza Aliyeva, Nazila Sadiq Huseynova, Ulviya Murshud Abilova, Famil Musa Chiragov
PCNCT-P-005	preparation and application of biocompatible carrier implant to be used in the controlled purchase of clindan Hesna Ural Kayalık, Çınar Erdem, Sema Çetin
PCNCT-P-006	preparation and aplication of biocompatible carrier implant to be used in the controlled purchase of vincristine Hesna Ural Kayalık, Şule Balcı, Sema Çetin
PCNCT-P-007	Effects of seed coating(by polymer A200) on germination speed of Astragalus adscendens under different moisture conditions and Planting depth in the Boroujerd region Hamidreza Mehrabi, Mandana Rezayee
PCNCT-P-008	Application of electrochemically-prepared polypyrrole as heavy metals sensor Lo Momath, Sall Lamine Mohamed, Diaw Karim Diagne Abdou, Niang Matar, Fall Mamadou, Sall Guingue Diariatou, Aaron Jacques Jean, Chehimi Mehdi Mohamed
PCNCT-P-009	Correlating Crystallization Kinetics and Rheological Properties of Polyethylene Using a Newly Developed Low-Field RheoNMR Combination Mürüvvet Begüm Özen, Karl Friedrich Ratzsch, Volker Raentzsch, Iakovos Vittorias, Nehir Kavak, Patrick Kurt Dannecker, Michael A. R. Meier, Manfred Wilhelm
PCNCT-P-010	Influence of Conformation of Hyaluronic Acid in Solution on Proton Mobility Gökce Mihlac, Davide Ret, Stefano Gentilini, Simone Knaus
PCNCT-P-011	Synthesis and monomer reactivity ratios of 3-(trimethoxysilyl) propyl methacrylate-co-N-vinyl pyrrolidone copolymer Mansor B. Ahmad, Ameen Hadi Mohammed, Nor Azowa Ibrahim, Norhazlin Zainoddin
PCNCT-P-012	Well-defined polyethylene-based graft terpolymers by combining nitroxide-mediated radical polymerization, polyhomologation and azide/alkyne "click" chemistry Nazeeha Suliman Alkayal, Hakan Durmaz, Umit Tunca, Nikos Hadjichristidis
PCNCT-P-013	Functionalization of nanocarbons for using as fillers with epoxy resin Sonam Tamang, Sabita Shreshtha, André Wutzler, Ralf Lach, Wolfgang Grellmann, Rameshwar Adhikari
PCNCT-P-014	Depth-sensing Macroindentation Test and Stepped Isothermal Method – Accelerated Assessment of the Local Retardation Behaviour of Thermoplastic Polymers Ralf Lach, Tobias Krolopp, Jan Schoene, Stephan Arndt, Beate Langer, Wolfgang Grellmann
PCNCT-P-015	Impact of Block Length and Temperature Over Self-assemling Behavior of Block Copolymers Samia A Kosa, Laila M. Al Harbi, Musa Kaleem Baloch, Irfan Ullah, Elsayed H. El Mossalamy
PCNCT-P-016	Ink Identification by ToF Secondary Ion Mass Spectroscopy Dimitre Karpuzov

PCNCT-P-017	Separation and Characterization of Branching of Water-Soluble Polyelectrolytes by Capillary Electrophoresis in the Critical Conditions Jean Baptiste Henri Lena, Gregory Thomas Russell, Patrice Castignolles, Joel Thevarajah, Alison Maniego, Marion Gaborieau
PCNCT-P-019	Composite of Polylactic Acid and Monoterpenic Phenols Elif Güneş
PCNCT-P-020	Synthesis, characterization and kinetics of adsorption of metal ions onto copoly(amide-thioamide)s Tabak Ghezalla, Nassima Khelllafi, Fatima Zohra Ramdane
Macromolecules & Nanotechnology	
MN-P-001	Preparation of nanometer CoFe2O4 by co-precipitation using different precipitants and its characterization Yasmina Bellal Hammiche, Laaldja Boukhobza Meddour, Amar Djadoun, Nadia Henda, Ahmed Ezzine Hassanine, Fateh Mernache, Mohamed Sennour, Marie Hélène Berger, Abdennour Meddour, Aline Auroux
MN-P-002	Preparation and characterization of cobalt spinel ferrite obtained by hydrothermal treatment Yasmina Bellal Hammiche, Laaldja Boukhobza Meddour, Amar Djadoun, Amel Benadda, Marie Hélène Berger, Mohamed Sennour, Aline Auroux
MN-P-003	In-Situ Preparation of Thermoset/Clay Nanocomposites by Thiol-Epoxy Click Chemistry Ozlem Purut, Mehmet Atilla Tasdelen
MN-P-004	Synthesis and characterization of nanocomposite materials based on poly(4-Vinylpyridine) and two different Algerian clays: via in situ polymerization Fayçal Dergal, Djahida Lerrari, Ali Mansri, Kheldoun Bachari
MN-P-005	Synthesis and Application of Anisotropic Particles Huarong Liu, Fangwei Wang, Ruikun Wang
MN-P-006	Effect of precursors influence on the distribution of active fractions of nanocatalysts Tokhir Khakim Rakhimov, Mukhtorjon Ganievich Mukhamediev
MN-P-007	Core shell carbon nanotubes/silver monohybrids coated with polypyrrole designed via interfacial chemistry of aryl diazonium salts and ultrasonic pathway Mekki Ahmed, Achour Sabrina, Ihdene Zaher, Mahmoud Rachid, Chehimi Mohamed.mehdi
MN-P-008	Large-Area Reduced Graphene Oxides with Excellent Thermal Conductivity and Electromagnetic Interference Shielding Effectiveness Soon Man Hong, Pradip Kumar, Chong Min Koo
MN-P-009	The Preparation of Some New 1H-Tetrazole Derivatives and Their Attaching on Nano Spheres Nurdan Kurnaz Yetim, Elvan Hasanoğlu Özkan, Murat Gümüş, Ali Dişli, Nurşen Sarı
MN-P-010	Dendrimers: Synthesis, Immobilization and Applications of Pesticide Nurdan Kurnaz Yetim, Nurşen Sarı
MN-P-011	Immobilizaiton of Acetylcholinesterase on Macromolecule for Pesticide Control Elvan Hasanoğlu Özkan, Nurdan Kurnaz Yetim, Murat Gümüş, Ali Dişli, Nurşen Sarı
MN-P-012	Synthesis of Novel Dendron Based on Ferrocene, Immobilization of Enzyme and First Investigation as Catalytic of Glucose in Artificial Urine Elvan Hasanoğlu Özkan, Nurşen Sarı
MN-P-013	Polymeric submicron particles via self-assembly and crosslinking of double hydrophilic poly(ethylene oxide)-b-poly(N-vinyl pyrrolidone) in aqueous solution Jochen Willersinn, Markus Antonietti, Bernhard V. K. J. Schmidt
MN-P-014	Synthesis, Characterization and Antimicrobial Evaluation of Silver nanoparticles embedded Alkyd resin derived from Neem seed oil Haruna Musa, Aminu Magaji
MN-P-015	New complex compounds of functionalized Dehydroacetic acid units and N,N dimethyl amino benzaldehyde: the synthesis, catecholase study and electrochemical properties Tabti Djitli Salima, Djedouani Amel, Djedouani Amel, Rahmouni Samra, Romdhane Samir
MN-P-016	Preparation and Evaluation of a Novel Bio-Based Waterborne Polyurethane Nanoparticles for Sustained Delivery of Raloxifene-Hydrochloride Mohammad Reza Nabid, Niloofar Babanejad, Ismail Omrani
MN-P-017	Polymerization of preorganized amphiphilic assemblies Saheli Chakraborty, Subramanium Ramakrishnan
MN-P-018	Antimicrobial and antifungal potential of ecofriendly synthesized silver nanoparticles using Lagenaria siceraria fruit Amara Dar, Parsa Dar, Usama Waqas
MN-P-019	Sensing of Crack and Crack Size in CNT Buckypapers Tulay Bal Demirci, Elif Avcu, Fatih Turan, Koray Gürkan, Yener Taşkın, Alaattin Aktaş
MN-P-020	Preparation and research of multiphase systems: sodium silicate solute/polymer component/nanoparticles Shukhrat Kurbanbaev, Sukhrob Telyaev, Olga Trunilina, Sirojiddin Mirzaev
MN-P-021	Thermal resistant cellulose fiber via graphene/polyvinyl phosphonic acid nanocomposite Zoha Nooralian, Mazeyar Parvinzadeh Gashti, Zahra Saeedi, Izadyar Ebrahimi
MN-P-022	Synthesis of Gold Nanoparticles with Core Cross-Linked Micelles Gökhan SOLMAZ, Gökhan Kocak, Vural Bütün
MN-P-023	In-Situ Preparation of Cr2O3 Nanoparticles by Photopolymerization Technique Hicret Kırtay, Nergis Arsu
MN-P-024	Polysiloxane Nanofilament Growth on Glass Esra Kasapgil, E. Gozde Atici, Ilke Anac, H. Yildirim Erbil
MN-P-025	Magnetically activated microcapsules – preparation and characterization Malgorzata Natalia Kostrzewska, Søren Hvilsted, Anne Ladegaard Skov
MN-P-026	Thermal Conductive Polymer Nanocomposites Containing Boron Nitride Tuba Erdogan Bedri

MN-P-027	Photochemically produced self-assembled Ag nanoparticles on calf thymus DNA chains Eyüp Metin, Gönül Saadet Batıbay, Neşe Atacı, Nergis Arsu
MN-P-028	Syntesis of Modified Multi Walled Carbon Nanotubes with Hydroxy Substituted Schiff Bases and Investigation of the Adsorption Properties Tulay Bal Demirci, Elif Avcu
MN-P-029	Amphiphilic Polymer-Enabled Nanometal Synthesis, Morphogenesis, and Colloidal Stabilization Paschalis Alexandridis, Toshio Sakai
MN-P-030	Poly(2-hydroxyethyl methacrylate) nanocomposite hydrogel filled with fluorescent halloysite nanotubes Bruno F Urbano, Libnny Belmar, Leandro Toledo, Susana Sánchez
MN-P-031	A new insight into producing electromagnetic shielding cotton via graphene spraying Zahra Saeidi, Mazeyar Parvinzadeh Gashti, Zoha Noralian, Izadyar Ebrahimi
MN-P-032	Photochemically Prepared Silver/Polymer Nanocoatings by using 2-(Carboxymethoxy)thioxanthone Yasin Ogün Beydoğan, Melisa Adıbelli, Nergis Arsu
MN-P-033	Synthesis and Characterization of a Polymer with Unsymmetrical Zinc Phthalocyanines as Terminal Group Yunus A Sütcüler, Tuba Ç Çanak, Altuğ M Sevim, Ersin I Serhatlı, Ahmet Gül
MN-P-034	Synthesis of a Novel Internal Emulsifier Derived from plant Oil for the Preparation of Bio-based Polyurethane Nanoparticles and Their Application in Coatings Mohammad Reza Nabid, Ismail Omrani, Niloofar Babanejad, Hasan Kashef
MN-P-035	Photochemical preparation of gold and platinum nanoparticles Gönül Saadet Batıbay, Nergis Arsu
MN-P-036	Polymer Supported Ligand for Visual Detection of Trace Levels of Hg(II) & Pb(II) Ions in Aqueous Media Roya Sedghi, Susan Kazemi, Bahareh Heidari
MN-P-037	Study of anti-cancer drug release (tamoxifen) of the nanofibers made of poly-caprolactone -chitosan Zahra Saeidi, Ali Ashjaran, Seyed Ali Reza Dabirsiyaghi
MN-P-038	Chiral Porphyrin Macrocyclic Catalysts for the Processive Oxidation of Polymer Substrates Shaji Varghese, Alan E. Rowan, Roeland J. M. Nolte
MN-P-039	One-step formation of w/o/w multiple emulsions stabilized by the mixture of biopolymers Soumia Seddari Kalache
MN-P-040	Biomineralization-inspired synthesis of calcium iodate/gelatin nanocomposites by gel growth approach Naeimeh Dehghan, Mazeyar Parvinzadeh Gashti
MN-P-041	The Use of Gold Nanoparticle Conjugated Porous Glycidyl Methacrylate Microparticles For Phenol Removal Eda Aşık, Ali Tuncel, Ayşegül Aşkın
MN-P-042	A homopolymer grafted-graphene oxide nanosheet in superior stress transfer in immiscible polymer blends Goutam Prasanna Kar, Suryasarathi Bose
MN-P-043	POLYMER BASED CARBON NANOTUBES AS GAS SENSOR: ACETONE and TOLUENE Elif Baysak, Muge Yucel, Umit Hakan Yildiz, Gurkan Hizal
MN-P-044	Electrospray Finishing of Alcohol Repellent Materials on Non-woven Fabrics and Investigation of Performance Properties Semih Özgeç, Ali Kireçci
MN-P-045	In-situ Preparation of Poly(2-ethyl-2-oxazoline)/clay Nanocomposites via Living Cationic Ring-Opening Polymerization Umut Ugur Ozkose, Cagatay Altinkok, Ozgur Yilmaz, Onur Alpturk, Mehmet Atilla Tasdelen
MN-P-046	Facile Modification Of Reactive Electrospun Nanofibers Using "Click" Chemistry Özlem Ipek Kalaoğlu Altan, Rana Sanyal, Amitav Sanyal
MN-P-047	Nano Silica-Polyurethane Composite Materials for Coating Application Mohammed A Bahattab, Jose M Martin Martinez
MN-P-048	Preparation of mixed polyacrylonitrile-polycaprolactone nanofiber mats for seperation purposes Sündüz Alemdar, Cansu Çelik, Nursel Pekel Bayramgil
MN-P-049	Silane Modified Surfaces for Biological Imaging by AFM Nilüfer Kıvılcım, Fatma Bilge Emre, Turgay Seçkin
MN-P-050	Hydrophilic Nano SiO2 – based Composite Hydrogels for Biomedical Purposes Özge Fatma Gökmen, Nursel Pekel Bayramgil
MN-P-051	Synthesis, Characterization and Photocatalytic Activity of Halloysite-TiO2 Nanocomposites Özlem Karahan, Nalan Bilgin Öncü, Alp Yürüm, Güliz Inan Akdemir, Cleva W. Ow Yang, Ismail Koyuncu, Yusuf Menceloğlu, Serkan Ünal
MN-P-052	Fluorescent electrospun nanofibers from dansyl-functional [poly(methyl methacrylate-co-(2-hydroxethylmethacrylate)]2-b-poly(ethylene glycol) terpolymer Hilal Kuday, Okan Günaydın, Mesut Görür
MN-P-053	Superior osseo activity on TiO2 via polymer brush-strontium coatings Sergio E Moya, Danijela Gregurec, Nikolaos Politakos
MN-P-054	Low Density Polyethylene Films Coated with Essential Oil Loaded Halloysite Nanotubes by Layer-by-Layer Technique for Antibacterial Surfaces Buket Alkan, Ekin Şehit, Cüneyt Erdinç Taş, Serkan Ünal, Fevzi Çakmak Cebeci, Yusuf Ziya Menceloğlu, Hayriye Ünal
MN-P-055	Study of effect of alumina on aluminium characteristic properties in composite alloys Amel Bourbia, Hayette Bedboudi, M. Yacine Dibili
MN-P-056	Preparation of PEG Modified Single-Walled Carbon Nanotubes: CNT and PEG Chain Length Effect Ayhan Ünlü, Mehdi Partovi Meran, Gürkan Hızal, Nilgün Yavuz, F. Seniha Güner
MN-P-057	Corrosion behaviour of DC magnetron sputtered aluminium coatings Bahroune Moufida, Mosbah Asma, Guenfoud Fatma, Saker Abdelhamid
MN-P-058	Preparation of Paper-based Biosensor for Detection of Phenol Using Laccase Hybrid Nanoflowers Cansu Ozkardas, Cevahir Altinkaynak, Serife Sacmaci, Nalan Ozdemir, Ismail Ocsoy

MN-P-059	Effect of the preparation method on structural properties of magnetite Laaldja Boukhobza Meddour, Yasmina Bellal Hammiche, Amar Djadoun, Nadia Henda, Ahmed Ezzine Hassanine
MN-P-060	Structure-property relationship in the PVC composites reinforced with carbon nanotubes Ali Reza Zanjanijam, Mohammad Bahrami, Morteza Hajian, Maryam Ramezani
MN-P-061	Modification of Halloysite Nanotubes with Styrene-Maleic Anhydride Copolymers Refik Arat, Nurseli Uyanık
MN-P-062	The effect of nanoclusters on the mechanical strength of dental composites Şeref Okay, Aslı Dörtler, Nadir Kiraz, Meltem Asiltürk
MN-P-063	Solid-State Synthesis of a New Core– Shell Chiral Nanocomposite of Poly [(±)-2-(sec-butyl) aniline] and Nanosilica- Supported Perchloric Acid as Dopant Solid Acid Ali Reza Modarresi Alam, Abdolkarim Farrokhzadeh
MN-P-064	The Effect of Different Hydrophobic Agents on Staining and Color Change of Dental Composites Aslı Dörtler, Şeref Okay, Nadir Kiraz, Meltem Asiltürk
MN-P-065	Multifunctional EPDM/PP-g-MA/Bioengineering Polyesters/Colloidal Poly(MA-alt-1-dodecene)-g-SiO2/Organoclays Nanocomposites by Reactive Extrusion Bayram Ali Göçmen, Zakir M. O. Rzayev, Deniz Demircan, Gunay Kibarer
MN-P-066	Electrochemical Preparation of Polyaniline nanoparticles Abdunnaser Mohamed Etorki, Ibrahime Shaban Shaban, Manal Ali Elharby
MN-P-067	In-Situ Preparation of Ag/Cu Alloy Nanoparticles by Photopolymerization Technique Hicret Kırtay, A. Nadir Sevinç, Nergis Arsu
MN-P-068	Heterogeneous Activation of Oxone by Graphene Oxide for the Photodegradation of Rhodamine B Under Visible Light Irradiation Hasan Yolcu, Rabah Boukherroub

Macromolecules in Biotechnology & Medicine

MBM-P-001	Applications of the functionalized liquid crystal double-emulsion hollow droplets prepared by microfluidics Soo Young Park, Jong Kyu Kim, Ju Hyun Jang, In Seok Huh
MBM-P-002	The Physical and Chemical Properties of some Strained Molecules Related to Biphenylene Ali Larkem, Hakim Hamada, Hamama Larkem
MBM-P-003	Impact of diesel exhaut particulates (DEPs) on the respiratory system Bouzid Mohamed, Djadi Amina, Bezzazi Boudjema
MBM-P-004	Synthesis of photopolymerizable macromers giving novel poly(β-amino ester)s for biomedical applications Ece Akyol, Duygu Avcı
MBM-P-005	Hyperbranched Glycerol Modified β-cyclodextrin: Effect of the Macromolecular Topology on the Encapsulation and Release of Phenolphthalein Yanyan Ru, Yunyun Lv, Yuhong Liu
MBM-P-006	Flurbiprofen Derivative Loaded Nanostructured Lipid Carriers: Synthesis of New Flurbiprofen Derivative Gulce Taskor, Nezire Saygili
MBM-P-007	Polyurethanes based on castor oil and chitosan as potencial biomaterials in tissue engineering Said José Arévalo, Claudia Alejandra Ramírez, Manuel F Valero, Luis E Diaz
MBM-P-008	Phosphonate/phosphonic acid end-modified poly(b-amino ester)s for pH-sensitive applications Mirac Tatlıyüz, Ece Akyol, Duygu Avcı Semiz
MBM-P-010	Novel bisphosphonate and bisphosphonic acid-functionalized methacrylates for dental applications Melek Naz Guven, Duygu Avcı Semiz
MBM-P-011	Structural Analysis of Octenyl Succinic Anhydride (OSA) Esterified Starch from Plants Senay Simsek, Kristin Whitney
MBM-P-012	In-vitro Application Studies of Breast Cancer Treatment Beads on Drug Loaded Stimuli-Responsive Microgel/Liposome Biomaterials Covered with Magnetic Nanoparticles which Provides a Way to Carry All System on a Target Damla Ülker, Vural Bütün
MBM-P-013	Enzyme Immobilization on Polymer-based Supports Dilek Nartop, Murat Güleç, Birtane Demirel, Elvan Hasanoğlu Özkan, Nurdan Kurnaz Yetim, Nurşen Sarı
MBM-P-014	Sol-Gel Coating For Inhibition Monomer Release From Acrylic Denture Base Selda Sezer, Ahmet Gultek, Ismail Hakki Uzun, Nilufer Tulin Polat, Turgay Seckin
MBM-P-015	Synthesis and characterization of poly(vinylphosphonic acid-co-acrylic acid) copolymers for application in bone tissue scaffolds Rebecca Dey, Xia Zhong, Peter Youle, Qi Guang Wang, Ian Wimpenny, Sandra Downes, Judith Hoyland, David Watts, Julie Gough, Peter Budd
MBM-P-016	Polyelectrolyte complexes of poly(4-vinylpyridine)-g-poly(ethyleneglycol) with c-myc antisense oligonucleotide as a potential non-viral vector Murat Topuzoğulları, Damla Gökkaya, Mehmet Murat Özmen
MBM-P-017	Construction of novel cholesterol biosensors by using dithienopyrrole type conducting polymer Huseyin Bekir Yildiz, Nilay Gazel, Semahat Kucukkolbasi
MBM-P-018	A novel application for polyDADMAC in biopharmaceutical production Almut Rapp, Sarah Koegler, Joerg von Hagen
MBM-P-019	Development of polyvinyl chloride/silver nanoparticle composites through different fabrication techniques and their performance as antimicrobial substrates Muhammad Anwaar, Emel Yilgör, Iskender Yilgör
MBM-P-020	CoFe2O4 -FRUCTOSE-METHOTREXATE BASED ANTICANCER DRUG DESIGN Idil Karaca Açarı, Süleyman Köytepe, Burhan Ateş, Sevgi Balcıoğlu, Turgay Seçkin, Ismet Yılmaz
MBM-P-021	The study of binding modes of Thioxanthone-Diamin with calf thymus DNA Rabia Halıcı, Eyüp Metin, Neşe Atacı, Nergis Arsu

MBM-P-022	Synthesis of Novel Thermosensitive Poly(Amine Functionalized Glycolide)-Poly(Lactide) - PEG Copolymers and Their Applications in Drug Delivery Systems Sezgi Erdoğan, Mehmet Onur Arıcan, Olcay Mert
MBM-P-023	Syntheses and nanoformulations of novel boron based poly(diisobutyl glycolide) homopolymers and their copolymers with lactide for drug delivery systems Mehmet Onur Arıcan, Aybüke Huri Kandıra, Gökhan Duruksu, Ufuk Yıldız, Asgar Kayan, Olcay Mert
MBM-P-024	Synthesis and Characterization of a New Class of Boron Based Asymmetric Poly(Substituted Glycolides) for Biomedical Applications Ayşenur Vardar, Mehmet Onur Arıcan, Ufuk Yıldız, Asgar Kayan, Olcay Mert
MBM-P-025	Synthesis of Tween-Based Polyurethane Adhesives Capable of Adhering in Moisture Environment Burhan Ates, Suleyman Koytepe, Merve Goksin Karaaslan, Unzile Kelestemur, Sevgi Balcioglu, Selam Gulgen
MBM-P-026	Synthesis and evaluation of bisphosphonate-containing poly(β-amino ester) macromers for biomedical applications Seçkin Altuncu
MBM-P-027	Hydrogels for biomedical applications using biodegradable oligomers of poly(lactic acid) Marcos Lopes Dias, Felipe Augusto Moro Loureiro, Iryna Grafova, Andriy Grafov
MBM-P-028	Synthesis and In vitro Evaluation of Poly(2-((2-aminoethyl)amino)ethyl methacrylate) as a Potential siRNA Delivery Agent Aykut Zelçak, Volga Bulmuş
MBM-P-029	Synthesis, Characterization and in vitro Evaluation of Arginine Polymers Damla Taykoz, Deniz Uğur, Volga Bulmuş
MBM-P-030	Immobilization of L-asparaginase onto PMMA Composites Activated with Starch Ahmet Ulu, Suleyman Koytepe, Burhan Ates
MBM-P-031	Preparation, characterization and antibacterial study of pHEMA (soft contact lens) hydrogel incorporated with boric acid Ahmet Ulu, Rukiye Keskin, Emre Birhanlı, Sevgi Balcıoglu, Suleyman Koytepe, Burhan Ates
MBM-P-032	Electrospun Biocompatible Core/Shell Structure Nanofibers with Potential Application for Local Cancer Therapy & Wound Healing Roya Sedghi, Alireza Shaabani
MBM-P-033	Zinc Strontium Chelated Chitosan - Brushite Composite Gel Fibers for Bone Tissue Engineering Eda Ayse Aksoy, Betul Suyumbike Yagci, Gulseher Manap, Sukru Ozturk, Kezban Ulubayram
MBM-P-034	The Stabilization of Camptothecin Anticancer Drug via Phosphazene Cored and Miktoarm Star-Shaped Polymeric Nanocarriers Duygu Çetin, Mehmet Onur Arıcan, Erdinç Doğancı, Nursel Olgaç, Mesut Görür, Faruk Yılmaz, Olcay Mert
MBM-P-035	Shape-persistent polymersomes possess tunable permeable membrane and functional-antennae as multifunctional devices for biomedical applications Mohamed A. Yassin, Dietmar Appelhans, Hanns Achim Temme, Brigitte Voit
MBM-P-036	Development of Tween 60/Boric Acid-based Antibacterial Polyurethane Biomaterials Including Hydrophobic Surface for Wound Dressing Application Merve Gökşin Karaaslan, Emre Birhanli, Selam Gülgen, Suleyman Koytepe, Burhan Ates
MBM-P-038	Purification of a phosphated biopolymer by selective ethanol precipitation in presence of SDS and sodium acetate Rodrigo Gabriel Simas, Raphael De Marco, Joaquin Cabrera Crespo, Mickie Takagi
MBM-P-039	Preparation of Theophylline Loaded Pectin-Zeolite Based Wound Dressing Materials Banu Arabacıoglu Kocaaga, Ozge Kurkcuoglu, Melkon Tatlıer, F. Seniha Güner
MBM-P-041	Biological properties evaluations for some partially alicyclic copolyimides with pendant carboxyl groups Dumitru Popovici
MBM-P-042	Polycationic Polymeric Brush Decoration on Polyurethane for Dual Effective Antibacterial Activity Merve Gultekinoglu, Yeliz Tunc Sarisozen, Ceren Erdogdu, Meral Sagiroglu, Yoo Jin Oh, Peter Hinterdorfer, Eda Ayse Aksoy, Kezban Ulubayram
MBM-P-043	The Comparison of Isothermal and Hydrolytic Degradability Properties of Non Aromatic Polyurethanes Prepared for Medical Applications Canbolat Gurses, Hakan Dastan, Merve Goksin Karaaslan, Selam Gulgen, Suleyman Koytepe, Burhan Ates
MBM-P-044	Computational and Experimental Approaches to Improve Minocycline Imprinted Hydrogels for Controlled Drug Release Büşra Eroğlu, Dilek Dalgakiran, Ozge Kurkcuoglu, F. Seniha Güner
MBM-P-045	Biochemical Characterization of Recombinant L-Arabinofuranosidase Emine Erdoğan Özgeker, Alper Akkaya
MBM-P-046	Bioconjugation of Microbial Rennet with Dextran Sulfate Mithat Celebi
MBM-P-047	Surface Modifiable Dendrimers as Efficient Drug Carriers of Sulfamethoxazole (SMZ) as Model Drug Mustafa Ulvi Gürbüz, Ali Serol Ertürk, Metin Tülü
MBM-P-048	Synthesis and anticancer effect of different monosaccharides-DOX conjugates on MCF-7 breast cancer cell Idil Karaca Açarı, Süleyman Köytepe, Burhan Ateş, Sevgi Balcıoğlu, Turgay Seçkin, Ismet Yılmaz
MBM-P-049	In vitro cytotoxicity of PEG modified single-walled carbon nanotubes Mehdi Partovi Meran, Ayhan Ünlü, Ebru Hacıosmanoğlu, Gürkan Hızal, Nilgün Yavuz, Pelin Deniz Akkuş, Özge Kürkçüoğlu, F. Seniha Güner
MBM-P-050	Sodium Alginate Films for Future Wound Dressing Applications, Effect of Crosslinking on Swelling, Mechanical Properties and Release of Curcumin Candan Kılınçkale, Funda Kaşıkçı, Zeynep Kurt, Hüceste Çatalgil Giz
MBM-P-051	The Release Of Curcumin From Alginate Beads In Different Media Esra Aksoy, Melisa Berberoglu, Cagla Ergun, Huceste Catalgil Giz
MBM-P-052	Delivery Kinetics and Preparation of Rosemary Essential Oil Formulation Microencapsulated with Alginate/Starch Incilay Gökbulut, Fatma Sezer Öztürk, Burhan Ates, Hakan Parlakpınar
MBM-P-053	Synthesis of Tetrazine Functional Hyperbranched Polymers Mustafa Yasin Ateş, Ahmet Çetinkaya, Muhammet Ubeydullah Kahveci

MBM-P-054	SURFACE MODIFICATION AND CHARACTERIZATION OF NANO SiO2 NANOPARTICLES FOR BIOLOGICAL APPLICATIONS Elif Ceyda Kanpara, Fatma Özge Gökmen, Nursel Pekel Bayramgil
MBM-P-055	Labeling of Proteins with Trans-cyclooctene for Tetrazine Ligation Ahmet Çetinkaya, Mustafa Yasin Ateş, Muhammet Ubeydullah Kahveci
MBM-P-056	In Situ Synthesis of Tetrazine Functional Polymers Sinem Sipahioglu, Sinem Sipahioglu, Mustafa Yasin Ates, Ahmet Cetinkaya, Muhammet Ubeydullah Kahveci
MBM-P-057	Targeted Delivery of Anti-Angiogenic Chemotherapy Agents Merve Karaçivi, Rana Sanyal
MBM-P-058	Phosphopeptide Enrichment From Human-Serum via Immobilized Metal Affinity Chromatography Using Phosphonic Acid Functionalized Capillary Monolith as a Stationary Phase in a Microfluidic System Kouroush Salimi, Fatos Cigdem Kip, Omur Celikbicak, Duygu Deniz Usta, Bekir Salih, Ali Tuncel
MBM-P-059	Evaluation of toxicity essential oils of aromatic plant (Rosmarinus Officinalis) on the pédofaune Karima Baba Aissa, Hadjer Rabah, Kamel Moussaoui, Zahr Eddine Djazouli
MBM-P-060	"Development Of Novel Bilayer Chitosan-nHap Composite Biomaterial For Guided Bone Regeneration" Sedef Tamburacı, Özge Karadaş, Engin Özçivici, Funda Tıhmınlıoğlu
MBM-P-061	Development, Characterization and In vitro cell culture studies of Zein Bilayer Composites as Bioactive Wound Dressing Material Seda Gunes, Sedef Tamburaci, Funda Tihminlioglu
MBM-P-062	Production And Characterization Of Novel Bilayer Chitosan-Zein Based Composites For Osteochondral Tissue Regeneration Sedef Tamburacı, Berivan Çeçen, Hasan Havıtçıoğlu, Funda Tıhmınlıoğlu
MBM-P-063	Effect of pH on Drug Release Kinetics for PHEMA/Itaconic Acid Imprinted Hydrogels Dilan Bozkurt, Dilek Dalgakiran, Seniha Guner
MBM-P-064	Metal ion carrying magnetic monodisperse-porous silica microspheres as an immobilized metal affinity chromatography (IMAC) sorbent for protein separation Duygu Deniz Usta, Kouroush Salimi, Ali Tuncel
MBM-P-065	The influence of the washing step on the drug loading performance of the p(hema) hydrogels Dilek Dalgakıran, Fatma Seniha Güner
MBM-P-066	The effect of using antibacterial agent and size of the inorganic micro fillers on flexural strength of the composite materials and their antibacterial activity Zerin Yeşil Acar, Nadir Kiraz, Meltem Asiltürk
MBM-P-067	New synthesis and reactivity of 4-hydroxy-6-methyl-2h-pyran-2-one Benmohamed Soumaya, Rachedi Yahia, Hamdi Maamar, Dumas Françoise
MBM-P-068	Development of polymeric organic capsules as a platform for opioids prolonged release Shumaila Razzaque, Tan Bien
MBM-P-069	Preparation of different biopolymer chitosan films with enzyme immobilization for biotechnological application Entsar I Rabea, Mohamed E. I. Badawy
MBM-P-070	The investigation of polymorphs and eutectic mixture of progesterone and estradiol benzoate loaded in silicone rubber by differential scanning calorimetry (DSC) Maryam Ramezani, Farhid Farahmandghavi, Mohammad Imani
MBM-P-071	Anti-angiogenic Drug Conjugated Linear Dendritic Triblock Copolymers as Micellar Drug Delivery Systems Burcu Sumer Bolu, Ece Gecici, Rana Sanyal, Rana Sanyal
MBM-P-072	The study of binding modes of 2,2'-Naftoin with ct-DNA Eyüp Metin, Betül Fatma Erkoç, Aslı Tuğla, Elif Özçelik, Nergis Arsu
MBM-P-073	Biodistribution of Radiolabelled Polymeric Nanoparticles Obtained by PISA Approach in HT1080 Tumor Xenografts Sadik Kaga, Nghia Truong, Danielle Senyschyn, John Quinn, Rana Sanyal, Lisa Kaminskas, Michael Whittaker, Thomas Davis

Energy, Optics, & Optoelectronics

EOO-P-001	Non-Covalently Bonded Polymers Induced by π-Stacking: Synthesis of New Tribenzopentaphenes Derivatives and Investigation of their Properties Bassam ALAMEDDINE, Christine Luscombe, Titus Jenny
EOO-P-002	Effect of the Addition of Additives on the Improvement of the Performance of Lead–Acid Batteries Foudia Malika, Zerroual Larbi, Deafa Imad
EOO-P-003	Effect of solution slow rate on growth and characterization of nanostructured ZnO thin film deposited using spray pyrolysis Chabane Sari Sidi Mohammed, Zargou Salem, Senoudi Rachida Assia, Aida Mohammed, Attaf Nasreddine, Hakem Ilhem Faiza
EOO-P-004	Alkyl-end-group Modification of Benzotriazole and Thiophene Containing Conjugated Polymers Şevki Can Cevher, Duygu Keleş, Şerife Özdemir Hacıoğlu, Levent Toppare, Levent Toppare, Levent Toppare, Levent Toppare, Ali Çırpan, Ali Çırpan, Ali Çırpan, Ali Çırpan
EOO-P-005	Air electrode based on conducting polymer for enhanced Li-air batteries Yong Joon Park, Seon Hye Yoon, Dae Ho Yoon
EOO-P-006	1,1'-Biferrocenylene Based D-A-D Oligomers Okan Demirel, Rochus Breuer, Mehmet Emin Cinar, Michael Schmittel, Turan Ozturk
EOO-P-007	[3+3] Imine and β-ketoenamine tethered fluorescent covalent-organic frameworks for CO2 uptake and nitroaromatic sensing Kaleeswaran Dhanaanjayan, Pratap Vishnoi, Ramaswamy Murugavel
EOO-P-008	Triphenylmine and Tetraphenylethylene Substituted Thienothiophene for OLED Applications Garen Suna, Sebahat Topal, Turan Öztürk
EOO-P-009	Boron and Tetraphenylethene Substituted Thienothiophene Possessing Molecules for OLED Applications Pelin Ulukan, Turan Öztürk
EOO-P-010	Greener Reaction for High Electron Mobility Copolymer: Naphthalene Diimide Copolymers via Direct Arylation Polycondensation with Controlled Molar Mass and High Performance Rukiya Matsidik, Hartmut Komber, Alessandro Luzio, Mario Caironi, Michael Sommer

EOO-P-011	Improvement of capacitive performances of supercapacitors by incorporation of exfoliated montmorillonite into conducting polymers Damia Amoura, Margarita Sánchez Jiménez, Francesc Estrany, Laid Makhloufi, Carlos Alemán
EOO-P-012	Synthesis and Optoelectronic Properties of Mono-Alkylated Thieno[3,2-b]Thiophenes and Their Polymers Koray Tansu Ilhan, Sebahat Topal, Turan Ozturk
EOO-P-013	Donor-Acceptor Systems Possessing Tetraphenylethylene Substituted Dithienothiophene-S,S-dioxides for OLED Applications Rengin Büşra Özek, Gözde Aydın, Turan Öztürk
EOO-P-014	preparation and caracterisation by X ray diffraction of $LiFexCo(1\text{-}x)O_2$ Ouartsi Ibticem
EOO-P-015	From Polymers to High-Capacity Anode Nanostructures for Lithium-ion Batteries Serap Hayat Soytas, Ali Ansari Hamedani
EOO-P-016	Ferrocene-Containing Polythiophene and Its Use as a Cathode Material for Li-ion Batteries Mesut Görür, Muhammet Aydın, Faruk Yılmaz
EOO-P-017	Polythiophene with Densely Populated Organic Robust Radicals as Cathode Material in Li-Ion Batteries Muhammet Aydın, Mesut Görür, Faruk Yılmaz
EOO-P-018	A post-functionalizable carbazole based low bandgap polymer for organic photovoltaic applications Deniz Kiymaz, Ceylan Doyranlı, Hakan Bilgili, Burak Gültekin, Pieter Verstappen, Jurgen Kesters, Wouter Maes, Ceylan Zafer, Dirk Vanderzande, Sermet Koyuncu
EOO-P-019	Highly Efficient Poly(fluorene phenylene) Graft Copolymer as a New Class of Binder for High-Capacity Silicon Anode in Lithium-ion Batteries Neslihan Yuca, Mehmet E. Cetintasoglu, Murat F. Dogdu, Huseyin Akbulut3, Sevcan Tabanli, Uner Colak, Omer Suat Taskin, Yusuf Yagci
	Polymer and Polymer-Based Membranes
PPBM-P-001	Surface modification of polyethersulfone/aminated polyethersulfone membranes by the self-assembly with antibacterial materials Chang Keun Kim, Lakwon Choi, Eun Yub Choi, Kwang Young Park
PPBM-P-002	Metal-Organic Frameworks: Coordination Polymeric Nanofibers Alper Inan, Nesrin Horzum
PPBM-P-003	Synthesis, Characterization and study of rheological behavior of copolymer hydrogels based on poly (N-isopropylacrylamide-co-acrylamide) Seddiki Nesrinne, Aliouche Djamel
PPBM-P-004	Formation of Cellulose Acetate Membranes: Influencing Structure and Performance Jakob Tolk, Annette Reiche
PPBM-P-005	Synthesis and characterization of highly fluorinated polytriazoles copolymers with controlled degree of sulfonation for proton exchange membranes Asheesh Singh
PPBM-P-006	Effect of trityl substituted triphenylamine group on gas transport properties Soumendu Bisoi, Susanta Banerjee
PPBM-P-007	Morphology and thermal stability of nanocomposite polymer membranes containing methylimidazolium ionic liquid Oscar Javier Osorio Pedroza, José Carlos Dutra Filhio, Victor Jayme Roget Rodriguez Pita, Marcos Lopes Dias
PPBM-P-008	Radiation grafting of crosslinked anion exchange membranes for bipolar membrane fuel cells Enver Güler, Selmiye Alkan Gürsel
PPBM-P-009	Preparation of Polysulfone-Modified Fiber Structure Sepiolite Hybrid Particles for Ultrafiltration Membrane Betül Hanbeyoğlu Yavaş, Nilgün Kızılcan, Birgül Benli
PPBM-P-010	The impact of graphene flake size on the properties of polymer/graphene formulations and coatings Oana Mihaela Istrate, Simon R. Gibbon, Jonathan Moghal, Peter M. Budd, Ian A. Kinloch
PPBM-P-011	Comparasion of Bioresorbable Adhesion Barrier Films which Produced from Polylactic Acid Mehmet Buğra Güner, Afife Binnaz Hazar Yoruç
PPBM-P-012	From cellulose tissue papers to nanoporous membrane actuators Huijuan Lin, John W. C. Dunlop, Jiayin Yuan
PPBM-P-013	Study of synthetics dyes adsorption onto poly (vinyl alcohol)/glutaraldehyde/β-cyclodextrin polymer membranes Djamila Ghemati, Djamel Aliouche
PPBM-P-014	Preparation and properties of graphene oxide/polyimide nanocomposites for simultaneous quantification of serotonin and dopamine Öznur Göngör, Aziz Paşahan, Süleyman Köytepe, Büşra Aksoy, Turgay Seçkin
PPBM-P-015	A polyimide film: Synthesis, Characterization and its sensitive biosensor application Süleyman Köytepe, Aziz Paşahan, Öznur Güngör, Büşra Aksoy, Turgay Seçkin
PPBM-P-016	Synthesis and Characterization of H_3PO_4 doped triazole containing proton conductive polymer for PEMFCs Önder Çelik, Tansel Şener, Ümit Kadiroğlu, Ali Ekrem Müftüoğlu, Faruk Yılmaz
PPBM-P-017	Lithium salt doped poly(MMA-co-HEMA) conductive polymer electrolyte for lithium ion batteries Ipek Karaaslan, Önder Çelik, Ümit Kadiroğlu
PPBM-P-018	PVDF Flat Sheet Membranes Comprising Antimicrobial N-Halamine Compounds Fatma Ozkan, Hasan Basri Kocer
PPBM-P-019	Glycerol Carbonate and Polyethylene Oxide Containing Borate Ester Based Polymer Electrolyte Membranes Esra Eren, Aysel Demirci, Sümeyye Köybaşı, Yunus Emre Özkara, Betül Karataş, Yunus Karataş
PPBM-P-020	Waterproof – Breathable Antibacterial Polyurethane Membranes Ahmet Aydın, Fatma Ozkan, Hasan Basri Kocer
PPBM-P-022	PS-b-PNIPAM block copolymers for thermoresponsive membrane applications Merve Cetintas, Marleen Kamperman

PPBM-P-024	Synthesis and Characterization of Imidazole Derivatives as Polymer Electrolyte Membranes For Fuel Cell Applications Aysel Demirci, Burak Gündüz, Sahl Sadeghi, Enver Güler, Selmiye Gürsel, Yunus Karataş
PPBM-P-025	Crosslinked proton-exchange membranes by radiation induced grafting of 4-vinylpyridine and divinylbenzene onto poly(ethylene-co-tetrafluoroethylene) films Sahl Sadeghi, Enver Guler, Ege Oyku Cerit, Selmiye Alkan Gursel
PPBM-P-026	Chitosan Based Nanocomposites for Bioactive Food Packaging Seda Gunes, Funda Tihminlioglu
PPBM-P-027	Synthesis and Characterization of Highly Branched, Functional Poly(Arylene Ether Sulfone)s for Industrial Wastewater Purification Membrane Applications Emine Billur Sevinis, Aybige Ozture, Ipek Sacligil, Turkan Ormanci Acar, Turker Turken, Fevzi Cakmak Cebeci, Derya Imer, Ismail Koyuncu, Yusuf Menceloglu, Serkan Unal
PPBM-P-028	Modification of Reverse Osmosis (RO) Desalination Membranes by Zwitterionic Silane Coatings Selda Erkoc Ilter, Serkan Guclu, Farzin Saffari, Derya Imer, Ismail Koyuncu, Serkan Unal, Yusuf Menceloglu
PPBM-P-029	A Novel UV- Curable Monomer for Coating Applications Yeşim Müge Şahin, Mustafa Hulusi Uğur
PPBM-P-030	Membrane filtration of Sudan orange G on a cellulose acetate membrane filter for separation–preconcentration and spectrophotometric determination in water, chili powder, chili sauce and tomato sauce samples Zeid A Alothman
PPBM-P-031	Preparation and characterizations of biopolymer chitosan-based membranes as edible coatings in food preservation technology Mohamed E I Badawy, Entsar I. Rabea
PPBM-P-032	Development and Characterization of Poly (Arylene Ether Sulphone) –Zeolite Beta Based Polymer Electrolyte Membranes for Direct Methanol Fuel Cell (DMFC) Applications Olga Ece Atlan, Erde Can, Nurcan Bac
PPBM-P-033	Comparative behavior of styrene polymerization with Cp2Ti Cl2 and Cp2Zr Cl2 based catalyst in presence of anionic and cationic clay as support Ahmed Benaboura, Khadidja Belalem
PPBM-P-034	Synthesis, Characterization and Dopamine Selective Properties Of Polyurethane Films Containing Chlorogenic Acid Ensar Erel, Merve Gökşin Karaaslan, Süleyman Köytepe, Burhan Ateş, Ismet Yılmaz
PPBM-P-035	Graphene Oxide-Polymer Based Membranes and Aerogels Öznur Doğan, Ayşe Elif Kıratlı, Sevil Demirci, Erhan Bat

Smart and Functional Polymers

SFP-P-001	Amphiphilic polypeptides containing a hydrophobic luminescent terminal: probing critical aggregation concentration and lowest critical solution temperature by aggregation-induced emission Jin Long Hong
SFP-P-002	Characterization, Synthesis of New Benzoylthiourea Polymers and Thermal Properties Gülşah Kurt Şahan
SFP-P-003	Elaboration in ionic liquids and characterization of polyaniline applied to nitrite ions detection Abdoulaye Diarisso, Modou Fall, Cheikh Ahmad Elkabir Lô, Noureddine Raouafi, Ramzi Zarrougui
SFP-P-004	Molecularly imprinted HPLC column salid-phase material design for Dermatan Sulfate/Chondroitin Sulfate Separation Zihni Onur Uygun, Ferhan Sağın, Burcu Okutucu, Şükriye Hacıkara
SFP-P-005	Electrochemical synthesis of poly (4-aminobenzenesulfonic acid) applied to lead-contaminated wastewater treatment Matar Niang, Mamadou Fall, Mohamed Lamine Sall, Momath Lo, Abdou Karim Diagne Diaw, Diariatou Gningue Sall
SFP-P-006	Functional UV Coating Composite For Steel Coating Hye Jin Yoo
SFP-P-007	Acid/base sensing layers on the surface of polypropylene fibers by grafting poly(acrylic acid)/poly(3,4-ethylenedioxythiophene) Xingping Wang, Si Meng, Wujun Ma, Zhe Zhou, Bin Sun, Zongyi Qin, Mengfang Zhu
SFP-P-008	Synthesis, modification and controlled release of poly(isobutylene vinly ether-co-itaconic anhydride Fatma Zehra Dikici, Ahmet Okudan, Makbule Bilge Sağkan
SFP-P-009	Induction of Chiral Structures to Conjugated Polymers Using Circularly Polarized Light Tamaki Nakano, Yue Wang
SFP-P-010	PDMS: a stimuli-responsive polymer for RFID sensing Kate E. Belsey, Adam V. S. Parry, Mohamed A. Ziai, Christina V. Rumens, Stephen G. Yeates, John C. Batchelor, Simon J. Holder
SFP-P-011	Synthesis of a Photo-Cleavable Core Prepared by Passerini Reaction and Preparation of an ABC 3-Mictoarm Star Polymer with This Core Gözde Deveci, Muhammet Ubeydullah Kahveci
SFP-P-012	Synthesis and Self-Assembly of High Molecular Weight Polystyrene-block-poly[2-(N-morpholino)ethyl methacrylate]: A Story about Microphase Separation, Amphiphilicity, and Stimuli-Responsivity Steffen Eggers, Felix Lauterbach, Volker Abetz
SFP-P-013	PEG-Based Reactive Copolymers via Azide-Alkyne Click Cycloaddition-Mediated Step-Growth Polymerization Mehmet Arslan, Özgül Gök, Rana Sanyal, Amitav Sanyal
SFP-P-014	Smart Windows Application of Carbazole and Triazine based Star Shaped Architecture Merve Guzel, Metin Ak
SFP-P-015	Shape-memory Polymers - How to Bring Hard and Soft Segments Together Hannes Schäfer, Katharina Koschek
SFP-P-016	Reversible ring opening reaction of a model 1,3-benzoxazine with thiols Tobias Urbaniak, Marc Soto, Katharina Koschek
SFP-P-017	Acryl amide based thermoresponsive copolymers: total control over the lower critical solution temperature Niklas Lucht, Steffen Eggers, Volker Abetz
SFP-P-018	"Studies On Poly(Acrylate)S Containing Pendant Ligand With Schiff Base Azo Compound And Its Divalent Metal Complexes" Keerthiga Rajendiran

SFP-P-019	Introducing upper critical solution temperature to polymer multilayer films Cansu Ustoglu, Eda Cagli, Irem Erel Goktepe
SFP-P-020	Castor oil-derived polyurethanes engineered by dynamic crosslinks for self-healing Marc Comí
SFP-P-021	"SYNHESİS,CHARACTERİZATİON AND ANTİBACTERİAL ACTİVİTY WİTH 8 -HYDROXY QUİNOLİNE OF POLY(ISOBUTYLENE VINLY ETHER -co-ITACONIC ANHYDRİDE) " Bilge Makbule Sağkan, Ahmet Okudan, Fatma Zehra Dikici
SFP-P-022	Controlled radical ring opening polymerisation for smart and degradable polyesters Jens Gaitzsch, Wolfgang Meier
SFP-P-023	Synthesis of stimuli responsible metallosupramolecular poly-L-lysine as smart polymer structure Imren Özcan, Süleyman Köytepe, Turgay Seçkin
SFP-P-024	Synthesis and characterization of new cationic photoinitiator Betül Fatma Erkoç
SFP-P-025	Synthesis and characterization of thermal stable metallo-supramolecular polymers bearing phosphazene unit Selda SEZER, Süleyman Köytepe, Ahmet Gültek, Turgay Seçkin
SFP-P-026	Effect of supramolecular assembly and hydrophilicity on supramolecular temperature sensor Kanykei Ryskulova, Patrice Woisel, Richard Hoogenboom
SFP-P-028	Hybrid microspheres from zirconium butoxide and an ionic-non ionic copolymer Bruno F Urbano, Leandro Toledo, Jorge Yañez, Cristian Campos, Bernabé L Rivas
SFP-P-029	Synthesis and Characterization of NIPAAm-based Responsive Thin Layers Anil Bozdogan, Esra Kasapgil, Ilke Anac
SFP-P-030	Poly(Ionic Liquid)s-Derived Nitrogen-Doped Porous Carbons for Environmental Remediation and Catalysis Jiang Gong, Huijuan Lin, Markus Antonietti, Jiayin Yuan
SFP-P-031	Functionalized Chitosan by RAFT for Pharmaceutical Formulations Catalina Natalia Yilmaz, Sinem Yaprak Karavana, Onur Yilmaz
SFP-P-033	Synthesis of alkoxysilane functional copolymers via RAFT polymerization technique Onur Yılmaz, Ersin Önem, Hasan Özgünay, Catalina Natalia Cheaburu Yılmaz, Hüseyin Ata Karavana, Ali Yorgancıoğlu
SFP-P-034	Release of Doxorubicin from Layer-by-Layer Films of Poly (2-Isopropyl-2-Oxazoline) Meltem Haktanıyan, Süleyman Atilla, Eda Çağlı, Irem Erel Göktepe
SFP-P-035	Synthesis and characterization of some polyurethanes bearing pyridine-metal complex unit Süleyman Köytepe, Birgül Köstek, Imren Özcan, Turgay Seçkin
SFP-P-036	Synthesis of novel supramolecular gels containing terpyridine-metal complexes bonding units Süleyman Köytepe, Birgül Köstek, Büşra Aksoy, Imren Özcan, Turgay Seçkin
SFP-P-037	Thermoresponsive PEGMA Polymer Nanoparticles: A New Colloidal Drug Carrier Mehmet Sahin Atas, Ayse Nur Ozkaya Balcı, Mustafa Selman Yavuz
SFP-P-038	New coatings based on fluoropolymer/ clay nanocomposites prepared by in situ polymerization: Synthesis and characterizations Raihane Mustapha, Karamane Mohamed, Yagçi Yusuf, Lahcini Mohamed, Ilsouk Mohamed, Atilla Tasdelen Mehmet
SFP-P-039	Solubility behavior of poly(oligo(ethylene glycol) methacrylate)s in water: Effect of different additives Maryam Bozorg, Birgit Fischer, Volker Abetz
SFP-P-040	"Synthesis, Modifications and Characterization of New -Type-TER-Polymer Containing Vinly Propionate " Şerife Civil, Ahmet Okudan, Bilge Makbule Sağkan, Fatma Zehra Dikici
SFP-P-041	Polypyrrole coating on cotton textile as adsorbent for methylene blue dye Mohamad Mohamad Ayad, Wael Amer, Sawsan Zaghlol, Boberc Boberc, Jaroslav Stejskal
SFP-P-042	Tailor-made Functionalized Polymers as a Basis for the Formation of Supramolecular Structures Elisabeth Wittenberg, Volker Abetz
SFP-P-043	A Novel Red to Transmissive Electrochromic Polymer Based on Phenanthrocarbazole Gizem Atakan, Görkem Günbaş
SFP-P-044	Functionalized Polydimethylsiloxanes for specific adhesion Robin Heedfeld, Xiaomin Zhu, Martin Möller
SFP-P-045	Introduction of Catalytical Property to the PAMAM type Dendrimers Metin Tülü, Kevser Harmandar, Mustafa Ulvi Gürbüz
SFP-P-046	Dispersible Multi-Walled Carbon Nanotubes via Non-Covalent Interaction with Functional Copolymers Busra Cengiz, Amitav Sanyal
SFP-P-047	Molecularly Imprinted Microspheres for the Adsorption of Chlorpyrifos Wei Yao, Yanjun Fang, Hong Yin, Zhixian Gao
SFP-P-048	Synthesis and Characterization of pH-Responsive Zwitterionic ABC Triblock Copolymers and Intermediary Layer Cross-Linked Micelles Agung Ari Wibowo, Gökhan Solmaz, Vural Bütün
SFP-P-049	Synthesis and Study of Chemical resistance of Short Bamboo fiber reinforced epoxy composites Anu Gupta
SFP-P-050	Microphase Separation in supramolecular poly(1-hexene) with self-complementary hydrogen bonding along the backbone Saeed Houshmandmoayed, Mohammad Hossein Jandaghian, Mostafa Ahmadi, Mohammad Mahdi Mortazavi
SFP-P-051	Chemical Preparation and Characterization of New Biodegradable Poly(ε-caprolactone) formed from Diosgenin and Cholesterol Sevinc Ilkar Erdagi, Erdinc Doganci, Faruk Yilmaz, Cavit Uyanik
SFP-P-052	Specific ion effects on the friction and adhesion for zwitterionic polymer brushes Yunlong Yu, G. Julius Vancso, Sissi De Beer
SFP-P-053	Semi-interpenetration with Conducting Polyanilines on Poly(N-isopropylacrylamide) Hydrogels by Radiation Induced Graft Polymerization: Synthesis, Characterization and Their Application in the Controlled Drug Delivery Systems Betül Taşdelen, Selçuk Poyraz, Aslıhan Koruyucu

SFP-P-054	Synthesis and Characterization of Self Healing Polydimethylsiloxane-graft-Poyurethane Ali Sabri Berkem, Turgut Nugay, Ilke Anac
SFP-P-055	Controlled release of Paclitaxel from multilayers of block copolymer micelles with temperature-responsive coronae and pH-responsive cores Eda Cagli, Irem Erel Goktepe
SFP-P-056	Highly fluorescent porphyrin-functional styrene copolymer nanofibers Sümeyra Bayir, Mesut Görür, Erdinç Doğancı, Tamer Uyar, Faruk Yılmaz
SFP-P-057	Synthesis, Kinetic, Equilibrium and Thermodynamic Studies on Adsorption of Cu(II) by Thiosemicarbazide Chitosan Mudasir Ahmad
SFP-P-058	Synthesis of Core Cross-Linked Micelles from a Triblock Copolymer Containing Poly(glycidyl metharylate) Functional Block Vural Bütün, Gokhan Solmaz, Gokhan Kocak
SFP-P-059	Polypyrrole functionalized by 4-amino-3-hydroxynaphthalene sulfonic acid for removal chromium hexavalent Mohamed Lamine Sall, Momath Lo, Matar Niang, Abdou Karim Diagne Diaw, Diariatou Gningue Sall
SFP-P-060	Benzoxazole Derived Triphenylamine Based Polyimides: Synthesis, Characterization and Photophysical Properties Humaira Masood Siddiqi, Asma Iqbal, Toheed Akhtar, O.ok. Park
SFP-P-061	Polysulfones with Pendant Benzoxazine Units by Click Chemistry Cemil Dizman, Çağatay Altınkök, Mehmet Atilla Taşdelen
SFP-P-062	Magnetically responsive controlled drug release of vinblastine-loaded chitosan particles Keng Shiang Huang, Chih Hui Yang
SFP-P-063	Polyethyleneimine modified Halloysite embedding p(Acrylamide) Cryogel Composites Sultan Butun Sengel, Selin Sagbas, Nurettin Sahiner
SFP-P-064	The Synthesis and Characterization of Self Healing Materials Based on Thioxanthone Disulfides Umut Koyunucu, Ömer Tahir Günkara, Nergis Arsu

Renewable Resources and Biopolymers

RRB-P-001	Potassium hydroxide pulping of rice straw in biorefinery initiative Md Sarwar Jahan, Fahmida Haris, Md Mostafizur Rahman, Purabi Rani Samaddar
RRB-P-002	Influence of Temperature and Itaconic Acid on Emulsion Copolymerization of Butyl Acrylate and Methyl Methacrylate in Semi-Batch Process Leonardo Zborowski Sobrinho, Daniela Beirao Porto, Jesus Roberto Taparelli, Lucia Helena Innocentini Mei
RRB-P-003	Synthesis of new resins based on epoxidized sunflower oil via acrylation reaction and study of their biodegradability Naima Belhaneche Bensemra, Ratiba Irinislimane
RRB-P-004	Improved anaerobic digestion in the thermophilic phase by a thermal pretreatment Kheiredine Bani, Derbal Kerroum, Benchikhlehocine Mosaab
RRB-P-005	The Kinetic of Biogas Production Rate from Dairy Wastewater in Batch Mode «Effect Inoculums to Substrate Ratio in thermophilic Phase » Kheiredine Bani, Benchikhlehoucine Mosaab, Derbal Karoum
RRB-P-006	Mechanical characterization of reversible hydration of PLA based materials María Martín, Freddys Beltrán, María Ulagares De La Orden, Joaquín Martínez Urreaga, Vicente Lorenzo
RRB-P-007	Mechanical properties of biocomposites based on sunflower oil and alfa fibers Naima Belhaneche Bensemra, Sihem Kadem, Ratiba Irinislimane
RRB-P-008	Characterization of new biodegradable blends based on ethylene-butyl acrylate copolymers with thermoplastic starch Jesus Luis Pablos Lagartos, Ana Morro, Fernando Catalina, Teresa Corrales, Irma Marín, Concepción Abrusci
RRB-P-009	Utilization of chemically treated waste rubber as ion exchangers for industrial waste water treatments Yassin Abdel Latif Aggour, Ayed Saad Al Shihri
RRB-P-010	Effect of soybean oil in crosslink density and dynamic mechanical properties of natural rubber compounds with silica Carlos Henrique Scuracchio, Aline Zanchet, Janaina Da Silva Crespo
RRB-P-011	Design, synthesis and characterization of novel high bio based content dual-cure systems: high solids and water-dispersible Ozlem Kubra AKDOGAN, Swaroop Shendre, Vijay Mannari
RRB-P-012	Biodegradable polyester synthesis from sugar beet pulp pectin Hande Barkan, Cüneyt Hüseyin Ünlü, Oya Galioğlu Atıcı
RRB-P-013	A lignin-derived chemical platform for biobased polymers via ADMET Andrea Hufendiek, Sophie Lingier, Laetitia Vlaminck, Filip E. Du Prez
RRB-P-014	Fundamental Properties of Cellulose Derivatives Important for Thermoplastic Processing Yu Bai, James H. Wang, Bing Zhou, Jun Kuang
RRB-P-015	Production of chitosan-based conductive hydrogel by photopolymerization technique for biomedical applications Neslihan Alemdar, Celil Ulutürk
RRB-P-016	Synthesis of polyhidroxyesters derived from 10-undecenoic acid. Modification with aminoacids Carmen Valverde
RRB-P-017	Polyaniline/wheat stalk dust composites for effective removal of arsenic (V) from aqueous solution Santosh Khanal, Indu Khadka, Rameshwor Adhikari
RRB-P-018	Screening of various moderately Bacillus strains for PHA production Meliha Fetahović, Pınar Çağlayan, Ayşe Oğan, Özkan Danış, Meral Birbir
RRB-P-019	Urban Carpentry Waste Based Natural Fibers for Preparation of Reinforced Composites with Thermoplastics Rom Nath Baral, Rameshwar Adhikari
RRB-P-020	Synthesis and Characterization of Pectin Based Hydrogel Films: Effect of pH and Drug Loading Method on Drug Release Kinetics Özde Zeynep Güner, Banu Arabacıoğlu Kocaağa, F. Seniha Güner
RRB-P-021	Conducting Polymer Nanocomposites using Multi-walled Carbon Nanotubes in a Biodegradable Polymer Matrix Kedar Nath Dhakal, Santosh Khanal, Ralf Lach, Wolfgang Grellmann, Gert Heinrich, Rameshwar Adhikari, Kedar Nath Dhakal, Kedar Nath Dhakal

RRB-P-022	In situ Synthesis of Poly(butylene adipate)/Moroccan Beidellite clay Bionanocomposites by Polycondensation Raihane Mustapha, Ilsouk Mohamed, Castelvetro Valter, Lahcini Mohamed, Bronco Simona, Rhouta Benaissa, Conzatti Lucia, Bianchi Sabrina
RRB-P-023	Drug-Loaded Pectin Based Wound Dressing Hydrogels: Effect of Crosslinker Ratio and Drug Amount on Release Profile Cansu Çam, Banu Arabacıoğlu Kocaağa, Fatma Seniha Güner
RRB-P-024	Isocyanate- and phosgene-free synthesis of polyhydroxyurethane thermosets by employing multifunctional bio-based and organic/inorganic cyclic carbonates and bio-based amine hardeners Hannes Blattmann, Rolf Mülhaupt
RRB-P-025	The Some of Physical, Chemical and Mechanical Properties of Polyurethane Foams-contained Potato Crust Wastes Polyol-I Tülay Gürsoy, Aycan Gür, Mehmet Hakkı Alma, Ertuğrul Altuntaş, Eyyüp Karaoğul
RRB-P-026	The Some of Physical and Termal Properties of Polyurethane Foams-contained Potato Crust Wastes Polyol-II Tülay Gürsoy, Aycan Gür, Mehmet Hakkı Alma, Ertuğrul Altuntaş, Eyyüp Karaoğul
RRB-P-027	Polysilicon Acetals via Silicon Acetal Metathesis Polymerization (SAMP) Ersen Göktürk, Ertuğrul Şahmetlioglu, Stephen Albert Miller
RRB-P-028	Environmentally degradable biohybrids of MMT with polypyrrolidones from microbial itaconic acid Mohammad Asif Ali, Nupur Tandon, Seiji Tateyama, Tatsuo Kaneko
RRB-P-029	Effect Of Modified Chitosan Biopolymer On Biodiesel Synthesis Zafer Ömer Özdemir, Halil Mutlubas
RRB-P-030	Renewable Biocomposite Production from Agricultural Waste Biomass Müge Sennaroğlu Bostan, Selim Ceylan
RRB-P-031	Improvement of Water Sensitivity and Mechanical Properties of Xylan-Based Polymeric Materials Merve Akkus, Necati Ozkan, Ufuk Bakir
RRB-P-033	Biodegradable Copolyester Composites Reinforced by Wheat Stalk Microcrystalline Cellulose Jyoti Giri, Ralf Lach, Hai Hong Le, Wolfgang Grellmann, Hans Joachim Radush, Goerg H Michler, Jean Marc Saiter, Sven Henning, Rameshwar Adhikari
RRB-P-034	Experimental and Computational Studies of Ring Opening Polymerization of α-angelica lactone with Zinc Acetate Sıla Gümüştaş, Armağan Kınal, Mehmet Balcan
RRB-P-035	Biopolymer based Composite Hydrogel as Food Packaging Material Nabanita Saha, Kezban Kezban Akkaya, Ramasubba Reddy Palem, Ruhan Benlikaya, Petr Saha
RRB-P-036	Super-anisotropy in cyanobacterial hydrogels of supergiant polysaccharide SACRAN Maiko Kaneko, Kittima Amornwachirabodee, Saranyoo Sornkamnerd, Tatsuo Kaneko
RRB-P-037	Investigating the effect of PEG and diisocyanates on physical and viscoelastic properties of poly L-(lactic-acid) blended with gelatin Babak Kaffashi, Nazanin Shakoury

Polymer Engineering, Processing, and Characterization

PEPC-P-001	Exploring the Influence of Electron Beam Irradiation on Performance Properties of Blends Based on Ethylene Vinyl Acetate (EVA) and Thermoplastic Polyurethane (TPU) Joyeeta Dutta, Kinsuk Naskar
PEPC-P-002	Synthesis and characterization of poly (orthophenylenediamine) catalyzed by Maghnite-H+ Belmokhtar Abdelkader
PEPC-P-003	Development and evaluation of the porous structure and its effect on the electrical conductivity of conductive polymer composites Abdelhafid Merzouki, Naceredine Haddaoui
PEPC-P-004	Effects of Cross-linking Reaction and Film Formulation on the Structural and Degradation Characteristics of Corn Starch and Corn Starch Based Blended Films Ilknur Gönenç, Ferhunde Us
PEPC-P-005	Synergistic Effects of Titanium Dioxide Nanoparticle and Polylactic Acid on Enhancing Degradation of Low Density Polyethylene Films Tatiya Trongsatitkul
PEPC-P-006	Synthesis, structure and thermal properties of a new porous building material containing powder waste tires and cullet Azzedine Ayadi, Younes Lamri, Ratiba Benzerga, Fayrouz Benhaoua, Abdlali Aouina, Laurent Legendre
PEPC-P-007	Electrochemical properties of a novel composite material based on polyaniline modified by manganese dioxide (PANI + MnO2) Souhila Abaci
PEPC-P-008	Synthesis and characterization of copper/ poly (5- aldehyde-2,2',5',2"-terthiophjene) composite material and its application for electrocatalysis Souhila Abaci, Naima Maouche
PEPC-P-009	Structural property and Mechanical behavior of Polypropylene –Elemental sulfur (S8) composites: Effect of sulfur loading S Vijay Kumar, Kishore Kumar Jena, Saeed Al Hassan
PEPC-P-010	Sulfur Doped Graphene Polymer Nanocomposites for Electromagnetic Interference Shielding Chong Min Koo, Faisal Shahzad
PEPC-P-011	Interfacial interactions in polyamide/graphene nanocomposites Thomas Raine, Oana Istrate, Bernadette Craster, Ian Kinloch, Peter Budd
PEPC-P-012	Investigation on the thermal behaviors of various polymer/CO2 systems using high-pressure DSC Xia Liao, Erbo Huang, Qi Yang, Guangxian Li
PEPC-P-013	Mechanical and Dynamic Mechanical Properties of Devulcanized Rubber and Recycled PP Blends Carlos Henrique Scuracchio, Pamela Sierra Garcia, Sandra Andrea Cruz
PEPC-P-014	Investigation of mechanical properties of sea-shell-CaCO3/LDPE composites Şerife Pınar Yalçın, Loïc Le Pluart, Daniel Chateigner, Sophie Eve
PEPC-P-015	Microwave assisted synthesis and characterization of sodium alginate-graft-poly(N-hydroxymethyl acrylamide) Alper Akın, Nuran Işıklan

PEPC-P-016	Aging of composite insulators under the influence of intense electric fields and pollution Mourad Houabes, Renaud Metz
PEPC-P-017	Comparative study between adiabatic reactor and isotherm reactor-Application for the synthesis of methanol Rachida Chemini, Hafidha Belakeb
PEPC-P-018	Thermal, mechanical, microstructural and morphological properties of side chain LCP-HDPE graft coproducts Behiye Öztürk Şen, Ugur Soykan, Sedat Cetin
PEPC-P-019	Synthesis of new epoxy curing agent with flame retardant poperties via thiol-ene route and application as coating Nedime Özdemir, Hüseyin Esen
PEPC-P-020	Effect of accelerated ageing in oven on mechanical properties of thermoplastic elastomers Douglas Naue Simões, Michele Pittol, Daiane Tomacheski, Vanda Ferreira Ribeiro, Ruth Marlene Campomanes Santana
PEPC-P-021	Synthesis of Imidacloprid-β-Cyclodextrin Inclusion Complex and Its Controlled Release Characteristics in Polypropylene Filament Yarns Ilhan Özen, Ahmet C Turan, Husnu K Gurakin, Enrico Fatarella
PEPC-P-022	Synthesis and characterization of conducting polymers and copolymers coated paper substrate and modified by metal nanoparticles Dehbia Oukil, Wezna Hamouma, Abderrazak Hamam, Razika Aitout, Laid Makhloufi
PEPC-P-023	In-situ Synthesized Fe3O4 Decorated Polyaniline/RGO Based Flexible PVDF Composites Film with Excellent Electromagnetic Interference Shielding Efficiency Ranadip Bera, Sarbaranjan Paria, Sumanta Kumar Karan, Amit Kumar Das, Anirban Maitra, Bhanu Bhusan Khatua
PEPC-P-024	The Effects of Aromatic Boronic Acid on Thermal Characteristics of Polymers Involving Ester Linkages Nehir Utku
PEPC-P-025	Poly(lactic acid)/Poly(ethylene glycol) Blends, Nanocomposites and Fibers Esra Ozdemir, Jale Hacaloglu
PEPC-P-026	Assessment of improved and optimized technological process for manufacturing polymer packagings for pharmaceutical and cosmetic industry Margarita Natova, Ivan Ivanov, Vassil Georgiev, Filip Ublekov, Hristo Penchev
PEPC-P-027	Usability and Performance of Magnetite Filler as Polymeric Ionizing Electromagnetic Radiation Shield Ezgi Eren Belgin, Gül Asiye Ayçık
PEPC-P-028	Electrospun Rice Grain-shaped TiO2 Mesostructures Sensitized by CdS Quantum Dots for Photovoltaic Application Shengyuan Yang, Hongjun Deng, Meifang Zhu
PEPC-P-029	Performance of non-halogenated polymerizable phosphorus based flame retardant in polystyrene Hilal Gündüz, Hüseyin Esen
PEPC-P-030	A conducting polymer and a calixarene derivative: A novel surface design for glucose detection Tugba Ceren Gokoglan, Saniye Soylemez, Melis Kesik, Hande Unay, Serkan Sayin, Ali Cirpan, Huseyin Bekir Yildiz, Levent Toppare
PEPC-P-031	Synthesis and Characterization of Fluorene Based Random Copolymer with Benzothiadiazole and Benzoselenadiazole Seda Kutkan, Seza Goker, Serife O. Hacioglu, Levent Toppare
PEPC-P-032	New Heating Technology for Polymer Processing Sascha Bach, Marcus Stein, Ronald Claus, Jens Peter Majschak
PEPC-P-033	Ultrahigh Thermally Stable Poly(ether ketone)/MWCNT Composites as an Efficient Electromagnetic Shielding Material With Improved Mechanical Properties Sampat Singh Chauhan, Mathew Abraham, Veena Choudhary
PEPC-P-034	An effective surface design based on a conjugated polymer and silver nanowires for the detection of paraoxon in tap water and milk Janset Turan, Melis Kesik, Saniye Soylemez, Seza Goker, Sahin Coskun, Husnu Emrah Unalan, Levent Toppare
PEPC-P-035	Investigation Parameter Effects on Tribological Performance of Short Carbon Fiber and Pumice Particle Reinforced PPS/PA66 BLEND Alp Eren Şahin, Sinan Yilmaz, Tamer Sinmazçelik, Taner Yilmaz
PEPC-P-036	Investigation of Mechanical Properties of Polysulfone Blends Reyhan Ozdogan, Mithat Celebi, Mehmet Arif Kaya, Ozgur Ceylan
PEPC-P-037	Removal of Cu2+, Ni2+ and Au2+ ions from acidic solutions using polypyrrole conducting polymer by electrochemical cementation process Houa Hammache Makhloufi, Boualem Saidani
PEPC-P-038	Synthesis and Characterization of Hexakis(propylamino)cyclotriphosphazene and Hexaphenoxycyclotriphosphazene and Uses on Polyolefins as Flame Retardants Muzaffer Kaan Karaöz, Yunus Karataş, Teoman Tinçer
PEPC-P-039	Preparation and characterization of abundant biomass reinforced PVC composites Amel Mohamed Ben Ali, Ali Gasmi, Pierre Magri, Amar Boukerrou, Nesrine Amouchi
PEPC-P-040	Functional PEG building blocks via copolymerization of ethylene carbonate and functional epoxides Gent Kapiti, Helmut Keul, Martin Möller
PEPC-P-041	Production and Characterization of Polyethylene/Organoclay Nanocomposite Films for Food Packaging Ceren Alpaslan, Sennur Deniz
PEPC-P-042	Dynamic rheological properties and phase separation in of polyethylene/ poly(1-hexene) blends Saeed Houshmandmoayed, Kiana Salamat, Mohammad Mahdi Mortazavi, Ehsan Nikzinat, Reza Rashedi, Hannan Sepahi, Mostafa Ahmadi
PEPC-P-043	Tea Mill Waste as Reinforcing Fillers in Thermoplastic Composites Mustafa Kuyumcu, Alper Kasgoz, Mehmet Atilla Tasdelen
PEPC-P-044	Investigation of mechanical properties of organic - inorganic hybrid reinforced composites Mehmet Mudu, Halil Demirer, Mustafa Öksüz
PEPC-P-045	Characterization of spray polymerization products varified by specific operating parameters in a spray tower Katharina Franke, Werner Pauer, Hans Ulrich Moritz

PEPC-P-046	SYNTHESIS OF HYDROGEL SUPPORTED NHC-Pd(II)-Pyridine CATALYTIC SYSTEMS AND SUZIKI-MIYAURA COUPLING REACTION KINETICS Cihangir Boztepe, Sedat Yasar, Asim Kunkul, Ismail Ozdemir
PEPC-P-047	Effects of Polyhedral Oligomeric Silsesquioxane Reinforced Polypropylene (PP) Nanocomposites On the Thermal, Mechanical, and Optical Properties of PP Sevil Kaynar, Nurseli Uyanık, Mükerrem Çakmak
PEPC-P-048	Halloysite Nanotube/Polyethylene Nanocomposites with Ethylene Scavenging and Barrier Properties as Food Packaging Materials Cüneyt Erdinç Taş, Buket Alkan, Mustafa Baysal, Serkan Ünal, Fevzi Çakmak Cebeci, Yusuf Ziya Menceloğlu, Hayriye Ünal
PEPC-P-049	Effect of Surface Treatment and Cure Time on the Shear Strength of an Polydimethyl Siloxane Adhesive on Single-lap Joins for Aerospace Applications Nalan Özbay, Onur Dinçer, Hakan Der, Yusuf Ziya Karabay
PEPC-P-050	Real-Time Study on The Optical Properties of Polymer complexes of Hydroxypropylcellulose and 4'-Hydroxy-4-biphenylcarbonitrile Esin Kaya, Enmin Wang, Mükerrem Çakmak
PEPC-P-051	Investigation of Mechanical Properties of Recycling Polyolefins and Poly(lactic acid) Blends Mehmet Mudu, Mithat Çelebi, Mustafa Öksüz, Çağdaş Aslan
PEPC-P-052	First Si/Ge/Sn Heterotrimetallic Dendrimers Si(CH2)2Sn[(CH2)4MPh3]34 (M= Ge, Sn): Synthesis and Characterization Yılmaz Aksu
PEPC-P-053	Synthesis and Characterization of Organotin Dendrimers up to 3rd Generation Sn(CH2)nSn[(CH2)4SnPh3]34 (n) 3, 4): A Convergent Approach Yılmaz Aksu
PEPC-P-054	Study of the influence of whiskers and/or glass fiber concentration on mechanical, structural and thermal properties of polypropylene-based composites Maria Sönmez, Laurentia Alexandrescu, Mihai Georgescu, Daniela Maria Stelescu, Mihaela Nituica, Anton Ficai, Denisa Ficai, Dragos Gudovan, Roxana Trusca, Dana Florentina Gurau
PEPC-P-056	Non-isothermal crystallization and melting behavior of polypropylene in the PP/PVB blends containing PP-g-MA Ali Reza Zanjanijam, Shokoufeh Hakim, Hamed Azizi
PEPC-P-057	Study of Paint Sludge Emissions from Polymeric Composites Hamed Ghorbanpoor, Esra Akkulak, Hazal Gergeroglu, Lale Cıvan, Macid Nurbas, Huseyin Avci
PEPC-P-058	Morphology development in the PP/ plasticized PVB blends: Influence of content of the compatibilizing agent Ali Reza Zanjanijam, Shokoufeh Hakim, Hamed Azizi
PEPC-P-059	Enhancement of the thermal stability of the plasticized poly(vinyl chloride) using single wall carbon nanotubes Ali Reza Zanjanijam, Mohammad Bahrami, Morteza Hajian, Maryam Ramezani
PEPC-P-060	Characterization and investigation of synthesized Self-Colored PET(polyethylene terephthalate) based on Naphthalimide dye Ehsan Zamani, Alireza Khosravi, Mehdi Rafizadeh
PEPC-P-061	Controlling the process of synthesizing Self-Colored PET (polyethylene terephthalate) in the reactor Ehsan Zamani, Alireza Khosravi, Mehdi Rafizadeh
PEPC-P-062	Use of Rubber and Sawdust Wastes via Sodium Pentaborate, α-Sepiolite and Magnesium Hydroxide in Application of Radiation Isolation of Composites Filiz Balkaya, Zübeyde Kıvrak, Hamed Ghorbanpoor, Macid Nurbas
PEPC-P-063	Halogen-free flame retardants for poly(butylene terephthalate) composite Ceren Yargıcı, Mustafa Sezer, Nurseli Uyanık
PEPC-P-064	Mechanism Study on Electron-beam Assisted Stabilization of Polyacrylonitrile Fibers for Shortening the Conventional Thermal Treatment Sejoon Park, Seung Hwa Yoo, Ha Ri Kang, Seong Mu Jo, Han Ik Joh, Sungho Lee
PEPC-P-065	Diffusion (DOSY) NMR for fast molecular weight analysis of polyethylene furanoate (PEF) polyesters Jan-Georg Rosenboom, Giuseppe Storti, Massimo Morbidelli

Polymer Physics

PP-P-001	Diffuse reflectance spectroscopy of γ-irradiated ultra-high molecular weight polyethylene Syed Asad Maqbool, Saqlain Saqib Mukhtar, Muhammad Zeeshan Khalid, Yaqoob Khan, Mansoor A. Baluch, Tariq Yasin, Yaqoob Khan, Malik Sajjad Mehmood
PP-P-002	Effects of Blend Composition and Crystallization Temperature on the Spherulitic Morphologies of Crystalline/Crystalline Poly(ethylene suberate)/Poly(ethylene oxide) Polymer Blends with Similar Thermal Properties Zhaobin Qiu
PP-P-003	Volume phase transitions in ionic hydrogels Demet Aktaş, Gökhan Uçar, Başar Turan
PP-P-004	Effective chain length of additives on electrical conductivity of poly(3,4-ethylenedioxythiophene):poly(styrenesulfonate) films Hyuck Sik Wang, Jae Hyun Jung, Seok Hyeon Kim, Kigook Song
PP-P-005	Study of Optical Properties and Electrical Conductivity of Polymer Latex/Carbon Nanotube Composites Okan Yıldız, Saziye Ugur
PP-P-006	One-dimensional crystallization of 20-nm-width polymer nanorods Jaime Martín, Aurora Nogales, Natalie Stingelin, Carmen Mijangos
PP-P-007	Motion of Impurity ions caused by Inhomogeneity of Microstructure of Polymer Solid Yuichi Anada
PP-P-008	Investigation on glass transition temperature of electrospun PVAc fibers Konrad Domanski, Ece Yapaşan, Tuğba Isık, Nesrin Horzum, Mustafa Muammer Demir
PP-P-009	Polyacrylamide- Graphene Oxide Composites: Optical and mechanical properties with various GO contents Gülşen Akın Evingür, Önder Pekcan
PP-P-010	Physical, Mechanical and Adsorption Kinetic Properties of Graphene Oxide-Chitosan Composites Gülşen Akın Evingür, Nilay Kahya, Semira Bener, Bedia Erim Berker

PP-P-011	The universality on the gelation of NIPA-GO composites Gülşen Akın Evingür, Batuhan Ünver
PP-P-012	The swelling effect on the elasticity of NIPA-GO Composites Gülşen Akın Evingür, Özgül Öztürk
Industrial Polymers	
IP-P-001	Changes in mechanical and antimicrobial properties of thermoplastic elastomers after natural ageing Daiane Tomacheski, Michele Pittol, Douglas Naue Simões, Vanda Ferreira Ribeiro, Ruth Marlene Campomanes Santana
IP-P-002	Precipitated Styrene/Butyl-Acrylate Emulsion Additive as Modifier on Some Properties of Acrylonitrile-Butadiene Rubber Vulcanizates Adel Abd Elrehiem Koriem, Abbas Abd Elkarim Yehia, Mohamed Yossef El Zayat, Ahmed Ismail Hussain
IP-P-003	Study about thermal stability of the new biopolymer used in drill in fluids Genaro José Bolívar, Manuel Antonio Mas, María Virginia Tortolero, Mayra Garcia
IP-P-004	A New Technology for the synthesis of some interesting polymers on the industry Hamama Boularaf Larkem, Lidya Larkem, Ali Larkem
IP-P-005	Study of effect of an oxidizing agent on the degradation of polypropylene bags submitted to thermooxydation Farida Djerada Larfi, Kamel Khalil Bentifour, Samia Kaddour
IP-P-006	Synthesis and Application of a New Polymeric Adsorbent for Dye Removal of Colored Wastewater Maedeh Vafaee, Mohammad Ebrahim Olya, Behnaz Akbari
IP-P-007	Mechanical Properties of Thermotropic Liquid Crysttalline Copolyesters with PA66 Blends Melek Bulut, Zekeriya Yerlikaya
IP-P-008	Performance of redispersible polymer powders commercially used in cement based masonry mortars Gozde Camli, Yasemin Yetimoğlu Balk
IP-P-009	Benzoin Derivatives as UV Absorbers Tuğçe Çinko, Nergis Arsu, Duygu Sevinç Esen, Deniz Er, Ebru Ergüven
IP-P-010	Physicochemical Research Of The Effects Of Polymers Having Antipilling Effect On Different Fabrics Burcu Büyükkoru, Ali Kara
IP-P-011	Physcochemical Applications Of The Interaction Between Magnetic Polymers And Dyes Burcu Büyükkoru, Ali Kara
IP-P-012	Selective removal of toxic cationic textile dyes from wastewater using disulfide-linked polymer networks Mehmet Sahin Atas, Halit Cavusoglu, Mustafa Selman Yavuz
IP-P-013	Synthesis of New Polymer Additives from Waste Plastics for Asphalt Binder Emel Başkent, Savaş Gürdal, Serenay Akyol, Melike Ocak, Gülşen Albayrak Arı, Muzaffer Yaşar, Ramazan Oğuz Canıaz, Refika Çetintaş, Serhat Arca, Elif Kocaman, Ziya Köstereli
IP-P-014	Polymerization Reactions of Elemental Sulfur Emel Başkent, Savaş Gürdal, Serenay Akyol, Gülşen Albayrak Arı, Muzaffer Yaşar, Ramazan Oğuz Canıaz, Refika Çetintaş, Serhat Arca, Elif Kocaman, Ziya Köstereli
IP-P-015	Synthesis and applications of a pH-responsive cyclopolymers containing the residues of the amino acid aspartic acid Hasan Ali Al Muallem, Zakariyah Abdulkareem Jamiu, Shaikh Asrof Ali
IP-P-016	The effects of waste polymeric materials on the rheological properties of autoclaved aerated concrete Gurhan Gereli, Taner Kavas, Utku Tiriç, Uğur Uzgan, Abdullah Doğan Soyal
IP-P-019	Surface Characterization of Silane Modified Quartz by AFM Fatma Bilge Emre, Nilüfer Kıvılcım, Turgay Seçkin
IP-P-020	Effect of etoxylated surfactant content on the emulsion polymerization and water resistance performance of poly (EA-co-MMA) latexes Ecem Temelkaya, Cansu Akarsu Dülgar
IP-P-022	The role of nano-ZnO as an activator on rheological and tensile properties of rubber blends based on BR/SBR Behnaz Akbari, Fereshteh Motiee, Sasan Mosavi
IP-P-023	Development of High Solid Containing UV Curable Paint and Varnish Systems for Wood Coatings Alev Tüzün, Halil Pekedis, Deniz Er, Ebru Erguven Cakmak
IP-P-024	Mechanical properties and morphological characteristics of recycled PET fiber- glass fiber - PP hybrid composites Orkun Kaymakçı, Hülya Kiraz
Porous Polymer and Gels	
PPG-P-001	Chitosan-based hydrogel composites prepared by UV polymerization for wound healing Hong Ru Lin, Tseng Ping Sue, Yiu Jiuan Lin
PPG-P-002	PU-hydrogel composites prepared byγ-ray irradiation for wound healing Hong Ru Lin, Ga Hwa Lee, Yiu Jiuan Lin
PPG-P-003	Preparation and Properties of the Novel Photoluminescent Hydrogel Based on Fluorene Group Wen Fu Lee, Wang Xun Wu, Wei Chih Chen
PPG-P-004	Influence of the Cross-linker Type on the Internal Structure of Poly(N-Vinylcaprolactam) Microgels Synthesized by Precipitation Polymerization Michael Kather, Stefan Chang, Miriam Al Enezy Ulbrich, Jörg Stellbrink, Andrij Pich
PPG-P-005	Evaluation of a bio-hydrogel composed of nanostructured cellulose crosslinked genepin as tissue engineering scaffold: Fabrication, properties and application Asabuwa Ngwabebhoh Fahanwi, Ufuk Yildiz
PPG-P-006	Stimuli-induced volume phase transition of cationic copolymeric hydrogels based on N,N-dimethylaminoethly methacrylate Büşra Yıldız, Nermin Orakdöğen
PPG-P-007	Structutral and crystallization studies of Well-defined PEG-Networks Muhammad Haris Samiullah, Detlef Reicher, Tatiana Zinkevich, Joerg Kressler
PPG-P-008	Hierarchical Porous Ovalbumin-Chitosan Cryogel Scaffolds for Tissue Engineering Applications Tuğçe Şen, Berkay Özçelik, Mehmet Murat Özmen

PPG-P-009	Hydrogel based nutritional supplement tablets for health care Lenka Jelinkova, Nabanita Saha, Petr Saha
PPG-P-010	Swelling and thermodynamic studies of poly(hydroxyethyl methacrylate)-based smart hydrogels formed at various polymer concentrations Ahmed Kasim Durmus, Nermin Orakdogen
PPG-P-011	Synthesis of a New Natural Chitosan Based Hyrogel and Its Usage in Drug Delivery Applications Demet Aydınoğlu, Merve Ünal
PPG-P-013	Sulphobetaine-Based Hydrogels: Synthesis, Characterization, Investigation of Their Stimuli-Responsive Behaviours Hüseyin Şahin, Mehtap Kılınç, Cansel Tuncer, Vural Bütün
PPG-P-014	From functional polymer particles to fibers as substrates for metal catalysis Franziska Exnowitz, Marco Albuszis, Hannes Alex, Dr. Norbert. Steinfeldt, Dr. Werner Pauer, Prof. Dr. Meifang Zhu
PPG-P-015	Superporous hydrogels based on chitosan / salicylaldehyde systems Iftime Manuela Maria, Luminita Marin
PPG-P-016	Fabrication of microporous poly(lactic acid) monoliths via thermally induced phase separation and solvent exchange Ozgun Can Onder, Emel Yilgor, Iskender Yilgor
PPG-P-017	Supramolecular luminescent chitosan gels Bejan Andrei, Pinteala Mariana, Barboiu Mihai, Marin Luminita
PPG-P-018	Oxidative Cross-Dehdyrogenative Coupling of Terminal Alkynes with Thiols Using Cu(I) Incorporated Schiff Base Network Polymer as Catalyst Yonca Alkan, Ömer Suat Taşkın, Barış Kışkan, Yusuf Yağcı
PPG-P-020	Sulfur-Bridged Nanoporous Polymer Networks for CO2 Capture Mehmet Sahin Atas, Halit Cavusoglu, Mustafa Selman Yavuz
PPG-P-021	Synthesis and characterization of imprinted fluorescent stimuli-responsive gels for selective analysis of uranyl ion Elif Gökçe Atçakan, Orhan Güney
PPG-P-022	Elaboration of nanocomposites poy(N-isopropylacrylamide)/clay, and study adsorption of cationic dey Seddiki Nesrinne, Aliouche Djamel, Ghemataoui Nouara
PPG-P-023	" Tannin-Isocyanate based rigid foams " Zeyneb Hamadi, Francisco Jose Santiago Medina, Rabah Chaid, Antonio Pizzi
PPG-P-024	"Novel Functionalizable Cryogels With Degradable Linkers For Drug Delivery Systems" Duygu Aydin, Mehmet Arslan, Amitav Sanyal, Rana Sanyal
PPG-P-025	Porous thermoplastic polyurethane films by breath figure approach Gizem Daban, Cem Bayram, Murat Demirbilek, Emir Baki Denkbaş
PPG-P-026	Protein Release from Redox Responsive Hydrogels Ismail Altınbaşak, Amitav Sanyal, Rana Sanyal
PPG-P-027	Effects of gel aging on synthesis of boride ceramics Ruixing Li
PPG-P-028	Facile Synthesis of Diene-Containing Hydrogels for Reversible Functionalization Laura Chambre, Suzan Gunbay, Amitav Sanyal, Rana Sanyal
PPG-P-029	HETEROGENIZATION OF HOMOGENOUS NHC-Pd(II)-Pyridine CATALYTIC SYSTEMS VIA RADICAL POLYMERIZATION: CATALYTIC ACTIVITIES FOR SUZIKI-MIYAURA COUPLING REACTIONS Cihangir Boztepe, Ismail Ozdemir, Asim Kunkul, Nevin Gurbuz
PPG-P-030	Release of Riboflavin from Interpenetrated Network Poly(NIPAAm-co-AAc) Hydrogel and Release Kinetic Studies Sanogo Brahima, Cihangir Boztepe, Mehmet Yuceer, Asim Kunkul
PPG-P-031	Preparation and Characterisation of Rigid Phenolic Foams Cláudio G. dos Santos, Tobias Böhringer, Victor A. Oliveira
PPG-P-032	Gold nanoparticles embeded disulfide-linked polymer networks Mehmet Sahin Atas, Halit Cavusoglu, Mustafa Selman Yavuz
PPG-P-033	Gold nanoparticles embedded disulfide-linked polymer networks Halit Çavuşoğlu, Mehmet Şahin Ataş, Mustafa Selman Yavuz
PPG-P-034	Synthesis and characterization of sodium alginate/poly(acrylic acid) interpenetrating network films and calcium ion-crosslinked sodium alginate films Seda Bekin Acar, Shokat Sarmad, Gülten Gürdağ
PPG-P-035	Macroporous hydrogels composed entirely of synthetic polypeptides: biocompatible and enzyme biodegradable 3D cellular scaffolds Steven J. Shirbin, Fatemeh Karimi, Nicholas J. Chan, Daniel E. Heath, Greg G. Qiao
PPG-P-036	Synthesis and characterization of chemically crosslinked hydrogels to enhance the water holding capacity of sandy loam soil Misbah Abdul Ghafoor, Ijaz Ahmad Bhatti, Tariq Aziz
PPG-P-037	Synthesis of Template hydrophobically associating Polyacrylamide/Na+ Montmorillonite Babak Kaffashi, Maryam Khak, Mahmood Hemmati
PPG-P-038	In situ conductive polymer synthesis in neutral, cationic, and anionic cryogels and their potential sensor applications Sahin Demirci, Sultan Butun Sengel, Nurettin Sahiner
PPG-P-039	Development of Graphene Oxide-Polymer Based Aerogels Öznur Doğan, Erhan Bat
Modeling and Simulation of Polymers	
MSP-P-001	On numerical instabilities in Brownian Dynamics Simulations of Polymer Flows in Microchannels Arpit Bansal, Prateek Kumar Jha
MSP-P-002	Molecular Dynamic Simulations of Biocompatible Carbon Nanotubes Pelin Deniz Akkus
MSP-P-003	Effect of reinforcing subgrade soil by geogrid Ninouh Tarek, Salhi Sadok

MSP-P-004	Computational Investigation of The Complete Reaction Path of Poly(L-lactide) initiated by Tin(II) 2-Ethylhexanoate with DFT methods Merve Güreşci, Sıla Gümüştaş, Mehmet Balcan, Armağan Kınal
MSP-P-005	Computational study on organization of PAA-Vancomycin assembly from electro-neutral complex to Nanoplex formation Suresh B Vepuri, Dhiraj R Sikwal, Rahul S Kalhapure, Sanjeev Rambharose, Mahmoud E Soliman, Chunderika Mocktar, Thirumala Govender
MSP-P-006	A DFT Study on the Stereoselective Propagation In Free Radical Polymerization of Acrylamides Gülru Kayık, Nurcan Şenyurt Tüzün
MSP-P-007	Mathematical kinetic modeling on Isoniazid release from Dex-HEMA-PNIPAAm polymeric nanogels Babak Kaffashi, Maryam Jafari

Bridging Continents & Bridging Molecules

MACRO 2016

World Polymer Congress Istanbul 17-21 July

PLENARY SPEAKERS

Krzysztof Matyjaszewski

Carnegie Mellon University, USA

Dr. Krzysztof Matyjaszewski is the J.C. Warner Professor of Natural Sciences at Carnegie Mellon University and served as the head of the Chemistry Department. He has written 17 books, 83 book chapters, over 850 peer-reviewed articles cited more than 68,000 times (one of the three most frequently cited polymer chemists in the world) as well as 48 US and 132 international patents. His research has impacted many, as over 100 postdoctoral fellows and 100 graduate students have been members of his research group. He is the editor of Progress in Polymer Science. Dr. Matyjaszewski's research interests include macromolecular engineering; synthesis of well-defined macromolecules via living and controlled polymerizations; homogeneous and heterogeneous catalysis; and the preparation of well-defined polymers and hybrids for optoelectronics, ceramics, and biomedical applications. He discovered Cu-mediated Atom Transfer Radical Polymerization, commercialized in US, Europe, and Japan. Dr. Matyjaszewski received the 2014 Japan National Institute of Materials Science Award, 2012 Maria Curie Medal, 2012 Prize of Société Chimique de France, 2011 Wolf Prize in Chemistry, 2009 Presidential Green Chemistry Challenge Award, and from the American Chemical Society: 2013 North America Science Award, 2011 Hermann Mark Award, 2011 Award in Applied Polymer Science, 2002 Polymer Chemistry Award, and 1995 Creative Polymer Chemistry Award. He was awarded with seven honorary degrees (Ghent, Lodz, Athens, Moscow, Toulouse, Pusan, Paris) and is a member of National Academy of Engineering, Polish Academy of Sciences, and Russian Academy of Sciences and a fellow of National Academy of Inventors.

Controlled Radical Polymerization: from Mechanisms to Applications

Recent years witnessed tremendous progress in various controlled radical polymerization procedures. They employ a dynamic equilibrium between growing free radicals and dormant species. This process enables taming uncontrolled free radical behavior by inserting a dormancy periods of a few seconds or minutes after ca 1 ms activity. Such an intermittent activation route extends life of propagating chains from ca. 1 s to several hours and days. It also permits precise synthesis of macromolecules with complex architecture from small functional molecules as building blocks. Some examples of new initiating and catalytic systems used at ppm amounts for ATRP (atom transfer radical polymerization) will be presented. They open avenues to new block, gradient, graft, star, brush and (hyper)branched functional (co)polymers as well as bioconjugates and organic-inorganic hybrids that find applications as various new advanced functional nanostructured materials.

Keywords: CRP, ATRP, advanced materials

Markus Antonietti

Max Planck Institute, Germany

Markus Antonietti studied Polymer Chemistry in Mainz and was infected with the "polymer virus" already those days. His Habilitation, also in Mainz in 1990, was dealing with microgels and the control of their nanostructure. Since 1993, he is Director of the Max Planck Institute of Colloids and Interfaces. He has published about 650 Papers, which were cited more than 50000 times. His work was honoured at diverse occasions, but he calls it most rewarding that 45 of his former coworkers hold leading professorships, worldwide. His current interests cover porous polymers and carbons, synthesis of novel polymers, green chemistry and Carbon Negative Polymer Materials on the base of biomass, energy storage and artificial Photosynthesis.

Polymer Ionic Liquids: From Green Polymer Chemistry to Stabilizers to Actuators

Markus Antonietti, Jiayin Yuan

Max Planck Institute of Colloids and Interfaces, Colloid Chemistry Department, Research Campus Golm , D.14424 Potsdam, Germnay
Email: office.cc@mpikg.mpg.de

Polymer Ionic Liquids (PILs) are polymerized version of Ionic Liquids. They are more than only polyelectrolytes: due to charge frustration, many of them are liquid in the bulk, are processable, and combine simply polymer properties with ionic liquid character.

In this talk I will present some of their special characteristics, talk about self-organization and delineate the extraordinary potential of PILs to stabilize very different kind of interfaces in a highly efficient fashion. I will also show that ionic liquid monomers might not necessarily stay exotic species, but can be made by principles of green chemistry from natural resources

In a second part of the talk, I will report on novel material processes enabled by PILs, namely porosion, stimulated carbonization, as well as some work on sensors and artificial muscles based on autoporous PIL membranes.

Literature

1. Poly(ionic liquid)s: Polymers expanding classical property profiles (Review), Jiayin Yuan, Markus Antonietti, POLYMER 52, 1469-1482 (2011)
2. Poly(ionic liquid)s: An update (Update-Review), Jiayin Yuan, David Mecerreyes, Markus Antonietti, Progress in Polymer Science 38, 1009-1036 (2013)

Keywords: Polymer Stabilizer, Ionic Liquid, Self Organisation, Actuator

Michael Buback (Bob Stepto Lecturer)
University of Goettingen, Germany

Michael Buback studied chemistry at the University of Karlsruhe, where he received his PhD in 1972. After Habilitation in 1978 he was awarded a Heisenberg Fellowship by the German Science Foundation (DFG). He became Professor for Applied Physical Chemistry at the University of Göttingen in 1981, Full Professor for Technical and Macromolecular Chemistry in 1995, and Member of the Göttingen Academy of Sciences in 2000. Michael Buback received the Dechema Award, the Bunsen-Denkmünze, and the Herman F. Mark Medal. Michael Buback has been a member of the Academy of Sciences in Göttingen since 2000. In January 2008 he accepted the position of the vice president of the Polymer Division of the International Union of Pure and Applied Chemistry (IUPAC). He has published over 300 peer-reviewed papers. His research interests cover the entire field of radical polymerization with a focus on detailed kinetic studies via pulsed laser initiation carried out in conjunction with highly time-resolved IR, near-IR, and EPR spectroscopy. Further activities address the kinetics and the phase behaviour of chemical processes in extended ranges of pressure and temperature. Special expertise centers around the quantitative monitoring, via online vibrational spectroscopy, of species occurring during chemical transformations at pressures up to 7000 bar.

Radical Polymerization Kinetics and Mechanisms studied by Pulsed Laser Initiation in Conjunction with Highly Time-Resolved EPR Spectroscopy

The accurate knowledge of the relevant rate coefficients, e.g., for propagation, kp, and termination, kt, is mandatory for the understanding and simulation of conventional and reversible deactivation radical polymerizations (RDRPs). Such information has become available by pulsed-laser polymerization (PLP) techniques, such as PLP–SEC. The most recent and powerful PLP method is SP–PLP–EPR, which combines instantaneous production of an intense burst of radicals by a single laser pulse (SP) with subsequent time-resolved detection of the type and concentration of radicals. EPR spectroscopy is unrivalled for such detailed investigations, as the reactive species, i.e., the radicals are directly monitored with a time resolution of microseconds.

Laser-pulse techniques have in common that the radical chain length, i, increases linearly with time t after pulsing. Thus SP–PLP–EPR is perfectly suited for investigations into chain-length-dependent (CLD) termination. The composite model introduced by the Russell group allows for the adequate representation of the termination rate coefficient, kti,i, for two radicals of identical chain length, i. From an SP–PLP–EPR trace, also kp may be deduced which is advantageous in case of slowly terminating monomers, where the PLP-SEC method may run into difficulties.

The SP–PLP–EPR technique also allows for quantitative kinetic studies into systems with more than one type of radicals being present. This is the case, e.g., with acrylates and acrylamide, where secondary propagating radicals are transformed into tertiary midchain radicals via an intramolecular 1,5-hydrogen transfer (backbiting) reaction. Both radicals have clearly different EPR spectra and kinetic properties.

The SP–PLP–EPR method is also perfectly suited for elucidating the kinetics and mechanism of RDRPs, such as RAFT and ATRP, as will be illustrated by studies into iron-mediated ATRP and into dithiobenzoate-mediated RAFT polymerization.

Keywords: Polymerization Kinetics, Time-Resolved EPR Spectroscopy, Pulsed Laser Initiation

Kristi S. Anseth
University of Colorado Boulder, USA

Kristi S. Anseth earned her B.S. degree from Purdue University in 1992 and her Ph.D. degree from the University of Colorado in 1994. She then conducted post-doctoral research at MIT as an NIH fellow and subsequently joined the Department of Chemical and Biological Engineering at the University of Colorado at Boulder as an Assistant Professor in 1996. Dr. Anseth is presently a Howard Hughes Medical Institute Investigator and Distinguished Professor of Chemical and Biological Engineering. Her research interests lie at the interface between biology and engineering where she designs new biomaterials for applications in drug delivery and regenerative medicine. Dr. Anseth's research group has published over 200 publications in peer-reviewed journals and presented over 180 invited lectures in the fields of biomaterials and tissue engineering. She was the first engineer to be named a Howard Hughes Medical Institute Investigator and received the Alan T. Waterman Award, the highest award of the National Science Foundation for demonstrated exceptional individual achievement in scientific or engineering research. In 2009, she was elected a member of the National Academy of Engineering and the Institute of Medicine. Dr. Anseth is also a dedicated teacher, who has received four University Awards related to her teaching, as well as the American Society for Engineering Education's Curtis W. McGraw Award. Dr. Anseth is a Fellow of the American Association for the Advancement of Science and the American Institute for Medical and Biological Engineering. She serves on the editorial boards or as associate editor of Biomacromolecules, Journal of Biomedical Materials Research — Part A, Acta Biomaterialia, Progress in Materials Science, and Biotechnology & Bioengineering.

Hydrogels as dynamic cell niches through photo-click and photo-clip reactions

Our group is interested in the development of biomaterials to serve as in vitro cell culture systems and decipher critical extracellular matrix (ECM) signals that are relevant in tissue development, regeneration, and disease. Specifically, we design synthetic ECM analogs that capture key features of the unique chemistry and physical properties of a cell's niche—an environment that is not only tissue specific, but can be strikingly heterogeneous and dynamic. Unique to our approach is the ability to create cell-laden matrices in three-dimensional space in which the matrix properties can be changed on demand —so-called 4D biology. Here, our group has focused on the development of photochemical reactions to create tunable cell-laden matrices, for example, the thiol-ene photo-click reaction and complementary photo-clip reactions to introduce and remove biological signals from a complex milieu. These photochemical reactions not only proceed rapidly and with high specificity, but are bio-orthogonal, spatiotemporally controlled, and cytocompatible. This talk will illustrate how we leverage these chemistries to create biologically responsive hydrogel matrices, and employ them to study the effects of matricellular signaling on diverse cellular functions and processes. For example, we exploit peptide-crosslinked PEG hydrogels to encapsulate human mesenchymal stem cells (hMSCs) and study how matrix density, degradability, elasticity, and adhesivity influence migration in real time. Beyond simply observing cells, we also apply microrheological techniques to measure local gel degradation, and reporter molecules to detect local cell activity in situ (e.g., protease activity, apoptosis). Finally, results will demonstrate that these photo-click and –clip reactions are compatible with protein encapsulation and conjugation while maintaining bioactivity for cellular signaling.

Keywords: hydrogels, biomaterials, photochemistry, thiol-ene, stem cells

Nikos Hadjichristidis
KAUST, Saudi Arabia

The research of Prof. Hadjichristidis focuses mainly on the synthesis of novel homopolymers, and copolymers with well-defined complex macromolecular architectures (star, comb, cyclic, dendritic, etc) by using anionic polymerization (AP) high vacuum techniques, as well as combination of AP with other polymerization methodologies (polyhomologation, ROP, ROMP, ATRP, etc). These polymers are ideal models for checking the theory, understanding and improving the performance of industrial polymers (e.g. polyethylene, polystyrene based thermoplastic elastomers) and are potential candidates for high-tech applications (e.g. nanolithography, drug delivery, high temperature membranes). He has published more than 450 scientific papers in referred scientific journals, 19 patents, is the editor of three books and author of one book on "Block Copolymers" (Wiley 2003). He has received many awards including: The American Chemical Society (ACS) National Award for Polymer Chemistry (2015); the ACS, Rubber Division Chemistry of Thermoplastic Elastomers Award (2011); The ACS, Polymeric Materials Science and Engineering (PMSE) Division Cooperative Research Award (2010); The International Award of the Society of Polymer Science, Japan (SPSJ, 2007); He has been elected for the Macro Group United Kingdom Medal for Outstanding Achievement for 2016. Social Media https://polymer.kaust.edu.sa.

Novel Strategies Towards Well-Defined Polyethylene-Based Complex Macromolecular Architectures

Access to well-defined (high degree of structural, molecular weight and compositional homogeneity) polyethylenes (PEs) and PE-based copolymers is necessary in order to correctly elucidate the structure-properties relationships and evaluate potential applications. Recently Shea, based on homologation,[1] an organic reaction that converts a reactant into the next member of the homologous series e.g. ethane to propane, developed a novel polymerization methodology[2] leading to perfectly linear PEs. The general reaction scheme involves the formation of an organoboron zwitterionic complex between a methylide (monomer) and a trialkylborane Lewis acid (activator/initiator) which breaks down by intramolecular 1,2-migration. As a consequence, the activated methylene group of methylide is randomly inserted one by one (C1 polymerization) into the three branches of the trialkylborane leading to a 3-arm PE star. The resulting star is subsequently oxidized/hydrolysed to give perfectly OH-end-capped linear PEs.

Our group is developing both novel borane initiators/ylide monomers and combines polyhomologation with other living or controlled/living polymerizations to synthesize model polymethylene (or PE)-based polymeric materials with complex macromolecular architectures.[2-8]

References
[1] Tufariello, J. J. et al., JACS 1966, 88, 4557; 1967, 89, 6804.
[2] Luo, J.; Shea, K. J. Acc. Chem. Res. 2010, 43, 1420.
[3] Zhang, H; Alkayal, N.; Gnanou, Y.; Hadjichristidis, N. Chem. Commun. 2013, 49, 8952.
[4] Zhang, H; Gnanou, Y.; Hadjichristidis, N. Polym. Chem. 2014, 5, 6431.
[5] Zhang, H; Alkayal, N.; Gnanou, Y.; Hadjichristidis, N. Rapid Commun. 2014, 35, 378.
[6] Zhang, Z.; Zhang, H.; Gnanou, Y.; Hadjichristidis, N. Chem. Commun. 2015, 51, 9936.
[7] Zhang, H.; Zhang, Z.; Gnanou, Y.; Hadjichristidis, N. Macromolecules 2015, 48, 3556
[8] Zhang, H.; Hadjichristidis, N. Macromolecules 2016

Keywords: complex macromolecular architectures, controlled/living polymerizations, polyhomologation

Scheme. PS: polystyrene; PB: polybutadiene; PI: polyisoprene; PCL: polycaprolactone; PEO: poly(ethylene oxide); PLA: polylactide

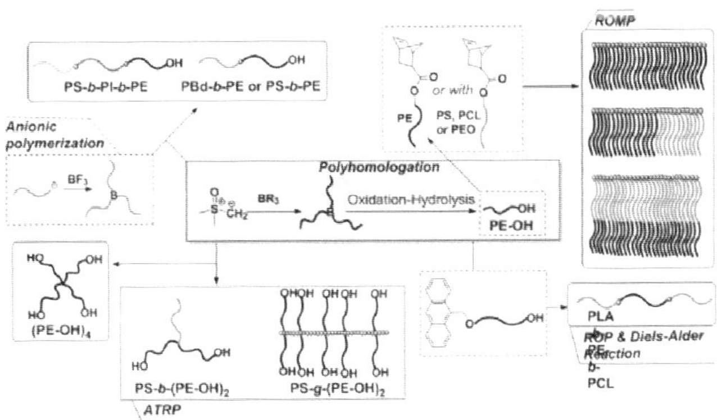

Niyazi Serdar Sariciftci

Johannes Kepler University, Austria

Niyazi Serdar Sarıçiftçi graduated from the Austrian St. George's College in Istanbul. He also studied classical piano at the Music Conservatory in Istanbul (1970-1980). Then he began studying physics at the University of Vienna (1980-1989). After obtaining the doctorate (1989), he conducted research on the 2nd Physical Institute of the University of Stuttgart, Germany (1989-1992). In 1992 he received the academic teaching license (venia docendi) by the Central Interuniversitary Commission (YÖK) in Ankara, Turkey. He then went to the Institute for Polymers & Organic Solids at the University of California, Santa Barbara, California, USA, where he worked for four years and, together with Alan J. Heeger (Nobel Prize in Chemistry, 2000) discovered and investigated the polymeric organic solar cells has (1992-1996). In April 1996, he accepted the appointment as Chair of Physical Chemistry at the Johannes Kepler University Linz. Since 1996 he gives lectures as a full professor at the JKU and is the head of the Institute for Physical Chemistry. In 2000 he was appointed founding director of the Linz Institute for Organic Solar Cells (LIOS) at JKU. Between 2003 and 2009 he was elected to the City Council of the City of Linz (SPÖ Group). Furthermore, Sarıçiftçi is a founding member of the Linz Circle. He is also a member of various associations and societies: Fellow of the Royal Society of Chemistry (FRSC), American Chemical Society (ACS), Materials Research Society (MRS), Austrian Physical Society (ÖPG), Austrian Chemical Society (GÖCH) and Fellow of SPIE. 2014, he was elected a corresponding member of the Austrian Academy of Sciences (AAS). Sarıçiftçi has published more than 500 scientific publications in scientific journals. He is one of the most cited scientists in his field. In a global ranking of the best materials scientists Sarıçiftçi was classified as 14th.

From Organic To Bioorganic Electronic Devices

Organic light emitting diodes (OLEDs), organic photovoltaic cells (OPVs) and organic field effect transistors (OFETs) are device elements for a future organic optoelectronics. Maturing from the academic research into the industrial development, such devices are entering the markets. Pure organic nanostructures and organic/inorganic hybrid nanostructures are comparatively studied for devices. This talk gives an overview of materials' aspect and devices with special emphasis on polymeric structures.

In order to account for a sustainable future, the application of biodegradable and biocompatible systems for organic optoelectronics are needed. The use of cheap electronic devices in a large scale will introduce a shift "from consumer electronics to consumable electronics". As such the contribution of electronic devices to urban waste is already increasing rapidly today. Therefore environmentally friendly materials are important to use. This is a next great challenge to material science in organic electronics. New developments of bio-inspired and/or bio-origin, bio-compatible materials are interesting. Such materials can also be used to interface the biological and biomedical research with the organic electronics field.

Keywords: Organic light emitting diodes, organic photovoltaic cells, organic field effect transistors

Takuzo Aida

University of Tokyo, Japan

Dr. Takuzo Aida was born in 1956. He received his Ph.D. in Polymer Chemistry from the University of Tokyo in 1984, and then began his academic career as an assistant professor at the same university on precision polymer synthesis. In 1996, he was promoted to full professor of the Department of Chemistry and Biotechnology, School of Engineering, the University of Tokyo. His research interests include optoelectronic soft materials, bioinspired macromolecules and materials including "Aqua Material", and molecular and biomolecular machines. He was appointed as a researcher for JST PRESTO Project for Fields and Reactions in 1996, and served as the director for JST ERATO AIDA Nanospace Project in 2000–2005 and then for JST ERATO-SORST Electronic Nanospace Project in 2005–2010. He is now the deputy director for Riken Center for Emergent Matter Science. He has received many awards including, as recent examples, American Chemical Society Award in Polymer Chemistry, Chemical Society of Japan Award in 2009, Purple Ribbon in 2010, and Fujiwara Prize and Alexander von Humboldt Research Award in 2011.

Stimuli-Responsive Smart Soft Materials

Machine technology frequently puts magnetic or electrostatic repulsive forces to practical use, as in maglev trains, vehicle suspensions or non-contact bearings. In contrast, materials design overwhelmingly focuses on attractive interactions, such as in the many advanced polymer-based composites, where inorganic fillers interact with a polymer matrix to improve mechanical properties. However, articular cartilage strikingly illustrates how electrostatic repulsion can be harnessed to achieve unparalleled functional efficiency: it permits virtually frictionless mechanical motion within joints, even under high compression. Here we describe a composite hydrogel with anisotropic mechanical properties dominated by electrostatic repulsion between negatively charged unilamellar titanate nanosheets embedded within it. Crucial to the behaviour of this hydrogel is the serendipitous discovery of cofacial nanosheet alignment in aqueous colloidal dispersions subjected to a strong magnetic field, which maximizes electrostatic repulsion6 and thereby induces a quasi-crystalline structural ordering over macroscopic length scales and with uniformly large face-to-face nanosheet separation. We fix this transiently induced structural order by transforming the dispersion into a hydrogel using light-triggered in situ vinyl polymerization. The resultant hydrogel, containing charged inorganic structures that align cofacially in a magnetic flux, deforms easily under shear forces applied parallel to the embedded nanosheets yet resists compressive forces applied orthogonally. We anticipate that the concept of embedding anisotropic repulsive electrostatics within a composite material, inspired by articular cartilage, will open up new possibilities for developing soft materials with unusual functions.

References

[1] Q. Wang et al., Nature 2010, 463, 339. [2] M. Liu et al., Nature Commun. 2013, 4, 2029. [3] M. Liu et al., Nature 2015, 517, 68. [4] Y.-S. Kim et al., Nature Mat. 2015, 14, 1002.

Keywords: Responsive materials, supramolecular chemistry, functional hydrogels

Bridging Continents & Bridging Molecules

MACRO 2016

World Polymer Congress Istanbul 17-21 July

AWARD LECTURERS

AWARD-O-001

Poly(2-oxazoline)s as basis for responsive polymers

Richard Hoogenboom

Ghent University

The living cationic ring-opening polymerization of 2-oxazolines has been studied in great detail since its discovery in 1966. The versatility of this polymerization method allows copolymerization of a variety of 2-oxazoline monomers to give a range of tunable polymer properties that enable, for example, hydrophilic, hydrophobic, fluorophilic, as well as hard and soft materials. However, this class of polymers was almost forgotten in the 1980s and 1990s because of the long reaction times and limited application possibilities. In the new millennium, a revival of poly(2-oxazoline)s has arisen because of their potential use as biomaterials and thermoresponsive materials, as well as the easy access to defined amphiphilic structures for (hierarchical) self-assembly.

Within this lecture I will provide a historical overview of my involvement in the development of poly(2-oxazoline)s, starting from the optimization and acceleration of the living cationic ring-opening polymerization of 2-oxazolines under microwave irradiation. This optimized polymerization procedure has then be applied for the development of highly tunable thermoresponsive polymers with lower critical solution temperature behavior in water as well as upper critical solution temperature behavior in ethanol-water solvent mixtures, the latter also leading to schizophrenic micellization. More recent work on supramolecular temperature sensors and poly(2-oxazoline) coated gold nanoparticles as logic gates will be highlighted too.

Keywords: poly(2-oxazoline), cationic polymerization, responsive polymer, thermoresponsive polymer

AWARD-O-003

Design, Synthesis, and Morphology of Ion-Containing Polymers

Moon Jeong Park

Department of Chemistry, Pohang University of Science and Technology, Pohang, Korea

Polymers containing charged species have been considered as the key materials in improving the efficiency of electrochemical devices. Although the synthesis and characterization of a wide variety of charge-containing polymers have been extensively reported over the last decade, quantitative understanding of the factors governing the transport properties of these materials is in its infancy. In this talk, I will present the ways to improve transport properties of various polymeric materials.1-4 Based on the current understanding of the diverse factors affecting the thermodynamics, morphologies and transport of charge-containing polymers, various strategies for accessing improved transport properties are elucidated.5-7 The creation of well-defined self-assembled morphologies in such materials is particularly highlighted as a novel prospective technique geared towards obtaining next generation devices with enhanced performance.8,9

[REFERENCES]

1. 1. S. Y. Kim, S. Kim, M. J. Park*, Nature Communications, 2010, 1:88.
2. S.Y. Kim, E. Yoon, T. Joo, M. J. Park*, Macromolecules, 2011, 44, 5289-5298.
3. O. Kim, S. Y. Kim, H. Ahn, C. W. Kim, Y. M. Rhee, M. J. Park*, Macromolecules, 2012, 45, 8702-8713
4. O. Kim, T.-J. Shin, M. J. Park*, Nature Communications, 2013, 4, 2208. DOI: 10.1038/ncomms3208.
5. G. Jo, H. Ahn, M. J. Park*, ACS Macro Letters, 2013, 2, 990-995.
6. S. Y. Kim, J. Lee, M. J. Park*, Macromolecules, 2014, 47, 1099-1108.
7. H. Kim, J. Lee, H. Ahn, O. Kim, M. J. Park*, Nature Communications, 2015 DOI: 10.1038/ncomms8278.
8. I. Y. Choi, J. Lee, H. Ahn, J. Lee, H. C. Choi, M. J. Park*, Angewandte Chemie International Edition, 2015, DOI: 10.1002/anie.201503332.
9. O. Kim, S. Y. Kim, J. Lee, M. J. Park* Chemistry of Materials, 2016, 28, 318-325.

Keywords: Ion-containing polymers, block copolymers, ionic liquids, phase behavior, conductivity

Bridging Continents & Bridging Molecules

MACRO 2016

World Polymer Congress Istanbul 17-21 July

INVITED SPEAKERS

Short Codes	
EOO	Energy, Optics, & Optoelectronics
IP	Industrial Polymers
MBM	Macromolecules in Biotechnology & Medicine
MN	Macromolecules & Nanotechnology
MSP	Modeling and Simulation of Polymers
PCNCT	Polymer Characterization by New and Combined Techniques
PE	Polymer Education
PEPC	Polymer Engineering, Processing, and Characterization
PP	Polymer Physics
PPBM	Polymer and Polymer-Based Membranes
PPG	Porous Polymer and Gels
RDPS-A	Recent Developments in Polymer Synthesis-A-Controlled/Living Polymerization
RDPS-B	Recent Developments in Polymer Synthesis-B-Complex Macromolecular Structures
RDPS-C	Recent Developments in Polymer Synthesis-C-Light-induced Reactions in Polymer Science
RDPS-D	Recent Developments in Polymer Synthesis-D-Condensation Polymerization and Thermoset
RRB	Renewable Resources and Biopolymers
SFP	Smart and Functional Polymers

Abstracts sorted alphabetically

EOO-I-001 [Energy, Optics, & Optoelectronics]

Polymer and Active Layer Design for Optimized Polymer-Fullerene Solar Cells

Barry C Thompson

Department of Chemistry, University of Southern California, Los Angeles, California, USA

Polymer-fullerene bulk heterojunction (BHJ) solar cells have consistently been at the forefront of the field of organic photovoltaics (OPV). The enduring vision of OPV is the promise of combining a simple, low-cost approach with an efficient, flexible, lightweight platform. While efficiencies have improved remarkably over the last decade through advances in device design, mechanistic understanding, and evolving chemical structural motifs, steps forward have often been tied to a loss of simplicity and a deviation from the central vision of OPV. Within the context of active layer optimization, our focus is to target high efficiency while maintaining simplicity in polymer design and active layer processing. To highlight this strategy, this talk will focus on our work on random poly(3-hexylthiophene) (P3HT) analogues and their application in binary and ternary blend polymer-fullerene solar cells. These random conjugated polymers are conceptually based on combining simple monomers strategically to influence polymer properties as opposed to the synthesis of highly tailored and synthetically complex monomers. Efforts toward further simplifying polymer synthesis via Direct Arylation Polymerization (DArP) will also be discussed. Finally, the ternary blend approach further exemplifies the focus on device simplicity by targeting efficiencies that are competitive with complex tandem solar cells, but within the confines of a single active-layer processing step.

Keywords: Solar Cell, Fullerene, Conjugated Polymer

EOO-I-002 [Energy, Optics, & Optoelectronics]

Semiconducting Polymers and Small Molecules for Transistors and Solar Cells

Iain McCulloch

SPERC, King Abdullah University of Science and Technology

The power conversion efficiency (PCE) of single junction organic solar cells has increased significantly during the last decade and now approaching the threshold considered necessary to commercialize the technology. During this period, the structural diversity of semiconducting donor polymers for solar cells has increased dramatically, enabling accelerated development of bulk heterojunction (BHJ) organic solar cells based on polymer donor materials and molecular fullerene derivatives. However both the fullerenes, and the low bandgap polymers typically suffer from low absorption coefficients due to weak oscillator strength. Our approach has been to use P3HT as a p-type hole acceptor, and design highly absorbing, low-bandgap n-type small molecules to replace fullerenes. These fullerene acceptors not only have weak absorption, but also poor tunability of absorption over the longer wavelengths of the solar spectrum; morphological instability in thin film blends over time; high synthetic costs and limited scope for synthetic control over electronic and structural properties. For these reasons, we have developed new, synthetically simple electron acceptor materials, based on rhodanine end groups, which have much larger absorption coefficients than fullerenes, coupled with high lying LUMO energy levels, to maximize cell voltages. In BHJ devices with P3HT donor polymer, the rhodanine molecules were demonstrated to outperform all fullerenes.

Keywords: photovoltaic, semiconductor, transistor

EOO-I-003 [Energy, Optics, & Optoelectronics]

Tuning the performance of conjugated polymers for transistor and solar cells applications

Martin Heeney

Department of Chemistry, Imperial College London, U.K.

Controlling the optical band gap and energy levels of conjugated polymers is fundamentally important for both transistor and solar cell applications. In this presentation I will discuss various synthetic approaches to tune optoelectronic properties in conjugated polymers, as well as discussing the performance of the resulting polymers in transistor and solar cell devices.

Keywords: Conjugated polymer, organic transistor, semiconducting polymer, organic solar cell

EOO-I-004 [Energy, Optics, & Optoelectronics]

New Polymers for Photovoltaics and NIR-II Imaging

Yingping Zou

College of Chemistry and Chemical Engineering, Central South University, Changsha 410083, China.

Although polymer solar cells have witnessed significant advances during the past few years, the efficiency and stability still need to be improved for large scale production and commercialization. Both materials and device designs play important roles in realizing such applications. Here, we will summarize our some recent work in new benzodifuran(BDF), thienobenzofuran (TBF) and benzodithiophene (BDT) based polymer photovoltaic materials.We synthesized a series of BDF, TBF and BDT based copolymers using different electron accepting unit. The optoelectronic properties were investigated. The most copolymers exhibit excellent photovoltaic performances with PCE up to 8.7% with a single layer device. As for as imaging, recent work have shown that in vivo fluorescence imaging with carbon nanotubes and quantum dots emitting in the second near-infrared (NIR-II, 1.0-1.7 μm) optical window can afford deep tissue penetration with high spatial resolution, owing to reduced scattering of long wavelength photons. Here, we synthesized a low-bandgap poly(benzodifuran/fluorothienothiophene) donor/acceptor copolymer exhibiting intrinsic fluorescence peaked at ~1047 nm in the NIR-II region. Non-covalent functionalization of the organic polymer with phospholipid-polyethylene glycol (DSPE-PEG) resulted in a water-soluble and biocompatible polymeric nanoparticle. Through antibody conjugation, molecular imaging of live cells was performed with conjugated polymer emitting > 1000 nm for the first time. The high quantum yield of the organic NIR-II agent allowed in vivo, deep-tissue and ultrafast imaging of mouse femoral artery.

Keywords: polymer solar cell, optoelectronic materials, device, NIR-II bioimaging

EOO-I-005 [Energy, Optics, & Optoelectronics]

Composites of Carbon Nanomaterials and Conjugated Polymers: Preparation and Properties

Jianxin Geng
Technical Institute of Physics and Chemistry, CAS, Beijing, China

Carbon nanomaterials including carbon nanotubes (CNTs) and graphene have attracted enormous attention among scientists due to their unique structures and unusual characteristics such as large specific surface area, high electrical and thermal conductivities, high mechanical strength, and prominent optical and electrochemical properties. In addition, their PI conjugated surfaces result in donor-acceptor feature of the composites of carbon nanomaterials and conjugated polymers. In our research, we have developed various methods including covalent and non-covalent approaches to modify CNTs and graphene using conjugated polymers. Poly(3-hexylthiophene) (P3HT) was grafted on the surfaces of single-walled CNTs (SWNTs) and multi-walled CNTs (MWNTs) through the amidation reaction between the terminal amine groups of P3HT chains and the carboxylic acid groups on the surfaces of CNTs. The composite films were demonstrated successful use in sensitive and selective electrochemical detection of metal ions (1,2). The photoinduced electron transfer from P3HT to carbon nanomaterials led to enhanced photoresponse and photothermal effect of the composites (3-5).

ncreasing the grafting density of P3HT chains on the surfaces of reduced graphene oxide (RGO) sheets resulted in red shifted absorption maximum, which was beneficial to efficient use of sunlight in optoelectronics (5,6). The electronic coupling between RGO and P3HT in the RGO-g-P3HTcomposite resulted in tuned photoluminescence (PL) emission of the RGO component, leading to resolved dual PL emissions of the RGO-g-P3HT composite (6).

References:

1. S. Yang, et al. *ACS Appl. Mater. Interfaces* 2014, 6, 7686.
2. S. Yang, et al. *RSC Adv.* 2014, 4, 25051.
3. J. Sun, et al. *J. Mater. Chem.* 2012, 22, 18879.
4. J. Sun, et al. *Chem. Commun.* 2013, 49, 5538.
5. D. Meng, et al. *Chem. Commun.* 2014, 50, 14345.
6. D. Meng, et al. *Chem. Sci.* 2014, 5, 3130.

Keywords: Carbon Nanotubes, Graphene, Poly(3-hexylthiophene), Photoinduced Electron Transfer, Photothermal Effect, Dual PL Emissions

EOO-I-006 [Energy, Optics, & Optoelectronics]

Conjugated polymers and their applications

Levent Toppare
Department of Chemistry & Department of Biotechnology & Department of Polymer Science and Technology & The Center for Solar Energy Research and Applications (GUNAM), Middle East Technical University, 06800 Ankara, Turkey

Conjugated polymers have gained great attention in recent years for their advantages over inorganic counterparts like low cost, easy structural modification and wide range of applications. Many types of conducting polymers with different architectures and functional moieties can be designed and synthesized to meet the requirements of their application areas. It is possible to use donor acceptor type polymers in different applications like electrochromic devices, organic light-emitting diodes (OLEDs), organic solar cells, biosensors and bioimaging.

Molecular design played a significant role for the rapid development in the field of these applications over the past decade. One of the most efficient strategies to achieve superior properties is the donor-acceptor approach; Combination of electron-rich donor and electron deficient acceptor moieties in the polymer backbone. Low-band-gap polymers having extended absorption in the longer wavelengths with an increased photovoltaic device efficiencies are obtained via this approach.

Different polymer scaffolds can be designed according to desired properties. Our team is interested in novel monomer and polymer design, synthesis for different applications. Herein, the recent studies of conducting polymer based electrochromic devices, organic light-emitting diodes (OLEDs), organic solar cells, biosensors and bioimaging will be demonstrated.

Keywords: conjugated polymers, electrochromic devices, organic light-emitting diodes, organic solar cells, biosensors

EOO-I-007 [Energy, Optics, & Optoelectronics]

Thienothiophenes (TT) and Dithienothiophenes (DTT) for Electronic and Optoelectronic

Turan Öztürk
Department of Chemistry, Istanbul Technical University, Istanbul, Turkey

Thienothiophenes (TT) 1 and dithienothiophenes (DTT) 2 possess two and three fused thiophenes, respectively.1-3 As the orientations of the rings vary depending on the location of the sulfur atom, four and six isomers for TTs and DTTs can be depicted, respectively. They are electron-rich heterocycles, which make them useful building blocks for the development of conjugated materials. They are particularly important for the development of numerous organic materials for electronic and optoelectronic applications such as thin film transistors (OTFTs), light emitting diodes (OLED), photovoltaic, energy storage, electrochromic devices and sensors.

Keywords: Thienothiophene, Dithienothiophene, Organic Materials

Figures

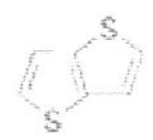

1

Thienothiophene (TT)

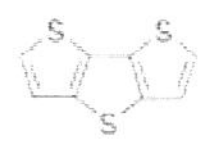

2

Dithienothiophene (DTT)

EOO-I-008 [Energy, Optics, & Optoelectronics]

Exceeding the Limits of Fullerene Solar Cells

Derya Baran[1], Thomas Kirchartz[2], Sarah Holliday[1], Andrew Wadsworth[1], Jeffrey Gorman[1], Christoph Brabec[3], James Durrant[1], Iain Mcculloch[1]
[1]Center of Plastic Electronics, Imperial College London, London, United Kingdom
[2]Julich Forschungszentrum GmbH, Julich, Germany
[3]Friedrich Alexander Univeristy Erlangen Nurnberg, Erlangen, Germany

Optimizing the energy levels at the donor-acceptor interface of organic solar cells has driven their efficiencies to above 10%. However, further improvements towards efficiencies of typical for the best inorganic solar cells are difficult because of empirical limits for the open-circuit voltage of polymer fullerene solar cells which hardly ever exceeds 1V due to voltage loss >0.6 eV (Figure 1). Here we show that this limit is overcome with a non-fullerene acceptor where a rhodanine flanked non fullerene small molecule acceptor FBR serves as acceptor with and a low band gap polymer PffBT4T-2DT as donor material. The combination of FBR and PffBT4T-2DT allows both a high open-circuit voltage of 1.12V and extremely low voltage loss ~ 0.47 V whilst keeping the external quantum efficiencies >55 % which could not been achieved with fullerene solar cells so far. These unconventional features result in photovoltaic performances as high as 7.8%. This study will inspire organic-photovoltaic community to focus on non-fullerene acceptors for further improvement of power conversion efficiencies without any complex multi-layer or -component device architectures. [1] W. Li, K. H. Hendriks, A. Furlan, M.M. Wienk, R. A. J. Janssen "High Quantum Efficiencies in Polymer Solar Cells at Energy Losses below 0.6 eV", JACS, Vol. 137, No. 6, (2015), pp 2231-34.

Keywords: Organic solar cells, non-fullerene acceptors, ternary solar cells,high efficiency

EOO-I-009 [Energy, Optics, & Optoelectronics]

Semiconducting Conjugated Copolymers for Optoelectronics: Effects of Side Chains and Molecular Weight

Věra Cimrová, Veronika Pokorná, Vagif Dzhabarov, Aleš Růžička, Anne Lutgarde Yonkeu, Drahomír Výprachtický

Institute of Macromolecular Chemistry, v.v.i., Academy of Sciences of the Czech Republic, Heyrovský Sq. 2, 162 06 Prague 6, Czech Republic

Semiconducting conjugated polymers are extensively studied due to their potential applications in optoelectronics, such as light-emitting diodes (LEDs), photovoltaic devices (PVs), organic field effect transistors (OFETs) and sensors. Low-bandgap donor-acceptor copolymers, which belong to the third generation of semiconducting polymers, are of interest due to their potential application namely in photovoltaics. Recently, we synthesized and studied novel donor-acceptor (D-A) copolymer series and new luminescent copolymers. The D-A copolymers contain highly planar 4,6-di(thien-2'-yl)thieno[3,4-c][1,2,5]thiadiazole (T) or its derivatives (electron-acceptor units) and various electron-donor units such as 9,9-bis(alkyl) fluorene, benzene, bithiophene or carbazole derivatives. Luminescent copolymers consist of 9,9-bis(alkyl)fluorene and bithiophene or carbazole derivatives. Copolymers with various side chain combinations and of various molecular weights were prepared and their photophysical, electrochemical, photoconductive or electroluminescent properties studied. The D-A copolymers exhibit interesting properties that are influenced by the nature and length of the alkyl chains attached to the thiophenes of T unit and also significantly by donor strength of the electron-donor unit. Absorption of thin films covers the entire visible spectral region extended up to NIR for some copolymers. Selected D-A copolymers were used in the blends with fullerene derivatives as active layers in photovoltaic devices. The effects of side chains, electron-donor strength and molecular weight on the photophysical, electrochemical, charge transport and photovoltaic properties were evaluated and are discussed. Luminescent copolymers were used as light-emitting materials in LEDs. Their electroluminescent properties are correlated with photoluminescent properties of thin films and discussed in relation to the copolymer structure, side chain lengths and molecular weight.
Acknowledgment: We thank the Czech Science Foundation for supporting this work with grant 13-26542S.

Keywords: conjugated polymers, photophysics, electrochemistry, electroluminescence, charge transport, photovoltaics

IP-I-001 [Industrial Polymers]

Polymer systems with latent reactivity for adhesive applications

Andreas Taden

Henkel AG & Co. KGaA, Adhesive Research

Polymers are crucial components in almost every adhesive in order to obtain strong cohesive strength. In a simplistic view high molecular weight and crosslinking contribute to resistance against mechanical fracture and separation. On the other hand an adhesive needs to be a liquid during handling, for dispensing and spreading/wetting on a given substrate. The same is true for coatings, sealants and matrix resins for composite materials. The hardening of an adhesive, which is also denoted as setting or curing, is therefore an important criteria and often determines its application. Fast curing is desired, but on the other hand typically limits the time for application, hampers the adhesive handling and involves reduced storage stability. In this respect latent reactive systems are highly demanded and can be regarded as important technology enablers.

In this talk two different approaches will be discussed. At first vinyl ether functionalized polyurethanes (VEPUs) will be presented as novel photopolymer systems with unique curing capabilities based on cationic ("living") polymerization and high crosslinking capabilities. As will be shown the materials can undergo a glass temperature induced vitrification, leading to dormant cationic polymerization centers and a highly desired latent reactivity for subsequent postcuring. Because the VEPUs are furthermore not sensitive towards oxygen inhibition and can polymerize in the absence of light after initiation ("dark cure") they enable manifold new application and handling possibilities.

Secondly, catalytic nanocapsules will be presented as colloidal concept towards thermolatent curing of thermosets. The respective polymer systems are characterized by an adjustable and very distinctive non-Arrhenius reaction kinetics which combines long handling times (also denoted as "pot life" or "open time") with fast cuing and short cycling steps.

Keywords: vinyl ether polyurethanes, photopolymers, cationic curing, catalyst encapsulation, thermosets

IP-I-002 [Industrial Polymers]

New Components for Dental Composites and Adhesives

Norbert Moszner

Research&Development, Ivoclar Vivadent AG, Schaan, Liechtenstein

Today in dentistry, for the restoration of anterior lesions, as well as for the supply of smaller and medium-sized defects in the posterior region, direct filling composites and enamel-dentin adhesives are used. Currently, restorative composites are light-curing hybrids consisting of a resin matrix, which is a mixture of various free-radically cross-linking dimethacrylates, inorganic fillers, such as milled radiopaque glass fillers or highly dispersed silica, a visible light photoinitiator system mainly based on a mixture of camphorquinone and a tertiary amine, and further additives. Enamel-dentin adhesives are used in restorative dentistry to achieve a strong bond between the filling composite and dental hard tissues. The main components of currently used enamel-dentin adhesives include strongly acidic adhesive monomers, such as methacrylate group-containing phosphoric acids, crosslinking dimeth-acrylates and the camphorquinone based visible-light photoinitiator. Finally, a mixture mainly of water and ethanol is used as solvent. Dental composites and adhesives are an interesting field of application of new chemical components, because on the one hand they enable significant improvements of important properties. On the other hand, the high price of new compounds is in general no killer criterion for their use in dental restorative materials. Improvements of restorative composites or adhesives can be achieved by the use of tailor-made components such as crosslinking and functionalized monomers, initiators, additives. Starting with the description of the state-of-the-art of methacrylate-based dental adhesives and composites, the following tailor-made compounds will be discussed:

- New crosslinkers for dental composites, e.g. alternatives to most frequently used Bis-GMA or TEGDMA, and polymer network modifier.

- New monomers for dental adhesive such as adhesive monomers or crosslinkers.

- Novel photoinitiators to substitute camphorquinone.

Keywords: Dental, Composites, adhesives, crosslinker, photoinitiator,

IP-I-003 [Industrial Polymers]

Photoinitiators – Key Components Enabling Novel Accesses to Crosslinked Industrial Polymers

Kurt Dietliker

Swiss Federal Institute of Technology Zurich (ETH-Z), Department of Chemistry and Applied Biosciences, Zurich, Switzerland

Radiation curing is a well-established technology with widespread uses in industry. Outstanding properties, such as control of the polymerization both in time and space, fast processes, high product quality and a favorable ecological footprint allowed for industrially attractive processes, spanning from well-established applications such as protective coatings, printing inks, adhesives, electronic materials to recent developments such as additive manufacturing.

A key component in all radiation curable formulations is the photoinitiator, which acts as trigger for the polymerization process upon activation with light. Research on these compounds is key for further improvements of existing technologies, as well as for the introduction of completely new applications. This paper illustrates with some examples how recent developments in photoinitiators are key drivers for novel industrial applications.

While radically and cationically curable formulations are well established in radiation curing, the recent introduction of photolatent bases made the vast field of materials cross-linkable under base catalysis accessible for this technology. Combination of inorganic precursors with such organic resins further allows the controlled design of organic-inorganic hybrid materials with outstanding properties.

The fast introduction of UV-LEDs as light sources is another driving force for new developments, both in photoinitiators and applications. While the photoinitiators have to be adapted to provide the necessary sensitivity with these narrow-band emitters, the unique properties of the LEDs allowed new applications not feasible with conventional radiation sources.

An important challenge for radiation curing is the polymerization of thick or highly filled materials, since the curable thickness depends on the penetration depth of the light. Work to overcome these limitations using various concepts is currently under way. Although not yet in use in industrial applications, a successful solution will propel radiation curing from a technology for the production of 2D articles of limited thickness to real 3D applications.

Keywords: photoinitiator, radiation curing, UV-LED, organic-inorganic hybrid materials, 3D applications

IP-I-004 [Industrial Polymers]

Trends in Compounding

Yakup Ulcer[1], Leonardo Bellomo[2]

[1]Management, Ravago Turkey, Istanbul, Turkey
[2]R&D Department, Ravago Petrokimya Üretim A.S.

Environmental concerns, Sustainability, Energy efficiency, Mobility and Internet of things are mega trends that are shaping our daily life. Compounding being one of the most dynamic and customer oriented sectors in Polymer Industry is naturally affected by these trends. In this paper selected recent developments in machinery and thermoplastic raw material technology in line with these megatrends will be discussed. On the machinery side emphasis will be given to environmental concerns, energy efficiency and industry 4.0. Light material formulations, Green materials (Natural fiber reinforced) and High mechanical strength/weight materials (Carbon fiber reinforced) will be covered on the material side.

Keywords: Energy efficient compounding, Intelligent production systems, Natural fiber reinforced polymers, Light Compounds, Carbon fiber reinforced polymers.

IP-I-005 [Industrial Polymers]

Novel Production of Aliphatic Polyketone in Water

Mustafa Yasin Sen

Research and Development Center, Kordsa Global, Kocaeli, Turkey

Polyamide 6.6, polyamide 6, polyethylene terephthalate (PET) and rayon are the most common synthetic fibers used in the production of tire cord fabrics. In addition to these polymers, p-aramid and polyethylene naphthalate (PEN) have recently found use in applications, where high performance is needed, due to their superior mechanical properties. However, because of the high cost of adhesive treatment process and low adhesion to rubber, which results in low fatigue resistance of the material, there is currently no commercially available fiber that has both good mechanical properties and good adhesion to rubber in cord fabric industry. It is predicted that aliphatic polyketone (poly-1-oxotrimethylene), which is formed by alternating copolymerization of ethylene and carbon monoxide, might fill the need in the industry due to its tenacity and modulus comparable to p-aramide and its good adhesion to rubber. Aliphatic polyketone can be produced effectively using a homogenous palladium catalyst system which involves:

- a palladium salt of a (preferably) carboxy acid (e.g. palladium acetate)
- a phosphorous-based bidentate ligand
- an anion of an acid, preferably an organic acid with $pKa \approx 2$
- an organic oxidant (optional) such as 1,4-benzoquinone.

A typical polymerization reaction is performed at 70-90 °C and 50-100 bar pressure in methanol. The use of sulfonated, water-soluble ligands in recent years rendered this process environmentally benign and economically attractive since they allowed the use of water as reaction medium. We herein present the preliminary results of ethylene-carbon monoxide polymerization in which a palladium complex of water soluble diphosphane was employed.

Keywords: Aliphatic polyketone, poly(1-oxotrimethylene), high-pressure polymerization, carbon monoxide, ethylene, propylen

MBM-I-001 [Macromolecules in Biotechnology & Medicine]

Engineered biointerfaces through tailormade surface-attached polymer networks – From to new diagnostic tools to implantable materials

Jürgen Rühe

Department of Microsystems Engineering, University of Freiburg, Germany

Surface-attached polymer networks can help to tailor the surface properties of a material precisely to the specific needs laid out by the medical application and to create a favorable interaction of the artificial material with the biological system. However, chemical tools are needed, which allow for the attachment of polymer molecules to the device surfaces to ensure sufficient stability of the coating. Our approach for such surface architectures is based on polymers bearing photochemically or thermally reactive groups. Upon activation these groups attach to neighboring C,H-groups via C,H insertion crosslinking (CHic). Surface-attached polymer networks can be generated through this process and deposited on different substrate materials ranging from biomaterials, polymers to glass or metal surfaces. We describe the useof these architectures to generate extremely protein repellent surfaces. These specifically designed surfaces can be used for a variety of biomedical applications. We use such coatings in blood contacting environments in the context of ventricular assist devices and for the reduction of scar formation after a glaucoma treatment. If further functional groups are incorporated, such layers can also take on an active role in the interaction with a biomedical environment and can be used to increase the sensitivity of biochips for DNA, RNA and protein or cell analysis. In the latter case we use such systems for the capture and analysis of circulating tumor cells (CTC). Additionally, we describe the development of new biochips which allow to diagnose HPV infections, mastitis in milk samples, or the generation of protein chips for cytokine analysis in the context of SIRS. In many cases the combination of a bioinert background and very specific probes are key to allow for a precise and reliable analysis especially in real serum samples.

Keywords: cell adhesion, biochips, surface-attached networks, CHic, biointerfaces

MBM-I-002 [Macromolecules in Biotechnology & Medicine]

Synthesis of Nature-inspired Biohybrid Architectures for Nanomedicine and Precision Sensing

Tanja Weil[1], Yuzhou Wu[1], Yu Tokura[2], David Ng[1], Seah Ling Kuan[1]

[1]Max-Planck Institute for Polymer Research, Mainz, Germany
[2]Institute of Organic Chemistry III/ Macromolecular Chemistry, Ulm University, Ulm, Germany

Proteins are sequence specific and geometrically defined macromolecules representing the central framework of all biological processes in Nature. Their precise physical architecture and consequent biochemical functions are unique and unrivalled in the synthetic world, providing an impetus for the incorporation of proteins into the development of contemporary hybrid materials. Unlike conventional polymers, their repertoire of chemical functionalities at discrete positions facilitates the grafting of designated synthetic moieties to achieve a nanoscale construct with exceptional macromolecular definition.[1] Through these synthetic appendages, supramolecular polypeptides, protein-polymer biohybrids and core-shell nanoparticles can be chemically programmed to possess new and improved physico-chemical properties while simultaneously exhibiting unique biological behaviour[2,3].

We have designed multifunctional copolymers that possess attractive features such as an accurately known length and a defined number of functional groups at distinct locations within the peptide backbone, low size-dispersity as well as intrinsic biocompatibility.1,3] Such copolymers allow the stabilization of fluorescent nanodiamonds[3] via multivalent interactions providing access to unique biosensors for in vitro and in vivo sensing and bioimaging applications.[4] In addition, multiple drug molecules with different modes of action as well as cell targeting entities were attached based on sophisticated peptide chemistry[5]. Such protein-derived hybrid particles have a great potential for in vivo therapy and pave the way to intelligent vehicles for nanomedicine applications.

1. D.Y.W. Ng, Y. Wu, S.L. Kuan, T. Weil. Acc. Chem. Res. 47, 12, 3471-3480 (2014).
2. D.Y.W. Ng, M. Arzt, Y. Wu, Seah Ling Kuan, M. Lamla, T. Weil. Angew. Chem. Int. Ed. 53, 1, 324–328 (2014).
3. Y. Wu, et al. Adv. Funct. Mater. 25, 42, 6576–6585 (2015).
4. Y. Wu, F. Jelezko, M. Plenio, T. Weil. Angew. Chem. Int. Ed. DOI: 10.1002/anie.201506556R1 (2016).
5. T. Wang, N. Zabarska, K. Monczak, M. Lamla, D.Y.W. Ng, S. Rau, T. Weil. Chem. Commun. 51, 63, 12552-12555 (2015).

Keywords: biohybrid copolymer, core-shell particles, nanodiamonds, drug delivery, precision sensing

MBM-I-004 [Macromolecules in Biotechnology & Medicine]

Robust Polymersomal Doxorubicin for Active Tumor Targeted Chemotherapy

Yan Zou, Weijing Yang, Zhiyuan Zhong, Fenghua Meng

Biomedical Polymers Laboratory, and Jiangsu Key Laboratory of Advanced Functional Polymer Design and Application, College of Chemistry, Chemical Engineering and Materials Science, Soochow University, Suzhou, 215123, P. R. China

The clinical success of many chemotherapeutics critically depends on the development of non-toxic and efficient tumor-targeting nanocarriers. However, despite of significant progress made in the past decades, current nanotherapeutics are challenged by their inadequate in vivo stability, drug leakage in circulation, low tumor accumulation, inefficient tumor cell uptake, or slow intracellular drug release. Recently, we have developed several novel types of disulfide-crosslinked degradable polymeric nanomedicines for potent and targeted cancer therapy in vivo. For example, based on a patented dithiolane-functionalized cyclic carbonate monomer, we obtained robust and active tumor-targeting polymersomal doxorubicin (Ps-Dox). Ps-Dox had a small size (~ 100 nm), excellent stability, high loading of DOX and fast redox-triggered intracellular drug release. The in vivo studies revealed that Ps-Dox functionalized with various peptides (e.g. cRGD, cNGQ, GE11) exhibited a long circulation time, high tolerated dose, superior targetability, and better therapeutic outcome in different tumor models, including ovarian cancer and non-small cell lung cancer, as compared to commercial pegylated liposomal DOX (Lip-Dox). Moreover, Ps-Dox caused significantly less side effects than Lip-Dox. Therefore, Ps-Dox has emerged as a better alternative to Lip-Dox and warrants further clinical translation.

Keywords: polymersomal doxorubicin, disulfide-crosslinking, targeting, cancer therapy

MBM-I-005 [Macromolecules in Biotechnology & Medicine]

Controlled and localized photopolymerization of molecularly imprinted polymer nanocomposites as plastic antibodies for bioimaging

Karsten Haupt, Maria Panagiotopoulou, Mariano Garcia Soto, Carlo Gonzato, Bernadette Tse Sum Bui

Sorbonne Universités, Université de Technologie de Compiègne, CNRS Institute of Enzyme and Cell Engineering

Molecularly imprinted polymers (MIPs) are synthetic antibody mimics that specifically recognize molecular targets [1,2]. They are highly cross-linked polymers synthesized in the presence of the target molecule acting as a molecular template. This templating induces three-dimensional binding sites in the polymer that are complementary to the template in size, shape and chemical functionality. The plastic antibody can recognize and bind its target with an affinity and selectivity similar to a biological antibody. We present here new approaches allowing for the synthesis of MIP by photoinduced living radical polymerization using a dendritic multiiniferter [3]. This results in protein-size, soluble MIP nanogels with a homogeneous size distribution. Their mean diameter is 17 nm and the average molecular weight 97 kDa, thus their size and density are very close to those of biological antibodies. The MIP nanogels show specific binding of their targets, small organic molecules or proteins, with a nanomolar affinity and a good selectivity. In addition, the direct coating of thin MIP films around fluorescent nanoparticles [4,5] by localized photopolymerization using the particles as individual internal light sources, will be described.

The use of these functional nanomaterials for cell and tissue imaging [5,6] by epifluorescence and confocal fluorescence microscopy will be demonstrated on fixated and living keratinocytes, and on tissue samples

[1] Arshady, R.; Mosbach, K., Makromol. Chem. 182, 687-692 (1981).
[2] Haupt, K. et al., Top. Curr. Chem. 325, 1-28 (2012).
[3] Çakir, P. et al. Adv. Mater. 25, 1048-1051 (2013).
[4] Beyazit, S. et al. Angew. Chem. Int. Ed. 53, 8919-8923 (2014).
[5] Panagiotopoulou, M. et al. (2016) Angew. Chem. Int. Ed. (in press, DOI: 10.1002/anie.201601122)
[6] Kunath, S. et al. Adv. Healthc. Mater. 4, 1322-1326 (2015).

Keywords: Localized photopolymerization, molecularly imprinted polymer, synthetic receptor, bioimaging, quantum dots, upconverting nanoparticles

QD-MIP imaging

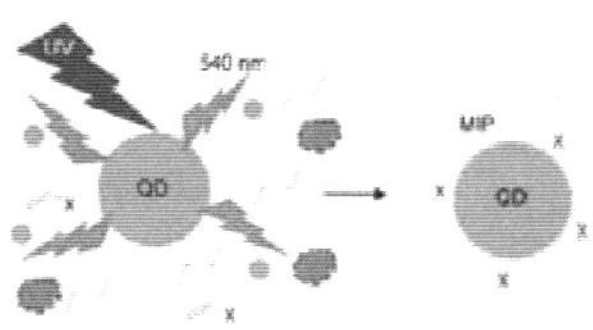

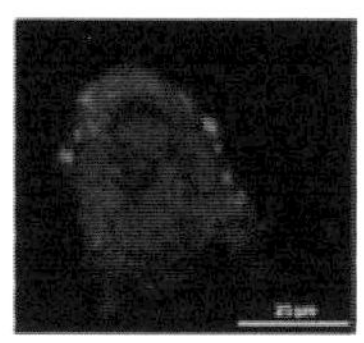

MBM-I-006 [Macromolecules in Biotechnology & Medicine]

Biomimetic Self-Assembled Peptide Nanostructures

Mustafa O Guler

Institute of Material Science and Nanotechnology, Bilkent University, Ankara, Turkey

Mimicking the structure and function of the biological materials through programmed self-assembly of small molecules provides us a wide variety of functionalities for life sciences applications. The self-assembly mechanism that forms the supramolecular aggregates involves noncovalent interactions such as hydrogen bonds, electrostatic and hydrophobic interactions. Diverse functional groups can be incorporated into nanostructures, for example bio-active peptide sequences and metal chelating groups as well as hydrophobic motifs that include alkyl chains, steroid rings, and aromatic systems. Here the potential impact of these nanostructures on bioactive functional materials will be discussed.

Keywords: Peptide, Nanofibers, Biomaterials, Regenerative Medicine

MBM-I-007 [Macromolecules in Biotechnology & Medicine]

Well-defined Fatty Acid Polymers as Potential Membrane Destabilizing Agents

Esra Aydınlıoğlu[1], Gürbüz Dursun[1], Ayşe Ezgi Şanal[1], Volga Bulmuş[2]

[1]*Biotechnology and Bioengineering Graduate Program, Izmir Institute of Technology, Urla, Izmir, 35430 Turkey*
[2]*Department of Bioengineering, Izmir Institute of Technology, Urla, Izmir, 35430 Turkey*

Non-viral transfection vehicles have been widely used in intracellular delivery of macromolecular therapeutics. Although they are less efficient in transfection ability, they are safer when compared with viruses and bacteria. Among non-viral vehicles, lipid-based systems usually have the highest transfection efficiency. Polymer-based systems, although they are in general less effective in transfections compared to lipids, are attractive alternatives as they can be easily endowed with desired properties including high stability and stimulus-responsive behavior.

We intended to synthesize fatty acid containing polymers as potential components of intracellular drug delivery systems combining the transfection efficiency of lipid-based systems with versatility of synthetic polymers. An unsaturated fatty acid, 10-undecenoic acid (UDA), was modified to form a methacrylate monomer (UDAMA). UDAMA was copolymerized with methacrylic acid (MAA) via RAFT polymerization to yield water-soluble, pH-responsive, amphiphilic, well-defined random copolymers. Polymerization of UDAMA and MAA using 4-cyano-4-phenylcarbonothioylthio pentanoic acid as a RAFT agent and 4,4'-azobis(4-cyanovaleric acid) as initiator yielded RAFT-mechanism controlled polymers (Table 1). The pH-dependent membrane-lytic activity of a number of random copolymers with varying UDAMA content and molecular weight was investigated via hemolysis assay. All polymers tested showed pH-dependent hemolytic activity. The activity at pH 5.5 varied between 40% and 80% depending on the molecular weight and UDAMA content of the polymers, while at pH 7.4 the highest hemolytic activity was only 14%. With increasing UDAMA content, membrane destabilization activity of polymers slightly decreased. Similarly, for polymers having the same composition, the hemolytic acitivity decreased with increasing molecular weight of the polymers. The DLS investigation showed pH-dependent self-assembly behavior of the polymers. The pH-responsive, fatty acid polymers are being investigated further in our laboratories as a potential endosome-disrupting agent.

The authors gratefully acknowledge the financial support of The Scientific and Technological Research Council of Turkey (TUBITAK) through the grant# 111T960.

Keywords: RAFT polymerization, endosome-disruptive polymers, intracellular drug delivery

RAFT-controlled polymerization conditions and properties of MAA-co-UDAMA polymers.

[MAA]/[UDAMA]/[RAFT]/[I]	Conversion % (by NMR)	Mn (by GPC)	PDI (by GPC)	UDAMA Content (mole %) of Polymers (by NMR)
[10]/[40]/[1]/[0.25]	30	3880	1,07	15
[20]/[80]/[1]/[0.25]	35	8360	1,28	15
[30]/[120]/[1]/[0.25]	44	14530	1,23	15

MBM-I-008 [Macromolecules in Biotechnology & Medicine]

Biodegradable Polyanhydrides from Radical Thiol-Ene Polymerizations: Surface Eroding Materials with Many Interesting Properties

Devon A. Shipp, Katie L. Poetz, Halimatu S. Mohammed, Brittany L. Snyder, Kelly R. Tillman

Department of Chemistry & Bimolecular Science, Clarkson University, Potsdam, United States

Polyanhydrides have found a niche in the degradable polymer field, largely because they often undergo surface erosion, but demanding synthesis conditions have held them back from widespread use. To address this problem, we have developed novel polyanhydrides that are based on radical thiol-ene 'click' polymerization, an easy to conduct step-growth mechanism of polymerization that can be applied to make materials that have relatively uniform network structure. Further, thiol-ene 'click' polymerization is robust, can be photo-, redox- or thermally-initiated and may use a wide variety of monomers. Thus, these have real potential in a variety of applications, including orthopedics and drug delivery. Discussed in this presentation will the be the synthesis of elastomeric and semi-crystalline polyanhydrides that have controllable degradation rates, and how this approach to network polyanhydrides can provide significant flexibility in tailoring characteristics such as crosslink density, functionality and hydrophilicity. In particular, it will be shown that polyanhydrides can be used in applications such as drug delivery, can also behave as shape memory polymers, exhibit self-healing properties and can be produced as nanoparticles.

Keywords: thiol-ene polymerization, polyanhydrides, click chemistry, surface erosion, drug delivery

MBM-I-009 [Macromolecules in Biotechnology & Medicine]

Surface-attached polymer networks for the detection of rare circulating tumor cells in whole blood

Marc Zinggeler, Frank Scherag, Thomas Brandstetter, Jürgen Rühe

IMTEK – Department of Microsystems Engineering, University of Freiburg, Freiburg, Germany

Circulating tumor cells (CTCs) can be found in patient´s blood at rather low concentrations of only 0.3-100 cells per mL. Nevertheless they are currently becoming more and more important as biomarkers for various cancers. In this work we present a novel dead-end affinity filtration system for the enrichment of circulating tumor cells (CTCs) from whole blood with high efficiency, selectivity and throughput. At the heart of the system lies an affinity membrane, which was prepared by chemical surface modification of a commercial micro-filter to exclusively allow for the specific adhesion of target cells. The high selectivity of the developed capture surface was confirmed both on the level of proteins and cells. The generated membranes can be employed in a simple dead-end filtration setup (e.g. applied as syringe filters) ideally suited for application in point of care settings. First filtration experiments with whole blood and breast cancer cells showed that the membrane is highly permeable for blood cells and allows the efficient capture of cancer cells at high filtration pressure. This promising combination is expected to yield high quality enrichments and could enable the reliable identification and characterization of CTCs directly on the membrane surface (e.g. by immunocytochemistry). We compare the obtained results to an alternative approach where we use a medical wire, which is to be inserted into the veins of patients and designed to capture CTC in vivo. The wire surface is coated with multilayer assembly which is strongly cell repellent, but contains cell-specific antibodies which allow the "fishing" of these cells from blood. In both cases the isolated tumor cells are then available for further downstream genetic analysis.

Keywords: circulating tumor cells, surface-attached polymer networks, C-H-insertion reactions

MN-I-001 [Macromolecules & Nanotechnology]

Single Molecule Properties of Covalently Responsive Polymers

Stephen Craig

Department of Chemistry, Duke University, Durham, NC

Mechanochemically active polymers provide the basis for new classes of stress-responsive and self-healing materials and soft devices. This talk will present single molecule force spectroscopy studies that quantify the covalent response of single polymer molecules to applied forces, yielding insights into the design of reactive polymer backbones and providing the basis for new approaches to responsive materials and chemically active soft devices.

Keywords: Single molecule force spectroscopy, mechanochemistry, reactivity

MN-I-002 [Macromolecules & Nanotechnology]

Simple processes for the preparation of superhydrophobic polymer surfaces

Çağla Koşak Söz, Emel Yılgör, Iskender Yılgör

Chemistry Department, Koç University, Istanbul, Turkey

Two simple processes; spin-coating, and doctor blade coating of silica/polymer dispersions are described for the preparation of superhydrophobic polymer surfaces. To demonstrate the versatility and broad applicability of the processes, polymeric surfaces modified included a thermoplastic resin, polystyrene and a thermoset, crosslinked epoxy resin. Micro/nano hierarchical nature of the surface topographies obtained were demonstrated by scanning electron microscopy, atomic force microscopy and white light interferometry studies. Formation of superhydrophobic surfaces were clearly demonstrated with very high static, advancing and receding water contact angles well above 150 degrees and contact angle hysteresis values of less than 10 degrees, similar to that of lotus leaves [1,2]. Roughness factor (r) and average surface roughness (Ra) values, which are critical in obtaining superhydrophobic surfaces were also determined for each polymeric system.

References:

[1] W. Barthlott, C. Neinhuis, Planta. 1997, 101(1), p.1-8
[2] M. Nosonovski, B. Bhushan, Current Opinion Colloid Interface Sci. 2009, 14(4), p.270-280

Keywords: superhydrophobic, thermoplastic, thermoset, polymer

Figure 1: Superhydrophobic epoxy resin surface with hierarchical micro/nano roughness

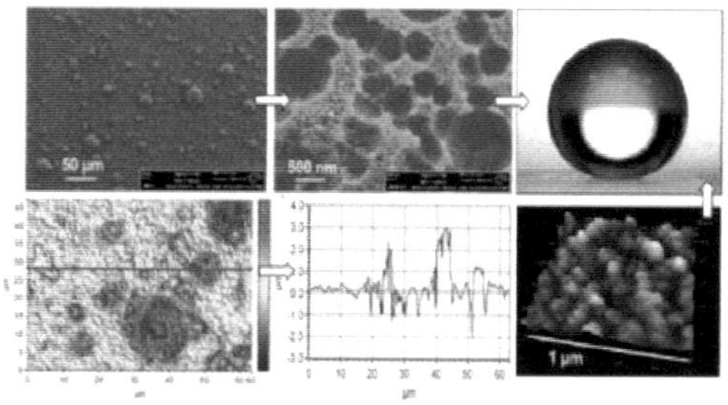

MN-I-003 [Macromolecules & Nanotechnology]

Nano-Patterned Polyelectrolyte Multilayers Spanning Square Centimetre Areas

Meirav Oded, Roy Shenhar

The Institute of Chemistry and the Center for Nanoscience and Nanotechnology, The Hebrew University of Jerusalem, Jerusalem, Israel

Over the last two decades, layer-by-layer (LbL) deposition has emerged as a cost-effective, versatile approach for the fabrication of functional polyelectrolyte coatings with alternating layers, with application ranging from catalysis, separation, and synthesis, through electrochemical and optical devices, to drug delivery and sensing. The depth-wise structure of such films was extensively studied, and the applicability of this technique to various types of surface geometries, including convex and concave substrates, was also demonstrated. Nonetheless, the vast majority of these works describe laterally homogeneous multilayers.

We present a non-lithographic approach for the creation of *nano-patterned* polyelectrolyte multilayers that span square centimetre areas. Our approach utilises the surface patterns afforded by block copolymer films as templates for the selective adsorption of polyelectrolytes. The presentation will discuss the fundamentals of selective polyelectrolyte adsorption on confined nano-domains and will elaborate on recent developments.

Keywords: Layer-by-Layer deposition, Block Copolymers, Self-Assembly, Polyelectrolytes

Shenhar abstract figure

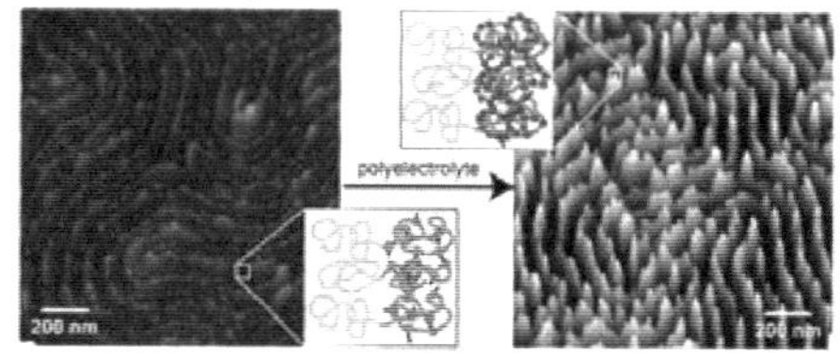

AFM images showing a block copolymer pattern before and after selective deposition of a polyelectrolyte (height scales: left image: 5 nm; right image: 15 nm).

MN-I-004 [Macromolecules & Nanotechnology]

Charge and surface fouling at the nanoscale: Isoelectric points of proteins by Atomic Force Microscopy

Julius G Vancso[1], Shifeng Guo[2], Xiaoying Zhu[2], Dominik Janczewski[3]

[1]*MESA+ Institute for Nanotechnology, University of Twente, the Netherlands*
[2]*Institute of Materials Research and Engineering A*STAR, Singapore*
[3]*Laboratory of Technological Processes, Warsaw University of Technology, Warsaw, Poland*

Biofilm forming, bacterial adhesion, and settlement of fouling species e.g. in marine environments depend on several physico-chemical parameters related to the surface of the substrates and the foulants. The first steps of fouling is often related to protein adsorption. Thus knowledge of the Coulombic charge and protein isoelectric point values are of pivotal importance to understand and control fouling. However, often only trace amounts of unknown (adhesion) proteins are available for characterization, which poses a challenge to conventional analytical techniques. Here we introduce a new method that requires the use of only minute quantities (on the order of hundreds to a few thousands) of macromolecules covalently immobilized to the surface of AFM colloidal probes to determine isoelectric points of proteins. We first validate our approach using well-known structures, and then study the surface charge of "footprint" adhesion proteins secreted by barnacle cyprid larvae of *Amphibalanus Amphitrite* during their search of settlement spots in sea water. We measure adhesion forces of the tip- immobilized proteins against positively and negatively charged robust and smooth surfaces, made by sequential deposition of polyelectrolytes, as a function of pH. The use of such reference substrates allowed us to study the influence of only one parameter (surface charge) while keeping the others (roughness, chemical composition) essentially unchanged. Examples for combating marine and bacterial fouling will be shown to illustrate possible applications.

Zhu, X., Jańczewski, D., Guo, S., Lee, S.S.C., Parra Velandia, F.J., Teo, S.L.M., He, T., Puniredd, S.R., Julius Vancso, G.
Polyion multilayers with precise surface charge control for antifouling (2015) ACS Applied Materials and Interfaces, 7 (1), pp. 852861.

Keywords: molecular isoelectric point, proteins, atomic force microscopy, nanoscale adhesion

MN-I-005 [Macromolecules & Nanotechnology]

Near Field Infrared Microscopy of Polymer Films and Boron Nitride

Gilbert Walker[1], Leonid Gilburd[1], Xiaoji Xu[3], Sissi De Beer[2], Kris Kim[1]

[1]*Department of Chemistry, University of Toronto, Toronto, Canada*
[2]*MSTP, University of Twente, Enschede, Netherlands*
[3]*Department of Chemistry, Lehigh University, Bethlehem, USA*

We present recent results on the mechanical and infrared properties of materials at nanoscale. The focus is will on block copolymers and materials that support surface phonon polaritons, including boron nitride nanotubes. Both single frequency images and infrared-infrared probe results will be provided, where the latter provide information on the coupling of surface phonon modes at nanoscale. The function of the infrared near field scanning instrument is quite simple to understand from a classical perspective: radiation is incident on the interface of a material and a sharp metal tip. The material response includes components that are both in- and out- of phase with the incident radiation – the out-of-phase part contains information on the absorption, and so it is amplified through a homodyning approach.

Keywords: infrared, scanned probe microscopy, imaging, block copolymers, boron nitride

MN-I-006 [Macromolecules & Nanotechnology]

Bi-phasic Hybrid Nanostructures for Functional Applications

Cenk Aktas[1], Salih Veziroglu[3], Ayman Haidar[2], Hashim Abdul Khaliq[2], Franz Faupel[1]

[1]*Christian Albrechts University-Kiel-Germany*
[2]*Saarland University-Homburg-Germany*
[3]*Meliksah University-Kayseri-Turkey*

Biphasic solid state composites of the type metal/metal oxide or element/element oxide can be synthesized in one pot chemical reactions using so called molecular "single source precursors". Due to their singular genesis these composites show peculiar hetero-structures based on core-shell hierarchies such as superlattices and composite nanospheres or nanowires. They exhibit superior or new functional properties compared to their individual constituent compounds. In the current work, we review in particular the synthetical and mechanistical approach of bi-phasic (Al/Al(2)O(3)) nanostructures such as nanospheres, nanowires and nanoloops using a single source precursor. Other bi-phasic materials of the general formula M/MO(x) (for example M = Ge, Sn, Pb) are addressed for comparison. The impact of different synthetical conditions as well as of modification of surfaces by organic polymers and their technological relevance are presented briefly. Additionally, functional applications of the prepared surfaces are explained with some outstanding case studies. These case studies are primarily concerned with their use as biomaterials and their application in medicine as well as with their use as thin films for optics and functional surfaces.

Keywords: Nanostructures, Biomedicine, Hydrophpobic, Omniphobic

MN-I-007 [Macromolecules & Nanotechnology]

Electropolymerization and characterization of polyaniline-titania nanocomposites for sensing applications

Ali Osman Sezer
Chemistry and Physics Department, California University of Pennsylvania, California, Pennsylvania, U.S.A.

Polyaniline-titania nanocomposite films were electrochemically deposited on ITO-coated glass substrates in a 1.0 M HCl solution containing 0.1 M aniline-HCl and 10 wt.% titania by potentiostatic polymerization of 1.0 V for 5 minutes with Pt as the counter electrode. The polymerization was carried out at 25 °C in an ultrasonic water bath. The sol-gel method was employed to synthesize titania nanoparticles with a 41 nm average size in the rutile phase confirmed by x-ray crystallography. SEM images revealed a dense, highly packed film formation on the substrate with nano-sized fibers with an average thickness of 100 nm. The fibers look entangled on a porous surface with clearly visible, spherical titania nanoparticles or clusters. Titania particles were visible both on surface and cross sectional SEM images with random distribution into the polymer matrix. FTIR and Raman spectra of deposited films confirmed the expected polymer structure as well as the presence of titania. XRD analysis of deposited films indicate lost crystallinity of polyaniline due to the presence to titania nano-particles. Electrochemical impedance spectroscopy study revealed low impedance of the nano-composite coated ITO substrates indicating semi-conductive nature of the composite material suitable for sensing applications. This talk will provide a review of conductive polymer-nanoparticle composites, their applications, and present the results of our study of polyaniline-titania nanocomposites.

Keywords: polyaniline, nanocomposite, electropolymerization, electrochemical impedance

SEM images of electrodeposited polyaniline-titania nanocomposite film on ITO-coated glass

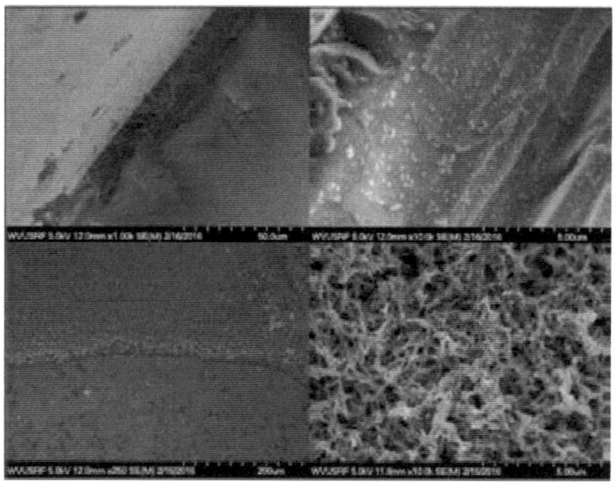

The SEM images show both surface and cross-sectional packing of the electrodeposited polyaniline-titania nano-composite. The random distribution of titania nanoparticles into the fibrous matrix of the host polyaniline in the doped form is clearly visible in these images.

MSP-I-001 [Modeling and Simulation of Polymers]

Simulation-Directed Discoveries of Concepts in Polymers: Primordial Nucleation, Translocation, and Charged Macromolecular Assemblies

Murugappan Muthukumar

Polymer Science and Engineering Department, University of Massachusetts, Amherst, MA, USA

Simulations of macromolecules by using a variety of computational algorithms and tools have generated a wealth of fundamental understanding of basic principles of macromolecular behavior. In addition, various simulations conducted by many scientists worldwide have aided discovery of new concepts, which control myriads of very complex phenomena exhibited by macromolecules. Such progress will be illustrated by considering the following major challenges in polymer science: (1) The primordial stage of polymer crystallization, (2) The effective charge of a macromolecule in polar solutions, (3) The translocation of a charged macromolecule, both synthetic and natural such as DNA, through protein channels and solid-state nanopores, and (4) Stability and kinetics of macromolecular assemblies such as viruses. Complementarity among various simulation methods, theory, and experiments will be emphasized.

Keywords: Polymer crystallization, nucleation, charge regularization, translocation, virus assembly

MSP-I-002 [Modeling and Simulation of Polymers]

Rubber Elasticity and Protein Behavior

Burak Erman

Chemical and Biological Engineering, Koc University, Istanbul, Turkey

Native proteins are polypeptides folded into a condensed state at a density of about 1.2-1.3 g/cm$_3$. Each atom of the protein in the condensed state exhibits large scale fluctuations about its mean position. The magnitudes of mean-square fluctuations vary in the range 0-5Å^2. Fluctuations about the mean positions may be approximated as Gaussians, thus leading to a model where each atom is attached to its center by a linear spring. This model, referred to as the Gaussian Network Model (GNM) of proteins in the literature, is in close resemblance to the molecular model of an amorphous, cross-linked rubbery elastic polymer. As in rubber elasticity, the amplitude of fluctuations of an atom varies inversely with the number of other atoms in its neighborhood. This imposes a strong relationship between protein structure and fluctuations and correlations of fluctuations. The function of proteins are determined to a large extent by the correlated fluctuations of its atoms.

The Onsager equation, $\langle \Delta Ri \Delta Rj \rangle = kBT\ (\partial Ri)/(\partial Fj) = (\Gamma^{-1})ij$ describes this important relationship between structure and function. Here, $\langle \Delta Ri \Delta Rj \rangle$ is the correlation between the fluctuations of atom i and j, kB is the Boltzmann constant, Ri is the equilibrium position of atom i, Fj is the equilibrium force on atom j, and Γ^{-1} is the ij›th element of the inverse of the spring constant matrix of the protein. This simple picture that has its origins in the theory of rubber elasticity has been useful in understanding the behavior of proteins.

Keywords: rubber elasticity, protein behavior, Gaussian Network Model

MSP-I-003 [Modeling and Simulation of Polymers]

Onsager Principle -A useful principle in soft matter dynamics

Masao Doi

Center of soft matter physics and its applications, Beihang University, Beijing, China

In the celebrated paper on the reciprocal relation for the kinetic coefficients in irreversible processes, Onsager extended Rayleigh's principle of the least energy dissipation to general irreversible processes. The principle has been shown to be useful in deriving basic equations which describe non-linear and non-equilibrium phenomena in soft matter. Here I will show that the principle is useful in getting an approximation solutions. Examples are given for contact line dynamics of liquid and gelation dynamics.

Keywords: soft matter, variational principle, diffusion, gel dynamics

MSP-I-004 [Modeling and Simulation of Polymers]

Directed Assembly of Polymeric Materials - Non-Equilibrium States and On-The-Fly Characterization of Assembly

Juan J De Pablo

Institute for Molecular Engineering - University of Chicago

This presentation will focus on the directed assembly of polymeric materials. Particular emphasis will be placed on the development of models capable of describing the dynamics of entangled polymers, and their use for design of assembly processes based on the use of surface patterns, topography, and solvent annealing strategies. The predictions of the models will be validated by comparison to experimental data, both at equilibrium and far from equilibrium. For equilibrium scattering measurements, we will present a new approach that relies on physics-based, evolutionary interpretation of data to extract information from multi-modal experiments. For dynamic measurements, we will present results of simulations at multiple levels of description, and compare these to results from rheological measurements and real-time AFM imaging of assembly processes.

Keywords: copolymer, directed assembly, x-ray scattering, equilibrium, non-equilibrium

MSP-I-005 [Modeling and Simulation of Polymers]

Thermodynamics and dynamics of polymers at soft interfaces

Paola Carbone

School of Chemical Engineering and Analytical Science, The University of Manchester, Oxford Road, M13 9PL, Manchester (UK)

The prediction of structural and dynamical properties of polymers at liquid/liquid interfaces is both a technological and a biological problem. Indeed, polymers are confined at or pass through liquid interfaces in many industrial processes, such as liquid/liquid extractions, solvent displacement methods, or emulsifications, and also when used for biological applications, such as drug nanocarriers, biocompatibilizers, or protective coatings. In these situations, the polymer chains can dissolve in one of the two solvents or reside at the interface, depending on their relative solubility. The latter case is quite common since, in order to lower the interfacial free energy, polymers can be adsorbed at the interface, behaving as surface active molecules. [1-3] When this situation occurs, the polymer chain is confined in a "soft slit" characterized by a solvent density that is lower than the bulk one. In this talk I'll present how computer modelling can help in predicting the thermodynamic stability and dynamics of homo and copolymers at soft interfaces. [4-7]

[1] L. Hu, M. Chen, X. Fang and L. Wu, Chem. Soc. Rev., 2012, 41, 1350–1362
[2] S. Sheiko et al., Nature Materials, 2013, 12, 735-740
[3] S. Nawaz, M. Redhead, G. Mantovani, C. Alexander, C. Bosquillon and P. Carbone, Soft Matter, 2012, 8, 6744–6754
[4] S. Nawaz, P. Carbone, J. Phys. Chem. B, 2011, 115, 12019–12027
[5] D. L. Cheung, P. Carbone, Soft Matter, 2013, 9, 6841– 6850
[6] T. Taddese, P. Carbone, D. L. Cheung, Soft Matter, 2015, 11, 81-93
[7] T. Taddese, D. L. Cheung, P. Carbone, ACS Macro Lett., 2015, 4, 1089–1093

Keywords: dynamical properties of polymers, liquid/liquid interfaces, modelling

MSP-I-006 [Modeling and Simulation of Polymers]

Glassy and nonlinear dynamics of polymers on random and patchy surfaces

Guido Raos

Department of Chemistry, Materials and Chemical Engineering "G. Natta", Politecnico di Milano, Milano, Italy

The behaviour of polymers on solid surfaces and under confinement is important for a number of reasons. My own interest in the subject originates from the need to understand the nature of the mechanical reinforcement of elastomers by solid nanoparticles, such as carbon black.[1] In the last few years, my coworkers and I have carried out several molecular dynamics simulations of both confined polymer melts and adsorbed polymer monolayers. The case of chemically heterogeneous surfaces, made up of a random mixture of weakly and strongly absorbing sites, is especially intriguing.[2] Due to energetic disorder, the slowing down of the polymer dynamics (measured by the activation energy for diffusion and the glass transition temperature) is maximum when the fraction of strongly interacting sites is f=0.75, instead of f=1.00 as might have been expected.[3,4] Finally, I will discuss our preliminary results on the dynamics of confined polymer melts, in which the spatial correlations in the chemical composition of the surfaces can be tuned to bridge random and patchy surface models (see Figure 1 and Figure 2).

References:

(1) V. A. Froltsov, M. Klüppel, G. Raos, *Molecular dynamics simulation of rupture in glassy polymer bridges within filler aggregates*, Phys. Rev. E **86**, 041801 (2012).

(2) G. Raos and T. J. Sluckin, *Pulling polymers on energetically disordered surfaces: molecular dynamics tests of linear and non-linear response*, Macromol. Theory Simul. **22**, 225-237 (2013).

(3) G. Raos and J. Idé, *Impact of interaction strength and surface heterogeneity on the dynamics of adsorbed polymers*, ACS Macro Lett. **3**, 721-726 (2014).

(4) R. Pastore and G. Raos, *Glassy dynamics of a polymer monolayer on a heterogeneous disordered substrate*, Soft Matter 11, 8083-8091 (2015).

Figure 1: Simulation of a polymer melt interacting with a random surface model (f=0.5).

Figure 2: Simulation of a polymer melt interacting with a patchy surface model (f=0.5).

Keywords: Molecular dynamics, polymer-surface interactions, glass transition, nanocomposites, rubber reinforcement

MSP-I-007 [Modeling and Simulation of Polymers]

Process-structure-property relationships for semicrystalline polymers: Crystallization from the melt and deformation mechanics

Gregory C Rutledge

Department of Chemical Engineering, MIT, Cambridge MA, USA

The properties and performance of semicrystalline polymers are dependent upon both the molecular scale architecture of the chains and the manner in which they are processed to form engineering materials. Much of the desirable behavior of these materials is a consequence of the semicrystalline morphology, in which crystallites are organized within an amorphous matrix, constrained by the connectivity of the chain molecules that compose both components of the resulting "nanocomposite". However, obtaining quantitative relationships between process, structure and properties by simulation requires realistic, chemical models. The kinetics of nucleation and growth of crystalline domains in polyolefins, in either the presence or absence of process flows, and the eventual realization of semicrystalline morphology continue to demand attention for both intellectual and commercial reasons. In this talk, we provide an update on the development of molecular simulations to illuminate the evolution, structure and properties of semicrystalline polyolefins from the amorphous melt. In the area of crystallization kinetics, nucleation and growth of crystallites in the quiescent melt have been extended to include the characterization of flow-enhanced nucleation of long alkanes. In the area of mechanics, large strain deformation of the lamellar stack model for a semicrystalline polymer provides insight into the complicated mechanisms of mechanical response. Yielding and the origins of toughness in polyolefins are evaluated at the molecular level.

Keywords: polyethylene, simulation, crystallization, nucleation, mechanics, plasticity

MSP-I-008 [Modeling and Simulation of Polymers]

Self-assembly of polymeric nanoparticles in flow

Johannes Heuser, Friederike Schmid

Institut für Physik, Johannes Gutenberg-Universität Mainz, Staudingerweg 7–9, 55128 Mainz, Germany

We present a method to incorporate hydrodynamic interactions into the successful dynamic density functional (DDFT) simulation method for polymers. This is done by coupling DDFT simulations to a Lattice-Boltzmann model that solves the Navier-Stokes equations in a consistent manner. As an application example, the method is applied to studying nanoparticle self-assembly from copolymer solutions in Poiseuille flow geometry. The shear induces both reversible and irreversible shape transformations in the nanoparticles and can hence be used to manipulate the final shapes of particles.

Keywords: self-assembly, dynamic density functional, nanoparticles

Figure

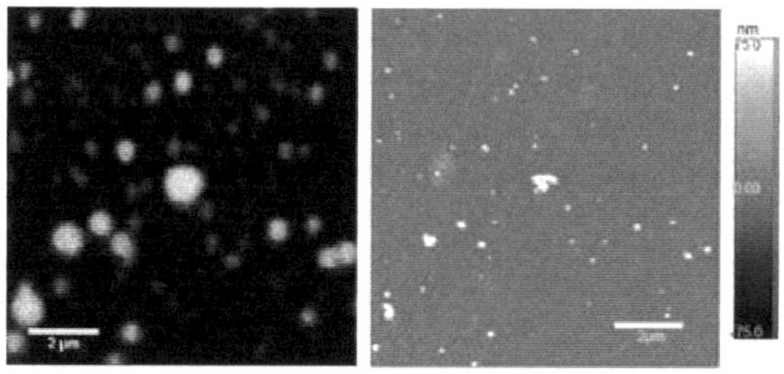

FLIM (left) - AFM (right) image of partially dried polystyrene-block-poly acrylic acid block copolymer vesicles on solid support. The different colors denote different average lifetimes of the probe dye calcein in different polymersomes, which are related to the different extent of drying according to vesicle size.

PCNCT-I-001 [Polymer Characterization by New and Combined Techniques]

Combined AFM - confocal fluorescence lifetime imaging microscopy (FLIM) to interrogate nanoenvironments in bacteria-sensing polymeric vesicles

Holger Schönherr, Stephan Wang, Sergey Druzhinin, Katrin Stephanie Tücking, Simon Haas, Cleiton Kunzler, Anna Schulte

Physical Chemistry I, Department of Chemistry & Biology, University of Siegen, Germany

Fluorescence lifetime imaging microscopy (FLIM) is an outstanding method for the analysis of fluorescently labelled species in biology, medicine and material science. By combining FLIM with atomic force microscopy (AFM), a powerful combination of complementary characterization methods can be realized that probes local properties in soft matter systems with an unprecedented range of information. Specifically, we report here on the analysis of the morphology and local properties of polymeric vesicles, which are developed in our group among others for theranostics (therapy and diagnostics) applications, i.e. for detection and treatment of bacterial wound infections, by combined AFM-FLIM. By identifying different excited state lifetimes of the local probe dye calcein, three microenvironments were confirmed in various enzyme-labile polymersomes of amphiphilic block copolymers: (1) solvated dye, (2) dye in the vesicle wall and (3) agglomerated dye. These experiments allow one to unequivocally differentiate vesicles from other polymer aggregates. Subsequently, the drying kinetics of the polymersomes was determined by analyzing the change of amplitudes of these excited state lifetimes.

It was found that the vesicles dried within minutes to hours, depending on the radius of the vesicles. The time constants for the drying process were determined, which were utilized to calculate the membrane wall thickness.

This demonstrates that it is possible to confirm the morphology of block copolymer vesicles by combined AFM-FLIM. Moreover, FLIM was applied as an effective method to determine the apparent diffusion constant of water vapor across the block copolymer membrane of vesicles. Finally, combined AFM-FLIM data from analyses of individual polymersomes will be reported. These experiments yielded information about the location of the probe dye ATTO 647N inside the vesicles, the number of dye molecules inside one vesicles and hence the local dye concentration, as well as the mechanical properties of the vesicle shell.

Keywords: atomic force microscopy, fluorescence lifetime imaging microscopy, polymersomes

PCNCT-I-002 [Polymer Characterization by New and Combined Techniques]

Characterization of Polyethylenes by Thermorheological Methods

Helmut Münstedt

Institute of Polymer Materials, University Erlangen-Nuremberg

Polyethylenes have the biggest market share among commercial polymeric materials. One reason for this success is their great variety of chemical and molecular structure. Ethylene and ethylene with various comonomers are used for polymerization. Beside the variation of molar mass and molar mass distribution common for polymeric materials, short-chain and long-chain branches can be introduced that significantly change processing and end-use properties of the different grades. The analysis of long-chain branched polyethylenes is still a challenging task, particularly because their composition can be rather heterogeneous. In this presentation it is shown how investigations of thermorheological properties can contribute to a characterization of various branched species. The rheological behavior of polymer melts distinctly depends on time and temperature. For many materials, time and temperature dependencies of rheological quantities are related to each other by the time-temperature superposition principle (TTS), which is based on the fact that all the relaxation times of a material change by the same factor only depending on temperature. Such a behavior is called thermorheologically simple. Polymers that do not fulfill TTS are denominated thermorheologically complex. Detailed information on these topics can be found, for example in the book "Deformation and Flow of Polymeric Materials" by H. Münstedt and F.R. Schwarzl, Springer Heidelberg (2014). It is demonstrated that short-chain branched polyethylenes behave thermorheologically simple, long-chain branched polyethylenes, however, thermorheologically complex. Furthermore, it is shown that the thermorheological complexity gives hints to long-chain branches in polyethylenes, which could not be detected by the classical characterization method of size exclusion chromatography combined with multi-angle laser light scattering (SEC/MALLS). For a more quantitative evaluation of the behavior of thermorheologically simple and complex polyethylenes, activation energies are considered. It is shown how activation energies determined from dynamic-mechanical measurements may contribute to an analysis of the molecular structure and the composition of polyethylenes.

Keywords: Thermorheological behaviour, activation energies, long-chain branching, polyethylenes, analysis of composition

PCNCT-I-003 [Polymer Characterization by New and Combined Techniques]

Micromechanics of Polymers: Electron Microscopic Methods to Investigate Micro- and Nanoscopic Processes of Deformation and Fracture

Sven Henning[1], Goerg H. Michler[2]

[1]*Fraunhofer Institute for Microstructure of Materials and Systems IMWS, Walter-Huelse-Str. 1, 06120 Halle (Saale), Germany*
[2]*Institute of Physics, Martin Luther University Halle-Wittenberg, 06099 Halle (Saale), Germany*

The main aims of micromechanical analyses are the identification of property determining micro- and nanostructures and the description of the principal deformation phenomena that occur under mechanical load, as there are, for instance, crazing, homogeneous deformation, shear band formation, micro- and nanovoid formation, fibrillation, etc. These processes can be correlated to a brittle, semiductile or ductile material behavior on the macroscopic scale. In that sense, the description of micromechanical phenomena can contribute to the formulation of material laws predicting the response of a material to mechanical load. Micromechanical processes are the link between the morphology (predicted by synthesis, composition, and processing), loading conditions and resulting mechanical properties. There are a number of different micromechanical techniques that can be adapted with respect to the actual problem. First of all, there are micromechanical in situ methods characterized by the application of tensile load, compressive or bending force using miniaturized testing devices inside the TEM, (E)SEM or AFM. Alternatively, microscopic analyses of deformation zones can be performed after mechanical testing ex situ with correlation to actual states of deformation. That approach also includes the inspection of fracture surfaces and, to some extent, problems of failure analysis. Thirdly, micromechanical techniques also include the application of miniaturized mechanical testing devices for small samples (e.g., micro-parts, biological samples, nanofibers) or for mechanical testing of small amounts of materials from new syntheses. Finally, the micromechanical approach allows the measurement of local mechanical properties in microscopic regions of interest. The presentation also covers different preparation routes with a special emphasis on preparation by means of (cryo-)ultramicrotomy. The advantages of the different techniques are illustrated by a number of examples covering semicrystalline polymers, block copolymers, composites and nanocomposites with different filler geometries (particles, layered structures, carbon nanotubes) and biological/biomedical materials (bone, electrospun nanofibers).

Keywords: Micromechanical processes, electron microscopic methods, deformation and fracture

PCNCT-I-004 [Polymer Characterization by New and Combined Techniques]

Precise characterization of polymers prepared by anionic polymerization and controlled radical polymerization

Taihyun Chang

Department of Chemistry and Division of Materials Science, POSTECH, Pohang, Republic of Korea

After Szwarc first demonstrated the mechanism of living anionic polymerization of styrene in 1956, anionic polymerization has been used widely to prepare well-defined polymers in molecular weight and chain architectures. The molecular weight distribution (MWD) of such living polymers was theoretically predicted by Flory even earlier in 1940. The experimental verification had to wait for 60 years until interaction chromatography (IC) and MALDI-MS became available due to the limited resolution of size exclusion chromatography (SEC) that has been used commonly for molecular characterization of synthetic polymers. In particular, the precise characterization of model branched polymers that are often prepared by anionic polymerization and subsequent linking of macromers for either backbone or branches is practically impossible by SEC. Let alone the intrinsic low resolution due to the serious band broadening, SEC separates polymers according to their hydrodynamic size and the size of branched polymers does not change as much as linear polymers with respect to their molecular weights. Atom transfer radical polymerization (ATRP) is a very useful method to obtain polymers with relatively narrow MWD and various chain-end functionalities (CEF). However, dead chains are always formed by suppressed but unavoidable termination reaction between free radicals while chains are growing that results imperfect CEF. Therefore, their MWD is intrinsically an overlapped distribution of both living and dead chains and the true MWD and CEF of polymers prepared by ATRP is yet to be addressed critically. The living and dead chains from ATRP can be separated by HPLC according to the CEF after the chain-end substitution reaction. In this presentation, precise molecular characterization of anionic-polymerized polymers and of living and dead chains of ATRP polymers will be discussed.

Keywords: Characterization, HPLC, anionic-polymerization, ATRP

PCNCT-I-005 [Polymer Characterization by New and Combined Techniques]

Investigation of Polymer Reactions by GPC/SEC with Mass Spectrometry

Peter Kilz[1], Christopher Barner Kowollik[2], Till Gründling[3]
[1]PSS Polymer Standards Service, POB 3368, 55023 Mainz Germany
[2]KIT, Macromolecular Chemistry, Karlsruhe, Germany
[3]BASF SE, Ludwigshafen, Germany

Size-exclusion chromatography (SEC, aka GPC) is the most versatile and widely used technique in macromolecular characterization. However, it suffers from limited selectivity which can be a major limitation in the analysis of complex macromolecules like copolymers, functionalized polymers and also in the determination of by-products which widely occur in any polymerization reaction and limit the efficiency of polymerization processes. As any separation technique, GPC/SEC cannot identify the separated species (even if they are well separated) without hyphenation with information-rich detection systems like mass spectrometry or NMR.

It has been shown that the combination of spectroscopic detection (e.g. FTIR) and size-separation can be efficiently used to elucidate polymerization pathways and identify multiple species simultaneously. Mass spectrometry is one of the most powerful spectrometric methods but has been difficult to interface with GPC/SEC systems (MALDI) or has been limited to the very low molar mass range (ESI).

The recent improvement in mass spectrometer design with efficient and gentle ionization and improved data analysis methods allows to apply these techniques to characterize complex macromolecules both from academia and industrial products alike by a hyphenation of GPC/SEC with MS devices. Research groups such as Karlsruhe Institut of Technology (KIT), National Institute of Standards and Technology (NIST), German Federal Institute for Materials Research and Testing (BAM), etc., have developed powerful analytical methods to successfully investigate polymers with mass spectrometry. The availability of GPC-MS software solutions designed for non-MS experts will enable each lab to take advantage of the new capabilities and obtain comprehensive information about new and complex macromolecules.

This presentation will present the latest findings and developments in overcoming the limitations of mass spectrometry with regard to synthetic polymers.

The high performance of modern MS instruments in combination with modern GPC-MS software empowers basically every lab to investigate not only the absolute molar mass, but also to disclose other properties, like:

- species with different end groups and cycles
- homopolymer contamination in copolymers
- functionalization efficiency and endgroup identity
- branch points in star and comb polymers

- analysis of chain termination in polymer synthesis
- information on chain degradation during aging/stress
- comprehensive out-of-spec tests

References
1. W.J. Simonsick et al., Rapid Comm. Mass Spec, 1993, 7, 853
2. C. Barner-Kowollik et al., Macromolecules, 2009, 42, 6366

Keywords: Size-exclusion chromatography, mass spectrometry, polymer reactions

Figure 1. WinGPC MS window: detector signals (top): RI (blue), MS TIC (red); bottom: mass spectrum at the elution volume of the green fraction marker with automatic assignment of charge-state, degree of polymerization and molar mas of end group

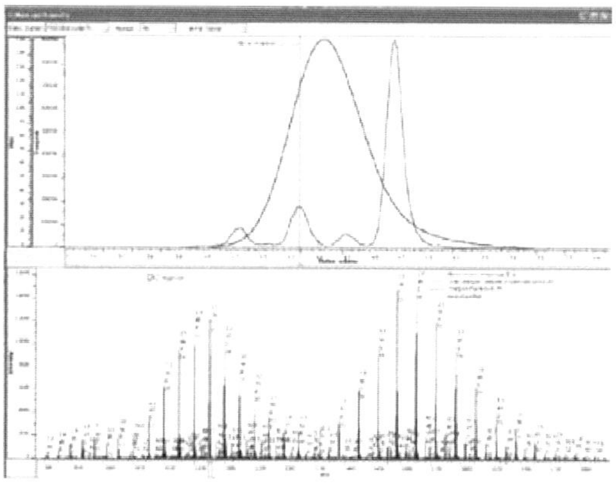

PCNCT-I-006 [Polymer Characterization by New and Combined Techniques]

Study of the effect of physicochemical parameters on chitosan-oleic acid nanoparticles size for encapsulation of anti-inflammatory hydrophobic drug

Paula Andrea Mendez[1], Gloria Vasquez[2], Carmiña Gartner[1], Betty Lucy Lopez[1]

[1] Grupo de Investigación Ciencia de los Materiales, Instituto de Química, Facultad de Ciencias Exactas y Naturales, Universidad de Antioquia, Calle 70 N° 52-21, Medellín, Colombia
[2] Grupo de Inmunología Celular e Inmunogenética, Facultad de Medicina, Universidad de Antioquia, Calle 70 N° 52-21, Medellín, Colombia

Self-assembled nanoparticles prepared from hydrophobic chitosan/oleic acid (Ch-OA) have showed antibacterial activity [1], efficacy in biomedical diagnostic [2] and as carriers for hydrophobic anticancer drugs [3]. In this study, a low molecular weight chitosan was modified with oleic acid with a degree of substitution of 12%. The self-aggregated nanoparticles were formed at low critical aggregation concentration of 0,025 mg/mL. The Ch-OA nanoparticles size was optimized by changing polymer concentration, pH, solvent, type and time of homogenization. The results showed that all these parameters affected the nanoparticle size. At the highest polymer concentration (1.2 mg/mL), the smallest size and the lowest polydispersity index were obtained as a result of stronger hydrophobic interactions in the nanoparticle core [4].

From Table 1, it can be observed that the nanoparticles size deceases in water, this behavior is associated with the change of chitosan conformation at higher pH, where the intermolecular interactions decrease [5]. Furthermore, by the magnetic stirring method, the particle size growth as the stirring time increases since prolonged stirring time allows the aggregation of larger number of polymer chains [6]. On the contrary, at higher sonication times the particle size decreases due to the shear forces, which break the largest particles and form smaller particles that consist of fewer number of Ch-OA chains [7].

Table 2 shows the effect of the organic solvent on the nanoparticles size of Ch-OA. The smallest particle size was obtained by using the fully water-soluble solvents (acetone and ethanol), since have faster diffusion rate which avoids the aggregation of the nanodroplets in the emulsion [8][9].

Table 1. Influence of the homogenization method on size and polydispersity index, in buffer and water

Homogenization method	Homogenization time	Particle size ± SD in buffer[a] (nm)	PdI ± SD	Particle size ± SD in water (nm)	PdI ± SD
Magnetic stirring	2 h	294.9±4.1	0.16±0.05	246.7±3.9	0.32±0.13
	10 h	292.1±1.0	0.33±0.15	271.9±45.2	0.45±0.14
	19 h	326.7±3.7	0.17±0.03	344.6±18.9	0.13±0.02
Sonication	2 min	331.5±5.3	0.24±0.11	339.0±4.2	0.12±0.03
	5 min	278.9±23.8	0.30±0.14	274.4±33.7	0.31±0.23
	8 min	261.1±11.3	1.39±0.39	258.1±10.7	0.98±0.70

[a] Acetic acid buffer pH 4.58

Table 2. Influence of the type organic solvent on size and polydispersity index

Solvent	Particle size ± SD in water (nm)	PdI ± SD
Blank	339.0±4.2	0.12±0.03
Acetone	325.5±8.7	0.08±0.02
Ethanol	351.9±7.1	0.11±0.02
Acetone : Ethanol 1:1 (v/v)	356.7±8.8	0.10±0.02
Ethyl acetate	603.7±78.5	7.57±5.34

The morphology of the particles was spherical, Figure 1. The self-assembled systems showed a drug loading about 7 μg.mL^{-1} and %EE close to 75.8% (±3.6) for Celecoxib, and stability to the store during four weeks. In vitro release study in PBS was done.

Figure 1. Morphology of the Ch-OA nanoparticles (a) TEM image and (b) SEM image of Ch-OA nanoparticles with concentration 1.2 mg/mL

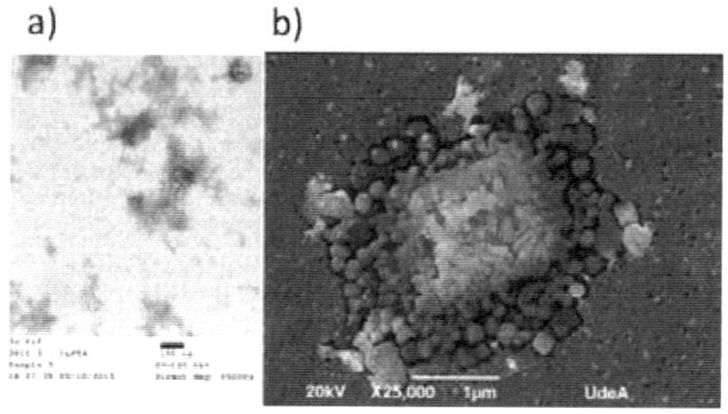

References

[1] L. Huang, X. Cheng, C. Liu, K. Xing, J. Zhang, G. Sun, X. Li, and X. Chen, "Preparation, characterization, and antibacterial activity of oleic acid-grafted chitosan oligosaccharide nanoparticles," *Front. Biol. China*, vol. 4, no. 3, pp. 321–327, May 2009.

[2] C.-M. Lee, D. Jang, J. Kim, S.-J. Cheong, E.-M. Kim, M.-H. Jeong, S.-H. Kim, D. W. Kim, S. T. Lim, M.-H. Sohn, Y. Y. Jeong, and H.-J. Jeong, "Oleyl-chitosan nanoparticles based on a dual probe for optical/MR imaging in vivo.," *Bioconjug. Chem.*, vol. 22, no. 2, pp. 186–92, Feb. 2011.

[3] J. Zhang, X. G. Chen, Y. Y. Li, and C. S. Liu, "Self-assembled nanoparticles based on hydrophobically modified chitosan as carriers for doxorubicin," *Nanomedicine Nanotechnology, Biol. Med.*, vol. 3, no. 4, pp. 258–265, 2007.

[4] Y.-Y. Li, X.-G. Chen, L.-M. Yu, S.-X. Wang, G.-Z. Sun, and H.-Y. Zhou, "Aggregation of hydrophobically modified chitosan in solution and at the air–water interface," *J. Appl. Polym. Sci.*, vol. 102, no. 2, pp. 1968–1973, Oct. 2006.

[5] T. López-León, E. L. S. Carvalho, B. Seijo, J. L. Ortega-Vinuesa, and D. Bastos-González, "Physicochemical characterization of chitosan nanoparticles: electrokinetic and stability behavior.," *J. Colloid Interface Sci.*, vol. 283, no. 2, pp. 344–51, Mar. 2005.

[6] N. Nazeri, M. R. Avadi, M. A. Faramarzi, S. Safarian, G. Tavoosidana, M. R. Khoshayand, and A. Amani, "Effect of preparation parameters on ultra low molecular weight chitosan/hyaluronic acid nanoparticles.," *Int. J. Biol. Macromol.*, vol. 62, pp. 642–6, Nov. 2013.

[7] E. Esmaeilzadeh-Gharedaghi, M. A. Faramarzi, M. A. Amini, A. Rouholamini Najafabadi, S. M. Rezayat, and A. Amani, "Effects of processing parameters on particle size of ultrasound prepared chitosan nanoparticles: an Artificial Neural Networks Study.," *Pharm. Dev. Technol.*, vol. 17, no. 5, pp. 638–47, 2012.

[8] M. G. Nava-Arzaluz, E. Piñón-Segundo, A. Ganem-Rondero, and D. Lechuga-Ballesteros, "Single emulsion-solvent evaporation technique and modifications for the preparation of pharmaceutical polymeric nanoparticles.," *Recent Pat. Drug Deliv. Formul.*, vol. 6, no. 3, pp. 209–23, Dec. 2012.

[9] K. C. Song, H. S. Lee, I. Y. Choung, K. I. Cho, Y. Ahn, and E. J. Choi, "The effect of type of organic phase solvents on the particle size of poly(d,l-lactide-co-glycolide) nanoparticles," *Colloids Surfaces A Physicochem. Eng. Asp.*, vol. 276, no. 1–3, pp. 162–167, 2006.

Keywords: Nanoparticles size, chitosan, oleic acid, celecoxib, anti-inflammatory.

PCNCT-I-007 [Polymer Characterization by New and Combined Techniques]

Magnetic Properties of π-stacked polymers

Young Wan Kwon[1], Chang Hoon Lee[2], Jung Il Jin[1], Min Ho Yang[3], Yves Henry Geerts[4], Tamaki Nakano[5]

[1]*KU-KIST Graduate School of Converging Science and Technology, Korea University, Seoul, Korea*
[2]*Department of Biochemical and Polymer Engineering, Chosun University, Korea*
[3]*Seoul Branch, Korea Basic Science Institute, Seoul, Korea*
[4]*Laboratoire de Chimie des Polymères, Université libre de Bruxelles, Bruxelles, Belgium*
[5]*Division of Macromolecular Science, Institute for Catalysis, Hokkaido University,Sapporo, Japan*

DNAs have been commonly regarded only as the important biomaterials connected to genetic information and hereditary. However, they are recently recognized as very important materials in nanobioscience and nanobiotechnology because of their unique structural and functional characteristics, for examples, self-assembly, molecular recognition, molecular wire, formation of metal nanoparticles, optics and optoelectronics, and organic molecular magnets, etc. [1~3].

In the earlier reports, we showed that for the first time the natural DNAs in dry state exhibit an intrinsic magnetism [4, 5] by relying on electron paramagnetic resonance (EPR) spectroscopy and superconducting quantum interference device (SQUID) measurement. Especially, organic based high-temperature ferromagnetic materials have been attempted by many researchers for lightweight, low-cost applications. In 1968, Itoh et al. proposed that large planar alternate hydrocarbons comprised of meta-substituted triplet diphenylcarbene moieties should have high spin ground state in accord with Hund's rule and spin polarization. However, the realization of a system exhibiting ferromagnetic ordering was not achieved. In our studies, the natural DNA reveals very discernible characteristic EPR peaks in three regions: a very broad signal of g>10 whose minimum appears at ca.1500G and the others of g~2.0 and 2.1 around 3500G. These peaks were explained by the presence of spins in morphologically different regions, i.e., ordered and disordered regions in the DNA.

In this presentation, magnetic properties of natural and modified DNAs including metal-complexed DNAs will be discussed. And the magnetic properties of helical structural based compounds like an organic discotic liquid crystals [6] and π-stacked polymers as mimics of DNA for the development of high temperature ferromagnetic materials will be also discussed.

These findings are expected to open a new research field as well as high technological applications of organic materials, for examples, in improved data storage and organic spintronics.

Acknowledgement: This work was supported by the Basic Science Research Program of the National Research Foundation of Korea (NRF) funded by the Ministry of Science, ICT and Future Planning (NRF-2013R1A1A2062323).

References

[1.] Y. -W. Kwon, C. H. Lee, D. -H. Choi, and J. -I. Jin, *J. Mater. Chem.* **19**, 1353-1380(2009).
[2.] Y.-W. Kwon, D.-H. Choi, J.-I. Jin, *Polymer J(Jpn.)* **44,** 1191-1208**(**2012).
[3.] Lee, C. H.; Kwon, Y.-W.; Jin, J.-I. In *Materials Science of DNA*; Jin, J. I., Grote, J. G., Eds.; (CRC press: Boca Ranton), 121(2012).
[4.] Lee, C. H.; Kwon, Y.-W.; Do, E. D.; Choi, D. H.; Jin, J.-I.; Oh, D. K.; Kim, J. *Phys Rev B* **73**, 224417(2006)
[5.] Kwon, Y.-W.; Lee, C. H.; Do, E. D.; Choi, D. H.; Jin, J.-I.; Kang, J. S.; Koh, E. K. *B Korean Chem Soc* **29**, 1233(2008).
[6.] C. H. Lee, Y. -W. Kwon, D. -H. Choi, Y. H. Geerts, E. -K. Koh, J. -I. Jin, *Adv. Mater.*, **22**(39) 4405–4409(2010**)**, .

Keywords: Ferromagnetic, DNA, Superconducting Quantum Interference Device (SQUID), Electron Paramagnetic Resonance (EPR), π-stacked polymers,

PCNCT-I-008 [Polymer Characterization by New and Combined Techniques]

Polymer Characterization with Compact NMR

Bernhard Bluemich, Kawarpal Singh, Yadollah Teymouri

Institut fuer Technische und Makromolekulare Chemie, RWTH Aachen University, Aachen, Germany

Conventional NMR instruments have large and fragile magnets, which are operated in dedicated laboratories. While compact instruments have been in use for over two decades for relaxation analyses of polymer materials, compact spectrometers are now available for chemical analysis on the workbench [1,2]. The features of compact NMR instruments are reviewed and their use for polymer analysis is illustrated with examples concerning aging phenomena [3] and composition of polymer materials as well as chemical analysis of raw rubber from different sources [4,5]. By NMR relaxometry the morphological changes of different polyethylenes upon solvent and temperature exposure can readily be characterized in the laboratory, and the rubber fraction in rubber-modified asphalt can be quantified nondestructively outside on the road. NMR spectroscopy has been shown to be capable of determining the wax fraction of factory-grade LDPE pellets with better accuracy than NMR relaxometry, and differences in NR from different plantations as well as SBR from different lots and different manufacturers are identified in ^{1}H and ^{13}C NMR spectra acquired with compact NMR instruments.

[1] B. Blümich, S. Haber-Pohlmeier, W. Zia, Compact NMR, de Gruyter, Berlin, 2014
[2] K. Singh, B. Blümich, NMR spectroscopy with compact instruments, Trends Anal. Chem. (2016) Doi:10.1016/j.trac.2016.02.014
[3] Y.Teymouri, R. Kwamen, B. Blümich, Aging and Degradation of LDPE by Compact NMR, Macromol. Mat. Eng., 300 (2015) 1063-1070
[4] K. Singh, B. Blümich, Desktop NMR Spectroscopy for Quality Control of Raw Rubber, Macromolecular Symposia xx (2016) xxx-xxx
[5] B. Blümich, K. Singh, W. Zia, NMR Spectroscopy and Nondestructive Testing of Rubber and Rubber Products, International Rubber Conference IRC 2015, Nürnberg, June 29 – July 2, 2015

Keywords: Compact NMR, polyethylene, rubber, ashalt, wax

PCNCT-I-009 [Polymer Characterization by New and Combined Techniques]

Structural characterisation of complex branched polymers to understand synthesis-structure-property relations

Robert G Gilbert[1], Hongyan Li[1], Wei Zou[1], Shiqing Zhou[2], Enpeng Li[3], Peng Zhang[3]

[1]*University of Queensland*
[2]*Ausinorigin Company*
[3]*Huazhong University of Science and Technology*

Structural characterization of complex branched polymers is fraught with difficulty: indeed, a full description requires an infinitely-hierarchical infinite-dimensional function. For special cases such as branched glucose polymers (e.g. starch and glycogen), a combination of two structural characterization techniques, multiple-detection size-exclusion chromatography and fluorophore-assisted carbohydrate electrophoresis, provides considerable structural information. Special mathematical models have been developed to treat the resulting data to obtain a small number of physically meaningful parameters. These are then used in a statistical methodology to develop mechanistically valid relations between synthesis, structure and properties. Examples of these will be given for developing materials with improved functional properties, and foods with improved digestibility, which is of special importance to public health, specifically diabetes, obesity and colo-rectal cancers.

Keywords: size exclusion chromatography, GPC, branched polymers, molecular structure, degradation

PE-I-001 [Polymer Education]

Unifying the polymer curriculum at universities: Pros and Cons

Patrick Theato

Institute for Technical and Macromolecular Chemistry, University of Hamburg, Hamburg, Germany

Polymer chemistry is part of almost every chemistry curriculum in the world. Thousands of students are taught polymer chemistry at universities annually. Those classes are often relying on or are based on selected textbooks, but the favored textbooks are varying from school to school and country to country. Further, more and more online video lectures are becoming available that allow students to master the fundamental aspects of polymer chemistry. Yet, how is it possible that in such an increasingly connected world, most classes taught at universities still remain as isolated castles of the past? Surely, every teacher has his or her own flavor in teaching a class and this is particularly true for the emphasis on certain topics in a polymer chemistry class. And similarly do students favor some topics over others. However at the end of the day, students are (or should be) able to know a common skillset in polymer chemistry. If so, then what is the common curriculum? Can we unify the polymer curriculum? And would such a unified curriculum make sense in polymer chemistry? Arguments for and against such a curriculum can and will be presented. As such, the intention of this talk is to trigger a serious discussion on future polymer chemistry curricula.

Keywords: teaching, polymer chemistry, curriculum

PE-I-002 [Polymer Education]

The Rise of Polymer Science Education in Belgium

Eric Goethals

Department of organic an macromolecular chemistry, Ghent University, Ghent, Belgium

The steady rising of educational programs of polymer science at the major Belgian Universities and the corresponding creation of high quality interdisciplinary research centers in the field are described. Belgium is worldwide the largest producer of polymers per capita. The production plants of commodity plastics are located mainly around the Antwerp harbor site. Specialty polymers are produced on many locations. Chemists and engineers with knowledge of polymer chemistry, physics and technology are therefore strongly needed. It took some time before the universities adapted their curricula to these needs. At present, courses on polymer science are given in all major universities on bachelor level, masters level and in "doctoral schools", the last ones usually in English. Additionally, courses on polymer science are given in the frame of "permanent education" programs. Further, several interdisciplinary associations on polymer-related topics such as recycling, biomedical polymers, functional polymer materials, sustainable materials, etc., have been created in which education and research are combined with industrial contacts.

Keywords: polymers, education, interdisciplinary courses

PE-I-003 [Polymer Education]

A Brief History of Chemistry and Polymer Science in Turkey

Teoman Tincer

Department of Chemistry and Polymer Science and Technology, Middle East Technical University Ankara

The earliest chemistry textbook was published by Derviş Ömer Şifai, a medical doctor, in Ottoman Empire in 1703. However, the first comprehensive work describing elements and some basic chemical reaction appeared in a mathematical journal published by Military Engineering School (Later became İstanbul Technical University) mostly referring the famous Swedish chemist J.J Berzelius in 1830. A well-educated scientist and engineer, Derviş Paşa, gave the first public conference on chemistry in the opening ceremony of Darül-Funün (The Door of Science, equivalent of today's University) on January 13th, 1863. Kimya-yı Tıbbı (Chemistry in Medicine) for the medicine students in Military Medicine Faculty was published by Dr. Md. Kırımlı Aziz Bey (later general, also one of leading founders of Turkish Red Crescent) in 1868. Late 19th century and in the beginning of 20th century German science became strongly influential in the Ottoman Empire compared. Among the German scientists Dozent Dr. Frizt Arndt (1885-1965) and his two co-workers, Dr. Kurt Hösch and Dr. Gustav Fester, are the most influential people in chemistry. Almost in the same years, Mendeleev's student Prof. Dr. Alibey Hüseyinzade Turan (1864-1940) joined İstanbul Darül-funün. He revised the periodic table, corrected the symbols of elements into Latin letters and reorganized periodic table according to his famous supervisor.

Prof. Dr. F. Arndt returned back to Turkey in 1933 and involved in reforms in higher education to convert the old fashioned Darül-fünun into modern university. During 1930's more than 100 German scientists immigrated to Turkey. Prof. Arndt contribution to chemistry in Turkey is invaluable.

Although WW II is full of tragedy, the scientific discipline and interest in research made İstanbul University worldwide known. At the end of WW II the second University was founded in Ankara. Some German scientist moved from Istanbul to Ankara in addition to some other European faculty members. Estonian Chemist, Prof. Dr. Adolf Gustav Parts (1904-1994) and Finnish chemist Prof. Dr. Yrgö Kauko (1886-1974) also joined newly established Ankara Science Faculty of Ankara University in 1946 and 1947, respectively. These two scientists separately supervised number of Ph.D thesis including Prof. Dr. Bahattin M. Baysal One (Parts's student, presently emeritus, 1922-) and Prof. Dr. Cemil Şenvar (1923- 2009, co-supervised by Prof. Kauku and E. Fischer). These two scientists were the first scientists who initiated polymer science in Turkey. Prof Baysal was then joined newly established Middle East Technical University (METU) in 1960 after his post-doctoral studies in USA. Later, he joined İstanbul Technical University (ITU) in 1978. Following postdoctoral studies in UK, Prof. Şenvar became the chairman of department of chemistry in newly established Hacettepe University (HU) in Ankara in 1966. Prof Şenvar then moved to Marmara University, İstanbul and continued teaching and research in polymer science after his official retirement.

METU and HU became the leader of polymer science and technology in Turkey. Polymer science were gradually developed in organic and physical chemistry chairs at ITU, Boğaziçi and İstanbul University. The first graduate interdisciplinary was established by METU in 1992 and then followed by ITU in 2000.

Keywords: polymer education, turkey, history

PE-I-004 [Polymer Education]

The Teaching and Learning of Organic Polymer Chemistry via the Open and Distance e-Learning Model at the University of South Africa

Gabriel Jeffrey Summers

Department of Chemistry, University of South Africa, PO Box 392, UNISA, Pretoria, 0003, South Africa

The University of South Africa (UNISA) is a comprehensive, dedicated open and distance e-learning (ODeL) institution of higher learning in South Africa. The tuition model at UNISA, one of the largest ODeL institutions in the world, transformed from the paper-based correspondence tuition and learning model, whereby well-edited standard study material was distributed to students via the post office mail system, to its current tuition model of open and distance e-learning. The Department of Chemistry at UNISA offers the module CHE4806: Organic Polymer Chemistry as one of the modules of the postgraduate qualification, BSc Honours in Chemistry and is the only postgraduate module in Polymer Science offered by the ODeL tuition model in South Africa. The flexible ODeL tuition model for the CHE4806 module is introduced via the blended learning format of providing students with printed and electronic versions of the study material and additional supporting academic references to textbooks, journal publications, different academic websites and other open source platforms. On-line tuition and academic resources available for the module includes the myUnisa management system, massive open on-line courses (MOOCS), digital open education resources (OER), the e-library and e-tutor systems and the general e-mail communication method. Face-to-face learner-instructor interaction and academic tutorial and group discussion classes are scheduled via the video conferencing method or via direct lecturer-student contact sessions at different UNISA regional centres across South Africa. The formative assessment strategy involves the on-line submission and marking of student assignments via the J-Router system which allows the rapid and efficient academic feedback to students during the academic year.

Keywords: Open and distance e-learning tuition model, polymer chemistry

PEPC-I-001 [Polymer Engineering, Processing, and Characterization]

Analysis and simulation of iPP morphology with fast temperature evolution at mold surface

Vito Speranza, Sara Liparoti, Roberto Pantani, Giuseppe Titomanlio

Department of Industrial Engineering, University of Salerno, Italy

A simple model describing morphology evolution during the injection molding process was developed along the years at the University of Salerno. The effect of flow on nucleation density and spherulite growth rate of an iPP was described in the model on the basis of a molecular stretch parameter, described by a nonlinear Maxwell model whose relaxation time was determined by the molecular stretch and by temperature, pressure and crystallinity (Pantani et al. 2015, AIP Conference Proceeding 1664 - 050006).

In this work, the predictions of the model were compared with the cross section morphology distributions of samples obtained by injection molding, coupled with fast evolution of mold cavity surface temperature (Liparoti et al. 2015 AIP Conference Proceeding 1695 – 020052). The molded material is a very accurately studied iPP as far as rheology, quiescent crystallization, effect of flow on nucleation, spherulitic growth rates and spherulitic/fibrillar transition. Mold surface temperature conditioning was obtained adopting a thin heating device layered on the cavity surface. Operative conditions are summarized in Table 1. It was found that pressure and temperature evolutions were accurately predicted by the injection molding simulation. Fig 1 shows, for temperature and pressure evolutions, a comparison between experimental and calculated values.

The thickness and the position of the fibrillar-shear layers were predicted by applying a criterion based on both the final viscous work and the molecular stretch distributions on the cross section. Fibrillar morphology is obtained when the viscous work and the molecular stretch are, simultaneously, above critical values. These values are respectively 20 GPa and 6. The identification of the shear layer on the cross section identifies also the thickness of the other layers since the skin layer and the spherulitic layers are adjacent to the shear layer. Table 1 also shows the experimental and calculated morphology distribution.

Keywords: injection molding, simulation, iPP, fast mold temperature evolution

Fig 1

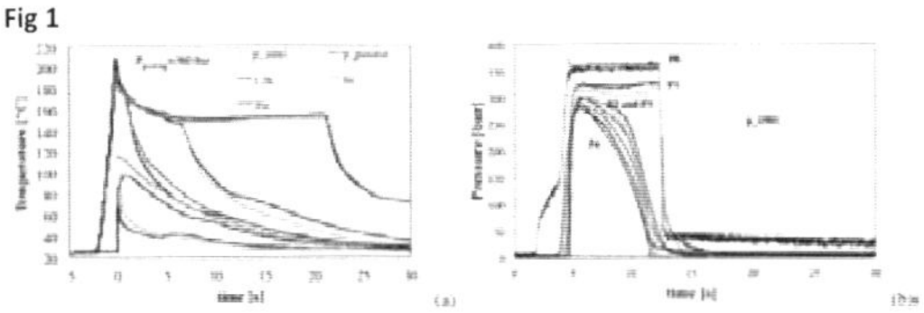

Experimental (black) and calculated temperature (a) and pressure (b) evolutions at different heating times of the mold cavity surface. Temperature evolutions obtained for p_steel and p_passive are also reported.

Table 1

Test run	P (W/cm²)	T_{heat} (°C)	t_h (s)	t_c (s)	P_{hold}	morphology distribution
p-steel	0	25	0	0	360	
p-passive	0	28	0	0	360	
p-1503	10	150	3	2	360	
p-1506	10	150	6	2	360	
p-15012	10	150	12	2	360	
p-15020	10	150	20	2	360	

Operating conditions for the experiments performed in this work (Tmelt=220°C; Volumetric flow rate=2.8 cm3/s), and morphology distributions as obtained from experiments and simulations.

PEPC-I-002 [Polymer Engineering, Processing, and Characterization]

Large scale R2R fabrication of piezoresistive films (Ni/PDMS) with enhanced through thickness electrical and thermal properties by applying magnetic field

Mukerrem Cakmak

Department of Polymer Engineering, The University of Akron, Akron OH, USA

The first successful development of a roll to roll (R2R) process that applies an external magnetic field to orient and organize magnetic nanoparticles along nanocolumns in the thickness direction of thin films to obtain high electrical and thermal conductivities in the thickness direction is reported. Utilizing a R2Rmachine that includes an in-line electromagnet, we orient and organize Ni nanoparticles in nanocolumns inside a flexible poly(dimethylsiloxane) matrix. In these films, the nanocolumns of Niparticles point in the magnetic field/thickness direction which leads to enhancement of the electrical and thermal conductivity in the thickness direction while maintaining optical transparency as the space between the nanocolumns is depleted of nanoparticles facilitating unimpeded light transmission. Exhibiting piezoresistivity, the electrical conductivity in these films increases by as much as 7 orders under moderate pressures. The thermal conductivity of the aligned composite films filled with 14 vol % Ni flakes was found to increase to 50 times the conductivity of the polymer matrix, or 13 times the conductivity of the non-aligned composite with the same concentration. This R2R method facilitates the manufacture of unique films with enhanced functional properties in the thickness direction to be used in a range of applications including Z direction heat spreaders, transparent switches, privacy protection screens and piezoresistive sensors.

Keywords: Roll to Roll manufacturing, piezoresistive films, pressure sensors, field assisted alignment of nanoparticles

PEPC-I-003 [Polymer Engineering, Processing, and Characterization]

Injection Molding of Slowly Crystallizing Polymers

Yakup Ulcer[1], Mukerrem Cakmak[2]

[1]*Ravago Istanbul, Turkey*
[2]*University of Akron, Akron, OH USA*

The effect of injection molding variables on the structure property relationship in injection molded slowly crysallizing polymers were studied. Poly ethylene 2,6 naphthalene dicarboxylate (PEN) and sydiotactic poly styrene (s-PS) were taken as model polymers. Characterization of structural gradients in both polymer showed the presence of a three layer structure consisting of an amorphous skin, a crystalline intermediate layer and an amorphous core. Major variables affecting the structure were found to be mold temperature and and mold geometry. Studies of annealed samples showed presence of stress induced nuclei that were not visible in the original samples. The experimental results were futher supported by mold filing and crystallization simulation results.

Keywords: Injection molding, (PEN), (s-PS)

PEPC-I-004 [Polymer Engineering, Processing, and Characterization]

Materials, process, and product design of thermoplastic polymers using hyphenated rheology

Dietmar Auhl

Aachen-Maastricht Institute for Biobased Materials, 6167RD Geleen, The Netherlands

Modern design loops with higher efficiency and shorter timeframes require comprehensive knowledge of structure-property-performance relations. It is well known that product properties and performance largely depend on material parameters and processing pathways. Examples are the influence of polymer architecture and composition, as well as temperature and flow history, on morphology and mechanical response since material functions can typically be designed via molar-mass, branching, and chemical structure distributions. In order to understand and tune material behavior and related processing effects with a fundamental macromolecular perception it often needs detailed information along broad length and time scales. Flow and phase behaviour are supremely examined by advanced and combined techniques in order to draw more concrete or extended conclusions, e.g. regarding multiple origins of stress or strain variations during melt flow. Hyphenated rheology methods like Rheo-Optics, Rheo-Spectroscopy or Rheo-Scattering with state-of-the-art instrumentation offer solutions for synchronized in-situ studies with different stimuli superposed with rheological flow. For example, investigations of corresponding material response and flow conditions in shear or elongation strongly benefit from additional analysis techniques like light imaging. Yet, especially also by closely linking experimental with theoretical studies and via computational tools beneficial progress has been achieved. Herewith, topics like molecular or polymer chain dynamics, crystal formation, orientation or distribution of phases and additives can be investigated under standard rheometric and complex flow scenarios equivalent to processing, e.g. non-isothermal and strong flow regimes. Example studies for materials, process, and product design of thermoplastic materials, but also model systems like polyethylene, polyisoprene as well as polylactide and liquid-crystalline polymers are discoursed. The above described developments and trends are further extended by interdisciplinary working groups both at academia and industry and grow to be more widely used as briefly highlighted.

Keywords: thermoplastic, rheology, material behavior

PPBM-I-001 [Polymer and Polymer-Based Membranes]

Ion Solubility, Diffusion, Permeation and Conductivity in Charged Polymers

Benny Dean Freeman

McKetta Department of Chemical Engineering, The University of Texas at Austin, Austin, Texas USA

Charged polymer membranes are widely used for water purification applications involving control of water and ion transport, such as reverse osmosis and electrodialysis. Additional applications, such as energy recovery ventilation and membrane-assisted capacitive deionization, rely on polymer membranes to control transport rates of water, ions, or both. Improving membranes for such processes would benefit from more complete fundamental understanding of the relation between membrane structure and ion sorption, diffusion and transport properties in both cation and anion exchange membrane materials. Ion-exchange membranes often contain strongly acidic or basic functional groups that render the materials hydrophilic, but the presence of such charged groups also has a substantial impact on ion (and water) transport properties through the polymer.

We are exploring the influence of polymer backbone structure, charge density, and water content on ion transport properties. Results from some of these studies will be presented, focusing on transport of salt, primarily NaCl, through various neutral, positively charged and negatively charged membranes via concentration gradient driven transport (i.e., ion permeability) and electric field driven transport (i.e., ionic conductivity). One long-term goal is to develop and validate a common framework to interpret data from both electrically driven and concentration gradient driven mass transport in such polymers and to use it to establish structure/property relations leading to rational design of membranes with improved performance.

Ion sorption and permeability data were used to extract salt diffusion coefficients in charged membranes. Concentrations of both counter-ions and co-ions in the polymers were measured via desorption followed by ion chromatography or flame atomic absorption spectroscopy. Salt permeability, sorption and electrical conductivity data were combined to determine individual ion diffusion coefficients in neutral, cation exchange and anion exchange materials. Manning's counter-ion condensation models and the Mackie/Meares model were used to correlate and, in some cases, predict the experimental data.

Keywords: ion transport, charged polymers, counterion condensation

PPBM-I-002 [Polymer and Polymer-Based Membranes]

Hydrocarbon, Solid Polymer Electrolytes for Acidic and Alkaline Electrochemical Devices

Steven Holdcroft, Andrew Wright, Thomas Skalski, Timothy Peckham, Thomas Weissbach, Benjamin Britton

Simon Fraser University

Full scale commercialization of Proton Exchange Membrane Fuel Cells (PEMFCs) is imminent yet fundamental knowledge in specialized areas is still required in order to resolve persistent technical barriers associated with high materials costs and limited reliability, versatility, and performance.

Perfluorinated proton-exchange polymers form the basis of standard high-performance PEMFCs but difficult synthetic chemistry hampers further materials development. Hydrocarbon proton-exchange materials, on the other hand, are founded on well-established and versatile synthetic chemistry that allows for rapid materials development, and offer a less expensive alternative to perfluorinated polymers. In the corollary case of Anion Exchange Membrane Fuel Cells, the search continues for an alkaline-stable, polymeric hydroxide-conducting medium. Solutions to these challenges require the undertaking of rigorous systematic studies on model polymers and representative materials of known and controllable molecular structure and preferred nano-morphology. In this presentation the evolution and properties of unique proton- and hydroxide-conducting polymers being developed at Simon Fraser University will be described.

References:

"Structurally-Defined, Sulfo-Phenylated, Oligophenylenes and Polyphenylenes", J. Am. Chem. Soc, 137 (2015) 12223-12226.

"Poly(phenylene) and m-Terphenyl as Powerful Protecting Groups for the Preparation of Stable Organic Hydroxides", Angewandte Chemie, 2016.

Keywords: Membranes, Fuel cells, electrolysers, ionic polymers

PPBM-I-003 [Polymer and Polymer-Based Membranes]

Integral Asymmetric Block Copolymer Membranes: Structure Formation and Preparation by Casting, Spinning, and Spraying

Volker Abetz[1,2]

[1]*Institute of Polymer Research, Helmholtz-Zentrum Geesthacht, Geesthacht, Germany*
[2]*Institute of Physical Chemistry, University of Hamburg, Hamburg, Germany*

Amphiphilic block copolymers are well-known to self-assemble into a rich variety of periodic structures mostly depending on their composition and molecular weight. However, also the conditions of the self-assembling process may influence the finally obtained structures, which may also be metastable. The tendency to self-assemble from solution can be combined with a quenching step in order to freeze an intermediate structure still far from equilibrium. A gradient such as the concentration gradient in an evaporating solution can initiate a directed self-assembly. The combination with a freezing step can lead to so-called integral asymmetric membranes.

In the first part of the lecture this process will be discussed in view of results obtained by scattering and microscopic techniques.

The second part will address questions concerning the control of pore sizes of such membranes and their preparation in different geometries. Both flat sheet and hollow fibre geometries will be discussed.

References:

7.02 - Block Copolymers in the Condensed State
A. Boschetti-de-Fierro, V. Abetz
Polymer Science: A Comprehensive Reference **7** **(Nanostructured Polymer Materials and Thin Films)** 3-44 **(2012)**

Characterization of Block Copolymer Self-Assembly: From Solution to Nanoporous Membranes
L. Oss-Ronen, J. Schmidt, V. Abetz, A. Radulescu, Y. Cohen, Y. Talmon
Macromolecules **45**, 9631-9642 **(2012)**

Evaporation-Induced Block Copolymer Self-Assembly into Membranes Studied by in Situ Synchrotron SAXS
C. Stegelmeier, A. Exner, S. Hauschild, V. Filiz, J. Perlich, S. V. Roth, V. Abetz, S. Förster
Macromolecules **48**, 1524-1530 **(2015)**

Isoporous Block Copolymer Membranes
V. Abetz
Macromolecular Rapid Communications **36**, 10-22 **(2015)**

Self-organized isoporous membranes with tailored pore sizes
S. Rangou, K. Buhr, V. Filiz, J. I. Clodt, B. Lademann, J. Hahn, A. Jung, V. Abetz
Journal of Membrane Science **451**, 266-275 **(2014)**

Tailored Pore Sizes in Integral Asymmetric Membranes Formed by Blends of Block Copolymers
M. Radjabian, V. Abetz
Advanced Materials **27**, 352-355 **(2015)**

Tailoring the Morphology of Self-Assembled Block Copolymer Hollow Fiber Membranes
M. Radjabian, J. Koll, K. Buhr, U. Vainio, C. Abetz, U. A. Handge, V. Abetz
Polymer **55**, 2986-2997 **(2014)**

A Facile Method to Prepare Double Layer Isoporous Hollow Fiber Membrane by in-situ Hydrogen Bond Formation in the Spinning Line
N. Noor, J. Koll, M. Radjabian, C. Abetz, V. Abetz
Macromolecular Rapid Communications **37**, 414-419 **(2016)**

Thin Isoporous Block Copolymer Membranes: It Is All about the Process
J. Hahn, J. I. Clodt, C. Abetz, V. Filiz, V. Abetz
ACS Applied Materials & Interfaces **7**, 21130-21137 **(2015)**

Keywords: block copolymer, self-assembly, blend, scattering, electron microscopy, ultrafiltration

PPBM-I-004 [Polymer and Polymer-Based Membranes]

New polymers and blends for block copolymer membranes

Suzana P Nunes, Klaus Viktor Peinemann, Yihui Xie, Haizhou Yu, Xiaoyan Qiu

King Abdullah University of Science and Technology (KAUST)

Block copolymers offer an excellent morphology control and versatile functionalization for membrane manufacture. Self-assembly and non-solvent induced phase separation has been demonstrated to be a successful strategy for preparation of asymmetric porous membranes in the ultrafiltration range[1]. More recently our group has been exploring new block copolymer systems, which could substantially extend the characteristics and application ranges of this class of membranes. We have demonstrated that blends of polystyrene-b-poly(4-vinyl pyridine) and polystyrene-b-poly(acrylic acid) can bridge the gap between ultrafiltration and nanofiltration, leading to the formation of pores selective for much smaller solutes than reported before when using block copolymers[2]. We have also invested our efforts on the synthesis of completely new copolymers for membrane preparation. We recently report the synthesis and application of block copolymer based on polysulfone, known to be mechanically more stable than polystyrene[3].

References:

1. S. P. Nunes, Macromolecules 2016
2. H. Yu et al., Angewandte Chemie Int. Ed. 2015, 54, 13937-13941
3. Y. Xie et al., Polymer Chemistry 2016

Keywords: Membranes, block copolymers, self-assembly

PPBM-I-005 [Polymer and Polymer-Based Membranes]

Electrospun Fiber Membranes for Clean Water Applications

Gregory C Rutledge[1], Fei Guo[1], Looh Tchuin Choong[1], Yi Min Lin[2]

[1]*Department of Chemical Engineering, Massachusetts Institute of Technology, Cambridge MA USA*
[2]*Department of Materials Science and Engineering, Massachusetts Institute of Technology, Cambridge, MA USA*

Water scarcity is a growing problem worldwide. Adequate supplies of fresh water are essential for growing populations and advancing economies, and are intimately intertwined with growing demands for energy and food production. Membrane technology offers a suite of solutions for many water resource problems, but its continued progress depends on development of more advanced and high-performing membrane materials. Fibrous membranes comprising fibers with diameters less than 1 micron can be produced by "electrospinning", a technology in which electrostatic forces are used to draw out exceptionally fine, and uniform, fibers. Nonwoven sheets of these novel fibers can be produced with a variety of chemistries and offer unique combinations of high specific surface area, high porosity, and small inter-fiber distances. As a result of these properties, such membranes can combine unusual liquid repellency with high porosity for membrane distillation applications, or can serve as microfilters for separation of oil-in-water emulsions. In this talk we examine some of the strengths and limitations of electrospun membranes for these applications. It is shown that the compressibility of the membranes can play a significant role in their performance. We also examine the interplay of fiber and emulsion chemistry on fouling in emulsion separations.

Keywords: Electrospinning, fiber, desalination, membrane distillation, emulsion, microfiltration

PPBM-I-006 [Polymer and Polymer-Based Membranes]

Bacteria encapsulated/immobilized electrospun polymeric nanofibrous webs for wastewater treatment

Tamer Uyar[1], Omer Faruk Sarioglu[1], Nalan Oya San[2], Asli Celebioglu[1], Turgay Tekinay[2]

[1]*Institute of Materials Science & Nanotechnology, Bilkent University, Ankara, Turkey.*
[2]*Life Sciences Application and Research Center, Gazi University, Ankara, Turkey*

Electrospinning is a higly promising fabrication technique for nanofibre/nanoweb production since it is simple, tuneable, versatile and cost-effective. Electrospun nanofibers and their nanofibrous webs have unique properties such as high surface to volume ratio and nanoscale porosity, hence they are promising for environmental applications as filtering and membrane materials. The integration of electrospun fibrous webs with specific bioremediating microorganisms can functionalise the webs for the filtration and purification purposes. In our studies, electrospun polymeric nanofibrous webs were found to be suitable matrices for immobilisation of microorganisms. We have immobilize and/or encapsulate variety of bacterial cells in electrospun nanofibrous webs for removal of textile dyes, heavy metals and surfactants from aqueous medium. Reusability tests revealed that bacteria immobilized/encapsulated electrospun nanofibrous webs can be used couple of times successively. In addition to the high efficiency of microorganism immobilized nanofibrous webs for remediation of different contaminants, their reusability makes them promising candidates for continuous remediation of contaminated water, which was observed in all of these studies. In short, our achievements reveal that bacteria immobilised electrospun nanofibrous webs could be quite applicable for water cleaning purposes due to their efficiency, versatility and reusability.

Keywords: electrospinning, nanofibers, bacteria, wastewater treatment, bioremediation

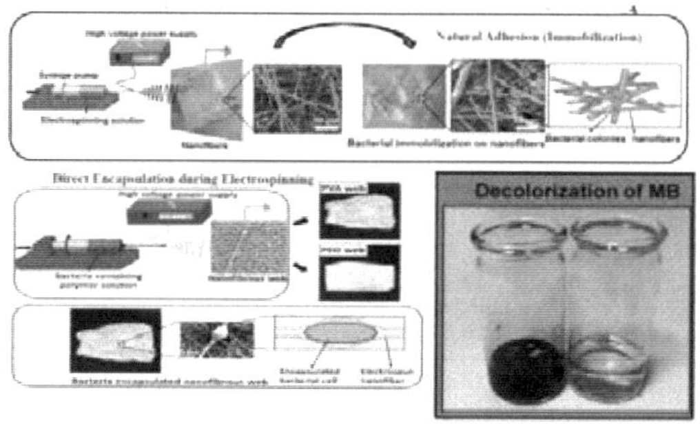

Fig 1. Bacteria encapsulated/immobilized electrospun polymeric nanofibrous webs for wastewater treatment.

PPBM-I-007 [Polymer and Polymer-Based Membranes]

On the Morphological Control and Transport Properties of Highly Tunable Block Ionomers as Membranes

Richard J Spontak[1], Kenneth P Mineart[1], Luca Ansaloni[2], Zhongde Dai[2], Joshua D Dickerson[3], Keara T Saud[3], Byongdu Lee[4], Liyuan Deng[2], May Britt Hagg[2]

[1]*Department of Chemical Engineering, North Carolina State University, Raleigh, NC, USA*
[2]*Department of Chemical Engineering, Norwegian University of Science & Technology, Trondheim, Norway*
[3]*Department of Materials Science & Engineering, North Carolina State University, Raleigh, NC, USA*
[4]*Advanced Photon Source, Argonne National Laboratory, Argonne, IL, USA*

Block ionomers, block copolymers possessing at least one charged block, are of increasing scientific and technological interest as amphiphilic materials capable of affording unique application opportunities. Previous studies have established that a midblock-sulfonated pentablock ionomer, which behaves as a thermoplastic elastomer, can be used in the design of ionic polymer-metal composites and photovoltaic elastomer gels. In this work, we demonstrate some of the promising gas-transport properties exhibited by this family of soft materials. One of the most interesting challenges generally associated with block ionomers is control over morphological development, which regulates solubility-driven swelling and continuous diffusive pathways and, hence, permeation efficacy. Using various solvent-related processes ranging from casting to annealing, we employ transmission electron microscopy/microtomography and synchrotron small-angle x-ray scattering to show that the morphological features of both equilibrated and nonequilibrated block ionomers are highly tunable. These features directly relate to the ability of these materials to absorb polar solvents and behave as physical hydrogels. An important aspect of these block ionomers is their ability to improve the permeability of humidified base and acid gases. Alone, they are capable of displaying an ultrahigh NH3/N2 selectivity of ~2000. Upon physical incorporation of low-molar-mass amines, these nanostructured materials likewise show promise for CO2 removal from mixed gases.

Block ionomer morphology

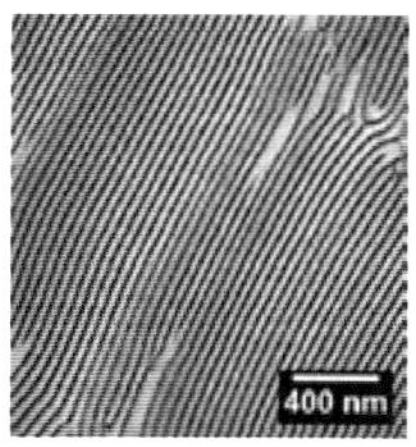

This lamellar morphology represents the expected equilibrium morphology in the midblock-sulfonated block ionomer used for gas-separation membranes.

Keywords: Block ionomer, nanostructured polymer, morphological control, solvent annealing, solvent selectivity, CO2 removal

PPBM-I-008 [Polymer and Polymer-Based Membranes]

Medium Cut-off Membranes for Hemodialysis - Closer to the Natural Kidney's Removal Function

Adriana Boschetti De Fierro[1], Manuel Voigt[1], Markus Hornung[2], Reinhold Buck[2], Michael Hulko[1], Lars Göran Nilsson[3], Werner Beck[4], Markus Storr[1], Bernd Krause[1]

[1]Research and Development, Gambro Dialysatoren GmbH, Hechingen, Germany
[2]Product Development, Gambro Dialysatoren GmbH, Hechingen, Germany
[3]Therapeutic Area, Gambro Lundia AB, Lund, Sweden
[4]Therapeutic Area, Gambro Dialysatoren GmbH, Hechingen, Germany

Renal replacement therapies (RRT) based on blood purification membranes currently cannot mimic the well-defined size exclusion filtration performed by the natural kidney. One of the restrictions is the limited selectivity achieved during membrane production by phase inversion technique. During production, stable and reproducible membranes are obtained, however at the cost of an inherent pore size distribution. One of the current gaps in hemodialysis treatment is the insufficient removal of the complete molecular weight spectrum of uremic toxins. Due to the lower selectivity of state-of-the-art synthetic membranes compared to the glomerular membrane, patients on hemodialysis have higher levels of middle and large molecular solutes in plasma. Clinical experience has been gathered with high permeability membranes such as the High Cut-Off membranes. Results from small pilot trials for time periods up to 12 weeks indicate that inflammatory state in patients might decrease with expanded toxin removal [Hutchison C. A. et al, J Am Soc Nephrol 19: 2008; 266A]. These results motivate that current efforts in membrane innovation are directed towards increased membrane permeability and enhanced removal of uremic toxins. The challenge in membrane manufacturing process is to increase the pore sizes without increasing the broadness of the pore size distribution. Controlled membrane manufacturing allows some improvement in this direction. The novel Medium Cut-Off membranes provide extended pore sizes with appropriate pore size distribution and permeability close to that of the natural kidney [Boschetti-de-Fierro A. et al, Sci Rep 5: 2015; 18448]. Their controlled characteristics allow for removal of an expanded range of uremic toxins compared to conventional high-flux membranes. Moreover, the protein layer formation during treatment restricts the removal of molecules above 3.5 nm in radius, optimizing removal of large uremic toxins while ensuring retention of albumin during treatment.

Keywords: hemodialysis, kidney, renal replacement

PPBM-I-009 [Polymer and Polymer-Based Membranes]

Membrane separations with PIMs and graphene

Peter Martin Budd
University of Manchester

Polymers of intrinsic microporosity (PIMs) are glassy polymers which possess high free volume and high internal surface area as a consequence of their relatively rigid, contorted macromolecular backbones. They comprise fused ring sequences interrupted by spiro-centres or other sites of contortion. PIM membranes are of interest for a variety of separation processes, including gas separations (e.g., carbon dioxide capture) and organophilic liquid separations (e.g., bioalcohol recovery). In gas separation, PIMs contributed to the revision of the upper bounds of performance by Robeson in 2008. If high free volume polymers are to achieve commercial success in membrane applications, the most significant issue to address is that of ageing, which leads to a reduction in permeability over time. A variety of strategies are being employed to tackle this problem.

Graphene and graphene oxide have also attracted attention for their potential in membrane applications. Graphene can be combined with PIMs to form mixed matrix membranes. The addition of small amounts of graphene can significantly influence the permeation properties and ageing behaviour of the polymer.

Keywords: PIM, graphene, membrane, carbon dioxide capture

PPBM-I-010 [Polymer and Polymer-Based Membranes]

Ionic liquids supported by porous and isoporous membranes for CO2/N2 gas separation applications

Judith Grünauer[1], Sergey Shishatskiy[1], Clarissa Abetz[1], Volkan Filiz[1], Volker Abetz[1], Volker Abetz[2]

[1]Institute of Polymer Research, Helmholtz-Zentrum Geesthacht, Geesthacht, Germany
[2]Institute of Physical Chemistry, University of Hamburg, Hamburg, Germany

The use of isoporous polystyrene-block-poly(4-vinylpyridine) diblock copolymer (PS-b-P4VP) and commercial polyacrylonitrile (PAN) membranes as matrix materials for ionic liquids (ILs) for gas separation tasks is reported and compared to an anodic alumina membrane. The compatibility of membrane components was investigated by means of porometry, tensiometry, and contact angle measurements. The CO2 gas diffusivity and solubility was determined by direct observation using a "time-lag" facility and a magnetic suspension balance. Overall liquid membranes supported by PS-b-P4VP and PAN achieved remarkably good results especially in combination with 1-ethyl-3-methylimidazolium dicyanoamide (EMIM DCA) and 1-ethyl-3-methylimidazolium tetracyanoborate (EMIM TCB) ionic liquids and are showing the way to attractive research and application opportunities.

Keywords: Ionic liquid, Supported liquid membrane, Isoporous membrane, Block copolymer, CO2 separation

PPBM-I-011 [Polymer and Polymer-Based Membranes]

Residual solvent effect on the CO2/CH4 and CO2/N2 separation properties of 6FDA-DAM membranes

Duygu Kahraman, Çiğdem Atalay Oral, Ş. Birgül Tantekin Ersolmaz

Department of Chemical Engineering, Istanbul Technical University, Istanbul, Turkey

The structure and performance of a polymeric membrane depend on the polymer structure as well as membrane fabrication method and processing conditions. The solvent type used in membrane fabrication, thermal treatment and aging conditions are among the parameters which considerably effect the structural, mechanical, thermal, and especially gas permeation properties. 6FDA based polyimides, containing bulky CF3 groups, show outstanding separation properties for CO2 mainly due to their wider intrasegmental packing. This study investigates the effect of thermal treatment conditions and residual solvent on the CO2 separation properties of dense 6FDA-DAM membranes. 6FDA-DAM is synthesized by the two-step solution imidization method. Membranes are prepared by casting-evaporation from solutions in three different solvents (NMP, DMF, THF), thermally treated at different temperatures, and characterized by WA-XRD, DSC, TGA, and gas (CO2, N2, CH4) permeation measurements at 35°C and 4 atm. The results show that post thermal treatment temperature and residual solvent affect the gas permeability coefficients significantly due to the plasticization effect of the solvent on the free volume and chain packing of the polymer which may be reversed to antiplasticization as the amount of residual solvent decreases. However, the effect of residual solvent varies for the two gas pairs studied. Drying near the Tg of the polyimide allows a significant selectivity enhancement for the CO2/CH4 gas pair over the membranes treated at lower temperatures while no significant change is observed for the CO2/N2 pair. These differences can be attributed to the variation of the influence of residual solvent for gases with different kinetic diameters and sorption characteristics. Implications of these results on 6FDA-DAM based mixed-matrix membrane formation will be discussed.

Abbreviations: 6FDA: 4,4 (Hexafluoroisopropylidene) diftalic dianhydride; DAM: 2,4,6-Trimethyl-m- phenylene diamine; NMP: N-Methylpyrolidone; DMF: Dimethylformamide; THF: Tetrahydrofuran

Keywords: Membrane, gas separation, residual solvent, permeability, polyimide

PPG-I-001 [Porous Polymer and Gels]

Porous polymers through emulsion templating

Michael S. Silverstein

Department of Materials Science and Engineering, Technion – Israel Institute of Technology, Haifa, Israel

PolyHIPEs are porous emulsion-templated polymers synthesized within high internal phase emulsions (HIPEs). HIPEs are highly viscous, paste-like emulsions in which the major, "internal" phase, usually defined as constituting more than 74% of the volume, is dispersed as discrete droplets within the continuous, minor, "external" phase. The surge in polyHIPE research and development parallels that of porous polymers in general and reflects the increasing number of potential applications (reaction supports, separation membranes, tissue engineering scaffolds, controlled release matrices, responsive and smart materials, and templates for porous ceramics and porous carbons).

Typical polyHIPEs have highly interconnected porous microstructures, low bulk densities, and the ability to rapidly absorb large quantities of liquid through capillary action. Surfactant-free polyHIPEs can be synthesized through the formation of nanoparticle-stabilized Pickering HIPEs. Various polyHIPE systems will be described and the relevant properties and applications discussed. The systems described will include hydrophobic polyHIPEs from water-in-oil emulsions; hydrophilic polyHIPEs from oil-in-water emulsions; biodegradable polyHIPEs; hydrophilic-hydrophobic bicontinuous polyHIPEs (hydrogel-filled hydrophobic polymers and rubber-filled hydrogels); shape memory polyHIPEs based on crystallizable size-chain polymers; polyHIPEs containing 85% water entrapped within an elastomeric matrix as individual micrometre-scale droplets; conductive polyHIPEs; hybrid polyHIPEs; and polyHIPE templates for porous ceramics and porous carbons. The profound effects that the polymerization mechanism and the locus of initiation can have upon the polyHIPE structure and properties will be described.

Keywords: porous, polyHIPE, templating

PPG-I-002 [Porous Polymer and Gels]

Injectable solid hydrogels constructed with peptide folding and solution assembly

Darrin Pochan

Materials Science and Engineering, University of Delaware, USA

Self-assembly of molecules is an attractive materials construction strategy due to its simplicity in application. By considering peptidic molecules in the bottom-up materials self-assembly design process, one can take advantage of inherently biomolecular attributes; intramolecular folding events, secondary structure, and electrostatic interactions; in addition to more traditional self-assembling molecular attributes such as amphiphilicty, to define hierarchical material structure and consequent properties. These self-assembled materials range from hydrogels for biomaterials to nanostructures with defined morphology and chemistry display for inorganic materials templating. The local nano- and overall network structure, and resultant viscoelastic and cell-level biological properties, of hydrogels that are formed via beta-hairpin self-assembly will be presented. Importantly, the hydrogels do not form until individual peptide molecules intramolecularly fold into a beta-hairpin conformation. Subsequently, specific, intermolecular assembly occurs into a branched nanofibrillar network. These peptide hydrogels are potentially excellent scaffolds for tissue repair and regeneration due to inherent cytocompatibility, porous morphology, and shear-thinning but instant recovery viscoelastic properties. Slight design variations of the peptide sequence allow for tunability of the self-assembly/hydrogelation kinetics as well as the tunability of the local peptide nanostructure and hierarchical network structure. In turn, by controlling hydrogel self-assembly kinetics, one dictates the ultimate stiffness of the resultant network and the kinetics through which gelation occurs. During assembly and gelation, desired components can be encapsulated within the hydrogel network such as drug compounds and/or living cells. The system can shear thin but immediately reheal to preshear stiffness on the cessation of the shear stress. Additionally, a new system comprised of coiled coil motifs designed theoretically to assemble into two-dimensional nanostructures not observed in nature will be introduced. The molecules and nanostructures are not natural sequences and provide opportunity for arbitrary nanostructure creation with peptides.

Keywords: hydrogel, peptide, self assembly, injectable solid

PPG-I-003 [Porous Polymer and Gels]

Nanostructural evolution of self-healing hydrogels formed in micellar solutions

Oguz Okay[1], Volkan Can[2], Zdravko Kochovski[3], Valentin Reiter[3], Nikolai Severin[3], Miriam Siebenbürger[2], Ben Kent[2], Justus Just[2], Jürgen P. Rabe[3], Matthias Ballauff[4]

[1]Department of Chemistry, Istanbul Technical University, Istanbul, Turkey, Helmholtz-Zentrum Berlin, Soft Matter and Functional Materials, Hahn-Meitner-Platz 1, D-14109 Berlin, Germany,
[2]Helmholtz-Zentrum Berlin, Soft Matter and Functional Materials, Hahn-Meitner-Platz 1, D-14109 Berlin, Germany,
[3]Humboldt-Universität zu Berlin, Department of Physics, Newtonstr. 15, D-14195 Berlin, Germany
[4]Helmholtz-Zentrum Berlin, Soft Matter and Functional Materials, Hahn-Meitner-Platz 1, D-14109 Berlin, Germany, Humboldt-Universität zu Berlin, Department of Physics, Newtonstr. 15, D-14195 Berlin, Germany

Understanding the nanoscale structure and dynamics of supramolecular hydrogels is essential for exploiting their self-healing mechanisms. In this talk, nanostructural evolution and self-healing mechanism of hydrogels formed from in-situ generated hydrophobically modified hydrophilic polymers and wormlike sodium dodecyl sulfate (SDS) micelles will be discussed [1-3]. A conformational transition in wormlike SDS micelles occurs upon addition of hydrophobic as well as hydrophilic monomers. Several hundred nm long SDS micelles completely disappear after the monomer addition, in favor of spherical micelles with a radius of 2.4 nm. After conversion of the monomers to hydrophobically modified polymer chains via micellar copolymerization, the spherical shape of the micelles remains intact but the radius increases to 2.8 nm. The interconnected spherical mixed micelles consisting of SDS and hydrophobic blocks of the polymer self-assemble to form a layered hydrogel structure. Self-healing response of the damaged hydrogel samples begins by reshaping the injured area into circular holes and ends by complete healing due to the intra- and interlayer mobility of the mixed micelles, respectively.

(1) Tuncaboylu, D. C.; Sahin, M.; Argun, A.; Oppermann, W.; Okay, O. Macromolecules 2012, 45, 1991-2000.
(2) Gulyuz, U.; Okay, O. Macromolecules 2014, 47, 6889-6899.
(3) Can, V.; Kochovski, Z.; Reiter, V.; Severin, N.; Siebenbürger, M.; Kent, B.; Just, J.; Rabe, J.P.; Ballauff, M.; Okay, O. Macromolecules 2016, 49, 2281–2287.

Keywords: Supramolecular hydrogels, self-healing, wormlike micelles.

PP-I-001 [Polymer Physics]

Static and Dynamic Bioinspired Material Systems

Andreas Walther

DWI - Leibniz-Institute for Interactive Materials

Biology is a source of inspiration for materials science by demonstrating macroscopic materials with advanced functionalities and excellent mechanical properties. However, although these rather static architectures stimulate large interest, even more thought-provoking are the dynamic, kinetically controlled processes and temporal evolution of structures in complex biological systems.

These are orchestrated through feedback loops and require energy input and dissipation to allow non-equilibrium materials and full spatiotemporal control.

In man made self-assemblies we have mastered to a large extent near-equilibrium structure formation in space and have gained an increasing understanding of how to construct very complex, hierarchically structured soft matter by using co-assemblies, competing interactions and hierarchical length scales. This has allowed to create real-life materials with unprecedented functionalities, inaccessible without delicate control over molecular interactions and sophisticated nano- and mesostructuration. The next step is to master temporal control in self-assemblies. This requires kinetic control of opposing reactions (built-up/destruction), internal feedback systems or the use of energy dissipation to sustain structures only as long as a chemical fuel is available. These approaches keep systems forcefully away from equilibrium and potentially allow neat access to temporal control.

In this talk I will present concepts for bioinspired materials formed in both static and dynamic conditions. The first part will deal with rather static, nacre-inspired high-performance nanocomposites, in which light-adaptive properties can be encoded through co-assembly and energy transfer approaches. The second part will focus on a platform concept, which allows to program self-assembling systems outside equilibrium with a lifetime by kinetic control of promoter/deactivator pairs and a simple internal feedback system. This will be showcased for different self-assembling systems.

References.

Self-Assembled Artificial Nacre: Angew. Chem. 2015, 54, 8653; Nat. Commun. 2015, 6, 5967;
Time Programmed Materials: Nano Lett., 2015, 15, 2213, Angew. Chem. 2015, 54, 13258; Emerging Area: Soft Matter 2015, 11, 7857.

Keywords: Self-Assembly, Colloids, Adaptive Materials, Non-equilibrium systems, Polymers, Peptides

Overview

Examples of static nacre-mimetics and dynamic time-programmed hydrogels

PP-I-002 [Polymer Physics]

Topological constraints do matter: polymer melts, elastomers, collapsed polymer globules, chromosome territories etc

Kurt Kremer

Max Planck Institute for Polymer Research, Mainz, Germany

The role of topological constraints on conformational as well as relaxational and dynamical properties of open linear and closed ring polymers as well as mixtures thereof is discussed. In the case of polymer melts the conformational statistics can be used to directly determine the entanglement molecular weight in excellent agreement to experiment. By manipulating the entanglements in long chain melts materials with new rheological properties can be achieved. For ring polymers the situation is completely different.

While linked rings act like DeGennes' Olympic gels, we find that non concatenated polymer rings segregate and form individual "collapsed" objects. I discuss some details of their conformations, which not only is related to one of the very basic problems in polymer science but also has far reaching consequences from the collapse of gels to chromosome territories.

References

J. D. Halverson, J. Smrek, K. Kremer and A. Yu Grosberg, From a melt of rings to chromosome territories: the role of topological constraints in genome folding. Rep. Prog. Phys. 77 (2014)
G. Zhang, T. Stuehn, K. Ch. Daoulas, K. Kremer, Communication: One size fits all: Equilibrating chemically different polymer liquids through universal long-wavelength description J. Chem. Phys. 142, 221102 (2015)

Keywords: entanglements, topology conservation, ring polymers, (non-) equilibrium, chromosome territories

PP-I-003 [Polymer Physics]

The principles of manipulating the phase transformations, solid-state order and properties of functional organic matter

Natalie Stingelin

Department of Materials, Imperial College London, London, UK

In the past decade, significant progress has been made in the field of organic functional materials with many new opportunities arising in the field of bioelectronics, energy harvesting and storage, organic electronics and photonics, etc. This progress has predominantly been due to important improvements of existing materials and the creation of a wealth of novel compounds. Many challenges, however, still exist. Intimate knowledge of relevant structure/processing/performance interrelations are required to further advance this interesting class of materials and to open new application platforms. Here, examples are given of how classical polymer science tools may be utilized to gain further understanding of functional organic matter and how the physical organization, from the molecular to the macroscale of functional organic matter such as polymer semiconductors can be controlled. To this end, we present a survey on the principles of structure development from the liquid phase of this materials family with focus on how to manipulate their phase transformations and solid-state order to tailor and tune the final 'morphology' towards technological and practical applications.

Keywords: functional organic materials, phase transformations, structure/property interrelationships

PP-I-005 [Polymer Physics]

Interface Driven Conformational Change and Aggregation in Proteins/Peptides

Mehmet Sayar[1], Cahit Dalgicdir[2], Beytullah Ozgur[2], Farhad Ramezanghorbani[2]

[1]Chemical and Biological Eng. & Mechanical Eng. Departments, College of Engineering, Koc University, Turkey
[2]College of Engineering, Koc University, Turkey

Conformation and assembly of proteins and peptides are not solely determined by their sequence, but are also strongly dependent on their environment. Interfaces, whether they are macroscopic (e.g. air/water interface, membrane), or molecular (e.g. surrounding molecules hydrophobic surfaces) strongly influence the conformation of proteins/peptides by enforcing strict partitioning of the hydrophobic/hydrophilic residues. In this study, by using molecular dynamics and enhanced simulation techniques, we investigate the conformational behavior of peptides in bulk water, at the air/water interface, and in the presence of other peptides. We demonstrate the role of the interface in altering both the preferred conformation and the self-assembly behavior of these peptides. By analyzing the potential of mean force curves between two peptides in different environments, we show that both the conformation and self-assembly behavior for such peptides strongly depends on the interplay between electrostatic forces, hydrophobic effect, and their strong tendency to form backbone hydrogen bonds.

These all atom analysis are typically rather limited in terms of time and length scale. Coarse grained (CG) models are used extensively to over come these barriers. However, with their reduced degrees of freedom validity of CG models is strictly restricted to the particular state that they are parametrized for. Development of CG models for proteins and peptides, which display large conformational changes in different environments, requires parameterization in more than a single state point. Here, we demonstrate how such transferable CG models can be developed by relying on properly chosen all atom reference states.

Keywords: Protein aggregation, interface, molecular dynamics, amyloid, peptide

PP-I-006 [Polymer Physics]

Self-Assembled Nanostructures of Polypeptide-Based Polymers

Jiaping Lin

School of Materials Science and Engineering, East China University of Science and Technology, Shanghai, China

Polypeptides, which bear rich conformations, are ideal building blocks for the self-assembled nanostructures with complicated morphologies. By designing polypeptides with complex topological architectures, such as amphiphilic block and graft copolymers, we prepare various nano-aggregates, such as vesicles and spindle-like micelles. In addition to those typical aggregates, hierarchical nanostructures are also expected to be obtained. One way to realize hierarchy in self-assemblies is to introduce additional polymers into the polypeptide-based systems. For example, mixtures of poly(γ-benzyl L-glutamate)-block-poly(ethylene glycol) (PBLG-b-PEG) and PBLG homopolymers are able to form super-helices with determined chirality. Another way to prepare hierarchical aggregates is to adopt step-wise self-assembly strategy. For example, we found that the anisotropic spindle-like micelles formed by polypeptide graft copolymers can serve as pre-assembled subunits, which can further self-assemble into hierarchical structures, such as nanowires. The mechanism underlying the second step of this novel self-assembly behavior was found to follow the rule of step-growth polymerization. The results obtained from our studies can enrich knowledge about the polymer self-assembly with hierarchy.

Keywords: polypeptide, self-assembly, copolymer, hierarchy

RDPS-A-I-002 [Controlled/Living Polymerization]

Click/transclick reactions in combination with CRP: "TAD's the spirit"

Filip Du Prez, Stef Vandewalle, Frank Driessen, Johan Winne

Department of Organic and Macromolecular Chemistry, Ghent University, Ghent, Belgium

By focusing on synthetic reactions that are highly specific and reliable, the concept of click chemistry has enabled a transfer of the creative power of chemical synthesis to a wide range of scientific research areas and applications. In this contribution, we introduce triazolinedione compounds (TAD), build up from an aza-moiety, as scalable and robust building blocks for click chemistry applications in the area of polymer synthesis (De Bruycker et al, Chemical Reviews, 2016). Examples of facile catalyst-free and ultrafast macromolecular functionalization, polymer-polymer linking and polymer cross-linking under ambient conditions, both in combination with copper-mediated CRP and RAFT, will be presented. An added feature to this powerful type of chemistry is the intense colour switch from the TAD compounds (red) to the corresponding adducts (colourless), providing a visual feedback system.

Moreover, triazolinediones, when combined with indole reaction partners, can also be 'unclicked' at elevated temperatures, and furthermore integrally transferred to an alternative reaction partner. This new concept of 'transclick' reactions has also been used to introduce thermoreversible links into materials, giving rise to dynamic bulk properties such as polymer network healing, reshaping and recycling.

In conclusion, we think that TAD-based click/transclick reactions, in combination with controlled radical polymerization techniques, can enable an unprecedented advancement in the field of macromolecular engineering.

Keywords: RAFT, ATRP, TAD-chemistry, click reactions, block copolymers

Figure 1

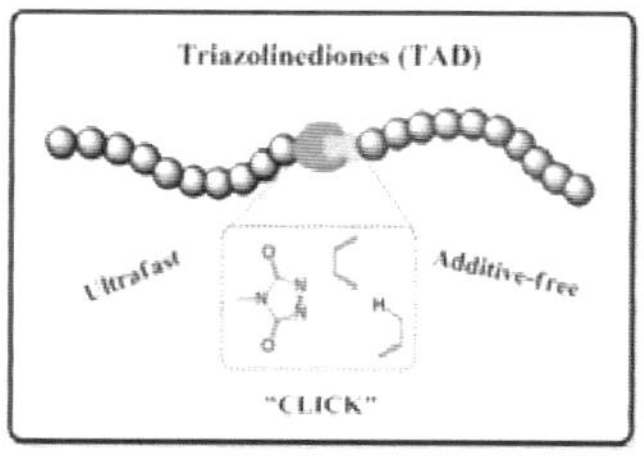

Combination of CRP and TAD-chemistry for the design of complex architectures.

RDPS-A-I-006 [Controlled/Living Polymerization]

Anionic Polymerization of Exomethylene Hydocarbon Monomers

Takashi Ishizone

Department of Organic and Polymeric Materials, Tokyo Institute of Technology, Tokyo, Japan

We have synthesized new exomethylene hydrocarbon monomers, α-methyleneindane (1)[1] and benzofulvene (2),[2] and examined the anionic polymerizability. Anionic polymerizations were carried out in an all-glass apparatus equipped with break-seals under high vacuum conditions (10-6 mmHg). The anionic polymerization of an exomethylene hydrocarbon, 1, quantitatively proceeded with sec-BuLi and potassium naphthalenide in 1,2-addition modes to form the polymers possessing cyclic structures vertical to main chain. No polymerization of 1 occurred with diphenylmethyllithium at −78 °C within 24 h, indicating the low anioic polymerizabiity comparable to styrene. The stability of the propagating carbanion of poly(1) was demonstrated by the quantitative initiation efficiency of the sequential copolymerization with styrene. On the other hand, well-defined polymers of benzofulvene, 2, were obtained by the initiation with not only sec-BuLi but also the low nucleophilic initiators including diphenylmethyllithium, diphenylmethylpotassium, and triphenylmethylpotassium. Even benzylmagnesium chloride (Grignard reagent) and potassium tert-butoxide (alkoxide) could initiate the polymerization of 2, while the initiation efficiency was far from quantitative. Furthermore, a well-defined block copolymer, poly(methyl methacrylate)-b-poly(2), was produced by the sequential copolymerization of 2 using living poly(methyl methacrylate) as a macroinitiator. The propagating carbanion of poly(2) could not initiate the polymeriation of styrene at all, indicating its low nucleophilicity. The polarized fulvene framework and π-stabilized indenyl propagating carbanion played very important roles to realize the high anionic polymerizability of 2. 1H and 13C NMR measurements confirmed that poly(2) obtained with sec-BuLi in THF possessed 59% of 1,4-structure and 41% of 1,2-structure, indicating that 2 acted as a conjugated 1,3-diene.

1) H. Ohishi, Y. Kosaka, K. Kitazawa, R. Goseki, S. Kawauchi, T. Ishizone, Macromolecules 48, 6900-6908 (2015).
2) Y. Kosaka, S. Kawauchi, R. Goseki, T. Ishizone, Macromolecules 48, 4421-4430 (2015).

Keywords: Anionic polymerization, Exomethylene, Hydrocarbon monomer, α-Methyleneindane, Benzofulvene, Living polymerization

figure 1(Ishizone)

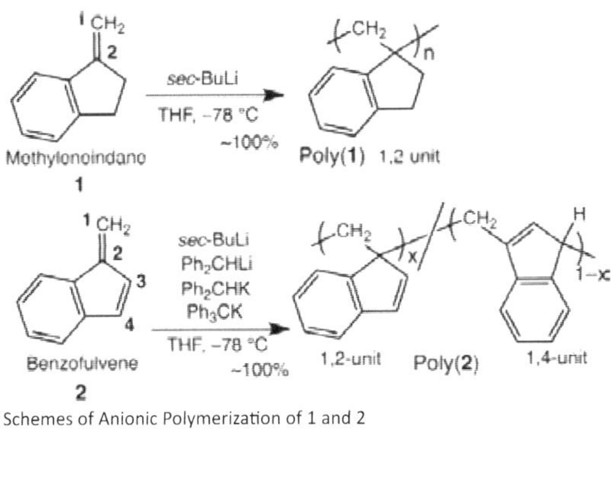

Schemes of Anionic Polymerization of 1 and 2

RDPS-A-I-008 [Controlled/Living Polymerization]

New Possibility of Living Anionic Polymerization: from Living Dendrimer to Sequence Controlled Polymer

Junpo He, Yunhai Yu, Chengke Qu

The State Key Laboratory of Molecular Engineering of Polymers, Department of Macromolecular Science, Fudan University, Shanghai, China

Anionic polymerization is a truly living polymerization system and has been widely used to prepare narrow disperse linear polymer. Recently, we have extended the capability of anionic polymerization to the preparation of living dendrimers, hyperbranched polymers, regio-specific and sequence-specific polymers. The syntheses of living dendrimers and hyperbranched polymers are based on an anionic inimer prepared through highly selective monoaddition reaction between carbanion and 1,3-bis(1-phenylvinyl)benzene (MDDPE). Copolymerization of the inimer with styrene through a self-condensing copolymerization process leads to the formation of living hyparbranched polymer[1-3]. Iterative polymerization of styrene and branching by the anionic inimer yield living dendrimer-like polymers[4,5]. We have also achieved control on regio-specificity and sequence-specificity of polymers through anionic polymerization of disubstituted 1,3-butadienes as template monomers[6,7]. The direction of monomer addition is affected by the substitution. These polymers were subject to hydrogenation and dehydrogenation, respectively, to prepare saturated polymers and substituted polyacetylenes.

References

1) Sun, W.; He, J.; Wang, X.; Zhang, C.; Zhang, H.; Yang, Y. Macromolecules 2009, 42, 7309.
2) Xie, C.; Ju, Z.; Zhang, C.; Yang, Y.; He, J. Macromolecules 2013, 46, 1473.
3) Ju, Z.; He, J. Chem. Comm. 2014, 50, 8480.
4) Zhang, H.; He, J.; Zhang, C.; Ju, Z.; Li, J.; Yang, Y. Macromolecules 2012, 45, 828.
5) Zhang, H.; Zhu, J.; He, J.; Qiu, F.; Zhang, H.; Yang, Y.; Lee, H.; Chang, T. Polym. Chem. 2013, 4, 830.
6) Zhang, Y.; Li, J.; Li, X.; He, J. Macromolecules 2014, 47, 6260.
7) Li, J.; He, J. ACS Macro Lett 2015, 4, 372.

Keywords: anionic polymerization, dendrimer, hyperbranched polymer, regio-specific polymer, sequence-specific polymer

RDPS-A-I-009 [Controlled/Living Polymerization]

Marrying Anionic Ring-Opening Polymerization with Chain-Growth Condensation Polymerization for the Synthesis of Aliphatic-Aromatic Copolyamides

Stephane Carlotti[1], Deniz Tunc[1], Camille Bakkali Hassani[1], Kevin Roos[1], Mikael Planes[1], Philippe Lecomte[2]

[1]University of Bordeaux, Bordeaux INP, CNRS, Bordeaux, France
[2]Center for Education and Research on Macromolecules, University of Liège, Liège, Belgium

The reduction of the energy consumption and the implementation of new technologies dedicated to the production of green energy for a sustainable development are nowadays major priorities. Lightweight materials with structural properties are expected to replace heavy metal components in order to reduce the consumption of energy and therefore the release of CO_2 in some cases. In composite materials, some polyamides are considered nowadays as an alternative to epoxy or polyester thermosets as far as their properties and processing can be adapted for the production of pieces at low price. This work presents a strategy based on the combination of Anionic Ring-Opening Polymerization (AROP) of ε-caprolactam and Chain-

Growth Condensation Polymerization (CGCP) of an aminobenzoate derivative for the synthesis of aliphatic-aromatic copolyamides. More generally, this approach is intended to demonstrate the possibility of marrying two chemistries and to synthesize new materials. Some tools to reach reversible (or not) crosslinked polyamide 6 in a controlled way will be also discussed.

Keywords: AROP, CGCP, polyamide, crosslinking

RDPS-A-I-010 [Controlled/Living Polymerization]

Organopolymerization by N-Heterocyclic Carbenes: Selectivity and Monomer Scope

Daniel Taton[1], Winnie Nzahou Ottou[1], Damien Bourichon[2], Camille Bakkali Hassani[1], Stéphane Carlotti[1], Anne Laure Wirotius[1], Jean Marc Sotiropoulos[2], Karinne Miqueu[2]

[1]*University of Bordeaux/CNRS/IPB - LCPO*
[2]*CNRS/University of Pau & des Pays de l'Adour - IPREM*

By virtue of their peculiar steric and electronic properties, N-heterocyclic carbenes (NHCs) have revolutionized organometallic chemistry as ligands for transition metals in the last two decades.[1] NHCs have also gained an increasing popularity as true organic catalysts in molecular chemistry and in metal-free polymer synthesis as well.[2-3] Cyclic esters have been the most investigated monomers in the context of NHC-organocatalyzed ring-opening polymerization (OROP). In this work, we will discuss our recent efforts to expand the monomer substrate diversity in NHC-mediated polymerization. In particular, NHC catalysis can be successfully applied to the OROP of N-activated aziridines, in presence of activated secondary amines as initiators.[4] NHCs can also be employed as initiators for the direct 1,4-conjugate addition organopolymerization of some (meth)acrylic monomers.2,3 However, minor modifications of the NHC structure can dramatically impact the reaction outcome. Thus, while reaction of 1,3-bis-(diiso-propyl)imidazol-2-ylidene (NHCiPr) with methyl methacrylate (MMA) affords an unprecedented imidazolium enolate cyclodimer (NHCiPr:MMA = 1:2), the tert-butyl derivative (NHCtBu) allows directly polymerizing MMA, evidencing the remarkable selectivity of NHCs.[5] The same NHCs have been further demonstrated to catalyze MMA polymerization, under the control of simple alcohols (ROH) as initiators.[6]

1. M. N. Hopkinson, F. Glorius, Nature 2014, 510, 485.
2. W. N. Ottou, H. Sardon, D. Mecerreyes, J. Vignolle, D. Taton, Prog. Polym. Sci. 2016, 56, 64.
3. M. Fèvre, J. Pinaud, Y. Gnanou, J. Vignolle, D. Taton, Chem. Soc. Rev. 2013, 42, 2142.
4. C. Bakkali-Hassani, E. Rieger, J. Vignolle, F. Wurm, S. Carlotti, D. Taton, in preparation.
5. W. Nzahou Ottou, D. Bourichon, J. Vignolle, A.-L. Wirotius, F. Robert, Y. Landais, J.-M. Sotiropoulos, K. Miqueu, D. Taton Chem. Eur. J. 2014, 20, 3989.
6. W. Nzahou Ottou, D. Bourichon, J. Vignolle, A.-L. Wirotius, F. Robert, Y. Landais, J.-M. Sotiropoulos, K. Miqueu, D. Taton Chem. Eur. J. 2015, 21, 9447.

Keywords: Organocatalysis, N-heterocyclic carbenes, ring-opening polymerization, methyl methacrylate, selectivity

RDPS-A-I-011 [Controlled/Living Polymerization]

Cyclopolymerization of 1,6-hepatadiynes: Synthesis, Mechanism and Application

Tae Lim Choi

Department of Chemistry, Seoul National University, Seoul, S. Korea

Olefin metathesis provides a versatile synthetic method to produce small, medium and large molecules. Over the past two decades, wide ranges of the molecules have been prepared by various types of olefin metathesis, such as cross metathesis (CM), ring-closing metathesis (RCM), ring-opening metathesis (ROM) reactions. Ru-based Grubbs catalysts have popularized the reaction because the catalysts are not just highly active but also very easy to handle with good functional group tolerance. Now the metathesis polymerization of olefin monomers is one of the most common ways to construct numerous polymer architectures and nanostructures. On the other hand, only a few study reported the polymerization of alkyne-containing molecules, which can efficiently produce conjugated backbone. So far especially in polymerization, olefins are the main functional group for the monomer. Here, we will describe alkyne polymerization by olefin metathesis using Grubbs catalysts. Also, our strategy to achieve controlled cyclopolymerization to prepare polymers with desired molecular weights and narrow polydispersity indices will be discussed. We studied the cyclopolymerization of 1,6-heptadiyne derivatives, which is another olefin metathesis polymerization, using Ru-based Grubbs catalysts. In the history of cyclopolymerization of diynes, these catalysts had shown poor activity and resulted in no polymer, whereas Mo-alkylidenes had achieved the efficient cyclopolymerization. We focused on the control of reaction conditions to improve the efficiency of polymerization utilizing common Grubbs catalysts, instead of the catalyst engineering. As a result, it was disclosed that severe dependence on various factors, such as the solvent, temperature, dissociative ligands, and even steric effect, could alter the efficiency of polymerization.

Keywords: Grubbs catatlyst, cyclopolymerization, conjugated polymer

cyclopolymerization

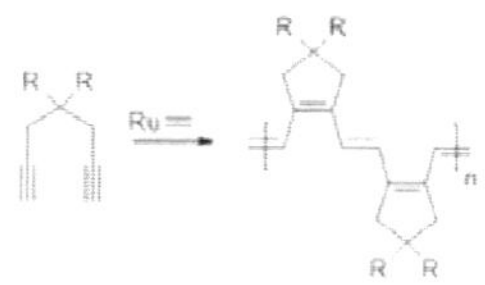

RDPS-B-I-001 [Complex Macromolecular Structures]

Highly aromatic hyperbranched polymers designed for film applications in opto-electronics

Brigitte Voit[1], Anton Kiriy[2], Robert Pötzsch[2], Qiang Wei[1]

[1]*Leibniz-Institut für Polymerforschung Dresden e.V. and Organische Chemie der Polymere and Center of Excellence cfaed, TU Dresden, Germany*
[2]*Leibniz-Institut für Polymerforschung Dresden e.V., Dresden, Germany*

The microelectronics industry and especially the organic/flexible electronics industry continue to demand new innovative polymeric materials. Besides the active (conducting and semiconducting) polymers, also the insulating materials in between, the dielectrics, as well as materials with special opto-electronic properties are highly important since they determine significantly the final performance of the device.

In the area of insulating materials high performance hyperbranched (hb) polymers (e.g. hb polyphenylenes, hb-PPh) offer significant potential in the application as gate dielectric material which can be adopted for high performance OFETs by design-matching with the respective semiconducting material. The branching provides the highly needed solubility and allows solution processing or even processing by printing techniques. For multilayer device preparation, we also developed suitable crosslinking strategies which can be thermally or photochemically induced.[1] Furthermore, hyper-branched polyvinylsulfides through thiol-yne addition of B3+A2 monomers [3] proved to be excellent candidates for high refractive index (HRI) materials well suited for 1D planar all-polymer photonic crystals and efficient light out-coupling from OLEDs [2,3]. Polymers with RI as high as 1.78 have been achieved when using a naphthaline-based B3 monomer and benzene-1,4-dithiol as A2. Furthermore, crosslinkable polycarbazoles have been developed through metal-free 1,3 dipolar cycloaddition reaction which can be potentially used as hole conductor/hole injection layer materials.

Acknowledgements: The work has been done in collaboration with Institute of Applied Photophysics (IAPP) TU Dresden (Reinecke, Will, Lenk) and Università di Genova (Comoretto) and is partially supported by Center of Excellence cfaed.

References:

[1] R. Pötzsch, B. Voit, Macromol. Rapid Commun. 2012, 33, 635.
[2] S. Gazzo, G. Manfredi, R. Pötzsch, Q. Wei, M. Alloisio, B. Voit, D. Comoretto, submitted J. Polym. Sci. Polym. Phys., 54, 2016, 73–80
[3] Q. Wei, R. Pötzsch, X. Liu, H. Komber, A. Kiriy, B. Voit, P.-A. Will, S. Lenk, S. Reineke, Adv. Funct. Materials, 2016, online 2016, dx.doi.org/10.1002/adfm.201504914

Keywords: Hyperbranched, polymers, optoelectronic

RDPS-B-I-002 [Complex Macromolecular Structures]

Topological Polymer Chemistry toward Macromolecular Graph Constructions

Yasuyuki Tezuka

Tokyo Institute of Technology

Topologically intriguing polymer chain architectures have continuously been an attractive research subject [1]. A class of *multicyclic* polymer topologies, including three subclasses of *fused*, *spiro* and *bridged* forms, are particularly unique not only from the topological geometry viewpoint but also from their programmed folding structures of biopolymer relevance. We have developed an *electrostatic self-assembly and covalent fixation* (ESA-CF) protocol [1], in which ion-paired polymer self-assemblies are employed as key intermediates. In addition to all three types of dicyclic constructions, i.e., θ (*fused*), 8 (*spiro*) and manacle (*bridged*) forms, a variety of tri-, tetra- and even hexacyclic polymer topologies of *spiro* and *bridged*-forms and three doubly-*fused* tricycle (b-, g- and δ-graph) forms, as well as a triply-*fused* tetracyclic and a quadruply-*fused* pentacyclic forms (unfolded tetrahedron-graph, and "shippo"-form, respectively) were subsequently constructed through the ESA-CF protocol in conjunction with a tandem alkyne-azide addition, i.e., *click*, and olefin metathesis, i.e., *clip*, reactions [2].

We have also completed the construction of a triply-*fused* tetracyclic macromolecular $K_{3,3}$ graph topology [3], which is known as a prototypical non-planar graph, and remarkably, moreover, has been identified in cyclic polypeptides (cyclotides) produced through the intramolecular S-S bridging with cysteine residues. An electrostatic polymer self-assembly consisting of a uniform-size dendritic polymer precursor having six cyclic ammonium salt end groups carrying two units of a trifunctional carboxylate counteranions was subjected to the covalent conversion under dilution. The $K_{3,3}$ graph product was subsequently isolated from the ladder form constitutional isomers by means of a recycling SEC technique.

[1] *Topological Polymer Chemistry: Progress of cyclic polymers in synthesis, properties and functions*, Y. Tezuka ed., World Scientific, Singapore, 2013. [2] Heguri H, Yamamoto T, and Tezuka Y, *Angwew. Chem. Int. Ed.*., **54**, 8688 (2015); [3] Suzuki T, Yamamoto T, and Tezuka Y, *J. Am. Chem. Soc.*, **136**, 10148 (2014).

Keywords: topological polymer chemistry, cyclic and multicyclic polymers, macromolecular graph constructions

RDPS-B-I-003 [Complex Macromolecular Structures]

From Simple Chemistries to Complex Functional Polymers

Patrick Theato

Institute for Technical and Macromolecular Chemistry, University of Hamburg, Hamburg, Germany

Properties of polymeric materials are inscribed in the chemical structure of the respective polymer. Over the years, we have introduced highly efficient post-polymerization modification protocols that allow the incorporation of multiple functional units onto a polymer chain. Noteworthy, modern syntheses allow for an ever increasing macromolecular molecular complexity. However, the achievable complexity comes at the price of relying on highly sophisticated reactions. We surely have also contributed to this development by establishing multi-component reactions in efficient post-polymerization modifications and examples will be presented. Yet, we want to ask the question of how complex a functional macromolecular chain can become when being prepared by only simple chemical reactions? This addresses not only the aspect of applicability but also of sustainability. We are convinced that *thinking outside the box* in polymer synthesis and processing can lead to new and promising approaches that enable precision polymer materials. Synthetic routes, possibilities, remaining challenges and opportunities will be discussed.

Keywords: polymer modification, multi-component reaction, thiol-ene reaction

RDPS-B-I-004 [Complex Macromolecular Structures]

Thiolactone Groups as Functional Handles: From Double End Group Modification to Macromolecular Lineups

Filip Du Prez, Frank Driessen, Steven Martens, Pieter Espeel

Department of Organic and Macromolecular Chemistry, Ghent University, Ghent, Belgium

The design and synthesis of tailor-made polymers for different applications requires a variety of synthetic methodologies to fine-tune the polymer structure and adjust its final properties. Introducing reactive functional handles in a subsequent post-polymerization modification (PPM) step, often via 'click'-type reactions, became a popular tool in macromolecular science. In order to increase the number of different functionalities in one polymeric unit, introduction of multiple functional groups per reactive handle can be a quite attractive feature.

A first part of this presentation will focus on the one-pot double modification of a variety of different polymers containing a thiolactone end group. For this end group modification, a functional amine opens the five-membered ring, releasing a thiol group. This group reacts in a subsequent orthogonal manner with the acrylate present in the same reaction medium, enabling the quantitative introduction of two distinct functionalities at the same reactive site. A series of different polymers containing a thiolactone end group were synthesized through CRP of a thiolactone-containing initiator or modification of commercial hydroxyl-functionalized polymers with an isocyanate-containing thiolactone. A library of different functionalities were introduced by reaction with different amine/acrylate combinations, introducing a broad range of different functionalities (Driessen, F et al Macromol. Rapid Commun., 2016).

This strategy was extended to the synthesis of a thiolactone-acrylate hetero-telechelic macromonomer. Upon addition of a functional amine, a multisegmented block copolymer was isolated with selected functionalities positioned at each segment connection and thus equally spaced along the polymeric backbone (Driessen et al, ACS Macro Lett., 2015).

Finally, these materials were applied in the design of glyco- and amphiphilic polymers and analyzed by SEC, DOSY-NMR, LCxSEC and DLS, revealing the enhanced properties of these unprecedented macromolecular structures.

Keywords: Complex macromolecular structures, thiolactone chemistry, precision synthesis, end group modification

Figure 1

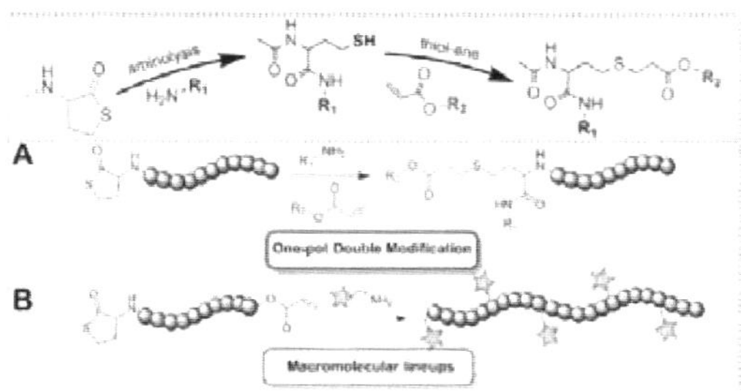

Amine-thiolactone-ene conjugation for (A) the double modification of polymer end groups and (B) the synthesis of macrolecular lineups.

RDPS-B-I-005 [Complex Macromolecular Structures]

Thiol-yne Click Chemistry for the Synthesis of Tough Materials

Andrew P Dove, Vinh X Truong, Laura Mcdougall, Craig A Bell, Anais Pitto Barry

Department of Chemistry, University of Warwick, UK

Click chemistry has transformed the synthesis and functionalisation of macromolecules on account of the simple and highly efficient nature of this class of reactions. among the wide range of reactions classified as click chemistry, the addition of thiols to unsaturated carbon-carbon bonds, so-called thiol-ene, or thiol-yne reactions in which an organic thiol can add via a nucleophilic or radical addition to an alkene or alkyne has provided a wide range of potential products that are of high interest in a number of areas. We have focussed on the development of the nucleophilic thiol-yne addition in which a thiol adds to an activate alkyne catalyzed by a simple base (for example triethylamine or aqueous base solution). These reactions have proven to be powerful for the functionalisation of polymer materials as well as the formation of polymers and polymer networks. In this presentation, the potential of this approach will be outlined for the synthesis of tough materials including hydrogels.

Keywords: thiol-yne, polymer synthesis, polymer functionlisation, gels

RDPS-B-I-006 [Complex Macromolecular Structures]

Inverted grafting-to PDMS block copolymers in a PDMS matrix key to superior aqueous lubrication

Søren Hvilsted[1], Irakli Javakhishvili[1], Katja Jankova[1], Troels Røn[2], Seunghwan Lee[2]

[1]Danish Polymer Centre, Department of Chemical and Biochemical Engineering, Technical University of Denmark, Kgs. Lyngby
[2]Department of Mechanical Engineering, Technical University of Denmark, Kgs. Lyngby, Denmark

Ability to control lubrication is essential to avoid damage of the underlying material and to ensure low energy dissipation in biological and man-made mechanical systems. Initially, our systematic approach to lower the surface friction of wet PDMS surfaces employed adhesion of various novel amphiphilic block or graft copolymers prepared by the controlled radical polymerization techniques ATRP and RAFT. The concept is relying on adsorption of the hydrophobic blocks onto the PDMS surface while the neutral or charged buoyant blocks should provide the lubricity in the aqueous phase. The systematic study concluded that only moderate lubrication of the PDMS surfaces could be achieved.

Alternatively anchoring hydrophilic polymer brushes onto surfaces by grafting seams as a powerful mean to render materials slippery in aqueous environment. Nevertheless, presently available approâches to graft polymer brushes on surfaces, "grafting-from" or "grafting-to", present severe restrictions in terms of practical and long-term applications.

Here we like to advocate the new concept: "inverted grafting-to" that forms hydrophilic polymer brushes by selective segregation of hydrophilic chains of an amphiphilic diblock copolymers, PDMS-*b*-PEG or PMDS-*b*-PAA, from a PDMS matrix. The block copolymers are prepared by "click chemistry" or ATRP by use of the macroinitiator concept, respectively. Excellent grafting stability and restoring capabilities are achieved as revealed even under harsh tribologial testing since the hydrophilic polymer brushes are generated from an internal source of the material. The film can easily be applied to elastomers, metals and ceramic substrates by spin- or drip-coating. The resulting sliding friction coefficients (μ) are 0.001 to 0.05 for soft contacts depending on substrate, load, counter surface, pH, and salinity. Here the hydrophilic PAA shows much superior lubricity compared to PEG, which is rationalized by larger reduction of total free energy of the former upon hydration.

Keywords: PDMS surface engineering, controlled radical polymerization, click coupling, aquous sliding experiments, water contact angle measurements

RDPS-B-I-007 [Complex Macromolecular Structures]

Building nanostructured materials from molecular bottlebrushes

Javid Rzayev

Department of Chemistry, University at Buffalo, SUNY, Buffalo, NY, USA

Bottlebrush copolymers are densely grafted macromolecules with a comb-like architecture. We have utilized multicomponent bottlebrush copolymers for the preparation of nanostructured materials via controlled cross-linking. End-reactive molecular bottlebrushes were prepared by a grafting from approach using controlled radical polymerization techniques. The synthesized copolymers were composed of polymethacrylate branches decorated with coumarin functionalities and poly(glycidyl methacrylate) end blocks, which were further modified to install olefinic groups. Orthogonal cross-linking of the end-reactive bottlebrushes by Ru-catalyzed cross-metathesis of olefinic groups and photodimerization of coumarin side chains produced well-defined brush-star and rigid brush-network architectures, which were easily converted to nanoporous materials by solvent removal. The influence of structural parameters on nanostructure formation was explored. The fabricated materials were characterized by electron microscopy, nitrogen adsorption and x-ray scattering methods.

Keywords: bottlebrush copolymers, nanoporous materials, controlled polymerization

RDPS-B-I-008 [Complex Macromolecular Structures]

Precision glycopolymers and their interactions with lectins

Remzi Becer

School of Engineering and Materials Science, Queen Mary University of London

Sequence controlled polymers have been attracting more and more attention to deliver the desired properties to the advanced materials by the help of their precisely controlled compositions and architectures. Development of various controlled radical polymerization techniques and "click" reactions provide a sufficient platform to prepare functional polymers. Understanding the specific multivalent carbohydrate-protein interactions is crucial to determine the structure-property relationships and to design accordingly the next generation of functional glycomaterials. Therefore, we investigate the structure-property relationships between the mammalian lectins and multivalent carbohydrate polymers, which may have applications for anti-adhesion therapy. Moreover, we have investigated the affinity of poly(mannose-methacrylate), helical glycocopolypeptides, gp120, start shaped glycopolymers, and cyclodextrin centered glycopolymers with a selected mannose binding lectin (DC-SIGN) that exists on dendritic cells, using SPR technique. Selected members of a glycopolymer library were used to demonstrate the interactions between DC-SIGN and mannose rich polymers. We extend this study to a broader set of polymers to examine the effect of chain length, end group, architecture, thermoresponsive block, and number of arms in the star shaped polymers on the lectin binding.

Keywords: glycopolymer, lectin, click reaction, controlled polymerization

RDPS-B-I-009 [Complex Macromolecular Structures]

Thiol-based Conjugations: A Versatile Tool for Fabrication and Functionalization of Polymeric Nanomaterials

Amitav Sanyal

Department of Chemistry, Bogazici University, Istanbul, Turkey

In recent years, 'click' chemistry has redefined our approaches towards functionalization of polymeric materials. Post-polymerization functionalization using 'click' reactions has emerged as a method of choice for obtaining functional materials due to their high efficiency. The rich toolbox of 'click' reactions enables facile multi-functionalization using combinations of different chemical transformations. In particular, thiol-based chemistries that proceed through either nucleophilic or radical pathway, provides a powerful tool for both fabrication and functionalization of polymeric nanomaterials. Soft nanostructures such as nanocoatings, nanofibers and nanoparticles find applications in various areas of biomedical sciences. Over the past years, we have developed several approaches towards synthesis of thiol-reactive polymeric materials. These range from soluble polymeric constructs to bulk materials such as hydrogels [1,2]. Also, reactive coatings that are several nanometer in thickness can be engineered on planar surfaces and curved interfaces. In particular, thiol-reactive polymers were 'grafted-from' silicon oxide and polymeric substrates, and 'grafted-to' magnetic nanoparticles [3,4]. These surfaces were derivatized with fluorescent dyes and ligands to attribute them with imaging and sensing capabilities. More recently, we have extended the methodology to fabricate nanofibers and nanoparticles that undergo facile multi-functionalization. The presentation will highlight how interplay of various chemistries can be weaved together to fabricate multi-functional polymeric nanostructures. Authors acknowledge The Scientific and Technological Research Council of Turkey (TUBITAK, Project No. 114Z139 and 104T307) for providing financial assistance for research.

References:

[1] Gok, O.; Kosif, I.; Dispinar, T.; Gevrek, T. N.; Sanyal, R.; Sanyal, A. Bioconjugate Chemistry 2015, 26, 1550-1560.
[2] Park, E. J.; Gevrek, T. N.; Sanyal, R.; Sanyal, A. Bioconjugate Chemistry 2014, 25, 2004-2011.
[3] Gevrek, T. N.; Bilgic, T.; Klok, H. A.; Sanyal, A. Macromolecules 2014, 47, 7842-7851.
[4] Arslan, M.; Gevrek, T. N.; Lyskawa, J.; Szunerits, S.; Boukherroub, R.; Sanyal, R.; Woisel, P.; Sanyal, A. Macromolecules 2014, 47, 5124-5134.

Keywords: 'click' chemistry, reactive polymers, hydrogels, nanofibers, nanoparticles

Thiol-ene Based Fabrication and Functionalization

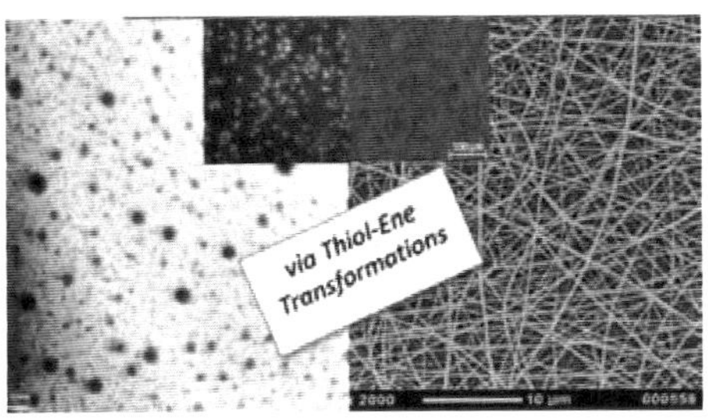

RDPS-C-I-001 [Light-induced Reactions in Polymer Science]

Photo-curable triazole-containing crosslinked networks from structure-property relationship to tough industrial thermosets

Christopher N Bowman, Han Byul Song, Austin Baranek, Matthew Mcbride, Abeer Alzahrani

Department of Chemical Engineering, University of Colorado, Boulder, USA

Photo-curable triazole-containing thermosets are mechanically robust, chemically stable, and extraordinarily tough polymer networks largely due to the stiff triazole linkage and the associated secondary interactions. The objective of this research is to study structure-property relationships of photo-initiated CuAAC (copper-catalyzed azide-alkyne cycloaddition) polymerization using easily synthesized azide and alkyne monomers for practical applications demanding a light-triggered, non-degradable, and tough thermosets. Five different structures of azide and alkyne containing monomers were synthesized utilizing facile and efficient synthetic procedures in order to perform systematic studies on kinetics and mechanics of photo-CuAAC polymerization. By varying the monomer structure, we demonstrated the photo-initiated CuAAC exhibited mobility limited polymerization kinetics in real-time FT-IR; however, the active copper species' catalytic activity after short light exposure enabled a considerable amount of dark polymerization, increasing conversion nearly 60% in the dark. CuAAC polymers presented significant increases of both Tg and toughness when compared with analogous control networks without triazoles when measure with DMA and mechanical testing. Furthermore, we show applications taking advantage of the mechanical properties including shape memory polymers, mechanically robust CuAAC foams, and CuAAC dental composites.

Keywords: CuAAC, photopolymerization, kinetics, mechanics, step-growth polymerization

RDPS-C-I-002 [Light-induced Reactions in Polymer Science]

In-situ preparation of metal/ polymer nanocomposite thin films by photopolymerization method

Nergis Arsu[1], Hicret Kirtay[1], Tugce Cinko[1], Gönül S Batibay[1], Tolga Ceper[1], Melisa Adıbelli[1], Tulin Ozbek Arasoğlu[2]

[1]Yildiz Technical University Chemistry Department, 34220 Esenler Istanbul, Turkey
[2]Yildiz Technical University Molecular Biology and GeneticDepartment, 34220 Esenler Istanbul, Turkey

Preparation of metal nanoparticles are still in progress to be the subject of intense research area due to unique properties which welcome wide range of potential applications from optics, biomedical and electronic fields.

The control of size, shape and distribution of metal nanoparticles takes very important part for preparation. Agglomeration is inevitable but reducing and getting good distiribution is important goal to achieve. There are several synthesis method including, chemical and photochemical. One pot preparation of metal nanoparticles and nanocomposites by using photochemical method were previously reported in the literature(Yağcı,Y et.al "Chem. Commun., 2008, 2771)

Photoinitiators play good role for photochemical way for preparation. There are mainly two type of photoinitiators, namely Type I and Type II. There are new generation of photoinitiators as well, one component Type II initiators which do not require any co-initiator to initiate the polymerization of monomers.

The most sensitive and significant route is to obtain in situ metal nanoparticles simultaneously with polymerization process and these plays a critical role on control of shape, size and distribution of particle.

In this study, thioxanthone and several derivatives of thioxanthone as one-component Type II initiator, and Type I initiators were used for photoreduction of metal ions to nanoparticles either in solvent or in polymer matrix by in situ photochemical method. Controlled size and well distribution of Nps accompanied for this synthesis. Antibacterial activities were also determined.

Keywords: photopolymerization, nanotechnology, nanocoatings, photoinitiators

RDPS-C-I-003 [Light-induced Reactions in Polymer Science]

One-Pot Three-Step Polymerization System Using Double Click Michael Addition and Radical Photopolymerization

M. Retailleau, A. Ibrahim, C. Croutxé Barghorn, X. Allonas

Laboratory of Macromolecular Photochemistry and Engineering, University of Haute Alsace, 3b rue Alfred Werner, 68093 Mulhouse, France

Introduced by Sharpless and al.1, the "click chemistry" is an attractive concept for polymer tailoring.2 Using this chemistry, polymers may be easily generated without solvents and heat while giving rise to high yields. Thiol-ene is a typical example of such a click chemistry. However, this strategy requires thiols which are difficult to use as starting materials. To overcome this issue, we found that using amines as nucleophile is possible without significantly affecting the thermo-mechanical properties.3 In this study, a click chemistry synthetic strategy based on aza-Michael addition and radical photopolymerization is proposed to generate a polymeric network via three time-controlled steps (Figure 1).4 The selection of primary diamines and diacrylates allows two consecutive aza-Michael reactions to occur. This reaction sequence affords the unique opportunity to interpose a radical photopolymerization reaction, enhancing the cross-link density. Consequently, the second aza-Michael addition appears as a valuable post-consolidation step of the polymer network. This kind of homogeneous 3D network with tunable properties is used in different examples to produce smart materials. As an example, this three-step approach was used to synthetize a new type of shape memory polymer (SMP). Photopolymers exhibiting excellent shape fixity and recovery within a narrow temperature range were achieved.5

References

[1] H.C. Kolb, M.G. Finn, K.B. Sharpless, Angew. Chem. Int. Ed. 40, 2004-2021 (2001).
[2] B.D. Mather, K. Viswanathan, K.M. Miller, T.E. Long, Prog. Polym. Sci. 31, 487-531 (2006).
[3] M. Retailleau, A. Ibrahim, C. Croutxé-Barghorn, X. Allonas, Prog. Org. Coat., (2016) under press
nas, C. Ley, D. Le
nas, RSC Advances
Addition

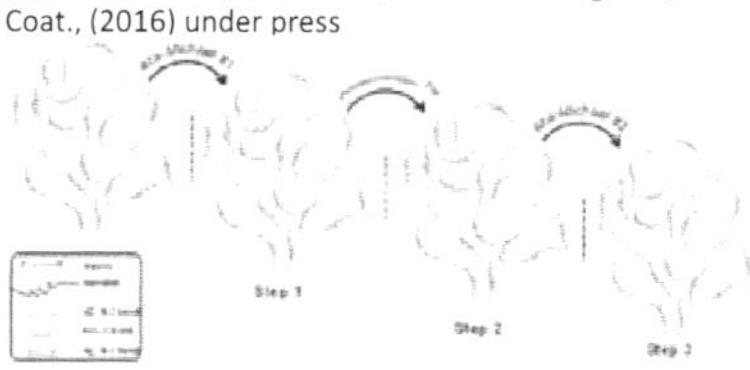

Schematic representation of the aza-Michael three-step process

RDPS-C-I-004 [Light-induced Reactions in Polymer Science]

Surface-attached polymer layers via C,H insertion crosslinking (CHIC)

Oswald Prucker, Jürgen Rühe

IMTEK – Department of Microsystems Engineering, University of Freiburg, Freiburg, Germany

The fabrication of modern microsystems often requires a precise control of the surfaces and interfaces of the devices. Many (bio)sensor formats also ask for surfaces which interact favourably with their biological environment and provide often delicate functionalities for sensing. It is obvious that these requirements ask for surface modification techniques that are simple and robust to perform and should be versatile in terms of the surface properties to be generated. The approach that we follow is based on polymers which carry photoreactive groups. Upon activation through light (or in some cases also heat) these groups attach to neighboring chains to form networks through C,H insertion crosslinking (CHic) reactions. Through an appropriate choice of the crosslinker unit the photochemical reaction can be carried out at a chosen wavelength. To this end more than a couple of dozen different systems (benzophenons, anthrachinons, diazocarboxyls, sulfonylazides etc.) have been studied. We describe a general model for the photocrosslinking process in such systems based on percolation theory, which allows a description of the crosslinking kinetics in detail. We show how these reactions can be used to generate surface-attached polymer networks with very precise thicknesses on a wide variety of substrates including typical oxide surfaces but also chemically rather inert polymeric substrates and even biological tissues. The techniques can also be combined with many methods for surface microstructuring. We will show examples of CHIC systems that are triggered photochemically at different wavelengths such that the layers are reliably generated within very short illumination times and we will also demonstrate how the technique can used to generate multilayer systems in a layer-by-layer fashion from different polymers. Two color lithography processes even allow the generation of undercut microstructures. Possible applications of such surface-attached polymer networks range from optical devices to biocompatible surfaces and various biochip formats.

Keywords: surface-attached polymer networks, C-H-insertion crosslinking, coatings

RDPS-D-I-001 [Condensation Polymerization and Thermoset]

The Most Versatile Polymer Chemistry, Polybenzoxazines: Various Preparation Methods and New Polymerization Schemes for Advanced Material Design

Hatsuo Ishida

Case Western Reserve University

Despite more than 70 years of the history of benzoxazine chemistry, the research activities on polybenzoxazines exist only in the past few dacades. It is the newest class of thermoset resins, and is one of only a few commercialized polymers in the past 40 years. Benzoxazine resins offer extremely versatile chemistry. Various structures can be synthesized by many commercially available raw materials. In addition to well-known Mannich condensation of phenol, formaldehyde and amine, there are other methods to prepare benzoxazine monomers. Some of the monomers prepared by these methods have never been reported to polymerize. In this lecture, these new polymerization schemes will be introduced to make benzoxazine chemistry more versatile than already existing vast potential to tailor desired properties. Despite active effort in understanding the polymerization mechanism, the benzoxazine initiation mechanism is still poorly understood. A new, generalized initiation scheme will also be introduced. Finally, using these newly understood polymerization schemes, it is possible to develop benzoxazines that polymerize at low temperatures. By combining smart chemistry with the intramolecularly hydrogen bonded structure, extremely thermally stable polymers that are readily processed can be developed. A representative reaction scheme is shown below.

Keywords: polybenzoxazine, benzoxazine, synthesis, polymerization mechanism, low polymerization temperature, high thermal stability

Scheme 1

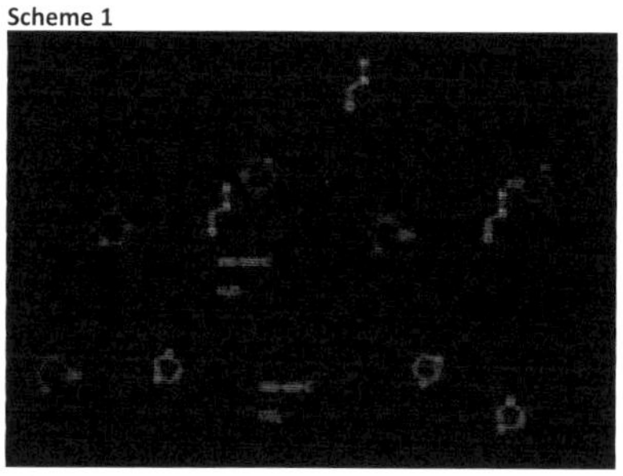

Scheme 1. Monomer synthesis (A), polymerization (B), and structural transformation upon heating (C1: Water formation and C2: Carbon dioxide formation).

RDPS-D-I-002 [Condensation Polymerization and Thermoset]

Metal free, superacid catalyzed step-growth polymerization: State of the art

Mikhail Zolotukhin

Instituto de Investigaciones en Materiales, Universidad Nacional Autonoma de Mexico

The report presents a general overview of novel polymer syntheses based on superacid catalyzed step-growth polymerizations of carbonyl compounds with aromatic hydrocarbons (polyhydroxyalkylations).1-3 The polymerizations performed at room temperature in the presence of Brønsted superacid highly chemoselective, tolerant of carbonyl, acetyl, N-oxime, carboxy, sulfoacid, methoxy, amino, bromomethyl groups demonstrate wide applicability and include simple reaction conditions, readily available starting materials and reagents and generation of an inoffensive by-product (water). It is often difficult to achieve high regiospecificity in Friedel-Crafts type reactions; and more surprising is the fact that polymer-forming electrophilic aromatic substitution reaction proceeds exclusively in para-position of the terminal phenyl groups. The demonstrated efficiency and orthogonality of some polymerizations shows it to be an essential addition to the family of polymer "click" reactions. Examples of new polymer preparations and the properties of the polymers thus obtained are presented. Polymers of complex topology (including 100% hyperbranched, semi-ladder and ladder polymers) will be also discussed. In contrast to the classical theory of step-growth polymerization, stoichiometrically imbalanced polymerizations of the carbonyl compounds with aromatic hydrocarbons gave for the first time Ultrahigh Molecular Weight (UHMW, Mw >2000 kDa) polymers. An origin of the non stoichiometric effect, unusual synthesis of linear, narrow polydispersity, UHMW polyme and applications of the non-stoichiometric polymerization for new polymer preparations will be demonstrated. On the whole, the results obtianed expand significantly the scope of both aromatic cation mediated and step-growth based polymer chemistries.

References

1. M. G. Zolotukhin, S. Fomine, R. Salcedo, L. Khalilov. Chem. Comm. 2004, 1030.
2. M. T. Guzman-Gutierrez, S. Fomine, M. G. Zolotukhin, et al., Macromolecules 2011, 44, 194.
3. A. R. Cruz, M. Carmen G. Hernandez, M. T. Guzmán-Gutiérrez, M. G. Zolotukhin, et al., Macromolecules, 2012, V. 45, 6774.

Keywords: polycondensation, superacid, polyhydroxyalkylation

RDPS-D-I-003 [Condensation Polymerization and Thermoset]

Synthesis of Conjugated Polymer by Alternating Intramolecular and Intermolecular Catalyst-Transfer Condensation Polymerization under Unstoichiometric Conditions

Tsutomu Yokozawa

Department of Materials and Life Chemistry, Kanagawa University, Yokohama, Japan

Catalyst-transfer condensation polymerization (CTCP) is a powerful methodology for synthesis of well-defined π-conjugated polymers with controlled molecular weight and low polydispersity. In CTCP with palladium (Pd) catalysts, *t*-Bu_3P ligated Pd complex is widely used as a catalyst for synthesis of well-defined π-conjugated polymers by polymerization of AB type monomers. We now report Suzuki-Miyaura coupling polymerization of A_2 and B_2 aromatic monomers with the *t*-Bu_3P ligated Pd catalyst. We first conducted polymerization of *p*-phenylenediboroinc acid ester and *p*-phenylene dibromide in the presence of a precatalyst for *t*-Bu_3PPd to obtain high-molecular-weight poly(*p*-phenylene) (PPP) with boronic acid ester moieties at both ends even by using excess amount of *p*-phenylene dibromide. This unstoichiometric polymerization is accounted for by successive substitution of *p*-phenylene dibromide via intramolecular transfer of the catalyst to afford always PPP oligomers and polymers with boronic acid ester moieties at both ends. Polymerization of other monomers will be also reported.

Keywords: Conjugated polymer, Catalyst-transfer condensation polymerization, Suzuki coupling, Pd catalyst, Unstoichiometric polycondensation

RDPS-D-I-004 [Condensation Polymerization and Thermoset]

Hyperbranched segmented polyurethaneureas: A versatile platform for the preparation of novel functional materials

Emel Yılgör, Çağla Koşak Söz, Iskender Yılgör

KUYTAM Surface Science and Technology Center, Chemistry Department, Koç University, İstanbul, Turkey

Hyperbranched segmented polyurethaneureas (HBPU) obtained by using A2 and B3 type monomers provide a versatile platform for the preparation of functional materials with a wide range of properties and applications. Physical and chemical properties of HBPUs obtained can be tailored by; (i) controlling the chemical composition of the polymer, (ii) choice of the polymerization method used, (iii) post-polymerization reactions, and (iv) use of additives. Critical influence of these factors on the degree of branching, type of the reactive end groups formed, polymer morphology, thermal and mechanical properties of the HBPUs obtained were investigated. HBPUs were prepared by using cycloaliphatic isocyanates (HMDI), polyether or PDMS soft segments with different molecular weights and various aliphatic amines. One of the critical factor, which had major influence on the topology, degree of branching and tensile properties of the HBPUs obtained was the polymerization procedure followed. As schematically shown in Figure 1, for a model system of 30 A2 and 30 B3 monomers, mixing A2 and B3 (A2+B3) or adding A2 over B3 (A2/B3) or B3 over A2 (B3/A2) during polymerization has a dramatic effect on the topology and degree of branching (DB) of the HBPU obtained. Tensile properties of HBPUs are also strongly influenced by the synthetic procedure used, as shown in Figure 2.

Keywords: hyperbranched, polyurethane, polycondensation

Figure 1

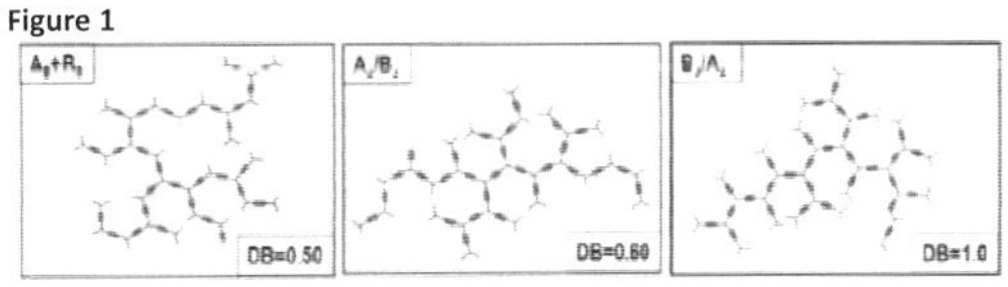

Influence of the polymerization procedure on the topology and degree of branching of model HBPUs obtained by the reaction of 30 A2 and 30 B3 monomers.

Figure 2

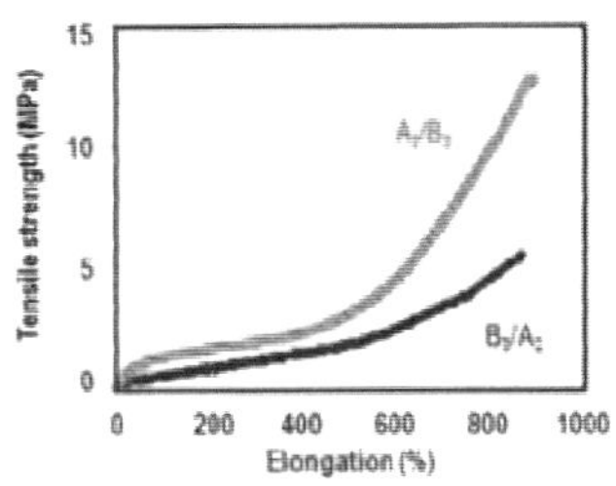

Influence of polymerization procedure on the tensile behavior of HBPUs with identical chemical compositions

RDPS-D-I-005 [Condensation Polymerization and Thermoset]

Comparison of In Situ to Ex situ Prepared Sol-gel Inorganic/Organic Hybrid Coatings

Mark Soucek

University of Akron

Several inorganic/organic systems will be presented from oil seed modified coatings to epoxide primers. Film properties in terms of tensile and in particular fracture toughness will be discussed with respect to aircraft primers. A comparison of in situ sol-gel based inorganic phase with ex situ SNAP colloids to similar sized particles without functionality. The fracture toughness was measure 3 different temperatures to reflect aircraft temperatures in flight and on the ground. In addition, a comparison of properties as a function of where the alkoxy silane was attached to the epoxide will be presented. The alkoxy silane was attached to either the epoxide end group or the internal hydroxyl group in the repeat unit. The end group was a replacement of one crosslinking mechanism for another whereas the alkoxy silane afford more of an opportunity to crosslink.

Keywords: sol-gel, hybrid, coatings

RRB-I-001 [Renewable Resources and Biopolymers]

Cellulose Swelling and Dissolution

Paschalis Alexandridis, Mohammad Ghasemi, Marina Tsianou

Department of Chemical and Biological Engineering, University at Buffalo, The State University of New York (SUNY), Buffalo, New York, USA

Efficient utilization of biomass is hindered by the recalcitrance to dissolution of cellulose, presenting crystallinity and extensive noncovalent interactions (e.g., hydrogen bonding and hydrophobic). Only a few solvent systems are effective for direct dissolution of cellulose, and these operate under rather strict conditions of composition and temperature. Moreover, the mechanism of cellulose dissolution from semicrystalline solid to homogenous polymer solution is not well-understood. We describe here the solvent-induced decrystallization, swelling, and dissolution of semicrystalline cellulosic fibers using a phenomenological model that is informed by experimental observations and polymer fundamentals. This model is based on the transport phenomena governing the dissolution of solid cellulose, e.g., solvent penetration, transformation from crystalline to amorphous domains, specimen swelling, and polymer chain untangling, and incorporates the thermodynamics and kinetics of dissolution. The model predicts the: (i) crystalline and amorphous cellulose and solvent concentrations as functions of time and position within the fiber; (ii) fraction of fiber dissolved and degree of crystallinity over time; (iii) fiber diameter and swelling rate over time; and (iv) total descrystallization time and total dissolution time. The effects of various parameters (such as fiber diameter, cellulose degree of crystallinity, solvent properties, temperature) on the kinetics of dissolution are analyzed, and the results are compared with experimental studies. The insights obtained from this analysis facilitate the design of efficient solvent systems and conditions for woody biomass processing.

Keywords: Biomass, biopolymer, cellulose, crystalline polymer, fiber, dissolution

RRB-I-002 [Renewable Resources and Biopolymers]

Fully bio-based Non-Isocyanate Polyurethanes (NIPU) via cyclic carbonate/amine route

Henri Cramail[1], Océane Lamarzelle[1], Lise Maisonneuve[1], Pierre Luc Durand[1], Etienne Grau[1], Guillaume Chollet[2]

[1]*University of Bordeaux, CNRS, Bordeaux INP*
[2]*ITERG*

Thermoplastic poly(hydroxyurethane)s (PHUs) raised industrial and academic research curiosity, since their synthesis is achieved via the ring-opening of bis-cyclic carbonates with diamines, enabling the replacement of phosgene and isocyanates employed in the classical polyurethane (PU) manufacture. Because of variations of fuel prices as well as environmental concerns, the use of building-blocks from renewable resources, such as vegetable oils, is on the rise. Combining PHUs synthesis and bio-based compounds, a large platform of fatty acid-based cyclic carbonates as poly(hydroxyurethane) precursors has already been synthesized through epoxidation/carbonation routes. However, such monomers exhibit a slow polymerization rate towards amines, due to the electron-releasing alkyl chains, which deactivate the cyclic carbonates. An alternative route consists in inserting a heteroatom nearby the cyclic carbonate to improve/activate its reactivity. Herein, the synthesis of new activated lipidic cyclic carbonates from glycerol carbonate and epichlorohydrin has been achieved, leading respectively to an ester or an ether linkage in β position of the carbonate. After kinetic investigations of the cyclic carbonate aminolysis on model compounds, the corresponding activated bis-cyclic carbonates were polymerized with two diamines and exhibited enhanced reactivities. A specific focus on the side reactions that could occur in both model reaction and polymerization is also discussed. On the other hand, a new route to access bio-based diamines using mild and green conditions has been set up through an optimization of aliphatic alcohol oxidation into the corresponding nitriles, followed by an hydrogenation. The resulting diamines were subsequently polymerized with activated cyclic carbonates in order to obtain fully bio-based poly(hydroxy urethane)s.

Keywords: poly(hydroxy-urethane)s, carbonates, bio-based amines

SFP-I-001 [Smart and Functional Polymers]

Contact Biocides Based on Poly(2-oxazoline)s and Poly(2-oxazine)s

Frank Wiesbrock, Klaus P. Luef, Martin Fimberger, Andrew M. Kelly

Polymer Competence Center Leoben, Leoben, Austria

Contact biocides are cationic polymers that kill microbes by mere contact. In contrast to disinfectants with comparably high vapor pressure at room temperature, they do not need require any maintenance or refill on regular basis. Like many other phenomena, also antimicrobial activity has its prominent role model in nature: Peptides with a high amount of cationic amino acids provide activity against bacteria and fungi. The cationic charge of the peptide qualifies it as successful competitor for interaction with the microbes' membranes; hydrophobic side-chains have been argued to synergistically enhance this effect. Based on a study on the use of partially hydrolyzed poly(2-oxazoline) s as antimicrobial contact biocides (*Macromol. Biosci.* 2013, *13*, 116-125), the next generation of contact biocides in the form of partially hydrolyzed poly(2-oxazoline)s and poly(2-oxazine)s have been developed. These works were particularly motivated by reducing the water solubility of these compounds and the amount needed to prepare polyolefin composites with surficial antimicrobial activity. Two approaches were followed, namely the cross-linking of copoly(2-oxazoline)s by means of the UV-induced thiol-ene reaction, and the synthesis and partial hydrolysis of higher homologues of poly(2-oxazoline)s, the poly(2-oxazine)s. In this lecture, the synthesis and acid-mediated partial hydrolysis of poly(2-oxazoline)s and poly(2-oxazine)s will be described. The structure of the corresponding copolymers, with special respect to amount of cationic charges and hydrophobic side-chains as well as the chain flexibility, will be correlated with the antimicrobial activity of surfaces containing these copolymers as additive (Figure 1). The antimicrobial activity was tested in full alignment with *ISO*

22196:2007 against *E. coli*, *P. aeruginosa*, *S. aureus*, and *C. albicans*. In addition, the correlation of the structures and the water solubility of the copolymers will be discussed due to the potential negative impact on the longevity of the antimicrobial properties.

Keywords: Poly(2-oxazoline)s, Poly(2-oxazine)s, Contact Biocides, Cationic Polymers

Figure 1

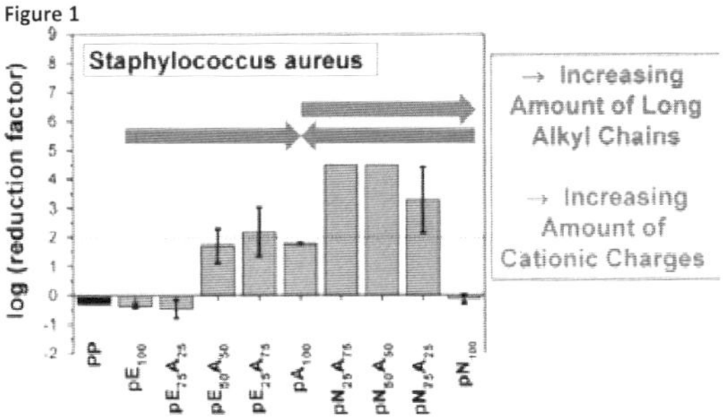

Antimicrobial activity of polypropylene compounds against S. aureus. The compounds contain the additives poly(2-ethyl-2-oxazoline) pE, poly(2-nonyl-2-oxazoline) pN or their partially hydrolyzed congeners in 5 wt.-% quantities.

SFP-I-003 [Light-induced Reactions in Polymer Science]

Exploiting Photochemical Selectivity in Macromolecular Synthesis

Christopher Barner Kowollik

Preparative Macromolecular Chemistry, Institut für Technische Chemie und Polymerchemie, Karlsruhe Institute of Technology, Karlsruhe, Germany

Installing chemical functionality at precise locations within complex macromolecules as well as onto three dimensional direct laser written constructs in the mildest possible fashion (visible light) and without any catalyst becomes possible with powerful light driven, quantitative ligation protocols. Advanced light induced processes enable the parallel encoding of multiple molecules onto predefined locations on 3D microstructures in one step as well as the wavelength dependent addressing of specific parts of a macromolecule for its highly orthogonal functionalization (λ-orthogonal photochemistry). Time permitting, the lecture will highlight how macromolecular ligations can be directed in their selectivity by the application of photonic fields, how light can be used as an orthogonal trigger in macromolecular chain folding, how the high efficiency of light driven chemistry can be exploited for the synthesis of sequence defined macromolecules or explore how diffraction unlimited laser lithography is possible based on photo-click chemistry concepts.

Keywords: Photochemical Selectivity, Orthogonal Photochemistry, Three Dimensional Microstructures, Surface Encoding, Sequence Defined Macromolecules, Visible Light Induced Ligation

SFP-I-002 [Smart and Functional Polymers]

Multifunctional Organic/Inorganic Hybrid-gels: from Polymer Design to Fiber Preparation and Applications

Meifang Zhu, Zhouqi Meng, Kai Hou, Peiling Wei, Mengge Xia, Yongtao Wu, Yan Zhang, Zhigang Chen

State Key Laboratory for Modification of Chemical Fibers and Polymer Materials, College of Materials Science & Engineering, Donghua University, Shanghai, China

Multifunctional organic/inorganic hybrid-gels have been designed as a novel platform for future biomedicine and intelligent sensing application. However, one of the formidable challenges in establishing these materials is to develop the safe, effective, high stability multifunctional hybrids for medicine and sensor such as tumor therapy and technology for multi-structure production. Here, a series of organic/inorganic functional hybrid-gels and gel-fiber have been prepared and characterized, including smart hydrogels, intelligent nanocapsules, and their application have also been investigated. Specifically, the hydrophilic CuS nanoplates/ W_8O_{49} nanowires and nanocapsules, can served as the safe and effective new photothermal agents and drug carriers, which could provide new opportunities for simultaneously photothermal-/chemo-therapy. Furthermore, a series of smart hybrid hydrogels and gel-fibers were developed with multiple sensitivity, and multi-functionalities. These materials offer prospects for new application in diverse fields including drug delivery, biomedical engineering and smart sensor.

Keywords: Organic/Inorganic hybrid-gels, smart hydrogel, hydrogel fiber, photothermal therapy

SFP-I-004 [Smart and Functional Polymers]

From light empowered to self oscillating hydrogel objects - on the rate and directionality control of microscopic morphing

Martin Moeller

DWI Leibniz-Institute of Interactive Materials and RWTH Aachen University, Aachen Germany

Natural as well as synthetic hydrogels are well studied examples for which a change in temperature or chemical potential can be transformed into a body shape deformation or morphing action. Certain requirements have to be fulfilled in order to exploit this for an active motor. One is a local energy input. Temperature effected motion requires a heat source within the hydrogel. Fast cooling can be achieved for small objects after the heating is ceased. The change in the state must be controlled in time and volume rigorously to allow a directed motion. Fast heat transfer accompanied by slower transport of water can cause a huge imbalance between the osmotic and elastic forces depending on the size and geometry of the object. This provides a means to control the motion. At the example of thermo-plasmonically heated microgels we will demonstrate that the design of such objects can enable rates higher than 2000 μm/s. In addition to the rate control also a path control can be achieved by modulation of the energy input, i.e. irradiation. This is essential to enable locomotion for small of small gel objects in water, i.e. at low Reynolds numbers. In order to cause a directed net movement, the morphing actuation must not follow a reciprocating forward and backward path. As a third point this refers to the requirement of a time gating in order to control that the first stroke (shrinkage upon heating) is followed in due time by a second stroke (expansion during cooling). In the case of a system that gains its energy by light absorption, this can be affected by modulation of the light intensity. A system that can move autonomously must, however, create its own pulse.

Keywords: Hydrogels, Morphing, self-sustaining

SFP-I-005 [Smart and Functional Polymers]

Functional polymer colloids prepared using radical thiol-ene/yne 'click' chemistry

Devon A. Shipp, Olivia Z. Durham, Hannah R. Norton, Dana V. Chapman

Department of Chemistry & Bimolecular Science, Clarkson University, Potsdam, United States

The use of thiol-ene and thiol-yne chemistries for polymerization and polymer modification are well understood and follow a step-growth mechanism, where polymer growth is dependent on the coupling of functional groups. These reactions are considered 'click' reactions, which can be defined as processes that are both facile and simple, offering high yields, no side products, easy product separation, and orthogonality with other chemistries. In addition, thiol-ene and thiol-yne chemistry offer the potential for the production of highly crosslinked functional materials with essentially 100% conversion. In this work we demonstrate the successful synthesis of crosslinked polymer particles using thiol-ene and thiol-yne chemistry in suspension, dispersion and emulsion polymerizations. The impact of various parameters on the development of thiol-ene and thiol-yne polymer particles will be explored. In addition, the self-limiting behavior of thiol-ene/yne reactions allows for successful synthesis of functional polymer colloids using off-stoichiometric amounts of monomers. This capacity to control functionality is illustrated through the creation of fluorescent polymer particles using both an in situ thiol-ene polymerization reaction with a vinyl chromophore as well as through post-polymerization modification via thiol-isocyanate chemistry. The development of 'clickable' and functional thiol-ene/yne polymer particles offers an expedient means to synthesize versatile and functional materials for a variety of potential applications, including paints and coatings, cosmetics, diagnostics, and drug delivery.

Keywords: thiol-ene/yne polymerizations, click chemistry, colloids, functional particles

SFP-I-006 [Smart and Functional Polymers]

Stimuli-Responsive Polymer-Based Sensors, Muscles, and Drug Delivery Platforms

Michael J Serpe

Department of Chemistry, University of Alberta, Edmonton, Alberta, Canada

The group's research is focused on the development of novel polymer-based materials for solving environmental and health-related problems. To solve these problems, the group primarily employs poly (N-isopropylacrylamide) (pNIPAm)-based spherical particles as the active component in our technologies. PNIPAm-based particles (nano or microgels, depending on their diameter) are extremely porous, and are fully water soluble and swellable. Additionally, pNIPAm-based nano/microgels are responsive to temperature, shrinking in diameter as the temperature is increased to >32 °C and reswelling when they are cooled to < 32 °C. Our group has exploited these properties for numerous applications. Today's talk will highlight the group's work on the development of these devices for sensing and biosensing, as muscles, and for controlled/triggered drug delivery.

Keywords: Stimuli responsive polymers, Structured materials, pNIPAm-based microgels

Bridging Continents & Bridging Molecules

MACRO 2016

World Polymer Congress Istanbul 17-21 July

ORAL ABSTRACTS

Short Codes	
EOO	Energy, Optics, & Optoelectronics
IP	Industrial Polymers
MBM	Macromolecules in Biotechnology & Medicine
MN	Macromolecules & Nanotechnology
MSP	Modeling and Simulation of Polymers
PCNCT	Polymer Characterization by New and Combined Techniques
PE	Polymer Education
PEPC	Polymer Engineering, Processing, and Characterization
PP	Polymer Physics
PPBM	Polymer and Polymer-Based Membranes
PPG	Porous Polymer and Gels
RDPS-A	Recent Developments in Polymer Synthesis-A-Controlled/Living Polymerization
RDPS-B	Recent Developments in Polymer Synthesis-B-Complex Macromolecular Structures
RDPS-C	Recent Developments in Polymer Synthesis-C-Light-induced Reactions in Polymer Science
RDPS-D	Recent Developments in Polymer Synthesis-D-Condensation Polymerization and Thermoset
RRB	Renewable Resources and Biopolymers
SFP	Smart and Functional Polymers

Abstracts sorted alphabetically

EOO-O-001 [Energy, Optics, & Optoelectronics]

Supercapacitor device performances of graphene/conducting polymers

Murat Ates

Department of Chemistry, Namik Kemal University, Tekirdag, Turkey

Conducting polymers are mostly used in energy-storage technology due to the high conductivity, and high electroactive sites. Graphene sheets are also used in energy storage technology due to the fast electron transfer, low resistivity, and fast ion movement, high conductivity, large surface area and electrochemical stability. Graphene oxide was reduced to graphene with sodiumborhydride (NaBH4). A few types of pi-conjugated monomers (Thiophene, 3,4-ethylenedioxythiophene, Pyrrole) were chemically synthesized as a nanocomposites of graphene/conducting polymers. The characterization of modified active materials were performed by FTIR-ATR, UV-vis, SEM-EDX, CV, and EIS techniques. The supercapacitor device performances of graphene/ conducting polymers were given by CV, CC, and EIS techniques.The highest energy and power density and capacitive retentions were presented in this study.

Keywords: Supercapacitor, graphene, nanocomposites, energy storage, conducting polymers, power density.

Figure 1

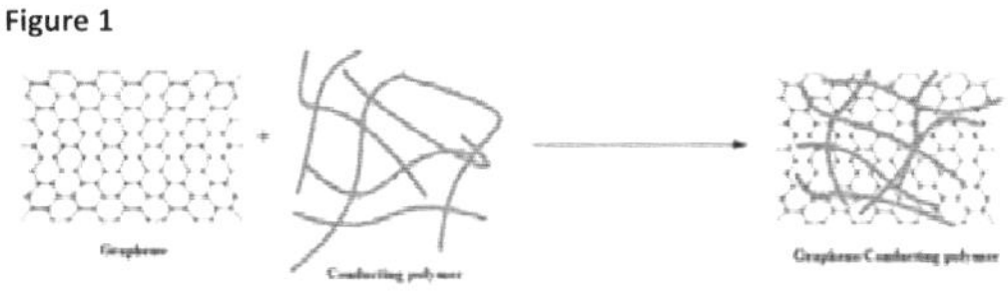

Illustration of graphene/conducting polymer nanocomposites

EOO-O-002 [Energy, Optics, & Optoelectronics]

Surface-attached polymer networks for planar optronic systems

Anne Katrin Schuler, Raimund Rother, Oswald Prucker, Claas Müller, Jürgen Rühe

IMTEK – Department of Microsystems Engineering, University of Freiburg, Freiburg, Germany

Optronic systems entirely produced from polymers may serve as sensor foils for many applications such as structural monitoring. It is crucial for such applications to control the optical properties of all materials in use but it is of equal importance to master all surfaces and interfaces involved in such systems. These can be numerous: Optical elements like light sources, wave guides or sensors need to be fixed on substrate or carrier foils, claddings need to wrap waveguides, connections join different elements and protective coatings shield all functional units from scratches or finger prints. We have developed chemistries to generate surface-attached polymer networks on chemically rather inert foils using polymers with thermally or photochemically reactive groups for C,H insertion crosslinking (CHIC) reactions. The presentation will describe how these materials are used to generate waveguides and other optically active units such that optical properties can be varied over a fairly wide range. In addition, we will show that C,H insertion reactions can also proceed across interfaces such that a strong adhesion between the various layers can be assured by basically generating monolithic multilayer assemblies. Aside from these practical implications we will also present theoretical studies that allow us to understand the crosslinking reactions and to predict the performance of the materials in such applications.

Keywords: Lamination, Surface-attached polymer networks, C-H insertion reactions, multilayer foils

Mulitlayer foil

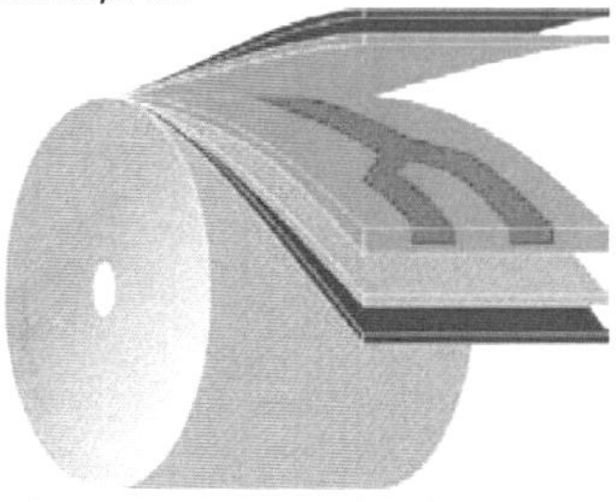

Schematic representation of a multilayer optical foil.

EOO-O-003 [Energy, Optics, & Optoelectronics]

Integration of triboluminescent crystals into transparent polymers

Mustafa M. Demir, Anıl Incel

Department of Materials Science and Engineering, İzmir Institute of Technology, İzmir, Turkey

Triboluminescence (TL) is known as the emission of light upon breaking of chemical bonds/interactions. The principle strategy for the produced TL is based on the light emission by mechanical forces for instance crushing, fracturing, smashing, and grinding. The development of TL materials utilizes the detection of any crack formation in real-time. Moreover, the fluorescent property of TL materials allows to determine the failure point. The integration of TL material into polymers has been emerged in a variety of stress based platform and the development of stress probes. In this study, one of the lantanide based TL material which is europium tetrakis (dibenzoylmethide) triethylammonium (EuD4TEA) was employed to develop EuD4TEA/transparent polymer film composites. Two different process were carried out: i) embedding and ii) surface impregnation. Three different transparent polymers were used: polymethyl metacrylate (PMMA), polystyrene (PS) and polyurethane (PU). Drop tower mechanism was designed for TL intensity measurement by the principle of transfering upon impact radiation to readable data via spectrophotometer. The obtained composite materials were also characterized by scanning electron and fluorescence microscopy techniques. The results show that [EuD4TEA] should be equal or more than 30% wt with respect to polmers to obtain readable signal for embedding process. However, even at 2.5% wt is enough to take TL response from surface impregnation method. Furthermore, PU based composite has tendency to deposit high amount crystal particles due to more microdomain segregation on film surface. Additionally, the stability of EuD4TEA/PU composites towards mechanical force is higher than other polymers.

Keywords: triboluminescence, optical nanocomposites, transparent polymers, mechanical sensors

EOO-O-004 [Energy, Optics, & Optoelectronics]

Dynamic Hologram Formation in Photorefractive Ferroelectric Liquid Crystal Blends

Takeo Sasaki

Tokyo University of Science

Ferroelectric liquid crystal blends composed of a smectic liquid crystalline mixture, a photoconductive chiral dopant, and an electron trap reagent exhibit significant photorefractivity together with rapid responses. As such, they allow the dynamic amplification of moving optical signals. In the present work, the photochemical durability of a photorefractive ferroelectric liquid crystal blend was investigated. A series of photoconductive chiral dopants was prepared and the durability of blends incorporating these dopants during laser irradiation was examined. In addition, the effect of the conduction of photogenerated ionic species on the photorefractivity decay was clarified.

Keywords: photorefractive effect, ferroelectric liquid crystals, hologram, optical amplification

EOO-O-005 [Energy, Optics, & Optoelectronics]

Cyano-functionalized polymer / barium titanate composites for organic electronics

Francesco Piana[1], Jiří Pfleger[1], Irina Yu Sapurina[2], Roman Jambor[3], Tomaš Řičica[3], Jan Macak[4]

[1]Institute of Macromolecular Chemistry, Academy of Sciences of the Czech Republic v.v.i., Heyrovsky Sq. 2, 162 06 Prague 6, Czech Republic
[2]Institute of Macromolecular Compounds, Russian Academy of Sciences, Bolshoy pr. 31, 199004, Saint-Petersburg, Russian Federation.
[3]Department of General and Inorganic Chemistry, Faculty of Chemical Technology, University of Pardubice, Studentská 95, Pardubice, Czech Republic.
[4]Center of Materials and Nanotechnologies, Faculty of Chemical Technology, University of Pardubice, Nam. Cs. Legii 565, Pardubice, Czech Republic.

The development of several devices in organic electronics requires performant dielectric layers, which allow for decreased operating voltage and have reduced electric leakage. Composite films prepared with barium titanate (BT) nanoparticles and high-k amorphous polymers fit perfectly these requirements. Synthesis of BT has been optimized to obtain regularly shaped small nanoparticles (diameter 40 – 90 nm) with limited tendency to aggregation. ζ-potential and QELS measurements revealed BT particles extremely stable in a water dispersion. Composite material of synthesized BT with poly(2-cyanoethyl vinyl ether) (CEPVA), a cyano-functionalized amorphous polymer with dielectric constant ~ 15, have been prepared by solution mixing. Morphology of BT, CEPVA and their composites has been investigated by SEM, DLS and DSC. Dielectric measurements performed at room temperature on films of micrometre scale thickness showed a dielectric constant between 35 – 40 in the range of 10^2 – 10^4 Hz, with only 30 vol.% content of BT in the composite and high permittivity of BT nanoparticles around 384 ± 33 calculated according to the Lichteneckers model.

Work supported by the NPU I program of the MEYS of the CR, Project POLYMAT, No. LO1507.

Keywords: Barium titanate, poly(2-cyanoethyl vinyl ether), composites, dielectric properties

EOO-O-006 [Energy, Optics, & Optoelectronics]

4-(Hydroxymethyl)-1,3-Dioxolane-2-on and Polyethylene Oxide Containing Borate Ester Based Polymer Electrolytes

Aysel Demirci, Esra Eren, Yunus Emre Özkara, Betül Karataş, Yunus Karataş

Department of Chemistry, Ahi Evran University, Kırşehir, Turkey

The technology evolves the need for energy storage is increasing every day. Lithium-ion and lithium-polymer batteries are the prominent and widely used types of batteries for energy storage. The use of polymers instead of the organic solvents seems to provide more reliable batteries. In addition, the problem of capacity fading with time in lithium-ion battery is much less significant for lithium-polymer batteries.

Besides the other common uses, borate esters have attracted attention in recent years as polymer electrolyte. There are examples of borate esters used as additives in especially poly ethylene glycol based polymers but their use as polymer electrolyte is rather limited [1].

In this study, new polymer electrolytes containing the structure of most commonly used commercial organic battery solvent component (ethylene carbonate) as the side group has been synthesized. The synergetic effect of the polymers and the organic solvents are here aimed. New polymer electrolytes using PEG derivatives with varying Mn values were prepared. The best ionic conductivity values were obtained with PEG-600. Using this electrolyte composition, different salts with varying ratios were also investigated for further optimization. The optimum room temperature ionic conductivity value was measured for the polymer electrolyte with (CF3SO2)2NLi salt and the ethylene oxide:lithium ratio of 24 as 4.09 x10-5 S/cm. Polymer electrolytes showed also improved lithium transference number and wide electrochemical window reference to Li+/Li.

Acknowledgements: This work was supported by Ahi Evran University (FBE. B2.16.001).

References:

[1] Wakihara, M.; Nakayama, M.; Kato, Y. "Lithium Ion Rechargeable Batteries", (Wiley) 2010, 213-255.

Keywords: polymer electrolyte, ionic conductivity, borate ester

Figure 1. The Arrhenius graph of temperature dependent ionic conductivity of borate ester based polymer electrolytes.

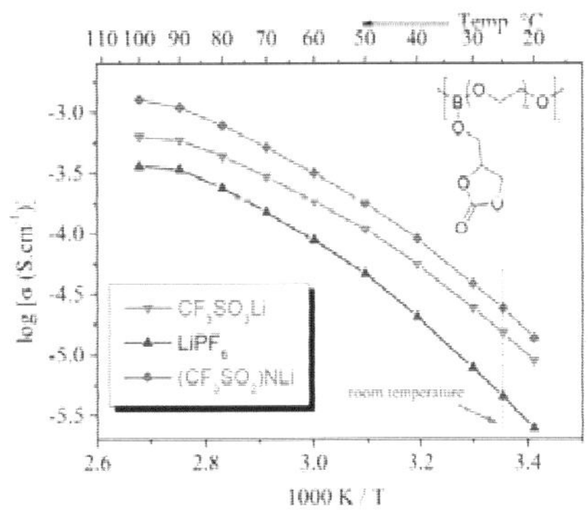

EOO-O-007 [Energy, Optics, & Optoelectronics]

Graphene based Electrodes for Fuel Cells and Li-ion Batteries

Selmiye Alkan Gürsel[1], Lale Işıkel Şanlı[1], Vildan Bayram[2], Begüm Yarar[2], Sajjad Ghobadi[2], Anaguli Abulizi[2], Adnan Taşdemir[2], Miad Yarali[2], Emre Biçer[1], Alp Yürüm[1], Elif Daş[3], Ayşe Bayrakçeken Yurtcan[4]

[1]*Sabanci University Nanotechnology Research & Application Center (SUNUM), Sabanci University, Istanbul, Turkey*
[2]*Faculty of Engineering &Natural Sciences, Sabanci University, Istanbul, Turkey*
[3]*Department of Nanoscience and Nanoengineering, Atatürk University, Erzurum, Turkey*
[4]*Department of Chemical Engineering, Atatürk University, Erzurum, Turkey*

Graphene has distinct properties such as high electrical conductivity, large surface area and enormous stability. In addition to its large variety of applications, graphene based materials are the most promising candidates for energy conversion and storage applications "Quesnel E., et al., 2D Mater. 2, 030204(2015)". Polymer electrolyte membrane (PEM) fuel cells are attractive for various applications due to their distinguishing features such as high power density and low temperature operation. Platinum (Pt) which is intensively used as catalyst in PEM fuel cells has not only high cost and performance problems but also has low abundance. Carbon black, the commonly used catalyst support, has several limitations causing degradation of catalyst activity and performance. In the present work, graphene nanoplatelets (GNP), reduced graphene oxide (rGO) and functionalized graphenes have been employed as the catalyst support. Graphene supported Pt nanoparticles were prepared by means of impregnation, supercritical carbon dioxide deposition and electrospinning/electrospraying methods. The preparation of highly dispersed and uniformly decorated Pt nanoparticles with a small particle size (3-5 nm) was achieved. Moreover, significantly better electrocatalytic activity and fuel cell performances for graphene based electrodes were obtained. Titanate nanotubes are promising materials owing to their special morphology which allows high ion mobility. Moreover, silicon (Si) has the highest theoretical capacity of about 4200 mAh/g and encapsulation with graphene can withstand the huge volume change upon lithiation and delithiation. In this study, rGO, nitrogen doped rGO and GNP were used as substrates to grow titanate nanotubes and Si nanoparticles (Figure 1). Synthesized nanocomposites were used as anode materials for LiBs and battery tests were conducted. We determined that the hybrid electrodes have promising energy storage performance and long durability properties. While the graphene substrate enhanced the electrical conductivity and structural integrity, synthesizing the active nanoparticles on graphene increased the cyclability of the electrode.

Keywords: graphene, electrode, fuel cell, battery

TEM image of Pt nanoparticles supported on graphene (left), SEM images of titanate nanotubes embedded on rGO (middle) and Si nanoparticles with N/rGO sheets as substrate (right)

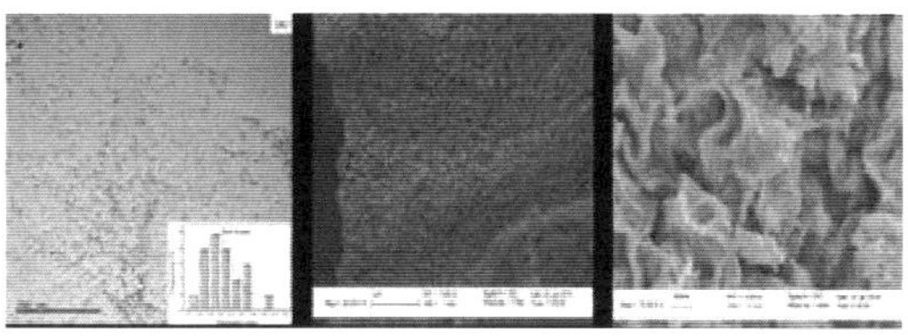

EOO-O-008 [Energy, Optics, & Optoelectronics]

Improvement of output performance of PVDF by incorporating Fe-oxides doped reduced graphene oxide: an efficient piezoelectric nanogenerator

Sumanta Kumar Karan, Bhanu Bhusan Khatua

Indian Institute of Technology Kharagpur, Kharagpur, India

Here, we explore the hybrid piezoelectric nanogenerator (HPNG) of non-electrically poled Fe-oxides doped reduced graphene oxide (Fe-RGO)/poly(vinylidene fluoride) (PVDF) nanocomposite film prepared by simple solution casting technique. The prepared HPNG exhibits an open circuit output voltage and short circuit current in the order of 5.1 V and 0.254 μA with superior power density (10.7 μW/cm3) which is much higher than pure PVDF at low pressure (≈ 12 kPa) by repetitive human finger imparting. The reason behind the greater output performance is that Fe-RGO can acts as effective nucleating agent which stabilize ≈ 99% relative proportion of electroactive polar γ-phase. Due to the electrostatic interaction among the different dipoles of PVDF and delocalized π-electrons and remaining oxygen functionality of Fe-RGO via ion-dipole and/or hydrogen bonding interaction, the electroactive polar γ-phase can be stabilized. The generated power from HPNG could instantly light up different color of LEDs without using any external energy storage devices. The HPNG could charge a commercial 2.2 μF capacitor up to 5.48 V within a very small time span of 110 sec, can be used to power up portable commercial electronics devices like wrist watch, calculator, speaker and mobile LCD screen. More importantly, HPNG retains its performance after 1600 compression cycles (≈500 sec, 10 weeks), signifying as a promising piezoelectric energy harvester that converts the small mechanical energy into electricity.

Keywords: Hybrid piezoelectric nanogenerator, PVDF, Fe-oxides doped RGO, Polar γ-phase, Self-Poled.

EOO-O-010 [Energy, Optics, & Optoelectronics]

End Group Effects on Molecular Order and Charge Transport in Conjugated Polymers

Ünsal Koldemir[1], Sreenivasa R Puniredd[3], Manfred Wagner[3], Sefaattin Tongay[5], Tracy D Mccarley[6], George Dimitrov Kamenov[7], Klaus Müllen[3], Wojciech Pisula[8], John R Reynolds[4], Ünsal Koldemir[2]

[1]*Research and Development Center, Hayat Kimya, Kocaeli, 41275 (Turkey)*
[2]*The George and Josephine Butler Polymer Research Laboratory, Department of Chemistry, Center for Macromolecular Science and Engineering, University of Florida, Gainesville, Florida 32611 (USA)*
[3]*Max Planck Institute for Polymer Research, Ackermannweg 10, 55128 Mainz (Germany)*
[4]*School of Chemistry and Biochemistry, School of Materials Science and Engineering, Center for Organic Photonics and Electronics, Georgia Institute of Technology, 901 Atlantic Drive, Atlanta, Georgia 30332 (USA)*
[5]*Department of Material Science, University of California Berkeley, Berkeley, California 94720 (USA)*
[6]*Department of Chemistry, Louisiana State University, Baton Rouge, Louisiana, 70803 (USA)*
[7]*Department of Geology, University of Florida, Gainesville, Florida 32611 (USA)*
[8]*Department of Molecular Physics, Faculty of Chemistry, Lodz University of Technology, 90-924 Lodz (Poland)*

Optimized microstructure through control of both intra- and inter-molecular interactions in organic semiconductors is critical in the context of enhanced charge transport for next generation organic field effect transistors (OFETs). These type of FETs offer low-cost, mechanical flexibility and easy to process at ambient conditions without compromising high performance. As such, we discuss the end capping protocols on the donor-acceptor alternating copolymers of dithienogermole (DTG) with 2,1,3-benzothiadiazole (BTD). As molecular order has been attributed as one of the main factors determining the charge carrier transport we probe the importance of end groups on the control of molecular order and microstructure. We have pursued a stoichiometric imbalance protocol for the post polymerization end capping reactions on DTG-BTD copolymer and further confirmed the presence of end groups via 1H-NMR, 2D NMR and MALDI-MS analyses. Further, we studied the molecular order by GI-WAXS, 2D-WAXS and AFM showing significant enhancement of polymer ordering in thin films on the DTG-BTD copolymer independent of the processing conditions. To this end, OFETs exhibited charge-carrier mobilities up to 0.60 cm2/Vs for the end-capped copolymer, which are an order of magnitude higher in comparison to the non-end capped analogue, which displayed a mobility of 0.077 cm2/Vs. In this work, we investigate that a simple synthetic approach, the introduction of end capping groups which remove reactive functionalities at the chain ends, can be and effective tool in the realization of next generation OFET and photovoltaic materials by promising better control of the polymer organization.

Keywords: organic field effect transistors, molecular packing, end capping reactions

EOO-O-011 [Energy, Optics, & Optoelectronics]

The effect of the triphenylamine unit on electrochemical behaviors of benzotriazole bearing conjugated polymers

Serife O. Hacioglu,[1] Sinem Toksabay,[1] Merve Sendur,[2] Levent Toppare[1,3,4,5]

[1]Department of Chemistry, Middle East Technical University, 06800 Ankara, Turkey
[2]Institut fur Physikalische und Theoretische Chemie, Universitat Regensburg, 93053 Regensburg, Germany
[3]Department of Biotechnology, Middle East Technical University, 06800 Ankara, Turkey
[4]Department of Polymer Science and Technology, Middle East Technical University, 06800 Ankara, Turkey
[5]The Center for Solar Energy Research and Application (G€UNAM), Middle East Technical University, 06800 Ankara, Turkey

Conjugated polymers have some impressive properties such as; flexibility, adjustable band gap and low cost [1] which attract great attention over the past two decades. and this enables them to use for different applications like electrochromic devices [2] organic light-emitting diodes (OLEDs) [3] and organic solar cells [4]. In literature, different conjugated polymers were designed by the combination of different donor and acceptor groups to enchance the mentioned behaviors. Herein, we designed and synthesized a series of benzotriazole (BTz) and triphenylamine (TPA)-based random copolymers. After successive synthesis, purifications and characterization of copolymers, electrochemical studies were performed via cyclic voltammetry(CV). During these studies three-electrode cell system was used consisting of an indium tin oxide (ITO) doped glass slide as the working electrode, platinum wire as the counter electrode, and Ag wire as the pseudo reference electrode.Spectroelectrochemical studies were performed using UV-Vis-NIR spectrophotometer to investigate maximum absorption wavelength, optical band gap and polaron/bipolaron formation. All electrochemical and spectroelectrochemical studies were performed in 0.1 M TBAPF$_6$/ACN solution. All polymers showed good solubility and high film quality. Poly4-(5-(2-dodecyl-7-methyl-2H-benzo[d] [1,2,3]triazol-4-yl)thiophen-2-yl)-N-(4-(5-methylthiophen-2-yl)phenyl)-N-phenylaniline has both n-type and p-type doping ability. Poly4-(5-(2-dodecyl-7-methyl-2H-benzo[d][1,2,3]triazol-4-yl)thiophen-2-yl)-N-(4-(5-methylthiophen-2-yl)phenyl)-N-phenylaniline has red color in its neutral state and multichromic in different oxidized states.

References:

[1]A. Heeger, Angew. Chem. Int. Ed. 2001, 40, 2591–2611
[2]A. Durmus, G. E. Gunbas, P. Camurlu, L. Toppare, Chem. Commun. 2007, 31, 3246–3248
[3] S. Setayesh, A. C. Grimsdale, T. Weil, V. Enkelmann, K. Mu¨llen, F. Meghdadi, E. J. W. List and G. Leising, J. Am. Chem. Soc., 2001, 123, 946
[4] C. Istanbulluoglu, S. Goker, G. Hizalan, S. O. Hacioglu, Y. Arslan Udum, E. D. Yildiz, A. Cirpan, L. Toppare, NewJ.Chem., 2015, 39, 6623

Keywords: triphenylamine, electrochemical behaviors, benzotriazole

EOO-O-012 [Energy, Optics, & Optoelectronics]

Resonance Energy Transfer from Quinolinone Modified Polystyrene-*block*-Poly(styrene-*alt*-maleic anhydride) Copolymer to Terbium Metal Ions

Drahomir Vyprachticky, Veronika Pokorna, Vera Cimrova

Institute of Macromolecular Chemistry, v.v.i., Academy of Sciences of the Czech Republic, Heyrovsky Sq. 2, 162 06 Prague 6, Czech Republic

Polystyrene-*block*-poly(styrene-*alt*-maleic anhydride) was synthesized by nitroxide mediated radical polymerization and modified with 7-amino-4-methylquinolin-2(1H)-one (I) and methanol. The formed *block* copolymer ligand contained a quinolinone fluorophore (Q) and carboxyl (III) or sodium carboxylate (IV) binding sites (Figure 1). The ligand-to-metal resonance energy transfer (RET) and ligand binding properties of [III-Tb^{3+}] and [IV-Tb^{3+}] complexes were investigated by steady-state and time-resolved luminescence spectroscopy in tetrahydrofuran/methanol and/or tetrahydrofuran/deuterated methanol mixtures and compared with those of a low-molecular-weight model ligand, i.e. the sodium salt of N-(4-methyl-2-oxo-1,2-dihydroquinolin-7-yl)succinamic acid (II). The long-lived emission intensities of Tb^{3+} at 490, 545, 585, and 620 nm corresponding to the $^5D_4$ - $^7F_6$, $^5D_4$ - $^7F_5$, $^5D_4$ - $^7F_4$, and $^5D_4$ - $^7F_3$ transitions, respectively, were strongly increased by the addition of ligands in the order [II-Tb^{3+}] << [III-Tb^{3+}] < [IV-Tb^{3+}]. The efficiency of energy transfer (E) was evaluated from the emission intensity of the donor (Q) in the presence or absence of the acceptor (Tb^{3+}) depending on the acceptor concentration and ligand neutralization. It was concluded that the macromolecular ligand structural properties (polymer coil and supramolecular structures, e.g. micelles) were responsible for the increase in RET. The time-resolved luminescence measurements revealed that the binding affinity of the ligands II, III, and IV increased in the order II < III < IV.

Acknowledgement: The authors thank the Czech Science Foundation for financial support (Grant number 13-26542S).

Keywords: Energy transfer, terbium luminescence, polymer complex.

Figure 1. The sodium salt of N-(4-methyl-2-oxo-1,2-dihydroquinolin-7-yl) succinamic acid (II) and poly(styrene-{alt}-maleic anhydride)-{block}-polystyrene modified with 7-amino-4-methylquinolin-2(1H)-one (I) and methanol yielding polymer (III)

EOO-O-013 [Energy, Optics, & Optoelectronics]

Solution Processed Orange Light Emitting Diodes Using Polyfluorene Derivative

Gönül Hızalan[1], Ozan Erlik[1], Ali Çırpan[2], Levent Toppare[3]

[1]Department of Chemistry, Middle East Technical University, Ankara, Turkey
[2]Department of Chemistry, Department of Polymer Science and Technology, The Center for Solar Energy Research and Application (GÜNAM), Department of Micro and Nanotechnology, Middle East Technical University, Ankara, Turkey
[3]Department of Chemistry, Department of Polymer Science and Technology, The Center for Solar Energy Research and Application (GÜNAM), Department of Biotechnology, Middle East Technical University, Ankara, Turkey

Poly((9,9-dioctylfluorene)-2,7-diyl-(4,7-bis(thien-2-yl)2-dodecyl-benzo[1,2,3] triazole)) was synthesized using benzene end cap. Molecular weights (Mn, Mw) of the polymer were determined as 22 kDa and 51 kDa, respectively. Optical properties of the polymer both in solution and in thin film were investigated with UV-Vis and photoluminescence spectroscopy. OLEDs were constructed with the ITO/ PEDOT:PSS/ Polymer/ Ca/ Al device architecture. I (current density)- V (voltage) characteristics of the devices were investigated using LabTracer program with Keithley. Electroluminescence and chromaticity studies were performed. Maximum EL intensity was obtained at 540nm with a shoulder at 580nm. Chromaticity values of 0.4569 – 0.5330 were obtained. Luminance voltage dependences were determined using Spectra Suit program with fiber optic by sweeping the voltage. Brightness of the device reached 14292cd/m^2 at 8 V, respectively. Device exhibited luminance efficiency values of 1.23cd/A at 6.0V.

Keywords: OLED, Polyfluorene, Conjugated polymers

EOO-O-014 [Energy, Optics, & Optoelectronics]

Polyfluorene-based Optoelectronic Device

Jinyi Lin[1], Bin Liu[2], Mengna Yu[2], Xuhua Wang[3], Linghai Xie[2], Wei Huang[2]

[1]Key Laboratory of Flexible Electronics (KLOFE) & Institute of Advanced Materials (IAM), Jiangsu National Synergistic Innovation Center for Advanced Materials (SICAM), Nanjing Tech University (NanjingTech), 30 South Puzhu Road, Nanjing 211816, China.
[2]Center for Molecular Systems and Organic Devices (CMSOD), Key Laboratory for Organic Electronics and Information Displays & Institute of Advanced Materials (IAM), Jiangsu National Synergistic Innovation Center for Advanced Materials (SICAM), Nanjing University of Posts & Telecommunications (NUPT), 9 Wenyuan Road, Nanjing 210023, China.
[3]The Blackett Laboratory, Imperial College London, Prince Consort Road, London SW7 2AZ, UK

Polyfluorenes (PFs) was an important light-emitting conjugated polymer for plastic devices. However, the intrinsic aggregation-induced green (g) emission is the most obstacle to extend their applications in deep-blue emitting optoelectronic devices. Herein, we make attempts to uncover the original of aggregation induced g-band emission and further construct novel high performance PFs via supramoelcular approach. The strong g-band emission in the photoluminance (PL) spectra of the supramolecular conjugated polyflurenes without steric group at 9-position (PFO, PPFOH and PDAF8-co-F8) film or solutions strongly supported the aggregation mechanism of green band emission. Two type aggregation mechanisms were first proposed in our works: fluorene dimer and fluorene distorted conformation. Subsequently, we rationally designed and prepared two novel polyfluorenes with the bulky groups at 9 sites (PDOPF) or carbazole (PDOPF-Cz). Interestingly, high quality PDOPF beta phase can be obtained in the film states attributed to the balance between steric interaction and Waals attractive forces, which is useful to improve spectral stability and device performance. What' more, for the stronger pendent carbazol π-π stacking interaction, PDOPF-Cz can self-assembled into chiral nanowire, which allow for intrachain photo-physical property in the film and gels states. Therefore, PDOPF and PDOPF-Cz had excellent spectral stability and deep-blue emission with high PLQY for high performance deep-blue PLED and laser. Therefore, supramolecular approach was an effective strategy not only to construct nanostructure but also obtain high performances polymer semiconductor and optoelectronic devices.

Keywords: Polyfluorenes, Supramolecular Approach, G-band Emission, Deep Blue Emission, Optoelectronic Device

Supramolecular Polyfluorene-based Optoelectronic Device

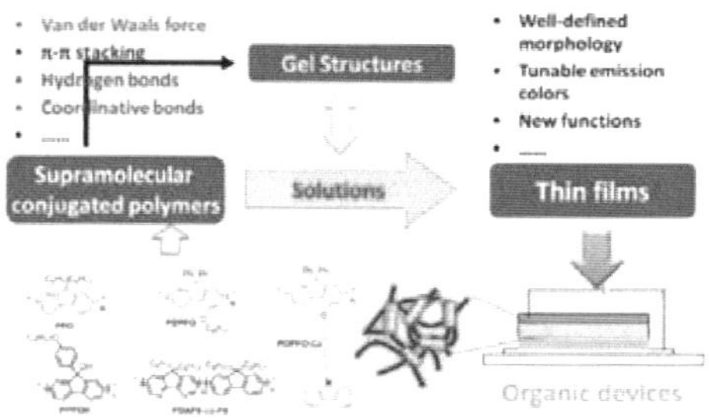

EOO-O-015 [Energy, Optics, & Optoelectronics]

Design of Squaraine-Based π-Conjugated Polymers For Organic Electronic: From Cross-Coupling Reactions to Metal-Free Polycondensation

Cyril Brochon, Guillaume Garbay, Eleni Pavlopoulou, Eric Cloutet, Georges Hadziioannou

Laboratoire de Chimie des Polymères Organiques (UMR 5629) Bordeaux INP/ CNRS - University of Bordeaux

Organic functional materials such as π-conjugated polymers are promising for the development of low-cost flexible optoelectronic devices. Most of the time, they are synthesized by reactions involving metallic catalysts. However, as theses metals can be considered as toxic for the final devices, the polymers need to be extensively purified [1]. π-conjugated materials prepared from condensation reactions in absence of metals constitute an interesting alternative to deal with this issue. Indeed, the only bi-product formed during the polymerization is water and the final materials can easily be purified. In this work, we are currently investigating the preparation of polysquaraines based material. Polysquaraines are donor-acceptor based polymers presenting near infra-red absorption and emission features. Hence they can be used in NIR fluorescent probes or dye-sensitized solar cells [2]. First, alternated squaraine-based π-conjugated polymers have been successfully obtained through "classical" polycondensation in presence of palladium and a metal free polycondensation. As expected, they exhibit interesting optical properties, such as a strong absorption in red and near infra-red and good emitting properties. Interestingly the metal free route gave rise to higher molecular weight which implies a shift of the maximum absorption to higher wavelength [3]. Furthermore, in addition of avoiding the presence of metals, metal-free polycondensation gives us access to original polymers. We succeeded in the synthesis of new polysquaraine exhibiting a broad emission without the need of mixing several compounds.

[1] Ö. Usluer, M. Abbas, G. Wantz, L. Vignau, L. Hirsch, E. Grana, C. Brochon, E. Cloutet, G. Hadziioannou, ACS Macro Letters, 2014, 3, 1134
[2] L.H. Liu, K. Nakatani, R. Pansu, J.J. Vachon, P. Tauc, E. Ishow, Advanced Materials, 2007, 19, 433
[3] J. Oriou, F. Ng, G. Hadziioannou, G. Garbay, M. Bousquet, E. Cloutet, C. Brochon, Polymer Chemistry, 2014, 5, 7100

Keywords: Semiconducting polymers, Squaraine-based polymers, metal free polycondensation, OLED active materials

EOO-O-016 [Energy, Optics, & Optoelectronics]

Electrochemical Coupling Layer by Layer Assembly of Multiscale Components for Fabrication of Optical Thin Films

Mao Li

State Key Laboratory of Polymer Physics and Chemistry Changchun Institute of Applied Chemistry, CAS, China

All solution processed technologies for transparent films are anticipated as the low cost next-generation electronics. Electropolymerization enables in situ preparation of uniform polymeric films from solution on conductive substrates. However, the electrochemical reactions usually result in conducting conjugated polymers with unavoidable highly doped state. In fact, the resulting conjugated polymers and highly doped states are often unnecessary for a wide library of building blocks. Herein, we present a quantitative (reaction-limited) electrochemical reaction for synthesis of low-doped film. The dimerization of bis-N-alkylcarbazole-substituted derivatives prevents the formation of a conducting conjugated polymer, which tends to lead to highly doped states. Our result represents the first example of preparation of a low-doped polymer and is a breakthrough in the broadening of the potential applications of electrochemical synthesis into the field of device fabrication.

Keywords: Layer by Layer Assembly, Electrochemical Synthesis, Carbazole, Optical Film, Device

EOO-O-017 [Energy, Optics, & Optoelectronics]

High-performance ternary composite electrode material based on polyaniline (PANI), molybdenum trioxide (MoO3) and graphene nanoplatelets (GNP) for next-generation supercapacitor prepared by sono-chemical method

Amit Kumar Das, B. B. Khatua

Materials Science Centre, IIT Kharagpur, Kharagpur, 721302, West Midnapore, West Bengal, India

Here we propose a facile and cost-effective route to synthesize a high-performance (i.e., having high energy and power density) composite electrode material (PMG composite) for next-generation supercapacitor through *In-situ* polymerization of aniline in presence of molybdenum trioxide (MoO_3) and graphene nanoplatelets (GNP). For structural characterization, Fourier transform infrared (FTIR) spectroscopy, wide angle X-ray diffraction (WAXD) and Raman spectroscopy were studied. Morphological analysis (field emission scanning electron microscopy (FESEM) and high resolution transmission electron microscopy (HRTEM)) revealed random adhesion of fibrillar PANI onto MoO_3/GNP composite. The prepared electrode material showed non-linear current (I)-voltage (V) relationship expressing its semiconducting nature. Most significantly, this PMG composite showed appreciably large specific capacitance of »593 F/g at the current density of 1 A/g with magnificent cyclic stability (92.4% retention of the original specific capacitance even after 1000 cycles) and 734 F/g at 10 mV/s scan rate in 1M H_2SO_4 electrolyte indicating the facilitated electrochemical reactions due to highly availability of the electrode-electrolyte interfacial area for the redox processes and conducting nature of the components, which are actually the outcome of synergistic contributions of the constituents of the ternary composite. Moreover, as the PMG composite shows excellent energy density along with high power density, it can successfully be implemented as the high-power energy resource in acidic electrolyte.

Keywords: composite, supercapacitor, polyaniline, graphene nanoplatelets, morphology

EOO-O-018 [Energy, Optics, & Optoelectronics]

Comparative Study for Monthly Average of Daily Global Radiation in Istanbul, Turkey

Zuhal Er, Inal Begum Turna
Physics Engineering Department, Faculty of Science and Letters,Istanbul Technical University, Istanbul,Turkey

The estimation of solar radiation at a given location is either done by related measurement devices or by developed models in the literature. The measurement equipment prices and maintenance costs are highly considerable for the developing countries, the solar radiation model developments are became an utmost important research area. In these premises, the first part of this study overarches the review of existing models which are reachable at the preparation time of this study. It has seen that reviewed and evaluated models are unique to invesment of the Global Radiation in Istanbul, Turkey. The other main objective of this study is to analyze the Global Radiation Models in Istanbul where is the part of the west side of Turkey. Furthermore, according to the diversity of solar energy potential of Turkey, the models are evaluated not among the entire field but restricted as one station. Therefore the evaluated model is selected as the territory of the study take place which is Istanbul station.The reviewed models in the literature are compared with graphical evaluations for Istanbul.The most appropriate model for Istanbul is defined by the commonly used statistical analysis methods which are mean percentage error (MPE), root mean square error (RMSE) and t-statistical analysis method. The results illustrate that, there are statistically significant relationships between relative global radiation and relative solar duration time. This is further clear by high values of coefficient of determination R2 (0,9873) across the variables. Consequently, the most applicable model for the station of the northern region of Turkey, Istanbul is enlighten.

Keywords: Global solar radiation, solar radiation distribution, solar irradiance, solar calculations, solar energy

EOO-O-019 [Energy, Optics, & Optoelectronics]

Highly Transparent and Flexible Biobased Polyimide/TiO2 and ZrO2 Hybrid Films with Tunable Refractive Index, Abbe Number, and Memory Properties

Tzu Tien Huang[1], Chia Liang Tsai[1], Seiji Tateyama[2], Tatsuo Kaneko[2], Guey Sheng Liou[1]

[1]Institute of Polymer Science and Engineering, National Taiwan University, Taipei, Taiwan
[2]Research Area of Energy and Environment, Japan Advanced Institute of Science and Technology (JAIST), 1-1 Asahidai, Nomi, Ishikawa 923-1292,Japan.

The novel biobased polyimide (**4ATA-PI**) and the corresponding PI hybrids of TiO_2 or ZrO_2 with excellent optical properties and thermal stability have been prepared successfully. The highly transparent **4ATA-PI** containing carboxylic acid groups in the backbone could provide reaction sites for organic-inorganic bonding to obtain homogeneous hybrid films. These PI hybrid films showed *tunable refractive index (1.60–1.81 for 4ATA-PI/TiO_2 and 1.60–1.80 for 4ATA-PI/ZrO_2)*, and the *4ATA-PI/ZrO_2 hybrid films revealed higher optical transparency and Abbe's number than those of 4ATA-PI/TiO_2 system* due to a *larger band gap of ZrO_2*. By introducing TiO_2 and ZrO_2 as electron acceptor into **4ATA-PI** system, the hybrid materials have a lower LUMO energy level that could facilitate and stabilize the charge transfer complex. Therefore, the memory devices derived from these PI hybrid films exhibited tunable memory properties from DRAM, SRAM, to WORM at different TiO_2 or ZrO_2 contents from 0 wt% to 50 wt% with a high ON/OFF ratio (10^8). In addition, the different energy level of TiO_2 and ZrO_2 revealed the specifically unique memory characteristics, implying the potential application of the prepared **4ATA-PI/TiO_2** and **4ATA-PI/ZrO_2** hybrids films in highly transparent memory devices.

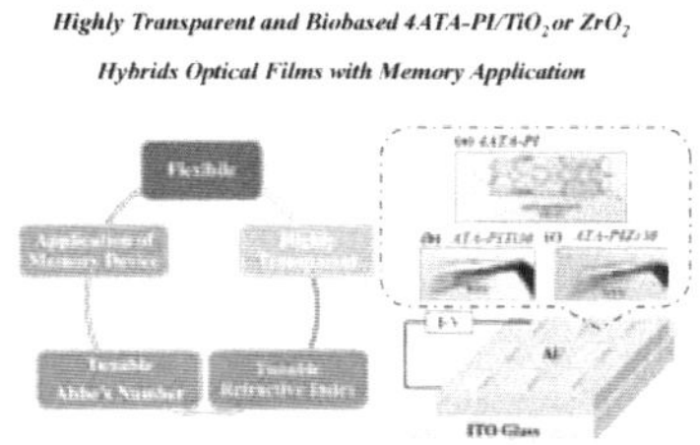

Keywords: Biobased Polyimides, Hybrid Material, Refractive Index, Resistive Memory

EOO-O-020 [Energy, Optics, & Optoelectronics]

A Novel Hybrid Type Conjugated Polymer Bearing Polyhedral Oligomeric Silsesquioxane Pendant Group

Salih Ertan, Atilla Cihaner

Department of Chemical Engineering and Applied Chemistry Atilim University, Ankara, Turkey

A new organic-inorganic hybrid monomer (POSSEDOT) was synthesized by using nano size support group Polyhedral Oligomeric Silsesquioxane (POSS). The chemical structure of the monomer was investigated by nuclear magnetic resonance (H-NMR) and high resolution mass spectroscopy (HR-MS). POSSEDOT monomer was polymerized electrochemically and the behavior of the polymer was investigated by cyclic voltammetry. In order to understand the effect of POSS unit on polymer chain, PEDOT polymer was studied under the same conditions. Spectroelectrochemical studies showed that incorparation of the POSS unit enchanced the percent transmittance and coloration efficiency (for-17 mC/cm2 film thickness, percent transmittance is about %68). The polymer film has highly transparent oxidized state and very bright blue color at its reduced state. In addition to that, electrochemical stability of the POSS containing polymer was superior to that of neat PEDOT.

Keywords: Electrochromism, Conducting Polymer, Polyhedral Oligomeric Silsesquioxane

EOO-O-021 [Industrial Polymers]

Facile Synthesis of Cyclotriphosphazene Based on Pyrole and Thiophene Monomers as Precursors for Novel Potential Conducting Polymers

Noreen Sajjad[1], Aeysha Sultan[2], Ferda Hacivelioğlu[3]

[1]*Department of Chemistry, University of Sargodha, Pakistan*
[2]*Department of Chemistry, University of Education, Pakistan*
[3]*Department of Chemistry, Gebze Technical University, Gebze, Turkey*

The organic conducting polymers (aka organic metals) are the material that combines the electrical properties of metals with the advantage of polymeric properties such as resistance to corrosion and chemical attack, low cost, light weight. Since the discovery of highly conducting polyacetylene[-Chiang, C. K.; Druy, M. A.; Gau, S. C.; Heeger, A. J.; Louis, E. J.; MacDiarmid, A. G.; Park, Y. W.; Shirakawa, H., J. Am.Chem.Soc,100, 1013-1015,1978.]. this field has grown tremendously.The organic conducting polymers so far synthesized are known for their high conducting stability and however these have limited practical applications. Since these polymers are associated with problems such as poor processability, insolubility and high manufacturing cost. Cyclotriphosphazene (CP) is an inorganic heterocyclic compound which is stable, nonflammable, and soluble in organic solvents. This unit, with high replaceable Cl atoms attached to phosphorus atoms can be easily tuned and functionalized to prepare a wide range of polymers. Furthermore, the functionalized CP cores are very stable and do not breakdown even under very aggressive chemical conditions [Cil E., Arslan M.,In Chimica Acta 362, 1421–1427, 2009]. This work presents our studies towards the synthesis of monomeric units that combine the superior properties of phosphazene and conducting properties conjugated system that would serve as precursor for development of polypyrole and polythiophenepolymers with the advantage of improved solubility, processability and electrical conduction for use in Hi-Tech applications.

Keywords: Organic conducting polymers, Phosphazene, Pyrole, Thiophene

EOO-O-022 [Energy, Optics, & Optoelectronics]

Highly Transparent to Truly Black Electroactive Ambipolar Devices Based on Polyamides and Viologen

Huan Shen Liu[1], De Cheng Huang[1], Yu Ruei Kung[2], Chyi Ming Leu[2], Guey Sheng Liou[1]

[1]*Department of Institute of Polymer Science and Engineering, National Taiwan University, Taipei, Taiwan(R.O.C)*
[2]*Material and Chemical Research Laboratories, Industrial Technology Research Institute, Hsinchu, Taiwan(R.O.C)*

Electrochromism is a phenomenon occurs in certain materials which we call these substances electrochromic materials (ECMs). Typically, ECMs will undergo a reversible color change if a specific electric potential is applied to them. Their color may have changed from the colorless to colored or vice versa, or even from one color to another color. Here we report an electroactive ambipolar system which combined electroactive polyamides with *N,N,N',N'*-tetraphenyl-p-phenylenediamine (TPPA) and tetraphenylbenzidine (TPB) units in the backbone and heptyl viologen (HV). Each material provides the three primary colors of one color, and arranges in groups. Through the collocation of counter electrode materials, more effectively reduces the overall operating voltage. Furthermore, the whole system is completely transparent in natural state and the transmittance at the redox state reduced to only 6% at both visible and NIR regions. The devices reveals a high ΔL* of 87 and a significant transmittance change of 90% at all visible light region. Thus, this system can be said to be the first electrochromic system can be from "highly transparent to truly black". This high-performance electroactive system with low driving voltage, high contrast ratio, and high absorbance at both visible and NIR regions can potentially use as technology to manipulate light, applications from shade material to insulation materials.

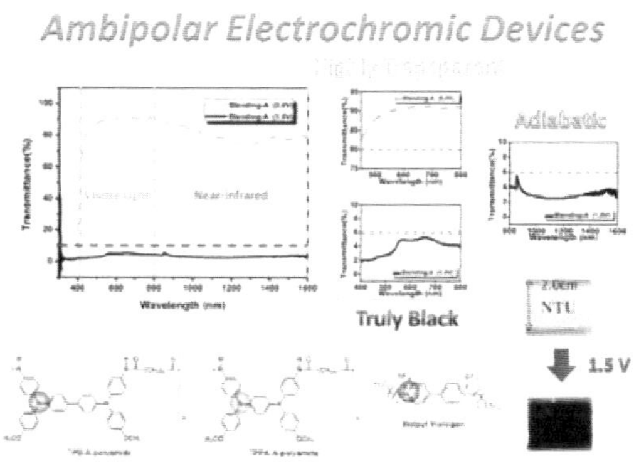

Keywords: Electrochromic device (ECD), Transmissive-to-black, Near-infrared (NIR), Polyamides, Viologen

EOO-O-023 [Energy, Optics, & Optoelectronics]

Syntheses, Electrochemical and Spectroscopic Properties of Fused-Thiophene Based Materials Possesing Benzonytrile Functional Group

Ilknur Demirtas[1], Erdal Ertas[1], Turan Ozturk[2]

[1]*Food Institute, TUBITAK Marmara Research Center, Kocaeli, Turkey*
[2]*Department of Chemistry, Istanbul Technical University, Istanbul, Turkey*

Fused thiophenes, formed by annulations of thiophene rings, have rigid structures with an extended π-conjugation, compare with thiophene [1]. Various studies have been performed to adjust thiophene to obtain advanced materials. Synthesis of fused thiophene rings is one of the methods for tuning the electronic properties of thiophene based materials [2]. Due to the unique features of thienothiophenes (TT), dithienothiophenes (DTT) and their polymers, they are used in designing and construction of magnetic and photosensitive receptors, electrochromic devices (ECD), organic light emitting diodes (OLED), field effect transistors (FET) and solar cells [3]. Theaimofthisstudyistodesignandsynthesizelowbandgapmaterialspossessing electron-rich thiophene based molecules having benzonitrile functional unit. Syntheses of the TT and DTT were performed through the ring closure reactions of mono and diketones using polyphosphoric acid. Introduction of two thienyl groups to the peripherals of the core TT moiety was achieved by Suzuki coupling. Conducting polymers of 1 and 3 were synthesized by potentiodynamic method and the resultant polymers were characterized via cyclic voltammetry (CV), Fourier transform infrared (FTIR) and UV-Vis spectroscopy. The monomer (Th2TT) 1 had absorption and emission maxima of 360 and 499 nm, respectively, with an optical band gap of 2.94 eV. The polymer 2 coated on ITO had a lower band gap, 1.85 eV. The monomer (DTT) 3 had absorption and emission maxima of 310 and 428 nm, respectively, with an optical band gap of 3.68 eV. Electropolymerization of 1 gave a polymer 4 with an optical band gap of 2.34 eV.

References:

[1] Kertesz, M.; Choi, C. H.; Yang, S. Chem. Rev. 2005, 105, 3448
[2] Wang, L.; Day, C. S.; Wright, M. W.; Welker, M. E. Beilstein J. Org. Chem. 2009, 5
[3] Dundar P.; Osken I.; Sahin O.; Ozturk T. Synthetic Metals 2012, 162, 1010-1015

Keywords: Low band gap materials, Thienothiophene, Dithienothiophene

Structures of TT, DTT derivatives and their polymers

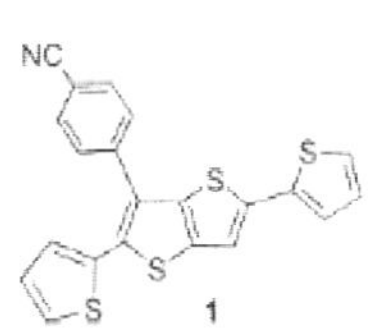

1

4-(2,5-di(thiophen-2-yl)thieno[3,2-*b*]thiophen-3-yl)benzonitrile

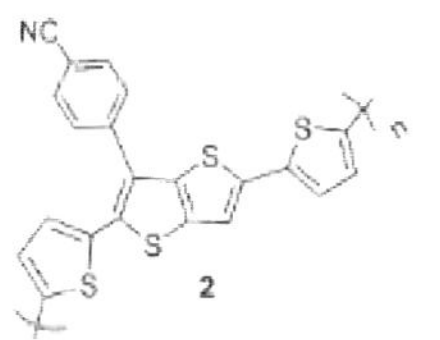

2

poly-(4-(2,5-di(thiophen-2-yl)thieno[3,2-*b*]thiophen-3-yl)benzonitrile)

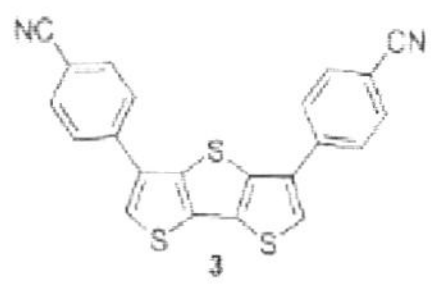

3

4,4'-(dithieno[3,2-*b*:2',3'-*d*]thiophene-3,5-diyl)dibenzonitrile

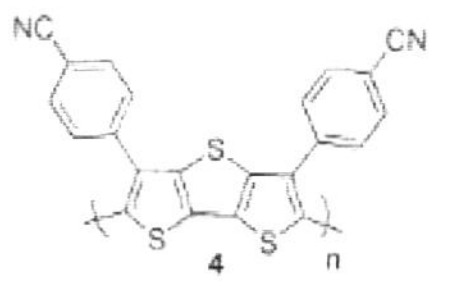

4

poly-(4,4'-(dithieno[3,2-*b*:2',3'-*d*]thiophene-3,5-diyl)dibenzonitrile)

EOO-O-024 [Energy, Optics, & Optoelectronics]

Conductive Polymer Nanotubes as Chemiresistive Sensors

Mehmet Can Zeybek[1], Gözde Özaydın Ince[2]

[1]*Department of Material Science and Engineering, Faculty of Engineering and Natural Science, Sabancı University*
[2]*Sabanci University Nanotechnology Research and Application Center (SUNUM)*

Conductive Polymers are one of the most attractive topics nowadays. Polypyrrole (PPy) among the conductive polymers has special place because of its high stability, good biocompatibility [1] and relatively high conductivity etc. Furthermore, sensor applications are one of the most extensive usage area of conductive polymers because of their suitable properties in this field.

In our study, vapor phase deposition technique oCVD is used for production of Polypyrrole Nanotubes. With processability at room temperature, highly-conformal deposition ability, applicability for unlimited samples from conductive to non-conductive, ordinary flat glass slide to samples with a very complex geometry on its surface; vapor phase deposition methods are suitable and quite promising instruments. In order to production of Polypyrrole nanotubes, hard template method has been conducted by using AAO template. During the polymerization of the pyrrole, antimony (V) chloride is used as dopant. For characterization studies, SEM, FTIR and 4-probe measurement is being used for morphological analysis, chemical analysis, and conductivity measurement respectively.

After nanotube production is completed, polypyrrole nanotube integrated sensor system is going to be used for detection of volatile organic compounds with high selectivity and sensitivity. For further improvement in sensitivity/selectivity we also aim to use some biomimetic geometry and hybrid material systems in the design of our sensor system.

Early diagnosing of health problems especially cancer in virtue of volatile biomarkers is a widely studied subject of today's world.[2] Usage of our candidate system has high potential to be utilized in early diagnosis of health problems by detection of volatile organic compounds which are the markers of some disease.

1. Balint, R., N. J. Cassidy, and S. H. Cartmell, Conductive polymers: Towards a smart biomaterial for tissue engineering. Acta Biomaterialia, 2014. 10(6): p. 2341-2353.
2. Peng, G., et al., Diagnosing lung cancer in exhaled breath using gold nanoparticles. Nat Nano, 2009. 4(10): p. 669-673.

Keywords: Chemiresistive Sensor, Conductive Polymer, Polypyrrole Nanotube

IP-O-001 [Industrial Polymers]

Mechanical Properties of New Trends Thermotropic Liquid Crystalline Copolyesters with PET Blends

Melek Bulut

Çanakkale Onsekiz Mart Üniversitesi, Acil Yardım ve Afet Yönetimi Bölümü, Çanakkale

Polymer blending is one of the most common methods in to adapt the properties of polymeric materials. Polymer blending based on thermoplastics and thermotropic liquid crystalline polymer (TP/TLCP) is a useful and attractive approach. Thermoplastics have the potential for high volume processing, low cost, and parts with superior impact resistance as compared to their thermosetting counterparts. But thermoplastics also have their disadvantages: creep, limited temperature capability, and limited processing techniques. Thermotropic liquid crystalline or mesomorfic copolyesters have recently attracted as high performance engineering polymers. This regularity of molecules cause high degree mechanical strength the direction of pull. These increases in mechanical properties can be obtained due to the inherent high orientation of LCPs in the melt state thereby easing the processability of TP. Moreover, the TP matrix can be reinforced due to the formation of TLCP fine fibrils, giving rise to the development of polymer composites that are commonly known as in situ composites. For last two decates, the rheological, morphological and mechanical properties of blends with thermotropic liquid crystalline copolyesters or other polymers as poly(hegzamethylene adipamide) (PA 66), poly(ethylene terephthalate) (PET), polycarbonate/(PC), poly(ether imide) (PEI), high molecular mass polyethylene (HMMPE) poly(trimethylene terephthalate) (PTT), poly(ethylene 2,6naphthalate) (PEN) carbon nanotube (CNT), acrylonitrilebutadiene styrene (ABS), polypropylene (PP) were studied. Anizotropic structure of this polymers cause high degree molecular orientation at during drawing. In this study, thermotropic liquid crystal copolymer nameds 67PHT and 60P40PET have been synthesized and blended with ındustrial polymer, poly(ethylene terephtalate) (PET). The Young's modulus values of the 67PHT/PET and 60P40PET/PET blends increase compare to pure PET when TLCP values increase. Tan δ values of 67PHT/PET and 60P40PET/PET blends show a single peak. This show that these blends are compatible and miscible.

Keywords: Blends, thermotropic liquid crystalline polymers, PET, mechanical properties

IP-O-002 [Industrial Polymers]

In-silico Screening of Green Plasticizers for PVC

Hüsamettin Deniz Özeren[1], Marcel Balçık[1], J. Richard Elliott[2], M. Göktug Ahunbay[1]

[1]*Department of Chemical Engineering, Istanbul Technical University, Istanbul, Turkey*
[2]*Department of Chemical and Biomolecular Engineering, University of Akron, Akron, OH, USA*

Polyvinyl chloride (PVC), is widely used industrial plastic, is processing with commonly phthalate derivatives plasticizers. Interest and research in new environment-friendly plasticizers has increased due to negative effect of phthalate derivatives on human health. The goal of the proposed research is to design "green" plasticizers for PVC to employ atomistic scale simulations. Our methodology is based on using molecular dynamics simulations to provide structure-property relationships for solution compatibility/stability and glass transition temperature that is most important descriptor of properties of amorphous polymers. The demonstration is based on computing/simulating glass temperature with various plasticizers in Table 1 derived from the list of the Department of Energy (DOE) of USA for sustainable building blocks that are important sources of non-petroleum based raw materials (Werpy, T. et al. *Top value added chemicals from biomass I. Results of screening for potential candidates from sugars and synthesis gas*; doe/go-102004-1992; 2004.) and PVC plasticizer uptake for these plasticizers. The methodology is based on molecular simulations. First, molecular models of the species in Table 1 and phthalate derivatives for comparison were constructed based on TRAPPE United-Atom approach and their geometries were optimized and electrostatic charges were evaluated. Next polymer matrices were constructed based on the generalized AMBER force field (GAFF). The glass transition temperature (Tg) and density of pure PVC were computed as 344 K and 1.38 g/cm^3, respectively, which are in very good agreement with the experimental data. The plasticizer uptakes were predicted through grand canonical Monte Carlo simulations at various temperatures. Finally, the variations in the polymer Tg were investigated and compared as a function of plasticizer type and content by simulating the pressure-volume-temperature (PVT) relationship through molecular dynamics simulations.

This work is part of the project supported by the Scientific and Technological Research Council of Turkey (TUBITAK) through the grant no. 114M178.

Keywords: Molecular simulation, Green chemistry, Thermodynamics, Polymer dynamics

Table 1. Biomass based plasticizer candidates

Succinic acid esters (Mono-methyl, di-methyl)
Fumaric acid ester (Di-methyl)
Levulinic acid ester (Mono-methyl)
Glycerol
Dioctyl adipate
Triethyl citrate
Oleic acid esters (Mono-methyl, di-methyl)

IP-O-003 [Industrial Polymers]

Single-Step Multiple-Colloid Coating for Multifunctional Biodefense Textiles: A Novel Application in Mosquito-Borne Disease Prevention

Nakarin Subjalearndee, Nipon Phanomkate, Varol Intasanta

National Nanotechnology Centre, National Science and Technology Development Agency, 111 Phahonyothin Road, Pathumthani, 12120, Thailand

A single-step coating was developed based on mixed-colloid systems for multifunctional biodefense applications in textiles. Among various preventive methods, mosquito control fabrics have been recommended by the World Health Organization for the prevention of mosquito-borne diseases such as malaria and dengue in rural parts of the world. While effective, their usage in low-income and highly infectious countries has been limited due to the embedded cost for logistics, cleaning and maintenance. Towards a novel application in smart and sustainable mosquito-borne disease prevention, this research pioneered the possibility of fabricating mosquito repellent textiles with antibacterial and easy care functions minimizing wash requirement while maximizing their life cycle, alleviating the burden of using such items in a rural and low-resource areas. Experimentally, water-based and multi-component formulations containing encapsulated citronella were prepared. Their stabilities were investigated and optimized via sizes and zeta potential of the starting materials and final mixtures. The stable formulations were applied on fabrics via conventional pad-dry-cure processes, with variation of weight of fabrics and % pick up. Our laboratory study resulted in unprecedented prototypes with multiple easy care properties—water repellent, anti UV, antibacterial and enhanced tensile strength. The laboratory samples were then evaluated against a standard bio efficacy method which unveiled satisfactory mosquito repellency. With the water-based formulations and under such conventional process, the successful coating steps represented a facile and scalable route to multifunctional mosquito repellent fabrics. Furthermore, our work extended to pilot coating under industrial circumstance which allowed for production scale multifunctional mosquito repellent fabrics which held promises for personal protection and defense against mosquito-borne diseases and biohazards.

Keywords: Colloid, nanocoating, mosquito, textiles, antibacterial, water repellent

IP-O-004 [Industrial Polymers]

Up-to Date Procedures for Polypropylene Production

Sonja Ketin[1], Rade Biocanin[2]

[1]*Environmental Engineering, Technical University, Novi Sad, Serbia*
[2]*Environmental, State University, Novi Pazar, Serbia*

All up-to-date methods to obtain polypropylene are based on the polymerization in the mass of monomers that in the reactor can be in liquid or gaseous state. The most important method of obtaining a polypropylene in the mass of a liquid monomer is Montell's (nowadays Basell) 'Spheripol's process. The reactor is a double loop pipe with screen where counter-flowing cooling water. The reactor is a relatively small volume and through it in the turbulent regime flow the suspension of the polymer powder in a liquid monomer. The velocity of flow in the reactor is large (about 10 m/s) *it contributes to the homogeneity of the product and allows the removal of large amounts of heat.* Production processes in the gas phase can be divided into three groups:

- the first group consists of " Unipol " process using fluidized layer which has been affirmed by the Union Carbide company
- the second group includes methods of obtaining in the horizontal reactor with a mixer where the most important is Amoco's " Innovene" process
- the third group includes methods of obtaining in a vertical reactor with a mixer where the most important is BASF's "Novolen" 'process.

A special procedure is "Borstar" which is a combination of existing (loop) reactors and reactors in the gas phase, which is developed by Borealis company. Novoline and Innovene process are similar. Mixer allows a good homogeneity and heat dissipation is performed by condensation of recycled propylene.

Keywords: polypropylene, production, polymerization

IP-O-005 [Industrial Polymers]

Optimization of Gas Phase Ethylene Polymerization: A Micro/Mini Reactor Study

Gözde Salkıç, Ertuğrul Erkoç

Department of Chemical Engineering, Bursa Technical University, Bursa, Turkey

Polyethylene is one of the most important materials in the polyolefin industry. Polyolefin polymerization is processed by using catalysts such as Ziegler-Natta, Metallocene and chromium-oxide types. Polymerization is started with breaking the double bonds by catalyst addition. In Ziegler-Natta catalysts consisting titanium alkyl bond, titanium ethylene alkyl is formed by this catalyst addition. High density polyethylene (HDPE), low density polyethylene (LDPE), ultra-high molecular weight polyethylene (UHMWPE) and medium density polyethylene (MDPE) are mostly used polyethylene types in industry. Although MDPE and HDPE are mostly produced in slurry processes, gas phase processes and patents utilizing fluidized bed reactors are also industrially available. However, there is no known UHMWPE production or any published patent for this type by gas phase polymerization. The main underlying reason for this is the fact that, olefin polymerizations are highly exothermic and efficient heat transfer currently cannot be realized in fluidized beds, causing hot spots and eventually, the deactivation of the catalysts. However, mass transfer from gas to liquid for solution process and additionally mass transfer from the liquid to the solid catalyst surface poses a problem for both of solution and slurry phase polymerizations. Moreover, for the said processes, environmentally hazardous solvent has to be separated from the product which brings additional installation and operation costs. In order to design microreactor system, a semi-batch mini-reactor setup was prepared to gather necessary knowledge in the presented study. Gas phase ethylene polymerization was performed by using activated Ziegler-Natta catalysts. Taguchi method was used to study the effect of important parameters such as temperature, pressure and catalyst-cocatalyst ratio on molecular weight and material properties of polyethylene.

Keywords: polyethylene, gas phase polymerization, microreactors

IP-O-006 [Industrial Polymers]

Isospecific polymerization of p-methylstyrene by means of a bis(phenolato)titanium based catalyst activated by MMAO: A systematic and kinetical study

Luis Valencia[1,2,3], Günter Reiter[4], Rolf Mulhaupt[1,2], Tianyu Wu[4], Pierre-J. Lutz[3]

[1]*Institute for Macromolecular chemistry, University of Freiburg, Freiburg, Germany*
[2]*Institut Charles Sadron, CNRS, Strasbourg, France*
[3]*Freiburger Materialforschungszentrum, University of Freiburg, Freiburg, Germany*
[4]*Physikalisches Institut, University of Freiburg, Freiburg, Germany*

The versatility of functionalization of Polyp-methylstyrene (PpMS), together with its lower density, higher Tg and lower melt viscosity in comparison to conventional Polystyrene have attracted the attention of research effort during the last decades.

On the other hand the isospecific polymerization of Poly(p-methylstyrene), early reported by Natta and coworkers [1], have been widely studied by means of different polymerization techniques [2-3], however a poor stereocontrol which have led to up to 40% of isotactic pentad fraction was in most of cases reported (pre-fractionation value), therefore the interesting features and capabilities of iPpMS have never been fully explored.

In this work we systematically study the isospecific polymerization of PpMS performed in the presence of the catalyst dichloro[1,4-dithiabutanediyl-2,20-bis(4,6-ditert-butyl-phenoxy)]titanium (A) activated by MMAO, introduced by Cappachione et al., [4-6].

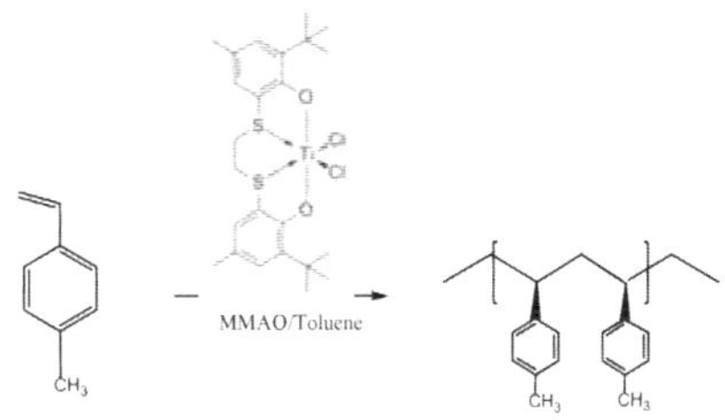

Scheme 1. Post-metallocene isospecific polymerization of *p*-methylstyrene

The produced polymers were highly isotactic, with more than 95% isotactic pentad fraction and narrow molecular weight distributions (see Table 1). The kinetic of formation, macro- and microstructure, thermal behavior and crystallinity of the resultant polymers was fully investigated. Furthermore a successful control of the molecular weight was achieved by adding a 1,9 decadiene as chain transfer agent leading to low molar mass vinyl terminated isotactic poly(p-methylstyrene).

Table 1. Homopolymerization of *p*-methylstyrene in the presence of A/MMAO

Run[a]	Temperature	P(m-mmm)-%[b]	P(m-m)-%[b]	P(m)-%[b]	Mw (x 10^4)[c]	Mw/Mn[d]
1	0 °C	97.00	97.85	98.92	3.4	1.25
2	20 °C	95.84	98.26	99.65	2.7	1.84
3	40 °C	96.78	99.31	99.65	6.1	1.24
4	60 °C	97.51	99.63	99.81	3.8	1.57

[a]*Ti = 5.84 umol; Al/Ti = 300; monomer = 22.77 mmol; toluene = 10 mL, 60 min reaction time*
[b]*proportions of mmmm pentad, mm triad and m dyad respectively, calculated by ^{13}NMR*
[c]*Weight-average molecular weight (g/mol)*
[d]*Weight-average molecular weight/number-average molecular weight*

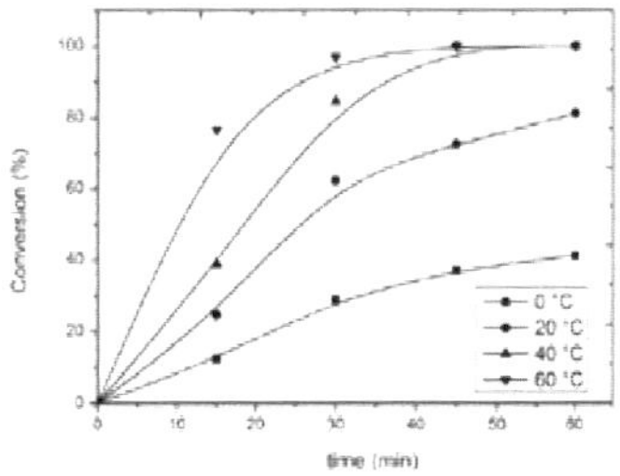

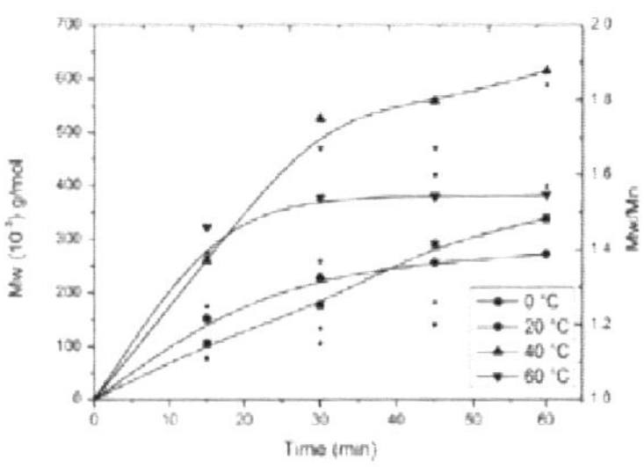

Figure 1. Conversion evolution and molar mass as a function of time

Acknowledgement

The authors thank specially the DFG (IRTG Soft Matter Science 1642) for financial support.

References

[1] Giulio Natta, Piero Pino, Paolo Corradini, Ferdinando Danusso, Enrico Mantica, Giorgio Mazzanti, G. M. (1955). Crystalline high polymers of a-olefins. *Journal of the American Chemical Society*, *77*, 1708–1710.

[2] N. Isihara, M. K. and M. U. (1988). Stereospecific polymerization of styrene giving the syndiotactic polymer. *Macromolecules*, *21*(12), 3355–3359.

[3] Li, B. Te, Wu, Y. X., Cheng, H., & Liu, W. H. (2012). Synthesis of linear isotactic-rich poly(p-methylstyrene) via cationic polymerization coinitiated with AlCl 3. *Polymer (United Kingdom)*, *53*(17), 3726–3734.

[4] De Carlo, F., Capacchione, C., Schiavo, V., & Proto, A. (2006). Reactivity of styrene and substituted styrenes in the presence of a homogeneous isospecific titanium catalyst. *Journal of Polymer Science, Part A: Polymer Chemistry*, *44*(4), 1486–1491.

[5] Buonerba, A., Fienga, M., Milione, S., Cuomo, C., Grassi, A., Proto, A., & Capacchione, C. (2013). Binary copolymerization of p -methylstyrene with butadiene and isoprene catalyzed by titanium compounds showing different stereoselectivity. *Macromolecules*, *46*(21), 8449–8457.

[6] Costabile, C., Capacchione, C., Saviello, D., & Proto, A. (2012). Mechanistic Studies on Conjugated Diene Polymerizations Promoted by a Titanium Complex Containing a Tetradentate [OSSO]-Type Bis(phenolato) Ligand. *Macromolecules*, *45*(16), 6363–6370.

Keywords: Poly(p-methylstyrene, Isospecific, Post-metallocene

IP-O-007 [Industrial Polymers]

Investigation of Parameters Affecting the Bitumen Modifications by Using Response Surface Method

Emel Başkent[1], Mustafa Hulusi Özkul[2]

[1]*Department of Polymer Science and Technology, Istanbul Technical University, Istanbul, Turkey*
[2]*Department of Civil Engineering, Istanbul Technical University, Istanbul, Turkey*

Bitumen, a binder material used in asphalt applications, is one of the important components that affects the service-life of asphalt. Bitumen chemistry, amount of Saturate-Aromatic-Resin-Asphaltene (SARA) fractions, and polymer addition affect the properties of bitumen. Besides warm mix additives providing the lower mixing temperatures during mixing of bitumen and aggregates. The characteristics of bitumen changes with the amount and type of additive and bitumen chemistry. In this study, understanding the simultaneous effect of variation of bitumen composition, amount of polymer and warm mix additive on the bitumen properties is aimed. For this purpose, SARA analysis is carried out for the same penetration value bitumen (Penetration 50/70) produced at İzmit, İzmir and Batman refineries in Turkey. Amount of asphaltenes, the heaviest and the most complex fraction of bitumen, is determined for the İzmit, İzmir and Batman bitumens as 23, 21, 34 (wt.)%, respectively. For the detailed characterization of the bitumen FTIR, DSC, TGA and GC-Simdis analysis are performed. Styrene-butadiene-styrene (SBS) copolymer and commercial paraffin wax as warm mix additive are used for modification of bitumens. The variables are determined as asphaltene content, amount of SBS and wax. The limits of asphaltenes, SBS and Wax are taken as 21-34, 2-5 and 1-3, (all in by mass) respectively. Design of experiment is formed by DesignExpert ® software followed by modification of bitumens that are analyzed with dynamic shear and bending beam rheometers, ductilometer and viscometer for determination of high temperature susceptibility, low temperature resistance, elastic recovery and workability, respectively. As a result of the measurements, the model of relationship between response functions and independent variables is constructed by using response surface method. Three-dimensional response surfaces and two-dimensional interaction curves are plotted by using model data that indicates the effect of each parameter on the properties of bitumen.

Keywords: Polymer modified bitumen (PMB), styrene-butadiene-styrene (SBS) copolymer, response surface method (RSM)

3D-Response Surfaces

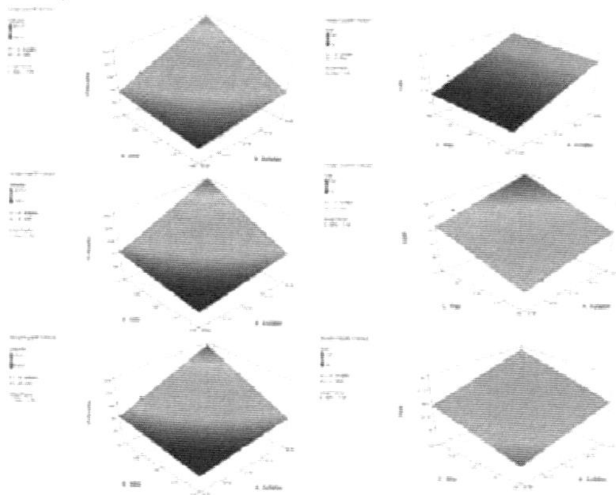

The trend of viscosity of bitumen, the meaning of workability, with the different amount of asphaltene and SBS content at constant wax addition (left 1-2-3). The change of dynamic shear rheometer, the meaning of deformation resistance at high temperatures, with the different amount of asphaltene and wax content at constant SBS addition (right 1-2-3).

IP-O-008 [Industrial Polymers]

Core-Shell Type Styrene-Acrylic Emulsions for Reduction of Formaldehyde Emission in Pigment Printing

Gökhan Akbulut, Sanem Şenler, Mesut Eren, Hakan Aşkun

Betek Boya, Istanbul, Türkiye

Pigment printing is the oldest and the easiest printing method with many advantages such as easy application onto various kinds of fibers, quality of the products and after fixation, being able to withstand washing processes. Gloss, color strength, penetration, soft touch and the fastness properties mainly determine the quality of the pigment print. This is affected by the choice of emulsion binder, thickeners and other additives, as well as amount and type of cross linkers. Binder should have low enough T_g to allow continuous film formation, but should be cross linked during the fixation to achieve fastness and other required properties. The N-methylol acrylamide (NMA) incorporated in the binder, up to 5 %, is the mostly employed cross linker. Since environmental regulations demand decreasing of the amount of formaldehyde which is released by NMA during thermal fixation process, we investigated the possibility of combination of easy film formation and toughness by employing core-shell type emulsion binders. Thus, in this study, a series of water based core-shell type styrene-acrylic emulsions have been synthesized, with core/shell (w/w) ratios of 20/80, 30/70 and 40/60. T_g of the core polymers were either +30 °C or +50°C, whereas the T_g of the shell polymers were -5 °C, -10 °C and -20 °C. The results revealed that, by additional variation of the amount of NMA in core or shell layers of the polymer, it was possible to prepare pigment printing pastes with 50-80 % less NMA, without loss of color strength and fastness properties. (A patent application has been made to European Patent Office, EP 15 175 437.1)

Keywords: Core-shell, pigment printing, formaldehyde reduction

IP-O-009 [Industrial Polymers]

Multifunctional polycarboxylate dispersants for the optimum soil anti-redeposition performance in new generation of powder detergents

Fatih Mehmet Erguney[1], Serap Ozay[1], Mehmet Gollu[1], Mehmet Atilla Tasdelen[2]

[1]Hayat Kimya R&D Center, Kocaeli, Turkey
[2]Department of Polymer Engineering, Yalova University

Polymeric detergent builders such as polycarboxylates are heavily used in powder laundry detergent formulations (>90% of all polymers)[1]. In the markets dominated by powder detergents, acrylicacid homopolymers are used as soil anti-redeposition agents and account for 2-5% and 10-20% of their weight and cost structure, respectively. They also help to the phosphate derivatives for Ca++ and Mg++ sequestering[2].

Some other polymers like polyalkyleneglycol and terephthalicacid based polyesters have gained importance in specialty applications (e.g. stain shield & soil release) due to remarkable improvement in the washing performance, but they are extremely expensive, so have not found extensive use, especially in developing markets. The polycarboxylates are indispensable to the industry, but there is still room to improve their soil anti-redeposition performance without increasing their cost structure, maybe by even improving their market competitiveness[3].

In this work, the most effort is given to disclosing the relationships between the molecular weight and polydispersity and their resulting soil anti-redeposition performance on different types of textiles such as cotton, polyacrylic, polyester and their hybrid compositions. In addition, some co-polymeric structures with phosphate and sulfate co-monomers have also been studied to verify any potential increase in synergistic washing performance of the surfactants (LABSA, SLES, Betaine, etc.).

The polyacrylates with 7.500-8.000Da molecular weight show much better performance on cotton while 15.000Da polymers are superior on polyacrylic and polyester based synthetic textiles. The polydispersity must be kept lower than 2, which is a difficult task in free radical polymerization especially when applied on an industrial scale. Furthermore, the best synergistic washing performance has been observed when both phosphoric and sulfonic acid, have been included in the copolymer structure in equal amounts with a 10 to 1 rate compared acrylicacid monomer. Finally, the polycarboxylate content can be kept around 1% in the detergent formulation by using such optimum polymers.

Keywords: acrylic acid dispersant, polymeric builder, powder detergent, calcium sequestering

IP-O-010 [Industrial Polymers]

Bridges between homogeneous and heterogeneous catalytic systems for polyolefins production: design and advanced understanding of supported aluminic activators

Regis M. Gauvin[1], Dominique W. Sauter[2], Nicolas Popoff[2], Muhammad A. Bashir[2], Kai C. Szeto[2], Laurent Delevoye[1], Tim Mckenna[2], Mostafa Taoufik[2], Christophe Boisson[2]

[1]UCCS, CNRS UMR 8181, University of Lille, Villeneuve d'Ascq, France
[2]C2P2, CNRS UMR 4265, CPE Lyon, University of Lyon 1, Villeurbanne, France

The development and understanding of supported group 4 catalysts for the production of polyolefins is a field attracting constant interest. In this context, both new characterization techniques and new preparation methods are key to further progress. In this contribution, we provide:

- 1) refined understanding of the structure of molecular and silica-supported methylaluminoxane (MAO) thanks to high-field solid-state NMR studies; We also investigate the origin of the impact of the support pretreatment temperature on the catalytic performances of the resulting systems in ethylene polymerization, in terms of nature and number of active sites.[1]

- 2) disclosure of a new single-site silica-supported activator, designed using a surface organometallic chemistry approach (Figure 1); this well-defined grafted aluminate ammonium species is analogous to established molecular activators, and we demonstrate that it efficiently activates metallocene complexes, thus achieving high activities in slurry polymerization of ethylene along with industrial-grade physical properties of the resulting polyethylene

These combined results provide elements that contribute bringing together homogeneous and heterogeneous catalysis in olefin polymerization.

[1] M. A. Bashir, T. Vancompernolle, R. M. Gauvin, L. Delevoye, V. Monteil, M. Taoufik, T. F. L. McKenna, C. Boisson, Cat. Sci. Tech., 2016, DOI 10.1039/C5CY01285F.
[2] D. W. Sauter, N. Popoff, M. A. Bashir, K. C. Szeto, R. M. Gauvin, L. Delevoye, M. Taoufik, C. Boisson, Chem. Commun. 2016

Keywords: supported catalysts, MAO, well-defined activators, polyolefins

Figure 1

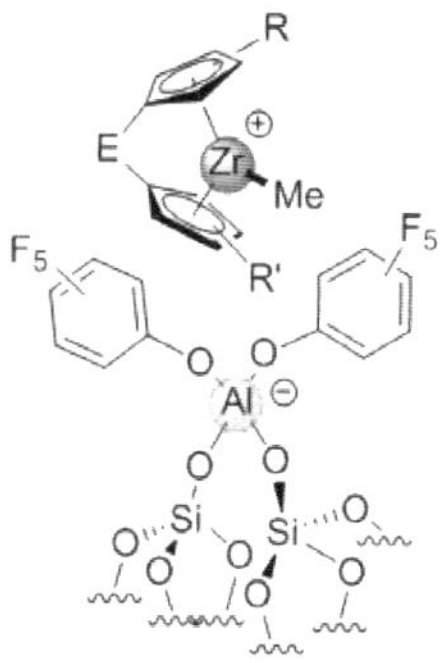

Well-defined single-site aluminate ammonium activator

IP-O-011 [Industrial Polymers]

Nanoreinforced Rigid Polyurethane-Organoclay Composite and Characterization Thereof

Orcun Yucel

R&D Materials Tech., Arcelik A.S., Istanbul, Turkey

Polymer-layered silicate nanocomposites have recently gained a great deal of attention for providing superior properties in comparison with neat and conventional filled polymer-composites. The properties include high dimensional stability, reduced gas permeability and increased mechanical and thermal performance [1–4]. Rigid polyurethane foams are widely used in global insulation market primarily for refrigeration and piping / household insulation purposes. However rigid PU foams have some disadvantages such as low aging performance, long-term thermal stability and flammability. To overcome these disadvantages, a great deal of effort has been devoted to the development of nanostructured polyurethane (PU)/montmorillonite (MMT) composites in recent years [4]. The effects of clay content (Cloisite 30B) and PU molecular structure on clay dispersion in rigid PU foam nanocomposites have been studied through ultrasonication process. The effect of nano-clay content on thermal and mechanical performance of rigid polyurethane foam was also investigated. By using ultrasonication process nano-clays were well-dispersed as determined by SEM analysis. Mechanical strenght was improved 15% with increased nano-clay content. In addition, thermal conductivity of rigid PU foam was also decreased.

Acknowledgements; The author thanks Arcelik A.S Company for support of the research.

[1] EP. Giannelis, J. Advan. Mat. (Weinheim, Germany) 1996, 8, 1:29.
[2] M. Alexandre, P. Dubois, J. Mat. Sci. Eng., 2000, 28(1–2):1.
[3] EP. Giannelis, R. Krishnamoorti, E. Manias, J. Adv. Poly. Sci., 1999, 138:107.
[4] X. Caoa, L. J. Lee, T. Widya, C. Macosko, J. Poly. 46, 2005, 775-783.

Keywords: polyurethane rigid foam, organoclay, insulation, nanocomposite

IP-O-012 [Industrial Polymers]

Fertilizer Release Profile of Pol(vinyl alcohol)/ Ammonium Sulfate Coated Nonwoven Structures

Ilhan Özen[1], Gamze Okyay[1], Sultan Karagöz[1], Abdullah Ulaş[2]

[1]Department of Textile Engineering, Erciyes University, Kayseri, Turkey
[2]Department of Soil Science and Plan Nutrition, Erciyes University, Kayseri, Turkey

The steady increase of world's population and environmental pollution and the gradual decrease of natural ressources has led to the "sustainable efficient agriculture" conception and the humankind has made efforts toward this end. Development of agrotextile materials is one of these efforts. Production of innovative agrotextile materials enabling reduced usage of water, fertilizer, and environmentally dangerous chemicals (pesticide) instead of conventional agrotextile materials which only protect agricultural crops from external conditions would be a very important progress in the field of sustainable efficient agriculture.

Fertilization is carried out once or twice throughout the whole agriculture season due to the easiness of application, although plants take the necessary nutrients slowly from the soil in time. This situation leads to fertilizer losses because of the external conditions. In order to avoid this from happening, generation of textile structures is aimed at in this study which enables the slow transfer of fertilizer to the plant in time. To this end, a nonwoven material was coated with a polymer/fertilizer mixture and cured under different conditions. The following materials were used: Poly(viniyl alcohol) (PVOH, 99% hydrolyzed, Mw: 146 000-186 000 g/mol) as matrix polymer, poly(ethylene terephthalate)/viscon (PET/CV) needle-punched nonwoven (130 g/m2), ammonium sulfate (AS, 21% nitrogen) as fertilizer, glutaraldehyde (GA, 25% aqueous solution) as crosslinker for the matrix polymer.

Coated structures were generated by using two different poly(viniyl alcohol) amounts (60 g/m2 and 120 g/m2), two different fertilizer loadings (27.5% and 55 %) and two different glutaraldehyde concentrations (PVOH/GA: 1/0.01 and 1/1). Fertilizer release profile of the PVOH/AS coated nonwoven structures was determined by conductivity and ammonium nitrogen measurements. In addition, gel fraction (degree of crosslinking), dynamic and equilibrium swelling ratios, water retention capacity and water diffusion of the generated materials have been revealed. Besides, morphological (SEM), thermal (DSC), and structural (FT-IR) properties were investigated.

Acknowledgment: This work was supported by the Scientific and Technological Research Council of Turkey (TUBITAK), Project Number: 115M718

Keywords: controlled release, poly(viniyl alcohol), ammonium sulfate, nonwoven

MBM-O-002 [Macromolecules in Biotechnology & Medicine]

Micellar Drug Delivery Systems

Rana Sanyal

Bogazici University, Department of Chemistry and Center for Life Sciences and Technologies, Bebek, 34342, Istanbul, TURKEY

Alongwith the potential cure they provide, most chemotherapy agents are inherently toxic to the entire body, since they aim to attack the fast reproducing cells. Though cancer cells are at the top of the reproduction list, many others such as bone marrow is also in the list, resulting in undesirable side effects. Packaging the chemotherapy agent to shield the body from the destroying effects of the drug till it reaches its destination is a universal solution to this obstacle. There are a variety of "packaging" options, among which nanomaterial based drug delivery systems are of great interest. In this talk, micellar carriers with the ability to carry hydrophobic chemotherapy agents will be explained. Two kinds of micellar constructs that are composed of dendritic and linear units, where the hydrophobic dendritic portion is capable of interacting with hydrophobic drugs will be discussed. The first carrier is composed of three blocks where the middle block, polyethylene glycol, provides the aqueous solubility. The two sides of the middle block is equipped with polyester dendrons, contributing to the lipophilic-hydrophilic balance of the overall system [Altin,Macromolecules,2010]. Increasing the concentration of this amphiphilic copolymer in aqueous media results in hydrogels [Kaga,Macromolecules,2015;JP-SPC,2016]. The number of dendritic surface groups has been a useful tool for investigation of the functionality in gels as well as soluble constructs. A library of constructs have been prepared with different dendron generations varying the number of drugs attached [Bolu,Mol Pharm,2016]. The second group of carriers are aiding the delivery of the hydrophobic drug without the benefit of the covalent linkage between the carrier and the drug. The micellar carriers have been characterized with SEC and DLS. Their stabilities, drug release profiles,cytotoxicities and cell internalization were investigated.

Acknowledgements: TUBITAK Project 107T623, Ministry of Development Project 2009K120520.

Keywords: drug delivery, micelle, nanomedicine

MBM-O-003 [Macromolecules in Biotechnology & Medicine]

Acid-disintegratable polymersomes for intracellular drug delivery

Guoying Zhang, Lei Wang, Yang Li

Department of Polymer Science and Engineering, University of Science and Technology of China, Hefei, China

Polymersomes self-assembled from amphiphilic polymers capable of synchronically loading of both hydrophilic and hydrophobic payloads have shown promising potential in drug delivery application. Herein, we report the fabrication of pH-responsive polymersomes via supramolecular self-assembly of amphiphilic diblock copolymers, poly(ethylene oxide)-b-poly(2-((((5-methyl-2-(2,4,6-trimethoxyphenyl)-1,3-dioxan-5-yl)methoxy)carbonyl)amino)ethyl methacrylate) (PEO-b-PTTAMA), which were synthesized via reversible addition-fragmentation chain transfer (RAFT) polymerization of a pH-responsive monomer (i.e., TTAMA) using a PEO-based macroRAFT agent. The resultant amphiphilic diblock copolymer then self-assembled into polymersomes consisting of hydrophilic PEO coronas and pH-responsive hydrophobic bilayers. The cyclic benzylidene acetals embedded in the hydrophobic polymersome bilayers were relatively stable under neutral pH, whereas they underwent hydrolysis to generate hydrophilic diol moieties when exposed to acidic pH milieu. By loading hydrophobic model drug (Nile red) and hydrophilic chemotherapeutic drug (doxorubicin hydrochloride, DOX·HCl) into the bilayer and aqueous interior of the polymersomes respectively, the subsequent release of both payloads was remarkably regulated by the solution pH values. In vitro experimental results revealed that the pH-responsive polymersomes were easily taken up by HeLa cells and were primarily located in the acidic organelles after internalization, where the pH-responsive cyclic acetal moieties were hydrolyzed and the embedded payloads were therefore released, allowing for on-demand release of the encapsulants. Besides, biomacromolecules (alkaline phosphatase, ALP) can also be encapsulated into the aqueous lumen of the polymersomes. Significantly, the pH-triggered degradation of polymersomes could also regulate the release of encapsulated ALP.

Keywords: Polymersomes, pH-responsive, amphiphilic diblock copolymer, drug delivery

Acid-disintegratable polymersomes for intracellular drug delivery

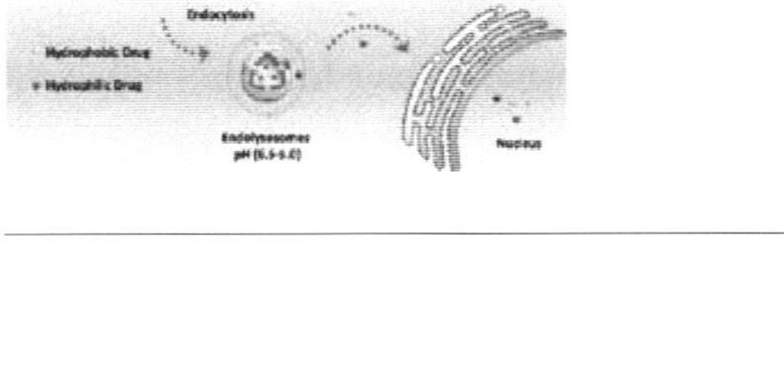

MBM-O-004 [Macromolecules in Biotechnology & Medicine]

Block Copolymer Micelles and Nanoparticles for Drug Delivery and Imaging

Paschalis Alexandridis[1], Yumiao Zhang[1], Jonathan F. Lovell[2]

[1]Department of Chemical and Biological Engineering, University at Buffalo, The State University of New York (SUNY), Buffalo, New York, USA
[2]Department of Biomedical Engineering, University at Buffalo, The State University of New York (SUNY), Buffalo, New York, USA

Block copolymers of the poly(ethylene oxide)-poly(propylene oxide) (PEO-PPO) family, commercially available as Pluronics or Poloxamers and approved for pharmaceutical use [Gu & Alexandridis, J. Pharm. Sci. 2004, 93 (6), 1454-1470; DOI: 10.1002/jps.20021], offer prime examples of self-assembling systems that find diverse applications. The presentation will utilize research findings from our group and others to highlight (i) fundamental aspects of Pluronic block copolymer micellization [Kaizu & Alexandridis, Colloids Surfaces A: Physicochem. Eng. Aspects 2015, 480, 203-213; DOI: 10.1016/j.colsurfa.2014.10.061] and gel (lyotropic liquid crystal) formation and structure [Sarkar et al., Macromol. Chem. Phys. 2012, 213 (23), 2514-2528; DOI: 10.1002/macp.201200438] in aqueous media, and (ii) applications of Pluronics in emulsion [Kaizu & Alexandridis, J. Colloid Interface Sci. 2016, 466, 138-149; DOI: 10.1016/j.jcis.2015.10.016], drug delivery [Zhang et al., Nature Commun. 2016, 7, in press], and imaging [Zhang et al., Nature Nanotech. 2014, 9 (8), 631-638; DOI: 10.1038/nnano.2014.130] formulations. Notable is the ability of Pluronic block copolymers to co-assemble with hydrophobic compounds in aqueous media and form "frozen" micellar nanoparticles having high cargo contents and excellent stability. Such frozen micelles incorporating naphthalocyanine dyes (Nanonaps) enabled real-time functional imaging of the gastrointestinal tract following oral administration in mice [Zhang et al., Nature Nanotech. 2014, 9 (8), 631-638; DOI: 10.1038/nnano.2014.130]. Further, Pluronic block copolymers, but not other surfactants, formed together with hydrophobic drugs (e.g., phylloquinone, cyclosporine, testosterone undecanoate, cabazitaxel) surfactant-stripped induced frozen micelles (ss-InFroMs) as a new class of drug delivery vehicles suitable for delivering a wide range of hydrophobic cargo [Zhang et al., Nature Commun. 2016, 7, in press]. By lowering the temperature to modulate the aqueous solubility of Pluronic, loose and free block copolymer molecules can be stripped away, leaving behind concentrated frozen micelles with extremely high drug-to-excipient ratios.

Keywords: Pluronic, poly(ethylene oxide), excipient, self-assembly, formulation, photoacoustic tomography

MBM O 005 [Macromolecules in Biotechnology & Medicine]

Polyhydroxyalkanoate-based diblock and triblock copolymers: Chemical composition and macromolecular architecture, an easy way to tune the nanoparticles properties for drug delivery applications

Ghislaine Barouti[1], Christophe Chassenieux[2], Sophie Guillaume[1]

[1]Institut des Sciences Chimiques de Rennes, Rennes, France
[2]Institut des Matériaux et Molécules du Mans, Le Mans, France

Poly(hydroxyalkanoate)s (PHAs) are renewable, biocompatible and biodegradable aliphatic thermoplastics which can be found as natural or synthetic polymers. These polyesters are promising biomaterials candidates (J.Nicolas et coll., *Chem. Soc. Rev.*, **2013**, 42, 1147).

Original PHAs based diblock and triblock copolymers, namely poly(β-malic acid)-(b)-poly(3-hydroxybutyrate) (PMLA-(b)-PHB) and PMLA-(b)-PHB-(b)-PMLA respectively, have been synthesized. The controlled sequential ring-opening polymerization of β-butyrolactone and benzyl β-malolactonate (MLABe) has been achieved through different catalytic approaches allowing the precise control of the hydrophilic weight fraction of the copolymers (G.Barouti et coll, *Polym.Chem.*, **2015**, 6, 5851-5859). Well-defined copolymers in terms of molar mass, dispersity, nature of end-chain groups and block structures were obtained (G.Barouti et coll., *Polym.Chem.*, **2015**, 6, 5414-5429; ibid manuscript submitted).

The control of the hydrophilic weight fraction ({f}: 11-82%) and the macromolecular architecture (diblock vs. triblock) of these copolymers provide a viable route to tune the physico-chemical properties of the resulting nanoparticles. Various self-assembled structures such as polymerzomes and micelles with tunable hydrodynamic radius (Rh: 16-300 nm), molecular weight and percentage of extension of the hydrophilic block (10-100%) were thus obtained as investigated by dynamic and static light scattering measurements.

Cell viability of these nanoparticles assessed from MTT assays on HepaRG cells evidenced no cytotoxic effect at different polymer concentrations (up to 88 µg/mL) (G.Barouti et coll., *Polym.Chem.*, **2015**, 6, 5414-5429). All these results demonstrated that these PHB-based copolymers are promising candidates as drug-delivery systems.

The Fondation pour la Recherche Médicale (FRM) is gratefully acknowledged for the financial support of this research.

Keywords: Poly(hydroxyalkanoate) (PHA), Poly(3-hydroxybutyrate) (PHB), Poly(malic acid) (PMLA), macromolecular architecture, nanoparticle, cytotoxicity

PHB based copolymers: From synthesis to well defined self-assembled structures

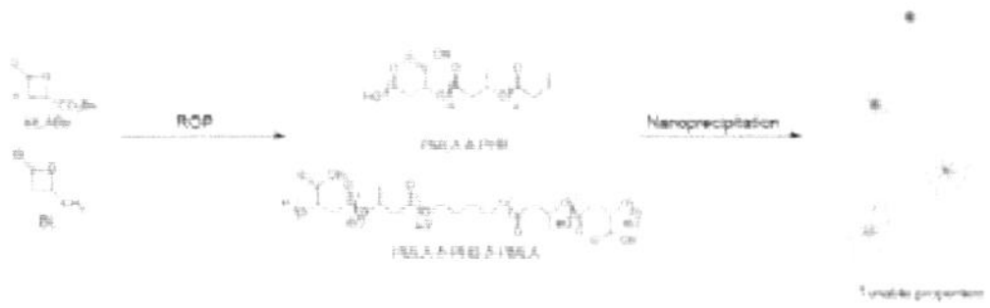

MBM-O-008 [Macromolecules in Biotechnology & Medicine]

Synthesis and Development of Microporous Organic Polymeric Capsules as a Drug Carrier

Shumaila Razzaque

Huazhong university of science and technology, Wuhan, China

Highly facile and commodious methodology was envisioned to synthesize the highly porous microspherical organic capsules (HMOCs). The presence of excessive carboxylic functionality on capsule surface reinforced its chemical and colloidal stability. On the basis of their hollow and porous architecture, they incorporate the macroscopic properties of nanocapsules as well as advantages of microscopic organic polymers. As the small pores in the range of few nanometers can affect matrix hydration, drug diffusion into the release medium and polymer degradation. Due to their porous structure they can effectively encapsulate the small molecules in nano-meter range and efficiently control their slow release. For instance, Doxorubicin, an extensively applied chemotherapeutic drug, has the dose-dependent toxic side effects, which limits it's efficacy of therapy applied to cancer patients that include bone marrow suppression and cardiotoxicity. Also, the stability of doxorubicin is a crucial issue, as doxorubicin metabolized very fast to inactive derivatives mainly by liver after an injection into the blood stream. To overcome such obstacles for its applicability, capsule systems are particularly appealing for controlled and long time delivery. Using capsules, a drug, nutrient, or healing agent is delivered without requiring modification and any external stimulus. These multifunctional HMOCs possessing zero order drug release kinetics and fluorescence properties, indicate their promising prospect for bio-medical applications.

Keywords: Hollow microporous organic capsules, Pore size, drug carrier, slow release

MBM-O-006 [Macromolecules in Biotechnology & Medicine]

Synthesis and characterization of psyllium-g-poly(acrylamide-co-acrylic acid) through microwave route for anticancer treatment

Pramendra Kumar[1], Deepak Kumar[2], Nida Khan[1], Jyoti Pandey[2]

[1]*Applied Chemistry Department, M J P Rohilkhand University, Bareilly (UP), India*
[2]*Applied Chemistry Department,BBAU, Central University, Lucknow(UP), India*

Microwave (MW) assisted synthesis of Psyllium (Psy) grafted poly (acrylamide-co-acrylic acid) (AAm-co-AA) a co-polymer of wide applications is being reported in the current work. Graft rapid co-polymerization of the psyllium (Psy) with acrylamide-co-acrylic acid (AAm-co-AA) under the influence of MW without any radical initiator/catalyst, resulted in grafting yields comparable to redox (ceric ammonium nitrate/Ascorbic acid) initiated in conventional heating graft co-polymerization. The effect of monomer concentration, initiator concentration, reaction time and temperature in terms of percent grafting has been investigated. The grafting efficiency up to ~40% was further increased when initiators and catalyst were used by MW heating. The effect of MW power and exposure time on the grafting yields also studied, both in presence and in the absence of redox initiator and catalyst were studied. Maximum grafting efficiencies achieved under MW ware 98 % in 90 seconds (in comparison to 54 % in 90 min by the conventional method). A representative MW grafted copolymer was characterized using IR, SEM, XRD, DSC and TGA, studies are further going for the potential application of colon cancer treatment.

Keywords: Psyllium, grafting, mocrowave, cancer

MBM-O-009 [Macromolecules in Biotechnology & Medicine]

Platelet adhesion and protein adsorption on heparin immobilized graphene enhanced poly(acrylonitrile-co-ethyl methacrylate) copolymer membranes

Tuğba Endoğan Tanır[1], Elif Vargün[2], Ayşegül Yıldız Ünal[3], Nesrin Hasırcı[4,5]

[1] *Middle East Technical University, Central Laboratory, 06800, Ankara, Turkey*
[2] *Mugla Sitki Kocman University, Department of Chemistry, 48000, Mugla, Turkey*
[3] *Mugla Sitki Kocman University, Department of Molecular Biology and Genetics, 48000, Mugla, Turkey*
[4] *Middle East Technical University, Chemistry Department, 06800, Ankara, Turkey*
[5] *BIOMATEN-CoE on Biomaterials and Tissue Engineering, 06800, Ankara, Turkey*

1. Introduction

Polyacrylonitrile (PAN) and its copolymers have been widely used in biosensor production and hemodialysis membranes. However, relatively poor hydrophilic surface of PAN may cause poor biocompatibility for blood contacting materials such as hemodialyzers and catheters. To improve hemocompatibility of these products, the surfaces can be modified by using some biomolecules like heparin which is a well-known anticoagulant [1]. In this study, poly(acrylonitrile-co-ethyl methacrylate) (P(AN-co-EMA)) copolymers were synthesized in two different molar ratios (92/8, P(AN92-co-EMA8) and 84/16, P(AN84-co-EMA16)) by emulsion polymerization. Reduced graphene oxide was added (at 0.5%wt) and composite membranes were prepared. Heparin was immobilized on the membranes and hemocompatibilities were examined in terms of platelet adhesion and plasma protein adsorption tests.

2. Experimental Procedure

Reduced graphene oxide (rGO) was synthesized by modified Hummers' method and characterized by XRD and TEM. rGO loaded P(AN-co-EMA) composite membranes were prepared and evaluated for their mechanical properties and hydrophilicity. Heparin was immobilized on the membranes using N-3-(dimethyl-aminoproply)-N-ethylcarbodiimide hydrochloride (EDC)

[2]. The membranes were analyzed by XPS to determine the degree of heparin attachment. The hemocompatibility of the membranes were analyzed by platelet adhesion and plasma protein adsorption studies.

3. Results and Discussion

Addition of rGO increased the mechanical properties of membranes. 16% and 36% increase in Young's modulus values were observed for P(AN92-co-EMA8) and P(AN84-co-EMA16) membranes, respectively.

Heparin was immobilized on membranes and these membranes were treated with platelet rich plasma (PRP) for 2 h. SEM results indicated that presence of heparin declined the platelet adhesion significantly and improved hemocompatibility of the membranes (Figure 1). Also, heparin immobilization has a strong effect on protein adsorption and the amount of adsorbed protein on heparinized membranes was less than the corresponding non-heparinized ones.

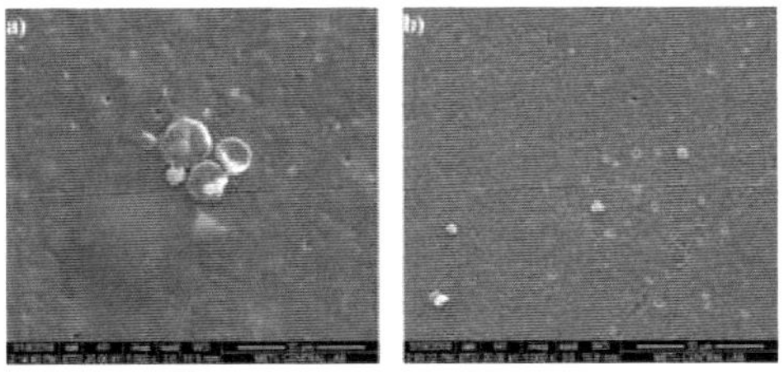

Figure 1 SEM micrographs after PRP incubation

a) P(AN92-co-EMA8) and b) Hep-P(AN92-co-EMA8)

4. References

[1] M.C. Yang, W.C. Lin, Journal of Polymer Research, 9, 201-206, 2002.
[2] T.Y. Liu, W.C. Lin, L.Y. Huang, S.Y. Chen, M.C. Yang, Biomaterials, 26, 1437-1444, 2005.

Keywords: polyacrylonitrile, heparin, membrane, hemocompatibility

MBM-O-010 [Macromolecules in Biotechnology & Medicine]

Measuring the Growth Kinetics of Layer-by-Layer Assemblies of Organic Nanoparticles and Polyelectrolytes

Maziar Mohammadi[1], Ali Salehi[2], Lucas Cygan[2], Ryan Branch[2], Cagri Besirli[3], Ronald Larson[2]

[1]*Department of Mechanical Engineering, University of Michigan, Ann Arbor, United States*
[2]*Department of Chemical Engineering, University of Michigan, Ann Arbor, United States*
[3]*Department of Ophthalmology and Visual Sciences, Kellogg Eye Center, University of Michigan, Ann Arbor, Michigan, 48105, USA*

We systematically measure the growth rate of layer-by-layer (LbL) assemblies of polyelectrolytes (PEs) with oppositely charged polystyrene (PS) nanoparticles (NPs) as a function of molecular weight (MW) of PEs, and NP size and charge. Both linear and exponential growth behaviors are observed, and under optimal conditions films of up to around 1 micron thickness can readily be assembled after 10 or so bilayers have been deposited. For many of the cases studied, an intermediate MW of PE leads to the fastest film buildup, for both cationic poly(ethyleneimine) deposited alternately with anionic PS NPs and for anionic poly(acrylic acid) deposited alternately with cationic PS NPs. The existence of an optimal MW suggests that growth rate is determined by a balance of thermodynamic factors including density of polymer bridges between particles, and kinetic factors, specifically the diffusivity of polymer in the film. The optimal MW, however, is very sensitive to the materials used. Finally, we characterize the surface morphology of the films with atomic force microscopy and scanning electron microscopy and find that the roughness increases less than linearly with film thickness. This work is expected to facilitate the design of thin films for emerging applications such as drug delivery.

Keywords: Growth kinetics, layer-by-layer (LbL) assembly, polyelectrolytes (PEs), nanoparticles (NPs), and molecular weight (MW).

MBM-O-011 [Macromolecules in Biotechnology & Medicine]

Synthesis of fluorescent peguilated lactide copolymers by Ugi four-component condensation

Marcos Lopes Dias[1], Luis Peña Icart[1], Lissette Aguero[3], Jose Ramon[3], Leonardo Andrade[4], Dionisio Zaldivar[2]

[1]*Universidade Federal do Rio de Janeiro, Instituto de Macromoléculas Professora Eloisa Mano, Rio de Janeiro, Brazil*
[2]*Universidad de la Habana, Facultad de Química, Havana, Cuba*
[3]*Universidad de la Habana, Centro de Biomateriales, Havana, Cuba*
[4]*Universidade Federal do Rio de Janeiro, Instituto de Bioquímica Médica, Rio de Janeiro, Brazil*

Ugi four-component condensation (UFCC) is a multicomponent reaction in which an amine, an aldehyde, a carboxylic acid, and an isocyanide react forming a bisamide in a one-pot reaction. In this contribution we report the use of UFCC for the synthesis of fluorescent Pegylated PLA block copolymers, and the application of these copolymers in the preparation of fluorescent microspheres. First functionalized poly(L-lactide) (end-activated PLA) was synthesized by L-lactide bulk polymerization using two coinitiators: salicylaldehyde and salicylic acid. Both PLACOOH and PLACOH presented a higher degree of crystallinity and Tg and Tm lower than that of non-functionalized PLA. Fluorescent block copolymers (FCP) was synthesized by UFCC. The method was also used to pegylate and to introduce fluorescein into the end-activated PLA. Two block copolymers were used: one using PEG–bisamine, carboxylic acid end-terminated PLA, aldehyde–fluorescein and tert-butyl isocyanide (FCP1); and another using PEG–bisamine, aldehyde-end-terminated PLA, acid fluorescein and tert-butyl isocyanide (FCP2). Pegylation by UFCC was confirmed by 1H-NMR analyses. Unimodal GPC peaks, relatively low polydispersities and higher molecular weight of FCPs as compared with the end-activated PLA confirmed also the formation of block copolymers. Fluorescent microspheres were then prepared from these copolymers by using a single emulsion–solvent evaporation methodology. Microparticles obtained by both copolymers FCP1 and FCP2 showed diameters with average sizes of around 13 mm, which allow application of these particles in parenteral, subcutaneous and intramuscular administration. The inclusion of fluorescein in these particles also could allow biodistribution studies. A kinetic study of drug release in vitro using HPLC showed a prolonged and controlled release of paclitaxel. Anticancer activity against large intestine cancer cell line (CACO2) was determined using the cell viability assay. Paclitaxel-loaded microspheres were able to inhibit cancer cell growth and colony formation.

Keywords: pegylation, PLA, fluorescein, copolymers, Ugi

MBM-O-012 [Macromolecules in Biotechnology & Medicine]

Surface-attached coatings on biological tissues and their potential in glaucoma treatment

Thananthorn Kanowijitsilp[1], Oswald Prucker[1], Alexandra Anton[2], Jürgen Rühe[1]

[1]*IMTEK – Department of Microsystems Engineering, University of Freiburg, Freiburg, Germany*
[2]*Eye Center, Medical Center - University of Freiburg, Faculty of Medicine, Freiburg, Germany*

Glaucoma is one of leading causes for blindness in the western world. The disease is related to a too high pressure in the eye. This pressure can be reduced if a small window into the eye is opened underneath a flap cut into the sklera. As long as this window stays open, fluid can flow out of the eye to reduce internal pressure. In many cases, however, scar formation leads to a too quick closure of the opening. Our approach to avoid scar formation is to coat the inner side of the flap with a thin layer of a hydrogel which repels proteins and, as a consequence, also strongly reduces the adhesion of tenon fibroblasts responsible for wound healing. So far, we have developed a technique to generate such layers on sklera tissue (obtained from pigs) using photoreactive polymers which upon illumination form bonds to neighboring chains as well as to the underlying collagenous tissue via simple C,H insertion crosslinking reactions (CHic).

These polymers are applied to the sklera as thin preprocessed pads using a carrier from which they can be released under moist conditions. The photoreactive group chosen for crosslinking and surface attachment reacts rather quickly if irradiated with UV light near the visible range. Even illumination times around one minute using lamps with rather low output are sufficient to link the gels to the tissues. These conditions did not damage the surrounding tissue to a noticeable extent. The layers on the tissue are stable in aqueous environments and repel tenon fibroblasts even for prolonged time spans.

Keywords: Surface-attached polymer networks, C-H-insertion crosslinking, glaucoma, sklera

MBM-O-013 [Macromolecules in Biotechnology & Medicine]

Multi-Functional Fluorescent Scaffold as a Multicolor Probe: Design and Application in Targeted Cell Imaging

Melis Kesik[1], Bilal Demir[2], Fırat Barış Barlas[2], Caner Geyik[3], Sevki Can Cevher[1], Dilek Odaci Demirkol[2], Suna Timur[4], Ali Cirpan[5], Levent Toppare[5]

[1]*Department of Chemistry, Middle East Technical University, 06800 Ankara, Turkey*
[2]*Department of Biochemistry, Faculty of Science, Ege University, 35100 Bornova-Izmir, Turkey*
[3]*Institute on Drug Abuse, Toxicology and Pharmaceutical Science, Ege University, 35100 Bornova-Izmir, Turkey*
[4]*Institute on Drug Abuse, Toxicology and Pharmaceutical Science and Department of Biochemistry, Faculty of Science, Ege University, 35100 Bornova-Izmir, Turkey*
[5]*Department of Chemistry and The Center for Solar Energy Research and Applications (GUNAM), Middle East Technical University, 06800 Ankara, Turkey*

Development of fluorescent probes have attracted great interest for diagnosis of cancer. The materials should be highly fluorescent, photostable, available for bioconjugation, have good biocompatibility, maximum spatial resolution, and minimal perturbation to biological systems. There are numerous organic and inorganic molecules; however, each class have excellent properties yet together with important drawbacks. This has motivated us to investigate a totally new strategy for the development of fluorescent probes for effective targeting and imaging. The newly proposed probe is developed as a multifunctional scaffold for the cellular imaging system. Such scaffold exhibits multicolor properties after combined with a targeting moiety (anti-CD44). The target design has to satisfy several requirements. Multiple materials are incorporated on the targeted probe showing their own benefits on the same platform. For this purpose, a fluorescent and a functional monomer, 3-(1H-phenanthro[9,10-d]imidazol-2-yl)phenol (PIP) and an antibody labeling kit (CF555) were merged on the same scaffold to develop the bioprobe. This proposed structure emits at dual wavelengths with no quenching in fluorescence. Also, pendant alcohol groups in the structure of PIP enable covalent attachment to labeled protein; CF555/anti-CD44 to enhance the biological activity and specificity towards the target. After combining with targeting moiety, the bioconjugate was characterized, tested for in vitro studies, and the cellular internalization was monitored in intact cells via fluorescence microscopy technique. Expression of the CD44 receptor in U87-MG cells (positive control) and HaCaT cells (negative control) was confirmed by flow cytometry. The results show that the bioconjugate, developed by this strategy, attached specifically to U87-MG cells with high efficiency. Such probe design for cellular imaging opens a new perspective for further improvement in fluorescence toward imaging.

Keywords: Fluorescent probe, cell imaging, multicolored, targeting

TOC

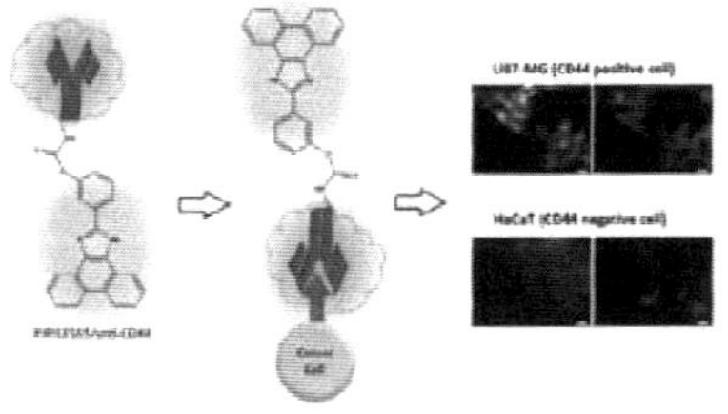

MBM-O-014 [Macromolecules in Biotechnology & Medicine]

Elastin-like polypeptide fusion of Interferon-alpha with enchanced pharmacokinetics profiles and superior antitumor efficacy

Jin Hu, Weiping Gao

Department of Biomedical Engineering, School of Medicine, Tsinghua University, Beijing, China.

Conjugating with poly(ethylene glycol) or human serum albumin is often used to increase the circulating half-lives of therapeutic proteins such as interferons, but these conjugation strategies usually lead to a heterogenous mixture of positional isomers or a fusion protein with unacceptably reduced bioactivity and low yield. To address these problems, we report a new and genetically encoded protein delivery platform, elastin-like polypeptide fusion (ELPfusion), to prolong the circulating half-life of a protein drug. We demonstrate proof-of-concept of ELPfusion by genetic fusion of an important therapeutic protein -interferon alpha (IFN-α)- at its C-terminus to an elastin-like polypeptide (ELP) to form a well-defined IFN-ELP fusion protein with high yield and retained bioactivity. Notably, the circulating half-life of the IFN-ELP protein (8.6 h) was 28-fold longer than that of IFN-α (0.3 h). In a murine cancer model, the IFN-ELP protein showed a 65-fold increase in tumor accumulation as compared to IFN-α 6 h post administration. Furthermore, it completely inhibited tumor growth without appreciable toxic effect, while IFN-α was just slightly effective at the same dosing. These findings show that ELPfusion may provide a new solution to precisely design protein conjugates with high yield, well retained bioactivity and significantly improved pharmacological profiles.

Keywords: interferon, elastin-like polypeptide, protein delivery, pharmacokinetics, anti-tumour effciency

The synthesis and in vivo antitumor efficacy of IFN-ELP

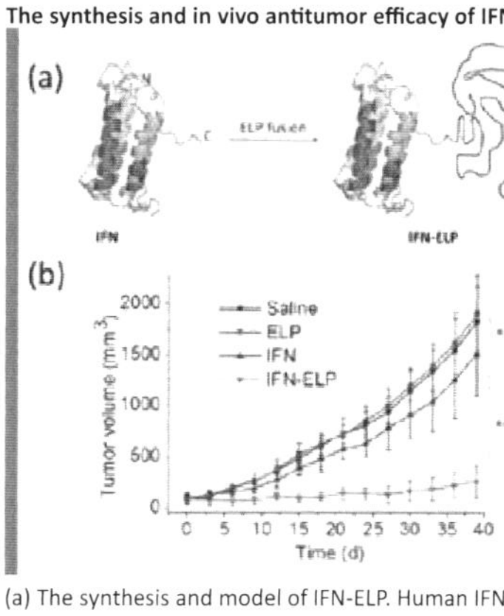

(a) The synthesis and model of IFN-ELP. Human IFN was genetically fused at its C-terminus to an ELP with the sequence of (VPGXG)90 to form an IFN-ELP fusion protein. (b) Inhibition of tumor growth post-treatment. ELP did not show any antitumor efficacy as compared to saline. IFN slightly inhibited tumor growth; in contrast, IFN-ELP almost completely inhibited tumor growth.

MBM-O-016 [Macromolecules in Biotechnology & Medicine]

How mucin structures the mucus barrier

Peter Charles Griffiths

University of Greenwich

Mucus is a dynamic gel that is continuously secreted, shed and digested. It is also a formidable barrier that protects many delicate surfaces from external insults, as well as limits the uptake of therapeutic modalities. The in-built mucus clearance mechanism removes putative drug carriers, such as nanoparticles, before the therapeutic agent reaches the underlying absorptive cells and enters the systemic circulation.

The structure and viscoelastic properties of mucus are determined by a matrix of glycosylated protein chains called mucins that trap and restrict foreign particles. In order to rationally design nanoparticles to penetrate this complex, heterogeneous matrix, a detailed understanding of its structure is required. Small-angle neutron scattering, rheology and spin-resonance spectroscopy has been used to quantify the structure and dynamics of mucus, of different sources, under a range of prevailing conditions. The mucus has been "deconstructed" - split into its constituent components - and their structure imposing characteristics quantified. One promising strategy to facilitate access to the underlying cell layer is a localised cleavage of the mucoglycoprotein substructures using proteases or other hydrophobic components. Changes in the same dynamic and structural properties induced by the addition of nanoparticles have been examined, coupling for the first time, a detailed insight into the structure-function of permeabilising agents. This insight will facilitate informed choice in the design of nanoparticles, and their surface decoration.

Keywords: mucin, mucus, scattering, rheology, spectroscopy

MBM-O-017 [Macromolecules in Biotechnology & Medicine]

New porous graphene oxide biomaterials with potential application in bone repair

Andreea Madalina Pandele, Sorina Alexandra Garea, Horia Iovu

University Politehnica of Bucharest

The idea of developing a biomaterial, appropriate for bone repair applications, with both osteoconductive and osteoinductive features is still a question of basic research.

Chitosan (CHT) is a promising material that encounters all the requirements for these applications, but as other polysaccharide shows low mechanical properties and poor procesability [P. A. L. Lima et. all, Mat. Sci. Eng. C-BIOMIM, 2013, 33, 3389]. To mitigate these drawbacks, graphene can be used as reinforcing agent. Assimilation of graphene into CHT matrix in order to expand the properties of the final material is a major challenge due to difficulty to achieve a good dispersion of the nanofiller within the biopolymer. In order to overcome this disadvantage the use of graphene oxide (GO), a graphene derivative, was proposed. The functional groups present on GO surface, epoxy(bridging oxygen atoms), carbonyl (C=O), hydroxyl (-OH) groups, bring an increase of the hydrophilic character to GO and favored interactions between the GO and polymer [W. Zhang et. all, Bulletion of Environmental Contamination and Toxicology, 2011, 87, 86]. The aim of this study is synthesis of composites scaffold based on CHT and different GO concentrations using freeze-drying method. Studies such as morphology, swelling, enzymatic biodegradation, biological tests and mineralization assays were performed.

Acknowledgement: This work was supported by a grant of the Romanian National Authority for Scientific Research, CNDI–UEFISCDI, project number PN-II-PT-PCCA-154/2012.

Keywords: graphene oxide, chitosan, bone repair

MBM-O-018 [Macromolecules in Biotechnology & Medicine]

Preparation and characterization of novel poly(proylene fumarate) (PPF) based scaffolds for bone tissue engineering

Erde Can[1], Görkem Cemali[2], Gamze Torun Köse[2]

[1]*Department of Chemical Engineering, Yeditepe University, İstanbul, Turkey*
[2]*Department of Genetic Engineering, Yeditepe University, İstanbul, Turkey*

In clinical treatment of bone damages, autologous and allogenous bone grafts or permanent prosthetic implants have been used. There are various limitations and disadvantages for these techniques and the creation of a biodegradable material which acts not only as a structural support to the damaged area but also acts as a scaffold that guides bone formation has emerged as a necessity to solve the problems associated with the present techniques. Polypropylene fumarate (PPF) is an unsaturated polyester which can be copolymerized with a vinyl monomer such as N-vinyl pyrolidone (VP) to give a biocompatible and biodegradable thermoset. There are several studies in literature on different compositions of PPF/VP and their composites with inorganic additives which gave promising results for their use as scaffolds for bone tissue engineering. In this study, the PPF was synthesized through a polycondensation reaction of propylene glycol and fumaric acid in an excess of propylene glycol. The PPF pre-polymer was then thermally cured with two different phosphonic acid based co-monomers, in the presence of benzoyl peroxide initiator, via radical polymerization at changing co-monomer weight ratios. The cure reactions were confirmed via FT-IR and DSC analysis and the degree of cross-linking was determined via solubility tests. Thermo-mechanical properties of the resulting polymers were determined via DMA, compressive properties were determined via compression tests, biodegradation rates in PBS buffer solution at 37°C were examined via both weight loss and pH measurements for 80 days. Properties like the equilibrium water content, and dynamic contact angle with water were also determined. In addition the biocompatibility of the resulting polymer samples was examined via MTS assay using human osteoblast cells. The effect of different co-monomers and the composition on various properties of the resulting PPF polymers were evaluated.

Acknowledgements: This project is funded by TUBITAK 1001 programme. (Project no: 114M195)

Keywords: Polypropylene fumarate, bone, tissue engineering, scaffold

Preparation of network polymers based on polypropylene fumarate and phosphonic acid based monomers

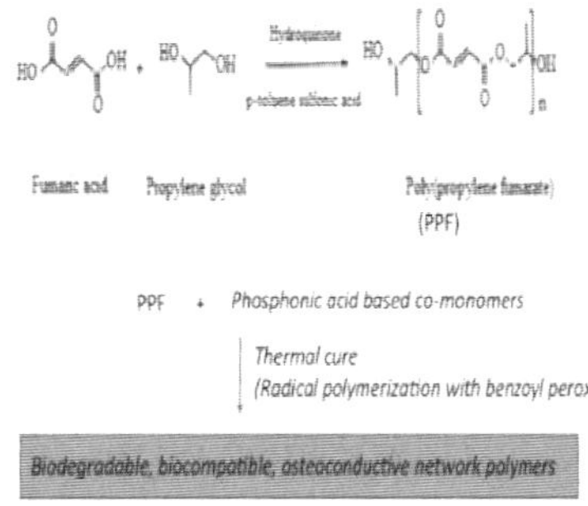

MBM-O-019 [Macromolecules in Biotechnology & Medicine]

Synthesis of Biodegradable PEG Based Poly(Substituted Glycolides)

Olcay Mert[1], Mehmet Onur Arıcan[2], Sezgi Erdoğan[2]

[1]*Department of Chemistry, Kocaeli University, Kocaeli, Turkey*
[2]*Department of Polymer Science and Technology, Kocaeli University, Kocaeli, Turkey*

Drug delivery is a new engineering technology that uses a simple method for administering a pharmaceutical compound with proper formulation to accomplish a better therapeutic effect than traditional methods in humans or animals. Biodegradable polymers can be designed as injectable systems to eliminate surgical procedures and provide efficient treatment directly at infectious site. PEG based polyesters such as poly(lactide) and poly(lactic-co-glycolic acid) (PLGA) are one of the most preffered injectable biodegradable polymers in biomedical field. In contrast, there have been limited reports that describe the use of glycolic acid derivatives as monomers for the preparation of glycolide family polymers. Therefore, synthesis, characterization and thermosensitive properties of PEG based novel substituted glycolic acid polymers were examined for the use of Drug Delivery Systems.

Acknowledgements. This study was granted by TUBITAK, Turkey, with project number 112T865.

Keywords: Drug Delivery, Substituted Glycolides, PEG

MBM-O-020 [Macromolecules in Biotechnology & Medicine]

Development of block copolymer based multifunctional nano-assemblies for a theranostic approach

Vusala Ibrahimova[1], Nathan D. Mcclenaghan[3], Sébastien Lecommandoux[2], Charles André Fustin[1]

[1]*IMCN, Université catholique de Louvain, Louvain-la-Neuve 1348, Belgium*
[2]*LCPO, Université de Bordeaux/CNRS, Pessac 33607, France*
[3]*ISM, Université Bordeaux/CNRS, 351 cours de la Libération, Talence Cedex 33405, France*

In our work, we developed PTS (aza-BODIPY) functionalized polypeptide based remote activated theranostic nano-assemblies [1] to be used in PDT. The singlet oxygen generation and fluorescence property of the aza-BODIPY provides simultaneous diagnosis and therapy of the tumor. Photodynamic therapy (PDT) is based on photoactivation of a photosensitizer (PTS) by near-infrared light to generate toxic reactive oxygen species (type I activation) or singlet oxygen (type II activation).[2] An asymmetric and multifunctional derivative of the aza-dipyrromethene boron difluoride chelate (aza-BODIPY) fluorophore was synthesized to be used as a photosensitizer due to its nontoxic properties, insensitivity to external biological environment,[3] high singlet oxygen generation and fluorescent quantum yield.[3b] To carry the photosensitizer to the tumor, amphiphilic block copolymers consisting of poly(γ-benzyl-L-glutamate) (PBLG, DP~50) and poly(ethylene oxide) (PEO, DP=45 and 113) chains, able to self-assembled into micelles and vesicles, will be used. We developed a synthetic strategy allowing covalent linkage of PTS to the amphiphilic block copolymers, thus preventing PTS leakage before the nano-assembly reaches the tumor site. Moreover, we investigated PTS activity dependence from concentration, morphology of the nano-assemblies and PTS localization on the nano-assemblies.

References

1. H. De Oliveira, J. Thevenot and S. Lecommandoux, Wiley Interdiscip. Rev. Nanomed. Nanobiotechnol. 2012, 4 (5), 525-46
2. H. Ding, H.Yu, Y. Dong, R. Tian, G. Huang, David A. Boothman, Baran D. Sumer and J. Gao, J. Control. Release, 2011, 156 (3), 276-80.
3. (a) A. Gorman, J. Killoran, C. O'Shea, T. Kenna, William M. Gallagher and Donal F. O'Shea, J. Am. Chem. Soc. 2004, 126 (34), 10619-31; (b) P. Batat, M. Cantuel, G. Jonusauskas, L. Scarpantonio, A. Palma, Donal F. O'Shea and Nathan D. McClenaghan, J. Phys. Chem. A, 2011, 115 (48), 14034-9.

Keywords: Polypeptide, self-assembly, photosensitizer, polymer nanoparticles, photodynamic therapy

Figure 1: Schematic representation of the project

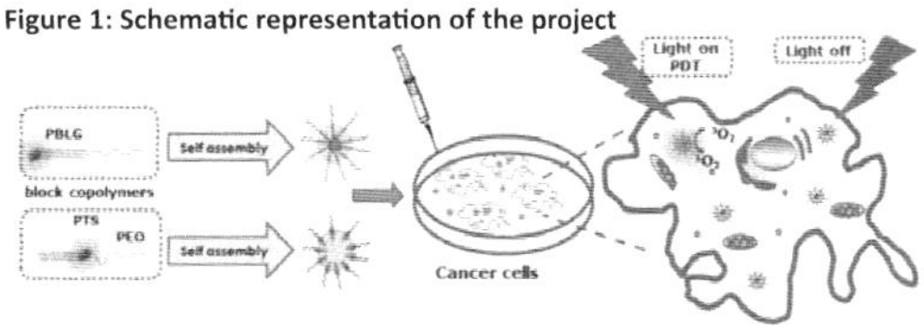

MBM-O-021 [Macromolecules in Biotechnology & Medicine]

Intrinsically antifouling ionically cross-linked silicones

Kate E Belsey[1], Viktorija Makarovaite[2], John C. Batchelor[3], Campbell W. Gourlay[2], Simon J. Holder[1]

[1]*Functional Materials Group, School of Physical Sciences, University of Kent, Canterbury, CT2 7NH, UK*
[2]*Kent Fungal Group, School of Biosciences, University of Kent, Canterbury, CT2 7NJ, UK*
[3]*School of Engineering and the Digital Arts, University of Kent, Canterbury, Kent, CT2 7NT, UK*

Biofilm growth (particularly that of *Candida albicans*) on silicone voice prostheses can have significant post-operative implications for patients, often requiring multiple further surgeries and frequent prostheses replacements. A number of modifications to silicone surfaces have been shown to reduce biofilm growth, but significant difficulties have been encountered with each potential approach [Talpaert et al., *J Med Microbiol.*, 2015, 64, (3)].

Here we present a new type of silicone elastomer, formed by the cross-linking reaction between an alkyl-halide terminated siloxane and the alkyl-amine branches of another siloxane. By modifying the ratios of these two components, and thus leaving different proportions of amine groups unreacted, it was possible to change the properties of the elastomer. The fewer the cross-linked groups, the greater the mobility of the chains within the siloxane and the greater the rearrangement potential. The distribution of the halide functional groups was seen (using SEM elemental mapping techniques) to be uniform throughout the polymer regardless of the proportions of components. Results of mechanical, calorimetric and wettability measurements will also be presented for these novel materials.

The elastomers were subsequently shown (both by halo and viability tests) to have inherent antifouling and antimicrobial properties with regards to *Candida albicans*, and preliminary data has been collected for other common biofilm components, such as *Streptococcus* and *Staphylococcus* species. The reduction in biofilm growth in comparison to standard poly(dimethyl siloxane) (PDMS) was seen to be proportional to the number of amine groups left unreacted within the network, likely due to the amine and ammonium groups being able to rearrange themselves towards the surface. The ease of synthesis of this elastomer together with its mechanical and intrinsic antifouling properties highlight its potential to be used in prostheses whilst also reducing some of the existing problems.

Keywords: Silicone, prosthetic, biofilm, antifouling

Figure 1

(a) schematic of the siloxane network, with free amine groups and ammonium cross-linking units; and (b) plot showing reduced biofilm growth with increased amine groups.

MBM-O-022 [Macromolecules in Biotechnology & Medicine]

Ultrafast tetrazine ligation for preparation bio-functional polymers

Muhammet Ubeydullah Kahveci[1], Maltish M. Lorenzo[2], Caitlin G. Decker[2], Samantha J. Paluck[2], Mustafa Yasin Ates[1], Ahmet Cetinkaya[1], Heather D. Maynard[2]

[1]Faculty of Chemical and Metallurgical Engineering, Department of Bioengineering, Davutpasa Campus, Yildiz Technical University, Esenler, 34220 Istanbul, Turkey

[2]Department of Chemistry and Biochemistry and California Nano Systems Institute, University of California, Los Angeles, 607 Charles E. Young Drive South, Los Angeles, California 90095-1569, United States

Demand on biologically functional polymers in biomedical applications such as biosensors, medical diagnostic devices, imaging and drug delivery systems have been getting more attention in recent years. The common approach for preparation of bio-active polymers is synthesis of a chemically functional polymers by controlled/living polymerizations and consequent modification of the polymers by ligation techniques. Super-fast inverse electron demand Diels-Alder (IEDDA) of tetrazine in ligation of biomolecules to polymers is expected to increase the yield of conjugation. Our recent studies involve preparation of homodimeric protein-polymer conjugates and hyperbranched multivalent protein-polymer conjugates. In the first study, tetrazine end-functionalized telechelic polymers were synthesized by controlled radical polymerization (CRP) and employed to generate T4 Lysozyme homodimers.1 Mutant T4 Lysozyme (V131C), containing a single surface-exposed cysteine, was modified with a protein-reactive trans-cyclooctene (T4L-TCO). RAFT polymerization yielded poly(N-isopropylacrylamide) (pNIPAAm) with an number average molecular weight of 2.0 kDa and a dispersity of 1.05. pNIPAAm was then modified at both ends by post-polymerization with 6-methyl tetrazine. For comparison, 2.0 kDa bis-tetrazine poly(ethylene glycol) (PEG) and 2.0 kDa bis-maleimide pNIPAAm were synthesized. Ligation of T4L-TCO to bis-tetrazine pNIPAAm or bis-tetrazine PEG resulted in protein homodimer in 38% yield and 37% yield, respectively, after only 1 hour, whereas bis-maleimide pNIPAAm resulted in 5% yield of dimer after 24 h (Figure). In the second study, which is still in progress, we have prepared tetrazine functionalized hyperbranched polymers by self-condensing vinyl polymerization of ATRP (SCVP-ATRP) and post-polymerization modification. The functional polymer will be utilized in a subsequent IEDDA with trans-cyclooctene labeled antibodies for preparation of multivalent antibody-polymer conjugates and evaluated for diagnostic capabilities in cancer.
This work has been financially supported by The Scientific and Technological Research Council of Turkey (TUBITAK, Project # 214M175)

Keywords: Tetrazine, trans-cyclooctene, protein-polymer conjugate, functional polymers, bioconjugation

Figure

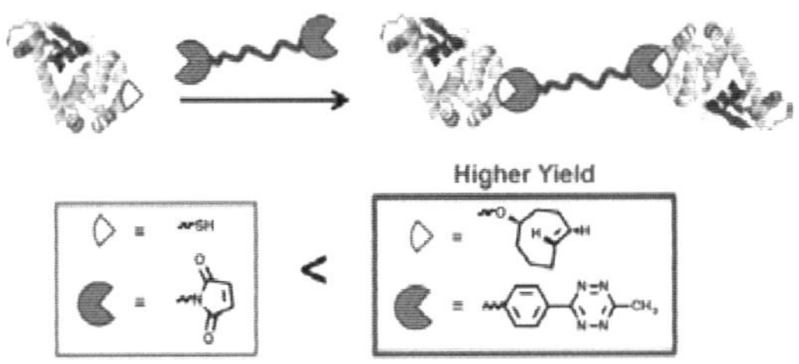

Preparation of homodimeric protein polymer conjugates by tetrazine chemistry

MBM-O-023 [Macromolecules in Biotechnology & Medicine]

Hydrophilic and Hydrophobic Balance in Tissue Adhesives

Sevgi Balcioglu, Burhan Ates, Suleyman Koytepe, Merve Goksin Karaaslan, Unzile Kelestemur, Selam Gulgen
Department of Chemistry, İnönü University, Malatya, Turkey

Biodegradable polymer adhesives should remain strongly attached to the underlying tissue in wet conditions[1]. Besides, tissue adhesives are required to perform a variety of functions: stop bleeding, seal leaks, be biodegradable, be biocompatible and/or facilitate healing [2]. One of the major problems of available tissue adhesives is poor adhesion in wet conditions. This problem is often caused by the partial hydrophilic structure of the adhesives. In order to provide adhesion, interaction is needed between adhesive and polar groups in the body. However, very high level of interaction limits the mechanic strength of adhesive. Therefore, polar-apolar group balance is necessary in the material. In our study, we used polar molecules to increase the interaction (PEG and chlorogenic acid (CLA)) as well as apolar molecules (tween 40, tween 60) to decrease high level of interaction in polymeric adhesives. Thus we planned to increase adhesion strengths in wet conditions. In the study, we carried out contact angle and adhesion tests to determine effect of hydrophilic-hydrophobic balance upon adhesion level of the materials.

Here, we synthesized the biocompatible and biodegradable polyurethane bioadhesives with good mechanical properties using 4,4'-methylenebis(cyclohexyl isocyanate) as aliphatic isocyanate source, PEG200, tween 40/60 and CLA as polyol source. The ratios of isocyanate/polyols were; 1:1 and PEG200/tween 40 or 60(T40/60)/ CLA were; 80:15:5, 80:10:10 and 80:5:15, respectively. The synthesized polyurethanes were characterized by FTIR, TGA, DTA, DSC and dynamic contact angle (Table 1). *In vitro* biodegradability and biocompatibility tests were performed and these materials were tested with adhesion strength test with aluminum substrates and bovine muscle tissue. According our adhesion test results with aluminum substrates, the highest adhesion was 466.8±49.6 kPa in PU-T40-CLA-15% (Figure 1) polymer for 24 hour. Furthermore, the highest adhesion was 18±2.8 kPa in PU-T40-CLA-15% for bovine muscle tissue in adhesion tests.

References

1. Mahdavi et al. "A biodegradable and biocompatible gecko-inspired tissue adhesive". *PNAS*, 2008, 105, 2307–2312.
2. Mehdizadeh et al. "Design Strategies and Applications of Tissue Bioadhesives". *Macromol Biosci*, 2013, 13, 271–288.

-This study was supported by grants from TUBITAK (TBAG-114Z591).

Figure 1.

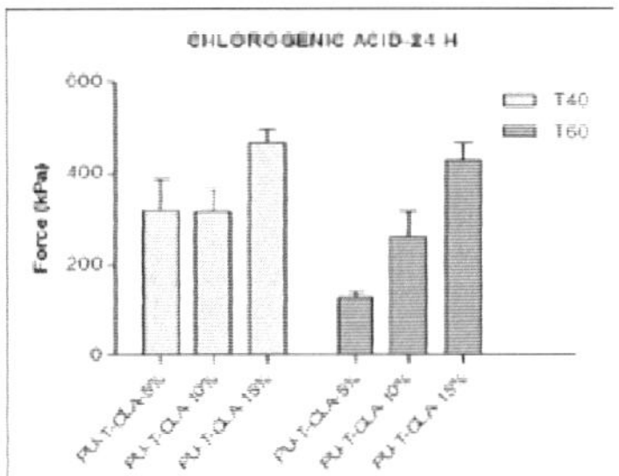

Adhesion strengths of synthesized polymers.

Table 1.

	T40	T60
PU-T-20%	63,38	66,15
PU-T-CLA-5%	59,67	64,97
PU-T-CLA-10%	57,59	60,77

PU-T-CLA-15%	53,59	59,08

Contact angles of synthesized polymers

Keywords: Tissue adhesive, tween, chlorogenic acid, hydrophilic, hydrophobic.

MBM-O-024 [Macromolecules in Biotechnology & Medicine]

Phosphonic Acid-Functionalized Acrylamides for Biomedical Applications

Hatice Betul Bingol[1], Ayse Altin[1], Tugba Bal[2], Sesil Agopcan Cinar[1], Seda Kizilel[2], Duygu Avci[1]

[1]Department of Chemistry, Bogazici University, Istanbul, Turkey
[2]Department of Chemical and Biological Engineering, Koc University, Istanbul, Turkey

Phosphorus-containing acidic monomers are able to interact with the inorganic phase of mineralized tissues such as enamel, dentin, and bone due to their affinity towards calcium ion. From this perspective, three phosphonic acid-containing acrylamide monomers were synthesized by the reaction of aminophosphonates with acryloyl chloride followed by the hydrolysis of phosphonate groups using trimethylsilyl bromide (J. Polym. Sci, Part A: Polym. Chem., 2015, 53, 2755–2767). As bone and dental tissues share some common structural features, these monomers are proposed to be used as a component in both self-etching dental adhesives and hydrogel scaffolds to be fabricated for mineralization of bone-like composites. To evaluate the potential of monomers to be used for self-etching dental adhesives, their interaction with hydroxyapatite, which is considered a model compound for bone tissue, were investigated by FTIR and XRD. Photopolymerization studies were performed by photo-DSC to explore the influence of monomer structure on polymerization rate and conversion. The monomers were found to be hydrolytically stable and their pH values were within the suitable range. To investigate the use of these monomers for tissue engineering applications, they were incorporated into hydrogel scaffolds. The hydrogels were characterized by swelling studies and SEM was used to explore their morphology. The phosphonic acid groups in the hydrogels acted as Ca^{2+} binding sites and induced mineralization upon immersion in simulated body fluid. Mineral deposits were observed in SEM micrographs, and energy dispersive X-ray spectroscopy was used to show that these minerals contain Ca and P. Ca^{2+} deposition assay was performed for the quantitative detection of Ca^{2+} in mineralized scaffolds. Viability of human dermal fibroblasts on hydrogels were also investigated. Altogether, these results suggest that the phosphonic acid-containing monomers are promising candidates to be used in fabrication of biomaterials for both dental and bone tissue engineering applications.

Keywords: phosphonic acid, photopolymerization, dental polymers, hydroxyapatite, hydrogels, mineralization

Phosphonic acid-functionalized biomaterials

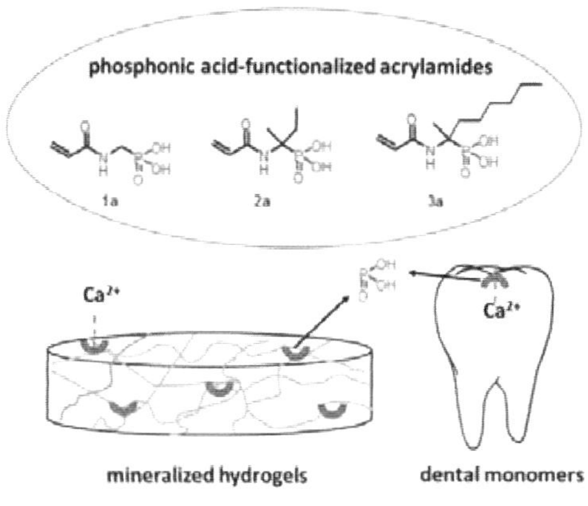

MBM-O-025 [Macromolecules in Biotechnology & Medicine]

Ti(IV) Attached-Polydopamine–Coated Magnetic SiO2 Microspheres for Enrichment of Phosphopeptides via Immobilized Metal Affinity Chromatography

Kouroush Salimi[1], Duygu Deniz Usta[2], Omur Celikbicak[3], Bekir Salih[3], Ali Tuncel[4]

[1]*Chemical Engineering Department, Hacettepe University, Ankara, Turkey*
[2]*Department of Medical Biology and Genetics, Gazi University, Ankara, Turkey*
[3]*Department of Chemistry, Hacettepe University, Ankara, Turkey*
[4]*Division of Nanotechnology and Nanomedicine, Hacettepe University, Ankara, Turkey*

Polydopamine (poly(dopa)) coated magnetic-monodisperse SiO2 microspheres functionalized with Ti(IV) groups Mag-SiO2@poly(dopa)-Ti4+ were first synthesized for the enrichment of phosphopeptides via immobilized metal ion affinity chromatography (IMAC). The as prepared materials were characterized using scanning electron microscopy (SEM), X-ray diffraction (XRD), vibrating sample magnetometer (VSM) and also using a surface area and pore size analyzer. The surface concentration of Ti(IV) groups and Ti(IV) functionalization conditions were optimized by protecting the saturation magnetization of starting material, Mag-SiO2@poly(dopa) microspheres. A good hydrophilic structure, high surface area, superior magnetic properties and high surface concentration of Ti(IV) chelates were achieved with the resulting microspheres. Mag-SiO2@poly(dopa)-Ti4+ microspheres were evaluated as IMAC sorbent for the enrichment of phosphopeptides from the digests of synthetic protein mixtures and human serum in batch-fashion operated based on magnetic separation principles. The selectivity of the method was investigated by matrix assisted laser desorption/ionization time-of-flight mass spectrometry (MALDI-TOF-MS). The detection limit for the enrichment of phosphopeptides by Mag-SiO2@poly(dopa)-Ti4+ microspheres from the tryptic products of β-casein was determined as 50 femtomols. Additionally, an ultrahigh selectivity in the mixture of β-casein and bovine serum albumin (BSA) prepared with different proportions (with the molar ratios of β-casein to BSA up to 1:500) was observed. Mag-SiO2@poly(dopa)-Ti4+ microspheres were also successfully used for phosphopeptide enrichment from human serum. The results indicated that as-prepared porous IMAC platform was a good candidate for the enrichment in the specific capture of phosphopeptides from a complex biological sample, human serum.

Keywords: SiO2 composite microspheres, Proteomics, Phosphopeptide Enrichment, Poly dopamine

MBM-O-026 [Macromolecules In Biotechnology & Medicine]

Area selective surface functionalization and enzyme immobilization in thiol-ene-epoxy microfluidic devices

Christian Hoffmann[1], Ines P Rosinha[3], Ulrich Krühne[3], Manuel Pinelo[2], John M Woodley[3], Anders E Daugaard[1]

[1]*Danish Polymer Centre, Department of Chemical and Biochemical Engineering, Technical University of Denmark, Søltofts Plads Building 229, DK-2800 Kgs. Lyngby, Denmark*
[2]*Center for BioProcess Engineering, Department of Chemical and Biochemical Engineering, Technical University of Denmark, Søltofts Plads Building 229, DK-2800 Kgs. Lyngby, Denmark*
[3]*CAPEC-PROCESS, Department of Chemical and Biochemical Engineering, Technical University of Denmark, Søltofts Plads Building 229, DK-2800 Kgs. Lyngby, Denmark*

The application of enzymes as green biocatalysts has demonstrated significant potential within a large variety of different industrial areas, such as the synthesis of chemical building blocks, pharmaceuticals and in food industry (DiCosimo et al. 2013). Enzyme immobilization is a well-known pathway within biocatalysis in order to improve the general stability of enzymes towards external conditions, for example temperature, pH and used solvents. The large variety of immobilization techniques, such as cross-linking of enzymes, entrapment or carrier binding has high potential for application in research scale, but also for industrial processes (Sheldon and van Pelt 2013). Binding to a carrier can be conducted either by adsorption or covalent bond formation on a solid support. In order to evaluate activity and stability of immobilized enzymes, microfluidic devices have drawn high attention. Their application capability directly as biocatalytic devices is an additional advantage. Recently, off-stoichiometric thiol-ene (OSTE) material was successfully exploited for the fabrication of microfluidic devices (Carlborg et al. 2011). This class of material provides beneficial mechanical properties combined with the possibility of tailored surface functionalization after device fabrication. In this study, we demonstrate the successful preparation of small scale microreactors from off-stoichiometric thiol-ene-epoxy (OSTE+) material. Subsequent area selective surface modification of remaining thiol groups on the microchip surface with a variation of allyl compounds via light induced thiol-ene reaction enabled the introduction of tailored functionality on the surface. By choosing various stencil masks for this procedure, different functional patterns could selectively be introduced on the surface. Epoxide functionality for example offered a pathway for covalently immobilization of horseradish peroxidase in desired areas of the microfluidic reactor. The impact of the enzymatically active pattern towards the overall conversion of the microreactor was evaluated by a model reaction in a continuous process.

Keywords: thiol-ene chemistry, surface functionalization, biocatalysis, enzyme immobilization

Figure 1

MBM-O-027 [Macromolecules in Biotechnology & Medicine]

Improving Antibacterial Efficacy of Chitosan Grafted Polyurethanes with Silver Nanoparticles

Eda Ayse Aksoy[1], Filiz Kara[2], Nesrin Hasirci[3], Serpil Aksoy[2]

[1]*Department of Basic Pharmaceutical Sciences, Faculty of Pharmacy, Hacettepe University, Ankara, Turkey*
[2]*Department of Chemistry, Faculty of Science, Gazi University, Ankara, Turkey*
[3]*Department of Chemistry, Faculty of Arts and Sciences, Middle East Technical University, Ankara, Turkey*

Biomaterials are essential substances used in the production of medical devices. In the cases where the sterilization is not proper, they carry some risk to cause infection at the implantation area. Therefore, modification of the surfaces of the biomaterials in the direction to prevent the attachment of bacteria can help to decrease infection problems. There are various modifications techniques, such as coating the surface with antibacterial agents like silver nanoparticles (AgNPs) or conjugation of antibacterial molecules such as chitosan (CH). The antibacterial spectrum of AgNPs is remarkably broad and includes both gram-positive and gram-negative bacteria. In this study, it was aimed to improve antibacterial efficacy of polyurethane (PU) membranes so that at the biointerfaces they will have enhanced biocompatibility. For this purpose, PU films were synthesized in medical purity by using toluene diisocyanate and polypropylene ethylene glycol without addition of any other ingredients. The samples were activated by glow-discharge oxygen plasma and CH was grafted on the films (PU-CH). PU-CH films were then treated with $AgNO_3$ and vitamin C solution to immobilize AgNPs (PU-CH-Ag). Properties of the surfaces of the films were investigated. XPS results showed the existence of Ag on the films at about 18%. SEM images identified the sizes of the silver particles ranging from 37 to 54 nm distributed on the PU-CH films (Fig. 1). Colony counting method results demonstrated PU-CH and PU-CH-Ag films had antibacterial effect against S. epidermidis after 24h inhibition. Bacteria suspension was tested as control. The obtained values were for the control and PU-CH films 6.5 and 4.8 log CFU/mL, respectively. By treatment with PU-CH-Ag films, all the bacteria were killed and no growth was observed. Results demonstrated that antibacterial efficacy of the interface of chitosan grafted polyurethane films were enhanced with the presence of silver nanoparticles.

Keywords: Surface Modification, Polyurethane, Chitosan, Silver Nanoparticles, Antibacterial Efficacy

Fig 1. SEM images of PU-CH and PU-CH-Ag (Scale bar 4µm)

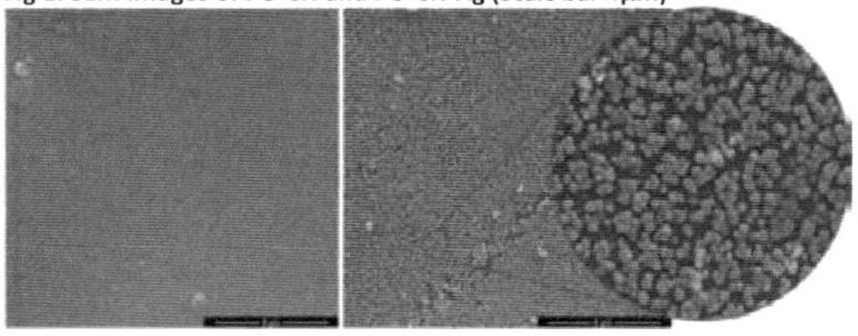

MBM-O-028 [Macromolecules in Biotechnology & Medicine]

New generation of antibiotics

Tarik Eren[1], Tanıl Kocagöz[2], Özlem Yalçın[3], Ceren Süer[1]

[1]Yıldız Technical University
[2]Acıbadem University
[3]Koc University

Infectious disease is a critically important global healthcare issue. For example, two million nosocomial infections occur in hospitals in the United States annually and add 5 billion dollars to healthcare costs. The rate of bacterial resistance, to even most powerful antibiotics, is increasing at an alarming rate. Hospitals are the main source and origin of multiple drug resistant bacteria and patients infected with these have clinical courses and they require prolonged hospital treatments. Pathogenic bacteria can easily deposit on the common objects such as door knobs, linen, public telephones, children's toys as well. When these surfaces are touched by infected people, then bacterial infections can be transferred. This is one of the primary pathways for the acquisition of nocosomial infectious pathogens. These ordinary materials are not antibacterial and they require modification. Increased effort has focused on materials that limit bacteria colonization, in other words, biofilm formation, on their surfaces in an effort to prevent the growth of bacteria and the transmission of infectious organisms. Polymeric antibacterial agents represent new developing direction in this field and offer great promise for enhancing activity and selectivity. Polymeric antibacterial agents represent new developing direction in this field and offer great promise for enhancing activity and selectivity. In this study, we synthesized and investigated the properties of phosphonium/pyridinium salts containing "well-defined" ROMP based polymers. Structure-activity relationship of the synthesized polymers has been investigated with real cell data, bacteria Staphylococcus aureus (ATCC 25923), Escherichia coli (ATCC 25922), Pseudomonas aeruginosa (ATCC 27853) and Mycobacterium tuberculosis (H37Ra). Biophysical technique (LIVE/DEAD Assay, Zeta potential etc) is also applied to get information about the mechanism in action.

Keywords: ROMP, antimicrobial polymers, phosphonium, pyridinium, hemolytic activity, structure-activity relationship

Figure 1

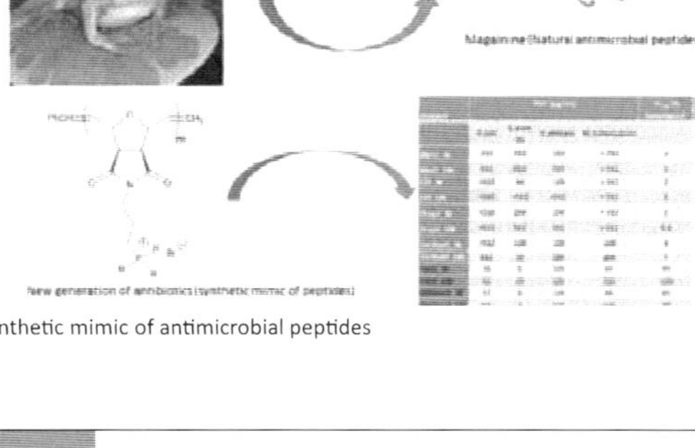

Synthetic mimic of antimicrobial peptides

MBM-O-029 [Macromolecules in Biotechnology & Medicine]

Development of Nanodroplets for Histotripsy-Mediated Cell Ablation

Yasemin Yuksel Durmaz[1], Eli Vlaisavljevich[2], Omer Aydin[2], Kuang Wei Lin[2], Adam Maxwell[3], Brain Fowlkes[2], Zhen Xu[2], Mohamed Elsayed[2]

[1]Department of Biomedical Engineering, Istanbul Medipol University, Istanbul 34810, Turkey
[2]Department of Biomedical Engineering, University of Michigan, Ann Arbor, MI 48109, USA
[3]Department of Urology, University of Washington, Seattle, WA, USA

This work focuses on the development of nanodroplets that can be combined with histotripsy to obtain an image-guided, targeted ultrasound ablation technique (nanodroplet-mediated histotripsy, NMH)[1]. Synthetic nanodroplets that encapsulate perfluorocabon (PFC) core transition upon exposure to ultrasound pulses into gas microbubbles, which will rapidly expand and collapse resulting in disruption of cells similar to the histotripsy process but at a significantly lower acoustic pressure. To be able to obtain PFC loaded nanodroplets, triblock copolymers composed of a hydrophilic poly(ethylene glycol) (PEG) block, a central poly(acrylic acid) (PAA) block, and a random copolymer of heptadecafluorodecyl methacrylate (HDFMA) and methyl methacrylate (MMA) as the hydrophobic block with different copolymer compositions were synthesized. Self-assembly of these copolymers into nanodroplets around PFC core was stabilized by shell crosslinking [2]. The encapsulation efficiency of different amounts of perfluoropentane (PFP) (1% and 2% v/v) and overall droplets stability were examined. The relationship between polymer composition and PFP content (0%, 1%, and 2% v/v) on the ability of the formulated nanodroplets to initiate a bubble cloud in agarose gel phantoms with similar mechanical and acoustic properties to tumor tissue in response to histotripsy pulse was also investigated. Furthermore, the effects of ultrasound frequency (345 kHz, 500 kHz, 1.5 MHz, and 3 MHz) on NMH was shown indicating that the NMH threshold was significantly below the histotripsy intrinsic threshold at all frequencies [3]. Finally, the effect of boiling temperature of PFC on NMH was investigated by comparing ablation characteristics of different encapsulated PFCs such as perfluorohexane (PFH) and PFP into nanodroplets [4].

[1] E. Vlaisavljevich et al., Theranostics 3, 802 (2013).
[2] Y.Y. Durmaz et al, Molecular Pharmaceutics 11, 3684 (2014).
[3] E. Vlaisavljevich et al Ultrasound in Medicine and Biology, 41, 2135 (2015)
[4] E. Vlaisavljevich et al., Ultrasound in Medicine and Biology, 42, 931 (2016)

Keywords: Self-Assembly, Nanodroplets, Perflurocarbons, Histotripsy, Mechanical cell ablation

Abstract figure-YY durmaz

MBM-O-030 [Macromolecules in Biotechnology & Medicine]

A New Microfluidic Extraction System Equipped with a Multiple Feeding System for Human Genomic DNA (hg-DNA) Isolation From Whole-Blood

Fatos Cigdem Kip[1], Eda Ogut[2], Duygu Deniz Usta[3], Erhan Senlik[2], Kouroush Salimi[1], Ali Tuncel[1]

[1]*Hacettepe University, Chemical Engineering Department, Ankara, Turkey*
[2]*Hacettepe University, Division of Nanotechnology & Nanomedicine, Ankara, Turkey*
[3]*Department of Medical Biology and Genetics, Gazi University, Ankara, Turkey*

A new *hg*-DNA microfluidic extraction system with a multiple feeding equipment was constructed using monodisperse-porous silica microspheres with bimodal pore-size distribution and with a mean size of 5.5 μm as the stationary phase. The microspheres were obtained by a newly developed staged-shape template hydrolysis-condensation protocol by using monodisperse-porous poly(methacrylic acid-co-ethylene dimethacrylate) microspheres as the seed material. The stationary phase was obtained by slurry packing of microspheres into the fused silica capillary tubing 300 μm in dimeter and 40 mm in length. Three feeding holes were separately used for feeding of adsorption, washing and elution buffers into the stationary phase using three separate microinjection pumps. Hence, the microextraction process can be completed in the continuous mode without pausing between adsorption, washing and elution stages. Human genomic DNA microextractions from whole blood were performed by controlling the flows of the adsorption, washing and desorption phases either in manual or computer controlled modes with RS232 interface. In the *hg*-DNA isolation experiments, 1X TE buffer containing 6M guanidium-HCl at pH 6.0, ethanol-water mixture (80/20 v/v) and 1X TE buffer at pH 8.5 were used as the binding, washing and desorption medium respectively using human whole-blood as the *hg*-DNA source. For this purpose, whole blood was lysed using Triton-X and Proteinase-K. In the *hg*-DNA isolation runs performed either in manual or computer controlled modes, the sample volume was changed between 10-50 μL by fixing the lysate/binding buffer volume ratio to 1/4. The isolation yields close to 40% w/w based on loaded hg-DNA was achieved. The results indicated that the microextraction system equipped with a multi-feeding system is promising particularly for the automatization of *hg*-DNA microextraction with low sample volumes.

This research was supported by a grant-in-aid for scientific research from the Scientific and Technological Research Council of Turkey (TUBITAK) under contract numbered 213S140.

Keywords: human genomic DNA ({hg}-DNA),DNA microextraction, microspheres, adsorption, sorbent, silica

MBM-O-031 [Renewable Resources and Biopolymers]

Riboflavin Release from Interpenetrated Network Poly(NIPAAm-co-AAc) Hydrogel and Release Data Modeling by using Response Surface Methodology and Artificial Neural Networks

Sanogo Brahima, Cihangir Boztepe, Mehmet Yuceer

Department of Chemical Engineering, Inonu University, Malatya, Turkey

Recent years, studies upon hydrogels in many fields and particularly in medical science for drug release are gained significant advance. In this study, interpenetrated network (IPN) poly(NIPAAm-co-AAc) based hydrogel was synthesized by two polymerizations methods namely emulsion and solution polymerization. NIPAAm (N-isopropylacrylamide) one of the most frequently employed temperature sensitive monomer and AAc (Acrylic Acid) a pH sensitive as comonomer were used for hydrogel synthesis. In this manner the synthesized hydrogel has both pH and temperature-sensitive behavior. The synthesized microgels by emulsion polymerizasyon method were used in solution polymerisation technique to obtain an IPN structure poly(NIPAm-co-AAc) hydrogel. After washing and drying, the synthesized hydrogel was loaded with a riboflavin drug, a B2 vitamin. As a function of time the release of riboflavin has been achieved in different pH and temperature environment. The obtained results show the key role played by pH and temperature effects on the riboflavin released. The determination of experimental conditions and analysis of drug delivery results were done by using response surface methodology (RSM)-central composite design of Design Expert-10 software (trial-version). Time (min), pH and temperature (°C) were used as model input variables. RSM has proposed a release maximum amount at pH=2.42, 51.58 oC and 345.82 min (Figure 1). In this work, Artificial Neural Networks (ANNs) in MATLAB was also used for release data modeling. The predictions from the ANN model, which associated input variables, produce results that show good agreement with experimental data compared to the RSM results. Table 1 gives the RSM and ANN performance in terms of mean squared error (MSE), mean absolute percentage error (MAPE, %), Root Mean Squared Error (RMSE) and the correlation coefficient (R) between the experimental data and neural network outputs.

Keywords: Hydrogels, IPN, modeling, ANN, RSM

Statistical calculations of ANN and RSM model

Model	R	MAPE (%)	RMSE	MSE
ANN (training data)	0.9953	6.4348	0.0188	3.5405x10-4
ANN (test data)	0.9894	8.3087	0.0274	7.5081 x10-4
RSM	0.8439	119.3	0.1334	0.0179

MBM-O-032 [Renewable Resources and Biopolymers]

A study of Dendritic Properties on Drug Solubility: Concentration, Generation Number and Surface Functionality

Metin Tülü, Ali Serol Ertürk, Mustafa Ulvi Gürbüz

Dept.of Chem. Faculty of Arts and Sciences, Yıldız Technical University, 34210 Istanbul, TR

Candesartan cilexetil (CC), prodrug of candesartan which is an angiotensin blocker (ARB) or a potent angiotensin II type 1 receptor (AT1), is useful in the treatment of hypertension, ventricular hypertrophy and diabetic nephropathy. The problem with CC is its poor solubility and, therefore, low bioavailability. The aqueous solubility of CC was measured in the presence amine, TRIS, and carboxyl terminated PAMAM dendrimers at 37 °C by using Higuchi and Connors method [1]. The effects of variables; such as concentration, generation size (G2-G4, and surface-group functionality of Ethylenediamine (EDA) core Poly(amidoamine) PAMAM dendrimers on the aqueous solubility of CC were investigated. Experimental measurements showed that there was reasonable increase in solubility of CC by the interaction of carboxyl and TRIS-terminated dendrimers. This increase was attributed to the surface functionality and the generation size. The CC-PAMAM dendrimer complexes were characterized with UV-VIS, ATR, and DTA techniques.

Acknowledgement: This research has been supported by Adıyaman University Scientific Research Projects Coordination Department. Project Number: MÜFMAP/2015-0009.

Reference: [1] Higuchi T, Connors KA. "Phase-Solubility Techniques." Advan. Anal. Chem. Instr. 1965; 4:117-212

Keywords: Solubilization; candesartan cilexetil; PAMAM dendrimers; Complexation

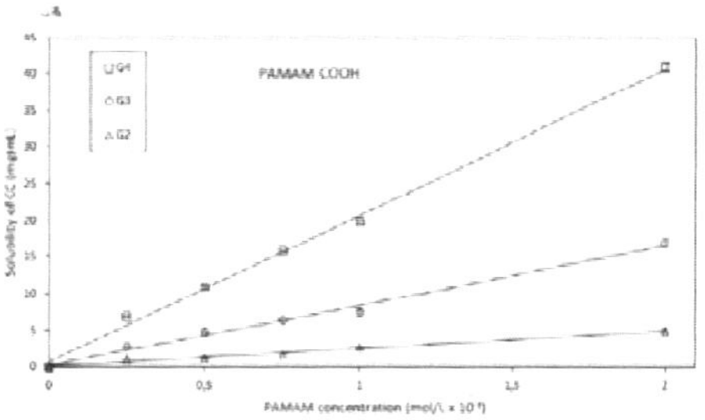

Figure 1. Solubility of Candesartan Cilexetil (CC) in the presence of increasing concentrations of (a) PAMAM-NH_2; (b) PAMAM-TRIS; (c) PAMAM-COOH

MN-O-001 [Macromolecules & Nanotechnology]

Bright and Patchy Polymersomes Enabled by a Symbiosis of Organic and Polymer Chemistry

Jens Gaitzsch[1], Vijay Chudasama[2], Giuseppe Battaglia[2], Wolfgang Meier[1]

[1]*University of Basel, Department of Chemistry, Klingelbergstrasse 80, 4056 Basel*

[2]*University College London, Department of Chemistry, 20 Gordon Street, WC1H 0AJ*

The chemical versatility polymers is often claimed to be a major benefit of polymersomes.[1] From the current polymers used, including a wide range of methacrylates, polypeptides, polypeptoids and polyesters many functionalities can already been accomplished.[2] However, even more potential can be unlocked if polymer chemistry is combined with organic chemistry to obtain polymersomes with enhanced propeties. Polymersomes usually have a surface without any domains, which represents a major drawback with respect to natural vesicles with their highly patterned surface. We accomplished stable domains on polymersomes by synthesising a miktoarm star terpolymer based on dibromomaleimide (see figure). Translated from organic chemistry, this core molecules proved to serve as a highly stable linker platform to form a miktoarm star terpolymer in a clean reproducible manner. By mixing this polymer with a diblock co-copolymer enabled the formation of polymersomes with stable domains (see figure). We were even able to regulate the size of the domains by ration of the diblock- and triblock copolymer.[3] Another important aspect is to efficiently label the vesicles with dyes or proteins. We tested a variety of ligation chemistries with the TAD chemistry being the most efficient.[4] This approach has just recently been translated from organic chemistry to polymer chemistry and our results show how well it can be applied to conjugate polymers.

Both results show that polymersomes benefit greatly if organic and polymer chemistry are combined to give the desired properties of the final vesicles.

1. Gaitzsch, J.; Huang, X.; Voit, B. Chem. Rev. 2016, 116, (3), 1053–1093.
2. Gaitzsch, J.; Karu, K.; Battaglia, G. Eur. Polym. J. 2015, 73, 447-454.
3. Gaitzsch, J.; Chudasama, V.; Morecroft, E.; Messager, L.; Battaglia, G. ACS Macro Lett. 2016, 351-354.
4. Gaitzsch, J.; Dalahaye, M.; Poma, A.; Du Prez, F.; Battaglia, G. Polym. Chem. 2016, DOI: 10.1039/c6py00518g.

Keywords: Polymersomes, Starpolymers, Block-Copolymers, Click-Chemistry, Amphiphilic Polymers

Figure 1

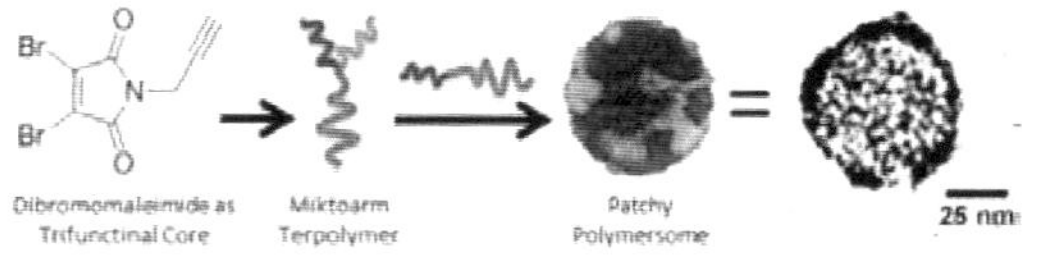

A dibromomaleimide core molecule is transformed to a miktoarm terpolymer in efficient reactions. The final star is then mixed with a linear amphiphilic block-copolymer to give patchy polymersomes (shown as symbol and TEM image)

MN-O-002 [Macromolecules & Nanotechnology]

Engineering Proteinosomes towards Biomimetic Protocells

Xin Huang

School of Chemistry and Chemical Engineering, Harbin Institute of Technology, Harbin, China

The mechanism of spontaneous assembly of microscale compartments is a central question for the origin of life, and has technological repercussions in diverse areas such as materials science, catalysis, biotechnology and biomedicine. Such compartments need to be semipermeable, structurally robust and capable of housing assemblages of functional components for internalized chemical transformations. In principle, proteins should be ideal building blocks for the construction of membrane-bound compartments but protein vesicles with cell-like properties are extremely rare. Here we present an approach to the interfacial assembly of protein-based micro-compartments (proteinosomes) that are delineated by a semi-permeable, stimulus-responsive, enzymatically active, elastic membrane consisting of a closely packed monolayer of conjugated protein–polymer building blocks. The proteinosomes can be dispersed in oil or water, thermally cycled to temperatures of 70 °C, and partially dried and re-inflated without loss of structural integrity. As a consequence, they exhibit protocellular properties such as guest molecule encapsulation, selective permeability, gene-directed protein synthesis and membrane-gated internalized enzyme catalysis.

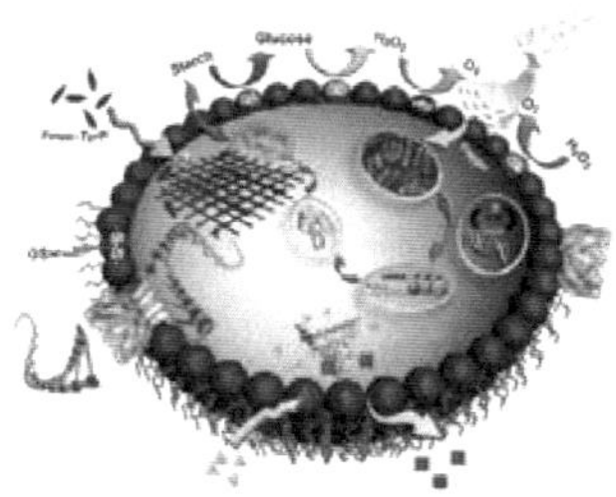

Designed proteinosome as a protocell model

References

[1] Huang, X. Patil, A. Li, M., and Mann, S.* *J. Am. Chem. Soc.* **2014**, *136*, 9225-9236.
[2] Huang, X., Li, M., and Mann, S.* *Chem. Commun.* **2014**, *50*, 6278-6280.
[3] Huang, X., Li, M., Green, D., Williams, D., Patil, A. and Mann, S.* *Nature Commun.* **2013**, *4*, 2239, doi:10.1038/ncomms3239.
[4] Gaitzsch, J. * Huang, X. * Voit, B. * *Chem. Rev.* **2016**, *116*, 1053-1093.

Keywords: Proteinosome, Protocell, Polymer capsules

MN-O-003 [Macromolecules & Nanotechnology]

Synthesis of glyconanoparticles with different morphologies and their interactions with DC-SIGN

Gokhan Yilmaz[1], Lea Messager[4], Daniel Mitchell[3], Giuseppe Battaglia[4], Caglar Remzi Becer[2]

[1]*Department of Chemistry, Warwick University, Coventry, UK*
[2]*School of Engineering and Materials Science, Queen Mary, University of London, London, UK*
[3]*Clinical Sciences Research Institute, Warwick Medical School, University of Warwick, Coventry, UK*
[4]*Department of Chemistry, University College London, UK*

Recent investigations in the glycopolymer technology have allowed the preparation of more complex and well-defined glyco-polymers/particles with several architectures from linear to globular structures (such as micelles, dendrimers and nanogels).[1,2] Over the last years, various self-assembled/decided nano-objects and scaffolds containing glycopolymers were designed to develop many biological and biomedical applications in diseases treatments.[3] Here, several types of amphiphilic block co-glycopolymers with optimal molecular weights and relatively narrow molecular weight distributions were synthesised via Single Electron Transfer Living Radical (SET-LRP) to prepare different glycopolymer nanostructures. These synthesised amphiphilic glycopolymers with the same number of mannose units self-assembled in water to generate glyconanoparticles with different morphologies such as spherical and worm-like micelles as well as spherical vesicles. The shape and size of these self-assembled glyconanopartices were characterized via Transmission Electron Microscope (TEM) and Dynamic Light Scattering (DLS). Finally, the interaction of these glyconanoparticles of different size and shape with dendritic cell-specific intercellular adhesion molecule-3-grabbing non-integrin (DC-SIGN) was monitored by SPR. The SPR results indicate that the size and shape of nanoparticles have significant effect on the binding level with DC-SIGN. According to the binding performance of these glycopolymer nanostructures, they can be potentially utilized in biomedical applications such as cell-targeted drug delivery and inhibition of viral infection such as HIV.

References

1) Yilmaz G., Becer C. R., *Frontiers in Bioenginn. and Biotech.*, 2014, **2**, 39.
2) Yilmaz G. Becer C. R., *Polymer Chemistry*, 2015, **6**, 5503.
3) V. Ladmiral, M. Semsarilar, I. Canton and S. P. Armes, *J. Am. Chem. Soc.*, 2013, **135**, 13574.

Graphical Abstract

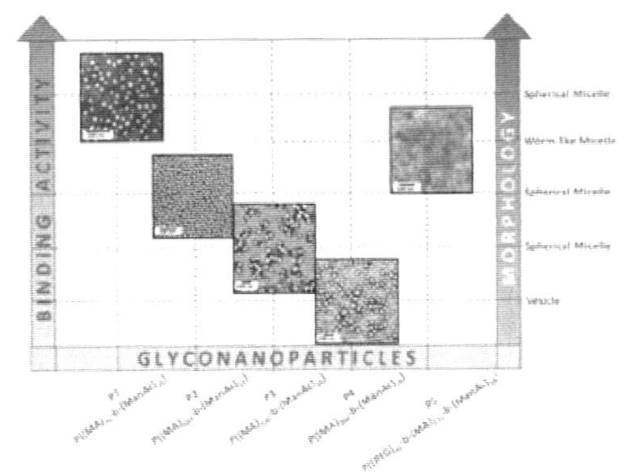

Keywords: well-defined glycopolymers, glyconanoparticles, binding ability, DC-SIGN

MN-O-004 [Macromolecules & Nanotechnology]

Supramolecular engineering of cellulose nanocrystals for responsive hybrid materials

Jani Markus Malho, Jérémie Brandt, Elisabeth Garanger, Gilles Sèbe, Sébastien Lecommandoux

Laboratoire de Chimie des Polymères Organiques (LCPO), CNRS UMR 5629, Université de Bordeaux, Bordeaux-INP, ENSCBP, 16 avenue Pey-Berland, 33607 Pessac Cedex, FRANCE

New concepts are evermore crucial for the generation of high performing hybrid materials from renewable and sustainable source materials. Biopolymers, particularly cellulose, have been studied extensively in the past due to their excellent mechanical properties in addition to their abundance, low density, high surface area and low-toxicity. Yet, their efficient utilization remains a challenge. Another interesting group of renewable materials are genetically engineered peptides and proteins, wherein recent developments in the designing and production have led to an expansion of available molecular level building blocks of high structural and functional precision. Genetically engineered peptides have been used extensively in biomedical field; still combinations of biopolymers and biosynthetic peptides are relatively rare, especially in the field of materials science. We present here a novel concept to combine biopolymers and genetically engineered elastin-like polypeptides (ELPs). ELPs are well known for their thermal responsiveness and mechanical properties, and thus desirable materials for biomedical applications. ELPs featuring a cysteine residue at the C-terminal end were grafted onto cellulose nanocrystals (CNCs) through the use of p-maleimidophenyl isocyanate (PMPI) as heterobifunctional cross-linker highly reactive towards hydroxyl groups that are present in large numbers on the surface of CNCs. PMPI provides maleimide anchor groups that are further reacted with the thiol-containing ELPs. Relatively straightforward, the method provides a promising approach for future generations of sustainable biohybrid materials for materials science applications.

Keywords: cellulose nanocrystals, genetically engineered peptides, functional polymers, supramolecular engineering, materials science, colloids

MN-O-005 [Macromolecules & Nanotechnology]

Nanoporous polymers as caging supports for catalytically active nanoparticles

Cafer T Yavuz[1], Jeehye Byun[1], Paul D Mines[2], Yuhoon Hwang[3], Dongah Ko[2], Hasmukh A Patel[4], Jeong Y Park[1], Jang W Choi[1], Henrik R Andersen[2]

[1]*Graduate School of EEWS, Korea Advanced Institute of Science and Technology, 291 Daehak-ro, Yuseong-Gu, Daejeon 34141, Republic of Korea*
[2]*Department of Environmental Engineering, Technical University of Denmark, Miljøvej, B113, DK-2800 Kongens Lyngby, Denmark*
[3]*Department of Environmental Engineering, Seoul National University of Science and Technology, 232 Gongneung-ro, Nowon-gu, Seoul 01811, Republic of Korea*
[4]*Department of Chemistry, Northwestern University, Evanston, IL, USA*

Nanoscale zero-valent iron (nZVI), with its reductive potentials and wide availability, offers degradative remediation of environmental contaminants. Rapid aggregation and deactivation hinder its application in real-life conditions. Here, we show that by caging nZVI into the micropores of porous networks, in particular Covalent Organic Polymers (COPs), we dramatically improved its stability and adsorption capacity, while still maintaining its reactivity[1]. We probed the nZVI activity by monitoring azo bond reduction and Fenton type degradation of the naphthol blue black azo dye. We found that depending on the wettability of the host COP, the adsorption kinetics and dye degradation capacities changed. The hierarchical porous network of the COP structures enhanced the transport by temporarily holding azo dyes giving enough time and contact for the nZVI to act to break them. nZVI was also found to be more protected from the oxidative conditions since access is gated by the pore openings of COPs.

We also previously reported[2] a surfactant-free method to produce Co_3O_4 nanocrystals with controlled sizes and high dispersity by caging templation of nanoporous networks. The morphologies of Co_3O_4 nanoparticles differ from wires to particulates by simply varying solvents. The composites of nanoparticles within network polymers are highly porous and are promising for many applications where accessible surface and aggregation prevention are important. The electrochemical performance of the composites demonstrates superior capacity and cyclic stability at a high current density (~980 mA h g^{-1} at the 60th cycle at a current density of 1000 mA g^{-1}). In a catalytic oxidation reaction of carbon monoxide, the composites exhibit a remarkable stability (in excess of 35 hours) and catalytic performance (T_{100} = 100 °C).

References:
1. ***J. Mater. Chem. A***, 2016, **4**, 632-639
2. ***J. Mater. Chem. A***, 2015, **3**, 15489-15497

Keywords: Nanoporous polymers, nZVI, Cobalt oxide nanoparticles, Catalyst

MN-O-006 [Macromolecules & Nanotechnology]

Novel self-assemblies via double-hydrophilic block copolymers in aqueous solution

Bernhard V. K. J. Schmidt

Max-Planck-Institute of Colloids and Interfaces, Potsdam, Germany

Self-assembly of block copolymers has a long tradition in polymer science, e.g. in solution or in bulk. Especially amphiphilic block copolymers have been utilized frequently in the formation of micellar or vesicular structures in aqueous solution. The self-assembly in these systems is based on hydrophobic interactions of the water-insoluble block. In recent years, double-hydrophilic block copolymers including one stimuli-responsive block were employed for self-assemblies in aqueous solution as well. Nevertheless, the underlying mechanism is the same as in amphiphilic block copolymer self-assembly as the polarity of one of the blocks is changed towards hydrophobic via external stimuli. In this contribution a different approach is presented. Double-hydrophilic block copolymers are utilized that do not show changes in water solubility in the observed temperature and pH range. The self-assemblies rely on the hydrophilic effect, which allows the formation of nano- to micro-scale aggregates in aqueous solution although completely water-soluble blocks are employed.[1] The self-assemblies are formed solely due to differences in the hydrophilicity of the blocks. To facilitate future applications in drug-delivery, biocompatible blocks are utilized, e.g. poly(*N*-vinylpyrrolidone), poly(ethylene oxide) or poly(saccharides). Due to the complete hydrophilic nature of the formed structures significant improved membrane permeability compared to amphiphilic systems is expected. The block copolymers were synthesized via reversible addition fragmentation chain transfer polymerization. To increase the stability of the formed assemblies, crosslinkable moieties were introduced into the block copolymers that do not alter the hydrophilic nature of the self-assembled structures. Subsequently, crosslinking reactions were performed and the formed structures analyzed via various techniques, e.g. dynamic light scattering, cryo scanning electron microscopy and confocal laser scanning microscopy. In summary, a new concept of block copolymer self-assembly based on completely water-soluble and biocompatible polymers will be presented.

[1] S.M. Brosnan, H. Schlaad, M. Antonietti, *Angew. Chem. Int. Ed.* 2015, *54*, 9715

Keywords: Block copolymers, double hydrophilic, RAFT polymerization, vesicles, self-assembly, biocompatible

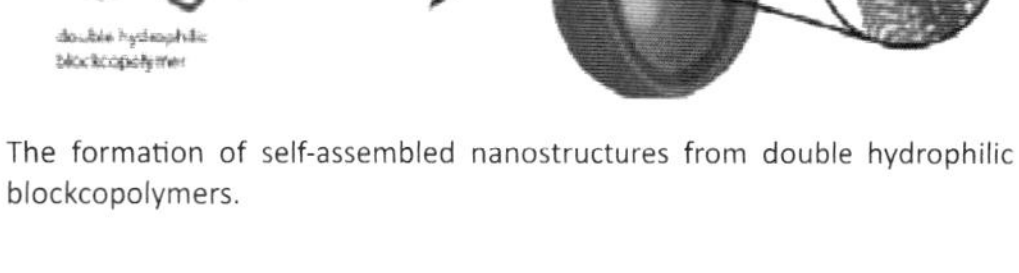

The formation of self-assembled nanostructures from double hydrophilic blockcopolymers.

MN-O-007 [Macromolecules & Nanotechnology]

CO2-Stimulable Block Copolymer Nanostructures

Qiang Yan, Bin Yan, Yue Zhao

Département de Chimie, Université de Sherbrooke

We report our studies of block copolymer (BCP) self-assembled nanostructures that can be stimulated by CO2 in aqueous solution to undergo a variety of deformations or morphological transitions. By introducing into BCP structures the proper functionalities that can be protonated by the presence of CO2, such as amidine, tertiary amine and benzimidazole, various types of covalent and supramolecular CO2-responsive BCPs were designed and synthesized. We explored the use of CO2 as a model physiological stimulus to induce and control changes of BCP self-assembled nanostructures for bio-mimicking. These include reversible expansion/contraction or dissociation/association, a sequence of transitions from microtubules to submicroscopic vesicles to nanosized micelles, and diversiform deformations driven by a restricted hydration effect (refs.1-4)

1. B. Yan, D. Han, O. Boissière, P. Ayotte, Y. Zhao, Soft Matter 2013, 9, 2011-2016.
2. Q. Yan, Y. Zhao, Angew. Chem Int. Ed. 2013, 52, 9948-9951.
3. Q. Yan, Y. Zhao, J. Am. Chem. Soc. 2013, 135, 16300-16303.
4. Q. Yan, H.J. Zhang, Y. Zhao, ACS Macro Lett. 2014, 3, 472-476.

Keywords: CO2 stimulation, block copolymer nanostructures, bio-mimicking

MN-O-008 [Macromolecules & Nanotechnology]

Fabrication of biodegradable Poly(lactic acid) stereocomplex microcapsules using Layer-by-Layer technique in non-aqueous medium

Li Zhao, Gleb B. Sukhorukov

School of Engineering and Materials Science, Queen Mary University of London

Drug delivery systems made of polyesters, which have excellent biodegradability and biocompatibility, are of extreme importance. In this work, Poly-L-(lactic acid) (PLLA) and Poly-D-(lactic acid)(PLLA) were synthesized and then assembled into PLA stereocomplex microcapsules in acetonitrile using Layer-by-Layer (LbL) technique with CaCO3 microparticles as the template. Proton nuclear magnetic resonance (1HNMR) and gel permeation chromatography (GPC) were applied to characterize the chemical compositions and molecular weight of synthetic PLA polymers. The chemical composition of the obtained microcapsules was confirmed by differential scanning calorimetry (DSC) analysis which helps to distinguish the change of melting point of stereocomplex polymers. Different deposition temperatures of LbL process were compared to investigate the best assembly condition for PLA microcapsules. PLA stereocomplex microcapsules with different number of multilayers were fabricated and imaged by scanning electron microscope (SEM). It was found that 30 layers are enough to construct an intact microcapsule structure. Intact Microcapsules were obtained after CaCO3 template removal by either addition of ethylenediaminetetraacetic acid (EDTA) into microparticles or dialysis of microparticles suspension in EDTA solution. Heating treatment was proven to effectively enhance the encapsulation capacity of the PLA microcapsules. Transmission electron microscope (TEM) was used to observe the internal morphologies as well as the thicknesses of PLA microcapsules. The biodegradability of PLA stereocomplex multilayers was studied by incubating the obtained films built up on planar substrates in different enzyme solutions and examining the decrease in thickness at different time intervals. In summary, a fully biocompatible process has been introduced to fabricate biodegradable PLA stercomplex microcapsules, which is believed to be fairly useful as biodegradable drug carriers.

Keywords: Poly(lactic acid), stereocomplex, microcapsules, Layer-by-Layer, non-aqueous, biodegradable

PLA stereocomplex microcapsules obtained using different methods of template removal

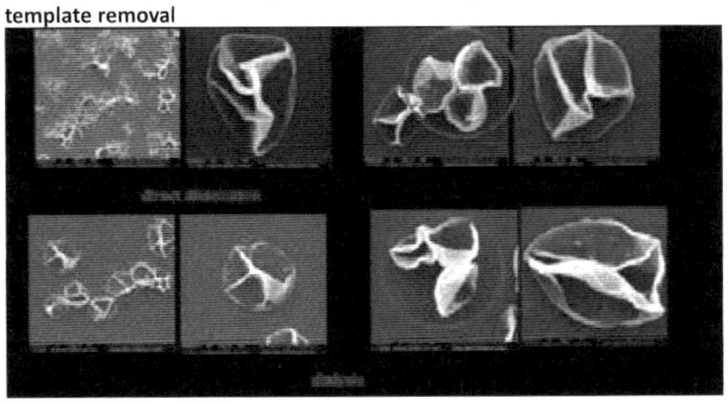

After the deposition of Poly(L-lactic acid) and Poly(D-lactic acid) on CaCO3 template, two different template removal methods were compared. The direct dissolution by adding ethylenediaminetetraacetic acid (EDTA) and the dialysis of particles in EDTA solution which is milder.

MN-O-009 [Macromolecules & Nanotechnology]

Synthesis and characterization of Au/Zirconium phosphate nanoplatelets for catalytic reduction of 4-nitrophenol

Mei Hui Tsai[1], Ying Hao Pai[1], Tsao Cheng Huang[1], Yi Hsiu Chung[2]

[1]*Department Chemical and Materials Engineering, National Chin-Yi University of Technology, Taichung 411, Taiwan (R.O.C.)*
[2]*Department Chemical Engineering, Tatung University, Taipei 104, Taiwan (R.O.C)*

Catalysts based on gold nanoparticles (Au NPs) have received much attention during the past two decades because of their unique catalytic properties in basic and applied research. There is great interest in the development of nanoparticles that combine multiple function or properties not obtainable in individual materials.α-zirconium phosphate(α-ZrP) nanoplatelets were initially prepared by thermal reflex method. And subsequently immobilization of Au nanoparticles onto α-zirconium phosphate nanoplatelets surface to prepared the catalyst of α-ZrP-Au nanocomposites. The nanopleatelets loading Au NPs needs prevent aggregation. The surface of the α-zirconium phosphate nanoplatelets was modified with organosilane molecules to generate an amine terminated surface. Au NPs to anchor on α-zirconium phosphate nanoplatelets surface through Au–N covalent bond formation. The characterization and structure was confirmed by Field Emission Scanning Electron Microscope(FE-SEM),Field Emission Transmission Electron Microscope(FE-TEM,)X-ray Diffractometer,(XRD),Solid State NMR Spectrometer(NMR), X-ray photoelectron spectroscopy(XPS),UV-visible spectrophotometer(UV-vis).It was demonstrated that α-ZrP-Au nanocomposites as a catalyst has excellent performance and stability for reduction 4-Nitrophenol(4-NPh) to 4-aminophenol(4-APh) within 30 sec.

Keywords: Au nanoparticles, α-ZrP, organosilane, catalyst, 4-Nitrophenol

MN-O-010 [Macromolecules & Nanotechnology]

Hydrogen-bonded layer-by-layer films of poly(2-alkyl-2-oxazoline)s

A Levent Demirel, Saman Hendessi, Pınar Tatar Güner, Annamaria Miko

Koç University, Istanbul, Turkey

Poly(2-alkyl-2-oxazoline)s (PAOX) form stable hydrogen-bonded aggregates with tannic acid (TA) in aqueous solutions at acidic conditions. Hydrogen bonds (H-bonds) are formed primarily between carbonyl groups of PAOX and hydroxyl groups of TA and are stable up to pH of 8.5. Using layer-by-layer (LbL) self-assembly, PAOX/TA pH-responsive multilayer films were prepared and characterized. The growth characteristics and pH-stability of various H-bonded LbL films prepared by different deposition methods (dipping, spinning) will be presented. pH-annealing (keeping the films in buffer solutions just below the critical disintegration pH) significantly slowed down the disintegration of H-bonded LbL films at basic pH indicating that the average number of H-bonds per chain increased during annealing. Ag nanoparticle (Ag-NP) loaded PEOX/TA LbL films were also investigated. The release of Ag+ ions from such films is important for antimicrobial functionality. Ag-NPs were first synthesized in aqueous solutions and stabilized by PEOX coating. PEOX coated Ag-NPs and TA were then used to form LbL films on solid substrates. The multilayers were characterized by scanning electron microscopy and UV-visible spectroscopy. Results showed that nanoparticles were distributed homogenously in the film and Ag+ was released into aqueous solutions. The potential of Ag-NP loaded LbL films for antimicrobial protection of bio-medical device surfaces will be discussed.

Keywords: Layer-by-layer films, H-bonded self-assembly, Antimicrobial coatings

MN-O-011 [Macromolecules & Nanotechnology]

Macromolecular Templates in Biomineral Formation

Marina Tsianou

Department of Chemical and Biological Engineering, University at Buffalo - The State University of New York (SUNY), Buffalo, NY, USA

Living organisms are able to form morphologically diverse minerals produced in a complex manner and in media of specified compositions. In most biomineralization processes many soluble biopolymers exist in minute quantities; these control the nucleation, growth, orientation, and polymorph selection. Our current group efforts are directed toward the crystallization of calcium-bearing minerals, calcium carbonate (the most abundant biomineral, and of high industrial relevance in a range of products), calcium oxalate (the major constituent of kidney stones), and calcium phosphate (the main component of bones and teeth). We establish that water-soluble macromolecules (polyelectrolytes, polypeptides, surfactants) mimic in an effective way the role of soluble biomineralization polymers and can control the final morphology and polymorphic occurrence by face-selective adsorption on crystals (thermodynamic regime) and/or by temporarily stabilizing 'less stable' polymorphs (kinetic regime). For example, in the case of calcium oxalate, we have demonstrated that anionic polyelectrolytes can be very effective toward kidney stone inhibition. Furthermore, we investigate the role of organic matrix as mediator of the mineralization and as provider of a structural framework for mechanical support. We employ hydrogels which serve as excellent models of the extracellular matrix microenvironment. Gel-mediated crystallization provides a great way to vary the local supersaturation through controlling the diffusion rate of the reacting ions. We study the nucleation, growth, type of polymorph and orientation of the calcium oxalate crystals as affected by the gel density, reservoir concentrations, molar stock solution ratio, and additives. Our findings and provide insights into the formation of organic-inorganic composite materials with the same specialization and complexity as those in nature.

Keywords: polyelectrolytes, hydrogels, biominerals

MN-O-012 [Macromolecules & Nanotechnology]

Self-assembled hydroxyapatite-chitosan nanocomposites: Promising materials for bone tissue engineering applications

Muhammad Anwaar, Emel Yilgör, Iskender Yilgör

KUYTAM Surface Science and Technology Center, Chemistry Department KOC University, Istanbul, Turkey

Preparation and characterization of self-assembled chitosan/nano-hydroxyapatite (CS/HA) nanocomposites will be reported. HA was synthesized through a sol-gel process in aqueous media with a Ca/P ratio of 1.67. We introduced formic acid as a new solvent to obtain homogeneous and stable dispersions of nano-sized HA particles in polymer solution. Self-assembling of HA nanoparticles during the drying of solvent cast films led to the formation of well-defined CS/HA nanocomposites (Figure 1-a). CS/HA nanocomposites with HA contents of 0, 5, 10 and 20 percent by weight were prepared and characterized by SEM, AFM, EDX, ATR-FTIR, XRD and TGA. SEM and AFM confirmed the presence of self-assembled islands of HA in the chitosan matrix. Complete degradation of the chitosan matrix in the nanocomposites during TGA experiments, followed by heat treatment at 1000 °C, led to the formation of nanoporous 3D films containing HA, tricalcium phosphate (β-TCP) and calcium pyrophosphate (Figure 1-b), which was confirmed by ATR-FTIR analysis. These 3D nanoporous HA/TCP composites can be considered as promising materials for bone tissue engineering applications since HA is known to promote adhesion, proliferation and osteogenic differentiation of osteoblast like cells [1].

References:

[1] Asaoka, T.; Ohtake, S.; Furukawa, K. S.; Tamura, A.; Ushida, T. Journal of Biomedical Materials Research Part A 2013, 101(11), 3295-3300.

Keywords: chitosan, hydroxyapatite synthesis, self-assembling, 3D porous scaffolds, bone tissue engineering

Figure 1. (a): CS/HA composites showing HA islands

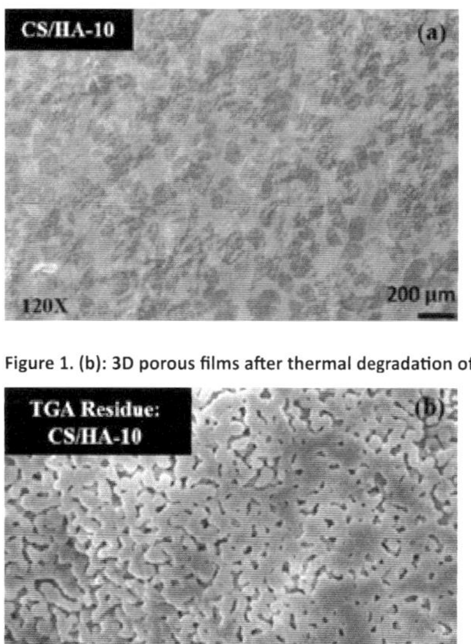

Figure 1. (b): 3D porous films after thermal degradation of CS

MN-O-013 [Macromolecules & Nanotechnology]

Direct Write Patterning of Multiple Proteins and Polymers at the Micrometer and Nanometer Scale

Erhan Bat[1], Heather D. Maynard[2]

[1]Department of Chemical Engineering, Middle East Technical University, Ankara, Turkey
[2]Department of Chemistry and Biochemistry, University of California, Los Angeles, USA

Electron beam lithography (EBL) is a powerful maskless lithography technique that allows generation of user defined complex patterns on surfaces at high resolution. Although a serial technique, EBL offers nanometer-scale alignment capability which enables interfeature spacings that are so small that different polymers or proteins may be touching or or arrayed one on top of the other. This allows generation of complex, multiplexed patterns or investigation of fundamentals of certain processes such as nanoparticle growth and protein-protein interactions. Direct patterning of multiple polymers, proteins, or particles on surfaces is of tremendous interest for a myriad of applications ranging from diagnostics and tissue engineering to biosensors. In this talk, several examples of multiplexed patterning of these materials by electrom beam lithography will be discussed.

Keywords: Electron beam lithography, micro and nanopatterning, multiplexing, glycopolymers, poly(ethylene glycol)

MN-O-014 [Macromolecules & Nanotechnology]

Simple green synthesis of CdTe/CdSe/ZnSe core-multi shell with reduced cytotoxicity for bio imaging

Ncapayi Vuyelwa[1], Oluwafemi Samuel Oluwatobi[2], Songca Pinda Sandile[3], Kodama Tetsuya[4]

[1]Department of Applied Chemistry, University of Johannesburg, P.O. Box 17011, Doornfontein 2028, Johannesburg, South Africa.
[2]Centre for Nanomaterials Science Research, University of Johannesburg, Johannesburg, South Africa.
[3]Department of Applied science, Walter Sisulu University Buffalo City Campus, East London, 5117, Eastern Cape, South Africa
[4]dGraduate school of biomedical engineering, Tohoku University, 4-1 Seiryo, Aoba Sendai, Japan, 980-8575

We herein report a simple, economical and green synthesis of highly water soluble and stable CdTe/CdSe/ZnSe multicore-shell nanoparticles (NPs) via a one pot synthetic route. The synthesis was carried out under ambient conditions in the absence of an inert environment and involved the use of K-2TeO3 and Na2SeSO4 as a stable tellurium and selenium precursor respectively. The temporal evolution of the optical properties and stability of the growing nanocrystals was monitored in detail by varying the refluxing time, pH and storing the NPs under ambient condition for several days. The as-prepared NPs were characterised using UV-Vis absorption and photoluminescence (PL) spectroscopy, Fourier transform infrared spectroscopy (FTIR), transmission electron microscopy (TEM), and high resolution transmission electron microscopy (HRTEM). The optical analyses showed an enhancement in the fluorescent intensity after the addition of the shell solution accompanied by red-shifting of the absorption and emission maximum. The stability study revealed an increase in the emission intensity as the ageing days increased. The TEM analysis showed that the materials are small, monodispersed, spherical and highly crystalline. The cytotoxicity of the NPs investigated on KM-Luc/GFP cell line at different concentrations showed a decrease in cell viability as the concentration of the NPs increased with the highest value of 88 % at 0.1µg/mL. Furthermore, the as-synthesised NPs prepared at longer reaction time showed lower cytotoxicity compared with those prepared at shorter reaction time. The results show variation in cell viability from 88 % to 57.6 % for the NPs produced at 7 h compared with those produced at 30 min as the concentration increased from 0.1 µg/mL to 60 µg/mL. The fluorescent imaging performed using confocal fluorescent microscope indicated that the as-synthesised NPs were taken up by the cells.

Keywords: Nanoparticles,Core-multi shell, Bio-imaging, Cytotoxicity

MN-O-015 [Macromolecules & Nanotechnology]

Synthesis and characterization of polyaniline/ SrTiO3 nanocomposite and its photocatalytic activity for degradation of methylene blue dye under visible light

Syed Shahabuddin, Norazilawati Muhamad Sarih, Sharifah Mohamad

Chemistry Department, Faculty of Science, University of Malaya, Faculty of Science, 50603 Kuala Lumpur, Malaysia.

Currently, the elimination of noxious chemicals from waste water is one of the most crucial concerns in pollution control from an ecological and biological point of view. The textile industries are the cause of worldwide pollution problem and contributes around 17-20% of the total industrial water pollution as per the World Bank estimate [1]. 3,7-Bis(dimethylamino)-phenothiazin-5-ium chloride, commonly known as methylene blue (MB) is a hazardous chemical compound commercially used for various applications such as textiles, papers, leathers, additives, laser printing, etc. and a major cause of water pollution. Amid the wide range of methodologies to degrade ominous waste materials, specially organic compounds, to less toxic materials, photocatalysis [2] has appeared as one of the most encouraging techniques since it provides a simple way of exploiting the energy of either natural sunlight or artificial illumination, such as ultraviolet light, microwaves, etc. Herein we highlights the enhanced photocatalytic degradation of methylene blue dye by polyaniline (PANI) based nanocomposites doped with SrTiO3 nanocubes synthesised via in-situ oxidative polymerization technique using ammonium persulfate(APS) as an oxidant in acidic medium [3]. The synthesised nanocomposites were characterized using Field emission scanning electron microscopy (FESEM), transmission electron microscopy (TEM), thermogravimetric analysis (TGA), X-ray diffraction (XRD), UV-vis spectroscopy, Brunauer-Emmett-Teller analysis (BET) and Fourier transform infrared spectroscopy (FTIR) measurements. The photocatalytic proficiencies of the photocatalysts were evaluated by degrading methylene blue (MB) under visible-light irradiation. The results revealed that SrTiO3 nanocubes doped polyaniline composites demonstrated higher degradation efficiencies as compared to undoped polyaniline. Furthermore, the effects of weight ratio of polyaniline to SrTiO3 on the photocatalytic activities were explored and results indicates that the nanocomposite P-Sr500, with 500 mg of SrTiO3 with respect to 0.4 M aniline, to be an optimal photocatalyst, with a 97% degradation efficiency after 90 minutes of irradiation under solar light.

Keywords: polyaniline, photocatalyst, nanocomposites, methylene blue, SrTiO3 nanocubes.

Schematic illustration for photocatalytic degradation of Methylene Blue

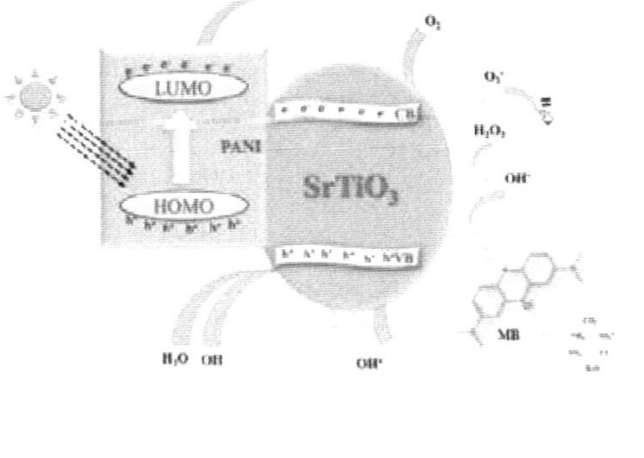

MN-O-016 [Macromolecules & Nanotechnology]

Simultaneous Enhanced SERS and Fluorescence Effects of Co-Encapsulated Au-Ag Alloy Nanoparticles and Silicon Quantum Dots within Polymer Nanoparticles

Noor Aniza Harun[1], Benjamin R. Horrocks[2], David A. Fulton[2]

[1]School of Fundamental Science, Universiti Malaysia Terengganu, 21030, Kuala Terengganu, Terengganu, Malaysia and Chemical Nanoscience Laboratory, School of Chemistry, Newcastle University, Newcastle upon Tyne, NE1 7RU, UK
[2]Chemical Nanoscience Laboratory, School of Chemistry, Newcastle University, Newcastle upon Tyne, NE1 7RU, UK

The development of multi-component polymer composite nanoparticles via co-encapsulation of Raman-active Au-Ag alloy nanoparticles and silicon quantum dots (SiQDs) within poly(divinyl)benzene by a single step miniemulsion polymerization is presented. The resulting polymer composite nanoparticles was successfuly characterized by dynamic light scattering, UV-Visible spectroscopy and transmission electron microscopy. Plasmon enhanced surface-enhanced Raman scattering and fluorescence of Raman-active Au-Ag alloy nanoparticles and silicon quantum dots encapsulated inside polymer composite nanoparticles was investigated by confocal microspectroscopy, where simultaneous and significant surface-enhanced Raman scattering and fluorescence enhancement up to five-fold were observed. These observations indicate that when the Raman-active Au-Ag alloy nanoparticles and silicon quantum dots were co-encapsulated within the polymer layer, the surface-enhanced Raman scattering and luminescence are enhanced, presumably on account of their close proximity of the Raman-active Au-Ag alloy nanoparticles to the silicon quantum dots. It was speculated that this polymer composite nanoparticles which offer dual optical characteristics associated with SERS of the Raman-active Au-Ag alloy NPs and metal-enhanced fluorescence effects associated with SiQDs could have potential in labelling applications.

Keywords: SERS, Metal-Enhanced Fluorescence, Silicon Quantum Dots, Miniemulsion Polymerization

MN-O-017 [Macromolecules & Nanotechnology]

Dispersion of nanoclay and its extend of compatbilization in immiscible natural rubber / nitrile rubber blends

Hanna J Maria, Sabu Thomas

International and Inter University Centre for Nanoscience and Nanotechnology, Mahatma Gandhi University, Priyadarshini Hills P. O. Kottayam, Kerala, India -686 560

A comparative study on the two methods for preparing highly dispersed O1Mt/polymer blend nanocomposites is done by the dry mixing and the solution mixing process and by doing dry mixing alone. In the combined method used, represented as solution mixing technique, clay particles are successfully delaminated into layers, and layer re-stacking is effectively prevented. HR-TEM and XRD results confirm that O1Mt layers achieve exfoliated or nearly exfoliated dispersion in all the NR/NBR blend nanocomposites. Compared with mechanically blended nanocomposites, NR/NBR/O1Mt nanocomposites prepared by solution mixing method exhibited comparatively high dynamic modulus, barrier and mechanical properties. This change is due to the good delamination of nanoclay layers and the consecutive dispersion and distribution which, may have created a good interaction between elastomer chains and the filler. This was confirmed by the morphological results obtained from their transmission electron microscopy (TEM) observations and the non-linear viscoelastic properties which showed payne effect.

Keywords: nitrile rubber natural rubber,immiscible,nanocomposites

MN-O-019 [Macromolecules & Nanotechnology]

Nano hybrid organic-inorganic polyhedral oligomeric silsesquioxanes (POSS) based nanocatalyst

Ali Akbari, Nasser Arsalani

University of Tabriz

INTRODUCTION: In recent years, due to higher stability and dispersibility of nanoparticles as well as the demand for higher activity and recyclability of catalysts, heterogeneous nano-catalysis has emerged as a sustainable and competitive alternative to conventional catalysis. The copper-catalyzed azide–alkyne cycloaddition (CuAAC) and Heck coupling reactions are two of the most attractive synthetic tools for convenient and reliable construction of 1,2,3-triazole derivatives and C-C bond formation respectively, whose subunits are often found in biologically active molecules. C. O. Kappe and E. Van der Eycken, Chem. Soc. Rev. 39 (2010) 1280 Silsesquioxanes, the three-dimensional oligomeric compounds with the general formula (RSiO1.5)2n, where R can be any functional group (alkyl, hydrogen, halogen, aryl, etc.), show unique properties. A number of silicon-based compounds have recently been developed as substrates for catalyst immobilization in a wide range of biochemical reactions. Among them, polyhedral oligomeric silsesquioxane (POSS) cage molecules, (RSiO1.5)8, have received a great deal of attention for the synthesis of organic-inorganic hybrid materials with precise control of nanoarchitecture. D. B. Cordes, P. D. Lickiss and F. Rataboul, Chem. Rev. 110 (2010) 2081 For the first time, we have designed and synthesized two types of POSS based nano-catalyst and used in click and heck reactions.: A. Akbari, N. Arsalani, M. Amini and E. Jabbari, J. Mol. Catal. A: Chem. 414 (2016) 47. A. Akbari, N. Arsalani, M. Amini and E. Jabbari (Submitted to Applied catalysis A: General, 2016)

CONCLUSION: In summary, we have synthesized two new different types of POSS-based nano-catalyst. These catalysts showed excellent mechanical and thermal properties. Superior catalytic activity, easy recovery and reusability are some of the advantages of these nano-catalysts in click and heck reactions.

Keywords: polyhedral oligomeric silsesquioxane, Nano-Catalyst, Organic Reactions

Click reaction

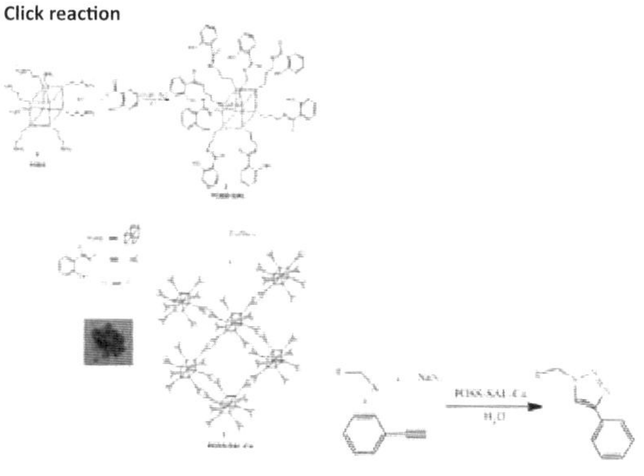

Heck Reaction

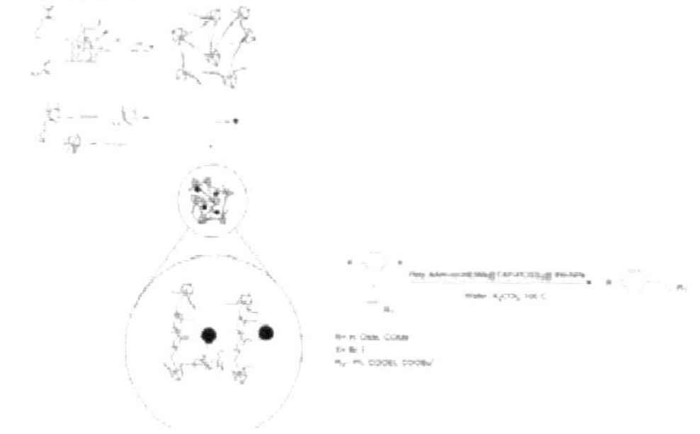

MN-O-020 [Macromolecules & Nanotechnology]

Controlled Synthesis of Silsesquioxane Based Hybrid Copolymers for Nanostructured Spheres and Fibers

Ali Demirci[1], Tamer Uyar[1], Tokuji Miyashita[2], Masaya Mitsuishi[2]

[1]Bilkent University, UNAM-National Nanotechnology Research Center
[2]Tohoku University, Institute of Multidiciplinary Research for Advanced Materials

Electrospining is a simple and versatile technique for generating nanostructured fiber materials, whose fibrous nanostructure are attractive for a range applications due to their intrinsically high porosities and large surfaces areas. Hybrid inorganic-organic materials have attracted much interest in recent years because they offer considerable opportunities to develop new nanostructured functional materials through the combination of inorganic and organic components. Polyhedral oligomeric silsesquioxane (POSS) is interesting hybrid macromolecules that has a nano-sized inorganic silica-like cage structure with a wide variety of organic groups. Hybrid organic-inorganic polymers based on POSS components are an important class of hybrid materials; they provide superior properties such as chemical resistance, thermal stability, mechanical strength, i.e. This study describes systematic studies for the development of hybrid network polymers based on POSS functional monomers and their use in nanostructured fiber and sphere materials. The use of POSS-based multifunctional building blocks in the controlled Piers-Rubinsztajn reaction enables various types of new hybrid polymer materials. The controlled copolymerization by varying total monomer concentration provide varying high-molecular-weigth products. While the electospining of their low-viscosity solutions of hybrid copolymer form nanostructured spheres, the higher viscosity solutions of hybrid copolymer form nanostructured fibers material. Based on these studies, a qualitative relationship between the molecular weight and the electrospun particle morphologies is discussed. Electrospinning at relatively low polymer concentrations produces polymer particles rather than fibers, by tailoring the solution properties, varying molecular weight, and electrospinning conditions, particles with different morphologies can be produced. Silsesquioxane based hybrid copolymer fiber and sphere nanostructured materials via electrospining offer tremendous routes to develop functional hybrid materials with uniform nanostructures.

Keywords: POSS, Hybrid, Electrospining, Nanofiber

MN-O-021 [Macromolecules & Nanotechnology]

Sodium Alginate Hydrogel Membrane Hybrided by Silver Nanoparticle as a Highly Effective Photocatalysis Microreactor for Large Scale Dye Wastewater Treatment

Xiaohuan Ji[1], Yibing Zhu[1], Yinghua Song[1], Peiwei Zhu[1], Xiaoze Jiang[1], Zhe Zhou[1], Bin Sun[1], Yushan Sun[2], Meifang Zhu[1]

[1]*State Key Lab for Modification of Chemical Fibers and Polymer Materials, College of Material Science and Engineering, Donghua University, Shanghai, 201620, China. E-mail: sunbin@dhu.edu.cn, zmf@dhu.edu.cn*
[2]*State Key Lab of Biobased Fiber Manufacturing Technology, China Textile Academy,Beijing, 100025, China*

Organic dyes are one of the most common harmful chemical compounds in wastewater to cause water pollution without adequate treatment, so rapid and convenient removal of organic dyes from wastewater has been a challenge faced by scientists. Silver nanoparticles (Ag NPs) as an efficient light harvesting photocatalysts are widely used. In order to solve the distribution and recycle problems of nano-sized photocatalysts, powdery silica, carbon and titania, fiber-like glass fiber or microsphere hydrogels have achieved varying degrees of success as supporting materials for various photocatalytic reactions, but it is still far from the real application, especially for the large scale water treatment. Herein, we provide a hydrophilic sodium alginate (SA) hydrogel membrane with monodisperse Ag NPs embedded in as a highly effective photocatalysis microreactor, which may find good potential application for large scale dye wastewater treatment. The membrane was prepared by dispersing the hydrophilic Ag NPs in the SA monomers with deionized water, and the crosslinking reaction was initiated in the presence of Ag NPs, so that the nanoparticles are simultaneously and homogeneously trapped in the polymer network. The membrane could absorb dye molecules, thus may act as the collector of dyes and microreactor for photo-degradation of dye. The photocatalysis reaction can be carried out by taking them out from the wastewater in a separate container under the irradiation of visible light. This has much sense for batch and efficient photocatalytic treatment. The membranes still keep good efficiency after five cyclic runs without any break of them, showing the stable reusability of photocatalysts in real application. This work provides a novel material, technology and idea to solve the big challenge for how to make nano-material with photocatalytic function to attain practical and industrial application in dye wastewater treatment.

Keywords: sodium alginate, silver nanoparticles, hybrided membrane, photocatalyst, dye wastewater treatment

The reuse and recycle process of Ag NPs/SA Membrane in large scale wastewater treatment

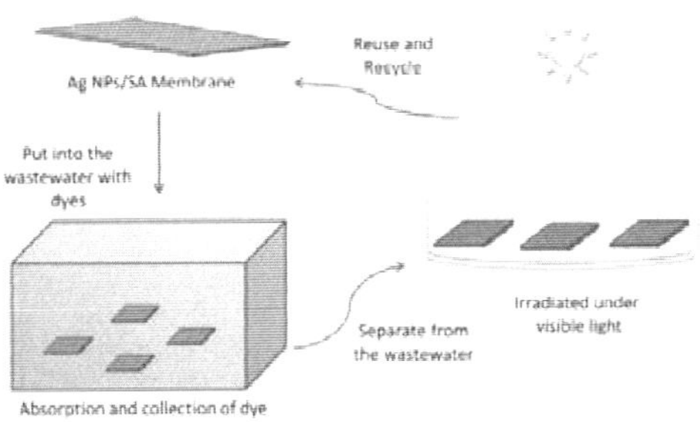

Scheme 1 The reuse and recycle process of Ag NPs/SA Membrane in large scale wastewater treatment

MN-O-022 [Macromolecules & Nanotechnology]

Superhydrophobic Polysiloxane Nanofilament Growth on Hydrophobic and Hydrophilic Polymer Coatings

Esra Kasapgil[1], Emine Gozde Atici[1], Ruhi Cicek[1], Ilke Anac[2], H. Yildirim Erbil[2]

[1]*Department of Materials Science and Engineering, Gebze Technical Univeristy, Kocaeli, Turkey*
[2]*Department of Chemical Engineering, Gebze Technical Univeristy, Kocaeli, Turkey*

Superhydrophobic polysiloxane nanofilaments were succesfully grown on non-activated hydrophobic and hydrophilic polymer coatings prepared by both the conventional emulsion and solution polymerizations using methyltrichlorosilane. Both liquid phase reactions using petroleum ether in an open system and gas phase reactions under controlled humidity environment were used in the nanofilament growth reactions. Wettability, morphology and transparency of the obtained polysiloxane nanofilaments were characterized by contact angle measurements, scanning electron microscopy, UV-vis transmittance tests respectively. Polysiloxane nanofilaments on polymer coatings (obtained by emulsion polymerization) having advancing contact angles between 166°-172° and 157°-169° were prepared via the gas phase reactions and the liquid phase reactions, respectively. Polysiloxane structures (in filament or spherical geometry) were grown on polymer coatings obtained by solution polymerizations during the solution phase reactions however, gas phase reactions were failed for the nanofilament growth on these surfaces. It is also shown that polysiloxane nanofilaments having advancing contact angle of 164° can be successfully prepared on polystyrene coating obtained by solution polymerization via the solution phase filament growth reactions. These results indicate that the presence of –OH groups on polymer coatings is not required for the growth of nanofilaments, only the presence of a water film on the polymer coatings having an adequate thickness is enough for the polysiloxane filament growth.

[1] G. R. J. Artus, S. Jung, J. Zimmermann, H. P. Gautschi, K. Marquardt, S. Seeger, Silicone nanofilaments and their application as superhydrophobic coating Adv. Mater. 18 (2006) 2758-2762.
[2] L. Gao, T.J. McCarthy, A perfectly hydrophobic surface J. Am. Chem. Soc. 128 (2006) 9052-9053.
[3] E. Kasapgil, E. G. Atici, R. Cicek, I. Anac, H. Y. Erbil, Superhydrophobic Polysiloxane Filament Growth on Non- Activated Polymer Coatings. Submitted to Applied Surface Science

Keywords: polysiloxane, nanofilament growth, emulsion polymer, superhydrophobicity, contact angle

Polysiloxane nanofilaments

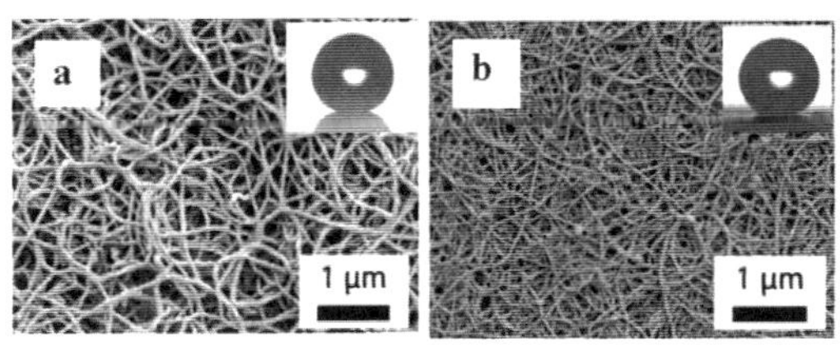

Polysiloxane nanofilaments prepared on butyl acrylate/methyl metharcylate/metharcylic acid (BA/MMA/MAA) (a) and polystyrene (b) coatings via solution phase reactions

MN-O-023 [Macromolecules & Nanotechnology]

Detection of Alzheimer`s Protein on Polycarbonate Nanopillared Films by Using Surface Enhanced Raman Spectroscopy

Sevde Altuntaş[1], Can Simon Pervane[2], Fatih Büyükserin[3]

[1]Micro and Nanotechnology Graduate Program, TOBB Univ. of Econ. & Technology, Ankara 06560, Turkey
[2]Institute of Complex Systems Simulation, University of Southampton, Southampton SO171BJ, United Kingdom
[3]Department of Biomedical Engineering, TOBB Univ. of Econ.& Technology, Ankara 06560, Turkey

Alzheimer`s disease is responsible for neuronal damage of cerebral cortex and hippocampus because of Amyloid β 1-42 protein accumulation. The protein detection studies have gathered around ELISA assay and PCR assays which are expensive, require specialized personnel and can contain complex protocols. On the other hand, Surface-enhanced Raman Spectroscopy (SERS) can potentially allow even single molecule detection in solutions or solid surfaces. In addition, SERS signal from a target molecule can be further increased by using nanopatterned surfaces when compared to smooth counterparts. Moreover, Raman signals are specific to molecules, so one can comment about content of a sample. In this context, our study proposes to fabricate diagnostic test models that utilize Au-coated nano patterned polycarbonate (PC) surfaces modified with Thioflavin - T to detect low concentrations of Amyloid-β 1-42 protein in water and artificial saliva medium by the enhancement of protein SERS signal. Nano patterned PC surface was fabricated by using nanoporous anodized aluminum films (AAM) as a mold. To AAM column structure, three different type PC arrays were fabricated to enhanced SERS signals. The PC films were then decorated with Au and Thioflavin –T for detection Amyloid-β 1-42 protein. The protein detection studies were conducted in water and artificial human saliva. SEM, SERS, FTIR, fluorescence microscopy and contact angle measurements were carried out for the characterization of different surfaces and further demonstration of the protein attachment.

In addition to experimental study, computational enhancement studies were carried out by using COMSOL Multiphysics. The results will be presented comparatively.

This work was supported by The Scientific and Technological Research Council of Turkey (TUBITAK) Grant No: 214Z167.

Keywords: SERS, Nanostructure, Enhancehment, Polycarbonate, Alzheimer, COMSOL

Figure 1

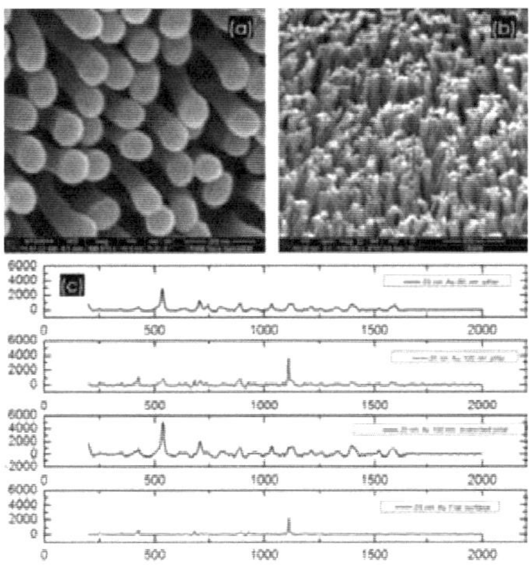

SEM images of nanopatterned surface (a), and branched nanopatterned surface (b). SERS spectrum of Thioflavin modified nanopatterned and flat PC surface.

MN-O-024 [Macromolecules & Nanotechnology]

Design of Virgin Coconut Oil Based Nanostructured Lipid Carriers to co-deliver phytochemical extracts

Rosnani Hasham[1], Nor Ayshah Rosli[2], Siti Hasyimah Suhaimi[2], Muhammad Fauwaz Puzi[2], Nor Azrini Azmi[2], Norhayati Mohamed Noor[1], Azila Abdul Aziz[1], Mohamad Roji Sarmidi[1], Ramlan Aziz[1]

[1]Institute of Bioproduct Development, Universiti Teknologi Malaysia, Johor, Malaysia
[2]Department of Bioprocess and Polymer Engineering, FCEE, Universiti Teknologi Malaysia, Johor, Malaysia

Virgin coconut oil (VCO) is the highest quality coconut oil that can be obtained from coconut fruit. The objective of the present study is to design functional nanostructured lipid carriers (NLCs) based on VCO able to co-encapsulate and co-deliver phytochemical extracts. This innovative approach could provide a significant advance in nanotechnology field and enhance the bioavailability of phytochemical extracts. In this study, the phytochemical extracts used were from four types of famous Malaysian plant medicines (Z. officinale, Z. zerumbet, O. stamineus and L. pumila). The nanocarriers were prepared using melt emulsification homogenization technique. The D-optimal design has been used to determine the optimum combination of different variables for delivering a desired response by using a minimum number of experimental runs. The obtained phytochemical extract loaded VCO based NLCs having diameters ranges of 100 to 200 nm with PDI value less than 0.3 were tested in vitro for their entrapment efficiency and skin permeation activity. Encapsulation of phytochemical extracts was found enhancing penetration across Strat-M membrane due to its nano-size and stability compared to non-encapsulated extracts. Taken together, the results indicated that the VCO based nanocarriers are highly efficient to co-encapsulate and co-deliver the phytochemical extracts, and suitable to be used in developing advanced bio-cosmetic preparations.

Keywords: Virgin coconut oil, Nanostructured lipid carriers, Phytochemical extract,

MN-O-025 [Macromolecules & Nanotechnology]

Non-stoichiometric complex coacervates

A. Başak Kayitmazer, Alaaddin Faruk Köksal, Elif Kılıç Iyilik

Department of Chemistry, Boğaziçi University, İstanbul, Turkey

Hyaluronic acid (HA) and chitosan (CH) can form nanoparticles, hydrogels, microspheres, sponges, and films, all with a wide range of biomedical applications. Turbidimetry, dynamic light scattering, light microscopy and zeta potential measurements show that the state of the dense phase depends on the molar ratio of HA carboxyl to CH amines, and is strongly dependent on their respective degrees of ionization. Due to the strong charge complementarity between HA and CH, electrostatic selfassembly takes place at very acidic pH, but is almost unobservable at ionic strength (I) greater than 1.5 M NaCl. All systems display discontinuity in the I-dependence of the turbidity, corresponding to a transition from coacervates to flocculates. An increase in either polymer chain length or charge density enhances phase separation. Remarkably, non-stoichiometric (not having a 1:1 charge ratio) coacervate suspensions form at zeta potentials far away from zero. This result is attributed to the entropic effects of chain semi-flexibility, charge mismatch between the two biopolymers. The disproportionation of Zhang and Shklovskii (Phys. A, 2005, 352, 216–238) also accounts for the existence of non-stoichiometric coacervates.

Keywords: polyelectrolytes, self-assembly, phase separation, biopolymers

MN-O-026 [Macromolecules & Nanotechnology]

Polymer Coated CdSe/CdS Nanocyristals: Synthesis And Optical Properties

Zeynep Eryılmaz, Vural Bütün

Eskişehir Osmangazi Üniversitesi

Polymers are one of the most commonly used stabilizer materials in wet chemistry due to their tunable and stimuli responsive properties that can be adjusted by varying their components, tail lenghts and sequences. This designable structures make possible to have water soluble, biocompatible and stable heterosystems for biological useage areas and colloidal systems [Pyun, J. and Emrick, T., Wiley-VCH Verlag GmbH & Co. KGaA, 2007]. Besides, during last two decades, size and shape dependent, fluorescent, semiconductor nanocrystals (NCs) have been getting more attention for both science and technical applications such as optoelectronics, photonics, catalysis, solar energy transformation, and biomedical because of their unique properties which cause to have efficient systems [Peng X. et. al., J. Am. Chem. Soc., 2003; Christianen P et. al., J. Am. Chem. Soc. Nano, 2014]. As much size, shape, composition, synthesis method as ligand surroundings that make NCs (photo)chemically stable, change the properties of NCs [Celso de Mello Donega´, Chem. Soc. Rev., 2011]. In this study, CdSe nanocrystals were synthesised via "Hot Injection Method" [Murray C. B., Noms D. J., Bawendi M. G., J. Am. Chem. Soc., 1993], CdSe/CdS hetero nanocrystals were synthesized via "Seeded Growth Method" and photoactive thiazolo-thiazole based polymers and copolymers were used to stabilize CdSe/CdS nanocrystals with different shapes in both hydrophobic and hydrophilic solvents. To have polymer coated nanocrystals "Ligand Exchange" and "Surface Initiated Polymerization" methods were used. Since pyridine is a well-known stabilizer for NCs 4-vinyl pyridine and pyridine end group monomers were used for further polymerization reactions. Optical properties and chemical characterizations of final hybrid structures were investigated for further dye sensitized solar cells and light emitting diode applications.

Keywords: Nanocyristals, Polymer coating, Hybrid nanostructures

Figure 1.

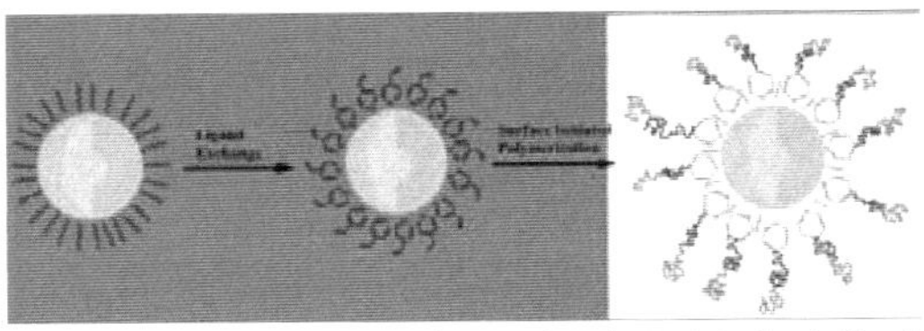

Illustration of polymer coated CdSe/ CdS NCs' synthesis via Surface Initiated Polymerization

MN-O-027 [Macromolecules & Nanotechnology]

Sustainable design of nanoparticles using the triazolinedione (TAD)-ene click reaction

Subrata Chattopadhyay, Filip Du Prez

Department of Organic and Macromolecular Chemistry, Ghent University

Green synthesis of nanoparticles is receiving great interest recently due to its capability to provide alternative, safer, energy efficient routes towards the sustainable design of particles [Chem. Soc. Rev. 2015, 44, 5778]. Two major directions recently evolved in this topic: (i) use of renewables as starting materials, (ii) use of robust, efficient and orthogonal click like reactions.

Our current research effort aims to combine these two directions, i.e. to establish green, robust and simple synthetic platforms to prepare colloidal nanoparticles from renewables. Recently, lots of our group activities focus on the exploration of triazolinedione (TAD) based reactions towards the preparation of materials [Chem. Rev. 2016, 116, 3919]. TAD chemicals, being one of the strongest known dienophiles and enophiles, react very fast with suitable 'ene' compounds. To expand the applicability of the TAD chemistry, in our current work we demonstrate the use of robust, efficient orthogonal 'TAD' based reactions towards the synthesis of nanopartices from different unsaturated renewables, such as plant oils, terpenes etc. Using the newly developed procedures, colloidal nanoparticles can be prepared from different renewables in a very efficient and sustainable way. These new procedures further provide an alternative platform for the future design of different types of nanoparticles, starting from unsaturated starting materials, for various targeted end applications.

Keywords: Plant oil nanoparticles, triazolinedione chemistry, lipid based nanoparticles, nanoprecipitation and nanoparticles from renewables.

Figure 1: Preparation of nanoparticles from renewables using triazolinedione (TAD)-ene reactions.

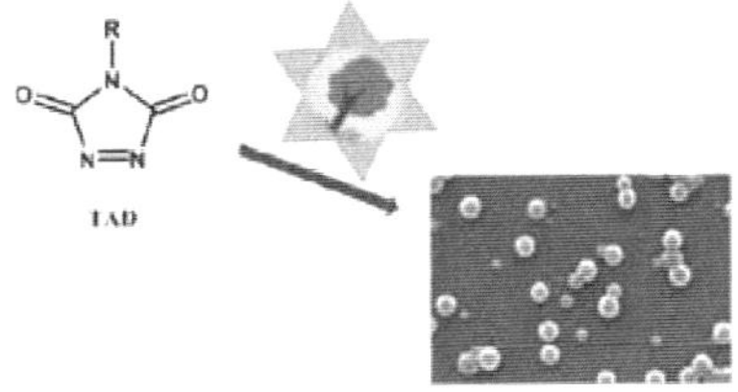

MN-O-028 [Complex Macromolecular Structures]

Synthesis and Surface Properties of PEG Based-Amphiphilic Star-shaped Polymers with POSS (Polyhedral Oligomeric Silsesquioxane) Core

Ahmet Üner[1], Erdinç Doğancı[2], Uğur Cengiz[3], Faruk Yılmaz[1]

[1]Department of Chemistry, Gebze Technical University, Kocaeli, Turkey
[2]Department of Chemistry and Chemical Processing Tech., Kocaeli Voc. Sch., Kocaeli University, Kocaeli, Turkey
[3]Department of Chemical Engineering, Çanakkale Onsekiz Mart University, Çanakkale, Turkey

Organic-inorganic materials have gained quite high interest due to the obtained properties derived from the synergies of organic-inorganic combination since last years, [1]. Polyhedral oligomeric silsesquioxanes (POSS), a class of cage molecules, can be incorporated into various polymeric systems by copolymerization, grafting or cross-linking techniques to obtain tailor-made polymers [2]. POSS molecules have a three dimensional structure with a common formula $(RSiO1.5)n$, $(n \geq 6)$. In essence, among the silsesquioxanes, octasilsesquioxanes have been mostly studied, where eight corners carring one organic group (R) makes them topologically ideal for the synthesis of unique polymeric materials. These eight corner groups provide not only various functionalities with high reactivity but also solubility in organic solvents [3]. Polymeric structures containing POSS have many promising properties such as enhanced oxidation resistance, mechanical, and thermal properties [4]. POSS has been also introduced to various polymeric structures to obtain amphiphilic molecules based on polyethylene glycol (mPEG) and polyacrylic acid (PAA) [5,6]. Here we report the design, synthesis, and self-assembly of POSS containing star-shaped amphiphiles in conjugation with different molecular weight of mPEG polymeric chains (Figure 1). POSS-PEG amphiphiles were synthesized via arm-first approach by using ring-opening polymerization (ROP) and "click" chemistry techniques. All the characterization results obtained are consistent with the chemical structures of the designed polymers. Micellar morphologies also successfully showed the self-assembly properties in transmission electron micsroscope (TEM) images.

Keywords: Amphiphilic polymer, star-shaped polymer, polymeric surfactant, POSS, self-assembly

Figure 1

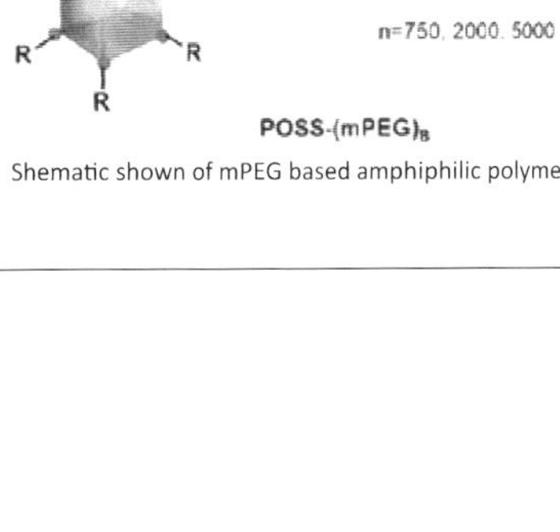

Shematic shown of mPEG based amphiphilic polymers with POSS core.

MN-O-029 [Macromolecules & Nanotechnology]

Magnetic Nanoparticles Surface Modified with Stimuli-Responsive Polymers: An Initial Study of Stimuli-Triggered Phase Transfer

Ersan Eyiler[1], Keisha B. Walters[2]

[1]*Department of Chemical Engineering, Cukurova University, Adana, Turkey*
[2]*Dave C. Swalm School of Chemical Engineering, Mississippi State University, Mississippi, USA*

Superparamagnetic iron oxide (Fe_3O_4) nanoparticles were modified with a pH and temperature responsive homo- and copolymers comprised of N-isopropylacrylamide (NIPAM), 2-(dimethylamino) ethyl methacrylate (DMAEMA), and itaconic acid (IA), via aqueous surface-initiated atom transfer radical polymerization (SI-ATRP). Neat and modified magnetic nanoparticles (MNPs) were characterized using Fourier transform infrared spectroscopy (FTIR), to discern chemical composition. UV-Vis, DSC, and DLS analyses were used to obtain the phase transitions in response to temperature. The LCST of PNIPAM-modified MNPs and PDMAEMA-modified MNPs were around 37 °C and 50 °C, respectively. The PIA-b-PNIPAM-coated MNPs were well-dispersed and stable in water and propanol. These nanocomposites could be used for separations, data coding, hyperthermic heating, and as (magnetofection) carrier for therapeutics, proteins, antigens, and DNA. Phase transfer of thermo-responsive polymer-grafted MNPs between aqueous and organic phases will be studied to store them in a desired solvent for a long time.

Keywords: Poly(itaconic acid), Poly(N-isopropylacrylamide), Stimuli-responsive, Nanoparticle, Phase transfer

MN-O-030 [Macromolecules & Nanotechnology]

Synthesis and characterization of eco-friendly bionanocomposites based on PVA and Zn/Al-LDHs

Ahmed Youssef

Packing and Packaging Materials, National Research Centre, Dokki, Giza, Egypt

Polyvinyl alcohol (PVA) nanocomposite was successfully prepared using Zn/Al (2:1) layered double hydroxide (LDHs) via solution casting method. Firstly, LDHs were prepared through co-precipitation technique. The morphology of the synthesized LDHs and PVA/LDHs nanocomposites were examined by transmission electron microscope (TEM), scanning electron microscope (SEM) and X-Ray diffraction (XRD). TEM and SEM results indicated that LDHs nanoparticles were clearly located in the polymer matrix, with good dispersion. While the XRD results were demonstrated that the exfoliated types of nanocomposites. Furthermore, the mechanical properties of the prepared nanocomposites were evaluated using dynamic mechanical analysis (DMA) and the PVA nanocomposites exhibited good mechanical properties compared with pure PVA. The thermal stability was studied by using thermal gravimetric analysis (TGA) and revealed good thermal stability. Moreover, the fire retardant of the prepared nanocomposites were assessed. The prepared PVA nanocomposites may be used as novel packaging applications.

Keywords: PVA, LDHs, nanocomposites, Exfoliation, Packaging

MN-O-031 [Macromolecules & Nanotechnology]

Nano lead oxide and epdm rubber composite for development of lightweight radiation shielding material: gamma irradiation

Ahmet Güngör[1], Habil Uzun[1], Ismail Kutlugün Akbay[1], Yurdaer Babuçcuoğlu[2], Tonguç Özdemir[1]

[1]*Mersin University Chemical Engineering Department, Mersin, TURKEY*
[2]*Mineral Research & Exploration General Directorate, Ankara TURKEY*

In this study, nano-sized lead II oxide particles were synthesized and was used to obtain a rubber composite material including EPDM and silane binding agent. In order to bind the synthesized PbO particles, surface modification with silane was used. The rubber compounds have been added into a kneader type mixer and homogeneous mixture has been achieved. Then, this homogeneous mixture transferred to a two roll mill and sheets of 2 mm-thickness has been prepared. Finally, the rubber sheet has been vulcanized for a period of 5 min at 180 °C at 20 MPa pressure. The composite materials including %1, %5 and %10 weight percent nano lead oxide was irradiated with doses of 80, 100 and 120 kGy for a maximum irradiation period of 248 days using a gamma ray source with a dose rate of 21.10 Gy/h. The mechanical, thermal properties of the irradiated material were investigated using DSC, DMA, TGA and Universal Testing. Modifications in thermal and mechanical properties of the composite were obtained.

Acknowledgment: This study was funded by The Scientific and Technological Research Council of Turkey (TUBITAK) under project number of 213M476.

Keywords: Lead (II) Oxide, Radiation Shielding, EPDM, Rubber, Gamma Radiation, Nanoparticle

Figure 1

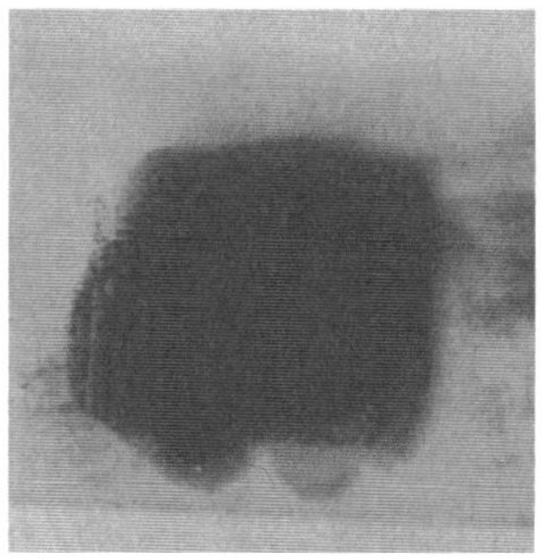

Nano-sized alpha lead (II) oxide nanoparticles

Figure 2

SEM image of the nano-sized beta lead (II) oxide nanoparticles

MN-O-032 [Macromolecules & Nanotechnology]

An advanced preparation and characterization of the PET/MgAl-LDH nanocomposites

Tsung Yen Tsai, Naveen Bunekar

Department of Chemistry, Master Program in Nanotechnology and Center for Nanotechnology, Chung-Yuan Christian University, Taoyuan, Taiwan, R.O.C.

A new nanocomposite poly(ethylene terephthalate)/sulfanilic acid salt-modified layered double hydroxide (PET/MgAl LDH-SAS) was synthesized by an in situ polymerization method using MgAl LDHSAS as a catalyst. This study employed a magnesium–aluminum layered double hydroxide (MgAl LDHs) with various particle sizes, which was modified by sulfanilic acid salt (SAS) through the hydrothermal process to form the MgAl LDH-SAS, and then intercalated by bis-hydroxy ethylene terephthalate (BHET) to form PET/MgAl LDH-SAS. Because of MgAl LDH-SAS, catalytic activity increased the polymerization rate, and nanocomposites also exhibited good thermal stability without the decomposition during the PET polymerization. The results indicated that the modified MgAl LDH-SAS improves the interlayer compatibility between the PET and MgAl LDH-SAS layers, thus making it easier for the oligomer to enter the gallery of MgAl LDH-SAS layers. Thus, the polymer chains can be included between the LDHs layers during the polymerization of the polymer matrix. The gas barrier, thermomechanical properties, and crystallization behavior of these new types of PET nanocomposites are investigated.

Keywords: Poly(ethylene terephthalate), Layered double hydroxide, Nanocomposite.

MN-O-033 [Macromolecules & Nanotechnology]

Preparation of Carbon Nanotube Decorated Fiber Reinforced Polymeric Composite Structures by Using Microwave Energy

Selçuk Poyraz

Department of Textile Engineering, Corlu Faculty of Engineering, Namik Kemal University, Corlu/Tekirdag, Turkey

As an attempt to tackle the potential drawbacks found in existing methods; a simple, facile, yet highly efficient and affordable nano-engineering method, i.e. the PopTube approach, is proposed to obtain fiber reinforced polymeric composites made up of fibrous substrates decorated with carbon nanotube (CNT) forest. This well established novel approach will be explored in details to unveil the "in-situ growth mechanism" of CNTs on various reinforcement fibrous substrates.

The novelty and main advantage of the PopTube approach lies beneath its simplicity. The in-situ CNT growth on any type and shape of engineering material, i.e. fibrous substrates, can be easily accomplished within very short time frames by using a standard kitchen microwave (MW) oven at ambient conditions; (I) without the need of any extra protection, expensive/complicated equipment, feeding gas stock or vacuum system, and (II) with the possibility of precise control on the process parameters such as MW time, power, the feeding stock ratio and so on.

The as-obtained CNT decorated fibrous substrates are meant to be used as reinforcing agents to prepare polymeric composites. The easily scalable and controllable nature of the PopTube approach has a great potential for the initiation of relevant industrial-level applications. The development and adaptation of this novel approach for different areas of science and industry will contribute both the discovery of new production technologies and the promotion of economy, in a positive manner.

The morphological features of the as-produced composites were characterized in details by using scanning electron microscopy (SEM) technique. Both the material characterization results and the versatile and easily controllable nature of the above mentioned process strongly support its promising success for the production of polymeric composites that could be effectively used for advanced engineering applications.

Keywords: Microwave energy, carbon nanotube, fibrous substrate, polymeric composite

MSP-O-001 [Modeling and Simulation of Polymers]

The Kinetic Mechanisms associated with the Phase Transition of a Biomimetic Polymer

Aviel Chaimovich[1], Christian Leitold[2], Kurt Kremer[1], Christoph Dellago[2]

[1]*Max Planck Institute for Polymer Research, Mainz, Germany*
[2]*Faculty of Physics, University of Vienna, Vienna, Austria*

It has been recently shown that a particular biomimetic polymer, whose beads are all uniformly identical, is capable of capturing the phase transition between the collapsed and expanded states of an arbitrary biological polymer [1,2]. Nevertheless, it appears that even for such a supposedly elementary scenario, the kinetic mechanisms cannot be clearly cast in terms of a reaction coordinate [3,4]. Consequently, we continue examining the kinetic mechanisms associated with the phase transition of this biomimetic polymer. Importantly, we systematically perturb the system's Hamiltonian by altering the energetics between a certain pair of beads, evaluating in turn if this perturbation corresponds in a peculiar way with a reaction coordinate.

[1] M. P. Taylor, W. Paul, and K. Binder. Physical Review E, 79(5):050801, 2009.
[2] M. P. Taylor, W. Paul, and K. Binder. The Journal of Chemical Physics, 131(11):114907, 2009.
[3] C. Leitold and C. Dellago. The Journal of Chemical Physics, 141(13):134901, 2014.
[4] C. Leitold, W. Lechner, and C. Dellago. Journal of Physics: Condensed Matter, 27(19):194126, 2015.

Keywords: Biomimetics, Kinetics, Molecular Modeling, Computational Physics

MSP-O-003 [Modeling and Simulation of Polymers]

Hierarchical multi-scale simulation approaches for properties of polymers at different interfaces

Gökhan Kaçar

Department of Genetics and Bioengineering, Trakya University, Edirne, Turkey

In this talk, we demonstrate the procedures we proposed to simulate polymeric materials either in bulk or at metal interfaces with coarse-grained Dissipative Particle Dynamics (DPD) method. Due to the simplicity in its parameterization, the method has become a popular tool to predict and to study the structure of soft matter. In DPD, all beads are treated with similar pure-liquid molar volumes. However, this assumption leads to severe limitations upon application. To overcome this drawback, we recently proposed a generalized parameterization procedure for variable DPD bead volumes. With this alternative parameterization, the structure of a cross-linked polymer, an epoxy, is studied. The simulations are further run for the polymer structure near a metal-oxide. Here, the DPD interactions associated with the soft polymer beads and the hard metal substrate are computed from a dual-scale approach where atomistic molecular dynamics (MD) simulations are coupled with DPD results. As a later step, the mechanical and the interfacial adhesion properties of the epoxy are obtained from a multi-scale approach. The multiscale approach involves fine-graining of the polymer mesoscopic structure and performing atomistic MD simulations. Mechanical properties such as Poisson's ratio, elastic modulus and glass transition temperature; and interaction energy between the polymer and the surface are computed from MD simulations and found to match well with the experimental measurements. In addition to the mechanical properties of polymers, we are interested in the physical properties of soft matter as well. Recently, we have been dealing with the implementation of hydrogen bonding to DPD method for physical behavior of liquid mixtures. In the final part, we will discuss our recent efforts to mimic hydrogen bonding in DPD and its application on self-healing polymers.

Keywords: Coarse-grained simulations, multi-scale modeling, polymers, surfaces and interfaces

MSP-O-004 [Modeling and Simulation of Polymers]

Simulation Study of Diblock Copolymer Crystallization in Thin Films

Chitrita Kundu, Ashok Kumar Dasmahapatra

Indian Institute of Technology Guwahati

The crystallization of diblock copolymer induced by thin films is explored by Dynamic Monte Carlo Simulation. We model double crystalline A-B diblock copolymer wherein crystallization driving force of A-block is higher than B-block (viz., melting point of A-block higher than that of the B-block). The block immiscibility has been modelled by a parameter, λ, which represents the strength of microphase separation and the crystallization driving force has been modelled by the parallel bond interaction energies, U_p (normalized by kT). Increase value of U_p (from 0 to 0.6) represents the transition from melt to crystal state. Film thickness is one of the most influential parameters of thin film crystallization which govern the rate of crystallinity as well as final crystal morphology. We study the non-isothermal crystallization of symmetric diblock copolymer for a series of film thickness (l) ranging from l = 4 to l = 20. With the increase of film thickness (l), the value crystallinity is reduced with the formation of thinner crystal. Similarly the transition points are also influenced by film thickness (l). The crystallization temperature of both blocks decreases with the increment of film thickness (l). Isothermal crystallization reveals that the transition kinetics follows different pathways with the variation of film thickness (l). The value of Avrami Index confirms the presence of homogeneous nucleation with two-dimensional crystal growth.

Keywords: Diblock Copolymer, Crystallization, Thin Film

MSP-O-005 [Modeling and Simulation of Polymers]

Thermodynamics of polymer nematics: comparing particle-based simulations with mean-field calculations

Kostas Daoulas, Cristina Greco, Kurt Kremer

Max Planck Institute for Polymer Research

Semiconducting liquid crystalline (LC) polymers are promising materials for organic electronics. Predicting their morphology on large-scales requires the implementation of mesoscale models. The statistical mechanics of these models is often addressed within mean-field (MF) approaches. MF methods are a basic tool of polymer physics due to computational efficiency and ability to straightforwardly access thermodynamic properties. For LC polymers, however, deviations between MF predictions and experimental results have been reported, e.g. for the molecular weight dependence of the nematic-isotropic transition. A question then arises: are the discrepancies due to the MF approximation or do they reveal fundamental limitations e.g. model transferability?

To address this issue, we consider a model representing polymer nematics as worm-like chains with soft anisotropic interactions and investigate phase behavior by (i) MF calculations based on partial enumeration of conformations, (ii) Monte Carlo simulations.

In the latter, accurate free energies are obtained via a special thermodynamic integration scheme which avoids thermodynamic singularities. Comparing the results of the two methods, we evaluate the effect of fluctuations and local correlations on the macroscopic behavior.

Keywords: liquid crystals, polymers, soft models, free energy calculations, self consistent field, organic electronics

MSP-O-006 [Modeling and Simulation of Polymers]

The Unusual Contrasting of the Same Group Element Chemical Behavior

Isa Degirmenci[1], Michelle L. Coote[2]
[1]Chemical Engineering Department, Faculty of Engineering, Ondokuz Mayıs University, 55139, Samsun, Turkey
[2]Research School of Chemistry, The Australian National University, Canberra, ACT, 2601, Australia.

Thiyl radicals are important intermediates in radical biochemistry, organic chemistry and polymer synthesis.[1] Whilst the chemistry of carbon- and oxygen- centred radicals have received considerable theoretical and experimental attention in the free-radical polymerization, the corresponding reactions of sulphur-centred radicals have generally been less well studied. The addition of thiyl radicals to C=C double bonds is of particular importance in radical thiol-ene reactions [2], and is known to be fast and often barrierless. There has been some debate about the role of polar effects in these reactions, with some experimental studies highlighting a correlation between reaction rate and the electron density of ene molecules, while others have highlighted the role of conjugation of the unsaturated double bond. Indeed, a recent computational study has suggested that a direct correlation between thiyl addition activation barrier and electron density or conjugation of olefins cannot be made directly.[3]

In this study, the reactivity differences of thiyl radical addition to C=C and C=S double bonds and their alkyl and alkoxyl radical analogues were studied via quantum chemical tools [4]. For this purpose, geometry optimizations and frequency calculations were carried out with MP2/6-31G(d) while the high level composite method G3(MP2)-RAD was chosen for improved single point energies. In this, the first theoretical study of thiyl radical addition to thioketones, we find that the chemical behavior of oxygen and sulphur are diametrically opposed, despite being in the same group of the periodic table. Moreover, the fast barrierless reactions of thiyl radicals with C=S bonds suggest that chain transfer processes play a role in the self-healing of materials based on thiruam disulfides.

Keywords: Self-healing polymers, radical thiol-ene reactions, quantum chemical calculations, G3(MP2)-RAD method

Reaction profile for the addition of thiyl radicals to thioketones and alkoxyl radicals to ketones

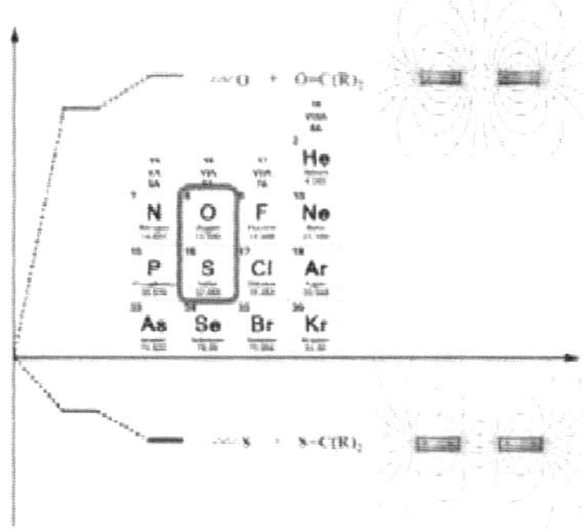

MSP-O-007 [Modeling and Simulation of Polymers]

Determining and modelling polymer crystal structures: the perspective of a polymer crystallographer

Stefano Valdo Meille, Antonino Famulari

Dipartimento di Chimica, Materiali e Ingegneria Chimica "G. Natta", Politecnico di Milano, Milano Italia

From its very beginnings, polymer crystallography has been extensively interrelated with molecular and crystal modelling. The relationship has never been unidirectional, as the two fields have substantially benefitted from the respective developments. Specifically, detailed molecular models have been and still are very often a prerequisite to solve polymer crystal structures. As the quality and the reliability of molecular models is increasing, so does their role in polymer crystallography. On the other hand experimental crystal structures, lattices and symmetries, not to speak of crystal morphologies and their evolutions represent essential testing grounds of modelling abilities. They can still hardly be predicted from molecular data, but rationalization of properties and of structural developments from known arrangements, is becoming on the other hand increasingly viable. Polymer crystal structures, as compared to low molecular mass structures, present often unique properties related to the unique features of polymer chains for which intramolecular properties tend to play a more significant role. Also the small size and the inherent anisotropy of crystals, the extensive role of disorder, the stability of mesophases, often dependent on molecular masses, the role of conformational chirality and of chain directionality in determining packing of chains, are very different than in low molar mass materials. In the present communication selected recent examples in the fields of conjugated polymers but also of more traditional polyolefins will be reported and discussed, especially with reference to the significance of so called "limited ordered" models and to the treatment of disorder in crystallographic and in modelling perspectives.

A. Buono, G. Raos, S. V. Meille et al.: Macromolecules, 43, 6772 (2010).
A. Buono, A. Famulari, S. V. Meille et al. Macromolecules, 44, 3681 (2011).
A. Famulari, S. V. Meille et al. J. Phys. Chem. B, 116, 14504 (2012)

Keywords: crystal structures, modelling, simulation

MSP-O-008 [Modeling and Simulation of Polymers]

Study on the physics of polymer crystallization based on the rod-coil multi-blocks chain model

Ping Tang, Faqiang Liu, Hongdong Zhang, Feng Qiu, Yuliang Yang

Key Laboratory of Molecular Engineering of Polymer, Fudan University, Shanghai, China

Only considering the long-chain characteristics of polymers can we understand the physics behind polymer crystallization. We aims to construct the microscopic molecular theory to describe polymer crystallization based on the more accurate chain model, which can both consider the long-chain conformation entropy and the segment orientation. By developing the self-consistent field theory for multi-block rod-coil copolymers as a model system, we can investigate the effect of chain structure such as the comonomer composition, comonomer sequence (such as polyolefins) on the crystallization degree and crystal structures. We also expect to simulate the chain folding mechanism to compare with the phenomenological crystallization model. The developed work will be helpful to understand the physical mechanism and essence of polymer crystallization and to guide the design of crystal structures.

Keywords: polymer crystallization, rod-coil multiblocks chain model, self-consistent field theory, chain folding

MSP-O-009 [Modeling and Simulation of Polymers]

Crystallization of Double Crystalline Binary Polymer Blends

Ashok Kumar Dasmahapatra

Department of Chemical Engineering, Indian Institute of Technology Guwahati, Guwahati - 781039, Assam, India

Polymer blends are known to be one of the exciting materials to prepare nanoscale devices via self-assembly. The thermodynamic and mechanical properties largely depend on their mutual miscibility. Miscible blends usually give a single glass transition temperature, whereas immiscible blends produce two distinct glass transition temperatures. Since the components in the blend are chemically dissimilar they phase separate via macrophase separation. We report dynamic Monte Carlo simulation results on the crystallization of A/B binary polymer blends with varying composition, wherein both the components are crystallizable. We model A-polymer as high melting component and hence its crystallization precedes the crystallization of B-polymer upon cooling from a homogeneous melt. The morphological development is controlled by the interplay between crystallization driving force (attractive) and de-mixing energy (repulsive) between the components. With increasing the composition of B-polymer, macrophase separation, crystallization and lamellar thickness follow a non-monotonic trend. This non-monotonic trend is attributed to the composition-heterogeneity in the blend. When one component is relatively less compared to the other, its mobility is reduced affecting transition temperature during crystallization. As a result, transition happens at a relatively lower temperature (viz., enhanced thermal driving force). Isothermal crystallization reveals that the crystallization behavior and crystal morphology strongly depends on the mode of cooling. Two-step, compared to on-step isothermal crystallization provides better crystalline structures.

Keywords: Binary blend, Crystallization, Monte Carlo Simulation

PCNCT-O-001 [Polymer Characterization by New and Combined Techniques]

About the thermo-mechanical Coupling of PET at the Yield-Point

Marcus Stein, Sascha Bach, Jens Peter Majschak

Chair of Processing Machines and Processing Technology, Technical University Dresden, Dresden, Germany

The deformation of material is generally coupled with a thermal reaction of the deformed material. Beside homogeneous temperature decrease at elastic deformation, plastic deformation during yielding and crack propagation lead to local dissipation of heat (1).

In this study we concentrate on the characterisation of the thermo-mechanical behaviour until yielding. At this characteristic stage of deformation a sudden local temperature rise above Tg can be observed for the investigated amorphous PET. Dependent on the maximum temperature and the temperature distribution in the necked area the deformation continues with progression of the neck or a crack is initiated immediately. These highly dynamic processes at yield are mainly determined by the deformation velocity and the shape of the test specimen. The influence of both were systematically investigated in this study.

For the tests a specimen according to the ISO 527 standard was used. The neck initialization was induced at a predetermined area. At this area the deformation and heat dissipation was observed by a high-speed IR camera. The self-similar specimen with various parallel length enabled to influence the active storage for elastic energy and thereby the available mechanical energy at yield. The data were analysed with the focus on the maximum temperature (Figure 1) and global energy parameters particularly mechanical work and dissipated heat at yield. The complete parameter set was statistically evaluated.

The result will be discussed under the model of the storage volume of elastic energy and will provide some new aspects for describing the complex interrelationships and processes at the yield point.

(1) Stein, M. & Schneider, K. (2016). Thermographic characterisation of the deformation und fracture behaviour of polymers with high time and spatial resolution. In W. Grellmann & B. Langer (Eds.), Deformation and Fracture Behaviour of Polymer Materials. Berlin: Springer (in progress).

Keywords: Heat Dissipation, High Speed Infra-Red Measurement, Elastic Energy Storage

Maximum temperature at yield depend on the parallel length of the test specimen for 100 mm/min stretching velocity

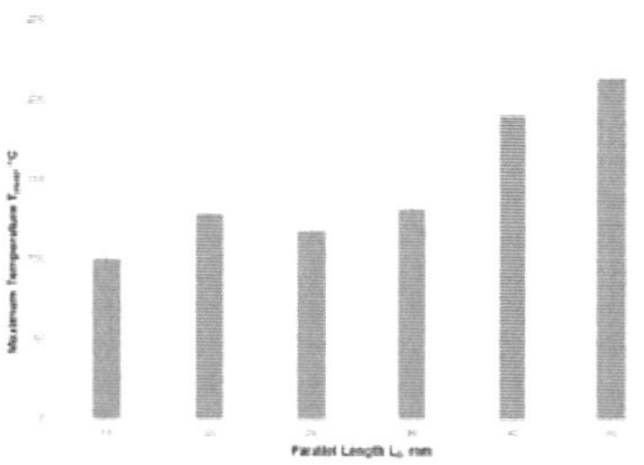

PCNCT-O-002 [Polymer Characterization by New and Combined Techniques]

New Applications for Science and Industry using a Universal Extensional- Fixture (UXF) with a counter rotation MCR Rheometer

Roman Fuehrer

Anton Paar GmbH, Graz, Austria

Extensional testing methods with controlled stress or strain play a significant role in polymer science and industry. These tests are performed in order to determine the mechanical coating strength and extensional rheology behavior, that can provide information about the inter particle molecular bonding, the degree of cross linkage or branching of polymer solids, melts or elastomers. Besides dynamic mechanical oscillatory testing (DMA) that can determine viscoelastic properties, the Universal Extensional Fixture (UXF) system can also be used for tensile testing, shrinkage testing and extensional rheological testing on thin films and fibers. Such combined testing system is used to investigate effects such as temperature stability, shrinkage, elongation, crystallizations, phase transitions, relaxations and the brittleness of films and fibers.

The UXF TwinDriveTM uses the counter rotation mode of Anton Paar MCR 702 to perform extensional Rheology. Since the drum is directly connected to the air bearing EC-motor, the UXF TwinDriveTM principle shows the best low torque sensitivity.

Shrinkage due to internal stresses and molecular orientations is a common problem in the production of polymeric films. The shrinkage is often visible not until months after manufacturing which makes it difficult to detect. We can accelerate this process through the use of elevated temperatures in the CTD oven or with solvents in the fluid immersion option with the UXF.

The influence of relative humidity on Films and Fibers can be investigated using the humidity option of the CTD. This feature can be extremely useful when considering poly condensation products such as polyamides that can be influenced from moisture on the modulus value and the glass transition temperature. The unique combination of special drums, TwinDriveTM rotation mode, immersion and humidity option and radiation/convection oven gives new applications for science and industry.

Keywords: Rheology, DMA, Extensional, TwinDrive

PCNCT-O-003 [Polymer Characterization by New and Combined Techniques]

The Use of Direct Pyrolysis Mass Spectrometry Technique to Investigate the Nanostructure of Polymer Nanoclay Composites

Jale Hacaloglu[1], Esra Özdemir[2]

[1]Department of Chemistry, Middle East Technical University, Ankara, Turkey
[2]Department of Polymer Science and Technology, Middle East Technical University, Ankara, Turkey

Polymer nanocomposites have attracted great attention due to the improvements in mechanical, thermal, barrier, durability, chemical stability, flame retardancy, scratch/wear resistance, biodegradability as well as optical, magnetic and electrical properties. Clays, one group of nano-fillers widely used for the preparation of polymer nanocomposites, are classified according to the degree upon which the clay layers are separated from one another. Composites vary between two main structures, intercalation and exfoliation. The morphologies of nanocomposites, pure clay or pure organoclay, the degree of intercalation/exfoliation are generally determined by two common techniques X-ray diffraction (XRD) analysis and transmission electron microscopy (TEM). Clays and organoclays show a characteristic peak, indicative of the platelet separation or d-spacing in clay structure, in XRD analysis due to their regular layered structures. The shift of characteristic peak to lower angles is an indication about intercalation, an increase in the basal spacing between clay layers. However, the disappearance of such a peak is not a conclusive evidence for a highly exfoliated structure in nanocomposite. A direct way to visually observe the nanostructure of nanocomposites and clay d-spacing is to the use of transmission electron microscopy (TEM) technique, which allows a qualitative understanding of the internal structure, exfoliation or spatial distribution of layers within the polymer matrix and their structural defects.

Recent studies in our group indicated a direct relation between the evolution and/or the decomposition of organic modifier and interlayer spacing between the clay layers during the pyrolysis of polymer nanoclay composites. As the interlayer spacing between the clay layers increases, the elimination of organic modifier takes place at lower temperatures. Furthermore, it has been determined that the direct pyrolysis mass spectrometry results not only give information on interlayer spacing but also on possible interactions between the organic modifier and the polymer.

Keywords: nanostructure, nanoclays, direct pyrolysis mass spectrometry

PCNCT-O-004 [Polymer Characterization by New and Combined Techniques]

Application of gel permeation chromatography in polymer analysis

Subin Damodaran, Bernd Wolf

Separations unit of Tosoh Bioscience GmbH, Griesheim, Germany

Consumers of polymers were attracted by its physical and chemical properties. These properties were tailored by molecular weight, presence of branching, molecular structure and shape of the polymers. For polymeric materials the molar mass and molar mass distribution plays a vital role in the determination of mechanical, bulk, and solution properties. These properties govern polymer processing and the end-use performance of a given material. Unlike small molecules, which have discrete molar mass distributions, synthetic polymers are typically composed of hundreds to thousands of chains of different molar mass that result in a distinctive molar mass distribution. The shape and breadth of a polymer's molar mass distribution will depend on the mechanism, kinetics and condition of the polymerization, and will dictate the end-use properties of the polymer. Polymer properties such as, hardness, tear strength, impact resistances, wear, brittleness, toughness, tackiness, etc., are important in determining the successes or failure of a given material. Gel permeation chromatography (GPC), also known as size exclusion chromatography (SEC) is a technique that separates the molecules according to their hydrodynamic volume and provides information on the molecular weight distribution of polymers. GPC/SEC can determine molar mass, short and long chain branching, crosslinking, tacticity and chemical heterogeneity of polymers.

Keywords: Gel permeation chromatography, Size exclusion chromatography, Polymers

PCNCT-O-008 [Polymer Characterization by New and Combined Techniques]

Dielectric and thermal properties of poly(ethylene oxide)-based polymer electrolytes

Chin Han Chan, Hans Werner Kammer

Faculty of Applied Sciences, Universiti Teknologi MARA, 40450 Shah Alam, Selangor, Malaysia

First part of the discussion is on systems with poly(ethylene oxide) (PEO) comprising the lithium perchlorate salt (LiClO4), and also pure low-molecular PEO. Long-ranging Coulomb interactions cause deviation from perfect behaviour. The accompanying osmotic pressure tends towards Debye-Hückel like behaviour. Under this general strategy, we discussed thermodynamic aspects of the relaxation process. Structure formation in these systems creating key quantity of length ratio (double layer/sample thickness). This rules the relaxation process via friction force or dissipation of energy. In this spirit the transport process was newly discussed and all properties can be understood comprehensively in phenomenological way. PEO-300K and PEO-1K with glass transition temperature (Tg) at around -56 °C and -60 °C, respectively. We notice that the influence of Tg of PEO-300K and PEO-1K on conductivity is insignificant. However, the melting temperature (Tm) of PEO-1K at round 37 °C exhibits great correlation to dielectric properties. Correlation of Tm on dielectric and transport properties is discussed as well.

In the second part of presentation, the influence of poly(methyl methacrylate) (PMMA) (miscible) (high Tg) or poly(methyl acrylate) PMA (miscible) (low Tg) on the conductivity of PEO after addition of Li salt is discussed. PEO, PMMA and PMA possess oxygen in their respective chemical structures, which may be able to coordinate with the Li salt added. The conductivity of PEO-salt mixtures is slightly higher than that of the miscible PEO/PMMA/salt or PEO/PMA/salt (dispersed PMMA or PMA in PEO phase). This may be due to reduced segmental motion cause by the increased Tg of the blend and a decrease in free ions in the amorphous phase of PEO as a small amount of the salt is solvated by PMMA in the blend. Therefore, the percolation path lies in the amorphous PEO rich phase of the blend.

Keywords: Dielectric impedance, transport coefficient, poly(ethylene oxide), solid polymer electrolytes

PCNCT-O-009 [Polymer Characterization by New and Combined Techniques]

Characterization of Hyaluronic Acid derivatives – Novel fundamental inside

Davide Ret[1], Gökçe Mihlaç[2], Stefano Gentilini[1], Simone Knaus[1]

[1]*Institute of Applied Synthetic Chemistry, Vienna University of Technology, Vienna, Austria*
[2]*Departmant of Chemistry, Istanbul Technical University, Istambul, Turkey*

Hyaluronic acid is a biomaterial used in medical and esthetic application such as eye surgery, viscosupplementation therapy of osteoarthritis, dermal filling, and tissue augmentation. To improve the chemical and physical properties in biological systems hyaluronic acid is chemically modified. In establishment of the relationship between different degrees of modification and mechanical properties as well as biocompatibility the accurate determination of the degree of substitution (DS) is of great importance. An effective and wide spread method for determining the DS is ^{1}H-NMR spectroscopy. This method gives accurate results only when the implied protons are mobile (Michel and Marguerite 2004). Hyaluronic acid proton mobility is affected by many parameters as: molecular weight, concentration, ionic force, pH and temperature. In the present work the determination of the DS was intensively studied and the relationship between particle size, conformation in solution and proton mobility was elucidated. It will be shown that accurate determination of the DS by ^{1}H-NMR spectroscopy for HA derivatives is possible (Figure 1) and that the same principle can be used for other glycosaminoglycans.

Acknowledgements : The authors gratefully acknowledge the financial support from Croma Pharma (Austria).

Keywords: Hyaluronic Acid, proton mobility, Degree Substitution, structure, NMR

Figure 1

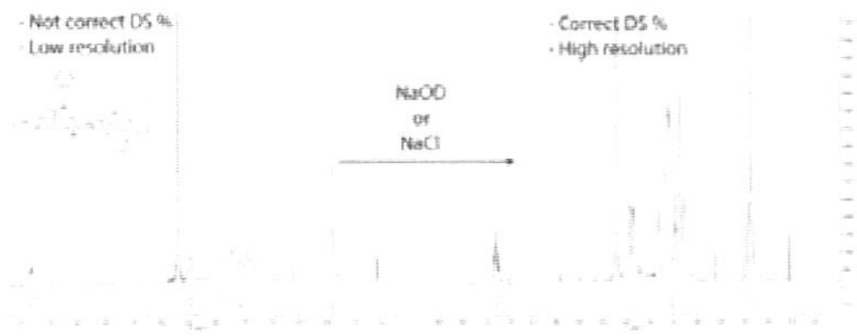

1H-NMR spectrum of HA benzylamine derivate acquired in D2O without and with NaOD or NaCl

PCNCT-O-010 [Polymer Characterization by New and Combined Techniques]

Characterization of Branching in Poly(Acrylic Acid) by 13C NMR Spectroscopy and Capillary Electrophoresis

Jean Baptiste Henri Lena[1], Gregory Thomas Russell[1], Alexander Goroncy[1], Patrice Castignolles[2], Joel Thevarajah[2], Alison Maniego[2], Marion Gaborieau[3]

[1]*Department of Chemistry, University of Canterbury*
[2]*Western Sydney University, Australian Centre for Research on Separation Science (ACROSS), School of Science and Health*
[3]*Western Sydney University, Molecular Medicine Research Group, School of Science and Health*

Poly(acrylic acid) (PAA) is a hydrophilic polymer with a wide range of applications, e.g. as a superabsorbent, water purifier, drug carrier and scale-control agent. Many characterisations of PAA have been carried out in this and in previous works, e.g. by ESI-MS, solution-state NMR, SEC, and capillary electrophoresis in critical conditions (CE-CC). Even if it is not always acknowledged, PAA is branched. The branching strongly influences the polymer's properties. Short branches influence physical properties such as melting point and glass transition temperature whilst long branches affect rheology. To provide a better understanding of the mechanism and kinetics of radical polymerization of acrylic acid that lead to branching, traditional and new methods of characterization are here employed. Namely, the degree of branching (DB) was determined by quantitative 13C solution-state NMR spectroscopy and the heterogeneity of branching was assessed by CE-CC. The aim of our research is to study the influence of different parameters in the synthesis of PAA on the structure of the polymer, including the presence of a chain transfer agent (CTA), the initial monomer concentration, and the temperature. It was found that the presence of a CTA reduces both the degree of polymerization and the DB, whilst also increasing the heterogeneity of branching. In the presence of CTA mid-chain radicals were found to transform into linear species by a "patching effect" or the frequency of the backbiting step was reduced. Consequently, the DB was considerably reduced. In this case, as both linear and branched chains were present, the heterogeneity of branching was higher. Temperature was also observed to increase DB, by increasing the frequency of backbiting.

Keywords: Capillary Electrophoresis, Solution-State NMR spectroscopy, Branching

PCNCT-O-011 [Polymer Characterization by New and Combined Techniques]

Structural characterization of synthetic polymers via mass spectrometry (MS)-based strategies

Esra Altuntas[1], Jan Jordens[1], Ynze Mengerink[1], Maarten Honing[2]

[1]*DSM Resolve, Geleen, The Netherlands*
[2]*Division of Bioanalytical Chemistry, Amsterdam Institute for Molecules Medicines and Systems (AIMMS), Faculty of Sciences, VU University Amsterdam, The Netherlands*

Structural characterization of synthetic polymers is an important part of polymer science due to the fact that the performance of the polymeric material is directly linked to its properties such as molar mass, architecture, functional and end groups. Mass spectrometry (MS)-based techniques play a crucial role in the characterization of synthetic polymers to fulfill the need for an absolute characterization method. MS is the most versatile and comprehensive method in OMICS sciences such as in proteomics and metabolomics. Sequence information of the full complement of biological samples can be obtained by the applications of MS and tandem MS (MS/MS) techniques together with the right sample preparation and separation methods. The control of polymer sequences and their accurate characterization is one of the significant challenges of current polymer science. Therefore, similar approach can be utilized for the structural characterization of synthetic polymers [1]. In this work, the detailed characterization of various synthetic polymers such as different polyamides (PA6 and PA66) was achieved via MS-based strategies applying as in top-down and bottom-up sequencing approaches in other OMICS fields. MS-based strategies allowed a relatively precise examination of the investigated polyamide structures. Tandem MS (MS/MS) experiments were performed in order to gain further structural information from the resulting fragmentation patterns and to understand the fragmentation mechanism of these polymers. The obtained information provided the necessary knowledge for the structural elucidation of this type of polymers in the future. Moreover, a new technique "Trapped Ion Mobility Mass Spectrometry" (TIMS) was applied to sequence various polymers. Polymers with a different amount of branching were also analyzed by TIMS and their structures were compared to determine the branching ratios.

1. E. Altuntas and U.S. Schubert, Analytica Chimica Acta, 808, 2014, 56–69.

Keywords: polymer characterization, mass spectrometry, structural elucidation

PCNCT-O-012 [Polymer Characterization by New and Combined Techniques]

A Novel Polymeric Radiation Shield reinforced with WO3: Design, Performance and Structural-Thermal Characterization

Ezgi Eren Belgin, Gül Asiye Ayçık

Department of Chemistry, Muğla Sıtkı Koçman University, Muğla, Turkey

During the last decades, composites are seemed to be good solution for a proper shielding design by possessing both filler and matrix properties. Polymer based composite shields are widely studied nowadays because of flexibility, lightness, cheapness and mechanic-chemical strength of the polymers. Thus in the study, production of a composite radiation shield with low density and high attenuation performance was aimed. Isophytalic unsaturated polyester with the density of 1.15 gcm^{-3} was used as the matrix polymer. Because of the instability and processing hardness of the pure tungsten the oxide form of tungsten with monoclinic crystal structure and 7.16 gcm^{-3} density was also used as filler material of the composite. Preparation of the composites was accomplished by free radicalic polymerization of the PES resin after dispersion of the filler particles within the prepolymerat different filler loading values.

Performance of the composites was investigated by gamma spectrometric methods for the three different radiation energy range of 0-2000 keV. The structure was cleared up by FTIR and SEM studies. Thermal properties were also investigated by TGA and DSC studies.

The attenuation performance of the composite providing best results approximately reached lead's performance for low radiation energies while for the highest energy it was reached to approximately 31% of the lead. But it was the most important result that the heaviest composite was approximately 6 times lighter than lead.By considering that the most common users of the radiation shields are the X-ray workers that are exposed to maximum 100 keV radiation energy this novel polymeric shields are remarkable candidates of being future's radiation shields with their nontoxic nature and lightness.

Keywords: Radiation shielding, composite shielding, polymeric composite

PCNCT-O-013 [Polymer Characterization by New and Combined Techniques]

Force spectroscopy with real-time modeling and analysis on soft biological and polymer materials

Rached Jaafar[1], Dalia Yablon[2], Nettikadan Saju[3], Christian Bippes[1], Patrick Frederix Patrick Frederix[1]

[1]*Nanosurf AG, Liestal, Switzerland*
[2]*SurfaceCharLLC, Sharon MA*
[3]*Nanosurf Inc., Woburn MA*

Atomic Force Microscopy (AFM) is one of the major enabling technologies paving the way to today's understanding of surfaces at the nanoscale. Single point measurements in the form of force spectroscopy, where force vs. tip-sample separation data is collected and then modeled with any number of contact mechanics models, can be especially useful for measuring properties such as material stiffness and adhesion. Current force spectroscopy approaches can be slow, accompanied by a tedious analysis that has to be applied to a very large data set that can make meaningful interpretation of the data challenging. We present force spectroscopy measurements that incorporate real-time quantitative measurements and analysis of adhesion and stiffness providing a rapid assessment of material properties. Measurements on biological systems and polymer blends reveal a quick, quantitative differentiation in stiffness and adhesion within the different components of the polymer blend and between the different biological samples. In addition to real-time analysis with adhesion-based contact mechanics models, analysis of the properties of interest can be further refined by assessing different parameters in the force curves such as such as bimodality, baseline correction, and splitting for improved assessment.

Keywords: atomic force microscopy, real-time analysis, material properties

PE-O-001 [Polymer Education]

Impact of Nanotechnology on Higher Education Curricula: Cases of "Materials Science" and "Polymer Science & Engineering" Curricula

Djafer Benachour

LMPMP; Department of Process Engineering, Faculty of Technology, Ferhat ABBAS University SETIF 1, Setif,Algeria

The fast and large development of nanotechnology over the last three decades, have led to: i) the production of nanomaterials that are now being used in different fields. These nanomaterials spread out to many scientific and technologic fields, notably in materials science, in catalysis and in medicine. For instance, smart nanocarriers for drugs delivery and targeting, ranging from liposomes, to polymeric nanospheres and micelles, are now used for cancer treatment. These drugs nanocarriers helped the emergence of what is now called "Nanomedicine". ii) the emergence of new techniques and technologies in chemical synthesis. For example, the production of a new type of catalysts that are much more efficient: nanocatalysts: this new type of catalysts, combined to microreactor technology, were behind the mutation of catalysis, from art to science. Many synthesis reactions are now performed with a 100% yield and with no side products... laboratory scale. These scientific and technological developments are having a strong impact, not only on chemical industries and chemicals production... but also on chemistry and chemistry related education curricula and research activities at different levels. This impact is clearly seen over the comparison of higher education curricula evolution of three major disciplinary fields: materials science. polymer science and chemical engineering. The evolution of such curricula from different leading higher education institutions, is followed over the last three decades. Nanotechnology related topics increased regularly, going from "discovery and optional" courses in the 1980's to "major and fundamental" teachings in this beginning of the XXIst century. For instance, polymer nanocomposites are now industrially produced and their teachings have become an important part in materials/polymer science curricula. Atomic force microscopy, a powerful and unavoidable technique for surface characterization at the nanoscale level, Is now available in most research laboratories...

Keywords: Nanotechnology; Education, Curricula, Materials, Polymers, Nanomaterials

PE-O-002 [Polymer Education]

Awareness on polymers among high school students

Emine Erdem, Hüseyin Kara

Hacettepe University Faculty of Education, Secondary Science and Mathematics Education, Chemistry Education, Beytepe/Ankara

The aim of this study was to raise the awareness of high school students on polymer materials, which are commonly encountered in everyday social life, and frequently discussed in academic contexts. We expect that this study will contribute to students' ability to "apply in everyday life the information learned in chemistry classes," which is one of the many important skills conferred by the education system. This pilot study was based on a descriptive survey method. The sample of the study consisted of 9th, 10th, 11th and 12th grade students taking chemistry classes an Anatolian Teacher High Schools, Social Sciences High Schools, Anatolian High Schools and Science High Schools in Turkey. Data collection was performed using the Opinion on Polymers Questionnaire prepared by the researchers as data collection tool. Data were analyzed using the content analysis technique. Categories were formed based on the data obtained through the Opinion of Polymers Questionnaire; these categories and their associated codes were then used to determine the frequency values of the concepts expressed by the students. Based on the study results, a general evaluation was made regarding the students' level of awareness on polymers in chemistry class, and a number of recommendations were also proposed.

Keywords: Polymer, Awareness, Chemistry Education

PEPC-O-001 [Polymer Engineering, Processing, and Characterization]

Twin-Screw Extrusion Isolates High Viscosity Polymers

Bernabe Quevedo Sanchez[1], Tople V. Nitin[2], Hemmer L. Gregory[2]

[1]Department of Technology and Innovation, Cartagena, Spain, SABIC
[2]Department of Technology and Innovation, Mt Vernon (IN), USA, SABIC

SABIC is a global leader in diversified chemicals headquartered in Riyadh, Saudi Arabia. We support our customers by identifying and developing opportunities in key end markets, such as construction, medical devices, packaging, agri-nutrients, electrical and electronics, transportation and clean energy. We employ more than 2000 scientists worldwide and operate in more than 50 countries. Fostering innovation and a spirit of ingenuity, we have 10,960 global patent filings. SABIC ranks as a top producer in polycarbonate, polyphenylene ether and polyether imide.

Production of resins typically requires the separation of the polymer from an organic solvent. Thermal-mechanical processors, such as agitated thin film evaporator (ATFE), twin-screw extruder (TSE) or a combination of the two, are the preferred equipment to isolate high viscosity resins.

Capacity expansion with attractive return on the investment required the minimization on equipment cost maintaining resin quality. For this reason, a novel extrusion process was recently developed to isolate polymers with viscosities up to 20,000 poise at 120 ºC above the glass transition temperature and solvent content up to 75% weight.

Process intensification to a one-step devolatilization extruder reduced equipment cost by 40% compared with ATFE's two-step process. Development of this technology required i) work at the laboratory scale to determine barrel configuration and a screw design that enhanced surface renewal and vapor disengagement; ii) optimization of operating conditions to produce resin with ppm levels of solvent; iii) scale-up demonstrating commercial feasibility of high viscosity polymers.

Long term stability in the extrusion operation has been demonstrated. Conditions must be within operability limits to avoid thermal-degradation of polymer and excessive gas velocities that may lead to vent plugging. In addition, the modular screw design enabled a wide range of production rates, from full capacity down to 30% of the maximum rate.

Keywords: extrusion, evaporation, degassing, devolatilization, process intensification

PEPC-O-002 [Polymer Engineering, Processing, and Characterization]

Rheological Behavior of HDPE/Natural Needle Like Clay Nanocomposites Under Shear and Extensional Flow

Vishwa Pratap Singh[1], Gurpreet Singh Kapur[2], Veena Choudhary[1]

[1]*Centre for Polymer Science and Engineering, Indian Institute of Technology, Delhi Hauz Khas New Delhi, India 110016*
[2]*Indian Oil Corporation Ltd. Research & Development Centre Sector 13, Faridabad, Haryana, India 121007*

In the present study, high density polyethylene (HDPE)-based composites containing varying amounts of sepiolite [1-10 wt%] and PE-g-MAH as compatibilizer [varying maleic anhydride content and MFI], were prepared by melt compounding using corotating twin screw extruder. First masterbatch of HDPE containing 20 wt% sepiolite was prepared which was then diluted to obtain the desired level of filler loading. The effect of sepiolite loading and compatibilizer type on the rheological behavior of HDPE-sepiolite nanocomposites was investigated with a strain-controlled rheometer. The melt-extensional properties such as melt strength and drawability were investigated by using Rheotens set up and extensional viscosity was examined by using two different methods i.e. Rheotens set up and capillary Rheometer. High shear viscosity and melt fracture analysis were also carried out using capillary rheometer. Incorporation of varying amounts of sepiolite enhanced the melt strength and extensional viscosity with slight reduction in drawability at high filler loading. Extensional viscosity calculated from Wagner's master curve using Rheotens data as well as calculated by modified Cogswell method using convergent flow analysis from capillary data exhibited good agreement. High shear viscosity increased upon increasing filler content with slight decrease in onset of melt fracture at 10 wt% sepiolite loading. These properties were further increased significantly with the addition of compatibilisers to HDPE/Sepiolite composites without affecting the shear viscosity. It was found that the storage modulus, loss modulus and complex viscosity in oscillatory rheometry were highest for nanocomposite prepared using 10% w/w sepiolite. The conclusion is that such composites with higher melt strength and extensional viscosity can be used in blow moulding applications.

Keywords: HDPE, Sepiolite, PE-g-MAH, Melt Extensional properties, Melt fracture

PEPC-O-003 [Polymer Engineering, Processing, and Characterization]

Optimization of injection molding process conditions while manufacturing of HDPE/TiO2 nanocomposites

A S Mohammad Sayem Mozumder[1], Hifsa Pervez[1], Abdul Hamid Ismail Mourad[2]

[1]*Chemical & Petroleum Eng Dept, UAE University*
[2]*Mechanical Eng Dept, UAE University*

This study presents the optimization of the injection molding process parameters during fabrication of titanium dioxide enriched high density polyethylene (HDPE-TiO2) nanocomposites. The mechanical properties of the nanocomposites including yield strength, Young's modulus and elongation were selected as the performance targets. Experimental runs were carried out based on the L4 Taguchi orthogonal array. Operating conditions such as barrel temperature, residence time and injection time were optimized using Grey relational analysis in which the degree of influence of each control factor was determined by the grey relational grade. The optimal process parameters were found 200°C - 50 min - 20 s on the basis of highest grade.

Keywords: Optimization, Polymer Processing, Injection Molding

EPC-O-005 [Polymer Engineering, Processing, and Characterization]

Graphene-Based Polymer Nanocomposites for Thermal Conductivity and Electromagnetic Interference Shielding Applications

Chong Min Koo

Korea Institute of Science and Technology

Highly thermally conductive materials with great electromagnetic interference (EMI) shielding efficiency have been attracted much attention for a wide range of applications in the modern high-power electronics, portable devices, and self-driving cars. The highly integrated and high-speed wireless communication devices suffer from significant heat generation and undesirable electromagnetic interference effect. The heat emission and EMI not only deteriorates the performance of the devices but also makes serious concern on harmful health problem to human. Due to unavailability of the dual function materials, unfortunately, there have been used multilayer films consisting of EMI shielding layer of metal or metal alloy, and thermally conductive materials layer of artificial graphite in the modern electronic devices, even though these multilayer sheets are ineffective and expensive. In this presentation, I would like to briefly demonstrate that graphene, and graphene-based polymer composites can be considered as one of the dual function materials with high thermal conductivity and EMI shielding effectiveness.

Keywords: Graphene, Nanocomposites, EMI shielding, Thermal conductivity

PEPC-O-006 [Polymer Engineering, Processing, and Characterization]

Effect of polymer on the existence of a critical thickness in coextruded multilayer films

Adrien Bironeau, Alain Guinault, Guillaume Miquelard Garnier, Cyrille Sollogoub, Gilles Régnier

Pimm, Ensam, Cnrs, Cnam, Paris, France

Coextrusion through a series of multiplying die elements enables the production of polymeric films containing thousands of layers of alternating polymers with individual layer thicknesses below 100 nm (Schrenk, USA Patent, 3,884,606, 20-May-1975). Previous studies have shown improved macroscopic properties (especially permeability) related to the nanoconfinement of the polymers within the layers (Ponting et al., Macromol. Symp. 2010, 294, 19-32; Wang et al., Macromolecules 2009, 42, 7055-7066). However, when a lower limit in thickness is reached, instabilities during process may occur and thus affect integrity and homogeneity of layers, resulting in altered final properties of the materials. However, an understanding of the reasons for these instabilities has not been achieved yet.

In this systematic study, the role of several processing and material parameters on the onset of instabilities and the possible existence of a critical thickness is discussed. Two couples of polymers were investigated: PMMA/PS and PMMA/PC. Films with 1025 to more than 8000 alternating layers, with different mass ratios and draw ratios, were fabricated in order to achieve the thinnest thicknesses possible (in theory, a few nm). Films were characterized by atomic force microscopy. A statistical analysis was used to estimate the sample size representativity and to determine the distribution in layer thicknesses and the stability of layers (Bironeau et al., J. Microsc., in press). Results confirm the existence of a critical thickness below which the layers spontaneously break-up. This critical thickness is reached whatever the processing parameters studied for a given couple and appears to depend on the polymers physical properties (such as molecular weight) and interactions with each other.

Keywords: Nanolayered film, Atomic force microscopy, Coextrusion, Critical thickness

PEPC-O-007 [Polymer Engineering, Processing, and Characterization]

Consolidation by Spark Plasma Sintering of Poly-Ether-Ether-Ketone

Jean Charles Sébileau[1], Sébastien Lemonnier[1], Elodie Barraud[1], Adele Carradò[2], Marie France Vallat[3], Michel Nardin[3]

[1]*French-German Research Institute of Saint-Louis, ISL, France*
[2]*Institut de Physique et Chimie des Matériaux de Strasbourg, IPCMS, France*
[3]*Institut de Science des Matériaux de Mulhouse, IS2M, France*

For decades, high-performance thermoplastics such as thermostables polymers have aroused considerable interest. In this class of polymers, polyetheretherketone (PEEK) has been extensively investigated thanks to unique combination of temperature stability, high wear resistance, stiffness, chemical resistance and high mechanical properties [1-3]. For these reasons, it is already well-known to be of great interest for engineering applications, particularly in the automotive, aerospace and biomedical industries [2, 3].

However, conventional processing like injection or extrusion molding applied to this category of polymers present several limitations. Indeed, the high viscosity in the molten state makes the shaping difficult and limits the use of fillers. In addition, the high temperature processing leads to some degradations and shrinkage problems during crystallisation due to the difference in density between amorphous and crystalline phases [4].

A solution to these problems is the use of uncommon method for the development of such materials. In this way, the Powder Metallurgy approach was considered and more precisely a consolidation process called "Spark Plasma Sintering" (SPS). Contrary to conventional techniques, its main interest is the processability at low temperature under the melting point. This way confers the capacity to keep a degree of crystallinity close to the native powder's one and hence the potential to increase the mechanical properties.

First, the effects of SPS parameters such as temperature, pressure, and dwell time on density, mechanical properties and structure of PEEK were investigated. Finally, the mechanisms of polymer's sintering were studied and a particular attention on the role of pressure's application on the chains mobility was considered.

References:

1. Attwood T.E. et al., Polymer, 1981, p. 1096-1103.
2. Díez-Pascual A.M. et al., Progress in Materials Science, 2012, p. 1106-1190.
3. Kurtz, S.M., PEEK biomaterials handbook, Plastics Design Library, 2012, Elsevier
4. Gottfried E., Matériaux polymères, 2000, Hermes Science Publications.

Keywords: Polymer processing, Spark Plasma Sintering, Polyetheretherketone

PEPC-O-008 [Polymer Engineering, Processing, and Characterization]

Structure- Property Relationship in Ethylene Octene Copolymer (EOC) - Polydimethyl Siloxane (PDMS) based TPEs for cable insulation applications

Padmanabhan Ramachandran, Kinsuk Naskar, Golok Nando

Indian Institute of Technology, Kharagpur, India

The present study focuses on thermoplastic elastomers based on ethylene octene copolymer (EOC) and poly dimethyl siloxane (PDMS) rubber prepared by melt mixing technique nd crosslinked using dicumyl peroxide (DCP). It is found that the addition of peroxide causes crosslinking in both the phases. However, crosslinking without affecting the crystallinity of the EOC polymer leads to tremendous improvement in the mechanical properties including the tensile strength which has improved nearly by 60 % for the 70:30 EOC: PDMS blends. For better understanding about the crosslinking characteristics of thermoplastic vulcanizates (TPVs), significant correlation has been made between the vulcanized network and the physico-mechanical properties. Further, the dynamic mechanical properties and creep behavior of these thermoplastic elastomers (TPEs) and TPVs have also been studied. It is inferred that the TPVs shows a 19 % decrease in the creep compliance, i.e. higher creep resistance compared to uncrosslinked blends. Subsequently, the morphology study of the blends by Atomic force microscopy (AFM) shows that, there is a decrease in the PDMS rubber domains from 0.8 µm to about 0.4 µm after vulcanization. From the thermal studies it was found that, through blending and subsequent crosslinking, the maximum degradation temperature of the EOC was increased from 492.1 °C to 514.8 °C. Ageing and reprocessing studies of the prepared TPVs also show better physico-mechanical properties even after reprocessing twice. All the radiation crosslinked blends exhibit lower dielectric constant, lower dielectric loss and higher electrical resistivity as compared to the virgin blends, which makes them an excellent material for high voltage cable insulating applications.

Keywords: Ethylene octene copolymer, blend morphology, Polydimethylsiloxane rubber, dynamic vulcanization

PEPC-O-009 [Polymer Engineering, Processing, and Characterization]

Novel High Performance TPV: a new generation thermoplastic elastomer for automotive applications

Kinsuk Naskar

Rubber Technology Centre, IIT Kharagpur, Pin- 721302, West Bengal, India

Thermoplastic vulcanisates (TPVs) are a special class of thermoplastic elastomers, produced by simultaneously mixing and cross-linking a rubber with a thermoplastic at elevated temperature. As a result a typical morphology is formed, where the cross-linked rubber particles are generally finely dispersed in a continuous matrix of thermoplastic. High performance TPVs or super TPVs are a new class of TPVs which exhibit high heat resistance as well as excellent oil resistance properties suitable for automotive under-the-hood applications. In the present work, a new super TPV based on carboxylated acrylonitrile butadiene rubber (XNBR) and polyamide (PA12) has been developed. TPVs of three different blend ratios (50:50, 60:40, 70:30) of (XNBR:-PA12) have been prepared at an elevated temperature in an internal mixer by using a fixed concentration of Trigonox 311, which is a novel cyclic mono-functional cross-linking peroxide. The final morphology of the TPVs varies from a co-continuous to a dispersed one depending on the blend ratio. TPV of 50:50 XNBR-PA12 shows the best mechanical properties and the superior thermal stability amongst all TPVs. Detailed thermal characterisation through DSC and DMA and rheological characterisation have been performed. The lowest tanδ of the TPV with a 50:50 blend ratio of XNBR-PA12 indicates the highest degree of crosslinking and this is also well supported by the overall crosslink density value of that TPV. Heat ageing and oil resistance studies have also been carried out to demonstrate the performance of these TPVs.

Keywords: Thermoplastic elastomer, XNBR rubber, Polyamide 12, dynamic cross-linking, peroxide, morphology

PEPC-O-010 [Polymer Engineering, Processing, and Characterization]

Determining the effect of width of waisted section on waisted tensile test specimen using HDPE non-linear elastic-plastic behaviour in FEA

Mohammad Taghipourfard[1], Mike Troughton[2], Jim Song[1], Amir Khamsehnezhad[2]

[1]Mechaincal Engineering, Brunel University, London, UK
[2]TWI Ltd,Granta Park, Great Abington, Cambridge, UK

The first plastic pipes were installed in the mid-1930s, with their usage increasing rapidly in the 1950s. Plastics have steadily replaced clay, copper, asbestos-cement, aluminium, iron and concrete pipes in various applications. Polyethylene is employed in about 20% of the plastic pipe applications [1,2]. High-density polyethylene (HDPE) pipes are used extensively for the transportation and distribution of natural gas, with over 80% of the new piping installations using HDPE.

There are a number of tests available for assessing the short-term performance of butt fusion welded joints in HDPE pipes. Previous work [3,4] has shown that the most discriminating short-term test is a tensile test a using waisted test specimen (Fig 1), such as those described in ISO 13953, EN 12814-2, EN12814-7 and WIS 4-32-08.

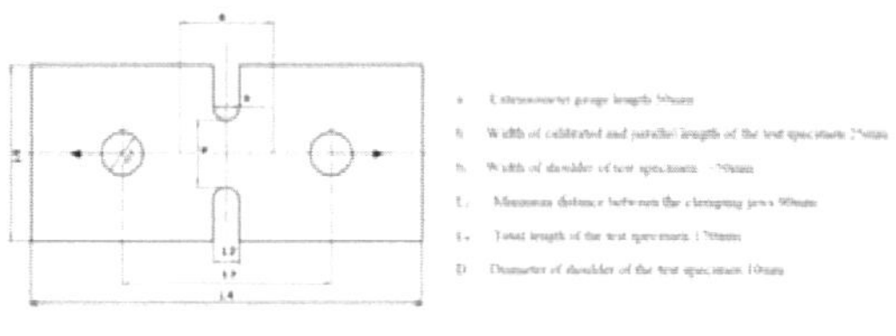

Figure 1, Waisted tensile test specimen

Most of these standards specify that the failure mode should be ductile, rather than brittle or mixed (Fig 2). However, the same specimen geometry has been used for all thicknesses. As the thickness of the specimen increases, the triaxiality factor also increases in the cross section, which reduces the ductility of the test.

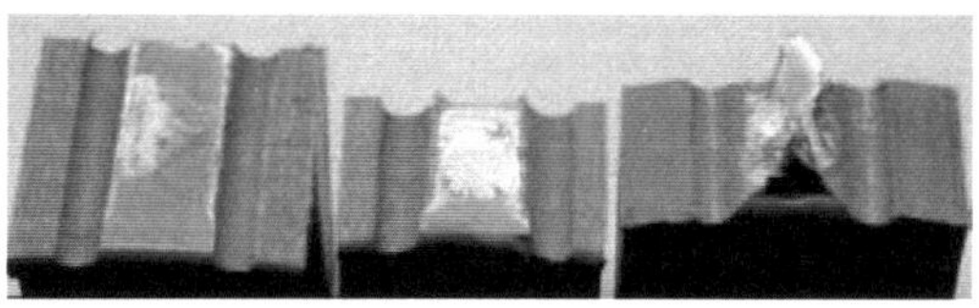

Figure 2, Three different types of failures in a waisted tensile test from left, brittle, mixed and ductile failure.

Effect of some parameters in different thicknesses on triaxiality factor, has been identified using a model which consists of linear elastic, non-linear deformation, hollomon's equation for plastic deformation and work hardening deformation behaviour of HDPE.

References;

1. Presentation by Duke Energy on the Use of PE Pipe in Non-Safety Related Service Water Piping at Catawba Nuclear Station,2006.
2. ASME Code Case N-755"Use of Polyethylene (PE) Plastic Pipe for Section ɩɩɩ, Division 1, Construction and section χɩ Repair /Replacement Activities", Revision 6a, September 12, 2006.
3. Brown C I and Troughton M J, 2004: 'Qualifying long-term performance of butt fusion welds in PE pipes from short-term tests'. Plastics Pipes XII Conference, Baveno, Italy, 19-22 April.
4. Chipperfield F A and Troughton M J, 1999: 'Comparison of short-term coupon tests for assessing the performance of butt fusion welds in polyethylene pipes'. 16th International Plastic Pipe Fuel Gas Symposium, New Orleans, USA, 1-4 November.

Keywords: HDPE, FEA, Polymer necking modelling

PEPC-O-011 [Polymer Engineering, Processing, and Characterization]

Flame retardands nanocomposites synergy effect of combining conventional antypirenes with carbon nanofillers

Ewa Kicko Walczak, Grażyna Rymarz

Polymers Department, Institute for Engineering of Polymers Materials and Dyes, Toruń, Poland

Introduction : The thermoset resins are proven construction materials for the technical and highly demanding applications of the transportation, electrical and building part industry. Heat stability, high thermal, low shrinkage, mechanical properties are typical for their type of polymers. Above applications also requires a good flame retardants(FR). Undertaken activities refer to official draft recommendations in UE states. This paper presents positive effect of reduced flammability of thermoset resins thanks to the use of nanocomposites containing multi-ingredient halogen-free flame retardants which combine phosphorus/nitrogen modifiers interacting with nanofillers: expandable graphite (EG) graphene (G), graphene oxide (GO), anthracite (AN).

Experimental part: The flame retardancy of modified polymers has been investigated by LOI analysis, TG; (TG-FTIR, TG-MS) and by using CC method. The fine-plates, phase morphology of nanocomposites was assessed by SEM.

Results overview: We confirm that nanocomposites formation is an important concept for the flame retardants industry. Laminates made of modified resins meet requirements LOI over 28-34 %, reduced 30-70% head release rate(HRR) by CC method. No adverse impact on strength properties.

Conclusion: A multi-ingredient combine of FR turned out to make significant progress in achieving a desired flammability. It should be assumed that the presence of conventional flame retardants led to a synergy effect promoting faster formation of a protection layer hindering oxygen flow-through resulting from the process of thermal destruction.

References:

[1] Bourbigot S., Duqusne S. (2007): J. Mater. Chem., 17, 2283.
[2] Kicko-Walczak E., Półka M. (2010): Przem. Chem., 89(1), 66.
[3] Katsoulis C., Kandare E., Kandola K. (2011): J. Fire Sci., 29, 361.
[4] Kicko-Walczak E., Rymarz G. (2013): Przem. Chem., 92(4), 504..
[5] Kicko-Walczak E., Rymarz G. (2015): JMEA, 5(9),510.

Keywords: Fire retardants (FR), halogen-free FR additives, flammability of thermoset resins, test methods,

PEPC-O-012 [Polymer Engineering, Processing, and Characterization]

Elaboration and characterization of nanostructured polyurethane rigid foams within the inclusion of O-MMT: impact and fire proof behavior

Zitouni Safidine, Yasmine Mahmoud

Macromolecular Chemistry Laboratory, Ecole Militaire Polytechnique, Bordj-El-Bahri, Algiers, Algeria

Abstract – the aim of the present work is the elaboration of nanostructured polyurethane rigid foams within the inclusion of O-MMT Nanoclay: PUR/O-MMT. The contents of O-MMT are 2, 3, 5 and 10wt %. The nature expected by these rigid foams is the fire andimpact proof behavior.The nanostructured materials were performed in two stages, firstly by the use of twin screw extruder (TSE) in contrarotative mode in which the blended polyols (CPP) were included between the MMT layers after treatment within the alkyl ammoniumC18. And secondly all pieces having parallelepiped shape are obtained by the combined processes:reaction injection molding (RIM) and controlled expansion. Both were performed by an appropriate montage.Various characterization methods toward reactants and Nano clays were used such as: FTIR, DSC, XRD and micro XRF. Adding the use of SEM and XRD behaviors for the final items. The normalized inflammability test according to the UL 94 and tensile strength for the solid specimens was also performed. From the results it's found that these new materials exhibit an enhancement of tenacity and notable flame retardant characterobviously for the threshold content which levels off at 2 wt. % of O-MMT.

Keywords: s: nanostructured, PUR rigid foam, TSE, RIM, impact strength, fire proof.

PEPC-O-013 [Polymer Engineering, Processing, and Characterization]

Study of thermal properties of UHMWPE/ MWCNTs nano composites: Effect of γ-rays modified MWCNTs incorporation

Saqlain Saqib Mukhtar[1], Syed Asad Maqbool[1], Muhammad Zeeshan Khalid[1], Bilal Ghafoor[3], Malik Sajjad Mehmood[1], Tariq Yasin[2]

[1]*Department of Basic Sciences, University of Engineering and Technology, 47050 Taxila, Pakistan*
[2]*Pakistan Institute of Engineering and Applied Sciences, 45650 Islamabad, Pakistan*
[3]*Materials Science and Engineering Department, Institute of Space Technology, 44000, Islamabad, Pakistan*

This experimental study was performed to examine the effects of γ-rays on the thermal properties of ultra-high molecular weight polyethylene (UHMWPE) nano composites with γ-irradiated and un-irradiated MWCNTs. UHMWPE was mixed with various concentrations (0.5% and 1% by weight) of γ-irradiated and un-irradiated MWCNTs through ball mailing process and then molded into sheets of micron sizes by using hot press (Gibitre instruments laboratory press). The composites were then subjected to different γ-dosed ranging from 25 kGy-to-150 kGy. In order to investigate the effect of irradiation on thermal properties differential scanning calorimeter (DSC) measurements were performed. Results obtained so far revealed that percentage crystallinity (%Xc) of UHMWPE increases with the increase in γ-dose and un-irradiated MWCNTs. In addition, no significant effect of aforementioned facts i.e. γ-dose and MWCNTs concentration had been observed on peak melting temperature (Tm). However, Crystalline lamellar thickness of UHMWPE/γ-MWCNTs decreases with absorbed dose while an increase had been observed for UHMPWPE/MWCNTs. We expect that results reported in this study will be useful for state-of-art industrial applications of UHMWPE e.g. orthopedic, thermal and radiation insulation shielding etc.

Keywords: Nano composites, Thermal Characterization, Radiation Processing of Polymer/Polymer Composites

PEPC-O-014 [Polymer Engineering, Processing, and Characterization]

Improving Mechanical and Thermal Properties of Poly(Lactic Acid) By Using POSS Cored Star-Shaped Poly(ε-Caprolactone) Polymers

Merve Dandan Doganci[1], Figen Aynali[2], Erdinc Doganci[3], Faruk Yilmaz[4], Guralp Ozkoc[5]

[1]*Department of Chemistry, Kocaeli University, Kocaeli, Turkey*
[2]*Department of Chemical Engineering, Gebze Technical University, Kocaeli, Turkey*
[3]*Department of Chemical Technology, Kocaeli University, Kocaeli, Turkey*
[4]*Department of Chemistry, Gebze Technical University, Kocaeli, Turkey*
[5]*Department of Chemical Engineering, Kocaeli University, Kocaeli, Turkey*

Poly (lactic acid) (PLA) is one of the most important and widely used biodegradable polymers in industrial fields.[1] However; pure PLA cannot provide the expected performance in the industrial applications due to its brittleness at room temperatures. In order to eliminate this disadvantage, PLA is toughened by blending with other flexible and biodegradable polymers which exhibit lower glass transition temperature. Poly (ε-caprolactone) (PCL), as a biodegradable polymer, is used for the toughness improvement of PLA as well as enhancing its features.[2] Besides being more flexible than PLA, hydrolysis of PCL is also slower compared with PLA.

In this work, polyhedral oligomeric silsesquioxane (POSS) cored star-shaped PCL (POSS-$(PCL)_8$) and miktoarm PCL-PLA ($(PLA)_4$-POSS-$(PCL)_4$) polymers with different chain lengths were synthesized. Then, the blends of PLA and star-shaped polymers were prepared in a twin screw extruder with different feeding ratios to increase the toughness of PLA. POSS is chosen as a core to the star polymers since it is compatible both with PLA and PCL together with its ability to increase the mechanical stability of PLA.[3,4]

References:

1. Garlotta, D. (2001), J Polym Environ, 9 (2), 63-84.
2. Patricio, T., Bartolo, P. (2013), Procedia Eng, 59, 292-297.
3. Turan, D., Şirin, H., Ozkoç, G. (2011), J Appl Polym Sci, 121, 1067-1075.
4. Lee, K. S.,Chang, Y. W. (2013), Polym Int, 62, 64-70.

Keywords: PLA, PCL, Miktoarm shape star polymer, Biodegredable polymer, Polymer Blends

Figure 1.

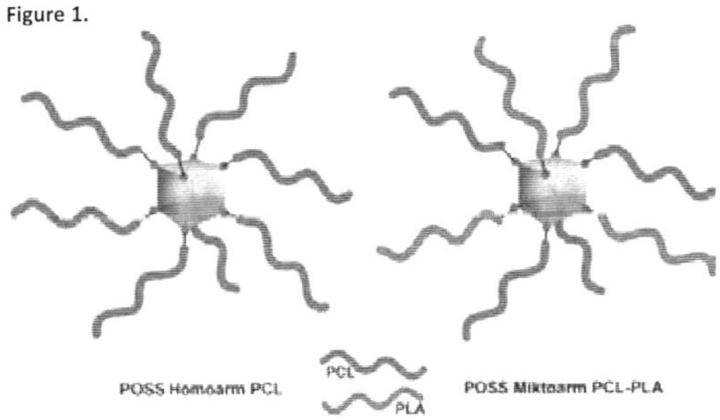

Schematic illustration of POSS-(PCL)8 and (PLA)4-POSS-(PCL)4 polymers.

PEPC-O-016 [Polymer Engineering, Processing, and Characterization]

Engineering at the Nanoscale: State of the Art, Challenges and New Opportunities

Sabu Thomas[1,2]

[1]*International and Inter University Centre for Nanoscience and Nanotechnology, Mahatma Gandhi University, Priyadarshini Hills P. O. Kottayam, Kerala, India -686 560*
[2]*School of Chemical Sciences, Mahatma Gandhi University, Priyadarshini Hills P. O. Kottayam, Kerala, India -686 560*

The talk will concentrate on various approaches being used to engineer materials at the nanoscale for various applications in future technologies. In particular, the case of clay, carbon nanostructures (e.g. nanotubes, graphene), metal oxides, bionanomaterials (cellulose, starch and chitin) will be used to highlight the challenges and progress. Several polymer systems will be considered such as rubbers, thermoplastics, thermosetts and their blends for the fabrication of functional polymer nanocomposites. The interfacial activity of nanomaterials in compatibilising binary polymer blends will also be discussed. Various self assembled architectures of hybrid nanostructures can be made using relatively simple processes. Some of these structures offer excellent opportunity to probe novel nanoscale behavior and can impart unusual macroscopic end properties. I will talk about various applications of these materials, taking into account their multifunctional properties. Some of the promising applications of clay, metal oxides, nano cellulose, chitin, carbon nanomaterials and their hybrids will be reviewed.

Keywords: nanostructures metal oxides, bionanomaterials

PEPC-O-017 [Polymer Engineering, Processing, and Characterization]

Effect of organoclays on the degradation of polyoxymethylene during melt processing

Alexis Ditta[1], Arnaud Travert[2], Loïc Le Pluart[1]

[1]*Normandie Univ, UNICAEN, ENSICAEN, CNRS, Laboratoire de Chimie Moléculaire et Thioorganique*
[2]*Normandie Univ, UNICAEN, ENSICAEN, CNRS, Laboratoire de Catalyse et Spectroscopie*

The improvement of Polyoxymethylene (POM) thermomechanical properties through the incorporation of nanofillers such as organoclays can be hindered by a detrimental effect of the latter on the thermal stability of the polymer[1-2]. In order to be able to develop polyoxymethylene-based nanocomposites, it is necessary to assess the relation between nanofillers chemical properties and the thermal stability of POM. For this purpose, polyoxymethylene nanocomposites containing various organoclays have been processed and characterized in this work. For a better understanding of the mechanisms involved, the degradation has been followed using temperature controlled Fourier Transform Infrared Spectroscopy in Attenuated Total reflection mode (FTIR-ATR). Eventually the organoclay surface properties have been characterized using pyridine as a probe molecule in order to quantify Lewis and/or Brønsted acidic sites types and the relationships between the nature of the organoclay and the thermal stability of nanocomposites have been established. This work confirms that organoclay containing nanocomposites are less thermally stable than neat polyoxymethylene. Some organoclays accelerate POM classical degradation by chain scission followed by depolymerization, whereas other induce a deprotection by cleavage of the ester bond of the acetyloxy end capping groups. The latter leads to a much faster degradation by unzipping depolymerization.We bring evidence that the degradation of the alkylammonium ions of the organoclay is not responsible for polyoxymethylene decreased thermal stability. Surface chemistry analyses of organoclays by probe molecules have shown that the presence of Brønsted acidic sites was responsible for the deprotection of chain ends, whereas Lewis acidic sites were only able to accelerate the native POM degradation mechanism. These acidic sites do not result from Hoffman elimination reaction but are originally present on the organoclays.

[1]Zeng *et al.* J. Phys. Chem. C 111, 13945–13950 (2007)
[2] Kongkhlang *et al.* Polymer 49, 1676–1684 (2008)

Keywords: Polyoxymethylene, nanocomposites, organoclay, thermal stability, degradation mechanism

Time-dependent IR spectra of neat POM during isothermal degradation at 200°C

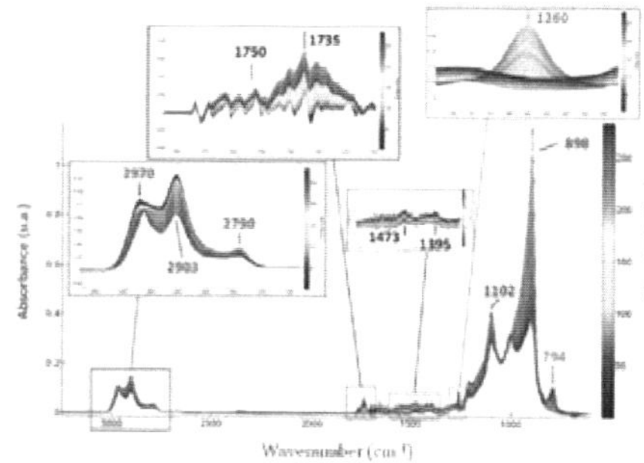

PEPC-O-018 [Polymer Engineering, Processing, and Characterization]

Composites based on amphiphilic block copolymers of PLLA, PEO and PPO blocks and nanohydroxyapatite

Lívia Mesquita Dias Loiola[1], Laura Alejandra Fasce[2], Bruna Antunes Más[3], Eliana Aparecida de Rezende Duek[3], Patricia María Frontini[2], Maria Isabel Felisberti[1]

[1]*Institute of Chemistry, University of Campinas – UNICAMP, P.O. Box 6154, 13083-970, Campinas, São Paulo, Brazil.*
[2]*Research Institute of Materials Science and Technology, INTEMA (UNMdP - CONICET), Av. J. B. Justo 4302, B7608FDQ, Mar del Plata, Argentina.*
[3]*Faculty of Medical Sciences, Pontifical Catholic University of São Paulo – PUC-SP, 18030-095, Sorocaba, São Paulo, Brazil.*

Composites which combine biocompatible polymers and hydroxyapatite are unique materials with regards to their mechanical properties and bioactivity in the development of temporary bone-fixation devices. Nanocomposites based on amphiphilic block copolymers of poly(L-lactide) (PLLA), poly(ethylene oxide) (PEO) and poly(propylene oxide) (PPO) - PLLA-*b*-P(EO-PO)-*b*-PLLA - and hydroxyapatite nanoparticles (nHAp at 5, 10 and 20 wt.%) were prepared by dispersion in benzene solutions, followed by freeze-drying and injection molding processes. The spherical nHAp nanoparticles (diameters of 78 ± 39 nm) are uniformly dispersed throughout the amphiphilic PLLA-*b*-P(EO-PO)-*b*-PLLA matrixes. The morphology of the copolymers of a hydrophilic and flexible PEO/PPO phase dispersed throughout a hydrophobic and hard PLLA matrix [1] was not changed with the addition of the nanofiller. The reinforce effect of the nanoparticles addition was evidenced by the increment of the storage modulus (*E'*) at temperatures above the glass transition of the polyether phase. Moreover, these particles acted as nucleating agents, inducing the earlier formation of PLLA nuclei, once the crystallization of PLLA blocks were shifted to higher temperatures for the nanocomposites than for the pure copolymers. The incorporation of nHAp increased both the reduced elastic modulus, *Er* (up to 66%), and the indentation hardness, *H* (up to 42%), compared with the pure copolymer matrixes. The mechanical properties of the studied materials are governed by the PLLA phase mass fraction and crystallinity degree. The non-cytotoxic character of PLLA-*b*-P(EO-PO)-*b*-PLLA was demonstrated thorough PicoGreen® assays [1], however the incorporation of 10 wt.% of nHAp did not caused an increment of alkaline phosphatase (ALP) activity of osteoblast cells after 14 days *in vitro* in comparison to pure copolymers. The nanocomposites are potential candidates for the design of biodegradable devices for temporary bone fixation.

[1] Loiola, L. *et al. European Polymer Journal*, 68 (2015) 618.

Keywords: Nanocomposite, amphiphilic block copolymer, nanohydroxyapatite

Figure 1:

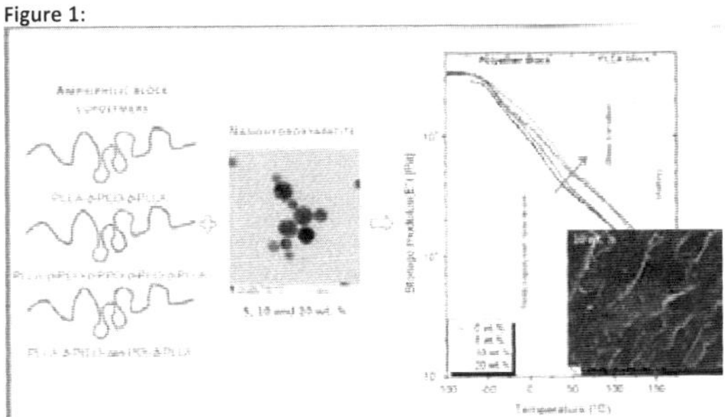

The amphiphilic block PLLA-b-P(EO-PO)-b-PLLA copolymers structures and the spherical nHAp nanoparticles are depicted. The reinforce effect of the nanoparticles addition is evidenced by the increment of the storage modulus.

PEPC-O-019 [Polymer Engineering, Processing, and Characterization]

Exploiting mild enzymatic hydrolysis to functionalize synthetic polymers

Enrique Herrero Acero[1], Barbara Zartl[2], Jennifer Ahrens[3], Alessandro Pellis[2], Caroline Gamerith[1], Sara Vecchiato[1], Andreas Ortner[1], Olivier Defrain[3], Bernhard Mueller[4], Georg M. Guebitz[4]

[1]*Polymers & Environmental Biotechnology; Austrian Centre of Industrial Biotechnology; Tulln; Austria*
[2]*Institute of Environmental Biotechnology; University of Natural Resources and Life Sciences, Vienna; Tulln; Austria;*
[3]*Textilcord Steinfort; Steinfort; Luxembourg*
[4]*Glanzstoff Industries; St, Poelten; Austria*

During the last years, significant progress has been achieved regarding the increase of the hydrolytic activity of certain enzymes towards synthetic polyesters. Particularly significant have been the molecular engineering efforts around cutinases in order to tune their PET hydrolysis activity [1,2]. Most of the research efforts have been directed to achieve complete enzymatic depolymerization leading to monomers for recycling purposes. However, the potential of enzymatic surface functionalization remained not fully exploited. Cutinases have been proven to be useful in the functionalization of sensitive polymers like PLA. Thanks to the new generation of more active enzymes, mild enzymatic hydrolysis generates new groups on the polymer surface in relatively short times suitable for the grafting of biologically active molecules [2] or loading of therapeutics. The use of harsh chemicals to achieve similar functionalization degree strongly damaged the polymer bulk structure decreasing the number-average molecular weights (Mw) from the initial 26.5x103 g/mol to 18.7x103 g/mol; whereas after 72 hours of enzymatic treatment the Mn remained 25.8x103 g/mol [2]. Under anhydrous conditions direct transesterification of acid containing substances is also possible, although the functionalization levels achieved are relatively modest [3]. Depending on the polymer properties in some cases partial hydrolysis is enough to modify the surface properties in a controlled way, like hydrophilicity or pore size leading to increase mass flux in membrane applications. During this presentation latest enzymatic mediated approaches for synthetic polymer surface functionalization developed in our group will be presented.

[1] Ribitsch, D. et al. Applied and Environmental Microbiology, 2015, 81 (11), 3586-3592.
[2] Wei, R. et al. Biotechnology and Bioengineering, 2016 DOI 10.1002/bit.25941
[3] Nyanhongo, G.S. et al. React. Funct. Polym., 2013, 73 (10) 1399-1404.
[4] Pellis, A.; et al. Biotechnology Journal, 2015, 10 (11), 1739-1749.

Keywords: enzymes, polymer, functionalization

PEPC-O-020 [Polymer Engineering, Processing, and Characterization]

Optical Biosensing Applications on Molecularly Imprintied Poly (Methacrylic Acid-co-Cyanurate Ester) Nanorods via Thermal Polymerization

Zehra Oluz[1], Tolga Yildirim[1], Martin Steinhart[2], Hatice Duran[1]

[1]*Dept. Mater. Sci. & Nanotechn. Eng.,TOBB University of Economics and Technology, Ankara, Turkey*
[2]*Institute für Chemie, Universitat Osnabrück, Osnabrück, Germany*

Thermal of methacrylic acid and cyanurate ester monomer were carried out in Anodic Alumina Oxide membranes (AAO). AAOs with diameters ranging from 35 to 280 and pore depth from 250 nm to 1 µm were used. Using L-phenylalanine anilide as template during polymerization led molecularly imprinted nanoparticles. Thermal polymerization offers an efficient control of solution propagation and grafting process, while molecularly imprinted polymers shows enhanced capacity as sensor. L-phenylalanine anilide imprinted copolymer nanrods were characterized by X-Ray photoelectron spectroscopy (XPS) and atomic force microscopy (AFM) for chemical structure and surface topography analysis. Performances of the particles were followed by surface plasmon resonance spectroscopy (SPR) after coating the final product on gold deposited glass substrate against four different analogous of analyte molecules: D-phenylalanine anilide, L-tyrosine, L-tryptophan and L –phenylalanine. SPR spectroscopy is a primary method and it allows real time imaging of sensing even at very low template concentrations. These characterizations indicated that Poly (methacrylic acid-co-cyanurate ester) do contain binding sites for L-phenylalanine anilide, and are highly selective for analyte molecule of interest.

This Project was supperted by TUBITAK (Project No. 112M804).

Keywords: anodic aluminum oxide, surface plasmon resonance, molecular imprinting

PEPC-O-021 [Polymer Engineering, Processing, and Characterization]

Simultaneously Improvement of the Mechanical Strength and Fractural Toughness Properties of Recycled Polycarbonate

Davut Demir[1], Nevin Gamze Karsli Yilmaz[2], Taner Yilmaz[1]

[1]*Kocaeli University, Mechanical Eng. Dept., Umuttepe Campus, 41380, Kocaeli/TURKEY*
[2]*Kocaeli University, Chemical Eng. Dept., Umuttepe Campus, 41380, Kocaeli/TURKEY*

Polycarbonate (PC) is a thermoplastic polyester, it is one of the most important engineering thermoplastic polymers and for this reason it is widely used for many applications. However, wide usage area of PC leads to come out the waste management problem of it. For overcoming this problem, recycling is a widely used method. On the other hand, deterioration in properties of recycled polymers and their inadequacy about performing the expected properties come up as another problem which should be overcome 1.

The mostly used method to overcome this disadvantage is to use chemicals (chain extenders) which contain two or more functional groups. These chemicals act as a bridge between polymer chains which shortened after degradation of recycling process. Thus, while molecular weight and viscosity of polymer increasing correspondingly mechanical performance of it enhances 2. However, increase in molecular weight, viscosity and correspondingly mechanical performance of polymer may not take place in its fracture toughness. For this reason, there is need to some other chemicals such as impact modifiers to enhance the toughness of polymer.

In this study, it was aimed to increase the mechanical strength and fractural toughness properties of recycled PC (r-PC) simultaneously. For this purpose, epoxy base chain extender and "ethylene, acrylic ester and glycidyl methacrylate terpolymer" based impact modifier was used. For evaluating the effects of chain extender and impact modifier on the flow properties of r-PC, melt flow index analysis was performed. Fracture toughness of the samples was evaluated by using essential work of fracture method. Finally, mechanical properties of samples were identified by using tensile test.

After the characterization tests, optimum chain extender-impact modifier loading level determined according to simultaneous improvement in mechanical strength and fracture toughness of r-PC result.

This study was supported by TÜBİTAK (Scientific and Technological Research Council of Turkey), Project Number: 115M142.

Keywords: Recycled polycarbonate (r-PC), chain extender, impact modifier, fracture toughness, tensile strength

PPBM-O-001 [Polymer and Polymer-Based Membranes]

Improved permeation flux, salt rejection and chlorine resistant properties of ZnO nanoparticles infused polymeric thin film RO desalination membranes

Aneela Sabir[1], Wail Falath[2], Karl Jacob[3], Muhammad Shafiq[1], Atif Islam[1], Muhammad Taqi Zahid Butt[4], Tahir Jamil[1]

[1]*Department of Polymer Engineering and Technology, University of the Punjab, Lahore, 54590 Pakistan*
[2]*School of Materials Science and Engineering, Georgia Institute of Technology, Atlanta, GA, USA*
[3]*Department of Mechanical Engineering, King Fahd University of Petroleum and Minerals, Dhahran, 31261, Saudi Arabia*
[4]*Faculty of Engineering and Technology, University of the Punjab, Lahore, 54590 Pakistan*

A novel thin film nanocomposite reverse osmosis (TFN-RO) membranes infused with zinc oxide nanoparticles (ZnO-NPs) and Pluronic F127 were prepared by dissolution casting methodology using single layer crosslinked PVA/DGEBA. Membrane characterization was conducted through ATR-FTIR spectroscopy, SEM, AFM, contact angle analysis and reverse osmosis (RO) permeation tests which determines functional group analysis, morphology and topography. It was found that that membrane surface became smoother, more hydrophilic, increased salt rejection, improved permeation flux and chlorine resistance after the successful infusion of ZnO-NPs. The chlorine exposure tests under accelerated conditions indicated that membrane chlorine stability has been enhanced effectively. TFN-RO membranes revealed more hydrophilicity and lesser surface roughness as compared to single layer crosslinked membrane. The nanopores were observered with no mottled surface observed in SEM micrographs. TFN4 composite membranes having ZnO-NPs and hydrophilic polymer exhibited a much higher water permeation flux (1.96m3/m2/day) and salt rejection 98.7% compared to the single layer crosslinked membrane.

Keywords: Polymeric Membrane, Reverse osmosis, Natural sea salt

PPBM-O-002 [Polymer and Polymer-Based Membranes]

Gas separation performance, sorption behavior and thermal transitions of semicrystalline PolyActiveTM

Md Mushfequr Rahman, Jelena Lillepärg, Jan Pohlmann, Silvio Neumann, Sergey Shishatskiy, Clarissa Abetz, Volker Abetz

Helmholtz-Zentrum Geesthacht, Institute of Polymer Research, Max-Planck-Straße 1, 21502 Geesthacht, Germany

Commercial multiblock copolymers of the series PolyActiveTM containing poly(ethylene glycol terephthalate) (PEGT) and poly(butylene terephthalate) (PBT) blocks are promising materials for separation of CO_2 from N_2. In this study single gas permeabilities of CO_2 and N_2 through three grades of PolyActiveTM composed of 77 wt% of PEGT and 23 wt% are investigated. These polymers have poly(ethylene glycol) (PEG) segments of 1500, 3000 and 4000 g/mol, respectively. The length of PEG segment dictates the melting and crystallization temperature of the PEG block which has a direct influence on CO_2 separation performance [M. M. Rahman et al. (2015) ACS Applied Materials & Interfaces, 7(23): 12289-12298]. The melting and crystallization of the PEG blocks is studied using high pressure differential scanning calorimetry in both CO_2 and N_2 gas environment. A substantial decrease of melting and crystallization temperature occurs with increasing CO_2 pressure. But no change occurs under N_2 pressure. The sorption of CO_2 is investigated using a magnetic suspension balance in the temperature range from 30 to 70 °C. The isotherms follow Henry's law when the poly(ethylene glycol) (PEG) blocks of PolyActiveTM are in a molten state. The sorption isotherms show a significant deviation from Henry's law when the PEG blocks are in a semicrystalline state. This deviation occurs because of the CO_2 induced melting point depression of PEG blocks which stems from the strong affinity of the polar ether oxygen towards the quadrapolar CO_2 [M. M. Rahman et al. (2016) Polymer, DOI: 10.1016/j.polymer.2016.04.024]. The obtained results provide a guideline for deeper understanding of sorption and transport behavior of CO_2 and N_2 through PolyActiveTM membranes.

Keywords: PolyActive, Gas sorption, Henry's Law, DSC

PPBM-O-003 [Polymer and Polymer-Based Membranes]

Propylene/Propane Plasticization in Unary and Binary Gas Sorption on 6FDA-Based Polyimide Membranes

Sadiye Halitoğlu Velioğlu, Göktuğ Ahunbay, Birgül Tantekin Ersolmaz
Department of Chemical Engineering, Istanbul Technical University, Istanbul, Turkey

Worldwide demand for propylene is expected to raise from 90 to 130 million tonnes by 2023. To meet this demand, the development of membrane processes is of interest in the petrochemical industry. Among the polymeric gas separation membranes, 6FDA (4,4'-(hexafluorisopropylidene) dipthalic anhydride) based glassy polyimides shine out with their high performance. However, many polyimides have limitations in their practical use due to plasticization phenomenon, which often results in unacceptable loss in propylene (C3H6)/propane (C3H8) selectivity. Propylene and propane, both being condensable, tend to plasticize polymeric membranes, even at partial pressures as low as 2 bars. In this study, molecular simulation techniques are used to estimate the degree of plasticization of three different 6FDA-based polyimide membranes induced by sorption of C3H8 and C3H6. The diamines considered here are 2,5-dimethyl-p-phenylenediamine (DPX), 2,4,6-trimethyl-m-phenylenediamine (DAM), and 4,4-oxydianiline (ODA). To reproduce the plasticization effect, sorption-relaxation cycles are applied until C3H6 or C3H8 concentration converges. The increase in the FFV of the resulting polymer structure is considered as the extent of plasticization. Results indicate that propane induced plasticization in all three polyimides increases with increasing backbone flexibility of the polyimide following the general trend in polymers. However, in propylene plasticization, 6FDA-DPX deviates from this trend due to high polarizability of propylene resulting from π-electron complexation as supported by the radial distribution function and isosteric heats of adsorption analyses. On the other hand, 6FDA-DPX, having a lower Tg than 6FDA-DAM, exhibits lower FFV increase due to the alternating electron rich and poor segments in the polyimide. This causes suppression of detrimental plasticization-induced matrix dilations of polyimide via inter-chain interactions such as charge transfer complexes or π-π stacking complexes. Binary mixture sorption simulations show that, FFV increase is lower than those in pure gas sorption simulations for 6FDA-DPX and 6FDA-ODA due to the change in sorption site preferences.

Keywords: Propylene/Propane separation, plasticization, Gas separation membranes.

PPBM-O-004 [Polymer and Polymer-Based Membranes]

High Thermally Stable Isoporous Integral Asymmetric Block Copolymer Membranes

Christian Höhme[1], Janina Hahn[1], Brigitte Lademann[1], Andreas Meyer[2], Barbara Bajer[1], Clarissa Abetz[1], Volkan Filiz[1], Volker Abetz[1,2]

[1]*Helmholtz-Zentrum Geesthacht, Institute of Polymer Research, Max-Planck-Str. 1, 21502 Geesthacht, Germany*
[2]*University of Hamburg, Institute of Physical Chemistry, Martin-Luther-King-Platz 6, 20146 Hamburg, Germany*

In consequence of an everincreasing demand for energy-efficient and sustainable technologies, ultrafiltration (UF) membranes attracted increasing attention e.g. in waste water treatment, dairy, and food industry, in recent years. For preparing membranes with a sharp pore size distribution, microphase separating block copolymers offer a tremendous potential due to their ability to selfassemble into highlyordered structures on the length scale of about 10 to 100 nm[1]. The combination of the selfassembly and the nonsolvent induced phase separation process (SNIPS), using e.g. polystyrene-*block*-poly(4-vinylpyridine) (PS-*b*-P4VP), leads to isoporous membranes. However, the T_g of polystyrene (100 °C)[2] limits the usability of PS-*b*-P4VP membranes e.g. for applications where membrane cleaning is realised via steam sterilisation.[3]

In this work, high-T_g poly(α-methylstyrene)-*block*-poly(4-vinylpyridine) (PαMS-*b*-P4VP) and poly(4-methylstyrene)-*block*-poly(4-vinylpyridine) (P4MS-*b*-P4VP) diblock copolymers with different molecular weights were synthesised via living anionic polymerisation. The ideal casting conditions (e.g. solvent composition, polymer weight fraction, evaporation time) for every block copolymer were determined and the block copolymers were used to prepare highly-ordered isoporous membranes using the SNIPS process. As indicated by the T_g's shown in Table 1 (measured by DSC) and emphasised by the SEM surface pictures (Figure 1), the developed membranes show a higher thermal stability compared to PS-*b*-P4VP membranes. Especially PαMS-*b*-P4VP membranes are high thermally stable up to 150 °C and therefore offer more possibilities for cleaning or postmodification steps.

Table 1. Glass transition temperatures of the individual blocks of the used block copolymers

	polystyrene	poly(α-methylstyrene)	poly(4-methylstyrene)	poly(4-vinylpyridine)
glass transition temperature [°C]	100	176	113	152

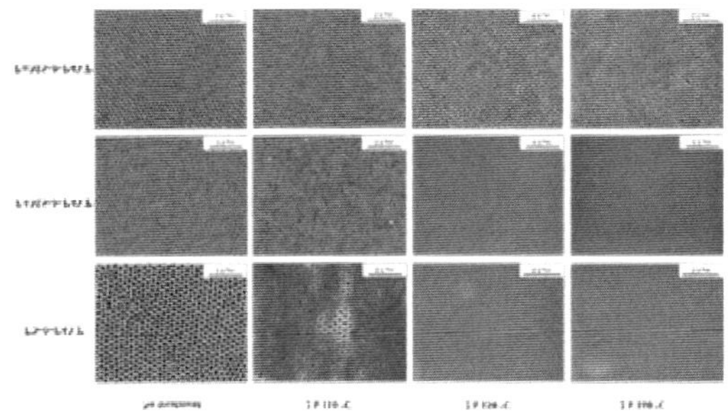

Figure 1. SEM surface images of different block copolymer membranes before and after thermal treatment

Furthermore, the properties of the casted membranes were analysed by water flux and retention measurements. The morphologies of the block copolymers in bulk state were investigated by transmission electron microscopy and small-angle X-ray scattering.

References

1. V. Abetz and A. Boschetti-de-Fierro, In *Polymer Science: A Comprehensive Reference*, ed. K. M. Möller, Elsevier, Amsterdam, 2012, pp. 3-44.
2. E. H. I. J. Brandrup, E. A. Grulke, A. Abe, D. R. Bloch, in *Polymer Handbook*, John Wiley & Sons, New York, 1999, vol. 4th edition.
3. C. Regula, E. Carretier, Y. Wyart, G. Gésan-Guiziou, A. Vincent, D. Boudot and P. Moulin, *Water Research*, 2014, **56**, 325-365.

Keywords: anionic polymerization, diblock copolymers, isoporous membranes, self-assembly, high thermal stability

PPBM-O-005 [Polymer and Polymer-Based Membranes]

PS-*b*-PNIPAM block copolymers for thermoresponsive membrane applications

Merve Cetintas, Marleen Kamperman

Laboratory of Physical Chemistry and Soft Matter, Wageningen University and Research Centre, Wageningen, The Netherlands

Responsive pores are receiving a great deal of attention in membrane systems for their advantages against fouling and easy cleaning properties[1]. Block copolymer self-assembly is of interest, because well-defined nanosized pores can be obtained[2,3] and stimuli-responsive functions can be incorporated[3,4]. In our study, we synthesized well-defined high molecular weight polystyrene-*b*-poly(N-isopropylacrylamide) (PS-*b*-PNIPAM) block copolymers by RAFT polymerization. The PNIPAM block in the copolymer acts as the temperature responsive part due to its lower critical solution temperature (LCST) of 32°C. In principle, the pore sizes of the membrane should increase at temperatures above the LCST of PNIPAM, due to collapsing of the PNIPAM chains. Membranes were prepared by (a) spin-coating on a porous substrate and solvent annealing to the desired morphology or by (b) using a phase inversion procedure, in which a viscous polymer solution was cast on a glass substrate and transferred to a water bath after a certain evaporation time. Morphological properties of the block copolymer membranes were investigated by a variety of techniques including Atomic Force Microscopy (AFM), Grazing Incidence Small Angle X-Ray Scattering (GISAXS) and Scanning Electron Microscopy (SEM). Membrane preparation using the phase inversion procedure resulted in nanoporous membranes and temperature dependent permeability experiments showed that our membranes have a thermoresponsive character. Reversibility tests proved that the membranes can be used repeatedly at temperatures between 20 and 50°C.

References:

[1] D. Bhattacharyya, T. Schäfer, S. R. Wickramasinghe, S. Dauner, Responsive Membranes and Materials, 2013, John Wiley & Sons Ltd., UK
[2] W. A. Phillip et al., Macrom. 2010, 43, 7763
[3] S. Rangou et al., J. Memb. Sci., 2014, 451, 266
[4] S. P. Nunes et al., ACS Nano, 2011, 5(5), 3516

Acknowledgements: This project is financially supported by Dutch Polymer Institute (DPI) under project #766.

Keywords: RAFT polymerization, block copolymer self-assembly, non-solvent induced phase inversion method, thermo-responsive nanoporous membranes

Figure. Membrane structure by SEM (left) Reversibility of the permeabilities versus temperature (right).

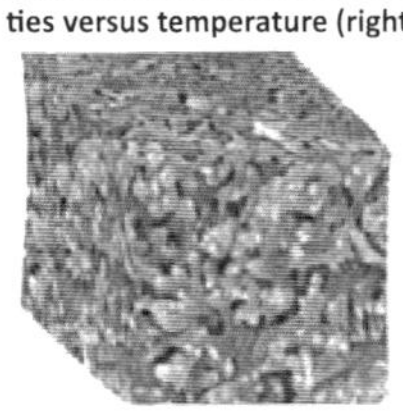

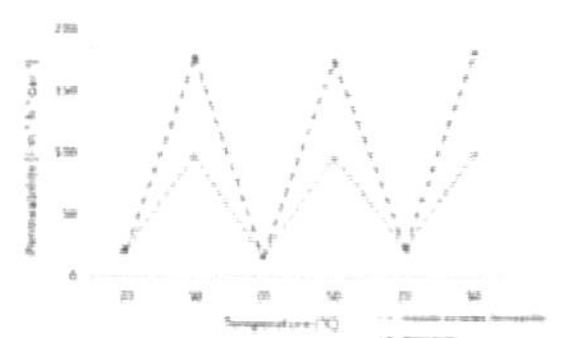

PPBM-O-006 [Polymer and Polymer-Based Membranes]

Microphase Separated Proton Exchange Membranes from Dual Electrospun Mats from Sepiolite-Based Nanocomposite Blends of Fully Sulfonated Poly(ether ketone)/ Non-sulfonated Poly(ether sulfone)

Maryam Oroujzadeh, Samal Babanzadeh, Shahram Mehdipour Ataei

Iran Polymer and Petrochemical Institute, P.O.Box 14965/115, Tehran, Iran

Adding nanoparticles to the polymeric matrix of proton exchange membranes improves mechanical strength, dimensional and thermal stability of the membrane, and also decreasing the water absorption and methanol permeability of prepared membranes.[X. Xu, L. Li, H. Wang, X. Li, RSC Advances, 2015, 5, 4934-4940]. However, this strategy suffers from some disadvantages. One of them is that nanoparticles act as a physical barrier against proton transportation and decrease proton conductivity of membrane. In this research two main purposes were followed: formation of hydrophilic/hydrophobic separated regions in the structure of proton exchange membrane by producing ionic paths in totally hydrophobic matrix from novel technique using dual electrospinning. [M. Oroujzadeh, S. Mehdipour-Ataei and M. Esfandeh, Int. J. Polym. Mater. 2015, 64, 279-286]. The other is introducing inorganic nanoparticles (Sepiolite) into the membranes structure in such a way that no negative effect on proton conductivity of resulted. Nanocomposite blends of fully sulfonated PEK and non-sulfonated PES were prepared from dual electrospinning process. Sepiolite was used as nanoparticle and just dispersed in non-sulfonated PES matrix to avoid barrier effects of nanoparticles against proton. As a novel and special technique, using dual electrospinning process and melting just non-sulfonated sepiolite dispersed fibers resulted membranes in which ionic paths (fibers of fully sulfonated PEK) were embedded in a sepiolite dispersed non-ionic matrix. The presence of these ionic channels was proved by sTEM microscopy. Thermal and mechanical properties and also water absorption, dimensional stability, and proton conductivity of these phase separated membranes were investigated in different sepiolite and sulfonated fiber contents. Membranes showed acceptable thermal and mechanical properties. In spite of the presence of fully sulfonated PEK fibers, water absorption and swelling ratios for these samples were lower than Nafion. The proton conductivity of membranes was excellent which was attributed to the presence of ionic paths in hydrophobic matrix.

Keywords: proton exchange membranes, electrospinning, poly(ether ketone), poly(ether sulfone), sepiolite.

Figure 1

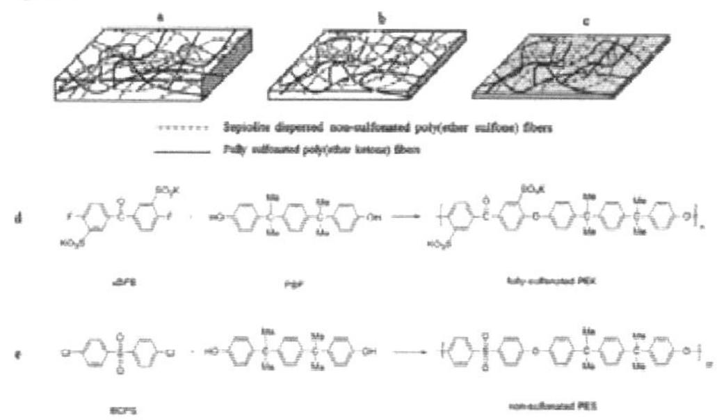

Schematic representation of prepared electrospun mats (a), mats after pressing (b), and membranes after hot-pressing (c), and also Synthesis of fully-sulfonated PEK (d) and non-sulfonated PES(e)

PPBM-O-007 [Polymer and Polymer-Based Membranes]

Thin film nanofiltration polyamide membranes for water purification using Na2SO4 draw solution

Tahir Jamil

Department of Polymer Engineering and Technology, University of the Punjab, Lahore, 54590 Pakistan

Thin film nanofiltration (TFNF) membranes, were fabricated using polyamide (PA) /asymmetric polysulfone substrates via dope formulations. The membrane permeation and Na2SO4 salt rejection performance was evaluated by using indigenously fabricated NF assembly. TFNF 4 was considered as an optimal membrane that exhibited a much higher permeation flux (35.4 L/m2.h) and salt rejection (95.6%) compared to the PSf membrane. Membranes were characterized by Fourier transform infra-red spectroscopy (ATR-FTIR), Scanning electron microscope (SEM), atomic force microscopy (AFM) and X-ray diffraction (XRD) analysis. These characterizations determine the successful interactions between polymer chains and other components and morphology. It was found that the membrane surface became smoother and more hydrophilic membrane. The SEM micrograph analysis showed average pore size < 100nm in TFNF4 membrane without any mottled surface.

Keywords: Thin filmpolymer membrane, Nanofiltration, Desalination

PPBM-O-008 [Polymer and Polymer-Based Membranes]

Porous polyisatinbiphenyl membranes for harsh environments

Bruno A. Pulido[1], Suzana P. Nunes[1], Mikhail G. Zolotukhin[2]

[1]*Biological and Environmental Science and Engineering Division (BESE), King Abdullah University of Science and Technology (KAUST), Thuwal, Saudi Arabia.*

[2]*Instituto de Investigaciones en Materiales (IIM), Universidad Nacional Autónoma de México (UNAM), Ciudad de México, México.*

A polymer with exceptional high stability at temperatures up to 400 C° in oxidative conditions was manufactured into porous membranes by phase inversion. The membranes were characterized by infra-red spectroscopy and thermal analysis. Pore diameter, porosity and morphology were studied by scanning electron microscopy. The pristine membranes show good chemical resistance, when exposed to common organic solvents, as well as pH=2 and pH=12 aqueous solutions. The polymer stability was increased by further crosslinking by two methods, one using different dibromides and a second by membrane modification with propargyl bromide followed by thermal treatment. The membranes can in this form be used in harsh environments and in applications like organic solvent nanofiltration.

Keywords: harsh environment membranes, organic solvent ultrafiltration, high temperature polymers

PPBM-O-009 [Polymer and Polymer-Based Membranes]

The promising chemical stable membranes developed through non-solvent induced phase separation (NIPS) and UV curing of acrylic functionalized polysulfone

Mian Farrukh Mehmood[1], Marco Sangermano[1], Paula Bosch[3], Mattia Giagnorio[2], Alberto Tiraferri[2]

[1]Dipartimento di Scienza Applicata e Tecnologia, Politecnico Di Torino, Torino, Italy

[2]*Dipartimento di Ingegneria dell'ambiente, del territorio e delle infrastrutture, Politecnico Di Torino, Torino, Italy*

[3]*Departamento de Química Macromolecular Aplicada, Instituto de Ciencia y Tecnología de Polímeros (ICTP), Consejo Superior de Investigaciones Científicas (CSIC), Madrid, Spain*

The acrylic functionalization of polysulfone (PSU) has been an attractive strategy to improve the chemical stability of polysulfone based membranes but this modification has also led to significant drop in the flux when compared with pristine polysulfone. [Ulbricht, Polymer - 47 (2006), 2217- 2262] [Hendrix, J. Membr. sci. 447 (2013) 212 - 221].

In order to overcome the aforementioned discrepancy; the current research was aimed to modulate the morphology and the performance of membranes in a way that a cross-linked network is developed for chemical stability and the pores morphology is retained for sufficient flux. For this purpose we incorporated two structurally different di acrylates, bisphenol A ethoxylate di acrylate (BPEDA, Mn=512) and poly (ethylene glycol) di acrylate (PEGDA, Mn=700), into 15% PSU (Mn=22,000) soln. in DMF. We have optimized all related parameters like percentage of acrylic functionality, total energy (UV) dose to ensure a cross linked network, wet thickness of the membranes and controlled non-solvent induced phase separation (NIPS). The thin films containing PSU, acrylic monomer and photo initiator, phenylbis (2,4,6-trimethylbenzoyl) phosphine oxide (BAPO), were casted on glass substrate, UV cured and subjected to NIPS.

The UV curing behavior, thermo – mechanical and morphological properties of membranes were fully investigated. The performance of membranes was measured in terms of contact angle, dead-end flux, qualitative solvent resistance and latex 20 nm particles rejection. The best formulation, 10 % BPEDA functionalized PSU, was found to have 241 L m-2 h-1 bar-1 flux, 92 % rejection of Latex 20 nm particles and resistant to DMF, THF, DMSO, acetone, toluene, and ethyl acetate after 120 hours of immersion. The developed membranes, as per best of our knowledge, exhibited performance for better than previously reported. The flux obtained is also higher than commercially available PVDF based solvent resistant membranes.

Keywords: Polysulfone, Acrylic functionalization, UV curing, phase separation

Representation of the procedure

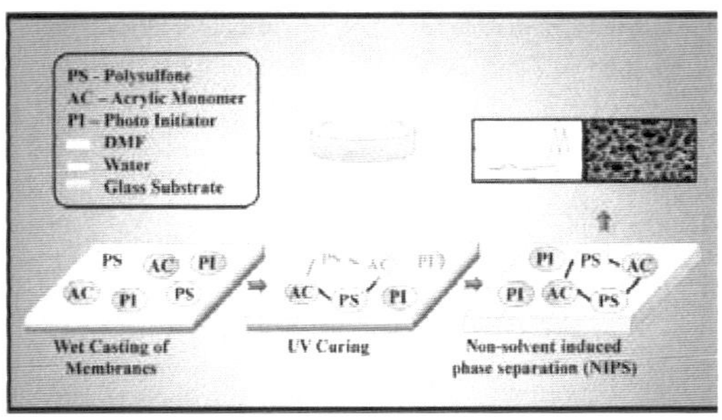

The figure carries the scheme representing the procedure, curves exhibiting bis phasic thermo-mechanical behavior and cross sectional morphology of a cross linked membrane

PPBM-O-011 [Polymer and Polymer-Based Membranes]

Gas diffusion as a probe for intrinsic microporosity

Alessio Fuoco[1], Elena Tocci[1], Peter M. Budd[2], Neil B. Mckeown[3], Karel Friess[4], Johannes C. Jansen[1]

[1]Institute on Membrane Technology, Consiglio Nazionale delle Ricerche, ITM-CNR, Via P. Bucci 17/C, 87036 Rende (CS), Italy
[2]School of Chemistry, University of Manchester, Manchester M13 9PL, UK
[3]School of Chemistry, University of Edinburgh, David Brewster Road, EH9 3FJ, UK
[4]University of Chemistry and Technology, Prague, Department of Physical Chemistry, Technická 5, 166 28 Prague 6, Czech Republic

Polymers of intrinsic microporosity (PIMs) form a novel class of polymers with promising performance in the field of gas and vapour separation membranes. They owe a series of remarkable properties to their rigid and highly contorted polymer backbone, which prevents efficient packing [1] and causes a high free volume and unusually high gas permeabilities. In the present paper, we will study the transport properties of PIMs in comparison with other polymers commonly used for gas separation membranes via detailed analysis of the diffusion behavior. Fig. 1 shows the diffusion coefficient of PIM-BTrip-TB [2] in comparison with the rubbery poly(ether-amide) PEBAX® 2533 [3] and another high free volume polymer Teflon AF2400 [4]. The figure confirms the high molecular sieve-like [5] size-selectivity for the larger gas molecules due to the extremely stiff polymer structure. It also evidences the presence of two distinct trends in the size-dependence of the diffusion coefficient for small (He and H2) and large molecules (N2, O2, CO2, CH4), which seems to be a typical characteristic of PIMs. In this work, we will discuss how this behavior is related to the polymer structure via experimental analysis of the diffusion coefficient and via molecular modelling of the polymer structure and its free volume distribution.

Acknowledgements: This research received funding from the EU's Seventh Framework Program, GA 608490, project M4CO2.

[1] NB McKeown, PM Budd; Macromolecules, 2010, 43, 5163–5176.
[2] I Rose et al.; ACS Macro Lett, 2015, 4, 912–915.
[3] P Bernardo et al.; Sep Purif Technol 2012, 97, 73–82.
[4] JC Jansen, K Friess, E Drioli; J Memb Sci, 2011, 367, 141–151.
[5] M Carta et al.; Science, 2013, 339, 303-307.
[6] V Teplyakov, P Meares; Gas Sep Purif, 1990, 4, 66–74.

Keywords: gas diffusion, Polymers of intrinsic microporosity, effective molecular diameter

Figure 1

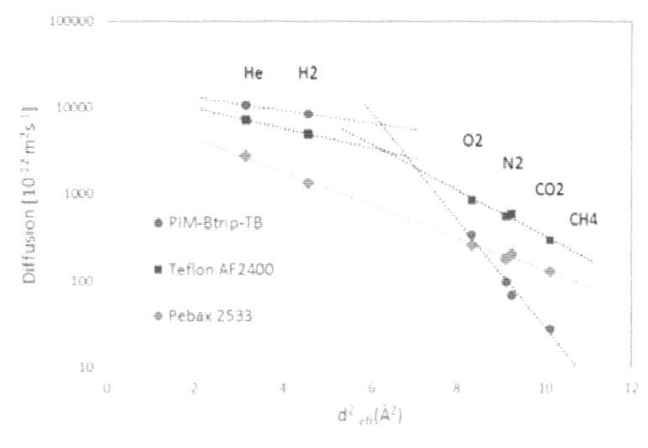

Diffusion coefficient in different polymers as a function of the effective molecular diameter of the gas [6].

PPBM-O-012 [Polymer and Polymer-Based Membranes]

Tailoring the selectivity of PIM-1 via chemical modification

Bekir Satilmis[1], Johannes C. Jansen[2], Alessio Fuoco[2], Peter M. Budd[3]

[1]Department of Chemistry, Faculty of Science and Arts, Ahi Evran University, Kirsehir 40100, Turkey.
[2]Institute on Membrane Technology (ITM-CNR), Via P. Bucci 17/C, 87036 Rende (CS), Italy.
[3]School of Chemistry, University of Manchester, Manchester M13 9PL, UK.

Polymers of intrinsic microporosity (PIMs) are composed of highly rigid and contorted structures that lead to inefficient packing. [Budd *et al.*, 2008, p. 230-231] Since their introduction, PIMs have received significant attention from the membrane research community. Particularly, PIM-1 has been studied extensively since it was found to surpass Robeson's upper bound for several gas pairs. [Robeson, 2008, p. 390-400] In this work, PIM-1 has been chemically modified (Figure 1) and the adsorption and separation properties of PIM-1 and modified PIM-1s have been examined. Hydrolysis of PIM-1 was performed in the presence of NaOH solutions (H_2O/ethanol). [Satilmis *et al.*, 2014, 52189-52198] The reduction of nitrile to primary amine was achieved using borane dimethyl sulphide complex, resulting in amine PIM-1. [Mason *et al.*, 2014, 1021-1029] The reaction of PIM-1 with ethanolamine and diethanolamine produced hydroxyalkylaminoalkylamide PIMs.[Satilmis *et al.*, 2015, 5663-5669] Volumetric gas sorption studies of modified PIMs showed that the sorption capacities of polymer altered, depending on the modification. Also, modified PIMs showed enhanced CO_2/N_2 ideal selectivities at 1 bar compared to parent PIM-1. Although chemical modifications reduced the permeability of the membranes, enhanced gas selectivities were also achieved in permeation studies. Enhanced H_2/CO_2 selectivity placed amine PIM-1 above the 2008 Robeson upper bound. The relationship between the degree of conversion and permeability of amine PIM-1 was studied in detail. The effect of temperature and pressure on the permeability of amine PIM was studied. Ethanolamine modified PIM showed size selective behaviour by enhanced H_2/N_2 and H_2/CH_4 selectivities, and it crossed the 2008 Robeson upper bounds. Dye adsorption studies revealed that chemical modification manipulated the interaction ability of PIM-1. PIM-1 showed high affinity for neutral dye. While hydrolysed PIMs showed high affinity for cationic species, amine and ethanolamine modified PIMs displayed high affinity for anionic dyes.

Keywords: Polymers of intrinsic microporosity, membranes, adsorption, permeation, separation

Figure 1

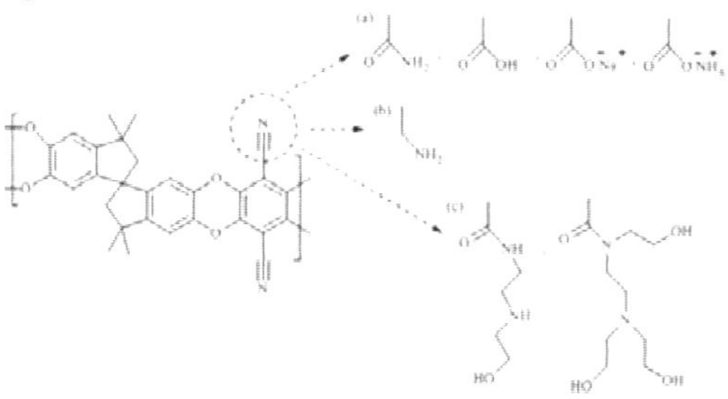

Polymers of intrinsic microporosity, PIM-1, (a) hydrolysed PIM-1s, (b) amine PIM-1, (c) ethanolamine and diethanolamine modified PIM-1s.

PPBM-O-013 [Polymer and Polymer-Based Membranes]

A novel method to determine the transport parameters of individual gas mixture components in polymers of intrinsic microporosity

Johannes Carolus Jansen[1], Marek Lanc[2], Alessio Fuoco[1], Elisa Esposito[1], Marcello Monteleone[1], Paola Bernardo[2], Karel Friess[2], Mariolino Carta[3], Neil B Mckeown[3]

[1]Consiglio Nazionale delle Ricerche, Institute on Membrane Technology ITM-CNR, Rende (CS), Italy.
[2]Department of Physical Chemistry, University of Chemistry and Technology Prague, Prague 6, Czech Republic
[3]School of Chemistry, University of Edinburgh, Edinburgh, United Kingdom

The present paper discusses a novel method for the continuous in-line measurement of the gas permeation through polymeric membranes, using a mass spectrometric gas analyser to determine the permeability and diffusion coefficients of the individual components in gas mixtures. Since slightly more than a decade, polymers of intrinsic microporosity (PIMs) are receiving a increasing attention from the scientific community for their potential use as novel high performance membrane materials. Transport of gases and vapours in dense polymeric membranes takes place according to the well-known solution-diffusion mechanism but transport of CO_2 and its mixtures in PIMs may show highly unusual behaviour. Reasons for the unusual transport properties may be the strong dual mode sorption behaviour or the interaction of the gases with the polymer matrix. In order to fully exploit these novel polymers as successful membrane materials, it is necessary to acquire a better understanding of their basic transport properties. The simplest method to determine the permeability, diffusion and, indirectly, the solubility coefficient of gases in dense polymeric membranes is the so-called time lag method. This measures the transient of the permeate pressure increase in a fixed-volume variable pressure setup. Unfortunately, this method is only suitable for pure gases. The aim of the present work is therefore to study the transient behaviour of gas mixtures, using a mass-spectrometric in-line gas analyser with high sampling rate to follow the permeate gas composition as a function of time in PIMs and other polymers. This allows the determination of the individual permeability and diffusion coefficients of each of the gas species in the mixture under different experimental conditions. By comparison with pure gases, it is also possible to identify positive or negative coupling effects.

Acknowledgements: This research received funding from the EU's Seventh Framework Program, Grant Agreement 608490, project M^4CO_2.

Keywords: PIM membrane, Permeation transient, Mass Spectrometer, Mixed gas permeation

PPBM-O-014 [Polymer and Polymer-Based Membranes]

CO2-Induced Plasticization in Copolyimides Containing a Sulfone Group

Marcel Balçık, Sadiye Velioğlu, Mehmet Göktuğ Ahunbay, Birgül Tantekin Ersolmaz

Department of Chemical Engineering, Istanbul Technical University, Istanbul, Turkey

The plasticization of the polyimide membranes, which are widely applied in gas separation processes due to their attractive permeability, selectivity and processability characteristics, is the main issue of consideration in the application in industry. The plasticization leads to reduction in selectivity of the polymeric material and also results unpredictable membrane properties. In order to investigate, the structure plasticization relationship, one homopolyimide, 6FDA-p-BAPS and two copolyimides,6FDA/BTDA-pBAPS and 6FDA-pBAPS/mPDA were selected to study on, based on our previous study (Velioglu et al., Journal of Membrance Science, 2015). in which the permeability coefficients of H_2, $O_2$2, He, CO_2, N_2, and CH_4 gases for more than 2200 possible copolyimide structures were estimated by the group contribution method. A polyimide with well-known plasticization properties, Matrimid, was studied in order to validate simulation methods. Atomic charges of polymers were calculated using B3LYP functional and ESP charge fitting. Polymer chains were modeled using Polymer Consistent Force Field (PCFF) while CO2 molecule was modeled by TRAPPE model. In order to investigate CO_2 induced plasticization of polymers, CO2 molecules were sorbed into polymeric structures using Grand Canonical Monte Carlo (GCMC) method. Series of NVT and NPT ensemble Molecular Dynamics (MD) runs were performed in order to equilibrate the resulting structure and Fractional Free Volumes (FFV) were calculated to investigate plasticization phenomenon. Open source LAMMPS software was used in simulations while FFV determination was carried out using open source Pore Blazer software. Comparison of plasticization of polymers were carried out using FFV increases and relation of plasticization with specific sites were investigated.

Keywords: Molecullar modeling, gas separation, polyimides, plasticization

PPBM-O-015 [Polymer and Polymer-Based Membranes]

Cellulose Membranes For Organic Solvent Nanofiltration

Faqih Muhamad Sukma, Pınar Zeynep Çulfaz Emecen

Department of Chemical Engineering, Middle East Technical University, Ankara, Turkey

Cellulose is an alternative polymer that can be used in Organic Solvent Nanofiltration (OSN) where its applications are hindered by the lack of suitable nonderivatizing solvents. Common solvent systems like NaOH/CS2 or NMMO are either toxic or unstable. Recent studies showed that there is an alternative way of dissolving cellulose using ionic liquids. The aim of this study is to process cellulose membranes for OSN via phase inversion using ionic liquids as solvents. 1-ethyl-3-methylimidazolium ([EMIM]OAc) is used as the ionic liquid to dissolve cellulose, while acetone is used as cosolvent and water as nonsolvent. Probe solutions in both ethanol and water are used to study the retention performance of membranes.

It is observed that increasing cellulose concentration in the membrane precursor solution, either by having more cellulose in the starting solution or by evaporating the acetone prior to coagulation, decreases membrane permeance and increases rejection of Bromothymol Blue and drying of the membranes further increases the retention. Solute-solvent-membrane interactions have an important role in determining the retention performance as a higher molecular weight dye (Brilliant Blue R, 826 Da) is retained less than Bromothymol Blue (624 Da) dissolved in ethanol, which may be related to the interaction of the ionizable groups in Brilliant Blue R with the membrane. Change of solvent also affects the rejection as was shown by higher value of Brilliant Blue R rejection in water compared to that in ethanol, probably due to the presence of a hydration shell around Brilliant Blue R in aqueous solution and its ionized groups. The membrane with the highest cellulose concentration of 20% which was prepared via phase inversion and dried after coagulation showed Bromothymol Blue (in ethanol) rejection of 94%, Brilliant Blue R (in ethanol) rejection of 40%, and Blue Dextran 5 kDa (in water) rejection of 94%.

Keywords: Organic solvent nanofiltration, Flat-sheet membranes, Cellulose

PPBM-O-016 [Polymer and Polymer-Based Membranes]

Phase Inversion Kinetics of Polymer-Ionic Liquid Solutions in Relation with Membrane Morphology and Macrovoid Formation

Elif Nur Durmaz, Pınar Zeynep Çulfaz Emecen

Department of Chemical Engineering, Middle East Technical University, Ankara, Turkey

Cellulose is known to be stable in many solvents due its inter- and intra- molecular hydrogen bonds. Dissolving cellulose with ionic liquids enables to produce membranes with phase inversion technique. Morphology of the resultant membrane depends on both thermodynamics of polymer-solvent-nonsolvent system and kinetics of the phase separation process. During the phase inversion, polymer solution becomes opaque and by measuring light transmission through the solution as a function of time, phase inversion rate can be measured.

In this study, three polymers and two solvent systems were chosen to investigate the phase inversion behavior in relationship with the resulting membrane morphology. Cellulose acetate and polyether sulfone (PES) are the polymers chosen to compare with cellulose. The ionic liquid, 1-ethyl-3-methylimidazolium acetate (EMIMAc) is the main solvent for all solutions, DMSO is added as co-solvent for some of them. Water and ethanol are the nonsolvents with different nonsolvent strength. Membranes are produced by immersion precipitation technique, at the same time phase inversion rate is measured. Performance of the obtained membrane is evaluated with probe rejection tests and SEM images are taken for the investigation of membrane morphology.

Phase inversion rates, SEM images and performance tests show that all cellulose membranes are very similar to each other, independent of the solvent or nonsolvent type. For cellulose acetate and PES membranes phase inversion rate is considerably higher than cellulose solutions. Similarly, morphologies of these membranes are very different. When EMIMAc is the only solvent and water is the nonsolvent, PES and cellulose acetate membranes have nodular morphologies quite similar to cellulose membranes with large circular macrovoids only near the bottom of the membranes. EMIMAc-DMSO binary solvent system changes the structure to typical asymmetric membrane morphologies. Moreover, use of ethanol as nonsolvent results in larger pores for cellulose acetate and PES membranes.

Keywords: Phase Inversion Kinetics, Cellulose Membrane, Ionic Liquid

PPBM-O-017 [Polymer and Polymer-Based Membranes]

Permeation and Separation Characteristics of IPA/Water Mixtures Through Poly(vinyl alcohol)-g-Poly(N-Hydroxymethylacrylamide) Copolymer Membrane by Pervaporation, Evapomeation and TDEV

Fatma Kurşun[1], Nuran Işıklan[2]

[1]*Chemistry Department, Kirklareli University, Kirklareli, Turkey*
[2]*Chemistry Department, Kirikkale University, Kirikkale, Turkey*

Pervaporation is one of the membrane separation processes, which has been most actively researched and has application in separation and purification of liquid mixture, particularly for dehydration of organic compounds or separation of azeotropic or close-boiling point mixtures [1]. Evapomeation is similar to pervaporation method, however, feed vapor is in direct contact with the membrane in evapomeation process. Therefore, the swelling of the membrane can be largely prevented [2,3]. Temperature difference controlling evapomeation (TDEV) is a new membrane separation method that temperature of the feed solution and membrane surrounding is different. It enables to improve a higher membrane performance in the permeation and separation for mixtures [4,5]. In this study, poly(vinyl alcohol)-g-poly(N-Hydroxymethylacrylamide) (PVA-g-PNHMAAm) membranes were synthesized as a membrane material. The grafted PVA membranes were prepared with a casting method and cross-linked by heat treatment in order to prevent water solubility. The permeation and separation characteristics for isopropyl alcohol (IPA)/water mixtures in the pervaporation, evapomeation and TDEV methods were investigated and compared. The effects of feed concentration and temperature on the membrane permeation rate and separation factor has been carried out. The Arrhenius activation parameters were calculated for pervaporation and evapomeation methods. It was observed that in the all methods, permeation rates increase while separation factors decrease with increasing amount of water in the feed and temperature. TDEV method for the separation of IPA/water mixtures through the grafted PVA membranes was found to be more effective than other methods. The best separation factor was found to be 1545 with a permeation rate of 0.0062 kg/m2.h. Keywords: Pervaporation, Evapomeation, TDEV, PVA, IPA, Seperation factor. Acknowledgments: The authors are grateful to the Kırıkkale University for financial support with the project number of 2012/18.

Keywords: Pervaporation, Evapomeation, TDEV, PVA, IPA, Separation Factor

PPBM-O-018 [Polymer and Polymer-Based Membranes]

Prepration of Styrene Based Membranes Including Substituted Anilidic and Amidic Acids, Characterization and Protein Adsorption

Deniz Baysal, Ramazan Coşkun, Ali Delibaş

Department of Chemistry, Bozok University, Yozgat, Turkey

Due to the high performance and low energy consumption, the use of membrane in separation technology is becoming more important every day. High permeability, high selectivity, mechanical strength, temperature resistance, chemical resistance and easy process ability increased interest to polymeric membranes [1,2].

In this study, maleic anhydride-styrene copolymer was prepared by using Bz_2O_2 initiator with radicalic copolymerization. Later, this synthesized copolymer modificated with different reagents such as aminobenzoic acid (BA), amino isophthalic acid (FA), iminodiacetonitrile (IDAN) and diaminomaleonitrile (MN). These functionalized styrene copolymers including anylidic and amidic acid were characterized with FTIR. Next step, membranes were prepared from functionalized copolymers using solution cast. The surface morphologies of prepared membranes were determined with SEM, the thermal stability was determined with TGA/DTA and additionally, glass transition temperatures (Tg) was determined with DSC. Mechanic stabilities were tested by tensile strength measurement. After, the gap volume and porosity of membranes were determined and swelling behavior was investigated in different pH and temperatures. The capacity of copper adsorption was determined. The study of protein adsorption was carried out. FTIR spectra results shown that synthesized copolymer was modified. Also, SEM analyses of prepared membranes surfaces were porous and the surface morphology was changed with using reactive groups. That from TGA/DTA analysis the thermal stability of membranes including anilidic acids was higher than including amidic acid group membranes. The membranes including amidic acid had higher Cu(II) capacity and protein adsorption.

Keywords: membranes, maleic anhydride-co-styrene, functionalization

Figure

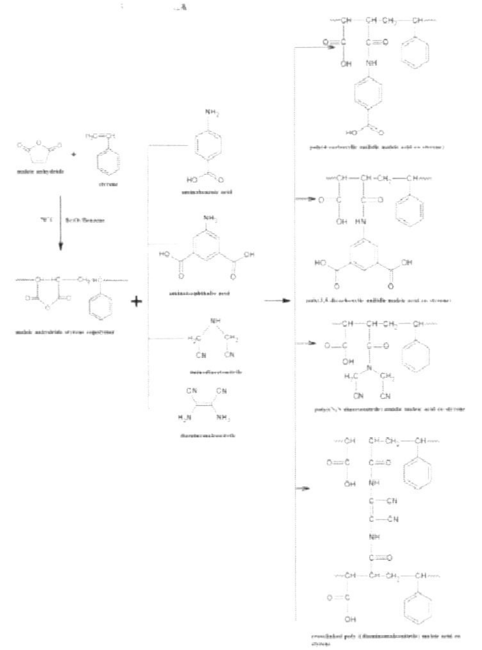

Functionalization schema of copolymer

PPG-O-001 [Porous Polymer and Gels]

Chemistry in the confined spaces of porous polymers

Cafer T Yavuz[1], Jeehye Byun[1], Nesibe A Dogan[1], Hasmukh A Patel[2], Damien Thirion[1], Saravanan Subramanian[1], Ercan Ozdemir[1]

[1]*Graduate School of EEWS, KAIST, Daejeon, Republic of Korea*
[2]*Department of Chemistry, Northwestern University, Evanston, IL, USA*

Porous polymers with robust organic structures through a covalent backbone feature permanent voids that can be used for chemical activities. We have developed a family of nanoporous (pore size < 100 nm) covalent organic polymers (COPs), which show significant capacities and selectivities for carbon dioxide. Surprisingly, azo (N=N) bearing COPs show lack of N_2-philicity by increasing temperature, in other words N_2-phobicity, leading to very high CO_2/N_2 selectivities. Under high pressures COP-1 shows a record high capacity of 5.6 g/g CO_2 uptake at 200 bar and 45 °C. COP-83 has a capacity of 5 mmol/g at 298 K and 1 bar, and COP-97 shows an uptake of 8 % (w/w) CO_2 in 2 minutes from a simulated flue gas mixture (CO_2 15%, H_2O 3.8%, He 81.2%, 40 °C, flow rate: 80 mL/min). More recently we established the first direct introduction of ethylene diamines on the walls of COPs through bromination intermediates. CO_2 uptake capacities multiply with the nitrogen content, up to an unprecedented four times of the starting porous polymer in dry flue gas conditions. Morphology control is another subject of interest and we made hollow spheres of COP-122 with nitrile pendant groups for further grafting chemistry. Our results point to a great potential in porous polymers for chemical conversions in their confined spaces since they are highly porous, inexpensive, physisorptive solids, which can be chemically modified with chemisorptive functionalities.

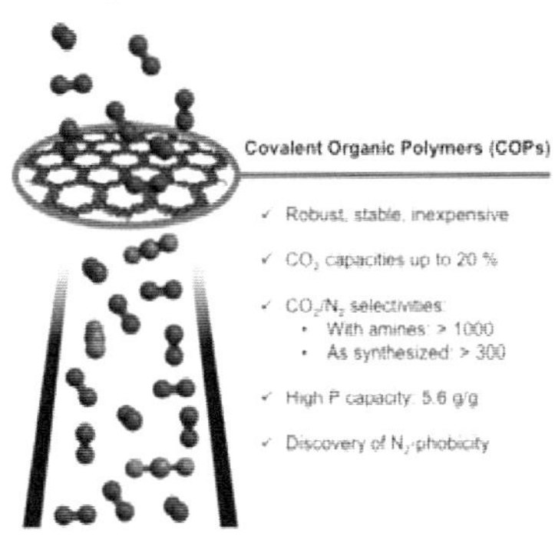

References:

1. **Chem. Mater.** 26 (23), 6729–6733 (2014)
2. **Chem. Eur. J.**, 20, 772-780, (2014)
3. **J. Mater. Chem. A**, 2, 12507-12512, (2014)
4. **Nature Commun.**, 4:1357, (2013)
5. **Adv. Funct. Mater.**, 23, 2270–2276 (2013)
6. **J. Mater. Chem.**, 22, 8431-8437 (2012)
7. **Chem. Comm.**, 48 (80), 9989–9991 (2012)
8. **Energy Environ. Sci.**, 4, 4528-4531 (2011)

Keywords: Porous Polymers, Carbon Capture, Nanomaterials

PPG-O-002 [Porous Polymer and Gels]

Open porous composites and ceramic derivatives by emulsion templating

Peter Krajnc[1], Irena Pulko[2], Robert Liska[3], Thomas Koch[3]

[1]*University of Maribor, Maribor, Slovenia*
[2]*Polymer Technology College, Slovenj Gradec, Slovenia*
[3]*Vienna University of Technology, Vienna, Austria*

Emulsion templating ia a technique, where internal phase droplets are utilised for pore templating within the continuous phase bulk. If the droplet phase is large by volume and poylmerisation of the monomers in the continuous phase is achieved, The resulting monolith of polymer is normally open porous, pores consisting of a larger series of pores resulting from droplets of internal phase being connected by a series of channels resulting of polymer film shrinkage during the polymerisation (Pulko I. et al, Macromol. Rapid Commun. 2012, 33, 1731). Introduction of metal oxide particles into continuous phase allows for the preparation of composite porous material while presenting a problem regarding emulsion stabilization. Careful optimization of emulsion preparation and subsequent thermal treatment can lead to corresponding ceramic material (Turnsek M. et al., J. Eur. Ceram. Soc. 2016, 36, 1045). Composite porous materials, based on methyl methacrylate and alumina and titania particles, will be presented. Furthermore, the influence of subsequent thermal treatment on the morphology of resulting ceramic material will be discussed.

Keywords: polyHIPE, emulsion templating, porous material, porous polymers, porous ceramic, poly(methyl methacrylate)

PPG-O-003 [Porous Polymer and Gels]

Emulsion-Templated Porous Polymer Scaffolds for 3D In-Vitro Cell Culture and Tissue Engineering

Ahmed M. Eissa[1], Neil R. Cameron[2]
[1]*School of Engineering, University of Warwick, Coventry, UK*
[2]*Department of Material Sciences and Engineering, Monash University, Clayton, Australia*

Highly porous polymeric materials can be created by the emulsion templating technique, whereby a high internal phase emulsion (HIPE) is created in which the continuous (non-droplet) phase contains polymerisable monomer(s). Polymerisation then results in a well-defined macroporous polymer (foam), known as a polyHIPE. The nature of the formation of polyHIPEs creates a highly porous, interconnected monolith structure, the architecture of which can be tightly controlled. These polyHIPE materials can serve as scaffolds for the culture of cells and, ultimately, tissues in 3D, allowing cells to mimic their native structure and function *in-vitro*. Applications of these polymeric scaffolds are foreseen in 3D cell culture, tissue engineering and regenerative medicine. However, one potential limitation of these polyHIPE materials as 3D cell scaffolds is surface chemistry. Cells *in-vivo* are typically surrounded by a complex extracellular matrix that contributes to cell anchorage and function. For example, it is well known that hepatocytes (liver cells) possess a galactose-binding cell receptor (asialoglycoprotein). The presence of galactose residues on the surface of polyHIPE scaffolds could improve hepatocyte adhesion and function while growing on the scaffold, leading to enhanced performance in liver tissue engineering applications.

This presentation will describe the surface functionalisation of polyHIPE materials by incorporating pentafluorophenyl acrylate (PFPA) into the initial styrene emulsion. Galactose-functionalised scaffolds are formed *via* displacement of the labile PFPA group by post-polymerisation treatment with 2'-aminoethyl-β-D-galactopyranoside (Fig. 1). Experiments with primary rat hepatocytes show that the galactose-functionalised scaffolds offer enhanced hepatocyte adhesion and functionality compared to the unfunctionalised/ glucose-functionalised scaffolds [A. S. Hayward, A. M. Eissa *et al.*, *Biomacromolecules* **2013**, 14 (12), 4271].

The surface functionalisation of the recently developed biodegradable polyHIPE scaffolds which are made by photopolymerisation of HIPEs containing multifunctional thiols and acrylates will also be presented. The resulting thiol-ene polyHIPE materials have been found to contain residual thiols which are reactive towards a range of (meth)acrylate, maleimide and allyl functionalities providing a facile route to functionalisation with biomolecules [C. R. Langford *et al.*, *Polymer Chemistry* **2014**, 5, 6200]. Initial cell culture experiments with human endometrial stem cells will furthermore be shown. Lithographic 3D printing approaches to create scaffolds with complex architectures and features are currently underway.

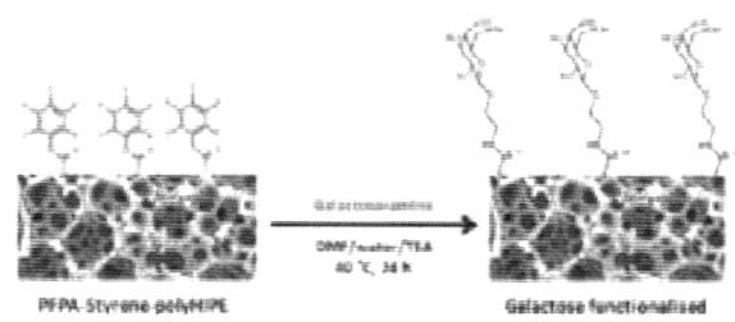

Figure 1. Schematic showing the surface functionalisation of a polyHIPE material.

Keywords: biomaterials, porous polymers, polyHIPEs, 3D cell culture, tissue engineering, regenerative medicine

PPG-O-004 [Porous Polymer and Gels]

Synthesis and characterization of new polymeric HPLC column

Arzu Sinem Satılmış[1], Musa Şölener[2], Osman Sermet Kabasakal[2]
[1]*Asos Mekatronik Otomasyon San ve Tic. Ltd. Şti, Ergene/Tekirdağ/Turkey*
[2]*Eskişehir Osmangazi University Chemical Engineering Department, Eskişehir,Turkey*

In this study, the purpose was the synthesis and characterization of polymeric column packing materials by using poly(GDMA-*co*-GDGDA) beads. For this purpose poly(glycidyl methacrylate) (poly(GMA)) particles were synthesized by dispersion polymerization method in monodisperse form as seed latex. The monodisperse poly(GMA) particles (2 mm in size) were directly used as the seed latex in the production of macroporous particles.

Monodisperse macroporous Poly(glycerol dimethacrylate-*co*-glycerol-1,3-diglycerolate diacrylate) (Poly(GDMA-*co*-GDGDA)) beads were obtained by multistage microsuspension polimerization technique by using poly(GMA) seed latex in an aqueous emulsion medium.

The physical characterizations of the polymeric particles were performed by scanning electron microscopy (SEM), surface area and pore size measuring (BET), and laser diffraction particle size analyzer.

Sodium azide (NaN_3) was attached on the monodisperse poly(GDMA-co-GDGDA) particles by using click chemistry method. Then, propiolic acid was attached on this poly(GDMA-co-GDGDA)-NaN_3 particles. By this way, the poly(GDMA-co-GDGDA)NaN_3-propiolic acid triple nitrogen rings were obtained.

The obtained polymeric structure was analyzed by HPLC using 4.6 x 150 mm sized stainless steel column and UV detector at 214 nm by injecting polar analyte.

Introduction

Porosity and homogeneous size distribution are important properties for high performance liquid chromatography (HPLC) column packing materials. Due to their hydroxyl functionality, glycerol dimethacrylate (GDMA), glycerol-1,3-diglycerolate diacrylate (GDGDA) or their derivatives are widely used for various applications (1, 2).

In this study, the purpose was the synthesis and characterization of polymeric column packing materials by using poly(GDMA-*co*-GDGDA) beads.

Materials-Methods

Synthesis of poly(GMA) seed latex

Primarily poly(glycidyl methacrylate) (poly(GMA)) particles were synthesized by dispersion polymerization method in monodisperse form as seed latex. For this purpose, polyvinylpyrrolidone (PVP) (PVP K-30) (0.45 g) was dissolved in ethanol (30 mL). Glycidyl methacrylate (GMA) (3 mL) and azobisisobutyronitrile (AIBN) (0.24 g) were added into the reactor. The polymerization was performed at 70 °C for 24 h in a shaking water bath. The obtained poly(GMA) latex was washed with ethanol and distilled water successively by centrifugation–decantation and finally redispersed in distilled water.

Synthesis of poly(GDMA-GDGDA) particles

Cyclohexanol (Cyc-OH) (3.1 mL) and dibutyl phthalate (DBP) (1.8 mL) were emulsified in water (60 mL) including sodium dodecyl sulfate (SDS) (0.15 g) by sonication. The poly(GMA) containing dispersion (0.2 g) was added to the emulsion to swell the particles with organic phase. The resulting dispersion was stirred at room temperature for 24 h. Separately, the GDMA (2.6 mL) and GDGDA (2.6 mL) monomers, isoamyl alcohol (IAA) (0.25 mL) and benzoyl peroxide (BPo) (0.12 g) were emulsified in water (60 mL) including SDS (0.15 g) by sonication. This mixture was added on the previous mixture and stirred at room temperature for more 24 h. Finally, poly(vinyl alcohol) (PVA) (0.8 g) dissolved in distilled water was added on the mixture. The polymerization of the swollen particles was conducted at 80 °C and 100 cpm shaking rate for 24 h. The poly(GDMA-co-GDGDA) particles were washed with ethanol and distilled water successively by centrifugation–decantation.

The physical characterizations of the polymeric particles were performed by scanning electron microscopy (SEM), surface area and pore size measuring (BET), and laser diffraction particle size analyzer. The SEM photographs of the poly(GDMA-*co*-GDGDA) beads are seen in figure 1.

Sodium azide (NaN_3) was attached on the monodisperse poly(GDMA-co-GDGDA) particles by using click chemistry method. Then, propiolic acid was attached on this poly(GDMA-co-GDGDA)-NaN_3 particles. By this way, the poly(GDMA-co-GDGDA)NaN_3-propiolic acid triple nitrogen rings were obtained.

Results and Discussion

The obtained polymeric structure was analyzed by HPLC using 4.6 x 150 mm sized stainless steel column and UV detector at 214 nm by injecting polar analyte. The HPLC spectra obtained for various flow rates are shown in figure 2. It was observed that the effect of pH on separation is not considerable as can be seen in figure 3.

References;

1) Saracoglu B; Uguzdogan E, Golgelioglu C, Tuncel A; *Ind. Eng. Chem. Res.* 2009, 48, 4844–4851.

2) Golgelioglu C, Bayraktar A, Celebi B, Uguzdogan E, Tuncel A; *J. Chromatogr. A.* 2012, 1224, 43–50.

Keywords: dispersion polymerization, multistage microsuspension polimerization, poly(GDMA-co-GDGDA) beads, high performance liquid chramotography, HPLC

Figure1

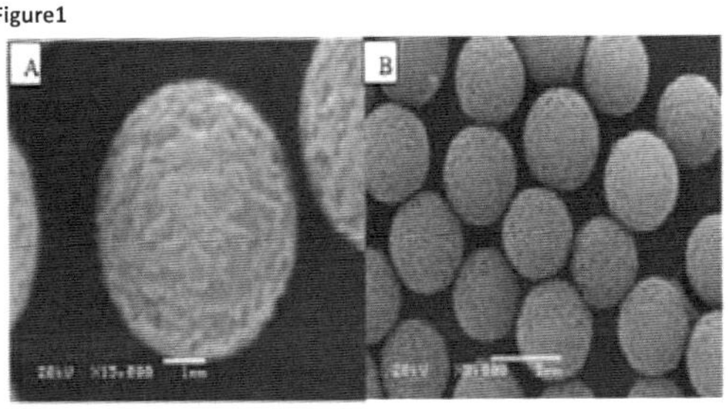

Poly (GDMA-co-GDGDA) particules SEM images. Magnification order: A) 15,000x B) 5,000X

Figure2

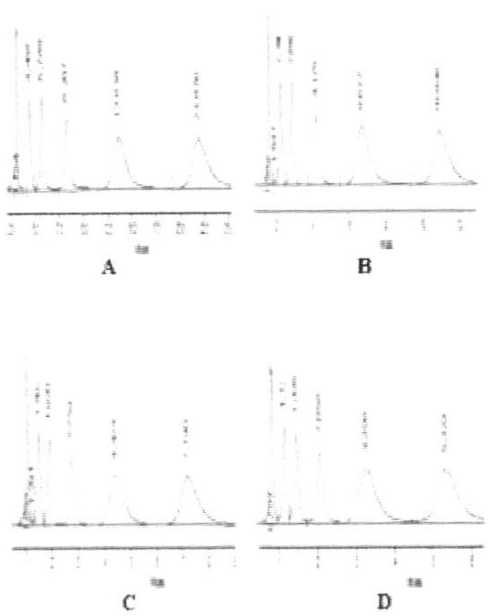

Separation behavior of the produced column for polar analytes on isokratic mode at various flow rates. Flow rates: A: 0.5; B: 1 C: 1.5; D: 2 mL/min. Conditions: Column size: 4.6x150 mm, Room temperature: (25 °C), UV detector: 214 nm, Mobile phase: (AcN/H2O) (%): 90/10, The order of elusion: 1. Toluene, 2. Acrylamide, 3. Thymine, 4. Thiourea, 5. Inosine, 6. Cyctosine.

Figure3

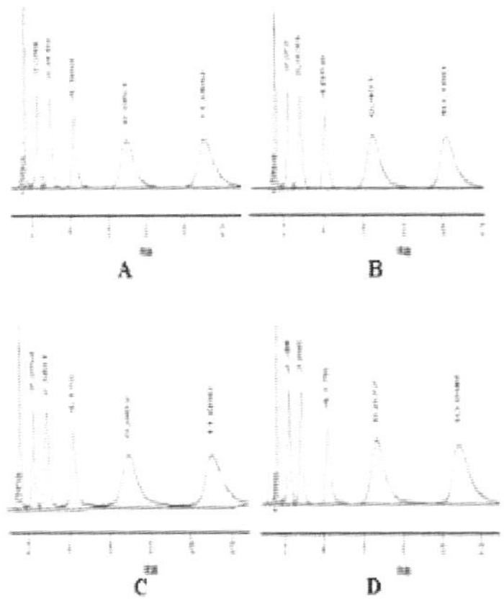

Separation behavior of the produced column for polar analytes on isokratic mode at different pH. pH: A: 3; B: 4 C: 5; D: Conditions: Column size: 4.6x150 mm, Room temperature: (25 °C), UV detector: 214 nm, Mobile phase: (AcN/H2O) (%): 90/10, Flow rate: 1 mL/min. The order of elusion: 1. Toluene, 2. Acrylamide, 3. Thymine, 4. Thiourea, 5. Inosine, 6. Cyctosine.

PPG-O-005 [Porous Polymer and Gels]

Self-structuring chitosan-cinnamaldehyde hydrogels

Luminita Marin

"Petru Poni" Institute of Macromolecular Chemistry, Iasi, Romania

Chitosan based hydrogels are a class of crosslinked materials intensely studied for their biomedical, industrial and environmental applications. Usually, they are prepared by crosslinking chitosan polyamine with various di-aldehydes, especially glutaraldehyde. On the other hand, the di-aldehydes possess a degree of toxicity which limits their use in biomedical applications. As alternative of di-aldehyde crosslinkers, we propose a dual crosslinking based on covalent and physical bonds, which can be reached exploiting the imine forming reversibility. In this context, this paper deals with the study of a series of chitosan/cinnamaldehyde hydrogels prepared by acid condensation reaction. The crosslinking of chitosan with cinnamaldehyde mono-aldehyde is an unusual pathway of hydrogel obtaining which imposes a deep structural characterization in order to understand the driving forces which govern the gelling process. The crosslinking process has been studied by 1H-NMR, HR-MAS, 13C-CP MAS and ATR-FTIR spectroscopy on hydrogels and corresponding xerogels. Further, X-ray diffraction demonstrated a three-dimensional nanostructuring due to the imine linkage reversibility with consequences on the micrometer-scale morphology and on the improvement of mechanical properties, as demonstrated by SEM and rheological investigation. The hydrogel morphology can be simple controlled by varying the crosslinking degree and time (Figure 1).

Keywords: chitosan, cinnamaldehyde, hydrogels, self-structuring, elastic network

Figure 1. Representative SEM images of hydrogels

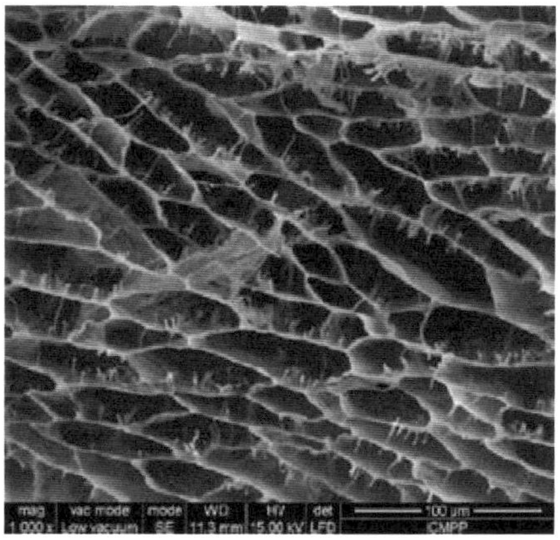

c) Hydrogel containing a 4/1 molar ratio of NH2/CHO funtional groups, crosslinked 1 weak

Figure 1. Representative SEM images of hydrogels

b) Hydrogel containing a 2/1 molar ratio of NH2/CHO funtional groups, crosslinked 1 weak

Figure 1. Representative SEM images of hydrogels

a) Hydrogel containing a 2/1 molar ratio of NH2/CHO functional groups, crosslinked 2 hours

PPG-O-006 [Porous Polymer and Gels]

Dual crosslinked chitosan based hydrogels with efficient antifungal activity

Daniela Ailincai, Luminita Marin, Mariana Pinteala, Mihai Barboiu

Petru Poni Institute of Macromolecular Chemistry, Iasi, Romania

Hydrogels are three-dimensional high-performance materials able to absorb large amount of water or biological fluids. Many polymers proved ability to form hydrogels by chemical or physical crosslinking. Among them, chitosan is by far the best choice due to its intrinsic outstanding biological properties: biocompatibility, biodegradability, hemostatic, hypolipidemic and hypoglycemic effect, antitumoral, antimicrobial and fungicidal activity. In order to combine the advantages of the physically crosslinked hydrogels with those of the chemically crosslinked ones, our aim was to obtain chitosan based hydrogels, dual crosslinked, by the use of an aldehyde which has the possibility to form covalent and H- bonds. Different hydrogels have been obtained by varying the ratio between NH2 and CHO functional groups in order to obtain hydrogels with different crosslinking degrees and in consequence, different morphologies and properties. FTIR and NMR spectroscopy demonstrated that chitosan gelation occurs by chemo-physically crosslinking, forming a supramolecular network. The X-ray diffraction demonstrated a three-dimensional nanostructuring due to the hydrophilic/hydrophobic segregation. SEM, rheological investigations and swelling ratio measurements have been used as complementary methods to characterize the obtained hydrogels. Their antifungal activity has been evaluated on *Candida* planktonic yeast and biofilms, and the obtained results recommend their use as materials for the treatment of the recurrent vulvovaginitis infections.

Keywords: chitosan, hydrogel, monoaldehyde, antifungal activity

Representative hydrogel images

PPG-O-007 [Porous Polymer and Gels]

Self-assembled networks composed of polymer and supramolecular chains

Olga Philippova, Andrey Shibaev

Physics Department, Moscow State University, Moscow, Russia

A combination of polymer and supramolecular chains is a promising way for the design of new smart materials. The present paper is devoted to the elaboration of self-assembled networks based on synthetic polymers and wormlike micelles of viscoelastic surfactant promising for the use in oil industry as one of the main components of fracturing fluids. The "living" character of wormlike micelles, which incessantly break and recombine, imparts to the network significant advantages over similar polymeric systems including the ability to self-heal and to completely recover the properties after the mechanical degradation at high shear rates as well as a high responsiveness to hydrocarbons. The latter is a key property for the application of these gels in petroleum industry, because when oil is pumped, the gel should be completely destroyed and do not to hinder the oil flow. It is shown that the proposed polymer/surfactant self-assembled networks demonstrate a very pronounced increase in viscosity as compared to pure component solutions. At the same time, the addition of n-alkanes leads to the drop of viscosity by up to 5 orders of magnitude thus inducing the transition of a gellike system to a fluid with low viscosity. SANS data evidence that this process is accompanied by the breaking of the wormlike surfactant micelles with the formation of microemulsion droplets, which results in the disruption of the whole network. Thus, in the prepared gels, the polymer component strengthens the network, whereas the supramolecular component imparts a high responsiveness to oil. Acknowledgment. The work was financially supported by Russian Science Foundation (project №15-13-00114).

Keywords: Polymer gels, responsive polymers, networks

PPG-O-008 [Porous Polymer and Gels]

Synthesis of metal entrapped stable Polymeric hydrogels for catalytic reduction of organic dyes

Mohammad Saleem Khan, Majid Khan, Luqman Ali Shah

National Centre of Excellence in Physical Chemistry, University of Peshawar, Pakistan

Stable Polymeric hydrogel of Acrylonitrile and 2-hydroxyethyl methacrylate were prepared through conventional free radical polymerization in water using N,N-Methylenebisacrylamide (MBA) as a crosslinker. The synthesized hydrogel then employed via in situ reduction to synthesize hybrid microgel of different metal salts like Nickel, Copper and Cobalt. The prepared metal entrappedhydrogel samples were characterized through FTIR, SEM and UV/VIS spectrophotometer. The results indicate that the hybrid hydrogels are stable for a long time. The synthesized hybrid materials show remarkable catalytic properties and used successfully for the degradation of different water soluble dyes.

Keywords: metal entrapped hydrogels, Acrylonitrile2-hydroxyethyl methacrylate, catalytic reduction, organic dye.

PPG-O-009 [Porous Polymer and Gels]

Visual Observation on the Foaming Process of Styrene-Methyl Methacrylate Copolymer / n-Pentane system Using Temperature Increment

Mostafa Salehi, Amin Parham, Mostafa Rezaei, Mahdi Salami Hosseini

Institute of Polymeric Materials, Department of Polymer Engineering, Sahand University of Technology, Tabriz, Iran

In this work, the synthesized styrene-methyl methacrylate copolymer particles with 70:30 weight percent (St70/MMA30) was used to study the visual foaming dynamics at non pressurized condition using temperature increment. First, all copolymer particles were impregnated by 10 wt% (gr/gr) of n-pentane using designed chamber consisted of the self-sealing observation cell equipped to the pressure and temperature controller. At the second stage, the impregnated particles were placed into the hot stage with controlled temperature and the foaming process was recorded by visualization equipment. The effect of foaming temperature on the expansion dynamics of copolymer particles was examined and it was found that expansion ratio was increased with increasing temperature as well as the foaming onset was conducted earlier at high temperature. Particle collapsing was also observed in foaming temperature above 140 °C.

Keywords: St-MMA Copolymer; n-Pentane; Foaming Dynamics; Visualization

PPG-O-010 [Porous Polymer and Gels]

Porous Polymers For CO2 Capture and Conversion

Onur Buyukcakir[1], Ali Coskun[2], Kyriaki Polychronopoulou[3]

[1]Graduate School of Energy, Environment, Water and Sustainability (EEWS), Korea Advanced Institute of Science and Technology (KAIST), Daejeon, South Korea
[2]Department of Chemistry, Korea Advanced Institute of Science and Technology (KAIST), Daejeon, South Korea
[3]Mechanical Engineering, Khalifa University, Abu Dhabi, United Arab Emirates

The increasing trend of CO_2 emission into the atmosphere along with its environmental effects positioned the CO_2 capture and conversion technologies as an integral part of today's energy industry. In this regard, porous polymers with high specific surface areas and distinct gas sorption properties emerged as an important and alternative class for CO_2 capture and utilization. In spite of their distinctive properties, however, because of their amorphous character predicting and controlling their surface area and pore morphology are highly challenging. In order to overcome these limitations, we have developed[1] a new strategy to introduce local-order into amorphous porous polymers using shape-persistent organic cage compounds as molecular building-blocks in the synthesis of porous cage frameworks (pCAGEs). These three-dimensional shape persistent molecules enabled us to tune extrinsic porosity of frameworks that allowed us to control the surface area of pCAGEs from 582 to 820 $m^2\,g^{-1}$ by simply increasing the size of the organic linker. Additionally, pCAGEs showed significantly high CO_2 uptake capacities (4.21 mmol g^{-1} at 1 bar, 273K) with prominent CO_2/N_2 selectivities. Furthermore, porous polymers offer a perfect platform for catalytic conversion of CO_2 into value added products. In this sense, we have introduced highly stable porous cationic polymers (PCPs) [2] incorporating charged bipyridinium subunits. PCPs showed excellent catalytic activity for the conversion of CO_2 into cyclic carbonates (up to 99% yield) with an unprecedented selectivity, thus introducing a metal-free two in one approach, that is the capture and conversion of CO_2 into industrially important products using a porous heterogeneous organocatalyst, which is a long-term goal for today's energy industry mainly due to economic and environmental benefits arising from the direct utilization of CO_2 as a sustainable source.

References:

1. Buyukcakir, O. et al., Chem. Mater., 2015, 27, 4149-4155.
2. Buyukcakir, O. et al., Chem. Commun., 2016, 52, 934-937.

Keywords: Porous Polymers, CO2 Capture and Conversion, Porous Cage Frameworks, Catalytic Conversion of CO2

Figure-1

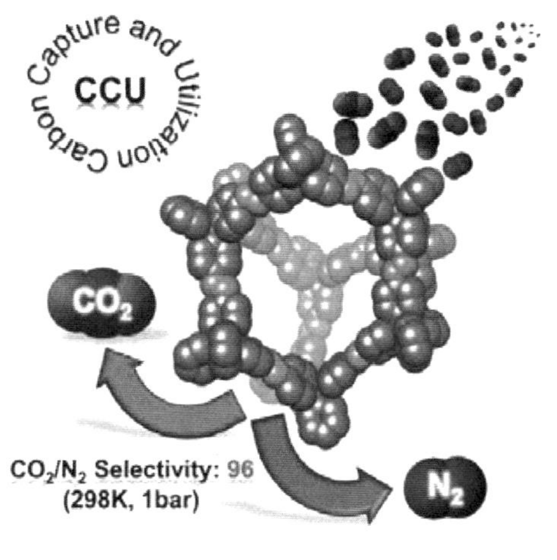

Rationally designed porous polymers for CO2 capture and conversion

PPG-O-011 [Porous Polymer and Gels]

Porous inverse vulcanised polymers for mercury capture

Douglas J Parker, Holly A Jones, Tom Hasell

Department of Chemistry, University of Liverpool, Liverpool, UK

Heavy metal contamination exists in the waste streams of many industries, and mercury is of particular concern for human health. Sulfur is an industrial by-product, removed as an impurity in oil-refining. This has led to vast unwanted stockpiles of sulfur, and resulted in low bulk prices. Sulfur is therefore a promising alternative feedstock to carbon for polymeric materials. Inverse vulcanisation (Chung et al., Nat. Chem., 2013, 5, 518) has made possible the production of sulfur polymers, stabilised against depolymerisation by crosslinking. Supercritical CO2 can be used to foam inverse vulcanised polymers – making them porous (Hasell et al., Chem Commum. 2016, DOI: 10.1039/C6CC00938G). These high sulfur-foams show excellent potential as low cost water filters to remove Hg. Alternative monomers for inverse vulcanisation, and how they can be used to reduce the cost and improve the properties of the resultant polymers, will be discussed.

Figure 1: a) Reaction of elemental sulfur and 1,3-diisopropenyl benzene produces a stable high sulfur polymer. b) SEM imaging of scCO2 foamed sulfur-DIB crosslinked polymer (50 wt% sulfur). The sample was exposed to CO2 at 28 MPa and 80 °C for 3 hours. The sample shows both closed cell and connected macropores. c) The percentage mercury remaining in solution after 3 hours exposure to each of the materials listed. Values are given as a mean of three repeats with standard deviation shown as error bars.

Keywords: Inverse vulcanisation, sulphur polymer, porous, mercury capture, supercritical

Figure 1

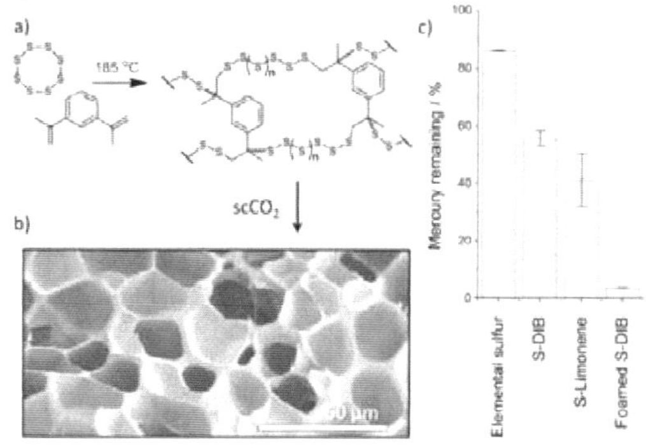

PPG-O-012 [Porous Polymer and Gels]

Microporous Graphene Frameworks For Gas and Energy Storage Applications

Ali Coskun
Graduate School of EEWS, Department of Chemistry, KAIST, Daejeon, Republic of Korea

Microporous polymers attracted[1] significant deal of attention in recent years due to their permanent porosity, chemical tunability, physicochemical stability and exceptional gas sorption properties.[2] Incorporation graphene nanoribbons (GNRs) or graphene sheets into microporous polymers, however, has been a significant challenge mainly due to their low solubility and high affinity to restack to form graphitic layers. Graphene is composed of single-atom-thick, sp^2 hybridized carbon atoms that are bonded together in a two-dimensional hexagonal honeycomb lattice, and presents high π-surface area, excellent chemical, thermal, mechanical stability along with superior thermal and electrical conductivities. Importantly, the resulting microporous polymers incorporating either graphene or GNRs are expected to inherit some of their unique properties. In order to prevent restacking and to take full advantage of high π-surface area, we introduced a permanent 'spacer' such as porosity between GNRs and graphene layers within the framework. We utilized both noncovalent[3] and covalent[4] approaches to form three-dimensional graphene frameworks. More recently, we also utilized[5] bottom-up approach to introduce graphene nanoribbons up to 2 nm in length and 1.1 nm in width into graphene nanoribbon frameworks (GNFs). These frameworks showed high thermal stability up to 400°C in air and exhibited BET surface areas up to 700 m2 g-1 along with promising gas separation properties.

[1] Patel, H. A.; Je, S. H.; Park, J.; Chen, D. P.; Jung, Y.; Yavuz, C. T.; Coskun, A., Nature Commun., 2013, 4, 1357.
[2] Talapaneni, S. N.; Hwang, T. H.; Je, S. H.; Buyukcakir, O.; Choi, J. W.; Coskun, A., Angew. Chem. Int. Ed., 2016, DOI: 10.1002/anie.201511553.
[3] Srinivasan, S.; Je, S. H.; Back, S.; Barin, G.; Buyukcakir, O.; Guliyev, R., Jung, Y.; Coskun, A., Adv. Mater., 2014, 26, 2725–2729.
[4] Song, K. S.; Coskun A., ChemPlusChem, 2015, 80, 1127–1132.
[5] Byun, Y.; Coskun, A., Chem. Mater., 2015, 27, 2576–2583.

Keywords: Carbon capture, graphene, porous polymers, gas sorption

PPG-O-013 [Porous Polymer and Gels]

Preparation and Antimicrobial Properties of Macroporous Polypeptide-based Cryogels

Mehmet Murat Ozmen[1], Steven J. Shirbin[2], Shu J. Lam[2], Nicholas J. Chan[2], Qiang Fu[2], Neil O'brien Simpson[3], Eric C. Reynolds[3], Greg G. Qiao[2]

[1]*Yildiz Technical University, Department of Bioengineering, 34220, Esenler, Istanbul, Turkey*
[2]*Polymer Science Group, Department of Chemical and Biomolecular Engineering, University of Melbourne, Parkville, Melbourne, Victoria 3010, Australia*
[3]*Melbourne Dental School and The Bio21 Institute of Molecular Science and Biotechnology, Oral Health CRC, The University of Melbourne, Parkville, VIC 3010, Australia*

Peptide-based antimicrobial hydrogels exhibit inherent antimicrobial activity upon physical contact between the gel and the bacteria (Chen, Adv. Mater. 2015, 27, 4, 648). These gels consist of a polycationic rich surface (i.e., polylysine), which results in disruption of the bacterial membrane. Macroporous hydrogels, so called cryogels, resulting from gelation at sub-zero temperatures, have attracted attention due to their superior swelling and mechanical (e.g., elasticity, toughness) properties. (Ozmen, Chem. Commun. 2015, 51, 98, 17479). Antimicrobial cryogels have the potential to exploit the important characteristics of these systems for biomedical and water purification applications (Ozmen, ACS Macro Lett. 2016, 5, 552).

In this study, novel cryogels were prepared at low temperatures via crosslinking of the amine residue of a polylysine-b-polyvaline copolymer with a glutaraldehyde cross-linker (Figure 1). The resulting cryogels were characterized to investigate their potential application as an antimicrobial material. The prepared cryogels exhibited excellent swelling properties and mechanical strength compared to their analogous hydrogel due to their macroporous internal morphology. Furthermore, these cryogels displayed inherent antimicrobial activity against E. coli resulting in a 95.6% reduction in the viable cell counts within a 1hr incubation. By comparing cryogel antimicrobial efficacy with that of the control traditional hydrogel, macropores were found to be crucial for the bactericidal activity observed.

Keywords: Antimicrobial, Cryogel, Macroporous, Polypeptide

Figure 1

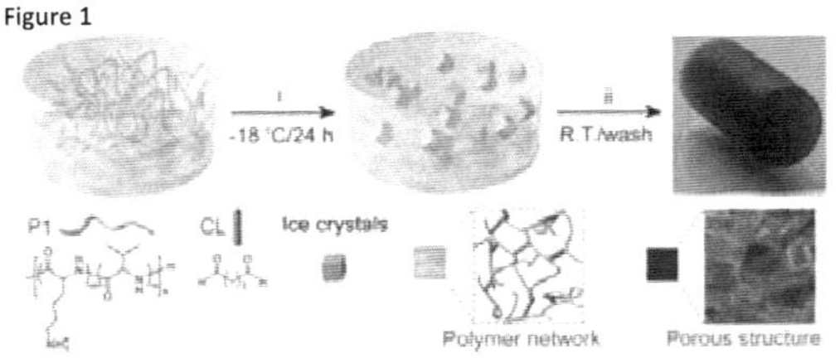

Preparation of PLL-b-PDLV block copolypeptide (P1) based cryogels: (i) Block copolypeptide (P1) and glutaraldehyde crosslinker (CL) dissolved in PBS, then frozen at -18 oC for 24 hrs (ii) Gel is thawed at RT and washed thoroughly in water.

PPG-O-014 [Porous Polymer and Gels]

Evaluations of molecular weight and reduction effects on Jeffamine-glutaraldehyde cryogels for potential application as scaffolds in cartilage tissue engineering

Ahmet Erdem, Asabuwa Ngwabebhoh Fahanwi, Ufuk Yıldız

Department of Chemistry, Kocaeli University, Kocaeli Turkey

Cryogels have demonstrated good candidacy as scaffolds for cartilage tissue engineering mainly due to their structural similarities with native cartilage tissue. Their close network structures of interconnected macroporous hydrophilic networks promotes mass transfer properties and provides increase surface area for cell attachment and proliferation. These macroporous interconnected networks of cryogels have been generated via different cryogelation techniques. In this study, Jeffamine based cryogels were synthesized via a lyophilization process by reacting glutaraldehyde with different molecular weight diamino polyether amines (Jeffamine ED Series) in an aqueous phase followed by a homogeneous crosslinking process using Jeffamine T400 as crosslinker. The obtained synthesized imine based cryogels were then successfully reduced to secondary amine functionalized cryogels using NaBH4. Molecular weight and reduction effects on the cryogels were characterized with FT-IR and SEM. In addition, swelling ratio, porosity, mechanical properties, degradation ratio, cytotoxicity and cryogel functionality change were investigated. The performed investigations depicted potential applications of such PEG aminated cryogels as suitable scaffolds for cartilage tissue engineering

Keywords: Jeffamine, Cryogels, lyophilization, Cytotoxicity, Tissue engineering

The reaction process of the prepared imine and reduced cryogels

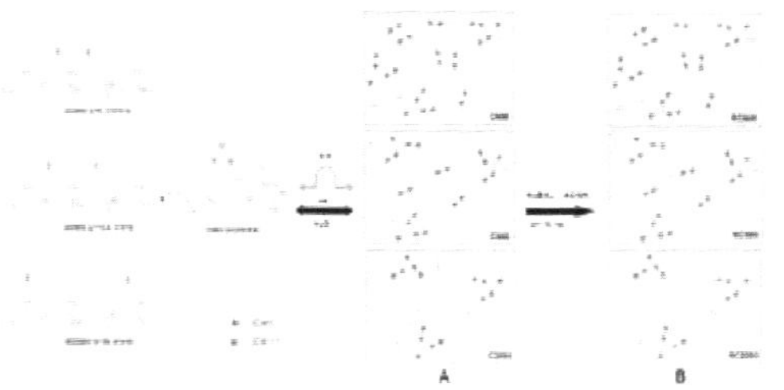

PPG-O-015 [Porous Polymer and Gels]

Dendron-polymer conjugates as precursors for designing functional hydrogels

Sadik Kaga, Rana Sanyal, Amitav Sanyal

Department of Chemistry, Bogazici University, Istanbul, Turkey

Hydrogels, resembling soft tissues in terms of physical structures and characteristics, have been widely used in tissue engineering [1], drug delivery [2] and biological sensing [3] applications in recent years. Functional properties of the hydrogels towards intended applications can be promoted by judicious design of these polymeric networks. The general physi-absorption and encapsulation methods have drawbacks for retention and release of encapsulated materials in a controlled manner [4]. Postgelation functionalization of the hydrogels is an efficient technique for covalent attachment of small molecules and biological components such as proteins, enzymes and growth factors into the hydrogel matrix. Dendritic constructs with multivalent and multifunctional characteristics possess advantages in either formation or post-modification of the hydrogels [5]. In the studies to be presented, as hydrogel precursors, dendron-linear polymer-dendron (DLD) conjugates were synthesized using alkyne [6], alkene [7] and methacrylate and alkyne [8] functionalized polyester dendrons and poly(ethylene glycol) with different lengths. Photo-cross-linking, Huisgen and thiol-ene click reactions were used for cross-linking to give functional hydrogels with high yields. Morphological characteristics and physical properties such as water uptake abilities and mechanical strengths of the hydrogels were evaluated. The leftover alkene and alkyne groups at the cross-linking sites of the network enabled post-modification of the hydrogels. As a representative of small molecule incorporation, dye molecules having thiol or azide groups were covalently attached to the residual functionalities using thiol-ene or Huisgen click reactions. Furthermore, biotin-streptavidine interaction was used following modification of the network by biotin azide for illustrating biomolecule incorporation. Tunable degree of cross-linking and density of functional groups were controlled by stoichiometric adjustment. Fluorescent microscopy was used for determination of post-modification degree. Photo-patterning and micromolding in capillaries methods were also utilized for the hydrogel synthesis using same hydrogel precursors to show post-modification of the hydrogel patterns for different applications.

Acknowledgements: This project has been funded by The Scientific and Technological Research Council of Turkey (TUBITAK, Project No: 108T898). Sadik Kaga is supported by the 2211-D grant program of TUBITAK.

References:

[1] Zhu, J., Marchant, R. E. Expert Rev. Med. Devices **2011**, 8, 607-626.
[2] Vashist, A., Vashist, A, Gupta, Y. K., Ahmad, S. J. Mater. Chem. B **2014**, 2, 147-166.
[3] Buenger, D., Topuz, F., Groll, J. Prog. Polym. Sci. **2012**, 37, 1678-1719.
[4] Silva, A. K. A., Richard, C., Bessodes, M., Scherman, D., Merten, O. W. Biomacromolecules **2009**, 10, 9-18.
[5] Kaga, S.; Arslan, M., Sanyal, R., Sanyal, A. Molecules **2016**, 21, 497.
[6] Altin, H., Kosif, I., Sanyal, R. Macromolecules **2010**, 43, 3801–3808.
[7] Kaga, S., Gevrek, T. N., Sanyal, A., Sanyal, R. J. Polym. Sci., Part A: Polym. Chem. **2016**, 54, 926–934.
[8] Kaga, S., Yapar, S., Gecici, E. M., Sanyal, R. Macromolecules **2015**, 48, 5106–5115.

Keywords: Dendron, polymer, hydrogel.

PP-O-001 [Polymer Physics]

Charging and Uncharging of Neutral Polymers in Solution Probed by SANS

Marina Tsianou, Ankitkumar I. Fajalia, Paschalis Alexandridis

Department of Chemical and Biological Engineering, University at Buffalo - The State University of New York (SUNY), Buffalo, NY, USA

Polymers in aqueous solutions are frequently used together with surfactants in order to provide structure, solubilization domains, and/or colloidal stability. Underlying these functions are inter- and intra-molecular interactions that depend on (i) the nature of the polymers, (ii) the type of surfactants, (iii) solvent quality (affected by, e.g., temperature, salinity, pH), and (iv) external stimuli (e.g., shear). The solution behavior and interactions of polymers associating with surfactants can be profoundly affected by the presence in water of polar organic solvents (cosolvents) which are ubiquitous in pharmaceutical, coating and home/personal care formulations. This presentation will analyze changes in the polymer conformation as affected by surfactant self-assembly and cosolvent effects, and as revealed from small-angle neutron scattering (SANS) measurements and other complementary techniques. Examples will be presented from poly(ethylene oxide) and cellulose ether solutions in the presence of ionic surfactants where contrast variation SANS has been instrumental in decoupling the structure of the polymer and that of the surfactant in the mixture. This study contributes fundamental insights on polymer and surfactant organization in solution, as well as new, quantitative information on systems that are widely used in practice.

Keywords: polymer, surfactant, neutron scattering

PP-O-002 [Polymer Physics]

Kinetics of thin polymer film rupture: model experiments for a better understanding of layer breakups in the multilayer coextrusion process

Guillaume Miquelard Garnier[1], Yakun Zhu[1], Jing Huang[1], Cyrille Sollogoub[1], Adrien Bironeau[1], Frederic Restagno[2]

[1]*PIMM lab, ENSAM/ le CNAM/ CNRS, Paris, France*
[2]*Laboratoire de Physique des Solides, CNRS/ Univ. Paris-Sud, Université Paris-Saclay, Orsay, France*

Multilayer coextrusion is an innovative process derived from classical co-extrusion which can be used to produce materials (especially films) made of thousands of alternating layers of two diffetent polymers, in which all the layers have nanometric thicknesses while the total thickness of the material is millimetric. Such materials showed lots of potential applications, due to the fact that many properties (mechanical, electrical conductivity, gas barrier) can be greatly improved by confinement of the polymers within the nanometric layers and the multiplication of the interfaces. However, it has been observed that the layers break up when trying to reduce the layer thickness down to a few nanometers, suggesting the existence of a critical thickness below which a spontaneous breakup occurs. The physical origin of the instabilities occurring during the multilayer coextrusion and leading to these break-ups, nonetheless, has not yet been understood, but can alter the final properties of the resulting material. In this study (Zhu et al. Polymer, 90, 2016, 156-164), we developed a model experimental setup allowing to follow the rupture kinetics of thin spin-coated polymer films for both bilayers systems and trilayers systems which occurs spontaneously when heated above Tg. We compare the results quantitatively to both theoretical and experimental work from the literature. We show in particular that the kinetics of bursting of a thin film embedded in another polymer matrix can be well described by a simple equation balancing the viscous and capillary forces. The role of viscosity, interfacial tension and thickness of the layers under static conditions is investigated. This kinetics was then compared to those of the instabilities occurring during multilayer coextrusion leading to the break-up of layers below a critical nanometric thickness. The results suggest that shear and elongation forces during the process actually stabilize the layers.

Keywords: dewetting, thin film, multilayer coextrusion, interfacial tension, viscosity

Experimental set-up and typical dewetting behavior for a trilayer

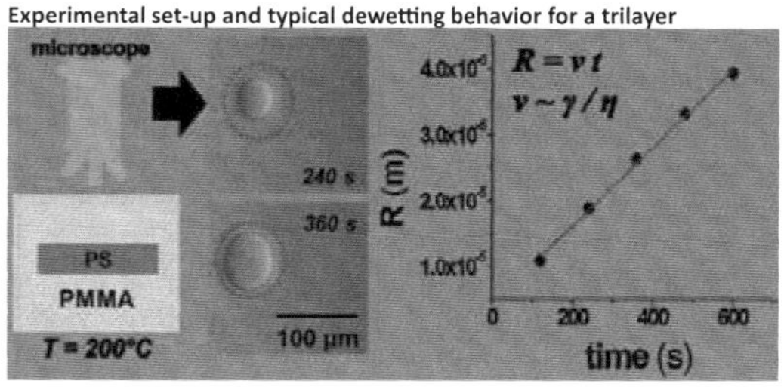

The thin PS film embedded in a PMMA matrix is heated and observed using an optical microscope. Holes growth is monitored over time. The measured speed is constant and varies as the interfacial tension over the viscosity of the matrix.

PP-O-003 [Polymer Physics]

Study and modeling of the coalescence of two different polymer particles PVDF-PMMA

Sara Aid[1], Anissa Eddhahaka[1], Daniel Froelich[2], Abbas Tcharkhtchi[1]

[1]*Laboratory PIMM, CNRS UMR 8006, Arts et Métiers ParisTech, Paris France*
[2]*Arts et Métiers ParisTech, Chambéry, France*

The present work focuses on the study of coalescence phenomenon of two different miscible polymers the ***PolyVinyliDene Fluoride** (PVDF)*, and *Poly (methyl methacrylate)* PMMA [H. Sasaki and al, 1995]. First the blend has been characterized by thermal and thermomechanical methods; its rheological characteristics have been studied at molten state. Then the coalescence tests have been performed on the grains of these polymers using a polarized light optical microscope combined by a hot stage, video camera and video capture software. Different tests were carried out on different pairs of PVDF/PVDF, PMMA/PMMA and PVD/PMMA. The effect of different parameters like temperature, particle size on the rate of coalescence was investigated. In order to follow the effect of the neighboring grains on the kinetic of coalescence of two grains a third grain has been put in contact with these two grains in different positions. These tests showed clearly the miscibility of PMMA and PVDF.

The second part of this study concerns the modeling of coalescence of the grains of different polymers. In previous works, our group studied the effect of viscosity, surface tension and relaxation time on coalescence phenomenon and used the Bellehumeur's model [C.T. Bellhumeur and al, 1996] in the case of two particles of the same polymer. We use the same model for the pair of PVDF/PMMA grains. In this case it is necessary to follow the effect of viscosity, surface tension and relaxation of each grain and also their miscibility on kinetic model.

Keywords: Coalescence, miscibility, modeling

PP-O-004 [Polymer Physics]

Rational manipulation of the microstructure of organic semiconductors by nanoscale confinement

Jaime Martín[1], Ruipeng Li[2], Aurora Nogales[3], Aram Amassian[4], Natalie Stingelin[1]

[1]*Department of Materials, Imperial College London, London, UK*
[2]*Cornell High Energy Synchrotron Source, Wilson Laboratory, Cornell University, Ithaca, USA*
[3]*Instituto de Estructura de la Materia IEM-CSIC, Madrid, Spain*
[4]*Division of Physical Sciences and Engineering, King Abdullah University of Science and Technology (KAUST), Thuwal, Saudi Arabia*

Typical devices manufactured from π-conjugated molecular materials, such as FET, OLED and OPV cells, are based on architectures where the semiconducting organic material is processed to have nanoscale dimensions. Therefore, it is to be expected that the material integrating such devices experiences some kind of finite-size-effect on the structure development during processing. However, little attention has been paid so far to confinement effects on the structural features of π-conjugated systems even though it is well known that microstructure (crystallinity, orientation, phase, etc) determines the devices performance of these materials. We have addressed this issue through an comprehensive study on the structure development of the model conjugated oligomer *p*-DTS(FBTTh$_2$)$_2$. *p*-DTS(FBTTh$_2$)$_2$ has proven to be among the most successful molecular chromophore structure known to date, delivering PCE efficiencies up to 9% in a single junction device [2]. As confining medium, we have selected the nanopores of Anodic Aluminum Oxide (AAO) templates[3,4], consisting of hexagonal arrays of cylindrical nanopores with diameters ranging from 25 to 400 nm.

The analysis of the microstructure of the material as a function of the temperature and degree of confinement revealed a notably different phase behavior as compared to its bulk counterpart. Deep depressions of crystallization temperature are measured, which evidence the change of nucleation mechanism from a typical heterogeneous mechanism of bulk materials to a homogeneous mechanism upon confinement. More striking is the discovery of new thermotropic liquid crystalline mesophase that is absent in the bulk material. Moreover, the strong nanoscopic confinement stabilizes **a new high-temperature crystal polymorph** that is characterized here for the first time. Thus, the complete microstructural behavior of *p*-DTS(FBTTh$_2$)$_2$ allowed us to resolve its **temperature-confinement phase diagram,** as well as to develop processing protocols that enable full control of the microstructure of *p*-DTS(FBTTh$_2$)$_2$ nanowires, i.e. on the crystal phase, crystal orientation, crystallinity, etc.

1. A. K. K. Kyaw, et al. Nano Lett. 2013, 13, 3796
2. Martín, J.*et al. Nature Commun.* **2014,** 5.
3. Martín, J. *et al. Phys. Rev. Lett.* **2010,** 104, (19), 197801.
4. Martín, J. *et al. Macromolecules* **2013**, *46, 1477*

Keywords: semiconducting polymers, nanotechnology, solar cells, confinement, conjugated polymers, crystallization

Figure 1

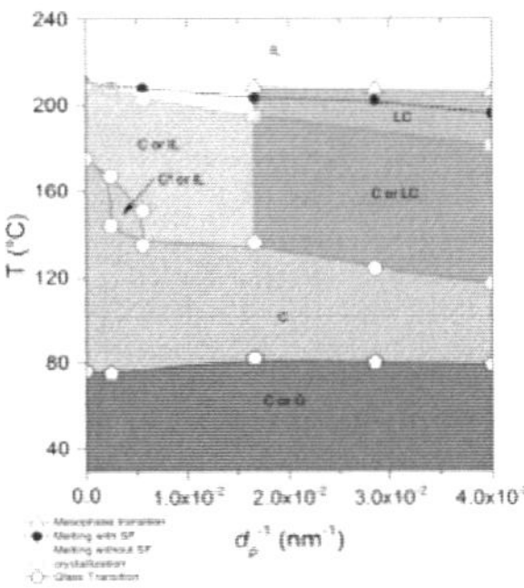

Temperature-confinement (inverse of the pore diameter) phase diagram for the *p*-DTS(FBTTh$_2$)$_2$ molecule. IL; isotropic liquid C; crystal, LC; liquid crystal, G: glass.

PP-O-005 [Polymer Physics]

High-performance polyethylene fibers "al dente": improved gel-spinning of ultra-high molecular weight polyethylene using vegetable oils

Raphael Schaller, Kirill Feldman, Paul Smith, Theo Tervoort

Department of Materials, ETH Zurich, Zurich, Switzerland

State-of-the-art high-performance polyethylene (HPPE) fibers have essentially the density of water, but combine a stiffness of about 110 GPa (half the value of steel) with a strength of about 3 GPa (more than double the value of steel). Due to these unprecedented properties, HPPE fibers play a key role in areas as diverse as superior bulletproof clothing to lifting a sunken ferry from a delicate ecosystem in front of the coast of Italy. We demonstrate, for the first time, solution-spinning of the HPPE fiber precursor from "green" solvents, such as vegetable oils, as opposed to the current industrial standard solvent, decaline, which is poisonous and flammable. Using these "green" solvents is not only environmentally friendly, but, additionally, leads to dramatically improved mechanical properties and/or economics of the final HPPE fiber (see Figure 1). The presented results offer an alternative efficient and environmentally friendly approach to the fabrication of HPPE fibers and films.

Keywords: Gel-spinning, polyethylene, mechanical properties

Figure 1

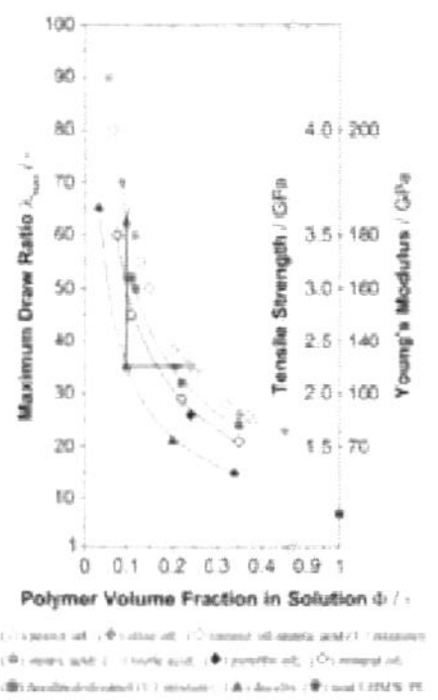

Maximum draw ratios and the corresponding tensile strength and Young's modulus values for UHMW PE gel-processed from different solvents as a function of polymer volume fraction. Colored lines are guides to the eye. The arrows in the graph at 10 vol% (typical concentration used in the industrial process) indicate the possibilities to improve either the final fiber properties (vertical arrow) or the economics of the process (horizontal arrow) by using vegetable oil based solvents (green line) instead of decalin (blue line). Note that the use of the left and right y-axis reflects the unique but non-linear relation between (maximum) draw ratio, tensile strength and Young's modulus of the final HPPE fiber.

PP-O-006 [Polymer Physics]

Particle sorption in shaken polymer films

Sissi De Beer, Bernard D Kieviet, Liz I. S. Mensink, G. Julius Vancso

University of Twente, Materials Science and Technology of Polymers

Mucus consists of sugar-chains in an aqueous liquid. It forms a soft, dilute polymeric barrier in e.g. the respiratory or intestinal system. This polymeric barrier acts as a filter allowing nutrient-uptake, while hindering the adherence and absorbance of bacteria and toxic materials. We aim at predicting particle-sorption in barrier-like solvated polymeric systems, such as gels and surface-anchored brushes, since it might ultimately aid in e.g. the design of efficient medication or help predict potential nano-toxicity of substances. We are specifically interested in the effect of out-of-equilibrium boundary conditions imposed by solvent flow and wobbling of the polymer film, because biological systems are constantly in motion.

Here, I present our first steps towards predicting out-of-equilibrium particle-sorption in polymer brushes. Using coarse-grained molecular dynamics simulations, we show that, depending on the shape, size and solvent-mediated particle-polymer interactions, particle sorption kinetics in brushes is slowed down in the presence of continuous solvent flow, while it is enhanced when the solvent is refreshed via a batch-wise, stop-and-go procedure. Imposing oscillatory motion to the polymer film also enhances sorption kinetics.

Keywords: polymer physics, polymer brush, molecular dynamics, particle polymer interaction

PP-O-007 [Polymer Physics]

Relevance of fractionation and non-covalent interactions to formation of a polymer-polymer complex

Adriana Sturcova, Jaroslav Kratochvil, Jiri Dybal, Antonin Sikora

Institute of Macromolecular Chemistry, Academy of Sciences of the Czech Republic, Prague, Czech Republic

If two polymers form a complex, such process may result in structures and materials suitable for technological and industrial applications. Complex formation is dependent on the composition of the two polymers, their concentration, chain length and chain stiffness, nature of the solvent, temperature, and pH. The process can be driven by Coulombic forces, hydrogen bonding interactions, van der Waals forces, charge transfer interactions and solvophobic interactions. Hydrogen-bonded complexes can also be viewed as a special class of miscible polymer blends. We have previously found that poly(N-methyldodecano-12-lactam) (PMDL) formed miscible blends with poly(4-vinylphenol) (PVPh) soluble in tetrahydrofuran.[1] The miscibility was brought about by hydrogen bonding of 2:1 and 1:1 stoichiometry between hydroxyl groups (PVPh) and carbonyl groups (PMDL). When PMDL and PVPh of higher molar mass were used, two types of a hydrogen-bonded associate formed in a single solution: a gel-like interpolymer complex and a soluble blend. The two types of associates were characterized by elemental analysis, differential scanning calorimetry, size exclusion chromatography, and thermogravimetric analysis. It was concluded that the association process was accompanied by fractionation into a high-molar-mass interpolymer complex and a low-molar-mass soluble blend.[2] FTIR spectroscopy and DFT calculations were used to test the usual assumption about ladder-like structure and cooperativity of hydrogen bonds in complexes and loose and random hydrogen-bond structure in soluble blends.[3] Conditions for preparation of the two types of associates were exactly the same, thus the role of chain length, hydrogen bonding, and solvophobic interactions can be assessed.

References:

1. J. Kratochvíl, A. Šturcová, A. Sikora, J. Dybal, *J. Polym. Sci. Part B: Pol. Phys.* 2011, 49, 1031
2. J. Kratochvíl, A. Šturcová, A. Sikora, J. Dybal, *Polym. Int.* 2014, 63, 1406.
3. A. Šturcová, J. Kratochvíl, A. Sikora, J. Dybal, *Eur. Polym. J.* 2014, 59, 200.

Keywords: interpolymer complex, hydrogen bonding, FTIR, DSC

PP-O-008 [Polymer Physics]

Morphological effects of microscale matrix-filler interactions in thermoplastic polymers

Antonis Gitsas[1], Thomas Gkourmpis[2], Davide Tranchida[3]

[1]*Innovation Headquarters, Borealis Polyolefine GmbH, Linz, Austria*
[2]*Innovation & Technology, Borealis AB, Stenungsund, Sweden*
[3]*Advanced Polymer Characterisation, Borealis Polyolefine GmbH, Linz, Austria*

For power applications good mechanical performance of flexible components needs to be combined with excellent dielectric properties. Since the production and processing is a very demanding process, tailored-made polymer design is required for matching the above requirements. The addition of high aspect ratio fillers offers novel design opportunities (Xu et al., Macromolecules 43, 2010, 5000) but simultaneously their dispersion adds an additional level of complexity to the system (Logakis et al. Chem. Phys. Let. 498, 2010, 125), in particular through the complex interaction between filler, matrix and the polymer amorphous and crystalline phases (Hallouet et al., J. Phys. D 42, 2009, 064004). Furthermore, morphological differences can be expected when examining different sample types (Laird et al., Macromolecules, 2013, 46, 2877). Therefore good knowledge of the thermoelectrical and morphological behaviour is key for the understanding of the long-range end-material performance (Almajid et al., Plast. Rub. Comp. 44, 2015, 50).

In this work selected polypropylene and polyethelene polymers polymers were employed as matrix for the dispersion ofhigh aspect ratio fillers. The dispersion quality and local morphology along with the related rheological, mechanical and electrical properties were investigated with the aid of X-ray scattering, dielectric spectroscopy, AFM, SEM, and mechanical probes. It was found that thin interfaces formed at the boundaries of the fillers, altering the local molecular dynamics as well as the macroscopic behaviour. Not only the filler design but also the matrix setup can be used for achieving seamless interfaces. This way, for example, the respective loading thresholds are reduced and the final electrical and thermal conductivity are increased. The analysis also shows that the extrusion process has a marked effect on the overall morphology: For high-load compositions we noticed a very good level of dispersion, a very smooth surface and a very smooth interface between filler-free and composite phases.

Keywords: polypropylene, morphology, mechanics, dielectrics, composites, interfaces

Interfaces around carbon black in polypropylene

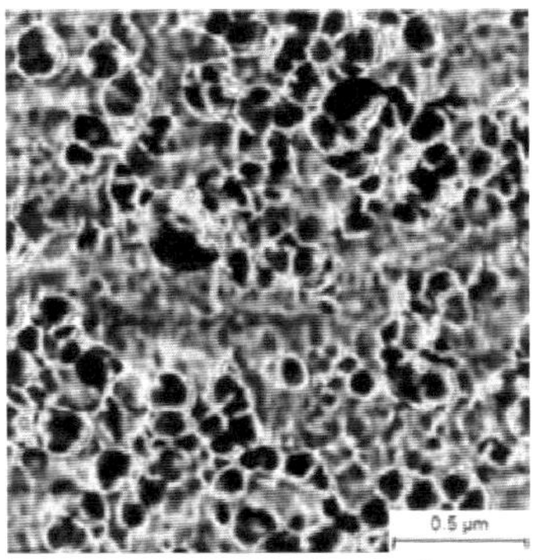

Contrast-enhanced AFM phase image of carbon black in polypropylene. The level of carbon black dispersion is very good and there appears to be a limited level of orientation. A thin homogeneous interface around the particles is clearly observed, indicating a nucleation effect of the filler.

PP-O-010 [Polymer Physics]

Two paradoxical phenomena of (smart) polymers in solvent mixtures: Effects of preferential adsorption and depleted depletion

Debashish Mukherji, Kurt Kremer

Max Planck Institute for Polymer Research

Smart polymers are a modern class of soft materials that show drastic changes in their structure, function, and stability by a slight change in external stimuli. This presentation will discuss two opposite (yet related) paradoxical phenomena of polymers that occur in mixed solvent environment. One of these phenomena is known as co-non-solvency. Co-non-solvency occurs when a polymer is added in two (perfectly) miscible and competing good solvents. As a result, the same polymer collapses into a globule within intermediate mixing ratios [1]. More interestingly, polymer collapses when the solvent quality remains good or even gets increasingly better by the addition of the better cosolvent [2]. This puzzling phenomenon, where the solvent quality is completely decoupled from the polymer conformation, is driven by strong local preferential adsorption of better cosolvent with the polymer [1,2]. Because a polymer collapses in good solvent, the depletion forces, that are responsible for standard poor solvent collapse, do not play any role in describing co-non-solvency [3]. However, a case where depletion forces do play a significant role is when a polymer swells in a mixture of two competing poor solvents. This swelling can be explained by the reduction of depletion induced attraction that is resulting from a delicate interplay between the local depletion effects and the bulk solution density. Furthermore, it will be presented that both these phenomena can be understood within universal (generic) concepts. Therefore, a broad range of polymers is expected to exhibit these phenomena and the specific chemical details do not play any role in understanding these complex conformational behaviors [4].

[1] Mukherji and Kremer, Macromolecules 46, 9158 (2013).
[2] Mukherji, Marques, and Kremer, Nature Communications 5, 4882 (2014).
[3] Oliviera, Netz, Mukherji, and Kremer, Soft Matter 11 8599 (2015).
[4] Mukherji, Marques, Stuehn and Kremer, Journal of Chemical Physics 142, 114903 (2015).

Keywords: Smart Polymers, Complex Mixtures, Solvation Thermodynamics, Generic Simulations

PP-O-009 [Polymer Physics]

Control over Microphase Separation, Crystallization and Molecular Orientation of All-Conjugated Diblock Copolymers

Yanchun Han

State Key Laboratory of Polymer Physics and Chemistry,Changchun Institute of Applied Chemistry, Chinese Academy of Sciences

All-conjugated block copolymers containing two different conjugated blocks can self-organize into well-defined structures at the nanometre scale both in solution and the solid state, driven by factors such as immiscibility or crystallizability difference between the blocks. In this paper, we investigate the influences of solvent solubility and the copolymer composition on the self-assembly of all-conjugated diblock copolymers poly(2,5-dihexyloxy-p-phenylene)-block-(3-hexylthiophene) (PPP-b-P3HT) thin films. The competition between microphase separation, break out crystallization and confined crystallization of the two conjugated blocks, the transformation between face-on and edge-on molecular orientation, and the self-epitaxial crystallization between the two blocks will be discussed in detail.

Keywords: Microphase Separation, Crystallization,Orientation,All-Conjugated Diblock Copolymers

PP-O-011 [Polymer Physics]

Dynamics of entangled Supramolecular Networks based on Poly (n-butyl acrylate-g-Ureidopyrimidinone)

Mostafa Ahmadi[1], Amir Jangizehi[1], Seyedreza Ghaffarian[1], Seyedhadi Goldansaz[2], Mostafa Shirani[1]

[1]Department of Polymer Engineering and Color Technology, Amirkabir University of Technology, Tehran, Iran
[2]Bio and Soft Mater (BSMA), Institute de la Matière Condensée et des Nanosciences (IMCN), Université Catholique de Louvain, Louvain-la-Neuve, Belgium

Complications due to polymer dynamics have been frequently avoided in studying supramolecular transient networks by utilizing chains shorter than entanglement molecular weight (Feldman et al. Macromolecules 2009, 42, 9072), while in most of the applications polymer chains are well-entangled (Ahmadi et al. Macromolecules 2015, 48, 7300). We have therefore used atom transfer radical polymerization (ATRP) to synthesize high molecular weight poly (n-butyl acrylate)s with low polydispersities and functionalized them with strong hydrogen bonded Ureidopyrimidinone (UPy) groups. Dynamic thermal and mechanical measurements revealed that samples with lower density of physical junctions expose similar behavior different from the ones with higher number of sticky groups. The latter had several melting peaks which are attributed to the diversity of association structures made by UPy stacks as confirmed by optical microscopy. Cylindrical stacking of UPy groups was more pronounced at higher fraction of sticky groups, which in turn led to considerably higher flow activation energies and larger relaxation times.

Keywords: Supramolecule, Hydrogen bonding, Entanglement

RDPS-A-O-001 [Controlled/Living Polymerization]

Polymer Electrolite Membrane Synthesis by ATRP and Iniferter Methods

Metin H Acar, Cuneyt E Tas, Mehmet Basdal, Buket Alkan

Department of Chemistry, Istanbul Technical University, Istanbul, Turkey

PVC based graft copolymers with AMPS monomer were synthesized by using two different graft copolymerization methods. Firstly, P(VC-g-AMPS) was synthesized by ATRP [1] using PVC as macroinitiator [2]. Secondly, PVC was reacted by sodium diethyl dithiocarbomate and PVC containing N,N-diethyl dithiocarbomate group (PVC-DDC) was obtained. Then, PVC-DDC polymer was used as macro-photoinitiator in order to synthesized P(VC-g-AMPS) via iniferter polymerization [3-4]. The obtained graft copolymers were characterized by using FT-IR, UV-Visible spectrophotometer, conventional and triple detector GPC and 1H-NMR. Thermal properties were measurements by DSC and TGA. Ion conductivity of the materials was investigated with the nyquist and magnetude results by using electrochemical impedance spectroscopy. After synthesized and characterized of each polymer, all materials were stirred throughout in water at 100 oC for two hours. A significant increase in the Tg value of PVC-g-AMPS by iniferter method was observed when it compared with PVC-g-AMPS via ATRP method. P(VC-g-AMPS) via iniferter method is shown higher ion conductivity (less resistance) than P(VC-g-AMPS) via ATRP method (Figure 1).

References:

[1] Wang, J. S. and Matyjaszewski, K.: Macromolecules, 1995, 28, 7901-7910
[2] Bicak, N. and Ozlem, M.: J. Polym. Sci. Part A: Polym. Chem., 2003, 41, 3457-3462
[3] Otsu, T. and Kuriyama, A.: Polymer J., 1985, 17, 97
[4] Guan, J. and Yang, W.: J. App. Polym. Sci., 2000, 77, 2569-2574

Keywords: PEM membrane, ATRP, Iniferter

Figure 1

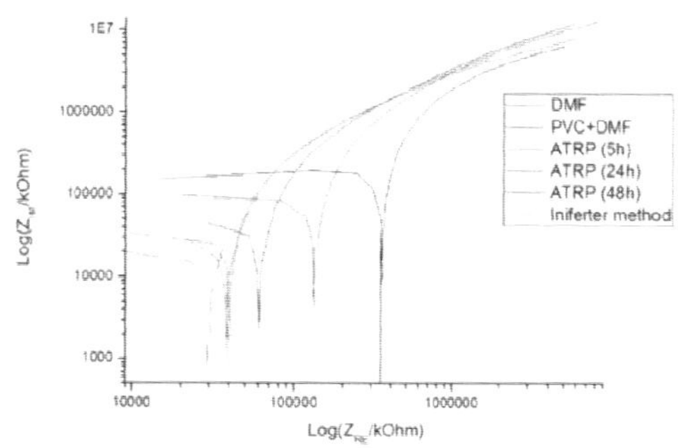

Nyquist graph of EIS measurements of P(VC-g-AMPS) for ATRP and iniferter method.

RDPS-A-O-002 [Controlled/Living Polymerization]

Anti-fog and anti-frost performance of zwitterionic polymer brushes prepared by SI-ATRP polymerization

Metwally Ezzat, Chun Jen Huang

Department of Biomedical Sciences and Engineering, National Central University, Jhong-Li, Taoyuan 320, Taiwan

Formation of fog and frost on transparent surfaces, which is strongly associated with surface physicochemical properties, lead to many problems in our daily life. In order to address them, superhydrophilic zwitterionic polymer brushes based on poly(sulfobetaine methacrylate) (pSBMA) and poly(sulfobetaine vinylimidazole) (pSBVI) have been prepared on glass substrates by surface-initiated atom transfer radical polymerization (SI-ATRP). To study the wettability effect of these brushes on anti-fog/ anti-frost properties, pSBMA and pSBVI polymer brushes with contact angles in the range from < 5° to 18° were prepared by controlling the thicknesses of polymer films. X-ray photoelectron spectroscopy (XPS), ellipsometry and atomic force microscopy (AFM) were applied to monitor interfacial elemental compositions, the thicknesses of the brushes and the roughness, respectively. The wettability of the polymer brushes was measured by water contact angle (CA) and their anti-fog efficiency was determined visually and tested quantitatively by UV visible light transmission. Consequently, the polymer films with contact angles of < 5° exhibit excellent anti-fogging and frost-resistance performance. The optical transmittance of modified substrates tested under both hot and cold fogging conditions was as high as that of unfogged bare glass. Additionally, no frost was formed on the coated substrates after storage in the refrigerator at -20 °C during experimental period. The results convincingly demonstrated that to obtain superior anti-fog and frost resistance capabilities, the substrates must exhibit a contact angle lower than 5°. Also, there is no comparable difference in the performance of both pSBMA and pSBVI polymer brushes, however the former is more desirable because SBMA monomer is commercially available and short polymerization time is needed to obtain superhydrophilicity.

Keywords: Polymer brushes, SI-ATRP, PSBMA, PSBVI, Anti-fog, Anti-frost

RDPS-A-O-003 [Controlled/Living Polymerization]

Recent developments in the field of light-controlled radical polymerizations

Qizhi Yang, Benjamin Cabannes Boué, Jacques Lalevée, Julien Poly

Mulhouse Institute of Materials Science, CNRS/UHA, Mulhouse

The introduction of light as an external stimulus in the mechanisms of controlled radical polymerizations has been extensively studied in recent years. This interest is mostly due to the supplementary advantages that the use of light can bring, such as temporal and spatial controls, which are two typical features of photopolymerizations. In this contribution, we will present two mechanisms recently developed in our group. The first one is a photocatalyzed atom transfer radical polymerization (ATRP) involving a copper-based complex with phenanthroline ligands. The photocatalysis cycle involves here the excited state derived from the activator form of the complex, which is original for a photocatalyzed Cu-based ATRP. The polymerization was made faster thanks to the additional regeneration of the activator enabled by reducing agents, being either external (additive such as an amine) or intrinsic (side group on a functional comonomer). The second example is based on a reversible addition-fragmentation chain transfer mechanism (RAFT). Before being considered as RAFT agents, several compounds such as xanthates or dithiocarbamates were already commonly used as photoiniferters. Although this dual reactivity has been already successfully exploited for the development of photoRAFT mechanisms, a RAFT agent with a structure especially designed for this purpose has not been reported yet. We will describe the use of such a RAFT agent. For both mechanisms, very efficient controls were reached (PDI values around 1.1) for methacrylic or acrylic monomers using cheap and soft LED light sources. Additionally, efficient temporal control of the polymerizations was achieved by simply switching on/off the light source.

Keywords: ATRP, RAFT, photocatalysis, light

photocatalyzed ATRP

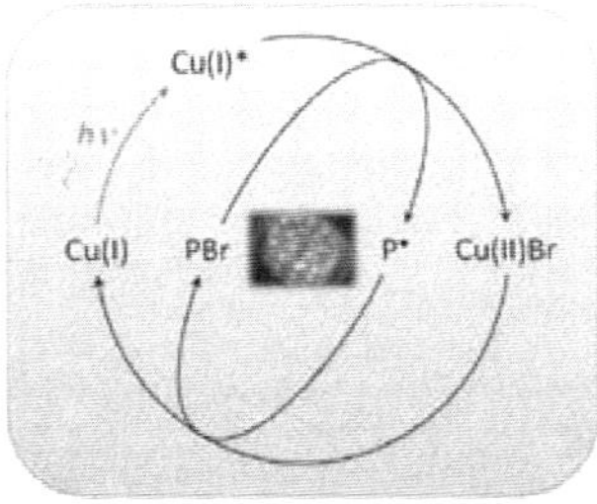

RDPS-A-O-004 [Controlled/Living Polymerization]

Polymerisation behaviour of 'switchable' RAFT agents

Christopher Michael Fellows[1], Sarah J Stace[1], Graeme Moad[2], Daniel J Keddie[3]

[1]*School of Science and Technology, The University of New England, Armidale, Australia*
[2]*CSIRO Manufacturing, Clayton, Australia*
[3]*School of Biology, Chemistry & Forensic Science, University of Wolverhampton, Wolverhampton, United Kingdom*

'Switchable' RAFT agents bear a Z group containing functionality that can be 'switched' to become more electron-withdrawing by coordination of Lewis base functionality (with, for example, a proton). In switched (protonated) form they provide effective control in reversible deactivation radical polymerization (RDRP) of more active monomers (MAM, e.g., acrylates, styrene). In unswitched form they are suitable for RDRP of less active monomers (LAM, e.g., N-vinylpyrrolidone) (1). We have been studying structure-activity relations in polymerisations carried out in the presence of N-pyridyldithiocarbamates, applying these RAFT agents to the preparation of MAM- LAM diblocks and triblocks (2,3), and evaluating methods for end group removal by various techniques including thermolysis. This presentation will outline some of our most recent work in this area.

(1) Benaglia, M.; Chiefari, J.; Chong, Y. K.; Moad, G.; Rizzardo, E.; Thang, S. H. J. Am. Chem. Soc. 2009, 131, 6914-6915.
(2) Stace, S. J.; Moad, G.; Fellows, C. M.; Keddie, D. J. Polym. Chem. 2015, 6, 7119-7126.
(3) Moad, G.; Keddie, D.; Guerrero-Sanchez, C.; Rizzardo, E.; Thang, S. H. Macromol. Symp. 2015, 350, 34-42.

Keywords: RAFT polymerisation, block copolymer, dithiocarbamate

Switchable RAFT agent

RDPS-A-O-005 [Controlled/Living Polymerization]

Reversible addition-fragmentation chain transfer polymerization of N-isopropylacrylamide and acrylic acid from nanocrystalline cellulose to prepare a dual sensitive material

Hossein Roghani Mamaqani, Marziyeh Haqani, Mehdi Salami Kalajahi

Sahand University of Technology

A shell of crosslinked poly (2-hydroxyethyl methacrylate) was applied at the surface of nanocrystalline cellulose (NCC) by distillation precipitation polymerization. Then, it was reacted with S-(thiobenzoylthioglycolic) acid to obtain NCC-RA. Homopolymers of N-isopropylacrylamide (NIPAAm) and acrylic acid (AA) as temperature- and pH-sensitive materials and also their dual-sensitive block copolymers were grafted at the surface of NCC-RA by an in situ reversible addition-fragmentation chain transfer polymerization via R-group method. Grafting of chain transfer agent and polymer chains at the surface of nanocrystals was studied by X-ray photoelectron (XPS), Fourier transform infrared, and Raman spectroscopies. Thermogravimetric analysis and XPS were also used to study the graft content of chain transfer agent and polymers. Sensitivity of the polymer-grafted NCCs to temperature and pH were studied by UV-visible spectroscopy. By the addition of poly(N-isopropylacrylamide) block to the poly(acrylic acid)-grafted nanocrystals, phase transition is observed at 32 °C. When the poly(acrylic acid) block is the outer layer, phase transition is observed at higher temperatures and pH-responsivity is observed at pH 7.0-10.0. Morphology of the neat and polymer-grafted NCCs was studied by scanning and transmission electron microscopies.

Keywords: Nanocrystalline cellulose, N-isopropylacrylamide, Acrylic acid, Block copolymer, Reversible addition-fragmentation chain transfer polymerization, Stimuli responsive.

RDPS-A-O-006 [Controlled/Living Polymerization]

Step by Step Investigation of Radical Polymerization Processes by a Combination of Electron Spin Resonance and Controlled Radical Polymerization Techniques

Atsushi Kajiwara

Department of Materials Science, Nara University of Education, Nara, Japan

The reaction steps at various stages of conventional radical polymerizations were able to be investigated by both steady-state electron spin resonance (SS ESR) spectroscopy in combination with atom transfer radical polymerization (ATRP) technique and with time-resolved ESR (TR ESR) spectroscopy. Initiation steps are usually too fast to observe by conventional spectroscopic technique. In this research, the initiation steps can be examined by TR ESR and SS ESR/ATRP combination method. The SS ESR/ATRP combination method can observe each initiation step by measurements of uniform oligomeric model propagating radicals. TR ESR spectra allowed direct observation of transient radicals (chain initiating radical) obtained by the addition reaction of a diphenylphosphinoyl radical to vinyl monomers (Figure 1). The addition rate constants (initiation rate constants, k2) and activation energies (Ea) of diphenylphosphinoyl radical to the (meth)acrylates, vinyl esters, and styrenes were determined based on the TR ESR spectra. A combination of steady-state ESR and TR ESR provides information on both the initiation and propagation processes in radical polymerization of various monomers. These research works are a part of an IUPAC project "Critically evaluated ESR (EPR) spectra of important polymerization-related radicals" that is just started at the end of 2015.

Keywords: Radical Polymerization, ESR/EPR, Elementary Processes, ATRP, Initiation Processes, kinetics

Figure 1

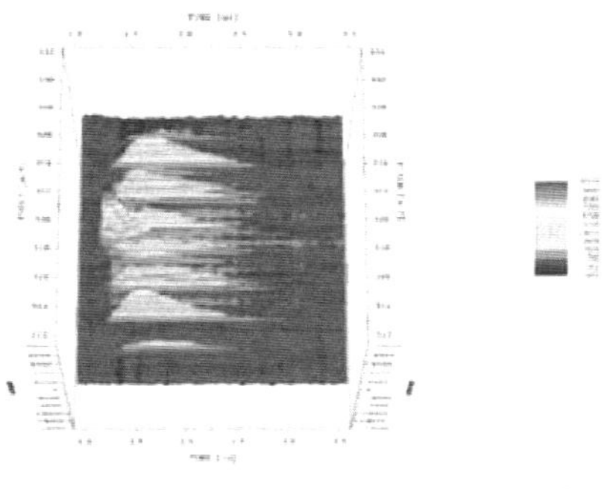

Figure 1 Time-resolved ESR spectra of Chain initiating radical of tBMA at 25 C. 3-axes are time, magnetic field, and ESR signal intensity.

RDPS-A-O-007 [Controlled/Living Polymerization]

Propagation rate coefficient for acrylamide radical polymerization in aqueous solution

Igor Lacik[1], Anna Chovancova[1], Lucia Uhelska[1], Calista Preusser[2], Robin A. Hutchinson[2], Michael Buback[3]

[1]Department for Biomaterials Research, Polymer Institute of the Slovak Academy of Sciences, Dúbravská cesta 9, 845 41 Bratislava, Slovakia
[2]Department of Chemical Engineering, Dupuis Hall, Queen's University, Kingston, Ontario K7L 3N6, Canada
[3]Georg-August-Universität Göttingen, Institut für Physikalische Chemie, Tammannstraße 6, D-37077 Göttingen, Germany

The understanding of the kinetics and mechanism of radical polymerization for water-soluble monomers polymerized in aqueous solution has significantly improved during the last decade thanks to the advent of pulsed-laser polymerization techniques resulting in accurate rate coefficients for each individual reaction step. Due to the presence of hydrogen bonds, the polymerization in aqueous solution exhibits features not encountered in organic solvents. For non-ionized monomers, the propagation rate coefficient, kp, significantly decreases toward higher monomer concentration in aqueous solution. This effect is assigned to interactions of the solvent environment, consisting of water and monomer, with the transition state structure for propagation [1]. The understanding of this feature allowed for the adequate description and modeling of several homo- and copolymerization systems [1].

This contribution summarizes our recent studies on kp for acrylamide (AAm) polymerized in aqueous solution employing pulsed-laser polymerization in conjunction with aqueous-phase size-exclusion chromatography [1]. The kp values were obtained at AAm concentrations from 1 to 30 wt.%, temperatures from 5 to 70 °C, and pulse repetition rates from 25 to 500 Hz. The effect of ionic strength between 0.001 and 1.0 mol/L and of pH up to pH 11 was also investigated. This broad range of experimental conditions resulted in a large set of kp values that enabled to understand (1) the propagation kinetics of AAm, and (2) the fluidizing action of water on the transition state structure and, based on that, to discuss this effect for the acrylate and methacrylate family of water-soluble monomers polymerized in aqueous solution.

This work was supported by the Slovak VEGA Grant Agency under the contract No. 2/0198/14 and by the Slovak Research and Development Agency under the contract No. APVV-14-0858.

[1] Lacík, I.; Chovancová, A.; Uhelská, L., Preusser, P., Hutchinson, R. A., Buback M. Macromolecules 2016, DOI: 10.1021/acs.macromol.6b00526, and references therein

Keywords: acrylamide, PLP-SEC, radical polymerization, propagation rate coefficient

RDPS-A-O-008 [Controlled/Living Polymerization]

Controlled living anionic polymerization of cyanoacrylates by frustrated Lewis pair based initiator

Rosa María Sebastián[1], Rubén Sáez[1], Jordi Marquet[1], Ciaran Mcardle[2], Fouad Salhi[3], Bernard Ryan[4], Cormac Duffy[4]

[1]Department of Chemistry, Science Faculty, Universitat Autònoma de Barcelona, 08193 Cerdanyola del Vallès, Barcelona, Spain
[2]Advanced Technology Group, Henkel Ireland Operations and Research Ltd, Tallaght Business Park, Dublin 24, Ireland
[3]Open Innovation Barcelona, Adhesive Technologies, AR Materials Science & Engineering Platform, Henkel KGaA & Co., Ed. Eureka, Campus UAB, 08193 Cerdanyola del Vallès, Barcelona, Spain
[4]AR Rapid Systems Platform, Henkel Ireland Operations & Research Ltd, Tallaght Business Park, Dublin 24, Ireland.

Cyanoacrylates are a family of strong fast-acting adhesives with industrial, medical, and household uses that are known as superglues. These monomers rapidly polymerize in the presence of nucleophiles, such as water, forming long strong chains, joining a large amount of different type of materials such as metals, ceramics, plastics, woods, rubbers, bones and skin. It is well known that the properties of a polymer are strongly influenced by its architectural features; the suitability of a polymer for a particular application is often determined by its specific structural and molecular parameters such as molecular weight, molecular weight distribution, the presence of functional groups and their spatial location among others. For most applications, the absence of control over these features is not a problem, however for sophisticated materials; polymers with defined molecular weight and narrow polydispersity are desirable or even imperative. Controlled living polymerization techniques allow the synthesis of polymers with tailored properties and architectures, affording industrial processes of interest for the design of high performance polymeric materials and composites for advanced technologies. Due to the fast reactivity of cyanoacrylates, their real controlled living polymerization has not yet been described. In this work we will present the use of a hydrogenated Frustrated Lewis Pair (FLP) as an initiator for the polymerization of cyanoacrylates, resulting in the formation of polymers with designed molecular weights, and polydispersities less than 1.1. The preparation of block copolymers of different cyanoacrylates has also been achieved. These results open an avenue for the modification of physical and mechanical properties of conventional polycyanoacrylates, being possible to find for them new high added value applications.

Keywords: cyanoacrylates, anionic living polymerization, frustrated Lewis pair, copolymers

Scheme

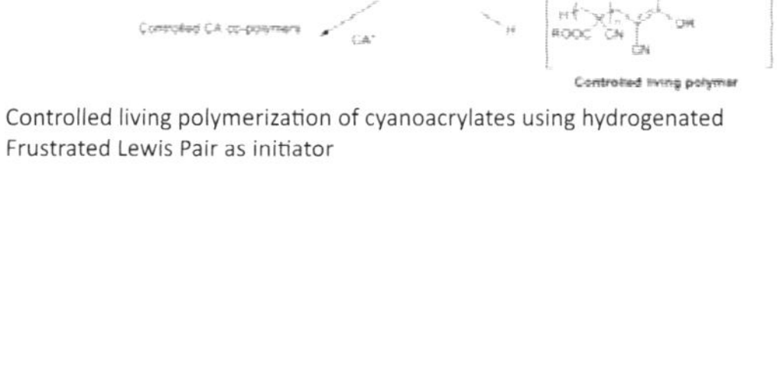

Controlled living polymerization of cyanoacrylates using hydrogenated Frustrated Lewis Pair as initiator

RDPS-A-O-009 [Controlled/Living Polymerization]

H-bonding donor and Brønsted acid co-catalysis in ring-opening polymerization

Zhenjiang Li, Xiaopei Li, Yaya Liu, Herui Sun, He Dong, Songquan Xu, Jiaxi Xu, Jiaqi Liu, Xianfu Pan, Kai Guo

State Key Laboratory of Materials-Oriented Chemical Engineering, College of Biotechnology and Pharmaceutical Engineering, Nanjing Tech University, Nanjing 211816, China

H-bonding donor (HBD) and Brønsted acid (BA) co-catalysis (HBD–BA) enabled ring-opening polymerization (ROP) of cyclic esters in solution at room temperature. HBD scaffolds with multiple H-bond donating sites in double, triple, and quadruple arrays complementary to the oxyanions of the corresponding BAs promoted the BA catalyzed cationic ROPs efficiently. Previously "non-polymerizable" monomers were made polymerizable by weak to medium strength BAs by HBD–BA strategy. Homopolymers and multi-block copolymers were prepared in controlled sequence, predicted molecular weights, and narrow dispersities. All the polymerizations were in controlled/living nature. The first H-bonding donor and Brønsted acid co-catalysis platform in controlled/living ring-opening polymerization is fulfilled.

Keywords: Brønsted acid, hydrogen bonding, ring-opening polymerization, organocatalysis

Figure 1. H-bonding donor (HBD) and Brønsted acid (AH) co-catalysis in ring-opening polymerization (ROP)

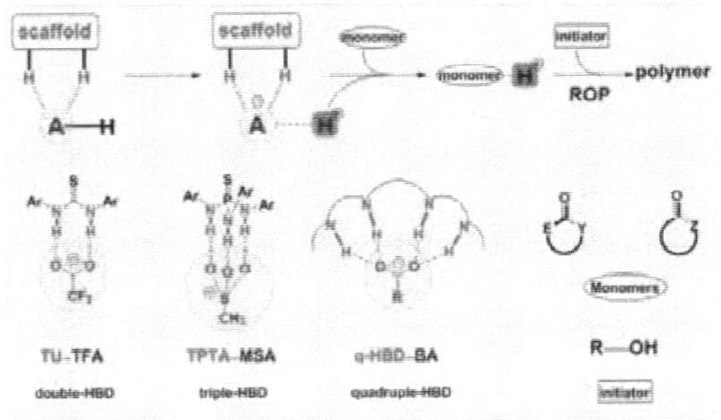

graphical abstract

RDPS-A-O-010 [Controlled/Living Polymerization]

Controlled Polymerization of Lysine Monomer

Youhua Tao, Xianhong Wang

Key Laboratory of Polymer Ecomaterials, Changchun Institute of Applied Chemistry, Chinese Academy of Sciences, China

L-lysine is a naturally available renewable resource, and along with dwindling petroleum and eventually depletion of fossil resources, the controlled synthesis of bio-based polymers derived from L-lysine has recently become a growing research focus [1-3]. ε-Poly-lysine (ε-PL) is an uncommon cationic, naturally-occurring homopolymer produced by the fermentation process. Due to its significant antimicrobial activity and nontoxicity to humans, ε-PL is now industrially produced as an additive, e.g. for food and cosmetics. However, the biosynthetic route can only make polymers with a molecular weight of about 3 kDa [4]. In this work, we report a new chemical strategy based on ring-opening polymerization (ROP) to obtain ε-PL from lysine. The 2,5-dimethylpyrrole protected α-amino-ε-caprolactam monomer was prepared through cyclization of lysine followed by protection. ROP of this monomer, followed by the removal of the protecting group, 2,5-dimethylpyrrole, ultimately yielded ε-PL with varying molecular weights.

References :

1. Chen, X. Y.; Lai, H. W.; Xiao, C. S.; Tian, H. Y., Chen, X. S.; Tao, Y. H.; Wang, X. H. Polymer Chemistry 2014, 5, 6495-6502.
2. Lai, H. W.; Chen, X. Y.; Lu, Q.; Bian, Z., Tao, Y. H.; Wang, X. H. Chemical Communications 2014, 50, 14183-14186.
3. Jia, F.; Chen, X. Y.; Zheng, Y.; Qin, Y. S.; Tao, Y. H.; Wang, X. H. Chemical Communications 2015, 51, 8504-8507.
4. Tao, Y. H.; Chen, X. Y.; Jia, F.; Wang, S. X.; Xiao, C. S.; Cui, F. C., Li, Y. Q., Bian, Z., Chen, X. S.; Wang, X. H. Chemical Science 2015, 6, 6385-6391.

Keywords: Controlled Polymerization, Lysine, ε-Poly-lysine

New chemosynthetic route to linear ε-poly-lysine

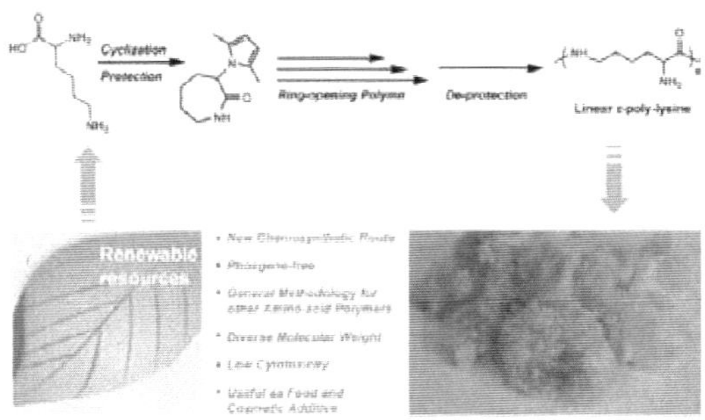

New chemosynthetic route to linear ε-poly-lysine

RDPS-A-O-011 [Controlled/Living Polymerization]

Hyperbranched polyethylene by living ethylene polymerization initiated by diimine and amino-imine nickel and palladium complexes

Robert Mundil[1], Soňa Hermanová[1], Laura Schlechte[2], Albena Lederer[2], Jan Merna[1]

[1]Department of Polymers, University of Chemistry and Technology, Prague, Czech Republic
[2]Leibniz Institute of Polymer Research, Dresden, Germany

The aim of this work is the preparation of model hyperbranched polyethylene (hbPE) with precisely controlled molar mass by living ethylene polymerization. For this purpose we synthesized a series of new nickel and palladium complexes bearing α-diimine and amino-imine ligands (Figure 1) that undergo chain walking mechanism during ethylene polymerization [Johnson, L. K. et al. J. Am. Chem. Soc. 1995, 117, 6414; Gao, H. et al. Chem. Commun. 2012, 48, 3312]. With a proper catalyst and suitable reaction conditions, an extensive chain walking can be reached, leading to hbPE [Guan, Z. et al. Science 1999, 283, 2059]. The effect of ligand structure (steric and electronic) on the extent of chain walking and the livingness of the polymerization was investigated. Highly branched PEs were obtained by nickel and palladium amino-imine complexes, however, their topology was linear. The only catalytic systems, which led to PE with branched (dendritic) topology, were based on Pd α-diimine complexes. Substitution of ligands by electron-withdrawing substituents did not lead to the increase in chain walking extent and to the formation of more dendritic PE as expected according to published data [Popeney et. al Organometallics 2005, 24,1145]. The polymerization strategy, which ensures fast initiation, for the preparation of hbPE with narrow molar mass distribution was developed. Block copolymers with blocks of different topology (topoblock copolymers) and narrow molar mass dispersity were prepared by Pd α-diimine catalyst.

This work was supported in frame of joint project by Czech Science Foundation (No. 15-5887J) and Deutsche Forschungsgemeinschaft (No. LE 1424/7-1).

Keywords: hyperbranched polyethylene, chain walking, nickel and palladium catalysts

Figure 1

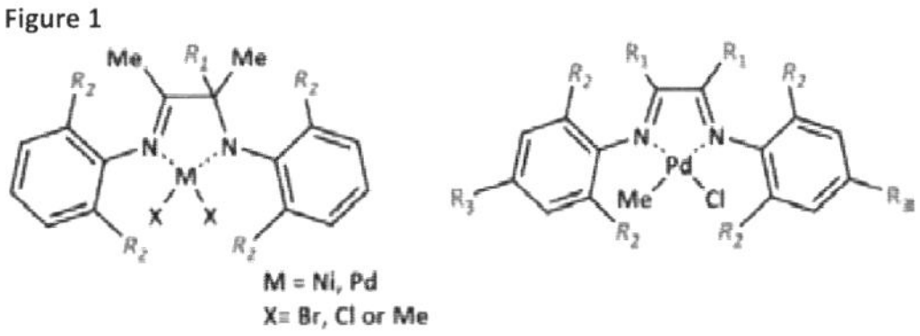

Catalysts used in this work

RDPS-A-O-013 [Controlled/Living Polymerization]

Ibuprofen imprinted ultrathin poly(N-(2-hydroxypropyl) methacrylamide) films

Tuncer Caykara

Faculty of Science, Department of Chemistry, Gazi University 06500Besevler, Ankara/Turkey

Surface-imprinted polymer films include the generation of cavities containing functional sites within a highly cross-linked polymer matrix. The building of these cavities is based initially on the formation of a pre-polymerization complex containing a template molecule and monomer. Molecularly imprinted polymerization of this complex then forms a defined spatial configuration. This highly cross-linked polymer film on a solid substrate is a rigid inflexible network. The template molecule is removed from the polymer matrix by extraction leaving behind the nanosized cavities with size and shape dimensions that match those of the template molecule(1-3). In this study, the surface-molecularly imprinted (MIP) poly(N-(2-hydroxypropyl) methacrylamide) [poly(HPMA)] films were prepared via the interface-mediated reversible addition-fragmentation chain transfer (RAFT) polymerization from 4-cyano-4-(propylsulfanylthiocarbonyl) sulfanyl pentanoic acid immobilized silicon disc using N-(2-hydroxypropyl) methacrylamide as the functional monomer, N,N`-methylene(bis)acrylamide as the cross-linking agent and ibuprofen as the template molecule. The highly cross-linked MIP layer (~12 nm) was homogeneously grafted on the silicon surface, which favors the fast mass transfer and rapid binding kinetics. Binding capacities of the MIP poly(HPMA) films were calculated from the root-mean-square (rms) roughness (nm) data obtained by atomic force microscopy measurements using in the Luzinov and Langmuir equations adopted for this study. The target binding assays demonstrate the desirable binding capacity and imprinting efficiency of the MIP poly(HPMA) films. This work provides a versatile approach for the quantitative determination of biomoleculs with low molecular weight on the surface-imprinted polymer films.

Acknowledgement. This research was supported by the Scientific and Technical Council of Turkey (TUBITAK), KBAG-114Z319.

Keywords: surface-molecularly imprinted, interface-mediated reversible addition-fragmentation chain transfer (RAFT) polymerization, poly(N-(2-hydroxypropyl) methacrylamide)

RDPS-B-O-001 [Complex Macromolecular Structures]

Sequence-control: an access to novel polymeric supramolecular nanostructures

Donatien De Rochambeau[1], Maciej Barlog[2], Johans J Fakhoury[1], Thomas Gw Edwardson[1], Robin S Stein[1], Hassan S Bazzi[2], Hanadi F Sleiman[1]

[1]Department of Chemistry, McGill University, Montreal, QC, Canada
[2]Department of Chemistry, Texas A&M University at Qatar, Doha, Qatar

Primary sequence allows proteins and nucleic acids to fold and self-assemble. The resulting 3D structures are responsible for the biopolymers' versatile functions such as catalysis, molecular recognition and transport. In order to mimic these properties, sequence control in synthetic polymers is of great interest. We present a library expansion of available non-natural monomeric units that may be used in sequence-defined polyphosphate synthesis previously reported by our group (1). Similar to automated solid-phase synthesis of oligonucleotides via phosphoramidite chemistry, non-natural monomers were coupled to a variety of nucleic acid strands and to each other in high yields. This method noticeably led to the discovery of novel "DNA-Teflon" polymer conjugates with remarkable properties (2).

Perfluorocarbons (PFC) are a promising class of molecules for medical applications: they are detectable through 19F nuclear magnetic resonance (NMR), have good affinity for cell membranes and bring unique supramolecular interactions into nanostructures. Modulating the PFC tail length of "DNA-Teflon" copolymers resulted in robust micelles that are traceable by 19F NMR and responsive to external stimuli at low concentrations. Strong fluorine-fluorine interactions in DNA duplexes allowed a remarkable increase in melting temperature and nuclease resistance. Finally, PFC insertion into siRNA was achieved, with the conjugates demonstrating efficient gene silencing, thus highlighting the potential application of these modifications in oligonucleotide therapeutics.

(1) Edwardson, T. G. W.; Carneiro, K. M. M.; Serpell, C. J.; Sleiman, H. F. Angew. Chemie. 2014, 126 (18), 4655.
(2) De Rochambeau, D, Barlog, M, Fakhoury, JJ, Edwardson, TGW, Stein, RS, Bazzi, HS, Sleiman, HF, Polym. Chem. 2016, accepted.

Keywords: Sequence-control, automated synthesis, DNA, Perfluorocarbon

Automated synthesis of sequence-controlled polymers

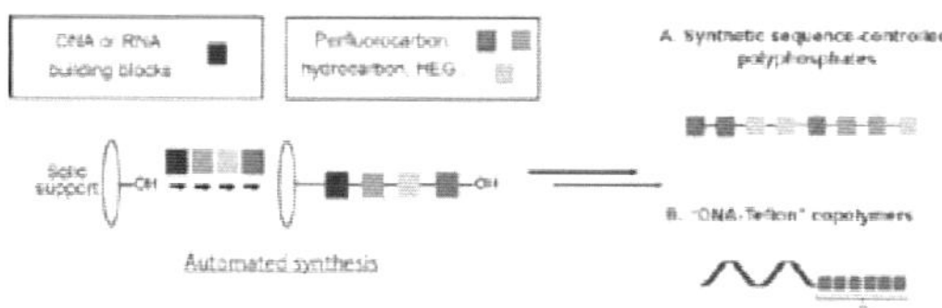

RDPS-B-O-002 [Complex Macromolecular Structures]

Main Chain Post-Functionalization Of Synthetic Polyesters Through Metal-Free Cycloaddition Reactions

Hakan Durmaz, Muge Butun Cetin, Cansu Esen, Ozgun Daglar, Gurkan Hizal, Umit Tunca

Department of Chemistry, Istanbul Technical University, Istanbul, Turkey

Novel polyesters including reactive alkyne that can act as a dipolarophile and a dienophile in the main chain have been prepared through the esterification reactions of acetylenedicarboxylic acid and various diols. Dipolar [3+2] and Diels-Alder [4+2] cycloaddition reactions have then been carried out in the main chain employing diverse dipoles and dienes. According to the experimental results, it was found that the dipolar cycloaddition reactions led to the complete functionalization of the main chain under moderately low temperatures (~40 °C) when compared to Diels-Alder cycloaddition reactions in which higher reaction temperatures (80-110 °C were necessary to reach the complete functionalization of the main chain. The ability of the main chain that can undergo hetero functionalization either by single cycloaddition reaction or sequential double cycloaddition reactions has also been studied.

Keywords: Polyester, Acetylenedicarboxylic acid, Dipolar cycloaddition, Diels-Alder cycloaddition

Schematic representation of main chain post-functionalization of synthetic polyesters via dipolar [3+2] and Diels-Alder [4+2] cycloaddition reactions

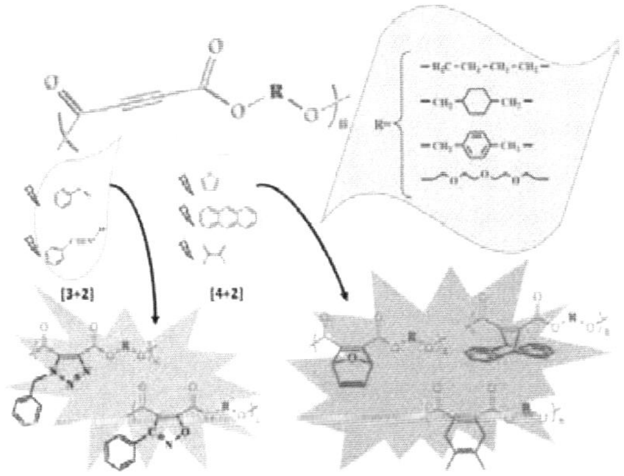

RDPS-B-O-003 [Complex Macromolecular Structures]

From star polymers to multifunctional dendrimers using thiolactone chemistry

Nezha Badi[1], Nazif Uğur Kaya[2], Filip E. Du Prez[3]

[1]Ghent university, department of organic and macromolecular chemistry, polymer chemistry research Group, Ghent, Belgium - University of Strasbourg-ECPM – institut Charles Sadron (CNRS), Strasbourg, France
[2]Ghent university, department of organic and macromolecular chemistry, polymer chemistry research Group, Ghent, Belgium & Polymer science & technology department, graduate school of science engineering & technology, Istanbul technical university, Istanbul, Turkey
[3]Ghent university, department of organic and macromolecular chemistry, polymer chemistry research Group, Ghent, Belgium

Thiolactone chemistry is a powerful tool for the design of complex multifunctional architectures (Espeel and Du Prez, Eur. Polym. J. 2015, 62, 247-272). In this study, this click-like synthetic strategy was conducted to prepare thermoresponsive star polymers as well as multifunctional dendritic structures. For these purposes, a thiolactone bearing an isocyanate functionality has been prepared in multi-gram scale. This molecule has been utilized for multiple step functionalization starting from a 4-arm-PEG (Scheme 1). Amine-thiol-ene conjugation was carried out in order to form the aforementioned structures. The efficient aminolysis of the thiolactone ring and the attachment of the generated thiol to the double bond of an acrylate derivative were completed in a one pot procedure. By choosing carefully the amine and/or the acrylate derivatives, multifunctional systems ranging from thermoresponsive star polymers to dendrimers have been prepared. The present strategy is an elegant and powerful way for preparing a wide range of multifunctional structures.

Keywords: thiolactone chemistry, click-like reaction, star polymers, dendrimer structures, multifunctional

scheme 1

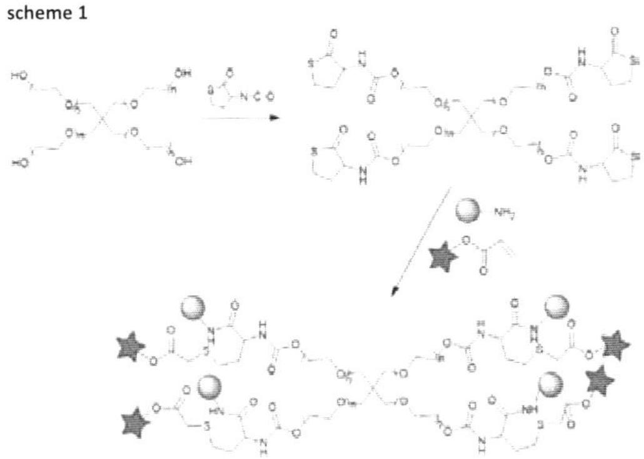

Scheme 1. General strategy used for preparing multifunctional macromolecular architectures.

RDPS-B-O-005 [Complex Macromolecular Structures]

Viscoelastic Properties of Peptide Nanofiber Gels via Microrheology

Deniz Kaya[1], Asli Gencaslan[2], Goksu Cinar[3], Mustafa Ozgur Guler[3]

[1]Physics Department, Akdeniz University, Antalya, Turkey
[2]Nuclear Sciences Research and Application Center, Akdeniz University, Antalya, Turkey
[3]National Nanotechnology Research Center, Bilkent University, Ankara, Turkey

Polypeptides are used extensively in regenerative medicine, drug delivery, biomimetic-bioinspired materials and supramolecular chemistry research fields. Amyloid inspired self-assembled peptide nanofibers can form gel network that can be used as extracellular matrix in biomedical applications. The viscoelastic properties of the gel network would affect the diffusion of macromolecules and yet the faith of the cell that they are supporting. For this reason, knowing about the viscoelastic properties of peptide gels is a very important feedback on the synthesis of new biomaterials. We measure the viscoelasticity of the peptide nanofiber gel system by passive microrheology technique by a fluorescent video microscope. The peptide macromolecules used in this study are positively and negatively charged. Upon mixing those under certain conditions would result in gel formation. When micrometer sized colloidal particles are embedded in the polypeptide gel network, the Brownian motion of the particles can be recorded via video microscopy and the complex viscoelastic modulus can be calculated from the experimental data. In our study, we used positively and negatively charged colloidal particles in separate experiments to investigate the possible specific interactions. The complex modulus values of the system are found to be comparable with the ones of similar polymeric systems. We show that microrheology technique can reveal the local behavior of the gel in detail and also the specific interactions are important in our peptide nanofiber gel system.
Acknowledgement: Authors would like to thank TUBITAK for the financial support of this research via grant number 114M143.

Keywords: Microrheology, peptide nanofiber gels, video microscopy, colloidal particles.

RDPS-B-O-006 [Complex Macromolecular Structures]

Controlling inter-lamellar spacing in periodically grafted amphiphilic copolymers

Sananda Chanda, Subramaniam Ramakrishnan

Department of Inorganic and Physical Chemistry, Indian Institute of Sience, Bangalore, India

Methods to tune lamellar domain spacing in block copolymers involves varying the lengths of the blocks, blending with homopolymers or introduction of small molecular additives (Rangarajan et al. *Macromolecules* **1997**, *30*, 494). Recently, we reported a new approach to prepare periodically grafted copolymers (PGAC) based on itaconic acid; the unique feature of this polymer is the presence of *exo*-chain double bonds which can serve as excellent Michael acceptors (Chanda et al. *Polym. Chem.*, **2015**, *6*, 2108). In our current study, we have examined a series of PGACs wherein pendant PEG segments of varying lengths (TREG, 350, 550 and 750) were grafted at periodic intervals along a long crystallizable alkylene (C-20) backbone. The immiscibility between the alkylene and PEG segments, and the strong propensity of the alkylene segments to crystallize, drive these PGACs to self-segregate by folding in a zigzag fashion (Roy et al. *Macromolecules* **2012**, *45*, 3063, Mandal et al. *J. Am. Chem. Soc.* **2014**, *136*, 2538), and subsequently organize into a lamellar morphology, which was evident from DSC, SAXS, WAXS and AFM studies; the inter-lamellar spacing was seen to increase linearly with PEG length. Co-grafted PGACs, prepared by grafting a mixture of two different PEGs (PEG 350 and PEG 750), also exhibited a lamellar morphology; interestingly, the inter-lamellar spacing was seen to depend only on the total PEG content, while the presence of a mixture of PEG lengths exerted little influence. However, a mixture of two homo-PGACs bearing pendant PEG350 and PEG750 segments, underwent macrophase separation suggesting that despite the central alkylene segment being the same, difference in the length of the pendant PEG segments alone is sufficient to induce immiscibility. This study provides a new design for controlling the dimensions of microphase-separated nanostructures at significantly smaller length scales (sub-10 nm) than is typically possible using block copolymers.

Keywords: grafted copolymers, itaconic acid, crystallize, self-segregate, folding, inter-lamellar spacing

Controlling interlamellar spacing in periodically grafted copolymers

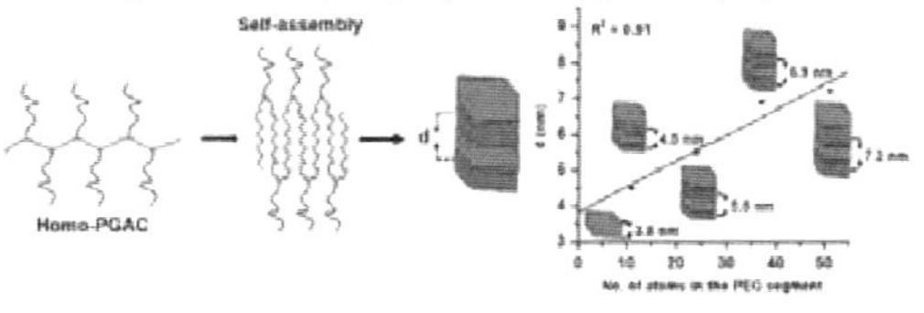

This schematic representation shows that periodically grafted copolymers self-segregate by folding in a zigzag fashion and subsequently organize into a lamellar morphology; interestingly, the inter-lamellar spacing is seen to increase linearly with PEG length

RDPS-C-O-001 [Light-induced Reactions in Polymer Science]

Remoldable Polymer Networks Using Light-responsive Anthracene Derivatives

Filip Du Prez, Jonas Van Damme, Otto Van Den Berg, Laetitia Vlaminck

Department of Organic and Macromolecular Chemistry, Ghent University, Ghent, Belgium

Reversible bonds in polymer materials enable a range of properties and functionalities for many applications. Nevertheless, the amount of suitable reversible bond forming chemistries is limited, especially when light is the preferred stimulus.

The anthracene [4+4] cycloaddition is a light driven reversible reaction that forms no significant side reactions.Therefore, it is more reliable than [2+2] cycloadditions when multiple dimerization/dissociation cycles are desirable.

We have synthesized a range of anthracene derivatives, having different substituents at the 9-position. The influence of the substituent on the absorption spectrum, photochemical dimerization and thermal dissociation has been studied. The substituents are not only shown to influence the absorption by mesomeric and inductive effects, enabling possible better spectral overlap with UV lamps, but also to shift the temperatures at which dissociation occurs. This enables the application driven selection of anthracenes for polymer use, having certain desirable dimerization/dissociation rates.

A selection of the synthesized derivatives were used to make remoldable networks that can self-heal upon sequential heating at different temperatures and irradiation, or reduce internal stress upon simultaneous stimulation.

Keywords: Anthracene compounds, remoldable networks, light-induced dimerization, coatings

Figure 1

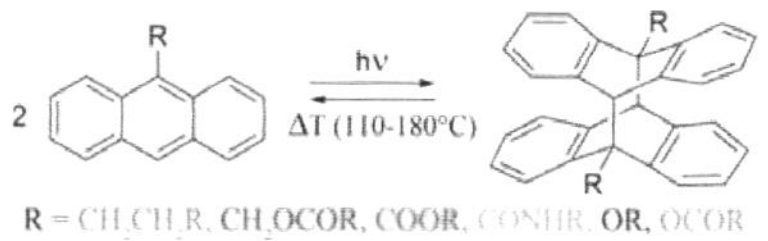

R = CH2CH2R, CH2OCOR, COOR, CONHR, OR, OCOR

Anthracene building blocks used in remoldable networks

RDPS-C-O-002 [Light-induced Reactions in Polymer Science]

New Photochemical Strategies for Macromolecular Syntheses

Gorkem Yilmaz, Yusuf Yagci

Istanbul Technical University, Faculty of Science and Letters, Chemistry Department, Maslak-Istanbul/Turkey

The application of photochemical redox reactions to the synthetic polymer chemistry brought significant advantages compared to the traditional strategies. For example, free radical promoted cationic polymerizations were successfully carried out in the presence of radical photoinitiators in extended wavelength regions. Similar photoinduced strategies were also applied to the other macromolecular synthesis methods, such as atom transfer radical polymerization (ATRP) and copper catalyzed azide-alkyne cycloaddition (CuAAC) reactions. These strategies provide spatial and temporal control over the processes. This abstract covers the recent studies on photo-induced syntheses of macromolecular structures.1-7

References.

(1) Arslan, M.; Yilmaz, G.; Yagci, Y. Polym. Chem. 2015, 6, 8168.
(2) Doran, S.; Yilmaz, G.; Yagci, Y. Macromolecules 2015, 48, 7446.
(3) Erdur, S.; Yilmaz, G.; Colak, D. G.; Cianga, I.; Yagci, Y. Macromolecules 2014, 47, 7296.
(4) Taskin, O. S.; Yilmaz, G.; Yagci, Y. ACS Macro. Lett. 2016, 5, 103.
(5) Yilmaz, G.; Iskin, B.; Yagci, Y. Macromol. Chem. Phys. 2014, 215, 662.
(6) Tasdelen, M. A.; Yilmaz, G.; Iskin, B.; Yagci, Y. Macromolecules 2012, 45, 56.
(7) Yilmaz, G.; Iskin, B.; Yilmaz, F.; Yagci, Y. ACS Macro. Lett. 2012, 1, 1212.

Keywords: Photo-induced Processes, Photopolymerization, CuAAC, ATRP

Scheme 1.

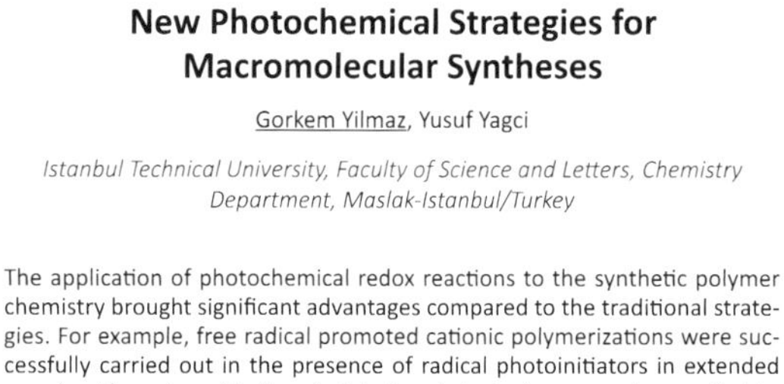

Photoinduced macromolecular synthesis

RDPS-C-O-003 [Light-induced Reactions in Polymer Science]

Comparative Study for Photopolymerization of Polyacrylamide Hydrogels using different UV light sources

Nur Farizah Binti Ayub[1], Jamarosliza Binti Jamaluddin[1], Mathias Ulbricht[2], Nadia Binti Adrus[1]

[1]Biopolymer Research Group, Department of Chemical & Energy Engineering, Universiti Teknologi Malaysia, 81310 UTM Johor Bahru, Johor, Malaysia
[2]Lehrshtul für Technische Chemie II, Universität Duisburg-Essen, 45141 Germany

The advantages of hydrogels photopolymerization using UV-mercury (UV Hg) light source are well known. Those include having the ability for in situ photo-patterning of hydrogels according to desired shape. However, the use of UV Hg curing system can produce harmful effect to the environment. Recent trends show that the UV LED light source is a green technology. Thus, it was a great worth to compare the photopolymerization of polyacrylamide (PAAm) hydrogels using either UV-mercury (300 < λ < 400 nm) or UV-LED (λ ~ 365 nm). First, we synthesized water soluble photoinitiator (WSPI) from the complexation of 2,2-dimethoxy-2-phenyl acetophenone (DMPA) and methylated-β-cyclodextrin (MβCD). The Fourier transform infrared (FTIR) and UV data have confirmed that WSPI possessed characteristic similar to DMPA (λ ~ 330 nm). Then, 2 wt% of WSPI was incorporated into monomer and crosslinker mixtures containing acrylamide (AAm; 15 wt %) and methylenebisacrylamide (MBAAm; 5 wt% relative to AAm) dissolved in distilled water. The pre-gel solutions were exposed to different UV curing system (Hoenle AG, Germany) at various irradiation times; i.e., from to 2.5 to 20 minutes. Regardless the light source used for photopolymerization, the optimum conversion (> 80 %) of the hydrogels was achieved at 15 min. Most interestingly, UV LED curable hydrogels showed comparably higher conversion than conventional UV Hg system. UV LED emitted monochromatic wavelength only and thus, this could be more efficient for hydrogel photopolymerization. The swelling and tensile properties of UV Hg and UV LED curable PAAm hydrogels are also compared. Importantly, this study has revealed that UV LED able to cure PAAm hydrogels without compromising its performance as compared to conventional UV Hg system. In addition, it is also envisioned that UV LED curing system can offer better options for curing of such temperature-sensitive hydrogels because it can be operated without heat generation.

Keywords: UV LED, UV Hg, polyacrylamide hydrogel, water soluble photoinitiator

RDPS-C-O-004 [Smart and Functional Polymers]

A Coumarin Bearing Amphiphillic Block Copolymer: Synthesis, Characterization and Micellization

Binnur Aydogan Temel, Belma Zengin Kurt, Gamze Camlik, Rabia Sare Yanikoglu

Department of Pharmaceutical Chemistry, Bezmialem Vakif University, Fatih, 34093 Istanbul, Turkey

In recent years, a number of strategies have been devoted in the use of nanoparticles for drug delivery applications. Drugs may be encapsulated, adsorbed or dispersed in nanoparticles, which can be colloidal-sized particles, possessing diameters ranging between 1 and 1000 nm.[1] A wide variety of nanoparticles composed of a range of materials including lipids, polymers and inorganic materials have been developed, resulting in delivery systems that vary in their physicochemical properties and thus their applications. Among them amphiphilic block copolymers with hydrophilic and hydrophobic segments have attracted a great deal of attention in terms of their ability to form various types of nanoparticles.[2-4] In this work, we describe the synthesis of a coumarin bearing amphiphilic block copolymer via a simple ring opening polymerization and click reaction. The structure of starting materials and corresponding copolymer is confirmed by ^{1}H NMR and GPC analyses. Micelle formation of amphiphilic copolymer is also investigated and characterized by DLS.

References:

1) Letchford, K., Burt, H., Eur. J. Pharm. Biopharm., 2007, 65, 259-269.
2) Adams, M. L., Lavasanifar, A., Kwon, G. S., J. Pharm. Sci., 2003, 92, 1343-1355.
3) Jin, Q., Maji, S., Agarwal, S., Polym. Chem., 2012, 3, 2785-2793.
4) Robb, M. J., Connal, L. A., Lee, B. F., Lynd, N. A., Hawker, C. J., Polym. Chem., 2012, 3, 1618-1628.

Keywords: Coumarin, amphiphilic, block copolymer, micellization

Scheme 1.

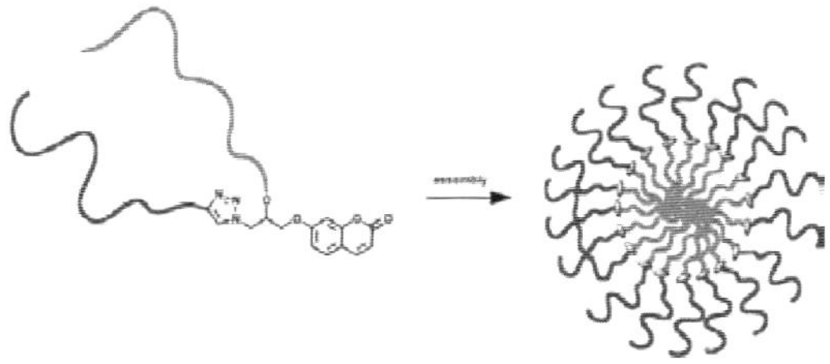

Schematic representation for the micellization of coumarin bearing amphiphilic block copolymer.

RDPS-C-O-005 [Light-induced Reactions in Polymer Science]

Light-Induced Grafting of Functional Polymers to Nanodiamonds

Kilian N. R. Wuest[1], Vanessa Trouillet[2], Anja S. Goldman[1], Martina H. Stenzel[3], Christopher Barner Kowollik[1]

[1]Preparative Macromolecular Chemistry, Institut für Technische Chemie und Polymerchemie (ITCP) and Institut für Biologische Grenzflächen (IBG), Karlsruhe Institute of Technology (KIT), Engesserstr. 18, 76131 Karlsruhe, Germany
[2]Institute for Applied Materials (IAM-ESS) and Karlsruhe Nano Micro Facility (KNMF), Karlsruhe Institute of Technology (KIT), Hermann-von-Helmholtz-Platz 1, 76344 Eggenstein-Leopoldshafen, Germany
[3]Centre for Advanced Macromolecular Design (CAMD), The University of New South Wales, Sydney NSW 2052, Australia

Nanodiamonds (ND) offer a high potential for drug delivery and imaging due to their unique properties. They are non-toxic, their surface can be readily functionalized and non-bleaching fluorescence can be introduced through vacancy centres in the carbon lattice. Moreover, NDs are accessible in large quantities and with low cost from the detonation of carbon containing explosives. However, for many applications the nanodiamond surface needs to be modified. For instance, a polymer coating can enhance their colloidal stability and introduce targeting functions. Herein, we report the light-induced grafting of polymers to NDs in the absence of any catalyst and at ambient temperature. Therefore, o-methyl benzaldehyde groups – that form o-quinodimethanes (photoenols) upon UV irradiation – were attached to the surface of NDs. The photoenol groups undergo Diels-Alder reactions with electron deficient double bonds, such as maleimides. A variety of maleimide end group functional polymers (poly(styrene, poly(N-isopropylacrylamide) and poly(2-(2′,3′,4′,6′-tetra-O-acetyl-α-D-mannosyloxy)ethyl methacrylate)) were grafted to NDs showing the modularity of this approach. The resulting polymer coated NDs were characterized by FT-IR, X-ray photoelectron spectroscopy, thermogravimetric analysis and dynamic light scattering. The effect of polymer chain length on grafting efficiency as well as the lectin binding behavior of the glycopolymer functional NDs were investigated. [Wuest, K. N. R.; Trouillet, V.; Goldmann, A. S.; Stenzel, M. H.; Barner-Kowollik, C. Macromolecules 2016, 49 (5), 1712–1721.]

Keywords: photochemistry, nanoparticles, nanodiamonds, grafting-to, glycopolymers

General schematic representation of the light-induced grafting of polymers to nanodiamonds.

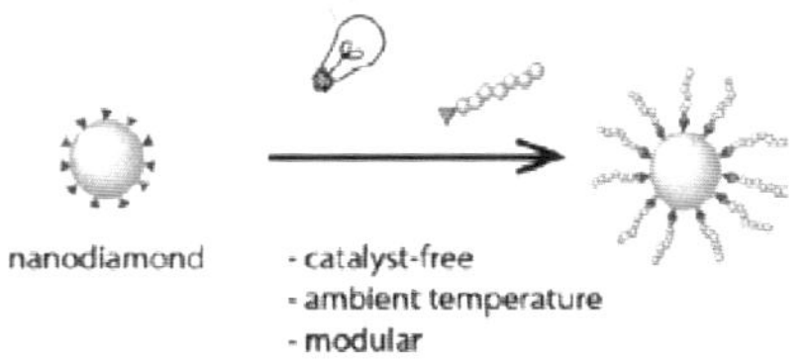

RDPS-C-O-006 [Light-induced Reactions in Polymer Science]

Light-triggered polymerization of dopamine through a photoinduced pH jump

David Perrot, Ahmad Ibrahim, Céline Croutxé Barghorn, Xavier Allonas

Laboratory of Macromolecular Photochemistry and Engineering, University of Haute Alsace, Mulhouse, France

Polydopamine is a bio-inspired polymer sharing similar properties to the adhesive secretions of mussels [1]. It has been reported that a thin, surface adherent coating, and multifunctional biopolymer-polydopamine (PDA) layer could be deposed on a wide range of inorganic and organic materials by self-oxidative polymerization of dopamine in an aqueous solution at basic pH [2]. Based on this approach, a new method for polydopamine deposition on different substrate was developed by the use of a photobase generator to phototrigger a pH jump [3]. In this study, we first investigated the influence of the PBG on the pH under irradiation. Then, dopamine photoinduced polymerization and its pH dependence were investigated by UV-visible spectroscopy. Contact angle measurement and SEM combined with EDX were implemented to investigate the polydopamine coating and the kinetics of film deposition. From these results, it turns out that the new photopolymerization method and the highly formulation stability could be of great interest for the development of functional coatings.

[1] H. Lee, S. M. Dellatore, W. M. Miller and P. B. Messersmith, Science, 2007, 318, 426-430.
[2] H. Lee, B. P. Lee and P. B. Messersmith, Nature, 2007, 448, 338-341.
[3] D. Perrot, C. Croutxé-Barghorn and X. Allonas, Polymer Chemistry, 2016, DOI: 10.1039/C6PY00356G.

Keywords: Photo triggered, Photobase Generator, pH jump, Polydopamine

photoinduced pH jump for dopamine polymerization

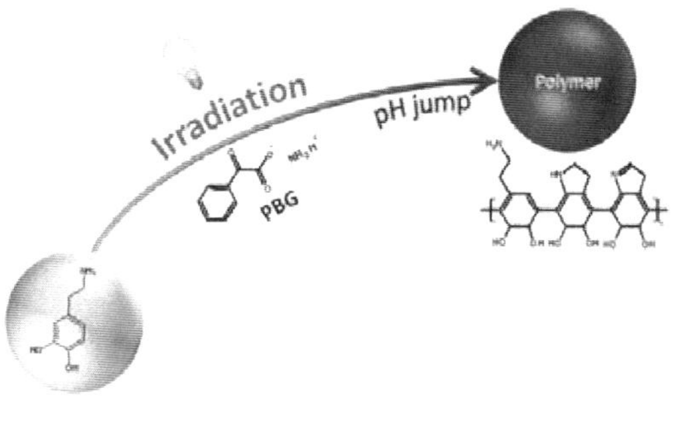

RDPS-C-O-007 [Light-induced Reactions in Polymer Science]

Hydrophobic coating based on fluorinated acrylate monomer using UV-LED polymerization

Siti Khairunisah Binti Ghazali, Nadia Binti Adrus, Jamarosliza Binti Jamaluddin

Biopolymer Research Group, Department of Bioprocess & Polymer Engineering, Faculty of Chemical & Energy Engineering, Universiti Teknologi Malaysia, Malaysia

Water repellent or hydrophobic coating offers various promising advantages such as anti-fouling and anti-corrosion surface properties. It is well known that fluorine based moiety can alter the surface wettability due to its low surface energy characteristic. In this study, UV-curable hydrophobic coating mixture was prepared based on aliphatic urethane acrylate oligomer, 2-ethylhexyl acrylate (2-EHA), methyl methacrylate (MMA), trimethylopropane triacrylate (TMPTA), heptadecafluorodecyl methacrylate (HDFMA) and hydroxyphenyl ketone photoinitiator (Chivacure 300). The reaction mixture was coated onto a glass substrate and cured via UV-LED irradiation. From the ATR-FTIR spectra, the C=C absorption peak at 1635 cm^{-1} was decreased after UV-LED irradiation, indicating the proceeding of polymerization. The effect of HDFMA content on the properties of UV-curable coating was investigated. In order to quantify the degree of hydrophobicity, water contact angle of the coating was determined. The addition of fluorinated acrylate monomer improved the water repellence behaviour and increased the water contact angle from 86.01° to 123.57°. As the HDFMA content increased, the surface tension of the coating also became lower, due to the presence of the fluorinated group. Moreover, the light transmittance of the coating layer was not significantly affected with increasing HDFMA content. It can be concluded that the fluorinated containing monomer gave the most contribution towards the hydrophobicity of the coating assisted by UV-LED curable system. Analogously, these results may attain various potential applications such as self-cleaning, optoelectronic and liquid-repellent coating. Importantly, the utilization of UV-LED in this research was a great initiative to develop green aspect for photopolymerization.

Keywords: UV-LED, curable coating, hydrophobic, heptadecafluorodecyl methacrylate

RDPS-D-O-001 [Condensation Polymerization and Thermoset]

Adapting Polybenzoxazines for Smart Applications

Baris Kiskan, Mustafa Arslan, Omer Suat Taskin, Yusuf Yagci

Department of Chemistry, Istanbul Technical University, Istanbul, Turkey

Polybenzoxazine is a new type of high performance phenolic thermoset having a wide range of interesting features and the capability to overcome several shortcomings of conventional novolac and resole type phenolic resins. They exhibit near zero volumetric change upon curing, low water absorption, for some polybenzoxazines Tg much higher than cure temperature, high char yield, no strong acid catalysts required for curing, release of no by-product during curing and also possess thermal and flame retarding properties of phenolics along with the mechanical performance (Ghosh et al., 2007, Prog. Polym. Sci., 32, 1344–1391). These polymers have interesting properties and potential for various applications. An overview is presented about polybenzoxazines highlighting the recent advances for smart applications ranging from porous materials to self-healing systems (Fig. 1).

Keywords: Polybenzoxazines, Smart Materials, Self-Healing, Ring-opening polymerization

Figure 1

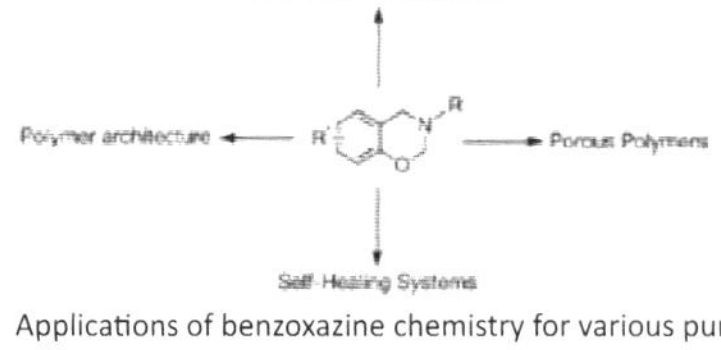

Applications of benzoxazine chemistry for various purposes.

RDPS-D-O-002 [Condensation Polymerization and Thermoset]

High molecular weight processable and thermally stable poly(cyclo)acetals

Sophie Lingier[1], Sil Nevejans[1], Pieter Espeel[1], Stefaan De Wildeman[2], Filip E. Du Prez[1], Andrea Hufendiek[1]

[1]Department of Organic and Macromolecular Chemistry, Polymer Chemistry Research Group, Ghent University, Krijgslaan 281 S4bis, 9000 Ghent, Belgium
[2]Biobased Materials, Faculty of Humanities and Sciences, Maastricht University, P.O. Box 616, 6200 MD Geleen, The Netherlands

Poly(cyclo)acetals are a class of polymers generally formed by the reaction of a tetravalent alcohol and a dialdehyde[1], which were first reported in 1912. Yet, only limited attention was given to this class of polymers despite their promising properties[2] In the course of the polymerization, cyclic acetal moieties are formed resulting in a rigid polymer backbone. Poly(cyclo)acetals show an overall high thermal stability and extraordinary mechanical properties. However, it was reported earlier that these polymers could only reach low molecular weights, were not soluble in common organic solvents and/or not processable.

To improve the processability of the poly(cylco)acetals, di(trimethylolpropane) (Di-TMP) instead of the commonly employed pentaerythritol was used successfully as tetravalent alcohol in the polyacetalization reaction with various commercially available dialdehydes.

In this way, processable poly(cyclo)acetals with high molar masses of up to 40 kDa were synthesized in a systematic manner. These polymers were thermally robust (degradation temperatures (Td) of up to 370°C) and exhibited glass transition temperatures (Tg) in the range from 20 to 110 °C, depending on the employed dialdehyde. The trends in Tg modulation were confirmed by a theoretical simulation. Moreover, the poly(cyclo)acetals showed favourable mechanical and optical properties, such as ductile behaviour and transparency.[3]

1. Capps, D.B, T.C. Corporation, Editor. 1956.
Cotting, J.-a. and A. Renner, Polycycloacetals containing epoxy groups. 1988. UCCLE-Bruxelles, 1958: Belgium.
2. Read, J., J. Chem. Soc., Trans., 1912(101): p. 2090-2094.
3. Lingier, S. et al., 2016 (Submitted)

Keywords: Poly(cyclo)acetal, processable, high molecular weight

Reaction of glutaraldehyde and Di-TMP in order to obtain a high MW and processable poly(cyclo)acetal

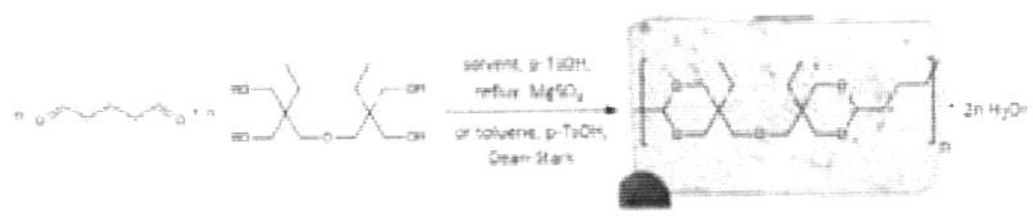

RDPS-D-O-003 [Condensation Polymerization and Thermoset]

Semi-crystalline Polyamide Thermosets with High Temperature Shape Memory Properties

Ming Li, Theo J Dingemans

Faculty of Aerospace Engineering, Delft University of Technology, Delft, The Netherlands

Polyamide thermosets based on PA10T (poly(decamethylene terephthalamide)) have been designed and synthesized for high-temperature shape memory applications. Copolyamides containing 5, 10 or 15 mol% of thermally crosslinkable PE (phenylethynyl) groups were used as melt processable precursors. All reactive PA10T derivatives were prepared via solution-polymerization and freestanding films were prepared using a high-temperature compression moulding method. Rheology experiments confirmed that the reactive copolyamides were close to being fully cured after 15 minutes at 350 °C. All fully cured thermoset films are stable up to 400 °C when tested using DMTA and this is the result of network formation. Crosslinking does not remove all crystallinity, in fact, crystallization of the polymer chains is strongly hindered by the crosslink chemistry, as indicated by the depression of Tm and ΔHm. The storage modulus (E') at the rubbery plateau above Tm is about 10-20 MPa, which enables sufficient stress storage for shape recovery. The shape memory properties of the semi-crystalline thermoset films were investigated using a rheometer in torsion mode. High-temperature dual-shape memory behavior with the Tm as switching temperature (>200 °C) was achieved for thermosets with 5 mol% and 10 mol% PE. Thermosets with 15 mol% PE showed two distinct recovery steps because the crystallinity is too low and cannot provide a crystal scaffold that locks the shape between Tg and Tm. The use of PE reactive functionalities allows us to control the degree of crystallinity in the final polyamide thermosets and tune the shape-memory behavior. PA-based thermosets with triple-shape memory and one-way reversible shape memory behavior can be designed with the Tg (~125 °C) as the second switching temperature.

Keywords: polyamide thermosets, semi-crystallinity, high temperature shape memory properties

RDPS-D-O-004 [Condensation Polymerization and Thermoset]

Novel thermoreversible and thermoadjustable blocking agents for triazolinediones: crosslinking on demand

Hannes A. Houck[1], Kevin De Bruycker[2], Stijn Billiet[2], Christopher Barner Kowollik[3], Johan M. Winne[2], Filip E. Du Prez[2]

[1]Polymer Chemistry Research Group and Laboratory for Organic Synthesis, Department of Organic and Macromolecular Chemistry, Ghent University, Krijgslaan 281 S4-bis, 9000 Ghent, Belgium & Preparative Macromolecular Chemistry, Insitut für Technische Chemie und Polymerchemie, Karlsruhe Institute of Technology (KIT), Engesserstraße 18, 76131 Karlsruhe, Germany
[2]Polymer Chemistry Research Group and Laboratory for Organic Synthesis, Department of Organic and Macromolecular Chemistry, Ghent University, Krijgslaan 281 S4-bis, 9000 Ghent, Belgium
[3]Preparative Macromolecular Chemistry, Insitut für Technische Chemie und Polymerchemie, Karlsruhe Institute of Technology (KIT), Engesserstraße 18, 76131 Karlsruhe, Germany

Sharpless' philosophy of utilizing only a few good reactions in as many processes as possible,[1] introduced the concept of 'click'chemistry, now one of the most powerful paradigms in material science.[2] Since then, numerous of covalent coupling reactions that meet the original -or extended[3]- definition have been intensively researched. Recently, we have developed a promising 'click'type platform, based upon the very reactive 1,2,4-triazoline-3,5-dione (TAD), a five membered cyclic azodicarbonyl component.[4] This TAD-based chemistry excels other familiar 'click'systems because it is additive-free, provides a useful color switch and shows ultrafast kinetics at ambient temperature.[5] In addition, a judicious choice of the complementary reaction partner offers the ability to select both an irreversible ((di)ene) or reversible (indole) bond formation. In particular, this reversible nature allows interchangeability of the clickable TAD group between two distinct moieties at elevated temperatures. However, the inherent reactivity and limited stability of TADs is as a major drawback. Therefore, resolving stability issues is considered a key next step in our futher development of this 'click'chemistry platform. To pursue this desire –and inspired by the concept of blocked isocyanates– different reversible TAD-based reactions have been investigated, resulting in the identification of indole components as suitable thermoreversible blocking agents for triazolinediones.

In this work, a variety of these indole components were synthesized and combined with TADs to afford quantitative, bench-stable adducts at room temperature.[6] It is shown that these adducts, depending on the indole substitution pattern, exhibit interesting reversibility properties in a well-defined, characteristic temperature range. Thus, indole moieties were proven to be efficient thermoadjustable blocking agents, which allow temporal control over TAD-based click reactions, as illustrated by the delayed crosslinking of one-component polyurethane systems.

[1]H.C.Kolb, M.G.Finn, K.B.Sharpless,Angew.Chem.Int.Ed.2001,40,2004-2021.
[2]W.Xi, T.F.Scott, C.J.Kloxin, et.al.,Adv.Funct.Mater.2014,24,2572-2590.

[3]C.Barner-Kowollik, F.E.Du Prez, P.Espeel, et.al.,Angew.Chem.Int. Ed.2011,50,60-62.
[4]S.Billiet, K.De Bruycker, F.Driessen,et.al.,Nat.Chem.2014,6,815-821.
[5]K.De Bruycker, S.Billiet, H.A.Houck, et.al.,Chem.Rev.2016,116,3919-3974.
[6]H.A.Houck, K.De Bruycker, S.Billiet, et.al.,2016,submitted.

Keywords: indoles, blocked triazolinediones, delayed crosslinking, 1K systems

RDPS-D-O-005 [Condensation Polymerization and Thermoset]

Sequential dual-curing processes based on thiol-click reactions

Xavier Fernández Francos[1], Ali Osman Konuray[1], Dailyn Guzmán[2], Alberto Belmonte[3], Silvia De La Flor[3], Angels Serra[2], Xavier Ramis[1]

[1]*Thermodynamics Laboratory ETSEIB, Universitat Politècnica de Catalunya, Barcelona, Spain*
[2]*Department of Analytical and Organic Chemistry, Universitat Rovira i Virgili, Tarragona, Spain*
[3]*Department of Mechanical Engineering, Universitat Rovira i Virgili, Tarragona, Spain*

Sequential dual-curing is an advatageous processing methodology based on the combination of two compatible polymerization reactions taking place sequentally and in a controlled way. This is achieved by the use latent catalysts that are activated selectively by stimuli such as UV light or heat, or by reaction kinetics. After the first curing stage, the intermediate material can be manipulated or transported safely before the activation of the second process, as in common B-stageable systems. The extent of both curing stages and the properties of the intermediate and final materials can be easily custom-tailored just by changing the monomer composition and functionality, making it possible to eliminate the risk over-drying present in common B-stageable systems and to meet a wide range of performance or processing requirements. Dual-curing processing of solventless formulations also minimizes VOC emission. Thiol-click reactions are attractive for dual-curing processing due to their orthogonality and the possibility of combining them with different polymerization reactions. Our recent advances in the use of thiol-click reactions in dual-curing processing with controlled curing sequence include: (a) thiol-ene/thiol-epoxy systems, using radical photoinitiator for the first reaction and a latent thermal catalyst for the second one [1] (b) thiol-epoxy systems with excess epoxy groups catalyzed using an imidazole catalyst with sequential curing controlled by kinetics [2] and (c) thio-Michael/thiol-epoxy formulations with kinetics control of the curing sequence using a moderately active tertiary amine or a photolatent base. The processing capabilities in terms of sequence control are discussed, as well as the wide range of intermediate and final properties that can be achieved starting from a reduced set of monomers.

References:

1. Guzman D, Ramis X, Fernandez-Francos X, and Serra A. RSC Advances 2015;5(123):101623-101633.
2. Fernandez-Francos X, Konuray A-O, Belmonte A, De la Flor S, Serra A, and Ramis X. Polymer Chemistry 2016;7:2280-2290.

Keywords: dual-curing, thiol-click reactions,latent catalysts, epoxy, acrylate

Sequential curing of a dual-curable formulation based on thiol-click reactions

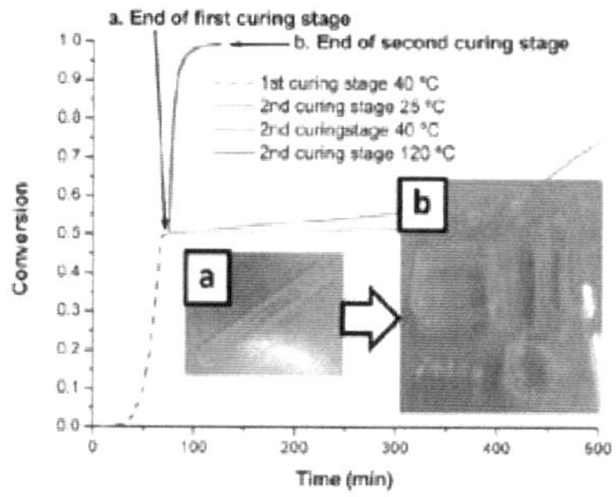

The figure illustrates the curing sequence of a dual-curable system with kinetics control. The first curing stage takes place at a moderate temperature, but the kinetics of the second curing stage are much slower, making it possible to handle or manipulate the material before its final application and activation of the second curing stage at an elevated temperature. The figure also illustrates the possible application of dual-curing processing for the production of complex shape objects starting from a simple temporary shape produced after the first curing stage that is modified and fixed after the second curing stage. The final material has shape-memory properties, therefore the produced final objects can find an application such as mechanical actuators (i.e. spring)

RDPS-D-O-006 [Condensation Polymerization and Thermoset]

New mechanism of liquid polysulfide cross-linking: a combined IR spectroscopy and rheology study

William Guérin[1], Vanessa Blasin-Aubé[2], Loïc Le Pluart[1], Arnaud Travert[2]

[1]*Normandie Univ, ENSICAEN, UNICAEN, CNRS, Laboratoire de Chimie Moléculaire et Thioorganique, 14000 Caen, France*
[2]*Normandie Univ, ENSICAEN, UNICAEN, CNRS, Laboratoire Catalyse et Spectrochimie, 14000 Caen, France*

Polysulfide rubbers are used in many fields (insulated glass, construction, automotive or aircraft industry) thanks to their chemical inertness and good thermal stability. These rubbers are obtained from liquid polysulfide (polyether oligomers), bearing thiol reactive groups [1]. Increase in oligomers molecular weight and crosslinking of the rubber are obtained by the formation of disulfide bonds through an oxidation reaction (Equation 1). While various oxidants can be used to perform this polymerization reaction, manganese dioxide is the most widely used thanks to its good reactivity and its low cost.

$2\ RSH + MnO_2 = RSSR + MnO + H_2O$ (Equation 1)

Despite its large use and important researches, the reaction mechanism is still in debate [2]. Improving the properties of polysulfide rubbers passes through a better knowledge of this mechanism. In this study, IR spectroscopy and rheology have been used in parallel to monitor in situ the cure of liquid polysulfide and perform a thorough kinetic study (Figure 1). Good agreement was found between the two methods of monitoring: the evolution of the concentration of characteristic species was correlated to the variation of the viscoelastic properties. Moreover, new species, such as manganese hydroxide (($Mn(OH)_2$), have been observed during the cure, showing that Equation 1 is insufficient to describe the reaction mechanism. Multivariate analyses on infrared spectra demonstrate the existence of a reactive intermediate. Overall, our results provide evidence of a complex mechanism involving at least two consecutive steps. It allows a better understanding of the reactivity of manganese dioxide as well as, ultimately, a better control of the properties of polysulfide rubber.

References :

[1] T. C. P. Lee, Properties and Applications of Elastomeric Polysulfides.

Editions iSmithers Rapra Publishing, 1999.
[2] Y. N. Khakimullin, V. S. Minkin, T. R. Deberdeev, Polysulfide Oligomer Sealants: Synthesis, Properties, and Applications, Appla Academic press, 2015.

Keywords: Polysulfide, rubber, cross-linking, manganese dioxide.

Figure 1

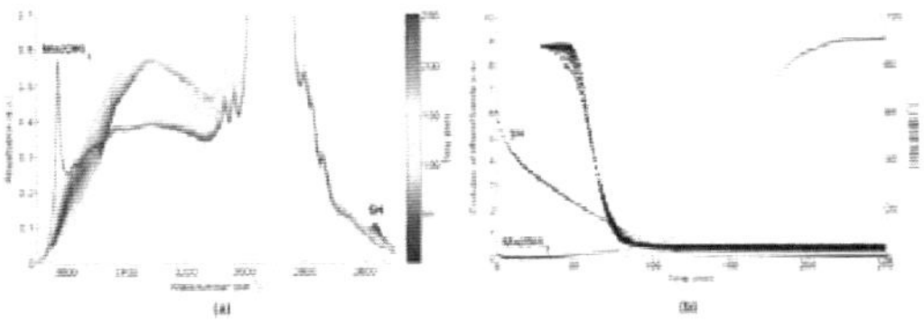

Infrared spectra (a) and evolution of characteristic infrared bands areas and of the loss angle (b) during the cure

RDPS-D-O-007 [Condensation Polymerization and Thermoset]

Controlled Polymerization of Conjugated Polymers: Towards All-conjugated (Co)polymers

Guy Koeckelberghs, Michiel Verswyvel, Pieter Willot, Tine Hardeman

KU Leuven, Leuven, Belgium

The discovery of the controlled polymerization of poly(3-alkylthiophene) s (P3AT) marked a milestone in the synthesis of conjugated polymers, as it enables the synthesis of P3ATs with predefined molar masses, low polydispersities and offers the possibility to construct all-conjugated block copolymers by successive monomer addition. The key to turn the polymerization into a controlled chain-growth polymerization is the fact that after the catalytic cycle, the catalyst remains associated with the π-conjugated backbone. However, the fact that the success of the controlled polymerization depends on the complexation of the catalyst to the growing π-conjugated polymer is also the weak point of this method, as this is system-dependent. A second consequence of using Ni(dppp) (dppp = 1,3-diphenylphosphinopropane) as catalytic entity in the polymerization of P3ATs is the fact that inherently HT-P3ATs is formed, even if a mixture of both isomers is used. (Random) copolymerization of different monomers is in most cases impossible. Finally, this also limits the possibilities for block-copolymer formation.. In this contribution, the scope of a new catalyst, Pd(RuPhos), of which the controlled nature of the polymerization does not depend on the association with the growing polymer chains, but originates from a desactivation of the monomer, is discussed. First, it is investigated whether the use of this catalyst indeed results in a controlled chain-growth polymerization of P3ATs, poly(fluorene)s and poly(3-alkylselenophene)s. Second, it is investigated whether the catalyst allows to tune the regioregularity in P3ATs. Third, (random) copolymerization of thiophene and fluorene will be discussed. Finally, it is attempted to prepare all-conjugated ABC block copolymers – a class of polymers that has never been prepared – and it is investigated whether this catalyst allows to vary the order of monomer addition.

Keywords: conjugated polymers, controlled polymerization, block copolymers, random copolymers

RDPS-D-O-008 [Condensation Polymerization and Thermoset]

Copolymerization of conjugated polymers: adding a new asset to their controlled synthesis

Tine Hardeman, Guy Koeckelberghs

Laboratory for Polymer Synthesis, Division of Polymer Chemistry and Materials, KU Leuven, Heverlee, Belgium

Since it was discovered that the polymerization of 2-bromo-5-chloromagnesio-3-hexylthiophene with Ni(dppp)Cl_2 was a chain-growth process, the field of conjugated polymers has evolved tremendously. The mechanism was unraveled and the obtained information was used to further expand the scope of this discovery. Not only different poly(thiophene)s, but also other conjugated polymers were synthesized with enhanced control. However, the synthesis of random copolymers remained a challenge. This is caused by the fact that the association of the catalyst to the polymer chain is essential to obtain a chain-growth polymerization. This association is different for each monomer, while a random copolymerization requires a similar association. Otherwise, the catalyst can be trapped by more electron-rich monomer sequences. In order to solve this problem, different strategies are considered. A first option is the use of biaryl monomers, which indeed resulted in a successful synthesis thiophene-phenylene copolymers with a random incorporation of the phenylene monomer. Second, the possibility to use a different type of reaction mechanism – not depending on catalyst association but nevertheless resulting in a controlled polymerization – is investigated. The Pd(RuPhos)-protocol was found to be a suited alternative, and a series of thiophene-fluorene copolymers could be synthesized. Finally, making alterations to the monomer is also a possibility. If one could tune the properties of both monomers so that their electronic properties become more similar, it is possible to avoid the catalyst being trapped in a more electron-rich site. The possible alterations of the monomer are (i) changing the halogen or (ii) changing the organometallic function. Both can have an effect on the properties of the monomer, but do not affect the building block which is eventually incorporated in the polymer chain. While the effect of changing the halogen proved to be limited, a change in the organometallic function shows promising results.

Keywords: Conjugated polymers, controlled polymerization, random copolymers

RDPS-D-O-009 [Condensation Polymerization and Thermoset]

Design and Synthesis of Novel Waterborne Polyurethane Dispersions for High Performance Coating and Adhesive Applications

Emine Billur Sevinis[1], Ayse Durmus[1], Nesibe Ayse Dogan[1], Nihan Ongun[1], Yusuf Menceloglu[1], Serkan Unal[2]

[1]Material Science and Nanoengineering Program, Sabanci University, Istanbul 34956, Turkey
[2]Nanotechnology Research and Application Center, Sabanci University, Istanbul 34956, Turkey

Waterborne polyurethane dispersions (PUDs) are a unique class of environmentally friendly polymeric coatings and adhesives due to their versatile, multi-component polyurethane chemistry in combination with their very low or zero volatile organic content (VOC). In addition to typical monomers such as building blocks for soft and hard segments in a polyurethane backbone, the chemical structure and nature of the ionic monomer used as the internal emulsifying agent plays a critical role not only in the final dispersion stability but also final coating and film properties obtained from these dispersions. This presentation will focus on the synthesis and characterization of novel PUDs with a focus on the comparison of different synthetic approaches (namely the prepolymer and acetone methods) and the effect of key structural parameters such as nature of the soft segment, hard segment content and chemical structure of the ionic monomer on the colloidal properties of final dispersions and physical properties of cast films obtained from them. Dynamic Light Scattering (DLS), solution rheology, Differential Scanning Calorimetry (DSC), Dynamic Mechanical Analysis (DMA) and stress-strain tests were employed for the characterization of final dispersions and their films. Lastly, the presentation will highlight examples of novel PUDs with self-healing, thermally active or reactive nature, designed and synthesized for functional coating and adhesive adhesives that can be readily employed by various industries.

Keywords: waterborne, polyurethane, coating, adhesive

RDPS-D-O-010 [Condensation Polymerization and Thermoset]

Effects of nanocomponent on the curing reaction and reaction-induced phase separation of epoxy composites

Wenjun Gan, Weizhen Li, Zonglian Xia

College of Chemistry and Chemical Engineering, Shanghai University of Engineering Science, Shanghai, China

Different kinds of Nanocomponents were introduced into the modified epoxy system to study their effects on the curing reaction and reaction-induced phase separation (RIPS) of modified epoxy systems by differential scanning calorimetry (DSC), Fourier transform infrared spectroscopy (FT-IR), optical microscopy (OM), transmission electron microscopy (TEM), scanning electron microscopy (SEM), and time-resolved light scattering (TRLS). Firstly, Multi-walled carbon nanotubes (MWCNTs) were coated with silica by a sol-gel method to improve interfacial bonding and dispersion of nanotubes in epoxy matrix. The effects of silica-coated multi-walled carbon nanotube (MWCNT@SiO2) addition on the curing behavior of epoxy resin, and on the physical and thermomechanical properties of epoxy composites, were studied. FT-IR measurements of different blends at different curing times indicate that the curing reaction is accelerated with the presence of MWCNTs and increases with the content of MWCNT@SiO2. DSC results confirm that the value of activation energy decreases with the introduction of MWCNTs in the order of MWCNT@SiO2 < r-MWCNTs < epoxy. Alternatively, silica nanoparticles were employed in polyetherimide (PEI) modified epoxy system and the effect of silica nanoparticles on the reaction-induced phase separation in thermosetting/thermoplastic (epoxy/PEI) blend was investigated. The activity energy of the curing reaction is decreased by existence of the silica nanoparticles. Depending on the specific interaction between the silica surface and the epoxy, the silica nanoparticles selectively distribute in the epoxy-rich domain, leading to a slow down of the diffusion of the epoxy molecules. The coarsening mechanism is diffusion-controlled, which enhances the viscoelastic effect and produces a more dynamic asymmetric epoxy/PEI system. The final morphology stops at an earlier stage of inverted phase structure (isolated epoxy-rich droplet and PEI-rich matrix).

Keywords: nanocomponent, epoxy, curing reaction, phase separation

RDPS-D-O-011 [Condensation Polymerization and Thermoset]

Transparent epoxy thermoset toughened by modified star block copolymer

Shankar Prasad Khatiwada[1], Shankar Prasad Khatiwada[2], Gert Gert Heinrich[1], Rameshwar Adhikari[3]

[1]*Leibniz-Institut für Polymerforschung, Hohe Straße 6, D-01069, Dresden, Germany*
[2]*Research Center for Applied Science and Technology (RECAST), Tribhuvan University, Kirtipur, Kathmandu, Nepal*
[3]*Central Department of Chemistry, Tribhuvan University, Kirtipur, Kathmandu, Nepal*

Epoxy resins are extensively used in various applications due to their excellent chemical resistance and mechanical performance. However, their brittleness is a major disadvantage which significantly limits their uses. In this work, we investigate the ways of improving thermal and mechanical properties of epoxidized polystyrene-polybutadiene (SB) star block copolymers modified thermoset system formed by diglycidyl ether of bisphenol-A (DGEBA) and diethyltoluenediamine (DETDA) hardener. The epoxidation of the copolymer was performed using meta-chloroperoxybenzoic acid (MCPBA) and characterized by Fourier transform infrared (FTIR) spectroscopy while the degree of epoxidation was determined by proton magnetic resonance (1H-NMR) spectroscopy. The mechanical behavior of the blends was analyzed by Shore hardness, impact and tensile tests. The blends were further characterized by differential scanning calorometry (DSC), thermogravimetric analysis (TGA) and dynamic mechanical analysis (DMA). It was found that the use of modified SB starblock copolymer containing high amount of styrene helped to maintain the rigidity of the blends while increasing their toughness. In spite of macrophase-separation of the blends, optical transparency was maintained.

Keywords: Epoxy resin, blends, star block copolymer, electron microscopy, FTIR spectroscopy

RDPS-D-O-012 [Condensation Polymerization and Thermoset]

Investigation on Morphology and Mechanical Properties of Nanostructured Epoxy Resin Blends

Rajesh Pandit[1], Ralf Lach[2], Wolfgang Grellmann[2], Goerg H. Michler[3], Sven Henning[4], Jean Marc Saiter[5], Albrecht Berkessel[6], Rameshwar Adhikari[1]

[1]*Central Department of Chemistry, Tribhuvan University, Kritipur, Kathmandu, Nepal*
[2]*Polymer Service GmbH Merseburg, Eberhard-Leibnitz-Strasse 2, Merseburg, Germany*
[3]*Institute of Physic, Martin Luther University, Halle-Wittenberg, Halle/Saale, Germany*
[4]*Fraunhofer Institute for Mechanics of Materials, Halle/Saale, Germany*
[5]*AMME- LECAP International Laboratory, Université de Rouen,Rouen, France*
[6]*Department of Chemistry, University of Cologne, Greinstraße 4, 50939 Cologne, Germany*

Styrene/butadiene (SB) block copolymers are characterized by nanostructures of comprising hard and soft parts with various molecular architectures. In this work, SB block copolymers were used to prepare blends with theremosetting resin, the former being obtained by the epoxidation of the block copolymers having various chain architectures using *meta*-chloroperoxybenzoic acid (MCPBA) as an epoxidizing agent. The thermoset system formed by a diglycidyl ether of bisphenol-A (DGEBA) epoxy resin and methylene dianiline (MDA) hardener was chosen to ensure the miscibility of most of the epoxidized styrenic block copolymer with the matrix. The epoxidation of the block copolymers was characterized by Fourier transform infrared (FTIR) spectroscopy while the morphology and mechanical behavior of the blends were analyzed using electron microscopy and recording microindentation technique respectively. The block copolymers with epoxy-philic and epoxy-phobic blocks segregated to form quite regular nanostructures within the epoxy matrix that persisted after curing. In this way, it was possible to design the nanostructured blends having tailored mechanical properties.

Keywords: epoxy resin, nanostructure, block copolymer, mechanical properties

RDPS-D-O-013 [Condensation Polymerization and Thermoset]

Synthesis and characterization of functionalized Graphene-oxide/Polyimide nanocomposites as potential candidates for optoelectronics

Asma Iqbal[1], Humaira Masood Siddiqi[1], Toheed Akhter[2], O. Ok. Park[2], Sung Woo Lee[2]

[1]Department of Chemistry,Quaid-i-Azam University,Islamabad, 45320Pakistan
[2]Department of Chemical and Bio molecular engineering, Korea Advance Institute of Science and Technology,Daehak-ro, Yuseong-gu, Daejeon,305-701, Republic of Korea.

In recent years, the use of graphene-based nano-fillers for synthesis of a new class of advanced polymer composites with unique mechanical and thermal, and optical properties has been greatly investigated. In this work, in order to achieve better dispersion and charge transfer properties in PI-GO composites, graphene oxide nanosheets were successfully surface modified by implanting imide linkage by using a triphenylamine based tetra-amine and pyromellitic dianhydride. The modified GO was then incorporated into the polyimide matrix. The polyimide was, in turn, synthesized from tetra-amine monomer and pyromellitic dianhydride by conventional two-step condensation polymerization process. The development of organic functional groups on graphene sheets greatly enhanced the forces of attraction between GO and polyimide by formation of a charge transfer complex between polyimide and imide moieties on GO. This enhanced forces of attraction between two components suppressed the aggregation of GO. The resultant composites showed excellent thermal stability and photophysical properties. In comparison to neat polyimide, their GO based composites demonstrated intense blue emission, bathochromic shift, and excitingly good PL quantum yields. Furthermore, in case of nanocomposites higher HOMO levels, as compared to neat PI, were found as shown in Fig1. These properties prove that our prepared nanocomposites are potential candidates for future flexible optoelectronic devices. This method could be of great interest for adjusting the electrochemical and photoluminescence properties for optoelectronics devices.

Keywords: triphenylamine, Surface modified GO, PI/MGO composites, blue light emitting, optoelectronics.

PL-emission of Neat Polyimide and its composites

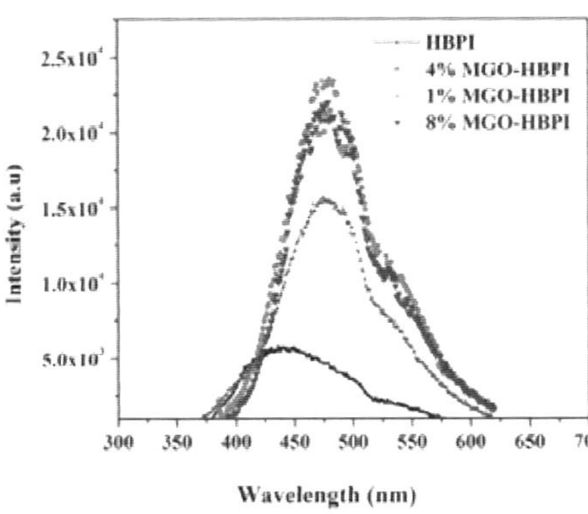

The blue light emission for neat polyimide and blue-green emission for PI-GO composites was observed with intense emission.

RDPS-D-O-014 [Condensation Polymerization and Thermoset]

Mussel-inspired, tannic acid based coatings modified with l-cysteine

Çağla Koşak Söz, Emel Yılgör, Iskender Yılgör

Chemistry Department, Koç University, Istanbul, Turkey

Mussels attach to solid surfaces in the sea through their adhesive (foot) proteins which are rich in 3,4-dihydroxy-l-phenylalanine (DOPA). The catechol unit in DOPA, which oxidizes to quinones is extremely reactive to both amines and thiols by Schiff base or Michael addition reactions, was found to be responsible for the adhesion capability of mussels [1]. In the past few years, researchers have been trying to mimic mussel foot proteins to fabricate multifunctional polydopamine based coatings [2]. Inspired by the mussel foot proteins, we prepared coatings using tannic acid (TA) and polyetheramines (PEA) instead of DOPA. TA, which is a polyphenol, is highly reactive towards amine or thiol groups. In our study, thin TA/PEA coatings were produced on glass slides. Effect of PEA structure and TA/PEA ratio on the properties of coatings were investigated. Coating surfaces were then modified by immobilization of l-cysteine through the Michael addition of thiol group to the phenolic sites. Surfaces obtained were characterized by ATR-IR spectroscopy, scanning electron microscopy (SEM), atomic force microscopy (AFM), white light interferometry (WLI) and advancing and receding water contact angle measurements. It was demonstrated that the glass slides coated with novel TA/PEA networks were further modified through the covalent attachment of zwitterionic l-cysteine aminoacid.

References:

[1] Lynge, ME. Nanoscale, 2011, 3, 4916–4928
[2] Messersmith, PB. Science, 2007, 318, 426- 430

Keywords: mussel, tannic acid, polyetheramine, coating, biomimetic

RDPS-D-O-015 [Condensation Polymerization and Thermoset]

Vinylogous Urethanes, a New Chemistry for High-Performance Vitrimers

Wim Denissen[1], Guadalupe Rivero[2], Renaud Nicolaÿ[3], Ludwik Leibler[3], Johan M. Winne[1], Filip E. Du Prez[1], Lucie Imbernon[1]

[1]Department of Organic and Macromolecular Chemistry, Ghent University, Belgium
[2]Instituto de Investigaciones en Ciencia y Tecnología de Materiales (INTEMA), Universidad Nacional de Mar del Plata, Mar del Plata, Argentina
[3]Matière Molle et Chimie, ESPCI ParisTech, Paris, France

Most covalent adaptable networks give highly interesting properties for material processing such as reshaping, recycling and repairing. Classical thermally reversible chemical cross-links allow for a heat-triggered switch between materials that behave as insoluble cured resins, and liquid thermoplastic materials, through a fully reversible sol–gel transition. In 2011, a new class of materials, coined vitrimers[1-2], was introduced, which extended the realm of adaptable organic polymer networks. Such materials have the remarkable property that they can be thermally processed in a liquid state without losing network integrity. This feature renders the materials processable like vitreous glass, without the need of moulds or precise temperature control. In this work[3], we present vitrimers based on transamination of vinylogous urethanes. These urethane-like chemical moieties emerge as interesting bonds for building vitrimers as they combine chemical robustness with rapid exchange kinetics. We show on model compounds that the exchange reactions is swift in a favorable temperature window and can even be fine-tuned by the presence of simple acids or base additives. Next, different kind of cross-linked materials, ranging from flexible elastomers to rigid thermosets, were made from bulk chemicals. These materials behave at room temperature as classical thermosets but become processable when heated as evidenced by stress-relaxation experiments. In addition, the networks are recyclable up to four times by consecutive grinding/ compression molding cycles without significant mechanical or chemical degradation.

(1) Montarnal, D.; Capelot, M.; Tournilhac, F.; Leibler, L. Science 2011, 334, 965.
(2) Denissen, W.; Winne, J. M.; Du Prez, F. E. Chem. Sci 2016, 7, 30.
(3) Denissen, W.; Rivero, G.; Nicolaÿ, R.; Leibler, L.; Winne, J. M.; Du Prez, F. E. Adv. Funct. Mater. 2015, 25, 2451.

Keywords: vitrimers, processable network, dynamic chemistry

TOC-figure

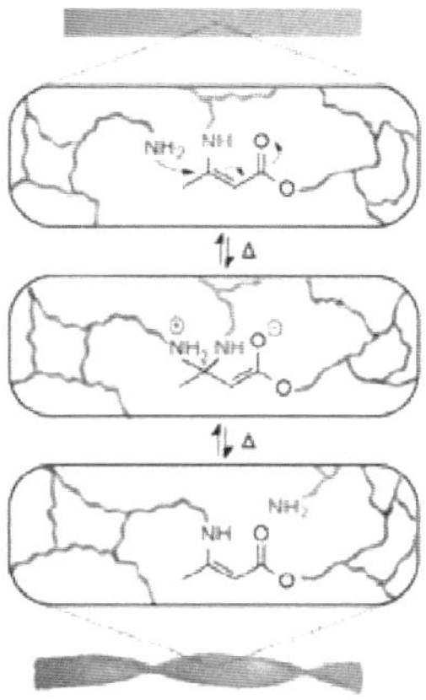

Polymer networks containing vinylogous urethanes that undergo swift exchange reactions upon heating, combines the mechanical properties of thermosets with a unique processability due to the thermally triggered exchange reactions.

RDPS-D-O-016 [Condensation Polymerization and Thermoset]

Solid-state polymerization and characterization of copolyamide based on adipic acid, 1,4-butanediamine, and 2,5-furan dicarboxylic acid

Yohana Kurnia Endah[1], Sang Hoon Han[3], Jae Hoon Kim[4], Nak Kyoon Kim[1], Woo Nyon Kim[2], Hong Shik Lee[1], Hyun Joo Lee[2]

[1]*University Of Science And Technology (UST), Daejeon, South Korea*
[2]*Korea Institute of Science And Technology (KIST), Seoul, South Korea*
[3]*Korea University, Seoul, South Korea*
[4]*Sungkyunkwan University, Suwon, South Korea*

2,5-Furandicarboxylic acid (FDCA) is a promising bio-based alternative material to terephthalic acid. In this paper, three types of poly(butylene adipamide) (PA-4,6) containing 10, 20, and 30 mol% of poly(butylene 2,5-furandicarboxylamide) (PA-4,F) were synthesized through the consecutive prepolymerization and solid-state polymerization (SSP). Incorporation of a 10 mol% PA-4,F component to PA-4,6 resulted in a slight increase of the intrinsic viscosity (IV) and glass transition temperature (Tg) after 12 h of SSP at 220 oC. By increasing the SSP temperature and reaction time, IV was proportionally increased. The highest IV value of 0.75 was obtained by 48 h of SSP at 240 oC. Whereas, an increase of the PA-4,F content to 20 mol% and 30 mol% gave rise to a decrease of the IV, Tg, and melting temperature, which interrupted the increase of SSP temperature. The thermal decomposition temperature of PA-4,F-incorporated polyamide was lower than that with PA-4,6 due to the lower thermal stability of the FDCA component.

Keywords: 2.5-furandicarboxylic acid (FDCA), polyamide, nylon, solid-state polymerization (SSP), PA-46, copolymer

RDPS-D-O-017 [Condensation Polymerization and Thermoset]

Room-temperature dual-curing processing of amine-acrylate-methacrylate mixtures

Ali Osman Konuray[1], Xavier Fernández Francos[1], Xavier Ramis[1], Gustavo González[1], Àngels Serra[2], Marco Sangermano[3]

[1]*Thermodynamics Laboratory, ETSEIB Universitat Politècnica de Catalunya, Barcelona, Spain*
[2]*Department of Analytical and Organic Chemistry, Universitat Rovira i Virgili, Tarragona, Spain*
[3]*Dept. of Material Science and Chemical Engineering, Politecnico di Torino, Torino, Italy*

Dual-curing is a controlled process methodology based on the combination of two sequential and compatible polymerization reactions that are activated by external stimuli such as UV light or heat. After the first curing stage, the stable intermediate material can be manipulated or transported safely before the activation of the second curing stage. The properties of the intermediate and final materials can be custom-tailored by changing the formulation composition and functionality of the monomers participating in the different curing stages, making it possible to meet a wide range of performance or processing requirements.

Two dual-curing schemes are proposed here which are based on the same curing steps: A self-limiting, click aza-Michael addition between acylates and amines followed by radical photopolymerization of excess acrylates [1] (first scheme) or methacrylates (second scheme). A significant advantage of these dual-curing schemes is that both curing stages are carried out at room or moderate temperature, reaching almost complete acrylate/methacrylate conversion. We investigate the effect of acrylate-amine stoichiometry, monomer structure and formulation composition on aza-Michael addition kinetics, viscoelastic properties and thermal degradation behavior. A wide range of physical properties can be achieved in both the intermediate and the final curing stages by selecting different monomers and by changing formulation composition. It is also shown how the aza-Michael reaction is selective towards acrylates, because of their better Michael acceptor characteristics in comparison with methacrylates [2]. This produces different network structure in the intermediate stage after the aza-Michael reaction using excess acrylates or methacrylates, and so the properties differ as well. The flexibility of this curing system makes possible the applicability of these products to various industrial needs.

1. Gonzalez G, Fernandez-Francos X, Serra A, Sangermano M, and Ramis X. Polymer Chemistry 2015;6:6987-6997.
2. Mather BD, Viswanathan K, Miller KM, and Long TE. Progress in Polymer Science 2006;31:487-531.

Keywords: acrylate, methacrylate, aza-Michael, dual curing, click reaction, radical polymerization

Selectivity of aza-Michael addition towards acrylates as shown by FTIR data

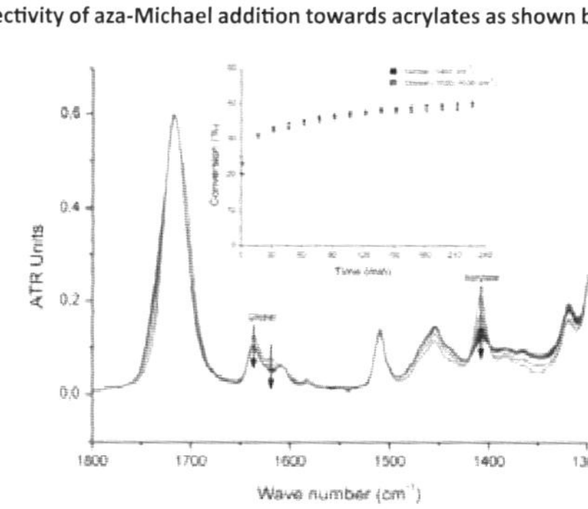

Evolution of FTIR spectra for 50MA_50Ac (i.e. a formulation containing acrylate and methacrylate groups in equal mass percentages) taken during aza-Michael addition for 4 hours. Spectra were taken every 15 minutes. Curves are normalized w.r.t. the carbonyl ester band at 1720 cm-1. Acrylate selectivity of aza-Michael addition is evident from the good fit between global conversions determined using two different FTIR absorption bands.

Conversion of acrylate and methacrylate groups and glass transition temperatures (measured with DSC) of the different products after 24 h of aza-Michael reaction (Stage 1) and subsequent photocuring (Stage 2).

Material	Acrylate conversion (%) (Stage 1)	Tg (ºC) (Stage 1)	Acrylate conversion (%) (Stage 2)	Global conversion (%) (Stage 2)	Tg (ºC) (Stage 2)
0MA_100Ac	86	-22.3	94	92	-9.2
25MA_75Ac	83	-37.5	97	95	8.8
50MA_50Ac	80	-46.4	99	89	26.3
75MA_25Ac	66	-67.4	89	84	52

This table shows the conversion and the glass-transition temperatures at the intermediate (i.e. after aza-Michael addition) and final stages (i.e. after photocuring). As can be seen, all materials achieve practically complete conversion at the final stage except 75MA_25Ac. In fact, this formulation also has the lowest conversion at the intermediate stage. As explained previously, the lower concentration of acrylates in this formulation slows down aza-Michael reaction rate and leads to lower acrylate conversions in the intermediate stage. Furthermore, its high methacrylate content results in a tighter polymethacrylate network which produces mobility restrictions for polymerization and inhibits complete conversion at the final stage. This sample was post-cured thermally to ensure complete conversion, however the improvement in Tg was practically insignificant. This shows the efficacy of our dual curing process. For each formulation, the final Tg is significantly higher than the intermediate Tg owing to the photo-polymerization of unreacted acrylate and methacrylate groups remaining at the intermediate stage, leading to a more densely crosslinked network structure. Finally, it can be noticed that the higher the dimethacrylate content, the higher is the final Tg, regardless of the intermediate Tg, indicating the dominant contribution of the polymethacrylate network in the viscoelastic properties of the materials.

RRB-O-002 [Renewable Resources and Biopolymers]

Process control in the production of engineered palm-based fibreboard

Khairiah Haji Badri[1], Azizah Baharum[2], Azlin Mohmad Azman[1]

[1]*School of Chemical Sciences and Food Technology, Faculty of Science and Technology, Universiti Kebangsaan Malaysia, 43600 Bangi, Selangor, Malaysia*

[2]*Polymer Research Center Faculty of Science and Technology, Universiti Kebangsaan Malaysia, 43600 Bangi, Selangor, Malaysia*

An investigation on the possibility of processing untreated empty fruit bunch (EFB) fibre into medium-density fibreboard (MDF) was conducted. A novel palm-based polyurethane prepolymer (pPU) was used as a binder. It was comprised of palm kernel oil based monoester polyol (PKO-p), 4,4-diphenylmethane diisocyanate (MDI) and polyethylene glycol (PEG200 molecular weight of 200 Da) in acetone. In considering the effect of state of matters, an optimization on fibre size and processing parameters were carried out. Three different fibre sizes were used; that were 250 to 500 µm (MDF S1), 500 to 1000 µm (MDF S2) and 1000 to 2000 µm (MDF S3). An improvement in the mechanical and thermal properties was observed as the fibre size decreased. MDF S1 showed the optimum flexural strength and modulus at 25.7 MPa and 1307 MPa respectively. A higher glass transition temperature (Tg at 77°C) was observed in MDF S1 compared to MDF S3 (Tg at 67°C). The morphological observation exhibited a much better matrix encapsulation in the form of interfacial adhesion between EFB and pPU occurred in MDF S1. The processing temperature was varied at 35°C (MDF T1), 45°C (MDF T2) and 55°C (MDF T3) with loading ratio of 65:35 (EFB fibre: pPU). Flexural test showed MDF T3 had the optimum strength and modulus at 31.2 MPa and 1405 MPa respectively. Bomb calorimety had supported the result with the least gross heat of 20.9 kJ/kg, showing MDF T3 is more stable as compared to MDF T1 and MDF T2. These findings were supported by the Fourier Transform Infrared Spectroscopy (FTIR) analysis exhibiting the formation of urethane backbone in the binder system. The engineered MDF was produced with excellent strength and thermal resistivity by using a smaller size EFB of 250 to 500 µm (MDF S1), compressed at 55°C (MDF T3) for 9 min.

Keywords: Medium-density fibreboard, Polyurethane prepolymer, Process control, Untreated empty fruit bunch

RRB-O-003 [Renewable Resources and Biopolymers]

Properties and Potential Applications of Modified Liquid Natural Rubber

Siti Fairus Mohd Yusoff, Ibrahim Abdullah

School of Chemical Sciences and Food Technology, Faculty of Science and Technology, University Kebangsaan Malaysia, 43600 UKM Bangi, Selangor, Malaysia

Natural rubber (NR) is a renewable material harvested from the sap of rubber trees, Hevea brasiliensis, and contains polyisoprene as its major component.[1] NR has been widely used in automobile and adhesive industries, owing to its superior tensile strength, excellent elasticity, high tear resistance, low permanent set, and waterproof nature. However, due to the large amount of double bonds in its chain structure, NR is susceptible to oxidative and thermal degradation, ozonolysis, and extreme weathering, thus limiting its applicability. Typically, NR has an average weight molecular weight (Mw) of ~2.5 × 10^6 that must be reduced during manufacturing processes. Low-molecular-weight NR can be easily modified into various useful products. One example is liquid natural rubber (LNR), which is a reduced form of NR with short polymeric chains and molecular weight lower than 10^5.[2] The short polymeric chain of LNR is more advantageous than NR because it makes chemical modifications possible and thus expands the applicability of LNR in various fields. In order to improve the thermal stability of LNR, chemical modification can be performed to produce highly saturated LNR. In this research, we have carried out number of modification onto LNR including hydrogenation, oxidation, hydroxylation and fluorination. The desired product of the modification reaction should withstand high temperature conditions and therefore could be used in various applications, such as in rubber blending and hydrophobic surfaces.[3]

References:

1. Piya-Areetham, P.; Rempel, G. L.; Prasassarakich, P. Polym. Degrad. Stab. 2014, 102, 112. DOI: 10.1016/j.polymdegradstab.2014.01.032
2. Daik, R.; Bidol, S.; Abdullah, I. MPJ. 2007, 2(1), 29.
3. Kongparakul, S.; Ng, F. T. T.; Rempel, G. L. Appl. Catal., A. 2011, 405, 129. DOI: 10.1016/j.apcata.2011.07.039

Keywords: natural rubber, modification, thermal, hydrophobic

RRB-O-004 [Renewable Resources and Biopolymers]

Renewable super-strong plastics derived from exotic amino acids

Tatsuo Kaneko, Maiko Kaneko Okajima, Seiji Tateyama

Energy and Environment Area, Japan Advanced Institute of Science and Technology, Ishikawa, Japan

Bio-based polymers obtained by a polymerization of biomolecules are indispensable for establishment of green sustainable society where bioplastics would be applied in various fields. Conventional bioplastics were composed of aliphatic polymers like poly(lactic acid)s having not very high mechanical strength. In order to solve this problem, we have prepared ultra-strong bioplastics are generated from fermented microbial monomers. An aromatic amino acid, 4-aminocinnamic acid, was prepared from a biomass using recombinant bacteriaxoti, and quantitatively photodimerized, and diacid- and diamino-monomers that were both characterized by a rigid α-truxillate structure were generated. These two monomers were polycondensed to create the polyamides with a phenylene cyclobutane repeating backbone such as poly*(4,4'-diyl-α-truxillic acid dimethylester) 4,4'-diacetamido-α-truxillamide* which was processed into amorphous fibers and plastic films having high transparency. The tensile strength, Young's modulus and transparency of the polyamide were 356±33 MPa, 11.4±1.6 GPa, and 93%, respectively, which were all higher than those of a widely-used transparent plastics. Especially mechanical strength of the fiber is 407 MPa at maximum higher than those borosilicate glasses, presumably due to the molecular spring function possibly showing tautomerization in the phenylene cyclobutanyl backbone, in spite of non-crystalline structure. In addition, the polyamide was efficiently degraded into monomer which was successfully re-polymerized in high yield. Thus we developed high-performance and recyclable super-engineering bioplastics which was first prepared and would be widely applied in materials of automobile parts, medical instruments, and electronic devices.

Keywords: Bioplastics, aromatic polymers, photoreaction, amino acids, microorganisms

renewable transparent film and fibers

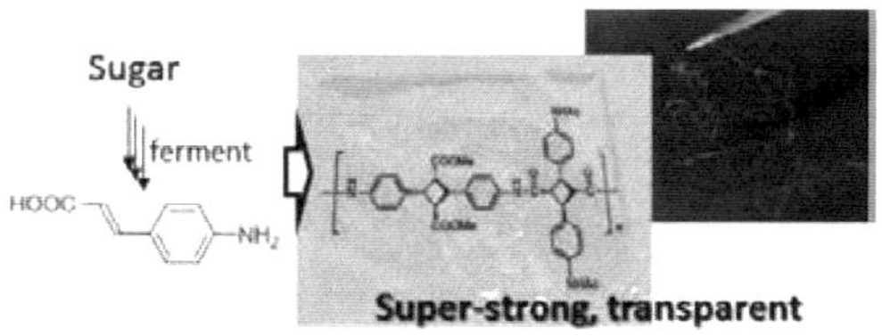

RRB-O-005 [Renewable Resources and Biopolymers]

Oxygen-generating photocrosslinkable hydrogel for cardiac progenitor cells survival under hypoxic conditions

Neslihan Alemdar[1], Jeroen Leijten[2], Gulden Camci Unal[3], Jesper Hjortnaes[4], Joao Ribas[3], Arghya Paul[3], Pooria Mostafalu[5], Akhilesh Gaharwar[3], Yiling Qiu[6], Sameer Sonkusale[5], Ronglih Liao[6], Ali Khademhosseini[3]

[1]*Department of Chemical Engineering, Marmara University, Istanbul 34722, Turkey*
[2]*Department of Developmental BioEngineering, MIRA Institute for Biomedical Technology and Technical Medicine, University of Twente, Enschede, The Netherlands*
[3]*Harvard-MIT Division of Health Sciences and Technology, Massachusetts Institute of Technology, 77 Massachusetts Avenue, Cambridge, MA 02139, USA*
[4]*Department of Cardiothoracic Surgery, University Medical Center Utrecht, Utrecht, Netherlands*
[5]*Department of Electrical and Computer and Engineering, Tufts University, Medford MA, 02115, USA*
[6]*Cardiac Muscle Research Laboratory, Cardiovascular Division, Department of Medicine, Brigham and Women's Hospital, Harvard Medical School, Boston, MA 02115, USA*

Providing sufficient amount of oxygen to cells is a critical issue, especially in the case of myocardial infraction, in which hypoxia induces ischemia of the myocardium, leading to cell death and tissue necrosis due to the lack of adequate oxygen. There is a vital need to develop and fabricate oxygen-generating biomaterials to decrease hypoxia-induced cell death for engineered cardiac tissues. Here, we report oxygen-generating hydrogels from methacrylated gelatin (GelMA) and calcium peroxide (CPO) to enhance the proliferation of cardiac side population cells (CSPs), one of the cardiac resident stem cell types, and minimized hypoxia-induced cell death. A range of oxygen-generation hydrogel was fabricated from CPO (0 to 3 wt%) and GelMA using photocrosslinking process. A sustainable release of oxygen was observed from CPO-GelMA hydrogel over a period of 5 days under hypoxic conditions (1% O2). The incorporation of CPO into the GelMA also increased the pore structure and swelling degree of the hydrogel network. A uniform distribution of CPO within GelMA matrix was confirmed by tracing the calcium content using Alizarin Red S. To evaluate the pro-survival role of the oxygen-generating hydrogel, CSP cells were encapsulated within CPO-GelMA hydrogel and cultured under hypoxic conditions (1% O2), mimicking the physiological conditions of myocardial ischemia. The results showed that GelMA with 2 and 3 wt% CPO significantly enhanced cell viability and minimized hypoxia-induced cell death compared to GelMA-only as evidenced by the increase in metabolic activity, proliferation capacity, cell viability and the decrease in LDH level which is a marker for cell damage. Overall, the results demonstrate that CPO-based oxygen-generating hydrogel could be efficiently used to provide sustainable and sufficient oxygen to cardiac cells under hypoxic conditions. Therefore, CPO-GelMA hydrogel can be a potential treatment for myocardial infarction.

Keywords: Hydrogel, biopolymer, hypoxic condition

RRB-O-006 [Renewable Resources and Biopolymers]

Multifunctional nanocellulose / paper composites

Mona Hassan Abdel Rehim[1], Magda Alsamahy[1], Salah Abuyousef[1], Maysa Mohram[2]

[1]*Packing and Packaging Department Materials, Division of chemical Industries Research, National Research Centre*
[2]*Department of Microbial Chemistry, Genetic Engineering and Biotechnology Division*

Cellulose, the major component of paper, can be obtained from plants. However, agricultural crop residues, as a source for cellulose, represent one of the most abundant organic materials all over the world with annual renewability especially for countries that do not have forests. In the past decade, research on nanostructures of cellulose has increased dramatically due to the potential applications in electronics, biosensors, and energy storage devices. Utilization of Nanocellulose (NC) in the paper modification, by either coating or addition, offers new favorable properties for the modified paper.

Modified paper sheets have been prepared from rice straw pulp and different ratios of Nanocellulose (NC). The nanomaterial was isolated from rice straw pulp by chemical method and characterized before addition. Moreover, effect of Chitosan, reduced graphene oxide on the modified paper was also investigated. Characterization of the obtained paper composites was carried out by FTIR, SEM, TGA and the gas barrier properties was also studied. The addition of the new materials to the paper sheets led to higher barrier and mechanical properties and eclectically conductive composite as well. The obtained results confirmed the utility of the modified paper sheets in hygienic, packaging and electronic devices.

Keywords: Agricultural residues, nanocellulose, barrier properties, electrical conductivity

RRB-O-008 [Renewable Resources and Biopolymers]

ADMET and TAD chemistry: A Sustainable Alliance

Laetitia Vlaminck, Kevin De Bruycker, Oğuz Türünç, Filip E Du Prez, Andrea Hufendiek

Department of Organic and Macromolecular Chemistry, Polymer Chemistry Research Group, Ghent University; Krijgslaan 281 S4-bis, B-9000 Ghent, Belgium

Acyclic diene metathesis (ADMET) has been known since the nineties, but the impact of the resulting polymers on the field of green chemistry, was certainly not anticipated. To open ways to tailor-made materials in the future, post-functionalisation methods of the available unsaturations are necessary. From '92 until now, some functionalisation methods have been described, with varying degrees of success.[1-3] Harsh conditions, multiple equivalents, long reaction times and/or low yields make the proposed reactions less interesting for industrial applications. A robust, additive-free, atom efficient and orthogonal procedure for the functionalisation of ADMET derived polymers is thus needed. (Figure 1)

We propose the use of triazolinediones as a functionalisation method for ADMET polymers. 1,2,4-Triazoline-3,5-dione (TAD) (Figure 2) exhibits a high reactivity toward both conjugated dienes and isolated alkenes in a Diels-Alder or an Alder-ene reaction, respectively. [4,5] These reactions reliably give a specific reaction outcome with olefinic reaction partners in almost any setting. On top of this, the reactions can be monitored visually, as all triazolinediones show a strong absorption in the visible light spectrum.

We demonstrated the successful post-functionalisation of ADMET derived polymers utilizing TAD chemistry. Post- functionalisation was performed in a quick and quantitative manner, on bio-based ADMET polymers with several polymer backbones and with a variety of TAD compounds. It was shown that the glass transition temperature of the polymers could be tuned by the degree of functionalisation. In a last step, the functionalized polymers were cross-linked by a bivalent TAD compound to obtain functionalized thermoset materials.

1. J. A. van Hensbergen *et al.*, Polym. Chem., 2014, 5, 6225-6235.
2. F. N. Führer *et al.*, Macromol. Chem. Phys., 2014, 215, 2268-2273.
3. M. Unverferth *et al.*, Polymer, 2014, 55, 5571-5575.
4. S. Billiet *et al.*, Nat. Chem., 2014, 6, 815-821.
5. K. De Bruycker *et al.*, Chem. Rev., 2016.

Keywords: ADMET, Triazolinediones, post-polymerization functionalization, crosslinking

Figure 1

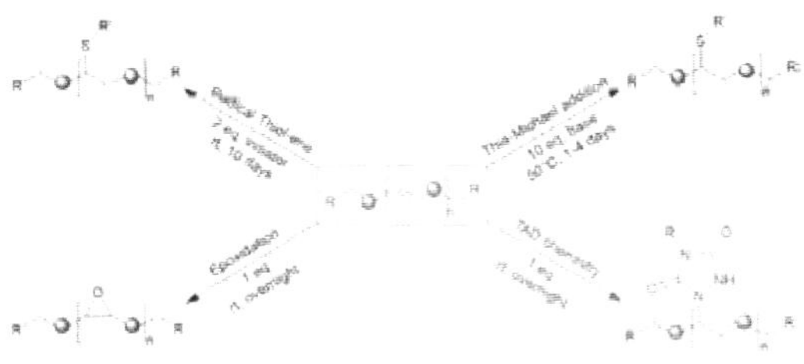

Overview of the possible post-polymerisation functionalisation methods

Figure 2

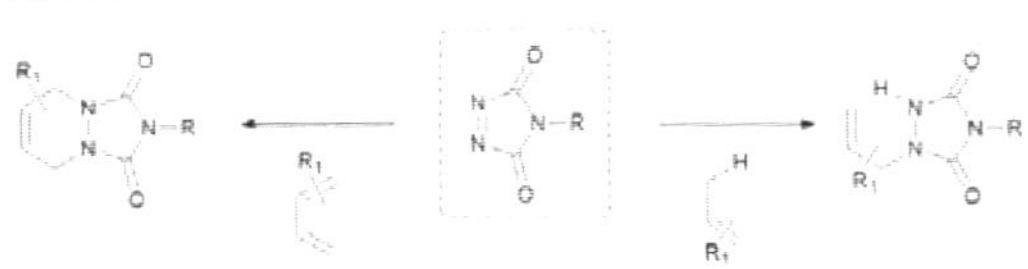

Diels-Alder (left) and Alder-ene (right) reaction of triazolinedione with dienes and isolated alkenes, respectively.

RRB-O-009 [Renewable Resources and Biopolymers]

Characterization of Sustainable Phenolic Resins Synthesized with Renewable Bioresources

Francisco Cardona, Thariq Mohamed Bin Hameed Sultan

University Putra Malaysia, UPM

Novel resole phenolic resins (PF) with improved fracture toughness and flexibility properties were synthesized and evaluated. A first modification consisted in the copolymerization of Phenol with a natural renewable component (cardanol from cashew-nut-shell-liquid, CNSL) during the synthesis of the liquid resole phenolic PF resins (CPF). An increases in the content of cardanol resulted in a proportional increases in the flexural strength and in the fracture toughness together with a decreases in the flexural modulus of the cured CPF/PF blended phenolic resins. It was observed a further increased plasticizing and toughening effect by the blending of the CPF/PF resins with propylene glycol (PG). Additionally, dissolved tannin (condensed wattle tannin) was incorporated into the CPF resins and the fracture toughness and flexibility values of the cured tannin-CPF resins were found to be proportional to and increasing with the tannin content. However, glass transition temperature (Tg), flexural stress, and flexural modulus values of the CPF resins decreased with the tannin content in the phenolic resins.

Keywords: Phenolic PF, Cardanol, Condensed tannin, DMA, Flexural properties, Fracture toughness

Figure 1.

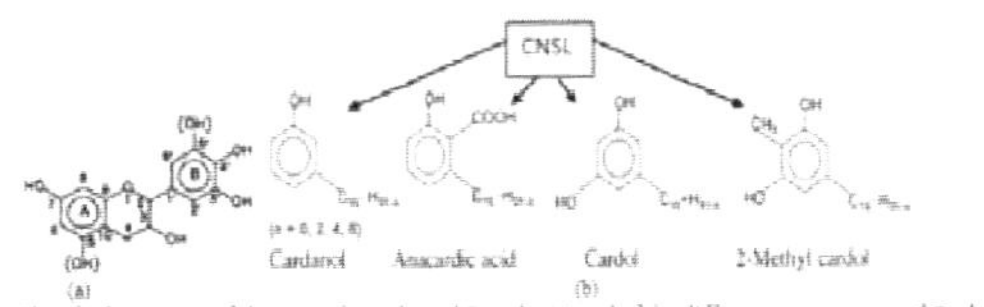

Chemical structure of the natural condensed Tannin (a) and of the different components of Cashew Nut Shell Liquid (CNSL) (b)

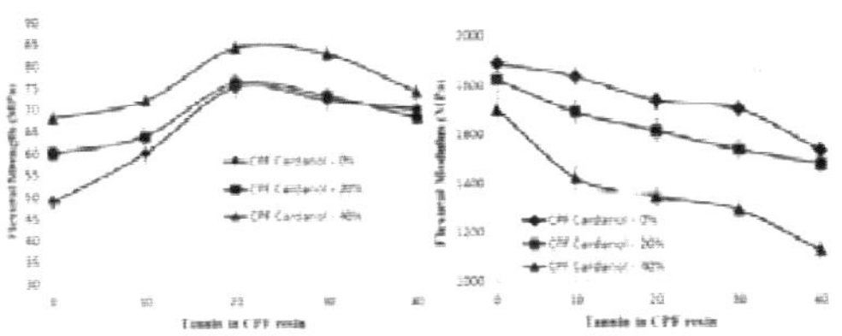

Flexural properties of the phenolic CPF resins with different cardanol content and vs the amount of dissolved tannin blended with the CPF resins. Prior to the flexural test the testing samples were cured at RT for 12 h followed by a post-curing at 80 °C for 4h.

Chemical composition of condensed wattle tannin (a) and of the components of Cashew Nut Shell Liquid (CNSL). Flexural properties of the blended CPF/PF phenolic resins with different cardanol content.

RRB-O-010 [Renewable Resources and Biopolymers]

Sustainable Benzoxazine and Sulphur copolymers as Li-S Battery Cathode Materials

Swapnil Shukla[1], Arnab Ghosh[2], Sagar Mitra[2], Bimlesh Lochab[1]

[1]Department of Chemistry, School of Natural Sciences, Shiv Nadar University, Uttar Pradesh, India
[2]Electrochemical Energy Storage Laboratory, Department of Energy Science and Engineering, Indian Institute of Technology, Bombay, Powai, Mumbai, India

Petroleum based sources are diminishing at a rapid rate which has given an urgent impetus for the exploration of alternate renewable sources as potential replacements. Same holds good for the rapidly expanding class of Polybenzoxazines (PBz), thermoset resins with superior properties viz. near zero shrinkage, high char yield, facile synthetic methodology, molecular design flexibility to name a few. In this paper we have explored the incorporation of sustainable benzoxazines as potential co-monomers for the entrapment of elemental sulphur as a potential Li-S battery cathodic material. Elemental Sulphur (S) annually (~70 M.T.) generated as a waste from the petroleum resources have been incorporated in the polybenzoxazine framework with high loadings of >90%. The key advantages include the utilization of renewable and waste materials as the feed stock available abundantly and a sustainable solventless, one pot synthetic route for the cathode preparation. In addition, the conventional drawbacks related to sulphur dissolution during the cycling process have also been addressed. Hence, the present work aims to address the drawbacks of Li-S battery system via a sustainable and scalable route.

Keywords: sustainable, renewable, polybenzoxazine, elemental sulphur, Li-S battery

RRB-O-011 [Renewable Resources and Biopolymers]

High molecular weight polyethylene furanoate (PEF) synthesis via ring-opening polymerization

Jan-Georg Rosenboom, Peter Fleckenstein, Giuseppe Storti, Massimo Morbidelli

Institute of Chemical and Bioengineering, ETH Zurich, Zurich, Switzerland

Polyethylene furanoate (PEF), the promising 100% renewables-based substitute for fossil-based polyethylene terephthalate (PET), has successfully been synthesized by polycondensation, where conversion rate and molecular weight are limited by required condensation byproduct removal. On the other hand, ring-opening polymerization (ROP) for PEF could become a faster synthetic route through a pseudo-living process. While recent efforts have resulted in only moderate molecular weights and yields, we present essential aspects of ROP to deliver sufficiently high absolute molecular weights and yields for commercial PEF applications such as bottles, textiles, medical grafts, etc.

Bio-based cyclic PEF monomers (cyOEF) have been synthesized through different routes such as cyclodepolymerization of short PEF oligomers in 2-methylnaphthalene at yields >70% with a ring-size distribution from dimer to heptamer. The residual 30% linears can be recycled to increase overall yield. Our search for better and "greener" solvents is ongoing, aided by solubility predictions from COSMO-RS that show several aromatics to possess promising solvent power and selectivities. Purification from residual linears via silica gel adsorption to yield >99% cycles is essential for ROP to deliver 1) sufficiently high molecular weights, 2) reproducible reaction control and 3) whiter products.

Cyclics melt up to 370°C, but can be plasticized with tetraglyme and undergo ROP catalyzed by tin-based initiators such as FDA-approved tin octoate, to form long PEF chains at yields of >95% and Mw > 60,000 g/mol, as required for PET bottles, within less than 20 minutes. Absolute molecular weight measurement is essential to avoid over-estimation, which we perform with laser scattering (MALS) and diffusion (DOSY) NMR. Compared with PET, ROP-based PEF shows a 10°C higher glass transition temperature, a 50°C lower melting point, and at least 5x higher oxygen diffusion barrier, which can be explained with a higher molecular rotational energy barrier derived from molecular dynamics simulations.

Keywords: polyethylene furanoate (PEF), renewable resources, biobased polymers, ring-opening polymerization, process development, 'green' bottle

High conversion high molecular weight synthesis of PEF via ROP

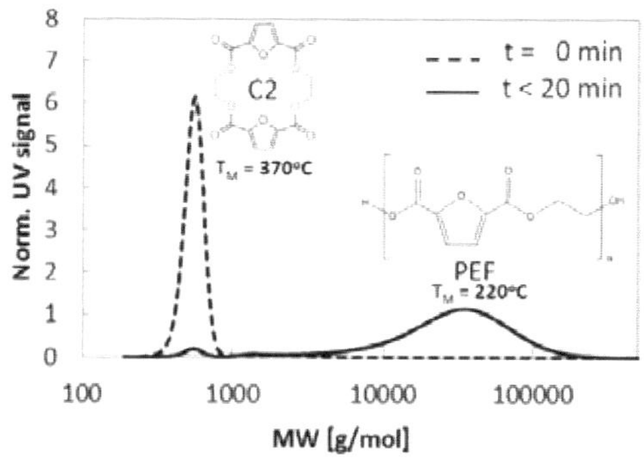

GPC / SEC trace of cyclic dimeric monomer (C2, dashed) undergoing ring-opening polymerization to form PEF polymer (continuous line)

Absolute molecular weights from MALS and DOSY

Sample	Mw (MALS) [g/mol]	Mw (DOSY) [g/mol]
PET standard 1	9900	10240
PET standard 2	37770	38540
PET standard 3	73230	77240
PET bottle	60900	61750
PEF product 1	19420	17670
PEF product 2	39910	41480
PEF product 3	72420	72270

Absolute weight-average molecular weights Mw derived for PET standards, PET bottles and synthesized PEF samples, measured with multi-angle light scattering (MALS) and diffusion (DOSY) NMR

Material Properties of PET and PEF

Material Property	PET	PEF
T_G [°C]	73	85
T_M (polymer) [°C]	260	215
T_M (cyclic dimer) [°C]	225	370
T_M (cyclic trimer) [°C]	320	270
O2 gas diffusion [mL/m^2/d]	1950	392
Rotational barrier [kJ/mol]	30	86

Comparison of essential material properties of renewables-based PEF from ROP vs fossil-based PET. T_G is glass temperature, T_M is melting temperature.

RRB-O-012 [Renewable Resources and Biopolymers]

Evaluation of Thermal, Adsorption and Microstructural Properties of Corn Starch Based Cross-linked Composite Films

Ilknur Gönenç[1], Ferhunde Us[2]

[1]*Directorate General of Food and Control, Ministry of Food, Agriculture and Livestock, Ankara, Turkey*
[2]*Department of Food Engineering, Hacettepe University, Ankara, Turkey*

In this study corn starch based and cross-linked biodegradable composite films plasticized were produced and effects of cross-linking agent and the addition of polymers that have good film forming ability such as methylcellulose, carboxymethylcellulose and polyvinyl alcohol on the thermal, adsorption and microstructural properties of corn starch based films plasticized with glycerol (GLY) or polyethylene glycol (PEG) were investigated. One glass transition temperature was obtained for the GLY plasticized films however dual glass transitions were obtained for PEG plasticized films from Differential Scanning Calorimetry analysis. Besides, glass transition temperatures increased with the addition of cross-linking agent. The moisture adsorption capacity of the films increased in the presence of GLY and decreased with crosslinking reaction. GAB model fitted well to the moisture adsorption isotherm data of the films and Mathematical fitting of adsorption data to GAB model gave the monolayer values an opportunity to evaluate the intensity of the interaction between the water and polymer matrix. The microstructural properties of the films were analyzed by X-ray diffraction, polarized microscopy, scanning electron microscopy (SEM) and methods. X-ray diffraction analysis indicated that the degree of crystallinity of the films decreased in the presence of cross-linking agent. The crystalline fractions were qualified with polarized light microscopy. Among the film samples only corn starch/methylcellulose based films displayed a double refraction pattern under polarized light. SEM observations showed the surface homogeneity of composite films. The surfaces of PEG plasticized films were found less homogeneous than GLY plasticized ones, and this might be due to the phase separation in the polymer matrix. Consequently, it was concluded that a significant improvement can be achieved in the physical and microstructural properties of starch-based films by blending and cross-linking.

Keywords: Biodegradable films, corn starch, cross-linking, differential scanning calorimetry, x-ray diffraction, scanning electron microscopy

RRB-O-013 [Renewable Resources and Biopolymers]

Novel thermosetting resins based on cardanol

Erde Can[1], Emre Kinaci[2], Giuseppe R. Palmese[2]

[1]*Department of Chemical Engineering, Yeditepe University, İstanbul, Turkey*
[2]*Department of Chemical and Biological Engineering, Drexel University, Philadelphia, PA, USA*

With the petroleum production facing exhaustion day by day, the use of renewable plant based feed-stocks for making polymers offer both economic and environmental advantages. One plant based resource for making polymers is the Cashew nutshell liquid (CNSL) which is a natural resin found in the honeycomb structure of the cashew nutshell. The main component of thermally treated CNSL is cardanol which is a phenol derivative mainly having a meta substutient of a C15 unsaturated hydrocarbon chain with one to three double bonds. In this study, thermosetting resins were prepared from both novel vinyl ester end epoxy monomers derived from cardanol. In the first part of the study, two commercially available CNSL resins, NC 514 with approximately two epoxies per molecule and LITE 2513HP with approximately one epoxy per molecule were methacrylated and the methacrylated NC-514 resin (NC514VE) was used as the thermosetting resin and the mono-functional methacrylated LITE monomer (LITEVE) was used as the reactive diluent in both conventional DGEBA vinyl ester resin (DGEBAVE) and in NC514VE resin at changing concentrations. Resin formulations of NC514VE using styrene as the reactive diluent were also prepared. The resins were cured in the presence of Trigonox A initiator and cobalt napthanate catalyst at room temperature and then post-cured at a higher temperature. In the second part of the study, the LITE 2513HP resin was subjected to epoxidation reaction from the unsaturation sites on the side chain in the presence of formic acid and hydrogen peroxide. The synthesized polyepoxy cardanol resin was cured with both petroleum based curing agents and cardanol based epoxy hardeners. Also for comparison, the cardanol based di-epoxy resin, NC514 and the petroleum based di-epoxy resin, DGEBA (EPON828) were thermally cured with the epoxy hardeners at stoichiometric ratios. Both the liquid resin and cured resin properties were determined and evaluated.

Keywords: cardanol, epoxy, vinyl ester, thermosetting resin

Cardanol based epoxy and vinyl ester resins synthesized in this study

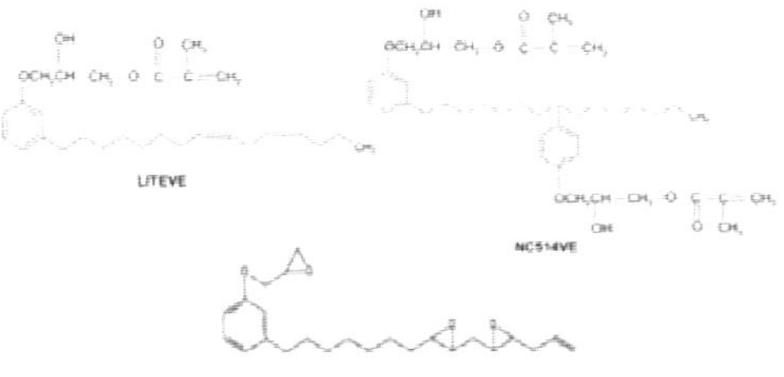

Side chain epoxidized LITE

RRB-O-014 [Renewable Resources and Biopolymers]

Syntheses of renewable polyamides with controlled optical activity from itaconic acid and amino acid

Mohammad Asif Ali, Seiji Tateyama, Tatsuo Kaneko

Energy and environment Area, japan advanced institute of science and technology, Ishikawa, Japan

The development of high-performance bio-based polyamide is produced from itaconic acid and amino acid, because engineering plastics from renewable resources are significant for the establishment of a sustainable society. Based on a previsouly-reported strategy of molecular design to produce the pyrrolidone ring composed of amine compound and itaconic acid, we here focus on bioavailable N-substituted heterocyclic diacid which is produced after the Michael addition in between the itaconic acid and amino acid, for the purpose of bio-based polyamide monomer syntheses. Then we prepare new renewable biopolyamides via salt mooner whichthermally convert into polyamide structure through melt condensation process. The polyamides with number-average molecular weights above 34 000 g/mol (polydispersity index: 2.0-3.4 (SEC)) are obtained and Tg values over 92 oC, which are higher than conventional polyamides (around 57 oC) were observed. The Young's modulus and mechanical strength of these polyamides also show higher values of 430-2800 MPa and 90-165 MPa, respectively. Furthermore, optical activity control of the diacids obtained from amino acid with D and/or L form enhances the thermal and mechanical properties under a certain molecular design due to cumulative effect of the crystallinity with new structure-property relationship. The polyamides became soluble in water by ring-opening reaction of the pyrrolidone, which led to environmental-corrosion by landfill or ultraviolet-irradiation.

Keywords: Itaconic acid, Amino acid, Nylon salt, Polyamides

RRB-O-015 [Renewable Resources and Biopolymers]

Polyglycolic Acid and Its Copolymers from Cationic Polymerization of Formaldehyde and Carbon Monoxide

Ersen Göktürk[1], Stephen Albert Miller[2]

[1]Department of Chemistry, Mustafa Kemal University, Hatay, Turkey
[2]Department of Chemistry, University of Florida, Gainesville/Florida, United States

Polyglycolic acid (PGA) is the simplest linear biodegradable polymer traditionally synthesized through anionic polymerization of glycolide with mostly using tin octanoate catalyst. Although this route has been the best option in terms of high yields and molecular weights, the need to use expensive glycolide monomer and toxic tin catalysts limit the industrial scale production of PGA. We present a new approach to synthesize PGA from cationic polymerization of formaldehyde from trioxane and carbon monoxide (CO). This method constitutes a cost-effective and efficient path for the synthesis of PGA and allows perfectly alternating copolymerization of formaldehyde and CO with 92% of yield. To improve on the physical properties of PGA, such as melting temperature and appearance, the method is used to make copolymers of PGA where some epoxide comonomers are included into the polymerization mixture to yield polyester-ether thermoplastics. By controlling the comonomer feed ratios and the polymerization temperatures, high quality PGA copolymers can be prepared. One of the PGA-based copolymers exhibited thermal properties similar to those of polypropylene, and we are suggesting that PGA copolymers may be replaced to this polyolefin in some applications.

Keywords: polyglycolic acid, formaldehyde, carbon monoxide, cationic polymerization

RRB-O-016 [Renewable Resources and Biopolymers]

Synergistic effect of nano reinforcement and cross-linking towards polyvinyl alcohol (PVA) properties

Amit Kumar Sonker[1], Naveen Tiwari[1], Rajaram Krishna Nagarale[2], Vivek Verma[1]

[1]Department of Materials Science and Engineering, Indian Institute of Technology, Kanpur, India
[2]Electro Membrane Processes Division, CSIR-Central salt and marine chemicals research institute, Bhavnagar, India

Here, for the first time we are reporting far increase in tensile strength of the PVA composite material. The obtained best value of 122 MPa and 14.6% swelling shows the excellent synergistic effect of both cross-linker and reinforcement material. The composite films were prepared by simple mechanical dispersion of reinforcement material, bacterial cellulose nano-whiskers (BCNW) into PVA solution followed by cross linking with diacids, succinic acid (SuA) and adipic acid (AdA). We have in detail evaluated the effect of aliphatic carbon chain length of cross-linker on thermal, mechanical and water uptake properties. Neat PVA had strength of 37.3 MPa. With 5% reinforced of BCNW i.e. without cross linking, it exhibited 97% increase i.e. 74.5 MPa. With cross-linking of 15 mmol of SuA and AdA PVA films for 2 hours had excellent thermal properties with swelling percentage of 19% and 26.6% and tensile strength of 103 MPa and 67 MPa, respectively. The best result obtained was for 5% BCNW-PVA films crosslinked with 15 mmol succinic acid. These results were explained on the basis of synergetic cross-linking and extended hydrogen bonding between PVA and reinforcement material. The BCNW-PVA cross-linked and cross-linked PVA films can be used as biological implants, bio-inspired materials, membrane for pervaporation and filtration system etc.

Keywords: Bacterial cellulose nano-whiskers, cross-linker length, swelling properties, mechanical properties, thermal properties

RRB-O-017 [Renewable Resources and Biopolymers]

Mechanical Behavior Approach of Biodegradable Materials Based on Polyvinyl Alcohol, Achira (*Canna edulis*) Starch and Collagen Recovered from Leather Shavings

Lauro Vladimir Valle Alvarez, Katherine Stephanie Encalada Flores, Valeria Estefanía Viteri Tapia, Francisco Javier Quiroz Chávez

Centro de Investigaciones Aplicadas a Polímeros, Escuela Politécnica Nacional, Quito, Ecuador

Tanning leather industry generates leather shavings as a result of the shaving process. This waste stream is usually disposed in landfills without previous treatment. Simultaneously, the huge increasing of post-consumed traditional plastics, which ones need large periods of time in order to get degraded, has induced the idea of developing an alternative eco-friendly material. In this context, this article proposes to draw upon leather shavings combined with renewable resources to obtain biodegradable plastics. This work aimed to formulate a thermoplastic biodegradable material by the combination of hydrolyzed collagen (HC) recovered from leather shavings and achira rhizome starch (AS). Both components were mixed with polyvinyl alcohol (PVA) and glycerol (GL) as plasticizer. Several formulations were prepared using an experimental mixture design in order to evaluate the interactions between the components. In this regard, 60% of the total mixture weight corresponded to PVA, while the rest of the three substances were added in different percentages. The range belonging to GL and AS, was from 15% to 20% whereas the interval of HC was ranked from 5% to 10%. The raw materials were mixed in a roll mill at 60°C. Afterwards, these blends were placed in a heated mold at 110°C under constant pressure of 5500 psi for 7 minutes after of which a translucent and flexible sheet was obtained. Mechanical study was conducted using ASTM D882 standard in order to evaluate the interaction between blend´s components and the influence of composition over the properties. The results showed that both HC and GL acted as plasticizers over the blend, resulting in a synergic effect that enhanced the percentage of elongation at break. On the other hand, Young´s modulus was directly proportional to starch quantity in the formulation, it means that the material stiffness increased with the increment of AS.

Keywords: biopolymers, polyvinyl alcohol, starch blends, recovered collagen

RRB-O-019 [Renewable Resources and Biopolymers]

Design of new camelina oil-based hydrophilic monomers for novel polymeric materials

Brindusa Balanuca, Raluca Stan, Horia Iovu

University POLITEHNICA of Bucharest

Vegetable oil represents the most important renewable raw material used in macromolecular field in order to replace the petroleum-based compound [S. Miao et all, Acta Biomaterialia, 2014, 10, 1692; N. B. Samarth and P. A. Mahanwar, Open Journal of Organic Polymer Materials, 2015, 5, 1]. In the current research study, camelina oil (CO), a novel and less explored source of polyunsaturated fatty acids, was involved in a synthesis strategy to design innovative hydrophilic monomers. Thus, through a rationally functionalization of CO hydrophobic skeleton, by grafting polymerizable methacylic groups and hydrophilic polyethylene glycol (PEG) units, different types of self-emulsifiable monomers were designed. The synthesis experiments were monitored through 1H-NMR, 13C-NMR and FT-IR spectrometry and the registered spectra indicate the presence of the attached groups. Surfactant behaviour was also determined. The capacity of the CO-based hydrophilic monomers to form polymeric networks was investigated using two different techniques: emulsion and bulk photo-polymerization respectively, in order to establish the appropriate curing procedure for the designed compounds. FT-IR spectrometry and thermogravimetric analysis indicate that polymeric materials were obtained using both emulsion and bulk photo-curing techniques. Thus, we note that the major advantage of the designed monomers stay in ability to exhibit a double role: monomer and surfactant favouring the green polymerization technique in water.

Acknowledgement: This work was supported by a grant of the Romanian National Authority for Scientific Research, CNCS-UEFISCDI, project number PNCDI II - 105 / 2011-10-10

Keywords: camelina oil, hydrophilic monomers, photo-polymerization

RRB-O-020 [Renewable Resources and Biopolymers]

Renewable Resource based Biocomposites from Natural Luffa Cylindrica Fiber and Poly(hydroxybutyrate-co-valerate) Biopolymer

Dildare Basalp, Funda Tıhmınlıoglu

Department of Chemical Engineering, Izmir Institute of Technology, İzmir, Turkey

This study is focused on the preparation and characterization of environmental friendly green biocomposites composed of totally renewable resources. Novel green biocomposites were fabricated from natural luffa cylindrical fiber and poly(3-hydroxybutyrate-co-3-hydroxyvalerate) (PHBV) by using melt mixing technique. PHBV is one of the bacterial synthesized PHA biopolymers used in various packaging and biomedical applications. Luffa cylindrica is a plant having several micro channels and cylindrical shape as cellulosic fibers. PHBV-luffa fiber (2-5-10-15 wt%) filled biocomposites were processed by melt rheometer. Impact modifier (IMP) (5 wt%) and aminopropyltriethoxy silane (AMPTES) coupling agent were used for improvement of compatibility and interfacial adhesion between PHBV and luffa fiber. The effects of modifier type and luffa fiber amount on the thermal, mechanical, thermomechanical, morphological properties and water sorption properties were investigated by using differential scanning calorimetry (DSC), thermogravimetric analyzer (TGA), tensile testing, dynamical mechanical analyzer (DMA), contact angle measurement and scanning electron microscopy (SEM).Tensile strength and tensile strain of the composites were improved by 108 and 310 % at 2 wt% AMPTES treated luffa fiber loading, respectively. Thermal stability of the composites was improved with luffa loading. The onset degradation temperatures of PHBV pure polymer film and PHBV consisting IMP and 10 wt% luffa composites were found as 266 and 274 oC, respectively. The derivative temperatures of PHBV pure polymer film and PHBV consisting 10 wt% AMPTES treated luffa composites were found as 301 and 306 oC, respectively. Scanning electron microscopy (SEM) results indicated good interfacial adhesion among the AMPTES treated composites. The decrease in water sorption and increase in contact angles values of luffa fiber filled composites were observed with IMP modifier and AMPTES treatment. As a result, AMPTES and impact modifier treated biocomposites indicated the improvement on mechanical, water sorption and surface properties due to the interfacial enhancement of PHBV-luffa composites.

Keywords: renewable biopolymer composite

RRB-O-021 [Renewable Resources and Biopolymers]

Mechanical Properties of Marine Sourced Hydroxyapatite/Polycaprolactone Electrospun Biocomposites

Yeşim Müge Şahin[1], Erdi Buluş[2], Gül Tosun[2]

[1]*İstanbul Arel University*
[2]*Fırat University*

As a novel alternative source of scaffold materials, a marine sourced bioceramic (MSB) was obtained by using a simple mechano-chemical conversion method. Different marine sourced calcium phosphate bioceramics [i.e. hydroxyapatite (HA), ß-tricalcium phosphate (ß-TCP), whitlockite, monetite and other phases], can be obtained via this new method, for bone tissue engineering [1,2]. The exact $CaCO_3$ content was determined by thermogravimetric analysis. The required volume of aqueous H_3PO_4 solution was calculated in order to set the stoichiometric ratio of Ca/P equal to 1.5 for ß-TCP and to 1.667 for HA. The temperature was set to 80 °C for 8 hours to complete the process. The precipitate was then dried and the resulting nano-powder was calcined in a high- temperature oven. For complete characterization of the obtained bioceramics, fourier transform infrared spectroscopy (FTIR) and x-ray diffraction (XRD) analyses were carried out.

Exhibiting high biocompatibility and mechanical properties, polycaprolactone (PCL) was chosen to prepare electrospun composites of marine sourced HA/PCL, for tissue engineering purposes. Three different compositions were used and optimum electrospinning conditions were adjusted to prepare these biomaterials. Morphological and mechanical analyses were carried out via scanning electron microscopy (SEM) and tensile testing respectively, for further biomedical applications.

REFERENCES:

1. Sahin Y.M.. Gunduz O., Bulut B., Ozyegin L.S.& Oktari, F. N. (2015). Nano-Bioceramic Synthesis from Tropical Sea Snail Shells (Tiger Cowrie-Cypraea Tigris) with Simple Chemical Treatment. methods, 23, 24.
2. Gunduz, O., Sahin, Y. M., Agathopoulos, S., Ben-Nissan, B., & Oktar, F. N. (2014). A new method for fabrication of nanohydroxyapatite and TCP from the sea snail Cerithium vulgatum. Journal of Nanomaterials, 2014, 1.

Keywords: Marine Sourced Hydroxyapatite, Mechano-chemical Conversion, Polycaprolactone, Elektrospun Biocomposites, Tissue Engineering.

RRB-O-022 [Renewable Resources and Biopolymers]

Thermobifida cellulosilytica cutinase as a powerful tool for the synthesis and functionalization of biobased polyesters

Alessandro Pellis[1], Enrique Herrero Acero[2], Alice Guarneri[3], Valerio Ferrario[3], Lucia Gardossi[3], Georg M. Guebitz[1]
[1]University of Natural Resources and Life Sciences, Institute of Environmental Biotechnology, Tulln an der Donau, Austria
[2]Austrian Centre of Industrial Biotechnology GmbH, Division Enzymes and Polymers, Tulln an der Donau, Austria
[3]Laboratory of Applied and Computational Biocatalysis, Dipartimento di Scienze Chimiche e Farmaceutiche, Università degli Studi di Trieste, Trieste, Italy.

Since several years, the polymer industry is under pressure to mitigate the environmental cost of petrol-based plastics. On that respect, biotechnology contributes to the gradual replacement of petrol-based chemistry and the development of new renewable products, leading to the closure of carbon circle. Enzymes add higher value to bio-based polymers by catalyzing not only their selective functionalization but also their synthesis under mild and controlled conditions. Enzymes belonging to the cutinase class demonstrated their ability both to hydrolyze and synthesize polyesters.1 In this work, the promising properties of cutinase 1 from *Thermobifida cellulosilytica* are described in terms of ability to catalyze the synthesis of aliphatic oligoesters in solvent-free conditions.2 Furthermore, the same cutinase catalyzes also the hydrolysis of various biobased polyesters including poly(L-lactic acid) and poly(ethylene furanoate), thus enabling the targeted introduction of chemical functionalities. Data on reactivity and structural features confirm the enormous potential of this enzyme for the green processing of polyesters.

1 Pellis A. et al. 2016, Trends Biotechnol., DOI: http://dx.doi.org/10.1016/j.tibtech.2015.12.009
2 Pellis A. et al. 2016, Catal. Sci. Technol., DOI: 10.1039/C5CY01746G

Keywords: Biobased polyesters, biocatalyzed synthesis, surface functionalization, cutinases, hydrolysis, polycondensation

RRB-O-023 [Renewable Resources and Biopolymers]

Blending of Polylactide with Different Elastomeric Materials for Toughening

Yelda Meyva[1], Cevdet Kaynak[2]

[1]*Natural and Applied Sciences, Polymer Science and Technology Department, Ankara, Turkey*
[2]*Faculty of Engineeering, Metallurgical and Materials Engineering Department, Polymer Science and Technology Department, Micro and Nano Technology, Ankara, Turkey*

Polylactide (PLA) is an important biopolymer used especially for biomedical and food packaging applications. The main problem to be used in other industrial applications is its very high level of brittleness. Therefore, the main purpose of this work is to compare rubber toughening influences of five different elastomeric materials. For this purpose PLA was blended by three ethylene copolymers, i.e. ethylene vinyl acetate (EVA), ethylene methyl acrylate (EMA), ethylene-n-butyl acrylate glycidyl methacrylate (EBA-GMA); and two thermoplastic elastomers, i.e. thermoplastic polyurethane (TPU), thermoplastic polyester (TPE). Blends having various amounts of elastomeric materials (5, 10, 15, 20 phr) are prepared by melt-mixing method in a laboratory scale twin-screw extruder. Specimens for testing and analysis are shaped in required dimensions by using a laboratory scale injection molder. Morphological analysis conducted by scanning electron microscopy (SEM) indicated that all elastomeric materials were immiscible forming round shaped domains in the PLA matrix, with certain level of agglomeration at higher contents such as 15 and 20 phr. "Strength and elastic modulus of the blends were determined by tensile and flexural tests, while toughness values were determined by using two different methods; (K_{IC}) and (G_{IC}) fracture toughness tests and Charpy impact toughness tests". It was observed that generally all elastomeric materials decreased strength and modulus values of PLA with a certain degree, while leading to significant increases in the toughness values due to their efficient rubber toughening mechanisms. "For example,(G_{IC}) fracture toughness of PLA increased by 90% and 130% with 15 phr TPU and TPE; while increases were as much as 158%, 136% and 133% with 20 phr EVA, EMA and EBA-GMA, respectively".

Acknowledgement: Authors gratefully acknowledge financial support from TUBITAK the Scientific and Technological Research Council of Turkey as part of the Project 113M586.

Keywords: Polylactide, Ethylene Copolymers, Thermoplastic Elastomers, Rubber Toughening

RRB-O-024 [Renewable Resources and Biopolymers]

Catalytic Transesterification of Maize Starch with Plant Oils: Sustainable Access to Termoplastic Starches

Zafer Söyler, Michael A. R. Meier

Institute of Organic Chemistry, Karlsruhe Institute of Technology, Karlsruhe, Germany

Starch is one of the major alternative renewable feedstocks that can be isolated from several raw materials including corn, wheat, pea and potato. There is considerably increasing interest for the preparation of industrial products from starch due to its low cost, eco-friendly and non-toxic nature. However, it has limited applications because of its inherent properties like low solubility, viscosity and brittleness. Chemical modification of starch is often necessary to enhance the thermal, mechanical and other properties,which opens various strategies to substitute the conventional synthetic plactics derived from petroleum feedstock[1]

In our work, we demonstrate a highly efficient and sustainable method for the homogeneous transesterification of maize starch with plant oils, such as olive oil and high oleic sunflower oil, in the presence of catalytic amounts of 1,5,7-triazabicyclo[4.4.0]dec-5-ene (TBD). Modified starches were highly soluble in common organic solvents and characterized by 1H NMR, 13C NMR, 31P NMR, ATR-FTIR, GPC, DSC, EA, TGA, water contact angle (WCA) and strain-stress test measurements. Degrees of substitution were calculated using the 31P NMR method[2] and resulted in DS of ~1.4. DSC analysis revealed thermal transitions (Tg) of the modified starches at around 80 °C. Hydrophobic surfaces were characterized by contact angle measurements. Mechanical properties were examined by strain-stress measurements and showed around 45% and 80% elongations for modified maize starch and modified amylose from corn, respectively. Solvent, oils and catalyst were recovered with an easy work-up and reused for the further modifications. The developed methodology represents significant step forward for the production of themoplastic starches with improved mechanical properties in a sustainable fashion.

References:

[1] B. Kaur, F. Ariffin, R.Bhat,and A. A.Karim, *Food Hydrocolloids*, 2012, vol. 26, no.2, 398–404.
[2] A. W. T. King, J. Jalomaki, M. Granström, D. S. Argyropoulos, S. Heikkinen. and I. Kilpelainen. *Anal. Methods*, 2010, 2, 1499-1505

Keywords: Catalytic transesterification, Maize starch, Plant oils, Sustainability, Strain-stress test

Modification of starch with plant oils

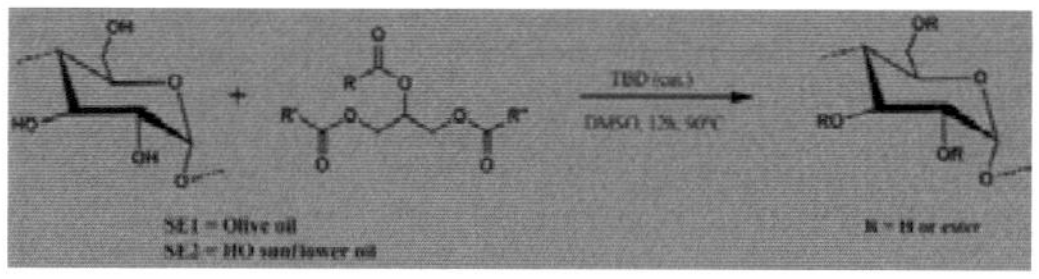

RRB-O-026 [Renewable Resources and Biopolymers]

Preparation and Characterization of *"Hibiscus Cannabinus"* (Kenaf) Grafted Natural Rubber Based Composites via Latex Route

Gunasunderi Raju[1], Manroshan Singh Jaswan Singh[2], Sasitaran Munuganan[1]

[1]*School of Distance Education, Universiti Sains Malaysia*
[2]*Malaysian Rubber Board*

"Hibiscus cannabinus" (kenaf) has recently been seen as a competitive reinforcing material in polymer composites. As Hibiscus cannabinus (kenaf) bast fibres are long, they were first processed using dry and wet grinding (ball mill) techniques to reduce the size and also to disperse the fibres in water. The resulting dispersions were then characterized for their particle size, zeta potential, rheology and morphology. They were then grafted onto epoxidized natural rubber latex (ENR25) using potassium persulfate as the reaction initiator. However, prior to that, the latexes were studied for their stability in the presence of the initiator concentration and reaction temperature and characterized via zeta potential, morphology and visual observations. The resulting kenaf/rubber composites were then characterized for their grafting efficiency using gel content and Fourier Transform Infra-Red techniques. From the dispersion studies, it was found that milling time affects not only the size of the particles, but also the morphology, shape, zeta potential and viscosity of the dispersion. Stability of nr latex was lower than ENR 25 latex, probably owing to the presence of non-ionic stabilizers due to the fact that the production of ENR 25 requires the use of strong acids. Nevertheless, the stability can be improved by adding more surfactant into the latex. Gel content and FTIR experiments confirmed the occurrence of grafting between kenaf fibres and the polymers.

Keywords: kenaf, grafting, natural rubber

SFP-O-001 [Smart and Functional Polymers]

CO2 induced UCST transition of amphiphilic triblock polymer and their self-assembly

Shaojian Lin, Patrick Theato

Institute for Technical and Macromolecular Chemistry, University of Hamburg

A well-defined temperature and CO2-responsive triblock polymer, poly((ethylene glycol)methyl ether-b-(acrylamide-co-acrylonitrile)-b-N,N-diethylaminoethyl methacrylate) (PEG-b-P(AAm-co-AN)-b-PDEAEMA), was synthesized via a reversible addition-fragmentation chain transfer polymerization (RAFT). It contains a temperature responsive segment based on P(AAm-co-AN), which exhibits an upper critical solution temperature (UCST) in water, and a CO2-responsive segment derived from PDEAEMA. Most importantly, the temperature responsive behavior (UCST) of this triblock polymer could be controlled by the amount of CO2 in solution, the UCST of the polymer solution decreased with increasing amount of CO2 in solution. The reason is that tertiary amine groups bearing on PDEAEMA are protonated to form ammonium bicarbonates resulting in their transition from hydrophobic to hydrophilic. Meanwhile, the prepared amphiphilic triblock polymer could self-assemble into vesicles in aqueous solution. These vesicles showed good stability in aqueous media even at high temperature and in the absence of CO2. After purging CO2 into solution, there is a transition from vesicles to micelles, which disassemble into unimers above UCST. The morphological transformations of these processes were confirmed by dynamic light scattering (DLS) and transmission electron microscopy (TEM).

Keywords: stimuli-responsive, Carbon oxide, temperature, UCST and self-assembly

SFP-O-002 [Smart and Functional Polymers]

A Photoresponsive Biomimetic Dry Adhesive Based on Doped PDMS Microstructures

M. Khaled Arafeh[1], Pamela Tannouri[2], Carlo Menon[2], Neil R. Branda[1]

[1]Chemistry Department, Simon Fraser University, Burnaby, BC, Canada
[2]School of Engineering Science, Simon Fraser University, BC, Canada

Inspired by the micro and nanoscale fibrillary structures found in many "Wall clambers" such as the gecko lizard, new artificial structured materials have been developed for use as dry adhesives. Unlike wet adhesive, these material do not need any moisture to bind and can work at a range of temperatures and pressures. The strong adhesive properties of such material make it difficult to detach easily without damaging the substrate surface which limits their reusability. Here we introduce light as an appealing stimulus to control the adhesive properties of PDMS micro-structured polymer by doping it with the photoresponsive chromophore. 0.25 wt% doping of PDMS with Spiropyran molecules allows photoinduced changes in the net charges at the surface of the polymer as Spiropyran molecules interconvert between neutral and zwitterionic isomers when exposed to visible and UV light. Exposing both nonstructured and microstructured doped films to UV light shows Increase in wettability and adhesion strength between film surface and glass tip. Further increase in the doping level doesn't lead to significant change which direct future research toward anchoring photochromic compound to the polymer backbone for more selective loading.

Keywords: smart materials, polymer, Dry Adhesive, Biomimetic, Photoresponsive polymers

SFP-O-003 [Smart and Functional Polymers]

Synthesis Pmma-Gold Core-Shell Based Metallodielectric Photonic Crystals As Substrates For Surface-Enhanced Raman Spectroscopy

Syara Kassim

School of Fundamental Science, Universiti Malaysia Terengganu, Kuala Terengganu, Malaysia

Bi-layered and multi-layered architectures embedded or enveloped with metal nanoparticles (NPs) have recently generated significant research attention due to their topological complexity and associated unusual optical properties often resulting from the activity of surface plasmon resonances, which originate from collective oscillation of their conduction electrons in response to optical excitation. Among those, core-shell(CS) of structures embedded with metal nanoparticles (NPs) usually silver or gold are of particular interest, which allow one to manipulate and use the metal NPs in a much easier and useful way.Opal-like assemblies of Au nanoshell exhibit the metal plasmonic resonances on the resulting optical properties of the materials. This paper demonstrates how the incorporation of functional Au nanoparticles onto the surface of poly(methyl methacrylate)(PMMA) spheres (as PMMA@Au core-shell particles) can be effectively utilised to improve the 3-D photonic crystal properties when they are immobilised on a glass substrate in the form of an ordered photonic crystal. Monodispersed PMMA spheres have been prepared by a modified emulsion polymerisation process and functionalised with Au nanoparticles by using citrate reduction method. The colloidal particles were then deposited onto a glass substrate by evaporation induced self-assembly and the optical properties of the resulting metallodielectric photonic crystal structures have been measured in reflection and transmission as a function of angle of incidence. The incorporation of metallic nanostructures such as Au, is interesting because such nanostructures possess surface plasmon resonances (SPRs),which manifest as additional absorption bands, when the incident photon frequency is resonant with the collective oscillation of the conduction electrons. Various alternative procedures for the fabrication of PhCs and MDPCs are described and preliminary results on the use of an Au-based MDPC for surface-enhanced Raman scattering (SERS) are presented. These preliminary results suggest perhaps a threefold increase of the Raman signal with the MDPC as compared to the bare PMMA PhC.

Keywords: photonic crystals, metallodielectric, surface-enhanced Raman scattering, polymer

SFP-O-005 [Smart and Functional Polymers]

Modified Smart Carbon Nanotube via Reversible-Deactivation Radical Polymerization

Nabendu B Pramanik, Souvik Ata, Nikhil K Singha

Rubber Technology Centre, Indian Institute of Technology Kharagpur, Kharagpur, India

Carbon nano-materials are one of the most interesting materials in this 21st century. Nanocomposites of polymer and carbon-nano tubes (CNTs) are used in different important applications. The main drawbacks of CNTs are their insolubility in suitable organic solvent and their incompatibility with most of the polymers. In this investigation CNT was modified via grafting of a polymethacrylate bearing furfuryl group via Diels-Alder (DA) reaction. In this case poly(furfuryl methacrylate) (PFMA) was prepared via reversible addition fragmentation and chain transfer (RAFT) polymerization. This RAFT polymer was grafted into the CNT surface via DA reaction in which the PFMA acted as dienophile and CNT surface as the diene. This was analysed via TGA, RAMAN, as well as XPS analyses. CNT was also modified to introduce brush-like polyacrylate bearing POSS unit via surface-initiated atom transfer radical polymerization (SI-ATRP). All these grafted CNTs were soluble in several organic solvents. This grafted CNT can have several potential applications.

Keywords: Reversible-Deactivation Radical Polymerization (RDRP), MWCNT, poly(furfuryl methacrylate) (PFMA), Diels-Alder, POSS

Figure 1

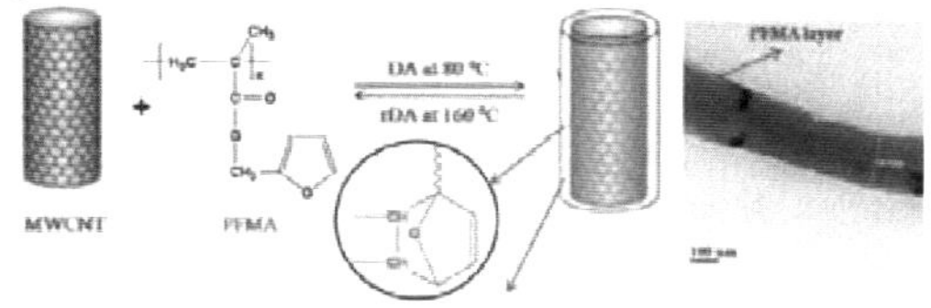

Schematic representation of direct functionalization of MWCNT via DA reaction between MWCNT and poly(furfuryl methacrylate) (PFMA).

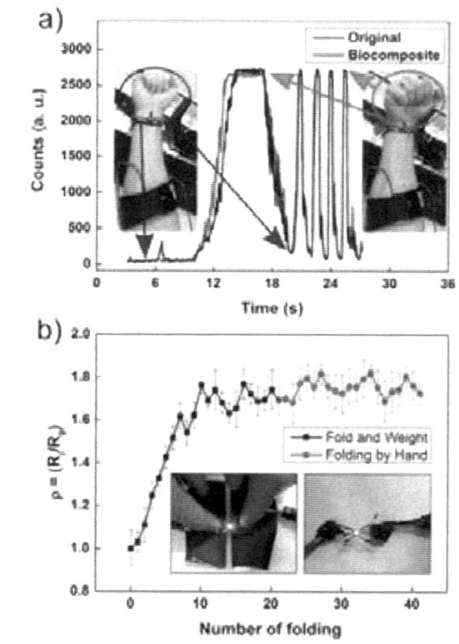

(a) Detection of muscle signals with the biocomposites and (b) demonstration of robustness of the biocomposites against severe folding events.

SFP-O-006 [Smart and Functional Polymers]

Graphene-polymer biocomposites for versatile high-performance flexible electronic applications

Ilker S. Bayer, Pietro Cataldi, Athanassia Athanassiou, Roberto Cingolani

Smart Materials, Istituto Italiano di Tecnologia

Bio-based foldable electrical conductors thanks to their biocompatibility, tunable biodegradation rate, and light-weight are emerging as new-generation components in surgery and food packaging. Moreover, they can considerably reduce non-degradable and potentially toxic electronic material waste mainly caused by the whirling race of continuous upgrades in portable electronic devices. Amongst biodegradable and biocompatible polymers, cellulose is the most abundant. Aside from its popular and wide use in paper-based printable electronics, cellulose is emerging as a key material for new potential applications such as functional flexible and foldable electronic biomaterial, not only as a substrate. Although cellulose fibers and nano-whiskers have been successfully used as reinforcement for various synthetic and bio-based polymers, to date, only a few studies focused on the development of functional composites combining continuous cellulose fiber networks with cellulosic thermoplastic polymers. To this end, we demonstrate that cellulose fiber networks can be sized or compounded with CA-graphene based composites, avoiding the use of any plasticizers, resulting in foldable, flexible and wear abrasion resistant bio-composite electrical conductors. The use of Graphene nano-platelet (GnPs) as the filler, results in robust, highly conducting, biocompatible and flexible nanocomposites, behaving like thin metal foils with excellent mechanical properties and bulk-like electric conductivity. In order to demonstrate their versatility, two applications were chosen both requiring isotropic electrical conduction, namely, (i) noninvasive electrodes for measuring small stochastic electrical signals generated by muscles (surface electromyography), and (ii) terahertz (THz) electromagnetic interference (EMI) shields. Contrary to conductive Xerox paper, the biocomposites feature isotropic bulk electrical conductivity and better water-proof resistance. Moreover, they demonstrate remarkable robust mechanical resistance against wear under 20 kPa applied pressure. Due to their light-weight, biodegradability, flexibility coupled with mechanical robustness and quasi-metal electrical conductivity, these bio-composites can be used in a variety of high-speed applications ranging from wearable electronics, electro-diagnostic medicine, aerospace and security.

Keywords: cellulose, biocomposites, graphene, electromyography, electromagnetic shileding

Robust Foldable biocomposites.

SFP-O-007 [Smart and Functional Polymers]

Flexible Composites for Multi functional Applications

Dipak Khastgir

Rubber Technology Centre Indian Institute of Technology Kharagpur

Flexible elastomers can be converted into multi functional composites through incorporation of appropriate Carbon fillers. However among all conductive fillers carbon based fillers are found to be most effective in imparting high conductivity coupled with improved mechanical properties to insulating polymer matrices. In fact carbon fillers are available in different forms carbon black powder, in fibrous form carbon fibre, carbon nano-tube, in the sheet form graphite and graphene. All these carbon fillers can modify both electrical and mechanical properties of the elastomeric composites. Judicial selection of elastomeric matrix and appropriate carbon can give rise to flexible conductive composites where electrical properties can be monitored through change in concentration of carbon filler. The present paper deals with development of conductive composites using different elastomers as matrix and different carbon fillers like carbon black, carbon fiber and carbon nano-tube as conductive additive to produce multi functional applications like Stress/strain sensor, Electromagnetic shielding Materials, Electrostatic charge dissipating materials and flexible electrode materials. The formation of continuous conductive network through distribution of conductive filler at and above critical concentration (percolation limit) is responsible for degree of electrical conductivity and its variation against applied stress/strain and variation of temperature. The formation and destruction of conductive networks of conductive particles in insulating polymer matrix under external applied stimuli is the main reason for its piezoelectric behavior and temperature response. The type of conductive networks formed in the polymer matrix depends both on matrix polymer as well as conductive filler. These composites can be efficiently used for stress/strain sensitive sensor, flexible electrode material for electrochemical synthesis and piezoelectric applications. Some such applications will be discussed.

Keywords: Carbon fillers,Elastomer Composites,Piezoelectric sensor,EMI Shielding,Electrode

SFP-O-008 [Smart and Functional Polymers]

Reversible stiffness photocontrollable materials

Jordi Royes Mir[1], Florence Benoit Marquié[1], Christophe Coudret[1], Véronique Pimienta[1], Clement Roux[1], Cazales Martine[2], Corinne Lorenzo[2], Anne Françoise Mingotaud[1]

[1]*Laboratoire des IMRCP, UMR 5623, Université Paul Sabatier, CNRS, Toulouse, France*

[2]*Equipe IP3D, ITAV, USR3505, CNRS, Toulouse, France*

Mechanical stimuli play a central role in many biological processes (Brafman, 2013); cells can sense their mechanical environment and adapt their genetic expression and functions. As a consequence, some stiffness modulable biomaterials powered by irreversible photoreactions have been applied to study the cellular response (Azagarsamy et al., 2014, Stowers et al., 2015). However, some cells, tissues or organs are continuously exposed to periodic mechanical forces that regulate cellular functions such as contraction, migration and proliferation (Wang et al., 2014). Therefore, irreversible biomaterials are not adapted to study dynamic processes. Recently, a few *smart* materials whose stiffness can be reversibly controlled, based on photoisomerization of azobenzene (Rosales et al., 2015) or spiropyran (Sun et al., 2016) moieties, have been developed. However, only minor changes in stiffness were achieved.

Exploiting the reversible formation and cleavage of multiple supramolecular interactions (Figure 1) we have prepared biocompatible photoresponsive hydrogels showing strong and reversible stiffness changes. The application of thiol-Michael addition to synthesize our *smart* materials has granted access to polymers with well-defined architecture with improved control over their properties. Moreover, using Doehlert design, we have adjusted the composition of our *smart* materials to yield transparent hydrogels mimicking the mechanical properties of commonly used agarose hydrogels. These properties make our reversible *smart* materials perfect candidates to study dynamic cellular functions triggered by reversible stiffness changes.

References:

Brafman D.A., Physiol. *Genomics*, 2013, 45, 1123-1135.
Azagarsamy, M.A.; McKinnon, D.D.; Alge, D.L.; Anseth, K.S., *ACS Macro Lett.*, 2014, 3, 515-519. Stowers, R.S.; Allen, S.C.; Suggs, L.J., *Proc. Natl. Acad. Sci. U.S.A.*, 2015, 112, 1953-1958.
Wang, J.; Lü, D.; Mao, D.; Long, M. *Protein Cell*, 2014, 5, 518-531.
Rosales, A.M.; Mabry, K.M.; Nehls, E.M.; Anseth, K.S., *Biomacromolecules*, 2015, 16, 798-806.
Sun, Z.; Liu, S.; Li, K.; Tan, L.; Cen, L.; Fu, G.; *Soft Matter*, 2016, 12, 2192-2199

Keywords: hydrogel, smart, stiffness, reversible, biomaterial, photoresponsive

Figure 1

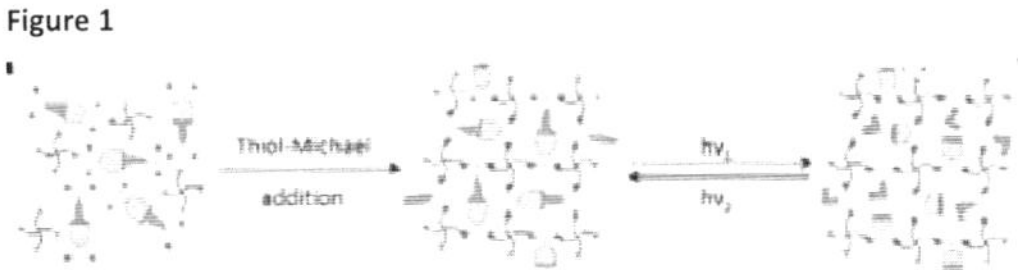

Preparation of {smart} materials with well-defined polymer architecture and reversible photoresponse due to the formation and cleavage of multiple supramolecular interactions.

SFP-O-009 [Smart and Functional Polymers]

Dextran hydrogels - from polysaccharides to *biocompatible* and *biodegradable* scaffolds

Nick Dibbert, Gerald Dräger, Andreas Kirschning

Department of Organic Chemistry, Leibniz Universität Hannover, Hannover

Developing new biomaterials containing a three-dimensional network based on biopolymers, polysaccharides respective, requires the merging of chemical knowledge to the field of tissue engineering. This fusion enables an analytical view into the modification of polysaccharides and enlarges the scope of applicable polysaccharides for biomedical research.

The focus of our research on polysaccharides like dextran and pullulan was based on the previous work of Kirschning *et al.* on hyaluronic acid and alginate, respective pure hyaluronic acid based hydrogels. This hydrogelation was driven via a condensation reaction between a hydrazide and an aldehyde of each functionalized biopolymer.

We applied this method of functionalization on different polysaccharides like pullulan, cellulose, chitosan, lentinan, lechinan and dextran. In the case of dextran we focused on the effect of different average molecular weights on functionalization as well as on hydrogelation. We could show that even after high modification degrees the dextran hydrogel was still degradable by dextranase.

Furthermore, we decorated the dextran proir to biological application with the RGDfK pentapeptide. The different methods that we used for the modification of dextran with RGDfK gave us the advantage to open up a new avenue towards the usage of our hydrogels as a drug delivery system.

Keywords: hydrogel, dextran, aldehyde, hydrazide, RGDfK, degradation

Concept of hydrogelation

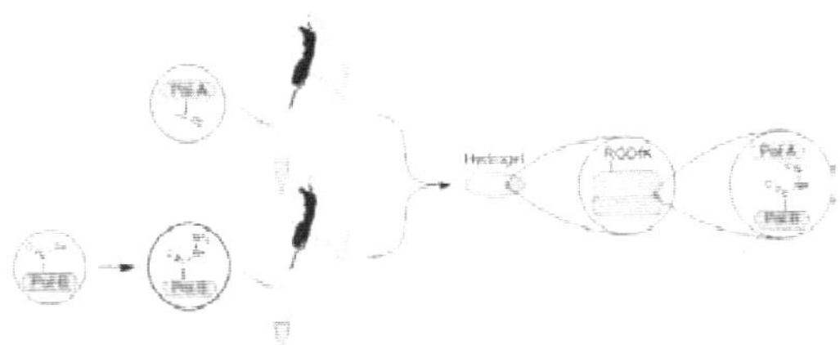

General concept of modification and hydrogelation of different polysaccharides

SFP-O-010 [Smart and Functional Polymers]

Redox-active poly(ferrocenylsilane)s as hydrogel actuators

Kaihuan Zhang, Mark Hempenius, Julius Vancso

Materials Science and Technology of Polymers, MESA+ Institute for Nanotechnology, University of Twente, The Netherlands

Stimuli-responsive polymer materials, capable of displaying large changes in response to minute external triggers, have attracted great interest as actuators for microfluidic control, drug delivery, artificial muscles, etc. Poly(ferrocenylsilane)s (PFSs) belong to a new class of redox-active materials which are composed of alternating ferrocene and silane units in the main chain and combine a high density of redox centers with excellent processability and redox characteristics.[1-3] Here we present a PFS-based hydrogel actuator (Fig.1a-c), prepared by a one-step thiol-Michael addition reaction, which is capable of generating a mechanical motion. This relatively tough hydrogel displayed a reversible color change and expansion-contraction upon oxidation and reduction under physiological salt conditions, even under high stress close to its rupture strain, which makes it suitable for artificial muscles in biological systems.

In addition to redox trigger, tributylphosphonium and alkanesulfonate anions were also introduced to the side group of PFS (Fig.1d), which result a poly(ionic liquid) with redox-controlled LCST-type behavior (Fig.1e, f). The LCST transition occurs in a wide temperature range depending on polymer concentration.[4] The aqueous solution also exhibits a redox-switchable LCST behavior. By using a thiol-Michal addition cross-linking chemistry (Fig.1g), redox-switchable shrinking and swelling was accomplished by utilizing the PFS-PEG-based hydrogel from reversible thermal stimuli (Fig.1h). This dual-responsive PFS-PEG-based hydrogel provides the opportunity for constructing a thermal actuator, which can be tuned between "ON" and "OFF" states by redox-stimuli.

References:

1. M. A. Hempenius, C. Cirmi, F. Lo Savio, J. Song, G. J. Vancso, *Macromol. Rapid Commun.* 2010, *31*, 772–783.
2. X. Sui, M. A. Hempenius, G. J. Vancso, *J. Am. Chem. Soc.* 2012, *134*, 4023–4025
3. K. Zhang, X. Feng, X. Sui, M. A. Hempenius, G. J. Vancso, *Angew. Chem. Int. Ed.* 2014, *53*, 13789–13793.
4. Y. Men, H. Schlaad, J. Yuan, *ACS Macro Lett.* 2013, *2*, 456–459.

Keywords: actuators, hydrogels, poly(ferrocenylsilane)s, poly(ionic liquid)s, redox chemistry

Figure 1

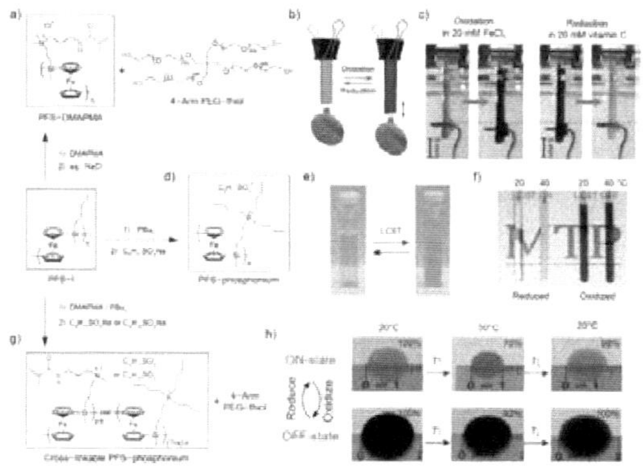

Figure 1. (a-c) PFS-PEG-based redox actuator; (d-f) PFS-phosphonium and redox-controlled LCST behavior in aqueous solution; (g,h) Cross-linkable PFS-phosphonium and hydrogels with redox-switchable thermal-actuation.

SFP-O-011 [Smart and Functional Polymers]

Development of Superhydrophobic Surfaces Based on Polyurethane Chemistry For Coating Materials

Seher Şenada, Sennur Deniz

Department of Chemical Engineering, Yıldız Technical University, Istanbul-Turkey

Superhydrophobic surfaces are created through the combination of surface roughening and lowering of the surface energy. It is well known that a fluorinated polymer has a low wettability due to its low surface energy [1-3]. This work based on the preparation of superhydrophobic surfaces was studied for coating applications. We investigated a simple method to prepare robust superhydrophobic surfaces. Polyurethane-based high-performance coating materials (FPU) were synthesized from fluorinated isocyanate and hydroxyl-end-capped solventless liquid oligoesters. Fluorinated polyurethane-silica nanocomposite (FPU-silica) coatings were prepared with different types of hydrophobic surface treated fumed nanosilica. The coating solutions of FPU-silica were prepared by mixing FPU with hydrophobic fumed nanosilicas having different surface groups in ethanol. For crosslinking, it was used a blocked isocyanate. Molecular structure of FPU-silica nanocomposite coating materials was characterized by ^{1}HNMR and FTIR spectrophotometer. Superhydrophobic properties of surfaces coated with FPU-silica nanocomposites were examined by water contact angle (WCA) measurements with a tensiometer instrument. In the absence of the silica particles, WCAs of non-fluorinated PU films having smooth top surface for advancing and receding were measured 100° and 48°, respectively. WCAs of FPU-silica coatings were found as 125° and 68° for advancing and receding cycle, respectively.

References:

[1] D. Wu, " Nature-Inspired Superlyophobic Surfaces ", Eindhoven University of Technology, ISBN: 978-90-386-1126-6,2007,pp.16
[2] D. Wu, W. Ming, R. A. T. M. van Benthem and G. deWith, "Superhydrophobic Fluorinated Polyurethane Films", Journal of Adhesion Science and Technology, 22 (2008) 1869–1881
[3]S.Deniz, B.Arıkan," Effect Of Silica Type On Superhydrophobic Properties Of PDMS-Silica Nanocomposite Coatings", International Journal of Engineering and Applied Sciences, Jan. 2016. Vol. 8. No. 01

Keywords: Superhydrophobic coating, Fluorinated polyurethane, Hydrophobic silica, Water contact angle.

SFP-O-012 [Smart and Functional Polymers]

Multi-arm Polymers for Conjugation and Release of Therapeutic Cargo

Özgül Gök[1], Pelin Ertürk[1], Burcu Sümer Bolu[1], Tuğçe Nihal Gevrek[1], Rana Sanyal[2], Amitav Sanyal[2]

[1]Department of Chemistry, Bogazici University, Istanbul 34342, Turkey
[2]Center for Life Sciences and Technologies, Bogazici University, Istanbul, 34342, Turkey

Recently, polymeric supports have been evaluated as smart platforms for improving stability and efficacy of therapeutic peptides and proteins, which are mostly subjected to proteolytic degradation in the blood [1]. Surrounding the small biomolecules with polymers can preserve their stability via the so called "umbrella-like effect", as well as provide them with an increased water solubility and improved therapeutic window. In this regard, poly(ethylene glycol) (PEG) based polymers have being widely utilized in form polymer-peptide conjugates [2]. Depending on the application, one may want to conjugate the protein through a non-reversible or a reversible conjugation. In this study, a multiarm polymer containing a thiol-reactive disulfide unit at its core was synthesized. This group provides a reversible conjugation for peptides/

proteins to polymers by reacting with their free thiol groups at known positions in their structure [3]. Design for multiarm polymer involves the growth of PEG constituted polymer chains from the periphery of polyester dendron [4]. Resultant water soluble and biodegradable multiarm polymer was conjugated to thiol-containing fluorescent dye and bioactive peptides such as glutathione. The studies establish that hydrophobic small molecules are easily solubilized through such a conjugation, and importantly, released under thiol-containing reducing environments, as encountered in cancerous tissues. Conjugation of an apoptotic peptide to this construct enabled the peptide to induce three-fold excess apoptosis compared to naked peptide. The design, synthesis of this novel multiarm polymeric construct, and in vitro studies highlighting its potential application as a delivery platform will be presented.
Acknowledgement: This project has been funded by The Bogazici University Research Fund (BAP, Project No:13B05D4).

Keywords: Dendron, polymer, polymer-peptide conjugate, bioconjugation.

SFP-O-013 [Smart and Functional Polymers]

Polyamidoamine/shellac: responsive hydrogel for controlled release applications

RK Dey[1], Arundhati Barik[2], Tanushree Patnaik[3]

[1]*Centre for Applied Chemistry, Central University of Jharkhand, Ranchi, India*
[2]*Department of Chemistry, Ravenshaw University, Cuttack, Odisha, India*
[3]*Department of Chemistry, Stewart Science College, Cuttack, Odisha, India*

Hydrogel (smart material) can control the drug releasing behavior by changing the gel struc-ture in response to changes in physiological environment. Fabrication of polyamidoamine-shellac based smart hydrogel material is relatively new in biomedical field of research. Shellac is a natural biocompatible polymer derived from secretion of lac insect. The combination of shellac with acrylic acid (AA) and polyamidoamine (PAA) in different ratios produce new set of biocompatible, biodegradable and pH responsive polymers (Figure-1). The in vitro cumula-tive drug release pattern is dependent upon ratio of PAA:acrylic acid content. The dynamic and equilibrium swelling properties of the polymeric materials were studied in simulated gas-tric and intestinal fluids. Very little amount of drug could be released from hydrogel at low pH (2.1); whereas, almost 96% of drug could be released when pH increases to 8.4. The hy-drogel material show good equilibrium swelling (1038%) in buffer medium. The controlled release kinetics of new set of polymers were studied using various mathematical models. The Fick's diffusion model shows a linearity until values of the swelling fraction Mt/Me ≤ 0.60 (first 60% of the dynamics water uptake); Mt and Me refers to the amount of buffer solution (drug) absorbed by the polymeric material at time 't' and in the equilibrium, respectively. The bio-degradation of materials were carried out in buffer medium and a higher enzymatic deg-radation was observed that attributed to the cleavage of ester bonds for hydrogels containing higher shellac content. The materials are found to be suitable for targeted drug delivery for colon specific applications.

Keywords: hydrogel, shellac, controlled release.

Scanning electron micrograph of shellac based hydrogel material.

SFP-O-014 [Smart and Functional Polymers]

Thermal Annealing of Polyelectrolyte Multilayers: An Effective Approach for the Enhancement of Cell Adhesion

Sergio E Moya[1], Nicolas Muzzio[2], Eleftheria Diamanti[2]

[1]*CIC biomaGUNE, Spain*
[2]*Inifta, Argentina*

Polyelectrolyte multilayers (PEMs) have many potential applications in tissue engineering and regenerative medicine. However, the softness of biocompatible PEMs results in limited cell adhesion. A novel strategy for the enhancement of cell adhesion on PEMs based on thermal annealing is presented here. The impact of thermal annealing at 37 ºC of poly-L-lysine (PLL) and alginate (Alg) polyelectrolyte multilayers on the adhesion of human lung cancer A549 and myoblast C2C12 cell lines is studied. Cell adhesion, measured by the projected average cell spreading and focal contact formation is remarkably improved for the annealed PEMs. The properties of the PEMs are characterized by means of the quartz crystal microbalance with dissipation (QCM-D), atomic force microscopy (AFM), atomic force spectroscopy (AFS), zeta potential and contact angle measurements after annealing at 37, 50 and 80 oC. After annealing, PLL/Alg PEMs become smoother displaying an increase in stiffness. Advancing contact angle in water varies from 36o before annealing to 93o, 77o and 95o after annealing at 37, 50 and 80 oC, respectively. Surface energy changes after annealing were calculated from contact angle measurements performed with organic solvents. Quartz crystal microbalance with dissipation, contact angle and fluorescence spectroscopy measurements show a significant decrease in the adsorption of the bovine serum albumin protein to the PEMs after annealing. Changes in the physical properties of the PEMs are interpreted as a result of the reorganization of the polyelectrolytes in the PEMs from a layered structure into complexes where the interaction of polycations and polyanions is enhanced.

Keywords: biopolymers, polyelectrolytes, termal annealing, cell adhesión, contact angle, antifouling

SFP-O-015 [Smart and Functional Polymers]

Multifunctional Polymersomes as Smart Nanodevices for Biomedical Applications: Probing Responsive Behaviour and Mechanical Properties

Banu Iyisan[1], Andreas Janke[2], Philipp Reichenbach[3], Lukas Eng[3], Dietmar Appelhans[2], Brigitte Voit[1]

[1]*Leibniz Institut für Polymerforschung Dresden e.V., Dresden, Germany; Organische Chemie der Polymere, Technische Universität Dresden, Germany; International Helmholtz Research School for Nanoelectronic Networks, Dresden, Germany*
[2]*Leibniz Institut für Polymerforschung Dresden e.V., Dresden, Germany*
[3]*Institut für Angewandte Photophysik, Technische Universität Dresden, Germany*

The demand for multifunctional nanovesicles is increasing greatly because of their high potential in various biomedical applications.[1-3] For this purpose, the previously established photo-crosslinked and pH responsive polymersomes decorated with adamantane and azide moieties[2] were taken as the basis of this study. Herein, this multifunctional system was immobilized on solid substrates by utilizing strong adamantane-β-cyclodextrin host-guest interactions which allowed to creating half-dome-shaped polymersome compartments on surfaces.[3] Besides, mechanical testing by force measurements with AFM was performed to gain more insight about the robust but also flexible nature of our multifunctional polymersomes. In addition, the pH responsive nature of the immobilized vesicles showed a stable and "breathable" membrane under different environmental conditions by in-situ AFM monitoring in the wet state. Another key feature of our approach was to perform Infrared light induced photochemical reactions at the membrane surface of immobilized polymersomes. To realize this, nitroveratryloxycarbonyl protected amine molecules as IR light sensitive photocleavable groups were introduced to the azido moieties of the polymersome surface before immobilization. This in turn leads to controlled photopatterned structures at the substrate and to freed amino groups on the polymersome surface which had been subsequently conjugated with a fluorescent dye. This approach mimics the concept of site-specific biomolecule binding which is highly interesting for the development of bio/chemo sensors. Overall, this responsive and multifunctional polymersomes are promising nanodevices to design smart and stable compartments on surfaces for various biomedical applications, e.g. making use of microfluidic devices.

References:

(1) Yassin, M. A.; Appelhans, D.; Wiedemuth, R.; Formanek, P.; Boye, S.; Lederer, A.; Temme, A.; Voit, B., Small 2015, 11, 1580-1591.
(2) Iyisan, B.; Kluge, J.; Formanek, P.; Voit, B.; Appelhans, D., Chem. Mater. 2016, 28, 1513-1525.
(3) Iyisan, B.; Janke, A.; Reichenbach, P.; Eng, L. M.; Appelhans, D.; Voit, B., ACS Appl. Mater. Interfaces 2016 (Submitted).

Keywords: polymersomes, IR light responsive, pH responsive, host-guest interaction

SFP-O-016 [Smart and Functional Polymers]

Thermal and Oxidation Responsive Polypeptide Materials

Zhibo Li[1], Xiaohui Fu[2], Yinan Ma[2], Chongyi Chen[2]

[1]*School of Polymer Science and Technology,Qingdao University of Science and Technology, Qingdao, China*
[2]*Institute of Chemistry, Chinese Academy of Sciences, Beijing, China*

A series of new functional amino acids were prepared in high yield via thiol-ene Michael addition between L-cysteine and monomethoxy oligo(ethylene glycol) (OEG) functionalized methacrylates (OEGxMA) and acrylate (OEGxA). Subsequent ring opening polymerization (ROP) of these NCA monomers gave a series of OEGylated poly-L-cysteine (poly-EGxMA-C or poly-EGxA-C) homopolypeptides. Depending on the length of OEG side chains, poly-EGxMA-C and poly-EGxA-C polypeptides displayed different solubility and secondary structure in water. More importantly, the obtained polypeptides can display reversible thermo-responsive properties in water when the x value is between 3 and 5. The oxidation-responsive behaviors of OEGylated poly-L-cysteine homopolypeptides, i.e., poly(L-EGxMA-C)n, were investigated. These poly-L-cysteine derivatives adopted mixed conformation in water, in which the β-sheet accounted for a significant proportion. Upon oxidation, the thioethers in polypeptide side-chains were converted to polar sulfone groups, which triggered the secondary structure transition from β-sheet preferred conformation to random coil. Accordingly, the increase of side-chain polarity together with conformation changes increased samples' water solubility and cloud point temperature. Using mPEG45-NH2 as macroinitiator, we synthesized PEG45-b-poly(L-EG2MA-C)22 diblock copolymer via ROP of L-EG2MA-C NCA. The PEG45-b- poly(L-EG2MA-C)22 was able to self-assemble into spherical micelles in aqueous solution, and the micelles could undergo an oxidation-triggered disassembly due to the oxidation-responsive thioethers. Such new class of oxidation-responsive polypeptides might provide a promising platform to construct inflammation targeting drug delivery systems.

Keywords: Thermal responsise, Polypeptide, Oxidation responsive, NCA

Illustration of thermal irreversible OEGlyated polycysteine.

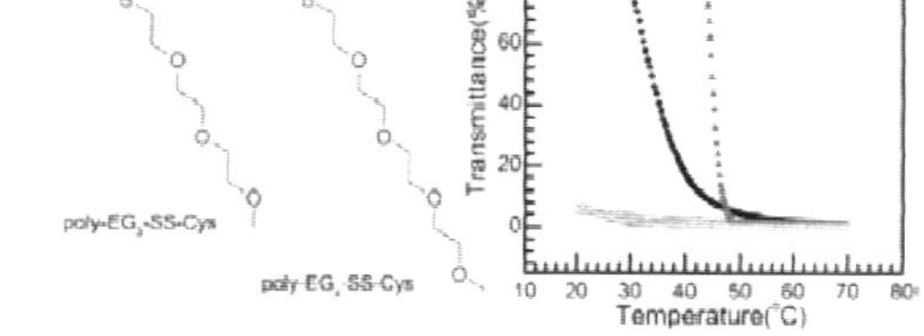

SFP-O-017 [Smart and Functional Polymers]

Encapsulation of Alkyd Resin in Urea Formaldehyde Resin for Self-Healing Coatings

Göksenin Kurt Çömlekçi, Sevgi Ulutan

Department of Chemical Engineering, Ege University, Izmir, Turkey

Self-healing materials have the ability to repair themselves upon exposure to external stimulus caused by several factors during their service life [Brown et al., 2003]. When the capsules embedded in coating are ruptured by damage, the self-healing mechanisms triggered through the release and reaction of the healing agent [Blaiszik et al., 2008]. In this research alkyd resin healing agents were encapsulated with poly(urea-formaldehyde) using in-situ polymerization technique at 55 °C and pH 3.5. The encapsulation parameters were stirring type and surfactant effects on capsule properties. Sonic homogenizator was considered versus mechanical mixer for stirring. Polyvinyl alcohol (PVA), tween 80 (T80) and ethylene maleic anhydride (EMA) were used as surfactants with different concentrations. The effect of PVA amount and stirring rate on microcapsule properties were carried out as well. The experiments were designed by using response surface method for optimum conditions. Formation of capsules was followed by optical microscope (OM). The encapsulation efficiency was estimated by evaluating soxhlet extraction data and the structures were investigated by means of FTIR spectroscopy. Capsule surface morphology was analyzed by using scanning electron microscope (SEM). Thermal gravimetric analyses were carried out to elucidate thermal stability of capsules. Experiments with sonic homogenizator and the surfactants except PVA resulted in nano sized capsules. As PVA concentration and stirring rate increased smaller spherical microcapsules were obtained while the capsule surface became smoother. As a result, alkyd resins encapsulated with poly(urea-formaldehyde) suitable for self-healing coating application were produced.

*This research is supported by TUBITAK project number 114M114 and partially by EBILTEM project number 2015/BIL/017.

Keywords: Microcapsule, Alkyd resin, Self healing

SFP-O-018 [Smart and Functional Polymers]

Magnetic polymer nanocomposite fibers for oil removal from stable water-oil emulsions

Suset Barroso Solares, Maryam Ghazi Zahedi, Javier Pinto, Despina Fragouli, Athanassia Athanassioua

Smart Materials, Nanophysics, Istituto Italiano di Tecnologia (IIT), via Morego 30, 16163 Genova, Italy.

Wastewaters containing emulsified oil are produced from several industrial processes (e.g. pharmaceutical, cosmetic, and food industries), presenting potential environmental and health risks. Therefore, the industrial and environmental demands emphasize the needs for innovative materials that can efficiently separate oil and water. Herein we present the fabrication of hydrophobic and oleophilic poly(methyl methacrylate)-based nanocomposite fibrous mats with magnetic properties, and their utilization for the oil removal from stable water-oil emulsions. The incorporation of ferromagnetic iron nanoparticles into the polymeric fibers increases the selectivity and oil removal performance of the fibers, as well as introduces magnetic actuation properties to the material. The functional mats exhibit oil absorption efficiencies up to 90%, independently from the oil content in water, which ranges from 5 to 30 v.%. The good performance achieved makes the presented material a promising candidate for water-oil emulsions separation.

Keywords: Wastewaters remediation, electrospun fibers, magnetic nanocomposite, stable emulsion separation.

SFP-O-019 [Smart and Functional Polymers]

Multi-template molecularly imprinted polymer: Synthesis, characterization and removal of selected acidic pharmaceuticals from wastewater

Lawrence Mzukisi Madikizela[1], Luke Chimuka[2]

[1]Department of Chemistry, Durban University of Technology, Durban, South Africa
[2]School of Chemistry, University of Witwatersrand, Johannesburg, South Africa

Removal of organics from wastewater offers a better water quality, therefore, the purpose of this work was to investigate the use of molecularly imprinted polymer (MIP) for the elimination of selected organics from water. A multi-template MIP for the adsorption of naproxen, ibuprofen and diclofenac was synthesized using a bulk polymerization method. A MIP was synthesized at 70°C by employing 2-vinylpyridine, ethylene glycol dimethacrylate, toluene and 1,1'-azobis-(cyclohexanecarbonitrile) as functional monomer, cross-linker, porogen and initiator, respectively. Thermogravimetric characterization indicated that the polymer backbone collapses at 250°C and scanning electron microscopy revealed the porous and roughness nature of the MIP after elution of templates. The performance of the MIP in aqueous solutions was evaluated by optimizing several adsorption parameters. The optimized adsorption conditions were 50 mg of MIP, extraction time of 10 min, a sample pH of 4.6 and the initial concentration of 30 mg/L. The imprinting factors obtained for naproxen, ibuprofen and diclofenac were 1.25, 1.42, and 2.01, respectively. The order of selectivity for the MIP was; diclofenac > ibuprofen > naproxen. MIP showed great swelling in water with an initial swelling rate of 2.62 g/(g min). The synthesized MIP proved to be able to adsorb naproxen, ibuprofen and diclofenac from contaminated deionized water, wastewater influent and effluent.

Keywords: Molecularly imprinted polymer, multi template, pharmaceuticals, adsorption

SFP-O-020 [Smart and Functional Polymers]

Design, synthesis and RAFT polymerisation of a quinoline-based monomer for use in metal-binding composite microfibers

Anna Isakova[1], Olga Efremova[2], Nikki Pullan[1], Larry Lüer[3], Paul D. Topham[1]

[1]CEAC, Aston University, Aston Triangle, Birmingham, B4 7ET, UK.
[2]Department of Chemistry, University of Hull, Hull, UK
[3]IMDEA Nanociencia, C/Faraday, 9, Cantoblanco, Madrid, Spain

Metal-binding polymer fibres have attracted major attention for diverse applications in membranes for metal sequestration from waste waters, non-woven wound dressings, matrices for photocatalysis, and many more. Here we present a straightforward solution to fabrication of polymeric fibres with a high loading of inorganic nanoparticles, starting from the design and synthesis of an 8-hydroxyquinoline-based zinc-binding styrenic monomer, QuiBoc. Its subsequent polymerisation by reversible addition–fragmentation chain transfer (RAFT) yielded well-defined polymers, PQuiBoc, of controllable molar masses (6 and 12 kg mol−1) with low dispersities (Đ, Mw/Mn < 1.3). Protected (PQuiBoc) and deprotected (PQuiOH) derivatives of the polymer exhibited a high zinc-binding capacity, as determined by semi-quantitative SEM/EDXA analyses, allowing the electrospinning of microfibres from a PQuiBoc/polystyrene (PS) blend without the need for removal of the protecting group. Simple "dip-coating" of the fibrous mats into ZnO suspensions showed that PQuiBoc/PS microfibres with only 20% PQuiBoc content had almost three-fold higher loadings of ZnO (29%) in comparison to neat PS microfibres (11%).

Keywords: RAFT polymerization, zinc oxide, microfibers, electrospinning,-composites, 8-hydroxyquinoline.

SFP-O-023 [Smart and Functional Polymers]

Thiolactones: Sustainable building blocks in advanced polymer design

Daniel Frank[1], Pieter Espeel[1], Sven Claessens[2], Edwin Mes[2], Tamara Dikic[2], Filip Du Prez[1]

[1]Polymer Chemistry research Group, Department of Organic and Macromolecular Chemistry, Ghent University, Krijgslaan 281 (S4bis), B-9000 Ghent, Belgium
[2]DOW Benelux BV, Herbert H. Dowweg 5, 4542 NM Hoek, Netherlands

Small-cyclic thiolactones are intermediates in organic synthesis, while γ-2-aminothiolactone analogues – derived from homocysteine – find applications in medicine. Implementation of larger size cyclic thiolactones in ring-opening polymerizations is impractical due to limited availability and high prices. Recent approaches demonstrate the successful use of D,L-homocysteine thiolactone based AB' monomers as versatile building blocks for thiol-ene polyaddition [1].

D,L-Homocysteine hydrochloride is a sustainable precursor for a platform of monomers that show superior advantages in polymer chemistry as "protected thiols" for thiol-ene polymerizations, while bearing the opportunity to introduce additional functionalities during *in-situ* activation using aminolysis (Figure 1). In comparison to 'common' thiol-ene compounds, bifunctional thiolactone AB' monomers have a virtually unlimited shelf-life under normal conditions, can contain various chemical functionalities and can be synthesized in multigram scale from readily available starting compounds.

A vast variety of research topics has been covered using thiolactone chemistry [2], including post-polymerization modification and sequence control. Various polymeric materials were made, such as hydrogels, polyamides as well as defined and sophisticated polymeric structures [3]. Applying carbamate-functionalized monomers, polyurethanes were synthesized through photoinitiated amine-thiol-ene polyaddition [1]. A potential application in coatings is being tested, using sustainable thiolactone monomers for scavenging applications.

1. Espeel, P., Goethals, F., Du Prez, F.E., Journal of the American Chemical Society 2011, 133(6), 1678-81.
2. Espeel, P., Du Prez, F.E. European Polymer Journal 2015, 62, 247-272.
3. Driessen, F., Herckens, R., Espeel, P., Du Prez, F.E. Polymer Chemistry 2016, 7(8), 1632-1641.

Keywords: Thiol-ene Polyaddition, Amine-thiol-ene Conjugation, Functional Materials, Homocysteine Thiolactone, Protected Thiols

Figure 1

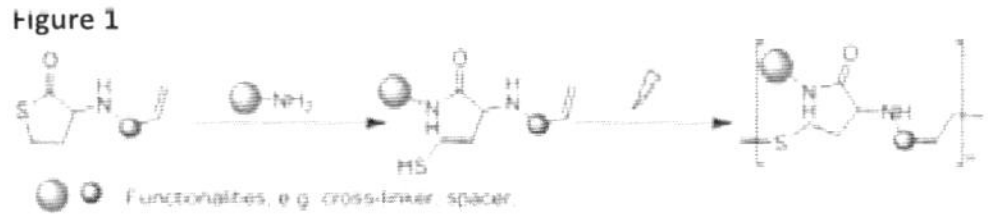

General scheme of the aminolysis and polymerization of AB' thiolactone monomers.

SFP-O-025 [Smart and Functional Polymers]

Temperature- and Voltage-Induced Ligand Rearrangement of a Dynamic Electroluminescent Metallopolymer

Demet Asil Alptekin[1], Jonathan A. Foster[2], Asit Patra[2], Xavier De Hatten[2], Jesus Del Barrio[2], Oren A. Scherman[2], Jonathan R. Nitschke[2], Richard H. Friend[3]

[1]*Department of Chemistry, Middle East Technical University, Ankara, Turkey*
[2]*Department of Chemistry, University of Cambridge, Cambridge, UK*
[3]*Department of Physics, University of Cambridge, Cambridge, UK*

Efforts are being made to design organic and inorganic hybrid structures that exhibit improved material functionality toward emissive materials, photovoltaic, sensing, data storage, nano-electronics and stimuli-responsive materials. Of particular interest are the synthetic polymers containing metal centres that offer a good strategy to construct low cost high performance optoelectronics and smart materials in the future. The presence of transition metal elements in a polymer chain offer an attractive platform for combining the chemical, electronic, magnetic, optical and redox properties of metal complexes with those of the organic materials.

A dynamic-covalent metal-containing polymer was synthesized by the condensation of linear diamine and dialdehyde subcomponents around copper(I) templates in the presence of bidentate phosphine ligands. In solution, the red polymers undergo a sol–gel transition upon heating to form a yellow gel, a process that can be either reversible or irreversible depending on the solvent used. When fabricated into a light-emitting electrochemical cell (LEC), the polymer emits infrared light at low voltage. As the voltage is increased, a blue shift in the emission wavelength is observed until yellow light is emitted, a process whichis gradually reversed over time upon lowering the voltage. The mechanism underlying these apparently disparate responses is deduced to be due to loss of the copper phosphine complex from the polymer.

Keywords: copper, metallopolymer, gel, electroluminescence

Temperature and Voltage Induced Electroluminescence Shift

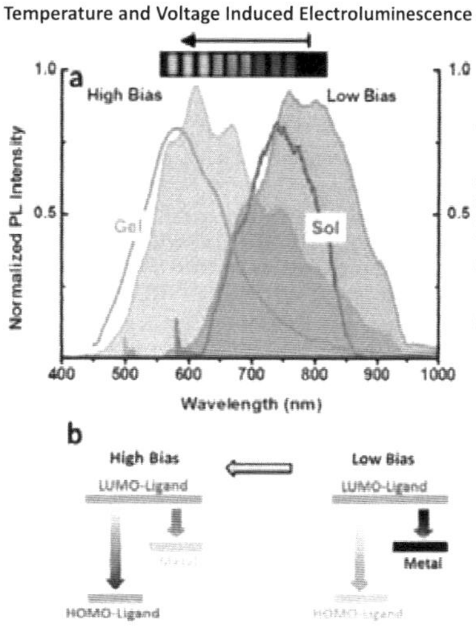

a) Photoluminescence spectrum of the polymer solution in DMF (red), corresponding gel (orange), and electroluminescence spectrum of the device at low bias (shaded red) and high bias (shaded orange). b) Schematic energy level diagram showing the dominant transition giving rise to electroluminescence emission for the device at low and high bias. (Figure is adapted from Demet Asil et. al., Angew. Chem. Int. Ed. 2014, 53, 8388–8391)

Bridging Continents & Bridging Molecules

MACRO2016

World Polymer Congress Istanbul 17-21 July

POSTER ABSTRACTS

Short Codes	
EOO	Energy, Optics, & Optoelectronics
IP	Industrial Polymers
MBM	Macromolecules in Biotechnology & Medicine
MN	Macromolecules & Nanotechnology
MSP	Modeling and Simulation of Polymers
PCNCT	Polymer Characterization by New and Combined Techniques
PE	Polymer Education
PEPC	Polymer Engineering, Processing, and Characterization
PP	Polymer Physics
PPBM	Polymer and Polymer-Based Membranes
PPG	Porous Polymer and Gels
RDPS-A	Recent Developments in Polymer Synthesis-A-Controlled/Living Polymerization
RDPS-B	Recent Developments in Polymer Synthesis-B-Complex Macromolecular Structures
RDPS-C	Recent Developments in Polymer Synthesis-C-Light-induced Reactions in Polymer Science
RDPS-D	Recent Developments in Polymer Synthesis-D-Condensation Polymerization and Thermoset
RRB	Renewable Resources and Biopolymers
SFP	Smart and Functional Polymers

Abstracts sorted alphabetically

EOO-P-001 [Energy, Optics, & Optoelectronics]

Non-Covalently Bonded Polymers Induced by π-Stacking: Synthesis of New Tribenzopentaphenes Derivatives and Investigation of their Properties

Bassam ALAMEDDINE[1], Christine Luscombe[2], Titus Jenny[3]

[1]Mathematics and Natural Sciences Department, Gulf University for Science & Technology, Kuwait
[2]Materials Science and Engineering Department, University of Washington, Seattle, USA
[3]Chemistry Department, University of Fribourg, Switzerland

We report the versatile synthesis of trapezoidal tribenzo[fj,ij,rst]pentaphene (TBP) derivatives 1a-h whose aromatic core induces a strong self-assembly by π- π interactions leading to the formation of non-covalent polymers. The synthetic strategy that we have developed offers the possibility to decorate the TBP core structure with various electroactive and/or solubilizing peripheral groups. Field effect transistor (FET) devices of the bis-diarylamine-TBP compounds were fabricated using spin coating under different concentration, spin speed, and solvent conditions. Emission spectra and surface investigation by atomic force microscopy (AFM) of 1a and 1b reveal the formation of non-covalently bonded aggregates induced by the strong π-π stacking of the aromatic core leading to island features. Investigation of the other derivatives 1c-h is underway.

Keywords: Organic Chemistry, Polymer Chemistry, Energy, Materials, and Nanotechnology for Resource Sustainability

EOO-P-002 [Energy, Optics, & Optoelectronics]

Effect of the Addition of Additives on the Improvement of the Performance of Lead–Acid Batteries

Foudia Malika[1], Zerroual Larbi[2], Deafa Imad[3]

[1]Foudia Malika (Laboratoire D'energétique Et Electrochimie Du Solide (Lees), Département De Génie Des Procédés, Faculté De Technologie, Université Sétif -1, Sétif 19000, Algeria
[2]Zerroual Larbi (Laboratoire D'energétique Et Electrochimie Du Solide (Lees), Département De Génie Des Procédés, Faculté De Technologie, Université Sétif -1, Sétif 19000, Algeria
[3]Drefa Imad (Laboratoire D'energétique Et Electrochimie Du Solide (Lees), Département De Génie Des Procédés, Faculté De Technologie, Université Sétif -1, Sétif 19000, Algeria

The lead–acid battery has many advantages, such as high power density, wide usable temperature range, complete recycling system and low price. For these reasons, the lead–acid battery has been used for numerous purposes. Therefore much research has been done to improve its capacity by adding positive plate additives.The objective of this work is to improve the electrical properties of lead-acid battery with the addition of additives in electrolyte and in the cured plates before oxidation. The results showed that the addition of surfactant (STTP) in sulfuric acid and 3% mineral additive (alumino-silicate) in the cured plates change the morphology and the crystallite size of PAM after oxidation. The discharge capacity increases with the decrease of the crystallite size and the resistance of the active mass. This shows that the addition of mineral additive and the surfactant additive to the PAM, the electrical performance and the cycle life of lead- acid battery are significantly increases. Key words: lead-acid battery, additives, positive plate, impedance (EIS).

Keywords: lead-acid battery, additives, positive plate, impedance (EIS).

EOO-P-003 [Energy, Optics, & Optoelectronics]

Effect of solution slow rate on growth and characterization of nanostructured ZnO thin film deposited using spray pyrolysis

Chabane Sari Sidi Mohammed[1], Zargou Salem[1], Senoudi Rachida Assia[1], Aida Mohammed[2], Attaf Nasreddine[2], Hakem Ilhem Faiza[3]

[1]Laboratory of Research on Macromolecules, Faculty of Sciences, University Abu Bekr Belkaid, Tlemcen, Algeria.
[2]Thin Films and Interfaces Laboratory, Faculty of Science, University of Constantine 1, Algeria.
[3]Materials Science and Engineering Department, Carnegie Mellon University 5000Forbes Avenue, Pittsburgh, PA 15213 USA.

Zinc Oxide (ZnO) is one of the promising II-VI semi conducting materials for technological applications (such as micro and optoelectronics, sensors, surface acoustics wave devices and solar cells) and thus plays an important role in several areas of physics, chemistry and materials science. It has been widely studied and continues to receive considerable attention, due to its low cost, nontoxicity, high stability and high optical transparency in the visible and near infrared region [1-6]. It has been well established that the optical and electronic properties of thin films are sensitive to minute variations in structural features that in turn sensitively depend on the fabrication conditions of thin film [7]. The economic viability of SPT derives from the advantage of inexpensive equipment (non-vacuum method), the ease of large area deposition and continuous fabrication by moving nozzle over a substrate. To achieve good quality sprayed thin films with the desired product composition and morphology, understanding of the role of process parameters such as substrate temperature, spray rate, solution concentration and, pressure of the carrier gas on the product characteristics is crucial [15]. In SPT, spray rate (Sf) optimization is a crucial step because it governs the preferred orientation, structural, optical and electrical properties [16-18]. The objective of the present work is to contribute to the better understanding of the role of process parameters on the physical properties of ZnO films fabricated by SPT. In particular, ZnO thin films were fabricated using SPT with systematically varied solution flow rate to establish the role of Sf on the optical, electrical, structural properties of ZnO films and to identify process conditions to maximize the figure of merit (ΦTC) of thin films.

Keywords: ZnO thin films, Ultrasonic spray, XRD, Optical transmission, Electrical properties

EOO-P-004 [Energy, Optics, & Optoelectronics]

Alkyl-end-group Modification of Benzotriazole and Thiophene Containing Conjugated Polymers

Şevki Can Cevher[1], Duygu Keleş[2], Şerife Özdemir Hacıoğlu[1], Levent Toppare[1], Levent Toppare[2], Levent Toppare[3], Levent Toppare[4], Ali Çırpan[1], Ali Çırpan[2], Ali Çırpan[4], Ali Çırpan[5]

[1]Department of Chemistry, Middle East Technical University, 06800 Ankara, Turkey
[2]Department of Polymer Science and Technology, Middle East Technical University, 06800 Ankara, Turkey
[3]Department of Biotechnology, Middle East Technical University, 06800 Ankara, Turkey
[4]The Center for Solar Energy Research and Application (GUNAM), Middle East Technical University, 06800 Ankara, Turkey
[5]Department of Micro and Nanotechnology, Middle East Technical University, 06800 Ankara, Turkey

This study focused on modifying the alkyl-end-group of benzotriazole and thiophene containing conjugated polymers. Introducing a moiety with structurally rigid and tendency to enhance the π-π stacking and also absorption in the high energetic region, influence the both electronic and optic properties of the conjugated polymers. Stille cross coupling chemical polymerization method was performed to synthesize the polymer contains: benzotriazole and bisthiophene. Characterizations of polymers were done in terms of molecular weight (GPC), absorption wavelengths (Uv-vis spec.), and electronic properties (CV). Broad absorption of polymer, which was located at λmax 582 and 533 nm, was found as great candidate for optoelectronic devices. However HOMO and LUMO levels play an important role for photovoltaic devices. The preliminary studies showed that low Voc and FF values as 0.4 V and 0.36 respectively. Without further optimizations Polymer:PCBM bulk heterojunction photovoltaic cell performance was at most 0.48% efficiency.

Keywords: polymer, solar cell, benzotriazole

EOO-P-005 [Energy, Optics, & Optoelectronics]

Air electrode based on conducting polymer for enhanced Li-air batteries

Yong Joon Park, Seon Hye Yoon, Dae Ho Yoon

Department of Advanced Materials Engineering, Kyonggi University, Korea

Currently, Li-air batteries are the most impressive next-generation battery systems, because their energy storage is expected to far exceed that achievable with state-of-the-art Li-ion batteries [1-3]. However, the physical realization of these devices continues to face several major challenges, such as their significant over-potential, low rate capability, and limited cyclic performance. In the aprotic Li-air cell, air electrode stores are charged by the reversible formation and dissociation of Li2O2. One of the serious problems affecting this mechanism is that side reactions, such as Li2CO3 formation and electrolyte decomposition, occur simultaneously during this process. So, side reaction suppression is a key point of research as regards improving the electrochemical performance of Li-air batteries. It is notable that the side reactions are promoted by carbon, a widely used air electrode base-material. In this work, conducting polymer is proposed as an air-electrode composite material to reduce the side reaction derived from carbon. The property of the air electrode based on conducting polymer was characterized and compared with that of the air electrode based on carbon.

Reference:

1. Padbury, R. & Zhang, X., J. Power Sources 196, 4436-4444 (2011).
2. Li, F., Zhang, T. & Zhou, H., Energy Environ. Sci. 6, 1125-1141 (2013).
3. Christensen, J., Albertus, P., Sanchez-Carrera, R. S., Lohmann, T., Kozinsky, B., Liedtke, R., Ahmed, J. & Kojic, A., J. Electrochem. Soc. 159, R1-R30 (2012).

Keywords: conducting polymer, air electrode, Li air batteries, side reaction

EOO-P-006 [Energy, Optics, & Optoelectronics]

1,1'-Biferrocenylene Based D-A-D Oligomers

Okan Demirel[1], Rochus Breuer[2], Mehmet Emin Cinar[2], Michael Schmittel[2], Turan Ozturk[1]

[1]*Department of Chemistry, Istanbul Technical University, Istanbul, Turkey*
[2]*Department of Chemistry, University of Siegen, Siegen, Germany*

Fused heterocyclic molecules with high π-delocalization, such as thienothiophenes (TT) and dithienothiophenes (DTT), have been attracting attention as building blocks for the construction of organic semiconductors for organic light emitting diodes (OLEDs), organic field-effect transistors (OFETs), organic solar cells (OSCs) and photovoltaic cells [1]. Compared to the interest in the conjugated polymers, the growing interest in the small conjugated molecules originates from their easier synthetic strategies, welldefined molecular structures, easy purification methods, better processability, and reproducibility [2].

The synthesis of materials composed of electron-donating (D) and -accepting (A) units offer the ability to manipulate the HOMO–LUMO energy levels. Triphenylamine (TPA) as a donor with a good charge transport property provides high hole mobility [3]. The other electron donor group is 1,1'-biferrocenylene (BFD), having a mixed-valence class III character for the BFD0/BFD+ redox stage [4]. On the other side, benzothiadiazole (BT) is a strong acceptor and extensively utilized for designing low HOMO–LUMO gap systems [1b].

Keywords: Semiconductors, Oligomers, OLED, Synthesis

Structures of D-A-D oligomers 1a-c – 4a-c.

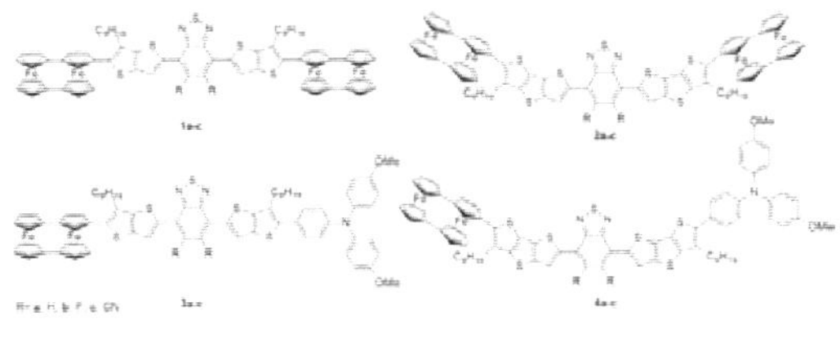

EOO-P-007 [Energy, Optics, & Optoelectronics]

[3+3] Imine and β-ketoenamine tethered fluorescent covalent-organic frameworks for CO2 uptake and nitroaromatic sensing

Kaleeswaran Dhananjayan, Pratap Vishnoi, Ramaswamy Murugavel

Indian Institute of Technology Bombay, Mumbai, India

Covalent-organic frameworks (COFs) comprise a unique class of purely organic materials with low mass density, high surface area, high thermal stability and long-range ordered architectures."O. M. Yaghi et al. *Science*, 2005, 310, 1166". COFs have gained significant attention in recent decades for their attractive design synthesis and widespread applications in gas storage and chemo-sensing "D. Jiang et al. *J. Am. Chem. Soc.*, 2013, 135, 17310". In this study a series of covalent organic frameworks with [3 + 3] structure motif have been synthesized from readily accessible building blocks via imine condensation, and the dependence of gas adsorption properties on the building block dimensions and functionalities was studied. The resulting imine and β-ketoenamine linked frameworks exhibit high surface area: the Brunauer–Emmett–Teller (BET) specific surface area up to 756 m2 g−1, and a Langmuir surface area up to 1515 m2 g−1. More importantly, the COFs exhibit moderate H2 (up to 1.15 wt %, 77 K, 1 bar) and excellent CO2 (up to 18.0 wt %, 273 K, 1 bar) uptake, which are comparable to other reported organic porous materials. Besides, due to the inherent fluorescent capability of triphenylbenzene, the COFs are endowed with fluorescence and fluorescence chemo-sensing ability for polynitroaromatic analytes "R. Murugavel et al. *J. mater. Chem. C*, 2015, 3, 7159".

Keywords: Covalent Organic Frameworks, Schiff base, Carbon dioxide capture and Sensing

Figure 1. (a) CO2 uptake and (b) fluorescence quenching of COFs by picric acid.

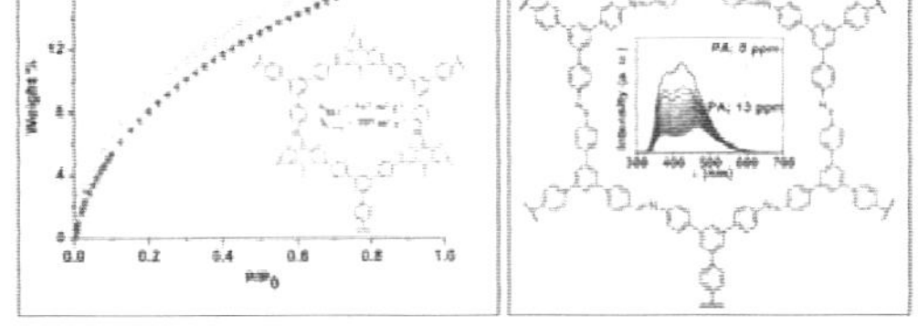

EOO-P-008 [Energy, Optics, & Optoelectronics]

Triphenylmine and Tetraphenylethylene Substituted Thienothiophene for OLED Applications

Garen Suna[1], Sebahat Topal[1], Turan Öztürk[2]

[1]*Istanbul Technical University, Department of Chemistry, 34469, Maslak, Istanbul*
[2]*TUBITAK UME, Chemistry Group Laboratories, PBox 54, 41470, Gebze-Kocaeli, Turkey*

Organic electronics is a fast developing research area, which already revolutionized the display market. Triphenylamine (TPA) is a well-known hole-transporting material but may suffers aggregation-caused emission quenching in solid state. Tetraphenylethene (TPE), on the other hand, is an archetypal luminogen that shows the phenomenon of aggregation-induced emission (AIE). One of the important advantages of using thiophene in designing organic materials is easy modification for tuning the electronic properties of the materials such as electron affinity, ionization potential,absorption and emission.1 Such tuning could also be performed by fusing thiophene rings,2 among which thienothiophenes (TT) play an important role as their core skeleton consists of two fused rigid thiophene rings to limit the rotational disorder between the rings and create a better π-conjugation.3,4 In this study, we have used the AIE effect of TPE molecule by incorporating TPE substituted phenyl groups to TT to preclude AQC leading to higher quantum yields in solid state, which was found to be 0.407. Further studies such as adding some acceptors is under investigation.

REFERENCES:

1. Roncali, J. Macromol. Rapid Commun. 2007, 28, 1761–1775.
2. Ozturk, T.; Ertas, E.; Mert, O. Tertrahedron 2005, 61,11055–11077
3. Skabara, P. J. In Handbook of Thiophene-basedMaterials; Perepichka, I. F., Perepichka, D. F., Eds.; John Wiley &Sons: Chichester, U.K., 2009; Chapter
4. Comel, A.; Sommen, G.; Kirsch, G. Mini-Rev. Org. Chem. 2004,1, 367–374.

Keywords: OLED, TPA, TPE, TT, AIE

Scheme 1

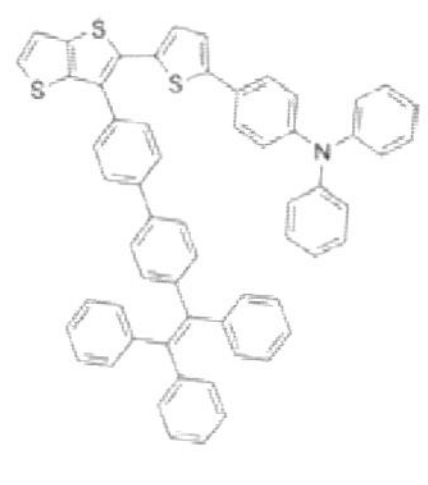

EOO-P-009 [Energy, Optics, & Optoelectronics]

Boron and Tetraphenylethene Substituted Thienothiophene Possessing Molecules for OLED Applications

Pelin Ulukan[1], Turan Öztürk[2]

[1]*Istanbul Technical University, Department of Chemistry, 34469, Maslak, Istanbul*
[2]*TUBITAK UME, Chemistry Group Laboratories, PBox 54, 41470, Gebze-Kocaeli, Turkey*

Organic light emitting diodes (OLEDs) are a great invention in display technology [1]. Since they show high color brightness, small film thickness and low production costs, they promise great developments in display market. OLEDs operate by converting electrical energy into light, a phenomenon called electroluminescence. The luminophores responsible for electroluminescent properties in these structures may show reduced emission in aggregate state compared to their solution states and consequently they experience emission quenching called as aggregation caused quenching (ACQ). There is another photophysical phenomenon known as Aggregation Induced Emission (AIE), which refers to the emission emerging from aggregate formation. Luminogens exhibiting AIE are termed as AIEgens. TPE (tetraphenylethene) is a kind of AIEgen with its central olefin stator and four phenyl rings. TPE possessing molecules start to emit light through aggregate formation. Molecules a and b, synthesized in our previous studies and having mesitylborane units as acceptors linked through electron rich thienothiophene (TT) π-conjugated spacers [5], demonstrated high quantum yields of 0.53 and 0.38 in solution, respectively, whereas they showed extremely low quantum yields in solid state owing to the ACQ. In this study, we have used the AIE effect of TPE molecule by incorporating TPE substituted phenyl groups to TT (c and d) to preclude AQC leading to higher quantum yields in solid state

REFERENCES:

1. C.W. Tang, S.A. Van Slyke, Appl. Phys. Lett. 51 (1987) 913 915.
2. a) T. Ozturk, Tetrahedron Lett. 37 (1996), 2821–2824. b) E. Ertas, T. Ozturk, Chem. Commun. (2000), 2039–2040. c) T. Ozturk, J. D. Wallis, Acta Crystallogr. Sect. C, 52 (1996) 2552–2554.
3. A. Wakamiya, T. Taniguchi, S.Yamaguchi, Angew. Chem. Int. Ed. 45 (2006) 3170-3173.
4. F. Turksoy, J. D. Wallis, U. Tunca, T. Ozturk, Tetrahedron 59 (2003) 8107–8116.
5. M. E. Cinar, T. Ozturk, Chem. Rev. 115 (2015) 3036–3140.

Keywords: Organic light emitting diodes (OLEDs), aggregation caused quenching (ACQ), Aggregation Induced Emission (AIE),

Figure 1. Molecules a–d

EOO-P-010 [Energy, Optics, & Optoelectronics]

Greener Reaction for High Electron Mobility Copolymer: Naphthalene Diimide Copolymers via Direct Arylation Polycondensation with Controlled Molar Mass and High Performance

Rukiya Matsidik[1], Hartmut Komber[2], Alessandro Luzio[3], Mario Caironi[3], Michael Sommer[1]

[1]*Institut für Makromolekulare Chemie, Universität Freiburg, Germany*
[2]*Leibniz Institut für Polymerforschung Dresden e.V., Dresden, Germany*
[3]*Center for Nanoscience and Technology @PoliMi, Instituto Italiano di Tecnologia, Milano, Italy*

Among the few structures with suitable n-channel properties available, naphthalene diimide (NDI) represents an impressive exception due to its high electron mobilities and unconventional morphological behaviour.[1-2] However, an unfavorable issue of NDI main-chain copolymers arises from the predominant preparation by Stille polycondensation involving highly neurotoxic organostannanes.[2] The most simple and straightforward way to construct conjugated polymers is the direct activation of a C–H bond. In direct arylation polycondensation (DAP), the synthesis of monomers with organo-metallic functional groups is redundant. This makes DAP superior compared to classical methods regarding simplicity, cost, environmentally friendly building blocks, scalability, and purity.[4] Here we establish a very simple and highly efficient DAP protocol for main-chain NDI copolymers in aromatic solvents with controlled molar mass and high performance.[5-6] Most critical is the suppression of two major termination reactions of NDIBr end groups: nucleophilic substitution and solvent end-capping by aromatic solvents via C-H activation. Indeed, these termination reactions are the hitherto unsolved issue associated with the copolymerization of the NDI unit and suitable C-H comonomers. The in situ solvent end-capping of a C-H reactive solvent (e.g. toluene) can be used to control molecular weight (MW) by varying monomer concentration, whereby end-capping is efficient and MW is low for low concentration and vice versa. Moreover, reducing C-H reactivity of aromatic solvent (e.g. mesitylene), leads to substantially increased MW. Chain perfection of obtained polymers is demonstrated by in detail NMR spectroscopy, which reveals chains to be fully linear and alternating. This is further confirmed by investigating the optical and thermal properties. Field-effect transistor (FET) electron mobilities are measured using off-centre spin-coating, with FET devices made from DAP synthesized polymers exhibiting better reproducibility compared to Stille controls.

1.Nature. 2009,457,679.
2.J. Mater. Chem. C, 2014, 2,3088.
3.Acc. Chem. Res. 2008,41,1110.
4.J. Am. Chem. Soc., 2015,137,6705.
5.ACS Macro Lett. 2015,4,1346.

Keywords: direct arylation polycondensation, naphthalene diimide, solvent C-H reactivity, solvent in situ end-capping, electron mobility

EOO-P-011 [Energy, Optics, & Optoelectronics]

Improvement of capacitive performances of supercapacitors by incorporation of exfoliated montmorillonite into conducting polymers

Damia Amoura[1], Margarita Sánchez Jiménez[2], Francesc Estrany[3], Laid Makhloufi[1], Carlos Alemán[4]

[1]*Laboratoire d'Electrochimie, Corrosion et de Valorisation Energétique (LECVE), Faculté de Technologie, Université de Bejaia, 06000 Béjaia, Algérie*
[2]*Departament d'Enginyeria Química, Escola Universitaria d'Enginyeria Tècnica Industrial de Barcelona, Universitat Politècnica de Catalunya, Comte d'Urgell 187, 08036 Barcelona, Spain*
[3]*Center for Research in Nano-Engineering, Universitat Politècnica de Catalunya, Campus Sud, Edifici C', C/Pasqual i Vila s/n, Barcelona E-08028, Spain*
[4]*Departament d'Enginyeria Química, E. T. S. d'Enginyers Industrials, Universitat Politècnica de Catalunya, Diagonal 647, 08028, Barcelona, Spain*

Conducting polymers have recently attracted considerable scientific and technological interest because of their interesting electrical, electrochemical, optical and magnetic properties. In the present work, the incorporation of exfoliated montmorillonite (MMT) into conducting polymers consisting in 3-layered films made of poly(3,4-ethylenedioxythiophene) (PEDOT) and poly(N-methylpyrrole) (PNMPy) arranged alternatively (prepared using the layer-by-layer electropolymerization technique) has been investigated and the structural, electrochemical and capacitive properties of the obtained material are explored using scanning electron microscopy (SEM), tapping-mode atomic force microscopy (AFM), cyclic voltammetry and chronopotentiometry. The influence of the clay on the properties has been investigated as function of the thickness of the internal and external PEDOT layers, which has been controlled through the polymerization time. The results show that the incorporation of MMT has an important impact for the activation of the electrochemical properties (i.e. electroactivity, electrostability and specific capacitance) of the layered conducting polymers. The ability to exchange charge reversibly and the electrochemical stability of films with MMT at the intermediate layer are higher than those of films without clay, these effects being more pronounced for films with micrometric PEDOT layers. The clay at the intermediate layer improves the specific capacitance of micrometric films and, most importantly, protects the capacitive behavior against electrochemical degradation in all cases.

Keywords: Conducting polymers, PEDOT, Polypyrrole, Montmorillonite, Supercapacitor

EOO-P-012 [Energy, Optics, & Optoelectronics]

Synthesis and Optoelectronic Properties of Mono-Alkylated Thieno[3,2-b]Thiophenes and Their Polymers

Koray Tansu Ilhan, Sebahat Topal, Turan Ozturk

Departmant of Chemistry, Istanbul Technical University, Istanbul, Turkey

In recent years, application area of conducting polymers has been continuously extending which gives rise to an impetus in both scientific researches and synthesis of brand new compounds by different research groups [1]. Organic light-emitting diodes (OLEDs), one of this application area of these kind of compounds, are going to improve the human lifestyle by offering a promising avenue to develop future energy saving solid-state lighting sources because of their intrinsic characteristics such as low driving voltage, high resolution, high brightness, large viewing angle, large color range, high contrast, less weight and size, and efficiency etc.. One of the most trend compounds on this area is thiophene based polymers with their wide range of applications. These compounds have not only chemical stability on chemical and electrochemical field but also applications on electronic and optoelectronic area due to their distinct electronic structures [2]. The simplest fused thiophenes having two annulated thiophene units are named as thienothiophenes (TTs) [3]. They are electron-rich structures which enable them to construct conjugated and low band gap organic semiconductors. Aim of this project is to synthesize new thienothiophene compounds containing various alkyl groups and their polymers, electronic and optoelectronic properties of which are anticipated to be promising for material chemistry. The syntheses of thieno[3,2-b]thiopenes having alkyl groups given in the literature involves four steps with undesired side products. Therefore, these compounds have not been investigated in terms of their applications. Our synthetic methodology with only two steps and high yields is an important contribution to the material chemistry.

REFERENCES:

1. Roncali, J.,1992, Chem. Rev., 92, 711-738.
2. (a) McCullough, R. D.1998, 10, 93-116.
(b) Roncali, J. 1999, J. Mater. Chem., 9, 1875-1893.
3. Cinar, E., Ozturk, T. 2015, Chemical Review, 115 (9), 3036–3140.

Keywords: Thiophenes, Thienothiophenes, Polymers, Solar Cells, Electropolymerization

Figure1

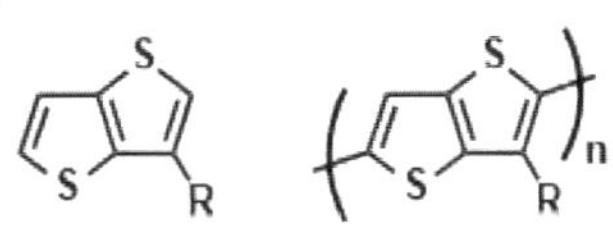

R= $-CH_3$, $-C_2H_5$, $-C_3H_7$, $-C_4H_9$, $-C_5H_{11}$, *iso*-C_5H_{11}
$-C_6H_{13}$, $-C_7H_{15}$, $-C_8H_{17}$, $-C_9H_{19}$

EOO-P-013 [Energy, Optics, & Optoelectronics]

Donor-Acceptor Systems Possessing Tetraphenylethylene Substituted Dithienothiophene-S,S-dioxides for OLED Applications

Rengin Büşra Özek[1], Gözde Aydın[2], Turan Öztürk[3]

[1]*Department of Chemistry, Istanbul Technical University, Istanbul, Turkey*
[2]*Department of Chemistry, Yildiz Technical University, Istanbul, Turkey*
[3]*Chemistry Group Laboratories, TUBITAK UME, Gebze-Kocaeli, Turkey*

The current researches in material chemistry demonstrate that conjugated organic molecules with donor–acceptor (D–A) units represent a relevant class of electro-conducting materials. These molecules have fine abilities to be used in organic light-emitting diodes (OLEDs) which are made by placing a thin film of them between two conductors of dissimilar work functions [1]. Additionally, dithienothiophenes (DTT) as being rich in electrons owing to three fused thiophenes are electron donors with attractive electrochemical and optical properties [2]. By courtesy of having extended π-conjugation and rigid structure, DTTs are considered as one of the most promising organic molecules in material science [3]. S,S-dioxide which has fluorescent property with, in general, an emission in the blue region. In spite of their ability to give high luminescence efficiency in dilute solutions, most of the organic molecules show relatively weak or no emissions in their solid states [4,5]. Thus, it is highly desirable to use electroluminescent materials possessing aggregation-induced emission (AIE) such as tetraphenylethylene (TPE), in order to overcome the aggregation quenching problem with achieving emission in solid state [5]. Inthisongoingresearch,itisaimedthatopticalandelectronicalpropertiesofDTT-S,S-O2 derivatives containing TPE units and π-spacer thiophenes will be investigated.

REFERENCES:

J. Morgado, N. Barbagallo, A. Charas, M. Matos, L. Alcácer, F. Cacialli, J. Phys. D: Appl. Phys., 36 (2003) 434–438.
T. Ozturk, E. Ertas, O. Mert, Tetrahedron, 61 (2005) 11055–11077. 3. M. E. Cinar, T. Ozturk, Chem. Rev., 115 (2015) 3036–3140.
4. I. Afonina, P. J. Skabara, F. Vilela, A. L. Kanibolotsky, J. C. Forgie, A. K. Bansal, G. A. Turnbull, I. D.
W. Samuel, J. G. Labram, T. D. Anthopoulos, S. J. Coles, M. B. Hursthoused, J. Mater. Chem. 20 (2010), 1112.
5. D. Jana, B. K. Ghorai, K. Tetrahedron, 53 (2012) 6838–6842.

Keywords: OLED, Donor-Acceptor Systems, Tetraphenylethylene, Dithienothiophene, aggregation-induced emission (AIE)

Target Molecule

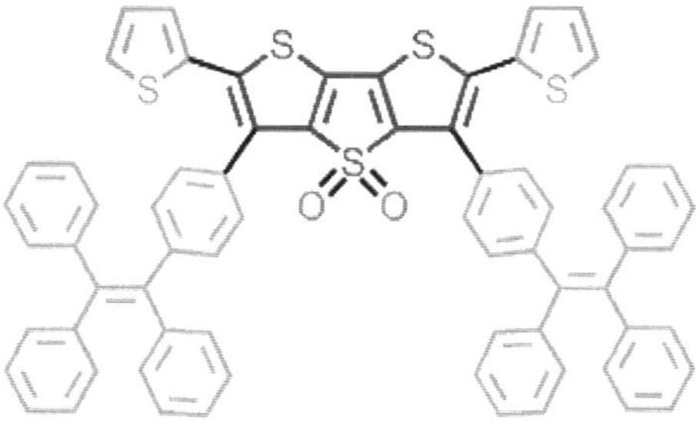

green: TPE, blue: DTT, orange: thiophene

EOO-P-014 [Energy, Optics, & Optoelectronics]

preparation and caracterisation by X ray diffraction of LiFexCo(1-x)O2

Ouartsi Ibticem

University 20Aout1955 Skikda

in this work we prepared samples from metalique oxydes of lithium, fer and cobalt which are used in rechargeable lithium batteries as cathodes. these oxydes are mixed according to their stoichiometric ratio then caractherized by X ray diffraction.

Keywords: lithium, batteries,cathodes

EOO-P-015 [Energy, Optics, & Optoelectronics]

From Polymers to High-Capacity Anode Nanostructures for Lithium-ion Batteries

Serap Hayat Soytas[1], Ali Ansari Hamedani[2]

[1]*Sabanci University Nanotechnology Research and Application Center, Sabanci University, Istanbul, Turkey*
[2]*Material Science and Nanoengineering, Sabanci University, Istanbul, Turkey*

Lithium-ion batteries are currently the most commonly used rechargeable energy storage systems due to their high energy density and specific energy capacity compare to other commercial battery technologies and have enabled important developments possible especially in consumer electronics since their first use in 1991. However, even the best battery systems available provide insufficient energy density and durability for transportation applications. Therefore, we need to develop new electrode materials that will provide long cycle life, high-energy capacity and power density. Silicon (Si) is one of the most promising anode materials for lithium-ion batteries because of its high theoretical specific capacity (4000 mAh/g) compared to the commercial graphite anodes (372 mAh/g). However, it experiences a high volume change during repeated charge-discharge cycles and rapidly loses its capacity. The use of nanosized Si helped to compensate for the high volume change during cycling. Si nanowires, nanotubes, nanoparticles and the like are some examples of nanosized Si, which showed promising performances in laboratory scale, but they have limited scalability because their production mostly requires chemical vapor deposition (CVD) relying on toxic silane gas or complicated experimental conditions. In this study, carbon nanofiber (CNF)/Si nanostructures are produced using affordable and easy-to-disperse Si precursors, instead of expensive and hard-to-disperse commercial Si nanoparticles. For this purpose, copolymers of acrylonitrile and alkoxysilanes are developed both as Si and CNF precursors. These newly developed copolymers are subjected to electrospinning – a simple and cost-effective method for nanofiber manufacturing – for the production of electrospun silane-containing nanofibers. Then, a single step heat treatment under autogenic pressure is employed for the production of silicon decorated CNF nanocomposite structures to be used as freestanding anodes in lithium-ion batteries.

Keywords: polyalkoxysilanes, electrospinning, lithium-ion battery

EOO-P-016 [Energy, Optics, & Optoelectronics]

Ferrocene-Containing Polythiophene and Its Use as a Cathode Material for Li-ion Batteries

Mesut Görür[1], Muhammet Aydın[2], Faruk Yılmaz[2]

[1]*Department of Chemistry, Istanbul Medeniyet University, Istanbul, 34700, Turkey*
[2]*Department of Chemistry, Gebze Technical University, Kocaeli, 41400, Turkey*

One of the recent challenges for scientific community is to develop fast-charging, lightweight, small, thin, and environmentally safe portable energy storage systems to supply increasing energy demands of growing portable electronics market [1]. The most common batteries used in portable consumer electronics and electrical vehicles are Li-ion batteries. These systems contain inorganic transition metal-oxide-based cathode-active materials, which depend on limited mineral resources. Therefore, scientists are in the pursuit of developing new electrode materials. In this sense, ferrocene (Fc) and its derivatives have been used as cathode-material in battery applications because of its air stability, excellent electrochemical response, and redox stabilities [3]. In this work, Fc-containing polythiophene polymer was prepared in a four-step synthetic procedure (Scheme 1). In the first step, compound 1 with two bromide functional groups was synthesized via esterification reaction of 3-thiophene acetic acid with 1,3-dibromo-2-propanol and then it was azidified by treating with sodium azide in DMF, yielding azide-functional thiophene derivative (2). 1,3-dipolar cycloaddition (click) reaction between 2 and ethynyl-Fc afforded Fc-functional thiophene derivative (3). Subsequently, 3 oxidatively polymerized with FeCl3 in CCl4 and redox-active polymer with Fc side-groups (P1). Finally, P1 was employed as the cathode material in the rechargeable Li-ion button cell assembly. The produced battery was then characterized for its charge-discharge cycle performances.

*The authors acknowledge TUBITAK for the financial support (Project No: 114Z302).

References:

1. M. Aydin, M. Gorur, and F. Yilmaz, React. Funct. Mater. Accepted Manuscript (doi:10.1016/j.reactfunctpolym.2016.02.008).
2. C. Su, Y. Ye, L. Xu, and C. Zhang, J. Mater. Chem., 22 (2012), 22658-22662.
3. C. Su, L. Wang, L. Xu, and C. Zhang. Electrochim. Acta, 104 (2013), 302-307.

Keywords: Organic radical battery, Nitroxy stable free radical, Redox-active polymer

Scheme 1

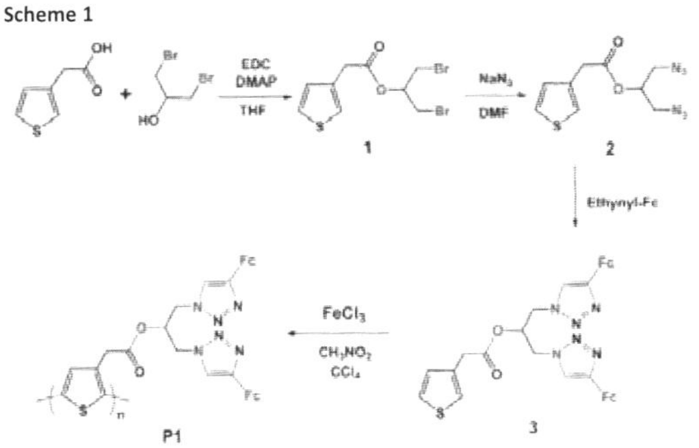

Synthesis of Fc-functional polythiophene (P1)

EOO-P-017 [Energy, Optics, & Optoelectronics]

Polythiophene with Densely Populated Organic Robust Radicals as Cathode Material in Li-Ion Batteries

Muhammet Aydın[1], Mesut Görür[2], Faruk Yılmaz[1]

[1]*Department of Chemistry, Gebze Technical University, Kocaeli, 41400, Turkey*
[2]*Department of Chemistry, Istanbul Medeniyet University, Istanbul, 34700, Turkey*

The ratio of portable equipment in the consumer electronics market have been increasing during the past decades. It necessitates the development energy storage devices with high energy density, power capacity as well as long cycle life. Current mobile devices generally includes conventional Li-ion batteries with Li insertion-based cathode materials (e.g. $LiCoO_2$ and $LiMn_2O_4$), relying on unrenewable mineral resources [1]. Organic radical batteries (ORB) have attracted great deal of scientific attention because of their fast redox kinetics, high power capacity and energy density. They are accepted as environmentally friendly materials since they do not contain any toxic heavy metals. Polymers with 2,2,6,6-tetramethylpiperidine-1-oxyl pendant groups have been widely used as cathode component in Li-ion rechargeable batteries [2]. In this work, brush-shaped polythiophene with pendant poly(2,2,6,6-tetramethylpiperidinyloxy-4-yl methacrylate) (PTMA) arms (P3) was prepared in a four-step synthetic procedure (Scheme 1). Bromide functional thiophene compound (1), which was obtained through a reaction of thiophene acetic acid with 1,3-dibromopropan-3-ol, was oxidatively polymerized in the presence of $FeCl_3$, yielding bromide functional polythiophene (P1). Then, P1 was used as a macroinitiator in the atom transfer polymerization (ATRP) of 2,2,6,6-tetramethyl-1-piperidin-4-yl methacrylate (TMPM) and the yielded PTMPM (P2) was subsequently oxidized with m-chloroperbenzoic acid (m-CPBA), giving polytiophene with pendant PTMA arms (P3). Finally, P3 was employed as the cathode material in the preparation of rechargeable Li-ion button cell. The produced batteries were then characterized for their charge-discharge cycle performances.

*The authors acknowledge TUBITAK for the financial support (Project No: 114Z302).

References:

1) G. Hauffman, A. Vlad, T. Janoschka, U. S. Schubert, J.F. Gohy, J. Mater. Chem. A, 2015,3, 19575-19581.
2) W. Guo, Y. X. Yin, S. Xin, Y. G. Guo, L. J. Wan, Energy Environ. Sci., 2012, 5, 5221–5225.

Keywords: Organic radical battery, Nitroxy stable free radical, Redox-active polymer

Scheme 1

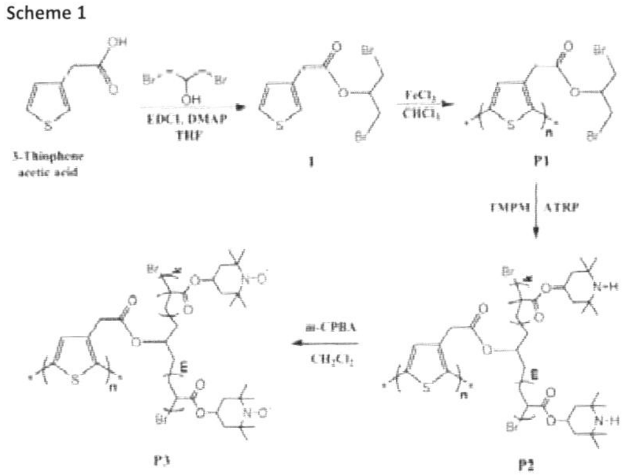

Synthetic route for the preparation of P3

EOO-P-018 [Energy, Optics, & Optoelectronics]

A post-functionalizable carbazole based low bandgap polymer for organic photovoltaic applications

Deniz Kiymaz[1], Ceylan Doyranlı[2], Hakan Bilgili[1], Burak Gültekin[1], Pieter Verstappen[3], Jurgen Kesters[3], Wouter Maes[3], Ceylan Zafer[1], Dirk Vanderzande[3], Sermet Koyuncu[4]

[1]*Ege University Solar Energy Institute, TR-35100, Izmir, Turkey*
[2]*Canakkale Onsekiz Mart University, Polymeric Materials Research Lab, TR-17020 Canakkale, Turkey*
[3]*Hasselt University, Institute for Materials Research (IMO), DSOS, Agoralaan 1 Bldg D, B-3590 Diepenbeek, Belgium*
[4]*Canakkale Onsekiz Mart University, Department of Chemical Engineering, TR-17020 Canakkale, Turkey*

The photovoltaic behavior of bulk heterojunction (BHJ) blends of conjugated polymers and (methano)fullerene derivatives has been subject of many fundamental studies in chemistry and materials sciences [1]. For a BHJ organic solar cell to work effectively, charge transfer (in the excited state) between the electron donating and electron accepting components in the photoactive layer must be very efficient [2]. Moreover, to obtain highly efficient solution-processed organic photovoltaic (OPV) devices, good film-forming properties must be combined with well-defined nanostructured active layer morphologies to fulfill the complex requirements of light absorption, charge separation and charge transport in confined geometries. Improving the interaction between the donor and the acceptor materials in an effort to control the morphology at the nano-scale (simultaneously trying to optimize the stability of the device) remains a fundamental challenge [3]. Herein, we report on the synthesis of a diketopyrrolopyrrole-carbazole (DPP-Cbz) based low bandgap polymer (BP-3) with post-functionalizable allyl side chain. Electrochemical and optical properties were investigated using cyclic voltammetry and UV-Vis absorption spectroscopy, respectively. We have achieved a high power conversion efficiency of over 6% by using a BP-3:P-C71BM (1:3) blend in ortho-dichlorobenzene (ODCB). This efficiency can possibly be further improved by attaching other functional moieties to the carbazole subunits.

We gratefully acknowledge the support from the Scientific and Technical Research Council of Turkey (TUBITAK) and the Research Foundation – Flanders (FWO); project number: 114Z064.

1. S. Gunes, H. Neugebauer, N. S. Sariciftci, Conjugated Polymer-Based Organic Solar Cells, Chem. Rev. 2007, 107, 1324–1338.
2. L. Lu, T. Zheng, Q. Wu, A. M. Schneider, D. Zhao, L. Yu, Recent Advances in Bulk Heterojunction Polymer Solar Cells, Chem. Rev. 2015, 115, 12666–12731.
3. F. Liu, Y. Gu, X. Shen, S. Ferdous, H. W Wang, T. P. Russell, Characterization of the Morphology of Solution-Processed Bulk Heterojunction Organic Photovoltaics, Prog. Polym. Sci. 2013, 38, 1990–2052.

Keywords: low band gap polymers, carbazole, diketopyrrolopyrrole (DPP), photovoltaic applications

EOO-P-019 [Energy, Optics, & Optoelectronics]

Highly Efficient Poly(fluorene phenylene) Graft Copolymer as a New Class of Binder for High-Capacity Silicon Anode in Lithium-ion Batteries

Neslihan Yuca[1], Mehmet E. Cetintasoglu[2], Murat F. Dogdu[2], Huseyin Akbulut[3], Sevcan Tabanli[4], Uner Colak[2], Omer Suat Taskin[5], Yusuf Yagci[6]

[1]*Enwair Energy Technologies Corporation & Istanbul Technical University, Energy Institute, Istanbul, 34469, Turkey*
[2]*Enwair Energy Technologies Corporation, Maslak, 34469, Istanbul, Turkey*
[3]*Istanbul Technical University, Faculty of Science and Letters, Department of Chemistry, Maslak, 34469, Istanbul, Turkey*
[4]*Istanbul Technical University, Faculty of Science and Letters, Department of Physics Engineering, 34469, Maslak, Istanbul, Turkey*
[5]*Enwair Energy Technologies Corporation, Maslak 34469 & Istanbul University, Institute of Marine Science and Management, Department of Chemical Oceanography, Vefa 34134, Istanbul, Turkey*
[6]*Istanbul Technical University, Faculty of Science and Letters, Department of Chemistry, Maslak, 34469, Istanbul, Turkey & King Abdulaziz University, Center of Excellence for Advanced Materials Research (CEAMR) and Chemistry Depart-ment, Faculty of Science, P. O. Box 80203, Jeddah 21589, Saudi Arabia.*

A novel approach involving the use of a fluorescent and ductile polymer for the high capacity Li-ion battery application is demonstrated. For this purpose, a conjugated polymer, poly(fluorene phenylene) graft copolymer, containing with lateral substituents, namely poly(ethylene glycol) (PEG) chains, as a potential building block for conjugation and electrolyte uptake was synthesized and characterized. The synthesis was achieved through Suzuki poly-condensation reaction in the presence of Pd(PPh3)4 catalyst by using independently prepared PEG functionalized di-bromo benzene in conjunction with dioctylfluorene-diboronic acid bis(1,3-propanediol) ester. The obtained conductive and flexible polymer was used as a binder for Si-anode with high performance. The observed full capacity of cycling of silicon particles, i.e., at C/3 with the capacity of ca. 605 mAh/g after 1000th cycle, confirms the high performance with-out any supplementary conductive additive. It was demonstrated that the designed and fabricated binder polymer with multi-functionality displays a high electronic conductivity, mechanical strength, high polarity, adhesion, and more importantly, the electrolyte uptake.

Keywords: li-ion battery, poly(fluorene phenylene), Suzuki coupling

Figure 1.

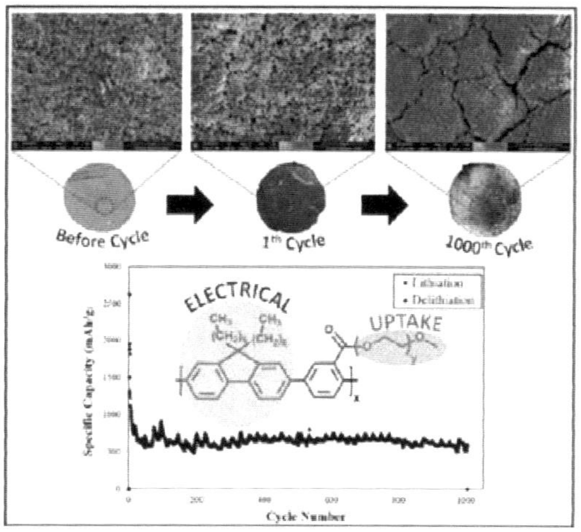

IP-P-001 [Industrial Polymers]

Changes in mechanical and antimicrobial properties of thermoplastic elastomers after natural ageing

Daiane Tomacheski[1], Michele Pittol[2], Douglas Naue Simões[1], Vanda Ferreira Ribeiro[1], Ruth Marlene Campomanes Santana[1]

[1]*Laboratory of Polymeric Materials, Federal University of Rio Grande do Sul, Porto Alegre, Brazil*
[2]*Softer Brasil Compostos Termoplásticos, Campo Bom, Brazil*

Nowadays, the spread of pathogenic bacteria have increased the search for products with antimicrobial properties. Thermoplastic elastomers (TPE) compounds are widely used for personal care objects, and can be produced with antimicrobial additives. The knowledge of how long the polymer maintain its bactericidal properties and how the loss occurs are important to ensure its efficacy during the operating life cycle. TPE samples were loaded with silver nanoparticles (NpAg, 0.5%) and titanium dioxide (TiO2, 4.0%); a compound with no antimicrobial additive (standard) was also tested. These samples were processed by extrusion and injection molded. Antimicrobial (antibacterial activity and fungal resistance) and mechanical properties of the samples were evaluated after 3 and 6 months of exposition to natural ageing. The changes on polymers surface were measured by FTIR-ATR. In six months of exposition, there was a reduction in 85% of tensile strength of standard sample, similar to sample loaded with NpAg (90%). TiO2 sample was more resistant, showing a reduction of 43% in tensile strength. The both metal loaded samples showed a decay of antimicrobial efficiency after 6 months of exposition, with a greater susceptibility of E. coli growth. It was notable that even after 6 months of exposition the TPE samples did not present a fungus mycelial growth. On the basis of the results obtained until now, it can be concluded that differences in antimicrobial properties of metal samples were due to changes on polymer matrices, and not the loss of additive due the leaching process. Samples with NpAg maintain their efficacy even after 6 months of natural ageing.

Keywords: Thermoplastic elastomers, nanosilver, titanium dioxide, antimicrobial

IP-P-002 [Industrial Polymers]

Precipitated Styrene/Butyl-Acrylate Emulsion Additive as Modifier on Some Properties of Acrylonitrile-Butadiene Rubber Vulcanizates

Adel Abd Elrehiem Koriem[1], Abbas Abd Elkarim Yehia[1], Mohamed Yossef El Zayat[2], Ahmed Ismail Hussain[1]

[1]*Department of Polymer and Pigment, National Research Center, Cairo, Egypt*
[2]*Department of Physics, Faculty of Science, Fayom University, Fayom, Egypt*

Nitrile Rubber (NBR) is commonly considered the work horse of the industrial and automotive rubber products. In this study styrene/butyle acrylate (SBt) emulsion was prepared using in-situ seed emulsion polymerization and precipitated in excess of methanol. The prepared (SBt) emulsion was used as modifier for acrylonitrile-butadiene rubber (NBR) in different doses (5 phr, 10 phr, 15 phr and 20 phr). The Rheometric characteristics and the physico-mechanical properties at room temperature and after aging of the vulcanizates, tensile strength, elongation at break and swelling in both toluene and motor oil, were studied. Also the morphology of the incorporated modifier and the dielectric properties of the vulcanizates were investigated. It was found that the dosage of 20 phr of the precipitated SBt emulsion has been incorporated successfully with (NBR) with the best physico-mechanical properties and dielectric properties as well.

Keywords: Acrylonitrile-butadiene rubber (NBR), Styrene/butyle acrylate (SBt), physico-mechanical properties, dielectric properties

Retained tensile strength of NBR vulcanizates containing styrene butyl acrylate copolymer versus aging time

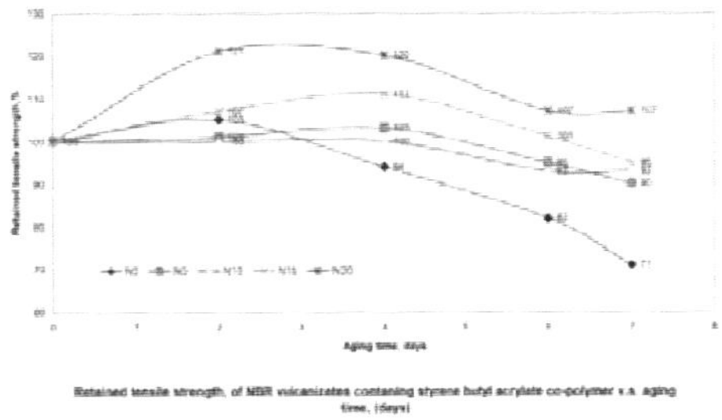

The retained values of tensile strength for all aging periods were calculated and illustrated as shown. The tensile strength was 71% for NBR without modification, however the value of tensile strength has been increased to 107% for the NBR loaded with 20 phr of styrene butyl acrylate co-polymer (SBt). The rubber sample containing 20 phr of the modified SBt has a great efficiency to protect the rubber vulcanizates against aging.

IP-P-003 [Industrial Polymers]

Study about thermal stability of the new biopolymer used in drill in fluids

Genaro José Bolívar[1], Manuel Antonio Mas[1], María Virginia Tortolero[1], Mayra Garcia[2]

[1]*PDVSA - INTEVEP*
[2]*Instituto Experimental Jardín Botánico "Dr. Tobías Lasser"*

In PDVSA-Intevep, we have formed a multidisciplinary team to assess the feasibility on the use of biopolymers from seaweeds because the sulfonate segments would promote the adequate thermo-resistance to the system. Therefore, in this paper we support the hypothesis about the use of biopolymer obtained from seaweed to design DRILL-IN fluids HPHT, as an alternative biopolymer used for maintain the thermal stability and salt resistance in compare with others polymers as PAM and Xanthan that have serious limitations.

Furthermore, viscosity studies for solutions λ-Carrageenan determine the thermal stability of the modules G 'and G'' in the range from 20 to 100 °C (68 – 212 °F), compared to those presented by xanthan gum solutions under the same conditions pH and ionic strength. Finally, the effects of ionic strength solutions for λ-carrageenan showed the opposite effect presented by xanthan gum solutions, where increased ionic strength of the environment promotes the decrease in viscosity values.

Keywords: Polysaccharide; Seaweed; Rhodophyceae; Carrageenans and Drill In

IP-P-004 [Industrial Polymers]

A New Technology for the synthesis of some interesting polymers on the industry

Hamama Boularaf Larkem[1], Lidya Larkem[2], Ali Larkem[1]

[1]Department of chemistry, faculty of Sciences, Annaba university. Annaba-Algeria
[2]Department of medicine, faculty of medicine, Annaba University. Annaba-Algeria

Recently it was found that compounds such as 1, 2 and 3 are very interesting for the synthesis of compounds 4 and 5 with new technology. As you know compound 1 was treated with NBS in DMF followed by second bromination in CCl4 to give2 which react with an excess of Na2 S or sulfuryl chloride on which in many steps produce compound 3.Compounds 4 and 5 are formed in good yield on which are necessary as a new polymer material could be used in the near future for many factories of the Industry.

Keywords: synthesis, bromomethylnaphthalene, polymers, sulfuryl chloride.

Synthesis of some new polymers

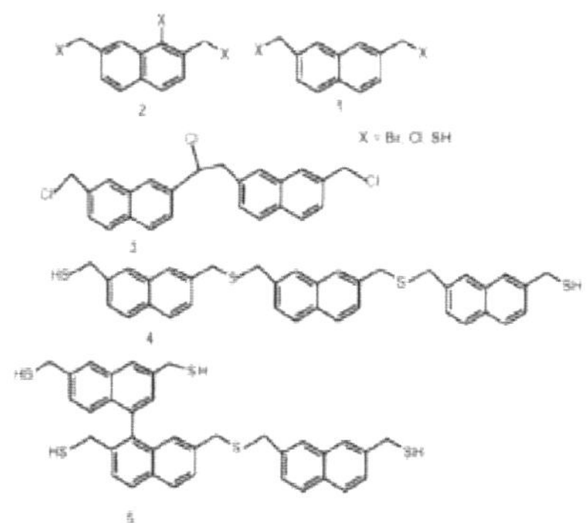

IP-P-005 [Industrial Polymers]

Study of effect of an oxidizing agent on the degradation of polypropylene bags submitted to thermooxydation

Farida Djerada Larfi[1], Kamel Khalil Bentifour[2], Samia Kaddour[3]

[1]Department of chemistry, A.Mira University, Béjaia,Algeria
[2]Sasace Industry, Algeria
[3]Department of Chemistry, Usthb University, Algiers, Algeria

This work focus on the effect of the rate of an oxidant agent on the degradation of polypropylene bags. This new class of plastics preserves the environment, since its degradation products are carbone dioxyde and water.

In this study, polypropylene bags in different percentage of oxidizing agent were prepared and submitted to thermo-oxidation at 80°C. Some samples were cut along the weft and the other samples were cut along the chain.

Each sample was characterised by FTIR, DSC and tensile tests. The results obtained showed that the degradation of the bags is depending on the percentage of oxidant agent and on the type of cutting.

Characterisation by FTIR showed the appearance of characteristic absorption bands of the carbonyl function in the bags containing the oxidizing agent. This is due to the split of macromolecular chain. The final products can be assimilated by mico-organisms.

Keywords: polypropylene, oxidizing agent, thermo-oxydation

FTIR spectra of weft the samples at different rates oxidant after expostion 36 days at 80 ° C in an oven

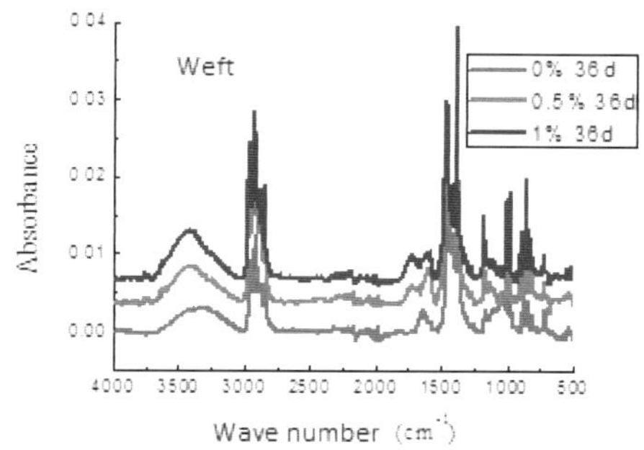

The FTIR spectra show an absorption band beteen 3100 and 3600 cm-1 corresponding to the elongation of O-H. another band is observed around 1700 cm -1 characteristic of carbonyl groups. The intensity of these bands increments with the rate of oxidant. The appearance of these bands is due to the split of macromolecular chains.

Elongation at break as a function of time

Time (Days)	Elongation at break (mm) 1% oxidant (chain)	Elongation at break (mm) 1% oxidant (weft)
0	14,71	12,5
8	14,68	11,16
15	14,11	11,02
26	11,92	8,74
29	9,93	7,91
33	8,79	7,19
36	7,88	6,94

The elongation at break decreases with time and is lower for weft samples. The degradation is better for this samples than for chain samples

IP-P-006 [Industrial Polymers]

Synthesis and Application of a New Polymeric Adsorbent for Dye Removal of Colored Wastewater

Maedeh Vafaee[1], Mohammad Ebrahim Olya[1], Behnaz Akbari[2]

[1]Department of environmental research, institute for color science and technology, Tehran, Iran.
[2]Department of Chemistry, Tehran North Branch, Islamic Azad University, Tehran, Iran

Azo dyes are regarded as xenobiotic compounds which their hazardous for the environment has been well-known [1]. Synthetic dyes are widely used in industries like textile, leather, food, cosmetics and papers [2, 3]. A large amount of dye has been lost in the manufacturing and utilizing process which lead to environmental pollution problems [4]. Numerous methods have been developed to treat colored effluents. Among those, some of them should be considered as chief methods; namely, coagulation, flocculation, biological treatment, ozone treatment, membrane filtration, colloidal gas aphrons, chemical oxidation, photocatalytic removal, and ion exchange [5–10]. In the present study, an attempt has been made to remove azo dye from an aqueous solution using polymeric adsorbent. Thus, the influence of Nano Poly Quaternary Ammonium Salt (PQAS) was carried out by the treatment and decolorization of wastewater. The polymeric adsorbent salt was synthetized in 3 steps, First, DE Isocyanate was reacted with N-methyl DE ethanol Amin. After preparing hydroxy ethyl mathacrylate, polymerized reaction was occurred. The key operating parameters such as adsorbent dosage, initial pH, dye concentration and temperature effect were explored within the experiments, the effect of time was a key concern. Furthermore, the effect of stirring on the adsorption of azo dye onto PQAS was examined. Moreover, the Langmuir model was used to quantify the amount of adsorption. The efficiency of the treatment and PQAS was evaluated via FT-IR spectroscopy and SEM techniques, respectively.

Keywords: Dye Removal, Poly Quaternary Ammonium Salt Adsorbent, Colored Wastewater

SEM

IP-P-007 [Industrial Polymers]

Mechanical Properties of Thermotropic Liquid Crysttalline Copolyesters with PA66 Blends

Melek Bulut[1], Zekeriya Yerlikaya[2]

[1]*Çanakkale Onsekiz Mart Üniversitesi, Acil Yardım ve Afet Yönetimi Bölümü, Çanakkale*
[2]*Kastamonu Üniversitesi, Eğitim Fakültesi, Kimya Bölümü, Kastamonu*

Liquid-crystal polymers are capable of forming regions of highly ordered structure while in the liquid phase. However, the degree of order is somewhat less than that of a regular solid crystal. Typically LCPs have a high mechanical strength at high temperatures, extreme chemical resistance, Blending is considered to be an effective medhod for the modification polymer properties. This strategy is usually cheaper and less time-consuming than the develop new polymers or new polymerizaton routes.Polymer blending based on thermoplastic and thermotropic liquid crystalline polymer (TP/TLCP) is a useful and attractive approach. In this study, thermotropicliquid crystalline polymer named 60P40PET has been synthesized an blended with PA66. Mechanical property of 60P40PET/67PHT blends have been investigated dynamic mechanical analysis (DMA). The single Tg values observed from the DMA shows that, the blends formed are homogeneous and miscible. In this study, thermotropic liquid crystal copolymers named 60P40PET have been synthesized and blended with PA66. Mechanical property of 60P40PET/PA66 blends have been investigated via static mechanical analysis (SMA) and dynamic mechanical analysis (DMA). The glass transition temperature, Tg was selected as the peak position of E'' when plotted as a function of temperature. Generally for incompatible system, the tanδ versus temperature curve shows two damping peaks corresponding to the glass transition temperatures of individual polymers. When blend components are compatible, a single peak is found for the combined processes. 60P40PET/PA66,blends show a single peak. It is said that these blends are compatible (According to the Fox mixing rule). Fibrillation of TLCP is the key factor for improving the mechanical properties of thermoplastics polymers. Mechanical properties change of 60P40PET/PA66 blends was performed between post-cold drawing and pre-cold drawing. The Young's modulus values of cold drawing 60P40PET/PA66 blends were higher. It is considered that cold drawing is a pro drawer

Keywords: Thermotropic liquid crystalline copolyesters, blend, PA66, compatibility

IP-P-008 [Industrial Polymers]

Performance of redispersible polymer powders commercially used in cement based masonry mortars

Gozde Camli, Yasemin Yetimoğlu Balk

Departman of R&D, KYK Construction Chemicals, Eskişehir, Turkey

Polymer powders is one of the most important component in cement based masonry mortar. Redispersible polymer powders (RDP) act as a second binder in the mortar, in combination with the cement. These mortars are prepared as a dry mixture. They, except for copolymer and ter-polymer, including cellulose ether, Portland cement and calcite, dolomite, silica sand.

The purpose of this study, in cement base masonary mortars of polymers commonly used in construction chemicals (Vinyl Acetate/Ethylene (VA/E) (RPP-3, RPP-6), Vinyl Acetate/Acrylic (VA/A) (RPP-2, RPP-4, RPP-5), Vinyl Acetate/Acrylic /Versatate (VA/A/VeoVa) (RPP-1)) investigate and compare performance. It was compared to result of capillary water adsorption, long term water absorption, flexural strenght, compressive strenght, and bond strenght.

One of the highlights basic properties of polymers there are water repellent. Therefore optimization study was performed in capillary water absorption test. Firstly, ratio of polymers was changed from 1 to 5%. Optimal polymer ratio was determined as 3%. This proportion of polymer was used in all subsequent tests. After ratio of polymers determined, the application of the mortar in the wall and taking into consideration the pot life, it worked fixed to water ratio.

Despite the pressure and bond strength results of the numerical difference not to be seen, approximately 35% difference between the flexural strengths were observed. While RPP-4 shows minimum flexural strength, RPP-6 shows maximum flexural strength. Although the observed differences in RDP of capillary water absorption rate and extended water absorption, VA/E and VA/A showed similar behavior and was found to have the best performance. As a result of studies, depending on the chemical structure and molecular size of the polymer it was observed to be highlighted in different performance.

Keywords: Redispersable Polymer Powders (RDP), Cement, Construction Chemicals

IP-P-009 [Industrial Polymers]

Benzoin Derivatives as UV Absorbers

Tuğçe Çinko[1], Nergis Arsu[1], Duygu Sevinç Esen[2], Deniz Er[2], Ebru Ergüven[2]

[1]*Department of Chemistry, Yildiz Technical University, Davutpasa Campus Istanbul 34220, Turkey*

[2]*Kayalar Kimya San. ve Tic. A.Ş. Kimya Sanayicileri Organize Sanayi Bölgesi Tem Yanyol F1 Blok, 34956 Tepeören/ Tuzla / İstanbul*

Benzoin and its ether derivatives have long been known to be efficient photoinitiators for free radical polymerization of vinyl monomers and have found wide application in the polymer industry. Benzoin after photoexcitation and efficient intersystem crossing populates the triplet state, which undergoes efficient α-cleavage (Norrish Type-I reaction) to generate benzoyl and α-hydroxy radicals. In the presence of vinyl monomers, these radicals can initate free radical polymerization. The major disadvantage of benzoin and its ether derivatives is the poor light absorption in the near-UV spectral region which is most attractive for industrial photopolymerizations. (Arsu et al., J.Polymer Science Part A-P.Chemistry, 2013, 51(8), 1865-1871)

Weather resistant system takes important part in the industrial applications especially their ability to withstand sunlight exposure. Thus, photooxidation reactions need to be stopped or slow down. The photostability of polymers is one of their most important properties.

To solve the problem of polymer stabilization, new benzoin derivatives namely 1,2-bis(3-(diethylamino)-5-hydroxyphenyl)-2 hydroxy etanon and 2-hydroxy-1,2-bis(2ydroxy -4,6-dimethoxyphenyl) etanon, were synthesized and characterized. Enhanced absorption properties of new benzoin derivatives in the visible region of electromagnetic spectrum allow them to use as UV absorbers in paint formulations. Several synthetic paint formulations were prepared and weathering studies were performed. Pendulum hardness, gloss and yellowing indices of coating were determined by using of various techniques.

Keywords: UV Absorber, Benzoin Derivatives, Polymer Stabilization

IP-P-010 [Industrial Polymers]

Physicochemical Research Of The Effects Of Polymers Having Antipilling Effect On Different Fabrics

Burcu Büyükkoru, Ali Kara

Sciences Faculty, Chemistry Department, Uludağ Univ., Bursa, Turkey

Pilling is a condition exhibited by certain fabrics made from spun yarns. Balls of lint are firmly attached to the fabric's surface and when the condition is severe, the fabric becomes visually unappealing and irritating to the touch. Spun fabrics made from 100% synthetics fibers, polyesters, acrylics, and blends of polyester/cotton are prone to pill. Abrasion and pilling resistance testing machine can be used for predicting a fabric's pilling tendency. Samples are rubbed against known abradants at low pressures and in continuously changing directions, and the amount of pilling is compared against standard parameters.

Finishing procedures have pronounced positive effect on fabric pilling. Many fabrics can be improved by selection of proper finishing conditions. Generally enzymes have been used. However, they are not effective on all fabrics. For this reason, antipilling polimers are needed.

Polymer is a large molecule (macromolecule) built up by the repetition of small chemical units. They can be produced with different characters according to the field of application. In this project, we need polymer which has an antipilling effect.

We will also examine how the polymer is connected to the fabric with physicochemical properties. After antipilling polymer is applied to the fabrics, yellowing, hydrophilicity and softness of the fabrics will be investigated. Also, the stability of the product will be observed.

Acknowlegement: I sincerely thank Dr. Ö. Nafi Yavuz who is the factory manager of the Rudolf Duraner for providing me the oppurtunity to work on this project. Martindale abrasion pilling tester and all fabrics can be used for the supply of Rudolf-Duraner. I am also grateful to this company.

References:

* [1] L. W. Man, Raymand, Mechanisms of pilling formation and reduction by attrition methods, The Hong Kong Polytechnic University Institute of Textiles & Clothing. (2009).

Keywords: pilling, abrasion, fabric

A schematic diagram for stages of pilling

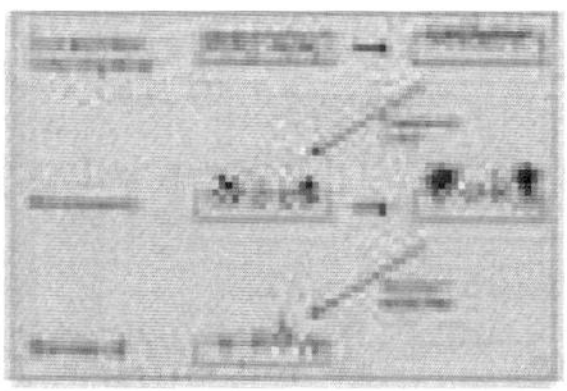

IP-P-011 [Industrial Polymers]

Physcochemical Applications Of The Interaction Between Magnetic Polymers And Dyes

Burcu Büyükkoru, Ali Kara

Faculty of Sciences, Chemistry Department, Uludağ Univ., Bursa, Turkey

Recently magnetic polymers have been used in a large scale. Magnetic micro and nano polymer particles with tailored surface chemistry using natural or synthetic polymers have been used in various fields of applications. The applications of microspheres became a new topic of research. They are composed of magnetic particles and polymer. The magnetic particles make the rapid separation possible in an external magnetic field. Also magnetic carriers with good properties are necessary and indispensable. Using magnetic polymer microspheres is an easy, cost effective and highly efficient method.

Dyes are used in many areas such as textiles, paper, plastics and so on. About 15% of dyes are lost during the dyeing process and released with the effluent. Generally, dyes are carcinogenic and mutagenic. So, it is important to remove them from waste water. Adsorption process has been extensively examined for the elimination of the dyes from waste water.

The application of dye adsorption process based on the use of magnetic polymer involves the following steps: (i) adsorption, (ii) magnetic separation of the adsorbent from waste water, (iii) adsorbent regeneration, and (iv) handling of the regeneration solution and the adsorbent.

Kinetic and isotherm studies of dye adsorption indicate which model is followed by the adsorption process. Also, the effect of solution, contact time and initial dye concentration on adsorption capacity have a great importance in this method. Physicochemical properties of the magnetic polymer can be characterized by SEM, TGA, XRD, CHN and FTIR.

References:

[1] A. Kara, E. Demirbela, N. Tekin, B. Osman, N. Besirli, Magnetic vinylphenyl boronic acid microparticles for Cr(VI) adsorption: Kinetic, isotherm and thermodynamic studies, Journal of Hazardous Materials. 286 (2015) 612–623.
[2] R. M. Kakhki, Application of magnetic nanoparticles modified with cyclodextrins as efficient adsorbents in separation systems, J Incl Phenom Macrocycl Chem. 82 (2015) 301–310.

Keywords: Magnetic polymer, microspheres, dye, adsorption

IP-P-012 [Industrial Polymers]

Selective removal of toxic cationic textile dyes from wastewater using disulfide-linked polymer networks

Mehmet Sahin Atas[1], Halit Cavusoglu[2], Mustafa Selman Yavuz[1]

[1]*Metallurgy and Materials Engineering Department, Selcuk University, Konya, Turkey*

[2]*Advanced Technology Research and Application Center, Selcuk University, Konya, Turkey*

Textile industry is the most prominent where effluents containing large amounts of dye residues have to be treated before being discharged into wastewater streams. If not done properly, water pollution with toxic dyes, metal ions and organic contaminants pose serious environmental hazards, especially for aquatic biosystems, where symbiotic processes are affected by the contaminant induced inhibition of the photosynthetic activity[1]. Unrefined dyes cause chemical and biological changes in our aquatic system. The presence of these compounds also make water pollution due to the direct destruction of metal ions and organic pollutants. Various techniques such as precipitation, membrane filtration, coagulation, electrochemical, ion exchange, chemical oxidation, adsorption have been used for the removal of dyes from wastewater[2]. Various kinds of adsorbents have been designed to remove dyes in water with considerable adsorption capabilities, such as carbon nanotubes, mesoporous silica, and zeolites. Recently, new porous covalent organic polymers (PCOPs) were tested as adsorbents for removal of dye effluents[3].

In this study, novel disulfidelinked PCOPs were synthesized by oxidative coupling procedure. We successfully and selectively removed the cationic toxic dyes including crystal violet, methylene blue and Rhodamine B from wastewater.

Acknowledgements: This study was carried out as a Master thesis by Mehmet Sahin Atas at the Graduate School of Natural and Applied Science at Selcuk University, Konya, Turkey. Academic Staff Training Program (ÖYP) (Project Number 2015-ÖYP-128), which authors gratefully acknowledge.

References:

[1]. Mahto, T. K.; Chowdhuri, A. R.; Sahu, S. K., Journal of Applied Polymer Science 2014, 131 (19), 40840.

[2]. Wang, L.; Wu, X.L.; Xu, W.H.; Huang, X.J.; Liu, J.H.; Xu, A.W., ACS Applied Materials & Interfaces 2012, 4 (5), 26862692.

[3]. Patel, H. A.; Yavuz, M. S.; Yavuz, C. T., RSC Advances 2014, 4 (46), 24320-24323.

Keywords: Adsorption, Cationic dyes, Crystal violet, Porous covalent organic polymers, Wastewater

IP-P-013 [Industrial Polymers]

Synthesis of New Polymer Additives from Waste Plastics for Asphalt Binder

Emel Başkent[1], Savaş Gürdal[2], Serenay Akyol[2], Melike Ocak[2], Gülşen Albayrak Arı[2], Muzaffer Yaşar[2], Ramazan Oğuz Canıaz[1], Refika Çetintaş[1], Serhat Arca[1], Elif Kocaman[1], Ziya Köstereli[1]

[1]*Turkish Petroleum Refineries Corporation, R&D Product Development Department, 41780, İzmit, Turkey*

[2]*Department of Chemical Engineering, Istanbul University, Istanbul, Turkey*

Inadequate recycling rate of wastes leading to increasing huge environment problems. Therefore studies on waste recycling are of paramount importance. In this research, chemical recycling of PET and obtaining new polymer materials from the recycled material as a bitumen additive is studied. Chemical recycling reactions that are carried out with PET flakes and ethylene glycol (PET/EG:0.2 wt/wt), in the presence of zinc acetate catalyst (PET/catalyst: 100 wt/wt), for 3 hours at 200 oC in nitrogen atmosphere. Water soluble (Bis(2-Hydroxyethyl) terephthalate (BHET) monomers) and water-insoluble (oligomers) fractions are separated at the end of the reaction followed by a polymerization reaction with dicarboxylic acids and BHET to get the polyester polyols (P.Polyols) with pre-determined physical and chemical properties as a bitumen additive. Two different acids, adipic and sebacic acids, are used with different diol:diacid ratios (1:1 to 1,5:1 by mass) to get the different chain lengths in the P.Polyols. Also, some hydroxyl groups are added with the BHET to enhance the branched structure while the reaction rate increases. Bitumen with/out P.Poliol, and also the synthesized P.Polyols are analyzed with the FTIR, GPC, DSC and TGA analysis. GPC results show that the increasing in the diol:diacid ratio decreases the molecular weight of P.polyol. Similarly, Tg values of polyols decreases with the increasing of both diol:diacid ratio and addition of hydroxyl group. When these P.polyols are added to bitumen, dynamic shear rheometer result that is critical for deformation resistance at high temperatures is increases like commercial Stiren-Butadien-Stiren (SBS) copolymer which is commonly used at state of the art to modify bitumen properties. Also, P.Poliols increases the elasticity of bitumen as four times higher than its base value. Moreover, P.Polyol does not increase the viscosity of bitumen as much as SBS which is an advantageous property regarding the workability of bitumen during paving.

Keywords: Waste Plastics, Poly(ethylene terephthalate), PET Recycling, Polymer Modifed Bitumen (PMB), Glycolysis, Polyester Polyol

IP-P-014 [Industrial Polymers]

Polymerization Reactions of Elemental Sulfur

Emel Başkent[1], Savaş Gürdal[2], Serenay Akyol[2], Gülşen Albayrak Arı[2], Muzaffer Yaşar[2], Ramazan Oğuz Canıaz[1], Refika Çetintaş[1], Serhat Arca[1], Elif Kocaman[1], Ziya Köstereli[1]

[1]*Turkish Petroleum Refineries Corporation, R&D Product Development Department, 41780, İzmit, Turkey*

[2]*Department of Chemical Engineering, Istanbul University, Istanbul, Turkey*

Elemental sulfur, a by-product of refineries and gas-processing plants, has been investigated for using in different industrial applications such as Li-S batteries, IR optical camera lenses and concretes. These research studies are set off by the storage problems comes with the high production capacities of sulfur production plants.

Therefore, this study is performed for the synthesis of valuable polymeric materials from elemental sulfur to use for different applications. For this purpose, copolymerization reaction is carried out with the styrene monomer in a two different ratios, 50 and 25%(wt.), and molten sulfur at reaction temperatures of 120, 130 and 160 oC for reaction times ranging from 45 to 180 minutes. Procedure for the removal of inhibitor which is found in styrene monomer is applied to monomer by using 2 M NaOH. After the purification step, monomers are charged into the molten sulfur at desired temperatures in nitrogen atmosphere.

Due to the nature of sulfur molecules, it may be in the form of orthorhombic (Sα), monoclinic (Sβ) or polymeric sulfur at different temperature. The transitions between the allotropes (orthorhombic ↔ monoclinic) decompose the sulfur structures into the reaction product and this can be seen by visually with time (A.M.O.Mohamed, ISBN-13:978-1-60427-005-1, 1966). Thus view of some samples for a period of 1- 10 and 20 days indicate the decomposition of sulfur molecules remaining unreacted in the reaction medium.

Reaction products are also characterized with the 1H-NMR and DSC analysis. Calibration curve of elemental sulfur is plotted for DSC analysis and the results show that copolymers obtained with 50%(wt.) styrene at 160 oC has no elemental sulfur at the end of reaction and H-NMR spectrums of these samples give the specific peaks for the polysulfide structures between the 4.0-4.5 ppm. Experimental results gave important insight about reaction kinetics of polymerization reactions of elemental sulfur.

Keywords: Elemental Sulfur, Copolymerization Reactions, Polysulfides

IP-P-015 [Industrial Polymers]

Synthesis and applications of a pH-responsive cyclopolymers containing the residues of the amino acid aspartic acid

Hasan Ali Al Muallem, Zakariyah Abdulkareem Jamiu, Shaikh Asrof Ali

Chemistry department, King Fahd University of Petroleum & Minerals, Dhahran 31261, Saudi Arabia

Reaction of diallylamine and dimethyl maleate afforded Michael addition product dimethyl N,N-diallylaspartate I which upon treatment with dry HCl and aqueous HCl gave its hydrochloride salt II and N,N-diallylaspartic acid hydrochloride III, respectively. The new monomers II and III underwent cyclopolymerization using various initiators to give respective cationic polyelectrolyte IV and polyzwitterionic acid (PZA) V. Acid hydrolysis converted polymer IV into V Under the influence of pH, water-insoluble V (ZH2±) has been equilibrated to water-soluble poly(zwitterion-anion) VI (ZH ± -), polydianion VII (Z=), and cationic polyelectrolyte VIII (ZH3+). The solution properties have been investigated in terms of degree of charge imbalances in polymer backbone. Apparent basicity constants of a carboxyl group and trivalent nitrogen in VII (Z=) have been determined. Evaluation of antiscaling properties of V using supersaturated solution of CaSO4 revealed ≈100% scale inhibition efficiency at a meager concentration of 20 ppm for a duration of 50 h at 40ºC. The composition and phase diagram of the aqueous two-phase systems of VII and polyoxyethylene (POE) have been studied for this new class of ionic polymers. Combined total mass fraction of less than 0.10 for the component polymers make these aqueous two-phase systems (ATPSs) viable for industrial applications. The transformation of water-soluble of VII to insoluble V makes it a suitable component for recycling ATPS as it can be recycled at a lower pH.

Keywords: pH- responsive polymers; cyclopolymerization; water-soluble polymers; Aqueous two-phase system; aspartic acid.

Scheme

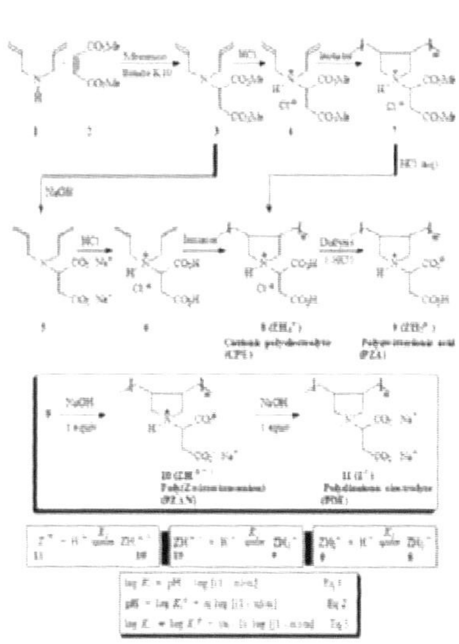

Synthesis of cyclopolymers having aspartic acid residues.

IP-P-016 [Industrial Polymers]

The effects of waste polymeric materials on the rheological properties of autoclaved aerated concrete

Gurhan Gereli[1], Taner Kavas[2], Utku Tiriç[1], Uğur Uzgan[1], Abdullah Doğan Soyal[1]

[1]AKG Gazbeton İşletmeleri San. ve Tic. A.Ş., Ar-Ge Merkezi, Işıkkent, İzmir, Turkey
[2]Afyon Kocatepe Üniversitesi, Mühendislik Fakültesi, ANS Kampüsü, Afyonkarahisar, Turkey

Autoclaved Aerated Concrete (AAC) is one of the most commonly used light-weight construction materials for contemporary buildings, especially due to its low density, unique thermal and breathing properties and high fire resistance. Moreover, such advantageous properties make it preferable for earthquake-resistant housing. On the other hand raw material mixture, mud rheology and production process are closely related with final product properties of it. Because of these facts, rheological properties of aerated concrete pastes are studied in order to assess effects of waste polypropylene (PP) and waste expanded polystyrene (EPS) on the technical properties of AAC. For this purpose, different cement types have been used. Viscosity, pH, thixotropy, shear stress, shear rate, shear thickening, shear thinning, and temperature of the pastes were measured.

Keywords: Autoclaved Aerated Concrete, Rheology, PP, EPS

IP-P-019 [Industrial Polymers]

Surface Characterization of Silane Modified Quartz by AFM

Fatma Bilge Emre[1], Nilüfer Kıvılcım[2], Turgay Seçkin[2]

[1]Department of Primary, İnönü University, Malatya, Turkey
[2]Department of Chemistry, İnönü University, Malatya, Turkey

Silicon dioxide (SiO_2) exists in three crystalline forms, respectively quartz, tridymite and cristobalite. SiO_2 is the main component in glass, brick and concrete and also forms the insulator in silicon devices. In this study, Quartz was etched with NaOH and NH_4OH and these quartzs were functionalized with (3-aminopropyl)triethoxysilane (APTES). This functionalization provided active amine groups on the quartz. Surface modification with namely polymers were carried out by surface free-radical polymerization and surface polymer grafting. There are three ethoxy ($-OCH_2CH_3$) in an APTES molecule, respectively. They can be hydrolyzed in aqueous environments or anhydrous organic solvents to form silanol groups (Si–OH) [1]. This group undergoes condensation with a hydroxylated surface via the hydrogen bond and then forms a siloxane (Si–O–Si) linkage over the treated surface. In addition, the condensation of neighbor APTES molecules also forms a polymer matrix linked by siloxane (Si–O–Si) bonds with a subsequent loss of water molecules by curing, leading to amino-terminated ($-NH_2$) terminated surfaces [2,3]. Amino modified SiO_2 surfaces are important for investigation of biological surfaces by AFM. Surface morphology of quartzs were examined by scanning electron microscopy and AFM. The polymers surface coating were explained elemental analyses, thermogravimetric analyses and ınfrared spectrometry.

[1] Xie, Y.,Hill,C.A.S.,Xiao,Z.,Militz,H.,Mai,C.,2010. Compos: PartA: Appl. Sci. Manuf. 41,806–819.
[2] Gunda, N.S.K.,Singh,M.,Norman,L.,Kaur,K.,Mitra,S.K.,2014. Appl.Surf. Sci.305, 522–530.
[3] Yadav,A.R.,Sriram,R.,Carter,J.A.,Miller,B.L.,2014. Mater. Sci. Eng. C: Mater. Biol. Appl. 35,283–290.

Keywords: Quartz, surface modification, APTES

An illustration of functionalization on quartz with APTES

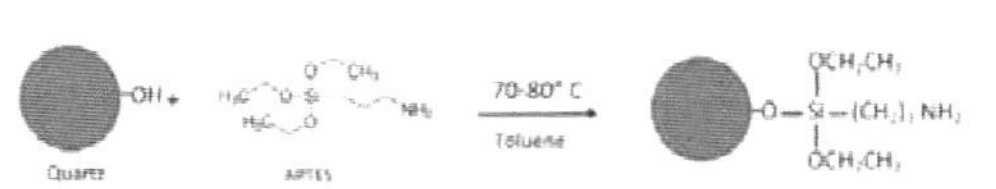

IP-P-020 [Industrial Polymers]

Effect of etoxylated surfactant content on the emulsion polymerization and water resistance performance of poly (EA-co-MMA) latexes

Ecem Temelkaya, Cansu Akarsu Dülgar

Organik Kimya San. ve Tic. A.Ş, Istanbul, Turkey

Impregnated papers are often used in short cycle presses. Impregnated papers are hardened under temperature and pressure. In commercial applications, emulsion acrylic co-polymers have the fastest growth in the market share. The interest in aqueous acrylic dispersions grew with the need to solve solvent emission problems.In this study, it has been focused on the effect of etoxylated surfactant content in the emulsion process and water resistance performance of poly (EA-co-MMA). The amount of surfactant has been studied at 0.24%, 0.48% and 0.72% based on total monomer composition. The materials used for this research were the monomers mentioned as ethyl acrylate (EA), methyl methacrylate (MMA), acrylic acid (AA), etoxylated surfactant, ammonium persulfate and deionized water. The aim of the study is to investigate the influence of surfactants on water resistance on acrylic co-polymers used for paper impregnation. The resultant polymers were applied on uncoated cellulosic paper in order to perform Cobb test.

Keywords: paper impregnation, emulsion, acrylate, water resistance

IP-P-022 [Industrial Polymers]

The role of nano-ZnO as an activator on rheological and tensile properties of rubber blends based on BR/SBR

Behnaz Akbari, Fereshteh Motiee, Sasan Mosavi

Department of Chemistry, Tehran North Branch, Islamic Azad University, Tehran, Iran

In this study, an attempt has been made to replace ZnO with nano-ZnO to prepare rubber compounds based on BR/SBR (Table.1) to minimize its pollution in the environment [1-4]. Seven rubber compounds have been made to investigate their rheological behavior, fatigue resistance, and tensile properties. The particle size of nano-ZnO was determined by using X-ray powder diffraction (XRD) and scanning electron microscopy (SEM). The compound containing 0.5 phr nano-ZnO as an activator has showed the maximum curing rate index (CRI) compared to other compounds. Improvement in modulus and tensile strength are observed. In addition, elongation at breaks was increased. The higher tensile strength for sample filled with 0.5 phr of nano-ZnO is illustrated in Table.2.There has not been any significant change in 100% tensile modulus for all the studied compounds. The sample containing 0.5 phr nano-ZnO enhanced the abrasion resistance, but reduced the wear rate of rubber compound. It appears that the compound containing 0.5 phr nano-ZnO gave better fatigue resistance. Likewise, thermo gravimetric analysis (TGA) showed two stages of sample degradation. As can be seen in the Fig.1, thermal stability of BR is less than SBR leading to fast degradation.The activation energy has been calculated to evaluate the conditions of sample degradation during thermal analysis tests. The results exhibited that the activation energy has not changed significantly for all seven samples. The results given in Table.2 showed a better distribution of nano-ZnO in comparison to the micro-sized ZnO. Finally, the use of nano-ZnO as an activator is a cost effective way to produce new blends of rubber compounds which is more environmentally friendly products.

Keywords: NanoZnO, Tensile, Abrasion

IP-P-023 [Industrial Polymers]

Development of High Solid Containing UV Curable Paint and Varnish Systems for Wood Coatings

Alev Tüzün, Halil Pekedis, Deniz Er, Ebru Erguven Cakmak

Kayalar Kimya Research and Development Center, Istanbul, Turkey

One of the major advantages of UV curing system is implemented without requiring additional solvents. Oligomers, monomers and photoinitiator like all components contribute to the final properties of the film formed by joining structure after curing is complete.[1] The main advantages of the UV curing systems are the following; the energy needed for the process is low, the polymerisation is very fast and highly efficient, no solvent is required, therefore the environmental pollution by VOC is avoided.[2-3] This high solids solvent-free or in the project that is a small amount of solvent can be sprayed that requires the use of UV curing lacquer and studies on the development of coating formulation has been performed. In this project, for the furniture industry wood painting systems with applications on MDF (medium density fibers) pigmented UV topcoat and undercoat have been studied. On the other hand, to be applied on types of wood (beech, oak, walnut, mahogany...) to create decorative and protective film without damaging the wood's natural appearance, UV sealers and UV topcoat systems have been performed by spray application. Generally, experiencing difficulties with spray application is possible since the viscosity of high solids systems are quite high. Therefore, low-viscos, high reactive oligomers with different structures (epoxy acrylate, polyester acrylates, polyether acrylates, amine modified acrylates, urethane acrylates, polyester...) have been performed depend on the reaction rate with controlling adhesion. The monomers and photoinitiators have been formulated in various options and quantity to obtain desired reaction speed and improvement in hardness.

Keywords: UV curing, high solids, VOC free, paint, varnish

IP-P-024 [Industrial Polymers]

Mechanical properties and morphological characteristics of recycled PET fiber- glass fiber - PP hybrid composites

Orkun Kaymakçı[1], Hülya Kiraz[2]

[1]*Arçelik A.Ş. Central R&D Department, 34950 Tuzla-Istanbul, Turkey*
[2]*Arçelik A.Ş. Washing Machine R&D Department, 34950 Tuzla-Istanbul, Turkey*

Glass fiber(GF) reinforced polypropylene(PP) composites are widely used in appliance and automotive industries due to their relatively high strength, high stiffness and improved dimensional stability. In addition, GF-PP composites offer good cost to performance ratio [1-2]. In this study, PET (Polyethylene terephthalate) flakes from post-consumer beverage bottles are added into PP matrix along with the glass fibers to develop eco-friendly, high performance hybrid PP composites as alternatives to conventional PP-GF composites. PET, glass fibers and PP matrix are compatibilized using maleic anhydride grafted PP coupling agents. In the extrusion process of the composites, PET flakes are stretched into the fiber form at optimized temperatures. Fiber form of the PET inside the GF-PP hybrid composites is investigated using Scanning electron microscopy (SEM). Along with the glass fibers, PET fibers in the composites reinforce the PP matrix improving the mechanical properties of the composites. In addition, it is found that, coupling agent has strong effect on the mechanical properties of the composites. Recycled PET-GF-PP hybrid composites can be a green and cost effective alternative to conventional glass fiber reinforced composites.

References:

[1] Karian, Harutun, ed. Handbook of polypropylene and polypropylene composites, revised and expanded. CRC press, 2003.
[2] Fu, S-Y., et al. "Tensile properties of short-glass-fiber-and short-carbon-fiber-reinforced polypropylene composites." Composites Part A: Applied Science and Manufacturing 31.10 (2000): 1117-1125.

Keywords: hybrid, PET, glass fiber, PP, composites, recycled

MBM-P-001 [Macromolecules in Biotechnology & Medicine]

Applications of the functionalized liquid crystal double-emulsion hollow droplets prepared by microfluidics

Soo Young Park, Jong Kyu Kim, Ju Hyun Jang, In Seok Huh

Department of Polymer Science and Engineering, Kyungpook National University

Uniform-sized cholesteric liquid crystal double-emulsion (CLCDE) hollow droplets were prepared by a microfluidic method. The CLCDE droplets in aqueous solution showed the unique defect structures and colors at the center which were dependent on the surface anchoring conditions. The CLCDE hollow droplets decorated with PAA-b-LCP (PAA: polyacrylic acid and LCP: side chain LC polymer) showed the structural change between parallel and perpendicular anchoring with respect to pH in aqueous solution. This structural change from parallel to perpendicular anchoring may cause color disappearance due to removal of the regular helical structure. Several enzymes (such as glucose oxidase, cholesterol oxidase, and urease) were immobilized to PAA on the CLCDE droplets. The enzyme-immobilized CLCDE hollow droplets were successfully demonstrated for biosensors with different colors inside the CLCDE hollow droplets for bar-code applications.

Keywords: liquid cryst, biosensor, microfluidics

MBM-P-002 [Macromolecules in Biotechnology & Medicine]

The Physical and Chemical Properties of some Strained Molecules Related to Biphenylene

Ali Larkem, Hakim Hamada, Hamama Larkem

Department of Chemistry, Annaba University. Annaba. Algeria

My research interests are chiefly concerned with polycyclic and heterocyclic systems related to biphenylene, employing methods of synthesis which usually involve, cyclo-addition reactions of transient intermediate (arynes, benzocyclobutadiene ect...), extrusion reactions from larger ring systems, and by using an Ullmann reaction coupling in order to synthesize some interesting compounds such as cyclo-octatetraene(COT) derivative (II) and (III).

All compounds have been characterized by alternative synthesis and spectral data.

Keywords: synthesis, polycyclic, heterocyclic, biphenylene, cyclo-octatetraene.

Synthesis of cyclo-octatetraene (II) and Structure (III)

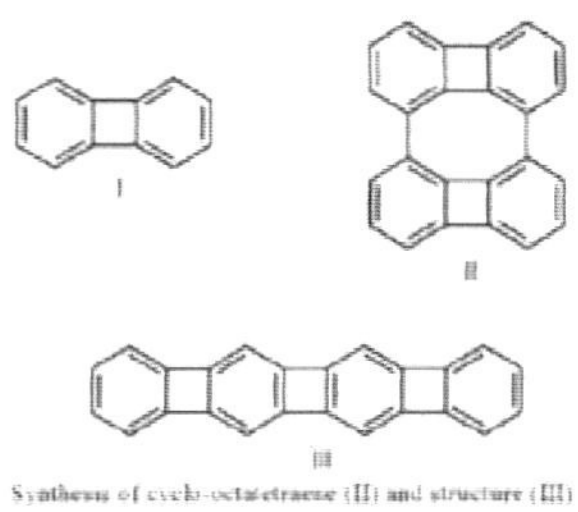

Synthesis of cyclo-octatetraene (II) and structure (III)

MBM-P-003 [Macromolecules in Biotechnology & Medicine]

Impact of diesel exhaut particulates (DEPs) on the respiratory system

Bouzid Mohamed[1], Djadi Amina[2], Bezzazi Boudjema[1]

[1]Research Unit: Materials, Processes and Environment, University M'Hamed Bougara, Frantz Fanon City, Boumerdes 35000 Algeria.
[2]Department of Research and Development, Health Technology of Algeria, Tipaza 42000, Algeria

What could be more precious than air? Yet, it is easily contaminated and then becomes the invisible agent of mass contamination as in cities, where highly carcinogenic diesel particles form up to 80% of all atmospheric particulate pollutants. The health impact of diesel exhaust particulates is well established, lungs being the most exposed; DEPs spread down the respiratory tract: in bronchi, bronchioles up to alveoli. In the present study, five scenarios regarding the deposition patterns of DEPs as reported by the literature are reviewed by means of Visioscopie. We shall introduce a sixth scenario we have observed in our experimental research: "the junction effect" occurring between the bronchiole walls and the alveolar ducts". This obstruction is the most noticeable one in lung pathologies.

Keywords: Diesel particulates, Air pollution, lung, Visioscopie

MBM-P-004 [Macromolecules in Biotechnology & Medicine]

Synthesis of photopolymerizable macromers giving novel poly(β-amino ester)s for biomedical applications

Ece Akyol, Duygu Avcı

Department of Chemistry,Bogazici University, Istanbul,Turkey

Photocrosslinkable, biodegradable and biocompatible materials are investigated widely due to their broad range of applicationssuch as drug-delivery vehicles and tissue engineering scaffolds. Among the biodegradable systems, poly(β-amino ester)s [PBAEs] have become important as biomaterials due to their pH sensitivity, biodegradability and high biocompatibilities. In this project, phosphonate-containing PBAE macromers were synthesized through the reaction of two primary amines (diethylaminoethylphosphonate, diethyl 6-aminohexylphosphonate) and four diacrylates (polyethylene glycoldiacrylate, 1,6-hexanedioldiacrylate, triethylene glycol diacrylate, trimethylolpropanetriacrylate) and photopolymerizedto generate phosphonate-functionalized PBAEs to investigate the effects ofthe phosphonate groups on polymers'degradability, mechanical properties and binding to tissues.

This work was supported by a grant from the Scientific and Technological Research Council of Turkey (TÜBİTAK) (114Z926).

Keywords: photopolymerizable,poly(β-amino ester)s,biodegradable

MBM-P-005 [Macromolecules in Biotechnology & Medicine]

Hyperbranched Glycerol Modified β-cyclodextrin: Effect of the Macromolecular Topology on the Encapsulation and Release of Phenolphthalein

Yanyan Ru, Yunyun Lv, Yuhong Liu

Department of Chemical Engineering, School of Chemical Engineering and Technology, Xi'an Jiaotong University,China

In the study, Hyperbranched Glycerol Modified β-cyclodextrin (CD-g-HPGs) with various arm length (10, 30, 50 or 80) and arm number (3, 7, 14 or 21) were synthesized and characterized by 1HNMR and Gel Permeation Chromatographic (GPC). CD-g-HPGs form molecular inclusion complexes with phenolphthalein (PP) in a state of aggregations. PP, as a small hydrophobic guest, is readily encapsulated in the cavity of β-CDs or the hydrophobic domains stacked by cavities of β-CDs. Longer arm length and denser arm number not only increase its loading capacity, but also have an interesting influence on the release behavior, that is, a longer arm length accelerates release, while a higher density of arm number slows release attributed to the synergy effect of aggregation and host-guest interaction between β-CD and PP. Due to regularly adjusted host-guest interaction, this method could be used to design efficient carriers and reasonable assemble systems containing β-CD.

Keywords: β-CD, hyperbranched glycerol, topology, encapsulation, release

Figure 1 Release of PP from CD-g-HPGs with different arm length

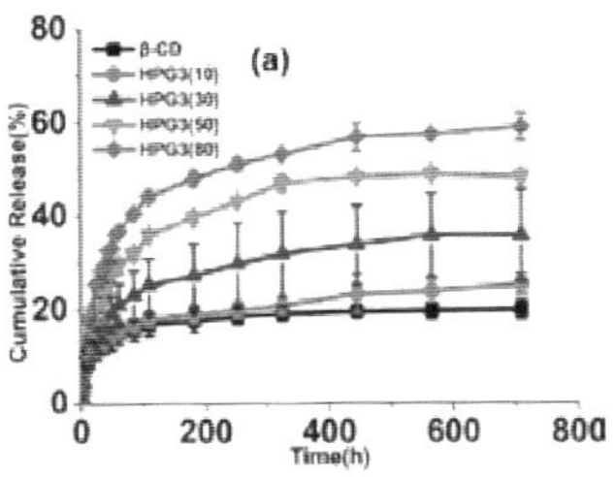

Figure 2 Release of PP from CD-g-HPGs with different arm number

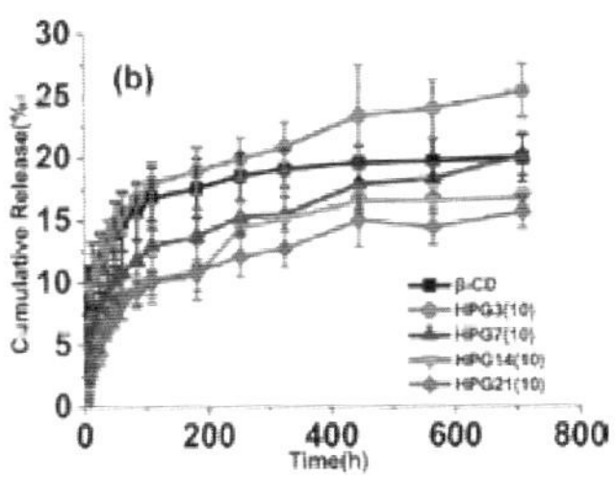

Figure 3 Aggregation and host-guest interaction between β-CD and PP

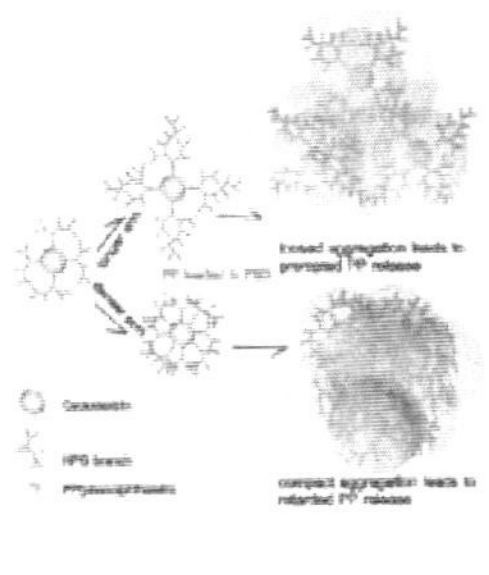

MBM-P-006 [Macromolecules in Biotechnology & Medicine]

Flurbiprofen Derivative Loaded Nanostructured Lipid Carriers: Synthesis of New Flurbiprofen Derivative

Gulce Taskor, Nezire Saygili

Department of Basic Pharmaceutical Sciences, Faculty of Pharmacy, Hacettepe University, Ankara, Turkey

The objective of this work was to formulate a nontoxic, lipid conjugated flurbiprofen derivative nanostructured lipid carriers with ability to reduce pain (an analgesic), to reduce fever (an antipyretic) and to reduce swelling and soreness (an anti-inflammatory agent) [1]. Using nanostructures to improve therapeutic profile of non-steroidal anti-inflammatory drugs (NSAID) is an important approach, especially concerning their gastric toxicity. Synthesis of new flurbiprofen derivative is significant due to their activities in biological systems with reduced side effects. Flurbiprofen chloride was reacted with amino alcohol to afford new flurbiprofen derivative (Scheme 1). Flurbiprofen and the derivative was formulated as nanostructured lipid carriers (NLCs) and characterized for size, zeta potential and morphology. The system were composed with polymeric-lipid hybrid drug molecules.

The reaction of flurbiprofen chloride with amino alcohol, in the presence of the weak nucleophilic base triethylamine and effective nucleophilic base dimethylamino pyridine gave the corresponding amino alcohol derivative of flurbiprofen. A significant reduction in toxicity is expected for NLCs in gastric mucosa compared to free flurbiprofen. This may be due to controlled release of flurbiprofen from the nanostructured lipid carriers, which decreases the mucosal contact to flurbiprofen.We describe the synthesis of new flurbiprofen derivative and the preparation of flurbiprofen derivative-nanostructured lipid carriers.

References:

1. Abdellatif, K. R. A., Chowdhury, M. A., Dong, Y., Das, D., Yu, G., Velázquez, C. A., Suresh, M. R., Knaus E. E., Bioorg.&Med. Chem. Lett. (2009), 19 (11), 3014-3018.

Keywords: non-steroidal anti-inflammatory drug, synthesis, nanostructured lipid carriers

Scheme1

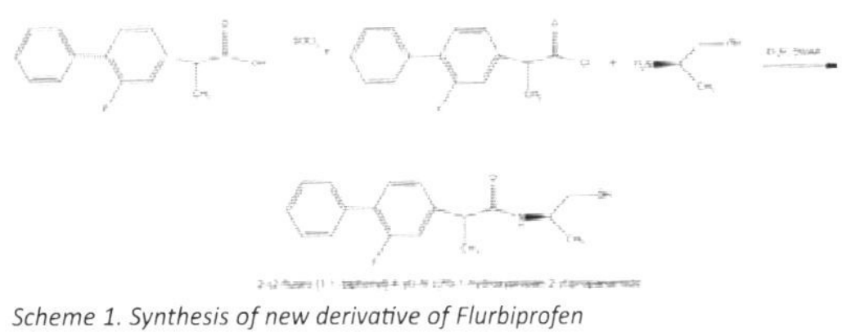

Scheme 1. Synthesis of new derivative of Flurbiprofen

MBM-P-007 [Macromolecules in Biotechnology & Medicine]

Polyurethanes based on castor oil and chitosan as potencial biomaterials in tissue engineering

Said José Arévalo[1], Claudia Alejandra Ramírez[1], Manuel F Valero[1], Luis E Diaz[2]

[1]*Deparment of chemical engineering, energy, materials and enviromental group, university of La Sabana, Chia, Colombia*
[2]*Deparment of agroindustrial production engineering, bioprospecting research group, university of La Sabana, Chia, Colombia*

In this investigation, materials derivate from chitosan and castor oil polyol were gotten with the aim of relating polymer properties with the functionality of polyols and chitosan concentration. Three different polyols derivated from transesterification reaction were mixed with chitosan to synthesize urethane bonds with the diisocyanate. Functional groups were studied by FTIR, identifying urethane bond peak as well as characteristics peaks from chitosan, which confirm the presence of both polymers in the materials. All polymers were thermal stable below 310 °C at TGA assays, ANOVA results show a decreasing trend with the addition of chitosan into the polymer matrix. Swelling behavior and in vitro degradation were evaluated in a period of 41 and 31 days respectively, showing a dependency between these properties. Swelling induces the main impact at degradation process above enzymatic (1,5 μg/ml of lysozyme and 37 °C). Statistical analysis didn't show a relationship between polyol functionality and chitosan concentration with those properties (R-square 31 % and 19,3 % for swelling and degradation respectively). At hardness assay, polymers with higher functionality had better results than their analog polymers with low hydroxyl index, hardness between 92 and 81 was gotten by polymers from P1, ANOVA showed a model where hydroxyl index was the variable with the greatest influence on this property with R-square of 91,1%. On the whole, materials keep the thermal and mechanical performance of polyurethanes but increase fluid exchange, due to hydrophilic behavior improvement showed in degradation analysis, those characteristics show potential applications as biomaterials. Polymers from P and P0 show potential on soft tissue and haemostatic application due to fluid exchange ability and it keep its structure, as well polymers from P1 have potential at hard tissue due to their mechanical properties.

Keywords: Polyurethane, Chitosan, polymer, castor oil, biomaterial

MBM-P-008 [Macromolecules in Biotechnology & Medicine]

Phosphonate/phosphonic acid end-modified poly(b-amino ester)s for pH-sensitive applications

Mirac Tatlıyüz, Ece Akyol, Duygu Avcı Semiz

Department of Chemistry, Bogazici University, Istanbul, Turkey

Stimuli-responsive polymer micelles or nanoparticles have been used as drug carriers for site-specific drug delivery. Poly(b-amino ester) (PBAE) s, cationic polymers which are soluble at an acidic pH but insoluble at the neutral pH, can be used for development of pH-sensitive drug delivery systems due to their high buffer capacity, biodegradability and biocompatibility. They are easily synthesized by Michael addition of difunctional amines to diacrylates, and the polymer molecular weight and chain end groups can be controlled by adjusting the amine/diacrylate monomer ratio. Here we present the synthesis of phosphonate/phosphonic acid end-modified PBAEs and determine their properties as drug delivery vehicles. Synthesis of polymers involved two steps: i) initial preparation of acrylate-terminated polymers from reactions of polyethylene glycol diacrylate and 1,6-hexanediol diacrylate with propyl amine and 5-amino-1-pentanol and ii) reaction of these polymers with a phosphonated amine to generate phosphonate end-modified PBAEs. The cleavage of the phosphonate groups using trimethylsilyl bromide gave phosphonic acid end-modified PBAEs. The influence of PBAE structure on pH sensitivity was determined. The average sizes of the polymer micelles under different pH conditions measured using DLS indicated their potential for intracellular drug delivery applications.

This work was supported by a grant from the Scientific and Technological Research Council of Turkey (TÜBİTAK) (114Z926).

Keywords: drug delivery, phosphonic acid, poly(b-amino ester)s, pH sensitive polymers

MBM-P-010 [Macromolecules in Biotechnology & Medicine]

Novel bisphosphonate and bisphosphonic acid-functionalized methacrylates for dental applications

Melek Naz Guven, Duygu Avcı Semiz

Department of Chemistry, Bogazici University, Istanbul, Turkey

A recent trend in the effort to increase interaction of dental monomers with dental tissue is to incorporate bisphonate/bisphosphonic acid groups in monomer structures, since bisphosphonates have strong affinity for hydroxyapatite. In this work, novel bisphosphonate and bisphosphonic acid-functionalized dimethacrylates were synthesized for dental applications (Figure 1). The reactions of two different bisphosphonated diamines with isocyanatoethyl methacrylate gave bisphosphonate-functionalized dimethacrylates which were then converted to the bisphosphonic acid-containing ones by the cleavage of bisphosphonate groups with trimethylsilyl bromide. Their copolymerizations with commercial dental monomers such as 2-hydroxyethyl methacrylate and triethylene glycol dimethacrylate were investigated with real time-FTIR. The interaction of bisphosphonic acid-containing monomers with hydroxyapatite were examined using XRD. The effect of monomer structure on polymerization and adhesive properties was investigated.

Keywords: bisphosphonic acid, bisphosphonate, dental adhesives, dental applications

Figure 1

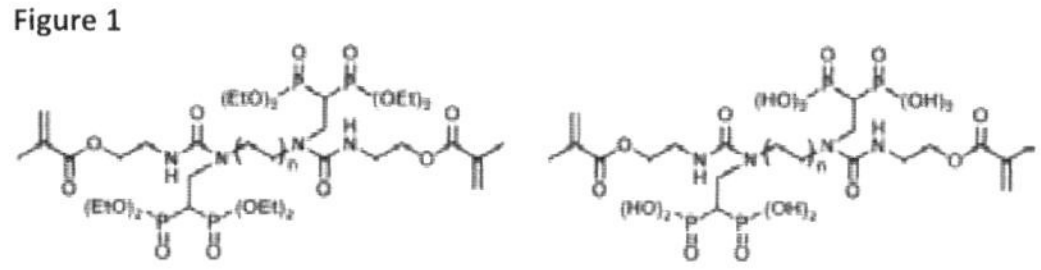

Structure of dental monomers

MBM-P-011 [Macromolecules in Biotechnology & Medicine]

Structural Analysis of Octenyl Succinic Anhydride (OSA) Esterified Starch from Plants

Senay Simsek, Kristin Whitney

North Dakota State University, Fargo, North Dakota, USA

Starch is the major carbohydrate reserve in plant tubers and seed endosperm where it is found as granules. Starch consists of two types of molecules, amylose and amylopectin. Both consist of polymers of α-D-glucose units in the 4C1 conformation. Starch has a wide variety of uses as an ingredient in different products. To improve starch functionality for specific applications, it is modified by physical, chemical and enzymatic processes. Octenyl succinic anhydride (OSA) esterified starch is one type of chemically modified starch obtained by esterification of OSA groups on repeating anhydroglucose units along the starch chains. Its preparation involves the reaction in aqueous medium under mild alkaline conditions with the starch in its granular form. The alkalinity helps to reduce hydrogen bonding between starch chains which consequently favors granular swelling and the diffusion of OSA molecules within the swollen starch granule Characterization of the fine structure of OSA starch would lead to a better understanding of functional properties. In this study, OSA rice and tapioca starches were analyzed using microscopy, liquid chromatography and nuclear magnetic resonance (NMR). Chain length distribution of amylopectin changed significantly ($P<0.05$) after OSA esterification. Weight averaged degree of polymerization (DPw) decreased significantly ($P<0.05$) from 16.47 to 13.29 and from 14.87 to 12.47 in native and OSA rice and tapioca starches, respectively. The chain length distribution of pure amylopectin fractions suggested that OSA groups were not present in the amylopectin portion of the starch. 1H-NMR analysis of pure amylose and amylopectin fractions indicated that OSA substitution was present only in amylose fractions of rice and tapioca starches. Esterification with 3% OSA results in starch that has OSA substituted mainly on amylose chains or possibly on amylopectin chains that have been hydrolyzed from the amylopectin molecules during esterification.

Keywords: octenyl succinic starch (OSA), starch, amylose, amylopectin

MBM-P-012 [Macromolecules in Biotechnology & Medicine]

In-vitro Application Studies of Breast Cancer Treatment Beads on Drug Loaded Stimuli-Responsive Microgel/Liposome Biomaterials Covered with Magnetic Nanoparticles which Provides a Way to Carry All System on a Target

Damla Ülker, Vural Bütün

Department of Polymer Science and Technology, Eskisehir Osmangazi University, 26480, Eskisehir, Turkey

'Microgels' which is a remarkable output of polymer chemistry and which has recently begun to be evaluated with liposomes, offers an opportunity to bring forward a new approach to the application of liposome [Hanus J. et. al. Langmuir, 2013]. Although there are many studies in the literature regarding liposomes, there is no study in the literature on liposomes embedded into multi-response microgels, covered with Fe3O4 and goal orientation and accumulation that respond to multi environmental conditions. In this study, multi-responsive microgels which were first introduced by our group in the literature were used in order to provide targeting and releasing properties to liposomes [Butun V. et. al. Langmuir, 2011; Tuncer C. et. al. J.Appl. Polym. Sci., 2015]. In the first stage, 5-Fluorouracil (5-FU) which is used clinically to treat breast, stomach and colon cancer, was loaded into liposomes in aqueous media. In the second stage, 2-(N-morpholino) ethyl methacrylate (PMEMA) based multi-responsive microgel hosting liposome-5-FU system was synthesized. In the third stage, a hybrid system that consists of liposome-5-FU loaded microgels were covered with magnetic-nanoparticles Fe3O4 in order to be able to direct these system by a magnetic field. Finally, this hybrid system was used in vitro studies as a drug carrier on a target and controlled drug release system in phosphate buffer solution.

Acknowledgement: The Scientific and Technological Research Council of Turkey is thanked for financial support (GR:115Z726 and 2211/C).

Keywords: magnetic liposome, multi-reponse microgels, drug carrying, targeted, release.

MBM-P-013 [Macromolecules in Biotechnology & Medicine]

Enzyme Immobilization on Polymer-based Supports

Dilek Nartop[1], Murat Güleç[2], Birtane Demirel[2], Elvan Hasanoğlu Özkan[3], Nurdan Kurnaz Yetim[4], Nurşen Sarı[3]

[1]Department of Polymer Engineering, Faculty of Technology, Düzce University, 81620 Konuralp, Düzce, Turkey
[2]Department of Chemistry, Faculty of Arts and Science, Nevşehir Hacı Bektaş Veli University, 50300 Nevşehir, Turkey
[3]Department of Chemistry, Faculty of Arts and Science, Kırklareli University, Kırklareli, Turkey
[4]Department of Chemistry, Faculty of Science, Gazi University, 06500 Teknikokullar, Ankara, Turkey

Nanoparticules involving metal ion play important roles in biotechnological applications such as enzymatic reactions in the human body [1]. Immobilization of enzymes onto polymeric platforms is the most useful strategy to improve the operational stability of biocatalysts [2]. The polymer-based macromolecules including metal atoms can be covalently linked with enzymes [3]. In this study, to investigate the effect of hydrogen and coordination covalent bonds on enzyme immobilization, we prepared two polymeric supports.

Firstly, a new polymer-based Schiff base and its Pt(IV) complex were synthesized by means of condensation reacting (aminomethyl)polystyrene (AMP), terephthaldehyde (TA) and 2-amino-4-methylphenol. Macromolecules were characterized by elemental analysis, of FT-IR, 1H-NMR, TGA and SEM analysis. Secondly, glucose oxsidase enzyme (GOx) were immobilized on the modified polymers with using the adsorption method. The effect of pH and temperature parameters on the free and immobilized GOx enzyme was examined. The optimum pH value for immobilizied glucose oxsidase enzyme on (AMP-TA-MAF) and (AMP-TA-MAF-Pt) were determined as 8. The optimum temperature values for immobilizied enzyme on (AMP-TA-MAF) and (AMP-TA-MAF-Pt) were found two optimum temperature 30 °C / 70 °C and 40 °C / 70 °C, respectively.

References:

1. N. Toshima, T. Yonezawa, New J. Chem. 22 (1998) 1179.
2. H.N. Bhatti, N. Saleem, Food Technol Biotechnol. 47 (2009) 331.
3. J. Huang, R. Zhao, H. Wang, W. Zhao, L. Ding, Biotechnol. Lett. 32 (2010) 817.

Keywords: Polymer-based macromolecules, glucose oxsidase enzyme, (aminomethyl)polystyrene

Figure

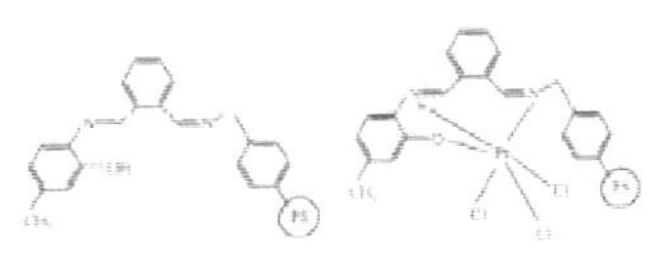

(AMP-TA-MAF) (AMP-TA-MAF-Pt)

Figure Structures of synthesized polymer-based macromolecules

MBM-P-014 [Macromolecules in Biotechnology & Medicine]

Sol-Gel Coating For Inhibition Monomer Release From Acrylic Denture Base

Selda Sezer[1], Ahmet Gultek[1], Ismail Hakki Uzun[2], Nilufer Tulin Polat[2], Turgay Seckin[1]

[1]*Department of Chemistry, Inonu University, Malatya, Turkey*
[2]*Department of Prosthodontics, Inonu University, Malatya, Turkey*

Acrylic based resins are widely used in dentistry. Polymethyl methacrylate (PMMA) based resins are commonly utilized in production of various dental prosthesis, prosthesis support materials, crones and orthodontic instruments and there isn't any ideal material that can replace these materials. Besides their advantages such as biocompatibility, aesthetics appearance, etc; acrylic resins have dimensional instability, water absorption, low mechanical properties and residual monomer content type disadvantages. During polymerization of acrylic resins, un-polymerized monomers stay in the material and they are released into the mouth in the course of time. These released compounds can cause allergic reactions in mucosa membrane or all over body. Amount of monomers released from acrylic resins varies with polymerization method, polymerization environment, degree of polymerization of monomer and biodegradation of dental materials inside the mouth. Another significant factor for residual monomer release is the surface condition of the denture base resin. In this study, leaching of monomer was investigated by coating the surface of polymethyl methacrylate based resins with (3-Glycidyloxypropyl)trimethoxysilane (GLYMO)- Tetraethyl orthosilicate (TEOS) using sol gel method. Organic-inorganic hybrid materials (ceromers) are profoundly suitable to be used for dental coatings. As coating material, ceromers are used to increase the rigidity and abrasion resistance of some plastics such as PMMA and to enhance life of these materials in humid environments. Coated and uncoated test specimens were held in 10% alcohol solutions for 1, 7, 14, 30 and 60 days. The surface characteristics of materials are examined by AFM, SEM and contact angle measurement. Released monomer from acrylic denture base was determinate by HPLC. Uncoated samples showed a fast monomer release within the first 14 days and monomer release rate got slower after this period, while no release was not observed for the coated samples after 60 days. A p value less than 0.05 was considered statistically significant.

Keywords: Sol-gel coating, monomer releasing, poly (methyl methacrylate), denture base

Releasing monomer from coated and uncoated acrylic denture base.

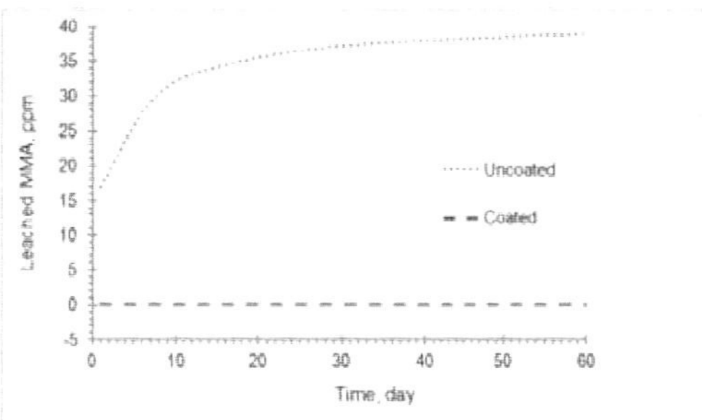

MBM-P-015 [Macromolecules in Biotechnology & Medicine]

Synthesis and characterization of poly(vinylphosphonic acid-co-acrylic acid) copolymers for application in bone tissue scaffolds

Rebecca Dey[1], Xia Zhong[1], Peter Youle[1], Qi Guang Wang[2], Ian Wimpenny[3], Sandra Downes[3], Judith Hoyland[2], David Watts[4], Julie Gough[3], Peter Budd[1]

[1]*School of Chemistry, University of Manchester, Manchester, M13 9PL, U.K.*
[2]*Centre for Tissue Injury and Repair, Institute for Inflammation and Repair, Faculty of Medical and Human Sciences, University of Manchester, Manchester, M13 9PL, U.K.*
[3]*School of Materials, University of Manchester, Manchester, M13 9PL, U.K.*
[4]*School of Dentistry and Photon Science Institute, University of Manchester, Manchester, M13 9PL, U.K.*

Poly(vinylphosphonic acid-*co*-acrylic acid) (PVPA-*co*-AA), has recently been identified as a possible candidate for use in bone tissue scaffolds (Bassi *et al.*, 2011). It is hypothesized that this copolymer can mimic the action of bisphosphonates, a class of drugs used to treat osteoporosis. Due to the strong binding of PVPA-*co*-AA to calcium in natural bone, it is thought that the copolymer can be internalized by resorbing osteoclasts, eventually leading to osteoclast apoptosis (Ghag *et al.*, 2014). The free radical polymerisation of acrylic acid (AA) and vinylphosphonic acid (VPA) has been investigated with varying experimental conditions (Dey *et al.*, 2016). A range of copolymers were produced with varying monomer feed ratios and their compositions were determined quantitatively using ^{31}P NMR spectroscopy. Monomer conversions were calculated using ^{1}H NMR spectroscopy and a general decrease was found with increasing VPA content. Titration studies demonstrated an increase in the degree of neutralisation as a function of VPA in the copolymer. However, a VPA content *ca.* 30 mol% was found to be the optimum for calcium chelation. The ability of the polymer to enhance matrix mineralisation and reduce bone resorption is closely related to its ability to chelate calcium from bone mineral. Therefore, this suggests that PVPA-*co*-AA, with a VPA content *ca.* 30 mol% is a promising candidate for biomaterials applications. Furthermore, assessment of cell metabolic activity showed that PVPA-*co*-AA had no detrimental effect on cells, regardless of copolymer composition.

Keywords: phosphonate-containing copolymer, tissue scaffold, calcium chelation

MBM-P-016 [Macromolecules in Biotechnology & Medicine]

Polyelectrolyte complexes of poly(4-vinylpyridine)-g-poly(ethyleneglycol) with c-myc antisense oligonucleotide as a potential non-viral vector

Murat Topuzoğulları, Damla Gökkaya, Mehmet Murat Özmen

Department of Bioengineering, Yildiz Technical University, Istanbul, Turkey

Polycations, polymers with positively charged repeating units are widely used in biomedical fields including drug and gene delivery applications. Polyvinylpyridines (PVP) are an important members of the polycations family. In the literature, there are many studies reporting the preparation of these polymers by quaternization of amines of pyridine rings in the polymer chain with alkyl halides2. However, there are limited studies on the preparation of quaternized PVP via grafting a polymer chain such as polyethylene glycol (PEG). Grafting a hydrophilic block to the PVP chain produces cationic copolymers suitable for interaction with genetic material to form polyelectrolyte complex (PEC) micelles. The PEG chains on the polymer are especially important in gene therapy studies since they prevent aggregation and non-specific interactions with blood proteins. The hydrophobic character and positive charge of the PVP chain directly affect the interactions inside the PEC micelle and the final structure of such micelles.

In this study, PVP was synthesized using RAFT polymerization and the resulting polymer was quaternized with methoxy-PEG-Br in different quarternization degrees. Polymers with different positive charges and PEG chain lengths were obtained. The resulting polymers were characterized using GPC, FTIR and NMR spectroscopy. Furthermore, the polymers were mixed with a 15-base antisense oligonucleotide by varying the ratios of the components to obtain PEC micelles. The effect of the positive charge, PEG chain length and molar ratio of the components on the size and charge of the formed micelles were investigated using dynamic and electrophoretic light scattering. The results revealed the successful formation of the PEC micelles and that these materials have potential as a non-viral vector in gene delivery applications.

References:

1-M. Topuzogullari, V. Bulmus, E. Dalgakiran, S. Dincer, (2014) Polymer, 55, 2, 525
2- A.San Juan, D. Letourneur, V.A. Izumrudov, (2007) Bioconjugate Chemistry, 18, 922.

Keywords: polyelectroyte complex, gene therapy, quaternization

MBM-P-017 [Macromolecules in Biotechnology & Medicine]

Construction of novel cholesterol biosensors by using dithienopyrrole type conducting polymer

Huseyin Bekir Yildiz[1], Nilay Gazel[2], Semahat Kucukkolbasi[2]

[1]*Department of Materials Science and Nanotechnology Engineering, KTO Karatay University, Konya, Turkey*
[2]*Department of Chemistry, Selcuk University, Konya, Turkey*

Cholesterol is a steroid, which is known as an important constituent in an organism. But the medical problems caused by the increase amount of cholesterol in the blood is also quite common. That's why, cholesterol determination is likely to be able to be carried out by an enzyme biosensor which has a fast and wide measurement range is the main idea of this study.

At this study, firstly prominent with its electron donating structure new generation conducting polymer 4-(4H-dithienol[3,2-b:2',3'-d]pyrrole-4)aniline is synthesized. Characterisation of polymer is controlled with 1H-NMR and 13C-NMR. Then glassy carbon electrode was modified with DTP(aryl)aniline by electropolymerization.

Ideal working conditions and performance factors of prepared enzym electrode was identified. Ideal working conditions which are optimum immobilization time, optimum pH value, optimum working potential, optimum temperature were determined. Performance factors were analysed under ideal working conditions of cholesterol oxidase enzym electrode. Working range was between 2 μM and 23,7 μM. LOD and LOQ values were found as 0,27 μM and 0,82μM respectively. Michealis-Menten constant, Km, to use for determining the enzym activity was calculated as 17,81 μM. Storage stability of enzyme electrode was observed as 1 week. Finally, under optimized conditions, the amount of cholesterol in spiked human serum samples and recovery experiments were conducted.

Keywords: Cholesterol, biosensor, cholesterol oxidase, conducting polymers, dithienopyrrole, electropolymerization

Figure 1

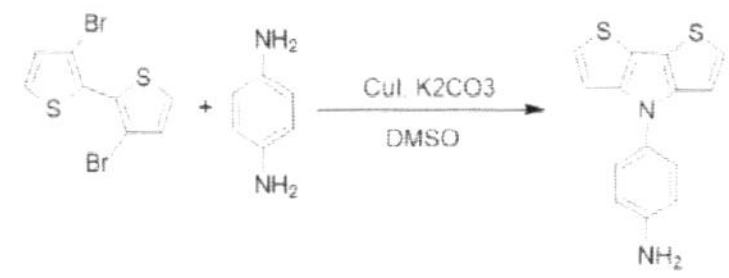

2 3 4

Figure 1. Schematic representation of 4-(4H-dithienol[3,2-b:2',3'-d]pyrrole-4)aniline synthesis

Schematic representation of 4-(4H-dithienol[3,2-b:2',3'-d]pyrrole-4)aniline synthesis

MBM-P-018 [Macromolecules in Biotechnology & Medicine]

A novel application for polyDADMAC in biopharmaceutical production

Almut Rapp, Sarah Koegler, Joerg von Hagen

Merck, Pharm Chemicals Solutions R&D, Darmstadt, Germany

PolyDADMAC (Polydiallyldimethylammonium chloride) is a well-known polymer that is water soluble and carries positive charges. Recently we have identified its tremendous benefit in therapeutic protein production as a flocculant. A recent trend in the manufacture of monoclonal antibodies (mAbs) is the use of flocculation and subsequent depth filtration instead of centrifuges to remove cells and cell debris during the downstream purification process. The use of polyDADMAC as a flocculant in water purification is well-known. Flocculation in cell culture harvest however faces tremendous challenges. While cells, cell debris and related cellular impurities need to be removed efficiently, a loss of antibody needs to be kept to an absolute minimum. It turns out that polyDADMAC with an average molecular weight of 300 000 g/mol is very well suited to flocculate cells and cell debris, but leaves monoclonal antibodies in solution. In combination with depth filters, cells and cell debris can be efficiently removed from the product stream. We demonstrate the effectiveness of polyDADMAC flocculation and its significant benefits over classical centrifugation, acid precipitation and other flocculation polymers such as polyethylenimines. The polyDADMAC flocculation mechanism was identified to follow the electrostatic patch model and we highlight the importance of correct dosing of the polymer. Since polymer toxicity is a primary concern for implementation of this process step, we have investigate the effective removal of residual polyDADMAC in the product stream to ensure that the therapeutic protein (mAb) is not contaminated with residual polyDADMAC.

Keywords: polyDADMAC, flocculation, biopharmaceutical production, downstream process

MBM-P-019 [Macromolecules in Biotechnology & Medicine]

Development of polyvinyl chloride/silver nanoparticle composites through different fabrication techniques and their performance as antimicrobial substrates

Muhammad Anwaar, Emel Yilgör, Iskender Yilgör

KUYTAM Surface Science and Technology Center, Chemistry Department Koc University, Istanbul, Turkey

Preparation of silver nanoparticle containing poly(vinyl chloride) (PVC) nanocomposites using three different methods, spin coating, cast molding and electrospinning will be reported. Silver nitrate ($AgNO_3$) was used as a precursor to generate silver nanoparticles (AgNP). Incorporation of AgNP into various polymers such as chitosan/PVC [1], gelatin [2], polyacrylonitrile [3], poly(lactic-co-glycolic acid) [4] and poly(vinyl alcohol) [5] have been reported. Dimethylformamide (DMF) can reduce silver nitrate to silver nanoparticles without the use of external reducing agents [4]. In this study PVC/AgNP nanocomposites were prepared by using electrospinning, cast molding and spin coating. 16 wt% solutions of PVC were prepared in DMF:THF (1:1) mixtures with $AgNO_3$ concentrations of 5 and 10% by weight on the weight of polymer. Nanocomposites obtained were characterized by ATR-FTIR, XRD, EDX, SEM and stress strain tests. As shown in Figure 1, zone inhibition tests performed on the substrates showed very high antibacterial activity against E coli in electrospun webs when compared with films prepared by cast molding.

References:

1. Dwivedi, P. et al., Int. J. Scientific and Research Publications, 2012, 2(7), 1-5.
2. Jeong, L., et al., Int. J. Molecular Sciences, 2014, 15(4), 6857-6879.
3. Mahapatra, A., et al., J. Appl. Polym. Sci., 2012, 124(2), 1178-1185.
4. Khalil, K.A., et al., Int. J. Electrochemical Science, 2013, 8(3), 3483-3493.
5. Tan, G., et al., Turk J Biol, 2016, 40, 1-9.

Keywords: Nano-composites, polyvinyl chloride, silver nano particles, electrospinning, antimicrobial materials

Figure 1. Results of zone inhibition tests against E coli in PVC/AgNP samples prepared by electrospinning and cast molding

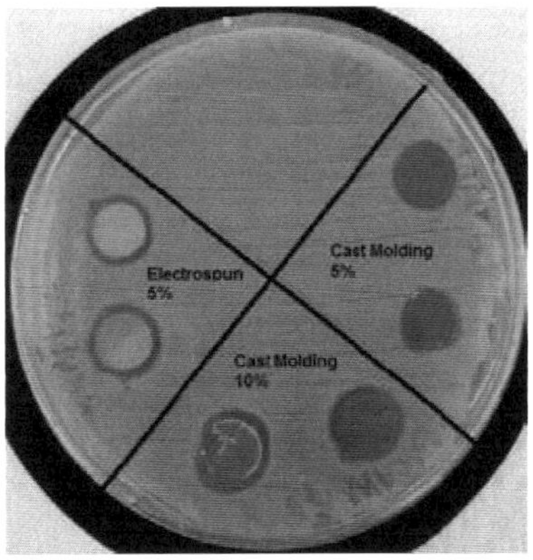

MBM-P-020 [Macromolecules in Biotechnology & Medicine]

CoFe2O4 -FRUCTOSE-METHOTREXATE BASED ANTICANCER DRUG DESIGN

Idil Karaca Açarı, Süleyman Köytepe, Burhan Ateş, Sevgi Balcıoğlu, Turgay Seçkin, Ismet Yılmaz

İnönü University, Faculty of Arts and Science, Chemistry Department, Malatya, Turkey

Recently spinel ferrite has conceived considerable attention due to extensive application in biomedical field such as contrast enhancing agent in mangnetic resonance imagine and an active agent in drug delivery. More importantly spinel cobalt ferrite has been proposed for biomedical application due to its large anisotropy compared to other ferrites [1,2]. Methotrexate (MTX) is one of the oldest and most widely utilized chemotherapy drugs, not only in anticancer therapy, but also in the treatment of autoimmune diseases and for the prevention of graft-versus -host disease after hematopoietic transplantation [3]. Recently, more effective faster and combine anticancer drug design has been extensively studied. Thus, the main objective in this study was basically to determine the cancer cell zone and MTX release in these areas. For this aim, $CoFe_2O_4$ nanoparticles had been synthesized using the hydrothermally technique. Then, prepared nanoparticles were modified with fructose for solubility and targeting of cancer cell. MTX was immobilized in fructose modified nanoparticles (Figure 1). Effects of synthesis conditions on the crystal structure, crystallite size, lattice parameter, microstructure and magnetic properties of the produced spinel ferrites and fructose modified nanoparticles were investigated by X-ray diffraction analysis (XRD), scanning electron microscope (SEM) and elemental analysis. The mean crystallite size of the prepared ferrite was in the range of 5–20 nm. The magnetic properties of nanoparticles and nanocomposite were investigated using a vibrating sample magnetometer (VSM). Anticancer effects of $CoFe_2O_4$-fructose-methotrexate based nanoparticles were investigated in vitro cell culture system on MCF-7 breast cell.

References:

[1] Gupta K, Gupta M. Biomaterials 2005; 26: 3992-4021.
[2] Kohler N, Sun C, Wang J, Zhang M. Langmiur 2005; 21: 8858-8864.
[3] Nevazhay D, Budzynska R, Kanska U, Jagıello M, Omar M.S, Boratynskı J, Opolskı A. Anticancer Research 2006; 26: 1135-1144.

Keywords: nanoparticle, cancer, drug design, MTX

Figure 1.

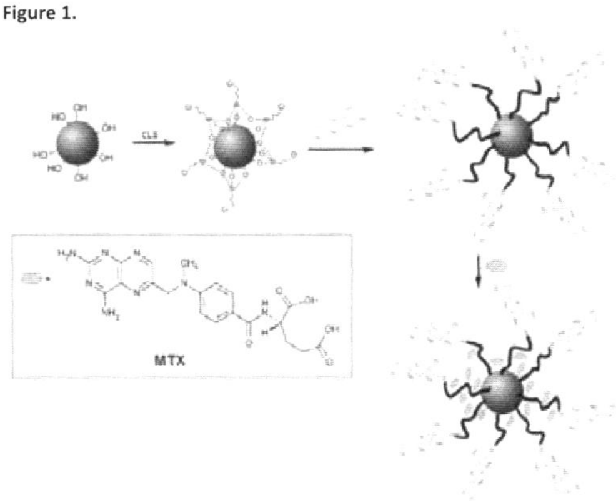

Synthesis of the CoFe2O4-fructose-methotrexate based nanoparticles.

MBM-P-021 [Macromolecules in Biotechnology & Medicine]

The study of binding modes of Thioxanthone-Diamin with calf thymus DNA

Rabia Halıcı, Eyüp Metin, Neşe Atacı, Nergis Arsu

Department of Chemistry, Yıldız Technical University, Istanbul, Turkey

The study of interactions between anticancer reagents or plausible (candidate) anti-cancer reagents and DNA is important in order to understand the mechanism of their anticancer activities. Therefore, it is necessary to introduce new compounds as anticancer drug candidate and more simple techniques. Therefore, it is necessary to introduce new compounds as anticancer drug candidate and more simple techniques. Among these techniques, fluorescence studies are a powerful way to investigate the dynamic properties of fluorescent–labelled molecules. (Qui *et al*, Biosens. Bioelectron, 2007, 22, 2629-2635)

Thioxanthone and its derivatives are important photoinitiators and they can be classed as Type II photoinitiators. Several thioxanthone derivatives e.g. mercapto, acetic acid, anthracene and polymeric, visible and water soluble ones were synthesized and used as photoinitiator for various formulations. (Arsu *et al*, Euro. Polymer J, 2010, 46, 1374-1379) Biochemical applications of thioxanthones in the last decade led important achievements especially for their antitumor activities and low toxicities. (Palmeira *et al*, Pharmacol., 2012, 83, 57-68) Expectations are the synthesis of new analogues with the thioxanthone scaffold in the field of anticancer therapy, with enhanced antitumor activity and without serious side effects. (Corbet *et al*, Inves. New Drugs, 1999, 17, 17-27)

2,7-Bis-(2-diethylamino-ethylamino)thioxanthen-9-one, Thioxanthone-Diamin (TX-DA), was synthesized as new thioxanthone derivative by using of modified synthesis method. (Mauss, 1948) In this study, to identify the interaction mechanism of TX-DA with DNA, UV–Vis absorption, fluorescence emission spectroscopy techniques and melting studies were used to distinguish the binding mode whether intercalative, electrostatic or groove.

Keywords: Thioxanthone diamin, binding mode, DNA, DNA binding, photoinitiator, TX-DA

MBM-P-022 [Macromolecules in Biotechnology & Medicine]

Synthesis of Novel Thermosensitive Poly(Amine Functionalized Glycolide)-Poly(Lactide) - PEG Copolymers and Their Applications in Drug Delivery Systems

Sezgi Erdoğan[1], Mehmet Onur Arıcan[1], Olcay Mert[2]

[1]Department of Polymer Science And Technology, Kocaeli University, Kocaeli, Turkey
[2]Department of Chemistry, Kocaeli University, Kocaeli, Turkey

Polyesters such as poly(lactide), poly(glycolide) and PLGA have been mostly chosen in biomedical field because of their features of biocompatibility and biodegradability over the last ten years. However, their high crystallinity, hydrophobicity and lack of functional group diversity in the back-bone prevent the methods to improve the mechanical, chemical and biological properties. Substituted glycolide monomers and their polymerizations have received increasing special attention because functional groups along the polymer chains adjust finely their properties such as drug permeability, hydrophilicity and biodegradability. Especially, it enables attaching drugs and bioactive reagents covalently onto the polymer. Here we describe the synthesis and thermosensitive properties of PEG based copolymers of lactide with amine-functionalized glycolide.

Acknowledgement : This study was granted by TUBITAK, Turkey, with project number 112T865.

Keywords: Functional Glycolide,Thermosensitivity, Drug Delivery Systems

MBM-P-023 [Macromolecules in Biotechnology & Medicine]

Syntheses and nanoformulations of novel boron based poly(diisobutyl glycolide) homopolymers and their copolymers with lactide for drug delivery systems

Mehmet Onur Arıcan[1], Aybüke Huri Kandıra[2], Gökhan Duruksu[3], Ufuk Yıldız[2], Asgar Kayan[2], Olcay Mert[2]

[1]Department of Polymer Science and Technology, Kocaeli University, Kocaeli, Turkey
[2]Department of Chemistry, Kocaeli University, Kocaeli, Turkey
[3]Center for Stem Cell and Gene Therapies, Research and Practice, Institute of Health Sciences, Stem Cell Department, Kocaeli University, Kocaeli, Turkey

Each day approximately twenty thousand people lost their lives in the world because of the fatal cancer diseases. Many cancer types (for example, prostate and colon in men, breast and cervix cancer in women, lung cancer in both sex) have a chance of treatment when they are diagnosed in early phase. In the treatment of primarily malignant cancer and other major diseases, drug delivery systems are now one of the most invested areas in biomedical and pharmaceutical industry. In this work, diflouro boron-based poly(diisobutyl glycolide) homopolymers and poly(diisobutyl glycolide-co-lactide) copolymers were synthesized via ring opening polymerization. The polymer structures were confirmed by NMR spectroscopy while the molar mass distributions were determined by GPC analysis. Preparation of nanoparticles were performed by both nanoprecipitation and single emulsion techniques with using poly(vinyl alcohol) or vitamin E TPGS as surfactants. All nano-formulations were characterized by dynamic light scattering (DLS). Paclitaxel anticancer drug was effectively loaded into nanocarriers and release studies from the nanoparticles were carried out with HPLC technique. Finally, morphology and toxicity analyses of paclitaxel loaded nanoparticles were performed in vitro using three different cell lines (MCF7, DU145, and Fibroblast).

Acknowledgement: We are grateful to TUBITAK (Project Number: 113Z263) for financial support.

Keywords: Early Detection, Drug Delivery Systems, Biodegradable Polymers

MBM-P-024 [Macromolecules in Biotechnology & Medicine]

Synthesis and Characterization of a New Class of Boron Based Asymmetric Poly(Substituted Glycolides) for Biomedical Applications

Ayşenur Vardar[1], Mehmet Onur Arıcan[2], Ufuk Yıldız[1], Asgar Kayan[1], Olcay Mert[1]

[1]Department of Chemistry, Kocaeli University, Kocaeli, Turkey
[2]Department of Polymer Science and Technology, Kocaeli University, Kocaeli, Turkey

Cancer, which causes the death of eight million people in the world ever year, is defined as uncontrolled growth and multiplication of cells. One of the most effective methods for treatment of cancer disease is drug delivery system. This study was conducted to ensure the monitoring response to treatment, goal oriented treatment and early diagnosis (theranostics). For this purpose, boron based asymmetric poly(isobutyl-methyl glycolide) (PIBMG) and poly(isobutyl-ethyl glycolide) (PIBEG) polymers were synthesized with melt polymerization under nitrogen atmosphere with using BF2dbmOEtOH as an initiator. New boron based biodegradable polymers were produced in various molecular weights and were characterized by spectroscopic, chromatographic, and thermal methods.

Acknowledgement: We are grateful to TUBITAK (Project Number: 113Z263) for financial support.

Keywords: Biodegradable Polymers, Asymmetric Poly(Substituted Glycolides), Theranostics

MBM-P-025 [Macromolecules in Biotechnology & Medicine]

Synthesis of Tween-Based Polyurethane Adhesives Capable of Adhering in Moisture Environment

Burhan Ates, Suleyman Koytepe, Merve Goksin Karaaslan, Unzile Kelestemur, Sevgi Balcioglu, Selam Gulgen

Department of Chemistry, İnönü University, Malatya, Turkey.

Tissue adhesives have been widely used in clinics as wound closure devices, hemostatic agents, and tissue sealants to prevent bleeding and promote tissue healing. Tissue adhesives should demonstrate suitable properties, such as tissue adhesiveness, biocompatibility, biodegradability and mechanic strength. However, the commercial tissue adhesives which have high biocompatibility genarally show poor mechanical and adhesion strength. In tissue adhesives, PEG based structures are commonly used because of biocompatibility, biodegradability and well tolerating in body. However, due to high hydrophilicity of PEG groups, a poor adhesion force is provided in moisture environment of the body. Therefore, in this study, amphiphilic polysorbate (tween (T)) molecules were used in synthesis of polyurethane-based tissue adhesives in order to decreace the interaction between body fluids and tissue adhesives. Furthermore, to increase mechanical strength and adhesion force, caffeic acid (CA) was used as crosslink agent and mussel bioinspired molecule. As experimental section, our work involves synthesis, thermal and morphological characterization, adhesion properties, in vitro biodegradation and in vitro biocompatibility properties of tween-CA based polyurethane bioadhesives. In synthesis design, PEG200 was used to increase biocompatibility and biodegradability, tween 20, 40, 60 and 80 were used to provide amphiphilic character to PUs, CA was used to increase mechanical strength in PU polymers. NPU-T20-CAs were produced with 4,4'-methylenebis(cyclohexyl isocyanate)(NPU), caffeic acid, PEG and tween 20,40,60,80 via polycondensation using THF as solvent and triethylamine as the catalyst. The ratios of NPU:PEG:tween:CA units in the PUs were 1:0.8:0.15:0.05, 1:0.8:0.10:0.10 and 1:0.8:0.05:0.15, respectively. According to adhesion test results, the highest adhesion strength was about 450 kPa of NPU-T20-CA-5% (Figure 1). The contact angles of NPU-T20-CAs were between 48-63°. Biodegrdability results of NPU-T20-CAs were about 10%. As a result, the synthesized adhesives may be used tissue adhesive to provide adhering in wet environment.

-This study was supported by grants from TUBITAK (TBAG-114Z591).

Keywords: Polyurethane, adhesive, tween

Figure 1. Adhesive strength of Tween/CA-based polyurethanes

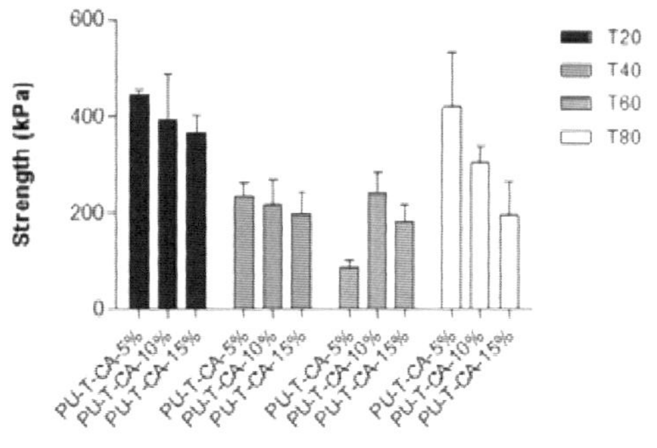

MBM-P-026 [Macromolecules in Biotechnology & Medicine]

Synthesis and evaluation of bisphosphonate-containing poly(β-amino ester) macromers for biomedical applications

Melek Naz Guven, Merve Seckin Altuncu, Duygu Avci

Department of Chemistry, Bogazici University, Istanbul, Turkey

Poly(β-amino ester)s (PBAEs) are a class of cationic polymers that have been developed to be used as biomaterials in drug delivery, gene delivery and tissue engineering applications due to their pH-sensitivity, biodegradability and high biocompatibility. They can be easily prepared by Michael addition reaction of primary or secondary amines to diacrylates without solvents or catalysts. In this work, six bisphosphonate-functionalized PBAE macromers (Figure 1) were synthesized. Bisphosphonate incorporation into PBAEs is novel and we expect improvement in relevant properties. The synthesis involved two steps: i) synthesis of bisphosphonate-containing diamines through the reaction of diamines (hexane-1,6-diamine, butane-1,4-diamine and 4,9-dioxa-1,12-dodecanediamine) with ethylidene bisphosphonate, ii) aza-Michael addition of bisphosphonated diamines to multifunctional acrylates such as poly(ethylene glycol) diacrylate and 1,6-hexanediol diacrylate. The macromers were photopolymerized to generate degradable networks to investigate the influence of variations of macromer structure on polymers' degradability and mechanical properties.

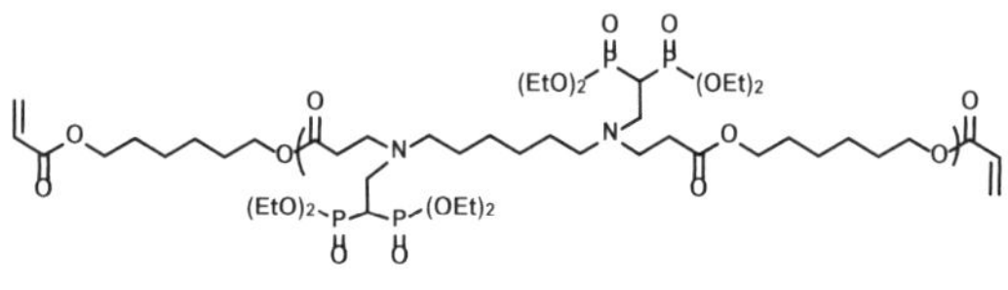

Figure 1. Structure of one of the macromers

This work was supported by a grant from the Scientific and Technological Research Council of Turkey (TÜBİTAK) (114Z926).

Keywords: bisphosphonate, poly(β-amino ester) macromers, biomedical applications

MBM-P-027 [Macromolecules in Biotechnology & Medicine]

Hydrogels for biomedical applications using biodegradable oligomers of poly(lactic acid)

Marcos Lopes Dias[1], Felipe Augusto Moro Loureiro[2], Iryna Grafova[3], Andriy Grafov[3]

[1]Instituto de Macromoléculas Professora Eloisa Mano (IMA), Universidade Federal do Rio de Janeiro, Rio de Janeiro, Brazil
[2]Faculdade do Centro Leste, Serra, Brazil
[3]Department of Chemistry, University of Helsinki, Helsinki, Filand

Hydrogels are three-dimensional structures with high hydrophilicity, formed by polymer networks capable of absorbing large amounts of water or biological fluids. Due to their high water content, porosity and soft consistency, they closely simulate natural living tissue, more than any other class of synthetic biomaterials. Hydrogels may be chemically stable or they may degrade and eventually disintegrate and dissolve, having applications as contact lenses, wound dressings, drug delivery systems, tissue engineering, and hygiene products. Recently, poly(lactic acid) (PLA) and polyethylene glycol (PEG) has been reported as interesting building blocks for biodegradable hydrogels capable of being used in many applications. In this contribution, an alternative route to produce ethylene glycol-lactic acid based hydrogels is reported. Oligomers of lactic acid (LA) was prepared by aminolysis reaction of commercial PLA with 1,6-hexamethylenediamine in chloroform solution using PLA:diamine mass ratio of 1:1, 1:2 and 3:1. Characterization of LA oligomers was carried out by spectroscopic techniques (FTIR, 1H NMR, 13C NMR), thermal analysis (TGA, DSC) and gel permeation chromatography (GPC). The oligomers showed presence of amide and amine end-groups and the molar mass dependence on PLA/diamine ratio. The hydrogels were produced by reaction of poly(ethylene glycol) diglycidyl ether (PEDGE) with the LA oligomers in different concentrations. The hydrogels were characterized according to their water absorption properties. Dependence of water retain capacity on the hydrogel composition is discussed.

Keywords: hydrogel, PLA, oligomer, poly(ethylene glycol) diglycidyl ether, aminolysis

MBM-P-028 [Macromolecules in Biotechnology & Medicine]

Synthesis and In vitro Evaluation of Poly(2-((2-aminoethyl)amino)ethyl methacrylate) as a Potential siRNA Delivery Agent

Aykut Zelçak, Volga Bulmuş

Department of Bioengineering, Izmir Institute of Technology, Izmir, Turkey

The effective and specific gene silencing ability of siRNA molecules has made them valuable tools as potential therapeutic agents. However, one of the biggest struggles in the development of siRNA based gene silencing therapies is the delivery of siRNA molecules. Diaminoethane motif bearing polymers have displayed superior properties in delivery of nucleic acids including DNA and siRNA.

In this study, a new spermine-like polymer bearing diaminoethane motifs -poly(2-((2-aminoethyl)amino)ethyl methacrylate)(P(AEAEMA))- was synthesized and evaluated as a potential siRNA carrier.

An amine containing monomer 2-((tert-butoxycarbonyl) (2-((tert-butoxy carbonyl)amino)ethyl)amino)ethyl methacrylate (BocAEAEMA) was synthesized according to the procedure reported previously (Kurtulus et al., 2014). RAFT polymerization was performed to prepare homo- and block co-polymers of BocAEAEMA. The synthesized polymers (P(AEAEMA)19, P(AEAEMA)41 and P(PEGMA)12-b-P(AEAEMA)32) were characterized via NMR and GPC.

The polymers at concentrations between 2.5μM and 20μM did not alter significantly the viability of Skov-3 cells after incubation for 24 h, as determined by MTT assay.

The ability of the BocAEAEMA polymers to form polyplexes with siRNA was investigated via gel electrophoresis. P(AEAEMA)19, P(AEAEMA)41 and P(PEGMA)12-b-P(AEAEMA)32 could efficiently form electrostatic complexes with siRNA at an N/P ratio of 5, 2 and 2, respectively.

siRNA protecting ability of P(AEAEMA)41 and P(PEGMA)12-b-P(AEAEMA)32 against serum nucleases was investigated by agarose gel electrophoresis. While naked siRNA degrades within 0.5 hours, both polymers could protect siRNA against serum components at least for 6 h.

The size and surface charge of the polyplexes were investigated by DLS and ELS, respectively. The diameter of the P(AEAEMA)41-siRNA complexes was found to be lower than 110 nm at all N/P ratios tested. In contrast, P(PEGMA)12-b-P(AEAEMA)32-siRNA complexes displayed aggregation tendency. All polyplexes displayed positive zeta potential.

The *in vitro* transfection efficiency of BocAEAEMA polymers is currently under investigation in our laboratory.

The authors acknowledge TUBİTAK for grant 115R301.

Keywords: cationic polymers, RAFT polymerization, gene therapy, siRNA delivery

MBM-P-029 [Macromolecules in Biotechnology & Medicine]

Synthesis, Characterization and in vitro Evaluation of Arginine Polymers

Damla Taykoz[1], Deniz Uğur[2], Volga Bulmuş[3]

[1]Department of Chemical Engineering, Izmir Institute of Technology, Urla, Izmir, 35430 Turkey

[2]Biotechnology and Bioengineering Graduate Program, Izmir Institute of Technology, Urla, Izmir, 35430 Turkey

[3]Department of Bioengineering, Izmir Institute of Technology, Urla, Izmir, 35430 Turkey

Arginine containing peptides are able to cross cellular membranes. This makes arginine containing peptides and polymers valuable tools for intracellular drug delivery systems. Considering the potential use of arginine-containing polymers in delivery of nucleic acid drugs into cells, we have synthesized an arginine containing methacrylate monomer, arginine methyl ester methacrylamide (AMME), using arginine amino acid (i.e. arginine methyl ester) having unprotected guanidine group as a starting compound, and polymerized via RAFT to directly obtain well-defined polymers of arginine amino acid. Both ^{1}H-NMR and mass spectrophotometer results showed that arginine monomer was synthesized successfully. Free radical polymerization results showed no control over the molecular weight of the synthesized polymers whereas in RAFT polymerization, molecular weight of the polymers was controlled over time with increasing monomer conversion. Linear increase in ln[M]0/[M] with polymerization time, and number average molecular weight (Mn) with monomer conversion indicated RAFT controlled polymerization of AMME monomer under the conditions tested. Figure 1 shows the ^{1}H-NMR spectrum of the RAFT-synthesized arginine polymers after dialysis. MTT experiments were performed using A549 human lung cancer cell line to investigate the cytotoxicity of low molecular weight (Mn: 14,000 Da; PDI: 1.21) and high molecular weight (Mn: 30,000 Da; PDI: 1.24) arginine polymers. The results showed that the toxicity of arginine polymers was observed at concentrations starting from 5 μM concentration as expected due to highly positively charged guanidine groups on arginine. Cell uptake studies at both 4°C and 37°C using arginine polymers with varying molecular weights are under investigation to understand the uptake mechanism of the polymers. Acknowledgment. The authors acknowledge Tübitak for grant#111T960.

Keywords: arginine, RAFT polymerization, intracellular drug delivery

Figure 1

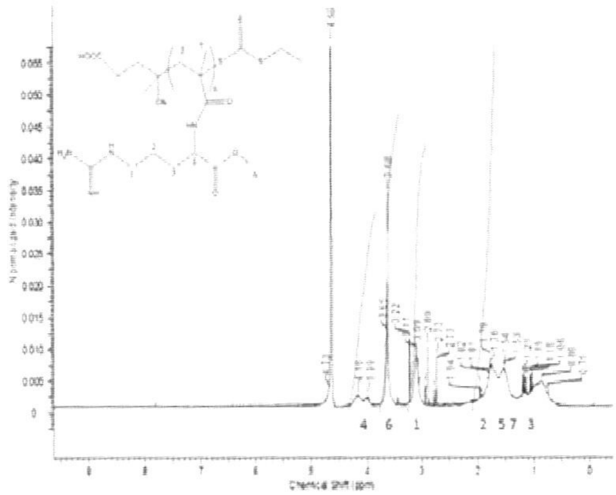

1H-NMR spectrum of the RAFT-synthesized arginine polymers after dialysis.

MBM-P-030 [Macromolecules in Biotechnology & Medicine]

Immobilization of L-asparaginase onto PMMA Composites Activated with Starch

Ahmet Ulu, Suleyman Koytepe, Burhan Ates

Department of Chemistry, Inonu University, Malatya Turkey

L-asparaginase (L-ASNase; known for its anti-leukemia properties) was chosen as a model enzyme to evaluate poly(methyl methacrylate) (PMMA)–starch composites as an immobilization matrix. The use of the starch for enzyme immobilization is preferable to other natural polymers because of the higher reactivity and porosity of starch and/or its lower cost [1]. In our works, the essential aim in the using of starch for the preparation of PMMA composites is the activation of PMMA with hydroxyl groups of starch for L-ASNase immobilization. PMMA activated with hydroxyl groups can also easily interact with L-ASNase due to the hydrophilic character. PMMA–starch composites were prepared by emulsion polymerization technique for L-ASNase immobilization as highly activated support. The pure PMMA and PMMA-starch composites were characterized as structural, thermal and morphological. PMMA-starch composites were found to have better thermal stability and more hydrophilic character than pure PMMA. After, L-ASNase was immobilized onto PMMA-starch composites contained the different ratio of starch (1, 3, 5, and 10 wt %). By immobilization, the temperature stability of the enzyme was improved and showed maximum activity at 45 ºC. pH dependent activities of the native and immobilized enzymes were also investigated, and it was found was 8.5 for both. Km value of immobilized enzyme decreased approximately 8 fold compared with the native one. Besides, native enzyme lost its activity completely within 30 days, while immobilized enzyme lost only 60 % of its activity in under same condition.

1. Ulu, A., Koytepe, S. and Ates, B. (2016). Synthesis and characterization of PMMA composites activated with starch for immobilization of L-asparaginase. Journal of Applied Polymer Science, 133, 40, 43421.

Keywords: PMMA-Starch composites, L-asparaginase, biodegradable, emulsion polymerization

MBM-P-031 [Macromolecules in Biotechnology & Medicine]

Preparation, characterization and antibacterial study of pHEMA (soft contact lens) hydrogel incorporated with boric acid

Ahmet Ulu[1], Rukiye Keskin[1], Emre Birhanlı[2], Sevgi Balcıoglu[1], Suleyman Koytepe[1], Burhan Ates[1]

[1]Department of Chemistry, Inonu University, Malatya Turkey
[2]Department of Biology, Inonu University, Malatya Turkey

Soft contact lenses are hydrogels, a cross-linked network of water-soluble monomers and have been used for vision correction for more than 45 years. The majority of soft contact lenses are poly(2-hydroxyethyl methacrylate) (pHEMA) hydrogels. Therefore, pHEMA-based hydrogels are extensively used as soft contact lenses due to their excellent biocompatibility and mechanical properties [1]. For lens uses, bacterial contamination has one of the most important risks. In order to reduce the bacterial-associated adverse events, we hypothesized that the incorporation of varying concentrations boric acid (BA) particles into contact lenses may enhance antibacterial properties of the lenses. Boric acid has been generally known killing of diverse organisms ranging from bacteria to rodents [2]. The present study includes polymerization of 2-HEMA by emulsion polymerization to produce hydrogel contact lenses incorporated with boric acid. The synthesized hydrogels were characterized by FTIR, TGA, DTA and DSC. Besides, the hydrogels were determined in terms of hydrophilic character, in vitro biocompatibility and biodegradability behaviors. The antibacterial effect all hydrogel was also examined using disk diffusion method. The results show the hydrophilic character and in vitro biodegradability increases with increasing hydrophilic groups in hydrogels contact lenses. The antibacterial data demonstrate that pHEMA/BA hydrogel contact lenses can effectively use to decrease the risk of bacterial-related adverse events for lens wearers.

1. Caló, E., Khutoryanskiy, V. V. (2015) Biomedical applications of hydrogels: A review of patents and commercial products. European Polymer Journal. 65, 252–267.
2. Siegel, E. and Wason, S. (1986). Boric acid toxicity. Pediatric Clinics of North America. 33, 2, 363–367.

Keywords: contact lens, pHEMA, boric acid, antibacterial

The FTIR spectra of pHEMA hydrogels containing different concentrations boric acid.

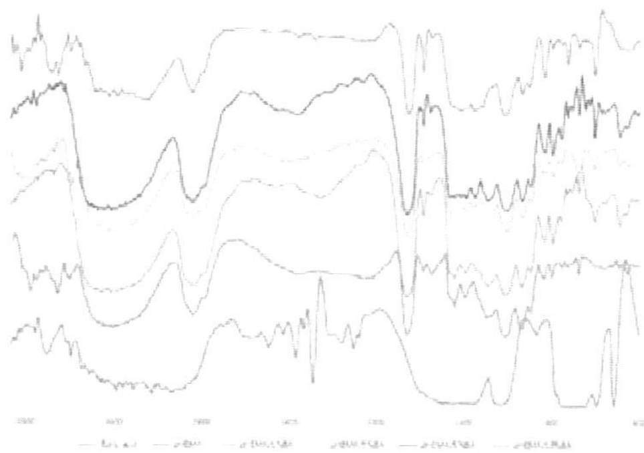

MBM-P-032 [Macromolecules in Biotechnology & Medicine]

Electrospun Biocompatible Core/Shell Structure Nanofibers with Potential Application for Local Cancer Therapy & Wound Healing

Roya Sedghi, Alireza Shaabani

Department of Polymer, Faculty of Chemistry, Shahid beheshti university

The first core/shell structure of Curcumin (Curcumin/poly (vinyl alcohol)–chitosan nanofibers (Cur/PVA–CS)) was prepared and properties of the system were characterized. For this purpose, the coaxial electrospinning technique was used to produce Cur/CS-PVA core-sheath nanofibers in which Curcumin was used as core of nanofibers and blend of chitosan and polyvinyl alcohol as shell layer. Chitosan was chosen as a one of the shell material of coaxial electrospinning because of its nontoxic, biocompatible, biodegradable, biorenewable and environmentally friendly properties [1]

Due to polycationic nature of CS, repulsive force, rigid chemical structures, powerful inter and intramolecular interactions electrospinning ability of CS restricted [2,3]. Because of biocompatibility, excellent mechanical properties, high hydrophilicity and fiber-forming ability of PVA, selected to blending with CS. Also Curcumin as a model of antimicrobial and anticancer drug loaded into nanofibers core to increase bioavailability, bioactivity and better in vitro controlled release. The results demonstrated that the Curcumin loaded in core/shell nanofibers have a great antimicrobial effect against more dangerous MRSA which is resistance to numerous antibiotics including methicillin, oxacillin and amoxicillin, thus making difficult to treat. Also the effect of these nanofibers on cancer cells viability which assessed by the XTT and shown the Curcumin loaded nanofibers were suitable for postoperative chemotherapy of breast cancers. The nanofibers had an excellent antimicrobial and potential for wound healing.

[1] S.-K. Kim, Chitin, chitosan, oligosaccharides and their derivatives: biological activities and applications, CRC Press, 2010.
[2] K. Desai, K. Kit, J. Li, S. Zivanovic, Biomacromolecules, 9 (2008) 1000-1006.
[3] H. Che, M. Huo, L. Peng, Q. Ye, J. Guo, K. Wang, Y. Wei, J. Yuan, Polymer Chemistry, 6 (2015) 2319-2326.

Keywords: wound healing, curcumin, Cur/PVA-CS nanofibers, core/shell

Figure 1.

SEM & TEM images of nanofibers.

MBM-P-033 [Macromolecules in Biotechnology & Medicine]

Zinc Strontium Chelated Chitosan - Brushite Composite Gel Fibers for Bone Tissue Engineering

Eda Ayse Aksoy[1], Betul Suyumbike Yagci[1], Gulseher Manap[2], Sukru Ozturk[3], Kezban Ulubayram[1]

[1]*Department of Basic Pharmaceutical Sciences, Faculty of Pharmacy, Hacettepe University, Ankara, Turkey*
[2]*Polymer Science and Technology Division, Institute for Graduate Studies in Science and Engineering, Hacettepe University, Ankara, Turkey*
[3]*Bioengineering Division, Institute for Graduate Studies in Science and Engineering, Hacettepe University, Ankara, Turkey*

The limited ability of the bone tissue to regenerate irregular large bone defects and the risks and limitations of autogenic and allogenic bone grafts have created the need for the development of tissue-engineered constructs. Natural and synthetic biodegradable polymers have long been processed to form sponges, 3-D constructs, micro and/or nanofiber meshes, and used as scaffolds for bone tissue engineering. However, analyzing the clinical results it is obvious that bone tissue scaffolds still need to be improved and tailored. In this study, it was aimed to prepare functional gel fiber scaffolds with enhanced osteoconductive properties. For this purpose, chitosan was combined with brushite (dicalcium phosphate dehydrate) which is an osteoconductive bioceramic with great potential of being a bioresorbable cement for bone regeneration. Chitosan-brushite composite gel fibers of variable diameters were prepared by wet spinning technique and these wet spun fibers were extended to woven gel structures by optimizing the processing parameters. The swelling ratio of up to 180 % was observed, which is superior for integration of the scaffold into irregular bone defect area. Enhancement of osteoconductive properties of chitosan-brushite fibers was carried out by zinc (Zn) and/or strontium (Sr) chelating which are both essential elements of the bone matrix. Fibers were characterized in terms of their bulk and surface properties, via chemical, thermal and morphological characterization techniques. In vitro cytotoxicity evaluations were carried out according to ISO10993-5 Indirect MTT cytotoxicity test. Prepared fibers showed promising results regarding the development of bone tissue gel fiber constructs.

Keywords: Bone Tissue Engineering, Wet Spun Fiber, Chitosan, Brushite, Zinc, Strontium

MBM-P-034 [Macromolecules in Biotechnology & Medicine]

The Stabilization of Camptothecin Anticancer Drug via Phosphazene Cored and Miktoarm Star-Shaped Polymeric Nanocarriers

Duygu Çetin[1], Mehmet Onur Arıcan[1], Erdinç Doğancı[2], Nursel Olgaç[3], Mesut Görür[4], Faruk Yılmaz[3], Olcay Mert[1]

[1]*Department of Chemistry, Kocaeli University, Kocaeli, Turkey*
[2]*Department of Chemistry and Chemical Processing Technologies, Kocaeli, Turkey*
[3]*Department of Chemistry, Gebze Institute of Technology, Kocaeli, Turkey*
[4]*Department of Chemistry, Istanbul Medeniyet University, Istanbul, 34700, Turkey*

Cancer is a dangerous disease produced by abnormal cells. Cancer cells can also easily diffused to one or many parts of the body. It was guessed that every year 14.1 million new cancer cases have occurred and it would probably rise to 19.3 million new cancer cases every year by 2025. Camptothecin (CPT) is a plant alkaloid isolated from bark and wood of the Chinese tree, Camptotheca acuminata, and it has been known as an effective drug for most of cancer therapy. CPT has two forms depending on the pH value, at pH=5 an active lactone form and inactive carboxylate form at higher than pH=7. Lactone form of CPT is a very important structural requirement for binding DNA topoisomerase I complex there by stabilizing it to destroy cancer cell. In the current study, we developed a nanocarrier system based on miktoarm star shaped poly(ε-caprolactone) (PCL)-poly(ethylene glycol) PEG polymers with phosphazene core, which keep the drug in lactone form. Preparation of nanoparticles were performed by both nanoprecipitation and single emulsion techniques with using poly(vinyl alcohol) (PVA) or vitamin E D-α-Tocopherol polyethylene glycol succinate (TPGS) as surfactants. All nano-formulations were characterized by dynamic light scattering (DLS). CPT anticancer drug was effectively loaded into nanocarriers, and loading ratios were found with HPLC technique.

Keywords: camptothecin, drug delivery, PCL-PEG

Camptothecin hydrolysis from active lactone to toxic carboxylate form

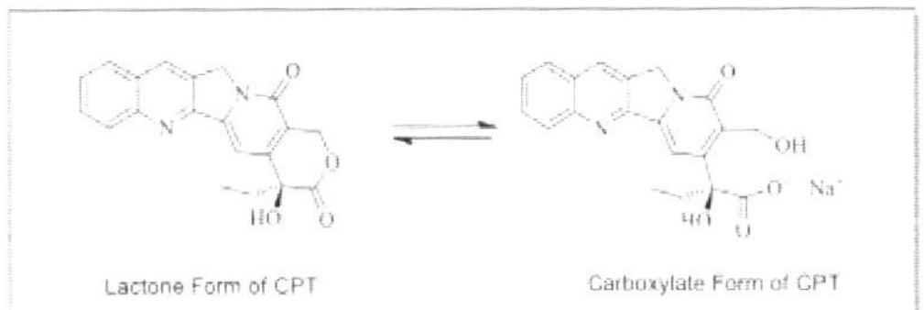

MBM-P-035 [Macromolecules in Biotechnology & Medicine]

Shape-persistent polymersomes possess tunable permeable membrane and functional-antennae as multifunctional devices for biomedical applications

Mohamed A. Yassin[1], Dietmar Appelhans[2], Hanns Achim Temme[3], Brigitte Voit[2]

[1]*Advanced Materials and Nanotechnology Laboratory, Centre of Excellence, National Research Center, Giza, Egypt*
[2]*Leibniz Institute of Polymer Research Dresden e. V., Dresden, Germany*
[3]*Institute of Immunology, Medical Faculty, Carl Gustav Carus Technical University Dresden, Dresden, Germany*

Self-assembly of amphiphilic block copolymers into three dimensional cavities, so called polymersomes has attracted increasing scientific interest in the past decade. This interest is derived from the numerous potential applications such as carriers for drug delivery and as nanoreactor for synthetic biology (1). Here, we present establishment of photo cross-linked polymersomes with pH-tunable membrane permeability. In particular, the polymersomes are based on poly(ethyleneglycol)-block-poly[2-(diethylamino)ethylmethacrylate-stat-2-hydroxy-4-(methacryloyloxy) benzophenone], PEG-b-P(DEA-stat-BMA). After short UV irradiation time, the polymersomes persist their vescular structures over the whole pH range. Furthermore they show impressive reversible swelling-shrinking behavior upon switching between acidic and basic pH. Such on-off switches allowed the polymersomes membrane to act as gate which traffic the release of their cargo such as doxorubicin. To impart smartness on polymersomes, the surface of polymersomes was decorated with functional folate and fluorescein antennae for the purpose of selective cell targeting and cell imaging (2). The polymersomes showed no cytotoxicity against the SBC-2 celles and the HEK 293 cells. Furthermore, the folate-antennae polymersomes were selectively taken up by tumor cells with about 4-fold higher than the healthy cells which in turn enhanced selective anti-tumor activity against the tumor cells.

[1] J. Gaitzsch, D. Appelhans, L. Wang, G. Battaglia, B. Voit, Angew. Chem., Int. Ed. 2012, 51(18), 4448
[2] M. Yassin, D. Appelhans, R. Mendes, M. Rümmeli, B. Voit, Chem. Eur. J. 2012, 18(39), 12227

Keywords: Polymersomes, Drug delivery, nanoreactors pH-tunable, pH-tunable, functionalization

MBM-P-036 [Macromolecules in Biotechnology & Medicine]

Development of Tween 60/Boric Acid-based Antibacterial Polyurethane Biomaterials Including Hydrophobic Surface for Wound Dressing Application

Merve Gökşin Karaaslan[1], Emre Birhanli[2], Selam Gülgen[1], Suleyman Koytepe[1], Burhan Ates[1]

[1]*Inonu University, Science Faculty, Department of Chemistry, Malatya, TURKEY*
[2]*Inonu University, Science Faculty, Department of Biology, Malatya, TURKEY*

Wound dressing biomaterials play an important role in the healing process on the surface of wound. Especially, wound dressing materials keep the moisture in healing environment, have an appropriate oxygen permeability and protect the wound mainly against microorganisms in the healing process [1]. In this context, the aim of study was the development of antimicrobial polyurethane(PU) based wound dressing materials that holds the desired level the fluid passage using tween/polyethlyenglycol (PEG)/boric acid as polyol resources. The following reactants were used in the syntheses of polyurethanes: non-aromatic 4,4'-methylenebis(cyclohexyl isocyanate), tween 60(T60), boric acid(B) and polyethylene glycol 200(PEG) (Figure 1). The ratios of PEG/tween 60/boric acid monomer units in the polyols were: 80/15/5, 80/10/10 and 80/5/15, respectively. The synthesized polyurethane based wound dressing materials were subsequently characterized by fourier transform infrared spectroscopy (FT-IR) (Figure 2) and thermal analysis techniques. Also, antibacterial and contact angle properties of tween/boric acid-based polyurethanes were determined. According to these results, it was determined that the contact angle increased with using the tween which have aliphatic end groups in structure. As a result, prepared biocompatible-antibacterial polyurethanes are good candidates for wound healing applications as potential dressing materials.

Reference:

[1] N. Liao, A.R. Unnithan et al.,Colloids Surf. Physicochem. Eng. Asp., 2015, 469:194–201.

Keywords: Polyurethane,wound dressing, antibacterial

Figure 1. Synthesis of PU-T60-Bs.

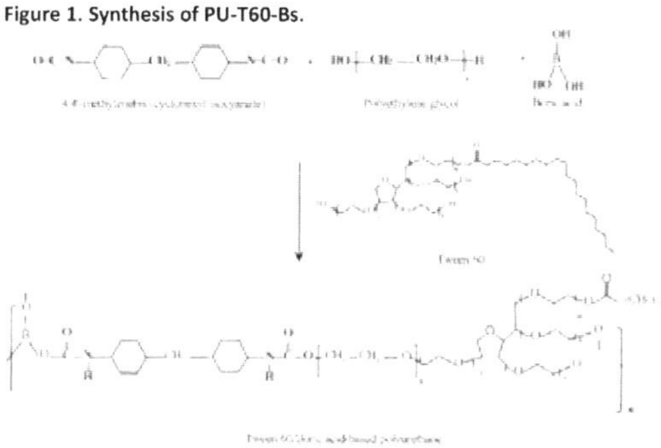

Figure 2. FT-IR spectra of PU-T60-Bs.

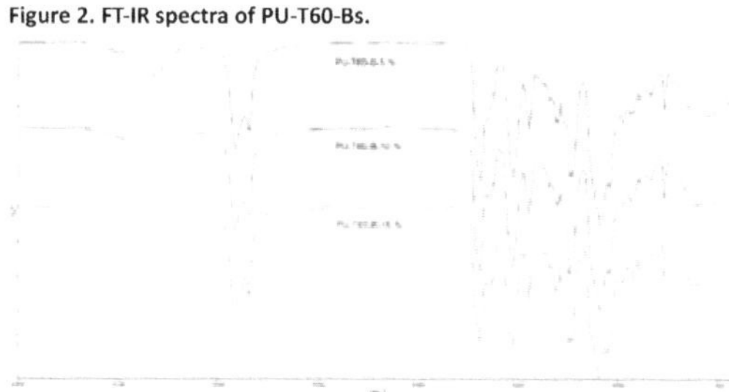

MBM-P-038 [Macromolecules in Biotechnology & Medicine]

Purification of a phosphated biopolymer by selective ethanol precipitation in presence of SDS and sodium acetate

Rodrigo Gabriel Simas, Raphael De Marco, Joaquin Cabrera Crespo, Mickie Takagi

Instituto Butantan

Polyribosyl-ribitol phosphate (PRP) is an exopolysaccharide produced by *Haemophilus influenzae* type b, used as antigen in a conjugated vaccine. Although its purification is well-established industrially, alternative methods are necessary in order to reduce the vaccine cost. The aim of this work was to improve a PRP purification process, focusing on ethanol precipitation and to evaluate the effects of SDS and sodium acetate (NaAc) on recovery and purity of the product. Concentrated cell-free broth containing PRP was obtained from *Haemophilus influenzae* culture and used to evaluate (a) the effect of SDS on product recovery and purity; (b) the combined effect of ethanol and NaAc addition in each precipitation step. The presence of SDS 0.5% resulted in protein and nucleic acids relative purity of 19 and 56 respectively, against 8 and 39 without SDS, both with similar PRP yield (72%). We also identified two important conditions in precipitation steps, in which, either no PRP or all PRP was precipitated. If no NaAc is added to the broth, PRP is completely soluble in media containing up to 40% ethanol. On the other hand, PRP is virtually insoluble when ethanol concentration is higher than 60%. In the second step, NaAc significantly reduced protein and nucleic acid precipitation due to salting in. Optimized condition resulted in protein and nucleic acid relative purity of 93 and 102 respectively, compared to 6 and 9 in the control. It implies that the use of SDS and NaAc combined was essential to improve PRP purity. In conclusion, we demonstrated that the new method promotes comparable polymer recovery and is more efficient in removing impurities, based mainly on differential ethanol precipitation.

Keywords: Biopolymer; purification; SDS, Sodium Acetate; Ethanol selective, Recovery

MBM-P-039 [Macromolecules in Biotechnology & Medicine]

Preparation of Theophylline Loaded Pectin-Zeolite Based Wound Dressing Materials

Banu Arabacıoglu Kocaaga, Ozge Kurkcuoglu, Melkon Tatlıer, F. Seniha Güner

Department of Chemical Engineering, Istanbul Technical University, Istanbul, Turkey

Hydrogels of polysaccharides promise as effective drug delivery matrices because of their unique advantages such as cell-cell recognition, transparency and permitting oxygen penetration. However, their disadvantages such as low mechanic strength and poor chemical stability inhibit their actual applications [1]. Pectin, an inexpensive, biocompatible polysaccharide, provides an acidic environment by acting as barrier against bacteria and has ability for binding to bioactive molecules [2]. Alumino-silicate based zeolites can be added to the pectin hydrogels to increase the mechanical properties. This study aims to develop pectin-zeolite based antimicrobial composite drug delivery matrix, focusing on two different drug loading systems; diffusion into membrane and diffusion into matrix. Different concentrations of cross-linker CaCl2 (5,7,9 and 10% w/w), Zn-zeolite (1,3 and 5% w/w) and drug theophylline (30 mg, 60 mg) were studied to obtain an effective drug release performance for the composite, where glycerol was used as plasticizer. The release studies in tris buffer pH 6.4 mimicking the biological environment, were performed by UV spectrophotometer at 271 nm. The release profile of theophylline was strongly affected by the loading method and calcium concentration. Formulation of variables is important to create optimum pore size during ionic gelation for the drug entrapment and release efficiency. Optimum conditions were achieved at 7% (w/w) CaCl2 and 3% (w/w) zeolite. Our results suggested that excess amount of zeolite particles may act as an obstacle on the diffusion path of drugs.

[1] Munarin, F., Tanzi, M.C., Petrini, P., 2012. International Journal of Biological Macromolecules, 51:681-689.
[2] Aguilar K.C., Tello F., Bierhalz A.C.K., Romo G., Florez H.E.M., Grosso R.F.C., 2015, Journal of Food Engineering, 154:17-24.

Keywords: pectin, wound dressing, zeolite, controlled drug delivery

Drug release profile of pectin-zeolite matrix

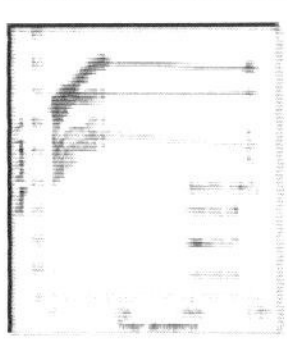

MBM-P-041 [Macromolecules in Biotechnology & Medicine]

Biological properties evaluations for some partially alicyclic copolyimides with pendant carboxyl groups

Dumitru Popovici

"Petru Poni" Institute of Macromolecular Chemistry, Iasi, Romania

The higher solubility and implicit the processability of partially alicyclic polyimides are dictated primarily by the interruption of linearity and conjugation along the macromolecular chain which lead to preventing the formation of the charge transfer complex with is responsible for the low solubility of aromatic polyimides - PI. Besides the well knows engineering uses of PI's, in last years it have begun to be successfully used in the biomedical applications.

Biological properties of some alicyclic copolyimides derived from 3,5-diaminobenzoic acid was investigated. The influence of carboxyl pendant group was investigated compared with polyimides based on 1,3 phenylenediamine, using the same echimolar mixtures of dianhydrides bicyclo[2.2.2]oct-7-ene-2,3,5,6-tetracarboxylic dianhydride and 4,4'-(hexafluoroisopropylidene)-diphthalic anhydride.

Bio- and hemo-compatibility has been evaluated base on contact angle of copolyimides films with three testing liquids, using van Oss, Good and Chaudhury theory. The presence of the pendant carboxyl groups increase the polarity of the films surface, that lead to an increasing of their hydrophilicity, lowering the free surface energy and preventing the fixation of various components of blood at the surface that can form clots. Antibacterial properties have been investigated using two standard bacteria: ATCC 25922 – *Escherichia Coli* and ATCC 25923 – *Staphylococcus Aureus*. After 24 hours of incubation the proliferation of the bacteria at the films surface were not observed, proving that these have a negative antibacterial response.

Controlled growth of osteoblasts and fibroblast cells was tested for both pristine films and for the reorganized films surface by rubbing in one direction with cotton velvet. Surface morphology for unmodified and modified films was studied using atomic force microscopy – AFM. All this results suggest that partially alicyclic copolyimides derived from 3,5-diaminobenzic acid are valuable candidates for biomedical applications such as implants, prostheses, catheters or septic environments.

Keywords: hemocompatibility, alicyclic copolyimides, contact angle, antibacterial, osteoblast, fibroblast

MBM-P-042 [Macromolecules in Biotechnology & Medicine]

Polycationic Polymeric Brush Decoration on Polyurethane for Dual Effective Antibacterial Activity

Merve Gultekinoglu[1], Yeliz Tunc Sarisozen[1], Ceren Erdogdu[2], Meral Sagiroglu[2], Yoo Jin Oh[3], Peter Hinterdorfer[3], Eda Ayse Aksoy[1], Kezban Ulubayram[1]

[1]*Department of Basic Pharmaceutical Sciences, Faculty of Pharmacy, Hacettepe University, Ankara 06100, Turkey*
[2]*Department of Pharmaceutical Microbiology, Faculty of Pharmacy, Hacettepe University, Ankara 06100, Turkey*
[3]*Institute for Biophysics, Johannes Kepler University, Linz A-4020, Austria*

Polyurethane (PU) is one of the most frequently used polymers for biomedical materials thanks to its excellent mechanical and biocompatible properties. However, antibacterial complications are still unresolved problem for PU. For this purpose, it was aimed to modify PU surfaces with antibacterial polycationic polyethyleneimine (PEI) molecules. PU consists of short chain (chain extender) and long chain diols with isocyanate domains in its chemical structure. In this study, polycationic PEI chains were grafted on to PU surface through isocyanate anchor sites. PEI chains that grafted on to PU surface had the brush-like conformation. Brush like character was varied with two different molecular weights PEI. In the further step, alkylation was performed to increase the surface charge and the antibacterial activity. PEI brushes have dual effective antibacterial activity due to dynamic bacteria repellent and contact killing ability of polycationic brushes. All the chemical reaction steps were examined with both spectroscopic and microscopic techniques. The sample groups exhibited at least 90% cell viability for L929 fibroblast and G/G uroepithelial cell lines. Also antibacterial test results revealed that increasing brush length and charge inhibited the bacterial attachment up to two orders of magnitude for *E. coli*, *K. pneumonia* and *P. mirabilis* bacteria species(Gultekinoglu, 2015). In addition force spectroscopy study results revealed that the bacteria-surface interactions are affected by the surface modifications in terms of PEI chain length, surface charge and topography at the single bacterium level(Gultekinoglu, 2016).

Acknowledgments: This study is financially supported by The Scientific and Technological Research Council of Turkey (TUBITAK)(Grant No.112M293).

References:

1. Gultekinoglu M., et al. "Designing of dynamic polyethyleneimine (PEI) brushes on polyurethane (PU) ureteral stents to prevent infections." Acta biomaterialia 21 (2015): 44-54.
2. Gultekinoglu M., et al. "Nanoscale characteristics of antibacterial cationic polymeric brushes and single bacterium interactions probed by force microscopy." RSC Advances 6.21 (2016): 17092-17099.

Keywords: polymer brush, polyurethane, polyethyleneimine, antibacterial

MBM-P-043 [Macromolecules in Biotechnology & Medicine]

The Comparison of Isothermal and Hydrolytic Degradability Properties of Non-Aromatic Polyurethanes Prepared for Medical Applications

Canbolat Gurses[1], Hakan Dastan[2], Merve Goksin Karaaslan[2], Selam Gulgen[2], Suleyman Koytepe[2], Burhan Ates[2]

[1]*Departmant of Molecular Biology and Genetics, İnönü University, Malatya, Turkey*
[2]*Departmant of Chemistry, İnönü University, Malatya, Turkey*

Today, polyurethanes have been used in many areas from industrial to biomaterial applications [1]. Since biodegradability is an essential parameter for biocompatibility, non-aromatic polyurethanes have been chosen in general. This work addresses the *in vitro* isothermal and hydrolytic degradation comparison of linear polyurethanes(LPU). The following reactants were used in the syntheses of polyurethanes: non-aromatic hexamethylene diisocyanide, quercetin and polyethylene glycol 200(PEG). The ratios of PEG–Quercetin(Q) monomer units in the polyols were 95:5, 90:10 and 85:15, respectively. The structures of polyurethanes were illuminated with FT-IR spectroscopy. Thermal properties of prepared polyurethanes were determined with DTA, TGA and DSC thermal analysis systems. The three essential mass losses of LPU-PEG-Qs have been seen in Figure 1A. The first mass loss occurs between 150-200°C and indicates the breakage of the cross-linking points in the polymeric structure. The second mass loss continues starting from 250°C to 305°C. In general, it is caused by the rupture of a polyphenol structure. The third mass loss occurs between 375°C to 435°C due to the heat-distortion of diisocyanate structure. A similar degradation curve is observed in all three polymers. After 445°C, the polymer is burned oxidatively. In addition, the comparison of isothermal and hydrolitic degration properties was carried out with FT-IR and weight loss, respectively. The hydrolytic degradation of polyurethane samples was carried out at 37±0.1 ºC for 4 weeks using phosphate buffer solution (pH=7.4±0.2). The mass losses on the fourth week for the materials were in the range of 15–19% depending on polyurethanes containing different ratios of quercetin (Figure 1B).

[1] McBane JE, Sharifpoor S, Cai K, Labow RS, Santerre JP. Biomaterials. 2011, 32(26):6034-44.

-This study was supported by grants from TUBITAK (2209-A).

Keywords: Linear Polyurethanes, Quercetin, Isothermal and Hydrolytic Degradability

Figure 1. A) Thermograms of LPU-PEG-Qs (5-10-15%) B) Hydrolitic degradations of LPU-PEG-Qs (5-10-15%) during 4 weeks

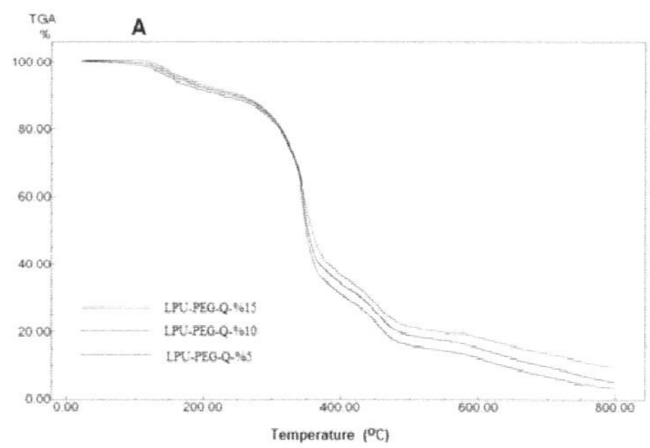

B

	Weight Loss (%)			
	1st Week	2nd Week	3rd Week	4th Week
LPU-PEG-Q 5%	16.38±1.77	14.99±0.54	15.85±0.13	16.31±0.54
LPU-PEG-Q 10%	16.29±0.44	17.44±0.06	17.49±0.26	18.18±0.24
LPU-PEG-Q 15%	13.85±0.48	14.65±0.46	14.34±0.14	16.29±0.29

MBM-P-044 [Macromolecules in Biotechnology & Medicine]

Computational and Experimental Approaches to Improve Minocycline Imprinted Hydrogels for Controlled Drug Release

Büşra Eroğlu[1], Dilek Dalgakiran[1], Ozge Kurkcuoglu[2], F. Seniha Güner[2]

[1]*Department of Polymer Science and Technology, Istanbul Technical University*
[2]*Department of Chemical Engineering, Istanbul Technical University, Istanbul, 34469, Turkey*

Molecular imprinting is a technique to create template-shaped cavities in polymer matrices, which gains memory for template molecule. This technique has been also suggested to promote controlled drug release [1]. The aim of this study was synthesizing antibiotic minocycline imprinted hydrogels and determining their drug delivery performances. Hydrogels were prepared with 2-hydroxyethylmethacrylate (HEMA) as backbone monomer, ethylene glycol dimethacrylate (EGDMA) as cross-linker, and acrylic acid (1.25%) as functional monomer, which is expected to play an essential role in effectively imprinting the minocycline. Different amounts of EGDMA (1-3%) were studied. The hydrogels were synthesized by thermal polymerization at 45°C for 24 h in the plate in the presence of the minocycline. Minocycline release performances were analyzed in 0.9% NaCl. The release profiles were evaluated using different kinetic models: zero-order, first-order, Higuchi and Korsmeyer-Peppas. Imprinted factors are also calculated. The imprinted factor was found between 0.44-0.735 for 1% EGDMA 1.188-3.054 for 2% EGDMA and 2.242-3.671 for 3% EGDMA. The results show that the best imprinting performance was obtained at 3% EGDMA concentration.

Full-atom molecular dynamics (MD) simulations were employed to study molecular interactions of the antibiotic and acrylic acid molecules in the pre-polymerization step; these interactions determine to a large extent the memory and controlled drug release performance of the hydrogel. 5 ns long MD simulations were performed using Materials Studio with COMPASS force field. MD simulation results suggested that acrylic acid molecules made stable H-bonding from several functional groups of the template, which suggested that acrylic acid was a suitable functional monomer in obtaining good cavities for hydrogel to recognize the drug in later usages and also in controlling drug release.

Reference:

1. Carmen Alvarez-Lorenzo, Fernando Yañez, Rafael Barreiro-Iglesias, Angel Concheiro. 2006. Imprinted soft contact lenses as norfloxacin delivery systems, Journal of Controlled Release, 113, 236–244.

Keywords: Molecular Imprinting, drug release, molecular dynamics

MBM-P-045 [Macromolecules in Biotechnology & Medicine]

Biochemical Characterization of Recombinant L-Arabinofuranosidase

Emine Erdoğan Özşeker, Alper Akkaya

Ege University, Faculty of Science, Biochemistry Department, Bornova, Izmir, Turkey

Enzymes are successfully used in many of the industrial process, such as food and paper industries. Enzymes, which are derived from microorganisms living under mild conditions, are not suited for industrial processes that require high temperature, acidic or basic reaction conditions. Enzymes, which are derived from microorganisms living under extreme conditions are needed for these industrial hard conditions1,2. α-L-arabinofuranosidases (E.C 3.2.1.55) belong to group of glycoside hydrolysis. This enzyme group hydrolyses 1,3 and 1,5 α-arabinosyl bounds of N-arabinose containing molecules. It is also an important part of a system which is responsible for complete hydrolysis of arabinoxylans. In this study, biochemical characterization of recombinant L-arabinofuranosidase has been purposed. For this purpose, effect of pH, temperature and mediator on enzyme activity were determined. Also the substrate specificity and the thermal stability of the enzyme were examined. Referring to the literature, it is seen that a small number of studies conducted with the enzyme. Therefore, this study will contribute to the literature. Alicyclobacillus acidocaldarius (DSM 446), which is a source of recombinant L-arabinofuranosidase, is acidothermophilic bacteria. As a result of characterization studies, it was determined that the enzyme kept its activity at high temperatures. The results are promising in terms of industrial applicability of this enzyme.

This study was supported by TÜBİTAK 1001 (Project Number: 214Z086) Scientific and Technological Research Projects Supporting Program.

References:

[1] D. C. Demirjian, F. M. Varas, C. S. Cassidy, Enzymes from extremophiles, Current Opinion in Chemical Biology, 5(2001), 144–151.
[2] B. Van den Burg, Extremophiles as a source for novel enzymes, Current Opinion in Microbiology, 6(2003), 213–218.

Keywords: L-arabinofuranosidase, recombinant protein, Alicyclobacillus acidocaldarius (DSM 446), enzyme characterization.

MBM-P-046 [Macromolecules in Biotechnology & Medicine]

Bioconjugation of Microbial Rennet with Dextran Sulfate

Mithat Celebi

Department of Polymer Engineering, Yalova University, Yalova, Turkey

Enzymes can be used wide range of applications industrial processes and products, such as biosensors, detergents, textiles, food industry, wastewater treatment and polymer synthesis. Microbial Rennet is used to cheese manufacture in food industry. Microbial Rennets from Mucor miehei, Endothia parasitica, Mucor pusilllus have been proved to be suitable for large-scale production. Microbial Rennet is an aspartic protease which cleaves 105 (phenyl alanine)-106 (methionine) peptide bond in kappa-casein (milk protein). Bioconjugation reactions have been used to provide new features to enzymes with biopolymers such as dextran, dextran sulfate, chitosan and cellulose derivatives etc. Dextrans are biodegradable and biocompatible polysaccharides containing repeating glucose units linked with each other from α-1,6 and variable positions of α-1,2, α-1,3, and α-1,4. Dextran sulfate is the polyanionic derivative of dextran. Dextran sulfate alters the activity of enzymes.

In this study, aldehyde dextran sulfate was synthesized by oxidation of dextran sulfate with sodium periodate. Dextran sulfate was conjugated to microbial Rennet via Schiff base formation with aldehyde groups in dextran sulfate chains and amine groups of the enzyme. Activities of the microbial Rennet and the synthesized bioconjugate were determined and compared at different pHs and temperatures. Activity of the conjugate was decreased according to free enzyme.

Keywords: bioconjugation reaction, microbial Rennet, Dextran Sulfate, milk protein

MBM-P-047 [Macromolecules in Biotechnology & Medicine]

Surface Modifiable Dendrimers as Efficient Drug Carriers of Sulfamethoxazole (SMZ) as Model Drug

Mustafa Ulvi Gürbüz[1], Ali Serol Ertürk[2], Metin Tülü[1]

[1]Dept.of Chem. Fac. of Arts and Sciences, Yıldız Technical University, 34210 Istanbul,TR.
[2]Dept. of Basic Pharm.Sciences, Faculty of Pharmacy, Adıyaman University, 02040 Adıyaman, TR.

Sulfamethoxazole (SMZ) is a sulfonamide and used widely in the treatment of bacteriostatic and urinary tract infections with trimethoprim as an antibiotic. The problem with SMZ is its poor water solubility, therefore, low bioavailability in clinical applications [1]. Current study shows the synthesis of TREN cored Amine, Tris, and Carboxyl terminated dendrimers with different generations (T2-T4). Synthesized dendrimers were characterized by 1H NMR, 13C NMR, ATR-FTIR, spectroscopic titrations, and evaluated as potential solubility enhancers and drug carriers of sulfonamides by taking SMZ as a model drug. The effects of dendrimer concentration, generation, and surface groups on the solubility of SMZ were also investigated. Results showed that the solubility of SMZ in the presence of PAMAM dendrimers improved significantly with an increasing generation size (T2-T4) and dendrimer concentration (0-2 mM). The role of PAMAM dendrimer in the solubility enhancement of SMZ was in the order of T4.NH2 > T4.COOH> T3.NH2 > T4.TRIS> T2.NH2 > T3.COOH> T3.TRIS > T2.COOH > T2.TRIS, and in the ranges of 5 to 45-fold with maximum SMZ loading 7 to 61 mole/mole per dendrimer molecule. In vitro release studies demonstrated that drug-dendrimer complexes at the end of 2 h. drug release (16-26%) was considerable slower than pure SMZ (38.8%).

Keywords: Dendrimers, Poly(amidoamine) PAMAM, Drug-delivery system, SMZ

Figure 1 SMZ Pamam

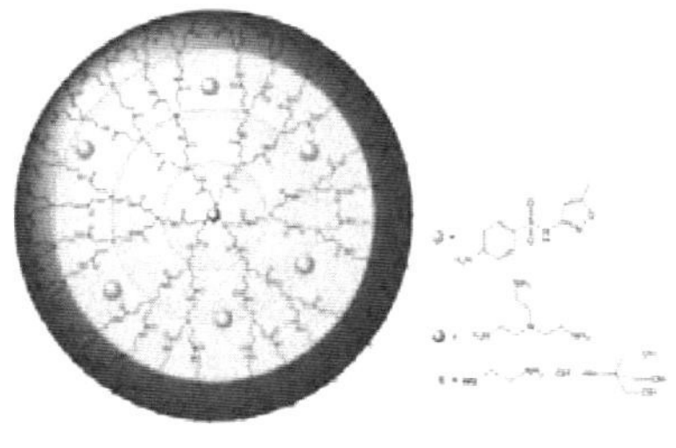

Figure 1. Representative illustration of the SMZ-PAMAM inclusion complex

MBM-P-048 [Macromolecules in Biotechnology & Medicine]

Synthesis and anticancer effect of different monosaccharides-DOX conjugates on MCF-7 breast cancer cell

Idil Karaca Açarı, Süleyman Köytepe, Burhan Ateş, Sevgi Balcıoğlu, Turgay Seçkin, Ismet Yılmaz

Faculty of Arts and Science, Chemistry Department, Inonu University, Malatya, Turkey

Cancer is a major public health problem despite the efforts that have been made in the search for novel drugs and treatments. Monosaccharides use for targeting and imaging of cancer cells [1]. The first anthracyclines were isolated from the pigment-producing Streptomyces peucetius early in the 1960s and were named doxorubicin. Nowadays, it is commonly used in the treatment of a wide range of cancers. The exact mechanism of action of doxorubicin is complex and still somewhat unclear, though it is thought to interact with DNA by intercalation and topoisomerase II inhibition [2, 3]. The aim of this study was to investigate effectively anticancer drug of different monosaccharites-DOX conjugate. Firstly different monosaccharites-DOX conjugates had been synthesized using the click chemistry Figure 1). Click chemistry refers to a group of reactions that are fast, simple to use, easy to purify, versatile, regiospecific, and give high product yields [4,5].Different monosaccharites-DOX conjugates were investigated by X-ray diffraction analysis (XRD), scanning electron microscope (SEM) and elemental analysis. In vitro anticancer activity on MCF-7 cells by the MTT (3-[4,5-dimethylthiazol-2-yl]-2,5-diphenyltetrazolium bromide) method.

References:

[1] Wang S, Yin D, Wang W, Shen X, Zhu J. J, Chen H.Y, Liu Z. Nature 2016; 6: 22727.
[2] Hynek D, Kreşcova L, Zitka O, Adam V, Trnkova L, Sochor J, Stiborova M, Eckschlager T, Hubalek J, Kizek R. Int. J. Electrochem. Sci. 2012; 7: 13-33.
[3] Minotti G, Menna P, Salvatorelli E, Cairo G, Gianni L. Pharmacol. Rev. 2004; 56: 185-229.
[4] Hein C. D, Liu X-M, Wang D. Pharm. Res. 2008; 25: 2216-2230.
[5] Zi C-T, Xu F-Q, Li G-T, Li Y, Ding Z-T, Zhou J, Jiang Z-H, Hu J-M. Molecules. 2013; 18: 13992-14012.

Keywords: Monosaccharides, cancer, drug design, doxorubicin

Figure 1.

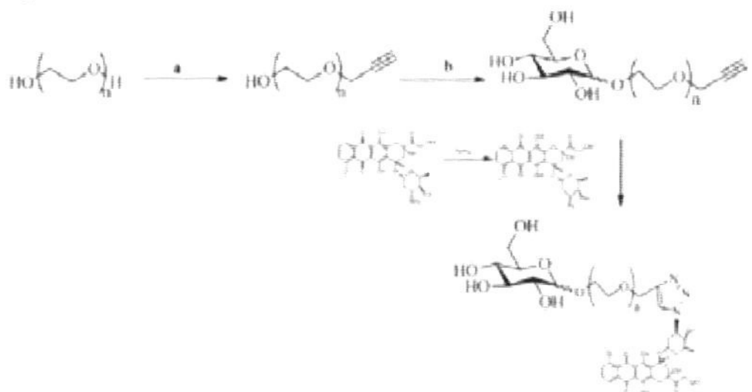

MBM-P-049 [Macromolecules in Biotechnology & Medicine]

In vitro cytotoxicity of PEG modified single-walled carbon nanotubes

Mehdi Partovi Meran[1], Ayhan Ünlü[2], Ebru Hacıosmanoğlu[3], Gürkan Hızal[4], Nilgün Yavuz[5], Pelin Deniz Akkuş[1], Özge Kürkçüoğlu[1], F. Seniha Güner[1]

[1]Department of Chemical Engineering, Istanbul Technical University, Istanbul, Turkey
[2]Department of Biophysics, Trakya University, Edirne, Turkey
[3]Department of Biophysics, Istanbul Faculty of Medicine, Istanbul University, Istanbul, Turkey
[4]Department of Chemistry, Istanbul Technical University, Istanbul, Turkey
[5]Department of Energy Science and Technology, Istanbul Technical University, Istanbul, Turkey

Carbon nanotubes (CNTs) due to their unique chemical and mechanical properties have been one of the most promising nanomaterials for a variety of biomedical applications. CNTs have the potential to be used both in in vivo imaging and drug delivery. The main obstacle in utilization of pristine CNTs is their scarce solubility in aqueous environment due to their graphitic. To overcome this problem, CNT surfaces should be modified by non-covalent coating or by covalent functionalization approaches. On the other hand, CNTs spontaneously internalized by a wide variety of cell types. Therefore, it is imperative to examine the toxicity of these carbon-based nanostructures.

In this study, single wall CNTs (SWCNTs) were non-covalently coated by various hydrophilic poly (ethylene glycol) (PEG) polymers for biocompatibility. We chemically linked PEG chains and pyrene molecules, which can strongly attach to SWCNT surface via π-π stacking. First, monocarboxylic acid-terminated PEG (PEG-COOH) was synthesized according to a procedure via reaction of Me-PEG with succinic anhydride. Subsequently, PEG-Pyrene was prepared via an esterification reaction of PEG-COOH with Pyrene Methanol. Finally, SWCNT and PEG-Pyrene heterogeneous mixture was sonicated in a low-power sonic bath and stirred vigorously in THF to gain PEG modified SWCNT. Various techniques including TGA, NMR and TEM, were employed to characterize each sample synthesized in the study. Interestingly, TGA analysis indicated that shorter SWCNTs have higher PEG chain density. In order to explain this observation, full-atom molecular dynamics (MD) simulations were performed, and interactions of SWCNT and different length pyrene-PEG chains were analyzed. Finally, we examined the cytotoxicity of different PEG-modified SWCNT samples in Human hepatoma cell (Hep3B).

Keywords: Carbon nanotubes, PEGylation, cytotoxicity, SWCNT

MBM-P-050 [Macromolecules in Biotechnology & Medicine]

Sodium Alginate Films for Future Wound Dressing Applications, Effect of Crosslinking on Swelling, Mechanical Properties and Release of Curcumin

Candan Kılınçkale, Funda Kaşıkçı, Zeynep Kurt, Hüceste Çatalgil Giz

Department of Chemistry, Istanbul Technical University, Istanbul, Turkey

Curcumin is an Indian spice, which is used as herbal medicine for treatment of several complaints such as rheumatism, diabetic ulcers, anorexia, cough and sinusitis. Recent researches show that curcumin has anti-inflammatory, anti-oxidant and anti-carcinogenic effects on human's health. It also accelerates the wound healing process in different stages. Plasticized alginate films are prepared and crosslinked by two-step contact with calcium ions. Firstly, films with low crosslinked degree (pre-crosslinked film) are prepared and on the second step additional crosslinking is performed with more concentrated CaCl2 solution. Swelling data is monitored as a function of time and the effect of two step crosslinking is discussed. Mechanical properties of films are determined and stress-strain curves are obtained. At the second step same alginate films are prepared by curcumin addition and films are pre-crosslinked. Swelling properties and mechanical properties of curcumin containing films and and curcumin release from the films are monitored with time.

Keywords: Alginate fims, Crosslinking, Swelling, Mechanical Properties, Release of Curcumin

MBM-P-051 [Macromolecules in Biotechnology & Medicine]

The Release Of Curcumin From Alginate Beads In Different Media

Esra Aksoy[1], Melisa Berberoglu[1], Cagla Ergun[2], Huceste Catalgil Giz[1]

[1]*Istanbul Technical University, Department of Chemistry, Istanbul, Turkey*
[2]*Bogazici University, Department Of Chemical Eng, İstanbul, Turkey*

Curcumin has been used for food coloring for a very long time and more recently a lot of research has focused on its antioxidant, anti-inflammatory and antitumor properties, however its insolubility in water limits its applications. According to the previously proposed recipe (Akkaya. NE) water soluble, Alginate –Curcumin complexes are prepared in this work. Complexes are encapsulated in alginate beads and release rates are monitored by online monitoring of polymerization (ACOMP) system by UV, viscosity and light scattering detectors. As a release medium, water, Tween80, gastric fluid and intestinal fluid are chosen. Release rates are compared according to the release medium. Pictures of beads in different media are shown below.

-Nil Erge Akkaya, Master Thesis, "Increasing the water solubility of curcumin, Preparation of Alginate- Curcumin complexes in US medium and bead production" ITU. 2014.

Keywords: Alginate, Curcumin, Release

Beads in gastric fluid

Beads in intestinal fluid

Beads in water

MBM-P-052 [Macromolecules in Biotechnology & Medicine]

Delivery Kinetics and Preparation of Rosemary Essential Oil Formulation Microencapsulated with Alginate/Starch

Incilay Gökbulut[1], Fatma Sezer Öztürk[1], Burhan Ates[2], Hakan Parlakpınar[3]
[1]Food Engineering, İnönü University, Malatya, Turkey.

[2]Department of Chemistry, İnönü University, Malatya, Turkey.
[3]Department of Pharmacology, Faculty of Medicine, Inonu University, Malatya, Turkey

Microencapsulation is a technique used for the entrapment of a thin film with coating materials in the food-purity of the liquid droplets, the solid particles or gas components (1, 2). In the food industry, this technique is used for various purposes such as protection against environmental factors of foodstuffs (humidity, temperature, air and light), prevention of evaporation losses, better preservation of the physical properties, the provision of targeted place and time release, masking undesirable flavors /odor and prevention of react with other components (3). Alginate is preferred due to which has neutral taste and smell, natural, renewable, inexpensive, highly hydrophilic nature, biocompatible, biodegradability, non-toxic and non- immunogenic properties as coating material (4). Rosemary (Rosmarinus officinalis L.) is extensively used for medicinal due to phenols located in structure and its essential oils. Rosemary essential oils have shown that antimicrobial, antioxidant, radical scavenger activities (5). Despite these positive features of rosemary essential oils, it is very difficult to storage at room temperature. Therefore, in this study, we prepared the microcapsules of rosemary essential oil using alginate/starch as coating agents by oil/water emulsion method (Figure 1A). Capsulation efficiency and delivery of essential oil from this microcapsule were investigated. Optimum microcapsules were prepared from 1% (w/v) sodium alginate, 7% (w/v) rosemary essential oil and cross-linked with 1.0% (w/v) $CaCl_2$. Capsulation efficiency was determined as 87.16% (Figure 1B). The delivery kinetic of rosemary essential oil from the microcapsules was followed for 240 minute and released essential oil ratio changed with time as linear, approximately 55%. As a result of, microencapsulation of rosemary essential may provide for elongated storage time.

-This study was supported by grants from TUBITAK (214O195).

Keywords: Encapsulation, alginate, starch

Figure 1. (A) Image of Alginate/Starch microencapsulate, (B) Encapsulation efficiency depend on concentration of rosemary essential oil

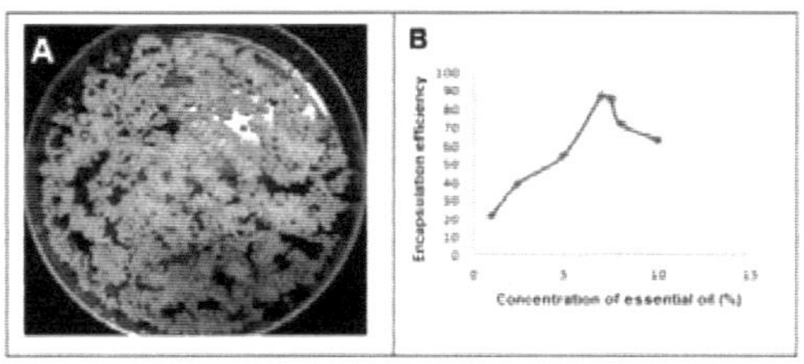

MBM-P-053 [Macromolecules in Biotechnology & Medicine]

Synthesis of Tetrazine Functional Hyperbranched Polymers

Mustafa Yasin Ateş, Ahmet Çetinkaya, Muhammet Ubeydullah Kahveci

Departmant of Bioengineering,Yildiz Technical University,Istanbul,Turkey

New ligation techniques, known as bioorthogonal reactions, that do not interfere with biological systems and allow selective modification in biological systems have been developed in recent years. They are not affected by complex biochemical reactions in biological environment, which consist of hundreds of functional groups. Furthermore, the reactions have not any adverse impact on biochemical reactions. (Sletten ve Bertozzi, 2009). Bioorhtogonal conjugation approaches have been getting increasing attention in recent years. Among the bioorthogonal ligations, tetrazine reactions are known as the fastest bioconjugation method (Knall ve Slugovc, 2013). Therefore, these reactions are widely employed in live cell and tissue imaging, PET techniques, proteomics and similar applications which requires efficient and fast binding. (Sletten ve Bertozzi, 2009). In this study, hyperbranched polymers with tetrazine functionalities have been prepared by combination of ATRP and self-condensing vinyl polymerization, and subsequent post-polymerization modification. Hyperbranched copolymers of poly(ethylene glycol)methacrylate (PEGMA) and N-methacrylamidosuccinimide (NMAS) were prepared and amino-functional tetrazine molecule (Tz-NH2) synthesized by well-established Lewis acid catalyzed reaction at high yields was incorporated onto the polymer via the succinimidyl ester moieties of the polymer leading to multifunctional hyperbranched polymers. The future studies involve use of the tetrazine functional polymers in targeting cancer and bacterial cells for the aim of diagnosis and treatment. In these studies, Inverse Electron Demand Diels-Alder (IEDDA) reaction between tetrazine-trans-cyclooctene will be employed as a fast and efficient conjugation.

This work was financially supported by The Scientific and Technological Research Council of Turkey (TUBITAK, Project # 214M175)

Sletten, E.M., Bertozzi, C.R. 2009. "Bioorthogonal Chemistry: Fishing for Selectivity in a Sea of Functionality". Angewandte Chemie-International Edition, 48 (38), 6974-6998.
Knall, A.C., Slugovc, C. 2013. "Inverse electron demand Diels-Alder (iEDDA)-initiated conjugation: a (high) potential click chemistry scheme". Chemical Society Reviews, 42 (12), 5131-5142.

Keywords: Tetrazine, trans-cycloctene, functional polymers, Inverse electron demand Diels-Alder

MBM-P-054 [Macromolecules in Biotechnology & Medicine]

SURFACE MODIFICATION AND CHARACTERIZATION OF NANO SiO2 NANOPARTICLES FOR BIOLOGICAL APPLICATIONS

Elif Ceyda Kanpara[1], Fatma Özge Gökmen[2], Nursel Pekel Bayramgil[1]

[1]Hacettepe University, Faculty of Science, Chemistry Department, 06800, Beytepe-Ankara
[2]Bilecik Şeyh Edebali University, Central Research Laboratory, 11210, Bilecik

Silica-polymer hybrid materials are the most widely studied inorganic or organic hybrid materials. The primary reason to choose these particles comes to have a lot of different application areas and easy particle synthesis. Chemical properties of silica surface are determined by both silanol and siloxane groups on the surface and inside. Silica particles, by the surface hydroxyl groups, are modified with different chemical (organic or inorganic) agents; thus provided opportunities for the expansion of applications of hybrid silica material. Surface modified silica particles are used successfully in many biomedical fields such as controlled drug release, tissue engineering, biosensors, and active body carrier for antimicrobial applications as well as cancer diagnosis. This work was planned to improve the usage success of modified silica nanoparticles in biomedical field. Firstly the activity of hydroxyl groups on the surface of nano SiO_2 were increased by acidic or basic reactions. Thereafter, the modification of hydrophilic polymers were performed covalently by using "grafting from" method. Modified nano SiO_2 particles were characterized by FT-IR, XRD, elemental analysis, TGA-DSC and SEM methods. The resulting final products were incorporated into the biodegradable films, and usage performance of them has been investigated as controlled drug delivery systems. We observed that the modified silica particles were dispersed homogeneously in polymeric films and bound strongly by polymer, hence to be improved the performance of drug release.

References:

1. Lee, S., L., Nur, H., And Wei, C., S. (July 2011). Synthesis and Characteriza-

tion of Acid Nodified Silica-Titania Aerogel as Oxidative-Acidic Bifunctional Catalyst. International Journal of Applied Physics and Mathematics, 43-47.
2. Ling, S., L., (2008). Sulfated Silica-Titania Aerogel as a Bifunctional Oxidative and Acidic Catalyst in the Synthesis of Diols. Journal of Non-Crystalline Solids,354, 3939-3943.
3. Sun, Xiaoxing. (2012). Mesoporous Silica Nanoparticles for Applications in Drug Delivery and Catalysis. Graduate Theses and Dissertations. Paper 12812.

Keywords: polymer, nanoparticles, drug delivery, biological applications, silica.

MBM-P-055 [Macromolecules in Biotechnology & Medicine]

Labeling of Proteins with Trans-cyclooctene for Tetrazine Ligation

Ahmet Çetinkaya, Mustafa Yasin Ateş, Muhammet Ubeydullah Kahveci

Yildiz Technical University, Davutpasa Campus, Faculty of Chemical and Metallurgical Engineering, Department of Bioengineering, Esenler, 34220 Istanbul, Turkey

Fast, accurate and reliable diagnosis and the development of diagnostic systems is one of the most important areas of the market volume about 19 billion dollars in health sector (Lesney,2001). In recent years, bioorhtogonal conjugation approaches, which do not interfere with biological systems, get increasing attention, and the number of studies for the establishment of a temporary modification made on technologies that allow targeting and drug delivery systems has increased considerably (Lee,2005; Duncan,2006). They are not affected by complex biochemical reactions in biological environment, which consist of hundreds of functional groups. Furthermore, the reactions have not any adverse impact on biochemical reactions. (Bertozzi,2009). There is currently strong interest in utilizing bioorthogonal chemistries to augment the detection of diseases both inside and outside of the body. The bioorthogonal cycloaddition reaction between tetrazine and trans-cyclooctene (TCO) is rapidly growing in use for molecular imaging and cell-based diagnostics (Haun,2014).

In this study, we synthesized an amine reactive TCO with succinimidyl ester moiety (TCO-NHS) and conjugated TCO-NHS to proteins (bovine serum albumin,BSA and lactate dehydrogenase, LDH) via NHS groups. For this purpose, cis-cyclooocten-4-ol was converted to trans-cyclooocten-4-ol (TCO) by photoisomerization at 254 nm in the presence of a sensitizer, methyl benzoate. NHS was incorporated to TCO by reacting with disuccinimidyl carbonate and characterized by 1H-NMR, Q-TOF-MS and FT-IR. Finally, TCO-NHS was utilized in labeling BSA and LDH. The labeled proteins will be conjugated to polymers containing tetrazine moieties after characterization. In the future studies, antibodies will be conjugated to polymers via tetrazine-TCO ligation and the system will be evaluated as a diagnostic platform. All this studies are in progress.

This work was financially supported by The Scientific and Technological Research Council of Turkey (TUBITAK, Project # 214M175)

Lee, J. 2005. ''Cell-specific targeting of nanoparticles by multivalent attachment of small molecules''. Nature, 23, 1418–1423

Keywords: Tetrazine, trans-cycloctene, protein labelling, bioconjugation, diagnosis, polymer

MBM-P-056 [Macromolecules in Biotechnology & Medicine]

In Situ Synthesis of Tetrazine Functional Polymers

Sinem Sipahioglu[1], Sinem Sipahioglu[2], Mustafa Yasin Ates[2], Ahmet Cetinkaya[2], Muhammet Ubeydullah Kahveci[2]

[1]Department of Bioengineering, Yildiz Technical University, Istanbul, Turkey
[2]Opticianry Programme, Beykent University,Istanbul,Turkey

A recently developed bioorthogonal conjugation approach, namely tetrazine ligation, has been getting remarkable attention beacuse of its super fast kinetic and selectivity. Intracellular small molecule imaging, genetically targeted protein tagging, postsynthetic DNA labeling, nanoparticle-based clinical diagnostics and in-vivo imaging are recent applications of tetrazine cycloadditions (Yang J. et al., 2012). Most of polymeric materials which used in the medical and biomedical fields are designed to be functional and tetrazine functional polymers have been prepared by post-polymerization techniques in the literature. The purpose of this study is direct synthesis of tetrazine molecules on polymer chains. The precursor polymer was sysnthesized by ATRP of poly(ethylene glycol) methyl ether methacrylate (PEGMA, Mn = 300 g/mol) and acrylonitrile (AN). The ratio of comonomers incorporated into the copolymer [P(PEGMAn-co-ANm)] was determined n:m = 2.28:1 by 1H-NMR. The well established tetrazine synthesis in which alkyl and aromatic nitriles are reacted with hydrazine in the presence of a Lewis acid and subsequent oxidation of yielding dihydrotetrazine was modified for synthesis of direct synthesis of tetrazine on polymer chains [P(PEGMA-co-AN-co-Tz)]. The nitrile group on the polymer was utilized as the alkyl nitrile and reacted with Boc-protected aminobenzyl nitrile under similar conditions leading to formation of tetrazine rings directly on the polymer chain. Among the partner molecule for tetrazine ligation is trans-cyclooctene (TCO) and this reaction is known as the fastest conjugation reaction under ambient conditions. Thus, conjugation of a TCO derivative with succinimidyl ester (TCO-NHS) to the tetrazine functional polymer [P(PEGMA-co-AN-co-Tz)] was followed by 1H-NMR spectroscopy and achieved successfully.

This work was financially supported by The Scientific and Technological Research Council of Turkey (TUBITAK, Project # 214M175)

Yang J., Karver M. R., Li W., Sahu S., Devaraj N. K., "Metal-Catalyzed One-Pot Synthesis of Tetrazines Directly from Aliphatic Nitriles and Hydrazine", Angew. Chem. Int. Ed., 51, 5222 –5225, 2012.

Keywords: Tetrazine, trans-cyclooctene, functional polymers, bioorthogonal conjugation

MBM-P-057 [Macromolecules in Biotechnology & Medicine]

Targeted Delivery of Anti-Angiogenic Chemotherapy Agents

Merve Karaçivi, Rana Sanyal

Department of Chemistry, Bogazici University, Istanbul, Turkey

The non-spesific biodistribution, low molecular weight and poor water solubility of chemotherapy drugs, result in several adverse side effects on healthy tissues.[1] To overcome all these toxic side effects in healthy tissues and increase drug circulation in plasma, many macromolecular drug delivery systems have been investigated in recent years.[2-3] One of the major problems in cancer therapy is excessive formation of blood vessels around the tumor.[4] To address this issue, water soluble polymeric carriers with methacrylate backbone that contain targeting groups have been designed and anti-angiogenic chemotherapy drug Combretastatin (CA4) was covalently attached to these copolymers. The resulting structures were characterized by GPC and H1-NMR. Cytotoxicity studies on human umbilical vein endothelial cells (HUVECs) were examined and cell internalization studies were performed. Finally, time dependent drug release studies were investigated under a variety of conditions.

Keywords: Macromolecular, Drug Delivery, Anti-angiogenic

MBM-P-058 [Macromolecules in Biotechnology & Medicine]

Phosphopeptide Enrichment From Human-Serum via Immobilized Metal Affinity Chromatography Using Phosphonic Acid Functionalized Capillary Monolith as a Stationary Phase in a Microfluidic System

Kouroush Salimi[1], Fatos Cigdem Kip[1], Omur Celikbicak[2], Duygu Deniz Usta[3], Bekir Salih[2], Ali Tuncel[4]

[1]*Chemical Engineering Department, Hacettepe University Ankara, Turkey*
[2]*Chemistry Department, Hacettepe University, Ankara, Turkey*
[3]*Department of Medical Biology and Genetics, Gazi University, Ankara, Turkey*
[4]*Division of Nanotechnology and Nanomedicine, Hacettepe University, Ankara, Turkey*

Phosphonic acid functionalized macroporous capillary monolith was obtained by the free-radical copolymerization of vinylphosphonic acid and ethylene dimethacrylate in a fused silica capillary 300 µm in diameter and 40 mm long. The poly(vinylphosphonic acid-co-ethylene dimethacrylate), poly(VPA-co-EDMA) monolith was derivatized with Ti°4+ ions using TiCl4 in an acidic aqueous medium. The bare and derivatized forms of capillary monolith were characterized by scanning electron microscopy, nitrogen adsorption-desorption method and micro-liquid chromatography in terms of morphology, porous properties and back-pressure flow rate relationship. Ti4+ functionalized capillary monolith was first tried as a stationary phase in a microfluidic system for phosphopeptide enrichment from synthetic mixtures and human serum. The synthetic phosphopeptide mixtures were prepared by the tryptic digestion of β-casein and BSA-β-casein mixture. The sample mixtures prepared with different BSA/β-casein molar ratios were fed into the Ti4+ functionalized-macroporous monolith in a microfluidic system using a micro-liquid chromatography pump at a flow rate of 5 µL/min. Aqueous ammonia solution was used as the eluent for the desorption of phosphopeptides selectively adsorbed onto the macroporous monolith. The selectivity of monolith was investigated by matrix assisted laser desorption/ionization time-of-flight mass spectrometry (MALDI-TOF MS). The detection limit for the enrichment of phosphopeptides by Ti4+ attached-poly(VPA-co-EDMA) monolith from the tryptic products of β-casein was 50 femtomole. In the case of β-casein and BSA mixtures as semi-complex samples (molar ratio of β-casein to BSA up to 1:500), the phosphopeptides were also successfully isolated. The phosphopeptides could be also isolated from human serum as a complex biological matrix. Four basic phosphopeptide structures were clearly identified in this sample by MALDI-TOF MS following to the enrichment in the microfluidic system

Keywords: monolith, phosphopeptide, vinylphosphonic acid

MBM-P-059 [Macromolecules in Biotechnology & Medicine]

Evaluation of toxicity essential oils of aromatic plant (Rosmarinus Officinalis) on the pédofaune

Karima Baba Aissa, Hadjer Rabah, Kamel Moussaoui, Zahr Eddine Djazouli

Department of Biotechnology, Saad Dahleb University of Blida, Road of Soumaa, BLIDA

In agriculture, the intentional use of plant protection products, which intended to protect plants against pests, diseases and weeds, (Calvet, 2003). However, the biological fertility has been neglected which often does not ensure the maintenance of soil organic matter and instead leads to a loss of biodiversity using pesticides (Robert, 2005).

In the context of elimination of harmful effects of plant health products on target, and non-target and soil quality. The agronomists orient towards a biological control method by use of natural substances of plant origin, however, the essential oils.

The objective of this work is to evaluate the toxicity of essential oil of Rosmarinus Officinalis one of Blida and the other of Tipaza on the abundance of pédofaune.

¬ From the results obtained, it shows a certain contrast in the effectiveness of treatments biological tested that boils in their repressive power on the abundance of populations nematode phytophagous and their attractive effect on mésofauniques communities.

¬ Treatment with essential oil of Rosmarinus Officinalis of the Blida region had a more attractive effect on population abundance mesofauna compared to biological treatment with Rosmarinus Officinalis in the region of Tipaza.

¬ After treatment, the effect of two essential oils formulated base Rosmarinus officinalis to 10 % show that oribates mites are the most affected because they recorded a decrease in abundance after 48 hours.

Keywords: essential oil, pédofaune, toxicity, Rosmarinus Officinalis

MBM-P-060 [Macromolecules in Biotechnology & Medicine]

"Development Of Novel Bilayer Chitosan-nHap Composite Biomaterial For Guided Bone Regeneration"

Sedef Tamburacı[1], Özge Karadaş[1], Engin Özçivici[1], Funda Tıhmınlıoğlu[2]

[1]*Izmir Institute of Technology, Department of Biotechnology and Bioengineering,Gulbahçe Campus, Urla, İzmir, Turkey*
[2]*Izmir Institute of Technology, Department of Chemical Engineering, Gulbahçe Campus, Urla, İzmir, Turkey*

Guided Bone Regeneration (GBR) is a process that used for the treatment of periodontal bone defects and it is performed with barrier membranes as bone grafts or bone substitutes. Recently GBR studies have focused on the development of resorbable polymeric polymer-based membranes with more attention than synthetic polymers because of their biodegradability. The aim of this study is proposed to fabricate novel bilayer membrane with microporous sublayer composed of chitosan and nano-hydroxyapatite particles (nHAp) and nanofiber chitosan/PEO upperlayer. Bilayer chitosan/nHAp composite was designed in order to support migration of osteogenic cells to the defect site with its sublayer and prevent epithelial fibroblastic cell growth impeding bone formation with its upper layer. Microporous and nanofiber layers were fabricated separately by freeze-drying and electrospinning methods. The effect of nHAp content on the morphological, mechanical and physical properties of the composites was studied by several characterization techniques using SEM, mechanical test, and micro Ct. In addition, *in vitro* biological activity of the composites were carried out with WST-1 assay for *in vitro* cytotoxicity and with rezasurin assay for osteoblast and fibroblast proliferation studies. The results showed that mechanical, pore size, protein adsorption and swelling increased with an increase of n-Hap. However, biodegradation of the composites decreased with rise of n-Hap content from 80% to 50% at 28 day incubation. Cell culture studies of the bilayer composite membranes showed high alkaline phosphatase activity and osteocalcin secretion of osteoblast cells on membranes. WST-1 assay results showed that fabricated bilayer membranes exhibited no cytotoxic effect and possess high cell proliferation on layers. Therefore, this study shows that the prepared membranes can be used as a potential membrane for guided bone regeneration applications.

Keywords: Barrier membrane, chitosan, guided bone regeneration, biomaterial

Bilayer structure

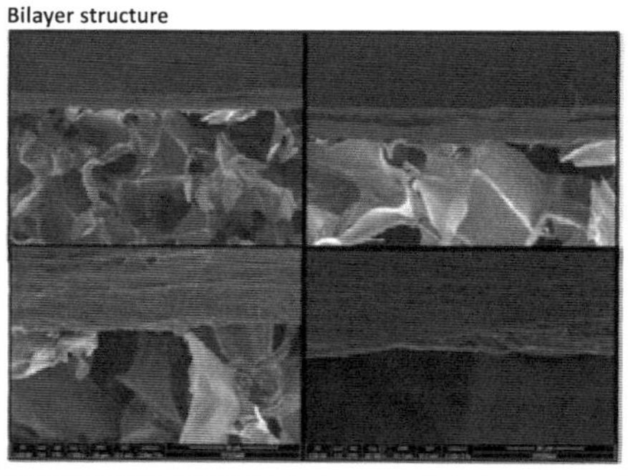

Figure shows the bilayer structure composed of microporous and nanofiber layers

Micro Ct

Micro Ct analysis of Chitosan-50% nHAp composite sublayer

MBM-P-061 [Macromolecules in Biotechnology & Medicine]

Development, Characterization and In vitro cell culture studies of Zein Bilayer Composites as Bioactive Wound Dressing Material

Seda Gunes[1], Sedef Tamburaci[1], Funda Tihminlioglu[2]

[1]Department of Biotechnology and Bioengineering, İzmir Institute of Technology
[2]Department of Chemical Engineering,İzmir Institute of Technology

In recent years, there has been a significant interest about functional bioactive materials for wound healing application.Wound dressing materials provide initial cell adhesion and promote wound healing process.Hypericum perforatum oil is used in several biomedical applications such as healing of wounds, peptic ulcer.Among several biomaterials, zein based composites have been showed a potent biomaterial for biomedical applications.Nanoclay nanocomposites have gained importance in order to improve mechanical, physical, thermal, and barrier properties of materials.The aim of this study is to develop and characterize Hypericum perforatum oil incorporated zein composites including nanoclay consisting of bilayer structures of film and nanofiber layer as a wound dressing material which covers the injured tissue, protects against infections and dehydration as well as gas permeability. Oil loaded zein films were successfully prepared by solvent casting method in different oil and nanoclay concentrations.Zein nanofibers were prepared by using electrospinning method.Bilayer structure was constructed by coating zein films with zein nanofiber. The films were characterized by the water vapor permeability, mechanical properties, TGA, swelling and surface hydrophobicity measurements.Fiber formation, diameter and bilayer morphology were examined by SEM.Antimicrobial test was carried by disc diffusion method against E.coli, S.aureus, C.albicans. In vitro cytotoxicity of monolayer film and bilayer films was performed on HS2 keratinocyte and fibroblast cells.Cell attachment and proliferation experiments were performed. H.perforatum oil incorporated zein composites showed antimicrobial effect. Oil addition slightly increased the water vapor permeability the films.Bilayer composite films were swollen much higher than one layer films. Nanofibers with homogenous fiber structure were observed by SEM analysis.The thermal properties were improved with nanoclay intercalation. Both prepared films and bilayer composites were found as cytocompatible and provided a good surface for cell attachment and proliferation.The results revealed that the H. perforatum incorporated zein based composites seems to be a potential and novel biomaterial for wound healing applications.

Keywords: Zein, plant essential oil, nanocomposites, bilayer, wound healing

MBM-P-062 [Macromolecules in Biotechnology & Medicine]

Production And Characterization Of Novel Bilayer Chitosan-Zein Based Composites For Osteochondral Tissue Regeneration

Sedef Tamburacı[1], Berivan Çeçen[2], Hasan Havıtçıoğlu[2], Funda Tıhmınlıoğlu[3]

[1]Department of Biotechnology and Bioengineering,Izmir Institute of Technology, Gulbahce Campus, Urla, Izmir, Turkey
[2]Institute of Health Science, Department of Biomechanics, Dokuz Eylul University, Inciralti Health Campus, Izmir, Turkey
[3]Department of Chemical Engineering, Izmir Institute of Technology, Gulbahce Campus, Urla, Izmir, Turkey

Osteochondral tissue is hard to regenerate after injuries or degenerative diseases. Traditional treatments have still disadvantages, such as donor tissue availability, donor site morbidity, implant loss, limited durability of prosthetics. Thus, recent studies has focused on tissue engineering strategies to regenerate the osteochondral defects. Scaffold designs have been developed from monolayer structures to bilayered scaffolds to repair the cartilage bone interface and to support each tissue separately. Chitosan is widely used as biomaterial for bone tissue regeneration due to its porous structure, gel forming and antibacterial properties, biocompatible structure, high affinity to in vivo macromolecules. Zein is the most widely employed plant protein with antibacterial activity and can be easily formulated into nanofibers with electrospinning method due to its solubility in organic solvent ethanol. Nanohydroxyapatite particles and POSS nanocage structures (hybrid silica-organic group) were used as reinforcements in polymer layers to mimic cartilage-bone tissue. This study is proposed to develop a bilayer scaffold produced via fabrication of two different layers with different polymer-inorganic composites in order to to satisfy the complex and diverse regenerative requirements of osteochondral tissue. Chitosan/nanohydroxyapatite microporous layer was designed for bone tissue and zein/POSS nanofiber layer was designed for cartilage tissue. Composites were characterized by SEM, mechanical test, swelling, biodegradation in order to determine morphological, physical, mechanical properties. Physical and morphological characterization results showed that bilayer scaffold has a structure composed of microporous and nanofiber layers joined at a continuous interface with good mechanical and morphological properties. Furthermore, *in vitro* cell culture studies has been performed with proliferation, fluorescence imaging, LDH and ALP activity assays using osteosarcoma and chondrosarcoma cell lines. *In vitro* studies showed that bilayer scaffold promoted cell adhesion and ALP and LDH activities for used cell lines, therefore, promises for osteochondral regeneration.

Keywords: Biomaterial, chitosan, zein, osteochondral tissue, bilayer, composite

Chitosan/nHAp porous layer

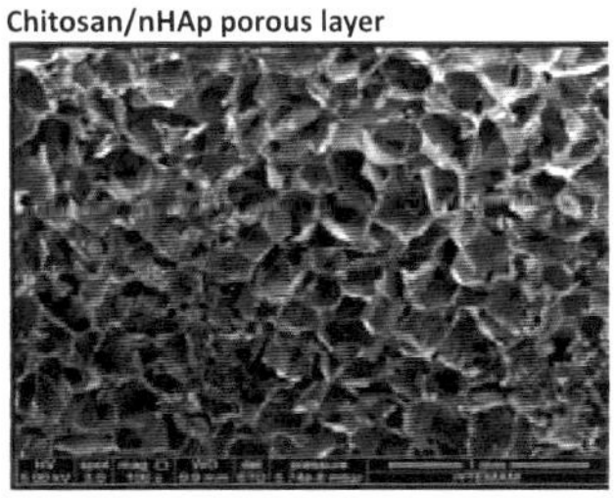

Scanning electron micrograph of Chitosan-40%nHAp composite layer

Zein/POSS nanofiber layer

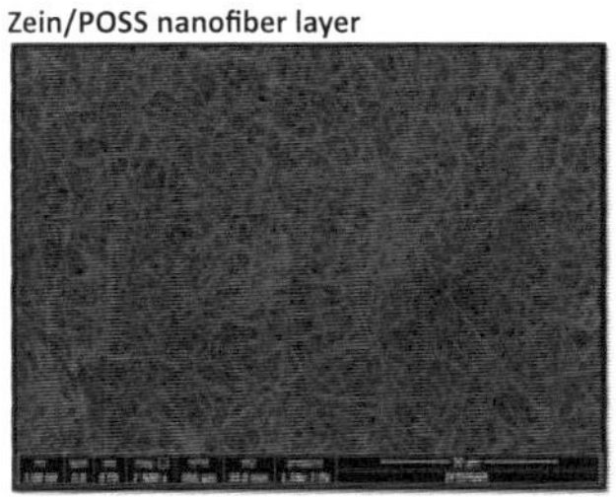

Scanning electron micrograph of Zein/POSS composite nanofiber layer

MBM-P-063 [Macromolecules in Biotechnology & Medicine]

Effect of pH on Drug Release Kinetics for PHEMA/ Itaconic Acid Imprinted Hydrogels

Dilan Bozkurt[1], Dilek Dalgakiran[2], Seniha Guner[1]

[1]*Department of Chemical Engineering,Istanbul Technical University, Istanbul, Turkey*

[2]*Polymer Science and Technology, Istanbul Technical University, Istanbul, Turkey*

Molecular imprinting is a special technique to produce polymers with specific target binding sites. Increasing knowledge and developing technology extend this technique into pharmacological, analytical, biological selective molecular recognition and controlled release systems where the technique has a great importance in transition through future drug technology [1, 2]. Properties of the release media is a limitation for the usage of the technique in drug release studies. pH is one of these factors, changing from acidic to basic values in different mediums [3]. Aim of this study is to investigate the effect of the pH on the release rate of the doxycyline hyclate antibiotic from molecularly imprinted poly(2-hydroxyethyl methacrylate) (PHEMA) hydrogels. The hydrogels were synthesized by thermal radicalic polymerization. Triethylene glycol dimethacrylate (TEGDMA) was used as crosslinker and itaconic acid (IA) as functional monomer. First, they were mixed with HEMA and doxycycline hyclate. After addition of thermal initiator 2,2'-azobis(2-methylpropionitrile), polymer mixture was degassed by using bubbling technique with nitrogen gas in order to remove oxygen. Polymerization mixture was transferred into glass mold and placed in vacuum oven at 45°C for 24 hours. FT-IR and DSC methods was used for characterization of hydrogels. Drug release kinetic studies were performed in acidic (pH=4.6), neutral (pH=6.8) and a basic (pH=7.5) mediums at 37 °C. The releasing profile of the hydrogels at different pH mediums were compared to each other.

References:

1. Samarth N.B, Kamble V., Mahanwar P.A, Ranez A.V., Abitha V. 2015 A historical perspective and the development of molecular imprinting polymer- A review, Chemistry International 1(4), 202-210.
2. Vasapollo G. et al. 2011. Molecularly Imprinted Polymers: Present and Future Prospective, International Journal of Molecular Sciences 12(9), 5908-5945.
3. Klein S. 2010. The Use of Biorelevant Dissolution Media to Forecast the In Vivo Performance of a Drug, The AAPS Journal, 12(3), 397-406.

Keywords: Drug release, molecular imprinted polymer, PHEMA

MBM-P-064 [Macromolecules in Biotechnology & Medicine]

Metal ion carrying magnetic monodisperse-porous silica microspheres as an immobilized metal affinity chromatography (IMAC) sorbent for protein separation

Duygu Deniz Usta[1], Kouroush Salimi[2], Ali Tuncel[3]

[1]*Department of Medical Biology and Genetics, Gazi University, Ankara, Turkey*
[2]*Chemical Engineering Department, Hacettepe University, Ankara, Turkey*
[3]*Division of Nanotechnology and Nanomedicine, Hacettepe University, Ankara, Turkey*

Magnetic monodisperse-porous silica microspheres ca 5 µm in size were obtained by a newly proposed "staged shape templated hydrolysis and condensation protocol". The magnetic silica microspheres were derivatized with N,N-iminodiacetic acid-glycidoxypropyltrimethoxysilane (IDA-GLYMO) complex via silane chemistry. The selected transition metal ions (i.e. Ni2+, Mn2 and Zn2+) were attached onto the IDA-GLYMO functionalized-magnetic silica microspheres via coordination complex formation between selected metal ion and carboxyl groups. Albumin, lysozyme and haemoglobin were selected as the model proteins isolated by the magnetic sorbents in batch fashion. The protein separation by immobilized metal affinity chromatography with the sorbents carrying different metal ions was invetigated at pH 7.0 phosphate buffer as the adsorption medium and pH 8.0 phosphate buffer with 0.5 M imidazole as the desorption medium. The effects of initial protein concentration, medium pH and sorbent concentration on the isolation yields of selected proteins were determined. Relatively higher isolation yields ranging between ca. 30-40 % were observed for albumin with Ni2+ functionalized magnetic microspheres. The results indicated that newly proposed magnetic sorbent exhibited an isolation performance comparable with the commercial IMAC resins for some common proteins.

This study is supported by Turkish Scientific and Technological Research Council, TUBİTAK by a project numbered 115M615, entitled "Synthesis of functionalized magnetic silica microspheres and determination of their biotechnological application performance".

Keywords: Protein Separation, Magnetic silica microspheres, IMAC

MBM-P-065 [Macromolecules in Biotechnology & Medicine]

The influence of the washing step on the drug loading performance of the p(hema) hydrogels

Dilek Dalgakıran[1], Fatma Seniha Güner[2]

[1]*Istanbul Technical University, Polymer Science and Technology, Maslak 34469 Istanbul*
[2]*Istanbul Technical University, Department of Chemical Engineering, Maslak 34469 Istanbul*

Molecular imprinted polymers have specific binding sites for a target molecule which has side groups interact with polymer functional groups. Template (target molecule) is present at the reaction mixture during polymerization and monomers make crosslink network and arrange around the template molecule. The template interacts with monomers by non-covalently or covalently [1, 2]. For this study, the non-covalent interaction has been achieved. Finally, the template is removed from polymer by a washing procedure for non-covalent binding. The washing procedure is important on the imprinting performance. It is carried out in order to maximize the specific interactions by discharge the template from cavities [3]. The aim of this study is to investigate the influence of the washing procedure on the molecularly imprinted polymer performance. The molecularly imprinted hydrogels were synthesized by radicalic polymerization. Itaconic acid and Acrylic acid as functional monomer and doxycyline as template molecule were added in hydroxyethyl methacrylate. The chemicals dissolved in each other by ultrasonic bath. Ethylene glycol dimethacrylate as crosslinker and finally 2,2'-azobis(2,4-dimethyl-pentanenitrile) as initiator was added to solution and allowed to mix. It was degassed with N2 bubbling in order to remove oxygen from polymer solution. The non-imprinted hydrogel was prepared similar to imprinted hydrogel procedure without adding doxycyline. The mold was placed in 45 ºC vacuum oven for 24 hours. Two step washing procedure performed. At the first step, four different solvent mixture (containing acetonitrile, acetic acid, methanol and oxalic acid) were used. At the second step, three solvents were used to investigate the effect of the washing solvents on the imprinted polymer drug loading and releasing performance. Drug loading tests were performed at 4ºC for 72 hours and imprinted factor (IF) values were compared to each other. The UV spectroscopy was used to detection of the drug after loading test at 273 nm.

Keywords: molecularly imprinted hydrogel, washing solvent, imprinting factor

MBM-P-066 [Macromolecules in Biotechnology & Medicine]

The effect of using antibacterial agent and size of the inorganic micro fillers on flexural strength of the composite materials and their antibacterial activity

Zerin Yeşil Acar[1], Nadir Kiraz[2], Meltem Asiltürk[1]

[1]*Akdeniz University, Faculty of Engineering, Department of Materials Science and Engineering, Antalya, Turkey*
[2]*Akdeniz University, Faculty of Science, Department of Chemistry, Antalya, Turkey*

Composite materials are one of the most widely used materials because of their adaptability to different applications such as dentistry. The advantage of composite materials is that they have unique properties in terms of mechanical, physical, and chemical aspects. Particularly, the antibacterial property that you acquire in terms of using them in different areas is of high importance. In this study, the effect of size of the inorganic micron sized fillers on the comosites' flexural strength has been evaluated. In addition, the antibacterial activity of zinc oxideas as an antibacterial agent was investigated on the composite material desired mechanical properties. The organic matrix is comprised of 2,2-Bis[4-(2-hydroxy-3-methacryloyloxypropyl)phenyl]-propane(BISGMA), urethanedimethacrylate(UDMA), and tri-ethyleneglycoldimethacrylate(TEGDMA) and small quantity of BOP as initiators. The inorganic fillers are consisted of commercial modified Aerosil R709 and M8000, M6000, M4000, M3500, and M3000 which were modified with methacryloxypropyltrimetoxysilane(MPTS). Aerosil R709 has nano size, while others have micron sizes. % proportions of nano size filler and micron size filler were 19 and 46 respectively. The maximum flexural strength values were obtained at the system that was included M8000. After that, zinc oxide was used as an antibacterial agent which was added in this system. For three point bending tests, 5 plate specimens, 2mmx-2mmx25mm, were prepared for each composite and tested at 0,75mm/min. Flexural strength (n=5) was measured by the method in ISO4049. Consequently, the mechanical values of the composite material decreased after zinc oxide was added.On the other hand, these systems showed antibacterial activity.

Keywords: Dental composite, antibacterial effect, ZnO

Figure 1. Flexural strength of the composites

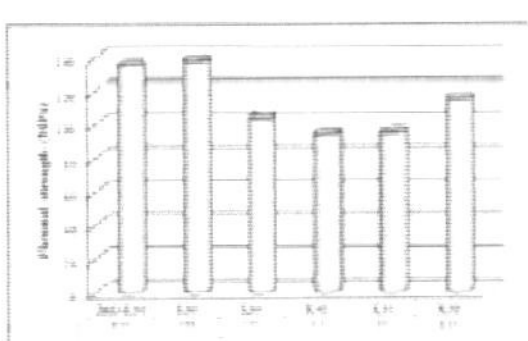

Figure 2.Images of the test plates used to detect the antimicrobial activity on S. mutans

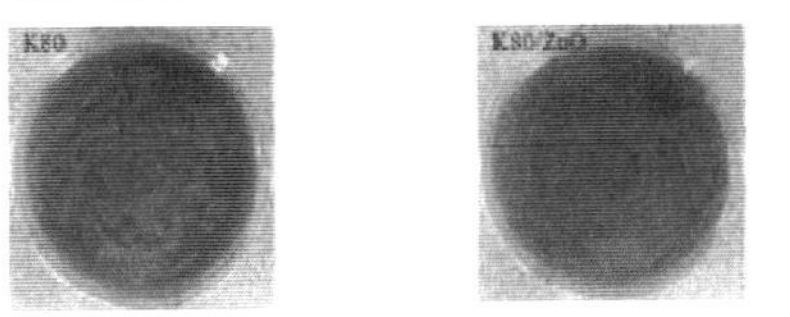

MBM-P-067 [Macromolecules in Biotechnology & Medicine]

New synthesis and reactivity of 4-hydroxy-6-methyl-2h-pyran-2-one

Benmohamed Soumaya[1], Rachedi Yahia[1], Hamdi Maamar[1], Dumas Françoise[2]

[1]*Faculty Of Chemistry, University Of Science And Technology Houari Boumediene Bp 32, El-Alia, 16111, Bab-Ezzouar, Algiers, Algeria.*
[2]*Université Paris Sud, Umr 8076 Bio Cis, Laboratoire De Pharmacognosie, Ipsit And Labex Lermit, Faculté De Pharmacie, 5, Rue Jean-Baptiste Clément, 92296 Chatenay-Malabry Cedex, France.*

Two new series of pyran- or coumarin-substituted thiazolyl-pyrazole-chromen-2-one derivatives 3-*1-[4-(4-Hydroxy-6-methyl-2-oxo-2H-pyran-3-yl)thiazol-2-yl]-3-methyl-1H-pyrazol-5-yl*-2H-chromen-2-one derivative a–d were efficiently synthesized under environmentally friendly reaction conditions through a convenient one-pot multicomponent reaction of a heterocyclic bromoacetyl derivative 4-hydroxy-6-methyl-2H-pyran-2-one and 3-(bromoacetyl)-4-hydroxy-6-methyl-2H-pyran-2-one with thiosemicarbazide and a substituted 3-(acetoacetyl) coumarin derivative a–d in anhydrous ethanol. The reaction proceeds through Hantzsch thiazole and Knorr pyrazole syntheses in refluxing ethanol. The key features of this approach are its operational simplicity, the quick access to the desired products, and the good to excellent yields. The structures of the newly synthesized compounds were established by IR spectroscopy, 1H, 13C, and 2D NMR spectroscopy, and mass spectrometry.

Keywords: Lactone, pyrone and coumarine,

MBM-P-068 [Macromolecules in Biotechnology & Medicine]

Development of polymeric organic capsules as a platform for opioids prolonged release

Shumaila Razzaque, Tan Bien
School of Chemistry, Huazhong university of science and technology. Wuhan, china

Pain is a major clinical issue which imposes substantial burden on society. Opioids are the most frequently prescribed medication for the treatment of moderate to severe pain. A regular administration of analgesics is necessary to ensure the normal daily life functioning by alleviating pain symptoms. Among various opioids, morphine is predominantly administered up to date. It is epidural opioid, considered as better quality analgesia. Despite of its effectiveness, the very short half-lives required the frequent dosing, thus paramount the inconsistent drug levels and enhanced the risk of undesirable therapeutic effects. Therefore, preferential drug formulations are mainly focused while designing a delivery system which proficiently controls the release on administration.To improve the therapeutic efficacy and enhance the physiological action of opioids as well as avoidance of adoption, polymer scaffold as drug carrier have attained a substantial attention. Polymer implants are capable of modulating controlled release thus providing constant and effective therapeutic drug levels and avoids the adverse effects. Among diverse polymer geometries, hollow capsules have recently received more attention. This interest is primarily fueled up due to their ability to confine the chemicals within their hollow cavities and their controlled release and avoid their ripening or poisoning. In this regard, fabricated hollow microporous organic capsules (HMOCs)showed the extraordinary high encapsulation efficiency of morphine and its controlled release. HMOCs release the drug at a rate which is proportional to the amount of drug remaining in its interior. Due to the presence of hollow and porous morphology they posses high SBET surface areas. Thus such a capsular system can be used as a storage device for numerous drugs that are non-polar as well as provide a necessary pathway for the controlled release. Compared to other inorganic capsules, HMOCs have better tolerance to acid/alkali and are more compatible with the organic molecules.

Keywords: Hollow microporous polymer (HMOCS), drug delivery system, opioid analgesics, controlled release, zero-order kinetics

Morphology

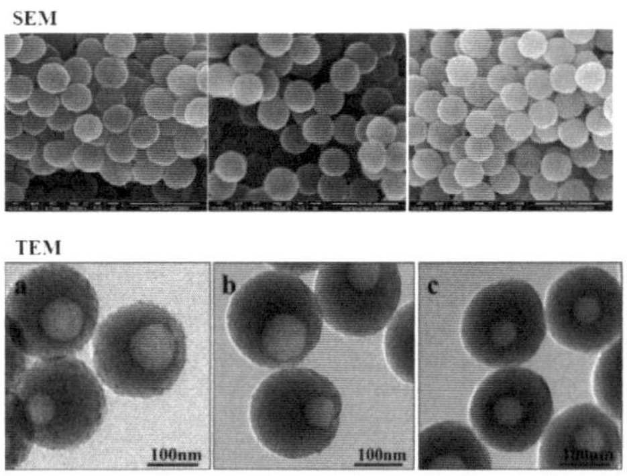

SEM and TEM micrographs, showing the morphology of HMOCs

schematics

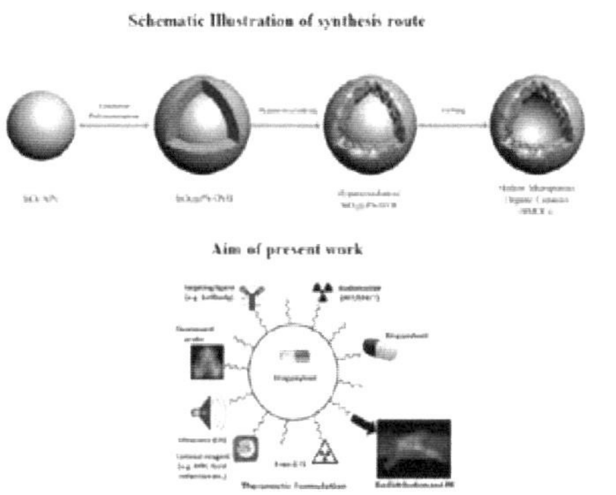

Scheme illustrating the synthetic methodology

Summarized tendency to drug release

Serial No	Sample	Encapsulation efficiency(%)	Release efficiency(%)	SBET surface area (m2/g)
1	HMOC-1	98	96	919
2	HMOC-3	95	95	794
3	HMOC-5	90	92	658
4	HMOC-10	88	90	584
5	HMOC-15	80	82	336

The tabular data explains about the tendency of a polymer to release the drug molecule, dependent on the pore size.

MBM-P-069 [Macromolecules in Biotechnology & Medicine]

Preparation of different biopolymer chitosan films with enzyme immobilization for biotechnological application

Entsar I Rabea[1], Mohamed E. I. Badawy[2]

[1]Department of Plant Protection, Faculty of Agriculture, Damanhour University, 22516-Damanhour, Egypt
[2]Department of Pesticide Chemistry and Technology, Faculty of Agriculture, 21545-El-Shatby, Alexandria University; Alexandria, Egypt

A biopolymer chitosan is an admirable material for immobilization of several enzymes, as it exhibits increase their stability compared to the free enzymes; hence it has been broadly used in electrochemistry and biosensors in recent years. It is a natural cationic co-linear polymer of ß-(1-4)-D-glucosamine and N-acetyl-β-(1-4)-D-glucosamine units, and is obtained by deacetylation of chitin, the second most abundant polysaccharide on the earth after cellulose. Composite biopolymers films were prepared by casting the mixture containing chitosan (2%), starch (2%) and sorbitol (1%) as a plasticizer. Chitosan-Ca and chitosan-Mg films were also prepared by the same composite polymers with ratio of 1:1 (chitosan:CaCl2 and chitosan:MgCl2) calculated to glucosamine unite of chitosan molecule. Carboxylesterase (CbE) that plays a significant role in numerous endogenous and exogenous ester-containing compounds was immobilized into the biopolymer films by crosslinking mechanism using glutaraldehyde. Concentration of glutaraldehyde showed a significant function in the immobilization procedure and the optimal immobilization was obtained with 30 mM glutaraldehyde after incubation for 24 h at 4ºC. The crosslinking mechanism conserved the enzyme entrapped into chitosan films with activity up to 87.39% for chitosan/starch-Ca film, 79.81% for chitosan/starch-Mg film and 71.14% for chitosan/starch film. Studies were done on the immobilized enzyme into chitosan films regarding the determination of the optimal pH, thermal stability, sensitivity to inhibitors, and kinetic parameters. The chitosan films degraded or chelated different concentrations of methomyl and chlorpyrifos pesticides however; the degradation percentage was increased with the films immobilized with CbE. Therefore, this study demonstrates that the chitosan/starch films and chitosan/starch-metal films provide excellent immobilization environment for CbE designed to use in coating of fresh fruit and vegetables containing low levels of organophosphorous and carbamate pesticides.

Keywords: Biopolymers, Chitosan films, Carboxylesterase, Immobilization, Biodegradation.

MBM-P-070 [Macromolecules in Biotechnology & Medicine]

The investigation of polymorphs and eutectic mixture of progesterone and estradiol benzoate loaded in silicone rubber by differential scanning calorimetry (DSC)

Maryam Ramezani, Farhid Farahmandghavi, Mohammad Imani

Polymer science Faculty, Iran Polymer and Petrochemical Institute, Tehran, Iran

The ability of some substances to exist in more than one crystalline form, known as polymorphism, is one of the most important solid state properties of drugs [1, 2]. Here, effects of weight fraction of the drugs and silicone rubber curing profile are analyzed and reported. In fact, probable changes occurring in crystalline and chemical structures of the progesterone (PRO) and estradiol benzoate (EB) as a function of heat treatment program were traced by differential scanning calorimetry (DSC). Heating programs were selected to imitate the curing process of silicone rubber which is usually used as a matrix to yield in implantable drug delivery devices releasing PRO and EB. Samples of the PRO and EB and also their mixtures with different weight ratios in silicone rubber were prepared. According to the DSC thermograms, different polymorphs were emerged by heating to melting point and cooling down cycles applied on PRO. In other words, α-PRO polymorph melted (at 131°C) at the first heating cycle but changed into β- form after first cooling showing melting point at 123.8 °C in the second heating cycle. DSC thermograms of the EB indicated its thermal degradation into estradiol (as base) by the same heating program. Investigation of mutual effects of the drugs on their crystalline structure in the mixtures containing 10, 30, 50, 70 and 90 wt.% EB revealed presence of significant interactions between PRO and EB. In fact, melting temperatures of the drugs shifted compared to the neat samples (0 and 100 wt.% EB). The interactions reached to the maximum level in the mixture containing 30 wt.% EB resulting in an eutectic mixture with a single melting point around 117°C.

References:

[1] Heinz et al, The Journal of pharmacy and pharmacology, 61, 971–88, 2009.
[2] Tripathi et al., AAPS PharmSciTech, 11, 1493–8, 2010.

Keywords: polymorphism, drug, silicone rubber, DSC,

MBM-P-071 [Macromolecules in Biotechnology & Medicine]

Anti-angiogenic Drug Conjugated Linear Dendritic Triblock Copolymers as Micellar Drug Delivery Systems

Burcu Sumer Bolu[1], Ece Gecici[1], Rana Sanyal[1], Rana Sanyal[2]

[1]*Bogazici University, Department of Chemistry, Istanbul, TURKEY*
[2]*Bogazici University, Center for Life Sciences and Technologies, Istanbul, TURKEY*

In this study, a family of alkyne functionalized polyester dendrons of generations ranging from G1 to G3 were coupled to linear poly(ethylene glycol) diazides through Huisgen type "click" chemistry and ABA type linear dendritic triblock copolymers were obtained.1 While taking advantage of PEG's hydrophilicity and anti-biofouling character, well defined hydrophobic polyester dendrons with alkyne functionalized peripheral groups render the final construct biodegradable and easily functionalizable. Henceforth anti-angiogenic drugs were conjugated to these dendron-linear polymer–dendrons (DPD) via "click" chemistry to generate a sturdy controlled drug delivery system. Moreover, once introduced into aqueous media hydrophobic-hydrophilic-hydrophobic character of the constructs enabled formation of flower-like micelles. In addition to generation dependent sub-nanomolar range critical micelle concentrations (CMC), resulting micelles had hydrodynamic diameters in nanoscale indicating that they are suitable for passive tumor targeting through EPR effect. Drug release and stability experiments highlighted an improved release profile at acidic pH and relative serum stability. Effectivity of anti-angiogenic drug was shown in vitro via Matrigel tube formation assays and for all constructs cellular cytotoxicity values were determined using HUVECs. Cell internalization was displayed by fluorescent dye doped micellar constructs. Finally this DPD based drug conjugated system provides a stable, tunable and attractive delivery platform for combating cancer by inhibiting angiogenesis.

Keywords: micellar drug delivery, polyester dendron, flower micelle, "click" chemistry, antiangiogenic

Synopsis

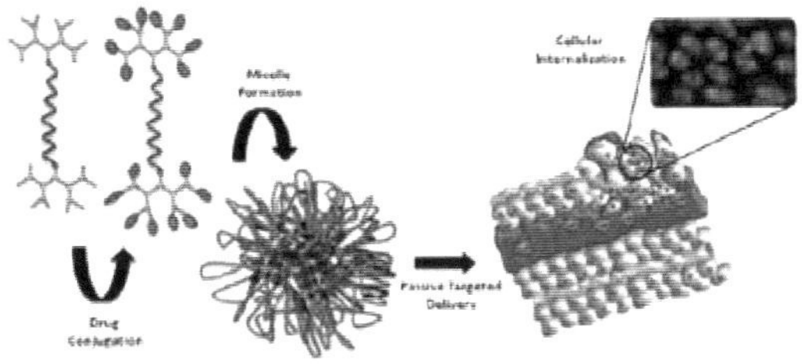

MBM-P-072 [Macromolecules in Biotechnology & Medicine]

The study of binding modes of 2,2'-Naftoin with ct-DNA

Eyüp Metin, Betül Fatma Erkoç, Aslı Tuğla, Elif Özçelik, Nergis Arsu

Department of Chemistry, Yıldız Technical University, Istanbul, Turkey

Small molecules can interact with DNA generally by non-specific outside edge binding, intercalation and groove binding. It is important to understand the binding modes of different molecules with DNA to understand the benefits and/or harms of the substances in living organisms. As an example, some anticancer drugs bind to DNA by intercalative interactions and inhibit the replication process of DNA and therefore these drugs (such as doxorubicin and daunorubicin) inhibits the uncontrolled cancer cell reproduction and tumor growth. (Blackburn *et al*, Reversible Small Molecule-Nucleic Acid Interactions, RCS Publishing, 2006, 3:342-379)

The study of interactions between anticancer reagents or candidate anti-cancer reagents and DNA is important in order to understand the mechanism of their anticancer activities. Therefore, it is necessary to introduce new compounds as anticancer drug candidate and more simple techniques. Among these techniques, spectroscopic studies are a powerful way to investigate the binding mode of small molecules with DNA. (Qui *et al*, Biosens. Bioelectron, 2007, 22, 2629-2635)

In this study, 2-hydroxy-1,2-di(naphthalen-2-yl)ethanone (2,2' Naftoin) was synthesized and characterized. In order to identify the interaction mechanism of 2,2' Naftoin with DNA, UV–Vis absorption spectroscopy, fluorescence emission spectroscopy techniques and melting studies were used to distinguish the binding mode whether intercalative, electrostatic or groove.

Keywords: naftoin, DNA binding, DNA, binding mode, intercalation, groove

2,2'-Naftoin Molecular Structure

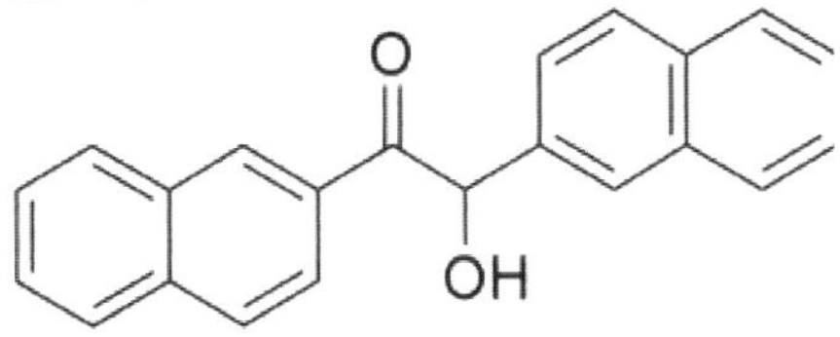

2-hydroxy-1,2-di(naphthalen-2-yl)ethanone (2,2' Naftoin)

MBM-P-073 [Macromolecules in Biotechnology & Medicine]

Biodistribution of Radiolabelled Polymeric Nanoparticles Obtained by PISA Approach in HT1080 Tumor Xenografts

Sadik Kaga[1], Nghia Truong[2], Danielle Senyschyn[3], John Quinn[2], Rana Sanyal[1], Lisa Kaminskas[3], Michael Whittaker[2], Thomas Davis[2]

[1]*Department of Chemistry, Bogazici University, Istanbul 34342, Turkey*
[2]*ARC Centre of Excellence in Convergent Bio-Nano Science & Technology, Monash Institute of Pharmaceutical Sciences, Monash University, Parkville, Melbourne, Victoria 3052, Australia*
[3]*Drug Delivery Disposition and Dynamics, Monash Institute of Pharmaceutical Sciences, Monash University, Parkville, Victoria 3052, Australia*

Since the discovery of the enhanced permeability and retention (EPR) effect, macromolecules with long blood circulation have become the primary substances for drug delivery due to improved accumulation in tumors through fenestrated blood vessels.[1] Likewise other biocompatible nanosized structures, polymeric nanoparticles have been widely used in drug delivery and diagnosis systems for cancer and cardiovascular disease mainly due to their biodistribution properties.[2] There are various types of techniques for the preparation of polymeric nanomaterials.[3] Block copolymers, having polymer blocks with different physical and chemical properties, have become commonly studied materials to generate polymeric nanoparticles via self-assembly.[6] Several research groups have focused on biodistribution of non-spherical nanoparticles with the diameter range about 5-25 nm.[4-6] They all concluded that non-spherical nanoparticles less likely accumulate in major organs (liver, spleen and lung) and more likely accumulate in tumor tissue compared to spherical ones. However biodistribution of non-spherical polymeric particles with larger diameter is unclear. To investigate biodistribution profile of non-spherical polymeric nanoparticles with diameter size around 60 nm compared to spherical ones, we used PISA (polymerization induced self-assembly) concept which allows generating non-spherical nanoparticles with desired larger diameters.[7,8] RAFT polymerization used for the synthesis of PGMA-b-POEGMA-b-PST block copolymer nanoparticles and they were tested for cytotoxicity in HT1080 cell line after inactivation of epoxide groups via ring opening reaction. Radiolabelling of sufficient epoxide groups on nanoparticle surfaces was followed by ring opening of remaining ones for biodistribution study in nude mice harboring HT1080 tumor xenografts. Size and morphology of nanoparticles showed significant effect on biodistribution patterns.

Keywords: PISA, Polymeric nanoparticles, biodistribution, tumor xenograft.

MN-P-001 [Macromolecules & Nanotechnology]

Preparation of nanometer CoFe2O4 by co-precipitation using different precipitants and its characterization

Yasmina Bellal Hammiche[1], Laaldja Boukhobza Meddour[1], Amar Djadoun[2], Nadia Henda[3], Ahmed Ezzine Hassanine[3], Fateh Mernache[4], Mohamed Sennour[5], Marie Hélène Berger[5], Abdennour Meddour[5], Aline Auroux[6]

[1]*L.mccco., Faculté de chimie, Usthb. BP32 El Alia, Bab Ezzouar, 16111, Alger, Algérie.*
[2]*L.geophysique, FSTGAT, USTHB. BP32 El Alia, Bab Ezzouar 16111, Alger, Algérie.*
[3]*Credec, Sonelgaz, Ouled Fayet*
[4]*Udec, Crnd, Comena, Draria, Alger, Algérie*
[5]*Mat, Centre des Matériaux, Mines-ParisTech, France.*
[6]*Cnrs, Université Claude Bernard, Lyon, France.*

The spinel structure has received extensive attention in numerous scientific and technological fields. The structural and textural properties of these materials are strongly influenced by the preparation methodology used in their synthesis and may influence the performances of these materials in their different fields of application [1-4]. This work is devoted to the investigation of the influence of the preparation process on the physical–chemical properties of a cobalt ferrite spinel applied. Samples of CoFe2O4 mixed oxide belonging to the inverted spinel type structure have been obtained by co-precipitation method using chloride, nitrate and citrate precursor procedures at 70°C. The obtained materials are calcined at 500°C. The bulk catalysts were characterized by ATG/ATD, FTIR, XRD, MEB, TEM-EDX, BET and TPR. In this work it is demonstrated that the chemical synthesis affect the crystal properties and cation distribution in the spinel structure, microstructure, surface area and reducibility. The infrared spectra of precursors, based on coprecipitated chloride salts at 70 ° C, are given in Fig (1). The TGA / DTA curves hydroxides and oxalate precursors, prepared at 70 ° C are shown in Fig (2). They are characterized by endothermic and exothermic peaks. The SEM/TEM photographs and the results of EDX analysis relating to catalytic systems, obtained at the calcination temperature 500 °C, are illustrated in Fig (3). Distinct morphologies were noted between solids based on hydroxides and the like based oxalate. The EDX analysis shows that the catalysts are composed only of the elements cobalt, iron and oxygen, no impurity was detected. Table (1) shows the crystallite sizes determined by XRD and surface areas determined by BET analysis. The three samples were subjected to TPR analysis, the resulting profiles are shown in Figure (4).

Keywords: spinel, nanomaterial, coprecipitation,

Figure.1: infrared spectra of precursors prepared at 70 ° C

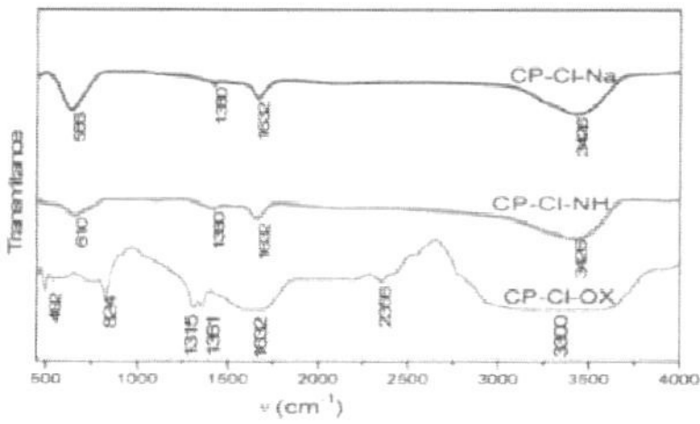

Figure 1 : infrared spectra of precursors prepared at 70 °C

The infrared spectra of precursors, based on coprecipitated chloride salts at 70 ° C, are given in Fig (1). Inspection of its spectra shows the presence of common bands at about 3500 cm -1, 1380 and 1632 cm-1 related to different water molecules of the vibration modes physisorbed adsorbed to the surface of solids. These bands are wider for the oxalate based precursor since the formed salt is in hydrate form. The spectrum of the CP-Cl-NH sample is characterized by the presence of metal hydroxide exhibiting broadband 610 cm-1 associated with vibrations of elongation of the metal hydroxide-network. The cobalt hydroxide Co(OH)2 is probably at the origin of this band. CP-Cl-Na solid product a sharp band at 586 cm-1 corresponding to the intrinsic vibration assigned to the spinel structure. Furthermore, the spectrum of the precursor CP-Cl-OX has a distinctly different profile; the presence of oxalate radical acts by banding, hydrated metal oxalate characteristics.

Figure.2: Profiles TGA / DTA precursor prepared by co-precipitation at 70 °C with NaOH (a), NH4OH (b) and H2C2O4 (c).

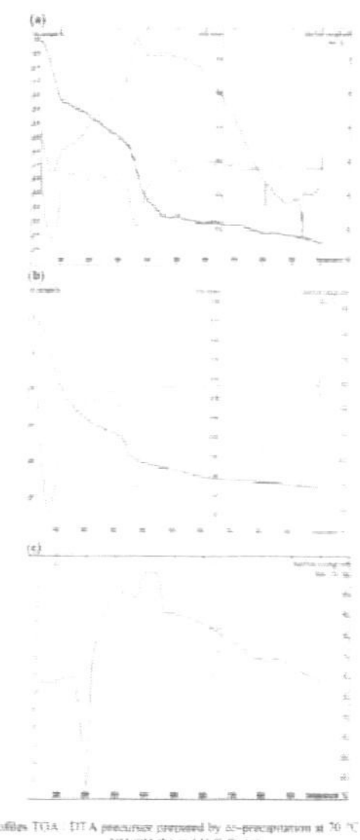

Figure 2 : Profiles TGA / DTA precursor prepared by co-precipitation at 70 °C with NaOH (a), NH4OH (b) and H2C2O4 (c)

The TGA / DTA curves hydroxides and oxalate precursors, prepared at 70 ° C are shown in Fig (2). They are characterized by endothermic and exothermic peaks suggesting changes detailed below: Bearings similar weight losses are observed in the profile of the two solid CP- Cl-Na and CP -Cl-NH (Fig (2): a and b) in the temperature range 50-200 ° C and 200-400 ° C. This analogy indicates that the thermal decomposition of species hydroxides similarly for CP-CP and Cl-NH-Cl-Na precursors. Fluctuations above 400 ° C can be attributed to loss of oxygen, assigning the evolution of an oxygen-rich spinel phase in a stoichiometric phase during the heat treatment. The oxalate precursor results of the TGA analysis (Fig (2):c) show that the total decomposition is also carried out in two stages, in the temperature range 150-200°C and 200-400°C with a total mass loss of 51% given to the conversion of the mixture of oxalates cash hydrated and CoO-Fe2O3 hydroxides.

Figure.3: SEM-EDX-TEM analysis of the samples prepared using chlorides precursor salts precipitated by the different agents (NaOH, NH4OH and H2C2O4 and calcined at 500 ° C.

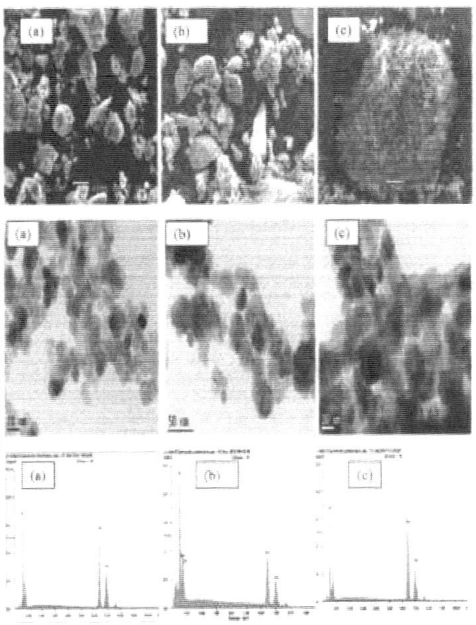

Elements	O	Fe	Co
%At Theorique	57.14	28.57	14.28
(a) CP-Cl-Na-500	45.92	35.41	18.66
(b) CP-Cl-NH-500	43.54	35.55	20.90
(c) CP-Cl-OX-500	59.08	26.77	14.15

Figure 3: SEM-EDX-TEM analysis of the samples prepared using chlorides precursor salts precipitated by the different agents (NaOH, NH_4OH and $H_2C_2O_4$ and calcined at 500 ° C

The SEM/TEM photographs and the results of EDX analysis relating to catalytic systems, obtained at the calcination temperature 500 °C, are illustrated in Fig (3). Distinct morphologies were noted between solids based on hydroxides and the like based oxalate. These are in the form of agglomerates size significantly larger and have a completely different aspect surface. The EDX analysis shows that the catalysts are composed only of the elements cobalt, iron and oxygen, no impurity was detected. However, atomic percentages were determined only in one point, which does not reflect the average stoichiometry of each sample.

Figure.4: Profiles TPR samples prepared using chlorides precursor salts precipitated by the different agents (NaOH, NH4OH and H2C2O4 and calcined at 500 ° C.

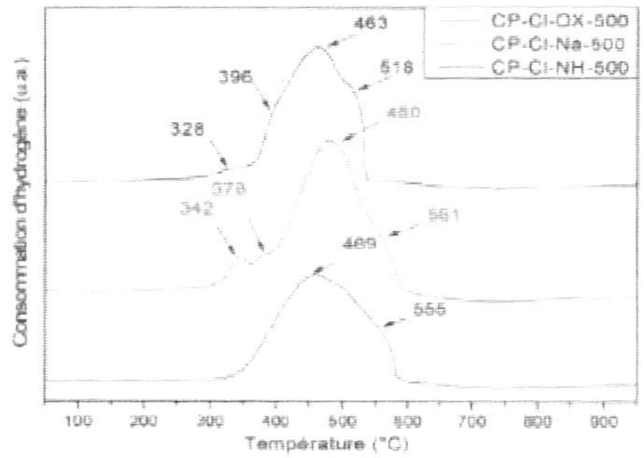

Figure 4 Profiles TPR samples prepared using chlorides precursor salts precipitated by the different agents (NaOH, NH_4OH and $H_2C_2O_4$ and calcined at 500 ° C

The three samples were subjected to TPR analysis, the resulting profiles are shown in Figure (4) The curves of samples CP-Cl-Na-500 and CP-Cl-OX-500 have four peaks reduction whose maxima are located at 342, 378, 480, 561 ° C and 328, 396, 463, 518 ° C respectively. The first step is attributed to the reduction of the Co3 + of the spinel structure in metallic Cobalt (Co 3+ +3/2 H 2 → Co0 + 3H +). The reduction peak CP-Cl-Na-500 is much more intense and appears at a temperature higher than that of CP-Cl-OX-500, indicating a more favorable formation of Co3+ ions during the coprecipitation based hydroxide. The CoO oxide of the spinel phase (CoO.Fe2O3) is reduced to metallic cobalt in the second block. The last two steps correspond to the successive reduction of the oxide Fe2O3 to Fe3O4 magnetite and Fe3O4 to iron metal. The TPR profile of the sample CP-Cl-NH-500 has a difference with the previous. Its shape is wider, which may indicate poor resolution reduction peaks. It has two maxima peaks at temperatures 469 and 555 ° C, respectively assigned to the sequences Fe Fe2O3 Fe3O4.

Table.1: crystallite sizes and specific surface samples prepared using chlorides precursor salts precipitated by the different agents NaOH, NH4OH and H2C2O4 and calcined at 500 ° C.

Simple	CP-Cl-Na-500	CP-Cl-NH-500	CP-Cl-OX-500
crystallite size (nm)	8.97	11.29	19.78
SBET (m2/g)	69.1	32.8	42.7

Table (1) shows the crystallite sizes determined by XRD and surface areas determined by BET analysis. From the results of XRD analysis, the samples have different crystallite size, the largest size is allocated to the oxatate sample. Similarly to the BET analysis, there is a clear influence of the precursor used in the synthesis of solids. Indeed, NaOH base obtained solid has the largest specific surface area.

MN-P-002 [Macromolecules & Nanotechnology]

Preparation and characterization of cobalt spinel ferrite obtained by hydrothermal treatment

Yasmina Bellal Hammiche[1], Laaldja Boukhobza Meddour[1], Amar Djadoun[2], Amel Benadda[1], Marie Hélène Berger[3], Mohamed Sennour[3], Aline Auroux[4]

[1]*Lmccco, Faculté de Chimie, Usthb. Bp32 El Alia, Bab Ezzouar, 16111, Alger, Algérie.*
[2]*Lgp, Fstgat, Usthb, Bp32 El Alia, Bab Ezzouar 16111, Alger, Algérie.*
[3]*Mat, Centre des Matériaux, MINES-ParisTech, France.*
[4]*Cnrs, Université Claude Bernard, Lyon, France.*

Ferrite materials are a large family of mixed oxides which have the potential applications in catalysis [1-2]. Several routes are used for the preparation of spinel ferrites [3-4]. In this study, a series of cobalt spinel ferrite compounds were prepared using coprecipitation method at room temperature followed by a hydrothermal treatment at 150°C for different times varied between 2 and 22 hours. The powders were characterized by Thermogravimetric and Differential Thermal Analysis (TGA/DTA), X-Ray Diffraction (XRD) and Temperature Programmed Reduction (TPR). The TGA/DTA (fig.1) results showed a thermal instability of the precursor and two selected samples, counting a weight loss attributed to a decomposition of hydroxide species, which involves the formation of impure spinel phase during the hydrothermal treatment. XRD study of products (fig.2) obtained at different times revealed the presence of mixture single oxides in addition to the spinel phase. The relative proportion of CoFe2O4 increases from 68,2% after a hydrothermal treatment of 2 hours to a maximum of 93% at a residence time of 16 hours in autoclave as reported in Table 1. The crystallite size and lattice parameters (Tab.1) were found to exhibit an irregular variation with increasing the synthesis time which may indicate a complex progress in the growth of the crystals. This behavior is also observed on the catalytic proprieties of the solids in the total oxidation of ethanol (fig.3). The comparison of the isoconversion temperatures leads to the conclusion that the most active sample is obtained at a hydrothermal treatment time of 5h, it is characterized by the lower crystallite size value and contains 76.1% of CoFe2O4 phase. A correlation between the catalytic activity and the TPR results (fig.4) is attempted, concluding that the lower the reduction temperature is, the lower are the isoconversion temperatures.

Keywords: mixed oxides,Ferrite materials, hydrothermal treatment, coprecipitation

Figure.1. TGA-DTA curves of the solids: precursor (Pr-CP); after 2hours hydrothermal treatment (HT-2h); after 8 hours hydrothermal treatment (HT-8h)

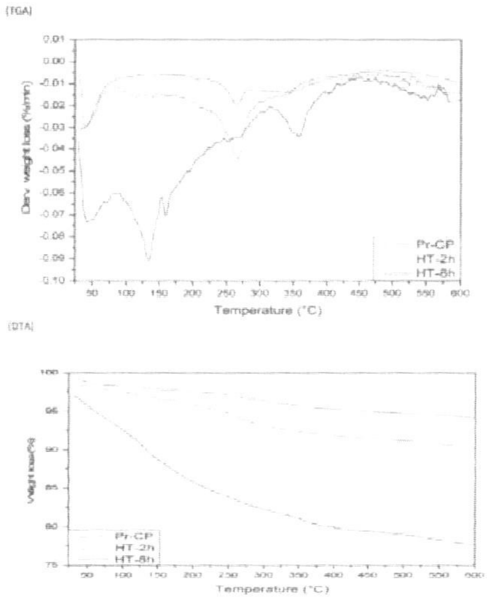

Figure. 2. XRD diffractograms of the hydrothermal prepared solids at 150°C for 2 hours (HT-2h) and 16 hours (HT-16h).

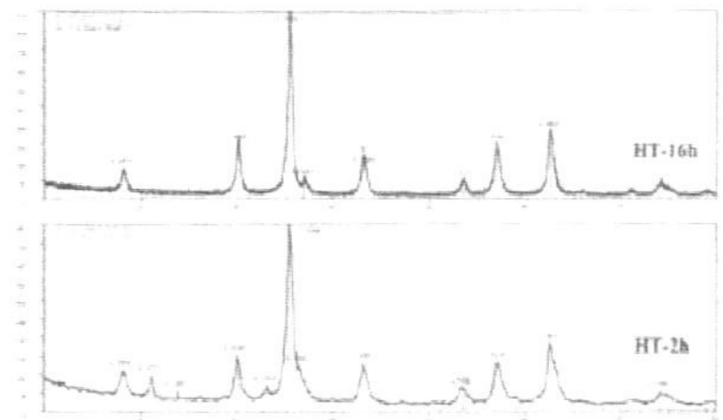

Figure.3. Crystallite size and isoconversion catalysts HT-2h and HT-5h.

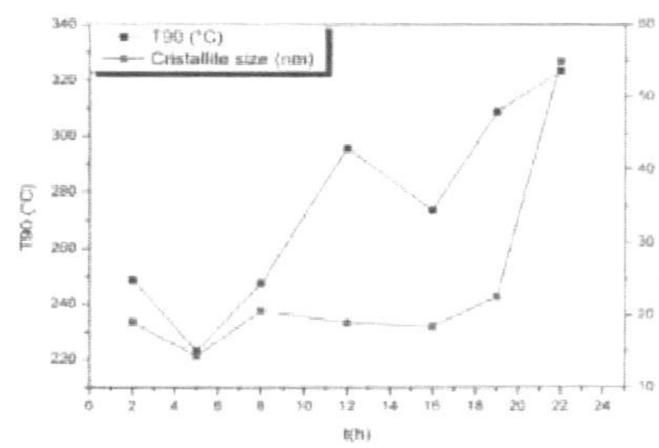

Figure.4. TPR profiles of two selected temperatures vs hydrothermal treatment time.

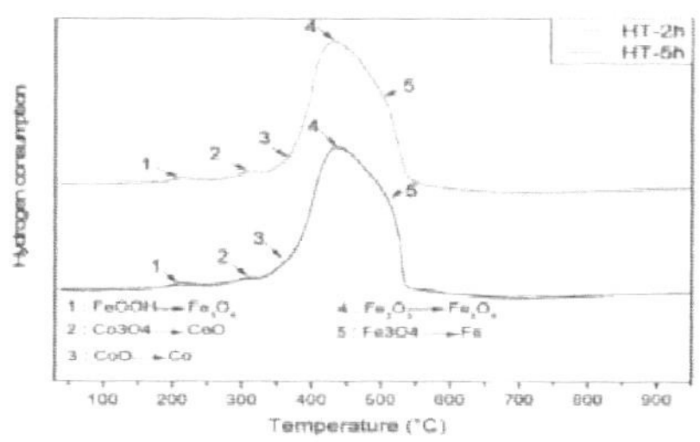

Table.1.Quantitative analysis and, XRD data and temperatures of 90% isoconversion levels of the samples prepared by hydrothermal treatment at different times.

Sample	HT-2h	HT-5h	HT-8h	HT-12h	HT-16h	HT-19h	HT-22h
% CoFe2O4	68.2	76.1	86.3	92.1	93.0	71.9	83.2
Cristallite size (nm)	18.96	14.23	20.53	18.87	18.31	22.49	54.86
L. Parameter (A°)	8.3860	8.3853	8.3490	8.3711	8.3686	8.3903	8.3405
T10 (°C)	177.5	177.5	199.6	224.8	211.7	210	211.7
T50 (°C)	235	209.5	235	271.8	252.8	292.8	308.5
T90 (°C)	248.7	223	247.7	295.7	273.4	308.5	323.3

MN-P-003 [Macromolecules & Nanotechnology]

In-Situ Preparation of Thermoset/Clay Nanocomposites by Thiol-Epoxy Click Chemistry

Ozlem Purut, Mehmet Atilla Tasdelen

Department of Polymer Engineering, Faculty of Engineering, Yalova University, TR-77100 Yalova, Turkey

Polymer/clay nanocomposites are one kind of composite materials containing of nanometer-sized inorganic nanoparticles, typically in the range of 1–100 nm, which are uniformly dispersed in and fixed to a polymer matrix. The inclusion of inorganic nanoparticles into a polymer matrix not only combines their properties but also brings advanced new functions that find applications in many industrial fields. Currently polymer/clay nanocomposites can be prepared by three ways such as solution mixing, melt blending, and in-situ polymerization. Recently, a highly efficient copper (I) catalyzed azide/alkyne cycloaddition "click" reaction, in which exfoliation is rooted in the functional groups of the intercalant that readily react with the antagonist groups of the preformed polymers has been developed. The quantitative efficiency of click reaction coupled with tolerance to a wide variety of functional groups and reaction conditions make this process highly attractive for the nanocomposite preparation.

Another important example of click chemistry reaction is thiol-epoxy coupling reaction, which can be conducted from commercial available and cheap epoxide and thiol compounds under ambient conditions. This chemistry is an efficient and simple tool for the synthesis of various polymers and produces a reactive secondary hydroxyl functionality that can be further utilized in another reactions.[2] Here, we report the synthesis of thermoset/clay nanocomposites by thiol-epoxy click reaction. This approach is conceived to greatly extend the synthetic capabilities of polymer/clay nanocomposites by careful choices of organic clays and polymers, and the optimization of the synthesizing process to deliver the biggest benefit.

This work was financially supported by Yalova University Research Fund (Grant No: 2015/YL/055)

[1] Tasdelen MA, Van Camp W, Goethals E, Dubois P, Du Prez F, Yagci Y. Macromolecules, 2008:41(16):6035-6040.
[2] Cengiz, N. Rao, J. Sanyal, A. Khan, A., Chem. Commun., 2013, 49, 11191

Keywords: polymer/clay nanocomposites, click chemistry, thiol-epoxy

Figure 1. Images of thermoset/MMT nanocomposites.

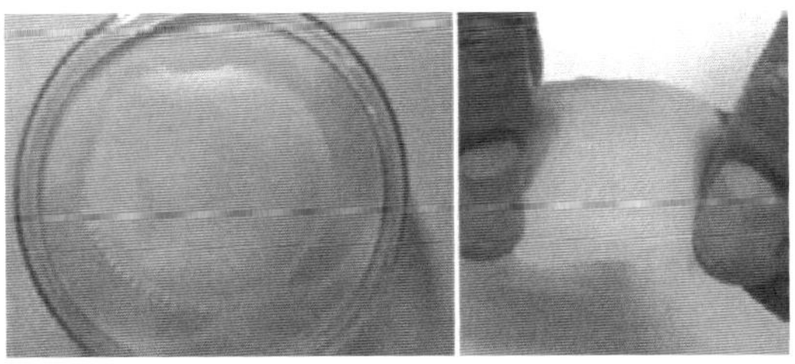

MN-P-004 [Macromolecules & Nanotechnology]

Synthesis and characterization of nanocomposite materials based on poly(4-Vinylpyridine) and two different Algerian clays: via in situ polymerization

Fayçal Dergal[1], Djahida Lerrari[1], Ali Mansri[2], Kheldoun Bachari[1]

[1]*Centre de Recherche Scientifique et Technique en Analyses Physico-Chimiques (C.R.A.P.C), Tipaza, Algeria*
[2]*Laboratoire d'Application des Electrolytes et des Polyélectrolytes Organiques (L.A.E.P.O), University of Tlemcen, Algeria*

A significant number of investigations have been reported on the elaboration and characterization of Polymer/Clay nanocomposites, via differents methods[1-5]. The poly(4-vinylpyridines) are an important matrix which exhibiting interesting properties due to the presence of the nitrogen atom in the pyridine ring [6].

The weakly basic nitrogen atom enables a variety of reactions on vinyl-pyridine polymer such as reactions with ac-ids, quaternization, and complexation of metals. New composites materials were successfully prepared by in-situ polymerization of 4-vinylpyridine in presence of two different types of Algerian modified clays (Maghnia and Mostaganem), noted (BC) and (MC), respectively, at different percentage of inorganics (1 wt%, 3 wt% and 5 wt%).

These materials were characterized using, Thermogravimetric analysis (TGA), Differential Scanning Calorimetry (DSC), Fourier Transformed Infrared (FTIR), Scanning Electron Microscopy coupled with Energy dispersive X-Ray Spectroscopy (SEM-EDX) and RMN-1H.

References:

[1] Nikolaidis A.K, Achilias D.S, Karayamnidis G.P, Effect of the types of organic modifier on the polymer-ization kinetics and the properties of poly(methylmethacrylate)/organomodified montmoriillonite nanocompo-sites. Eur.Polym. j.48, 240-251.
[2] Ouaad K, Djadoun S, Vincent L, Sbirrazzuoli N, (2013), Elaboration and thermal behavior of nanocomposites based on poly(ethylmethacrylate) and an Algrian bentonite prepared via in situ polymerization initiated by Ni(II) α-benzoinoxime complex, Thermo Acta, 30-36.
[3] F.Dergal, A.Mansri and L.Billon (2013), Characterization by TGA and Uv-visible of New Pigment Materials Containing Mica, P4VP and D&C Red 6 Dye, Chem Eng trans, 32, 2131-2136.
[4] L. Ghannam, M. Bacou, H. Garay, M. Shanahan, J. François, L. Billon, Elastomer monolayers adsorbed on mica surfaces by nitroxide-mediated polymerization. Polym. 45, 7035 –7045.
[5] L.Ghannam, J. Parvole, G. Laruelle, J. Francois, L. Billon, (2006), Surface-initiated nitroxide-mediated polymeri-zation: a tool for hybrid inorganic/organic nanocomposites "in situ" synthesis, Polym. Inter. (Review). 55, 1199.
[6] A. Mansri, K.I. Benabadji, J. Desbrières, J. François, (2009). Chromium removal using modified poly(4-vinylpyridinium) bentonite salts. Desal. 245, 95–107.

Keywords: Nanocomposite, P4VP, Clay.

MN-P-005 [Macromolecules & Nanotechnology]

Synthesis and Application of Anisotropic Particles

Huarong Liu, Fangwei Wang, Ruikun Wang

CAS Key Laboratory of Soft Matter Chemistry, Department of Polymer Science and Engineering, University of Science and Technology of China, Hefei, China

Anisotropic particles with morphological and/or compositional anisotropies (also termed Janus particles) have tremendous potential use in scientific research and frontier technology, such as biomedical applications, sensors, photonic crystals, optical devices, colloidal stabilizer and so on. To meet the need of different applications, various anisotropic particles with different sizes, nature and morphologies have been synthesized via seeded emulsion polymerization or seeded dispersion polymerization, sometimes as well combining with sol-gel method and the synthetic approach of inorganic nanoparticles by controlling the nature and the amount of the seeds, the nature and the amount of the second monomers, the type and the concentration of emulsifier or dispersant, the absorbed dose and dose rate, temperature and other factors. These various anisotropic particles can be used as solid surfactants to stabilize emulsions and further to fabricate hierarchical microspheres or porous materials. Meanwhile, some raspberry-like particles can be used to form transparent film with fire retardant properties or to build superhydrophobic surfaces with a high adhesive force to water, and thus to form superhydrophobic surface having antibacterial ability after being further decorated by Ag nanoparticles.

Acknowledgment: We thank the National Natural Science Foundation of China (Project No. 21074122 and 51373160) for financial support.

Keywords: Anisotropic particles, Pickering emulsions, Hierarchical structures, Seeded polymerization

MN-P-006 [Macromolecules & Nanotechnology]

Effect of precursors influence on the distribution of active fractions of nanocatalysts

Tokhir Khakim Rakhimov, Mukhtorjon Ganievich Mukhamediev

National University of Uzbekistan

The influence of various factors on boundaries of nanoproperties studied by method of calculation of the boundary dimensions in disperse systems [1] which opens up new prospects for the synthesis of nanosystems with specified characteristics. Whether the media influences the nature of the active fractions to size, or active particles are the same size regardless of the nature of the carrier? The latter is the most probable, especially if used as carrier sufficiently inert materials like carbon fibers. In this paper we studied the palladium nanocatalysts on different carbon fibers: Mtilon and Karbopon. Synthesized a two series (100 samples), differing mean particle sizes and their size dispersion. The samples were subjected to a soft recovery by hydrogen. We measured the initial rate of reaction and the initial values of the activity. A calculation of the values of the boundaries of the activity [2]. To do this, set the value consistently lower and upper activity boundaries (Fig. 1). As shown by the data presented, contrary to expectations, the nature of the support has a significant impact on the size of the boundary of the active fractions. Karbopon nanocatalysts have to be observed not only in the direction of displacement of the active particles with a smaller diameter, but also a significant narrowing of the interval of diameters within which nanoparticles retain activity. Recovery results in a continuous narrowing of the interval, and consequently the total activity becomes lower with increasing content of the recovered phase, possibly a metal lattice. Mtilon possible to obtain the supported catalysts with a broad size distribution, but the activity exhibit unusually large particle sizes close to their micron size. Perhaps the active particles are in two non-contiguous areas sizes are very small and quite large, between which there are low-active fraction of particles.

Keywords: nanocatalysts, palladium, boundaries sizes, carbon monoxide, oxidation, carbon fibers

Figure 1.

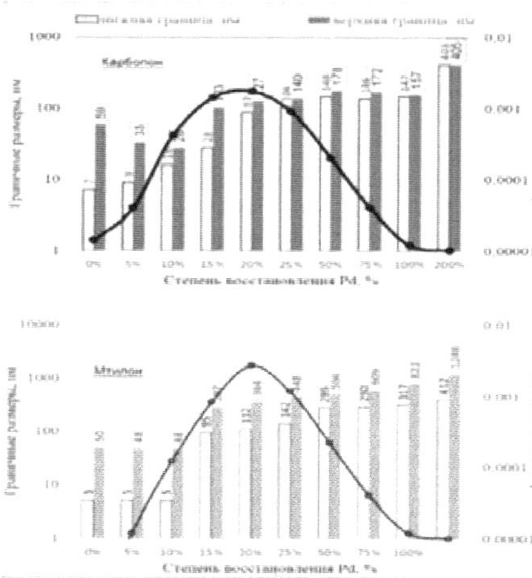

The lower and upper boundaries of the active fractions and the initial value of activities at various palladium nanocatalysts UVM in CO oxidation.

MN-P-007 [Macromolecules & Nanotechnology]

Core shell carbon nanotubes/silver monohybrids coated with polypyrrole designed via interfacial chemistry of aryl diazonium salts and ultrasonic pathway

Mekki Ahmed[1], Achour Sabrina[1], Ihdene Zaher[1], Mahmoud Rachid[1], Chehimi Mohamed.mehdi[2]

[1]Ecole Militaire Polytechnique, BP 17, Bordj El Bahri, 16111 Alger Algérie
[2]Institut de Chimie et des Matériaux Paris Est – Equipe Systèmes Polymères Complexes, UMR 7182 CNRS–Université Paris Est Créteil, 2-8 rue Henri Dunant, 94320 Thiais, France.

A simple and versatile synthesis process of novel composites consisting of multi-walled carbon nanotubes (MWCNTs), silver nanoparticles and polypyrrole were carried out through two facile steps. Firstly, the surface of MWCNTs was modified via an in-situ diazonium generation process, this functionalization helped to easily disperse the MWCNTs in aqueous solutions, then after, without using any supplementary oxidant or reduction agent, an in-situ pyrrole polymerization were conducted onto the top of surface modified MWCNTs accompanied with a simultaneous reduction of Ag nanoparticles by application of ultrasound irradiations waves at room temperature. Polypyrrole layer formation and Ag nanoparticles incorporation on the surface of MWCNTs were confirmed by FTIR, Raman and XRD. The SEM and TEM images revealed highly dispersed surface modified MWCNT/Ag/PPy nanocomposites with nanotubular morphologies having an average diameter of 15-30 nm. The electrical conductivity of the formed nanocomposite was found to be higher than of pyrrole/silver and nanocomposite based crude MWCNTs. The prepared nanocomposite will certainly find several applications in different area such as gas sensing, catalysis and electronic devices.

Keywords: aryl diazonium, carbon nanotube, polypyrrole, silver nanoparticles, nanocomposites, ultrasonic irradiation.

MN-P-008 [Macromolecules & Nanotechnology]

Large-Area Reduced Graphene Oxides with Excellent Thermal Conductivity and Electromagnetic Interference Shielding Effectiveness

Soon Man Hong, Pradip Kumar, Chong Min Koo

Korea Institute of Science and Technology

This presentation demonstrates that reduced large-area graphene oxides (rLGO) with maximum surface area of 1592 μm2 work were fabricated through a cost-effective chemical reduction process at low temperature. The product revealed large electrical conductivity of 243±12 S cm−1 and thermal conductivity of 1390±65 W m−1 K−1, values much superior to those of a conventional reduced small-area graphene oxide (with electrical conductivity of 152±7.5 S cm−1 and thermal conductivity of 900±45 W m−1 K−1). The rLGO thin film also exhibited not only excellent stiffness and flexibility with Young modulus of 6.3 GPa and tensile strength of 77.7 MPa, but also an efficient electromagnetic interference (EMI) shielding effectiveness of ~ 20 dB at 1 GHz. The excellent performance of rLGO is attributed to the fact that the larger area LGO sheets include much fewer defects that are mostly caused by the damage of graphene sp2 structure around edge boundaries, resulting in large electrical conductivity. The manufacturing process of rLGO is an economical and facile approach for the large scale production of highly thermally conducting graphene thin films with efficient EMI shielding properties, greatly desirable for future portable electronic devices.

Keywords: Graphene, Polymer Composite, EMI shielding, Thermal Conductivity

MN-P-009 [Macromolecules & Nanotechnology]

The Preparation of Some New 1H-Tetrazole Derivatives and Their Attaching on Nano Spheres

Nurdan Kurnaz Yetim, Elvan Hasanoğlu Özkan, Murat Gümüş, Ali Dişli, Nurşen Sarı

Department of Chemistry, Faculty of Science, Gazi University, 06500 Teknikokullar, Ankara, Turkey

Tetrazoles which represent a significant class of heterocycles have attracted considerable interest in recent years because of their wide utility [1]. Tetrazoles have been successfully used in various material sciences and synthetic organic chemistry as analytical reagents and synthons [2]. So, synthesis of this heterocyclic nucleus is of much current importance. Also, in recent years, the chemistry of polymer-based macromolecules has attracted considerable attention due to their applications, including as stabilizers in dispersions and drug delivery carriers [3]. Therefore, it is very important the preparation of various spheres for use in various fields. As far as we can see from the literature, no studies have been carried out on nanoparticles attached 1H-tetrazole derivatives. This study aimed to fill in this gap. Structure of the synthesized compounds were illuminated by spectroscopic methods. The 13C-NMR and 1H-NMR spectra data of the Schiff bases are also in accordance with the proposed structures. SEM images of modified polymers were not markedly different from those of (2AEPS). According to GPC, modified polymers have a very narrow Mn (PDI: 1.20, 1.10, and 1.08 for 2AEPS-(o-F-Tet-1H), 2AEPS-(m-F-Tet-1H) and 2AEPS-(p-F-Tet-1H), respectively.

References:

1. G. D. Çelik, A. Dişli, Y. Öner, L. Açık, Med Chem Res, 2013, 22, 1470.
2. F. Abrishami, M. Ebrahimikia, F. Rafiee, Appl. Organometal. Chem. 2015, 29, 730.
3. E. Bozkır, N. Sarı, H. Öğütcü, Journal of Inorganic and Organometallic Polymers and Materials, 2012, 22, 1146.

Keywords: Acetylcholinesterase, Nanoparticule, Tetrazole

Synthesis Rotation of 1H-tetrazole derivatives attached on nanoparticles

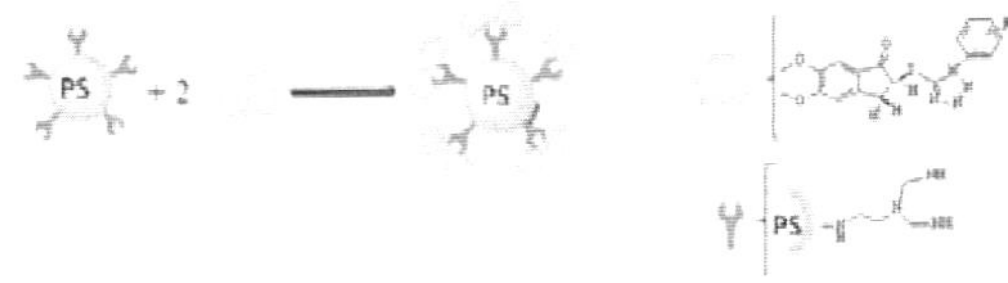

MN-P-010 [Macromolecules & Nanotechnology]

Dendrimers: Synthesis, Immobilization and Applications of Pesticide

Nurdan Kurnaz Yetim, Nurşen Sarı

Department of Chemistry, Faculty of Science, Gazi University, 06500 Teknikokullar, Ankara, Turkey

It is known that, Acetylcholinesterase (AChE) is the primary target of inhibition by organophosphorus compounds. So, exposure to even small amounts of organophosphorus compounds can be fatal.OPs play an important role in increasing agricultural productivity due to their high insecticidal activity[1]. Nowadays, monitoring of AChE activity has been employed as a mean for qualitative or quantitative measurement of organophosphates[2]. Therefore, the detection of OPs has become increasingly necessary and it is important to develop access method for the rapid detection of OPs.It is known various solid supports for immobilization investigation.But, investigations are not common use of dendrimers as support.Dendrimers are monodisperse synthetic polymers with a regular and highly branched three-dimensional structure.Dendrimers are "hot topics" of nanotechnologies like carbon nanotubes.The most important characteristics of dendrimers are their highly functionalized perfect surface for enzyme immobilization[3]. In this study, new support material existing in the structure of the dendrimer was prepared and used for the determination of the pesticides. Immobilized AChE may use for determination of phosmet insecticide from colour change or spectral shift.As a result of this study, the amount of pesticides used against codling moth in agricultural land has been identified as qualitative thanks to prepared immobilized material support.

References:

[1]M.Wei,J.Wang, Sensor.Actuat.B-Chem.211(2015)290–296.
[2]R.R.Dutta,P.Puzari, Biosens.Bioelectron.52(2014)166–172.
[3]A.Kuranl,A.Salimi, Biosens.Bioelectron.50(2013)186–193.

Keywords: Dendrimer, AChE, Immobilization

Second generation dendrimer including azomethine

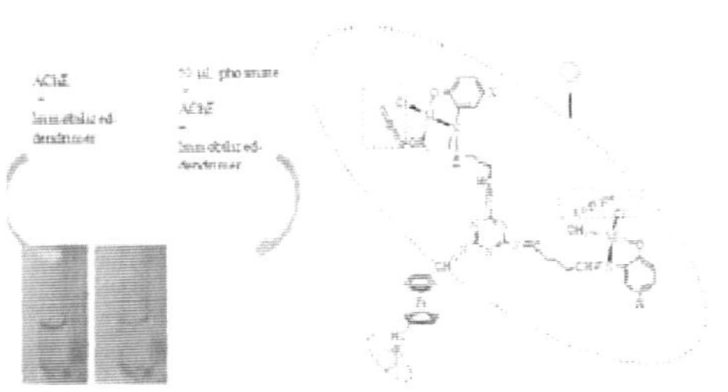

MN-P-011 [Macromolecules & Nanotechnology]

Immobilizaiton of Acetylcholinesterase on Macromolecule for Pesticide Control

Elvan Hasanoğlu Özkan, Nurdan Kurnaz Yetim, Murat Gümüş, Ali Dişli, Nurşen Sarı

Department of Chemistry, Faculty of Science, Gazi University, 06500 Teknikokullar, Ankara, Turkey

Organophosphorus (OP) compounds, commonly used as insecticides [1]. The inhibition of acetylcholinesterase (AChE) activity by OP compound is an irreversible process. In this process, AChE is inactivated. (AChE) plays a crucial role in synaptic transmission by controlling the action of acetylcholine. Reduction in the density of (AChE) leads to failures of neuromuscular transmission. Acetylcholinesterase (AChE) is a serine protease enzyme. The inhibition of AChE catalytic activity by OP is caused due to phosphorylation of serine residue[2]. In this work, (AChE) immobilized on nanoparticles attached tetrazol (Fig. 1) for practical determination Phosmet (N-(Mercaptomethyl) phthalimide S-(O,O-dimethyl phosphorodithioate)). Phosmet is organophosphate insecticide used on plants and animals. In this work, AChE immobilization has been successfully fabricated for the detection of pesticide. Especially, as a result of research on a phosmate insecticide, our developed immobilization strategy would be advantageous for various industrial applications.

This work was supported by the Gazi University Research Fund (05/2013-03 and 05/2014-02).

References:

1. A. Gahlaut, A. Gothwal, A. K. Chhillar, V. Hooda, Journal of Applied Biosensor, 2012, 1, 1-8.
2. S. Z. Ul Hassan, J. Militky, American Journal of Analytical Chemistry, 2012, 3, 93-98.

Keywords: AChE, Nanoparticule, Hazardous Materials, Tetrazole

AChE immobilized on nanoparticles attached tetrazole

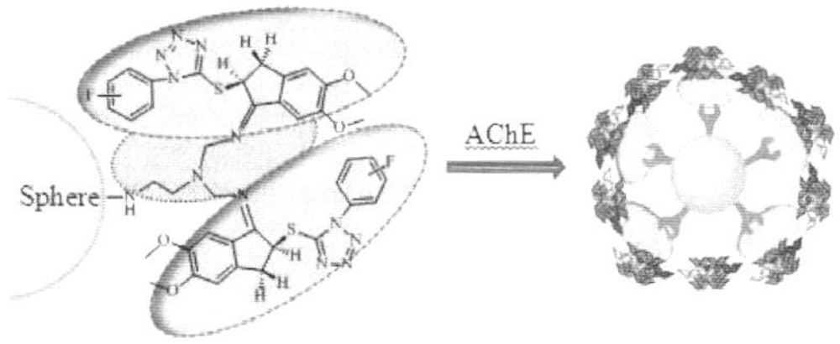

MN-P-012 [Macromolecules & Nanotechnology]

Synthesis of Novel Dendron Based on Ferrocene, Immobilization of Enzyme and First Investigation as Catalytic of Glucose in Artificial Urine

Elvan Hasanoğlu Özkan, Nurşen Sarı

Department of Chemistry, Faculty of Science, Gazi University, 06500 Teknikokullar, Ankara, Turkey

Over the past three decades, new materials have been prepared for biotechnological applications. The technology, hard or soft materials are remarkable. As soft materials, dendrons have been used in biosensor[1]. In biotechnological applications, immobilization is importance on support of enzymes, proteins and cells. Aromatic dendritic materials are one from used as support in enzyme immobilization.For enzyme immobilization, supports involving ferrocene and its derivatives are important due to high chemical stability, which are electron transfer mediators. E.t.m. is important for reductase enzymes. GOx is an reductase enzyme and is a well-studied, commercially available redox enzyme in medical and industrial applications for glucose sensing[2].

In this work, dendrons were prepared and then GOx immobilized on them. Results showed that, interaction enzyme-substrate in dendrons including Pt(II) and Pd(II) is better than FcMTpal. Furthermore, first use was investigated on determination of glucose in artificial ürine of immobilized-dendron. FcMTpal-AHPt-GOx was used reusability 10 times and the residual activity was about 83.09 % its initial activity.

References:

1.P.Díez,C.G.Piuleac, M.R.Paloma, S.Romano, M.Gamella, R.Villalonga, J.M.Pingarrón, Anal Bioanal Chem.2013,405,3773.
2.M. Şenel and N. Cevdet, Synthetic Met.2012,162,688.

Keywords: Dendron, Artifical urine, GOx

Synthesis of extended second generation dendrons

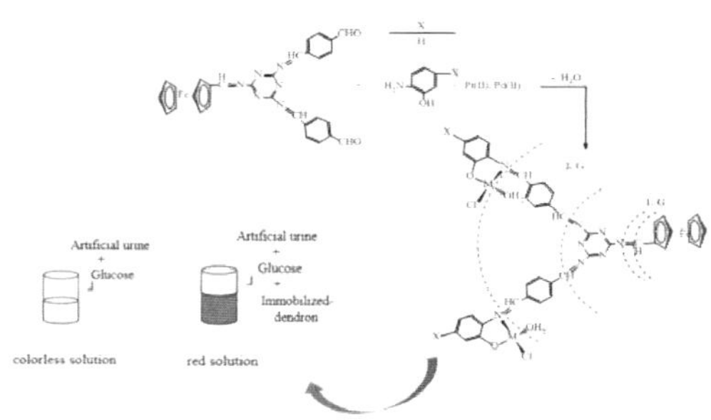

MN-P-013 [Macromolecules & Nanotechnology]

Polymeric submicron particles via self-assembly and crosslinking of double hydrophilic poly(ethylene oxide)-b-poly(N-vinyl pyrrolidone) in aqueous solution

Jochen Willersinn, Markus Antonietti, Bernhard V. K. J. Schmidt

Department of Colloid Chemistry, Max-Planck Institute of Colloids and Interfaces, Potsdam, Germany

The synthesis of novel vesicular systems bearing a water permeable shell with adjustable pore sizes is of high interest in biomedical and polymer science. Vesicles derived from amphiphilic block copolymers lack of this ability due to the hydrophobic character of the inner membrane. In contrast to amphiphilic block copolymers, the use of purely hydrophilic block copolymers enables an access to the desired particle features. However, the self-assembly of these block copolymers to suprastructures in aqueous media is strongly dependent on the difference in hydrophilicity[1] of the different blocks and yet only few examples of self-assembly of purely hydrophilic block copolymers were reported.[2,3] Here, we present an access to purely hydrophilic submicron particles based on poly(ethylene oxide)-*b*-poly(*N*-vinyl pyrrolidone) diblock copolymers. Block copolymers were synthesized with reversible deactivation radical polymerization (RDRP) using a PEO-xanthate as macro RAFT/MADIX chain transfer agent. With increasing polymer concentration in aqueous solutions, the block copolymer is able to self-assemble into spherical structures with apparent hydrodynamic diameters in the range between 200 nm and 2 μm. The self-assembly can be further improved by copolymerization with a more hydrophilic monomer,*N*-vinyl imidazole. Submicron particles obtained from PEO-*b*-P(VP-co-VIm) can be preserved by crosslinking with 1,4-diiodo tetraethylene glycol, forming imidazolium units as linkages. These crosslinked particle solutions can be highly diluted. Furthermore, almost quantitative formation of submicron particles with an average diameter of 200 nm can be afforded by self-assembly in the polar organic solvent DMF. Particles formed and crosslinked in DMF can be easily transferred to an aqueous system by dialysis. These particles show a promising ability to incorporate various molecules, such as dyes.

(1) Taubert, Andreas, et al. *Chem. Comm.*, 2004, *19*, 2170-2171.
(2) Casse, Olivier, et al. *Macromolecules*, 2012, *45*, 4772-4777.
(3) Brosnan, Sarah M., et al. *Angew. Chem. Int. Ed.*, 2015, *54*, 9715-9718.

Keywords: double hydrophilic, self-assembly, RAFT polymerization, vesicles

DLS size distribution comparison

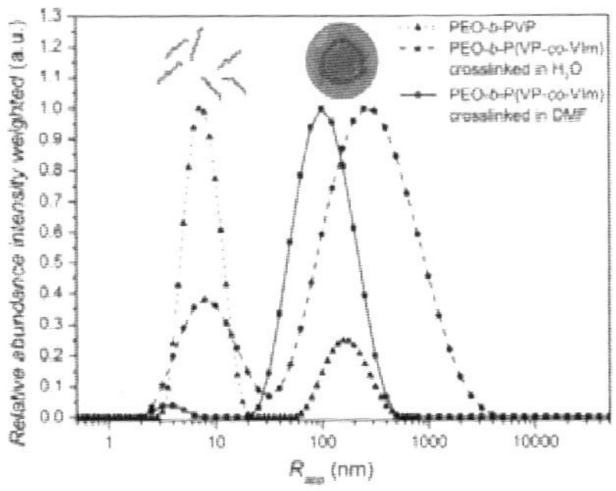

Size distributions of self-assembled block copolymers obtained via DLS

MN-P-014 [Macromolecules & Nanotechnology]

Synthesis, Characterization and Antimicrobial Evaluation of Silver nanoparticles embedded Alkyd resin derived from Neem seed oil

Haruna Musa[1], Aminu Magaji[2]

[1]*Department of Pure and Industrial Chemistry, Bayero University Kano, Kano State, Nigeria*
[2]*Department of Chemistry, Sule Lamido University Kafin-Hausa, Jigawa State, Nigeria*

AgNP was synthesized by reducing 0.1M C7H5O2Ag with 2ml of neem leaf extract, in an eco-friendly, cost effective process. AgNP was characterized by UV-visible spectrophotometer at maximum absorption of 435nm and SEM which reveals the spherical shape of the nanoparticles. The AgNP was tested to have strong zone of inhibition in the antibacterial and antifungal isolates used by disc diffusion method. Neem seed oil (NSO) was synthesized to alkyd resin by alcoholysis and esterification reaction methods and was characterized by FTIR, acid value, solubility and viscometric tests. Disc diffusion method was also used in the antimicrobial screening of the neem seed oil and its respective alkyd resin. The alkyd resin of Neem seed oil was used in the paint formulation with %PVCs of 3%, and is classified as automotive clearcoat. Chemical resistance, scratch resistance, light fastness, and drying schedule tests were conducted on the AgNP incorporated paint which was found to exhibited excellent properties than the conventional paint. Antimicrobial evaluation of the paint incorporated with AgNPs reveals more inhibition zones than a conventional one.

Keywords: AgNP, C6H5O2Ag, PVC%, NSO, NSOR, Paint

MN-P-015 [Macromolecules & Nanotechnology]

New complex compounds of functionalized Dehydroacetic acid units and N,N dimethyl amino benzaldehyde: the synthesis, catecholase study and electrochemical properties

Tabti Djitli Salima[1], Djedouani Amel[2], Djedouani Amel[3], Rahmouni Samra[4], Romdhane Samir[5]

[1]*Faculte des Sciences et Technologie, Universite Mohamed el Bachir El Ibrahimi, El Anasser, 34000, Bordj Bou Arreridj, Algeria*
[2]*Laboratoire de Physicochimie Analytique et Cristallochimie des Materiaux Organometalliques et Biomoleculaires, Universite Constantine 1, 25000, Constantine, Algeria*
[3]*Ecole Normale Superieure de Constantine, Ville Universitaire Ali Mendjeli, 25000, Constantine, Algeria*
[4]*Faculte des Sciences, Universite Sétif –el baz- 19000, Sétif, Algeria*
[5]*Département de Physique, Faculté des Sciences de Tunis. Campus Universitaire 2092 El Manar Tunis, Tunisie*

In this study, the treatment of α,β-unsaturated carbonyl compounds, obtained by the reaction of dehydroacetic acid (DHA) and heteroaromatic aldehyde results in the formation of a heterocyclic compound, named 4-hydroxy-3-[(2E)-3-(4- (dimethylamino)phenyl)prop-2-enoyl]-6-methyl-2H-pyran-2-one (HL) in good yield that depending upon the reaction conditions. The ligand HL behaves as a monoanionic bidentate OO chelating agent. Its structure was characterized using the usual spectral characterization such as electronic absorption, by FT-IR, and 1H NMR in addition to mass and elemental analyses. M(II) coordination compounds of Cu, Co and Ni with a this ligand (HL) derived from the condensation of DHA with aromatic amine were synthesized and characterized by several techniques, including elemental analysis, UV-Vis, FT-IR, EPR and 1HNMR spectral studies. The analytical values indicated that the water and ammoniac fractions coordinate to the metal ions. Based on these studies, the general formulae [M(L)2(H2O)] (M(II) = Cu) and [M(L)2(NH3)4] for (M(II) =Ni and Co) are proposed for the complexes. Electrochemical behavior of all these complexes has been investigated by cyclic voltametry on glassy carbon electrode in acetonitrile at 100 mV/s scan rate. This study indicate that the reduction process corresponding to CuII/CuI is electrochemically quasi-reversible in complex of copper, and irreversible reduction process for NiII/NiI and CoII/CoI in complexes of nickel and cobalt respectively. Cathecolase study of HL reveals an important result for the oxidation of catechol.

Keywords: α, β –unsaturated ketones, Electrochemistry, DRX, catechol, cyclic voltametry

MN-P-016 [Macromolecules & Nanotechnology]

Preparation and Evaluation of a Novel Bio-Based Waterborne Polyurethane Nanoparticles for Sustained Delivery of Raloxifene- Hydrochloride

Mohammad Reza Nabid, Niloofar Babanejad, Ismail Omrani

Department of Polymer, Faculty of Chemistry, Shahid Beheshti University, Tehran, Iran

In the present study, a nanoparticulate system based on bio-based waterborne polyurethane for the sustained delivery of raloxifene-hydrochloride, a drug with poor bioavailability, is described. Polyurethane (PU) was synthesized from bio-based polyol, which was derived from sunflower oil. Nowadays production of PUs from bio-based polyols has attracted much attention. In approaching efficient PUs, which are highly robust as well as non-toxic, preparing these biodegradable polymers from bio-based materials is a good idea. Having considered the biocompatibility and biodegradability of bio-based PUs and their wide applications in biomedical science, in the current work, we synthesized a novel waterborne bio-based PUs from biodegradable polybutylene adipate and bio-based segments; and investigated their potential for the delivery of raloxifene hydrochloride. The nanoparticles were evaluated for in vitro release. The drug release last up to two weeks and was mainly driven by diffusion-based mechanism. The structure of bio-based PU was characterized using 1HNMR, FT-IR, and TGA. In addition, the size and morphology of nanoparticles were investigated by DLS and SEM, respectively. Degradation of nanoparticles during the release of raloxifene hydrochloride was investigated applying SEM and caused mainly through the hydrolysis of its carrier. The bio-based PU has potential as a useful carrier for sustained release of raloxifene hydrochloride.

Keywords: Bio-based, Polyurethane, Raloxifene hydrochloride, Nanoparticles, Sustained Release

SEM images of degraded NPs, First day (a), 5 (b), 10 (c) and 14 (d) days incubation in 0.1 % polysorbate 80 in water, at 37 ºC).

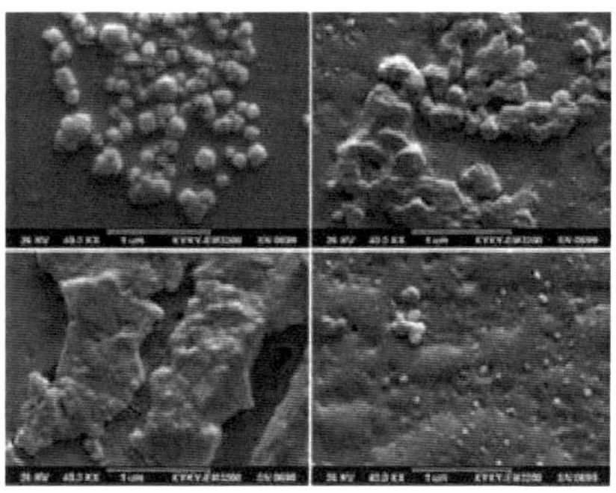

The synthesis route of Bio-based PU

MN-P-017 [Macromolecules & Nanotechnology]

Polymerization of preorganized amphiphilic assemblies

Saheli Chakraborty, Subramanium Ramakrishnan

Department of Inorganic and Physical Chemistry, Indian Institute of Science, Bangalore, India

Amphiphiles are an interesting class of molecules that spontaneously self-assemble to generate a wide variety of elegant nano-structures depending on the relative volume fractions of mutually immiscible segments and these have potential applications as templating media in nanotechnology, nano-containers for reactions, drug delivery vehicles and so on. Since the amphiphilic assemblies are dynamic in nature, reactive amphiphiles with polymerizable group either at the hydrophobic tail or at the hydrophilic head are generally used to make the nano-structures more robust (Jin et al. *Acc. Chem. Res.*, 2001, *34*, 973). The objective of our study is to design and synthesize an itaconic acid-based polymerizable amphiphile containing cetyl chain and polyethylene glycol (PEG) chain as hydrophobic and hydrophilic units, respectively; wherein the polymerizable unit is located at the interface of these two segments. Preorganization of these amphiphiles in water forces the collocation of the sterically congested itaconate double bond and increases its polymerizibility when compared to polymerization in solution. Polymerization of these assemblies generated different types of robust nano-structures depending on the initial monomer concentration and these are found to be stable against dilution. The amphiphile has also been investigated as a *'surfmer'* in emulsion polymerization of styrene to generate PEGylated polystyrene beads. We have further extended our idea of polymerizable nano-assembles from solution to the solid state with the intention of exploring 2-D polymerization to generate sheet-like structures. In my presentation I will present some of our recent findings.

Keywords: Amphiphile, Preorganization, Emulsion polymerization, 2-D polymerization

Preorganization of polymerizable amphiphiles

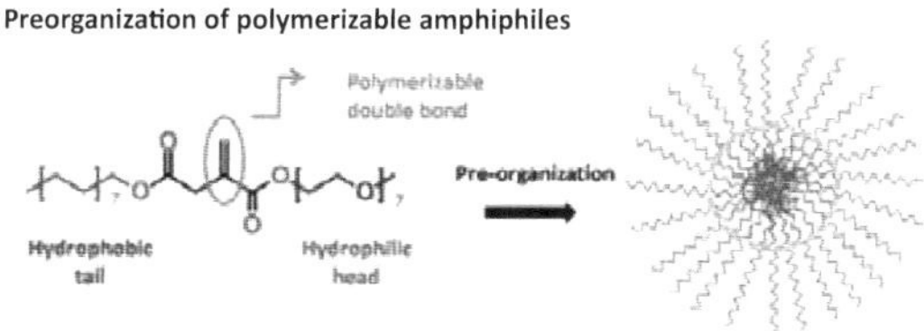

MN-P-018 [Macromolecules & Nanotechnology]

Antimicrobial and antifungal potential of ecofriendly synthesized silver nanoparticles using Lagenaria siceraria fruit

Amara Dar[1], Parsa Dar[2], Usama Waqas[3]

[1]*Center for Undergraduate Studies,University of the Punjab,Lahore.Pakistan*
[2]*Department of Chemistry,Quaid-e-Azam University,Islamabad,Pakistan*
[3]*Government College University Lahore,Pakistan*

The field of nanotechnology is developing and influencing human life very rapidly. Using plants for the synthesis of nanoparticles, is evolving as a new branch of nanotechnology. In plants, wide-ranging bio-molecules act as reducing and capping agents, therefore, increases the reduction rate and stabilization of nanoparticles. Present study deals with the green synthesis of silver nanoparticles using methanolic extract of Lagenaria siceraria fruit mesocarp and its various fractions. The formation and characterization of silver nanoparticles is established by UV-Visible spectroscopy, Fourier transform infrared spectroscopy (FTIR), Scanning electron microscope (SEM), Energy-dispersive X-ray spectroscopy (EDX) and X-ray diffraction (XRD). The antimicrobial assay is conducted against bacterial strains including Pseudomonas aeruginosa, Escherichia coli, Staphylococcus aureus and fungal strain including Candida albicane by agar well diffusion method.

Keywords: Silver Nanoparticles, Lagenaria siceraria fruit, Green Synthesis, Antibacterial, Antifungal.

MN-P-019 [Macromolecules & Nanotechnology]

Sensing of Crack and Crack Size in CNT Buckypapers

Tulay Bal Demirci[1], Elif Avcu[1], Fatih Turan[2], Koray Gürkan[3], Yener Taşkın[2], Alaattin Aktaş[2]

[1]*Department of Chemistry, Istanbul University, Turkey*
[2]*Department of Mechanical Engineering, Istanbul University, Turkey*
[3]*Department of Electric and Electronic Engineering, Istanbul University, Turkey*

Buckypaper is a very important material containing entangled networks of carbon nanotubes formed by van der Waals interactions. In this study, crack and crack size in a buckypaper will be investigated by electric resistance method. Buckypapers will be fabricated from the MWCNT (multi walled carbon nanotubes) by using vacuum filtration method. After fabricating, the initial resistance of buckypaper, R0, will be measured. Then, different cracks, those of whose size gradually increased, will be cut by using a sharp tool. The ratio of relative change in resistance (ΔR/R0) in cracked buckypaper will be measured. From this relative changing, the crack and crack size in a buckypaper will be predicted and we will try to find a correlation between crack size and relative change in resistance. This study has been sponsored by The Scientific and Technological Research Council of Turkey (TUBITAK), (Project No: 213M601).

Keywords: Carbon Nanotubes, Buckypaper, Resistance

MN-P-020 [Macromolecules & Nanotechnology]

Preparation and research of multiphase systems: sodium silicate solute/polymer component/ nanoparticles

Shukhrat Kurbanbaev[1], Sukhrob Telyaev[2], Olga Trunilina[2], Sirojiddin Mirzaev[2]

[1]*Higher technical school of fire safety of the Ministry of Internal Affairs of the Republic of Uzbekistan*
[2]*Institute of Ion-Plasma and Laser Technology Academy of Sciences of Uzbekistan*

It is well known that materials not in reaction with liquid glass (sodium silicate) are less than those that enter into chemical interaction with it. Whereby new substances are formed, presenting both practical and theoretical interest [1,2]. In this study, on the first stage, to obtain a bonding agent with good adhesive properties and strength properties, studies were conducted on mixtures, sodium silicate solute with various resinous polymeric substances and polymer emulsions, by means of simple mechanical mixing.

Prepared mixtures have following composition: sodium silicate + aminoformaldehyde resin, sodium silicate + polyvinylacetate, sodium silicate +acrylic emulsion.

When preparing mixtures of above composition, observed good mixability of components. Obtained uniform "mixtures" of components.

epending on the density of sodium silicate and chemical composition of the organic component (on content of the main substance), as well as on volumetric ratio of components, separation time on organic - and inorganic phase made from 3-4 minutes to several days.

Further, on resulting mixtures added the following particles: microcrystalline cellulose, vermiculite mineral fine powders, nanoparticles - SiO2. Studied rheology and thermophysical characteristics of obtained systems.

Reference :

1. Mirzayev S.Z., Kurbanbayev Sh.E., Muzafarov U.T. INTUMESCENT FLAME RETARDANTS: Structure and thermal properties // The book, Germany. LAP LAMBERT Academic Publishing, Saarbrücken, 2015. -129 P.
2. O'zbekiston Respublikasi Intelektual mulk agentligi. RASMIY AXBOROT-NOMA. Tashkent 2015 12(176). IAP 2014 0242. Institute of Ion-Plasma and Laser Technology Academy of Sciences of Uzbekistan, UZ. Composition for the manufacture of multi-stage fire-retardant coating. 17 P.

Keywords: polymer multiphase systems, sodium silicate, nanoparticles

MN-P-021 [Macromolecules & Nanotechnology]

Thermal resistant cellulose fiber via graphene/ polyvinyl phosphonic acid nanocomposite

Zoha Nooralian[1], Mazeyar Parvinzadeh Gashti[2], Zahra Saeedi[3], Izadyar Ebrahimi[4]

[1]*Zoha Noralian.: Yadegar-e-Imam Khomeini (RAH) Branch, Islamic Azad University, P. O. Box 18155/144, Tehran, Iran*
[2]*Mazeyar Parvinzadeh Gashti.: Yadegar-e-Imam Khomeini (RAH) Branch, Islamic Azad University, P. O. Box 18155/144, Tehran, Iran*
[3]*Zahra Saeedi.: Yadegar-e-Imam Khomeini (RAH) Branch, Islamic Azad University, P. O. Box 18155/144, Tehran, Iran*
[4]*Izadyar Ebrahimi.: Yadegar-e-Imam Khomeini (RAH) Branch, Islamic Azad University,P. O. Box 18155/144, Tehran, Iran*

In this study, graphene were stabilized on a cotton surface using vinyl phosphonic acid monomer to produce polyvinylphosphonic acid/graphene nanocomposite. The influence of graphene and synthesized polymer on the performance of the cellulose fiber was investigated using a FTIR spectroscopy, thermogravimetric analyzer and scanning electron microscope. The possible interactions between graphene, polyvinylphosphonic acid, and cotton functional groups at the surface were evaluated by FTIR spectroscopy. The results indicate that the graphene/polyvinylphosphonic acid nanocomposite is able to improve the thermal stability and increase the functionality of cotton. SEM clearly showed graphene distribution on the cotton surface.

References:

1. S. Thakur, N. Karak, Alternative methods and nature-based reagents for the reduction of graphene oxide: A review, Carbon 94 (2015) 224-242.
2. A. Zurutuza, C. Marinelli, Challenges and opportunities in graphene commercialization, Nat. Nanotechnol. 9 (2014) 730–734.
3. B. Fugetsu, E. Sano, H. Yu, K. Mori, T. Tanaka, Graphene oxide as dyestuffs for the creation of electrically conductive fabrics. Carbon 48 (2010) 3340–3345.
4. X. Liu, Z. Qin, Z. Dou, N. Liu, L. Chen, M. Zhu, Fabricating conductive poly(ethylene terephthalate) nonwoven fabrics using an aqueous dispersion of reduced graphene oxide as a sheet dyestuff. RSC Adv., 4 (2014) 23869–23875.
5. M. Shateri-Khalilabad, M. E. Yazdanshenas, Fabricating electroconductive cotton textiles using graphene. Carbohyd. Polym. 96 (2013) 190–195.

Keywords: graphene, cotton, polyvinylphosphonic acid, nanocomposite, Thermal resistant

sem

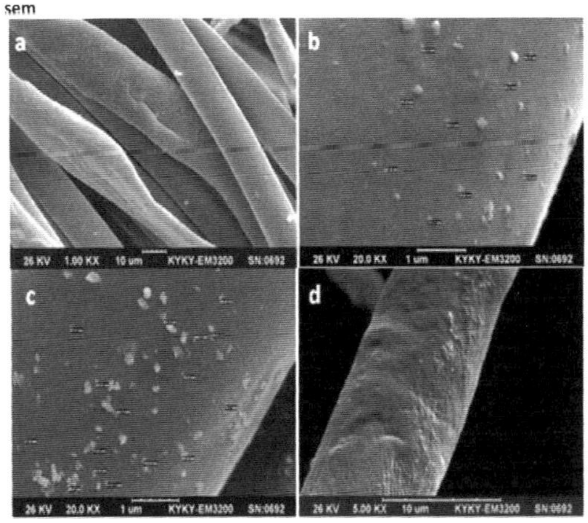

SEM micrographs of: (a) Untreated cotton, (b) Cotton coated with 10 cycles of 10 mg/ml FGN/CTAB colloidal dispersion and polyvinylphosphonic acid, (c) Cotton coated with 10 cycles of 20 mg/ml FGN/CTAB colloidal dispersion and polyvinylphosphonic acid, (d) Cotton coated with polyvinylphosphonic acid.

MN-P-022 [Macromolecules & Nanotechnology]

Synthesis of Gold Nanoparticles with Core Cross-Linked Micelles

Gökhan Solmaz[1], Gökhan Kocak[2], Vural Bütün[1]

[1]*Department of Polymer Science and Technology, Institute of Science, Eskisehir Osmangazi University, Eskisehir, Turkey*
[2]*Department of Chemistry, Faculty of Arts and Science, Eskisehir Osmangazi University, Eskisehir, Turkey*

Nanometals have attracted great attention in the fields of physics, chemistry, and biotechnology owing to their potential technological uses in opticals, medicine, catalysts, and electronic [Henglein A. Chem Rev 1989, Schmid G. Chem Rev 1992, Roucoux A. et al. Chem Rev 2002]. Additionally, stabilization of metal nanoparticles is one of the most important problems in nanoscience and nanotechnology. Synthesis of metal nanoparticles using polymers as a stabilizing agent is very common.

In this study, an amphiphilic poly[(ethylene glycol)methyl ether]-block-poly(glycidyl methacrylate) diblock copolymer (MPEG-*b*-PGMA) was first synthesized via atom transfer radical polymerization (ATRP). ts core-shell micellar solutions (5 and 10 mg L^{-1}) were then prepared in acetone/ water (1:9 v/v) mixture by PGMA block forming micellar core. Core cross-linked (CCL) micelles were synthesized by cross-linking the epoxy functional group of poly(glycidyl methacrylate) block with three different cross-linkers; ethylenediamine, hexamethylenediamine and cysteamine. Systematically, the degree of cross-linking was studied to see its effect on the sweeling behaviour of micellar structure. Their responses to solution conditions were investigated with dynamic light scattering (DLS) studies. These CCL micelles were also successfully used in the synthesis and stabilization of gold nanoparticles (AuNPs) in aqueous media. Transmission electron microscopy (TEM) images showed well-dispersed AuNPs. Nanoparticles were formed within the MPEG-*b*-PGMA spherical CCL micelles. CCL micelles found to be good stabilizer for the NPs in aqueous media.

Acknowledgments: We are grateful for the financial support of Eskisehir Osmangazi University (ESOGU). This work was supported by the Commission of Scientific Research Projects of ESOGU (Grant Numbers 201119006 and 201319A205).

Keywords: Core cross-linked micelle, Glycidyl methacrylate, Diblock copolymer, Gold nanoparticle, ATRP

MN-P-023 [Macromolecules & Nanotechnology]

In-Situ Preparation of Cr2O3 Nanoparticles by Photopolymerization Technique

Hicret Kırtay, Nergis Arsu

Department of Chemistry, Yildiz Technical University, 34220 Istanbul Turkey

Chromium (III) oxide has attracted much attention recently because of their importance both in science and technology. Special attention has been focused on the formation and properties of chromium (III) oxide (Cr_2O_3) which is important in specific applied applications such as in high temperature resistant materials, corrosive resistant materials, liquid crystal displays, green pigments, catalysts, coating materials for thermal protection and wear resistance and antibacterial activity (Jaswal et.al. Oriental J. Chemistry, 2014, 30, 559-566). There have been some ways to obtain Cr_2O_3 nanoparticles, including microwave plasma, decomposition of chromium (III) nitrate solution, laser induced deposition, sonochemical reaction, gas condensation and so on (Z. Pei, Y. Zhang, Material Letters, 2008, 62, 504-506). Nanocomposite materials have been produced via free radical polymerization of acrylic formulations with noble metals such as gold and silver (Yağcı. Y. Sangermano Rizza, G., Chem. Commun, 2008, 2771-2773). The aim of this study was to synthesize chromium(III) oxide nanoparticles in reproducable size and shape by the means of photopolymerization techniques in situ. In-situ photochemical reduction in the presence of various photoinitiators from metal (Cr) salts to metal oxide (Cr_2O_3) nanoparticles was achieved and morphological properties were determined by using SEM techniques.

Keywords: chromium (III) oxide nanoparticle, photopolymerization, polymer.

MN-P-024 [Macromolecules & Nanotechnology]

Polysiloxane Nanofilament Growth on Glass

Esra Kasapgil[1], E. Gozde Atici[1], Ilke Anac[1], H. Yildirim Erbil[2]

[1]*Department of Materials Science and Engineering, Gebze Technical University, Kocaeli 41400, Turkey*
[2]*Department of Chemical Engineering, Gebze Technical University, Kocaeli 41400, Turkey*

Fluorine free, transparent, superhydrophobic polysiloxane nanofilament were succesfully grown on glass using methyltrichlorosilane (MTCS) both via liquid phase reactions in an open system and gas phase reactions under controlled humidity environment. The glass substrates were activated by piranha solution prior to reactions. Wettability, morphology and transparency of the obtained polysiloxane nanofilaments were characterized by contact angle measurements, scanning electron microscopy, UV-vis transmittance tests respectively. Superhydrophobic polysiloxane nanofilament coatings on glass were obtained for the first time in an open system using standard petroleum ether giving advancing contact angle values from 156 up to 168°. Surfaces having contact angle hysteresis as low as 9° were obtained by varying the reaction time and MTCS concentration in petroleum ether solution. Relative humidity (RH) of the medium was found to be the most important factor in the gas phase reactions where 38 % RH was found to be the optimum value for the uniform growth of polysiloxane nanofilaments on glass.

[1] G. R. J. Artus, S. Jung, J. Zimmermann, H. P. Gautschi, K. Marquardt, S. Seeger, Silicone nanofilaments and their application as superhydrophobic coating Adv. Mater. 18 (2006) 2758-2762.
[2] L. Gao, T.J. McCarthy, A perfectly hydrophobic surface J. Am. Chem. Soc. 128 (2006) 9052-9053.
[3] E. G. Atici, E. Kasapgil, I. Anac, H. Y. Erbil, Methyltrichlorosilane Polysiloxane Filament Growth on Glass Using Low Cost Solvents and Comparison with Gas Phase Reactions. Submitted to Thin Solid Films

Keywords: Polysiloxane, nanofilament growth, methyltrichlorosilane, superhydrophobicity

Polysiloxane nanofilament on glass

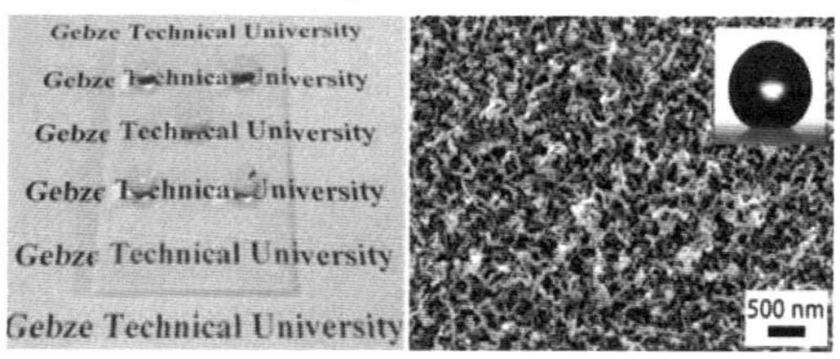

MN-P-025 [Macromolecules & Nanotechnology]

Magnetically activated microcapsules – preparation and characterization

Malgorzata Natalia Kostrzewska, Søren Hvilsted, Anne Ladegaard Skov

Danish Polymer Center, Department of Chemical and Biochemical Engineering, Technical University of Denmark, Kgs. Lyngby, Denmark

The utilization of magnetic nanoparticles (MNPs) in hyperthermia for cancer treatment, based on heat dissipation mechanism of MNPs in an alternating magnetic field has inspired us to design magnetically activated microcapsules with the purpose of remotely controlling poly(dimethylsiloxane) (PDMS) cross-linking reaction.

A PDMS cross-linker and various types of MNPs have been encapsulated within a polymeric shell by a phase separation. Scanning electron microscopy (SEM), thermogravimetric analysis (TGA) and nuclear magnetic resonance (NMR) were employed to investigate the morphology and the composition of the microcapsules. Moreover, the magnetic microcapsules were exposed to the alternating magnetic field at different frequencies.

Magnetic microcapsules containing the MNPs were successfully prepared without compromising the impermeable nature. The nanoparticles are mainly located in the core of the capsule remaining the shell's structure intact (Figure 1). Subjecting the mixture of the magnetic microcapsules and reactive (vinyl-terminated) PDMS has led to increase of the mixture temperature meaning that the heat generation is observable even at low concentration of the MNPs. This results in collapse of the microcapsules and allows for controlling the cross-linking reaction between the encapsulated cross-linker and the reactive PDMS compound.

Figure 1. Microscopic images of the microcapsules with magnetic nanoparticles. Figure 1a presents SEM images of dried microcapsules, whereas Figure 1b shows microcapsules in a water solution. Distribution of the magnetic nanoparticles (Figure 1c) was investigated by EDS analysis. Purple color represents iron atoms.

Keywords: microcapsules, alternating magnetic field, remote release

Figure 1

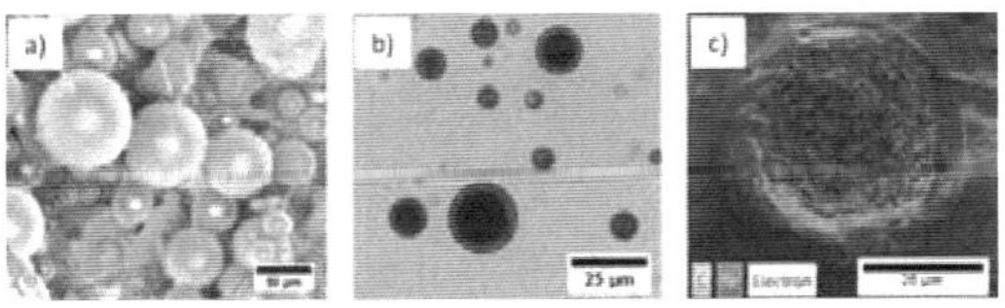

The figure should appear at the end of the abstract.

MN-P-026 [Macromolecules & Nanotechnology]

Thermal Conductive Polymer Nanocomposites Containing Boron Nitride

Tuba Erdogan Bedri

Institute of Chemical Technology, TUBITAK Marmara Research Centre, Kocaeli, Turkey

Materials that are light, cost-effective, easily handled and suitable component design with desired thermal and electrical properties, namely polymeric matrix nanocomposites, are regarded as new generation materials for electronic/microelectronic technology components. Hexagonal crystalline boron nitride (h-BN) has high thermal conductivity as well as electrical insulating properties where carbon nanotube and graphene cannot provide. Therefore, it attracts attention as a strategic material for electric-electronic applications (1-4).

The purpose of the work is to synthesize new generation polymeric matrix nanocomposites having hexagonal boron nitride (h-BN), and having high thermal conductivity, to determine the thermal, electrical, morphological and mechanical properties, to develop thermal pad for the aimed electronic component. In the design of thermally conductive nanocomposite, in order to increase the inter-surface interactions between the polymer and the nanoparticle and to provide a homogeneous distribution of the nano-filling in the matrix, we will first produce the h-BN nanosheet (Figure 1) with different exfoliation methods and then modify the surface. The polymeric matrix will be produced with surface polymerization using controlled polymerization approach. With this approach which will allow the formation of covalent bonds between the matrix and filling, the nanocomposite will be handled in different complex shapes.

The preparation of thermally conductive films of the synthesized boron nitride nanolayered composites will be handled by solvent casting method. Investigation of surface morphologies, which arise by controlled changing of the parameters, namely molecular weight, composition, and nanoparticle amount, will be realized by Scanning Electron Microscope (SEM). The thermal and thermo-mechanical properties of novel thermally conductive polymeric nanocomposites with h-BN nanosheet will also be determined for their performances in electronic applications.

Keywords: Hexagonal boron nitride, boron nitride nanosheet, thermally conductive nanocomposite, controlled radical polymerization, thermal interface materials

Figure 1

SEM micrograph of h-BN nanoparticle

Figure 2

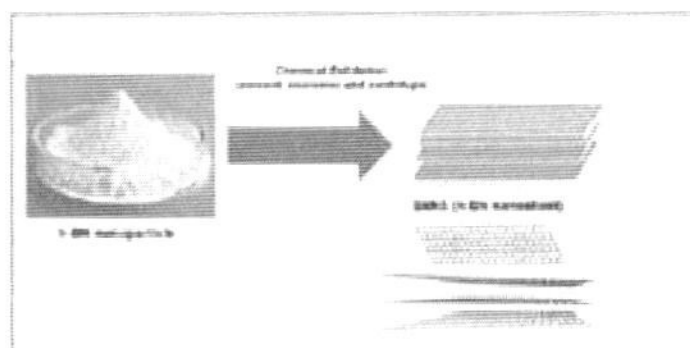

Exfoliation of h-BN

MN-P-027 [Macromolecules & Nanotechnology]

Photochemically produced self-assembled Ag nanoparticles on calf thymus DNA chains

Eyüp Metin, Gönül Saadet Batıbay, Neşe Atacı, Nergis Arsu

Department of Chemistry, Yıldız Technical University, Istanbul, Turkey

The size and distribution controlled nanoparticle synthesis has still receiving tremendous interest due to their enhanced optical, electronic, transport and mechanical properties. The noble metal nanoparticles, especially Au and Ag NPs have attracted great interest due to their remarkable physical and chemical properties and also potential application in various fields such as catalysis, nanoelectronics, sensing and biomedicine. (Nie and Emory, Science, 1997, 275, 1102) Recently, one dimensional nanomaterials are also gaining great interest and various methods have been employed for fabrication of 1-D nanomaterials. DNA has been used as a template to build inorganic nanostructures. (Kundu and Liang, Adv. Mater., 2008, 20, 826) The DNA molecule has two binding sites; first is the negatively charged phosphate groups, and the other is the aromatic base molecules. (Braun *et al*, Nature, 1998, 391, 775)

In literature there are several reports for the assembly of NPs onto DNA with an organized structure within nanoscale.

In this study, photochemical formation of self-assembled Ag NPs on DNA chain were achieved through in-situ reduction of Ag ions in the presence of 2-mercaptothioxanthone as photoinitiator (Arsu *et.al* Macromolecules 2003, 36, 2649-2653) by using photoirradiation technique. Ultraviolet irradiation technique has been used to prepare NPs with higher speed.

The synthesized nanomaterials were characterized by UV-Vis spectroscopy, SEM analysis and Mean Particle Size, Zeta Potential and Polydispersity Index of nanoparticles were measured by Zeta Sizer Nano ZS.

Keywords: DNA, DNA and silver, silver nanoparticles, DNA and Ag NPs, photoinitiator, 2-mercaptothioxanthone

MN-P-028 [Macromolecules & Nanotechnology]

Syntesis of Modified Multi Walled Carbon Nanotubes with Hydroxy Substituted Schiff Bases and Investigation of the Adsorption Properties

Tulay Bal Demirci, Elif Avcu

Department of Chemistry, Istanbul University, Istanbul, Turkey

Adsorption process is one of the conventional methods for the removal of heavy metals from aqueous solutions. In this work, absorbtion capacities of the modified novel multi wall carbon nanotubes (MWCNT-f) will be studied. Carboxylated multiwalled carbon nanotubes (MWCNT-COOH) will be functionalized with hydroxy substituted schiff bases. Then functionalized multiwalled carbon nanotubes (MWCNT-f) will be used as adsorbents of heavy metals from aqueous solutions. Effects of pH, concentration of the heavy metals, amount of adsorbents and contact time will be investigated. The products will be characterized by Fourier Transform Infrared Spectroscopy (FT-IR), Thermogravimetric Analysis (TGA), Scanning Electron Microscope (SEM),UV-Vis Spectrophotometer and Atomic Absorption Spectrometry (AAS) or ICP-MS. At the end of the study, we aim to synthesize a novel nano adsorbent for heavy metals.

Keywords: CNT, Nano Adsorbents, Schiff Bases, Heavy Metals

MN-P-029 [Macromolecules & Nanotechnology]

Amphiphilic Polymer-Enabled Nanometal Synthesis, Morphogenesis, and Colloidal Stabilization

Paschalis Alexandridis[1], Toshio Sakai[2]

[1]Department of Chemical and Biological Engineering, University at Buffalo, The State University of New York (SUNY), Buffalo, New York, USA
[2]Department of Chemistry and Material Engineering, Shinshu University, Nagano, Japan

Nano-size metals exhibit novel optical and catalytic properties, and attract considerable interest in a range of applications, e.g., photonics, diagnostics, sensors. The morphology (size and shape) of the nanoparticles and their surface/colloidal properties are very important in the various applications. A patented methodology [Alexandridis & Sakai, United States Patent 7,718,094, 2010] for the synthesis in aqueous media of metal nanoparticles with controlled size and shape and exceptional colloidal stability is reviewed [Alexandridis, Chem. Eng. Technol. 2011, 34 (1), 15-28; DOI: 10.1002/ceat.201000335]. This methodology is based on polymers that can exhibit multiple functions, i.e., reduction, shape/size control, surface modification and colloidal stabilization, on the basis of the polymer intramolecular and supramolecular organization. In addition to being water-based, this methodology requires no additional reductant and no external energy input, and employs commercially available polymers, e.g., poly(ethylene oxide)-containing Pluronics or Poloxamers, resulting in low cost, potential environmental benefits, and facile integration in medical applications (several Poloxamers are approved for pharmaceutical use). Recent results are presented on the use of PEO-based polymers with different end-functionality for gold [Sakai et al., J. Colloid Interface Sci. 2013, 394, 124-131; DOI: 10.1016/j.jcis.2012.12.003] and silver [Sakai et al., Colloids Surfaces A: Physicochem. Eng. Aspects 2015, 487, 84-91; DOI: 10.1016/j.colsurfa.2015.09.058] nanoparticle synthesis and stabilization. In addition to PEO-containing polymers, metal nanoparticles can synthesized and stabilized by other functional polymers such as poly(N-vinyl-2-pyrrolidone) (PVP) and poly(L-proline) [Gkikas et al., J. Polym. Sci. Part A: Polymer Chemistry 2013, 51 (6), 1448-1456; DOI: 10.1002/pola.26517].

Keywords: Nanoparticle, metal, dispersion, poly(ethylene oxide), polymer template

MN-P-030 [Macromolecules & Nanotechnology]

Poly(2-hydroxyethyl methacrylate) nanocomposite hydrogel filled with fluorescent halloysite nanotubes

Bruno F Urbano, Libnny Belmar, Leandro Toledo, Susana Sánchez

Polymer Department, University of Concepción, Concepción, Chile.

Hydrogels have been widely used in biomedicine because of their similarity to biological tissues, biocompatibility, viscoelasticity, etc. and have been applied in the fields of hygiene products, regenerative medicine, controlled release of drugs, repair wounds, and supports for cells, etc. Despite these advantages, the hydrogels have disadvantages associated with the low mechanical resistance as consequence of the water absorption. A strategy to improve the properties of the hydrogels and/or to incorporate new functionalities, consists in reinforce them by incorporating nanoparticles within the hydrogel network, resulting in a hydrogel nanocomposite. In this research was obtained a nanocomposite hydrogels based on poly (2-hydroxyethyl methacrylate) (HEMA) and halloysite nanotubes (HNT). The HNTs are nanomaterials, morphologically similar to the multi-walled CNTs, are available in a natural deposits, they have a high mechanical resistance and not cytotoxic. The HNT were labeled with fluorescein iso-thiocyanate (FITC) to provide fluorescent properties. Then, hydrogels were prepared by polymerization in-situ in a mixture of 2-hydroxyethyl methacrylate (1.8 and 3.7 mM), HNT-FITC (0.5, 1.5, and 3.0 wt-%) and initiator. Labeled nanotubes and hydrogels were characterized using various techniques for example: TGA, TEM, SEM, FTIR, DSC, among others. We observed that those nanocomposites prepared with lower concentration of monomer (1.8 mM) exhibited significant changes on thermal, morphological and rhodamine B release varying the filler content. The opposite was observed on hydrogels prepared with 3.8 mM of HEMA. We attributed the differences of the HNT effect to polymer-nanoparticles interaction. While a more concentrated monomer solution results in a polymer where the polymer-polymer interaction predominates over polymer-nanoparticles interaction, in a less concentrated polymer, the polymer-HNT interactions are stand out leading to a more accentuated changes on the properties.

FONDECYT Regular No 1140454 y a CIPA-CONICYT Regional R08C1002.

Keywords: hydrogel, nanocomposite, halloysite nanotubes, fluorescent nanoparticles

Figure 01: Fluorescence microscopy image of the nanocomposite hydrogel PHEMA-HNTFluo 3.0 wt-%.

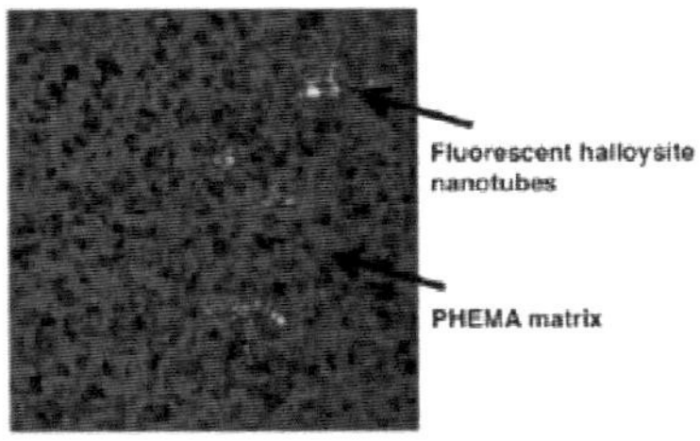

Figure 1

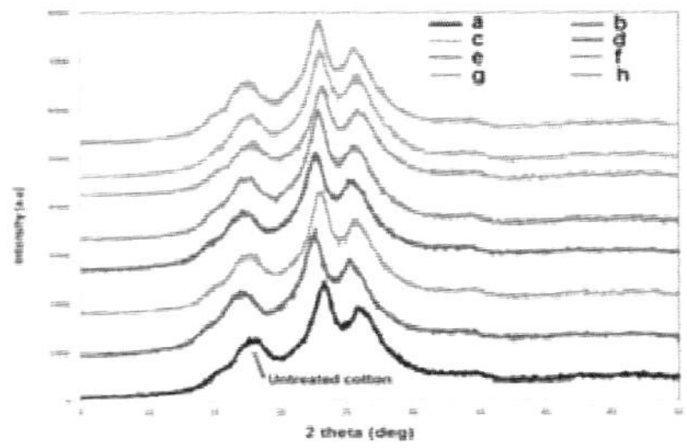

XRD spectra of different samples:(a) Untreated cotton, (b) Cotton coated with polyvinylphosphonic acid, (c) Graphene/polyvinylphosphonic acid/cotton nanocomposite prepared with 2 coating cycles of 10 mg/ml FGN/CTAB colloidal dispersion and polyvinylphosphonic acid, (d) Graphene/polyvinylphosphonic acid/cotton nanocomposite prepared with 5 coating cycles of 10 mg/ml FGN/CTAB colloidal dispersion and polyvinylphosphonic acid, (e) Graphene/polyvinylphosphonic acid/cotton nanocomposite prepared with 10 coating cycles of 10 mg/ml FGN/CTAB colloidal dispersion and polyvinyl phosphonic acid, (f) Graphene/polyvinylphosphonic acid/cotton nanocomposite prepared with 2 coating cycles of 20 mg/ml FGN/CTAB colloidal dispersion and polyvinylphosphonic acid, (g) Graphene/polyvinylphosphonic acid/cotton nanocomposite prepared with 5 coating cycles of 20 mg/ml FGN/CTAB colloidal dispersion and polyvinylphosphonic acid, (h) Graphene/polyvinylphosphonic acid/cotton nanocomposite prepared with 10 coating cycles of 20 mg/ml FGN/CTAB colloidal dispersion and polyvinylphosphonic acid.

MN-P-031 [Macromolecules & Nanotechnology]

A new insight into producing electromagnetic shielding cotton via graphene spraying

Zahra Saeidi, Mazeyar Parvinzadeh Gashti, Zoha Noralian, Izadyar Ebrahimi

Young Researchers and Elites Club, Yadegar-e-Imam Khomeini (RAH) Branch, Islamic Azad University, P. O. Box 18155/144, Tehran, Iran

In this study, we produced electromagnetic shielding and electro-conductive cellulose fibers via graphene spraying. Polyvinyl phosphonic acid was used as stabilizing agent to embed graphene on cellulose. The influence of graphene content and stabilizer on the electromagnetic shielding and electro-conductivity of the cellulose fiber were investigated using network and four-point probe electrical conductivity analyzers according to AATCC Test Method 76-2005 and ASTM D4935-99, respectively. Results showed that the electrical conductivity and EMI shielding of cotton is highly dependent on the graphene content and it increases the electrical conductivity properties of cotton due to intrinsic properties. Our simple spray coating is a novel promising method for utilizing highly electro-conductive cellulosic substrates.

Keywords: grapheme, electroconductive cellulose, polyvinylphosphonic acid

MN-P-032 [Macromolecules & Nanotechnology]

Photochemically Prepared Silver/Polymer Nanocoatings by using 2-(Carboxymethoxy) thioxanthone

Yasin Ogün Beydoğan, Melisa Adıbelli, Nergis Arsu

Department of Chemistry, Yildiz Technical University, Istanbul, Turkey

Preparation of metal nanoparticles are still in progress to be the subject of intense research area due to uniques properties which welcome wide range of potential applications from optics, biomedical and electronic fields. Among the metal nanoparticles silver (Ag) has very important place in the field of plasmonics and it interact very strongly with visible light. The control of size, shape and distribution of metal nanoparticles takes very important part for preparation.

The photochemical synthesis method has attracted much attention by offering several advantages e.g. convenient, fast and space selectivity. Photopolymerization method widely used for the preparation of nanocomposite materials since allows the fast formation of polymer network into thin films with novel properties. Photoinitiators play good role for photochemical way for preparation. There are mainly two type of photoinitiators, namely Type I and Type II. There are new generation of photoinitiators as well, one component Type II initiators which do not require any co-initiator to initiate the polymerization ofmonomers.(Aydin *et al.*, Macromolecular Rapid Communications, 2003, 24(12),718-723) In this study a rapid method was developed to obtain Ag/ polymer nanocomposite material by which silver salt is reduced and polymer matrix is simultaneously generated by the means of UV light.

In this study, acetic acid derivative of thioxanthone namely 2-(Carboxymethoxy)thioxanthone acid which classed as one component Type II photoinitiator was used for photoreduction of $AgNO_3$ salt to silver nanoparticles (AgNp) either in solvent or in polymer matrix by in situ photochemical method. Controlled size and well distribution of AgNps accompanied for this synthesis. UV-Vis spectroscopy and SEM pictures of coatings were confirmed Ag Nps formation.

Keywords: photoinitiator, photopolymerization, nanoparticles

MN-P-033 [Macromolecules & Nanotechnology]

Synthesis and Characterization of a Polymer with Unsymmetrical Zinc Phthalocyanines as Terminal Group

Yunus A Sütcüler[1], Tuba Ç Çanak[2], Altuğ M Sevim[2], Ersin I Serhatlı[2], Ahmet Gül[2]

[1]Polymer Science and Technology, Istanbul Technical University, Istanbul,Turkey
[2]Chemistry Department, Istanbul Technical University, Istanbul,Turkey

Phthalocyanines (Pcs) are well-known 18-π-electron aromatic molecules with unique physicochemical properties that make them interesting building blocks to be used in different fields of science and technology[1]. Polystyrene(PS) is made from petroleum, a non-sustainable, non-renewable, heavily polluting and fast-disapperaing commodity. PS polymer is known to be a clear brittle plastic that is synthesized by a free radical polymerization [2]. Recently phthalocyanines(Pc) conjugated with well-defined polymers is gaining huge interest due to their unique electronic and optical properties that stem from Pc macrocycle[3]. Conjugation of the macrocycles with various polymer chains enhances their processability, provides a means to control the molecular arrangement of the phthalocyanine rings, improves their solubility in common organic solvents.

Cu(I)-catalyzed azide-alkyne cycloaddition (CuAAC) click reaction which was initially postulated as a powerful tool for organic synthesis, has been extensively applied as a successful polymer functionalization technique. This study was undertaken to prepare polystyrene and unsymmetrical zinc phthalocyanine and involve azidation of polystyrene, performing click reaction between azidated polystyrene and asymmetric zinc phthalocyanine. Polystyrene was synthesized by using ATRP method and azidation was done in a step by using sodium nitrate. Azidated polystyrene and asymmetric zinc phthalocyanine were bonded with each other via Huisgen 1,3-dipolar cycloaddition. All the preparation steps of novel polystyrene-zinc phthalocyanine composite was investigated with various spectroscopic techniques such as 1H-NMR, FTIR and UV-Vis spectroscopy.

[1] Lezznoff CC., Lever ABP, Phthalocyanines Properties and Applications, Vol 1, (1989), Weinheim: VCH.
[2] Akovali G., "Handbook of Composite Fabrication", Rapra Technology Ltd., United Kingdom, 2001.
[3] McKeown NB. Phthalocyanine-containing polymers. J Mater Chem 2000;10: 1979-95.

Keywords: Polystyrene, Pthalocyanine, Composite,

Structure of Polystrene-Phthalocyanine composites

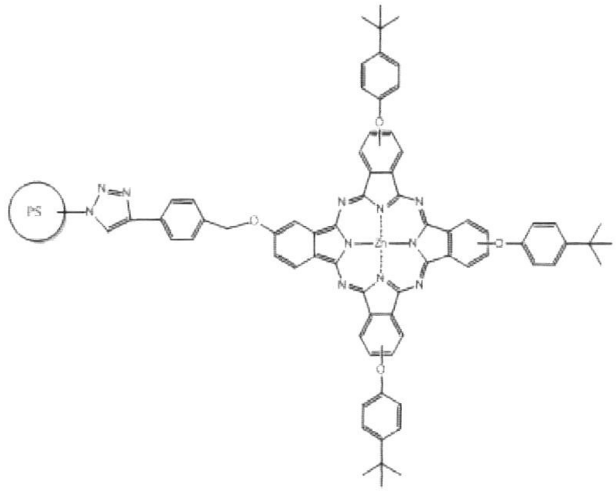

MN-P-034 [Macromolecules & Nanotechnology]

Synthesis of a Novel Internal Emulsifier Derived from plant Oil for the Preparation of Bio-based Polyurethane Nanoparticles and Their Application in Coatings

Mohammad Reza Nabid, Ismail Omrani, Niloofar Babanejad, Hasan Kashef

Department of polymer, Faculty of chemistry, Shahid Beheshti University, G.C, 1983963411, Tehran, Iran

A renewable, green and scalable dihydroxy acid was synthesized from plant oil and applied as a chain extender in the preparation of bio-based polyurethanes. In order to prepare dihydroxy acid, epoxidized plant oil was ring-opened with methanol followed by saponification. Chemical structure of dihydroxy acid verified by FT-IR and 1HNMR spectroscopies. Acetone process was used as a main method for the synthesis of waterborn polyurethane nanoparticles from polybuthylene adippate and dihydoxy acid. Particle size and morphology of nanoparticles were investigated by Dynamic Light Scattering (DLS) and Scanning Electron Microscopy (SEM), respectively and confirmed the narrow distribution with small particle diameters of nanoparticles (50-100 nm). The stable dispersion of nanoparticles applied for the preparation of films and subsequent characterizations of their thermal and mechanical properties. The obtained results corroborated that these kinds of novel bio-based polyurethane nanoparticles would be extensively efficient and applicable in coatings.

Keywords: Plant Oil, Bio-based Polyurethane, Nanoparticles, Coating, Mechanical Properties

The synthesis route of nanopolyurethane

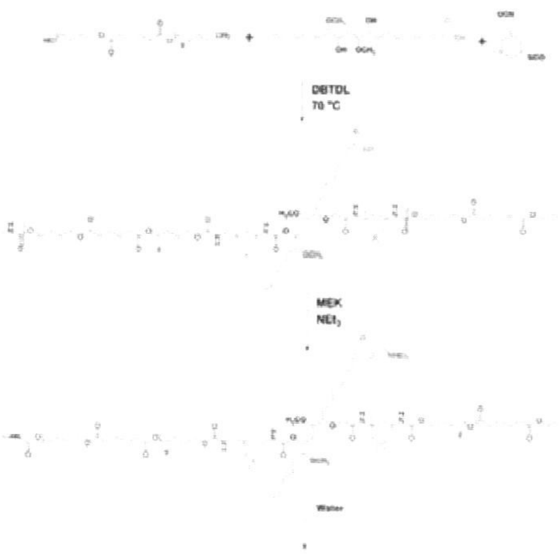

PBA and Plant Oil based Nanopolyurethane Dispersion

MN-P-035 [Macromolecules & Nanotechnology]

Photochemical preparation of gold and platinum nanoparticles

Gönül Saadet Batıbay, Nergis Arsu

Department of Chemistry, Yıldız Technical University, Istanbul, Turkey

Several strategies including chemical, photochemical, radiation and thermal methods were reported for the preparation of metal – polymer nanocomposites. The photochemical synthesis method has attracted much attention by offering several advantages e.g. convenient, fast and space selectivity. Therefore photopolymerization method widely used for the preparation of nanocomposite materials since allows the fast formation of polymer network into thin films with novel properties. Usually photopolymerizable formulations consist of photoinitiators, oligomers, reactive diluents and additives. Among those constituents photoinitiator play an important role for final properties of the cured coating as well as oligomers and reactive diluents. Photoinitiators can be classed as cleavable and abstractable type. Cleavable initiators called as Type I and they are less prone to oxygen inhibition. Type II initiators hydrogen abstracted type, generally needs a hydrogen donor such as amines and acids to produce initiating radicals. Among them thioxanthones are the most important Type II initiators (Yagci,et.al, The Royal Society of Chemistry, 2008, 2771–2773.) In this study, in-situ photochemical reduction from noble metal (Au,Pt) salts or mixtures to metal nanoparticles is aimed. The effect of polymer matrix will be investigated, and the effect of change in proportion of different crosslink density for various acrylates and methacrylates on nanoparticle size, shape and distiribution will be examined. Moreover the effect on stabilization of metal and metal oxide nanoparticles by different photoinitiators which will be used and its structures will be investigated.

Keywords: nanoparticles, gold-platinum alloy, gold nanoparticles, platinum nanoparticles, in-situ

MN-P-036 [Macromolecules & Nanotechnology]

Polymer Supported Ligand for Visual Detection of Trace Levels of Hg(II) & Pb(II) Ions in Aqueous Media

Roya Sedghi, Susan Kazemi, Bahareh Heidari

Department of Polymer, Faculty of chemistry, Shahid Beheshti University

In this study, we anchored dithizone on synthesized poly(t-butyl acrylate N,N methylenebisacrylamide -vinyl pyridine) beads and attempted to use it in the visual detection of trace levels of Hg(II) and Pb(II) ions. In this manner, the focus of this work is the preparation and characterization of poly(t-butyl acrylate- N,N-methylenebisacrylamide -vinyl pyridine) supported dithizone and discusses the applicability of these sorbents in the adsorption and visual detection of Hg(II) and Pb(II) ions trace levels in aqueous media. To the best of our knowledge, this is the first time that the polymeric nanocomposite loaded indicator chromogens and applied as a highly selective and sensitive chemosensosr for ultratrace levels detection mercury and lead ions from aqueous media. Detection of heavy metal ions requires simple, rapid, selective and sensitive detection methods. In this regard optical chemosensors based on in-situ detection, facility and continuous approaches in the environment samples is in great interest. On the basis of the results revealed in this work, the proposed mercury sensor based on the polymeric nanocomposite loaded indicator chromogens has many advantages including: fast response time, low cost, highly selective and sensitive method, easy preparation, wide dynamic range and low detection limit. The developed method was applied in order to high enrichment factor because of several heteroatoms in polymer backbone and is suitable for the determination of ultra trace Hg ions in real samples. The prepared chemosensor nanocomposite was characterized using different techniques such as SEM, TGA, XRD and FT-IR.

Keywords: sensor, visual, nanocomposite

Figure 1.

From left to right: The morphology of TiO2 nanoparticles, MAPTMS modified TiO2 nanoparticles and TiO2/poly(t- butylacrylate-co-methylene bis acrylamide-co-2-vinylpyridine)

MN-P-037 [Macromolecules & Nanotechnology]

Study of anti-cancer drug release (tamoxifen) of the nanofibers made of poly-caprolactone -chitosan

Zahra Saeidi[1], Ali Ashjaran[1], Seyed Ali Reza Dabirsiyaghi[2]

[1]*Young Researchers and Elites Club, Yadegar-e-Imam Khomeini (RAH)-shahre-rey Branch, Islamic Azad University, P. O. Box 18155/144, Tehran, Iran*
[2]*Member of the Department of Pharmaceutics, College of Pharmacy Azad University, Tehran, Iran*

Breast cancer is a kind of cancer that begins than breast tissue. that in a kind of it, breast skin is quite involved. than symptoms of the breast cancer is a dimple on the skin and peeling of skin and on the chest and..... Natural polymers in terms of the environmental characteristics and their compatibility with the human body have vast applications in the industry: cosmetics, wound dressing, delivery of medicine scaffold, and so on. In this study, the cancer drug (tamoxifen) used to complete the PCL-chitosan nanofibers made of a biocompatible polymer packag that deliver drugs will be examine.

For this purpose, first the nanofibers made of PCL-chitosan-containing drugs (tamoxifen) with optimal concentration of the drug was produced by electrospinning. The surface morphology of nanofibers microscopy (SEM) were studied. Using infrared spectrometer instrument (FTIR) evaluated the drug in nano-fibers. The effects of drug release and antimicrobial from nanofibers with standard (AATCC100) measurement and determined. According to antimicrobial test the nano-fiber manufacturing against Gram-positive and Gram-negative bacteria, *Escherichia coli* and *Staphylococcus Cocos* considerable influence along.

Based on the results of tests it was found that nano-fiber PCL-chitosan-containing drug with a ratio of 70-30 and a concentration of 1 gr and 0.3 gr, speed And release better itself show compared to concentrations and the other ratios. The test results of drug release indicate a total release rate was high.

Keywords: electrospinning, poly-caprolactone, chitosan, tamoxifen

1

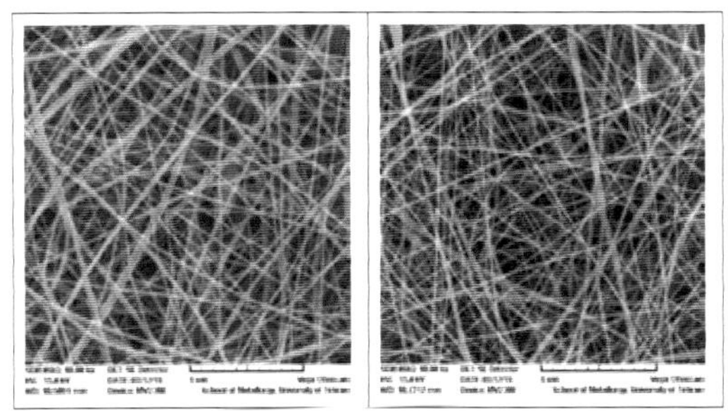

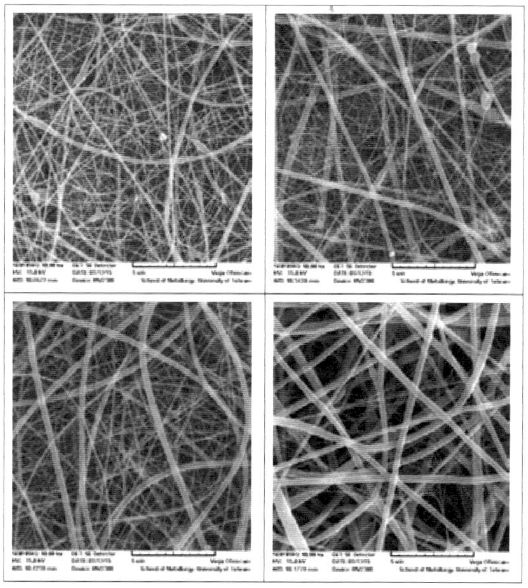

2

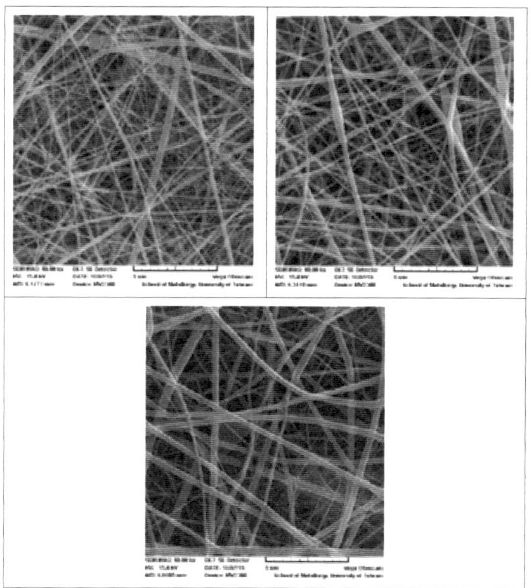

MN-P-038 [Macromolecules & Nanotechnology]

Chiral Porphyrin Macrocyclic Catalysts for the Processive Oxidation of Polymer Substrates

Shaji Varghese, Alan E. Rowan, Roeland J. M. Nolte

Institute for Molecules and Materials, Radboud University Nijmegen, The Netherlands.

Processive catalysis is ubiquitous in nature and processivity is utilized as a way to increase the efficiency and fidelity of a process. DNA polymerase is a nice example of a processive biocatalyst; it assembles nucleotides in a processive manner to synthesize new strands of DNA.(1) We have previously shown that an artificial processive catalytic system based on a porphyrin macrocycle (Fig.1) was able to thread onto polymers (e.g. polybutadiene) and epoxidize the double bonds of this polymer in a "processive" manner, that is, by carrying out several rounds of catalysis without detaching from the polymer chain. (2) However, the potential of chiral processive catalyst for the synthesis chiral polymers was not explored so far. Recently, we were able to synthesize a series of chiral porphyrin macrocycle for enantioselective processive catalysis. This chiral processive catalyst is currently under study with the objective to convert polyalkene into tacticity regulated chiral polymers. This has a great potential for future applications in fields ranging from the intelligent tailoring of polymers to data storage and even molecular computing.(3)

References:

(1) van Dongen, S. F. M.; Elemans, J. A. A. W.; Rowan, A. E.; Nolte, R. J. M. Angew. Chem. Int. Ed. 2014, 53 (43), 11420.
(2) Thordarson, P.; Bijsterveld, E. J. A.; Rowan, A. E.; Nolte, R. J. M. Nature 2003, 424 (6951), 915.
(3) Varghese, S.; Elemans, J. A. A. W.; Rowan, A. E.; Nolte, R. J. M. Chem Sci 2015, 6 (11), 6050.

Keywords: Catalysis, Polymers, Macrocycles

Fig.1: Processive catalysis exerted on a synthetic polymer chain by a porphyrin-based catalyst.

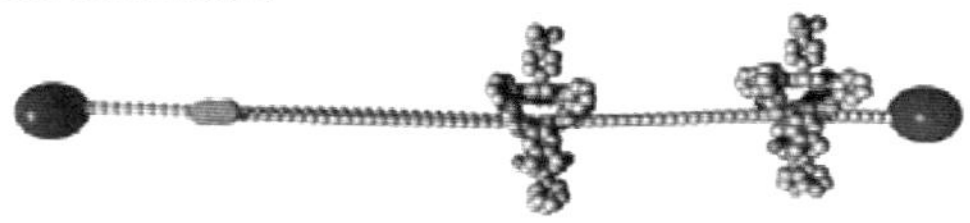

MN-P-039 [Macromolecules & Nanotechnology]

One-step formation of w/o/w multiple emulsions stabilized by the mixture of biopolymers

Soumia Seddari Kalache, Nadji Moulai-Mostefa

Materials and Environmental Laboratory Faculty of Sciences and Technology University of Medea, Ain D'Heb, 2001 Medea, Algeria

Summary: Multiple emulsions are complex systems in which the droplets of the dispersed phase themselves contain small dispersed droplets [1]. There are two types of multiple emulsions: water-in-oil-in water (W/O/W) and oil-in-water-in-oil (O/W/O). Due to their multi-compartment structure, these materials have many applications in the pharmaceutical, cosmetic and food fields [2].

These emulsions are thermodynamically unstable. However, they can be kinetically stable over a significant period, which determines their lives. Biopolymers such as proteins and polysaccharides are good examples of natural surfactants commonly found in the food industry and can be used in the pharmaceutical industry [3].

The objective of this study is to formulate multiple W/O/W emulsions by a mixture of natural biopolymers namely (sodium caseinate) NC and XG (xanthan gum) using the direct process.

The stability of double emulsions prepared was conducted for 20 days at various storage temperatures. The effect of pH formulation on the stability of emulsions prepared was studied. Measurements of the apparent viscosity of each formulation were performed and compared with other work. In parallel with these measures, microscopic was performed about different systems formulated.

The results obtained showed that the emulsions made by the SN / GX mixture kept good stability for one month. This stability is a function of pH of external phase. The optimal pH for the formulation of these double emulsions is 5 or 5.5.

References:

[1] Garti, N. (1997) Double emulsions - scope, limitations and new achievements. Colloids and Surfaces.123-124, 233.
[2] Colmenero, F.J. (2013) Potential applications of multiple emulsions in the development of healthy and functional foods. Food Res. Int. 52, 64-74.
[3] Eleonore, B., Ghozlene, M., Veronique, R., Grossiord, J.L., Agnelya, F. (2012) proteins, polysaccharides, and their complexes used as stabilizers for emulsions: alternatives to synthetic surfactants in the pharmaceutical field. Int. J. Pharm. 436, 359-378.

Keywords: multiple W/O/W emulsions, Xanthan Gum, Sodium caseinate, stability, viscosity, Morphology.

MN-P-040 [Macromolecules & Nanotechnology]

Biomineralization-inspired synthesis of calcium iodate/gelatin nanocomposites by gel growth approach

Naeimeh Dehghan, Mazeyar Parvinzadeh Gashti

Young Researchers and Elites Club, Yadegar-e-Imam Khomeini (RAH) Branch, Islamic Azad University, Tehran, Iran

Diffusion of inorganic salts in gels is a simple, inexpensive, and versatile technique for the synthesis of inorganic/organic hybrid nanocomposites. This paper introduces a novel method for producing calcium iodate/gelatin bionanocomposites using single diffusion of calcium ions in gelatin hydrogel at ambient temperature. Scanning electron microscopy clearly demonstrated nanostructures of calcium iodate after synthesis in gelatin. X-ray diffraction study stated the crystalline peaks related to calcium iodate powders produced in gel. This study demonstrates that gelatin hydrogel can be used for crystallization of calcium iodate by producing nanosheets. Such a nanocomposite may enable new future options for antimicrobial, pharmaceutical and biocompatible composites for use in tissue engineering applications.

Keywords: Biomimetic, Calcium iodate, Nanocomposites, Gelatin

fig.1.

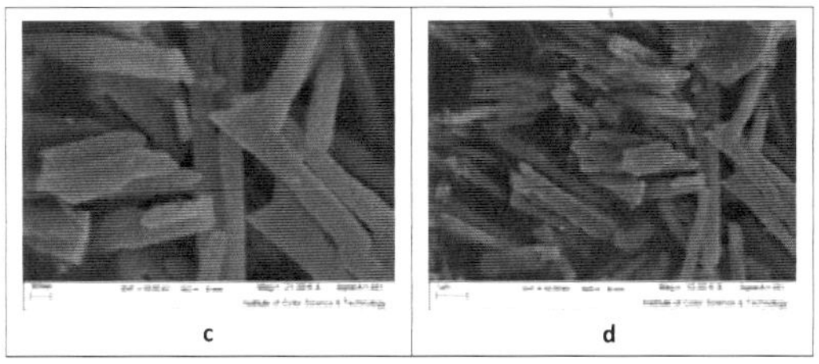

fig. a,b) gelatin-free calcium iodate with different magnification
c,d) calcium iodate/gelatin composite crystals with different magnification

fig.2.a

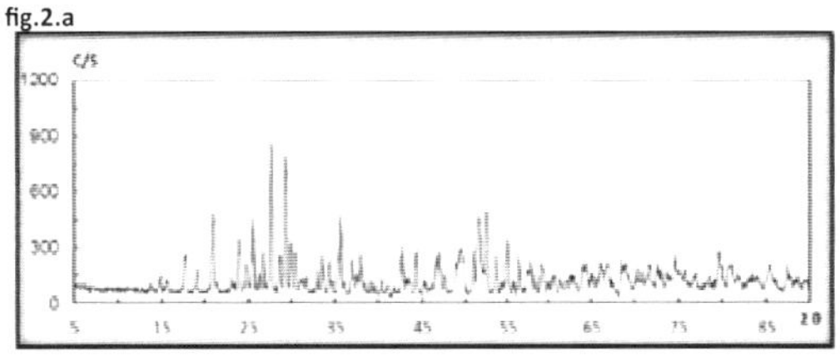

fig.2.b

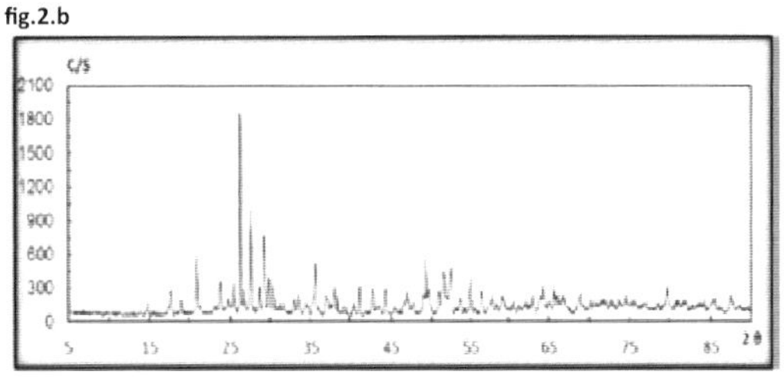

Fig. 2: The XRD pattern of calcium iodate crystals (a) pure calcium iodate crystals; (b) calcium iodate/gelatin composites synthesized in hydrogel.

fig.3

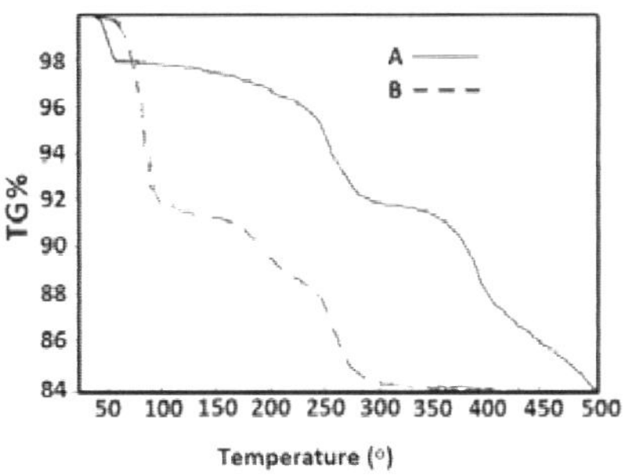

Fig. 3: The thermal degradation behavior of the crystals synthesized in this study: a) gelatin-free calcium iodate, b) calcium iodate /gelatin nanocomposites synthesized in gel.

MN-P-041 [Macromolecules & Nanotechnology]

The Use of Gold Nanoparticle Conjugated Porous Glycidyl Methacrylate Microparticles For Phenol Removal

Eda Aşık[1], Ali Tuncel[2], Ayşegül Aşkın[3]

[1]*Department of Polymer Science and Technology, Osmangazi University, Eskisehir, Turkey*
[2]*Department of Chemical Engineering,Hacettepe University,Ankara, Turkey*
[3]*Department of Chemical Engineering, Osmangazi University, Eskisehir, Turkey*

Gold nanoparticle conjugated porous polymeric microparticles can be used for in many waste water treatments. Many studies show that active phenol compounds, including 4 nitro phenol, have harmful effects on the environment, humans and all ecosystems. For this reason, many methods including biological, electrochemical, sonochemical etc. are developed and used for the reduction of toxic phenol compounds. However; desired efficiency for phenol removal haven't been reached, yet.

It is possible to develop a system with high surface area and high efficiency via gold nanoparticle conjugated polymeric support materials. In this study, gold nanoparticle conjugated porous glycidyl methacrylate (GMA) microparticles were synthesized, characterized and used for phenol degradation studies.

Porous microparticles were synthesized with suspension polymerization method. GMA monomer and ethylen dimethacrylate (EDMA) cross linker were mixed in toluen and benzoil peroxide (BPO) under certain temperature and shaking conditions, and microparticles were synthesized. After particle synthesis, microparticles were aminated with ethylen diamine. On the other hand, gold nanoparticles were synthesized with trisodium citrate and HAuCl4 by using Turkevich method. Gold nanoparticles were conjugated to the microparticles under room temperatures. With presence of NaBH4, microparticles have been added to 4-Nitro Phenol solution. Samples were taken from the solution at different time intervals and measured via spectrophotometer.

Results show that, our particles can be used for phenol removal. Reduction process was very successful in different phenol concentrations and with different particle amounts. This particles can be used in continuous flow chemistry systems due to their short reduction time. For the future studies, a continuous flow phenol reduction system setup can show that even with short retention times, particles would be able to reduce phenol concentration to an acceptable level.

Keywords: Gold Nanoparticle, Porous Microparticle, Glycidyl Methacrylate, Phenol

MN-P-042 [Macromolecules & Nanotechnology]

A homopolymer grafted-graphene oxide nanosheet in superior stress transfer in immiscible polymer blends

Goutam Prasanna Kar, Suryasarathi Bose

Department of Materials Engineering, Indian Institute of Science, Bangalore 560012, India

A novel strategy was adopted to compatibilize immiscible polymer blend. PMMA as mutually miscible polymer with PVDF and ABS polymer was chosen as compatibilizer (J. Machado and C. Lee, Polymer Engineering & Science, 1994, 34, 59-68). The interfacial location of PMMA in PVDF/ABS blend was proven by high resolution SEM morphology. The concentration dependent of PMMA in compatibilization and evolution of morphology in PVDF/ABS blend was assessed. This strategy was further improved with free-radical polymerization of methyl methacrylate (MMA) in presence of graphene oxide (GO) nanosheet. High mechanical performance of GO was exploited to enhance mechanical properties of PVDF/ABS blends. A dramatic increment of ca. 84% in the Young's modulus and ca. 124% in the yield strength was observed in the presence of PMMA-g-GO with respect to the neat blends. A simultaneous increment in both the strength and the modulus was observed in the presence of PMMA-g-GO whereas, only addition of GO resulted in a moderate improvement in the yield strength. This study reveal that a homopolymer-grafted-GO can results better stress transfer between immiscible polymer blends.

Keywords: Compatibilization, PVDF, ABS, Graphene oxide

MN-P-043 [Macromolecules & Nanotechnology]

Polymer Based Carbon Nanotubes ss Gas Sensor: Acetone and Toluene

Elif Baysak[1], Muge Yucel[2], Umit Hakan Yildiz[3], Gurkan Hizal[4]

[1]*Istanbul Technical Universiy, Istanbul, Turkey*
[2]*Izmir Institute of Technology, İzmir, Turkey*
[3]*Izmir Institute of Technology, İzmir, Turkey*
[4]*Istanbul Technical Universiy, Istanbul, Turkey*

Carbon nanotubes (CNTs) have great attention among the world since they have excellent mechanical, thermal and electrical properties which make them a very versatile material ranging from electronics, biotechnology, energy conversion and sensor systems. CNTs are classified as single walled and multi walled CNTs due to their wall numbers. Furthermore, single walled CNTs show higher electrical conductivity values which make them more usable and valuable in sensor systems. Conductivity of the CNTs can be enhance physically or chemical modification with metals, DNA or organic molecules and polymers.

In this study, pyrene end group functionalized polymers are prepared by controlled polymerization systems. As a result, pyrene bearing Pyr-PS-PS, Pyr-PEG-PEG, Pyr-PEG-PS, Pyr-PEG and Pyr-PCL branched and linear polymers are synthesized and they are physically bonded to single wall carbon nanotubes to monitor the resistivity change when exposed to toluene and acetone vapors.

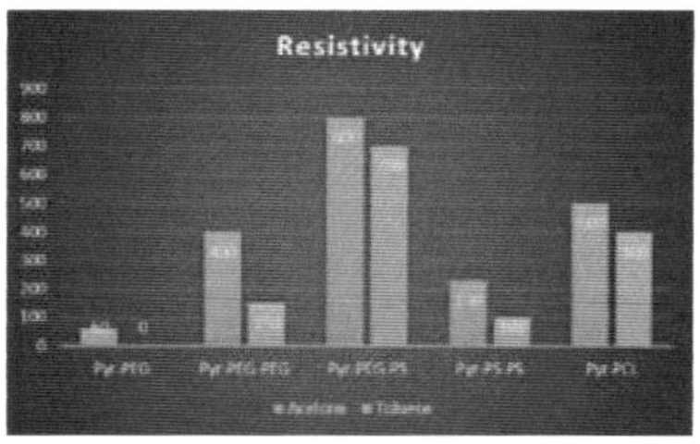

Figure 1. Resistivity Responses of the Polymer-CNT Compositions to Acetone and Toluene

[1] Ndiaye, N., Bonnet, P., Pauly, A., Dubois, M., Brunet, J., Varenne, C., Guerin K., Lauron, B.: *J. Pyhs. Chem*, **2013**, *117*, 20217

[2] Durmaz, H., Dag, H., Tunca, U., Hizal, G.: *J. Polym. Sci. Part A: Poly. Chem.*, **2012**, *50*, 2406

Keywords: polymer, CNTs, gas sensor

MN-P-044 [Macromolecules & Nanotechnology]

Electrospray Finishing of Alcohol Repellent Materials on Non-woven Fabrics and Investigation of Performance Properties

Semih Özgeç[1], Ali Kireçci[2]

[1]Semih ÖZGEÇ, Textile Engineering Department, Gaziantep University, Gaziantep 27310, Turkey
[2]Ali KİREÇCİ, Textile Engineering Department, Gaziantep University, Gaziantep 27310, Turkey

Electrospraying method was used for different application, especially for finishing application of textile materials. There are several methods for wet finishing treatments of fabrics such as padding, foaming, spraying and kiss-roll application. However each method may have its own advantages and drawbacks. This study aim the characterization of electrospray method for application of alcohol repellent finishing for nonwoven fabrics.

Electrospray is a method of generating liquid atomization with electrical forces. In electrospray, a liquid is passing through the nozzle. The charged liquid in the nozzle becomes unstable as it is forced to hold more and more charge. Soon liquid can keep a much electrical load which reaches a critical point. The nozzle head has highly charged droplets. The droplet size depends on a lot of parameters such as applied electrical charge, surface tension of finishing materials, distance of collector to nozzle head, viscosity, conductivity and the others [1,2]. The charged droplets accelerate through the collector surface. Thus, the surface of any object between the nozzle and collector would be covered with atomized solution.(See Figure 1).

This study aims the determination of appropriate electrospraying parameters to obtain the best finishing results. Therefore different alcohol repellent materials, which are produced CHT, Pulcra Chemicals and Rudolf, were electrosprayed onto several nonwoven and woven fabrics at different solution and process parameters. Nonwoven fabrics is 100% polypropylene and they have different weights such as; 35gram/m2, 40gram/m2 and 45gram/m2. The woven fabrics which made of %100 polyester and 65% Polyester %35 cotton fabric. Nonwoven fabrics are used at medical textiles for single-use, however woven fabrics are suitable for several use in this study. The basic criteria were to obtain the smaller size and homogenized distribution of particle. All the samples were tested by standard of AATC 193. Table I shows some results.

Keywords: Electrospraying, Nanotechnology, Alcohol Repellent, Non-woven, Performance properties

Figure 1.

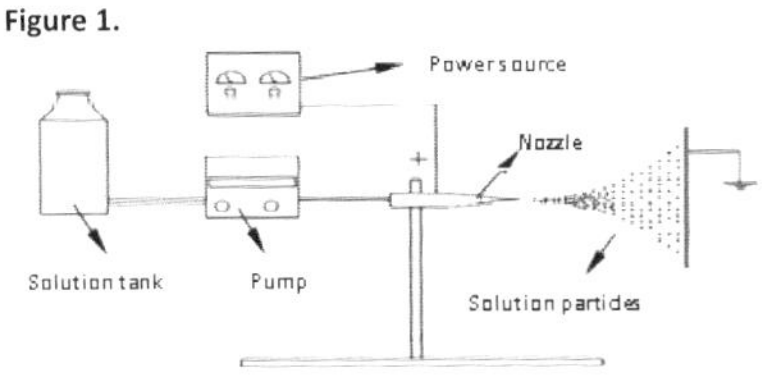

Electrospray Set-up

Table 1.

				Alcohol Repellent Test Results (Limited)					
			Non-woven Fabrics		Woven Fabrics				
		Padding		Elect-rospraying		Padding		Elect-rospraying	
	Test Method	40gsm	45gsm	40gsm	45gsm	65/35% Pes/Co	100% PES	65/35% Pes/Co	100%PES
CHT	AATCC 193	8	8	6	6	8	7	6	5
Pulcra	AATCC 193	6	5	4	4	7	6	4	3
Rudolf	AATCC 193	6	6	5	5	8	7	5	5

Test Results According to AATCC 193:2005(Limited)

MN-P-045 [Macromolecules & Nanotechnology]

In-situ Preparation of Poly(2-ethyl-2-oxazoline)/clay Nanocomposites via Living Cationic Ring-Opening Polymerization

Umut Ugur Ozkose[1], Cagatay Altinkok[2], Ozgur Yilmaz[3], Onur Alpturk[4], Mehmet Atilla Tasdelen[5]

[1]Department of Chemistry, Faculty of Science and Letters, Istanbul Technical University, Istanbul, Turkey & Materials Institute, Marmara Research Center, TUBITAK, Gebze, Kocaeli, Turkey
[2]Department of Polymer Engineering, Faculty of Engineering, Yalova University, Yalova, Turkey & Materials Institute, Marmara Research Center, TUBITAK, Gebze, Kocaeli, Turkey
[3]Materials Institute, Marmara Research Center, TUBITAK, Gebze, Kocaeli, Turkey
[4]Department of Chemistry, Faculty of Science and Letters, Istanbul Technical University, Istanbul, Turkey
[5]Department of Polymer Engineering, Faculty of Engineering, Yalova University, Yalova, Turkey

The present study reports for the first time the preparation of poly(2-ethyl-2-oxazoline)/clay nanocomposites (PEtOx/MMT). The tosyl-functionalized montmorillonite clay is utilized as initiator for the living cationic ring-opening polymerization of 2-ethyl-2-oxazoline, which triggers the delamination of silicate layers in the polymer matrix and leading to nanocomposite formation. The living nature of the polymerization has been confirmed by kinetic studies. The morphology and thermal properties of nanocomposite have been evaluated using X-ray diffraction (XRD), transmission electron microscopy (TEM), differential scanning calorimeter (DSC) and thermogravimetric analysis (TGA). All nanocomposite samples have a mixed exfoliated/intercalated silicate layers and an improved thermal stability compared to pure poly(2-ethyl-2-oxazoline).

Keywords: cationic ring-opening polymerization, living polymerization, nanocomposites, organoclay, polyoxazolines

In-situ preparation of PEtOx/MMT nanocomposites via living cationic ring-opening polymerization.

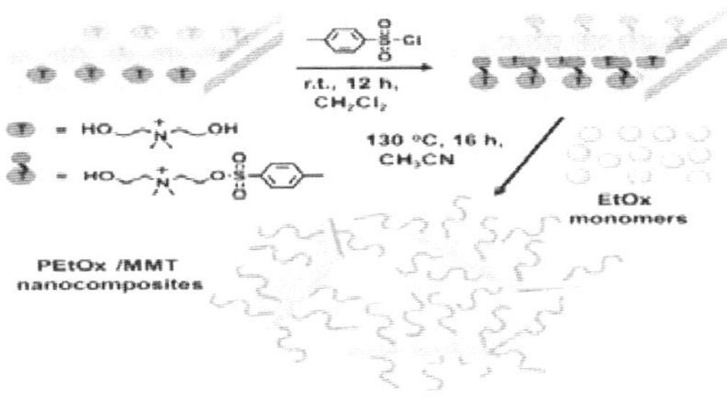

MN-P-046 [Macromolecules & Nanotechnology]

Facile Modification Of Reactive Electrospun Nanofibers Using "Click" Chemistry

Özlem Ipek Kalaoğlu Altan, Rana Sanyal, Amitav Sanyal

Department of Chemistry, Bogazici University, Istanbul, Turkey

Polymeric nanofibers have gained great attention lately for biomedical applications due to their unique properties like high surface area, porosity, physical resemblance to the extracellular matrix and diversity of composition [1]. The modification of the nanofiber surfaces with various (bio)molecules also contribute to enhance their biocompatibility, cell-adhesivity and stimuli-responsivity for drug delivery [2]. Recently, 'click' chemistry has become a promising functionalization method due to its advantageous properties such as mild reaction conditions, high efficiency and orthogonality to many reactions. Recent studies by Becker and coworkers have showed that 'click' chemistry is a powerful method for the functionalization of nanofibers [3,4]. In our study, we investigated the synthesis of reactive nanofibers with various clickable groups at their chain ends using solution electrospinning method. We synthesized different types of reactive polymers using ring opening polymerizations with a variety of clickable groups. The solutions of these polymers in suitable solvents were electrospun into reactive nanofibers. The obtained clickable nanofibers were easily functionalized with various (bio) molecules bearing complementary clickable groups at ambient conditions. Fabrication of novel nanofibers and their orthogonal functionalization will be presented.

Authors acknowledge financial support from The Scientific and Technological Research Council of Turkey (114Z139).

References :

1. Yoo, H.S., Kim, T.G., Park, T.G. Advanced Drug Delivery Reviews, 61, 10331042, 2009.
2. Pham, Q.P., Sharma, U., Mikos, A.G. Tissue Engineering, 12, 1197-1211, 2006.
3. Zheng, J., Hua, G., Yu,J., Lin, F., Wade, M.B., Reneker, D.H., Becker, M.L. ACS Macro Letters, 4, 207-213, 2015.
4. Zheng, J., Kontoveros, D., Lin, F., Hua, G., Reneker, D.H., Becker, M.L., Willits, R.K. Biomacromolecules, 16, 357-363, 2015.

Keywords: electrospinning, nanofibers, click chemistry, reagent-free bio(-functionalization)

MN-P-047 [Macromolecules & Nanotechnology]

Nano Silica-Polyurethane Composite Materials for Coating Application

Mohammed A Bahattab[1], Jose M Martin Martinez[2]

[1]National Of Petrochemical Technology Center, King Abdulaziz City For Science And Technology
[2]Adhesion & Adhesives Laboratory, University Of Alicante

Polyurethane can be tailored to meet the highly demand of modern technologies such as coatings, adhesives, fibers, foams and thermoplastic elastomers. Thermoplastic polyurethane are commonly used as transparent coating and commonly modified by adding fillers to impart viscosity, improve mechanical properties and modify their rheological and surface properties. Thermoplastic polyurethanes (TPU) are multi-phase segmented polymers that exhibit a two-phase microstructure, which arises from the incompatibility between the soft and hard segments. The hard rigid segment segregates into a glassy or semicrystalline matrices, and the polyol soft segments form amorphous or rubbery matrices in which the hard segments are dispersed. Several kinds of silica have been added to improve different polyurethanes properties. In this study, three different sizes, 20 nm, 50 nm and 100 nm manufactured by NYACOL Company and two modified surface hydrophobic fumed silica (AEROSIL R 9200) and hydrophilic modified fume silica (AEROSIL R 7200) manufactured by Degussa company were added to a TPU formulations. The effect of addition of nanosilica was evaluated by different techniques such as TEM, DSC, TGA, DMA and Contact angle. Effect of nanosilica edition on mechanical, thermal and morphology properties will be discussed.

Keywords: Thermoplastic polyurethanes, Nano silica, Rigid segment, Soft segments

MN-P-048 [Macromolecules & Nanotechnology]

Preparation of mixed polyacrylonitrile-polycaprolactone nanofiber mats for seperation purposes

Sündüz Alemdar, Cansu Çelik, Nursel Pekel Bayramgil

Department of Chemistry, Hacettepe University, Ankara, Turkey

Today nanofibers are preferred in many biomedical application due to the high surface area to volume ratio that they have. Among the methods used to obtain nanofibers, electro-spinning process comes first. Polyacrylonitrile (PAN) derived from acrylonitrile monomer, is a particularly important commercial polymer used in the production of carbon fibers [1]. Meanwhile polycaprolactones, unlike the polymers of the polyester family, can be degraded by enzymatic and microbial environment. However, biodegradation of the PCL takes place in a long and slow process. PCL can be used as homopolymer or as copolymer, or as the blend with another polymer. It has a very wide range of applications ranging from the pharmaceutical production to tissue engineering applications to medical devices to the textile industry to the paint and to the adhesives. [2-4]. PCL is a biocompatible polymer is thus increased PCL's biological uses. In drug carrier systems, some features of PCL such as easy incorporation of active pharmaceutical ingredients to porous PCL structure, high release rate of active ingredients in the body, complete discharging of active ingredient from the body and hold active ingredient stabile leads to increase its uses [5,6]. Hydrophobic nature of the microspheres produced in the micro and nano scale prevents the active substance dissolved in water dispersion. In addition, the nano-sized drug loaded PCL spheres are making drug release only on the diseased areas in the body, they are selectively-targeted feature. Thus, the smart drug production with current issues of today's pharmaceutical industry can be made with the PCL.. The surface, structure and thermal characterization of nanofibers were performed by using SEM, XRD, FT-IR, and TGA-DSC methods. Applications for environmental or biomedical fields, nanofiber mats were subjected to tensile-elongation tests for the mechanical strength of the most important parameters that affect the use of performance nanofibers.

Keywords: Nanofiber, Polyacrylonitrile, Polycaprolactone, Electrospinning

MN-P-049 [Macromolecules & Nanotechnology]

Silane Modified Surfaces for Biological Imaging by AFM

Nilüfer Kıvılcım[1], Fatma Bilge Emre[2], Turgay Seçkin[1]

[1]*Department of Chemistry, İnönü University, Malatya, Turkey*
[2]*Department of Primary,Faculty of Education, İnönü University, Malatya, Turkey*

It is clear that the major impediment to successful AFM imaging of biological materials is sample preparation. This is a particularly difficult issue as every sample requires a unique approach. The objective with any sample is to achieve a firm adhesion to a substrate of sufficiently minimal topography that the topography of the sample is easily discriminated. This fixing to the surface is to guarantee that the probe tip does not push the sample around during imaging. Fort his reason surface modified materials by using quartz was established. Quartz is the second-most-abundant mineral in Earth's continental crust. Its crystal structure is a continuous framework of SiO_4 silicon–oxygen tetrahedra, with each oxygen being shared between two tetrahedra, giving an overall chemical formula of SiO_2. The silanol functionality on surface has been icreased by an etching pretreatment with hydroxides [1]. Silanization and derivatization of quartz were carried out in Petri dishes that had been washed in "piranha bath" (3.5 % H_2O_2 in 18 M H_2SO_4), followed by rinsing with water and acetone[2]. The following steps were carried out at room temperature unless stated otherwise. Quartz was washed once with concentrated HCl/HNO_3 (3:1) and five times for 1 min with destilled water in an ultrasonic bath (50 kHz). They were etched with trifluoroacetic acid for 90 min and stored in vacuum over solid KOH for at least 10 h. Coverslips were then silanized with CLS (2 % in 95 % aqueous acetone) for 3 min followed by washing with acetone (12 times, 5 min each). The surface morphology was examined by scanning electron microscopy and AFM. Elemental analyses, thermogravimetric analyses and ınfrared spectoscopy were conducted to explain the surface characteristics.

[1] Gunda, N.S.K.,Singh,M.,Norman,L.,Kaur,K.,Mitra,S.K.,2014. Appl.Surf. Sci.305, 522–530.
[2]Yin, R.,Ottenbrite,R.M., Siddiqui, J.A., 1996, Polym.Prep., 37, 751.

Keywords: AFM, Quartz,Surface modification

Reaction for the fonctionalization of quartz with APS

MN-P-050 [Macromolecules & Nanotechnology]

Hydrophilic Nano SiO2 – based Composite Hydrogels for Biomedical Purposes

Özge Fatma Gökmen[1], Nursel Pekel Bayramgil[2]

[1]*Central Research Laboratory, Bilecik Seyh Edebali University, Bilecik, Turkey*
[2]*Department of Chemistry, Hacettepe University, Ankara, Turkey*

Gels are described as cross-linked polymers swellable in the solvent medium. When the solvent medium is selected as H2O, gel is called as hydrogel. Gels can absorb H2O up to thousands of times its own weight. Three-dimensional crosslinked structure is formed in a chemical reaction to obtain physical hydrogels via hydrogen bonds and van der Waals interactions [1,2]. Recent studies on strengthening the polymeric hydrogels and to give them different functionalities are focused on combining nanoparticles with hydrogels. Nanoparticles which are polymeric, ceramic, metallic, and carbon-based are incorporated in hydrogel structure to produce nanocomposite hydrogels which have excellent facilities and set-changeable functionalities [3,4]. The final product has shown superior physical, chemical, electrical and biological properties.

In our study, to prepare nanocomposite hydrogels, we aimed to use poly(N-vinylpyrrolidone) (PVP) as polymer which is known as highly popular in biology, medicine and pharmaceutical areas, with hydrophilic nano-SiO2. Physical and chemical properties of nanocomposite hydrogels were characterized by using FT-IR, XRD, SEM, TGA and DSC analysis techniques. The results were compared between PVP including nano-SiO2 and hydrophilic nano-SiO2. When the hydrophilic nano-SiO2 was added to nanocomposite hydrogels, there was provided a significant improvement in physical properties. Which nanocomposite hydrogel is more resistant to thermal and mechanical maintenance, the usability of these hydrogels for drug delivery studies were discussed.

References:

[1] Hoffman, A.S. J. Controll. Release 1987, 6, 297-305.
[2] Brannon-Peppas, L.; Peppas, N.A. J. Controll. Release 1991,16, 319-330.
[3] Gaharwar, A.K..; Peppas, N.A.; Khademhosseini, A. Biotechnol. Bioeng. 2014, 111, 441-453.
[4] Gaharwar, A.K..; Rivera, C.; Wu, C.J.; Chan, B.K..; Schmidt, G. Mater. Sci. Eng. C. 2013,33,1800–1807.

Keywords: Composite, hydrogel, hydrophilic nanosilica.

MN-P-051 [Macromolecules & Nanotechnology]

Synthesis, Characterization and Photocatalytic Activity of Halloysite-TiO2 Nanocomposites

Özlem Karahan[1], Nalan Bilgin Öncü[1], Alp Yürüm[1], Güliz Inan Akdemir[2], Cleva W. Ow Yang[2], Ismail Koyuncu[3], Yusuf Menceloğlu[2], Serkan Ünal[1]

[1]*Nanotechnology Research and Application Center, Sabanci University, Istanbul, Turkey*
[2]*Materials Science and NanoEngineering Program, Sabanci University, Istanbul, Turkey*
[3]*Environmental Engineering, Istanbul Technical University, Istanbul, Turkey*

Among various semi-conducters TiO_2 has been extensively used in a wide range of applications (gas sensor, solar cell, batteries and so on) because of its high photochemical stability, low cost and nontoxicity. Halloysites (HNTs) are two-layered aluminosilicate nanotubes with a huge potential as catalyst support because of their inherent hollow structure, large surface area, and abundance of surface hydroxyl groups. In this study, two HNT/TiO_2 nanocomposite structures were prepared using two different initial HNT agglomeration sizes (appx. 5 and 63 micron) by depositing anatase-TiO_2 on the HNT surface using titanium isopropoxide under hydrothermal treatment (Wang, ACS App. Mat. & Int., 2011, 3, 4154,-4158; Li, J. Mater. Chem. A, 2013, 1, 8045-8054). These HNTs can be directly used to attach TiO_2 nanoparticles because of the multitude of surface hydroxyl groups. Therefore, the combination of TiO_2 and HNTs is promising to simultaneously possess excellent photocatalytic activity and adsorptivity. Prepared HNT/TiO_2 nanocomposites were characterized using XRD, TEM, SEM, FTIR and Raman spectroscopies. While pore size distributions, surface areas and pore volumes were measured by BET, phase analysis was performed by applying Rietveld refinement method on the XRD data. FTIR results revealed the presence of hydrogen bonding between TiO_2 and the outer surface of HNTs. These hybrid nanostructures were tested for their photocatalytic activities for decomposition of methyl orange, Reactive Red 120 and Reactive Orange 16 dyes. HNT/TiO_2 nanocomposite structures exhibited significantly higher photocatalytic activity compared to HNTs alone, as demonstrated by UV-Vis spectrometric measurements, which was attributed to the presence of well-dispersed TiO_2 nanoparticles of controlled size on the HNTs surface, as evidenced by HRTEM analysis. This research is supported by The Scientific and Technological Research Council of Turkey (TUBITAK) grant number 113Y350.

Keywords: Halloysit nanotubes, TiO2 nanoparticles, Hydrothermal method, photocatalyst

MN-P-052 [Macromolecules & Nanotechnology]

Fluorescent electrospun nanofibers from dansyl-functional [poly(methyl methacrylate-co-(2-hydroxethylmethacrylate)]2-b-poly(ethylene glycol) terpolymer

Hilal Kuday[1], Okan Günaydın[2], Mesut Görür[3]

[1]Department of Organic Chemistry, Istanbul Yeni Yuzyil University, Istanbul, Turkey
[2]Department of Chemistry, Gebze Technical University, Kocaeli, Turkey
[3]Department of Chemstry, Istanbul Medeniyet University, Istanbul, Turkey

Electrospinning is the most efficient method for generating sub-micron polymer fibers from various materials. Electrospun nanofibers have found wide variety of applications, including tissue engineering, filtration, enzyme immobilization, and sensors, due to their porous structures with high surface area [1].

Poly(ethylene glycol) (PEG) and Poly(2-hydroxyethyl methacrylate) (PHEMA) are a well-known water soluble polymers with good biocompatibility which are used in different applications, such as medicine, industrial, biosensor, drug-delivery systems [2,3]. PMMA could be copolymerized with HEMA to tailor the hydrophilicity and wettability of PMMA, which can then be used in special applications, particularly, in solid-state pulsed dye lasers and in biomedical fields [4]. The dansyl functional group and its derivatives have been drawing wide attention because of their extensive applications [5]. The fluorescence from the dansyl group is relatively insensitive to quenching by oxygen and can be used as a specific fluorescence derivative reagent. Dansyl group is therefore an important fluorescence label used mainly in the studies and applications concerning separation, detection and conformation of amino acids, proteins and in the determination of clinical samples [6]. In this work, dansyl-functional [Poly(MMA-*co*-HEMA)]2-*b*-PEG terpolymer was synthesized in a three-step procedure, including atom transfer radical polymerization (ATRP) and nucleophilic substitution reaction. Nanofibers were produced from solution by electrospinning. Chemical structures of the polymers were confirmed by FT-IR and 1H-NMR techniques and relative average molecular weight was determined using gel permeation chromatography (GPC). The morphological, and photochemical parameters of these fibers were obtained by scanning electron microscope and time resolved fluorescence spectrophotometer.

Keywords: Electrospinning, dansyl, terpolymer, fluorescence, ATRP

Figure 1: Dansyl-functional [Poly(MMA-{co}-HEMA)]2-{b}-PEG terpolymer

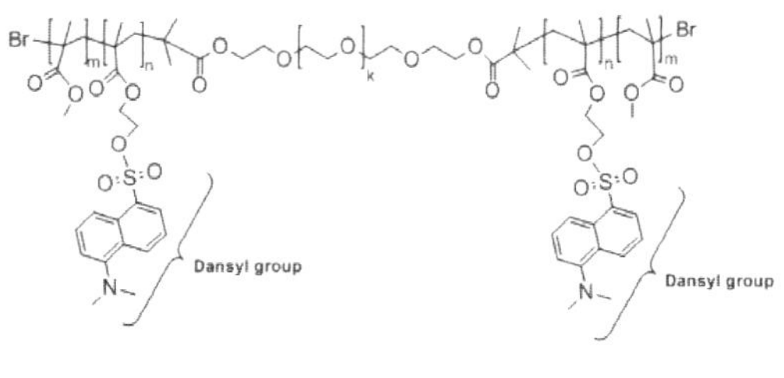

MN-P-053 [Macromolecules & Nanotechnology]

Superior osseo activity on TiO2 via polymer brush-strontium coatings

Sergio E Moya, Danijela Gregurec, Nikolaos Politakos

CIC biomaGUNE, Spain

Strontium, an essential element in bone remodeling cycle, has large perspectives for application in therapies for bone- related diseases and replacements but however its use is limited use due to toxic effects and unwilling side effects due to its biodistribution in the body. In this study we show a remarkable positive effect of Sr+2 on the osteoblast cells by confining its action to the implant site using polyacrylyc acid (PAA) brush at the interface. This approach enables Sr+2 activity on the cells beeing in direct contact with the titanium surface. The concentration of strontium can be varied through the grafting density of PAA to the titania surface. At higher grafting density brushes offer more carboxyl groups for Sr binding. Initial adhesion of osteoblast is superior on these structures compared to bare TiO2. Moreover, these coatings show ability to promote bone tissue mineralization and retaher rapid mineralization is observed at higher Sr2+ concentration. The synergistic effect of the polymer coating enhancing biocompatibility and the osseoactive strontium is confirmed with increased alkaline phosphatase activity, the main gene marker of the mineralized extracellular matrix.

Keywords: polyelectrolyte brushes, surface coatings, titania, osseointegration

MN-P-054 [Macromolecules & Nanotechnology]

Low Density Polyethylene Films Coated with Essential Oil Loaded Halloysite Nanotubes by Layer-by-Layer Technique for Antibacterial Surfaces

Buket Alkan[1], Ekin Şehit[1], Cüneyt Erdinç Taş[1], Serkan Ünal[2], Fevzi Çakmak Cebeci[1], Yusuf Ziya Menceloğlu[1], Hayriye Ünal[2]

[1]Material Science and Nanoengineering Department, İstanbul, Turkey
[2]Nanotechnology Research and Application Center, İstanbul, Turkey

Antibacterial polymeric materials that can inhibit the growth of pathogenic microorganisms on their surfaces hold great promise to prevent and mitigate bacterial infections. In this study, antibacterial essential oils were integrated onto polymeric film surfaces by using nanoencapsulation and Layer-by-Layer (LbL) thin film coating techniques. Halloysite nanotubes (HNTs) which are natural clay nanoparticles with hollow tubular structures were utilized for the encapsulation and sustained release of essential oils. Essential oil loaded HNTs were coated onto the surface of low density polyethylene (LDPE) films by using dip LbL assembly where multilayers of chitosan and HNTs were deposited as the cationic and anionic layers, respectively. Furthermore, spray LbL approach was performed on larger films surfaces to demonstrate scalability. The distribution of HNTs on the LDPE film surfaces that were obtained by dip and spray LbL was investigated with Scanning Electron Microscopy and the correlation between the layer number and the film thickness was determined by surface profilometry. Lastly, LBL coated HNT/PE films were shown to reduce the viability of *Aeromonas Hydrophila* up to 90%. Polymeric films developed in this study have a great potential for food packaging applications as they are composed of natural and non-toxic components which do not pose any risks to human health and yet present strong antibacterial activity against pathogenic microorganisms.

Keywords: Low Density Polyethylene, Halloysite Nanotube, Layer-by-Layer, Antibacterial Polymeric Material

MN-P-055 [Macromolecules & Nanotechnology]

Study of effect of alumina on aluminium characteristic properties in composite alloys

Amel Bourbia[1], Hayette Bedboudi[2], M. Yacine Dibili[3]

[1]*Departement of Physics, Preparatory School For Science And Technique (Epst)- Annaba, Algeria*
[2]*2Laboratory of Magnetism And Spectroscopy of Solids (Lm2s), Department of Physics, Faculty of Science, Badji-Mokhtar University, Pb 12 Annaba, 23000 Algeria*
[3]*Laboratory of Magnetism And Spectroscopy Of Solids (Lm2s), Department of Physics, Faculty of Science, Badji-Mokhtar University, Pb 12 Annaba, 23000 Algeria*

This article focuses on crystalline microstructure and intrinsic hardness properties of Al-Al2O3 alloys solidified by an electromagnetic high frequency (hf) melting process from cold compacted mixtures of pure aluminium and alumina (α-Al2O3) powders. The crystalline microstructure characteristics were investigated by X-ray diffraction (XRD) spectra measures, optical micrography observations. The hardness was evaluated by Vickers microhardness tests. It was found that the crystalline microstructure of the as-melted Al-Al2O3 alloys appear as a mixture of the fcc Al solid solution and the rhomboedric α-Al2O3 alumina phases. The surface morphologies exhibit a refined texture. The measured microhardness Vickers Hv0.2 gives a constant value of around 300 MPa. This observed strengthening of Al by Al2O3 in Al-Al2O3 composite alloys is essentially due to fine Al2O3 particles reinforcement of Al matrix.

Keywords: Elaboration processes, Al alloys, Microstructure, Strengthening.

MN-P-056 [Macromolecules & Nanotechnology]

Preparation of PEG Modified Single-Walled Carbon Nanotubes: CNT and PEG Chain Length Effect

Ayhan Ünlü[1], Mehdi Partovi Meran[2], Gürkan Hızal[3], Nilgün Yavuz[4], F. Seniha Güner[1]

[1]*Department of Biyophisics, Trakya University, Edirne, Turkey*
[2]*Department of Chemical Engineering, Istanbul Technical University, Istanbul, Turkey*
[3]*Department of Chemistry, Istanbul Technical University, Istanbul, Turkey*
[4]*Department of Energy Science and Technology, Istanbul Technical University, Istanbul, Turkey*

Carbon nanotubes (CNTs) have stimulated a great interest due to their unique properties as one of the most promising nanomaterials for a variety of biomedical applications. CNTs have the potential to be used not only in imaging but also for drug delivery. The application of carbon nanotubes in many different fields, e.g. in materials chemistry or in bionanomedicine, does require their surface derivatization to overcome the low solubility in organic or aqueous media. To overcome this problem, CNT surfaces should be modified by noncovalent coating or by covalent functionalization approaches. Compared to the chemical functionalization, non-covalent functionalization has two main advantages: (i) it could be operated under relatively mild reaction conditions, and (ii) the perfect graphitic structure of CNTs could be maintained. In light of the unique properties of carbon nanotubes and the hydrophilic and flexible ability of PEG chains, we examined noncovalent binding of different methoxy PEG to SWCNT using THF and Dichloromethane as solvent. Fabrication of the PEG-modified SWCNT was planned in several stages. First, monocarboxylic acid-terminated PEG (PEG-COOH) was synthesized according to a procedure via reaction of Me-PEG with succinic anhydride. Subsequently, PEG-Pyrene was prepared via an esterification reaction of PEG-COOH with Pyrene Methanol. Finally, SWCNT and PEG-Pyrene heterogeneous mixture was sonicated in a low-power sonic bath and stirred vigorously in THF to gain PEG modified SWCNT. Three methoxy PEG with different molecular weight were noncovalently loaded via π-stacking onto side walls of SWCNTs. Various techniques including TGA, NMR and TEM, were employed to characterize each sample synthesized in the study. TGA analysis indicated that shorter SWCNT has higher PEG chain density. Moreover, noncovalent binding efficiency significantly increased when THF was used as solvent.

Keywords: CNT, PEG-modified CNT, SWCNT, Noncovalent functionalization

PEG Modified Short and Long CNTs

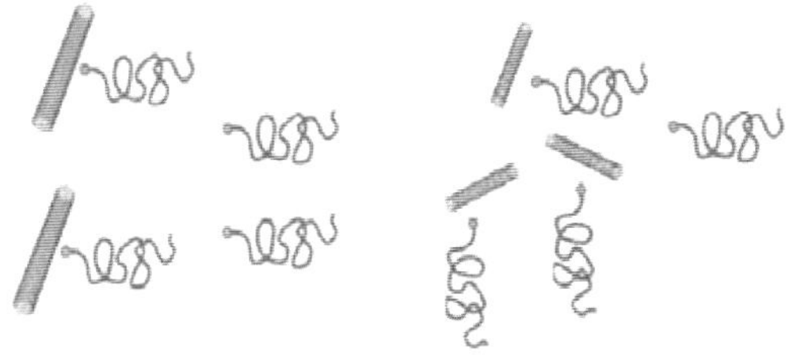

a) PEG Modified Long CNTs b) PEG Modified Short CNTs

MN-P-057 [Macromolecules & Nanotechnology]

Corrosion behaviour of DC magnetron sputtered aluminium coatings

Bahroune Moufida, Mosbah Asma, Guenfoud Fatma, Saker Abdelhamid

Departement of CPST, Nationa and Superior School of Mines, Annaba, Algeria

Protection of steel parts from corrosion is of ground importance for the industry as mechanical resistance is to be maintained. Corrosion protection by metal coatings using wet processes is still widely applied. However, such processes are not environmentally friendly because of the required disposal of heavy metal containing effluents. This is particularly true for hard chromium plating. Physical vapour deposition (PVD) offers a good alternative to replace the electrolytic deposits of hard chromium. For that purpose, base aluminium coatings (Al-M) were deposited by d.c magnetron sputtering and investigate in order to answer the mechanical and physico-chemical requirements. The objective of this study is to elaborate architecture multilayers nanostructured with base aluminium coatings. First of all, we elaborated and characterized the binary layers on base of Aluminium (Al-M: M = Mo, Mn and Mg). Afterward, classification by mean corrosion resistance, are done for finding the potential candidates as a sacrificial anode for steels of construction.

Keywords: Cathode sputtering magnetron, thin films, nanostructure, corrosion, aluminium alloys.

MN-P-058 [Macromolecules & Nanotechnology]

Preparation of Paper-based Biosensor for Detection of Phenol Using Laccase Hybrid Nanoflowers

Cansu Ozkardas[1], Cevahir Altinkaynak[2], Serife Sacmaci[1], Nalan Ozdemir[1], Ismail Ocsoy[2]

[1]Department of Chemistry, Faculty of Science, Erciyes University, Kayseri, Turkey
[2]Department of Analytical Chemistry, Faculty of Pharmacy, Erciyes University, Kayseri, Turkey

Enzymes as macromolecules have wide applications areas such as chemistry, biochemistry, biomedical, industry etc. However, free forms of enzymes have a short life time and this situation limites applications of enzymes in many areas. Also separation of free enzyme and reuse of it is difficult. So, it is necessary to increase the efficiencies, activities, stabilities and recoveries of enzymes in order to use them as biocatalysts. Recently, Zare and co-workers reported an elegant approach for the synthesis of immobilized enzymes in the form of nanoflower with highly enhanced catalytic activity and stability [1]. These flower-like hybrid nano structures are developed and used for various applications. Phenol compounds have high toxicity to the living organisms and unfortunately they are widely distributed as environmental pollutants and these compounds can cause several human health problems such as cancer, nervous and respiratory system problems, abnormal function of thyroid etc [2]. So, sensitive, rapid and on-site detection of these compounds are very important.

In this study, laccase-Cu2+ hybrid nanoflowers were prepared (Figure 1) and some characteristics of them were determined. And these prepared nanoflowers were incorporated onto a paper and then tested for phenol detection as paper-based biosensor.

The synthesis of nanoflowers were accomplished using a described method before [3,4]. The structure of the synthesized nanoflowers were confirmed by FT-IR, XRD and EDX. The catalytic activity of synthesized nanoflowers were evaluated. Then the synthesized nanoflowers were injected onto the paper, dried in the air at room temperature. In a typical experiment, different concentrations laccase nanoflowers and phenol solutions were tested. Finally, we demonstrated a paper-based biosensor for visual detection of phenol.

Keywords: Enzyme immobilization, hybrid nanoflower, laccase, phenol, paper-based

Figure 1: SEM image of synthesized Laccase nanoflower

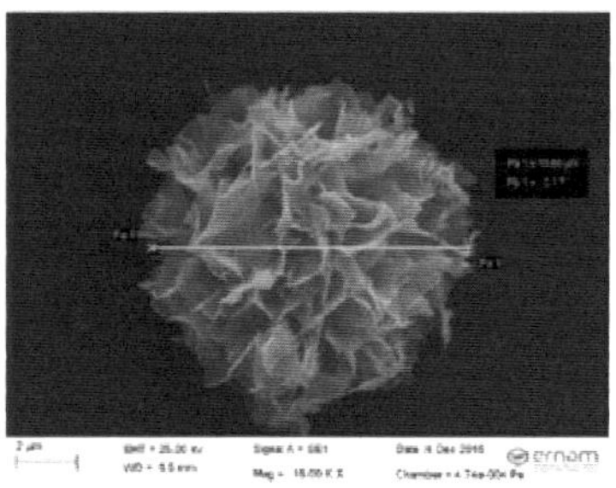

MN-P-059 [Macromolecules & Nanotechnology]

Effect of the preparation method on structural properties of magnetite

Laaldja Boukhobza Meddour[1], Yasmina Bellal Hammiche[1], Amar Djadoun[2], Nadia Henda[3], Ahmed Ezzine Hassanine[3]

[1]L.mccco., Faculté de chimie, Usthb. BP32 El Alia, Bab Ezzouar, 16111, Alger, Algérie.
[2]Lgp, Fstgat, Usthb, BP32 El Alia, Bab Ezzouar 16111, Alger, Algérie
[3]Dcemp, credec, rue ouled fayet, Alger, Algérie

This work is devoted to the influence of preparation process on the physical-chemical properties of a magnetite Fe3O4. The samples were synthesized by co-precipitation and hydrothermal methods, applying chloride precursor procedure and using sodium hydroxide as precipitating. The obtained compounds were characterized by X-ray diffraction (XRD), nitrogen adsorption desorption isotherms (BET/BJH), electron microscopy (SEM/EDX) and temperature programmed reduction (TPR) techniques. In this work, it is demonstrated that the chemical synthesis affect the crystal properties, microstructure, surface area and reducibility. The diffractograms shown by XRD (fig.1), exhibit diffraction lines corresponding to the spinel phase Fe3O4, no secondary phase is detected, but the presence of small amounts of other species cannot be totally excluded. Crystal growth of magnetite appears to depend strongly on the method of preparation as indicated by the calculated values of the crystallite size, lattice parameters and specific surface reported in table-I

Keywords: magnetite, spinel, nanomaterial,

Fig.I. Diffractograms of magnetite obtained by co-precipitation and hydrothermal treatment.

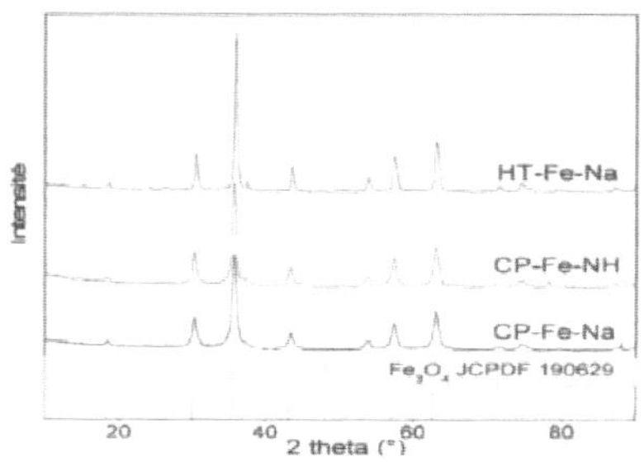

Figure 1: diffractograms of magnetite obtained by co-precipitation and hydrothermal treatment.

Table I: Cristallite sizes, lattice parameters and specific areas of de magnetite.

Samples	Fe3O4 JCPDF190629	CP-Fe-Na	CP-Fe-NH	HT-Fe-Na
Cristallites sizes (nm)	-	13.05	14.41	24.75
Lattices parameters (A°)	8.396	8.320	8.394	8.320
Surface BET(m2/g)	-	85.9	67.8	50.7

This table shows the crystallite sizes, the lattice parameters and specifics surfaces of magnetite samples synthesized by co-precipitation and hydrothermal

MN-P-060 [Macromolecules & Nanotechnology]

Structure-property relationship in the PVC composites reinforced with carbon nanotubes

Ali Reza Zanjanijam[1], Mohammad Bahrami[2], Morteza Hajian[2], Maryam Ramezani[1]

[1]Iran Polymer and Petrochemical Institute, Tehran, Iran
[2]Faculty of Chemistry, University of Isfahan, Isfahan, Iran

Carbon nanotubes possess extraordinary mechanical properties such as unique flexibility, high stiffness and tensile modulus. This combination of properties makes the nanotubes ideal reinforcement for preparation of multifunctional composites [1, 2]. Polyvinylchloride (PVC) is a polymer which is employed in many versatile applications such as automobile industry, toys, electrical wire insulation, and etc [3]. This work reports preparation of the PVC/ single wall carbon nanotubes (SWCNT) composites via plastisol curing method and investigates morphology-mechanical properties relationship in the fabricated composites. SEM micrographs for the samples containing 0.25-0.75 wt.% SWCNT showed homogeneous dispersion of the nanotubes in the PVC and strong interfacial interactions between polymer and filler. However, at 1 wt. %, agglomeration of the nanotubes was observed. The mechanical tests revealed that in the range of 0.25–0.75 wt.% SWCNT, there is a continuous increase in tensile strength, Young's modulus and elongation at break of the composites. It is ascribed to the high aspect ratio and intrinsic properties of the SWCNT, and good impregnation of the nanotubes with the PVC chains that provide simultaneous improvement in stiffness and toughness of the composites. Incorporation of higher contents of the SWCNT caused a decreasing trend in mechanical properties. It is due to agglomeration of the nanotubes and inappropriate impregnation of the nanofiller with PVC (as observed in the SEM). In fact, the agglomerates act as imperfections and have a negative effect on the mechanical properties of the samples. It could be concluded that mechanical properties of the plasticized PVC increases with addition of the SWCNT, as long as complete impregnation of the nanofiller with polymeric matrix is ensured.

References:

[1] Li et al, Int. J. Solids. Struct., 40, 2487 (2003).
[2] Chang et al, J. Mech. Phys. Solids, 51, 1059 (2003).
[3] Patrick, Practical Guide to PVC. Rapra Technology Limited, (2005).

Keywords: Composites, Polyvinylchloride, carbon nanotubes

MN-P-061 [Macromolecules & Nanotechnology]

Modification of Halloysite Nanotubes with Styrene-Maleic Anhydride Copolymers

Refik Arat[1], Nurseli Uyanık[2]

[1]ITU Graduate School of Sci.Eng.&Techn., Chemistry Dept., Maslak 34469 Istanbul/Turkey
[2]ITU Faculty of Sci. and Letters,Chemistry Dept., Maslak 34469 Istanbul/Turkey

Incorporation of nanoclay materials into polymer matrix is a big challenge to obtain polymer nanocomposites. Modification can be used as a method to overcome that topic. The aim of this study is adding of halloysite nanotubes (HNTs) to polystyrene (PS) without using any compatibilizer for preparing PS/HNTs nanocomposites. In this study, HNTs were modified with styrene-maleic anhydride copolymers (SMA) to improve their dispersion in PS matrix. Modification of HNTs was prepared through the esterification reaction by using p-tolueno sulfonic acid (PTSA) as the catalyst between SMA and –OH groups of HNTs'. Firstly, SMA was dissolved in THF, and then introduced into a glass flask. Certain amount of HNTs was added into the flask (the weight ratio of SMA/HNTs was chosen 1:1). 4wt% PTSA of total amount of component was added and the solution refluxed for 24 h under nitrogen atmosphere. After the reaction, the product was isolated and purified by centrifugation/redispersion process with THF solution (for 10 min at 10,000 rpm, 3 times) to remove the excess SMA. And then, the product was dried at 60 °C for 24 h to remove the solvent residues and allow the further condensation take place. Modified HNTs with SMA was characterized by fourier transform infrared spectroscopy (FTIR), thermal gravimetric analysis (TGA), differential scanning calorimetry (DSC) and X-ray diffraction analysis (XRD). Thermal and XRD analysis results showed that SMA was grafted to HNTs, magnificently. All these analysis results were supported by FTIR analysis.

Keywords: Halloysite nanotubes, Styrene-maleic anhydride copolymer, Modification of halloysite nanotubes, Polystyrene/halloysite nanotubes nanocomposite

Reaction schema of HNTs modification

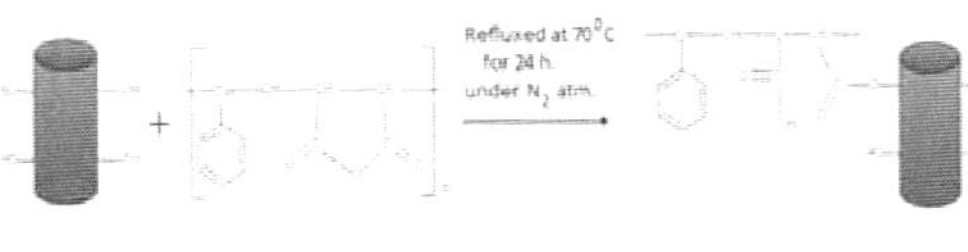

MN-P-062 [Macromolecules & Nanotechnology]

The effect of nanoclusters on the mechanical strength of dental composites

Şeref Okay[1], Aslı Dörtler[1], Nadir Kiraz[1], Meltem Asiltürk[2]

[1]Department of Chemistry, Akdeniz University, Antalya, Turkey
[2]Department of Material Science and Engineering, Akdeniz University, Antalya, Turkey

In recent years, dental composites have started to replace dental alloys because of aesthetic concerns. Polymer matrices, containing nano- and micro-sized filler particles are used to obtain composites having good mechanical strength. Nanoparticles provide translucency and natural tooth color as well as good polishability and wear resistance. However, high loading of nanoparticles into polymer resins is limited due to the large surface area to volume ratio. Structures known as clusters have been previously developed to solve this problem. In clusters, each nanoparticle maintains its form and optical properties. As they are entrapped in polymer matrix in the form of agglomerates, they are bigger in size, smaller in volume and can be incorporated into composites more easily.

In this study, mixtures of diurethane dimethacrylate(DUDMA), triethyleneglycol dimethacrylate(TEGDMA), ethoxylated bis phenol A dimethacrylate (bis-EMA) and bis phenol A diglicidyl dimethacrylate (bis-GMA) with camphorquinone were prepared. Dimethylaminoethyl methacrylate (DMAEMA) was used as co-initiator. Each mixture was loaded with 80 wt% of silanated filler particles of various sizes and clusters.

In order to prepare clusters, nanoparticles were dispersed in polymer resins, light-cured and then crushed into smaller pieces. Prepolymerized particles were ground by ball-milling and clusters below 25 microns were produced. The produced clusters were incorporated in the composites. Flexural and compressive strength test were applied to the prepared composites. The effect of cluster amount on mechanical strength was examined and the results were evaluated.

Keywords: nanoclusters, dental composites, light-cured resins, radical polymerization

MN-P-063 [Macromolecules & Nanotechnology]

Solid-State Synthesis of a New Core– Shell Chiral Nanocomposite of Poly [(±)-2-(sec-butyl) aniline] and Nanosilica- Supported Perchloric Acid as Dopant Solid Acid

Ali Reza Modarresi Alam, Abdolkarim Farrokhzadeh

Department of Chemistry, Faculty of Science, University of Sistan and *Baluchestan, Zahedan, Iran.*

In continuation of our recent works on the application of heterogeneous reagents and solid acids, "1. Modarresi-Alam, A. R., Zafari, S. and Miandashti, A. R. Polym. Adv. Technol. 2015, 26, 645–657. 2. Modarresi-Alam, A. R., Amirazizi, H. A., Movahedifar, F., Farrokhzadeh, A., Asli, G. R. and Nahavandi, H. J. Mol. Struc. 2015, 1083, 17–26. 3. Movahedifar, F. and Modarresi-Alam, A. R. Polym. Adv. Technol. 2016, 27, 131-139. 4. Farrokhzadeh, A. and Modarresi-Alam, A. R. J. Solid. State. Chem. 2016, 237, 258-268." herein we wish to report the synthesis of a core-shell nanocomposite of poly [(±)-2-(sec-butyl) aniline] (with the bulky chiral substituent) in solid-state. Poly [(±)-2-(sec-butyl) aniline]/silica-supported perchloric acid nanocomposites were synthesized by the simple combination of poly[(±)-2-sec-butylaniline] base (PSBA) and the silica- and nanosilica-supported perchloric acid (NSSPA) as dopant solid acid in solid-state. The morphology and size of samples by the scanning electron microscopy (SEM) and transmission electron microscopy (TEM) were studied and was confirmed core-shell structure. Furthermore, their coating were investigated by X-ray photoelectron spectroscopy (XPS), SEM-map and energy-dispersive X-ray spectroscopy (EDX). The XPS and Elementary Analysis (CHNS) results confirm nigraniline oxidation state and complete doping for nanocomposites (about 75%). The conductivity of samples was (≈ 0.05 S/cm) in agreement with the percent of doping obtained of the XPS analysis. Also, contact resistance was determined by circular-transmission line measurement (TLM). The structure of silica was studied by 29Si-NMR. The key benefits of this work are the preparation of core-shell chiral nanocomposite without application of any template under green chemistry and solid-state condition with the delocalized polaron structure, high conductivity, the improvement of the processability by inclusion of the 2-sec-butyl group and the use of dopant solid acid (NSSPA) as dopant.

Keywords: Chiral conducting nanocomposite, core–shell, poly [(±)-2-sec-butylaniline], nanosilica-supported perchloric acid, solid-state.

TEM images of various sites of nanocomposite PSBA/NSSPA.

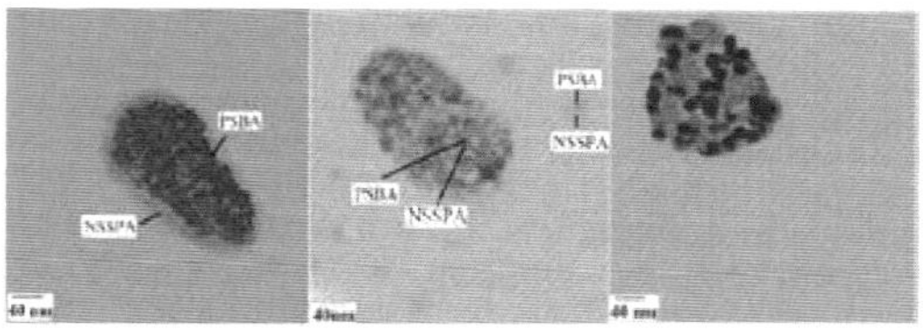

MN-P-064 [Macromolecules & Nanotechnology]

The Effect of Different Hydrophobic Agents on Staining and Color Change of Dental Composites

Aslı Dörtler[1], Şeref Okay[1], Nadir Kiraz[1], Meltem Asiltürk[2]

[1]Department of Chemistry, Akdeniz University, Antalya, Turkey
[2]Department of Material Science and Engineering, Akdeniz University, Antalya, Turkey

In recent years, one of the most popular research fields is dental composites. It has been considered very important that the aesthetic dental restorations have natural appearance, because dental composites are prefered not only for restoring teeth, but also for having aesthetic appearance. Permanant stains on dental composites and changes in color affect the aesthetic appearance. So, in this study, increasing the resistance of material to stains and color changes over time was aimed, by incorporating hydrophobic agents into the dental composite. Staining tests were applied to determine the color change in the composites.

Butyl methacrylate, t-butyl methacrylate, i-butyl methacrylate, lauryl methacrylate and hexyl methacrylate were used as hydrophobic agents in diurethane dimethacrylate (DUDMA) - bis phenol A glicidyl dimethacrylate (bis-GMA) - triethyleneglycol dimethacrylate (TEGDMA) composite resins to prevent staining and color change. Camphorquinone(CQ) was used as a photoinitiator in the composite resin. Silanated ceramic particles of various sizes were added into the monomer/hydrophobic agent mixture to produce composites. Disc-shaped specimens (n=4), which were 15mm in diameter and 1mm-thick, were prepared by using stainless steel molds. The specimens were kept in coffee solution for 6 hours and color change was checked every hour. The color change which occured during staining test was measured with Vita Easyshade Advance color measurement device. The measured values(L, a, b) were used to calculate *dE*, which shows the change in color. The same test was applied to some commercial products. The results of experimental and commercial products were compared and evaluated.

Keywords: staining test, hydrophobic agents, color change measurement, dental composites, light-curing resins

MN-P-065 [Macromolecules & Nanotechnology]

Multifunctional EPDM/PP-g-MA/Bioengineering Polyesters/Colloidal Poly(MA-alt-1-dodecene)-g-SiO2/Organoclays Nanocomposites by Reactive Extrusion

Bayram Ali Göçmen[1], Zakir M. O. Rzayev[1], Deniz Demircan[2], Gunay Kibarer[2]

[1]The Institute of Science and Engineering, Division of Nanotechnology and Nanomedicine, Hacettepe University, Beytepe, 06800 Ankara, Turkey
[2]Department of Chemistry, Faculty of Science, Hacettepe University, Beytepe, 06800 Ankara, Turkey

Multifunctional polymer blend nanocomposites (NCs) consisting EPDM elastomer as a matrix polymer, bioengineering polyesters (PLA and PCL), PP-g-MA compatibilizer and covalently encapsulated colloidal copolymer-g-silica nanoparticles (NPs) as reactive compatibilizer -nanofillers, and organoclays (reactive ODA-MMT and non-reactive DMDA-MMT) nanofillers were fabricated in melt by a one-step reactive extrusion using a Rondol twin-screw compounding line (diameter of twin-screw = 21 mm and L/D ratio = 25:1). The chemical and physical structures, morphology, thermal behaviors, mechanical and rheological properties as well as effects of origin organically and polyester, and colloidal copolymer-g-silica (NPs) on the main properties of the nanocomposites were estimated by FTIR, solid state 13C and 29Si NMR, XRD, SEM-TEM, DSC-TGA, and dynamic rotary rheumatic analysis methods. The effects of bioengineering polyesters and their molecular mass, origin of organoclays, colloidal copolymer-g-silica on the thermal, mechanical and rheological properties and morphology of NCs were evaluated. Moreover, colloidal copolymer-g-silica NPs as reactive compatibilizer-nanofiller play an important role in the formation well dispersed morphology due to affective in situ physical and chemical interfacial interactions in multifunctional polymers/nanofillers blends. A basic copolymer-g-SiO2 encapsulated NPs as reactive compatibilizer-nanofiller was synthesized by sol-gel method. The XRD patterns and reflection parameters, as well as DSC analyses indicate predominantly semi-crystalline structures with predominantly amorphous areaes of NCs due to colloidal structure of used copolymer-g-SiO2 NPs. SEM and TEM morphological structures strongly depend on the copolymer-g-SiO2 NPs and origin of loaded organoclay and biopolymer. The results of the mechanical (shear stress) and rheological (viscosity) properties of NCs confirmed that structural factors such as chemical and physical interfacial interactions play an important role in the formation of self-assembled nanostructures that singificantly improved the thermal, mechanical and rheological parameters of nanocomposites compared with pristine EPDM due to transport of elestomer matrix polymer to thermoplastic nanocomposite forms during reactive extrusion in situ processing.

Keywords: EPDM, PP-g-MA, Organosilica NPs, Bioengineering Polyesters, Reactive extrusion

MN-P-066 [Macromolecules & Nanotechnology]

Electrochemical Preparation of Polyaniline nanoparticles

Abdunnaser Mohamed Etorki[1], Ibrahime Shaban Shaban[2], Manal Ali Elharby[2]

[1]*Department of chemistry, Unversity of Tripoli, PO-BOX-13340*
[2]*Department of Environmental Science and Engineering, Libyan Academy, Tripoli. Libya*

The polyaniline (PANI) nanoparticles were synthesized by electrochemically method. Firstly the aniline monomer was distilled under reduced pressure, aniline was dissolved in absolute ethanol at different mol ratios: (1:1),(1:2),(1:3),(1:4),(1:5) and (1:6). Electropolymerization of polyanline nanoparticles was performed by the pulsed potentiostatic method. A freshly cleaved screen printed carbon electrode was used as a working electrode. The polyaniline nanoparticles were characterized by scanning electron microscope, X-ray Diffraction, Fourier transfer infrared spectroscopy, Tunneling electron microscopy and cyclic voltammetry. The electrochemical performances of the samples were tested. The results showed that the dimeters of polyaniline nanoparticles were 30-140nm. The SEM and TEM results showing that spherical nanoparticles were formed. The effect of different non-solvent mediums (ethanol, methanol), scan rate, pH and amplitude of the sonication to size of polyaniline nanoparticles was also studied.

Keywords: polyaniline nanoparticles, electrochemical polymerization, nanochemistry, cyclic voltammetry

MN-P-067 [Macromolecules & Nanotechnology]

In-Situ Preparation of Ag/Cu Alloy Nanoparticles by Photopolymerization Technique

Hicret Kırtay, A. Nadir Sevinç, Nergis Arsu

Department of Chemistry, Yıldız Technical University, Istanbul, Turkey

Metal nanoparticles have attracted much research interest due to their unusual chemical and physical properties, such as catalytic activity, novel electronics, optics, and magnetic properties, and they have potential applications in solar cells and biosensors[1].

Alloy nanoparticle systems have been found to exhibit optical limiting properties due to surface plasmon resonance and have been used in biodiagnostic applications. Alloy nanoparticles are materials used to tune the position of surface plasmon resonance, and thus help to produce materials for use in nonlinear optical applications.. The position of surface plasmon resonance for Ag NPs is about 450 nm. The position of surface plasmon resonance for Cu NPs is 570 ~ 580 nm. The prediction of the optical properties of such alloy systems is desirable if they are to be used in the design of optical devices. The aim of this study was to synthesize Cu/Ag nanoparticles by in- situ photopolymerization techniques. in reproducable size and shape. In-situ photochemical reduction in the presence of various photoinitiators from metal (Cu and Ag) salts to metal (Cu/Ag) nanoparticles was achieved either in polymer matrix or in solvent Morphological properties were determined by using SEM techniques.

Keywords: Ag/Cu alloy, nanoparticles, polymerization

MN-P-068 [Macromolecules & Nanotechnology]

Heterogeneous Activation of Oxone by Graphene Oxide for the Photodegradation of Rhodamine B Under Visible Light Irradiation

Hasan Yolcu[1], Rabah Boukherroub[2]

[1]*Institut de Recherche Interdisciplinaire (IRI, USR CNRS 3078), Université Lille 1, Parc de la Haute Borne, 50 Avenue de Halley, BP 70478, 59658 Villeneuve d'Ascq, France & Education Faculty, Kafkas University, Kars/36000/Turkey*
[2]*Institut de Recherche Interdisciplinaire (IRI, USR CNRS 3078), Université Lille 1, Parc de la Haute Borne, 50 Avenue de Halley, BP 70478, 59658 Villeneuve d'Ascq, France*

Synthetic dyes is big challenge for the environmental pollution as a consequence of the more common using many industrial areas such as textiles, paper, leather, food, and cosmetics. And they also have chemical stability, toxic and carcinogenic. Hence, search effective way to remove dyes from the water is significant scientific research field. Using some reactive radicals, such a *OH, *O2-, *OOH and SO4-* have emerged as a promising method for degration azo dyes. Recently, sulfate radical SO4•– has gain significance for degradation of azo dyes. This sulfate radical is a very strong one-electron oxidant with a high redox potential (2.5–3.1 V) and has a longer half time compared to •OH (1.8–2.7 V). One advantages of this method that avoids the problem of required as ambient oxygen is the oxidant such as hydrogen per-oixde and ozone. In this paper reports on the positive effect of graphene oxide (GO) in the heterogeneous activation of oxone for the photocatalytic degradation of rhodamine B under visible light irradiation. The effects of operational parameters such as the amount of oxone, GO, dye concentration, on photo mineralization have been analyzed.

Keywords: graphene oxide, ohotodegradation, oxone

MSP-P-001 [Modeling and Simulation of Polymers]

On numerical instabilities in Brownian Dynamics Simulations of Polymer Flows in Microchannels

Arpit Bansal, Prateek Kumar Jha

Department of Chemical Engineering, Indian Institute of Technology Roorkee, India

Brownian Dynamics simulations of polymer chain flowing in a microchannel are performed to explore the numerical instabilities occurring at high flow strengths. Polymer chains are modeled using the bead-spring model comprising of Fraenkel/harmonic springs for chain elasticity, Lennard-Jones potential for polymer-solvent interaction, and the flow effect included as a passive convective force. Flow profile is assumed to be parabolic and flow strength is characterized by Peclet number (Pe) defined as Pe=dv_o/D where d is the diameter of channel, v_o is the flow velocity at the center of channel and D is the diffusion coefficient. Numerical instabilities occurring at high flow strengths give rise to non physical behavior, such as chain stretching beyond the maximum possible length or beads collapsing into each other. We observe that there is a critical Pe value beyond which numerical instabilities appear in Brownian dynamics simulations, the magnitude of which varies with simulation time step. An alternative numerical scheme is proposed to overcome these issues, wherein chain elasticity is incorporated as a bond length constraint and a continuous tension is assumed to act along the chain segments. Newton's method is used to calculate tension values of chain segments while respecting the bond constraint. The results of our simulations appear in accord with experimental results and provide a more realistic description of polymer chains than the Brownian Dynamics simulations.

Keywords: Brownian Dynamics simulations, numerical instabilities, high flow strengths, non physical behavior, continuous tension

Typical simulation snapshots.

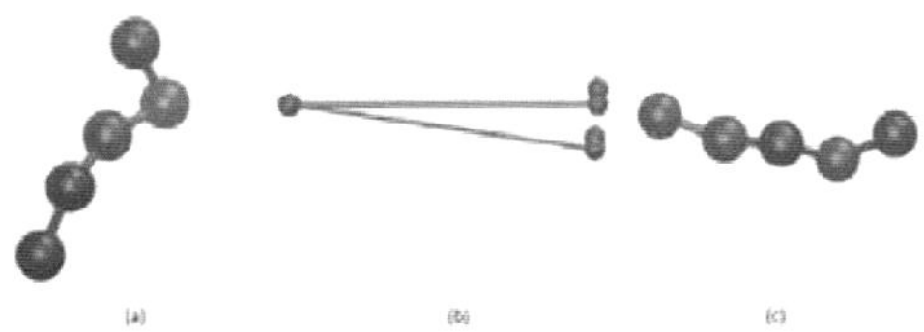

Figure (a): Structure of polymer when under no flow using Brownian Dynamics Simulations. Figure (b): Structure of polymer when under high flow strength using Brownian Dynamics Simulations (showing non physical behavior with beads collapsing and chains stretched beyond the maximum possible length). Figure(c): Structure of polymer when under high flow strength using Continuous Tension in Brownian Dynamics Simulations (non physical behavior is not observed).

MSP-P-002 [Modeling and Simulation of Polymers]

Molecular Dynamic Simulations of Biocompatible Carbon Nanotubes

Pelin Deniz Akkus

Department of Chemical Engineering, Istanbul Technical University, Istanbul, Turkey

Carbon nanotubes (CNTs) are attractive nanomaterials in theranostic cancer therapies due their uniform structure offering excellent physical, chemical and mechanical properties that enable both characterizing and preventing metastatic effects [1]. Poly(ethylene) glycol (PEG) coated CNT complexes are promising theranostic agents especially due to their hydrophilic properties masking these nanocomposites from the immune system, therefore making these biocompatible. These agents can travel in the blood stream for long periods with their valuable cargos such as therapeutics until they reach their target. The purpose of this study is to investigate non-covalent functionalization of single wall CNTs (SWNTs) with PEG chains using full-atom molecular dynamics (MD) simulations in water. 4 nm length-0.7 nm radius SWNT and pyrene functionalized PEG chains of different lengths (Mw =2000 and 5000) were modeled and geometrically optimized using Materials Studio v8.0. MD simulations at 300K were performed using COMPASS force field in periodic boundary conditions, and stability of interactions between free pyrene molecules and the SWNT, and between pyrene-PEG chains and the SWNT were monitored. According to our results, pyrene molecules made stable π-π stacking interactions with SWNT, while scanning a large surface area on SWNT. The effect of fluctuations of long hydrated PEG chains on the strength of π-π interactions between pyrenes and SWNT was apparently insignificant. Nonetheless, fluctuations of PEG chains increased the mean-square displacements of pyrene molecules on the SWNT surface. Our results suggested that biocompatibility of these SWNTs can be maintained in water without damaging their surface and mechanical properties, which are critical for their diagnostic and therapeutic capabilities.

[1] Kievit, F., M. and Zhang, M., 2011. Adv Mater 23(36):H217-47

Keywords: Molecular Dynamics Simulations, Carbon Nanotubes, Polymers

MSP-P-003 [Modeling and Simulation of Polymers]

Effect of reinforcing subgrade soil by geogrid

Ninouh Tarek, Salhi Sadok

Laboratory LGCA Department of Civil Engineering University Of Tebessa Algeria

A geogrid is geosynthetic material (polymer) used in civil engineering and public works (earthworks, road, railway, airport, retaining walls and extra-steep slopes). Geogrids are a great solution to the strengthening and stabilization of infrastructure Projects, because they reduce largely, the formation of cracks and ruts in asphalt pavements. And they are very effective in soft and unstable soils. Geogrids considerably reduce the cost of road project, capable of reducing the required thickness of granular material. They are designed to withstand short-term high dynamic loads (the trucks passing on the highway) or during longer periods of loads. The tensile strength is substantially equal in the longitudinal and transverse directions. Their geometry allows the particles to be caught in the meshes thus increasing soil reinforcement. This research work deals with the numerical simulation of multilayer asphalt pavement structure under traffic loads. The subgrade of pavement reinforced with geogrids. The purpose of this reinforcement is to improve the bearing capacity of the subgrade. We used numerical simulations to analyze the process of improving the soil bearing carrier to reduce the thickness of the pavement layers. This decrease in thickness of the pavement layers will reduce the consumption of aggregates and also environmental protection.

AASHTO «American Association of State Highway and Transportation Officials ", 1993.
Austin, D. N., and Coleman, D. M., "A field Evaluation of Geosynthetic-Reinforced Haul Roads Over Soft Foundation Soils," Proceedings of the conference Geosynthetics, British Columbia, Canada, vol. 1, pp. 65-80, 1993..
Bathurst R. J., Palmeira, E.M., Stevenson P.E., Tatsuoka F., Zornberg J.G., "Advances in Geosynthetics Materials and Applications for Soil Reinforcement and Environmental Protection Works". EJGE, 2008.
Burd, H. J., and Brocklehurst, C. J. "Finite Element Studies of the Mechanics of Reinforced Unpaved Roads," Proceedings of the Fourth International Conference on Geotextiles, Geomembranes and Related Products, Balkema, Vol. 1, The Hague, Netherlands, pp. 217-221, 1990.
BURMISTER D.M. "The theory of the stress and displacements in layered systems and applications of design of airport runway". Proceding of the Highway Research Board, 23. 1943,pp. 126-148.

Keywords: Geogrid, Modeling, Pavement, Reinforcement, Subgrade

MSP-P-004 [Modeling and Simulation of Polymers]

Computational Investigation of The Complete Reaction Path of Poly(L-lactide) initiated by Tin(II) 2-Ethylhexanoate with DFT methods

Merve Güreşci, Sıla Gümüştaş, Mehmet Balcan, Armağan Kınal

Department of Chemistry, Ege University, İzmir, Turkey

Polyester was first approved for use as biodegradable suture in the 1960s. Since then, various products based on cyclic ester have been produced as biocompatible materials. Polylactide is one of the most prominent examples, which has been being developed for use as a commercial polymer for packaging, film and fiber productions [1].

Ring opening polymerization (ROP) is one of the most widely used and effective technique for obtaining polyesters with high molecular weight. In the literature, two methods of ROP mechanism are proposed: the activated monomer mechanism and the coordination-insertion mechanism. The coordination-insertion is the best method due to its advantages such as easy control of molecular weight, the lower risk of side reactions and the higher molecular weight obtained [2]. Although many there are lots of experimental and theoretical studies about ROP of lactide, its mechanism has not been fully understood.

This study presents a theoretical study of the coordination-insertion ROP mechanism of L-lactide initiated by Sn(Oct)2 in the presence of methanol as a co-initiator. The DFT-B3LYP and M06-2X methods were utilized to determine the geometries, energies and vibrational frequencies of all stationary points along with polymerization reaction path in gas phase and in toluene medium. Initial activation barriers of the ROP reaction were predicted by B3LYP and M06-2X methods as 26.28 and 22.86 kcal/mol, respectively. GAUSSIAN 09 program suite was employed throughout the study [3].

References:

1. Middleton J.C. and Tipton A.J. Biomaterials 2000, 21: 2335-2346
2. Ryner M., Stridsberg K., Albertsson A.C., Schenck H. Von and Svensson M., Macromolecules, 2001, 34: 3877-3881
3. Frisch, M. J. et al., Gaussian 09, Revision A.1, Gaussian, Inc., Wallingford CT, 2009

Keywords: density functional theory, coordination-insertion mechanism, cyclic esters

MSP-P-005 [Modeling and Simulation of Polymers]

Computational study on organization of PAA-Vancomycin assembly from electro-neutral complex to Nanoplex formation

Suresh B Vepuri, Dhiraj R Sikwal, Rahul S Kalhapure, Sanjeev Rambharose, Mahmoud E Soliman, Chunderika Mocktar, Thirumala Govender

Discipline of Pharmaceutical Sciences, University of KwaZulu-Natal, Westville, Durban, South Africa

Nanotechnology provided a tremendous platform to create various Nano Drug Delivery Systems (NDDS) to bring significant advancements in the disease treatment strategies. At our laboratory we design, synthesize, characterize and evaluate the NDDS for effective antibiotic delivery against Staph. aureus and MRSA. We also characterize the intermolecular forces that govern the formation, stability and drug delivery performance of NDDS by using molecular and material modelling tools. In this paper we like to present the details of in silico study on our recently reported nano drug delivery system produced by the polyelectrolyte complex of poly (acrylic acid) (PAA) and vancomycin (VCM). The essential electro neutral complex formation and subsequent nano aggregation for the nanoplex production was successfully explained by using molecular modelling study. The coordination number for the VCM(dimer) Zij=5.665, which is equivalent to approximately 11VCMmolecules and 22 cationic units combined clearly explained the ability of VCM to form electro neutral complex with poly ionic PAA by electrostatic interactions. Further, the warranted electro neutral complex structure within the accessible PAA molecular volume for VAM accommodation was explained based on the experimental evidence for VCM occurrence as a dimer. Later, the unit cell dimensions for primitive nanoplex packing model obtained from molecular dynamics simulation confirmed the formation of hexagonal like particles, as observed in the SEM study. Overall, the molecular modeling study of the electro neutral complex, followed by self-associated aggregation was well correlated with experimental observations and also supported the concept behind the construction of VCM–PAA nanoplex drug delivery system.

Keywords: Poly acrylic acid, vancomycin, electroneutral complex, nanoplex and molecular modelling

MSP-P-006 [Modeling and Simulation of Polymers]

A DFT Study on the Stereoselective Propagation In Free Radical Polymerization of Acrylamides

Gülru Kayık, Nurcan Şenyurt Tüzün

Department of Chemistry, Istanbul Technical University, Istanbul, Turkey

In this study, the stereoselective radical propagation reactions of N,N-dimethylacrylamide (DMAAm), N-methyl-N-phenylacrylamide (MphAAm) and N,N-diphenylacrylamide(DPAAm) monomers were investigated in their free radical polymerization with density functional theory (DFT) calculations. The bulkiness of substituent, temperature and the polarity of the solvent were considered to elucidate their effect on the stereospecific addition modes. For this purpose, model reactions at dimeric stage have been considered. The experimental stereospecificity (J. Polym. Sci. Part A: Polym. Chem. 2009, 47, 1192-1203) in these monomers could be reproduced by DFT calculations at the B3LYP/6-31+G(d,p) level of theory. Gauche conformation was favored in the pro-meso and pro-racemo additions of the considered monomers in general. Further refinement of the quantitative trend in m:r ratios was possible with implicit solvent (IEFPCM) calculations. The experimentally observed effect of the polarity of the solvents (toluene, THF, chloroform and 2-propanol) on the stereospecificity could be reproduced both qualitatively and quantitatively in the calculations. In the experimental study, tartrate molecules have been used as stereospecifity inducer in DMAAm. The tartrate as tacticity inducer has changed the addition orientation in DMAAm. In this study, we have modelled pro-meso and pro-racemo addition modes of DMAAm monomer in the presence of tartrate to account on the experimentally observed change in stereospecificity from iso to syn. Calculations have shown that although gauche orientation is preferred without tartrates in pro-meso addition, the presence of tartrates enforces a synperiplanar conformation. In pro-racemo case, trans preference was conserved. It is concluded that the selectivity in the addition modes are explained via interplay between the steric effects and the hydrogen bonding interactions.

Keywords: DFT, radical polymerization, stereoselectivity, acrylamide

Pro-meso and pro-racemo addition modes

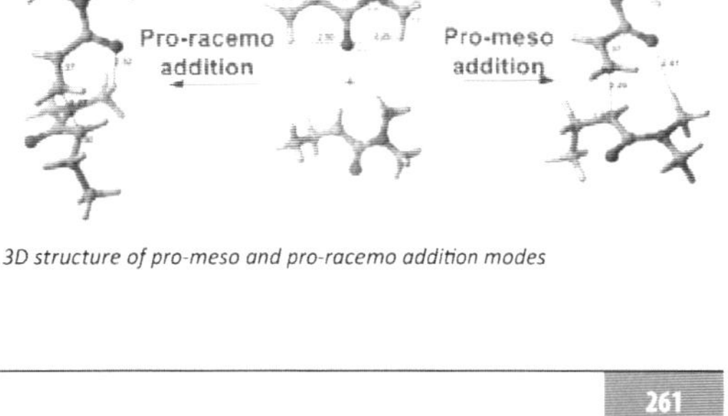

3D structure of pro-meso and pro-racemo addition modes

MSP-P-007 [Modeling and Simulation of Polymers]

Mathematical kinetic modeling on Isoniazid release from Dex-HEMA-PNIPAAm polymeric nanogels

Babak Kaffashi, Maryam Jafari

Department of Polymer Engineering, School of Chemical Engineering, College of Engineering, University of Tehran, Tehran, Iran

The quantitative calculation of release data is more facil when mathematics come to help. A mathematical modeling could aid optimizing and amending a delivery system design. The aim of this study was to find out the isoniazid release kinetic in a controlled release drug process. In this work, the degradable temperature sensitive Dex-HEMA-PNIPAAm nanogels synthesized by the UV polymerization were loaded by Isoniazid. The Isoniazid release amounts taken from in vitro studies at two different temperatures, below and above the LCST, were mathematically modeled to investigate the kinetic of drug release. The Math inquiry of release phenomenon of Isoniazid makes it easier to predict and recognize the influence of delivery the device laying out parameters on the release kinetic formulation. The modeling was performed using the model dependent methods, such as zero order, first order, Higuchi, Korsmeyer- Pepas, Hixon and Crowel. The best fitted model showing the highest determination coefficient (R2) was the Korsmeyer-Pepas which means that the predominant release mechanism is controlled by the diffusion process.

Keywords: release kinetic; modeling, zero order, first order, Higuchi method

PCNCT-P-001 [Polymer Characterization by New and Combined Techniques]

Characteristics and synthesis of fluorene copolymer thin films via plasma polymerization

Utku Özenir

Canakkale Onsekiz Mart University, Çanakkale / Turkey

In this study, fluorene copolymer thin films were synthesized via double decharging technique. The reactor consists of three cylinder-shaped hollow electrodes (inner radius 2 cm and length 10 cm) and two quartz tubes between those (inner radius cm). In the reactor, there are two anodes, one DC glow decharging electrode as the common cathode and another two anodes of which generate pulsed decharge.

The initial distance between the cathode and the first anode is 5 cm and between the anodes, 15 cm. Fluorene copolymers were generated by tiyophene, anilyn and pirol.

Thin film copolymers were synthesized on si wafer and substrates; molecular structure via xray photo electron spectroscopy (xps), Fourier transform infrared (ftr), ultraviolet visible part (uv-vis) spectrums and morphologies were detected via electro scanner microscope (sem).

It was observed that copolymers were generated through xps, ftir and uv-vis spectrums; monomers were disintegrated under high plasma energy. In addition to this, uv-vis energy band space (Eg) was monitored to be in a range between 1.8-2.5 eV. and in an ascending order, followed as fluorene-pirol, fluorene-tiyofene and fluorene-anilin.

Detected via SEM, the thin film surface was smooth under high levels of energy, whereas it was composed of irregular particles in sub-micron levels under low energy.

Keywords: Plasma

PCNCT-P-002 [Polymer Characterization by New and Combined Techniques]

Polyether Polyols as GPC Calibration Standards for Determination of Molecular Weight Distribution of Polyether Polyols

Mohd Azmil Mohd Noor[1], Vahid Sendijarevic[2], Hazimah Abu Hassan[1], Ibrahim Sendijarevic[2], Tuan Noor Maznee Tuan Ismail[1], Seng Soi Hoong[1], Nurul 'ain Hanzah[1], Razmah Ghazali[1]

[1]*Advanced Oleochemical Technology Division, Malaysian Palm Oil Board, Selangor, Malaysia*

[2]*Troy Polymers, Inc., Michigan, United States of America*

Average molecular weights (Mn, Mw and Mp) are important characteristics of oligomers and polymers, and therefore there is a need to have a precise and reliable determination method. A Gel Permeation Chromatography (GPC) coupled with a single refractive index detector was used to determine the molecular weight distributions of commercial polyether polyols calibrated against a series of polyether polyols with known molecular weights and low polydispersity. Results of these GPC analyses were compared to the ones calibrated against the commercially-available polystyrene (PS) standards. The number-average molecular weights (Mn) obtained with GPC using polyether polyols calibration were closer to the theoretical values than the Mn obtained using PS as calibration standards. Hence, these GPC analyses using polyether polyols as calibration standards can provide reliable determination of molecular weight distribution of polyether polyols and can be potentially applied to natural oil-based polyols, including palm oil-based polyols.

Keywords: gel permeation chromatography, polyether polyols, molecular weight distribution, calibration standards, polystyrene

PCNCT-P-003 [Polymer Characterization by New and Combined Techniques]

Determination of linear programmed temperature gas chromatography retention indices for some plasticizers

Farhi Halaimia[1], Imane Lakehal[2]

[1]*Halaimia farhi, Department of chemistry, Badji Mokhtar University, Annaba, Algeria*

[2]*Lakehal Iman, Department of chemistry, Badji Mokhtar University, Annaba, Algeria*

A method which allows the determination of retention indices in programmed temperature gas chromatography (P.T.G.C) ic described. The method, based on cubic spline interpolation, does not require the presence of a complete homologous series of n-alkanes in the sample. Its application to some plasticizers, eluted on two columns of different polarity, shows that the indices obtained behave more linearly than those determined by the classical method.

Keywords: Gas Chromatography, Retention indices, Temperature programming, Cubic spline model

PCNCT-P-004 [Polymer Characterization by New and Combined Techniques]

Sorbent for concentrating lead(II)

Rafiqa Alirza Aliyeva[1], Nazila Sadiq Huseynova[2], Ulviya Murshud Abilova[1], Famil Musa Chiragov[1]

[1]*Chemistry department, Baku State University, Baku, Azerbaijan*
[2]*Faculty of Pharmacy, Azerbaijan Medical University, Baku, Azerbaijan*

In this paper, was obtained by a new polymeric sorbent containing fragment of p-sulfoaniline. This invention can be used in analytical chemistry for the hydrometallurgical extraction of Pb(II) and its concentration. Essence of the invention is concluded in that copolymer of malein anhydride-styrene modification with p-sulfoaniline. Whereas, reaction is conducted in water phase anhydride groups consist in copolymer hydrolyzes. In system from interaction phormaldehide and amine is formed unstable carbonylamin, which interacts with carbocsyl group consist in macromolecule with the result that amino fragment falls into macromolecule. The composition of obtained sorbent has been studied by the IR-spectroscopy method. The Announced sorbent possesses high sorption characteristic to Pb(II). As follows, sorption capacity to lead ion in water solution 497 mg /g and degree of sorbtion -96%. By the potenthiometric method has been determined constant ionization of sorbent and constant stability complexes formed by the ions Pb(II). Exploring the various factors affecting the concentration was found optimal conditions for sorption of Pb(II). In static condition has been researched influence pH an ionic strength liquid phase to sorption ion Pb(II) by the synthesized sorbent. This sorbent adsorb Pb(II) at pH 6 and the install time, sorption equilibrium does not exceed 120 minutes. Number of adsorbed Pb(II) was determined by the difference of the metal content before and after sorption. It has been influence sorption processes from concentration of ion Pb(II) and time. It has been research influence equally capacity and concentration of acid on desorption sorbed ion Pb(II) from polymer sorbent. The experiment shows that 2M HCl desorbs Pb(II) ion from the polymer adsorbent better than others. Above this concentration desorption degree does not increase.

Keywords: concentration,lead, sorbent, sopolimer, p-aminoaniline

PCNCT-P-005 [Polymer Characterization by New and Combined Techniques]

preparation and application of biocompatible carrier implant to be used in the controlled purchase of clindan

Hesna Ural Kayalık[1], Çınar Erdem[2], Sema Çetin[2]

[1]*Ufuk University, Vocational School of Health Services, Ankara, Turkey*
[2]*Kirikkale University, Faculty of Science and Literature, Department of Biology, Kirikkale, Turkey*

PURPOSE: To make controlled-release profile of Clindamycin phosphate active ingredients that are used in the symptomatic treatment of infections caused by bacteria, while poly hydroxy ethyl methacrylate (pHEMA) developed as a biomaterial can be used in the controlled drug release system.

MATERIALS-METHODS: pHEMA and P(HEMA-MMA) hydrogels were prepared with UV photopolymerization method in the form of cylinder. Biocompatible is improved by adding the PEG and serum albumine to the structures of both gels. It was done evaluation of FTIR, SEM, DSC, contact angle for the characterization of the prepared copolymers. In order to detect swelling behavior, solvent molecules transfered to the swellable hydrogel. The emission profiles have been done with the spectrophotometric method.

CONCLUSION: It is concluded that carrier implant p(HEMA-MMA) obtained by adding MMA to biocompatible hydrogel pHEMA showed more biocompatible compared with pHEMA It was observed that 100, 250, 500ul Clindamycin sulphate adding to p(HEMA-MMA) release studies provided the best results as spectrophotometric with 400 nm wavelength. It was decided that p(HEMA-MMA) hydrogel could be used in the controlled-release studies of Clindamycin sulphate, the active ingredients of drugs used in the symptomatic treatment of infections

Comments: It was observed that p(HEMA-MMA) gel provided the best results in the use of controlled drug release systems in the studies by adding MMA to the structure of pHEMA. It was decided that pore size of the gel surface can be adjusted with the addition of MMA and drug release can be maintained for a longer period.

Keywords: pHEMA, p(HEMA-MMA), Characterization, Controlled release

PCNCT-P-006 [Polymer Characterization by New and Combined Techniques]

preparation and aplication of biocompatible carrier implant to be used in the controlled purchase of vincristine

Hesna Ural Kayalık[1], Şule Balcı[2], Sema Çetin[2]

[1]*Ufuk University, Vocational School of Health Services, Ankara, Turkey*
[2]*Kirikkale University, Faculty of Science and Literature, Department of Biology, Kirikkale, Turkey*

PURPOSE: To make controlled-release profile of Vincristin sulphate active ingredients of drugs used in the treatment of acute leukemia, while poly hydroxy ethyl methacrylate (pHEMA) developed as a biomaterial can be used in the controlled drug release system.

MATERIALS-METHODS: p(HEMA-MMA) hydrogel was prepared with UV photopolymerization method in the form of cylinder. Biocompatiible is improved by adding the PEO and serum albümine to the structures of p(HEMA-MMA). It was done evaluation of FTIR, SEM, DSC, contact angle for the characterization of the prepared copolymers. In order to detect swelling behavior, solvent molecules transfered to the swellable hydrogel. The emission profiles have been done with the spectrophotometric method.

CONCLUSION: It is concluded that carrier implant p(HEMA-MMA) obtained by adding MMA to biocompatible hydrogel pHEMA showed more biocompatible compared with pHEMA. It was observed that 100, 250, 500ul Vincristin sulphate adding to p(HEMA-MMA) release studies provided the best results as spectrophotometric with 450 nm wavelength. It was decided that p(HEMA-MMA) hydrogel could be used in the controlled-release studies of Vincristin sulfate, the active ingredient of drugs used in the treatment of acute leukemia.

Comments: It was observed that p(HEMA-MMA) gel provided the best results in the use of controlled drug release systems in the studies by adding MMA to the structure of pHEMA. Pore size of the gel surface can be adjusted with the addition of MMA and can be formed more rigid structure. Due to gel can swell up in a limited way, it was decided that constant release could be done in the work of drug release. Controlled release experiments of hydrogel installed Vincristin sulphate is preliminary work. The success obtained in the study is instructive for animal experiments.

Keywords: pHEMA, p(HEMA-MMA), Characterization, Controlled release, vincristine.

PCNCT-P-007 [Polymer Characterization by New and Combined Techniques]

Effects of seed coating(by polymer A200) on germination speed of Astragalus adscendens under different moisture conditions and Planting depth in the Boroujerd region

Hamidreza Mehrabi, Mandana Rezayee

Department of range management, Borujerd Branch, Islamic Azad University, Borujerd, Iran

The coated seed process is from amplifier ways that stick various materials on The outer surface of the seeds that minimize the negative environmental effects and increase the ability of Plant establishment. This study was done to assess the effects of coated seed on the germination speed of Astragalus adscendens in different conditions of drought stress and planting depth as it was conducted with completely randomized factorial design with four replications. treatments of covering material was used in Four Treatments: Non coating (NC), mineral- polymer based coating (CC), organic- polymer based coating (OC) polymer A200-based coating (HC); treatment of moisture percent used in three levels of dried soil content, treatments of planting depth in two surfaces of planting and three times of the seed diameter was 9%, 14% and 21 % respectively. During the test, it was evaluated the germination speed attribute. the main results showed that moisture treatments and planting depth at surface of 1% (P <0/01) was significant and has no significant effect of treatment materials. Also In examining of the interaction between type of covering material and soil moisture were not observed significant differences for germination speed between covering treatments and controls covering, But there was significant difference between treatments in 9% and 21%. Although In examining the triple interaction, increasing moisture and planting depth enhanced the speed of germination process, But it was not significant statistically, while it has made important differences in terms of description;Because it had not growth in the moisture level of 9% and shallow cultivation (high stress).But treatment of covered materials growth has developed significantly, so it can be useful in enhancing plant performance

Keywords: : seed coating, sowing depth, seed germination, treatment, soil moisture

Fig 1: mutual effect of seed coating material and moisture percentage on speed germination of astragalus adscendens.

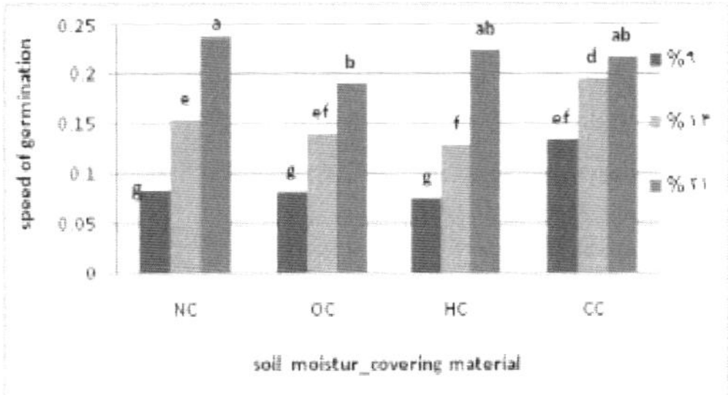

The mutual effect consideration of seed coating material and soil moisture showed germination speed was partly more 14% moisture level in all seed coating treatments in 21% moisture level. But increasingly difference has observed between 9 and 21% moisture treatment. Over all meaningful difference in point of germination speed haven't observed between seed coating material and wetness treatment (NC), (Fig 1).

Fig 2: mutual effect of soil moisture percentage sowing depth and coating material on speed of germination

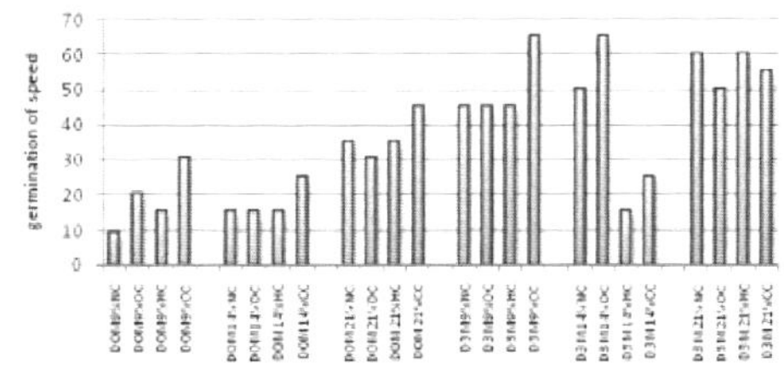

germination of speed calculator under triple mutual effect condition of soil moisture percent treatment, sowing depth and the kind of Seed coating material in 9% moisture level and wetness treatment (NC). In this point partly, have been different with other seed coating treatments. But this difference hasn't meant statistically. The most germination speed has happened in hydrogen material treatment (HC) with sowing depth of triple seed diameter in 21% moisture level to 36% degree. In this point meaningful difference hasn't observed between wetness treatment (NC) and other treatment(Fig2).

Table2: Average comparison of main effect treatment on speed of germination Aatragalus adscendens by using of Donkon,s multi

Treatment	Adjectives	speed of Germination
Moisture	9% dry soil weight	0/0935c
	14%	0/1540b
	21%	0/2167a
Sowing depth	surface	0/084417b
	Triple seed diameter	0/225083a
Seed coverage	NC	0/158000a
	OC	0/137000a
	HC	0/142333a
	CC	0/181667a

Between two moisture levels 9 and 21% have observed meaningful difference in 5 hundredth level bases on comparison of main effect average by using of Donjon's multi scope test. This result showed generation has had additive trend by degree increase in soil moisture. While 0/0935 hundredth reached to the less degree in 9% moisture level and the highest level 9% equal to 0/2167 hundredth in 21% moisture level. In point of generation speed between two surface cultivation depth (0/084417) triple seed diameter (0/225083) have observed meaningful difference in 5 hundredth level. But between seed coating treatments haven't observed meaningful difference base on germination speed (Table 2).

PCNCT-P-008 [Polymer Characterization by New and Combined Techniques]

Application of electrochemically-prepared polypyrrole as heavy metals sensor

Lo Momath[1], Sall Lamine Mohamed[1], Diaw Karim Diagne Abdou[1], Niang Matar[1], Fall Mamadou[1], Sall Guingue Diariatou[1], Aaron Jacques Jean[2], Chehimi Mehdi Mohamed[3]

[1]*Université Cheikh Anta Diop, Faculté des Sciences et Techniques, Laboratoire de Chimie Physique Organique et d'Analyses Environnementales, Dakar, Sénégal.*
[2]*Université Paris Est Marne La Vallée, Laboratoire Géomatériaux et Environnement, Marne la Vallée, France.*
[3]*Paris Est, ICMPE (UMR7182), CNRS, UPEC, F-94320 Thiais, France.*

Polypyrrole-modified electrodes obtained in the presence of a chelating salt were used for the detection and removal of heavy metal ions. Polypyrrole was electropolymerized on ITO substrates in aqueous media and doped with benzene sulfonic acid. The resulting films were electroactive and homogenous. A redox couple related to doping/undoping processes), with well-defined peaks and a regular growth were observed. The heavy metals detection ability of the films were tested by se films were tested by Differential Pulse Anodic Stripping Voltammetry (DPASV). The analytical response of the films of the films in contact with heavy metal ions showed a good selectivity and Cu2+ and Pb2+ limits of detection of about 1.16 and 3.92 µgL-1, respectively. These detection limits are lower than the acceptable limits in a drinkable water. These results are comforted par spectroscopic methods. Fluorescence spectroscopy analysis showed Stern-Volmer constant values of about 173000 and 90000 for Cu2+ and Pb2+, respectively. UV visible absorption spectroscopy determination of the ions concentration in the aqueous solution showed that after 240 minutes, almost all the Cu2+, are removed from the solution, in contrast to Pb2+. Furthermore, by XPS investigation of the polymer films, it was found that the Cu2+ peak areas were higher than those of Pb2+, indicating a lower entrapped amount of the latter ions. Finally, FTIR analysis revealed that the organic dopant (benzene sulfonic acid) is responsible for the detection and elimination of metal ions throughout the polymer matrix. It appeared that the detection and elimination by polypyrrole films were more efficient in the case of cupric ions compared to lead ions.

Keywords: Polypyrrole, DPASV, heavy metals

PCNCT-P-009 [Polymer Characterization by New and Combined Techniques]

Correlating Crystallization Kinetics and Rheological Properties of Polyethylene Using a Newly Developed Low-Field RheoNMR Combination

Mürüvvet Begüm Özen[1], Karl Friedrich Ratzsch[2], Volker Raentzsch[1], Iakovos Vittorias[3], Nehir Kavak[4], Patrick Kurt Dannecker[4], Michael A. R. Meier[4], Manfred Wilhelm[1]

[1]*Karlsruhe Institute of Technology (KIT), Institute for Chemical Technology and Polymer Chemistry (ITCP), Karlsruhe, Germany*
[2]*University of Freiburg, Freiburg Materials Research Center (FMF), Freiburg, Germany*
[3]*Basell Polyolefine GmbH, Frankfurt, Germany*
[4]*KIT, Institute of Organic Chemistry (IOC), Karlsruhe, Germany*

Polyethylene (PE) along with its copolymers is used in a wide variety of consumer products. During thermoplastic processing, parameters such as molecular weight (*Mw*) and its distribution, branching properties and processing conditions, e.g cooling and deformation rate, play crucial roles in the sensitivity of polymer crystallization, which has a strong impact on the mechanical properties of the product, to deformation [ref. 1]. Low-field (time domain) NMR is a useful tool to quantitatively investigate and monitor the molecular dynamics in the phases of a crystallizing polymer [ref. 2]. Shear induced crystallization experiments can be run in a laboratory rheometer [ref. 3]. In order to gain a deeper understanding of the material properties a promising idea was to combine rheology with NMR measurements. Therefore the newly developed unique low-field RheoNMR setup consists of a permanent 0.7 T magnet (30 Hz proton resonance) combined with a commercial high-end strain-controlled rheometer (Rheometrics / TA ARES) to enable simultaneous measurements of the full rheological shear behavior (*G'*, *G''*, LAOS, *I3/I1*, FT-Rheology [ref. 4]) and development of crystallinity [ref. 5]. Additionally, this technique allows correlating macroscopic material properties and microscopic molecular dynamics. To demonstrate capabilities of this new technique, studies on the crystallization of PE as a function of *Mw*, crystallization temperature, additives and applied shear are presented.

References:

[1] E. Piorkowska, G. C. Rutledge, *Handbook of Polymer Crystallization*, Wiley, Canada, 2013.
[2] A. Maus, C. Hertlein, K. Saalwächter, *Macromol. Chem. Phys.*, 207, 1150-1158 (2006).
[3] T. Dötsch, M. Pollard, M. Wilhelm, *J. Phys.: Condens. Matter*, 15, 923-931 (2003).
[4] K. Hyun, C. O. Klein, M. Wilhelm, K. S. Cho, J.G. Nam, K. H. Ahn, J. S. Lee, R. H. Ewoldt, G. H. McKinley, *Prog. Polym. Sci.*, 36, 1697 (2011).
[5] V. Räntzsch, M. Wilhelm, G. Guthausen, *Magn. Reson. Chem.*, DOI 10.1002/mrc.4219 (2015).

Keywords: Low-field RheoNMR, crystallization, polyethylene

Figure 1. Illustration of the new combined low field RheoNMR method.

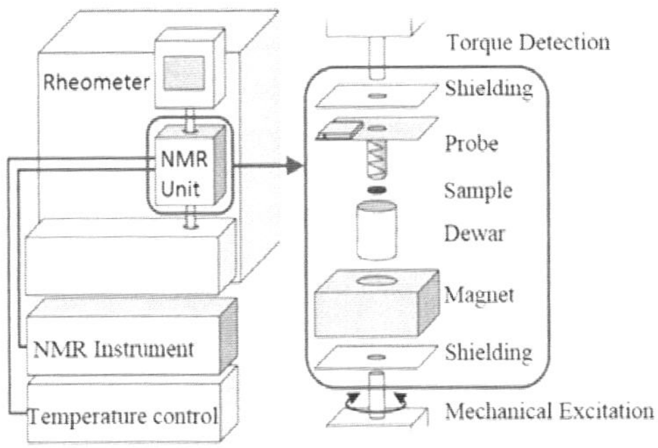

PCNCT-P-010 [Polymer Characterization by New and Combined Techniques]

Influence of Conformation of Hyaluronic Acid in Solution on Proton Mobility

Gökce Mihlac[1], Davide Ret[2], Stefano Gentilini[2], Simone Knaus[2]

[1]Departmant of Chemistry, Istanbul Technical University, Turkey
[2]Instute of Applied Synthetic Chemistry, Vienna University of Technology, Austria

Hyaluronic acid (HA) is a linear polysaccharide consisting of repeating units of D-glucuronic acid and N-acetyl glucosamine. Due to its high capacity for holding water and high viscoelasticity, HA has an outstanding position among biological materials. These properties make it suitable for various medical and pharmaceutical applications such as intradermal micro implants, viscosupplementation therapy for osteoarthritis treatment, eye surgery, tissue augmentation and dermal filling. It is very common to modify hyaluronic acid to improve its mechanical properties and prolong the duration of HA in vivo (Khabarov et al. 2014; Kenne et al. 2013). ^{1}H-NMR spectroscopy is very useful method to characterize HA derivatives (Schanté et al. 2011). However, as all polyelectrolytes HA chains interact with each other even at dilute solution, this behaviour produces not only viscous solutions at low concentration but also affects proton mobility and consequently the intensity of signals in ^{1}H-NMR spectra. The purpose of the study is to find a correlation between proton mobility and conformation of HA in solutions. For this aim, HA solutions of different ionic strength and pH were investigated by ^{1}H-NMR spectroscopy, rheological and static light scattering measurements. It has been shown that proton mobility strongly depends on conformation of HA in solution.

Acknowledgements: The authors gratefully acknowledge the financial support from Croma Pharma (Austria)

Keywords: Hyaluronic acid, conformation in solution, proton mobility, 1H-NMR

PCNCT-P-011 [Polymer Characterization by New and Combined Techniques]

Synthesis and monomer reactivity ratios of 3-(trimethoxysilyl) propyl methacrylate-co-N-vinyl pyrrolidone copolymer

Mansor B. Ahmad, Ameen Hadi Mohammed, Nor Azowa Ibrahim, Norhazlin Zainoddin

Department of Chemistry, Faculty of Science, Universiti Putra Malaysia, 43400 UPM Serdang, Selangor, Malaysia

Copolymerization of of 3-(trimethoxysilyl)propyl methacrylate (TMSPMA) with N-vinyl pyrrolidone (NVP) was carried out by free radical polymerization in dry benzene at 70 °C using benzoyl peroxide (BPO) as initiator. The resulting copolymer was characterized by FTIR and 1H-NMR and its thermal properties were determined by Differential Scanning Calorimetry (DSC) and Thermogravimetric Analysis (TGA). The copolymer composition was determined by elemental analysis. The monomer reactivity ratios were computed by linearization methods proposed by Fineman-Ross and Kelen-Tudos. The intersection method proposed by Mayo-Lewis and nonlinear method proposed by curve fitting procedure at lower conversion. The reactivity ratios obtained r1 (TMSPMA) 2.78 and r2 (NVP) 0.14 showed richer content of TMSPMA than NVP in the copolymer. The distribution of monomer sequence along the copolymer chain was calculated using a statistical method based on the reactivity ratios.

Keywords: 3-(Trimethoxysilyl) propyl methacrylate, N-vinyl pyrrolidone, reactivity ratios, sequence distribution

PCNCT-P-012 [Polymer Characterization by New and Combined Techniques]

Well-defined polyethylene-based graft terpolymers by combining nitroxide-mediated radical polymerization, polyhomologation and azide/alkyne "click" chemistry

Nazeeha Suliman Alkayal[1], Hakan Durmaz[2], Umit Tunca[2], Nikos Hadjichristidis[1]

[1]Physical Sciences and Engineering Division, Catalysis Center, Polymer
[2]Department of Chemistry, Istanbul Technical University, Maslak 34469, Istanbul, Turkey.

Novel well–defined polyethylene–based graft terpolymers were synthesized via the "grafting onto" strategy by combining nitroxide-mediated radical polymerization (NMP), polyhomologation and copper (I)-catalyzed azide-alkyne cycloaddition (CuAAC) "click" chemistry. Three steps were involved in this approach: (i) synthesis of alkyne-terminated polyethylene-b-poly(ε-caprolactone) (PE-b-PCL-alkyne) block copolymers (branches) by esterification of PE-b-PCL-OH with 4-pentynoic acid; the PE-b-PCL-OH was obtained by polyhomologation of dimethylsulfoxonium methylide to afford PE-OH, followed by ring opening polymerization of ε-caprolactone using the PE-OH as macroinitiator, (ii) synthesis of random copolymers of styrene (St) and 4-chloromethylstyrene (4-CMS) with various CMS contents, by nitroxide-mediated radical copolymerization (NMP), and conversion of chloride to azide groups by reaction with sodium azide (NaN3) (backbone) and (iii) "click" linking reaction to afford the PE-based graft terpolymers. All intermediates and final products were characterized by high-temperature size exclusion chromatography (HT-SEC), Fourier transform infrared spectroscopy (FTIR), proton nuclear magnetic resonance spectroscopy (1H NMR) and differential scanning calorimetry (DSC).

Keywords: NMP, polyhomologation, CuAAC

graphical abstract

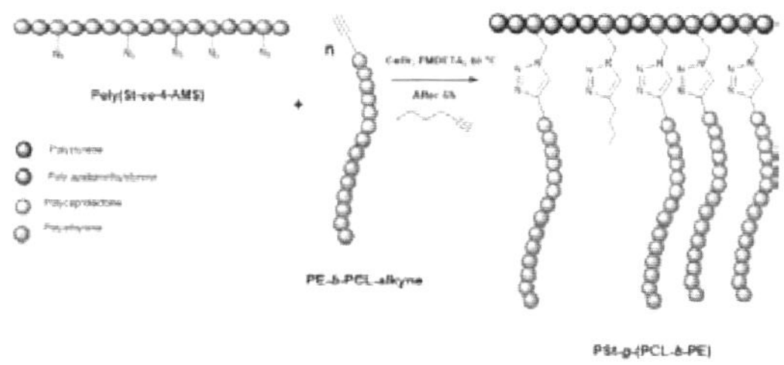

PCNCT-P-013 [Polymer Characterization by New and Combined Techniques]

Functionalization of nanocarbons for using as fillers with epoxy resin

Sonam Tamang[1], Sabita Shreshtha[2], André Wutzler[3], Ralf Lach[3], Wolfgang Grellmann[3], Rameshwar Adhikari[1]

[1]Central Department of Chemistry, Tribhuvan University, Kirtipur, and Nepal Polymer Institute (NPI), P. O. Box 24411, Kathmandu, Nepal
[2]Central Department of Chemistry, Tribhuvan University, Kirtipur, Kathmandu, Nepal
[3]Polymer Service GmbH Merseburg, Eberhard-Leibnitz-Str. 2, D-06217 Merseburg, Germany

Generally, the polymers are insulators. However, those can be made to behave electrically or thermally conducting by doping with different fillers such as nanocarbons (NCs).In this regard, epoxy surface coatings are among the most widely used industrial finishes and provide superior adhesion, flexibility and corrosion resistance when applied to metallic substrates.In these applications, polymeric materials with optimum electrical and mechanical properties are desirable for several practical applications. In this work, we investigate the nanocomposites of a thermosetting resin (amine cured epoxy resin, EP) with different forms of nanocarbons (NCs) such as carbon nanotubes, graphite and nanodiamondswith the aim of developing strong and electrically conducting coating materials. For this purpose, the NCs were functionalized to different extent and introduced into EP matrix via solution blending. The materials were characterized by different spectroscopic and microscopic techniques as well as mechanical tests. The electrical and mechanical properties of the composites will be discussed in correlation with the observed materials morphology. The materials thus developed are expected to be useful for applications as coating materials and sensors.

Keywords: Nanocarbons, Epoxy Resin, Sensors

PCNCT-P-014 [Polymer Characterization by New and Combined Techniques]

Depth-sensing Macroindentation Test and Stepped Isothermal Method – Accelerated Assessment of the Local Retardation Behaviour of Thermoplastic Polymers

Ralf Lach[1], Tobias Krolopp[1], Jan Schoene[2], Stephan Arndt[1], Beate Langer[3], Wolfgang Grellmann[4]

[1]Polymer Service GmbH Merseburg, Eberhard-Leibnitz-Strasse 2, 06217 Merseburg, Germany
[2]Institut für Polymerwerkstoffe e.V., Eberhard-Leibnitz-Strasse 2, 06217 Merseburg, Germany
[3]University of Applied Science Merseburg, Department of Engineering and Science, Eberhard-Leibnitz-Strasse 2, 06217 Merseburg, Germany
[4]Martin Luther University Halle-Wittenberg, Centre of Engineering, 06099 Halle/Saale, Germany

The prototype of a non-commercial depth-sensing macroindentation testing system with temperature chamber and temperature controlling unit for contact heating/cooling was upgraded in terms of control and software in such a way, that it allows application of the Stepped Isothermal Method (SIM). The SIM is a single-specimen method to analyse the long-term creep behavior less time-consuming, for which the temperature is raised stepwise at constant load and kept constant for a given time for each step, where the changes in indentation depth are measured as a function of time in this case. After correction of the thermally induced change in indentation depth and mathematical procedures a master curve is constructed along a logarithmic time axis. The method was successfully applied to the assessment of the long-term creep behaviour of different thermoplastic polymers such as PP homo- and copolymers, PE-HD, PET as well as PVDF. The results obtained were compared with results of the SIM using tensile testing and the short-term creep test using the depth-sensing macroindentation testing system. Because SIM using macroindentation analyses the mechanical performance of polymer surfaces it has been proved as a feasible tool to be sensitive to material changes due to physical and chemical aging phenomena.

Keywords: long-term creep behavior, depth-sensing macroindentation testing, Stepped Isothermal Method, physical and chemical aging, master curve

PCNCT-P-015 [Polymer Characterization by New and Combined Techniques]

Impact of Block Length and Temperature Over Self-assemling Behavior of Block Copolymers

Samia A Kosa[1], Laila M. Al Harbi[1], Musa Kaleem Baloch[2], Irfan Ullah[2], Elsayed H. El Mossalamy[3]

[1]King Abdulaziz University, Faculty of Science, Department of Chemistry, Jeddah, Saudi Arabia
[2]Department of Chemistry, University of Sargodha Sub-campus Bhakkar, Bhakkar, Pakistan
[3]Banha University, Faculty of Science, Department of Chemistry, Banha, Egypt

Self-assembling behavior of block copolymers having water soluble portion as one of the block plays key role in the properties and applications of the copolymers. Therefore, we have synthesized block copolymers of different block length and investigated their self-assembling behavior with reference to concentration and temperature using surface tension and conductance measurement techniques. The results obtained through both the techniques, concluded that critical micelles concentration (CMC) was decreased from 0.100 to.078 g/dl with the increase in length of water insoluble block and 0.100 to 0.068g/dl for the increase temperature. The ΔGmic was also decreased with the increase in temperature of the system; concluding that the micellization process was encouraged with the increase in temperature and block length. However, ΔHmic values were highest for short block length copolymer. The surface excess concentration obtained from surface tension data concluded that it was highest for short block length and vice-versa and was increased with the increase in temperature of the system. Whereas minimum area per molecules was highest for high molecular weight copolymers or having longest water insoluble block and low for high temperature.

Keywords: Copolymers, ΔG mic, surface tension, ΔH mic

PCNCT-P-016 [Polymer Characterization by New and Combined Techniques]

Ink Identification by ToF Secondary Ion Mass Spectroscopy

Dimitre Karpuzov

Department of Chemistry, University of Alberta, Edmonton, Canada

Optical methods are often unreliable in identification of ink layering on paper. The extreme sensitivity of Time-of-Flight SIMS makes the technique attractive for this purpose.

Secondary ion mass spectra of different inks over blank paper were acquired, and the characteristic ion masses corresponding to each color were identified. Some of these masses were then selected to collect images from overlapping writings with different color inks. The ion images can clearly show the order of ink layering on the paper. The method can be useful in forensic studies.

Keywords: ink, characterization, SIMS

PCNCT-P-017 [Polymer Characterization by New and Combined Techniques]

Separation and Characterization of Branching of Water-Soluble Polyelectrolytes by Capillary Electrophoresis in the Critical Conditions

Jean Baptiste Henri Lena[1], Gregory Thomas Russell[1], Patrice Castignolles[2], Joel Thevarajah[2], Alison Maniego[2], Marion Gaborieau[3]

[1]*Department of Chemistry, University of Canterbury, Christchurch, New Zealand*
[2]*Australian Centre for Research on Separation Science (ACROSS), School of Science and Health (SSH), Western Sydney University, Parramatta, Australia*
[3]*Molecular Medicine Research Group, School of Science and Health (SSH), Western Sydney University, Parramatta, Australia*

Water-soluble polyelectrolytes are an important type of polyelectrolytes and have a wide range of applications: as surface coatings, superabsorbants, adhesives and drug carriers. When poly(acrylic acid) (PAA) is produced by radical polymerization, the macroradical undergoes inter- and intramolecular (chain) transfer to polymer. As a result, long- and short-chain branches (LCBs and SCBs) are present. Over recent years, free-solution capillary electrophoresis (CE) has been applied to separate polyelectrolytes according to their branching topology,[1] and to quantify their heterogeneity of branching[2]. For homopolyelectrolytes such as poly(sodium acrylate) (PNaA) above 10 monomer units in size, the analysis occurs in the critical conditions. This does not refer to a separation mode but rather to an analogy with liquid chromatography conditions in which polymers are not separated according to their molar mass. It was observed that the electrophoretic mobility of PNaA decreases when the degree of branching increases. In the case of branched PNaA, two populations are observed in electropherograms.[3] The aim of this study was to identify the two peaks using the influence of ionic strength on electrophoretic mobility of an electrolyte, both as applied previously and by improving the separation selectivity (i.e., having both peaks completely resolved) by adding transition metals and/or altering the buffer concentration. Moreover, this study was useful to understand better the mechanism of separation of branched species by CE.

1. Maniego, A. R.; Ang, D.; Guillaneuf, Y.; Lefay, C.; Gigmes, D.; Aldrich-Wright, J. R.; Gaborieau, M.; Castignolles, P., Analytical and Bioanalytical Chemistry 2013, 405 (28), 9009-9020.
2. Thevarajah, J. J.; Sutton, A. T.; Maniego, A. R.; Whitty, E. G.; Harrisson, S.; Cottet, H.; Castignolles, P.; Gaborieau, M., Analytical Chemistry 2016, 88 (3), 1674-1681.
3. Lena, J.-B.; Goroncy, A. K.; Thevarajah, J.; Maniego, A. R.; Russell, G. T.; Castignolles, P.; Gaborieau, M., in preparation 2016.

Keywords: Capillary Electrophoresis, Branching, Polyelectrolytes, Poly(acrylic acid), Selectivity, Separation

PCNCT-P-019 [Polymer Characterization by New and Combined Techniques]

Composite of Polylactic Acid and Monoterpenic Phenols

Elif Güneş

Department of Chemical Engineer, Anadolu University,Eskişehir,Turkey

In recent decades, biodegradable and renewable polymers have intensively attracted attention due to their huge influence on reducing the environmental pollution. Polylactic acid is one of the most studied biodegradable polymers mainly because of its high mechanical properties and the easy production from its monomer. The high biocompatibility and biodegradability of PLA make it a very good candidate for many applications.In spite of the advantages, its range of application is still limited by its poor mechanical properties and its low thermal stability. So development of new methods which can improve the properties.In our study, monoterpenic phenols polymers are being used to improve the PLA properties and expand the application areas. Extruder is being used to produce composite from PLA and various compositions of monoterpenic phenols polymers. The physical and chemical properties of PLA will monitor under the influence monoterpenic phenols.The optimum concentration will be find, mechanical tests DSC (Differential Scanning Calorimetry) -TGA to (Thermogravimetric analysis) -FTIR (Fourier transform infrared spectroscopy), DMA (Dynamic Mechanical Analysis) analysis will be applied on composite materials which are prepared. Morphological features will monitor with SEM (Scanning electron microscope) analysis.

Keywords: polymer composite,polylactic acid,monoterpenic phenols

PCNCT-P-020 [Polymer Characterization by New and Combined Techniques]

Synthesis, characterization and kinetics of adsorption of metal ions onto copoly(amide-thioamide)s

Tabak Ghezalla, Nassima Khelllafi, Fatima Zohra Ramdane

Laboratory of macromolecular chemistry and thio-organic chemistry,Departement of polymers, Faculty of chemistry USTHB BP32 el alia Algiers Algeria 32 el Alia Bab ezzouar Alger

Series of bisthioamide-amines bearing flexible chain or aromatic cycle were prepared by reaction of bisdithioesters(1,2) and an excess of diamines at 45°c in solvent toluene. The copoly(amide-thioamide)s were synthesized by polycondensation of bisthioamide-amines (3) and terephtaloyl dichlorure in N methyl pyrolidone. The copoly(amide-thioamide)s (4), were obtained in moderate to good yields (30-45%) and possessed inherent viscosities in the range of 0,4-2,28 dL/g. All these polymers were amorphous except the (4d) which present a tendency to form liquid cristal mesophases and could be easily dissolved in amide-type polar aprotic solvent. They showed good thermal stability with glass transition temperature Tg from 109°c -209°C in nitrogen atmosphere. An increase of diamine length result in an decrease of Tg. An increase of the aromatic contents leads to formation of polymers with high molecular weight, an increasing of Tg, cristallinity and thermal stability of the copolymer samples and a reduction of their solubility. Copoly(amide-thioamide)s and polythioamides are utilized for adsorption of metals ions: Cu(II), Cd(II) and Zn(II) at different PH and their kinetics at 298K were studied.The optimum PH for quantitative adsorption onto coPATA is 10 for all these heavy metals, PH=4 for Cu, PH=8 for Cd and PH=10 for Zn onto PTA

(1) Adsorption kinetic data follows pseudo- second-order kinetic model.
(2) Adsorption onto PTA, or coPATA can be expressed better with Langmûr type adsorption isotherm.
(3) The experimental isotherms obtained were of L type.
(4) Adsorption selectivity for Cd(II) at PH=10 in the ternary system Cu(II), Cd(II) and Zn(II) onto PTA. Formation of a metal-polymer nanocomposite during contact of a mixed solution of cadmium dichloride and coPATA or PTA.

Keywords: bis thioamide-amines, copoly(amide-yhioamide)s, adsorption,kinetic,isotherm.

PEPC-P-001 [Polymer Engineering, Processing, and Characterization]

Exploring the Influence of Electron Beam Irradiation on Performance Properties of Blends Based on Ethylene Vinyl Acetate (EVA) and Thermoplastic Polyurethane (TPU)

Joyeeta Dutta, Kinsuk Naskar

Rubber Technology Centre, IIT Kharagpur, Kharagpur, West Bengal, India

Novel blends based on Ethylene vinyl acetate/Thermoplastic Polyurethane (EVA/TPU) at different ratios were prepared via melt blending technique. EVA/TPU 80/20 blend was observed to exhibit optimum tensile strength and elongation at break. Various models have been applied to predict the change in modulus with blend composition and Veenstra B model was found the best to correlate the experimental modulus values. All the blends show a single Tg and a shift in Tg with change in blend composition was observed through Differential scanning calorimetry (DSC) and Dynamic mechanical analysis (DMA), indicating technological compatibility. The optimized blends were further modified with the help of low energy electron beam irradiation in a cost effective and less hazardous way for technological (cable sheath) applications. The samples were irradiated by using a 2.5 MeV electron beam accelerating energy over a dose range from 25 to 200 kGy. The blends exhibit drastic improvement in mechanical properties with increasing radiation dose upto an optimum dose. Modification of the blends via EBR enhances the elastic recovery of the blends resulting significant improvement in tension set behaviour. Scanning electron microscopy (SEM) study was performed to explore the changes in morphology before and after irradiation. Thermogravimetric analysis (TGA) suggests that irradiation induced crosslinks also help to improve the thermal stability of the blends to some extent. All the irradiated blends have higher electrical resistivity than the blends without irradiation and the volume resistivity increases upto 150 kGy. The samples were found to exhibit remarkable improvement in oil resistance property after irradiation which is more prominent in EVA/TPU 70/30 blends.

Keywords: EVA, TPU, Electron beam irradiation, Morphology, Modeling and Thermal properties

PEPC-P-002 [Polymer Engineering, Processing, and Characterization]

Synthesis and characterization of poly (orthophenylenediamine) catalyzed by Maghnite-H+

Belmokhtar Abdelkader

Université Mustapha Stambouli de Mascara, Faculté des Sciences et de la Technologie,Algerie

The synthesis and charaterization of a poly (orthophenylenediamine) is described. This compound is synthesized catalyzed by Maghniyte-H+. Here we report the synthesis of a poly (orthophenylenediamine) obtaine

Keywords: charcterization, polymerization, Maghnite-H+, poly (orthophenylenediamine)

PEPC-P-003 [Polymer Engineering, Processing, and Characterization]

Development and evaluation of the porous structure and its effect on the electrical conductivity of conductive polymer composites

Abdelhafid Merzouki, Naceredine Haddaoui

Department of Process Engineering, Faculty of Technology, University Ferhat Abbas Setif-1

An attempt is made to develop a porous structure in a conductive polymer composite with a large surface area which will be used in contact with the surrounding fluid medium.

We studied the influence of the incorporation of a blowing agent, azobisformamide, on the development of a porous structure in conductive polypropylene-based composites and its effect on the conductivity of the resulting porous materials. We present the evolution of the parameters related to the development of the porous structure; namely conductivity, density, extent of porosity, azobisformamide efficiency, and material conductivity regression as a function of the blowing agent concentration for a fixed filler concentration of two types of conductive fillers, Carbon Black and Acetylene Black. The results showed that porosity is developed starting from a concentration of 0.8 phr of the blowing agent in the composites filled with 14.5 V % of Carbon Black or Acetylene Black.

Keywords: Polymer Composite, conductivity, carbon black, porosity, azobisformamide.

PEPC-P-004 [Polymer Engineering, Processing, and Characterization]

Effects of Cross-linking Reaction and Film Formulation on the Structural and Degradation Characteristics of Corn Starch and Corn Starch Based Blended Films

Ilknur Gönenç[1], Ferhunde Us[2]

[1]Directorate General of Food and Control, Ministry of Food, Agriculture and Livestock, Ankara, Turkey
[2]Department of Food Engineering, Hacettepe University, Ankara, Turkey

The objective of this study was to investigate structural and degradation characteristics of corn starch, corn starch/methylcellulose, corn starch/carboxymethylcellulose and corn starch/polyvinyl alcohol blend films cross-linked with glutaraldehyde and plasticized with glycerol (GLY) and polyethylene glycol (PEG). Fourier Transform Infrared (FTIR) spectrums of the produced film samples were obtained for the determination of structural characteristics. Degradation characteristics of the films was also examined. FTIR spectrums of the films were examined and frequency deviation values (which are evidence of hydrogen bond formation) were detected among film components. In the composite film samples containing GLY, more hydrogen bond formation was determined compared to the ones containing PEG. In the films which are not cross-linked, the OH tension vibration band was observed at 3331.5 - 3456.9 cm-1 while it switched to a higher frequency interval of 3346.9 - 3456.9 cm-1 in cross-linked films. The OH tension vibration frequency values of cross-linked films switching to higher frequencies were evaluated and it was concluded that hydrogen bonds in the structure decreased due to cross-link formation. Generally higher degradation properties are desired for biodegradable films. Depending on the formulation and presence of cross-linking agent, solubility values of the film samples varied significantly. PEG plasticized films showed higher solubility values than GLY plasticized ones. Considering the type of polymer, corn starch/carboxymethylcellulose containing films showed the highest solubility values. Although solubility of the films increased with time, a significant decrease was observed as a result of cross-linking.

Keywords: Biodegradable films, Glutaraldehyde, Corn starch, Fourier Transform Infrared, Structural properties, Degradation of edible films

PEPC-P-006 [Polymer Engineering, Processing, and Characterization]

Synthesis, structure and thermal properties of a new porous building material containing powder waste tires and cullet

Azzedine Ayadi[1], Younes Lamri[1], Ratiba Benzerga[2], Fayrouz Benhaoua[1], Abdlali Aouina[1], Laurent Legendre[2]

[1]*Research unit: materials, processes and environment URMPE, University of Boumerdes, Algeria*
[2]*IETR, IUT of Saint-Brieux, University of Rennes1, France*

In this work, we studied the effect of powder waste tires on the structure of a new material developed from glass waste. The addition of powder waste tires selected for this study was 5, 10, 15, 20 and 30 wt % in relation to the foam glass load. The structural alteration of the samples prepared by sintering was studied by varying the temperature from 700 °C to 900 °C with a heating rate of 6.5 °C / min. Physico-chemical analysis techniques such as SEM-EDS, porosity, thermal insulation, and differential thermal analysis were exploited for the characterization of this porous material.

The results of the microstructure clearly show that the addition of powder waste tires contributes to the formation of a cellular structure and increases the porosity of the material which gives a low thermal conductivity. Increasing the size and the pore distribution is explained by the phenomenon of coalescence between particles of raw materials produced during sintering. The differential thermal and gravimetric analysis showed that the powder waste tires is involved in the formation of the porous structure by the release of carbon dioxide, which favors their use as foaming agent. The product obtained has excellent thermal properties (λ = 0.027 W / m °C), which favors its use in the construction industry as insulating material. The reuse of this waste has a very great range; it is not only limited to energy saving in the manufacture of building materials on an industrial scale, but it also contributes considerably in the environmental protection while being accentuated as a partner against pollution.

Keywords: cullet; waste, tires, porosity; Thermal, environment.

PEPC-P-007 [Polymer Engineering, Processing, and Characterization]

Electrochemical properties of a novel composite material based on polyaniline modified by manganese dioxide (PANI + MnO2)

Souhila Abaci

Département de Génie des Procédés. Faculté de Technologie.Université Ferhat-Abbas Sétif-1, Algeria

The manganese dioxide (MnO2) is very important material in reason of its use as active electrode in the dry cells such as: lithium, alkaline, and zinc chloride cells, which are based on the procedure developed in years of twenties. Several works showed that the manganese oxide can be employed in several applications such as materials of electrode, oxygen combustible batteries and as composite material for energy storage, photovoltaic cells and light-emitting diodes. The MnO2 exists under several chemical and electrochemical varieties.

We present in this work an electrochemical and spectroscopic characterization of a novel composite material based of polyaniline (PANI) and MnO2. The (PANI+MnO2) was fabricated electrochemically by cyclic voltammetry (CV) to form a thin film of polyaniline in presence of MnO2 particles in the same solution. Then, the characterization of the obtained composite material was carried out by CV, electrochemical impedance spectroscopy (EIS) and UV-visible spectroscopy.

The cyclic voltammograms shows three redox couples characteristic of various redox states of the material composite (PANI+MnO2). Also, an increase in the currents of the peaks of oxidation and reduction during cycling was observed. This increase of current density exhibits the formation of conducting polymer at the electrode surface. The impedance measurements show that the diagram presents a semi circle in high frequencies and gives the charge transfer resistance of the polymer. This resistance varied with the content of MnO2 incorporated in the film. This phenomenon of transfer was followed by diffusion process and a capacitive blocking behavior at low frequencies. UV-visible spectrum displays two adsorption bands, characteristic π-π^* transition in the polymer chain of novel composite material. As a result, the new electrochemical, optical and electronic properties of the PANI+MnO2 are interesting and they can be use as modified electrode in the field of elecyrochemistry applications.

Keywords: Composite material, polyaniline, MnO2, electropolymerization.

PEPC-P-008 [Polymer Engineering, Processing, and Characterization]

Synthesis and characterization of copper/ poly (5- aldehyde-2,2',5',2"-terthiophjene) composite material and its application for electrocatalysis

Souhila Abaci, Naima Maouche

Département de Génie des Procédés. Faculté de Technologie.Université Ferhat-Abbas Sétif-1, Algeria

Electrochemical modification of electrodes surfaces with electrodeposition of conducting polymers plays an important role in electroanalysis and electrocatalysis [1-3]. Among of these conducting polymers; oligothiophenes which are the most studied and synthetized, in raison of their chemical stability, and electrochemical properties.

In this work copper/poly (5-aldehyde-2,2',5',2"-terthiophene (3TCHO)) composite material (Cu/3TCHO) have been prepared electrochemically. The poly(3TCHO) films were electropolymerized by cyclic voltammetry from deaerated solution of CH3CN which contained (10-2 M) of monomer (3TCHO) and 0.1 M TBAP as supporting electrolyte, by cycling the potential between 0V and 1.5V (versus SCE) at 50 mVs−1. Then the formed film was conditioned in metallic solution of copper.

The resulting materials were characterized by cyclic voltammetry, X-ray Photoelectron Spectroscopy (XPS), and Scanning Electron Microscopy (SEM). The electrochemical analysis and XPS measurements show that copper exists in the final products. A SEM image shows the copper particles dispersion which are spherical in nature and seem to be microsized. Finally, the composite material was tested for the electrocatalysis of ascorbic acid and it's exhibited a best signal compared with the simple film of poly(3TCHO). This work provides a simple route for the synthesis of Cu/polyoligothiophenes microcomposites materials.

References :

[1] G. Inzelt, M. Pineri, J.W. Schultze, M.A. Vorotyntsev, Electrochim. Acta. 45 (2000) 2403.
[2] M.E.G. Lyons (Ed.), Electroactive Polymer Electrochemistry. Part 2, Plenum Press,
NewYork,1996.
[3] Naima MAOUCHE, Saloua CHELLI, Belkacem NESSARK, Salah AEIYACH. Physics
Procedia 2 (2009) 1241.

Keywords: Composite materials; Electropolymerization; 5-aldehyde-2,2',5',2"-terthiophene; Ascorbic acid.

PEPC-P-009 [Polymer Engineering, Processing, and Characterization]

Structural property and Mechanical behavior of Polypropylene –Elemental sulfur (S8) composites: Effect of sulfur loading

S Vijay Kumar, Kishore Kumar Jena, Saeed Al Hassan

Department of Chemical Engineering, The Petroleum Institute, Abu Dhabi, UAE

Elemental sulfur is currently produced on the level of 70 million tons annually by petroleum refining, majority of which is used in the production of sulfuric acid, fertilizer and other chemicals. Still, over 6 million tons of elemental sulfur is generated in excess, which creates exciting opportunities to develop new chemistry to utilize sulfur as a feedstock for polymers. Development of new polymer composite materials using sulfur is not widely explored and remains an important challenge in the field. Utilization of sulfur as filler in the polymer matrix could be an interesting study. This work is to presents the possibility of utilizing elemental sulfur as reinforcing fillers in the polypropylene matrix. The physical, mechanical and morphological properties of the newly developed composites were studied according to the sulfur loading. In the sample preparation, four levels of elemental sulfur loading (5, 10, 20 and 30 wt. %) were designed. Composites were prepared by the melt mixing process by using laboratory scale mini twin screw extruder at 180°C for 15 min. The structure and crystallization behavior of composites was investigated by Raman, FTIR, XRD and DSC analysis. It was observed that sulfur interfere with the crystalline arrangement of polypropylene and depresses the crystallization. Tensile strengths and tensile modulus of the composites were slightly decreased with increasing in filler loading, however, percentage of elongation improved by more than 350% compared to neat polypropylene. The effect of sulfur on the morphology of polypropylene was studied with TEM and SEM techniques. Microscope analysis revels that sulfur is homogeneously dispersed in polymer matrix and behaves as single phase arrangement in the polymer. The maximum elongation for the polypropylene can be achieved by adjusting the sulfur loading in the polymer. This study reviles the possibility of using elemental sulfur as a solid plasticizer in the polypropylene matrix

Keywords: Polypropylene, Elemental sulfur, Crystallization, Morphology, mechanical properties

XRD of Polypropylene-sulfur nanocomposites

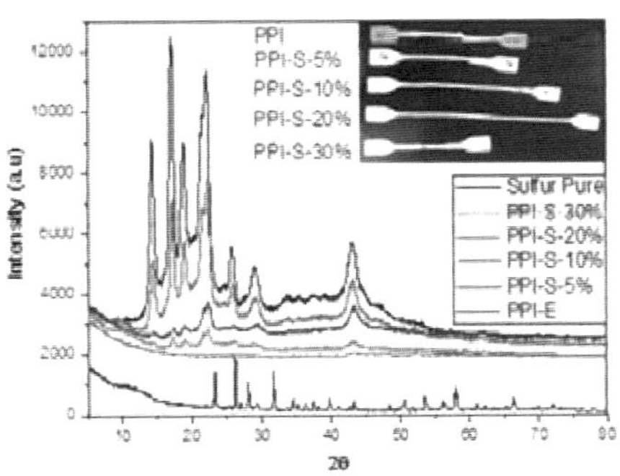

Figure XRD results for neat PPI, elemental sulfur and PPI-sulfur composites shows the decrease in crystallinity with the sulfur loading., insight picture shows digital images of the specimens of polymer composites subjected to mechanical test. The sulfur filled materials showed higher elongation with increasing amount of sulfur up to 20 weight percent.

PEPC-P-010 [Polymer Engineering, Processing, and Characterization]

Sulfur Doped Graphene Polymer Nanocomposites for Electromagnetic Interference Shielding

Chong Min Koo, Faisal Shahzad

Materials Architecturing Research Center, Korea Institute of Science and Technology, Republic of Korea

In this presentation, for the first time, we demonstrate a simple and straightforward method to improve not only electrical conductivity and permittivity but also electromagnetic interference (EMI) shielding effectiveness of reduced graphene oxide (rGO)/polystyrene (PS) nanocomposites through sulfur doping. Sulfur-doped reduced graphene oxide (SrGO) with thiophene-like structure, synthesized through a simple heating process of a mixture of graphene oxide and sulfur powder, had almost three times larger electrical conductivity (1095 S m−1) than undoped rGO (395 S m−1). The SrGO/PS nanocomposite showed not only 150% larger electrical conductivity and 60% larger dielectric properties, but also 22% larger EMI shielding effectiveness (24.5 dB) at 18 GHz than rGO/PS nanocomposite at the same loading level of 7.5 vol%. Considering the simplicity and effectiveness, sulfur doping of graphene is expected to be used as a versatile method to improve EMI shielding efficiency of graphene/polymer nanocomposites.

Keywords: EMI Shielding, Graphene, Nanocomposite

PEPC-P-011 [Polymer Engineering, Processing, and Characterization]

Interfacial interactions in polyamide/graphene nanocomposites

Thomas Raine[1], Oana Istrate[2], Bernadette Craster[3], Ian Kinloch[2], Peter Budd[1]

[1]School of Chemistry, University of Manchester, Manchester, M13 9PL, UK
[2]School of Materials Science, University of Manchester, M13 9PL, UK
[3]TWI Ltd., Granta Park, Great Abingdon, Cambridge, CB21 6AL, UK

Recently, a number of groups have investigated the incorporation of graphene and graphene oxide into polyamide, a class of thermoplastic that has a plethora of uses. However, often little discussion of the polyamide/graphene compatibility is provided. In this work, we employ twin screw extrusion to create polyamide/graphene nanocomposites at 1, 5, 10 and 20 wt% and evaluate the polyamide/graphene compatibility through SEM, TGA and DSC. From SEM, the exfoliation and dispersion of graphene in the polyamide matrix appears to be excellent, however, extensive voiding is observed in some cases. Through premixing powdered polyamide and graphene prior to extrusion, voiding can be reduced, although widespread voiding is still observed at higher loadings. TGA suggests that dispersion of graphene in the samples is good, regardless of the processing method. DSC suggests that samples created through premixing and then extrusion exhibit a higher degree of crystallinity, possibly due to better intimate mixing of the samples leading to myriad nucleation sites.

Keywords: Polyamide, Graphene, Nanocomposite, Thermoplastic, Extrusion, Characterization

PEPC-P-012 [Polymer Engineering, Processing, and Characterization]

Investigation on the thermal behaviors of various polymer/CO2 systems using high-pressure DSC

Xia Liao, Erbo Huang, Qi Yang, Guangxian Li

College of polymer science and engineering, Sichuan University, Chengdu, China

Supercritical carbon dioxide ($scCO_2$) can dissolve considerably in polymer, thus, can induce crystallization and depress the glass transition temperature (Tg), crystallization temperature (Tc), and melting temperature (Tm). The extent of the depression in Tg, Tc, and Tm depends on pressure or the concentration of gas in polymer matrix. Knowledge of the thermal behavior of a given polymer/CO_2 system is an important parameter in the development of a polymer foam fabrication process. Such Tg–pressure, Tc–pressure, and Tm–pressure data are also required to characterize the optimal temperature–pressure window within which a gas separation membrane should be used and for the extraction of an unreacted species or impurities. In this work, a high-pressure differential scanning calorimetry (HP-DSC) was used to establish the Tg–pressure, Tc–pressure, and Tm–pressure profiles in polymer/CO_2 systems at pressures higher than those previously reported. This study is the first to report the Tm measurements of polylactide (PLA) and polypropylene (PP) in CO_2 at pressures up to 15 and 30 MPa, respectively. We clearly observed the glass transitions of polystyrene (PS) and polycarbonate (PC) in CO_2 at pressures up to 5 and 6 MPa, respectively, in the DSC traces. Because of the dissolved CO_2, the Tg and Tm values of the polymers decreased; however, the increasing anti-plasticization effect of gas played a role at elevated pressures, and the Tm values of PLA and PP remained almost constant above a certain pressure. The Tc of PP in CO_2 followed the same trend. These very high-pressure DSC data will be effectively used for various applications such as foam and supercritical technology.

Keywords: Supercritical carbon dioxide, high-pressure DSC, thermal behavior, polymer

PEPC-P-013 [Polymer Engineering, Processing, and Characterization]

Mechanical and Dynamic Mechanical Properties of Devulcanized Rubber and Recycled PP Blends

Carlos Henrique Scuracchio[1], Pamela Sierra Garcia[2], Sandra Andrea Cruz[3]

[1]Department of Materials Engineering, Federal University of São Carlos
[2]Program in Nanosciences and Advanced Materials, Federal University of ABC
[3]Department of Chemistry, Federal University of São Carlos

The aim of this work is comprehend the improvement in the toughness of immiscible polymer blends developed with post-consumer polymers. For this purpose, scrap tire rubber devulcanized by microwave (GTR) was embedded in the recycled polypropylene (PPr - matrix phase), in order to make the thermoplastic more ductile [1]. In this work, the microwave devulcanization method was applied to the vulcanized rubber to obtain a material able to flow [2,3]. Also, these rubbers were submitted to different exposure periods to the microwave. According to a previous work [3], long microwave exposure periods resulted in many structural changes, such as the decrease of the insoluble phase content, breaking of chemical bonds and increase in the fluidity of rubber. Regarding the polymer blends, dynamic thermomechanical analysis (DMA) indicated that the long exposure periods to microwave interferes positively in the adherence of the rubber with the PPr (Table 1). In this case, the analysis revealed markedly that the glass transition temperature (Tg) values of the most devulcanized GTR were shifted to higher values and the Tg of the PPr shifted to a lower temperature. These results indicate a certain degree of compatibility between the two phases. Probably some chemical interactions may have occurred at the interface. The interface can also have influenced the deformation mechanism of each sample. Blends composed of the most devulcanized GTR exhibited high toughness (Table 2) and high elongation at break (Figure 1).

Aknowledgements: The author would like to thank FAPESP (process 2010/15799-6) and Braskem S.A. for the financial support.

Keywords: polymer blends, recycled rubber, devulcanization, dynamic mechanical characterization

Figure 1

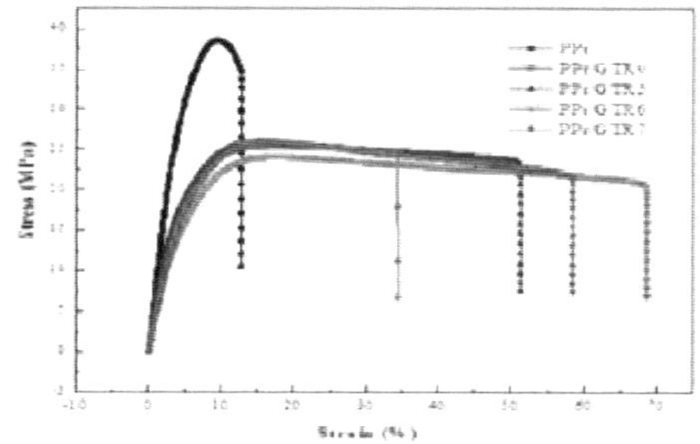

Stress-strain curves obtained from tensile tests of the polymeric blends

Table 1

Samples	Tg - GTR (ºC)	Tg - PP (ºC)
PPr	--	6.14
PPr/ GTRv	-46	6.0
PPr/ GTR5	-42.0	4.9
PP/GTR6	-36.9	4.7
PP/GTR7	-30.1	2.3

Glass transition temperature (Tg) of each phase

Table 2

Samples	Toughness (J.m-3.104)
PPR	377.0 ± 73.9
PPr/GTRv	787.6 ± 63.7
PPR/ GTR5	1,160.5 ± 231.4
PPR/ GTR6	1,477.9 ± 274.1
PPR/ GTR7	1,294.6 ± 181.7

Toughness values of each sample

PEPC-P-014 [Polymer Engineering, Processing, and Characterization]

Investigation of mechanical properties of sea-shell-CaCO3/LDPE composites

Şerife Pınar Yalçın[1], Loïc Le Pluart[2], Daniel Chateigner[3], Sophie Eve[3]

[1]Harran University, Faculty of Arts and Science, Department of Physics, Osmanbey Campus, Şanlıurfa, Turkey
[2]Normandie Univ, ENSICAEN, UNICAEN, CNRS, Laboratoire de Chimie Moléculaire et Thioorganique,Caen, France
[3]Normandie Univ, ENSICAEN, UNICAEN, CNRS, Laboratoire de Cristallographie des matériaux, Caen, France

Mollusk shells are fascinating examples of high performance organic/inorganic bio composite materials that have been the subject of numerous studies1. The investigation of mollusk shell microstructures is of interest in many fields of science2,3. Calcium carbonate (CaCO3) has been commonly used as a filler to toughen polymers because of its reducedcost, abundance and potentialities as industrial applications4-7.

In this study, calcium carbonate particles extracted and ground from three different sea shells (crepidula, oyster and scallops) were studied by Scanning electron microscopy and have unveiled complex microstructures (Figure 1). They were then used as fillers in polyethylene.The structure and mechanical properties of the composites have been evaluated. Comparisons with commercial calcite and synthesized aragonite, two of the allotropic forms of CaCO3, have been performed. Tensile tests and dynamic mechanical analysis (DMA) show improvements in tensile strength and torsion modulus when scallops-based fillers are used. It is concluded that the incorporation of these calcitic biosourced fillers is able to increase the mechanical properties of the polymers in the same way conventional CaCO3 does.

Keywords: LDPE, Mollusk Shell, SEM, DMA, Mechanichal properties

PEPC-P-015 [Polymer Engineering, Processing, and Characterization]

Microwave assisted synthesis and characterization of sodium alginate-graft-poly(N-hydroxymethyl acrylamide)

Alper Akın, Nuran Işıklan

Department of Chemistry, Kırıkkale University, Kırıkkale, Turkey

Graft copolymerization is the easiest modification method for natural polymers to produce new biomaterials with specific properties. In the last two decades, the graft copolymers of some important natural polymers especially alginate, chitosan and starch have been synthesized and characterized [1,2-3]. Under microwaves irradiation, initiators produce faster free radicals to initiate and propagate the grafting reactions effectively by conversion of microwave energy to heat energy. Hence higher grafting yields are obtained in comparison with the conventional grafting method used by the thermal heating [4]. Modification of sodium alginate (NaAlg) was carried out using N-hydroxymethyl acrylamide (HMAAm) as a monomer and potassium peroxodisulfate (KPS) as an initiator under microwave irradiation and nitrogen atmosphere. The effect of reaction conditions such as irradiation time, concentrations of HMAAm, KPS, NaAlg as well as microwave power and temperature on grafting and grafting efficiency has been investigated. Maximum grafting and grafting efficiency has been observed at 1 hour of exposure time, 0.435 M of HMAAm concentration, 500 W of microwave irradiation power, 0.0089 M of AIBN concentration, 0.5 g/dL of NaAlg concentration and 60 oC of reaction temperature, respectively. The grafted copolymer has been characterized by FTIR, DSC, TGA, XRD and SEM analyses. It has been found that the grafted copolymer is thermally more stable than ungrafted alginate.

Acknowledgments The authors are grateful to the Kırıkkale University for financial support with the project number of 2016/038.

REFERENCES:

[1] A.V. Singh, L.K. Nath, M. Guha, Carbohyd. Polym. 86 (2011) 872-876.
[2] N. Işıklan, F. Kurşun, M. İnal, Carbohyd. Polym. 79 (2010) 665-672.
[3] V. Singh, D.N. Tripathi, A. Tiwari, R. Sanghi, Carbohyd. Polym. 65 (2006) 35-41.
[4] V. Singh, P. Kumar, R. Sanghi, Prog. Polym. Sci. 37 (2012) 340-364.

Keywords: Graft copolymerization, N-hydroxymethyl acrylamide, Sodium alginate, Microwave, Potassium peroxodisulfate

PEPC-P-016 [Polymer Engineering, Processing, and Characterization]

Aging of composite insulators under the influence of intense electric fields and pollution

Mourad Houabes[1], Renaud Metz[2]

[1]Department of Electrical Engineering, Annaba University, Annaba, Algeria
[2]LCC Laboratory, University of Montpellier II, Montpellier, France

The objective of this work is to make an appropriate choice of insulating polymers can withstand several constraints: thermal, electrical, mechanical, chemical and climate. In fact, the main concern is the appearance of partial discharges in the volume formed by the insulating material (polyamide loaded to 50% by weight of glass fibers). These discharges create premature aging of the product under the combined action of a corrosive atmosphere (salt spray) and an electric field. The experimental part is concerned, first of all, the measurement of the flashover voltage, the activity of partial discharges propagating at the surface of the insulation and breakdown voltage in alternating voltage.

To do this, samples polymers which have planar surfaces or surfaces provided with protuberances were molded in order to compare their dielectric behavior in the presence of polluted salt spray such atmospheres; these measures are necessary to make an appropriate choice of material to integrate into the connector.

Thereafter, the study and the optical and electrical characterization of creeping discharges, propagating in the surface of various polymeric materials, in pulse voltage for both polarities (positive and negative) and two types of electric field (normal and tangential) are presented for the case of solid / air interface. Measuring the final length of the sliding discharges and the visualization of the current waveform are the two main parameters for distinguish characteristics of materials used. All outcome measures were used to select the cycloaliphatic epoxy resins as potential candidates for the design of a new generation of connectors. In addition, the simulation of the electric field, with the Flux 2D / 3D software, the electrical connector (TTDC 45501FA) helped lead to the proposal of a connector with a new prototype geometry capable of operating at 52kV.

Keywords: connector, insulating, polymer, electrical field, pollution, aging

PEPC-P-017 [Polymer Engineering, Processing, and Characterization]

Comparative study between adiabatic reactor and isotherm reactor-Application for the synthesis of methanol

Rachida Chemini, Hafidha Belakeb

Faculty of mechanical engineering and engineering processes, University of sciences and technology Houari Boumediene, Algiers, Algeria

The economic importance of methanol generated much attention to develop processes in general and technology of reactor in particular. There are many types of reactors which can be used in the synthesis of methanol. The difference between them are the nature of catalyst used, as well as the evacuation mode of heat.The catalytic reactor can function according to two modes:

- Adiabatic reactor with catalytic beds superposed with intermediate cooling by injection of emergency coolant (quench).

- Isotherm reactor with fixed bed multitubular filled with catalysts and inert ceramic particles. The tubes are cooled externally by fluid coolant.

The objective of this work is to compare the two technologies of reactors adiabatic and isotherm, in order to carry out the reaction of methanol synthesis.

The adiabatic reactor of methanol synthesis is favorable at low pressures. The reaction takes place in the presence of solid catalyst CuO/ZnO/AlO3 and the rate is extremely dependent on temperature. The adiabatic reactor gives a conversion rate 14.8% carried out in the first catalytic bed, because the reaction rate of hydrogenation is maximum in agreement with the profile of temperature and ratio CO2/CO of the feed. It is interesting to work at temperatures higher or equal to 246°C because the production of methanol becomes apréciable.

The isotherm reactor requires a number of 8892 tubes and a heat transfer surface 6841m2, in order to evacuate all the heat of the hydrogenation reaction, which is exothermic. The mole fraction of methanol at the exit of isotherm reactor is more significant than that of the adiabatic reactor, in agreement with the conversion rate obtained in these reactors. The conversion rate in the isotherm reactor reaches the value 27.19%. However, the isotherm reactor gives a more significant production of methanol and a higher catalyst lifetime.

Keywords: Adiabatic reactor, Isotherm reactor, Methanol synthesis, Simulation and modelisation, conversion, reaction rate

PEPC-P-018 [Polymer Engineering, Processing, and Characterization]

Thermal, mechanical, microstructural and morphological properties of side chain LCP-HDPE graft coproducts

Behiye Öztürk Şen[1], Ugur Soykan[2], Sedat Cetin[1]

[1]Department of Chemistry, Abant Izzet Baysal University, Bolu, Turkey
[2]Department of Food Quality Control and Analysis Programs, Yenicaga Yasar Celik Vocational High School, Abant Izzet Baysal University, Bolu, Turkey

In this study, high density polyethylene (HDPE) has been reinforced with a side chain LCP, poly(p-benzophenoneoxycarbonylphenyl methacrylate), poly(BPOCPMA), by graft copolymerization of p-benzophenoneoxycarbonylphenyl methacrylate (BPOCPMA) onto it. Thermal, mechanical and morphological behavior of the grafted products and microstructural changes (in unit cell parameters) of HDPE matrices were investigated. BPOCPMA used in the productions was synthesized by the reaction of p-methacryloyloxybenzoyl chloride (MBC) with 4-hydroxybenzophenone in xylene in the presence of pyridine. The yield was 57%. MBC was previously obtained by refluxing the solution of methacryloyloxybenzoic acid in thionyl chloride [1]. The graft copolymerization of BPOCPMA (5, 10, 15, 20, 30, 40% of the reaction mixture) onto HDPE were performed by bulk melt polymerization in vacuum at constant temperature (140°C) for 1h with the use of dicumyl peroxide. The crystalline melting temperature of HDPE varied consistently with poly(BPOCPMA) content, and increased from 131°C up to 134-135°C in some products. Significant expansions in the unit cell dimensions, a, b and c, and ab basal area of the orthorhombic structure, and in particle size of HDPE crystals in consistence with the graft content were recorded. The tensile properties increased considerably by grafting. The maxima, 93% increase in tensile strength and 42% in Young's modulus were achieved with the copolymer involving about 12% poly(BPOCPMA). The improvements could be explained by the advances in the orientation and alignment of HDPE chains in the products, conduced by the poly(BPOCPMA) graft units having the potential of arrangement in high degree of long-range order. SEM analyses of the coproducts showed completely homogeneous structures without any phase separation. The tensile and impact fractured surfaces displayed a gradual transition from ductile fracture to brittle nature dominated at high poly(BPOCPMA) percentages.

[1]- A. V. S. Sainath, A. K. Rao and A. V. R. Reddy, J Appl Polym Sci., 75, 465 (2000).

Keywords: LCP, HDPE, orthorhombic, bulk melt polymerization

PEPC-P-019 [Polymer Engineering, Processing, and Characterization]

Synthesis of new epoxy curing agent with flame retardant poperties via thiol-ene route and application as coating

Nedime Özdemir, Hüseyin Esen

Polymer Engineering Department, Yalova University, Yalova, Turkey

The necessities in human life enforced researches to invent and serve new materials with improved properties. One such example can be given as polymeric materials which are not only regarded as versatile material in transportation, but also with enhanced flame retardant properties. Regulations in almost all unions or countries aim to reduce both human and capital loss by setting the limits of fire retardancy at predefined limits. For instance, use of halogenated compounds in daily life articles made from polyesters, epoxies, etc. is deliberately reduced day by day due to the release of toxic and corrosive gases during fire. Other non-halogenated inorganic materials added to the formulation to impart necessary flame retardant properties ends up with increased article densities. Thus, alternative compounds, such as phosphorus based flame retardants have been extensively studied due to their preferred properties. Our group aimed in this study to obtain a curing agent with flame retardant properties. This compound is obtained through thiol-ene reaction and planned to be used in curing both commercial and renewable epoxy resins. Thiol-ene chemistry was selected because of its high selectivity and efficiency. This curing kinetics and flame retardance properties of this new epoxy curing agent is investigated.

Keywords: Epoxy curing agent, flame retardance, thiol-ene

PEPC-P-020 [Polymer Engineering, Processing, and Characterization]

Effect of accelerated ageing in oven on mechanical properties of thermoplastic elastomers

Douglas Naue Simões[1], Michele Pittol[2], Daiane Tomacheski[1], Vanda Ferreira Ribeiro[1], Ruth Marlene Campomanes Santana[1]

[1]Laboratory of Polymeric Materials, Federal University of Rio Grande do Sul, Porto Alegre, Brazil
[2]Softer Brasil Compostos Termoplásticos, Campo Bom, Brazil

Thermoplastic elastomers (TPE) compounds are used in a wide range of applications, like automotive industry, wire and cables revetment, medical devices, footwear, personal products, home appliances and others. These materials are susceptible to microbial attack, causing cracking in polymer chain and contributing to dissemination of diseases. Metal oxides like zinc oxide (ZnO), and others can be added to TPE composition to prevent attack by microorganisms. It is important to know the influence of these additives in the properties of the compounds, both at the beginning and during the life cycle. TPE samples were loaded with different commercial zinc oxide, symbolized as WR and Pe in proportion of 1%, 3% and 5% weight. A compound with no antimicrobial additive (standard) was also tested. These samples were processed by extrusion followed by injection molded. Mechanical properties of the samples were evaluated after 168 hours of exposition to accelerated ageing in oven at 105 °C. The changes on polymers surface were measured by FTIR-ATR. Differences in tensile strength, elongation and hardness were observed when compared compounds properties before and after thermal aging. WR1 reduced in 4.7% in tensile strength at rupture, standard increased 3.4% and WR3 decreased 1.1% in elongation. Also, Pe3 decreased 0.8% and WR1 increased 0.6% in elongation.

Keywords: Degradation, properties, thermoplastic elastomer, zinc oxide

PEPC-P-021 [Polymer Engineering, Processing, and Characterization]

Synthesis of Imidacloprid-β-Cyclodextrin Inclusion Complex and Its Controlled Release Characteristics in Polypropylene Filament Yarns

Ilhan Özen[1], Ahmet C Turan[2], Husnu K Gurakin[2], Enrico Fatarella[3]

[1]Department of Textile Engineering, Erciyes University, Kayseri, Turkey
[2]Setas Technology Center, Cerkezkoy, Tekirdag, Turkey
[3]Next Technology Tecnotessile, Prato, Italy

Recently, there has been a growing interest in developing pesticide controlled release technology. Much of the impetus has resulted from demands for enhanced agricultural production at lower levels of environmental risk. The main problem with conventional pesticide applications is using greater amounts of pesticides over a long period of time than what is actually needed, possibly leading to crop damage and environmental contamination. The immensity of the problem is apparent when considering the total pesticide application and the annual run-off for drainage areas. Agricultural chemical leaching and subsequent pesticide transport to non-target environments can be greatly reduced and/or possibly eliminated by controlled-release systems based on polymeric materials. The purpose of controlled release systems is to allow the automatic release of the agent to the target at a controlled rate and to maintain its concentration in the system within the optimum limits over a specified period of time, thereby providing great specificity and persistence without diminishing efficiency. In line with this target, inclusion complex of imidacloprid with β-cyclodextrin was synthesized. Then, the imidacloprid-β-cyclodextrin (IMI-β-CD) inclusion complex was incorporated into polypropylene matrix polymer and polypropylene filament yarns were produced. In the framework of this study, the following materials were used: Imidacloprid ($C_9H_{10}ClN_5O_2$, molecular weight: 255.66 g/mol, Confidor® SC 350 containing 350 g/l imidacloprid, Bayer Cropscience, Germany) as insecticide, technical grade β-cyclodextrin ($C_{42}H_{70}O_{35}$, molecular weight: 1134.98 g/mol, Nearfesh TS Pwd, Nearchimica spa Italy) as complexing agent and homopolymer polypropylene (Moplen HP500N, MFI230oC/2.16 kg: 12 g/10 min Tm: 161 oC, LyondellBasell Industries, Italy) as polymer matrix.

The release characteristics of imidacloprid were monitored by high-performance chromatography (HPLC) measurements. Besides, the physical and thermal properties of the inclusion complex (IMI-β-CD) were determined by Fourier transform infrared (FT-IR) spectroscopy, thermogravimetric analysis (TGA), and differential scanning calorimetry (DSC).

Keywords: Imidacloprid, Cyclodextrin, inclusion complex, polypropylene, controlled release

PEPC-P-022 [Polymer Engineering, Processing, and Characterization]

Synthesis and characterization of conducting polymers and copolymers coated paper substrate and modified by metal nanoparticles

Dehbia Oukil[1], Wezna Hamouma[2], Abderrazak Hamam[3], Razika Aitout[4], Laid Makhloufi[5]

[1]Wezna Hamouma
[2]Dehbia Oukil
[3]Abderrazak Hamam
[4]Razika Aitout
[5]Laid Makhloufi

This work is focused on the development of conducting polymers and copolymers films such as polypyrrole (PPy), polyaniline (PANI) and (PANI/PPy) on a flexible and insulating substrate which is paper. The chemical polymerization of PPy and PANI onto paper substrate is carried out using $FeCl_3$ as an oxidant and pyrrole and aniline as monomers. Different parameters were optimized (monomer concentration, duration of the experiment, nature of supporting electrolyte, temperature, etc...). The copolymer PANI/PPy is made by electrochemical synthesis of polyaniline onto the polypyrrole film previously deposited by chemical oxidation. The electrochemical reactivity of the obtained electrodes is tested by cyclic voltammetry technique (CV) and electrochemical impedance spectroscopy (EIS) in sulfuric acid solution.

Then, the modification of the polymers and copolymers surface by incorporation of metal nanonoparticles (Fe, Ni, Cu, Co) is conducted by cyclic voltammetry technique and by applying a galvanostatic procedure from $Fe(NO_3)_2$, $Ni(NO_3)_2$, $Cu(NO_3)_2$, $Co(NO_3)_2$ solutions. Surfaces characterization has been carried out using scanning microscope (SEM) coupled with energy dispersive X-ray analysis (EDX), and X-ray diffraction (XRD). The analysis showed the presence of the different metal nanoparticles in the polymers and copolymers films with dimensions less than 50 nm.

Keywords: Polypyrrole, Polyaniline, Copolymers, Paper substrate, metal nanoparticles

PEPC-P-023 [Polymer Engineering, Processing, and Characterization]

In-situ Synthesized Fe3O4 Decorated Polyaniline/ RGO Based Flexible PVDF Composites Film with Excellent Electromagnetic Interference Shielding Efficiency

Ranadip Bera, Sarbaranjan Paria, Sumanta Kumar Karan, Amit Kumar Das, Anirban Maitra, Bhanu Bhusan Khatua

Materials Science Centre, Indian Institute of Technology Kharagpur, Kharagpur, India

Polyaniline and chemically reduced graphene oxide (RGO) were mutually decorated by in-situ synthesized ferrosoferric oxide (Fe3O4) during the polymerization of aniline. Fe3O4 decorated polyaniline and RGO (PFR) were homogeneously dispersed in poly(vinylidene fluoride) (PVDF) matrix through solution blending process to develop a flexible PVDF-PFR composites. Polyaniline and RGO form three dimensional continuous conductive networks throughout the insulated PVDF matrix and make it highly electrically conductive material. The highly conductive flexible PVDF-PFR composite shows an excellent electromagnetic interference shielding effectiveness (EMI SE) property due to the presence of polyaniline-RGO conductive network and ferromagnetic Fe3O4 nanoparticles. Presence of PFR enhances the permittivity and permeability of the composites (0.5 mm thickness) which facilitates to achieve high EMI SE value (≈ −21.4 dB) at very low loading (5 wt.%) of RGO with a conductivity value of ≈1.10×10-1 S.cm-1 in the composites. The conductive network of RGO and polyaniline scatter the internally reflected microwaves and plays an effective role in absorption of microwave.

Keywords: EMI shielding, Conductivity, Microwave absorption, Reduced graphene oxide, In-situ Fe3O4.

PEPC-P-024 [Polymer Engineering, Processing, and Characterization]

The Effects of Aromatic Boronic Acid on Thermal Characteristics of Polymers Involving Ester Linkages

Nehir Utku

Middle East Technical University

Several studies in the literature were concentrated on the use of boron compounds as an additive in polymer matrixes to improve thermal, mechanical characteristics and flame retardancy of polymers. The early work in our group showed that the improvements were mainly due to condensation and/or trans-esterification reactions that may take place between the boron compounds in the presence of an OH, COO or CONH group along the polymer chain. Boronic acid groups considered as condensed phase flame retardants, release water on thermolysis, thereby leading to the formation of boroxines or boronic anhydrides. They become good char-yielding compounds and in the condensed phase, by assisting intermolecular interactions form a cross-linked boroxine network through the loss of water.

In this study, composites of benzene-1,4-diboronic acid (BDBA) and polymers involving ester group along the main change such as poly(lactic acid), PLA, or along the side chain such as poly(methyl methacrylate), PMMA, with or without methyl tallow bis-2-hydroxyethyl ammonium modified montmorillonite, Cloisite 30B were prepared and characterized. The effects of the amount of aromatic boronic acid, the method of blending process on morphology, thermal and mechanical characteristics and flame retardancy were investigated systematically with the use of TEM, SEM, XRD, DSC, TGA and applying mechanical and flame retardancy tests.

Thermal behaviors of the composites were also elucidated by direct pyrolysis mass spectrometry, (DP-MS) technique in addition to DSC and TGA. As a consequence, the thermal degradation mechanism and the thermal degradation products as a function of temperature were determined. Thus, products due to possible chemical interactions between the aromatic boronic acid, the polymer matrix and the organic modifier were investigated.

Keywords: Boron Compounds, Poly(lactic acid), Poly(methyl methacrylate), Poly(caprolactone), Polymer Composites, Thermal Characterization

PEPC-P-025 [Polymer Engineering, Processing, and Characterization]

Poly(lactic acid)/Poly(ethylene glycol) Blends, Nanocomposites and Fibers

Esra Ozdemir[1], Jale Hacaloglu[2]

[1]Polymer Science and Technology Department, Middle East Technical University, Ankara, Turkey
[2]Department of Chemistry, Middle East Technical University, Ankara, Turkey

Poly(lactic acid) (PLA), among all bio-based biodegradable polymers, is a promising polyester that has drawn significant attention due to not only its comparable properties with conventional polymers but also relatively better properties than other biodegradable polymers. However, although PLA has significant environmental friendly properties like biocompatibility, it still shows limited thermal, mechanical and barrier properties for many applications. Mechanical and thermal properties of PLA can be improved by copolymerization with other monomers, blending with other polymers, plasticization using biocompatible plasticizer and incorporation of filler materials. Among these, being a simple and more economic way, blending of PLA with various polymers has been investigated by many authors. Poly(ethylene glycol), (PEG), due to its miscibility, biodegradability, and food contactable application, is almost the most suitable material which has been used as an impact modifier for PLA.

In this work, morphological, and thermal characteristics of PLA/ PEG blends (90/10, 85/15 and 80/20 wt%), their nanocomposites involving Cloisite 30B and their electrospun nanofibers prepared by solution mixing or melt blending techniques were investigated by direct pyrolysis mass spectrometry, DPMS, in addition to XRD, TEM, SEM, TGA and DSC techniques. DPMS results indicated cleavage of ether linkages of PEG chains by COOH end groups of PLA during the blending and/or pyrolysis processes. The extent of these interactions depends on the preparation technique, the composition of the blend, the presence of nanoclay and leads to a decrease in the thermal stability of PEG chains. However, these interactions also generated a cross-linked structure increasing mainly the thermal stability of PLA chains.

Keywords: biodegradable polymers, poly(lactic acid), poly(ethylene glycol), polymer blends, nanocomposites, electrospun fibers

PEPC-P-026 [Polymer Engineering, Processing, and Characterization]

Assessment of improved and optimized technological process for manufacturing polymer packagings for pharmaceutical and cosmetic industry

Margarita Natova[1], Ivan Ivanov[2], Vassil Georgiev[3], Filip Ublekov[4], Hristo Penchev[4]

[1]Department of Physico-Chemical Mechanics of Institute of Mechanics of Bulgarian Academy of Sciences, Sofia, Bulgaria
[2]Himicheski produkti OOD, Yambol, Bulgaria
[3]Department of Phosphorus-Containing Monomers and Polymers of Institute of Polymers of Bulgarian Academy of Sciences, Sofia, Bulgaria
[4]Department of Conjugated Polymers of Institute of Polymers of Bulgarian Academy of Sciences, Sofia, Bulgaria

During polymer injection molding, flow lines and weld lines arise at the molded parts caused by disturbed melt flow when it crosses different parts of the equipment. Surface inhomogeneities are clearly seen in areas where injection molding delivers discolored lines, caused by a misalignment of the polymer structure elements inside the flow line. The differences in weld lines performance and in distantly located from them parts lead to instability of the properties. They cause crude esthetic perception and significantly reduce the mechanical features. In order to improve the mold construction, to stabilize the melt flow and to equalize dimensions of the products a new approach for pulse injecting has been applied during production of polymer packagings. Conventional tests for measuring mechanical performance of processed products are not applicable due to appearing of discrete inhomogenical zones of very small dimensions that are difficult to examine by standardized macroscopic test methods. To overcome this testing inconvenience, changes in visual surface roughness and mechanical performance of products were assessed by physico-mechanical METHODS: melt flow indices, DSC, X-Ray, nano- and macromechanical testing. As extensive methods applied to different parts of produced goods, defective and non-defective ones, they give a quantitative measurement of the polymer structure for assessing different parts of the products structure. Based on the obtained results a reasoned assumption has been made for the assessment of polymer processing effect after application of the pulse injection molding. It is proved that enhanced polymer performance is due to the fact that pulse injecting intensifies the molecular diffusion across the weld lines, improves the orientation of macromolecular chains and reduces the weld line weakness without any visible marks. Experimental data can be used for assessment of dissimilarity in different areas of the products and classification before exploitation of polymer packagings for pharmaceutical and cosmetic industry.

Keywords: pulse injection molding, flow- and weld lines in products, physical and mechanical test methods, assessment of technological process

PEPC-P-027 [Polymer Engineering, Processing, and Characterization]

Usability and Performance of Magnetite Filler as Polymeric Ionizing Electromagnetic Radiation Shield

Ezgi Eren Belgin, Gül Asiye Ayçık

Department of Chemsitry, Muğla Sıtkı Koçman University, Muğla, Turkey

Radiation exposure of human is increased by development of nuclear technology in the modern era. Consequently the need for radiation shields at different application areas is raised. Elemental lead and heavy concretes are the most widely used shielding materials. But because of their high heaviness the application areas of them remains limited especially when a mobile or wearable shielding is needed. In addition high toxicity of elemental lead threatens human health and environment. Therefore the need for light weight and non-toxic radiation shields enhanced and improvement of superior radiation shields became an important challenge.

In this study a novel polymeric shielding material is produced and shielding performance of the material was tested by using gamma spectrometer for three different ionizing electromagnetic radiation (IEMR) regions as low (0-500 keV), intermediate (500-1100 keV) and high (>1100 keV). Magnetite (Fe_3O_4) was used as filler material while isophytalic polyester was used as composite matrix. The IEMR attenuation ratios, linear attenuation coefficients, mass attenuation coefficients and half value layers of the composites and also the most widely used shielding material lead were determined, Table 1.

Table 1. The radiation shielding performances of produced composites, pure polyester and lead.

Low IEMR Energy (0-500 keV)	Material Type	Polyester+Fe_3O_4	Polyester	Lead
	IEMR attenuation rate (%)	51.50	18.22	98.18
	Mass attenuation coefficient ($cm^{-1}g^{-1}$)	0.87	0.18	0.61
Intermediate IEMR Energy (500-1100 keV)	Material Type	Polyester+Fe_3O_4	Polyester	Lead
	IEMR attenuation rate (%)	18.14	13.07	60.91
	Mass attenuation coefficient ($cm^{-1}g^{-1}$)	0.20	0.12	0.08
High IEMR Energy (>1100 keV)	Material Type	Polyester+Fe_3O_4	Polyester	Lead
	IEMR attenuation rate (%)	14.75	8.83	43.92
	Mass attenuation coefficient ($cm^{-1}g^{-1}$)	0.15	0.08	0.05

The performance of the composites was reached to 52, 30 and 34% of lead's performance at low, intermediate and high IEMR energy regions, respectively, Table 1.

Keywords: Polymeric shielding, radiation shielding, composite radiation shielding

PEPC-P-028 [Polymer Engineering, Processing, and Characterization]

Electrospun Rice Grain-shaped TiO2 Mesostructures Sensitized by CdS Quantum Dots for Photovoltaic Application

Shengyuan Yang, Hongjun Deng, Meifang Zhu

State Key Laboratory for Modification of Chemical Fibers and Polymer Materials, College of Materials Science and Engineering, Donghua University, Shanghai 201620, China

Development of Quantum Dot-sensitized Solar Cells (QDSCs) with intermediate efficiency is of great interests among global researchers. Here a rice grain-shaped TiO_2 mesostructure was produced by electrospinning a polymeric solution containing titanium isopropoxide, polyvinyl acetate and acetic acid in N,N-dimethyl acetamide followed by calcination at 500 °C. CdS quantum dots were then deposited on the mesoporous 'rice grains' by successive ionic layer adsorption and reaction (SILAR) method. Promising photovoltaic performance of an overall power conversion efficiency of ~ 1% was obtained in the presence of a methanol-free polysulfide electrolyte under standard 1-sun illumination of 100 mW/cm^2. We attribute the superior performance to the compromise of the high surface area of spherical nanoparticles and the directed electron transport capability of continuous nanofibers due to the unique inter-connected rice grain-like structure. Without any elaborate and complicated fabrication procedures typically involved in dye-sensitized solar cells (DSCs), the present methodology is also believed to provide a mass production way for alternative low-cost solar cells in view of the fact that both electrospinning and SILAR are generally considered as simple and scalable processing techniques.

Keywords: Electrospinning, Nanofibers, Quantum Dot-sensitized Solar Cells

PEPC-P-029 [Polymer Engineering, Processing, and Characterization]

Performance of non-halogenated polymerizable phosphorus based flame retardant in polystyrene

Hilal Gündüz, Hüseyin Esen

Polymer Engineering Department, Yalova University, Yalova, Turkey

According to the projections made by Plastic Research and Development Foundation (PAGEV) about the plastic manufacturing revealed that Turkey is expected to hold the first position within two years. This argument may be supported with number of small and middle sized firms that reach to 5000 and the 18 firms that are included within the Turkey's biggest 500 firm list. Within manufacturing, thermoplastics such as polyethylene (with its various types), polypropylene, polystyrene, polyurethane and thermoset systems such as unsaturated polyester resins were used. The formulation of the final plastic product has different kinds of additives to enhance the properties of the final product with respect to its final application. The report named World Fire Statistics published by International Association of Fire and Rescue Services (CTIF) in 2012 revealed the 1 % of Gross Domestic Product of the world has been spent to the direct or indirect losses caused by fire and related events. European statistics reported 17000 human death and a couple of more injuries in 2009. Therefore, not only the European countries but also all the countries in the world have attempts to lower or minimize the human and capital loss. Within this concept lowering the flame spread rate is the key point to achieve human evacuation and to gain time for extinguishing. This concept can be performed by adding flame retardants in to the daily used plastic products. In our project we plan to synthesize phosphorous based flame retardant and to integrate into polystyrene. This flame retardant will be synthesized from commercial phosphorous flame retardant after two step chemical modification. This new flame retardant can copolymerize with styrene. Copolymerization of this chemical with styrene will be characterized and flame retardant properties will be investigated.

Keywords: DOPO, Flame retardant, polystyrene

PEPC-P-030 [Polymer Engineering, Processing, and Characterization]

A conducting polymer and a calixarene derivative: A novel surface design for glucose detection

Tugba Ceren Gokoglan[1], Saniye Soylemez[1], Melis Kesik[1], Hande Unay[2], Serkan Sayin[3], Ali Cirpan[1], Huseyin Bekir Yildiz[3], Levent Toppare[4]

[1]*Department of Chemistry, Middle East Technical University, Ankara, Turkey*
[2]*Department of Polymer Science and Technology, Middle East Technical University, Ankara, Turkey*
[3]*Department of Material Science and Nanotechnology Engineering, KTO Karatay University, Konya, Turkey*
[4]*The Center for Solar Energy Research and Application (GUNAM), Middle East Technical University, Ankara, Turkey*

In this study, a novel amperometric glucose biosensor based on a conducting polymer and a calixarene was developed. Conducting polymer of (2-(2-octyldodecyl)-4,7-di(selenoph-2-yl)-2H-benzo[d][1,2,3]triazole)) (SBTz) was used as the immobilization matrix for biomolecule deposition to achieve an effective surface design to detect glucose. After successful deposition of SBTz on graphite electrode, a newly synthesized calixarene and gold nanoparticle (AuNP) mixture were used for improving biosensor characteristics. Glucose oxidase (GOx) was immobilized on modified electrode surface. Amperometric detection was observed at -0.7 V vs Ag reference electrode in phosphate buffer solution (50 mM, pH 6.5). Moreover, the constructed biosensor was characterized using scanning electron microscopy (SEM) technique to investigate the surface morphologies. KMapp and sensitivity values were calculated as 0.025 mM and 102 μA/mM cm2, respectively. The proposed biosensor showed a wide linear range between 0.005 mM and 0.5 mM glucose concentration with a low limit of detection (LOD) of 0.004 mM as S/N = 3. Finally, the biosensor was tested for the detection of glucose in beverage samples successfully.

Keywords: Conducting Polymer, Calixarene, Glucose Biosensor, Glucose Oxidase, Gold Nanoparticle, Amperometric Biosensor

PEPC-P-031 [Polymer Engineering, Processing, and Characterization]

Synthesis and Characterization of Fluorene Based Random Copolymer with Benzothiadiazole and Benzoselenadiazole

Seda Kutkan[1], Seza Goker[1], Serife O. Hacioglu[1], Levent Toppare[2]

[1]*Department of Chemistry, Middle East Technical University, Ankara, Turkey*
[2]*The Center for Solar Energy Research and Application (GUNAM), Middle East Technical University, Ankara, Turkey*

A novel donor-acceptor type random copolymer (PBThBSeFL) based on fluorene unit was synthesized via Pd (0) catalyzed Suzuki polycondensation reaction. Benzothiadiazole and benzoselenadiazole were used as the acceptor units. The characterization of the monomers was done using 1H and 13C Nuclear Magnetic Resonance spectroscopy. The weight average and number average molecular weight of the polymer were calculated using GPC. Electrochemical characterization of PBThBSeFL was performed in 0.1 M TBAPF6/ACN supporting electrolyte/solvent couple at a scan rate of 100mV/s in order to determine the redox behavior of the polymer. PBThBSeFL had doping/dedoping potentials around 1.53 V / 1.27 V respectively. Cyclic voltammetry studies at different scan rates (100 mV/s, 200 mV/s and 300 mV/s) showed the increase in the current density with respect to scan rate and this linear relationship proved that PBThBSeFL was well adhered to the ITO coated glass and a non-diffusion controlled mass transfer occured in the redox process of the polymer. Since PBThBSeFL had only p-dopable character, HOMO was calculated from cyclic voltammetry study but LUMO energy level was calculated from HOMO and optical band gap that was found from the spectroelectrochemistry studies. Corresponding HOMO and LUMO energy levels were -6.05 eV and -3.84 eV with an optical band gap of 2.21 eV. PBThBSeFL showed a maximum absorption around 475 nm which is consistent with its yellowish-orange color in its neutral state. Upon applied potential PBThBSeFL was oxidized, polaron/bipolaron regions were produced and PBThBSeFL revealed gray and greenish-gray colors in the oxidized states. Kinetic studies were performed to calculate the optical contrast and switching time. PBThBSeFL showed percent transmittance of 70% at 1625 nm, 31% at 700 nm and 27% at 475 nm, respectively. According to these results PBThBSeFL has promising properties such as good solubility, chemical stability and lower band gap than its analogues in the literature.

Keywords: Donor-Acceptor type copolymers, low band gap, fluorene, benzothiadiazole, benzoselenadiazole

PEPC-P-032 [Polymer Engineering, Processing, and Characterization]

New Heating Technology for Polymer Processing

Sascha Bach, Marcus Stein, Ronald Claus, Jens Peter Majschak

Chair of Processing Machines and Processing Technology, Technical University Dresden, Dresden, Germany

A number of processes for polymer processing like extrusion, injection moulding, sealing or thermoforming are using the temperature dependent flow and mechanical behaviour of polymers. In all these processes a heating system is necessary to achieve the required material properties. Thus deviations of the heating parameters have negative impact on the product quality directly. Therefore the poster will present an innovative heating technology, called cera2heat, providing new opportunities in flexible and accurate heating. The system is based on ceramic thick-film technology. The current status of development is a matrix-heater with 8 x 8 independent heat and sensor circuits formed as pixel each 5 x 5 mm for contact heating of sheet material. The matrix heater works like a common TFT, but with a defined temperature field instead of a visible picture (Figure 1). Each circuit is temperature controlled by a very fast multi-channel closed-loop system. The control accuracy is better than plus/minus 1 K. Due to the very small thermal mass of the system it needs less than 1 s to reach the set temperature. These advantages are used in the thermoforming process the starting application for this technology. The individual heating enables to use the temperature-dependent forming behaviour of the polymers by applying a defined inhomogeneous temperature field. That allows to controls the wall thickness distribution by applying a defined temperature distribution on the sheets. It has been proven for commonly used packages and an increase of wall thickness up to 240 % was observed. That means a down-gauging potential of the initial sheets up to 30 %. The technology also enables large forming degrees without any plug assistance. Using this technology completely new approaches and concepts can be introduced for thermoforming and other polymer processes. It additionally provides a new degree of freedom for simulation based process optimization.

Keywords: Heating Technology, Thermoforming, Temperature fields, Process Control

Matrix-heater with 8x8 individual controllable heat circuits (left) and thermographic picture of exemplary temperature field with 6 heated pixels (right)

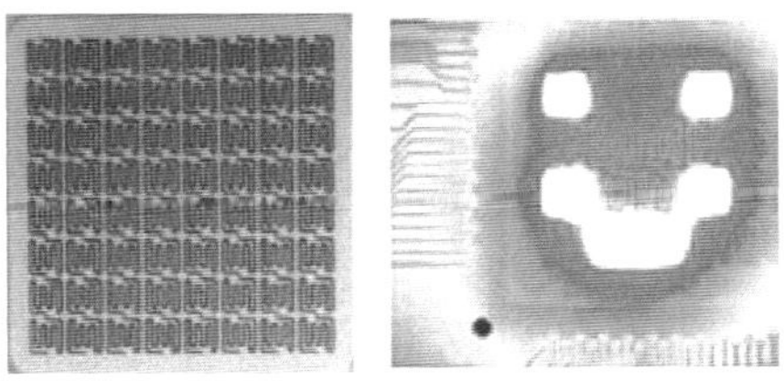

PEPC-P-033 [Polymer Engineering, Processing, and Characterization]

Ultrahigh Thermally Stable Poly(ether ketone)/MWCNT Composites as an Efficient Electromagnetic Shielding Material With Improved Mechanical Properties

Sampat Singh Chauhan[1], Mathew Abraham[2], Veena Choudhary[1]

[1]Centre for Polymer Science & Engineering, Indian Institute of Technology Delhi, Hauzkhas New Delhi 110016, India
[2]Gharda Chemicals Ltd. Dombivili East, Mumbai 421203, India

Poly(ether-ketone) [PEK]/multiwalled carbon nanotube [MWCNT] composites having varying amounts of MWCNTs were prepared via industrially compatible melt blending approach to investigate the capability of such composites as an effective electromagnetic radiation shielding material (ERSM) in the frequency range of 26.5-40 GHz (Ka band). In order to design this type of material (ERSM), electrical conductivity was targeted as key property using a conducting inclusion; carbon nanotubes (CNTs). Electrical conductivity increased from ~10-13 S/cm for neat PEK to ~10-5 S/cm for 3.2 vol % (~ 5 wt %) CNTs loaded PEK which was close to the requirement for microwave shielding. Nanocomposites are shown to exhibit upto -38 dB of total shielding effectiveness (SET) at 3.2 vol % loading in frequency range of 26.5-40 GHz (Ka band). This formulation of nanocomposites (3.2 vol % loading) shows shielding effectiveness due to absorption (SEA) of ~ -34 dB (>99.99 % attenuation) with very less reflection loss of ~ -4 dB in the prescribed frequency range. Achievement of such high absorption dominant shielding effectiveness (-38 dB) along with high thermal stability (T0.1 ~ 581 oC, char residue ~ 60% at 800 oC) and mechanical strength (tensile strength ~ 116 MPa & tensile modulus ~ 5.2 GPa) substantiate its capability as light weight, thermally stable and mechanical strong EMI shields for demanding applications such as aerospace, defense and space applications.

Keywords: Electrical conductivity, ERSM, Microwave shielding, Reflection loss, Shielding effectiveness

PEPC-P-034 [Polymer Engineering, Processing, and Characterization]

An effective surface design based on a conjugated polymer and silver nanowires for the detection of paraoxon in tap water and milk

Janset Turan[1], Melis Kesik[2], Saniye Soylemez[2], Seza Goker[2], Sahin Coskun[3], Husnu Emrah Unalan[3], Levent Toppare[4]

[1]Department of Polymer Science and Technology, Middle East Technical University, Ankara, Turkey
[2]Department of Chemistry, Middle East Technical University, Ankara, Turkey
[3]Department of Metallurgical and Materials Engineering, Middle East Technical University, Ankara, Turkey
[4]The Center for Solar Energy Research and Application (GUNAM), Middle East Technical University, Ankara, Turkey

Over the last decades, pesticides have been extensively used in agriculture, industry and chemical warfare leading to contamination of agricultural products and aquatic systems. These pesticide residues may cause severe health defects in nervous system hence their detection gained high priority. In this study, an amperometric butyrylcholinesterase (BChE) biosensor was constructed for the analysis of an organophosphorus pesticide; paraoxon in tap water and milk. For this purpose, a conjugated polymer; poly(5,6-bis(octyloxy)-4,7-di(thieno[3][3,2-b]thiophen-2-yl)benzo[c][1,2,5]oxoadiazole) (PTTBO) was coated on graphite electrode and used as an immobilization platform. The obtained surface has a hydrophobic nature that provides physical interaction between enzyme molecules and polymer. Then, silver nanowires (AgNWs) were used in biosensor construction to improve electron transfer rate, to achieve a large specific surface area, an effective biomolecule absorption and biocompatibility. With this novel surface design, rapid, highly sensitive and stable amperometric sensors were obtained. Fabricated biosensor showed two linear ranges between 0.5–8 μM and 10–120 μM with a low detection limit of 0.212 μM when butyryl thiocholine iodide was used as the substrate. Surface modifications were monitored using scanning electron microscope (SEM) and cyclic voltammetry (CV) techniques. The biosensor was tested for paraoxon detection in milk and tap water based on the inhibition of the enzyme molecules, where recovery tests proved the applicability of the designed system.

Keywords: Amperometric biosensor, Conducting polymer, Silver nanowire, Pesticide biosensor, Butyrylcholinesterase

PEPC-P-035 [Polymer Engineering, Processing, and Characterization]

Investigation Parameter Effects on Tribological Performance of Short Carbon Fiber and Pumice Particle Reinforced PPS/PA66 BLEND

Alp Eren Şahin, Sinan Yilmaz, Tamer Sinmazçelik, Taner Yilmaz

Mechanical Engineering Department, Kocaeli University, Kocaeli, Turkey

In many engineering applications, polimeric materials are increasingly finding place as a result of technological development and requirement, owing to the need to develop new and light materials. Hence, using of polimer matrix based composites is very common application for meeting the requirements. The composites can be produced with particle, fibre or particle/fibre (hybrid) reinforcement. Although these composite samples fullfil the requirement of advance material technology, they have some handicaps inabilities for some engineering applications. One of these handicaps is low wear resistance and using of these materials in application subject to wear. Wear is loss of material from the workpiece surface undesirably as a result of relative movement and wear resistance of polimeric materials is weak when compared to conventional materials. In this study, (phenylene sulfide) (PPS)/polyamide6,6 (PA66) blend matrix were reinforced with carbon fibre and pumice particles (in two different sizes <100μm and >100μm) and pumice/carbon fibre(hybrid) reinforcements. Investigation of adhesive wear behaviour of composite samples were tested with ball on disc method under two different load and three different rotational speed of the disc. According to test results, while increasing of load increased wear rate of composite samples were achieved. But it is found that rotating speed has no significant effect on wear rate of composite samples. Among the all experimental parameters the best wear resistance behaviour showed by hybrid samples. The study described in this paper is supported by a Scientific and Technological Research Projects Program grant from The Scientific and Technological Research Council of Turkey (Project No: 114M785).

Keywords: Tribology, PPS/PA66 Blends, Pumice, Carbon Fibre

PEPC-P-036 [Polymer Engineering, Processing, and Characterization]

Investigation of Mechanical Properties of Polysulfone Blends

Reyhan Ozdogan, Mithat Celebi, Mehmet Arif Kaya, Ozgur Ceylan

Department of Polymer Engineering, Yalova University, Yalova, Turkey

Polysulfone (PSf) has been used in many engineering applications due to excellent mechanical and thermal properties. PSf is a hydrophobic polymer owing to chemical structure of their backbones. PSf has been especially used in polymeric membranes because of having excellent film forming properties and resistant to extreme pHs and temperatures. On the other hand, PSf has some disadvantages such as high raw material price and cost of processing. PSf has been blended with low cost biopolymers to decrease the cost and enhance some properties of the membrane such as hydrophilicity. In this study, PSf was blended with biodegradable polymer which is aliphatic polyester based from renewable resources in solution casting. The effect of blend ratios were investigated on film characteristics. Mechanical strength was examined via tensile testing machine. Biodegradation rate of samples were determined by petri dish screen method and hydrophilicity of the films were examined by contact angle microscopy.

Keywords: biodegradability, blend, hydrophilicity, mechanical property, polysulfone, solution casting

PEPC-P-037 [Polymer Engineering, Processing, and Characterization]

Removal of Cu2+, Ni2+ and Au2+ ions from acidic solutions using polypyrrole conducting polymer by electrochemical cementation process

Houa Hammache Makhloufi, Boualem Saidani

Laboratoire d'Electrochimie, Corrosion et de Valorisation Energétique (LECVE), Faculté de Technologie, Université de Bejaia, 06000 Béjaia, Algérie

Intrinsic conducting polymers with conjugated double bonds have attracted much attention as advanced materials. Polypyrrole (PPy) is especially promising for commercial applications because of its good environmental stability, facile synthesis and high conductivity. In this work, the electrochemical cementation process approach was used to recover Cu2+, Ni2+ and Au2+ metal ions from acidic solution using PPy films. PPy was prepared by electrochemical deposition on iron substrate from aqueous oxalic acid solution at different current densities. The Cu2+, Ni2+ and Au2+ concentration time-dependent was followed during cementation with atomic absorption spectrophotometry. The resulting PPy/Cu, PPy/Ni and PPy/Au films electrodes were characterized by scanning electron microscopy and EDX analysis. It is shown that, both the initial concentration of metals ions and the thickness of the PPy films play an important role on the rate of the Cu2+, Ni2+ and Au2+ cementation process. The rate constant, k, of the process was found to increase both with increasing concentration of metal ions and PPy electropolymerization charge. The highest removal efficiency was achieved for Cu2+ and Au2+ ions.

Keywords: Conducting polymers, Cementation, Polypyrrole, Cu2+, Ni2+, Au2+

PEPC-P-038 [Polymer Engineering, Processing, and Characterization]

Synthesis and Characterization of Hexakis(propylamino)cyclotriphosphazene and Hexaphenoxycyclotriphosphazene and Uses on Polyolefins as Flame Retardants

Muzaffer Kaan Karaöz[1], Yunus Karataş[2], Teoman Tinçer[1]

[1]*Department of Polymer Science and Technology, Middle East Technical University, Ankara, Turkey*
[2]*Department of Chemistry, Ahi Evran University, Kırşehir, Turkey*

Most of the phosphazene studies depend on the integration of fundamental research and application areas. In some of these types of investigations, there exist many examples in which phosphazene molecules are used as a model for discovering new reactions. In research area of polyphoshazenes, reactions of cyclic phosphazenes are used in general to model the high molecular weight derivatives. Hexachlorocyclotriphosphazene (trimer) is one of the well-known and most widely studied example among phosphorus and nitrogen compounds. Different kinds of trimers have been synthesized as a result of substitution reactions of chlorine atoms and flame retardant phosphazene derivatives can be modelled by replacing the chlorine atoms with proper side groups [1].

In this study, hexaphenoxycyclotriphosphazene (a) and hexakis(propylamino) cyclotriphoshazene (b) are synthesized via the reactions of hexachlorocyclotriphosphazene with propyl amine and phenol respectively (Figure 1). The structures of the trimers are confirmed with 1H, 31P and 13C NMR spectrums and other basic spectroscopic techniques.

The synthesized cyclic phosphazene derivatives will be used to prepare blends. They will be investigated for their mechanical and thermal properties. Furthermore, flame retardancy tests will be performed. Beside the fact that aromatic structures contribute the flame retardancy, the aliphatic groups are known to be compatible with polyolefins. The contributions of these two factors are aimed to be studied and compared systematically here.

References:

[1] M. Gleria, R. De Jaeger, Phoshazenes: A Worldwide Insight, Nova Science Publishers, 2004.

Keywords: hexachlorocyclotriphosphazene, hexakis(propylamino)cyclotriphosphazene, hexaphenoxycyclotriphosphazene, flame retardancy

Synthesis of cyclic phosphazenes used as flame retardants: (a) hexaphenoxycyclotriphoshazene, (b) hexakis(propylamino)cyclotriphosphazene

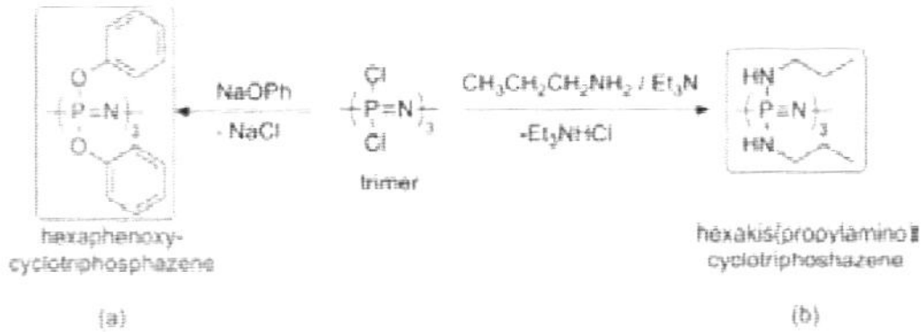

PEPC-P-039 [Polymer Engineering, Processing, and Characterization]

Preparation and characterization of abundant biomass reinforced PVC composites

Amel Mohamed Ben Ali[1], Ali Gasmi[2], Pierre Magri[3], Amar Boukerrou[4], Nesrine Amouchi[1]

[1]Département de Technologie, Université 20 août 1955-Skikda 21000- Algeria
[2]Département de Physique, Université Badji Mokhtar, Annaba 23000- Algeria
[3]Laboratoire LCP-A2MC, Université de Lorraine, 1 boulevard Arago, Metz, France
[4]Department de Genie des procédés, Université de Abderrahmane Mira, Bejaia 06000,-Algeria

This work aims to study the effects of fiber surface treatment on the properties of the composites. Untreated and treated fibers with varying amounts (ranging from 5-20 wt %) were mixed with PVC matrix using two-roll mill followed by compression molding. Morphology and fiber-matrix interfacial adhesion, thermal and mechanical properties were examined. The results show that, the surface treatments can improve interfacial adhesion between these two phases. Treated fibers improve mechanical properties of the composites. Young's modulus, tensile strength and elongation have been increased compared to untreated fibres. Whereas, the thermal stability, as measured by thermogravimetry (TGA) methods, indicated that the addition of natural fibres into polymer matrix induced an increase of the onset degradation temperature of the composite compared to the PVC virgin matrix, except of the composite at fiber loading of 20 wt %, for the thermograms DTG of composites, we find that the speeds decomposition of different composites developed during the first stage of decomposition are lower than that of virgin PVC (F0). It can be said that the incorporation of cellulosic filler significantly retards the decomposition of PVC and plays a role of an inhibitor of thermal degradation.

Keywords: PVC, Surface treatments, Thermal properties, Mechanical properties.

PEPC-P-040 [Polymer Engineering, Processing, and Characterization]

Functional PEG building blocks via copolymerization of ethylene carbonate and functional epoxides

Gent Kapiti, Helmut Keul, Martin Möller

Dwi - Leibniz Institute for Interactive Materials

Using a mono-mode microwave reactor, copolymers of ethylene carbonate (EC) and functional epoxides (e.g. tert-butly glycidyl ether(TBGE) or allyl glycidyl ether (AGE)) were obtained by reaction of mixtures of the monomers or sequential polymerization of first EC followed by the functional epoxide using CsOH as initiator. Depending on the reaction conditions different monomer sequences within poly(EO-co-TBGE/AGE) copolymers were obtained with molecular weights of 1500 and 2500 g/mol and dispersities in the range of Đ = 1.06-1.17. In order to determine the monomer sequence within the narrowly distributed copolymers kinetic studies were performed using ^{1}H- and ^{13}C NMR spectroscopy, size exclusion chromatography, and MALDI-ToF as analytical tools. Transcarbonylation, transalkylation and carbonate fragmentation were identified as fundamental reactions for the formation of carbonate free products with variable microstructure.

Keywords: synthesis, functional polymers, sequence control

graphical abstract

PEPC-P-041 [Polymer Engineering, Processing, and Characterization]

Production and Characterization of Polyethylene/ Organoclay Nanocomposite Films for Food Packaging

Ceren Alpaslan, Sennur Deniz

Department of Chemical Engineering, Yıldız Technical University, İstanbul, Turkey

One of the most useful polymers which are used in the packaging industry is polyethylene.However it is not desirable in the food packages through its high gas permeability.For this reason barrier featured polyethylene films can only be produced in multilayer extruder machines by blown extrusion technique.Although multilayer extruder machines require high cost investment monolayer and barrier,featured polyethylene films can be produced by constituting polyethylene nanocomposites instead of this technique [1].When the dimension of material is diminished in nanoscale it can show different properties which have in macroscopic size.It is known that,nanoparticules using is a good filler materials which improve product properties and shelf life of foods in food packaging production [2].

The aim of this study is to produce nanocomposite polyethylene films, which can be especially used in the food packaging industry.In this study the polymer films were consisted of low density polyethylene and linear low density polyethylene as polymer matrix,montmorillonite modified organoclays as gas permeability filler.They were prepared with the laboratory type film device which is integrated to the twin screw compounder.Moreover,another goal of this study is to apply production techniques which can also be suitable for the industrial scale.For increasing interfacial interactions resulting polyethylene/clay nanocomposites, the clay layers within the polymer matrix to ensure dispersion of nanometer level is required to use a compatibilizer. Therefore, maleic anhydride grafted polyethylene was used as compatibilizer. The polyethylene granules were prepared by using different ratio of organoclay(%1,3,5) by twin screw extruder.Polyethylene/organoclay nanocomposite films were prepared with the laboratory type film device which is integrated to the twin screw compounder and they were characterized in terms of thermal,morphological,gas permeability and mechanical properties.All test results were compared with sample of reference film which is not including organoclay and were investigated the effects of organoclay amount.At the end of the tests,optimum organoclay amount was determined.

Keywords: Polyethylene film, organoclay, gas permeability, twin screw extruder, food packaging, nanocomposite

PEPC-P-042 [Polymer Engineering, Processing, and Characterization]

Dynamic rheological properties and phase separation in of polyethylene/ poly(1-hexene) blends

Saeed Houshmandmoayed[1], Kiana Salamat[2], Mohammad Mahdi Mortazavi[3], Ehsan Nikzinat[4], Reza Rashedi[1], Hannan Sepahi[1], Mostafa Ahmadi[4]

[1]Jam Petrochemical Co
[2]Chemical Engineering Dept. Islamic Azad University, Tehran North Branch, Tehran, Iran
[3]Polymerization Engineering Department, Iran Polymer and Petrochemical Institute (IPPI), Tehran, Iran
[4]Department of Polymer Engineering and Color Technology, Amirkabir University of Technology, Tehran, Iran

Polyethylene (PE)/ poly(α-olefin) blends are among compatible blends with versatile applications. In this study the rheology, morphology, and interfacial interaction of several PE/ Poly(1-hexene) (PH) blends in different percentages of PH are investigated. The SEM images showed a droplet-matrix structure in all range of compositions and the size of dispersed phase increased proportionally with PH content. The miscibility of blends at various compositions is evaluated by viscoelastic parameters determined by dynamic oscillatory shear in linear region viscoelastic region. The complex viscosity of samples at various fraction of PH showed the log-additivity mixing rule behavior in low frequencies and positive-negative deviation behavior (PNDB) at high shear rates. The deviation of Cole– Cole plots from the semi-circular shape, and the tail in relaxation spectrums at high relaxation times are the evidences of two phase heterogeneous morphology. The morphological properties could be correlated to the linear rheological properties using well-known emulsion model. Samples expose significant strain-hardening behavior in nonlinear extension tests that is attributed to the high-molecular weight PH component.

Keywords: Polymer blend, rheology, morphology

PEPC-P-043 [Polymer Engineering, Processing, and Characterization]

Tea Mill Waste as Reinforcing Fillers in Thermoplastic Composites

Mustafa Kuyumcu[1], Alper Kasgoz[2], Mehmet Atilla Tasdelen[1]

[1]Department of Polymer Engineering, Faculty of Engineering, Yalova University, 77100 Yalova, Turkey
[2]Department of Polymer Engineering, Faculty of Engineering, Yalova University, 77100 Yalova, Turkey, Department of Chemical Engineering, Faculty of Engineering, Istanbul University, Avcılar, Istanbul, Turkey

Tea mill waste is a reusable lignocellulosic material that is manufactured about 30.000 tons annually from state-owned companies. There are a few valuation studies including energy fuel production, natural fertilizer, antioxidant, adsorbent, bio-sorbent, biocomposite and panel manufacturing from tea mill waste in literature. [1-3]

In this study, tea mill waste fiber was compared against wood flour for their ability to act as reinforcing fillers in melt-blended composites with polypropylene as the matrix polymer. The composites were prepared by compounding using a twin-screw extruder and test specimens for mechanical properties were prepared by hot-press method. Various feeding ratios between 5, 15, 30 and 50% were investigated to determine the influence of tea mill waste fiber concentration on the mechanical properties of the composites. Based on mechanical and thermal analyses results, some property differences were observed compared to wood flour-based composites, the tea mill waste fiber-polypropylene composites offered cost advantages.

References:

1- Klyosov, A. A. Wood-Plastic Composites, 2007.
2- Schneider, J. P., Myers, G. E., Clemons, C. M., English, B. W. Journal of Vinyl and Additive Technology, 1, 2, 103–108, 1995.
3- Donmez Cavdar, A., Kalaycioglu, H., Mengeloglu, F., Journal of Reinforced Plastics Composites, 30, 10, 833-844, 2011.

Keywords: tea mill waste, wood composites, recycling

PEPC-P-044 [Polymer Engineering, Processing, and Characterization]

Investigation of mechanical properties of organic - inorganic hybrid reinforced composites

Mehmet Mudu[1], Halil Demirer[2], Mustafa Öksüz[2]

[1]School of Enginnering, Yalova University, Yalova Turkey
[2]Faculty of Technology, Marmara University, İstanbul Turkey

Composite materials are composed of two or more same or different group materials to obtain better properties. Nowadays, developments of material science are recognized for hybrid composites than old times. Because of the economical reasons we can utilize inorganic materials like wallostonite or ground hazelnut shell could also be used as a filler material in plastic composites. In Turkey the hazelnut shell is being produced substantially but unfortunately it is generally used for heating in houses. the utilisability of ground hazelnut shell in PP matrix composites was investigated with respect to its relative hardness, characteristic colour and availability. In this study 10 % ground hazelnut shell and 10% wollastonite by weight were mixed with polypropylene matrix. 3, 6, 9 % SEBS and SEBS-g-MA were also added to this mixture. Then, the tensile, bending, izod impact, hardness, melt flow index tests were conducted on specimens produced by means of injection molding method. It was observed that values for E-Modulus, yield strength, shore D hardness decreased whereas the values for elongation at yield and break increased due to addition of SEBS and SEBS-g-MA to wollastonite and hazelnut shell filled PP composites

Keywords: Composite materials, hybrid, mechanical properties

PEPC-P-045 [Polymer Engineering, Processing, and Characterization]

Characterization of spray polymerization products varified by specific operating parameters in a spray tower

Katharina Franke, Werner Pauer, Hans Ulrich Moritz

Institute for Technical and Macromolecular Chemistry, University of Hamburg, Hamburg, Germany

Spray drying is a process mainly used in the pharmaceutical, chemical and in the food industry to produce particles with definite properties for various applications [1]. The spray polymerization is a technical development of spray drying. Spray polymerization is a fusion of polymerization and spray drying. Due to the phenomenon of the wet-bulb temperature of a droplet, the polymerization enthalpy is removed by evaporation of the dispersion medium. Thus, the droplet temperature can be assumed as isothermal. After shell formation the droplet temperature drastically increases and the ongoing polymerization leads to the final particle. Due to the process intensification the spray polymerization is interesting for industrial use. The constant and controllable particle morphology plays an important role for industrial applications. Just like the product morphology during spray drying [2], properties and morphology in the spray polymerization can be controlled by operating parameters such as gas flow rate, gas stream composition and temperature. These parameters affect the spray characteristics and, in particular, the later product properties, like stickiness, porosity, surface texture or morphology, flowability, storage stability. E.g. Krueger investigated the product morphology during co-current flow of the tower reactor [3]. Results for countercurrent operation are not published yet. This study shows that the process temperature has an influence on both, particle size distribution and inner as well as upper surface morphology in a countercurrent operation mode of the spray reactor. Further, we are able to show a dependence of the morphology on the gas flow rate, gas composition and countercurrent flow rate. The above mentioned process parameters are also analyzed regarding their influence on product characteristics such as molecular weight, residual monomer content and residual moisture. Figure 1 shows the technical drawing of the tower reactor used in this study and an example of a powdery product.

Keywords: Spray polymerization, tower reactor, polymer particles, particle morphology, particle formation

Figure 1

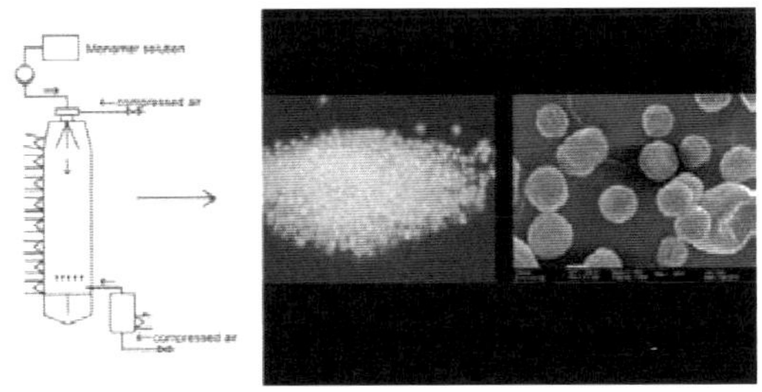

Technical drawing of the tower reactor (left) and the dry powdery polymer product (middle) and a SEM picture of the polymerized particles (right).

PEPC-P-046 [Polymer Engineering, Processing, and Characterization]

Synthesis Of Hydrogel Supported NHC-Pd(II)-Pyridine Catalytic Systems and Suziki-Miyaura Coupling Reaction Kinetics

Cihangir Boztepe[1], Sedat Yasar[2], Asim Kunkul[1], Ismail Ozdemir[2]

[1]*Department of Chemical Engineering, Inonu University, Malatya, Turkey*
[2]*Catalysis Research and Application Center, Inonu University, Malatya, Turkey*

Hydrogels are three-dimensional crosslinked polymeric materials that can absorb and hold large amounts of water without undergoing dissolution. The ionic charge content is important for water absorption. The hydrogel swelling capacity is increase with increasing 2-Acrylamido-2-methylpropane sulfonic acid (AMPS) content because AMPS includes hydrophilic groups, such as –COOH, –NH, –SO2, and –OH [1]. The Suzuki reaction is usually carried out with a homogeneous Pd catalyst but, in order to use the Suzuki reaction in a wide range of industrial applications, a heterogeneous Pd catalyst must be developed [2].

In this study, novel two NHC-Pd-pyridine polymerizable carbene complexes having catalytic activity, have been prepared. Hydrogel supported NHC-Pd(II)-pyridine heterogeneous catalytic systems are prepared via free radical polymerization of acrylamide (AAm, as monomer), AMPS (as co-monomer), N-N' Methylenebis acrylamide (MBA) and a novel monomer of NHC-Pd(II)-pyridine. The effects of AMPS content to Suziki reaction rate are investigated. So, poly(AAm-NHC-Pd(II)-pyridine) and poly(AAm-AMPS-NHC-Pd(II)-pyridine) systems are prepared.

The heterogeneous catalytic reaction rate of poly(AAm-AMPS-NHC-Pd(II)-pyridine) system is higher than poly(AAm-NHC-Pd(II)-pyridine) system.

Keywords: Hydrogel, Heterogeneous Catalyst, N-Heterocyclic carbene, Suzuki-Miyaura reaction.

Figure 1. Synthesis of imidazole based polymerizable NHC-Pd(II)-pyridine complex

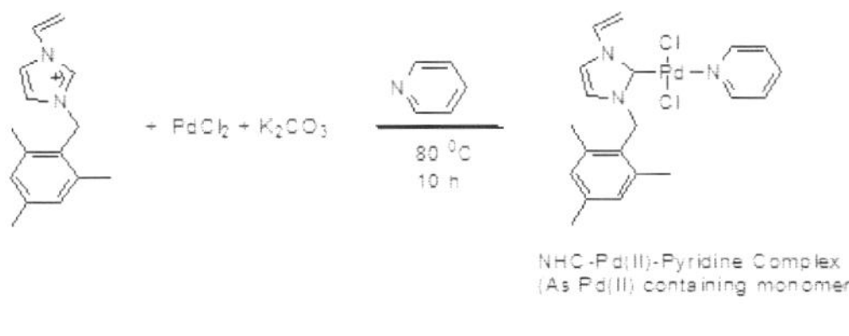

Figure 2. Suzuki-Miyaura coupling reaction.

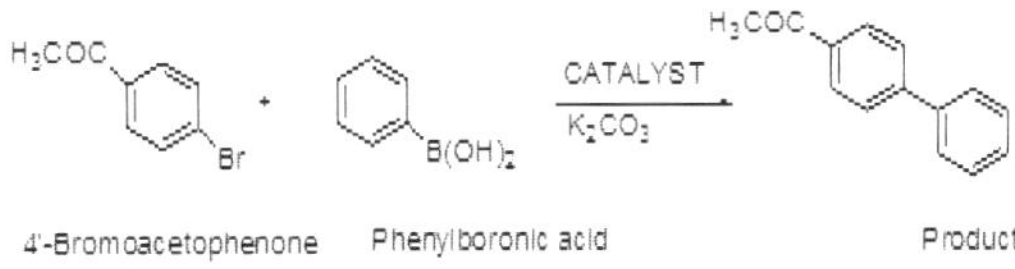

Figure 3. Conversion in the Suzuki reaction of bromoacetophenone with phenylboronic acid

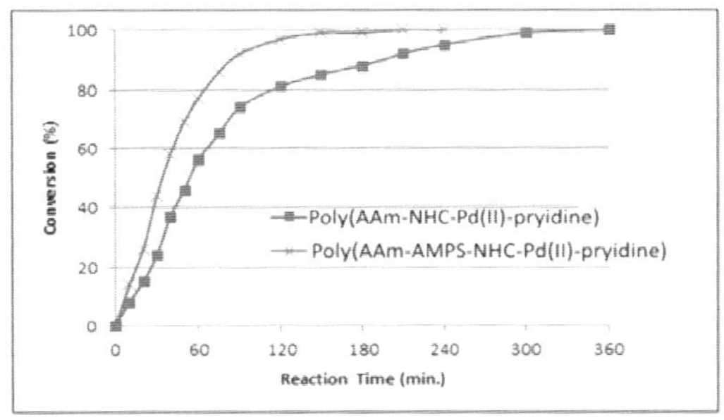

(Reaction conditions: 10 mmol bromoacetophenone (1 equiv.), 20 mmol K2CO3 (2 equiv.), 15 mmol PhB(OH)2 (1.5 equiv.), water =200 mL, 0.1 mol% of Pd, T = 80 °C)

PEPC-P-047 [Polymer Engineering, Processing, and Characterization]

Effects of Polyhedral Oligomeric Silsesquioxane Reinforced Polypropylene (PP) Nanocomposites On the Thermal, Mechanical, and Optical Properties of PP

Sevil Kaynar[1], Nurseli Uyanık[2], Mükerrem Çakmak[3]

[1]*ITU Graduate School of Sci.Eng.&Techn., Polym.Sci.&Techn. Dept., Maslak 34469 Istanbul/Turkey*
[2]*ITU Faculty of Art and Sci.,Chemistry Dept., Maslak 34469 Istanbul/Turkey*
[3]*The Univ.of Akron, Dept.of Polym.Eng., Akron, 44325-0301, Ohio/USA*

The influence of methyl-POSS on the properties of PP was studied, in this work. 1, 3, 5, 8, 10 wt-% POSS containing PP nanocomposites (PP/POSS NC) were prepared by melt blending via extrusion. The prepared PP/POSS NCs and neat PP were investigated by thermally and rheologically. These NCs and PP pellets were also subjected to melt spinning process by using capillary rheometer with six take up speed and three plunger speed to produce fibers. Draw down ratio (DDR=Ddie2/Dfiber2) of fibers were calculated and the effects of draw down ratio on the strength, and polymer chain orientation were investigated by mechanically, optically and morphologically. Rheological studies of PP/POSS NC pellets were operated by using parallel plate rheometry at three different temperatures. It was observed that complex viscosity decreases with increasing frequency, both storage modulus and loss modulus progressively increase with increasing frequency. Complex viscosities showed large increase with reduction of temperature at high frequency. Thermal behaviors of NC pellets were characterized by DSC and TGA. The enthalpy of fusion decreased and the degredation temperature of NCs increased with increasing of POSS content. Mechanical properties showed that POSS composition and DDR of the NCs effected the tensile properties of these materials. The modulus increased with increasing of take up speed and decreased by increasing of POSS content. Morphological characterizations were made by using POM and SAXS. Birefringence data of fibers calculated by using POM images and DDR of samples. For all fibers, increasing of DDR lead to increase birefringence and increasing of POSS content caused the decrease of birefringence values with the comparison of neat PP. SAXS pattern clearly showed that isotropy of fibers increased with increasing of POSS content and increasing of take up speed increased anisotropy and the fiber orientation.

Keywords: PP, POSS, Nanocomposites, Birefringence

PEPC-P-048 [Polymer Engineering, Processing, and Characterization]

Halloysite Nanotube/Polyethylene Nanocomposites with Ethylene Scavenging and Barrier Properties as Food Packaging Materials

Cüneyt Erdinç Taş[1], Buket Alkan[1], Mustafa Baysal[1], Serkan Ünal[2], Fevzi Çakmak Cebeci[1], Yusuf Ziya Menceloğlu[1], Hayriye Ünal[2]

[1]*Material Science and Nanoengineering Department, Sabancı University, İstanbul, Turkey*
[2]*Nanotechnology Research and Application Center, Sabancı University, Istanbul, Turkey*

Packaging materials with ethylene scavenging properties can increase storage life of fruits and vegetables by absorbing the ethylene gas which is a plant hormone responsible for the ripening processes. Barrier properties can provide impermeability to gases such as oxygen, carbon dioxide and water vapor which lead to deterioration of food by increasing their respiration rate. Polymeric films featuring both ethylene scavenging and barrier properties were prepared by the incorporation of halloysite nanotubes (HNTs) into low density polyethylene (LDPE) in order to develop packaging materials that can improve the quality and shelf life of fresh food products. HNTs which are natural clay nanoparticles with hollow nanotubular structures were utilized as nanofillers that absorb the ethylene gas secreted by food samples and at the same time create a torturous path within the polymer matrix that hinders the diffusion of gas molecules. HNT/LDPE nanocomposite films were prepared by melt compounding of HNTs and LDPE at different loading ratios using a twin screw extruder followed by blown film extrusion. The effect of HNT loading ratio on the dispersion quality of nanoparticles, thermal/mechanical properties and morphology of prepared LDPE/HNT films was investigated. Ethylene scavenging capacity of HNTs and HNT/LDPE nanocomposite films were determined by gravimetric analysis and prepared films were shown to have a significantly larger ethylene scavenging capacity than neat LDPE films. Barrier properties of nanocomposite films were investigated by their oxygen and water vapor transmission rate measurements. Ethylene scavenging and barrier properties of nanocomposite films were also evaluated on food samples. HNT/LDPE nanocomposite films were demonstrated to slow down the ripening process of bananas significantly and retain the firmness of tomatoes longer compared to neat LDPE films. Furthermore, due to enhanced water vapor barrier properties, strawberries packaged with HNT/LDPE films presented significantly lower weight loss compared to strawberries packaged with neat LDPE films.

Keywords: Low Density Polyethylene, Halloysite Nanotube, Nanocomposite Films, Polimeric Food Packaging Materials

PEPC-P-049 [Polymer Engineering, Processing, and Characterization]

Effect of Surface Treatment and Cure Time on the Shear Strength of an Polydimethyl Siloxane Adhesive on Single-lap Joins for Aerospace Applications

Nalan Özbay, Onur Dinçer, Hakan Der, Yusuf Ziya Karabay

Advanced Materials Technologies Division, TUBITAK SAGE, Turkey

Alloys composed mostly of aluminium have been very important in aerospace manufacturing indusry. The appropriate treatment of the adherend aluminium surfaces prior to bonding and the thickness of the adhesive it is to be used is very important for durable bonds which are capable of taking structural loads and can be created due to a careful consideration that must be given the way in which the joint is formed. For this purpose, it is necessary to search further into the knowledge and characterisation of the mechanical properties of this type of joint depending on the technical parameters. One of the most relevant parameters is the surface finish of the substrates, as this has an important influence on the mechanical properties of the joint. Furthermore, the relationship between cure time and adhesion is also important for searching mechanical properties. In this present research, the shear strength of Polydimethyl siloxane adhesive bonded aluminium alloys for the aerospace industry, were investigated. The lap shear strength tests of adhesive on aluminium joints were performed by changing the variables affecting the strength of single lap joints, such as surface treatment and cure time. Aluminium single lap joints (SLJs) were fabricated and tested to assess the adhesive (structural one-component polydimethyş siloxane adhesive sealant) performance in a joint. The surface treatments were abrasion, passivation and cleaning by a solvent. The results of increase in surface roughness and cure time will be presented.

Keywords: Aluminium and alloys, polydimethyl siloxane, Surface roughness, Lap-shear

PEPC-P-050 [Polymer Engineering, Processing, and Characterization]

Real-Time Study on The Optical Properties of Polymer complexes of Hydroxypropylcellulose and 4'-Hydroxy-4-biphenylcarbonitrile

Esin Kaya[1], Enmin Wang[2], Mükerrem Çakmak[2]

[1]*Department of Science Education, Faculty of Education, MuşAlparslan University, Muş 49250, Turkey*
[2]*Polymer Engineering Department, University of Akron, Akron, OH, 44325-0301, USA*

In recent years, it has been recognized that molecular interactions between different materials through non-covalent interactions are advantageous for the design of functional molecular aggregates such as liquid crystals as well as crystalline materials. Hydrogen bonding, as a typical non- covalent interaction, has been identified as a new strategy for manufacturing polymer complexes [1,2]. We prepared polymer complexes of Hydroxypropylcellulose (HPC)/ 4'-Hydroxy-4-biphenylcarbonitrile with different compositions by attaching the 4'-Hydroxy-4-biphenylcarbonitrile molecules to HPC via hydrogen bonds. The films of these polymer complexes were obtained by blade casting from Dimethylformamide (DMF) solutions and simutaneously characterised using differential scanning calorimetry, FTIR spectroscopy and polarising optical microscopy. A real-time measurement system was used to track the real-time development of the optical parameters both in-plane and out-of-plane birefringence, together with the weight, thickness and the surface temperature of the casting films during the drying process. At the end of drying, optical anisotropy was observed in the thickness direction as indicated by the out-of-plane birefrignence value whereas the level of this anisotropy decreased and ultimately disappeared when the component of 4'-Hydroxy-4-biphenylcarbonitrile is increasing (Fig.1). The in-plane birefringence remains zero regardless of the polymer complexes, indicating that the molecular chains remain randomly oriented in the film plane throughout the drying procedure [3].

References:

1. T. Kato, N. Mizoshita, K. Kanie, Macromol. Rapid Commun. 22 (2001) 797.
2. X. Han, S. Zhang, R. A. Shanks, D. Pavel, Reactive Functional Polymers 68(2008) 1097.
3.F. Jiang, M. Cakmak Polymer 68 (2015) 168

Keywords: Hydroxypropylcellulose, birefringence, real time characterization, Solvent-cast film

Fig.1

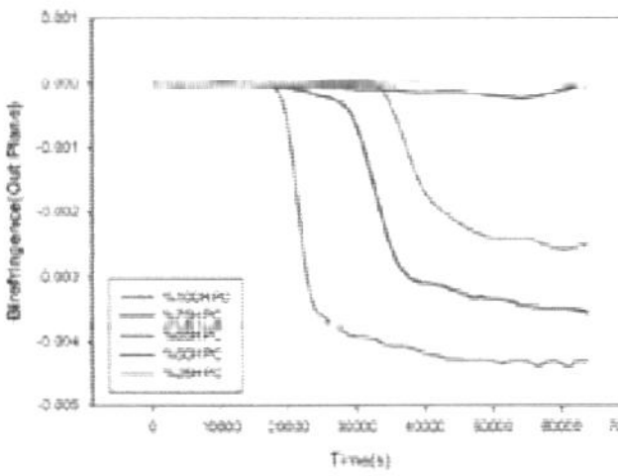

Correlation between percent HPC in cast films and out-of-plane birefringence of Polymer Complexes

PEPC-P-051 [Polymer Engineering, Processing, and Characterization]

Investigation of Mechanical Properties of Recycling Polyolefins and Poly(lactic acid) Blends

Mehmet Mudu[1], Mithat Çelebi[1], Mustafa Öksüz[2], Çağdaş Aslan[1]

[1]*School of Engineering, Yalova University, Yalova,Turkey*
[2]*Faculty of Technology, Marmara University, Göztepe, Turkey*

Population increase worldwide together with the need of people to adopt improved conditions of living led to a dramatical increase of the consumption of polymers. Polyolefins (LDPE, HDPE, PP) are a major type of thermoplastic used throughout the world in such applications as bags, toys, containers, pipes (LDPE), houswares, industrial wrappings and film, gas pipes (HDPE), film, battery cases, automotive parts, electrical components (PP). Recycling polyolefins and bioplastics can be play important role in order to supply this plastic demand. Poly(lactic acid) (PLA) is aliphatic polyesters derived from corn and sugar beets, and degrades into nontoxic compounds in landfill. PLA has some negative properites such as brittleness, low viscosity, low thermal stability, high moisture sensitivity, medium gas barrier properties, and low solvents resistance (e.g., against water). PLA blends can be prepared to enhance its properties. Recycling polyolefins and PLA were prepared with maleic anhydride grafted polyolefins as the compatibilizer to evaluate the effectiveness of PLA as a modifier for polyolefin. Because of the significant difference in polarity between PLA and polyolefin blends of these components results in imcompatible systems with poor physical properties. PLA has the ability to be stress crystallized, termally crystallized, impact modified, filled and processed in most polymer processing equipment. PLA also has excellent organo-leptic characteristic and is suitable for food contacts and related packaging. In this study, PLA blends were prepared with recycling polyolefins using twin screw extruder. Test samples were prepared using injection moulding. The tensile, bending, izod, impact, hardness, melt flow index tests were performed.

Keywords: blend, recycling, polyolefin, mechanical properties, poly(lactic acid)

PEPC-P-052 [Polymer Engineering, Processing, and Characterization]

First Si/Ge/Sn Heterotrimetallic Dendrimers Si(CH2)2Sn[(CH2)4MPh3]34 (M= Ge, Sn): Synthesis and Characterization

Yılmaz Aksu

Department of Materials Science and Engineering, Faculty of Engineering, Akdeniz University, Antalya, Turkey

In the past decade, dendrimers attracted enormous attention due to their unique chemical properties arising from their special and well-defined architecture1 and their potential application in catalysis, medicine, material sciences, or host guest chemistry. A further challenge was the synthesis of heterometallic dendrimers, which were suggested to improve specific chemical and physical properties of the homometallic dendrimers. However, preparative methods for the synthesis of metallodendrimers containing different metals, in particular main group metals, are hardly developed. The only examples reported up to now are silicon-based dendrimers with alternating direct bonded silicon and germanium atoms as well as peripherally substituted carbosilane dendrimers containing lithium and organotin fragments.

Recently, we succeeded in the synthesis of the tin analogues [1] of the organosilicon and organogermanium dendrimers already known. In the meantime, we developed a method to synthesize the first all-tin dendrimer by the divergent [2] as well as by the convergent [3] route and consequently prepared a series of novel tin-based dendrimers.[4] In this paper we describe a simple method for the synthesis of hybrid metallodendrimers of group 14 elements, resulting in the isolation of the first Si/Ge/Sn heterotrimetallic dendrimer. All compounds were characterized by multinuclear NMR studies (1H, 13C, 119Sn), mass spectrometry (MALDI-TOF, EI), and elemental analyses.

[1] Schumann, H.; Wassermann, B. C.; Schutte, S.; Velder, J.; Aksu, Y.; Krause, W.; Radüchel, B. Organometallics 2003, 22, 2034.
[2] Schumann, H.; Aksu, Y.; Wassermann, B. C. J. Organomet. Chem.2006, 691, 1703.
[3] Schumann, H.; Aksu, Y.; Wassermann, B. C. Organometallics 2006, 25 (14), 3428.
[4] Schumann, H.; Aksu, Y.; Schutte, S.; Wassermann, B. C.; Mühle,S. H. J. Organomet. Chem. 2006, 691, 1417.

Keywords: Dendrimer, Metallodendrimer, Tin-Based Dendrimer, Organotin-dendrimer, MALDİ-TOF

PEPC-P-053 [Polymer Engineering, Processing, and Characterization]

Synthesis and Characterization of Organotin Dendrimers up to 3rd Generation Sn(CH2) nSn[(CH2)4SnPh3]34 (n) 3, 4): A Convergent Approach

Yılmaz Aksu

Department of Materials Science and Engineering, Faculty of Engineering, Akdeniz University, Antalya, Turkey

An increasing expectance for materials with new properties focused interest on polymer research, from the traditional inear polymers to the more promising hyperbranched and dendritic polymers. In contrast to pure organic dendrimers, the number of metallodendrimers, especially those consisting of covalently bonded metal-carbon moieties, is very limited, mainly due to the lack of suitable reagents and synthetic methods. Only a very few main group metal based dendrimers have been described so far. Recently we have been successful in the synthesis of organotin dendrimers analogous to the well-known organosilicon and organogermanium dendrimers. On the basis of the results of our previous work[1-3] in this area, we focused our intention on developing a method to prepare organotin dendrimers by the convergent way. In this paper, we report the preparation of dendritic organotin dendrons and their utilization for the convergent synthesis of organotin dendrimers. 1H, 13C, and 119Sn NMR spectroscopic as well as MALDI-TOF mass spectrometric investigations evidenced that all synthetic steps proceeded efficiently without steric hindrance and competitive reactions.

References:

[1] Schumann, H.; Wassermann, B. C.; Schutte, S.; Velder, J.; Aksu, Y.; Krause, W.; Radüchel, B. Organometallics 2003, 22, 2034.
[2] Schumann, H.; Aksu, Y.; Wassermann, B. C. J. Organomet. Chem. 2006, 691, 1703.
[3] Schumann, H.; Aksu, Y.; Schutte, S.; Wassermann, B. C.; Mühle,S. H. J. Organomet. Chem. 2006, 691, 1417.

Keywords: Dendrimer, Metallodendrimer, Tin-Based Dendrimer, Convergent, Organotin, MALDİ-TOF

PEPC-P-054 [Polymer Engineering, Processing, and Characterization]

Study of the influence of whiskers and/or glass fiber concentration on mechanical, structural and thermal properties of polypropylene-based composites

Maria Sönmez[1], Laurentia Alexandrescu[1], Mihai Georgescu[1], Daniela Maria Stelescu[1], Mihaela Nituica[1], Anton Ficai[2], Denisa Ficai[3], Dragos Gudovan[4], Roxana Trusca[5], Dana Florentina Gurau[1]

[1]*Rubber Research Department, National Research & Development Institute for Textiles and Leather–Division: Leather and Footwear Research Institute, Bucharest, Romania*
[2]*Department of Science and Engineering of Oxide Materials and Nanomaterials, University POLITEHNICA of Bucharest, Romania*
[3]*Department of Inorganic Chemistry, Physical Chemistry and Electrochemistry, University POLITEHNICA of Bucharest, Romania*
[4]*Department of Chemistry and Engineering of Organic Substances and Petrochemistry, University POLITEHNICA of Bucharest, Romania*
[5]*S.C. METAV R&D S.A., Bucharest, Romania*

Fiber-reinforced polymer composites are increasingly demanded for industrial applications due to their relatively simple and fast processability, resistance to impact and delamination, low moisture adsorption as well as due to the possibility of infinite raw material storage. These properties provide composites with many possibilities of application in various areas such as the automotive, aerospace, construction, defense, transportation and marine industries [1-2]. According to theories, short fiber-reinforced polymer composites show improvement of mechanical and thermal properties while reducing thermal expansion coefficient and wear resistance [3], compared to traditional fibers (carbon and glass fibers). This is related to the fact that whiskers have a much higher resistance and rigidity due to their almost perfect crystalline structure as well as to the very high length/diameter ratio. In the last few years, inorganic whiskers such as $CaCO_3$[4], Al_2O_3, SiC [5] and K_2Ti_6O13[6] have been synthesized and introduced in the composite development process using various types of polymer matrixes (thermo-rigid, thermoplastic, etc.). The aim of this paper was to investigate the influence of the concentration of whiskers and/or mineral fibers treated with organosilanes on mechanical, structural and thermal behavior of composites based on polypropylene. Composites based on Tipplen K948 polypropylene (PP) reinforced with whiskers based on silicon carbide (SiCw) and/or glass fibers functionalized with 5% dichlorodimethylsilane (FG-DcDMS) and in the presence of low amounts of polypropylene-graft-maleic anhydride (PP-g-MA) as compatibilizer were obtained using a twin screw extruder granulator. The percentage of SiCw reported to the amount of polymer mixture (PP + PP-g-AM) was 3% and that of FG-DcDMS was the same. Composites obtained were analyzed structurally by FTIR, mechanically (tensile strength and three-point bending), as well as in terms of thermal stability under loading (HDT) using QUALITEST HDT1 equipment.

ACKNOWLEDGEMENTS: This research project PNCDI II – 168/2012, "Hybrid composite materials with thermoplastic matrices doped with fibers and disperse nanofillings for materials with special purposes", was financed by UEFISCDI.

Keywords: polypropylene, coupling agents, extrusion, thermal resistance, mechanical behavior

PEPC-P-056 [Polymer Engineering, Processing, and Characterization]

Non-isothermal crystallization and melting behavior of polypropylene in the PP/PVB blends containing PP-g-MA

Ali Reza Zanjanijam[1], Shokoufeh Hakim[1], Hamed Azizi[2]

[1]*Engineering Faculty, Iran Polymer and Petrochemical Institute, Tehran*
[2]*Processing Faculty, Iran Polymer and Petrochemical Institute, Tehran*

Considering application of plasticized polyvinylbutyral (PVB) film in safety glasses of automobiles and buildings, it is estimated that about 100000 tons of PVB is consumed annually. As a result, large amounts of used safety glasses (that contain PVB film) are disposed off in landfills [1]. Due to high price and environmental concerns, it is necessary to reuse waste PVB film in suitable applications. Regarding good mechanical properties of the PVB film, it appears that there is a possibility for using waste PVB in polymer blends. In this study, different PP/waste PVB blends were prepared and influence of the PVB and compatibilizer on crystallization and melting behavior of PP was investigated. The DSC measurements showed that addition of the PVB had a decreasing effect on melting peak temperature (Tm) of the PP crystals (Figure 1). This decrease was more remarkable for PP/PVB30 blends compared to PP/PVB15 ones, because it is due to migration of PVB plasticizer to PP phase during processing of the blends. This phenomenon was confirmed by changes in glass transition temperatures of the phases in dynamic mechanical analysis. On the other hand, the compatibilizer had trivial influence on Tm of the PP. The cooling scan for uncompatibilized blends demonstrated that Tonset and Tc for PP crystals didn't change upon addition of the PVB. Incorporation of PP-g-MA in the blends affected nonisothermal crystallization of PP and increased considerably aforementioned temperatures. It is ascribed to improvement of interfacial adhesion between phases due to role of the compatibilizer. Therefore, size of the PVB particles reduces and a more homogenous dispersion could be obtained in PP matrix. The fine PVB particles could act as heterogeneous nucleation agents causing an increase in crystallinity [2].

References

[1] Tupy et al, J. Appl. Polym. Sci, 2014.
[2] Yang et al., J. Appl. Polym. Sci, 2011.

Keywords: polyvinylbutyral, polypropylene, blend, crystallization

Melting behavior of the a)PVB, b)PP, c)PP/PVB15, d) PP/PVB15/3com, e) PP/PVB15/5com, f)PP/PVB15/10com, g)PP/PVB30, h)PP/PVB30/5com, and i) PP/PVB30/10com

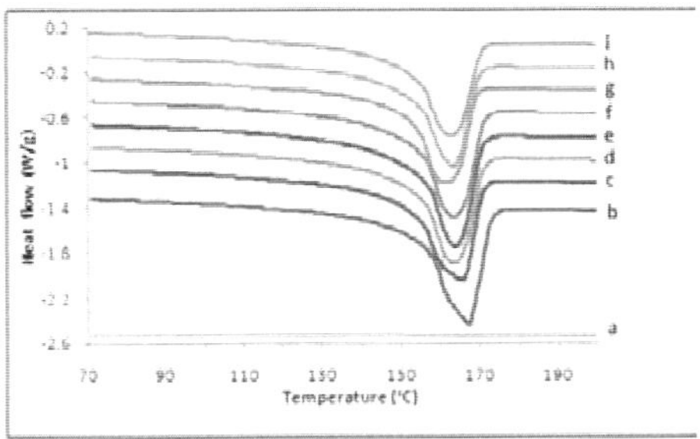

PEPC-P-057 [Polymer Engineering, Processing, and Characterization]

Study of Paint Sludge Emissions from Polymeric Composites

Hamed Ghorbanpoor[1], Esra Akkulak[2], Hazal Gergeroglu[3], Lale Cıvan[4], Macid Nurbas[5], Huseyin Avci[2]

[1]*Department of Polymer science and Technology, Eskisehir Osmangazi University, Eskişehir, Turkey*
[2]*Department of Metallurgical and Materials Engineering, Eskisehir Osmangazi University, Eskişehir, Turkey*
[3]*Department of Nanotechnology and Nanoscience, Eskisehir Osmangazi University, Eskişehir, Turkey*
[4]*Department of Biotechnology and Biosafety, Eskisehir Osmangazi University, Eskişehir, Turkey*
[5]*Department of Chemical Engineering, Eskisehir Osmangazi University, Eskişehir, Turkey*

Waste paint sludge is one of the most hazardous wastes in the segment industry, especially in automobile manufacturing which menace human health and environment. Nowadays, by waste generation increasing, various ways for re-using and recycling wastes also increases. For instance, Segala (2003) burned waste to co-processing in a cement kiln. Another research showed that paint sludge can be used in producing cement as ingredient materials row (Gautam et al., 2010). In addition, we already applied paint sludge in mortar as aggregate and observed significantly increase of volume in concretes comparing to control specimens. Paint sludge is divided into two categories: water-based and solvent-based (Salihoglu et al., 2016). In this study we investigated solvent-based paint sludge. The aim is study of paint sludge effects as reinforced materials in polymeric composites. In fact, because of toxicity and dangers of paint sludge, we evaluated emission of them into water when composites encounter with water for a long time. We prepared several specimens with different paint sludge ratio into epoxy resin matrix. In this experiment, UV-Vis was used to determine emissions of paint sludge into water after 24 hours. In conclusion, waste paint sludge has very good potential to use in polymeric composite materials because of not showing any emission into water in the first 24 h. To better understand the emission of paint sludge, we are keeping specimens into water for a longer time to further study.

Keywords: paint sludge, hazardous waste, waste management, automotive paint waste

PEPC-P-058 [Polymer Engineering, Processing, and Characterization]

Morphology development in the PP/ plasticized PVB blends: Influence of content of the compatibilizing agent

Ali Reza Zanjanijam[1], Shokoufeh Hakim[1], Hamed Azizi[2]

[1]*Engineering Faculty, Iran Polymer and Petrochemical Institute, Tehran, Iran*
[2]*Processing Faculty, Iran Polymer and Petrochemical Institute, Tehran, Iran*

Polymer blending has been attracted a lot of interest in academics and industry in recent years, because it offers an appropriate way to combine useful properties of different polymeric matrices [1]. However, immiscible polymer blends show poor interfacial adhesion which leads to unfavorable performance. Therefore, compatibilizers are added to the blends to improve their compatibility. Concentration of the compatibilizer plays an important role in behavior of the blends [2].

In this study, influence of the compatibilizer content on behavior of the PP/ plasticized PVB blends is probed. According to the results, saturation of interface with compatibilizer and micelle formation in the matrix occurs at certain content of the maleated PP. This content was about 3 and 5 wt.% for the PP/ PVB15 and PP/PVB30 blends, respectively. Actually, above the critical content, excess compatibilizer migrates to the PP matrix. This in turn decreases viscosity and elasticity of the matrix and hence a lower shear stress is imposed on the dispersed phase particles leading to increase in average particle size [3, 4]. Furthermore, due to this phenomenon, interparticle distance between dispersed phase particles was decreased by adding maleated PP up to the critical content, beyond which it increased. The interparticle distance has an important role in the mechanical properties of the blends.

Also, decrease in ratio of intensity of the hydroxyl peak to that of the peaks of butyral ring (IO-H / IC-O-C) in the FT-IR spectra revealed that a part of the OH groups in the PVB chain were reacted with MA groups of the compatibilizer and converted to ester linkage.

References:

[1] Al-Malaika et al, (2005), Polym. Degrad. Stab,90(2): 197-210.
[2] Fortelny, I. (2000) J. Macromol. Sci. Phys. B., 39(1): 67-78.
[3] Zhao et al, (2014) RSC Advances, 4 (103): 59302-59309.
[4] Entezam et al, (2012) Macromol. Mater. Eng., 297 (4): 312-328.

Keywords: blend, compatibilizer, polypropylene, polyvinylbutyral

PEPC-P-059 [Polymer Engineering, Processing, and Characterization]

Enhancement of the thermal stability of the plasticized poly(vinyl chloride) using single wall carbon nanotubes

Ali Reza Zanjanijam[1], Mohammad Bahrami[2], Morteza Hajian[2], Maryam Ramezani[1]

[1]Iran Polymer and Petrochemical Institute, Tehran, Iran
[2]Faculty of Chemistry, University of Isfahan, Isfahan, Iran

Poly (vinyl chloride) (PVC) is a thermoplastic that possesses widespread applications in plastic industry due to its inherent properties, such as superior mechanical and high chemical and abrasion resistance. Preparation of the PVC plastisol and its curing is an industrial method for fabrication of the PVC films. PVC suffers from low thermal stability. Therefore, various additives such as thermal stabilizers and nanofillers are used to solve this problem [1, 2]. This study presents the effects of the single wall carbon nanotubes (SWCNT) on thermal degradation of the plasticized PVC films. From the TGA curves (Figure 1), it could be seen that addition of very low amount of the SWCNT led to significant improvement in the thermal stability of the plasticized PVC. In other words, onset temperature of decomposition (Tonset) and decomposition temperature at maximum rate of the weight loss (Tmax) of the composites shifted to the higher temperatures and weight loss decreased notably. Also, the residual mass at 600 °C increased remarkably. The higher thermal stability of the composites is due to the high thermal conductivity and barrier effect of the carbon nanotubes. The high thermal conductivity helps to dissipate heat within the polymer matrix. On the other hand, the barrier effect of the nanotubes can hinder diffusion of gaseous decomposition products throughout the composite. These results indicate that carbon nanotubes can effectively improve thermal stability of the PVC.

References:

[1] Gong et al., Polym. Degrad. Stab., 84, 289 (2004).
[2] Liu et al., Polym. Degrad. Stab., 92, 503 (2007).

Keywords: Polyvinylchloride, thermal properties, Composites, carbon nanotubes

PEPC-P-060 [Polymer Engineering, Processing, and Characterization]

Characterization and investigation of synthesized Self-Colored PET(polyethylene terephthalate) based on Naphthalimide dye

Ehsan Zamani, Alireza Khosravi, Mehdi Rafizadeh

Amirkabir University of Technology (Tehran Polytechnic),Tehran, Iran

Synthesize and characterization of a dye based on naphthalimide has been described. Acenaphtene was used as a first material and four steps of nitration, oxidation, imidation and reduction were carried out to synthesize 4-amino-N-propanoic acid-1,8-naphthalimide as a fluorescent and polymerizable dye. In addition, the synthesized dye was added during polymerization of polyethylene terephthalate (PET). The wavelength of maximum absorbance and maximum intensity of fluorescence of dye were at 420nm and 523nm by utilizing UV-visible and fluorometry, respectively. The wavelength of maximum absorbance and maximum florescence intensity of polymer were at 410nm and 519 nm, respectively. UV-visible spectrum indicated that, 98-99 percent of dye was reacted chemically with polymer chains. The results indicated that a self-colored PET with appropriate properties have been prepared. The viscometer showed inherent viscosity was 0.556cm3/g and molecular weight was obtained 35156 g/mol using the viscometer technique.

Keywords: Self-Colored PET, Naphthalimide, Polyethylene terephthalate

PEPC-P-061 [Polymer Engineering, Processing, and Characterization]

Controlling the process of synthesizing Self-Colored PET (polyethylene terephthalate) in the reactor

Ehsan Zamani, Alireza Khosravi, Mehdi Rafizadeh

Amirkabir University of Technology (Tehran Polytechnic),Tehran, Iran

The polymerizable dye was synthesized based on naphthalimide with acenaphthene as a first material. After processing of four processes such as nitration, oxidation, imidation and reduction, 4-amino-N-propanoic acid-1,8-naphthalimide was synthesized as a fluorescent and polymerizable dye. In the next step, the synthesized dye should be add during polymerization of polyethylene terephthalate (PET). The first process of synthesizing PET is esterification. In the esterification process, the temperature of reactor was increased to 242°C and the pressure was controlled over 4.5 bar. The progress of reaction is evaluate by the amount of produced water as a by-product, which showed that the progress of reaction is near complete(96percent). After more than 2 hours, the esterification was complete and the pre-polymer of PET, Bis-hydroxyl ethyl terephthalate (BHET), was synthesized. The polycondensation of reaction was started by adding synthesized dye, thermal stabilizer and catalysis at 282°C and low vacuum conditions. Polycondensation reaction continued for 3 hours to obtain the desired polymer. The variation of temperature and pressure during esterification and trans-esterification of self-colored PET are controlled at accurate amounts.

Keywords: Pre-polymer, Self-colored PET, Polyethylene terephthalate

PEPC-P-062 [Polymer Engineering, Processing, and Characterization]

Use of Rubber and Sawdust Wastes via Sodium Pentaborate, α-Sepiolite and Magnesium Hydroxide in Application of Radiation Isolation of Composites

Filiz Balkaya[1], Zübeyde Kıvrak[1], Hamed Ghorbanpoor[2], Macid Nurbas[1]

[1]Department of Chemical Engineering, Eskisehir Osmangazi University, Eskisehir, Turkey
[2]Department of Polymer Science and Technology, Eskisehir Osmangazi University, Eskisehir, Turkey

The effect of radiation has been increasing on human health and environment in parallel with technological development and use of electronic devices. Radiation not only damages human health and environment but also it is effective for miles. For instant, Chernobyl disaster in Ukraine is the simplest example which was also observed in Black Sea.Therefore, we need to use composite materials with radiation absorption capacity from devices and radiation-emitting facilities. Protection of radiation has become imperative to take the necessary protective measures. In this study, we applied rubber and sawdust wastes via sodium pentaborate, α-sepiolite and magnesium hydroxide to undrestand effect of rubber and sawdust wastes in radiation absorption of composites. We prepared several composite panels with different rubber and sawdust ratio into epoxy resin matrix. Fig.1 shows all the composite panels. Radiation was measured with ion chamber meter device in order to measure gamma and x-ray radiation rate. Results diplayed that 30%rubber powder + 30% (sawdust-urea formaldehyde) + 15% α-sepiolite + 15% Sodium pentaborate + 10% Magnesium hydroxide into epoxy resin was the best efficiency ratio. In conclusion, use of rubber and sawdust wastes via sodium pentaborate, α-sepiolite,and magnesium hydroxide in application of radiation isolation of composites can be affected by absorbtion.

Keywords: Rubber Waste, Sawdust Waste, Radation Absorption, Waste Management

Fig.1

Composite panels images

PEPC-P-063 [Polymer Engineering, Processing, and Characterization]

Halogen-free flame retardants for poly(butylene terephthalate) composite

Ceren Yargıcı[1], Mustafa Sezer[2], Nurseli Uyanık[3]

[1]*Graduate School of Science Engineering and Technology, Istanbul Technical University, Istanbul, Turkey*
[2]*Arcelik Research and Development Center, Istanbul, Turkey*
[3]*Chemistry Department, Istanbul Technical University, Istanbul, Turkey*

Fire hazards arise from polymeric materials are of particular concern among government regulatory bodies, consumers, and manufacturers. Additive flame retardants are widely used in polymers, but in recent years, halogenated flame retardants are not preferable because of reaction mechanism of polymer under fire cause lots of harmful impact on environment like formation of smoke, toxic gasses and corrosive decomposition products [Mai, Y.,W., 2013]. In this study, the halogen-free flame retardant additives, boron phosphate (BP), triphenyl phosphate (TPP) and zinc borate (ZB) were employed with low-loading level in neat PBT and 30% glass fiber (GF) reinforced PBT. PBT composites were prepared by using a co-rotating twin screw extruder and PBT samples were shaped by injection molding machine. Limiting oxygen index (LOI) test and glow wire test (GWT) for flammability properties were conducted for PBT composites. Thereafter, the mechanical performance of these compounds were also performed. TPP additive was the most effective one in all flame retardants which were investigated in this research. TPP passed both 750 0C and 850 0C GWT and obtained highest LOI test result when 10 % TPP was added to PBT matrix, LOI increased to 27.2 from 22.5. Moreover, it was also found that when 5% BP with 5% ZB were used together, they showed a significant synergistic effectiveness in terms of flammability. Addition of 30 % glass fiber to PBT matrix increased mechanical properties of composite considerably. Tensile strength value of PBT increased to twice with 30 % glass fiber reinforcement and raised to 105 MPa from 51 MPa, flexural strength value increased to 157 MPa from 76 MPa, Izod (notched) impact strength increased to 10.6 kJ/m2 from 5.2 kJ/m2. This high increase was seen all glass fiber reinforced flame retardant composites. Furthermore, TPP additive affected the toughness and increased impact strength of PBT composites.

Keywords: PBT, flame retardant, composite, glass fiber

PEPC-P-064 [Polymer Engineering, Processing, and Characterization]

Mechanism Study on Electron-beam Assisted Stabilization of Polyacrylonitrile Fibers for Shortening the Conventional Thermal Treatment

Sejoon Park, Seung Hwa Yoo, Ha Ri Kang, Seong Mu Jo, Han Ik Joh, Sungho Lee

Carbon Composite Materials Research Center, Institute of Advanced Composite Materials, Korea Institute of Science and Technology

Carbon fibers (CFs) are attractive materials as reinforcing materials in various high performance structural composites from sporting goods to aircraft parts because they possess high strength and modulus per unit weight. Although a number of precursor materials such as pitch, rayon, and lignin were investigated for their commercialization, polyacrylonitrile (PAN) based CFs have dominated the high performance CF market due to their superior mechanical properties. The manufacturing process of PAN based carbon fiber (CF) can generally be divided into three processing steps: 1) spinning and stretching, 2) stabilization (200~300 °C) and 3) carbonization (≥1000 °C). A number of experimental parameters associated in each step has been extensively optimized to enhance the mechanical properties of the CFs. Among these three steps, the stabilization is the most crucial step as a determinant of not only high performance but also of processing cost. The conventional stabilization is carried out by passing the fibers through a series of ovens with gradual temperature gradient (200~300 °C) for a few hours, which is the most time and energy consuming step. Effects towards shortening stabilization time were made. As one of the methods to shorten such lengthy step with expensive energy consumption, electron beam irradiation on PAN precursor fibers has been employed prior to short thermal treatment. In conclusion, with short treatment of electron beam (200 sec), the irradiated fibers were successfully stabilized with short time of thermal treatment (30 min). In the presentation, we focus the study of how the electron beam treatment enables rapid stabilization by presenting several characterizations of irradiated fibers.

Keywords: Carbon fiber, electron beam irradiation, rapid stabilization, polyacrylonitrile

PEPC-P-065 [Polymer Engineering, Processing, and Characterization]

Diffusion (DOSY) NMR for fast molecular weight analysis of polyethylene furanoate (PEF) polyesters

Jan-Georg Rosenboom, Giuseppe Storti, Massimo Morbidelli

Institute of Chemical and Bioengineering, ETH Zurich, Zurich, Switzerland

Polyethylene furanoate (PEF), the promising 100% renewables-based substitute for fossil-based polyethylene terephthalate (PET), presents challenges to established molecular weight analysis methods such as size exclusion chromatography (SEC) due to its need for solvents such as hexafluoroisopropanol (HFIP), experimental care and special HFIP-resistant equipment. In this work, we present diffusion (DOSY) NMR as an alternative and quick method to measure molecular weight of PEF.

DOSY applies pulsed field gradients to spatially label molecules and derive diffusion coefficients from a fit to the signal decay of their diffusion-based limited spin-refocusing. The system can be readily calibrated using PET standards of known absolute molecular weights and the DOSY procedure is easily accessible with pre-installed double stimulated echo pulse (dstegpd2) sequences, which enable very reproducible analysis through convection compensation. DOSY has successfully been applied to derive number-average (Mn) and weight-average (Mw) molecular weights for PEF and PET, which match closely the absolute molecular weight predictions from SEC with multi-angle light scattering (MALS). Optimizing the procedure, we successfully applied a range of 0.4 – 30 mg/mL of polymer in 0.75 mL deuterated trifluoroacetic acid (TFA-d), where Brownian motion regime without molecular interaction was always confirmed. At higher concentrations, sufficient signal-to-noise with less scans allows DOSY analysis time to be as short as 1 min without significant loss in accuracy (<5%). Alongside molecular weight analysis with DOSY, PEF synthesized through ring-opening polymerization exhibits separate chemical shifts of cyclic monomers and linear PEF, which allows for simultaneous conversion analysis using simple 1H NMR peak integration.

The simple and robust setup, reliable molecular weight and conversion evaluation besides the usually 1H NMR derived product and impurity identification, as well as the short analysis time make DOSY NMR an attractive technique to be used in polymer reaction monitoring in industry and academia.

Keywords: molecular weight analysis, diffusion, DOSY, NMR, polyethylene Furanoate, PEF

Molecular weights derived from DOSY NMR compared with those from SEC-MALS

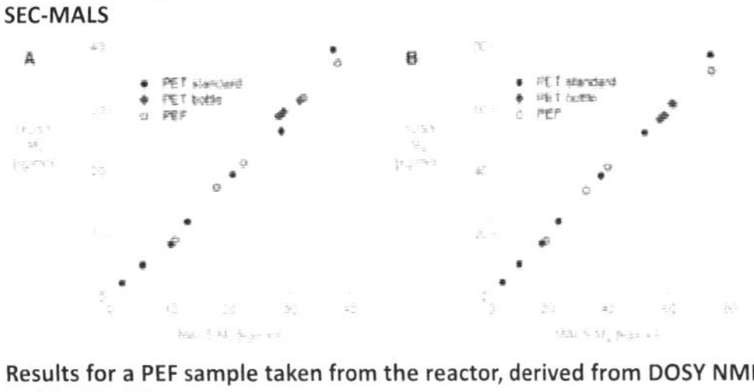

Results for a PEF sample taken from the reactor, derived from DOSY NMR and SEC-MALS

	DOSY NMR	SEC-MALS
Conversion	43.4%	42.3%
Mn [g/mol]	16600	16100
Mw [g/mol]	32200	31900

PPBM-P-001 [Polymer and Polymer-Based Membranes]

Surface modification of polyethersulfone/ aminated polyethersulfone membranes by the self-assembly with antibacterial materials

Chang Keun Kim, Lakwon Choi, Eun Yub Choi, Kwang Young Park

School of Chemical Engineering & Materials Science, Chung-Ang University, Seoul, Korea

Ultrafiltration membranes composed of polyethersulfone (PES) and aminated PES (PES-NH2) were fabricated by the nosolvent induced phase separation process and then the zinc oxide (ZnO) nanoparticles were assembled onto their surface. To assemble ZnO nanoparticles onto the PES/PES-NH2 membrane surface, thionyl chloride terminated ZnO nanoparticles were reacted with the amine groups in PES-NH2. The hydrophilicity and water flux of the membranes increased with increasing PES-NH2 content, and a reduction in solute rejection was not observed when membranes contains less than or equal to 20 wt% of PES-NH2. No antibacterial activity was observed for PES/PES-NH2 membranes, while PES/PES-NH2 membranes assembled with more than or equal to 0.8 wt% of ZnO nanoparticles displayed antibacterial activity (=6.1). Membranes having antibacterial activity, high water flux, and improved fouling resistance without a loss in solute rejection could be fabricated by assembling ZnO nanoparticles grafted onto the PES/PES-NH2 membrane surface.

Keywords: ultrafiltration, self-assembly, antibacterial activity, antifouling

PPBM-P-002 [Polymer and Polymer-Based Membranes]

Metal-Organic Frameworks: Coordination Polymeric Nanofibers

Alper Inan[1], Nesrin Horzum[2]

[1]Department of Chemistry, Izmir Institute of Technology, Izmir, Turkey
[2]Department of Engineering Sciences, Izmir Katip Çelebi University, Izmir, Turkey

Coordination polymers (CPs) have in the recent years emerged as a class of magnificent inorganic-organic hybrid materials due to their featuring various architecture, having tunable pore size, desirable framework and high thermal stability [1]. The fabrication of coordination-polymeric nanofibers is commercially important aspect of polymer science because of their ease of functionalization, having high surface area to volume ratio and easy material combination of nanofibers in terms of their potential applications in sensing, purification, separation, catalysis and energy. In this study, the generation of novel coordination polymers constructing with two central component: a series of transition metals as connector and 1, 3, 4-thiadiazole derivatives as linker in the form of nanofibers is reported. Novel coordination polymers have been synthesized by mixing 2, 5-dimercapto-1,3,4- thiadiazole (DMTD) with biologically important transition metal ions (Ag1+, Cu2+, Au3+, Ti+4, V+5), worked up into nanofibers by electrospinning and characterized by Fourier Transform Infrared Spectroscopy (FTIR), X-ray diffraction (XRD), Nuclear Magnetic Resonance (NMR), Gel Permeation Chromatography (GPC), Thermogravimetric analysis (TGA). Scanning electron microscopy (SEM) and Brunauer-Emmett-Teller (BET) analysis were used to characterize the structure, surface area and pore-size determination of nanofibers. Electrochemical performance of the nanofibers was studied by Cyclic Voltammetry (CV).

Reference:

[1] Kitagawa, S., Kitaura, R., Noro, S. Angew Chem Int Ed Engl. 2004, 43, 2334-75.

Keywords: metal-coordination polymer, electrospinning, hybrid nanofibers

PPBM-P-003 [Polymer and Polymer-Based Membranes]

Synthesis, Characterization and study of rheological behavior of copolymer hydrogels based on poly (N-isopropylacrylamide-co-acrylamide)

Seddiki Nesrinne, Aliouche Djamel

Laboratoire De Traitement Et Mise En Forme Des Polymères,Faculté Des Sciences, Université De Boumerdes- Algérie

In this work, a series of copolymers hydrogels based on the poly (N-Isopropylacrylamide) was synthesized through a radical mechanism of polymerization initiated by persulphate ions at room temperature, N, N' - methylene bisacrylamide (BIS) was used as crosslinking agent. The copolymers were characterized by FTIR spectroscopy, thermal analysis (DSC) and dynamic rheology. Results indicated that the strong interaction in the hydrogels resulted in the formation of a more stable copolymer. The single Tg in sample suggested that the two polymers in the hydrogel have good miscibility. Dynamic rheology study showed a linear behavior, indicating a stable structure and a good mechanical strength.

Keywords: copolymer, hydrogels, poly (N-isopropylacrylamide-co-acrylamide), rheological behavior

PPBM-P-004 [Polymer and Polymer-Based Membranes]

Formation of Cellulose Acetate Membranes: Influencing Structure and Performance

Jakob Tolk, Annette Reiche

Sartorius Stedim Biotech GmbH, Göttingen, Germany

Regenerated cellulose (RC) micro- and ultrafiltration membranes are promising constituents for filtration and purification steps of the biopharmaceutical downstream process since they give a superior performance compared to other membranes. They are highly hydrophilic, low adsorptive for proteins and surfactants and exhibit an excellent chemical resistance. Furthermore, they are gamma sterilizable and hardly susceptible to fouling. Many of the advantages of RC membranes lay in its chemically inert nature, which makes fabrication of cellulosic membranes difficult. To overcome this, the membranes are usually manufactured by saponification of a cellulose acetate (CA) precursor membrane. On an industrial scale, this precursor membrane can be produced by casting a multi-component polymer solution on a continuous stainless steel belt, which runs through several evaporation stages. During this path, the polymer solution demixes due to the solvent evaporation and the desired membrane structure is formed. The aim of the current work is to examine the influence of the casting solution composition and main process parameters on the structure and the performance of the cellulose acetate precursor and the final product made of regenerated cellulose. For this purpose, several membrane samples were made from different casting solutions under various conditions using a small-scale pilot plant.

The resulting membranes were characterized regarding their performance and their mechanical stability before and after saponification. Additionally, the crystallinity, which has a significant impact on the chemical accessibility and the mechanical properties of the polymers, was determined for both CA- and RC-membranes by differential scanning calorimetry and x-ray scattering, respectively.

The results show a significant impact of the casting solution composition on the structure and the mechanical stability of the produced membranes as well as their shrinking during the saponification process. Thus, the current project forms an ideal basis to optimize existing production processes and to develop innovative new products.

Keywords: cellulose acetate, membranes, microfiltration

PPBM-P-005 [Polymer and Polymer-Based Membranes]

Synthesis and characterization of highly fluorinated polytriazoles copolymers with controlled degree of sulfonation for proton exchange membranes

Asheesh Singh

Materials Science Centre, Indian Institute of Technology, Kharagpur 721302, India

A series of new highly fluorinated sulfonated polytriazole copolymers (PTFQSH-XX) were prepared by the click polymerization of 1,1,1,3,3,3-hexafluoro-2,2-bis(4-(prop-2-ynyloxy)phenyl)propane (TF), 4,4′-diazido-2,2′- stilbene disulfonic acid disodium salt ("S") and 4,4-bis[3-trifluoromethyl-4(4-azidophenoxy)phenyl]biphenyl (QAZ). The degree of sulfonation (DS) of PTFQSH-XX was varied by changing the molar ratio of the sulfonated diazide ("S") to the non-sulfonated diazide (QAZ). The structures of the polytriazoles were confirmed by FTIR and NMR (^{1}H, ^{13}C and ^{19}F) spectroscopy. The presence of hexafluoroisopropylidene groups in the copolymers contributed to high thermal and oxidative stability and low swelling of the membranes. TEM micrographs of the polytriazole membranes showed a good phase separated morphology with ionic clusters in the range of 3–75 nm. The proton conductivities of the fluorinated membranes were found in the range of 27–136 mS cm-1 at 80 °C in water.

Keywords: Click polymerization, Polytriazole copolymers, Proton conductivity, TEM micrographs

PPBM-P-006 [Polymer and Polymer-Based Membranes]

Effect of trityl substituted triphenylamine group on gas transport properties

Soumendu Bisoi, Susanta Banerjee

Materials Science Centre, Indian Institute of Technology, Kharagpur- 721302, India

Researches on polymer membranes are of great technological interest for a range of molecular separations due to their potential for enhanced energy efficiency and fast-growing separation technology by economically and environmentally over other competing technologies. The advancement of the membrane based separation technology depends on the availability of the superior membrane materials with high gas permeability and gas selectivity. In this scenario aromatic polyamides (PA) can be used as a membrane material. A series of new processable PA membranes containing trityl group substituted triphenyl amine (TPA) unit for gas separation were prepared from a newly synthesized diamine [Bis-(4-amino-phenyl)-(4-trityl-phenyl)-amine] monomer with four different diacid monomers. The synthesized PAs showed high thermal stability (up to 371 °C in air, for 10% weight loss) and tensile strength up to 74 MPa. The permeability of four different gases CH4, N2, O2 and CO2 through these PA membranes was investigated at 35 °C under a defined applied upstream pressure. PAs showed high gas permeability (PCO2 up to 141 and PO2 up to 33 barrer) and high permselectivity (PCO2/PCH4 up to 44.33 and PO2/PN2 up to 8.74). The gas transport properties have been correlated with the polymer structures with respect to their experimental and simulated molecular modelling fractional free volume (FFV) and intersegmental distance (dsp).

Keywords: Triphenylamine moiety, gas permeability, fractional free volume.

Permeability/selectivity trade-off map for O2/N2 separation

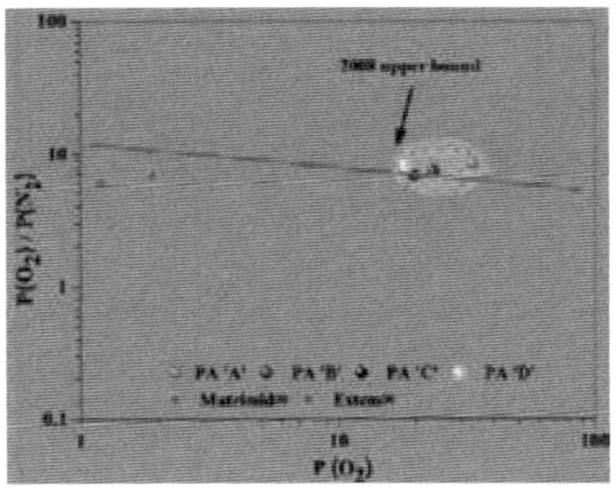

PPBM-P-007 [Polymer and Polymer-Based Membranes]

Morphology and thermal stability of nanocomposite polymer membranes containing methylimidazolium ionic liquid

Oscar Javier Osorio Pedroza[1], José Carlos Dutra Filhio[2], Victor Jayme Roget Rodriguez Pita[2], Marcos Lopes Dias[2]

[1]*Servicio Nacional de Aprendizaje (SENA) Centro Industrial y de Desarrollo Tecnológico (CIDT), Barrancabermeja, Colombia*
[2]*Universidade Federal do Rio de Janeiro (UFRJ) Instituto de Macromoléculas Professora Eloisa Mano (IMA), Rio de Janeiro, RJ., Brazil.*

Composite membranes were prepared by electrospinning/casting method aiming fuel cell application. Non-woven nanofiber mats of poly(styrene sulfonic acid-co-maleic anhydride) (PSSAMA)/poly(vinyl alcohol)(PVA)/ionic liquid (LI) blends were prepared by electrospinning using water as solvent. The mass ratio PSSAMA/PVA= 2/3 was employed and 1-Buthyl-3-methylimidazolium trifluoromethanesulfonate (BMIM TfO) was used as LI with concentrations of 10, 30 and 50 wt% in relation to the polymers. The parameters used in the electrospinning of the nanofibers were: needle-collector distance = 17 cm, voltage= 15 kV and solution flow rate= 0.25 mL/h. Composite membranes were prepared by casting a polycarbonate (PC) methylene chloride solution over the PSSAMA/PVA/BMIM TfO nanofibrous mats. Scanning electron microscopy (SEM) showed that these composite membranes are dense structures formed by nanofibers which act as composite filler and proton conducting elements. SEM micrographs revealed also that the addition of the smaller amount of BMIM TfO (10%) to the blend decrease the average diameter of the fiber up to 76 nm when compared with fibers prepared without any LI (average diameter= 123 nm). Thermogravimetric analysis showed that nanofibers on set degradation temperature was not altered up to 30 w% of LI (around 259°C), but it increases to 291°C when 50 wt% of BMIM TfO was added to the fibers.

Keywords: Composite membrane, Ionic liquid, Electrospinning, Poly(styrene sulfonate), Policarbonate, Fuel cell

PPBM-P-008 [Polymer and Polymer-Based Membranes]

Radiation grafting of crosslinked anion exchange membranes for bipolar membrane fuel cells

Enver Güler[1], Selmiye Alkan Gürsel[2]

[1]*Sabancı University Nanotechnology Research and Application Center (SUNUM), Istanbul, Turkey*
[2]*Faculty of Engineering and Natural Sciences, Sabancı University, Istanbul, Turkey*

In order to maintain an appropriate hydration of membranes and an effective operation, polymer electrolyte membrane fuel cells often require elaborate water management, which dramatically increases the complexity and cost of the fuel cell system. Bipolar membrane fuel cells, which have been very recently recognized, present an entirely different approach to membrane hydration by exploiting the concept of bipolar membranes containing both proton exchange membranes and anion exchange membranes. A novel design such that PEM on the anode side and AEM on the cathode side will provide self-humidification at the interface of the composite membrane structure to ensure high ionic conductivity (Figure 1). Besides, non-platinum catalysts will be allowed due to inherent faster oxygen reduction kinetics on an alkaline cathode. This makes AEM the key component of the bipolar membrane fuel cells. However, the further development of AEMs is still blocked by their low anionic conductivity compared to PEMs. An ideal AEM should maintain high ionic conductivity and good mechanical properties. However, it is not straight-forward to simultaneously provide these properties. Membranes having high ionic conductivity usually exhibit a declined dimensional stability and mechanical properties or vice versa. To prevent this trade-off in this work, a crosslinking strategy using a tertiary diamine such as 1,4-Diazabicyclo[2.2.2]octane has been applied to simultaneously provide high ionic conductivity and good mechanical properties in the design of AEMs. Prior to crosslinking, the ETFE films were irradiated and grafted with vinyl monomers to form versatile AEMs for bipolar membranes. Compared to conventional crosslinkers, the use of a tertiary diamine was recognized to obtain high performance AEMs in bipolar membranes.

Keywords: radiation grafting, anion exchange membrane, bipolar fuel cell, crosslinking

Schematic of a bipolar membrane fuel cell containing a proton exchange membrane (PEM) and a crosslinked anion exchange membrane (AEM).

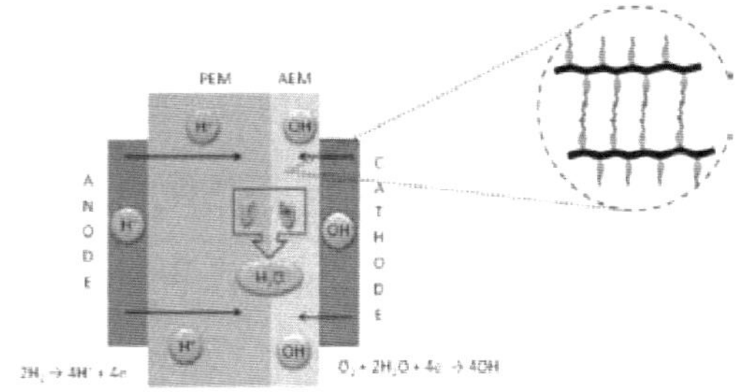

PPBM-P-009 [Polymer and Polymer-Based Membranes]

Preparation of Polysulfone-Modified Fiber Structure Sepiolite Hybrid Particles for Ultrafiltration Membrane

Betül Hanbeyoğlu Yavaş[1], Nilgün Kızılcan[1], Birgül Benli[2]

[1]*Istanbul Technical University, Faculty of Science, Department of Chemistry, Maslak, 34469, Turkey.*
[2]*Istanbul Technical University, Department of Mineral Processing Engineering, Maslak, 34469,Turkey.*

Polysulfone (PSU) is a widely used amorphous engineering thermoplastic polymer with extraordinary chemical, thermal and mechanical properties. These properties might be listed as thermal stability, mechanical strength, high rigidity, stiffness, excellent resistance to hydrolysis, and acids and bases, resistance to creep, oxidative resistance and an extensive operative range of temperature and pH. Owing to these properties, polysulfones are used in the fields of materials science, biology and polymer science and have been the basis of numerous applications.. Up to this point, PSU membranes are prepared for the applications such as filtration, hemodialysis, bioartificial organs, gas separation, water purification systems, and fuel cells, etc. Sepiolite is a natural clay mineral. It is a member of layered silicates; especially it is a magnesium hydrosilicate. Its unique properties of high surface area, fibrous structure, porosity, crystal morphology and composition, surface activity, production of stable suspensions of high viscosity at low concentrations leads to many applications. The aim of this study is to investigate the optimum conditions for the preparation of sepiolite added polysulfone composite fabrication for the application of ultrafiltration membranes from casting solutions containing PSU and fibrous sepiolite. To obtain compatible surfaces, it is known that the achievement of high clay-solvent interaction and well dispersion during in-situ polymerization is important. Sepiolite was utilized as both raw material and enriched with minerals (Ag, Cu). Synthesized composites were characterized by FTIR, NMR. XRD and SEM measurements and they are going to be used as membrane fabrication.

Keywords: Polysulfone, membrane, sepiolite, in-situ polymerization, minerals

Figure1.

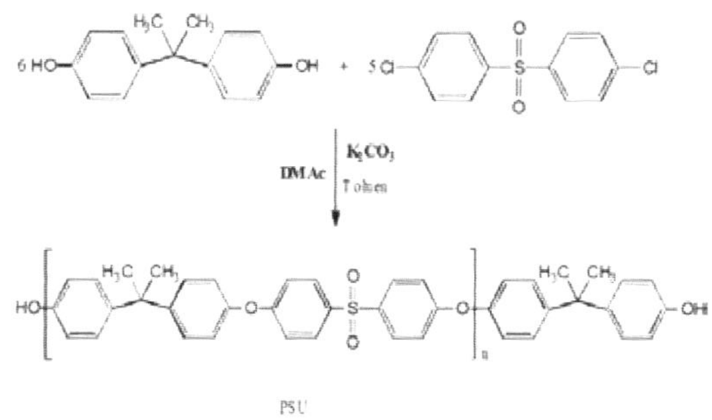

PSU Formation

PPBM-P-010 [Polymer and Polymer-Based Membranes]

The impact of graphene flake size on the properties of polymer/graphene formulations and coatings

Oana Mihaela Istrate[1], Simon R. Gibbon[2], Jonathan Moghal[3], Peter M. Budd[4], Ian A. Kinloch[1]

[1]*School of Materials, University of Manchester, Manchester, United Kingdom, M13 9PL*
[2]*AkzoNobel Research Development & Innovation, PO Box 221, Northallerton, North Yorkshire, DL7 7BJ, UK*
[3]*CROWN Technology, Downsview Road, Wantage, OX12 9BP, United Kingdom*
[4]*School of Chemistry, University of Manchester, Manchester, United Kingdom, M13 9PL*

The coating industry has been constantly growing and adapting over the past few decades. Recently, graphene based coatings have become a field of great importance due to graphene's excellent conductivity, high surface area and impermeability which make this material ideal for anti-corrosive barrier coatings (Bohm, 2014; Dumée et al., 2015; Nine, Cole, Tran, & Losic, 2015; Raju et al., 2014; Ramezanzadeh, Niroumandrad, Ahmadi, Mahdavian, & Moghadam, 2016; Su et al., 2014). Polymer/graphene nanocomposite suspensions were prepared using graphene nanoparticles, XGnP® 5M, 15M and 25M (XG Sciences, Inc., Lansing, Michigan, USA) with different flake sizes (i.e., 5 µm, 15 µm and 25 µm), and a thermosetting matrix, i.e., BPA free coating, Gold Lacquer (AkzoNobel, Packaging Coatings Ltd., Birmingham, UK). To better disperse the graphene flakes into the lacquer, a solvent, propylene glycol monomethyl ether acetate (Sigma-Aldrich Co Limited, UK), was used. The solvent was completely evaporated before the polymer/graphene nanocomposite suspensions were characterized and further used. The polymer/graphene formulations were subjected to rheological characterization using a TA Discovery HR-3 rheometer with parallel-plate geometry (60 mm diameter and 1 mm gap). Preliminary tests show that the size of the graphene flakes reflects in the storage modulus at low and at high frequencies. The relative viscosity measured during steady shear rate flow (Figure 1), using the same graphene flake concentration, decreased with increasing flake size. The polymer/graphene formulations were used to coat steel surfaces (Crown Packaging UK Plc). Preliminary results show that the size of graphene flakes used influences the coating resistance of the materials.

Keywords: polymer/graphene formulations, coatings, corrosion, barrier

Figure 1. Relative viscosity versus graphene flake size

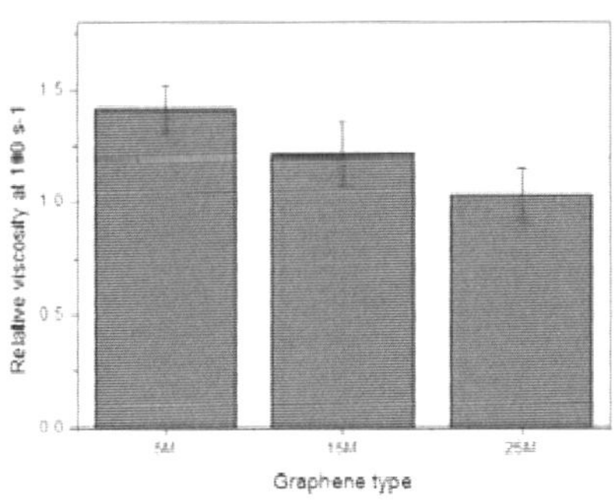

PPBM-P-011 [Polymer and Polymer-Based Membranes]

Comparasion of Bioresorbable Adhesion Barrier Films which Produced from Polylactic Acid

Mehmet Buğra Güner, Afife Binnaz Hazar Yoruç

Department of Metallurgical and Materials Engineering, Yıldız Technical University, Istanbul, Turkey

Polylactic acid (PLA) is one of the most developed materials which is the strongest canditate for replacing petroleum based materials. Because of environmental issues, usage of eco-friendly materials like PLA is increasing dramatically. On the other hand biomedical uses of PLA draw attention from scientists, due to excellent bioresorbable structure. One of the usage areas of PLA is adhesion prevention barriers in human body. Adhesions are very harmful and unwanted situations that trigerred after surgeries or other foreign body. Patients, who faces adhesions, suffers unendurable pains and repeatative surgeries are going to be necessity. For preventing adhesions, phsical seperation via using membrans is very commonly using strategy. Also PLA membranes have advantages that comes from familiar structure of lactic acid to human body. Lactic acid is a natural product of human metabolism and medical devices from lactic acid don't act as foreign body to trigger adhesions. Also controllable degredation kinetics of PLA, give options to produce situation spesific end products. Control of degredation time, gives possibility to membrans to degrade after completely recovery of wound and eliminate from human body as CO_2 and H_2O via breathe.

In this study, bioresorbable PLA was synthesized using lactide monomers. Synthesis was carried out with crystalline and mechanically strong L-lactide and fastly degredable D, L-lactide with %70, %30 molar ratios, respectively. $Sn(Oct)_2$ was used as a initiator for ring opening polymerization and reaction temperature has been fixed to 130-140°C degrees. After synthesis, chemical and thermal characterization has been done. Than, adhesion barrier films were produced with solvent casting method and degredation and mechanical properties of films was investigated. Finally investigated properties of synthesized polymers and produced films were compared to commercially available equivalent products.

Keywords: Polylactic Acid, Ring Opening Polymerization, Bioresorbable, Adhesion Barrier

PPBM-P-012 [Polymer and Polymer-Based Membranes]

From cellulose tissue papers to nanoporous membrane actuators

Huijuan Lin[1], John W. C. Dunlop[2], Jiayin Yuan[1]

[1]Department of Colloid Chemistry, Max Planck Institute of Colloids and Interfaces, Potsdam, Germany
[2]Department of Biomaterials, Max Planck Institute of Colloids and Interfaces, Potsdam, Germany

In this poster, we designed a hybrid porous membrane actuator based on a cationic PIL poly[3-cyanomethyl-1-vinylimidazolium bis(trifluoromethanesulfonyl)imide], poly(acrylic acid) and cellulose tissue paper (termed PIL-PAA/tissue paper).[1] This membrane was fabricated by coating a solution of PIL and PAA in DMF onto tissue paper, followed by drying and then treated with aqueous ammonia solution to produce a gradient crosslinking degree and porous membrane. The presence of the pores and the structural gradient enabled membranes to serve as fast and sensitive actuators.[2]

Compared to the native PIL-PAA membrane which is too fragile to be processed upon drying, the hybrid porous membrane based on PIL-PAA/tissue paper exhibits superior mechanical properties that can be easily folded into artificial flowers in drying state (Figure 1a). Besides, the hybrid membrane shows reversible actuation in acetone vapor and air. In acetone vapor, a fresh prepared membrane readily bent to a loop with a bending angle of 180o within 6 s. Upon pulling back to air, it recovered to flat state and further bent to opposite direction with -180o. Afterwards, the curving membrane again was put back into acetone vapor with a bending angle recovery from -180o to 180o, then turned to -180o upon back to air (Figure 1b). This is ascribed to the diffusion of acetone vapor into the gradient pores along the membrane cross-section, which creates asymmetric stress in the membrane and thus triggers the movement. Moreover, the composite membranes show high stability without obvious decrease of bending angle after 5 cycles within 70 s. These membranes can also be designed as S-shape deformation and responsive textile (Figure 1c).

Keywords: large deformation, tissue paper, poly(ionic liquid), porous actuator

Figure 1

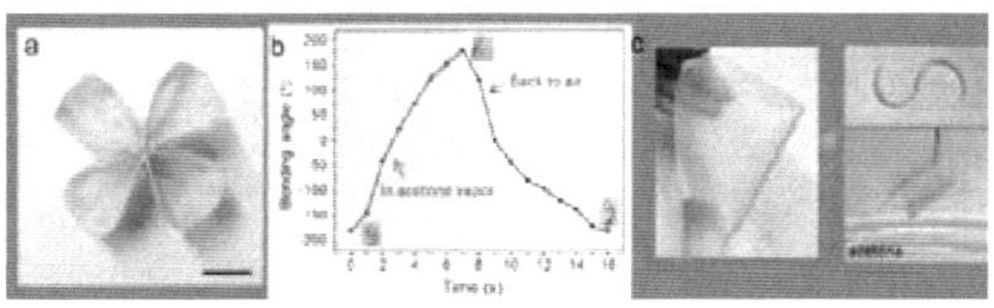

Figure 1. (a) Photograph of a paper flower folded from the hybrid membrane, scale bar: 2 cm. (b) plot of bending angle (o) of the membrane actuator against time in acetone vapor and air. (c) The design of the membranes as S-shape deformation and responsive textile.

PPBM-P-013 [Polymer and Polymer-Based Membranes]

Study of synthetics dyes adsorption onto poly (vinyl alcohol)/glutaraldehyde/β-cyclodextrin polymer membranes

Djamila Ghemati, Djamel Aliouche

Laboratory of Fibrous Polymers Treatment and Forming. F.S.I, M'Hamed Bougara University, Boumerdes, 35000, Algeria.

Among the numerous techniques of dye removal, adsorption on modified polymers is a procedure of choice; due to easy handling and removal efficiency [1,2]. Hydrophilic polymer materials are widely used in the adsorption processes for their high resistance to organic solvents. Polyvinyl alcohol (PVA) is a water-soluble synthetic biopolymer, biodegradable, and non-toxic [3]. However, to improve the functionality of PVA, this polymer is often combined with nanoparticles and other synthetic and natural polymers, such as polyacrylic acid, chitosan, cyclodextrins, and gelatin [4,5]. In this work, Crosslinked poly(vinyl alcohol)/glutaraldehyde (PVA/GA) membranes were prepared, and attempts to obtain hydrophilic crosslinked PVA membranes were made by adding various amounts of β-cyclodexrin, which is a typical cyclic oligosaccharide able to form inclusion complexes with organic host molecules (host-guest complexes). Thus, membranes of poly (vinyl alcohol)/glutaraldehyde/β-cyclodextrin (PVA/GA/β-CD) were synthesized. The membranes were characterized by infrared spectroscopy (FTIR) and swelling measurements. The ability of cyclodextrin to include a wide variety of chemicals was also exploited for the dye adsorption to show the potentialities of the membranes in the textile liquid waste processing. Adsorption of reactive, methyl orange and methylene blue dyes on (PVA/GA/β-CD) membranes was consequently studied using UV-Vis spectroscopy at wavelengths of 547, 463 and 660 nm respectively. Adsorption reached equilibrium after 24 hours.

Results indicated that there is no covalent bond formation between PVA and β-CD; the β-cyclodextrin is completely mixed into the PVA matrix polymer. The adsorption capacity increase with increasing amount of cyclodextrin; the maximum adsorption capacity was obtained with 8% β-CD. Therefore, change in adsorption capacities may be due to dye structure effect, the negative value of free energy indicated the spontaneous nature of adsorption.

Keywords: poly (vinyl alcohol), β-cyclodextrin, dyes, adsorption, swelling

PPBM-P-014 [Polymer and Polymer-Based Membranes]

Preparation and properties of graphene oxide/polyimide nanocomposites for simultaneous quantification of serotonin and dopamine

Öznur Göngör, Aziz Paşahan, Süleyman Köytepe, Büşra Aksoy, Turgay Seçkin

İnönü University, Faculty of Arts and Science, Chemistry Department, Malatya, Turkey

Serotonin (5-hydroxytryptamine, 5-HT) is a monoamine neurotransmitter of enormous biological importance widely distributed in the central nervous system, deficiency of which leads to mental disorders such as Alzheimer's disease, mental retardation, infantile autism, sleep disorders and depression [1-3]. Dopamine (DA) is one of the most significant catecholamines in the brain and belongs to the family of excitatory chemical neurotransmitters. Uncommon levels of DA may result in drastic neurological disorders such as schizophrenia, Alzheimer's, Parkinson's disease and addiction [4-5]. The prepared graphene oxide/polyimide nanocomposite films were characterized for their structure, morphology, and thermal behavior by Fourier transform infrared spectroscopy (FTIR), scanning electron micrograph (SEM), X-ray diffraction (XRD), and thermal analysis (DTA/TGA/DSC) techniques. The distribution of graphene particles inside the nanofibrous composites was investigated by scanning electron microscopy. It was found that the graphene particles were well dispersed in the polyimide matrix. The graphene oxide -containing polyimides were formed by casting the film on the electrode surface. The graphene oxide/polyimide nanocomposite electrode showed efficacy, excellent stability and reproducibility in simultaneously detecting of DA and 5-HT in the presence of a high concentration of ascorbic acid (AA). Moreover, a high concentration of AA was not the only influence on the oxidation signals of DA and 5-HT. And interference materials such as AA, 5-HT had negligible effect on the DA response, demonstrating the high selectivity of the proposed sensor. The proposed sensor demonstrated selectivity, and sensitivity towards DA and 5-HT.

References:

[1] I.P. Kema, W.G. Meijer, G. Meiborg, B. Ooms, P.H.B. Willemse, E.G.E. de Vries, Clinical Chemistry, 47(10), 2001, 1811-1820.
[2] A. Andrews, Neurotransmissions 16, 2000, 18-21.
[3] X. Jiang, X. Lin, Analytica Chimica Acta 537, 2005, 145–151
[4] R.M. Wightman, L.J. May, A.C. Michael, Anal. Chem. 60, 1988, 769A–779A.
[5] D. Merims, N. Giladi, Parkinsonism Relat. D 14, 2008, 273–280.

Keywords: Amperometric sensor, serotonin, dopamine, graphene oxide/polyimide nanocomposite

PPBM-P-015 [Polymer and Polymer-Based Membranes]

A polyimide film: Synthesis, Characterization and its sensitive biosensor application

Süleyman Köytepe, Aziz Paşahan, Öznur Güngör, Büşra Aksoy, Turgay Seçkin

İnönü University, Faculty of Arts and Science, Chemistry Department, 44280, Malatya, Turkey

Dopamine (DA) is one of the most significant natural catecholamine neurotransmitters, which is very important in human brain and body. Therefore, more attentions have paid in the research community for the rapid and accurate determinations of DA concentration, since extreme abnormalities of DA caused clinical diseases such as Parkinson's Alzheimer's and schizophrenia[1-4]. Electrochemical methods are widely used for determination of DA than that of traditional methods due to low detection limit, very fast response time, high sensitivity and simplicity.

A novel PI film as dopamine selective film was synthesized and its characteristic properties were evaluated. The structure, morphology, and thermal behavior of PI film were characterized by Fourier transform infrared spectroscopy (FTIR), scanning electron microscope (SEM), X-ray diffraction spectroscopy (XRD), and thermal analysis techniques. Furthermore, it was also examined whether this PI film can be used as a membrane for the rapid and accurate determination of DA. The PI film was formed on bare Pt electrode using a simple drop-casting method. Electrochemical measurement results illustrated that the PI electrode completely eliminated the unwanted responses from uric acid (UA) and ascorbic acid (AA) (Figure 1). As a result, it has been claimed that PI film could be used as selective polyimide membrane for the rapid and accurate determination of DA.

References:

[1] M.Wightman, L.J. May, A.C. Michael, Anal. Chem. 60 (1988) 769-779.
[2] B. Fang, H. Liu, G. Wang, Y. Zhou, S. Jiao, X. Gao, J. Appl. Polym. Sci. 104 (2007) 3864–3870.
[3] S. Reddy, B.E.K. Swamy, H. Jayadevappa. Electrochimica Acta 61 (2012) 78-86.
[4] H.Y. Wang, Y. Sun, B. Tang, Talanta. 57 (2002) 899–907.

Keywords: Polymeric membrane, polyimide, dopamine, sensor

Figure 1

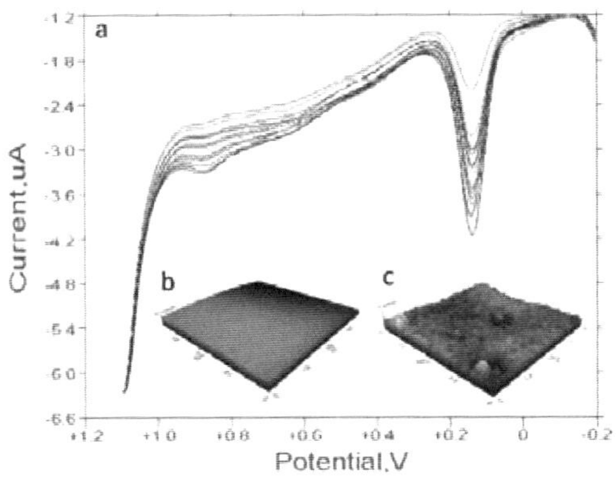

(a) DPVs of the PI electrode to increasing concentration of DA in the presence of interferences. AFM image of bare electrode (b) and PI electrode (c).

PPBM-P-016 [Polymer and Polymer-Based Membranes]

Synthesis and Characterization of H3PO4 doped triazole containing proton conductive polymer for PEMFCs

Önder Çelik[1], Tansel Şener[2], Ümit Kadiroğlu[1], Ali Ekrem Müftüoğlu[3], Faruk Yılmaz[4]

[1]*Department of Chemistry, Kocaeli University 41380, Kocaeli, Turkey*
[2]*TDinamik Energy Industrial Trade Inc., 34692, İstanbul, Turkey*
[3]*Department of Chemical Engineering, Yildiz Technical University, 34220, İstanbul, Turkey*
[4]*Department of Chemistry, Gebze Technical University, 41400, Kocaeli, Turkey*

Polymer electrolyte membrane fuel cells (PEMFCs) have been considered promising power sources for vehicles and electronic devices due to their high efficiency and nonpolluting properties. Perfluorosulfonic acid membranes, such as Nafion®, have been extensively studied because of their chemical and electrochemical stability; however, their development and commercialization are limited to operating temperatures well below 100 oC because the proton transport mechanism relies on the presence of water.[1] A strategy to develop proton conductive membranes for use in low humidity conditions has employed basic N-heterocycles (imidazole, pyrazole and triazole), allowing Grotthuss proton conduction (i.e. non-vehicular mechanism), in which protons move from site to site without the assistance of the vehicular diffusion mechanism. The nitrogen atoms in the heterocycles may act as donors and acceptors in proton transfer reactions, however, there is also an intermolecular proton transfer mechanism controlled by hydrogen bond breaking and formation. [2] In this study, we synthesized H3PO4 doped poly(styrene-co-vinylbenzylhydroxytriazole) copolymer via nitroxy mediated polymerization (NMP) and click reaction. Firstly, poly(styrene-co-vinylbenzylchloride) was synthesized by nitroxy mediated polymerization. Then copolymer was modified with sodium azide for click reaction with an alkyne functional moiety to obtain poly(styrene-co-vinylbenzylhydroxytriazole) copolymer. After purification, resulting copolymer was characterized by FT-IR, 1H-NMR, 13C-NMR, GPC, DSC, TGA. Proton conductivity of H3PO4 doped copolymer was found as 3.6x10-3S/cm.

Keywords: PEMFCs, conductive polymer, NMP, click chemistry

PPBM-P-017 [Polymer and Polymer-Based Membranes]

Lithium salt doped poly(MMA-co-HEMA) conductive polymer electrolyte for lithium ion batteries

Ipek Karaaslan, Önder Çelik, Ümit Kadiroğlu

Department of Chemistry, Kocaeli University, 41380, Kocaeli, Turkey

Lithium ion conducting polymer electrolytes are materials of great scientific and technological importance. Traditional Li-ion battery technologies, based on liquid electrolytes, suffer from unsatisfactory safety level. Polymer electrolyte-based, all solid state Li-ion batteries are regarded as better option and are believed to replace traditional technologies in the majority of prospective applications.[1] This is mainly motivated by their advantages over conventional liquid electrolytes, such as flame resistance, flexibility and low cost in design, functioning as a separator.[2] To achieve this goal, however, a scientific and technological breakthrough in the field of lithium conducting polymeric materials is strongly desired.

In this study we synthesized lithium salt doped Poly(methylmethacrylate-co-hydroxyethylmethacrylate) via nitroxy mediated polymerization. After synthesis resulting polymer was precipitated in methanol for removing and unreacted products and unwanted oligomers. After purification process resulting polymer was characterized by FT-IR, 1H-NMR, 13C-NMR, GPC, DSC, SEM and TG-DTA. Then various lithium salts doped in the polymer and resulting product used as electrolyte in Li-Ion battery.

References :

[1] Shaowei,F, et al., Electrochimica Acta, 93 (2013) 254-263
[2]Wetjen,M, et al.,Electrochimica Acta, 87 (2013) 779– 787

Keywords: Li-Ion batteries, conductive polymer, polymer electrolyte, NMP

PPBM-P-018 [Polymer and Polymer-Based Membranes]

PVDF Flat Sheet Membranes Comprising Antimicrobial N-Halamine Compounds

Fatma Ozkan, Hasan Basri Kocer

Department of Fiber and Polymer Engineering, Bursa Technical University, Bursa, Turkey

Polyvinylidene fluoride (PVDF) membranes have been widely used in water treatments due to their excellent chemical resistance, good thermal stability, easy processing and mechanical strength. However, their high hydrophobic nature causes fouling and shortens their use-life. One of the major reasons of fouling is biofouling caused by the colonization of microorganisms on the surface of membranes. In general, to remove this biofouling periodic back-washings are performed with chemicals such as sodium hypochlorite those deteriorate the membrane structure. By adding N-halamine polymers into the PVDF membrane structure holding the halogens occur during the back-washing cycles, microorganism colonization could be decreased and the use-life of membranes could be lengthened by reducing the number of harmful washing cycles. In this study, N-halamine polymers were synthesized and compounded into PVDF polymers to produce antimicrobial membranes. The effect of the compounded polymers was determined by the investigation of mechanical, thermal, and morphological properties of the produced membranes. The porosity and pore size of the membranes were determined by pure water permeation under pressure and BSA rejection tests. The permeability tests with bacterial solutions were also performed to investigate the biofilm formation on the membranes.

Keywords: Membrane, Antimicrobial, Water treatment

PPBM-P-019 [Polymer and Polymer-Based Membranes]

Glycerol Carbonate and Polyethylene Oxide Containing Borate Ester Based Polymer Electrolyte Membranes

Esra Eren, Aysel Demirci, Sümeyye Köybaşı, Yunus Emre Özkara, Betül Karataş, Yunus Karataş

Department of Chemistry, Ahi Evran University, Kırşehir, Turkey

The need for energy storage is increasing every day as the technology progress. Lithium-ion and lithium-polymer batteries are the prominent and widely used types of batteries for energy storage. The use of polymers instead of the organic solvents results in more reliable batteries. In addition, the problem of capacity fading with time in lithium-ion battery is much less significant for lithium-polymer batteries.

Besides the other industrial use, borate esters have attracted attention in recent years as polymer electrolyte. These examples include the use of borate esters as additives in especially poly ethylene glycol based polymers. However, limited number of studies exists where borate esters are directly used as the polymer electrolyte [1].

In this study, new polymer electrolyte membranes containing glycerol carbonate units, which contains the structure of most commonly used organic battery solvent (ethylene carbonate), has been synthesized. For this purpose, PEG-1000 and glycerol carbonate were reacted with boric acid in a proper stoichiometric ratio to maintain the crosslinking and the polymer electrolyte membranes were prepared as shown in Figure 1. The electrochemical properties of the polymer electrolyte membranes were investigated using different lithium salts (LiClO4, CF3SO3Li and (CF3SO2)2NLi). The optimum room temperature ionic conductivity were measured for the membrane with (CF3SO2)2NLi as 4.09 x10-5 S/cm.

References:

[1] Wakihara, M.; Nakayama, M.; Kato, Y. "Lithium Ion Rechargeable Batteries", (Wiley) 2010, 213-255.

Keywords: polymer electrolye membrane, borate ester, ionic conductivity

Figure 1: The representative syntheses of the polymer electrolyte membranes with a photo of a membrane sample.

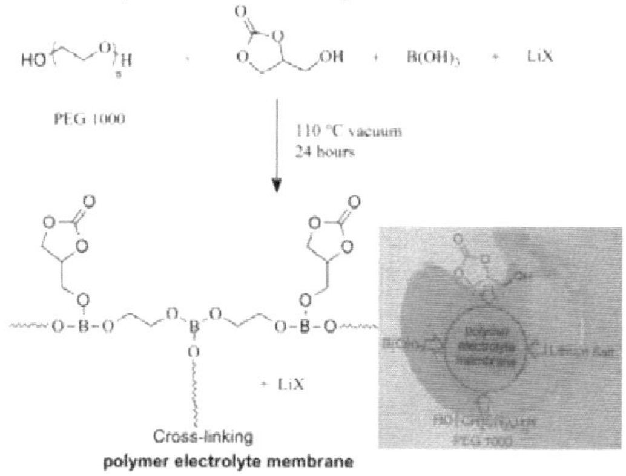

PPBM-P-020 [Polymer and Polymer-Based Membranes]

Waterproof – Breathable Antibacterial Polyurethane Membranes

Ahmet Aydın, Fatma Ozkan, Hasan Basri Kocer

Department of Fiber and Polymer Engineering, Bursa Technical University, Bursa, Turkey

Waterproof and vapor permeable membranes are commonly being used in protective textiles and medical products. These usage fields often require antibacterial properties to reduce the risk of cross contamination through direct contact or air born pathways. This property can be provided to polyurethane formulas by introducing antibacterial Quaternary Ammonium Salt (QAS) diols. In this study, nonporous hyrophilic antibacterial polyurethane membranes, that allow the passage of water vapor molecules through their polymer chains by the solution diffusion mechanism, were produced with the addition of QAS diols. QAS diols used in this study were very appropriate to produce breathable membranes due to their hyrophilic backbones which can make hydrogen-bonding with water vapor molecules, their ability to make covalent bond to the structures, and having very simple structures. Polyurethanes synthesized with the addition of various amounts of QAS diols, were used for production of membranes. The effect of QAS diols addition was determined by the investigation of mechanical, thermal, and morphological properties of the produced membranes. The effect of morphological structures on the membrane water and vapor permeability and on the antibacterial properties was determined.

Keywords: Polyurethane, Membrane, Antibacterial

PPBM-P-022 [Polymer and Polymer-Based Membranes]

PS-b-PNIPAM block copolymers for thermoresponsive membrane applications

Merve Cetintas, Marleen Kamperman

Laboratory of Physical Chemistry and Soft Matter, Wageningen University and Research Centre, Wageningen, The Netherlands

Responsive pores are receiving a great deal of attention in membrane systems for their advantages against fouling and easy cleaning properties[1]. Block copolymer self-assembly is of interest, because well-defined nanosized pores can be obtained[2,3] and stimuli-responsive functions can be incorporated[3,4]. In our study, we synthesized well-defined high molecular weight polystyrene-b-poly(N-isopropylacrylamide) (PS-b-PNIPAM) block copolymers by RAFT polymerization. The PNIPAM block in the copolymer acts as the temperature responsive part due to its lower critical solution temperature (LCST) of 32°C. In principle, the pore sizes of the membrane should increase at temperatures above the LCST of PNIPAM, due to collapsing of the PNIPAM chains. Membranes were prepared by (a) spin-coating on a porous substrate and solvent annealing to the desired morphology or by (b) using a phase inversion procedure, in which a viscous polymer solution was cast on a glass substrate and transferred to a water bath after a certain evaporation time. Morphological properties of the block copolymer membranes were investigated by a variety of techniques including Atomic Force Microscopy (AFM), Grazing Incidence Small Angle X-Ray Scattering (GISAXS) and Scanning Electron Microscopy (SEM).

Membrane preparation using the phase inversion procedure resulted in nanoporous membranes and temperature dependent permeability experiments showed that our membranes have a thermoresponsive character. Reversibility tests proved that the membranes can be used repeatedly at temperatures between 20 and 50°C.

Acknowledgements: This project is financially supported by Dutch Polymer Institute (DPI) under project #766.

References:

[1] D. Bhattacharyya, T. Schäfer, S. R. Wickramasinghe, S. Dauner, Responsive Membranes and Materials, 2013, John Wiley & Sons Ltd., UK
[2] W. A. Phillip et al., Macrom. 2010, 43, 7763
[3] S. Rangou et al., J. Memb. Sci., 2014, 451, 266
[4] S. P. Nunes et al., ACS Nano, 2011, 5(5), 3516

Keywords: RAFT polymerization, block copolymer self-assembly, non-solvent induced phase inversion method, thermo-responsive nanoporous membranes

Membrane structure by SEM (left) Reversibility of the permeabilities versus temperature (right)

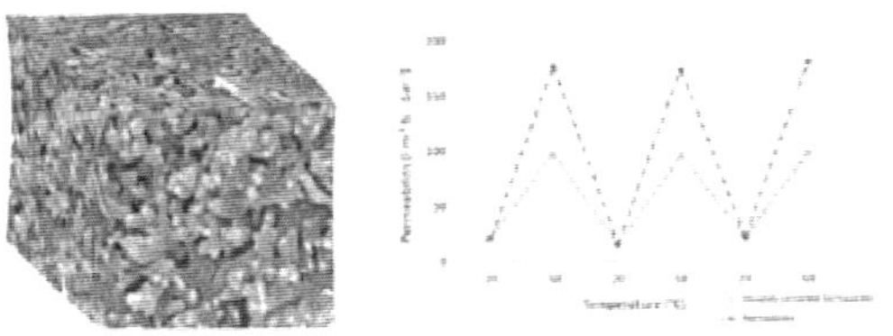

PPBM-P-024 [Polymer and Polymer-Based Membranes]

Synthesis and Characterization of Imidazole Derivatives as Polymer Electrolyte Membranes For Fuel Cell Applications

Aysel Demirci[1], Burak Gündüz[1], Sahl Sadeghi[2], Enver Güler[3], Selmiye Gürsel[2], Yunus Karataş[1]

[1]Department of Chemistry, Ahi Evran University, Kırşehir, TURKEY
[2]Faculty of Engineering &Natural Sciences, Sabancı University, Istanbul, TURKEY
[3]Nanotechnology Research & Application Center (SUNUM), Sabancı University, Istanbul, TURKEY

Fuel cells today are the most widely studied as the power source on electrochemical technologies. Polymer Electrolyte Membrane (PEM) fuel cells with their higher electrical power density and easy interoperability as a clean technology in especially electric vehicles are attracting great attention among the other types of fuel cells. The production of PEMs with low cost and high performance at elevated temperatures (above 100°C) are of special importance. There is therefore a need to synthesize membranes able to work effectively in such a system at temperatures above 100°C and low humidity conditions.

In this study, the syntheses of several imidazole derivatives have been targeted to serve as monomer for PEMs (Figure 1). For this purpose, imidazole derivatives bonded to a vinyl group with varying aliphatic chains (x = 1, 2, 3) were designed and synthesised. They were purified and characterized by proper methods. Subsequently, the synthesized monomer(s) were employed in radiation induced grafting for membrane production Radiation grafting is a versatile method for the preparation of cost competitive proton conducting membranes. In this method poly(ethylene-alt-tetrafluoroethylene) films were irradiated by gamma ray with a suitable dose to generate radicals on the polymer backbone. Successively, graft polymerization proceeded through an oxygen free solution of monomer(s) and suitable solvent(s) by immersing the irradiated film. Resultant copolymer films were impregnated with phosphoric acid to introduce proton conductivity. Conjugation of phosphoric acid with nitrogen in imidazole ring provides suitable condition for proton conduction at elevated temperatures, while ETFE backbone provides required mechanical stability for the membrane.

Keywords: Imidazole, Polymer electrolyte membrane, Fuel cell

PPBM-P-025 [Polymer and Polymer-Based Membranes]

Crosslinked proton-exchange membranes by radiation induced grafting of 4-vinylpyridine and divinylbenzene onto poly(ethylene-co-tetrafluoroethylene) films

Sahl Sadeghi[1], Enver Guler[2], Ege Oyku Cerit[1], Selmiye Alkan Gursel[1]

[1]*Faculty of Natural Science and Engineering, Sabanci University, Istanbul, Turkey*

[2]*Sabanci University Nanotechnology Research and Application Center (SUNUM), Sabanci University, Istanbul, Turkey.*

Energy supply plays a crucial role in economic, environmental and social development. A sustainable society requires energy resources which are economically affordable, environmentally friendly, and as little as possible relying on depletable or foreign resources. Fuel cell technology among other types of renewable energy resources has already proven its capabilities. Although this technology is utilized in many different sectors, yet there are possibilities to improve performance and efficiency of fuel cell systems. One of the promising areas of research in this field is high temperature polymer electrolyte membrane fuel cells (HT-PEMFC) operating at temperatures above 100°C. HT-PEMFC is very desirable due to improved electrode kinetics, enhanced ionic conductivity and reduced humification. Thus, the development of high performance membranes is of special importance for this application.

In this work, we study the effect of divinylbenzene (DVB) crosslinker on the physicochemical properties of irradiation graft polymerization of 4-vinylpyridine and poly(ethylene-co-tetrafluoroethylene) (ETFE) films for HT-PEMFC application. The process commences with cutting ETFE films in desired dimensions, and later on irradiating them with gamma rays by suitable dosage to generate radicals on the polymer backbone. Afterward, graft polymerization proceeds through an oxygen free solution of monomer(s) and suitable solvent(s) by immersing the irradiated film. The resulted film from polymerization is washed, dried and doped with phosphoric acid. Conjugation of phosphoric acid with nitrogen in pyridine ring provides a suitable condition for proton conduction at elevated temperatures, while ETFE backbone provides required mechanical stability for the membrane. In order to further improve the mechanical properties of membrane, DVB is introduced during the polymerization to strengthen its mechanical properties due to the crosslinked rigid network (Figure 1).

Keywords: Proton conducting membrane, Radiation induced grafting, 4-vinylpyridine, ETFE film, Phosphoric acid doping, High temperature PEM fuel cell

Figure 1

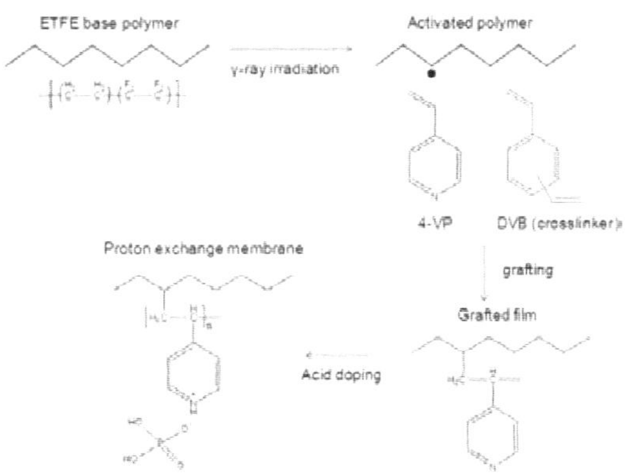

Mechanism of preparation of crosslinked phosphoric acid doped poly(4-VP) grafted ETFE membrane

PPBM-P-026 [Polymer and Polymer-Based Membranes]

Chitosan Based Nanocomposites for Bioactive Food Packaging

Seda Gunes[1], Funda Tihminlioglu[2]

[1]*Department of Biotechnology and Bioengineering, İzmir Institute of Technology*

[2]*Department of Chemical Engineering, İzmir Institute of Technology*

In recent years, bioactive packaging takes increasing attention due to its potential to provide quality and safety benefits. Bioactive packaging prolongs shelf-life, enhances safety, protects from oxidation while maintaining food quality. Various antimicrobials have been incorporated into films and essential oils (EOs) are frequently studied due to their antimicrobial, antioxidant property. As one of the biopolymers, chitosan displays biological activities such as antibacterial, antioxidant, antifungal, biodegradable and biocompatible. Nowadays, polymer/clay nanocomposites have been widely studied as a packaging matrix. Intercalated nanocomposites improve mechanical, thermal, rheological, release and barrier properties compared to neat organic polymers. The objective of this study is to develop a novel bioactive packaging system and characterize chemical, physical, mechanical, and antimicrobial properties. Hypericum perforatum oil incorporated chitosan nanocomposite films were successfully prepared by solvent casting method in different oil (0.25; 1% v/v) and nanoclay concentrations (1; 3 w/v). FTIR analysis, mechanical test, XRD analysis were performed in order to characterize the prepared films. In addition, the antimicrobial test was performed by disc diffusion method and the growth inhibition effects of the films including different amount of H. perforatum oil were investigated on Escherichia coli and Staphylococcus aureus. FTIR analysis showed that oil was successfully incorporated into chitosan films. XRD results demonstrated that clay content had a good dispersion in chitosan matrix. Nanoclay addition showed intercalated structure in pure chitosan films whereas exfoliated structure in chitosan-oil incorporated films. Nanoclay addition improved its physicochemical and mechanical properties of chitosan films. Antimicrobial test showed that H.perforatum oil incorporated nanocomposite films have antimicrobial effect on both microorganisms. Bioactive packaging materials based on renewable sources, such as chitosan and essential oils, presents a potency for the development of improved ecofriendly materials. These findings showed that oil incorporated chitosan based nanocomposites could be an alternative for the application in active food packaging applications.

Keywords: Chitosan, nanocomposites, Essential oil, bioactive packaging, food preservation, edible films

PPBM-P-027 [Polymer and Polymer-Based Membranes]

Synthesis and Characterization of Highly Branched, Functional Poly(Arylene Ether Sulfone)s for Industrial Wastewater Purification Membrane Applications

Emine Billur Sevinis[1], Aybige Ozture[1], Ipek Sacligil[1], Turkan Ormanci Acar[3], Turker Turken[3], Fevzi Cakmak Cebeci[1], Derya Imer[3], Ismail Koyuncu[3], Yusuf Menceloglu[1], Serkan Unal[2]

[1]*Material Science and Nanoengineering Program, Sabanci University, Istanbul 34956, Turkey*

[2]*Nanotechnology Research and Application Center, Sabanci University, Istanbul 34956, Turkey*

[3]*Environmental Engineering, Istanbul Technical University, Istanbul, Turkey*

Polysulfone based membranes are commonly employed for ultrafiltration and nanofiltration purposes in wastewater recovery in the industry. Despite their high overall performance, current membrane solutions still have performance limitations especially when operating under high pH and temperature. In this study, highly branched poly(arylene ether sulfone)s (PES) were synthesized using the A_2+B_3 polymerization approach to develop membranes

with enhanced thermo-mechanical properties and purification performance. In addition to short A_2 and B_3 monomers such as 4,4'-dichlorodiphenyl sulfone (DCDPS) and 1,1,1-Tris(4-hydroxyphenyl)ethane (THPE), respectively, previously synthesized A_2 oligomers with varying lengths were utilized to tailor the degree of branching and the average distance between branch points. Additionally, various strategies were developed to introduce functional groups such as acrylate or silane functionality on chain ends of final branched PES products, which were characterized by FT-IR and NMR spectroscopies, Size Exclusion Chromatography (SEC), Dynamic Mechanical Analysis (DMA), Differential Scanning Calorimetry (DSC) and stress-strain tests. Finally, the effect of the degree of branching and the type of functional end-groups was investigated on the performance of ultrafiltration and nanofiltration membranes prepared from these branched, functional PES in an effort to enhance their performance in industrial wastewater purification applications.
This research is supported by The Scientific and Technological Research Council of Turkey (TUBITAK) grant number 113Y350.

Keywords: Polysulfone, Branched, Nanofiltration, Ultrafiltration

PPBM-P-028 [Polymer and Polymer-Based Membranes]

Modification of Reverse Osmosis (RO) Desalination Membranes by Zwitterionic Silane Coatings

Selda Erkoc Ilter[1], Serkan Guclu[3], Farzin Saffari[3], Derya Imer[3], Ismail Koyuncu[3], Serkan Unal[2], Yusuf Menceloglu[1]

[1]*Materials Science and NanoEngineering, Sabanci University, Istanbul, Turkey*
[2]*Nanotechnology Research and Application Center, Sabanci University, Istanbul, Turkey*
[3]*Environmental Engineering, Istanbul Technical University, Istanbul, Turkey*

Polyamide thin film composite (PA-TFC) reverse osmosis (RO) membranes are commonly used for water desalination in industry because of their superior properties. However, their operational performances are seriously affected by the trade-off between water permeability and salt rejections, biofouling and chlorination. Therefore, various strategies have been applied to enhance the performance of RO membranes and among them, surface modifications and incorporation of nanoparticles have been the most effective ones. In this study, various zwitterionic silane compounds were synthesized for the treatment of commercial PA-TFC desalination membrane (Hydranautics SWC5 40) surface in order to enhance its performance. Surface modification of membranes was simply based on the alkoxysilane condensation reaction including different sulfobetain alkoxysilane compounds which were obtained by the reaction of the ""*tert*""-amine functional alkoxysilanes with 1,3-propane sultone and 1,4-butane sultone (P Wang et al., Journal of Membrane Science 2015, 492, 547-55, N. Kim et al., Journal of Membrane Science 2007, 300, 224-231). Commercial composite RO membranes were treated with zwitterionic silane coupling agents by spray coating of their aqueous solutions at three different concentrations: 1.0, 1.5, and 2.0% (w/v). Surface characterizations of the modified membranes were performed by SEM, ATR-FTIR spectroscopy, water contact angle (WCA), zeta potantial and optic profilometer measurements. In addition, membrane performances were evaluated by the permeability, salt rejection and biofouling tests. Surface characterizations showed that the membrane surfaces were successfully coated with zwitterionic silane which resulted in an increase in the surface hydrophilicity and significant changes in the surface roughness and zeta potentials. Significant improvements in the water flux and salt rejection performances were also observed upon surface treatment with certain solution concentrations of sulfobetain silane compounds.

This research is supported by The Scientific and Technological Research Council of Turkey (TUBITAK, project number 113Y376).

Keywords: PA TFC membrane, reverse osmosis membrane, zwitterionic silane coupling agent, surface modification

PPBM-P-029 [Polymer and Polymer-Based Membranes]

A Novel UV- Curable Monomer for Coating Applications

Yeşim Müge Şahin[1], Mustafa Hulusi Uğur[2]

[1]*İstanbul Arel University*
[2]*Marmara University*

Anthraquinone including natural components are used in the fields of medicine, pharmacy due to their antimicrobial properties. These components can be either isolated from several different types of plants or be synthesized. In the proposed study a novel allylic antraquinone molecule was synthesized as a UV-curable monomer. Chemical structure of the UV- Curable monomer was clarified by fourier transform infrared spectroscopy (FTIR) and nuclear magnetic resonance spectroscopy (H1-NMR and C13-NMR). Exhibiting good cure response and low viscosity, polyester tetraacrylate was chosen to obtain UV-curable films where surface hardness, chemical resistance and good adhesion is needed. Fotopolimerization of monomer with polyester tetraacrylate is conducted with Darocur 1173 and N-Vinylpyrrolidone (NVP) as fotoinitiator and reactive diluent respectively. Four different compositions were prepared in order to obtain the UV-curable films and their gel content were identified. For complete characterization of the obtained films, FTIR, X-ray diffraction (XRD) and NMR analyses were conducted. These films were prepared purposefully to improve the antimicrobial property of a coating and can further find applications in particularly in vacuum coating, laminating adhesives or in food packaging systems.

Keywords: UV-curable systems, Antraquinone monomer, Antimicrobial, Coating

PPBM-P-030 [Polymer and Polymer-Based Membranes]

Membrane filtration of Sudan orange G on a cellulose acetate membrane filter for separation–preconcentration and spectrophotometric determination in water, chili powder, chili sauce and tomato sauce samples

Zeid A Alothman

King Saud Universitry

A simple membrane filtration procedure for separation–enrichment of Sudan orange G is presented. The method is based on the adsorption of Sudan orange G on a cellulose acetate filter and its elution from the membrane with 10 mL of ethanol. Sudan orange G in the eluent was determined by UV–visible spectrophotometry at 388 nm. The effect of analytical conditions, including pH, flow rates and eluent, sample volume, type of membrane for quantitative preconcentration and separation of Sudan orange G were examined. The influences of matrix components on Sudan orange G recoveries were studied. The preconcentration factor was 125. The detection limit was 4.9 µg L−1. The relative standard deviation was 4.3%. The presented procedure was applied to chili powder, chili sauce, tomato sauce, powdered beverage and water samples.

Keywords: Sudan orange G, Separation, Preconcentration, Spectrophotometry

PPBM-P-031 [Polymer and Polymer-Based Membranes]

Preparation and characterizations of biopolymer chitosan-based membranes as edible coatings in food preservation technology

Mohamed E I Badawy[1], Entsar I. Rabea[2]

[1]Department of Pesticide Chemistry and Technology, Faculty of Agriculture, 21545-El-Shatby, Alexandria University, Alexandria, Egypt
[2]Department of Plant Protection, Faculty of Agriculture, Damanhour University, 22516-Damanhour, Egypt

Chitosan is a reactive functional biopolymer, which gives possibilities for chemical modifications to generate new properties and functions. Biocompatibility, biodegradability, non-toxicity, and potential biological activities make this polymer with its advantageous derivatives for many applications in different fields. Chitosan has the ability to form thin membranous films because of its high hydrophilicity and good film-forming characters (homogeneous matrix, stable structure, good permeable, good water barrier, and mechanical properties). Therefore, the current study investigates the preparation of different chitosan biopolymer membranes. The membranes were prepared by casting the composite polymers of 2% chitosan with gelatin or starch (2 and 5% for each polymer) with different concentrations of glycerol and sorbitol (1, 2, 5, and 10%, w/w) as plasticizers. Effect of gelatin, starch, and plasticizers on the membrane properties was determined by measuring their physical properties, and the antioxidant activity. The appearance and cross-sectional characteristics by scanning electron microscopy (SEM) showed good crosslinking and compatibility, generating the desired miscibility between the biopolymers. The results suggested that the incorporation of gelatin or starch in blending with plasticizers into chitosan membranes improved the swelling, water vapor permeability (WVP), and the antioxidant activity of the membranes. This study showed the benefits of the combination of such components into chitosan membranes and the potential for using them in different areas of applications such as dehydration of some organic solvents, separation of organic mixtures, ultrafiltration of biomaterials, protein adsorption/separation, and in food packaging.

Keywords: Biopolymer membranes, Microstructure, WVP, Antioxidant activity.

PPBM-P-032 [Polymer and Polymer-Based Membranes]

Development and Characterization of Poly (Arylene Ether Sulphone) –Zeolite Beta Based Polymer Electrolyte Membranes for Direct Methanol Fuel Cell (DMFC) Applications

Olga Ece Atlan, Erde Can, Nurcan Bac

Department of Chemical Engineering, Yeditepe University, Istanbul, Turkey

Fuel cells are electrochemical instruments used to convert chemical energy to electrical energy. Direct Methanol Fuel Cells (DMFCs) appear to be one of the most promising candidates among other fuel cell types. DMFCs eliminate the need for a methanol steam reformer and they have a similar working principle like Proton Exchange Membrane Fuel Cells (PEMFC). To date, commercially available perfluoro sulfonic acid membranes (eg. Nafion®) are widely used as polyelectrolyte membranes, although they have certain disadvantages such as low proton conductivity at high temperatures, high methanol cross over and high costs. In this study, polysulfones were sulfonated using trimethylsilyl chlorosulfonate as the sulfonating agent. Composite membranes were prepared by adding zeolite H beta into sulfonated polysulfone (sPSU), in order to improve mechanical properties, thermal stability. Additional benefits of composite membranes are increasing proton conductivity and decreasing methanol crossover and water uptake. First composite membranes were prepared by solution casting method using solutions of commercial PSU (Udel 1700) with different zeolite loadings (2,5,10,20,30 wt.%) which were synthesized with various Si/Al (20,30,50) ratios. The thermomechanical properties of composite membranes were analyzed by DMA, and proper Si/Al ratios and loadings were determined. According to pre-determined zeolite ratios and loadings, composite membranes were casted with sPSU. The formation of zeolite H beta was confirmed by FTIR spectroscopy, XRD and SEM analysis and its thermal degradation profile was determined by TGA. The thermal transitions, thermomechanical properties and thermal stabilities of composite membranes were evaluated using DSC, DMA and TGA. The proton conductivity and water uptake measurements were also carried out. In addition, SEM was used to characterize the distribution of the zeolite particles in the PSU polymer matrix. The prepared sulfonated polysulfone-zeolite H beta composite membranes show potential to be used in DMFC applications.

Keywords: fuel cell membranes, fuel cell, composite membranes, composite, inorganic additive, zeolite

PPBM-P-033 [Polymer and Polymer-Based Membranes]

Comparative behavior of styrene polymerization with Cp2Ti Cl2 and Cp2Zr Cl2 based catalyst in presence of anionic and cationic clay as support

Ahmed Benaboura, Khadidja Belalem

University of Sciences and Technologie Houari Boumedienne Faculty of chemistry B.P. 32 El-alia Bab Ezzouar Algiers Algeria

We report here our preliminary investigations on the in situ preparation of polystyrene using cationic and anionic clays as supports for Cp2Ti Cl2 and Cp2Zr Cl2 catalytic systems. The influence of the mineral charge in terms of polymerization yield, morphology of the polymers and stereoselectivity of the catalytic systems has been evaluated. The anionic filler has a strong influence on the stereoregularity of the obtained polymers with both catalysts. The structure of the obtained polymers has been characterized by Fourier transform infrared spectroscopy (FTIR) and wide angle X-ray scattering (WAXS). Thermal properties have been studied by thermogravimetric analysis (TGA) and differential scanning calorimetry (DSC).

Keywords: styrene, stereospecific polymerization, nanofiller,

PPBM-P-034 [Polymer and Polymer-Based Membranes]

Synthesis, Characterization and Dopamine Selective Properties Of Polyurethane Films Containing Chlorogenic Acid

Ensar Erel, Merve Gökşin Karaaslan, Süleyman Köytepe, Burhan Ateş, Ismet Yılmaz

Department Of Chemistry, Inonu Universty, Malatya Turkey

Dopamine is one of the significant natural catecholamine neurotransmitters, which is very important in human brain and body metabolisim [1]. Therefore, more attentions have paid in the research community for the rapid and accurate determinations of dopamine concentration in body fluids [2]. Polymeric membranes and electrochemical amperometric techniques have been successfully used to elimination of the interferant species for determination of dopamine [3] In this study, a chlorogenic acid-containing polyurethane film as dopamine selective film was synthesized and characterized. The following reactants were used in the syntheses of polyurethanes: aromatic 4,4'-diisocyanodiphenylmethane, chlorogenic acid and polyethylene glycol 200(PEG). The ratios of PEG–chlorogenic acid monomer units in the polyols were: 99:1, 97:3 95:5, and 90:10, respectively. Structural characterizations of polyurethanes were characterized by FTIR, DSC, DTA, TGA and dynamic contact angle measures. Furthermore, polyurethane film was formed by dropping of polyurethane solution on bare Pt electrode to prepared dopamine selective electrode. The voltametric results indicate that the polyurethane based electrodes can be used as a sensor for determination of dopamine with the good sensitivity, selectivity and high reproducibility (Figure 1).

Keywords: sensor, polymer, dopamin, Chlorogenic Acid

Figure1.

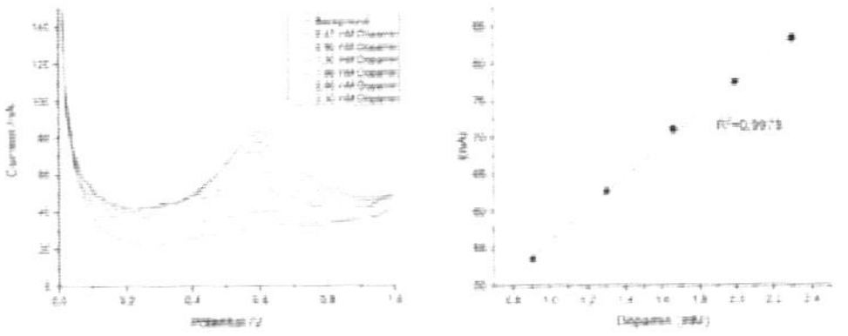

DPVs of the polyurethane electrode to increasing concentration of dopamine in the presence of interferences and calibration curve of the polymer electrodes

PPBM-P-035 [Polymer and Polymer-Based Membranes]

Graphene Oxide-Polymer Based Membranes and Aerogels

Öznur Doğan, Ayşe Elif Kıratlı, Sevil Demirci, Erhan Bat

Middle East Technical University

Graphene is a single atom thick 2D material, which consists of carbon atoms arranged in a honeycomb lattice. Isolation of graphene by Geim and Novoselov in 2004 set ground for graphene and graphene based materials to become the focus of interest of many research area such as sensors, polymeric nanocomposites, membranes and biomaterials. Graphene oxide, which is a derivative of graphene, is easy to process due to its dispersibility in polar solvents. It can be obtained by chemical exfoliation and oxidation of graphite. Due to hydroxyl, carboxylic acid and epoxide groups in its structure, surface modifications can be performed and if required, it can be reduced to graphene by using thermal and chemical methods. In this study, polymeric membranes and aerogels based on graphene oxide-polymer hybrids are prepared. Graphene oxide was synthesized by Tour Method and ATR-FTIR, UV-Vis, TGA, SEM, TEM and AFM analyses were conducted for characterization. Carboxylic acid groups of the graphene oxide were reacted with 1,3-diaminopropane to attach amine groups to the edges of graphene oxide then further reacted with alpha-bromoisobutyryl bromide to synthesize an ATRP initiator. ATR-FTIR, XPS and TGA analyses were performed on the products to confirm the functionalization. Poly(glycidyl methacrylate-co-styrene) or poly(oligoethylene glycol) methacrylate(POEGMA) were grown from graphene oxide macroinitiator via surface-initiated ATRP. Then, graphene oxide, polymer functionalized graphen oxide, and amine functionalized graphene oxide were used in different proportions to produce chemically crosslinked aerogels. POEGMA-graphene oxide was used as a surfactant in preparing polyether sulfone membranes via breath figure technique. The changes in the morphologies of the graphene oxide based aerogels and membranes were examined by SEM analysis.

Keywords: Graphene oxide, aerogel, surface-initiated polymerization, ATRP, membranes

PPG-P-001 [Porous Polymer and Gels]

Chitosan-based hydrogel composites prepared by UV polymerization for wound healing

Hong Ru Lin[1], Tseng Ping Sue[1], Yiu Jiuan Lin[2]

[1]Department of Chemical and Materials Engineering, Southern Taiwan University of Science and Technology, Tainan 710, Taiwan
[2]Department of Nursing, Chung Hwa University of Medical Technology, Tainan 717, Taiwan

Wound healing involves a complex series of cellular and extracellular matrix events. The effects and functions of cells involved in wound healing processes are modified and regulated in vivo by several growth factors which are secreted at the wound site. Among them, epidermal growth factor (EGF) receives much attention in healing the wound. In this study, UV polymerization was used to prepare the hydrogels composed of Chitosan, acrylic acid (AAc) and 2-Hydroxyethyl methacrylate (2-HEMA). EGF was incorporated in the hydrogels to stimulate the wound healing. The results indicate that hydrogels after UV polymerization has a swelling ratio of 4.32 (w/w), a water content of 80.6%, a water vapor transmission rate of 1732 (g/m2/day), and a compressive strength of 0.34 MPa with 53.4% compressive strain, and 15% acceleration thrombus formed rate was obtained. Chitosan-PAA-pHEMA hydrogel also has antibacterial properties in Staphylococcus aureus, Escherichia coli, and Pseudomonas aerus. The release of EGF reaches 50% in the first day and the effect would continue for 7 days. The result of cell culture indicates all the prepared hydrogels are non-toxic material which may help cell adhesion and proliferation. The animal test indicates Chitosan-PAA-pHEMA hydrogel incorporated EGF could accelerate wound close up rate. The Chitosan-PAA-pHEMA hydrogel prepared in this study may be a potential candidate used as a wound dressing material.

Keywords: Chitosan, Acrylic acid, 2-Hydroxyethyl methacrylate, Hydrogel, Epidermal growth factor, Wound dressing

Wound closure

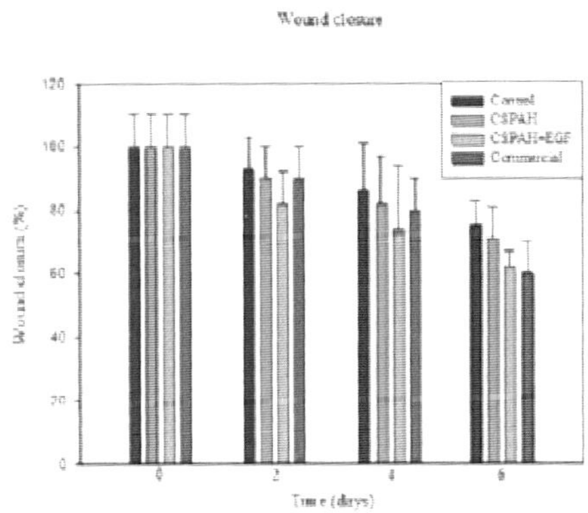

Comparison of wound healing using different dressings

PPG-P-002 [Porous Polymer and Gels]

PU-hydrogel composites prepared byγ-ray irradiation for wound healing

Hong Ru Lin[1], Ga Hwa Lee[1], Yiu Jiuan Lin[2]

[1]Department of Chemical and Materials Engineering Southern Taiwan University of Science and Technology, Tainan 710, Taiwan
[2]Department of Nursing Chung Hwa University of Medical Technology, Tainan 717, Taiwan

Growth factor receives much attention in healing the wound. However, most of the growth factors have short duration of retention at wound sites due to their high susceptibility to in vivo proteolysis. A number of controlled delivery vehicles such as heparin-binding have been developed to overcome their shortcomings, however, an ideal method for their administration remains unclear. In this study, Polyurethane (PU)/hydrogel composites were fabricated for wound repair applications. The hydrogel is a copolymer of thermosensitive N-isopropyl acrylamide (NIPAAm) and acrylic acid (AAc). γ-ray irradiation was employed to copolymerize NIPAAm with AAc and to graft hydrogel onto porous PU simultaneously. Fibroblast growth factor-2 (FGF-2) was incorporated into the composite to facilitate the wound healing. The physical properties of composites were characterized, their in-vitro release of FGF-2, and in-vivo tests were conducted. The results indicate that porous PU not only provides moderate mechanical strength of composite but also minimizes the water vapor transmission due to its hydrophobic nature. Thermosensitive hydrogel could absorb most of the wound exudates due to its high water uptake ability. Since its thermosensitive property, PU/hydrogel composite is easier to strip off than that of commercial wound dressing, this prevents from the second injury of wound when replacing the wound dressing. In vivo results show that the PU/hydrogel composite incorporating FGF-2 could accelerate wound healing and reduce the scar formation.

Keywords: Polyurethane; N-isopropyl acrylamide; Acrylic acid; Hydogel; Fibroblast growth factor-2; γ-ray; Wound healing

Histological examination

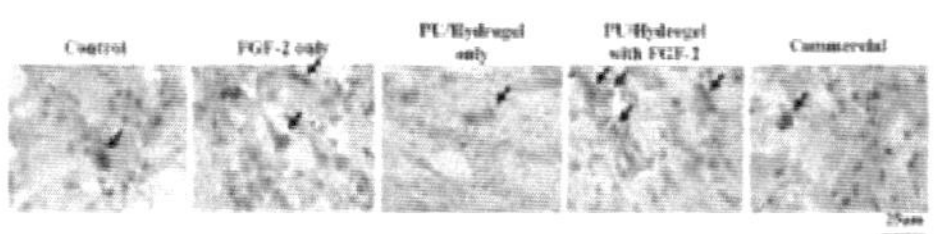

Histological examination of wound repair in each group on day 2 after initial wounding. Arrows indicate the vascularization formation.

PPG-P-003 [Porous Polymer and Gels]

Preparation and Properties of the Novel Photoluminescent Hydrogel Based on Fluorene Group

Wen Fu Lee, Wang Xun Wu, Wei Chih Chen

Department of Chemical Engineering, Tatung University, Taiwan

Photoluminescent hydrogel based on fluorene group was developed in this research. The optical and other properties of the hydrogels were investigated and compared with hydrogel without fluorene group. At first, bisphenoxy ethylacrylate fluorene (BPEAF) was synthesized and characterized via FT-IR,1H-NMR,13C-NMR, 1H-1H COSY,13C-1H HSQC and element analysis. Then N-isopropyl acrylamide was polymerized with BPEAF as crosslinker to prepare a series thermosensitive hydrogels. The influence of BPEAF content on the swelling behavior, mechanical properties, and drug release behavior of the hydrogels was investigated. The results showed that the swelling ratios of the hydrogels decreased with increasing of the BPEAF content, this is because the introduction of nonpolar flurorene structure into the hydrogels decreased the hydrophilicity of the hydrogels compared with those hydrogels not containing flurorene structure. In caffeine release, the fractional release of the caffeine in the hydrogels diminished with the swelling ratio of the hydrogels. In mechanical properties, the shear modulus(G) and elastic modulus (E) significantly increased with the content of the BPEAF due to the rigidity of the fluroene structure, as well as with the concentration of the BPEAF. The influence of BPEAF content on the photoluminescent intensity of the hydrogels were also investigated.

Keywords: Hydrogel, Photoluminescent, Bisphenoxy ethylacrylate fluorene, thermosensitive.

fig1

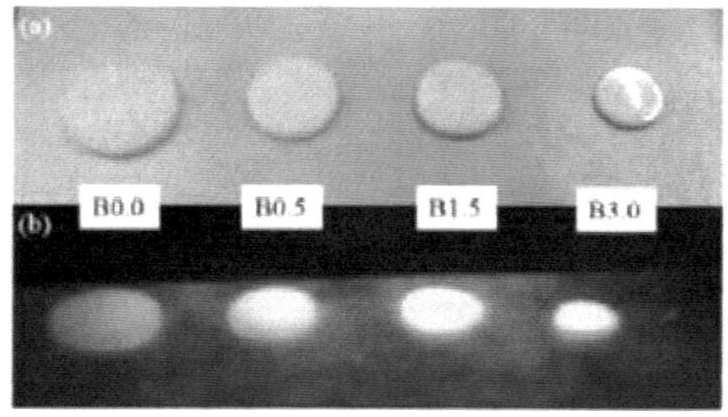

Figure1. Photoluminescent intensity of the hydrogels containing various BPEAF contents under (a) non UV-irradiation and (b) UV-irradiation.

PPG-P-004 [Porous Polymer and Gels]

Influence of the Cross-linker Type on the Internal Structure of Poly(N-Vinylcaprolactam) Microgels Synthesized by Precipitation Polymerization

Michael Kather[1], Stefan Chang[2], Miriam Al Enezy Ulbrich[1], Jörg Stellbrink[2], Andrij Pich[3]

[1]Institute of Technical and Macromolecular Chemistry, RWTH Aachen University, Aachen, Germany
[2]Juelich Centre for Neutron Science JCNS and Institute for Complex Systems ICS, Forschungszentrum Juelich GmbH, Juelich, Germany
[3]DWI - Leibniz Institute for Interactive Materials, Aachen, Germany

Microgels are a three dimensional network of physically or chemically cross-linked polymer particles with a size range of 10 to 10,000 nm that are able to disperse and swell in a solvent.[1] These kind of particles have generated an increasing interest in the past decades, due to their stimuli responsive behavior. Especially microgels based on *N*-isopropylacrylamide (NIPAm) and *N*-vinylcaprolactam (VCL) have gained special interest due to their volume phase transition temperature (VPTT) of 34 °C (NIPAm[2]) respectively 32 °C (VCL[3]) close to the human body temperature and their biocompatibility. In this work we address the influence of the chemical structure of cross-linking agent on the internal structure of the microgels synthesized by precipitation polymerization. In previous work we showed that use of *N,N'*-methylenebis(acrylamide) (BIS) as cross-linking agent in the precipitation polymerization of VCL leads to microgels with a heterogeneous distribution of cross-links (densely cross-linked core and loosely cross-linked coronae).[4] In present study we synthesized microgels using a new cross-linking agent that consists of two N-vinylcaprolactam molecules connected by a diethyl ether bridge and applied it in the precipitation polymerization of VCL. Series of PVCL microgels with variable content of cross-linking agent were synthesized. The polymerization kinetics as well as properties of microgels were studied and compared to analogue samples synthesized with BIS. The experimental data show that due to the comparable reactivity of new cross-linking agent and VCL, the distribution of the cross-links in microgels becomes more homogeneous, temperature-induced swelling degree increases and size distribution of microgels becomes narrower.

(1) Fernandez-Nieves, A. et al. *Microgel Suspensions: Fundamentals and Applications*; 2011; Vol. 1.
(2) Karg, M.; Hellweg, T. *Curr. Opin. Colloid Interface Sci.* 2009, *14* (6), 438–450.
(3) Gao, Y.; Au-Yeung, S. C. F.; Wu, C. *Macromolecules* 1999, *32* (11), 3674–3677.
(4) Schneider, F. et al. *Langmuir* 2014, *30* (50), 15317–15326.

Keywords: Microgel, Cross-linker, Internal Structure, Reaction Calorimetry, Small Angle Neutron Scattering (SANS)

Particle size distribution of microgels synthesized with BIS and BVE

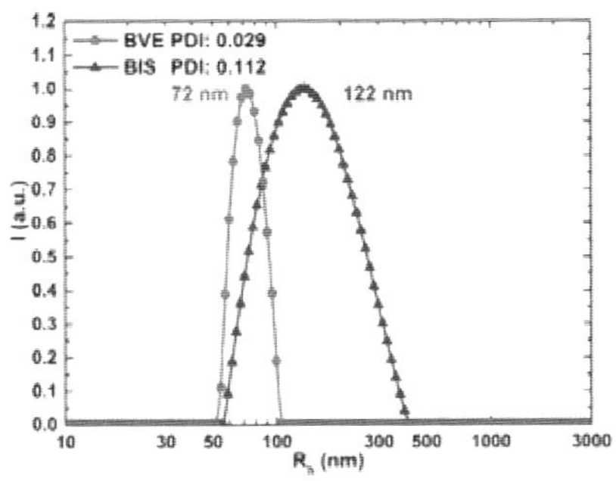

PPG-P-005 [Porous Polymer and Gels]

Evaluation of a bio-hydrogel composed of nanostructured cellulose crosslinked genepin as tissue engineering scaffold: Fabrication, properties and application

Asabuwa Ngwabebhoh Fahanwi, Ufuk Yildiz

Department of Chemistry, Kocaeli University, Kocaeli, Turkey

Developments and applications in tissue engineering rotates around the use of 3D scaffolds to work as templates for cellular manoeuvres to repair, reassemble and regenerate damaged tissues (Ko, Sfeir, & Kumta, 2010). Bio-hydrogels are widely utilized in drug delivery and tissue engineering due to their 3D porous nature which mimics biological tissue (Yang, Bakaic, Hoare, & Cranston, 2013). However, crosslinking of such hydrogel network structures suitable for these applications requires crosslinker molecules which are non-toxic, biocompatible and biodegradable. In this study, bio-hydrogel films of cationic nanocrystalline cellulose (CNCC) have been developed via in situ crosslinking with genepin, whose biocompatibility is essential for tissue engineering applications. The physicochemical, morphological and mechanical properties of CNCC films were evaluated using FTIR, SEM, NMR and rheological testing. The results depicted crosslinking interactions of intermolecules occurred during the fabrication process. The films demonstrated relative good mechanical strength and swelling properties. Cytocompatibility and biodegradability were investigated by a series of in vitro experiments. The obtained results revealed that CNCC films had good biocompatibility. In addition, the CNCC films were degradable in vitro and the degradation rate could be controllable by tailoring the crosslinking content. Hence, CNCC films possess promising prospective applications in biomedical fields.

Keywords: Bio-hydrogel, Cellulose, Genepin, tissue engineering

PPG-P-006 [Porous Polymer and Gels]

Stimuli-induced volume phase transition of cationic copolymeric hydrogels based on N,N-dimethylaminoethly methacrylate

Büşra Yıldız, Nermin Orakdöğen

Istanbul Technical University,Department of Chemistry,Soft Materials Research Laboratory, Istanbul, Turkey

Hydrogels are hydrophilic three-dimensional polymeric networks composed of interlocked polymer backbones, which have the ability to respond to external stimuli such as tempera-ture, pH, ionic strength, specific chemical compounds, and electrical or magnetic fields de-pending on the nature of the pendant groups along the polymer chains. It is desirable to find a way to produce responsive hydrogels with any predetermined value of pH at the point of the swelling transition to broaden their usage in the potential applications in various fields of life science and technology. In this respect, homopolymer poly(N,N-dimethylaminoethyl methacrylate) (PDMAEMA) is especially interesting because of its dual temperature- and pH-dependent solubility. In the present work, P(DMAEMA-DMAPMA) copolymeric hydro-gels were synthesized by free-radical solution copolymerization of monomers DMAEMA, N-[3-(dimethylaminopropyl)]methacrylamide in water to explore their swelling properties, the mechanical behavior as well as their robustness to large mechanical deformations by uniaxial compression testing.

The uniaxial compression measurements were conducted on cylindrical P(DMAEMA- DMAPMA) copolymeric hydrogel samples both after their preparation and after their equi-librium swelling in water. The results showed that the elastic modulus of P(DMAEMA- DMAPMA) hydrogels both after preparation and after equilibrium swelling in water change depending on the comonomer DMAPMA content used in the preparation. The effect of pH on the equilibrium swelling ratio of P(DMAEMA- DMAPMA) hydrogels was also investi-gated in buffer solutions and all P(DMAEMA- DMAPMA) hydrogels showed a pH-dependent swelling transition. The compression test results showed that the modulus of co-polymer hydrogels, both G0 and G, decrease continuously as the comonomer DMAPMA content increased and this result is attributed to the effective crosslink density distribution within the polymer matrix and the role of cross-links.

Acknowledgement: This work was supported by Istanbul Technical University Research Fund.

References :

1. Samchenko Y, Ulberg Z, Korotych O, (2011) Multipurpose smart hydrogel systems. Ad-vances in Colloid and Interface Science 168.1-2, 247-262.
2. Li C, Xu L, Zhai M, Peng J, Yang C, Li J, Wei G (2009) Swelling behavior of amphiphilic gels based on hydrophobically modified dimethylaminoethyl methacrylate. Polymer 50:4888-4894.
3. Li H, Yew YK (2009) Simulation of soft smart hydrogels responsive to pH stimulus: Ionic strength effect and case studies. Materials Science and Engineering: C 29:7, 2261-2269.
4. Gupta P, Vermani K, Garg S (2002) Hydrogels: from controlled release to pH-responsive drug delivery. Drug Discovery Today 7:10, 569-579.
5. Orakdogen N (2013) Investigation of mechanical and thermodynamic properties of pH-sensitive poly(N,N-dimethylaminoethyl methacrylate) hydrogels prepared with different crosslinking agents. Polym. Eng. & Sci. 53:(4) 734-743.

Keywords: Stimuli-responsive hydrogels,swelling,elasticity

PPG-P-007 [Porous Polymer and Gels]

Structutral and crystallization studies of Well-defined PEG-Networks

Muhammad Haris Samiullah[1], Detlef Reicher[2], Tatiana Zinkevich[2], Joerg Kressler[1]

[1]*Institute of Chemistry, Martin Luther University Halle-Wittenberg, Halle, Germany*
[2]*Institute of Physics, Martin Luther University Halle-Wittenberg, Halle, Germany*

Polyethylene glycol (PEG) networks with a well-defined structure were synthesized using copper(I)-catalyzed azide–alkyne cycloaddition (CuAAC). Network structures of two different type were prepared (i) by the combination of two tri-arm star PEG oligomers together and (ii) by connecting tri-arm PEG star units with bifunctional linear PEG oligomers of different molar masses.

PEG oligomers were end-group functionalized with azide and alkyne moeities by reacting them with sodium azide and propargyl bromide respectiviely. On the other hand, catalytic system of $CuSO_4$ and sodium ascorbate in aqueous solution was used for network formation process. Modification of precursors and the formation of network were verified by 13C-MAS NMR and FTIR spectroscopy. Change in the crystallization behavior of PEG chains within the network structure was investigated by Differential Scanning Calorimetry (DSC) experiments. A comparative study showed a significant change in the crystallization of PEO chains before and after the network formation process

Characterization of the network topology, particularly the estimation of the amount of free ("dangling") chain ends and network imperfections like double links were detected for both types of samples and studied by 1H-Double Quantum (DQ) NMR Spectroscopy.

Keywords: Polymer networks, network topology, Double Quantum (DQ) NMR

PPG-P-008 [Porous Polymer and Gels]

Hierarchical Porous Ovalbumin-Chitosan Cryogel Scaffolds for Tissue Engineering Applications

Tuğçe Şen[1], Berkay Özçelik[2], Mehmet Murat Özmen[1]

[1]*Yildiz Technical University, Department of Bioengineering, 34220, Esenler, Istanbul, Turkey*
[2]*Commonwealth Scientific and Industrial Research Organisation (CSIRO), 3168, Clayton, Victoria, Australia*

For tissue engineering applications, the pore structure of scaffolds is critical for cell attachment, proliferation, nutrient transfer, oxygen diffusion to cells and vascularization. Macroporous scaffolds can be prepared by a number of methods including phase separation, gas formation, and porogen leaching. Recently, macroporous hydrogels, so called cryogels, were prepared by cryogelation which is as a facile and attractive way to produce scaffolds (Ozmen, Chem. Commun. 2015, 51, 98, 17479). In addition, the use of hierarchical scaffolds enhances nutrient transfer and vascularization. While there are many studies on hierarchical porous synthetic cryogels, there are limited studies where natural polymers were used to prepare hierarchical porous cryogels (Dinu, Carbohyd. Polym. 2013, 94, 170).

In this study, we produced hierarchical porous ovalbumin-chitosan scaffolds to mimic hierarchical morphologies of tissues by combining cryogelation with phase separation. Initially, ovalbumin and chitosan were dissolved in aqueous acetic acid solutions at predetermined ratios; subsequently n-butanol was added to enable phase separation. Control samples were prepared without n-butanol addition. Glutaraldehyde was added to the mixtures as a crosslinker and the polymer solution was transferred to a syringe and stored at -20°C for 24h. The resulting cryogel scaffolds were taken out and immersed in acetone for solvent leaching. The obtained scaffolds were characterized by SEM, swelling and mechanical measurements. As can be seen in Figure 1, while the internal morphology of the n-butanol added scaffolds have a hierarchical structure, the control samples possess monotype porosity. Characterization and evaluation of the produced hierarchical porous scaffolds demonstrate their suitable properties for potential tissue engineering applications.

Keywords: scaffolds, cryogels, macroporous, hierarchical

Figure 1

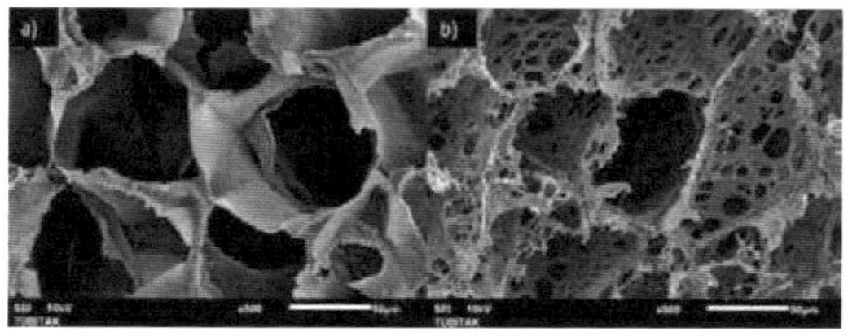

SEM images of a) traditional cryogel (control sample) with monotype pore structure b) hierarchical porous cryogel formed by cryogelation and phase separation

PPG-P-009 [Porous Polymer and Gels]

Hydrogel based nutritional supplement tablets for health care

Lenka Jelinkova, Nabanita Saha, Petr Saha

Centre of Polymer Systems, University Institute, Tomas Bata University in Zlin, třída Tomáše Bati 5678, 760 01 Zlin, Czech Republic

"Nutritional supplements" are commonly termed as "dietary supplements" which are in fact primarily supplements of essential nutrients such as vitamins and minerals, including related compounds such as omega-3 fatty acids and fiber etc. for the use of human beings to supplement the normal diet by increasing the total dietary intake. Until now, many known active ingredients and formulations are reported which are used as food supplements. They are usually formulated in the form of capsules or pastilles, sachets, ampoules of liquid drops or other simple forms of liquids and powders, but there are no information about the availability of betaglucan/agar hydrogel based dietary supplements. Thus, objective of this work was to prepare *hydrogel based nutritional supplement tablets (HNSTs)*, which are not only supporting and strengthening the immunity system of the human body, but also biocompatible and biodegradable. Moreover, nutritional supplements are sometime necessary to maintain a healthy life. Here, the HNSTs were prepared using nature based ingredients by means of a mixture of edible, biocompatible, and biodegradable polymers in addition to some supporting active substances (e.g. Chlorella Kessleri, Mumio etc.) to supply vitamins, minerals and other important nutrients. A series of recipes was prepared to achieve a suitable polymeric hydrogel matrix with active substances. Properties of these HNSTs were characterized using both dried and lyophilized samples. In order to compare their internal structures scanning electron microscopy was used, which confirmed that HNSTs are basically porous (see Fig.1). Swelling behaviour, water adsorption and desorption analysis reveal, that the prepared HNSTs i.e. *nutritional supplement tablets* are principally hydrogel. Futher, antioxidant activity and other interesting properties of some selected HNSTs were investigated, they will be discussed during poster presentation.

Acknowledgement: This work was supported by the Ministry of Education, Youth and Sports of the Czech Republic - NPU Program I (LO1504).

Keywords: antioxidant activity, dietary supplements, hydrogel, HNST, porous

SEM image of hydrogel based nutritional supplement tablet (HNST)

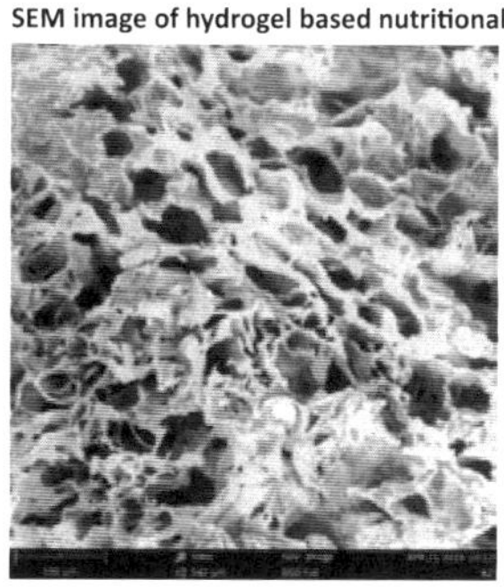

PPG-P-010 [Porous Polymer and Gels]

Swelling and thermodynamic studies of poly(hydroxyethyl methacrylate)-based smart hydrogels formed at various polymer concentrations

Ahmed Kasim Durmus, Nermin Orakdogen

Department of Chemistry, Istanbul Technical University, Istanbul, Turkey

One of the most significant intelligent or smart hydrogels finding an ever greater number of biomedical applications, such as bio-compatible materials, corrective contact lenses, coating for catheters, corneal and cardiovascular implants is derived from the polymers of a hydrophilic monomer, 2-hydroxyethyl methacrylate (HEMA) [1,2]. Because of its tunable chemical and three-dimensional crosslinked network structure, chemical and biochemical stability, non-toxicity, excellent biocompatibility and physicochemical properties similar to those of living tissues, poly(2-hydroxyethyl methacrylate) PHEMA has been widely used as the backbone for synthesizing stimuli-responsive hydrogels [3]. The presence of a pendant hydroxyl group and a carbonyl group on each repeating unit makes the polymers of a hydrophilic monomer, HEMA compatible with water, and the hydrophobic a-methyl group and backbone impart hydrolytic stability to the polymer and support the mechanical strength of the resulting network matrix [4].

In this work, a series of PHEMA-based hydrogels was prepared at various monomer concentrations but at a fixed crosslinker ratio. The properties of the resulting hydrogels was investigated by systematic variation of the monomer feed composition to evaluate their physico-chemical properties in relation to the concentration and the type of the groups incorporated into the polymer matrix. The swelling and elasticity tests were carried out on PHEMA hydrogel samples formed under various gel preparation concentrations. Since the mechanical properties of hydrogels are influenced by a number of factors such as the type of crosslinker, the water content, the preparation temperature, the initial monomer concentration, the effective crosslink density, the presence of ionic groups, and the swelling capacity of hydrogel matrix, the compressive stress-strain measurements were performed to determine the compressive modulus of PHEMA hydrogels. The elasticity results showed that the mechanical strength of PHEMA hydrogels increased greatly by increasing the gel preparation concentration.

This work was supported by Istanbul Technical University Research Fund.

References :

1. Wichterle O., Lim D. (1960) Hydrophilic gels for biological use. Nature, 185 (4706), 117–118.
2. Omidian H., Park K., Kandalam U., Rocca J. G. (2010) Swelling and Mechanical Properties of Modified HEMA-based Superporous Hydrogels. J. Bioact. Compat. Polym., 25, 483–49.
3. Lou X., and Coppenhagen C. (2001) Mechanical characteristics of poly(2-hydroxyethyl methacrylate) hydrogels crosslinked with various difunctional compounds. Polym. Int., 50, 3, 319-325.
4. Han Y. H., Lee E. M., Ji, B. C. (2008) Mechanical properties of semi-interpenetrating polymer network hydrogels based on poly(2-hydroxyethyl methacrylate) copolymer and chitosan. Fibers and Polymers, 9, 393-399.
5. Kizmaz B. and Orakdogen N. (2014) Compression modulus and equilibrium swelling studies of poly(2-hydroxyethyl methacrylate) hydrogels formed at various polymer concentrations" Soft Materials, 12, 2, 237-246.

Keywords: smart hydrogels, 2-hydroxyethyl methacrylate, polymer concentration, swelling, elasticity

PPG-P-011 [Porous Polymer and Gels]

Synthesis of a New Natural Chitosan Based Hyrogel and Its Usage in Drug Delivery Applications

Demet Aydınoğlu[1], Merve Ünal[2]

[1]Department of Food Process Technologies, Armutlu Community College, Yalova University,Yalova/Türkiye
[2]Department of Polymer Engineering, Yalova University, Yalova, Türkiye

The issue of controlled drug delivery systems has been drawing scientists' attention for over six decades. In particular, the investigations have been focused on polymeric materials, especially hydrogels. So far, several polymeric hydrogels were studied and investigated their drug delivery rates and mechanisms. Although some of them were succesful in some respects, for the others, it was a problem that their insufficient amount of delivering or delivery rates, as well as their inability of degredation. As it is well known that, in drug delivery applications, it is very important to deliver at intended amount of drug to the specific region of the body. We aimed to synthesize a new, totally natural hydrogel composites, composed of chitosan crosslinked with genipine and a kind of microalge, spirulina, in order to improve the drug delivery properties of current hydrogels. To examine of delivery mechanism of these hydrogels, 5 FU (Fluorouracil) which is one of the drug active agents that used in cancer treatment was chosen. Both drug amounts and delivery mechanisms were searched systematically by changing composition such as genipine and spirulina ratios of the hydrogel The results show that chitosan hydrogels immobilized with spirulina increase both of the drug amount loaded into and released from the hydrogels, as well as decrease the delivery rates compared with the hydrogel without spirulina. For example the addition of spirulina into the hydrogel system at a constant genipine ratio cause increase in drug delivery almost to threefold. On the other hand the mechanic strength analysis of these hydrogels are on the stage of carrying out. According to findings, we suppose that the hydrogels in present study can be candidate for drug delivery applications, especially when it is taken into account of their improved properties and totally natural character of constituents.

Keywords: hydrogel, drug delivery, spirulina, chitosan

PPG-P-013 [Porous Polymer and Gels]

Sulphobetaine-Based Hydrogels: Synthesis, Characterization, Investigation of Their Stimuli-Responsive Behaviours

Hüseyin Şahin[1], Mehtap Kılınç[1], Cansel Tuncer[2], Vural Bütün[1]

[1]Department of Polymer Science and Technology, Eskisehir Osmangazi University, Eskisehir, Turkey
[2]Department of Chemistry, Faculty of Arts and Science, Eskisehir Osmangazi University, Eskisehir, Turkey

Polymer gels are water insoluble, cross-linked, three-dimensional, hydrophilic polymeric structures which can absorb water at least 20% of its dry mass. Since polymeric gels have similarity to living tissue, contain a high proportion of water, very soft and elastic material, pH-, temperature-, salt-responsive, they are widely used in biomaterials field.

Swelling-shrinking behaviors of acrylate and methacrylate based hydrogels show sensitivity to solution pH. In the literature, there are various hydrogel synthesis with tertiary amine methacrylate based monomers such as 2-(N-dimethylamino)ethyl methacrylate (DMA) monomer. However, its lack of biocompatibility is a limitation for their biotechnological applications. Thus, such system should be converted to biocompatical structures. In this study, firstly, DMA monomer was reacted with 1,3 propane sultone to get biocompatable betainized form (β-DMA). β-DMA monomer was then used in the synthesis of betaine-hydrogels. This hydrogel had swelling and shrinking behavior depending on ionic strength of the medium. Furthermore, its hydrogels were also sythesized with different cross-linking degrees and swelling behaviours were investigated at different salt concentration. In further studies, these hydrogels will be used as host or adsorbant for various metal cations. Various metal nanoparticles including cobalt, gold, platinum, silver will also be synthesized in the hydrogel structure. Characterization of β-DMA monomer and β-DMA hydrogels were carried out by using 1H-NMR, SEM and TEM instruments, respectively.

Keywords: polymeric gels, hydrogel, nanoparticles, sulphobetaine-based hydrogels, biocompatible gel

PPG-P-014 [Porous Polymer and Gels]

From functional polymer particles to fibers as substrates for metal catalysis

Franziska Exnowitz[1], Marco Albuszis[1], Hannes Alex[2], Dr. Norbert. Steinfeldt[2], Dr. Werner Pauer[1], Prof. Dr. Meifang Zhu[3]

[1]*Institute of Technical and Macromolecular Chemistry, University of Hamburg, Hamburg, Germany*
[2]*Leibniz-Institute for Catalysis, Rostock, Germany*
[3]*College of Materials Science & Engineering, Donghua University, Shanghai, China*

Over the last decade the interest in nano structured catalysts such as with carrier metals has increased significantly. We present reactive uniform crosslinked highly porous surface functionalized polystyrene-divinylbenzene microspheres as well as fibers, which can contain alkynes and azides. The azide containing polystyrene-divinylbenzene microspheres were further functionalized with amino alkynes or thiourea. These particles were loaded with Platinum nanoparticles by adsorption of H2PtCl6 followed by reduction of Platinum with sodium borohydride, amino borane or ethylene glycol. Independent on the applied amine and reduction reagent small Pt nanoparticles with sizes between 1 and 5 nm were mainly formed. The composite material was applied as catalyst for the selective hydrogenation of 1-iodo-4-nitobenzene to 4-iodoanilin in continuous operation mode. At a flow rate of 0.125 mmol/min 80% selectivity of 4-iodoanilin at > 98 % conversion was obtained with the amine containing microsphere (T = 110°C, PH2 = 25 bar). Fibers with a functional surface could be produced through electrospinning of functional seed particles. The seeds were monodisperse and had a diameter between 1.4 – 3.2 μm depending on the ratio of the functionalized groups. The seeds were swollen for at least 12 h in different solvents depending on the functional groups. Afterwards the solution was spun. The fibers were collected on a plate or on a collector and analyzed via SEM, IR and Elemental Analysis. The IR-results show both of the functional groups in the fiber matt. The SEM-pictures show a porous structure throughout the fibers. In the next step, the synthesized fibers will be also loaded with catalytically active metal nanoparticles and applied in the reaction mentioned above. The formed composite materials will be characterized by different methods in order to establish correlations between structures and properties as well as the catalytic performance of these materials.

Keywords: functionalized microspheres and fibers, electrospinning, catalytically active metal nanoparticles, catalytic performance

PPG-P-015 [Porous Polymer and Gels]

Superporous hydrogels based on chitosan / salicylaldehyde systems

Iftime Manuela Maria, Luminita Marin

Petru Poni Institute of Macromolecular Chemistry, Iasi, Romania

Hydrogels are a class of materials with impressive applications in biomedical field. Among them, chitosan hydrogels own an important place due to the chitosan intrinsic properties as biocompatibility and biodegradability, hemostatic, hypolipidemic, hypoglycemic, antitumoral, antimicrobial and fungicidal activity – to mention only some. Usually, the chitosan hydrogels are obtained by acid condensation reaction of amine groups of chitosan with di-aldehyde crosslinkers. In such a way, chitosan backbones are random linked together giving a network. Our study revealed another pathway of chitosan crosslinking, using mono-aldehydes which covalently link to the chitosan backbones, causing their straightening, which further facilitate a hydrophilic/hydrophobic segregation forming a supramolecular network. This concept was proved using salicylaldehyde as chitosan crosslinker. To prove the driving force in crosslinking process, a model compound was synthesised too, and its supramolecular self-assembling was investigated by single-crystal X-ray diffraction. FTIR and NMR spectroscopy demonstrated the physical-chemical crosslinking and X-ray diffraction confirmed a three-dimensional self-assembly. The hydrogels have a superporous morphology and elastic networks with a high degree of swelling. Besides, anti-pathogenic activity was detected.

Keywords: chitosan, salicylaldehyde, hydrogels, supramolecular, superporous

SEM image of salicylaldehyde-chitosan superporous hydrogel

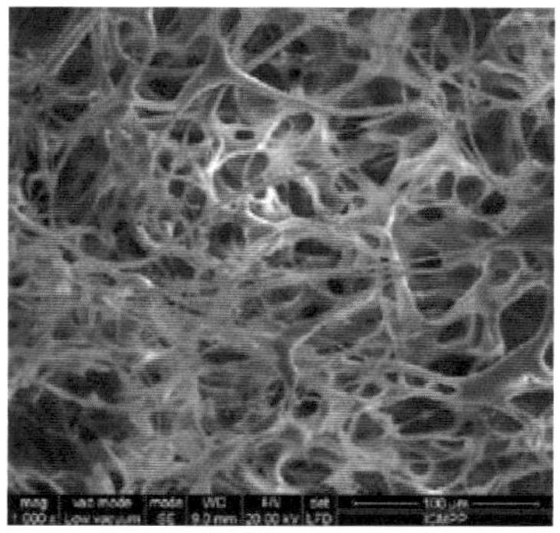

PPG-P-016 [Porous Polymer and Gels]

Fabrication of microporous poly(lactic acid) monoliths via thermally induced phase separation and solvent exchange

Ozgun Can Onder, Emel Yilgor, Iskender Yilgor

Koc University

Microporous poly(lactic acid) (PLA) monoliths were prepared by thermally induced phase separation followed by solvent exchange and vacuum drying. Tetrahydrofuran/water solvent system was used for the induction of liquid-liquid phase separation of PLA solution. PLA gels obtained were mechanically stabilized by replacing THF/water solvent mixture with a nonsolvent. Characterization of PLA monoliths was performed by scanning electron microscopy, mercury intrusion porosimetry and differential scanning calorimetry analysis. Effects of fabrication parameters on the morphology and pore structure were systematically examined. The parameters investigated included; (i) the polymer concentration (6, 7, 8 and 10 wt%), (ii) THF/water ratio (84:16, 86:14, 88:12 w/w), (iii) nonsolvent type (ethanol, methanol, ethanol/water, isopropyl alcohol), and (iv) quench temperature (24 °C, 4 °C and -20 °C). As demonstrated in Figure 1 for the PLA concentration, results show that, degree of porosity and the morphology of the pores such as the pore size and connectivity can be controlled by tuning the fabrication parameters.

Keywords: Poly(lactic acid), porous, thermally induced phase separation, morphology

Figure 1. Effect of PLA concentration on pore size and distribution.

PPG-P-017 [Porous Polymer and Gels]

Supramolecular luminescent chitosan gels

Bejan Andrei, Pinteala Mariana, Barboiu Mihai, Marin Luminita

Petru Poni Institute of Macromolecular Chemistry, Iasi, Romania

Supramolecular hydrogels are an important versatile class of materials with potential applications in tissue engineering, biomaterials, controlled release and others. Incorporation of heavy metal luminescent ions in supramolecular hydrogels is thought to be an interesting method to generate new optical materials and impressive scientific effort has gone into this direction. Our research activity in the chitosan hydrogel domain revealed the possibility to prepare supramolecular hydrogels by using mono-aldehydes in proper reaction conditions as dual physical-chemical crosslinkers. Starting from this premises, we considered interesting to prepare supramolecular luminescent hydrogels based on chitosan and a luminescent aldehyde. To do this, a D-A heterocycle aldehyde with green light emission and good quantum yield has been used as crosslinker to give luminescent hydrogels (Figure 1b). The FTIR and NMR investigation proved a covalent crosslinking degree while the wide angle X-ray diffraction indicated their supramolecular nature. They have a porous morphology and elastic behavior. The hydrogels have interesting luminescent properties, varying under the nature of the external stimuli, as metal ions (Figure 1).

Keywords: chitosan, luminescent aldehyde, supramolecular, luminescence

Hydrogel sample

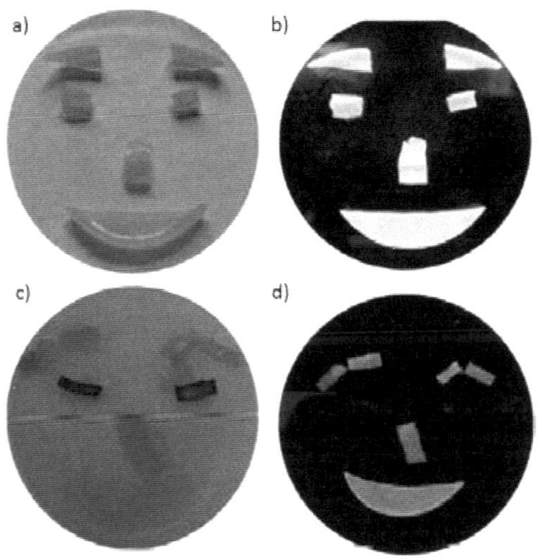

a) Image of the untreated hydrogel b) Untreated hydrogel under UV light c) Hydrogel samples treated with different heavy metal solutions d) Treated hydrogel samples under UV light

PPG-P-018 [Porous Polymer and Gels]

"" Oxidative Cross-Dehdyrogenative Coupling of Terminal Alkynes with Thiols Using Cu(I) Incorporated Schiff Base Network Polymer as Catalyst ""

Yonca Alkan, Ömer Suat Taşkın, Barış Kışkan, Yusuf Yağcı

Department of Chemistry, Istanbul Technical University, Istanbul, Turkey

Alkynes are an important class of organic molecules because of their versatile applications in materials science and organic synthesis as building blocks. Sulfur-containing compounds have also a crucial role in synthetic organic chemistry since sulfur moiety acts as an important auxilary functional group. The discovery of a new catalyst to perform a coupling reaction between an alkyne and a thiol in better conditions gains a considerable attention.

Despite the fact that homogenous catalysts are more reactive than heteregenous catalysts, they have disadvantages such as removal from the reaction medium and reusability. For that reason, design of an heteregenous catalysis have taken increasingly more attention due to its possible advantages such as reusability, waste minimization derived from reaction workup and help to the development of green chemistry concept.

The reactivity of homogenous catalysts result from their high interaction with its substrates since they are in the same phase and the less reactivity of heterogenous catalysts result from their limited surface area which is in contact with its substrates. In this point, use of porous materials serves a very important advantage. As the pore size decreases, surface area increases which results in more reactive sites capable of catalytic activity. When compared with the other porous polymers, microporous polymers are unique in that they exhibit permanent porosity with extremely high BET surface area.

In this study, a microporous polymer was synthesized and Cu(I) ions were incorporated inside the pores. Then, direct oxidative cross-dehydrogenative coupling reactions of various terminal alkynes and thiols and polymerization reactions between bifunctional alkynes and bifunctional thiols by using this copper-incorporated heterogenous catalyst were investigated. The products were characterized by NMR studies and the molecular weight characteristics of the polymers were investigated by GPC measurements.

Keywords: microporous polymers, catalysis, heterogenous catalysis, oxidative cross-dehydrogenative coupling

Coupling Reaction Between Thiophenol and Phenylacetylene

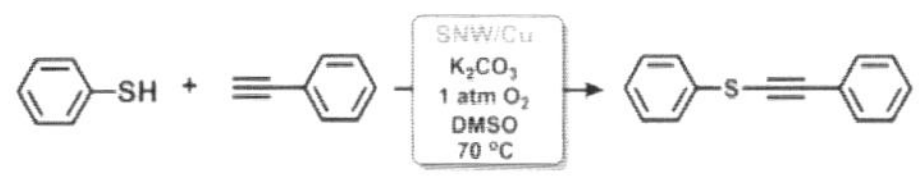

PPG-P-020 [Porous Polymer and Gels]

Sulfur-Bridged Nanoporous Polymer Networks for CO2 Capture

Mehmet Sahin Atas[1], Halit Cavusoglu[2], Mustafa Selman Yavuz[1]

[1]*Metallurgy and Materials Engineering Department, Selcuk University, Konya, Turkey*
[2]*Advanced Technology Research and Application Center, Selcuk University, Konya, Turkey*

As the global energy crisis becomes progressively acute, the dependence on natural gas has been economically attractive from the environmental view point. During the past few years, the increase in burning fossil fuels and petroleum products has led to the higher concentrations of CO2 in the atmosphere[1]. Effective strategies of CO2 remediation have to include capture and utilization in close connection with each other. To begin such an endeavour, capturing of CO2 generated by large point sources (e.g. fossil-fuel-fired power gasification plants) is essential[2]. Several porous materials such as mesoporous silica, layered materials, zeolites, activated carbon, metal–organic frameworks (MOFs), covalent-organic frameworks (COFs), and porous covalent-organic polymers with excellent CO2 adsorption capacities have been reported. Among them, MOFs, COFs, and porous polymers gained substantial attention due to their high surface area, ability to be functionalized, thermal stability, recyclability, and selectivity for CO2 over other gases[3]. The present work focuses on the removal performance of CO2 by adsorption process utilizing the highly stable sulfur-bridged porous covalent organic polymers (PCOPs). The research results showed that this highly stable PCOPs as adsorbant has a high adsorption capacity for CO2 system. Optimal CO2 adsorption capacities at 273 and 298 K were 5.97 cm3/g and 5.76 cm3/g.

This study was carried out as a Master thesis by Mehmet Sahin Atas at the Graduate School of Natural and Applied Science at Selcuk University, Konya, Turkey. Academic Staff Training Program (ÖYP) (Project Number 2015-ÖYP-128), which authors gratefully acknowledge.

References:

[1]. Patel, H.A. and C.T. Yavuz, Faraday Discuss, 2015.
[2]. Pandolfi, J.M., Conolly, S.R., Marshall, D.J., Cohen, A.L. Science, 2011. 333(6041): p. 418-22.
[3]. Patel, H.A., Karadas, F., Byun, J., Park, J., Deniz, E., Canlier, A., Jung, Y., Atilhan, M.,Yavuz, C.T. Advanced Functional Materials, 2013. 23(18): p. 2270-2276.

Keywords: porous polymers, carbon dioxide, adsorption

PPG-P-021 [Porous Polymer and Gels]

Synthesis and characterization of imprinted fluorescent stimuli-responsive gels for selective analysis of uranyl ion

Elif Gökçe Atçakan, Orhan Güney

Department of Chemistry, Istanbul Technical University, Istanbul, Turkey

Stimuli-responsive gels are crosslinked polymer networks which respond to external stimuli such as pH, temperature, ionic strength, electric field, light, pressure and magnetic field [1]. The responses of the gel to these external stimuli result in changing of phase, shape, volume and optical properties. Molecular imprinting is a sophisticated technique to create molecular recognition system in three-dimensional cross-linked polymers, namely molecularly imprinted polymers. Imprinted gels are synthesized in the presence of target molecules or ions, functional monomers and cross-linking agents by covalent, non-covalent or semi-covalent methods. Binding sites with molecular recognition properties are formed after template molecules have been removed from the polymerized material, leaving behind cavities for the subsequent rebinding process that are complementary in size and shape to the target molecules or ions [2].

In this study, ion-imprinted fluorescent stimuli-responsive gel was synthesized in the presence of uranyl ion, methacrylic acid and 9-vinylcarbazole as a ternary complex, N-isopropyl acrylamide as a matrix forming component and N,N-methylenebis acrylamide as a crosslinker. The dual stimuli- responsive properties of the gels were characterized depending on composition of solvent mixture and temperature. By obtaining the fluorescence spectra of carbazole moiety in slab gel as a function of uranyl ion concentration, the sensing character of fluorescent gel for the determination of uranyl ion was elucidated both in swollen and shrunken states.

References:

[1] Orhan Güney, Erdal Serin "Stimuli-responsive molecularly imprinted hybrid polymer gel as potential system for controlled release" J. Appl. Polym. Sci. (2016) 133: 42913
[2] F. Karagöz, O. Güney, "Development and characterization of ion-imprinted sol-gel derived fluorescent film for selective recognition of mercury(II) ion", J. Sol-Gel Sci. Technol. (2015) 76:349–357.

Keywords: Stimuli-responsive gel, Molecular imprinting, Uranyl analysis

PPG-P-022 [Porous Polymer and Gels]

Elaboration of nanocomposites poy(N-isopropylacrylamide)/clay, and study adsorption of cationic dey

Seddiki Nesrinne, Aliouche Djamel, Ghemataoui Nouara

Laboratory Processing and Forming Polymers. Faculty of Engineering Science. University M'Hamed Bougara Boumerdes, 35000.Algérie

The main objective of this work is to show the increasing importance of the chemical modification of polymer hydrogels to obtain a material sensitive to external stimuli, with a high capacity to absorb liquids and structural stability improved by introducing charges minerals (phyllosilicates) to a nanometer level.

We first synthesized synthetic clay-based nanocomposites (Laponite) to strengthen the structure of hydrogels sensitive to temperature stimuli. Characterization was done by FTIR, XRD, and by transmission electron microscopy (TEM); thermal stability was studied by thermogravimetric analysis (TGA), the adsorption of the dyes was tested by adsorption of cationic dyes. The characterization confirmed the new structures nanocomposite hydrogels. The thermal stability was greatly improved and nanocomposites obtained appear homogeneous and have very interesting unique properties.

Keywords: nanocomposite, hydrogels, poly (N-isopropylacrylamide), crosslincker, adsorption, cationique dey

PPG-P-023 [Porous Polymer and Gels]

" Tannin-Isocyanate based rigid foams "

Zeyneb Hamadi[1], Francisco Jose Santiago Medina[2], Rabah Chaid[1], Antonio Pizzi[2]
[1]URMPE, University M'hamed Bougara Boumerdes, rue de l'Indépendance, 35000 - Boumerdès, Algérie
[2]ENSTIB-LERMAB, Nancy-University, 27 rue du Merle Blanc, Bp1041, 88051 Epinal cedex 9, France

New Tannin-Isocyanate rigid foam were prepared using Toluene diethylene isocyanate (less 13%).

Scanning electron microscopy (SEM) showed the addition of castol oil/ ethylene glycol yielded a more ordred microstruscture of the foam and a smaller cell size.

Density, thermal conductivity and mechanical property of the foam were mesured. the result show that this materiel has a good mechanical property Young'smodulus 24±3 Mpa and compressive strang between 2.57-2.63Mpa.

hematerielswerealsofoundtopresentgoodresistancetostrangacidandbases. thermal conductivity measured in this workare (0.065 $Wm^{-1} K^{-1}$ for densitie 0.16 g cm^{-3}). Thermal coductivity are fully comparable with those of phenolic material. we show that this materiel can remplace the synthetic materiels in different applications.

Keywords: Tannins foams, mechanical properties, thermal conductivity

Density,porosity, compressive strangths and thermal conductivity of tannin-isocyanate foam

Sample	Bulk-density (gr/cm3)	Porosity (%)	Thermal Conductivity (W/mK)	compressive strangth (Mpa)	Young'smodulus (Mpa)	% open cell
Tannin-isocyanate foam	0.251	82.4	0.0652	2.57±0.03	24± 3	94.42

PPG-P-024 [Porous Polymer and Gels]

Novel Functionalizable Cryogels With Degradable Linkers For Drug Delivery Systems

Duygu Aydin[1], Mehmet Arslan[2], Amitav Sanyal[1], Rana Sanyal[1]

[1]*Department of Chemistry, Bogazici University, Istanbul, Turkey*
[2]*Department of Polymer Engineering, Yalova University, Yalova, Turkey*

Hydrogels are chemically or physically crosslinked three dimensional networks that are constructed using hydrophilic monomers or polymers that can absorb high amount of water while retaining the internal network structure and viscoelastic behavior [1,2]. They have been widely evaluated as hydrophilic matrices for fabricating controlled release systems for drug delivery, biomolecular immobilization, implant coatings, contact lenses and wound healing dressings. Functionality, reactivity and porous structure of gels are also important parameters especially for these type of applications. It is often beneficial to have chemically crosslinked functionalizable porous hydrogels [3]. These porous gels are usually synthesized using cryogelation technique with reactive monomers or macromers so that they can be functionalized with molecules of interest with ease [4,5].

In this study, poly(ethylene glycol) methacrylate-based cryogels containing activated carbonate groups as amine reactive probes have been fabricated. Degradable monomers based on NHS-activated carbonate group were utilized. The resulting cryogels were then characterized by their physical properties ie. water uptake and surface morphologies and then conjugated to amine-containing molecules. Efficient functionalization of these reactive cryogels was emphasized by conjugation of chemotherapy drugs. Finally, the time dependent releases of conjugated molecules were investigated under physiological pH conditions.

References:

[1] Hoffman, A. S.; *Adv. Drug Delivery Release* **2012**, *64*, 18-23.
[2] Gupta, P.; Vermani K.; and Garg S.; *Drug Discovery Today*, **2002**, *7 (10)*, 569-579.
[3] Altin, H.; Kosif, I.; Sanyal, R.; *Macromolecules.* **2010**, *43*, 3801-3808.
[4] Polizzotti, B. D.; Fairbanks, B. D.; and Anseth, K. S.; *European Journal of Pharmaceutics and Biopharmaceutics*, **2000,** 50 (1), 27–46.
[5] Dispinar, T.; Van Camp, W,; De Cock, L. J.; De Geest B. G.;, Du Prez F. E.; *Macromolecular Bioscience*, **2012**, 12, 383-394.

Keywords: Gels, carbonate, reactive hydrogels

PPG-P-025 [Porous Polymer and Gels]

Porous thermoplastic polyurethane films by breath figure approach

Gizem Daban[1], Cem Bayram[2], Murat Demirbilek[2], Emir Baki Denkbaş[1]

[1]*Department of Bioengineering, Hacettepe University, Ankara, Turkey*
[2]*Advanced Technologies Application and Research Center, Hacettepe University, Ankara, Turkey*

Porous surfaces and bulk structures with interpenetrating pore networks are essential in biomaterial studies, aiding the nutrient transport as well as the cell infiltration. Biomedical grade thermoplastic polyurethane (TPU) has extensively used in biomaterials science thanks to its biocompatibility and especially, superior mechanical properties. One of the featured use of TPU is blood vessel prostheses, which requires both biocompatibility and high durability. Porous TPU films were prepared by several methods in the literature such as electrospinning of nonwoven mesh, porogen leaching, gas foaming and phase separation. Breath Figure (BF) is a biomimetic method of water droplets that are seen inside of bottle walls, polymer thin layers. Francois et. al used that method for the first time to obtain porous thin layers in 1994 (Francois et al., 1994). Today there are lots of possible ways of producing porous thin layer, yet BF is the possible cheapest method since it does not require any expensive and complex equipment. There is an another and the most important advantage of BF is "contamination free nature" which is the dream of surface chemistry. Pore arrays produced through droplets condensed on cold substrate surface, two possible ways of preparation procedures. One is the direct method, there should be moist gas flow while solvent evaporates, and the other one is the static method, which we use in the study, evaporation of solvent is completed spontaneously without any gas flow in closed system. In our study porous TPU membranes were produced and characterized in terms of chemical and structural properties. Polymer concentration, casting thickness, humidity and moist exposure time were considered as effective parameters. Additionally, membranes were tested with human umbilical vein endothelial cells (HUVEC) for the in vitro biological performance.

References:

Widawski, G., Rawiso, M., and Francois, B. (1994) Nature, 369, 387–389.

Keywords: breath figure, polyurethane, porous membrane

Porous TPU membrane

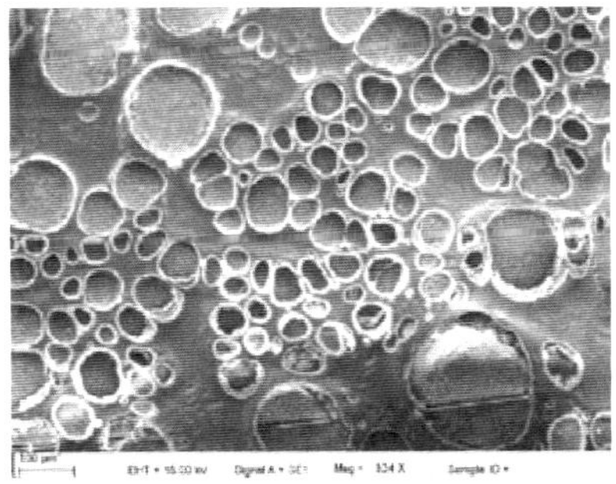

PPG-P-026 [Porous Polymer and Gels]

Protein Release from Redox Responsive Hydrogels

Ismail Altınbaşak, Amitav Sanyal, Rana Sanyal

Department of Chemistry, Boğaziçi University, Istanbul, Turkey

Hydrogels are three-dimensional crosslinked networks of hydrophilic polymers that can be prepared with tunable mechanical strength and water uptake capacity. In last few decades, design of responsive hydrogels that responds to specific environmental stimuli has drawn a lot of attention. Recently, redox-responsive hydrogels have attracted a tremendous amount of attention due to their high sensitivity and rapid response to thiols, which are biologically relevant stimuli.[1] Redox-responsive hydrogels have been prepared using different chemistries such as amide bond formation,[2] radical polymerization,[3] and oxidation of thiol groups on the polymer into disulfide bonds.[4] Although click reactions such as azide-alkyne cycloaddition, nucleophilic thiol-ene, radical thiol-ene and Diels-Alder reactions are the most popular techniques that used in the fabrication of hydrogels,[5]sensor technology, and tissue engineering. Increasing demand of these materials for such advanced applications has necessitated the development of hydrogels with complex chemical compositions such as incorporating small molecules and biomolecules that provide the functional attributes. This Focus Review highlights the tremendous impact of click chemistry on the design, synthesis, and functionalization of hydrogels. The high efficiency and fidelity of the click reactions have enabled rapid and modular synthesis of hydrogels with near-ideal network structures. Efficient incorporation of biomolecular building blocks, such as peptide sequences either during or after the fabrication of hydrogels, have been achieved through various click reactions. Utilization of these efficient reactions has led to the fabrication of many stimuli-responsive or 'smart' hydrogels in recent years.", "author" : [{ "dropping-particle" : "", "family" : "Yigit", "given" : "Sezin", "non-dropping-particle" : "", "parse-names" : false, "suffix" : "" }, { "dropping-particle" : "", "family" : "Sanyal", "given" : "Rana", "non-dropping-particle" : "", "parse-names" : false, "suffix" : "" }, { "dropping-particle" : "", "family" : "Sanyal", "given" : "Amitav", "non-dropping-particle" : "", "parse-names" : false, "suffix" : "" }], "container-title" : "Chemistry, an Asian journal", "id" : "ITEM-1", "issue" : "10", "issued" : { "date-parts" : [["2011", "10", "4"]] }, "page" : "2648-59", "title" : "Fabrication and functionalization of hydrogels through \"click\" chemistry.", "type" : "article-journal", "volume" : "6" }, "uris" : ["http://www.mendeley.com/documents/?uuid=e2d12591-c8db-49d7-8959-4f5becfe4fdc"] }], "mendeley" : { "formattedCitation" : "[5]", "plainTextFormattedCitation" : "[5]", "previouslyFormattedCitation" : "(Yigit et al. 2011 there are very few examples of redox-responsive hydrogel system that is fabricated using these click chemistries.

Diels-Alder click chemistry provides a robust methodology that does not require utilization of any catalyst or photochemical irradiation and generally involves diene and dienophiles as functional groups that are quite stable. In this study, disulfide bond containing redox-responsive hydrogels have been prepared using the Diels-Alder click chemistry. Hydrogels containing different amount of crosslinker were synthesized via the Diels-Alder reaction between furan bearing Poly(ethylene glycol)-based polymer and disulfide containing bismaleimide crosslinker. Furan bearing polymer was synthesized via single electron transfer living radical polymerization technique using poly(ethylene glycol) methacrylate and furfuryl methacrylate. Rheological analysis of different hydrogels with various crosslinker densities was done and then their degradation rates were investigated in phosphate buffer and reducing agent containing solutions. Fluorescein isocyanate-labeled bovine serum albumin was loaded into hydrogels physically. Protein release studies for hydrogels with different crosslinking density were carried out in the presence and absence of reducing agent to demonstrate superiority of present construct.

References:

[1] Chen X, Zhou Y, Peng X, Yoon J. Fluorescent and colorimetric probes for detection of thiols. Chem Soc Rev 2010;39:2120–35. doi:10.1039/b925092a.

[2] Wang P, Zhang Y, Cheng L, Liu W. The Unusual Mechanical Evolution of Biodegradable Double Hydrogen Bonding Strengthened Hydrogels in Response to pH Change. Macromol Chem Phys 2015;216:164–71. doi:10.1002/macp.201400456.

[3] Yoon JA, Kamada J, Koynov K, Mohin J, Nicolaÿ R, Zhang Y, et al. Self-Healing Polymer Films Based on Thiol–Disulfide Exchange Reactions and Self-Healing Kinetics Measured Using Atomic Force Microscopy. Macromolecules 2012;45:142–9. doi:10.1021/ma2015134.

[4] Hisano N, Morikawa N, Iwata H, Ikada Y. Entrapment of islets into reversible disulfide hydrogels. J Biomed Mater Res 1998;40:115–23.

[5] Yigit S, Sanyal R, Sanyal A. Fabrication and functionalization of hydrogels through "click" chemistry. Chem Asian J 2011;6:2648–59. doi:10.1002/asia.201100440.

Keywords: Hydrogel, Redox responsive, Protein release

PPG-P-027 [Porous Polymer and Gels]

Effects of gel aging on synthesis of boride ceramics

Ruixing Li

School of Materials Science and Engineering, Beihang University, Beijing, China

Sol-gel method is an effective way for low temperature synthesis of oxide ceramics. In this study, ZrB_2 are synthesized by a sol-gel route using zirconium n-propoxide, boric acid, acetic acid and D-fructose. D-Fructose acts as both a modifier and a carbon source for carbothermal reduction reaction. For comparison purposes, both nascent state gel and aged gel were used to clarify the aging-time-dependent mechanism for ZrB_2 synthesis. As a result, a single phase ZrB_2 powder with a uniform size and shape distribution can be obtained from the aged gel with a boron and carbon to zirconium molar ratio of 3.5 ~ 4 and 7, respectively, reduced at 1550 °C for 2 h.

Keywords: aging, sol-gel, synthesis

PPG-P-028 [Porous Polymer and Gels]

Facile Synthesis of Diene-Containing Hydrogels for Reversible Functionalization

Laura Chambre[1], Suzan Gunbay[2], Amitav Sanyal[1], Rana Sanyal[3]

[1]Department of Chemistry, Bogazici University, Bebek 34342, Istanbul, Turkey
[2]Department of Chemistry, Dublin Institute of Technology, Dublin 2, Ireland
[3]Center for Life Sciences and Technologies, Bogazici University, Istanbul, 34342, Turkey

Hydrogels are three-dimensional cross-linked network of hydrophilic polymers that have wide implementation in biological applications. Porous nature of these materials enables loading and release of drugs and biomolecules of interest into the hydrogel matrix. Such non-covalent modifications often lead to undesirable burst release. As an alternative, chemical modification of hydrogels allow robust and stable conjugation with control over subsequent release. So far, hydrogel have been mostly functionalized via azide/alkyne,[1] radical thiol-ene,[2] and Michael addition reactions.[3] The Diels-Alder (DA) cycloaddition provides an attractive alternative towards functionalization of hydrogels since it proceeds under benign conditions.[4] Oftentimes no catalyst is required and the reaction can also proceed in aqueous environment. Herein, we report the facile synthesis of diene-containing hydrogels and their subsequent functionalization via the DA click reaction. By adjusting the monomer ratio in the hydrogels, their swelling profile and porosity, as well as the reactive group density were altered. Release of attached molecules from these hydrogels via the retro Diels-Alder reaction was studied.

References:

[1](a) Altin, H.; Kosif, I.; Sanyal, R. Macromolecules, 2012, 43, 3801-3808
(b) Uttamapinant, C.; Tangpeerachaikul, A.; Grecian, S.; Clarke, S.; Singh, U.; Slade, P.; Gee, K. R.; Ting, A. Y. Angew. Chem., Int. Ed. 2012, 51, 5852–5856.
(c) DeForest, C. A.; Polizzotti, B. D.; Anseth, K. S. Nat. Mater. 2009, 8, 659–664.
[2] Beria, L.; Gevrek, T. N.; Erdog, A.; Sanyal, R.; Pasini, D.; Sanyal, A. Biomater. Sci., 2014, 2, 67
[3] Park, E.J.; Gevrek, T. N.; Sanyal, R.; Sanyal, A. Bioconjugate Chem., 2014, 25 (11), 2004–2011.
[4]Koehler, K. C.; Alge, D. L.; Anseth, K. S.; Bowman, C. N.; Biomaterials, 2013, 34(16), 4150-4158

Keywords: Hydrogel, Diels-Alder, Reversible Functionalization

PPG-P-029 [Porous Polymer and Gels]

Heterogenization Of Homogenous Nhc-Pd(II)-Pyridine Catalytic Systems Via Radical Polymerization: Catalytic Activities For Suziki-Miyaura Coupling Reactions

Cihangir Boztepe[1], Ismail Ozdemir[2], Asim Kunkul[1], Nevin Gurbuz[2]

[1]Department of Chemical Engineering, Inonu University, Malatya, Turkey
[2]Catalysis Research and Application Center, Inonu University, Malatya, Turkey

The synthesis of NHC-PdCl2-pyridine has been the focus of considerable attention due to their stability towards moisture, air, and heat and useful catalytic properties. The Suzuki reaction is usually carried out with a homogeneous Pd catalyst but, in order to use the Suzuki reaction in a wide range of industrial applications, a heterogeneous Pd catalyst must be developed [1]. Hydrogels are three-dimensional crosslinked polymeric materials that can absorb and hold large amounts of water without undergoing dissolution. Hydrogel catalyst supports have several advantages, such as big surface area, fast recovery and the simple recycling of the catalysts by filtration, which in turn prevents the contamination of the ligand and decreases the environmental pollution caused by residual metals in the waste [2, 3].

In this study, novel two NHC-Pd-pyridine polymerizable carbene complexes having catalytic activity, have been prepared.

These complexes (monomers) were polymerized with Acrylamide (AAm) and N,N'-Methylenebisacrylamide (MBA) by radical polymerization in DMSO. Ammonium peroxydisulfate (APS) and TEMED used as initiator and accelerator, respectively. Synthesized heterogeneous polymeric catalysts named as Catalyst 1 and Catalyst 2 were used in Suzuki-Miyaura coupling reactions of arylboronic acid and various aryl bromides. We investigated the effects to catalytic performance of these catalyst types in the Suzuki- Miyaura cross-coupling reactions and illustrated in Figure 2. Acrylamide based polymeric NHC-Pd-pyridine catalysts exhibited good recyclability and could be reused at least five times without remarkable loss of its catalytic activity.

Keywords: Hydrogel, Heterogeneous Catalyst, N-Heterocyclic carbene, Suzuki-Miyaura reaction.

Suzuki- Miyaura coupling reactions of aryl boronic acid and various aryl bromides with the Heterogeneous catalysts at 80 0C in water.

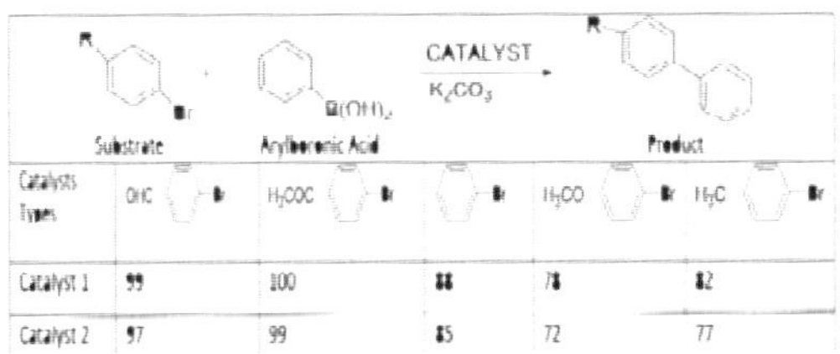

Reaction conditions: bromobenzene (0.25 mmol), phenylboronic acid (0.375 mmol, 1.5 eq.), K2CO3 (0.75 mmol, 3 eq.), H2O (5 mL), and 1 equimolar % Pd catalyst, 3 hours.

Synthesis of imidazole and benzimidazole based polymerizable NHC-Pd(II)-pyridine complexes.

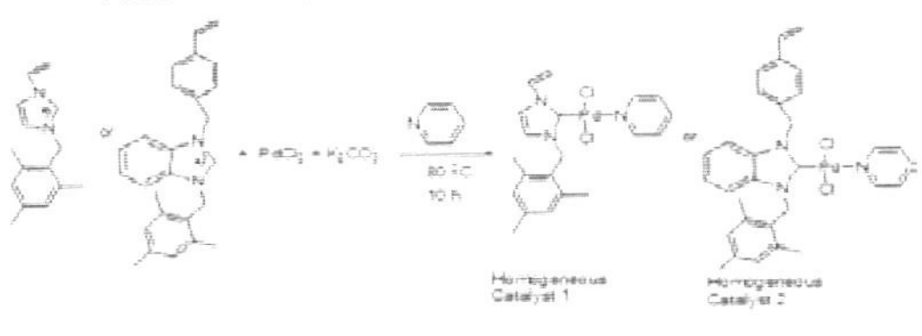

PPG-P-030 [Porous Polymer and Gels]

Release of Riboflavin from Interpenetrated Network Poly(NIPAAm-co-AAc) Hydrogel and Release Kinetic Studies

Sanogo Brahima, Cihangir Boztepe, Mehmet Yuceer, Asim Kunkul

Department of Chemical Engineering, Inonu University, Malatya, Turkey

The objective of this work is to make the release of riboflavin (B2 vitamin) from synthesized interpenetrated network (IPN) poly(NIPAAm-co-AAc) hydrogel and to investigate the kinetics of drug release by models taken in the literature. A pH and temperature sensitive IPN poly(NIPAAm-co-AAc) hydrogel was synthesized by microemulsion polymerization in first step then by solution polymerization method interpenetrated network structure hydrogel was obtained. During two days a synthesized dry poly(NIPAAm-co-AAc) IPN was immersed in riboflavin 200 ppm solution for drug loaded by swelling. The amount of loaded drug obtained was 21.33 mg/g. Then the release as a function of time of riboflavin experiments were realized at physiological temperature 37 °C, in pH=2, pH=4 and pH=7 solutions. Released riboflavin's amount was measured by UV-vis spectrophotometer at 445 nm wavelength. The drug release from IPN poly(NIPAAm-co-AAc) hydrogel results show that the drug releases were dependent on pH and temperature. The release amounts obtained at pH=2, pH=4 and pH=7 were 62.45%, 53.87% and 69.57%, respectively. The profiles of release for riboflavin are shown in Figure 1. The kinetic studies of drug release were made on different models taken in the literature and the best results were found on diffusion approch model (Eq.1) presented in this work. Sequential Quadratic Programming (SQP) technique in MATLAB software was used for parameters estimation of the model (Table 1). Table 2 gives the model performance in terms of the correlation coefficient (R), root mean squared error (RMSE) and mean absolute percentage error (MAPE, %) between the experimental data. Diffusion Approach: $M_t = a\cdot\exp(-k\cdot t)+(1-a)\cdot\exp(-k\cdot b\cdot t)$ (1) Where; M_t: Concentration of drug released at time t; a,b and k: Model parameters; NIPAAm: N-isopropylacrylamide; AAc: Acrylic Acid

Keywords: Kinetic studies, IPN, Drug release, Parameter estimation

Profiles of Riboflavin Release from IPN poly(NIPAAm-co-AAc) at 37 °C

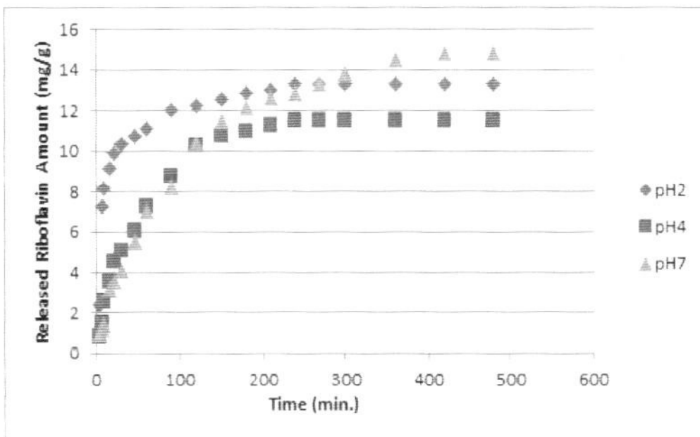

Parameter estimation results

Model Parameters	pH=2	pH=4	pH=7
a	0.4796	0.3508	0.4779
k	0.1526	0.0323	0.0156
b	0.0095	0.0485	0.0861

Statistical evaluation of goodness of fit

	R	RMSE	MAPE(%)
pH=2	0.9912	0.5056	2.6390
pH=4	0.9962	0.3543	2.1478
pH=7	0.9982	0.3221	1.7997

PPG-P-031 [Porous Polymer and Gels]

Preparation and Characterisation of Rigid Phenolic Foams

Cláudio G. dos Santos[1], Tobias Böhringer[1], Victor A. Oliveira[2]

[1]*Department of Chemistry, Federal University of Ouro Preto, Ouro Preto, Brazil*
[2]*Department of Materials and Metallurgical Engineering, Federal University of Ouro Preto, Brazil*

Resols were synthesized by reacting mixtures of phenol and formaldehyde using basic catalysts. These resols were used in formulations with blowing agents and surfactants to prepare phenolic resins in the form of cellular materials, referred to as phenolic foams. The relative proportions of blowing agents and surfactants had an influence on the properties of the resulting foams. Curing efficiency of the resins were evaluated by the acetone soluble content method and the results indicated curing indices above 98%.

Infrared spectroscopy showed the typical bands expected for phenolic resin structures and the foams had densities of 399 kg/m^3 on average. Such value of density is typical of foams classified as structural foams. The product was also characterized by thermal analysis, which showed that the foams are thermally stable until around 400 °C, when they start to degrade at a higher rate, leaving about 2% of solid residue.

SEM analyses of the foams confirmed the cellular structure of the material with a randomly distribution of open and closed pores with 4 μm average diameters. Mechanical tests of the foams resulted in compressive strengths of 1.9 MPa on average, which supports their application of structural foams.

The porous nature and the outstanding compressive strength of the foams suggest they could be carbonized and used as adsorption material that might be employed, for instance, to remove contaminants from stream waters. Such possibility is currently being investigated and shall be presented in a future work.

Keywords: phenolic resins, strutuctural foams, porous materials, characterization

Microstructure of the Phenolic Foam

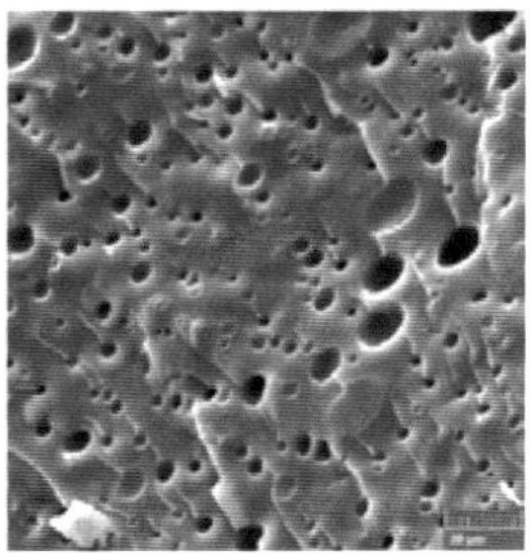

SEM image of the phenolic foam showing the porous structure

Phenolic Foam

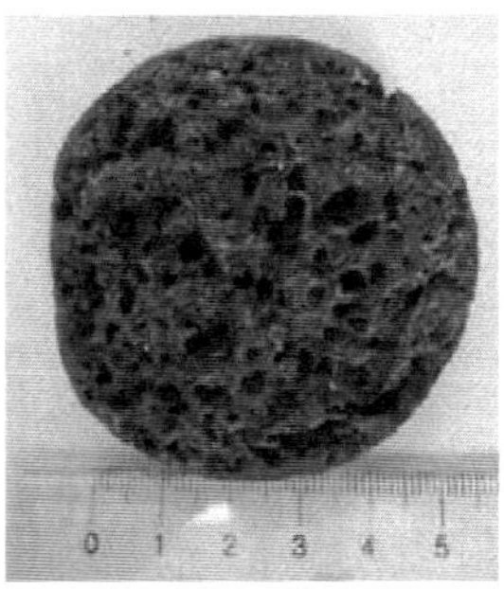

Typical image of the phenolic foams prepared

PPG-P-032 [Porous Polymer and Gels]

Gold nanoparticles embeded disulfide-linked polymer networks

Mehmet Sahin Atas[1], Halit Cavusoglu[2], Mustafa Selman Yavuz[1]

[1]*Metallurgy and Materials Engineering Department, Selcuk University, Konya, Turkey*
[2]*Advanced Technology Research and Application Center, Selcuk University, Konya, Turkey*

Gold (Au) nanoparticles decorated on suitable supports hold unique advantages as heterogeneous catalysts under mild conditions even at ambient temperature with high reaction rates. Although there are many reports about supported Au catalysts, which showed high catalytic activity for different systems including oxidations, hydrogenation, cyclization, rearrangement, C–C coupling reactions, the unstablility of the supports having these nanoparticles brings limitations to their uses[1]. Covalent organic frameworks (COFs) are lightweight and porous materials well-known for their applications in gas storage, catalytic supports, semiconductive and photoconductive devices. In addition to other porous materials such as charcoal, dendrimers, polymers, mesoporous silica, zeolites, MOFs, metal doped COFs have also been applied as catalysts for various organic reactions[2]. Despite the limited surface area of COFs, nanoparticle decorated COFs have been used as catalyst for Suzuki coupling, C–H activation, Knoevenagel condensation, nitro reduction, glycerol oxidation, etc[3]. Thus, in order to overcome these issues regarding the optimal interaction, we believe that the synthesis of new materials having strong interactions between support and loaded nanoparticles with high stability in aqueous–acidic–alkaline media is desperately needed[4]. Herein, a simple and efficient method for the preparation of gold nanoparticles supported on COP-65 porous materials is reported. In this study, Au nanoparticles were stabilized with tri-dentate material (3SH) and then the porous network was synthesized by oxidative coupling reaction in order to generate Au nanoparticles embeded disulfide linked polimeric COP65 networks. This novel material was characterized by using UV, TEM, SEM, solid-state C-NMR and FTIR.

Acknowledgements: This work was supported by Selcuk University Scientific Research Fund (BAP).

References:

[1].Sakaushi,K.;Hosono,E.;Nickerl,G.;Gemming, T.; H Zhou,; S Kaskel,; J Eckert,;Nature communications 4(2013), 1485
[2].Corma,A. and Garcia,H., Chem. Soc. Rev., 37(2008), 2096.
[3].Proch,S.;Herrmannsdörfer,J.; Kempe,R.; Kern,C.; Jess,A.;Seyfarth, L. andSenker, L.,Chem.–Eur. J., 14(2008), 8204
[4]. Pachfule, P.;Kandambeth,S.;Díaz,D.D.;Banerjee, R.,Chemical Communications, 50(2014), 3169-3172.

Keywords: Gold nanoparticles, Covalent organic polymers, catalysts

PPG-P-033 [Porous Polymer and Gels]

Gold nanoparticles embedded disulfide-linked polymer networks

Halit Çavuşoğlu[1], Mehmet Şahin Ataş[2], Mustafa Selman Yavuz[1]
[1]Metallurgy and Materials Engineering Department, Selcuk University, Konya, Turkey
[2]Advanced Technology Research and Application Center, Selcuk University, Konya, Turkey

Gold (Au) nanoparticles decorated on suitable supports hold unique advantages as heterogeneous catalysts under mild conditions even at ambient temperature with high reaction rates. Although there are many reports about supported Au catalysts, which showed high catalytic activity for different systems including oxidations, hydrogenation, cyclization, rearrangement, C–C coupling reactions, the unstablility of the supports having these nanoparticles brings limitations to their uses[1]. Covalent organic frameworks (COFs) are lightweight and porous materials well-known for their applications in gas storage, catalytic supports, semiconductive and photoconductive devices. In addition to other porous materials such as charcoal, dendrimers, polymers, mesoporous silica, zeolites, MOFs, metal doped COFs have also been applied as catalysts for various organic reactions[2]. Despite the limited surface area of COFs, nanoparticle decorated COFs have been used as catalyst for Suzuki coupling, C–H activation, Knoevenagel condensation, nitro reduction, glycerol oxidation, etc.[3]. Thus, in order to overcome these issues regarding the optimal interaction, we believe that the synthesis of new materials having strong interactions between support and loaded nanoparticles with high stability in aqueous–acidic–alkaline media is desperately needed[4]. Herein, a simple and efficient method for the preparation of gold nanoparticles supported on COP-65 porous materials is reported. In this study, Au nanoparticles were stabilized with tri-dentate material (3SH) and then the porous network was synthesized by oxidative coupling reaction in order to generate Au nanoparticles embedded disulfide linked polimeric COP-65 networks. This novel material was characterized by using UV-Vis, TEM, SEM, solid-state C-NMR and FTIR analyses.

Acknowledgements: This work was supported by Selcuk University Scientific Research Fund (BAP).

References:

[1].Sakaushi,K.;Hosono,E.;Nickerl,G.;Gemming, T.; H Zhou,; S Kaskel,; J Eckert,;Nature communications 4(2013), 1485
[2].Corma,A. and Garcia,H., Chem. Soc. Rev., 37(2008), 2096.
[3].Proch,S.;Herrmannsdörfer,J.; Kempe,R.; Kern,C.; Jess,A.;Seyfarth, L. andSenker, L.,Chem.–Eur. J., 14(2008), 8204
[4]. Pachfule, P.;Kandambeth,S.;Díaz,D.D.;Banerjee, R.,Chemical Communications, 50(2014), 3169-3172.

Keywords: Gold nanoparticles, covalent organic polymers, porous materials

PPG-P-034 [Porous Polymer and Gels]

Synthesis and characterization of sodium alginate/poly(acrylic acid) interpenetrating network films and calcium ion-crosslinked sodium alginate films

Seda Bekin Acar[1], Shokat Sarmad[2], Gülten Gürdağ[3]

[1]*Department of Polymer Engineering, Faculty of Engineering, Yalova University, 77100, Yalova, Turkey*
[2]*Department of Engineering Sciences and Mathematics, Luleå University of Technology, 971 87, Luleå, Sweeden*
[3]*Department of Chemical Engineering, Faculty of Engineering, Istanbul University, 34320, Avcilar, Istanbul, Turkey*

Hydrogels are three-dimensional hydrophilic networks that are capable of absorbing large amounts of water or biological fluids. Such hydrogels can be composed of homopolymers or copolymers and are not soluble in water because of their crosslinked structure. Sodium alginate (SA) based hydrogels have gained importance recently due to their gel-forming abilities with polyvalent metal ions or glutaraldehyde that find applications in various fields [1,2].

In this study, physically and chemically crosslinked gel films based on SA were prepared. In the first part, linear SA/SA2 films and Ca2+-crosslinked SA-Ca/SA2-Ca films were prepared by solution casting and solvent evaporation technique. To determine the effect of the crosslinking on the film properties, two types of films were fabricated: SA films at two different amounts, namely thicknesses, of SA (SA and SA2) and their Ca2+-crosslinked counterparts (SA-Ca or SA2-Ca). In the second part, structural and thermal characterization of glutaraldehyde (GA) crosslinked sodium alginate/poly(acrylic acid) (SA-AA-GA) interpenetrating polymer network (IPN) films were studied. Characterizations of the films were performed by Fourier transform infrared spectroscopy (FTIR), X-ray diffraction (XRD), differential scanning calorimetry (DSC), and thermogravimetric analysis (TGA). Sodium alginate (SA) and SA-AA semi-IPN films were also prepared for comparison. The effect of initial concentration of acrylic acid on thermal properties of SA-AA semi-IPN and SA-AA-GA full-IPN films was studied [3].

References:

[1] J.P. Zhang, Q. Wang, X.L. Xie, X. Li, and A.Q. Wang, J. Biomed. Mater. Res. B, 92B, 205 (2010).
[2] N.M. Mahmoodi, B. Hayati, and M. Arami, Ind. Crop. Prod., 35, 295 (2012).
[3] S. Bekin, S. Sarmad, G. Gürdağ, Sensors and Actuators B 202 (2014) 878-892, 2014.

This study was supported by Scientific and Technical Research Council of Turkey (TUBITAK) with a grant number: MAG/110M444.

Keywords: sodium alginate-based gels, poly(acrylic acid), interpenetrating polymer network.

PPG-P-035 [Porous Polymer and Gels]

Macroporous hydrogels composed entirely of synthetic polypeptides: biocompatible and enzyme biodegradable 3D cellular scaffolds

Steven J. Shirbin, Fatemeh Karimi, Nicholas J. Chan, Daniel E. Heath, Greg G. Qiao

Department of Chemical and Biomolecular Engineering, University of Melbourne, Victoria 3010, Australia

Synthetic polypeptides are a class of bio-inspired polymers with well demonstrated biocompatibility, enzyme biodegradability and cell adhesive properties, making them promising materials for the preparation of macroporous hydrogels as 3D cellular scaffolds. Despite this, synthetic polypeptides have been scarcely utilized in the fabrication of macroporous hydrogels. In all cases, the synthetic polypeptide is only one component of the gel, where crosslinkers and other polymers are incorporated into the final gel structure, potentially altering its intended properties and making it difficult to ascertain the true effect of the synthetic polypeptide on the properties of the gel. Therefore, to effectively study the suitability of synthetic polypeptides as a material for 3D cell culture applications, and to better utilize the beneficial properties owing to these materials, the preparation of macroporous hydrogels composed entirely of synthetic polypeptides is desirable.

Three-dimensional macroporous hydrogels composed entirely of biocompatible and enzyme biodegradable synthetic polypeptides have thus been prepared. Under cryoconditions, macroporous hydrogels in the form of macroporous cryogels were prepared using a single copolymer component through direct EDC/sulfo-NHS zero-length coupling between poly(L-glutamic acid) (PLG) and poly(L-lysine) (PLL) residues on a PLG-*r*-PLL random copolypeptide chain. The resulting macroporous cryogels were found to contain large interconnected pores (≥100 μm) highly suitable for tissue engineering applications. Tuning the relative ratios of the amino acid components could result in cryogels with very different pore structures, swelling and mechanical properties, suitable for developing gels for a range of possible soft tissue engineering applications. These cryogels were shown to be enzymatically biodegradable and demonstrated excellent biocompatibility, cell attachment and cell proliferation profiles with mammalian fibroblast (NIH-3T3) cells; demonstrating the appeal of these novel cryogels as highly suitable cellular scaffolds.

Keywords: Polypeptides, macroporous hydrogel, cryogel, tissue engineering, biocompatible, biodegradable

Figure 1. Macroporous hydrogels composed entirely of synthetic polypeptides as suitable 3D cellular scaffolds

PPG-P-036 [Porous Polymer and Gels]

Synthesis and characterization of chemically crosslinked hydrogels to enhance the water holding capacity of sandy loam soil

Misbah Abdul Ghafoor[1], Ijaz Ahmad Bhatti[1], Tariq Aziz[2]

[1]*Department of Chemistry, University of Agriculture, Faisalabad, 38040, Pakistan*
[2]*2Institute of Soil and Environmental sciences, University of Agriculture, Faisalabad, 38040, Pakistan*

Drought is one of the major abiotic stresses affecting crop growth. A number of approaches are being used for increasing water use efficiency in Agriculture. The problem is more evident in soil having sand as major fraction and less frequent irrigation. A strategy is needed to improve water holding capacity of this type of soil. Keeping in view, present study has been carried out to synthesize superabsorbent hydrogels based on acrylic acid and acrylamide. The hydrogels were synthesized by solution polymerization using N- metoxy methyl methacrylamide as cross linker and potassium persulphate (KPS) as initiator. Synthesized hydrogels were characterized by Fourier transform infra red (FTIR) spectroscopy. Surface morphology of hydrogels before and after swelling was observed by Scanning electron microscopy (SEM). Some hollow spots were noted on the hydrogel surface before swelling. These were probably the sites for water absorption. An increment in the amount of crosslinker during synthesis resulted in increased water holding capacity upto some extent but further increase resulted in decreased water absorbency. Swelling capacity was observed to be increased by increasing initiator. The synthesized hydrogels absorbed and retained a large quantity of water for a long period of time which can make it suitable to improve sandy loam soil quality and water holding capacity. Further, pot culture and field trials are needed to evaluate beneficial effect of these hydrogels in improving water use efficiency and crop yield.

Keywords: Superabsorbent, Crosslinked, Swelling, Sandy loam, Crop

PPG-P-037 [Porous Polymer and Gels]

Synthesis of Template hydrophobically associating Polyacrylamide/Na+ Montmorillonite

Babak Kaffashi, Maryam Khak, Mahmood Hemmati

Department of Polymer Engineering, School of Chemical Engineering, College of Engineering, University of Tehran, Tehran, Iran

In comparison with the pure polyacrylamide, the hydrophobically modified polyacrylamides (HM-PAM) present enhanced rheological properties, and thermal stability due to the hydrophobic association via intermolecular hydrophobic interactions. On the other hand, the HM-PAM prepared with a template exhibits stronger association ability to form a long hydrophobic block along the polymer chains. Montmorillonite (MMT) is an inexpensive inorganic layered material which can make unique structure in polymers with a good heat tolerance. Therefore, its nanocomposites have been studied extensively in the recent years. The purpose of this research is to improve the thermal stability of template hydrophobically modified acrylamide based copolymers through the synthesis with Na-MMT. We used the inverse miniemulsion polymerization to synthesize the AM/DDPAB copolymers in the presence of poly(acrylic acid) as a template. In the present research, similar procedure was also used with adding 1% of Cloisite Na+ as a nanoclay. By using a magnet stirrer, a predefined amount of clay was dispersed in the aqueous phase and then the reaction was started in a stainless steel Buchi reactor equipped with a mechanical stirrer and a heating system. The rheological measurements and flow curve, were processed using a rotational rheometer at a temperature range from 10 to 90°C. The concentrations of the prepared samples were 5000 ppm. We compared the viscosity of all samples at different temperatures at this concentration. The decrease in solution viscosity was reported with increasing temperature for aqueous solutions of hydrophobically associating water-soluble polymers.

Keywords: template, hydrophobically modified, rheology modifiers, copolymerization

Effect of temperature on viscosity of the synthesized samples with concentration of 5000 ppm

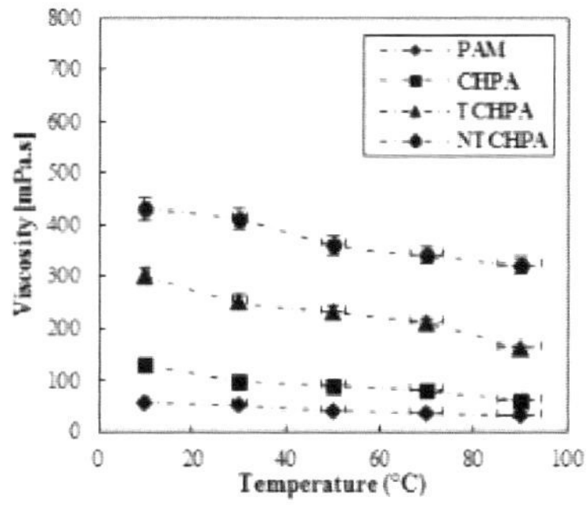

Structural parameters of the samples

SAMPLE	Mwx10-6
Polyacrylamide	3.4
Cationic Hydrophobically Acrylamide	3.1
Template Cationic Hydrophobically Acrylamide	3.3
Nano Template Cationic Hydrophobically Acrylamide	3.2

PPG-P-038 [Porous Polymer and Gels]

In situ conductive polymer synthesis in neutral, cationic, and anionic cryogels and their potential sensor applications

Sahin Demirci, Sultan Butun Sengel, Nurettin Sahiner

Department of Chemistry, Canakkale Onsekiz Mart University, Canakkale, Turkey

Conductive polymers such as p(aniline) (p(An)), p(pyrrole) (p(Py)), and p(thiophene) (p(Th)) were synthesized within the network of superporous cryogels with various charges using their corresponding monomers. The cryogels network templates from 2-hydroxyl ethyl methacrylate (2-HEMA) with neutral, 4-vinyl pyridine (4-VP) as cationically, and 2-acrylamido 2-methylpropane sulfonic acid (AMPS) as negatively charged monomers were used in the responsive cryogel synthesis under cryogenic conditions. FT-IR, TGA, and SEM techniques were employed for the characterization of bare cryogels and cryogel-conductive polymer composite. The I-V graphs of the synthesized bare cryogels and cryogel-conductive polymer composites were constructed from the current-voltage measurements using computer controlled electrometer. Potential sensor application of the prepared cryogel-conductive polymer composites against various dye molecules such as methylene blue (MB), methyl orange (MO), and organic pollutants such as 2-nitrophenol (2-NP), 4-nitrophenol (4-NP), and even against some herbicides such as paraquat and mecoprop were tested. Additionally, the prepared cryogel-conductive polymer composites were exposed to HCl and NH3 vapor for potential gas sensor applications.

Keywords: Cryogel/hydrogel, conductive polymers, conductive IPN composites, sensor.

PPG-P-039 [Porous Polymer and Gels]

Development of Graphene Oxide-Polymer Based Aerogels

Öznur Doğan, Erhan Bat

Middle East Technical University

Graphene is a single atom thick 2D material, which consists of carbon atoms arranged in a honeycomb lattice. Isolation of graphene by Geim and Novoselov in 2004 set ground for graphene and graphene based materials to become the focus of interest of many research area such as sensors, polymeric nanocomposites, membranes and biomaterials. Graphene oxide, which is a derivative of graphene, is easy to process due to its dispersibility in polar solvents. It can be obtained by chemical exfoliation and oxidation of graphite. Due to hydroxyl, carboxylic acid and epoxide groups in its structure, surface modifications can be performed and if required, it can be reduced to graphene by using thermal and chemical methods. Aerogels have open pores, large surface areas and low density and can be used in supercapacitors, electro- and photocatalysis, as adsorbants etc. They are formed by removing water and exchanging it by air in the hydrogels while preserving the porous structure. In this study, it is aimed to produce and characterize graphene oxide-polymer based aerogels. Graphene oxide was synthesized by Tour Method and ATR-FTIR, UV-Vis, TGA, SEM, TEM and AFM analyses were conducted to confirm the synthesis. Carboxylic acid groups of the graphene oxide were reacted with 1,3-diaminopropane to attach amine groups to the edges of graphene oxide then further reacted with alpha-bromoisobutyryl bromide to synthesize an ATRP initiator. ATR-FTIR, XPS and TGA analyses were performed on the products to confirm the functionalization. Poly(glycidyl methacrylate-co-styrene) was grown on graphene oxide surfaces via surface-initiated ATRP. Then, graphene oxide, polymer functionalized graphen oxide, and amine functionalized graphene oxide were used in different proportions to produce chemically crosslinked hydrogels. These hydrogels were then dried to obtain graphene oxide based aerogels using lyophilization method. The changes in the morphologies of the graphene oxide based aerogels were examined by SEM analysis.

Keywords: Graphene oxide, aerogel, surface-initiated polymerization, ATRP, hydrogel

PP-P-001 [Polymer Physics]

Diffuse reflectance spectroscopy of γ-irradiated ultra-high molecular weight polyethylene

Syed Asad Maqbool[1], Saqlain Saqib Mukhtar[1], Muhammad Zeeshan Khalid[1], Yaqoob Khan[1], Mansoor A. Baluch[1], Tariq Yasin[2], Yaqoob Khan[3], Malik Sajjad Mehmood[1]

[1]*Department of Basic Sciences, University of Engineering and Technology, 47050 Taxila, Pakistan*
[2]*Pakistan Institute of Engineering and Applied Sciences, 45650 Islamabad, Pakistan*
[3]*National Center for Physics, 44000, Islamabad, Pakistan*

The present experimental study was performed to investigate the optical properties of γ-irradiated ultra-high molecular weight polyethylene (UHMWPE) by using Diffuse Reflectance Spectroscopy (DRS). Micron size sheets of UHMWPE were prepared by using hot press with controlled heating and cooling protocol and then γ-irradiated at various doses ranging from 25-150 kGy. The samples were tested for reflectance measurements in the wavelength range from 200nm to 1200 nm, and Kubleka-Munk function F(R) = ([(1-R)]^2)/2R obtained from these measurements was subsequently used to obtain the optical properties including Urbach energy (Eu), direct and indirect energy band gaps for pristine and irradiated samples. Results obtained so far revealed that Eu has direct relation with absorbed dose even at low doses i.e. ≤100 kGy (low in the sense because above these doses saturation in cross linking density occurred within polyethylene). Similar trends were also observed for direct and indirect energy band gaps. We hope that reported results in this study might be a first step toward the new and novel approach for non-destructive radiation dosimetry applications.

Keywords: Diffuse Reflectance Spectroscopy, Erbuch Energy, Energy band gaps, UHWMPE

PP-P-002 [Polymer Physics]

Effects of Blend Composition and Crystallization Temperature on the Spherulitic Morphologies of Crystalline/Crystalline Poly(ethylene suberate)/ Poly(ethylene oxide) Polymer Blends with Similar Thermal Properties

Zhaobin Qiu

State Key Lab of Chemical Resource Engineering, Beijing University of Chemical Technology, Beijing, China

In this work, we studied the effects of blend composition and crystallization temperature on the spherulitic morphologies of melt-miscible crystalline/ crystalline polymer blends of poly(ethylene suberate) (PESub) and poly(ethylene oxide) (PEO) with similar thermal properties. The spherulites only from the major component were found in the 85/15 and 15/85 samples (PESub and PEO in weight ratio). The spherulites from both of the components were observed in the mediate blend compositions from 70/30 to 30/70; moreover, the nonbanded PEO spherulites continued to crystallize inside the banded PESub spherulites after they contacted each other, thereby forming the interpenetrating spherulites in most cases. Depending on crystallization temperature, the 70/30 blend exhibited two different crystalline morphologies. The formation of interpenetrating spherulites occurred, when crystallization temperature was 42.5 oC and below. However, a novel crystalline morphology, i.e., the surrounding spherulites, was observed, when crystallization temperature was 45 oC and above. The novel crystalline morphology was observed for the first time in miscible crystalline/crystalline polymer blends. The features of the surrounding spherulites were summarized, and the conditions for the formation of them were also discussed. The effect of crystallization temperature on the formation of the interpenetrating spherulites or the surrounding spherulites was further discussed on the basis of the diffusion length in the blends.

Keywords: Polymer blends, Crystallization, Morphology, Miscibility

Figure 1

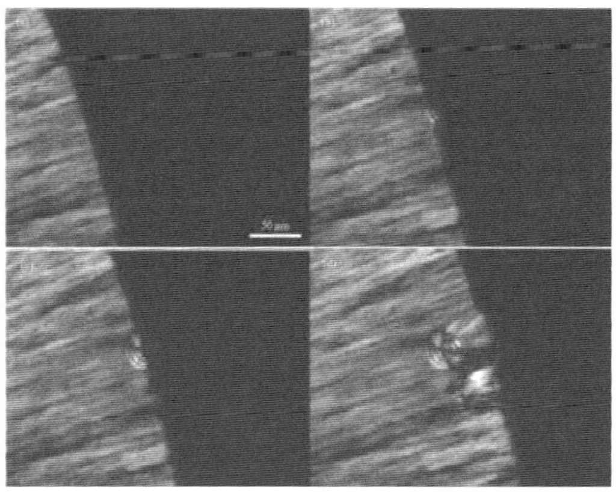

The formation process of interpenetrating spherulites in a 30/70 PESub/PEO blend crystallized at 47.5 oC at different crystallization times: (a) 360 s, (b) 366 s, (c) 370 s, and (d) 379 s (same magnification with bar = 50 μm).

PP-P-003 [Polymer Physics]

Volume phase transitions in ionic hydrogels

Demet Aktaş, Gökhan Uçar, Başar Turan

Istanbul Technical University Department of Physics Maslak Istanbul Turkey

Steady-state fluorescence (SSF) technique was employed for studying swelling and drying of poly(acrylamide-co-acrylic acid) (P(AAm-co-AAc)) composite gels for low and high pH. Disc shaped composite gels were prepared by free-radical crosslinking copolymerization of P(AAm-co-AAc). N, N'- methylenebis (acrylamide) (BIS) and ammonium persulfate (APS) were used as crosslinker and initiator, respectively. Pyranine 4 (4sPy) was introduced as a fluorescence probe. P(AAm-co-AAc) composite gels dried before using for swelling experiments. Fluorescence intensity of 4sPy was monitored during in situ swelling and drying processes of composite gels. It was observed that fluorescence intensity values decreased as swelling is proceeded. Li-Tanaka equation was used to determine the swelling time constants, Tτ and cooperative diffusion coefficients, D from intensity variations during the swelling processes. It was also observed that fluorescence intensity values increased as drying is proceeded. Gravimetrical and volumetrical experiments were also performed. It was shown that swelling time constants, Tτ decreased, diffusion coefficients for swelling process, DF increased and diffusion coefficients for drying process DG decreased as the pH values are increased.

Keywords: Swelling, drying, fluorescenc e, pH

PP-P-004 [Polymer Physics]

Effective chain length of additives on electrical conductivity of poly(3,4-ethylenedioxythiophene):poly(styrenesulfonate) films

Hyuck Sik Wang, Jae Hyun Jung, Seok Hyeon Kim, Kigook Song

Department of Advanced Materials Engineering for Information and Electronics, Kyung Hee University, Yongin, South Korea

Poly(3,4-ethylene dioxythiophene) (PEDOT) doped with Poly(styrenesulfonate) (PSS) is quite promising as a transparent electrode material owing to its enormous advantages, such as its high electrical conductivity, optical transparency, and excellent mechanical flexibility. Commercially available PEDOT:PSS has a conductivity of less than 1 Scm-1, which is lower than that of ITO by about three orders of magnitude. In 2002, Kim and co-workers (Synth. Met. 126, 2002, 311) reported the conductivity enhancement of PEDOT:PSS by adding a polar organic solvent, such as DMSO, DMF or THF into PEDOT:PSS aqueous solution. Since then, there are a lot of research efforts on enhancing its electrical conductivity to the level of ITO through chemical and/or physical processing. We investigated the effect of the different concentrations and molecular weight of the additive on the conductivity and other properties of PEDOT:PSS composite thin film using polyethylene glycols (PEGs) and water-soluble polymers. The electrical conductivity of the secondary dopant /PEDOT:PSS thin films was measured by the four-probe method. The conductivity enhancement depends on both the molecular weight and concentration of additives used. With the increase of additive concentrations, the electrical conductivity sharply increased to reach a maximum and then slowly decreased down. It is proposed that three mechanisms contribute to improved conductivity; conformational changes in the polymer from coil to expanded, phase separation of PEDOT:PSS grains and charge screening. Detailed studies on the structures and morphologies of these modified PEDOT:PSS thin films, in comparison to that of PEDOT:PSS without additive were characterized primarily by FTIR, Raman, XPS, and AFM to investigate the underlying mechanisms..

Keywords: conductive polymer, PEDOT:PSS, secondary dopant, additive, electrical conductivity

PP-P-005 [Polymer Physics]

Study of Optical Properties and Electrical Conductivity of Polymer Latex/Carbon Nanotube Composites

Okan Yıldız, Saziye Ugur

Department of Physics Engineering, Istanbul Technical University, Istanbul, Turkey

In the present work, the effects of multiwalled carbon nanotubes (MWCNT) on the optical, film formation and electrical conductivity properties of polystrene (PS) latexes were investigated by using the photon transmission technique and electrical conductivity measurements. At first, the PS latex dispersion was mixed with different amounts of MWCNTs, varying in the range of (0-20) wt%. Then, these composite films were prepared from these dispersion by drop casting method on glass substrates and drying in an oven at 45 0C. After drying, these composite films were separately annealed above the glass transition temperature (Tg) of PS, ranging from 100 to 250 0C, for 10 min. The optical and film formation properties of PS/MWCNT films were monitored by measuring transmitted light intensity, Itr after each annealing step. For (0-7.5) wt% MWCNT content films, Itr intensity increased with annealing temperature. However, above 7.5 wt% MWCNT no change was observed in Itr with annealing. The increase in Itr during annealing was explained by void closure and interdiffusion processes[1], respectively. The corresponding activation energies were measured and found to be independent of MWCNT content in the film. SEM images demonstrated that PS particles undergo complete film formation with annealing forming a continuous phase on the top surface of the film and MWCNTs sink to the bottom of the film as bundles. The surface conductivity of composites annealed at 250 0C showed a substantial increase above 1.4 wt%MWCNT which was defined as the percolation threshold of conductivity[2]. Above this fraction, the connection of the MWCNT bundles in the bottom forms a conductive network.

[1] S. Ugur, A. Elaissari, and O. Pekcan, J. Colloid Interface Sci. 263, 674 (2003).
[2] M. Sahimi, Applications of Percolation Theory London: Taylor and Francis (1994).

Keywords: Polymer,Latex,Carbon,Nanotube,Composites

PP-P-006 [Polymer Physics]

One-dimensional crystallization of 20-nm-width polymer nanorods

Jaime Martin[1], Aurora Nogales[2], Natalie Stingelin[1], Carmen Mijangos[3]

[1]Department of Materials, Imperial College London, London, UK
[2]Instituto de Ciencia y Tecnología de Polímeros, CSIC. Madrid, Spain
[3]Instituto de Estructure de la Materia, CSIC, Madrid, Spain

Crystallization of polymers under confinement has emerged as a new tool to face classical problems of the crystallization of polymers, such as the early stages of the liquid-solid transformation or the presence of mesophases at the crystal growth front. Moreover, these investigations are essential to gain control on the functionality of polymer nanostructures, because mechanical, optical, and electrical properties are directly correlated with the microstructure. Here we show a new crystallization strategy that allow the fabrication of completely oriented crystallization of ultra-small polymer rods -20 nm in diameter- via the induction of heterogeneous nucleation centers exclusively at one of the tips of nanorods, while the formation of additional nucleation centers along the nanrod can be ruled out due to temperature conditions. Then, crystals unidirectionally propagate along the nanorod, which leads to nanorods with high crystallinity and uniaxial orientation. Interestingly, the new strategy also allows studying for the first time the kinetics of the unidirectional polymer crystal growth under confinement. Thus, we show that the crystallization of the polymer nanorods shows the time dependency of a diffusion process, i. e. $t^{0.5}$. Moreover, the linear crystallization rate value of the confined flat on lamellae is of 2.7×10^{-7} cm/s. Since the polymer nanorods crystallize exclusively through of the growth of crystals heterogeneously nucleated at one of their tips, and just a single flat on lamella can be accommodated in a 20 nm in diameter pores due to space reasons, what we present in this paper may be the study of the crystallization of a single lamella within each pore, which would mean that single-crystalline nanorods are obtained from the melt.

Keywords: Crystallization, Nanotechnology, Confinement, Physics

Unidirectional, Dynamic and Isothermal crystallization of PEO in 20 nm pores

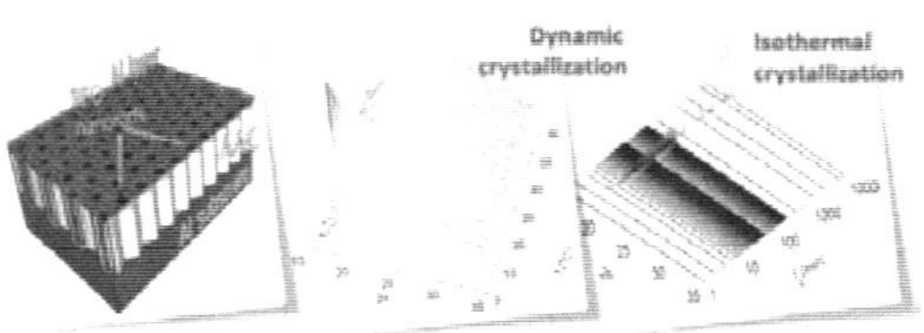

PP-P-007 [Polymer Physics]

Motion of Impurity ions caused by Inhomogeneity of Microstructure of Polymer Solid

Yuichi Anada

Department of Business and Information Systems, Hokkaido Information University, Ebetsu, Japan

In the non-crystalline part of melt-crystallized polymers, free spaces with inhomogeneous size exist between crystallites. This inhomogeneity of free spaces results from a thermal motion of the polymer molecules with amorphous conformation under the spatial restriction made by crystallite morphology. The impurity ions mixed in processing in the crystalline polymers move through these inhomogeneous free spaces. Consequently, this motion of ions reflects the inhomogeneous microstructure of non-crystalline part. In fact, in previous study of our laboratory, plural kinds of motion of impurity ions, which are related with the inhomogeneity of the non crystalline part, are observed through the electric modulus as a function of frequency in low frequency region below 10 Hz. Furthermore, the conductivity relaxation times of the ionic motion analyzed for poly(vinyl chloride) plasticized with dioctyl phthalate, isotactic polypropylene, low density polyethylene and poly(ethylene telephthalate) reflect the difference of the microstructure of non-crystalline part among these polymers. In the present study, the ionic motion of impurity ions in poly(vinylidene fluoride) (PVDF) is investigated as continuation of this study. Electrical measurement was made by the method of broadband dielectric spectroscopy by using the apparatus of Thermal analysis DEA2970 of TA instrument at frequencies between 0.01 and 10000 Hz. The interdigital sensor was used as an electrode. The thickness of the film of PVDF is 0.1mm. Frequency dependence of the loss permittivity of PVDF shows the characteristic of dc conduction below about 10000Hz in frequency at 162 degree centigrade. The electric modulus-frequency curve shows the characteristic of single conductivity relaxation. Namely, this characteristic shows the Debye type dispersion. The conductivity relaxation time analyzed from this curve is 0.00003 s at 162 degree centigrade. In a plot of conductivity relaxation time of PVDF together with the other polymers against permittivity shows a tendency downward with increasing permittivity.

Keywords: ionic motion, poly(vinylidene fluoride), broadband dielectric spectroscopy, electric modulus, conductivity relaxation time

PP-P-008 [Polymer Physics]

Investigation on glass transition temperature of electrospun PVAc fibers

Konrad Domanski[1], Ece Yapaşan[2], Tuğba Isık[2], Nesrin Horzum[3], Mustafa Muammer Demir[2]

[1]University of Cambridge
[2]Izmir Institute of Technology
[3]Izmir Katip Çelebi University

Glass transition temperature (*Tg*) is a second order transition that may be seen in semicrystalline and amorphous polymers. *Tg* has been examined in the literature for thin polymeric films. Since these films are heavily affected by substrates, the preparation of thin films free from substrate is intended as a model system in this study. Electrospun fibers appears out to be a potential technique for the formation of such model system. Electrospun fiber mats contain submicron diameter nonwoven films. Polyvinyl acetate (PVAc) has *Tg* value close to room temperature (45 °C). We focused on *Tg* measurements for electrospun fibers usinf Differential Scanning Calorimeter and their comparison with bulk materials to understand the presence of any confinement phenomenon. The electrospun fibers were prepared at different average fiber diameters (AFD). Results showed that fibrous materials show greater Tg than the corresponding bulk materials and thin films.

Keywords: Confinement effect, Electrospinning, Glass Transition Temperature

PP-P-009 [Polymer Physics]

Polyacrylamide- Graphene Oxide Composites: Optical and mechanical properties with various GO contents

Gülşen Akın Evingür[1], Önder Pekcan[2]

[1]Faculty of Engineering, Piri Reis University, Tuzla, İstanbul
[2]Faculty of Engineering and Natural Sciences, Kadir Has University, Cibali, İstanbul

Polyacrylamide (PAAm) - Graphene Oxide (GO) composites were prepared by free radical crosslinking copolymerization. Optical and mechanical properties of the composites were performed by fluorescence, ultra violet spectroscopy (UV), and compression measurements, respectively. Firstly, the gelation, energy band gap of PAAm- GO composites were characterized by the steady state fluorescence and UV technique, respectively. Critical exponents of gelation process were modeled by Classical and Percolation theory, and also energy band gap of the composite was calculated. Secondly, the compression elastic modulus was calculated by the theory of rubber elasticity. Lastly, the results show that GO can be well dispersed in the PAAm- GO composites and consequently contributed to a significant improvement of their optical and mechanical properties, respectively.

Keywords: composites,graphene oxide, fluorescence,energy band gap, elasticity

PP-P-010 [Polymer Physics]

Physical, Mechanical and Adsorption Kinetic Properties of Graphene Oxide-Chitosan Composites

Gülşen Akın Evingür[1], Nilay Kahya[2], Semira Bener[3], Bedia Erim Berker[2]

[1]*Piri Reis Universitty, Tuzla, Istanbul-TURKEY*
[2]*Istanbul Technical University, Maslak, Istanbul-TURKEY*
[3]*Kultur University, Istanbul-TURKEY*

Chitosan is polysaccharide that composed of D-glucosamine and N-acetyl-D-glucosamine monomers linked with beta-1, 4 glycosidic bonds. Due to its biocompatibility and biodegradability, chitosan biopolymer was chosen in present study. Recently, graphene oxide have been used as filler compound for polymeric systems. Graphene oxide (GO) alters thermal, mechanical, electrical characteristics of materials. In this study, chitosan and graphene oxide-chitosan composite films were prepared. Then the films were cross-linked in 1% and 20% (w/v) sodium sulfate solutions. Characterization of chitosan and GO/chitosan composites was investigated with FTIR, thermogravimetric analysis and mechanical test, respectively. It was shown that thermal strength of the films increases with various contents of graphene oxide and sodium sulfate don't have any considerable effect on thermal behavior of chitosan films. Also mechanical properties of films was analyzed with tensile testing machine. Graphene oxide has an impact on modulus of chitosan films with considering sodium sulfate in elastic region. Pyranine is a fluorescent dye and it can easily soluble in water. Adsorption was chosen as an application area for sodium sulfate crosslinked graphene oxide-chitosan composite films. Pyranine dye adsorption experiments were performed by immersing chitosan films into pyranine aqueous solution. Dye adsorption to the composite films was followed by fluorescence spectroscopy. Adsorption kinetic models was applied to experimental results.

Keywords: chitosan, graphene oxide, fluorescence, thermogravimetric,adsorption,elasticity

PP-P-011 [Polymer Physics]

The universality on the gelation of NIPA-GO composites

Gülşen Akın Evingür, Batuhan Ünver

Piri Reis University

Graphene oxide (GO) has been used to prepare Poly (N-isopropylacrylamide) (PNIPA)-GO composites with improved to the universality on the gelation of the composite by a free radical copolymerization with various GO contents in the range between 0-50 μl. GO is a significant filler for the enhancement of electrical, mechanical and thermal properties of composite materials. PNIPA gel is a typical temperature-sensitive gel exhibiting volume phase transition. The temperature sensitivity of PNIPA gels has been studied both for its fundamental interest and technological application. Gelation process of NIPA-GO composites was characterized by Steady State Fluorescence Spectroscopy. The gelation and its universality were monitored and tested as a function of GO contents. The exponents α and β for the weight average degree of polymerization, and the gel fraction, near the percolation threshold were calculated.

Keywords: Graphene oxide, gelation, Poly (N-isopropylacrylamide), composites

PP-P-012 [Polymer Physics]

The swelling effect on the elasticity of NIPA-GO Composites

Gülşen Akın Evingür, Özgül Öztürk

Piri Reis University

N-isopropyl acrylamide (NIPA) has been widely investigated as it shows attractive volume exchange through absorbing or excluding water, besides its biocompatibility with a high content of water. However, this hydrogel often shows a poor mechanical performance. NIPA -Graphene oxide (GO) composites are prepared via free radical cross-linking copolymerization with different amounts of GO varying in the range between 0 to 40μl per 1 mole NIPA. The elasticity of NIPA-GO composites is performed by compressive test. It is determined that GO content affects elasticity of them, revealing a change in the modulus and resilience of the composites.

Keywords: N-isopropyl acrylamide (NIPA), composites, graphene oxide,elasticity

RDPS-A-P-001 [Controlled/Living Polymerization]

RAFT Polymerization of Styrene mediated by Naphthyl Functionalized RAFT Agents

Gabriel J Summers, Thembinkosi S Mdletshe, Carol A Summers

Department of Chemistry, University of South Africa, PO Box 392, Pretoria, 0003 South Africa

Reversible addition-fragmentation chain transfer (RAFT) polymerization provides an efficient polymer synthesis method for the preparation of polymers with predictable number average molecular weight, narrow molecular weight distribution, well-defined chain end functionality and polymer topology [1]. Three novel naphthyl functionalized RAFT agents, ethyl 1-naphthylmethyl trithiocarbonate, butyl 1-naphthyl trithiocarbonate and dodecyl 1-naphthyl trithiocarbonate, have been prepared and employed in RAFT polymerization reactions. The RAFT polymerization of styrene mediated by the naphthalene functionalized RAFT agents with AIBN as initiator at 75 °C proceeds via controlled/living polymerization processes to produce quantitative yields of naphthalene functionalized polymers with the naphthyl group introduced at the α-terminus of the polymer chain. The different RAFT agents display good control of the polymerization process at different RAFT agent concentrations. Polymer kinetics measurements show that the polymerization reaction follows first order rate kinetics with respect to monomer consumption and the number average molecular weights increase linearly with the monomer conversion to produce polymers with narrow molecular weight distributions. Percentage monomer conversion for each reaction was monitored by ^{1}H NMR spectroscopy. The naphthalene chain transfer agents and the functionalized polymers were characterized by ^{1}H NMR spectroscopy, FTIR spectroscopy, size exclusion chromatography, ultra-violet visible spectroscopy and fluorescence spectroscopy

Keywords: RAFT polymerization, Naphthalene, Polystyrene

RDPS-A-P-002 [Controlled/Living Polymerization]

RAFT Polymerization of Styrene mediated by Diphenyl Functionalized RAFT Agents

Gabriel Jeffrey Summers, Teboho Simon Motsoeneng, Carol Ann Summers

Department of Chemistry, University of South Africa, Pretoria, South Africa

Reversible addition-fragmentation chain transfer (RAFT) polymerization provides an efficient polymer synthesis method for the preparation of polymers with predictable number average molecular weight, narrow molecular weight distribution, well-defined chain end functionality and polymer topology [1]. Three novel diphenyl functionalized RAFT agents, benzhydryl 2-(ethylsulfanylthiocarbonylsulfanyl)propanoate, benzhydryl 2-(butylsulfanylthiocarbonylsulfanyl)propanoate, benzhydryl 2-(dodecylsulfanylthiocarbonylsulfanyl)propanoate, have been prepared and employed in RAFT polymerization reactions. The RAFT polymerization of styrene mediated by the diphenyl functionalized RAFT agents with AIBN as initiator at 80 0C proceeds via controlled/living polymerization processes to produce quantitative yields of the diphenyl functionalized polymers with the diphenyl groups introduced regiospecifically at the α-terminus of the polymer chain. The different RAFT agents display good control of the polymerization process at different RAFT agent concentrations. Polymer kinetics measurements show that the polymerization reaction follows first order rate kinetics with respect to monomer consumption and the number average molecular weights increase linearly with the monomer conversion to produce polymers with narrow molecular weight distributions. Percentage monomer conversion for each reaction was monitored by 1H NMR spectroscopy. The diphenyl functionalized chain transfer agents and the functionalized polymers were characterized by 1H NMR spectroscopy, FTIR spectroscopy, size exclusion chromatography, ultra-violet visible spectroscopy and fluorescence spectroscopy.

Keywords: RAFT Polymerization, diphenyl functionalized RAFT agents, Polystyrene

RDPS-A-P-003 [Controlled/Living Polymerization]

One-step synthesis of a mid-chain functional macrophotoinitiator of a polystyrene-poly(ε-caprolactone) diblock copolymer via simultaneous ATRP and ROP using a dual-functional photoinitiator

Mustafa Degirmenci, Mustafa Durgun, Zafer Uyar

Harran University, Science Faculty, Department of Chemistry, Sanliurfa 63190, Turkey

Structurally well-defined block copolymers are an important class of polymeric materials owing to their technological applications and extraordinary properties. Recently, one-step processes using dual initiators successfully synthesized well-defined block copolymers through several combinations of polymerization techniques, such as NMP with ROP, ATRP) with NMP, RAFT with ROP as well as ATRP with ROP. In this study, we have synthesized a well-defined mid-chain functional macrophotoinitiator of a polystyrene-poly(ε-caprolactone) diblock copolymer (PSt-PI-PCL) using a novel dual-functional photoinitiator via the simultaneous atom transfer radical polymerization (ATRP) and ring-opening polymerization (ROP) in one step at 110 °C. The dual-functional photoinitiator (Br-PI-OH), possessing a bromine group on one end and a hydroxyl group on the other end, was synthesized by reaction of 2-hydroxy-4'-(2-hydroxyethoxy)-2-methylpropiophenone (HO-PI-OH)) with 2-bromopropanoyl bromide. The two distinct polymerization reactions initiated by Br-PI-OH proceeded in a controlled manner without mutual interference. The reaction pathway for this process is shown in Figure 1.

Keywords: Atom transfer radical polymerization (ATRP), Block copolymer, Dual Photoinitiator, Macrophotoinitiator, One-step polymerization, Ring-opening polymerization (ROP).

Figure 1.

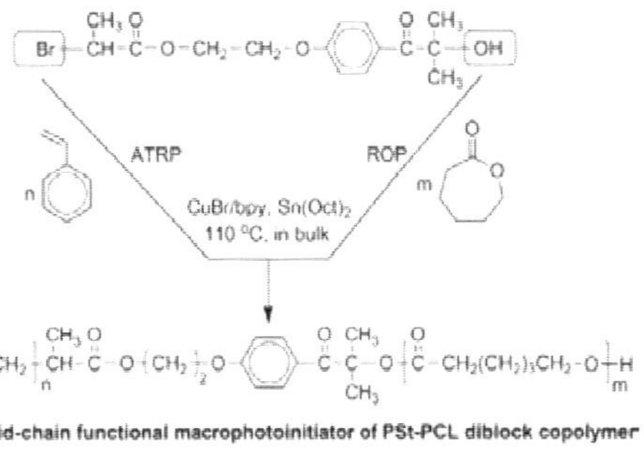

One-step synthesis of mid-chain functional PSt-PCL diblock copolymer

RDPS-A-P-004 [Controlled/Living Polymerization]

Synthesis of an A2B2 Type Star-shaped Copolymer by Combination of ROP, Click Chemistry and Photoinitiated Polymerization

Zafer Uyar, Mustafa Değirmenci, Bahattin Abacı
Department of Chemistry, Harran University, Şanlıurfa, Turkey

In this study, an A2B2 type star-shaped copolymer was synthesized by combination of ring opening polymerization (ROP), click chemistry and photoinitiated free radical promoted cationic polymerization. First a new well-defined benzoin group end-functionalized poly(ε-caprolactone) macrophotoinitiator [(PCL)2-(PI)2] was synthesized by the combination of ring opening polymerization (ROP) and click chemistry. The ROP of ε-CL monomer in bulk at 110 °C, by means of a dihydroxyl functional initiator, namely 2,2-Bis(bromomethyl)-1,3-propanediol, in conjunction with stannous-2-ethylhexanoate, (Sn(Oct)2), yielded a well-defined two-armed PCL with two bromine end-chain groups [(PCL)2-Br2]. Simple azidation of the bromine groups at the chain ends gave bisazido functionalized poly(ε-caprolactone) [(PCL)2-(N3)2]. Separately, an acetylene functionalized benzoin photoinitiator (PI-alkyne) was synthesized as a precursor from benzoin and propargyl bromide. Then the click reaction between (PCL)2-(N3)2 and this precursor was performed by Cu(I) catalysis. The spectroscopic studies revealed that low-polydispersity macrophotoinitiator with two poly(ε-caprolactone) chains and two benzoin photo functional groups at the chain ends (PCL)2-(PI)2) was obtained. In the last step, the obtained macrophotoinitiator was used as a precursor in photoinitiated free radical promoted cationic polymerization to synthesize an A2B2 type star-shaped copolymer consisting of poly(ε-caprolactone) (PCL, as A block) and poly(cyclohexene oxide) (PCHO, as B block).

Keywords: Star-shaped polymers, ROP, Click Chemistry, Caprolactone, Photoinitiated polymerization

RDPS-A-P-005 [Controlled/Living Polymerization]

Structure-Acitivity Relationships in Squaramide Catalyzed Ring Opening Polymerization of Lactide: The Effect of Tertiary Amine Group

Ali Rostami, Shaghayegh Ahmadi, Elahe Sadeh

Department of Chemistry, Shahid Beheshti University, Tehran, Iran

Inspired by synergistic cooperation of multitude functional groups present in effective and specific enzyme-mediated catalysis, devising bifunctional organocatalysts enabling dual activation of substrates has been receiving a significant attention. Of the existing controlled/living polymerization techniques, organocatalytic ring-opening polymerization (ROP), pioneered by Hedrick and co-workers (Angew. Chem., Int. Ed., 2001, 40, 2712–2715) is characterized by its ability to produce biodegradable polymers with exceptionally low polydispersity and excellent end-group fidelity that are entirely free from metal residues suitable for biomedical and microelectronic applications. Squaramide is a unique hydrogen bond donor motif that is commonly used as a synthetic anion receptor due to its special recognition properties (Angew. Chem. Int. Ed. 2011, 50, 2059–2062). Since the seminal work of Rawal in 2008 (J. Am. Chem. Soc., 2008, 130, 14416–14417), squaramides with different skeletons have been synthesized and applied as efficient bifunctional organocatalysts in various asymmetric transformations. Owing to their duality, rigidity, increased canted H-bond angle and acidity, squaramides often exhibit faster reaction rates and higher degree of stereo-control over thiourea counterparts (Chem. Eur. J. 2011, 17, 6890–6899). Here, a series of tertiary aminosquaramide bifunctional organocatalysts carrying different alkyl groups were synthesized and further applied to the ring opening polymerization of L-lactide. The formed polylactides exhibit narrow polydispersity and high end-group fidelity featuring a living polymerization process. Details of aminosquaramides preparation, their structure-activity relationship towards ROP, kinetic measurements and chain-extension experiments will be discussed.

Keywords: Ring-Opening Polymerization, Aminosquaramides, Hydrogen Bonds, Bifunctional Organocatalysis, Polylactide

Graphical Abstract

RDPS-A-P-006 [Controlled/Living Polymerization]

Simplifying the synthesis of polypeptoids

Afroditi Doriti[1], Sarah Mae Brosnan[1], Helmut Schlaad[2]

[1]*Department of Colloid Chemistry, Max Planck Institute of Colloids and Interfaces, Potsdam, Germany*
[2]*Institute of Chemistry, University of Potsdam, Potsdam, Germany*

Polypeptoids, i.e. poly(*N*-alkylglycine)s, are promising materials for pharmaceutical applications as they are biocompatible and can be degraded by reactive oxygen species under physiologically relevant conditions. Additionally, compared to their isomers, polypeptides, they are soluble in a larger variety of solvents. However, the tedious synthesis and the sensitivity of *N*-alkylglycine *N*-carboxyanhydride (NCA) monomers hinders their more widespread use. Herein, we report a method that simplifies their synthesis, avoiding the direct use of NCA. Polysarcosine is synthesized using the air and moisture stable *N*-phenoxycarbonyl-*N*-methylglycine (Poc-Sar), which reacts to form the respective Sar-NCA *in situ*. The formation of the NCA and subsequent polymerization, initiated by a primary amine, is accelerated by a non-nucleophilic tertiary amine base (see Scheme). The method produces polysarcosines with molar masses of 3.6-20 kg/mol and dispersities of ~1.1 within 24 hours.

Keywords: polypeptoids, polysarcosine, NCA

Polysarcosine synthesis from the air and moisture stable N-phenoxycarbonyl-N-methylglycine (Poc-Sar)

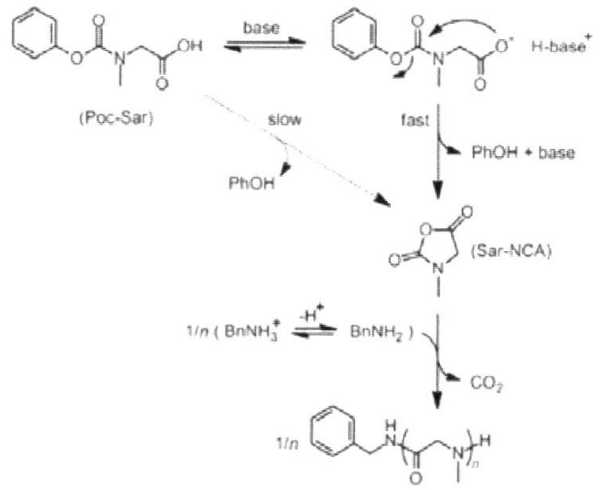

RDPS-A-P-007 [Controlled/Living Polymerization]

Surface Properties of Silicon Disc Coated with Fluoro-Based Polymers

Mehtap Evci, Tuncer Caykara

Department of Chemistry, Gazi University, Ankara, Turkey

Due to their wide range of applications such as waterproof textiles, separation membranes, self-clening property, corrosion resistance provided by oleophobic surfaces have a great interest for the scientific and the industrial communities. The hydrophilic property of a surface enhanced through chemical modification reduces surface energy. Fluorine containing polymers are the best candidates to synthesis materials which exhibit such behaviour because of the presence of strong and highly polar C-F bond and their low surface energy In addition, they exhibit excellent resistance to heat, chemicals, fuel and oil. [1-4]. In the literature, the reversible addition-fragmentation chain transfer polymerization (RAFT) which is an example of living/controlled radical polymerization methods of fluoroacrylates is very limited. In this study, poly(2,2,3,3,4,4,5,5,6,6,7,7-Dodecafluoroheptyl acrylate), poly(3,3,4,4,5,5,6,6,7,7,8,8,9,9,10,10,11,11,12,12,12-Heneicosafluorododecyl acrylate) and poly(pentafluorophenyl acrylate) were synthesised via the interface-mediated RAFT polymerization. Then, the silicon discs were coated with fluoro-based polymers by using spin coating technique. The surfaces coated with fluoro-based polymers subjected to annealing at 120 ° C and freeze drying at -37 ° C. After these processes, the surface compositions and other characteristic features of the fluoro-based polymer films were investigated by elipsometry, atomic force microscopy, X-ray photoelectron spectroscopy, grazing angle Fourier transform infrared spectroscopy and contact angle measurements.

[1] J.S. Liang, D.Z. Wang, K.Y. Liu, L. Chen, Preparation of stable superhydrophobic film on stainless steel substrate by a combined approach using electrodeposition and fluorinated modification, Appl. Surf. Sci. 293 (2014) 265–270.
[2] C.H. Xue, Y.R. Li, P. Zhang, J.Z. Ma, S.T. Jia, Washable and wear-resistant superhydrophobic surfaces with self-cleaning property by chemical etching of fibers and hydrophobization, ACS Appl. Mater. Interfaces 6 (2014) 10153–10161.

[3] J. Neil, G.M. Shirtcliffe, M.I. Newton, Y. Zhang, Superhydrophobic copper tubes with possible flow enhancement and drag reduction, ACS Appl. Mater. Interfaces 6 (2009) 1316–1323.
[4] S. Taleb, T. Darmanin, F. Guittard, Elaboration of voltage and ion exchange stimuli responsive conducting polymers with selective switchable liquid-repellency, ACS Appl. Mater. Interfaces 6 (2014) 7953–7960.

Keywords: Surface properties, fluoro-based polymers, living/controlled radical polymerization

Preparation of the surfaces coated with fluoro-based polymers

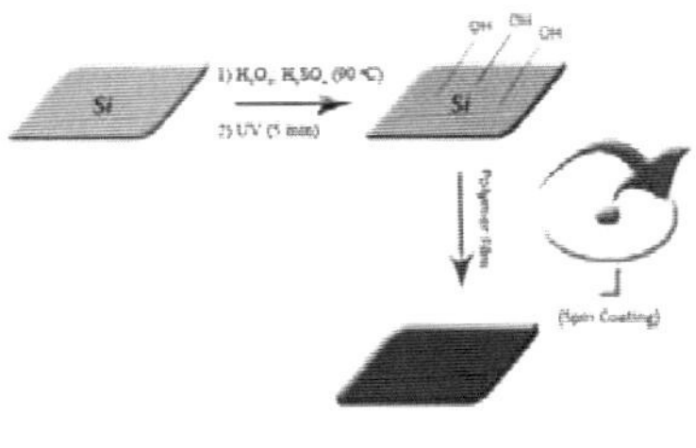

RDPS-A-P-008 [Controlled/Living Polymerization]

Preparation of Polyhedral Oligomeric Silsesquioxane (POSS) Materials with Fluorogenic Boronic Acid

Ahmet Gultek, Selda Sezer, Turgay Seckin, Fadime Nilufer Kivilcim

Department of Chemistry, Inonu University, Malatya, Turkey

Boronic acid containing polymers have many potential applications such as electrolytes (blue emissive), self-healing materials, fire retardants and precursors for functional polyolefines. Boronic acid containing macromolecules are among the stimuli responsive polymers and they have drug release properties under the influence of external field. They are used in medicinal applications due to self-assembly properties occurring by external forces and they are especially used in diagnosis of diabetes and regulation of blood glucose or glycoprotein levels. Although it is known that these type of polymers have many potential applications, it is very important to control topology, molecular weight and functionality of these polymers. Also it is necessary to control chemical composition and physical nature of the polymers for various applications. Fluorogenic polymers are mainly importanat to determine the drug releasing region where targeting is more important of than the releasing. In this study, amino functional POSS-(NH2)8 was used as a core material to prepare the targeted material Poly (POSS-(APBA-PA)) was prepared starting with 3-(2-azido-acetylamino) phenyl boronic acid (APBA) and propargyl acrylate (PA). Topology of the final material was controled by the combination of the functionality and molecular weight of the polymer. We found that morphology of the material that is stable in physiological conditions. Stimuli-response properties of fluorogenic groups to cis-diol were studied by altering properties such as shape, permeability or the number of functional groups.

Keywords: Fluorescent properties, boronic acid, POSS, ATRP, functional polymer

Synthesis of POSS-(APBA-PA) hybrid polymer

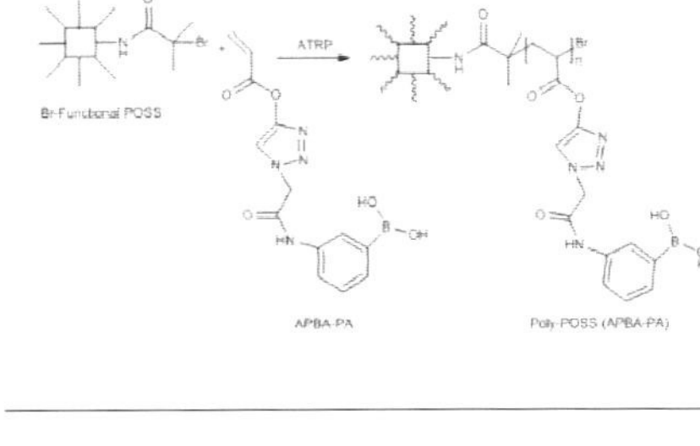

RDPS-A-P-009 [Controlled/Living Polymerization]

Preparation of stereoblock olefin copolymers and their model blends with high impact polypropylene

Anatolij Sokolohorskyj[1], Pavla Doubravová[1], Zdeněk Buráň[2], Jan Merna[1]

[1]Department of Polymers, University of Chemistry and Technology, Prague, Czech Republic
[2]Polymer Institute Brno, Unipetrol RPA, s. r. o., Brno, Czech Republic

Our aim is to prepare a model block copolymer *i*PP-*block*-EPR and its blends with *i*PP and EPR, which can be helpful for understanding of multiphase system like high impact polypropylene (HI PP). For this purpose three types of catalysts which enable stereospecific and simultaneously living polymerization of propylene were chosen and synthesized. Modified α-keto-β-diimine nickel complexes (1) [Bazan, *Angew. Chem. Int. Ed.*, 2009, 48, 1089] were prepared, however, they were found only slightly isospecific for propylene polymerization even below -30 °C. Amino-bis(phenolate) zirconium complexes (2) afforded highly isotactic PP with very narrow dispersity ($Đ$ = 1.02) as well as random ethylene-propylene copolymer (EPR, $Đ$ = 1.78). Despite short block copolymer *i*PP-block-PE (10 kg/mol *i*PP block + 8 kg/mol (PE block) was already reported [Busico, *Macromolecules*, 2004, 37, 8201], our attempt to prepare high molar mass *i*PP-*block*-EPR by this catalyst failed. We prepared the most defined *i*PP-*block*-EPR copolymers using α-diimine nickel complex (3) for which chain-walking isomerization (CHW) is a typical undesired side reaction way causing a formation of irregular PP structure. When CHW is suppressed (below -50 °C) the asymmetrical complex (3) can produce moderately isotactic PP [Coates, *J. Am. Chem. Soc.*, 2009, 127, 13770]. Taking an advantage of CHW the second block ("EPR-like") can be prepared just by switching the polymerization temperature to 0 °C with no need to use ethylene comonomer. *i*PP-*block*-EPR copolymers with Tm ≈ 130 °C and Tg ≈ -40 °C were prepared having molar mass 20 – 155 kg/mol. Model blends of these block copolymers with HI PP were prepared in micro-extruder and their morphology and mechanical properties were compared to commercial *i*PP and HI PP samples.

This work was supported from specific university research (MSMT No 20/2016) and from TA CR (TA03011394).

Keywords: living/controlled polymerization, high impact polypropylene, stereoblock copolymers, blends

Catalysts used for polymerization

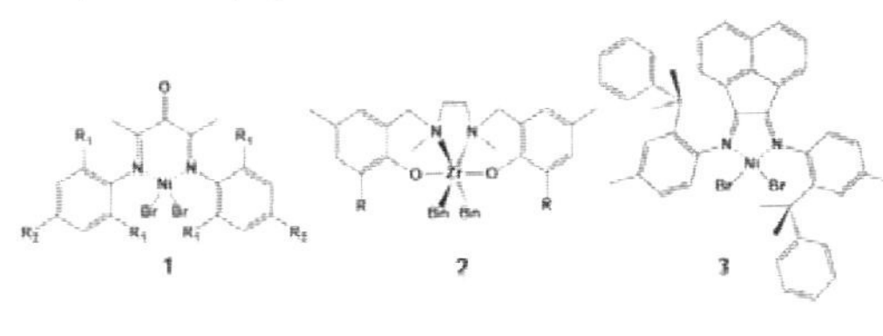

RDPS-A-P-010 [Controlled/Living Polymerization]

Hydrophilic Pentablock Copolymers Synthesized by One-Pot RAFT Polymerization of Methacrylates

Povilas Radzevicius, Medeina Steponaviciute, Tatjana Krivorotova, Ricardas Makuska

Department of Polymer Chemistry, Vilnius University, Vilnius, Lithuania

It was shown recently that RAFT polymerization is well suited for the synthesis of high-order multiblock copolymers by one-pot sequential monomer addition, where each step is taken to near full monomer conversion without intermediate purification. In the present study, one-pot methodology was used for the synthesis of a series of hydrophilic pentablock copolymers with rather long blocks (up to DP 100) differing by the nature of the monomers as well as succession of their blocks.

Hydrophilic pentablock copolymers containing the blocks of pHEMA, pBMA, pDMAEMA, or pPEOMEMA were synthesized by one-pot RAFT polymerization using difunctional RAFT CTA, which allowed the extension of growing chains into two directions and the synthesis of pentablock copolymers in three steps. Conversion of the monomers was monitored by 1H NMR spectroscopy, and molecular weight of the copolymers was determined by SEC with triple detection. Polymerization of every monomer was carried out up to very high conversion (usually, over 98 mol.%). Dispersity Đ of the pentablock copolymers was very low, usually 1.1-1.2.

Pentablock copolymers containing the blocks of pHEMA, pDMAEMA and pPEOMEMA are well soluble in water, while those containing the blocks of pHEMA and pBMA could be solubilized in organic solvents only. Solution behaviour and aggregation of hydrophilic pentablock copolymers was studied by 1H NMR spectroscopy, DLS, wetting angle measurements, DSC and TGA. The copolymers containing pPEOMEMA blocks with rather long PEO side chains represent molecular brushes while those containing pDMAEMA blocks are cationic in acidic media and possess lower critical solution temperature (LCST). It was shown that LCST transition of multiblock copolymers was at higher temperature, compared to random copolymers of the same composition, and formation of micelle-like structures in their solutions was less pronounced.

Acknowledgements. Financial support from the Research Council of Lithuania under the project MIP-54/2015 is gratefully acknowledged.

Keywords: RAFT, One-Pot, Pentablock Copolymers, Solution Behavior

RDPS-A-P-011 [Controlled/Living Polymerization]

Star Polymers; Synthesis, Characterization and Applications

Cansel Tuncer, Vural Bütün

Department of Chemistry, Faculty of Arts and Science, Eskisehir Osmangazi University, Eskisehir, Turkey

In recent years, branched polymers as compared to linear analogs have great attention because of their unique properties. One special type of branched polymer is star (co)polymer. Star polymers are macromolecules formed by connecting many linear polymer to a central core [Gao H. et. al. Macromolecules, 2008]. Due to their compact structure, they have unique properties, such as lower melting points and solution viscosity than linear analog (co)polymers [Bywater S. Physical Chemistry, Springer, 1979]. Their various application studies were reported in the literature (e.g. as drug carriers, gene delivery agents [Liu T. et. al. Biomaterials, 2014] and dispersants for carbon nanotubes [Motoyanagi J. et. al. Chem. Lett., 2013]. In this study, we report the synthesis, characterization and applications of novel water-soluble star copolymers. These star copolymers with well controlled molecular weights and branching have been successfully synthesized via group transfer polymerization without any macrogelation. Additionally, derivative star copolymers were also prepared and used as stabilizers in the synthesis of nearly monodisperse pH responsive poly[2-(*N*-diethylamino)ethyl methacrylate] microgels via emulsion polymerization. These derivative polymers were also useful in the production of Au-nanoparticle (AuNP) dispersions. These nanometal dispersionss were found to have good catalytic activity. Finally, because of having cavities and functional groups, branched polymers pulled methyl orange dye molecules from the aqueous phase to the organic phase and they encapsulated dye molecules. This study was conducted utilizing host-guest relationship between the copolymer and the dye molecules. Copolymer structures were characterized by using GPC and ^{1}H NMR spectroscopy. TEM-SEM and DLS were used for the characterization of microgels. UV-vis spectrophotometer was used for catalytic activity studies of polymer stabilized AuNP dispersions.
Acknowledgement: Eskisehir Osmangazi University Scientific Research Projects Commission (ESOGU BAP) is thanked for financial support (Grand Number: 201319033).

Keywords: Star polymer, microgel, Au-nanoparticle, encapsulation, host-guest relationship

RDPS-A-P-012 [Controlled/Living Polymerization]

Synthesis of Highly Crosslinked Polymers and Their Usage as Dye Adsorbent Surfaces

Mehtap Kılınç[1], Cansel Tuncer[2], Vural Bütün[1]

[1]Department of Polymer Science and Technology, Eskisehir Osmangazi University, Eskisehir, Turkey
[2]Department of Chemistry, Faculty of Arts and Science, Eskisehir Osmangazi University, Eskisehir, Turkey

Waste water of textile dyes contains various contaminants including chemical complex dye mixtures, salts, acids, bases, and sometimes consists of chlorinated organic compounds from heavy metals. Thus, paint removal is one of the main problems in the treatment of such waste water [Santos S.C.R. et. al. J. Hazard. Mater., 2008]. Activated carbon is the most widely used adsorbent for this purpose. However, due to its high costs, there is increasing interest in the use of low-cost adsorbents for the removal of dyes from waste water [Şener S. Chem. Eng. J., 2008]. Polymeric adsorbent material plays an important role in such waste water applications [Rao G.V.R. et. al. Chem. Mater., 2002].

In our study, the new type of intensive cross-linked polymers (ICL) were synthesized from a tertiary amine methacrylate based monomers and used as a dye adsorbent. Water-soluble and anionic dye methyl orange was used as a model dye in this study. Systematically, the effect of pH change on dye adsorption, by varying both the amount of polymer at fixed dye concentration and the dye concentration on a fixed amount of polymer were examined. The adsorption capacity of related polymer was calculated accordingly. Adsorption isotherms were analyzed to understand dye-polymer interactions. From related isotherm, kinetic adsorption rate of the polymer-dye system and thermodynamical parameters (ΔH, ΔS and ΔG) were calculated. Mw, Mw/Mn values of ICL polymers were determined by using gel permeation chromatography. Comonomer ratios of ICL polymers were calculated by comparing the related signals in their proton NMR spectra. UV-vis spectrophotometer was used in the calculation of adsorption capacity of ICL polymers.

Acknowledgement: Eskisehir Osmangazi University Scientific Research Projects Commission (ESOGU BAP) is thanked for financial support (Grand Number: 2016-1048).

Keywords: Polymer-dye interaction, dye absorban surfaces, highly cross-linked polymer, group transfer polymerization (GTP).

RDPS-A-P-013 [Controlled/Living Polymerization]

Synthesis and Solution Behavior of a Novel Polymer of Phosphorus Containing

Yasemin Samav[1], Vural Bütün[2]

[1]The Program of Chemistry, Bilecik Seyh Edebali University, Vocational School of Higher Education, Bilecik, Turkey,
[2]Department of Chemistry, Eskisehir Osmangazi University, Eskisehir, Turkey

Phosphorus-based polymers have found applications such as adhesion promoters for paints and adhesives, and complexing agents to recover metal ions from the environment and industrial liquids [Gefter Y. L. et al. Organophosphorus Monomers and Polymers; New York, 1962, Weil E. D., Encycl. Polym. Sci. Eng. 1987, Kaiser J. et al. Makromol. Chem. 1992]. One of the methods of incorporating phosphorus functionality into polymers is the polymerization or copolymerization of phosphorus containing vinyl monomers such as vinyl and allyl phosphonates [Kim Y. et al. Polymer 2003, Ebdon J. R. et al. Polym. Degrad. Stab. 2000]. Reversible Addition Fragmentation Transfer (RAFT) is one of the most robust controlled/living radical polymerization techniques [Moad G. et al. J. Chem. 2005] and allows for the controlled polymerization of such phosphonate-based monomers.

In this study, a novel water soluble block copolymer poly(2-N-morpholinoethyl methacrylate)-b-poly(2-methacryloyloxy ethyl phosphate) (PMEMA-b-PMMEP) was synthesized via RAFT in water using PMEMA-CTA as macro raft agent. MMEP monomer was synthesized with 2-hydroxyethyl methacrylate (HEMA) and triethylamine at 0 oC under nitrogen. PMEMA-CTA and PMMEP-CTA were synthesized in DMF using cyanoisopropyl dithiobenzoate (CIDB) as chain transfer agent. Polymers were characterized by using GPC and 1HNMR spectroscopy. The pH titration of PMMEP enabled determination of the dissociation constant of the phosphonic diacid group. pKa1 and pKa2 values were determined to be 2.5 and 7.1 from titration curve, respectively. Isoelectric point of PMEMA-b-PMMEP block copolymer was found to be 5.8 by zeta potentiometer.

PMEMA-b-PMMEP block copolymer formed micelles by the addition of Na2SO4 or at elevated temperature in the basic aqueous solution via formation of the micellar core with MEMA. Characterization of micelles was performed by DLS and TEM studies.

Keywords: RAFT polymerisation, phosphorus-containing polymer,

RDPS-A-P-014 [Controlled/Living Polymerization]

Synthesis, Characterization and Antibacterial Activity of Reduced Graphene Oxide-Poly(n-butylacrylate-block- n-vinylpyrrolidone) Nanocomposite Films

Elif Vargün, Cihan Açık, Emine Sonay Elgin

Department of Chemistry, Mugla Sitki Kocman University, Mugla,Turkey

Poly(N-vinyl-2-pyrrolidone), PVP, and its copolymers have been attracting much attention due to their hydrophilic nature, low toxicity and complexing ability. PVP has been used in various biomedical applications such as contact lenses, blood plasma expander, drug delivery and wound dressings. The antimicrobial property of the polymers plays an important role in medical devices, health care products, hospital and dental equipments etc [1]. Recently, several studies have demonstrated that graphene oxide (GO) and reduced graphene oxide (rGO) exhibit strong antibacterial activity [2]. Hence, in the present study, poly(n-butylacrylate-block-n-vinylpyrrolidone), P(BA-b-NVP) amphiphilic block copolymers were synthesized by atom transfer radical polymerization (ATRP) and then the rGO nanosheets were introduced the copolymers. Antibacterial activities of the resulting nanocomposite films were evaluated. The synthesis strategy is based on the preparation of hydrophobic poly(n-butyl acrylate) as a macroinitiator (PBA-MI) and then the hydrophilic N-vinyl-2-pyrrolidone monomer is polymerized as a second block by ATRP. Amphiphilic block copolymers were characterized by FTIR and ^{1}H-NMR spectroscopy (Figure 1). The molecular weight and its distribution of copolymers were determined by gel permeation chromatography. Differential scanning calorimetry analysis confirmed the phase separated morphology of the block copolymers and indicated two separate Tg's of block copolymers. The rGO was synthesized by modified Hummers' method and characterized by X-ray diffraction (XRD) and transmission electron microscopy (TEM). Various amounts of rGO were introduced into the copolymer films and the antibacterial activity of composite films was screened by agar well diffusion method against *Escherichia coli*.

References:

1. P Hemalatha, M K Veeraiah, S Prasannakumar, K V Anasuya, International Journal of Research in Engineering and Technology, 3(3) 56-64, 2014.
2. S. Liu, T. H. Zeng, M. Hofmann, E. Burcombe, J. Wei, R. Jiang, J. Kong and Y. Chen, ACSNano, 5(9), 6971–6980, 2011.

Keywords: ATRP, amphiphilic block copolymer, reduced graphene oxide

1H-NMR Spectrum of poly(n-butylacrylate-block-n-vinylpyrrolidone), P(BA-b-NVP)

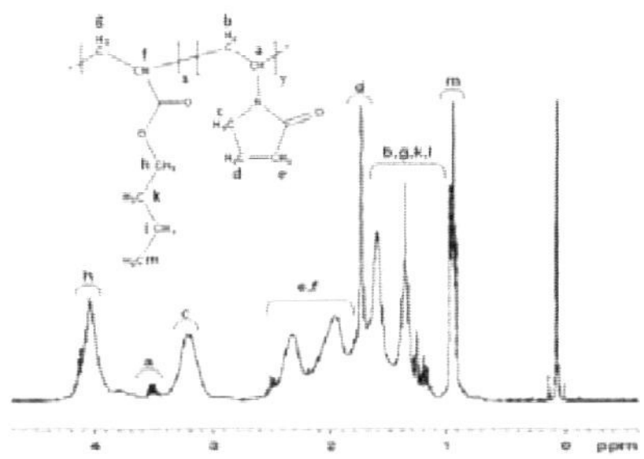

RDPS-A-P-015 [Controlled/Living Polymerization]

α-Diimine binuclear Ni-based catalysts for ethylene polymerization: effect of *ortho*-substituent

Mostafa KHOSHSEFAT[1], Saeid Ahmadjo[1], Seyed Mohammad Mahdi Mortazavi[1], Gholamhossein Zohuri[2], Majid Zahmaty[3], Khosrow Valieghbal[3]

[1]*Department of Catalyst, Polymer and Petrochemical Institue, Tehran, Iran*
[2]*Department of Chemistry, Faculty of Science, Ferdowsi University of Mashhad, Mashhad, Iran*
[3]*Centeral Laboratory of Jam Petrochemical Co., Assaluyeh, Iran*

Two binuclear α-diimine nickel (II) catalyst (A) *2,6-diisopropyl-C6H2-N=C-(C10H6)-C=N-C6H4-N=C-(C10H6)-C=N-2,6-diisopropyl-C6H2NiBr2* and *2,6-di-isopropyl-C6H2-N=C-(C10H6)-C=N-C6(CH3)4-N=C-(C10H6)-C=N-2,6-diisopropyl-C6H2NiBr2* (B) were prepared and used in ethylene polymerization. To investigate ortho-substituent effect (methyl) on catalyst behavior and produced polyethylene (PE) properties, the experiments were performed under similar condition and methylaluminoxane (MAO) was used as cocatalyst. Catalyst B (1645 g PE/mmolNi. h.) showed higher catalytic performance in comparison catalyst A (1073 g PE/mmolNi. h.). The vinyl content and branching of resulted PE in presence of cat A were 9.1 and 78 (in 1000 carbon atom), however, these quantities decreased to 7.2 and 59 for cat B, respectively. These observations assigned to lower steric and electronic effects of substituents onthe ortho-position of middle aryl ring (hydrogens in cat A) which can cause dissociation of the chain along with vinyl-end and reinsertion of the macromonomers lead to branches producing [Capacchione et al., 2015]. Besides, methyl groups in cat B caused retarding β-hydrogenelimination and chain transfer due to reducing monomer diffusion from ortho-position tometallic center [Merna et al., 2009]. These mechanistic reasons reduce vinyl and branching content in cat B. DSC analysis showed a peak along with a broad shoulder for both synthesized polymer using cat A and B. The broadening transition has been attributedto a sequence of melting followed byrecrystallization steps of less ordered domainswith the variable amounts of chainbranching [Pourmahdian et al., 2011]. The crystallinity contents and melting points were 24.7% and 123.0 °C for cat A and 41.3% and 112.2 °Cfor cat B. Based on aforementioned reason for branching, cat B which has methyl groups on ortho position instead of hydrogens in cat A, produce less branching with higher crystallinity.

Keywords: polyethylene, catalytic polymerization, binuclear catalyst, structure dependence, post metallocene

Scheme

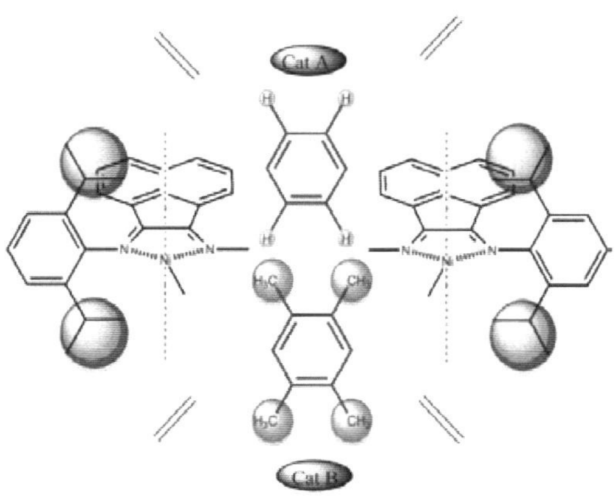

Schematic presentation of molecular structures of catalyst A and B.

RDPS-A-P-016 [Controlled/Living Polymerization]

Towards "Ultimate Control": Controlled Radical Polymerization in Pores Towards Length, Sequence and Tacticity Defined Macromolecules

Hui Chun Lee, Bernhard V. K. J. Schmidt, Markus Antonietti

Max Planck Institute of Colloids and Interfaces, Department of Colloid Chemistry, Research Campus Golm, Potsdam, Germany

This research aims to develop a facile approach towards the synthesis of well-defined polymers in several aspects, such as molecular weight, composition and tacticity. Biology demonstrates that an universal strategy towards well defined polymers is the proper design of confined space as reaction environment via specific molecular interaction. Consequently, an approach combing confined porous textures and reversible deactivation polymerization techniques is proposed to advance the synthetic polymer chemistry, i.e. a connection of metal-organic frameworks (MOFs) and atom transfer radical polymerization (ATRP). MOFs are formed by metal ion and/ or linking ligands and are known to contain diverse porous structures1-3 with high specific surface area, different sizes, shape and surface functionality, serving as an unique niche for host-guest interaction.4-6 In polymer science, ATRP and activators regenerated by electron transfer atom transfer radical polymerization (ARGET ATRP) are applied to conduct and further control the polymerization process with respect to molecular weight, end functionalities and architecture.7, 8 Interestingly, through the "host-guest" synergistic feature reflected by intrinsically composed framework the MOFs, more than as templates, can also serve as universal reactors for comprehensive polymer synthesis, such as polyisoprene (PI) and poly(4-vinylpyridine) (P4VP), which used to be hardly controlled and achieved by traditional ATRP/ ARGET ATRP. To simplify the synthesis process, the ATRP-oriented MOFs, the initiator-functionalized ones, are designed and fabricated to facilely realize an all-embraced MOF system for versatile polymerization and application. Combining the concept of biological processes that are based on confined space with controlled polymerization processes, ultimate control on polymers with desired chemical and physical properties, such as length, sequence and stereo specificity can be as realized, in a facile method (as shown in Scheme 1).

References:

1. O. M. Yaghi, M. O'Keeffe, N. W. Ockwig, H. K. Chae, M. Eddaoudi and J. Kim, Nature, 2003, 423, 705.
2. S. Kitagawa, R. Kitaura and S.-i. Noro, Angew. Chem., Int. Ed., 2004, 43, 2334.
3. G. Fe´rey, Chem. Soc. Rev., 2008, 37, 191.
4. J. C. Hulteen and C. R. Martin, J. Mater. Chem., 1997, 7, 1075.
5. D. J. Cardin, Adv. Mater., 2002, 14, 553.
6. T. Uemura, N. Yanaia and S. Kitagawa, Chem. Soc. Rev., 2009, 38, 1228.
7. K. Matyjaszewski, Macromolecules, 2012, 45, 4015.
8. K. Matyjaszewski and J. Xia, Chem. Rev., 2001, 101, 2921.

Keywords: ATRP, ARGET ATRP, MOFs, tacticity control, PI, P4VP

Scheme 1

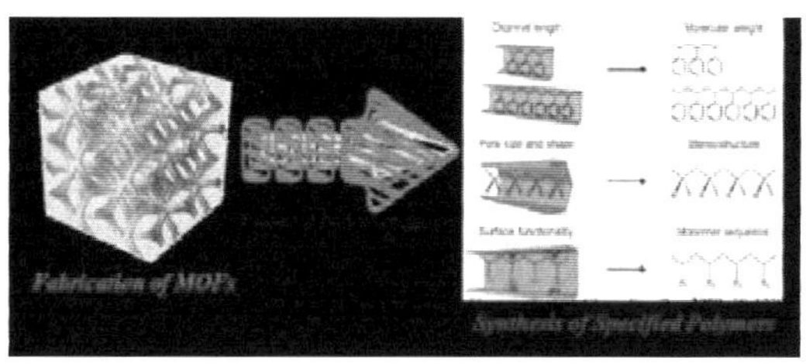

The executive flow chart of this research: starting from fabrication of MOFs, polymerization in the as-synthesized MOF, and characterization of the synthesized polymers.

The ATR-FTIR profiles of the PS@MOFs

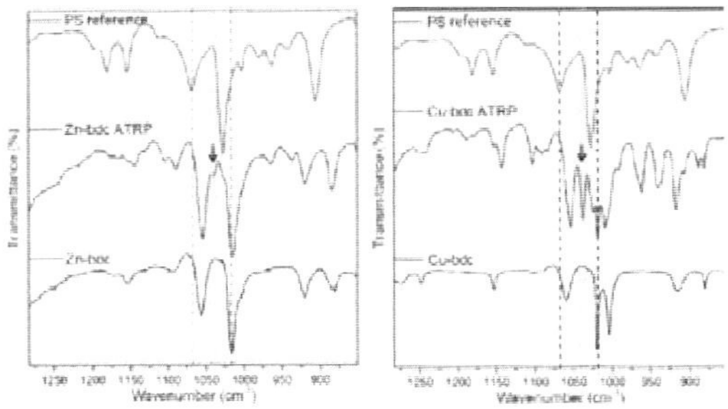

The ATR-FTIR profiles of the PS@MOFs comparing to the original bdc MOF and PS from bulk solution: the Zn-bdc ATRP (left) and the Cu-bdc ATRP (right).

The SEM images of the synthesized initiator-functionalized MOFs

the Zn-Brbdc (left) and the Cu-Brbdc (right)

RDPS-A-P-017 [Controlled/Living Polymerization]

TBD catalyzed ring-opening polymerization in microreactor

Ning Zhu[1], Xin Hu[2], Weiyang Feng[1], Kai Guo[1]

[1]College of Biotechnology and Pharmaceutical Engineering, Nanjing Tech University, Nanjing, China
[2]College of Materials Science and Engineering, Nanjing Tech University, Nanjing, China

Aliphatic polyesters have drawn much interest in biomedical application with outstanding biodegradability and biocompatibility. Organocatalyzed ring-opening polymerizations (ROP) have been the alternative powerful methodology to synthesize metal-free biomaterials, since the pioneering work of Hedrick et al was reported in 2001. Although great progress has been made in this topic, the challenges still remained including but not limited to long-time process, low level control of monomer sequence and stereoselectivity. Microreactor exhibited huge surface-to-volume ratio and continuous flow characteristic. By transferring polymerizations into microreactor, remarkable benefits could be achieved that cannot be done in batch. The idea was raised to view the challenge by combing organocatalysis and microreactor. 1, 5, 7-triazabicyclododecene (TBD) was chosen as model organocatalyst to mediate ROP of ε-caprolactone (CL) in the presence of benzyl alcohol. Living/controlled nature of continuous flow polymerization was demonstrated. The molecular weight of poly(ε-caprolactone) (PCL) was exactly controlled by changing the flow rate. Due to the improved mixing and heat/mass transfer efficiency, the apparent polymerization rate constants were increased in the assembled PTFE tubular reactor. This work might find its application in sequence controlled/defined polymer fabrications.

Keywords: TBD, microreactor, ring-opening polymerization, poly(ε-caprolactone)

RDPS-A-P-018 [Controlled/Living Polymerization]

Synthesis and Characterization of Poly(*n*-butyl acrylate)-*b*-Poly(4-vinyl pyridine) Block Copolymers

Hümeyra Mert[1], Nicolay V. Tsarevsky[2], Krzysztof Matyjaszewski[3]

[1]Department of Polymer Engineering, Hitit University, Çorum, Turkey
[2]Department of Chemistry, Southern Methodist University, Dallas, USA
[3]Department of Chemistry, Carnegie Mellon University, Pittsburgh, USA

Well-defined poly(vinyl pyridine) (PVP) block copolymers have attracted much interest, due to their potential applications, such as coordination reagents for transition metals, molecular imprinting templates or materials for membranes [1]. Atom transfer radical polymerization (ATRP) is an efficient technique enabling the synthesis of well-defined block copolymers with controlled molecular weight, narrow molecular weight distribution and high degrees of chain end functionalization [2, 3]. However, the polymerization of 4VP by ATRP is challenging because both 4VP and P4VP are strongly coordinating compounds that can compete with the ligand in the ATRP catalyst for binding with the metal center. Recently, it has been demonstrated that ATRP of 4VP could be achieved in a controlled fashion when relatively large amount of Cu(II) chloride (deactivator) in conjunction with a ligand forming stable and active copper complexes, tris(2-pyridylmethyl)amine (TPMA), was used [4]. In the present study, poly(*n*-butyl acrylate)-*b*-poly(4-vinyl pyridine) (P*n*BA-*b*-P4VP) diblock copolymers with different compositions and controlled molecular weights were synthesized via sequential copper-mediated ATRP of BA followed by chain extension of the produced macroinitiators with 4VP. Chain-end functionality of diblock copolymers was confirmed by efficient and clean chain extension polymerizations. These novel block copolymers were characterized by ^{1}H NMR, GPC, SEM, and DSC analyses. DSC measurement on the P*n*BA-*b*-P4VP block copolymers exhibited two glass transitions, indicating that the resulting block copolymers are phase separated.

References:

1. Rivas B L.; Maturana H A.; Pereira E. Angew Makromol Chem, 1994, 220, 61.
2. Wang, J-S.; Matyjaszewski K. J Am Chem Soc, 1995, 117, 5614.
3. Kato, M.; Kamigaito, V.; Sawamoto, M.; Higashimura, T. Macromolecules, 1995, 28, 1721.
4. Pietrasik, J.; Tsarevsky, N V.; European Polymer Journal, 2010, 46, 2333.

Keywords: ATRP, 4-vinyl pyridine, block copolymer

Synthetic protocol for P{n}BA-{b}-P4VP Block Copolymers

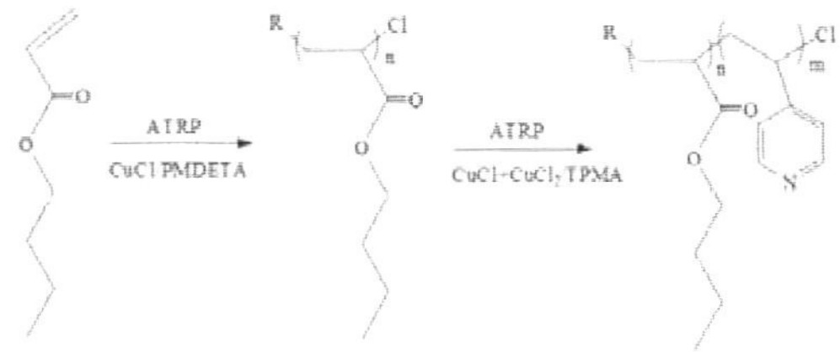

PMDETA: N,N,N',N'',N''-Pentamethyldiethylenetriamine
TPMA: Tris(2-pyridylmethyl)amine

RDPS-A-P-019 [Controlled/Living Polymerization]

Synthesis and characterization of antibacterial polysulfone via photoiniferter method

Serhat Oran, Mehmet Atilla Tasdelen

Yalova University Central Campus, 77200 Yalova

Microbial infection caused by bacteria is a serious health problem to the human population. In recent years, considerable attentions have been devoted to investigations of novel antibacterial materials that have been used for their ability to kill many microorganisms. The antibacterial materials not only comprise metal ions such as silver ions, but also have quaternary ammonium salts to enhance the antibacterial effects.Polysulfones (PSU) are the family of thermoplastics that have excellent thermal and chemical stability, high radiational degradation resistance, transparency, high rigidity, mechanical strength and good electrical properties. Also, polysulfones have good film properties therefore these polymers have been used for many applications that medical devices (haemodialysis), food processing, feeding systems, automotive, electronics, water distillation, gas separation and fuel cells. The chemical modification of polysulfones, especially cloromethlylation and quarternization, is a subject of considerable interest from recent investigations. Functional polysulfones obtained by cloromethlylation reaction and quaternization with ammonium groups will be display antibacterial properties that can be used for several applications, e.g. biomaterials, medical devices and membranes.

In this thesis, firstly, commerically polysulfones is functionalized by cloromethlylation route. Then, these polysulfones react with sodium diethyl dithiocarbamate to obtain functional polysulfones bearing iniferter groups at side chain. After that, these iniferter groups can be activated with UV light to graft 2-(dimethylamino)ethyl methacrylate onto the polysulfone backbone. Finally, quarternization experiment proceed with methly iodide to synthesize corresponding graft copolymer containing quetarnary ammonioum salts. The thin films of this polymers were prepared by a dry-cast process using dimethyl sulfoxide as solvent. The structure and properties of obtained materials were characterized by spectroscopic (FT-IR and 1H-NMR), chromarographic (GPC), thermal (DSC) and antibacterial analyses. Compared to neat graft copolymers, the quaternized graft copolymers display a slight antibacterial effect.

Keywords: İniferter, polysulfone, graft copolymer

RDPS-A-P-020 [Controlled/Living Polymerization]

Modification of Cell Culture Dishes by Consecutive Grafting of Poly(AAm)/Poly(NIPAAm) for Cell Sheet Recovery

Murat Barsbay, Olgun Güven

Department of Chemistry, Hacettepe University, Beytepe, 06800, Ankara

This study investigates the RAFT mediated and radiation induced grafting of poly(acrylamide) (PAAm) from polystyrene (PS) cell culture petri dishes and subsequent chain extension of the grafts by a temperature-responsive polymer, poly(N-isopropylacrylamide) (PNIPAAm), via the living feature of PAAm chains provided by the RAFT chain-end moieties. The thickness of the thermoresponsive PIPAAm layer is a key factor affecting the hydration features and hence the cell sheet recovery characteristics [1]. In order to confine the thickness of PNIPAAm within the desired limits we have performed controlled free-radical graft polymerization, namely RAFT polymerization, of PNIPAAm from PS petri dish [2]. Characterization of the synthesized copolymers by ATR-FTIR spectroscopy, X-ray photoelectron spectroscopy, elemental analysis, thermogravimetric analysis and contact angle measurements revealed the grafting of double polymeric layer from PS dishes. The basal hydrophilic PAAm component in the double layer is expected to promote the hydration of the upper PIPAAm layer and to induce rapid cell detachment compared to conventional temperature-responsive cell culture surfaces.

Grafting studies were carried out by using e-beam treatment. FTIR analysis revealed new absorption peaks following the grafting, indicating a successful modification. The elemental surface compositions calculated from XPS spectra and the results of elemental analysis performed to investigate the fullest extent of the samples indicated grafting of PAAm and PNIPAAm from the PS petri dishes. The decomposition pattern and thermal stability of the samples investigated by the thermogravimetry (TG) analysis also indicated the attachment of PAAm and PNIPAAm chains to PS substrate. By measuring the contact angle (CA) of a water droplet deposited on the surfaces of pristine and grafted PS surfaces, the changes in hydrophilicity due to the grafting of PAAm and PNIPAAm chains were determined. We believe that this technique could be widely useful for constructing various types of polymer-based biomaterial surfaces.

Acknowledgements: M.B. appreciates the financial support of the Scientific and Technological Research Council of Turkey (TUBITAK, Project Number: 113Z889)

References:

[1] Fukumori, K., Akiyama, Y., Kumashiro,Y., Kobayashi, J., Yamato, M., Sakai, M., Okano, T., Macromol. Biosci. 10 (2010) 1117–29.
[2] Barsbay, M., Güven, O., Stenzel, M.H., Davis, T.P., Barner-Kowollik, C., Barner, L., Macromolecules 40 (2007) 7140-7147.

Keywords: RAFT polymerization, surface modification, grafting

Schematic expression of the synthesis steps

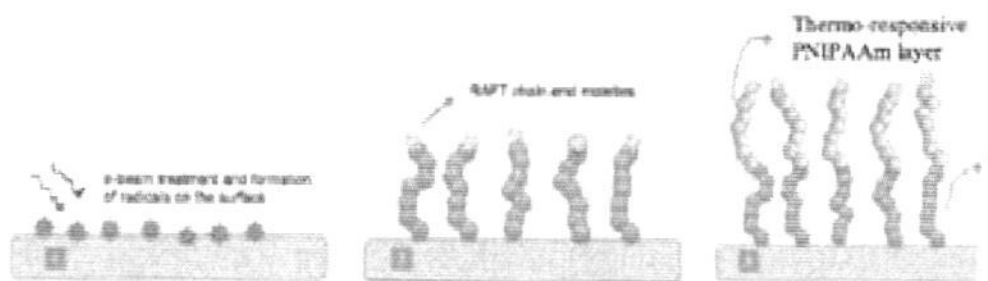

1. Formation of surface-radicals by e-beam treatment, 2. RAFT mediated grafting of poly(acrylamide) (PAAm) from polystyrene cell culture petri dish, 3. Chain extension of the grafts by a temperature-responsive polymer, poly(N-isopropylacrylamide) (PNIPAAm)

RDPS-A-P-021 [Controlled/Living Polymerization]

Well-defined Proton Exchange Membranes with enhanced performance via RAFT polymerization

Gökçe Çelik, Murat Barsbay, Olgun Güven

Department of Chemistry, Hacettepe University, Beytepe, 06800, Ankara

The replacement of the state-of-the-art and expensive perfluorinated membranes such as Nafion, Flemion, and Aciplex with cost-effective proton exchange membrane alternatives is one of the main challenges for commercialization of polymer electrolyte fuel cell technology. Poly(ethylene-alt-tetrafluoroethylene) (ETFE) has been identified as a promising alternative base polymer film to prepare low-cost polymer electrolyte membranes because of its advantageous characters like superior mechanical properties and high resistance to radiation-induced damage [1,2].

The radiation-induced grafting of styrene into ETFE film and further chemical modification steps (sulfonation) allows functionalization of the base material and introduction of the desired property (proton conductivity) for preparing a fuel cell membrane [3]. However this simple conventional method suffers from one simple flaw: The molecular weight and the dispersity of the grafted chains cannot be controlled. Predetermined molecular weights and low dispersities of grafts can be achieved by grafting of monomer onto base polymer under living/controlled free radical polymerization (CRP) conditions. Among the CRP methods, Reversible Addition Fragmentation-Chain Transfer (RAFT) is of particular interest as a very wide range of functional monomers can be polymerized in a controlled manner under non-demanding reaction conditions (e.g., tolerance to oxygen and low temperatures) [4].

The present study deals with the RAFT mediated radiation-induced (0.032 kGy/h, 60Co) grafting of styrene on ETFE films followed by the sulfonation of the polystyrene grafts. The effect of monomer concentration, absorbed dose and RAFT agent concentration on the grafting were investigated. The synthesized films and sulfonated membranes were characterized by ATR-FTIR, XPS, DSC, AFM and TGA methods. Cross-sections of these films were investigated by SEM- EDS to evaluate the grafting profile. The ion-exchange capacities of the sulfonated films were quite promising. This study show that RAFT mediated grafting has a great potential in the synthesis of fuel cell membranes with better performance.

References:

[1] S.A. Gürsel, L. Gubler, B. Gupta, G.G Scherer, Adv. Polym. Sci., 2008, 215, 157.
[2] L. Gubler, S.A. Gürsel, G.G. Schere, Fuel Cells, 2005, 5, 317.
[3] M.M. Nasef, E.S.A. Hegazy, Prog. Polym. Sci., 2004, 29, 499.
[4] M. Barsbay, O. Güven, Radiat. Phys. Chem., 2009, 78, 1054.

Keywords: RAFT polymerization, Proton Exchange Membrane (PEM), grafting

RDPS-A-P-023 [Controlled/Living Polymerization]

Synthesis and application of magnesium and zinc complexes with imino(phenoxide) ligands in L-lactide polymerization

Marcos Lopes Dias[1], Alana L. C. Oliveira[1], Leonardo C. Ferreira[1], Marco A. Chaer Nascimento[2], Rodrigo S. Bitzer[2], Iryna Grafova[3], Andriy Grafov[3]

[1]*Instituto de Macromoléculas Professora Eloisa Mano (IMA), Universidade Federal do Rio de Janeiro, Rio de Janeiro, Brazil*
[2]*Institute of Chemistry, Universidade Federal do Rio de Janeiro, Rio de Janeiro, Brazil*
[3]*Departmant of Chemistry, University of Helsinki, Helsinki, Finland*

Aliphatic polyesters such as polylactides (PLA) and their copolymers have attracted considerable attention due to their biocompatibility and biodegradability. Ring opening polymerization (ROP) of lactides (LA) using metal complexes as initiators or catalysts is an efficient way to produce PLA. Ligands coordinated to the metal center may significantly influence the polymerization reaction and the final polymer properties. The use of ROP to produce these polymers offers a high degree of control over the reaction, polymer molecular weight and tacticity. In this work, new zinc and magnesium complexes with different phenoxy-imine ligands were used as L-lactide polymerization initiators, in order to obtain the final polymer containing non-toxic metal residues [1,2,3]. Phenoxy-imine ligands were synthesized from diisopropylaniline and the aldehydes, viz. 2-hydroxybenzaldehyde and 2-hydroxy-2,4-di-t-butyl-benzaldehyde and characterized by FTIR and 1H-RMN. The ligands were used for synthesis of zinc and magnesium complexes by reaction with diethylzinc and ethylmagnesium bromide, respectively, in the presence of n–butanol. Chemical structure of the complexes in question was confirmed by FTIR and 1H NMR spectroscopies. Ligand and complex geometries have been optimized at the DFT level of theory using either the B3LYP or PBE0 hybrid functionals along with the triple-zeta LACV3P** basis set. The molecular structures of the complexes have been authenticated by DFT vibrational analysis. Our calculations have shown that both Zn complexes display trigonal planar coordination geometry, as expected for three-coordinated Zn(II). The complexes were then tested for L-lactide polymerization at molar ratios LA/complex of 1/500, 1/1000, and 1/2500. The reaction was carried out for 2 h at 180°C. The study revealed a good activity of the zinc complex in ROP, particularly, when increasing the initiator concentration in the reaction. Polymer yield increased with the decrease of LA/Zn ratio, when 90-92% conversion was attained. The PLLA obtained possess high stereoregularity, molecular weight, and Tm.

Keywords: ROP, Zinc, magnesium, phenoxy-imine ligands, polylactide

RDPS-A-P-024 [Controlled/Living Polymerization]

The First Successful Cyclopolymerization of 1,6-Heptadiynes by Grubbs 1st Generation Catalyst

Cheol Kang, Eun Hye Kang, Tae Lim Choi

Department of Chemistry, Seoul National University, Seoul, Korea

One of powerful ways to synthesize conjugated polymers is cyclopolymerization. Using olefin metathesis catalysts with diyne monomers, polyacetylene backbone can be easily constructed. Especially, cyclopolymerization of 1,6-heptadiynes give 5-membered rings bearing pi-conjugated system. And due to their planarity, the polymer have highly long effective conjugated length affording narrow band-gaps. Early studies of cyclopolymerization was developed by using Ziegler catalyst, MoCl5/WCl5, and Schrock catalysts. After then, modified Grubbs catalyst was used for cyclopolymerization with controlled molecular weights. More recently, our group developed living cyclopolymerization system using fast-initiating Grubbs catalyst with coordinating solvents or ligands. Meanwhile, there have been reports that side products are generated when cyclopolymerization is conducted with Grubbs catalysts. Our recent mechanistic studies have proven that this side reaction proceeds not by olefin metathesis process but by the action of decomposed Grubbs catalyst which mediates [2+2+2] alkyne trimerization. Especially, the side reaction was highly dominant when Grubbs first generation catalyst (G1), which is why the cyclopolymerization by G1 was never been reported. Herein, we report the first example of cyclopolymerization catalyzed by G1. By using acidic additives, we could successfully synthesize polymers with suppressing the generation of side products. Various monomers having different functional groups afforded polymers with moderate to good efficiencies. Through mechanistic studies, we concluded that the additive helps phosphine dissociation of Grubbs catalyst especially for conjugated Ru-carbene which is crucial for propagation. Also, it was confirmed that the additive prevents the deformation of propagating species and suppresses the [2+2+2] alkyne trimerization.

Keywords: Cyclopolymerization, Grubbs catalyst, Conjugated polymer

RDPS-A-P-025 [Controlled/Living Polymerization]

Exploration of microstructure of olefin block copolymers made by combination of chain-walking and chain-shuttling reactions in ethylene polymerization

Mostafa Ahmadi[1], Mahsa Karimi[2], Saeid Ahmadjo[2], Majid Zahmati[3], Davoud Jafarifar[3], Mohammad Mahdi Mortazavi[2]

[1]*Department of Polymer Engineering and Color Technology, Amirkabir University of Technology, Tehran, Iran*
[2]*Polymerization Engineering Department, Iran Polymer and Petrochemical Institute (IPPI), Tehran, Iran*
[3]*Central laboratory, Jam Petrochemical Co., Assalouyeh, Iran*

Olefin Block Copolymers (OBCs) have received new interests both in academia and industry due to their appealing features in synthesis, characterization and properties. They are originally made by copolymerization of ethylene and α-olefins using two catalysts with highly different comonomer affinities in presence of a chain shuttling agent (CSA) (Arriola et al. Science 2006, 312, 714). However both rac-ethylene bis-(indenyl) zirconium dichloride and α-diimine nickel(II) bromide catalysts have shown reversible chain transfer to diethyl zinc, providing the potential for production of OBCs from single ethylene stock (Xiao et al. Macromolecules 2009, 42, 1834). In this contribution we have synthesized and revealed the blocky structure of the so made OBCs by characterization of molecular weight (MWD) and chemical composition distributions (CCD). While the former catalyst makes linear segments the latter provides branched blocks through chain walking reaction and successive blocks are formed by consecutive shuttling through CSA. The broad MWD made by combination of two catalysts turned into narrower distribution in presence of CSA and the thick lamellae dispensed into much narrower segments as revealed by successive self-nucleation annealing (SSA) measurements.

Keywords: Chain walking polymerization, Chain shuttling polymerization, Olefin block copolymer

RDPS-A-P-026 [Controlled/Living Polymerization]

Effect of reversible transfer reaction on comonomer incorporation in coordinative chain transfer polymerization

Mostafa Ahmadi[1], Ali Ebrahimi[1], Mahsa Karimi[2], Saeid Ahmadjo[2], Majid Zahmati[3], Khosrow Valieghbal[3], Mohammad Mahdi Mortazavi[2]

[1]*Department of Polymer Engineering and Color Technology, Amirkabir University of Technology, Tehran, Iran*
[2]*Polymerization Engineering Department, Iran Polymer and Petrochemical Institute (IPPI), Tehran, Iran*
[3]*Central laboratory, Jam Petrochemical Co., Assalouyeh, Iran*

Coordinative chain transfer polymerization (CCTP) is a new living polymerization system based on reversible transfer reaction between a chain transfer agent (CTA) and a coordination catalyst (Valente et al. Chem Rev. 2013, 113, 3836). Simultaneous introduction of two catalysts with significantly different comonomer selectivity in such a system in presence of a comonomer gives rise to new olefinic multi-block copolymers (OBCs) with crystallizable homopolymer hard segments and amorphous copolymer soft blocks (Arriola et al. Science 2006, 312, 714). The complexity of the active species formed in the presence of two catalysts and a chain transfer agent can give rise to unexpected reactivities (Zinck, Polym Int, 2016, 65, 11). We have therefore studied two catalysts with different comonomer affinities (bis-(indenyl) zirconium dichloride and bisiminopyridine iron dichloride) in copolymerization of ethylene and 1-hexene in presence of different concentration of CTA to separately investigate the effect of reversible chain transfer reaction on comonomer incorporation. Both catalysts are also utilized simultaneously in presence of different levels of CTA to produce OBC. Comonomer incorporation and its distribution are studied by normal DSC and successive self-nucleation annealing (SSA) measurements. Thermal properties reveal surprising effects of CTA on elaboration of comonomer incorporation. The observed behavior is tried to be interpreted using modeling of microstructure and kinetics using the method of moments.

Keywords: Coordinative chain transfer polymerization, chain shuttling polymerization, Copolymerization

RDPS-A-P-027 [Controlled/Living Polymerization]

RAFT Polymerization of Styrene mediated by Diphenyl Functionalized RAFT Agents

Teboho Simon Motsoeneng, Gabriel Jeffrey Summers, Carol Ann Summers

Department of Chemistry, University of South Africa, Florida, Souty Africa

Reversible addition-fragmentation chain transfer (RAFT) polymerization provides an efficient polymer synthesis method for the preparation of polymers with predictable number average molecular weight, narrow molecular weight distribution, well-defined chain end functionality and polymer topology [1]. Three novel diphenyl functionalized RAFT agents, benzhydryl 2-(ethylsulfanylthiocarbonylsulfanyl)propanoate, benzhydryl 2-(butylsulfanylthiocarbonylsulfanyl)propanoate, benzhydryl 2-(dodecylsulfanylthiocarbonylsulfanyl)propanoate, have been prepared and employed in RAFT polymerization reactions. The RAFT polymerization of styrene mediated by the diphenyl functionalized RAFT agents with AIBN as initiator at 80 0C proceeds via controlled/living polymerization processes to produce quantitative yields of the diphenyl functionalized polymers with the diphenyl groups introduced regiospecifically at the α-terminus of the polymer chain. The different RAFT agents display good control of the polymerization process at different RAFT agent concentrations. Polymer kinetics measurements show that the polymerization reaction follows first order rate kinetics with respect to monomer consumption and the number average molecular weights increase linearly with the monomer conversion to produce polymers with narrow molecular weight distributions. Percentage monomer conversion for each reaction was monitored by 1H NMR spectroscopy. The diphenyl functionalized chain transfer agents and the functionalized polymers were characterized by 1H NMR spectroscopy, FTIR spectroscopy, size exclusion chromatography, ultra-violet visible spectroscopy and fluorescence spectroscopy.

Reference:

1. Chiefari J, Chong YK, Ercole F, Krstina J, Jeffery J, Le TPT, Mayadunne RTA, Meijs GF, Moad G, Rizzardo E and Thang SH, Macromolecules 31: 5559-5562 (1998)

Keywords: RAFT Polymerization, diphenyl functionalized RAFT agents, Polystyrene.

RDPS-A-P-028 [Controlled/Living Polymerization]

Novel Poly(methylmethacrylate)-Based Two-Armed Macroinitiators via ATRP: Synthesis and Characterization

Çiğdem Yağcı[1], Murat Mısır[2], Ahmet Bilgin[1], Nagihan Acıerik[3]

[1]*Kocaeli University*
[2]*Ahi Evran University*
[3]*Karabük Universtiy*

Atom Transfer Radical Polymerization (ATRP) is one of the remarkable processes by which precise polymer synthesis may be achieved giving well-defined polymeric architectures, inaccessible by other methods [1]. Macrointermediates such as macroinitiators, macromonomers, and macrocrosslinkers are very important in polymer modification leading to block and graft copolymers. The methyl methacrylate, which was one of the green and environment-friendly materials with the advantages of good chemical stability, fine solubility, high-temperature resistance, high degree of transparency, ease of processing, etc. and widely used in the preparation of optical limiting materials, is selected as the polymerized monomer [2]. In this work a series of two-armed polymethylmethacrylate macroinitiators were synthesized via ATRP of methylmethacrylate (MMA), as a monomer in presence of CuBr/PMDETA catalyst system at 90 °C. For this purpose [(4-nitrophenyl)-imino]bis(ethane-2,1-diyl)bis[2-bromo-2-methylpropanoate) via conversion of the hydroxyl end groups of two-armed N-(4-nitrophenyl) diethanolamine to two bromine-ended initiator for ATRP of MMA by esterification with 2-bromopropionyl bromide. All the novel compounds have been characterized by FT-IR, NMR and GPC, DSC, TGA techniques.

1. J. L. Hedrick, M. Tröllsas, C. J. Hawker, B. Atthoff, H. Claesson, A. Heise, R. D. Miller, D. Mecerreyes, R. Jerome, Ph. Dubois, Macromolecules, 31, 8691, (1998).
2. N. Qiu, D. Liu, S. Han, X. He, G. Cui, Q. Duan, Journal of Photochemistry and Photobiology A: Chemistry, 272, 65 (2013).

Keywords: Atom transfer radical polymerization, methyl methacrylate, controlled living radical polymerization

Scheme 1

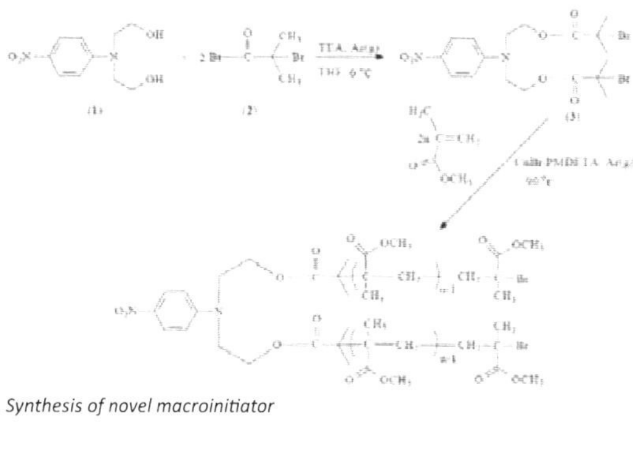

Synthesis of novel macroinitiator

RDPS-A-P-029 [Controlled/Living Polymerization]

Synthesis of Linear and Star Shaped Block Copolymers with ε-caprolactone and L-lactide by using Bismuth(III)Acetate Catalyst

Dildare Basalp, Funda Tıhmınlıoğlu

Department of Chemical Engineering, Izmir Institute of Technology, İzmir, Turkey

The tailoring of molecular structure of biopolymers by means of various co-initiators has attracted a great importance in biomedical applications. Enhancement of degradation and physical properties of polylactide (PLA) or poly(ε-caprolactone) (PCL) homo and copolymers can be optimized by incorporation of branch structures. This study is proposed to synthesize linear and star shaped PLA-PCL copolymers with variable architectures. Block copolymers were synthesized by successive two step ring opening polymerization of L-lactide and ε-caprolactone at 140 °C by using Bismuth(III)acetate (Bi(III) Ac) catalyst. Firstly, di, tetra and hexa OH functional PCL prepolymer were synthesized by using ethylene glycol, pentaerythritol, dipentaerythritol and myo-inositol as co-initiators, respectively. The ε-caprolactone monomer/catalyst ratios were kept constant as 500 for Bi(III)Ac. CI/I ratios were taken as 20 and 100. Characterization of synthesized polymers were carried out by using Nuclear Magnetic Resonance Spectroscopy (NMR), Differential Scanning Calorimetry (DSC), Thermal Gravimetric Analyzer (TGA), Gel Permeation Chromatograpy (GPC) and Matrix Assisted Laser Desorption/Ionization-Time of Flight (MALDI-TOF) Mass Spectrometry. Tetra-hydroxy functional PCL-PLLA copolymer structure was proved by 1H NMR spectra as seen seen in Figure 1. The synthesized copolymers by using Bi(III)Ac catalysts were found to be biocompatible according to MTT assay by using 3T3 cell.

Keywords: PLA, PCL, copolymer, ring opening polymerization

RDPS-A-P-030 [Controlled/Living Polymerization]

Graft copolymers from commercial chlorinated polypropylene via Cu(0)-mediated ATRP

Gokhan Acik, Mehmet Atilla Tasdelen

Department of Polymer Engineering, Faculty of Engineering, Yalova University, 77100 Yalova, Turkey

Polyolefins are the most widely used of commercial polymers finding many applications in packaging, toys, building, automotive, transport, medical and other industries.[1] They are generally made of propylene and ethylene monomers that composed of only carbon and hydrogen atoms.[2] One important characteristic of these polymers is their hydrophobicity, which can be considered to be the key factor limiting the paintability, wettability, and adhesion properties. To overcome these problems, it is necessary to modify the polyolefins. In this study, the commercially available chlorinated polypropylene has been used as a macroinitiator for the Cu(0)-mediated atom transfer radical polymerization of methyl methacrylate and tert-butyl acrylate to obtain well-defined graft copolymers. The relatively narrow molecular weight distribution in the graft copolymers and linear kinetic plots indicated the controlled nature of copolymerization reactions. Both FT-IR and 1H-NMR studies confirmed that the graft reactions had taken place successfully. After the graft copolymer formation, tert-butyl groups of poly(tert-butyl acrylate) side chains were completely converted into poly(acrylic acid) chains to attain corresponding amphiphilic graft copolymer.

Keywords: atom transfer radical polymerization (ATRP), chlorinated polypropylene, graft copolymers, polyolefins

Graft copolymerization of MMA and tBA via Cu(0)-mediated ATRP using PP-Cl as macroinitiator.

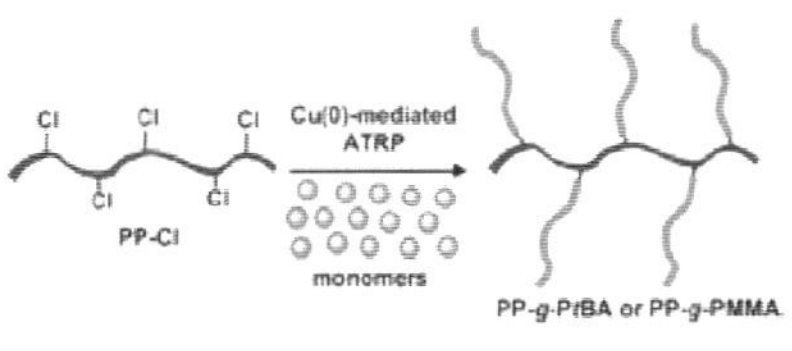

RDPS-A-P-031 [Controlled/Living Polymerization]

Synthesis and Characterization of New Olefinic Block Copolymers

Mahsa Karimi[1], Ali Ebrahimi[2], Mostafa Ahmadi[2], Saeid Ahmadjo[1], Majid Zahmati[3], Khosrow Valieghbal[3], Mohammad Mahdi Mortazavi[1]

[1]Polymerization Engineering Department, Iran Polymer and Petrochemical Institute (IPPI), Tehran, Iran
[2]Department of Polymer Engineering and Color Technology, Amirkabir University of Technology, Tehran, Iran
[3]Central laboratory, Jam Petrochemical Co., Assalouyeh, Iran

Olefinic multi-block copolymers (OBCs) are made by simultaneous application of two catalysts with significantly different comonomer selectivities in copolymerization of ethylene and α-olefins, in presence of a chain shuttling agent (CSA) (Arriola et al. Science 2006, 312, 714). Growing chains experience consecutive periods of living states on either catalysts and dormant states attached to the CSA. By each time cross-shuttling between two different catalysts, a new block with different comonomer content is formed. We have utilized two catalysts with different comonomer affinities; bis-(indenyl) zirconium dichloride and bisiminopyridine iron dichloride (which have proved the ability of catalyze chain growth on zinc (van Meurs et al. JACS, 2005, 127, 9913) in copolymerization of ethylene and 1-hexene in presence of different concentration of $ZnEt_2$ as a new system for production of OBC. While the former catalyst has high comonomer affinity, the latter is expected to incorporate negligible comonomer units, giving rise to production of successive amorphous and crystalline blocks along the chain. Comonomer incorporation and its distribution are studied by normal DSC and successive self-nucleation annealing (SSA) measurements. Monomer consumption profiles reveal higher initial activities compared to the case without CSA. The broad MWD made in presence of two catalysts turns much narrower upon introduction. Dynamic moduli at low temperatures, as measured by DMTA, reveal the fingerprint of blocky structure as already reported for OBCs. The obtained results propose a new viable method for production of new OBC products.

Keywords: Olefinic Block Copolymers, chain shuttling, matallocene, iron catalyst

Chain shuttling system for production of multi-block copolymer.

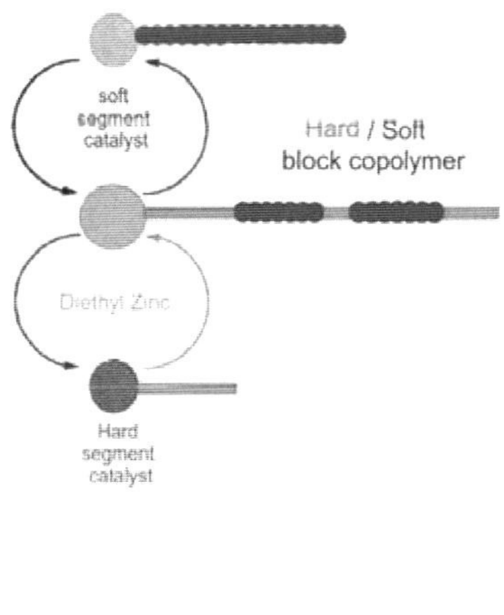

RDPS-A-P-032 [Controlled/Living Polymerization]

Polymer Bottlebrushes with Complex Architecture as Templates for Nanoparticle Synthesis

Yaoming Zhang, Joanna Pietrasik, Krzysztof Matyjaszewski

Institute of Polymer and Dye Technology, Lodz University of Technology

Well-defined polymer bottlebrushes were synthesized and used as templates for Au@carbon nanoparticles synthesis. The complex designed architecture of the template enables the control of nanoparticles geometry, tunable through the composition of side chains of bottlebrushes. Bottlebrushes were synthesized by atom transfer radical polymerization. Through control over the degree of polymerization of the backbone and side chains, the well-defined polymer bottlebrushes with different aspect ratios were obtained. The synthesis of the template was a multi-step reaction. First, poly(tert-butyl acrylate) (PtBA) was grafted from the backbone of poly(2-(2-bromopropionyloxy) ethyl methacrylate) (PBPEM), to obtain the bottlebrush PBPEM-g-tBA. Then PBPEM-g-tBA was used for side-chain extension with block copolymer of polyacrylonitrile-b-polystyrene (PAN-b-PS) by click reaction. The appropriate sequence of particular block allowed Au@carbon nanoparticles to be synthesized. The PtBA could serve as a template for loading gold precursor after hydrolysis to poly(acrylic acid) (PAA). PAN and PS are used as carbon source and stabilizer, respectively. Such materials could find potential applications as e.g. highly efficient catalyst for fuel cells.

Keywords: polymer bottlebrush, template, nanoparticle

RDPS-A-P-033 [Controlled/Living Polymerization]

Ethylene polymerization using a binary catalytic system; activity and microstructure study

Mostafa Khoshsefat[1], Saeid Ahmadjo[1], Mohammad Mahdi Mortazavi[1], Gholamhossein Zohuri[2], Meisam Haghpanah[2]

[1]Department of Catalyst, Polymer and Petrochemical Institue, Tehran, Iran
[2]Department of Chemistry, Faculty of Science, Ferdowsi University of Mashhad, Mashhad, Iran

There are three main methods which can be used for combination of different types of catalysts including catalyst alloys, hybrid catalysts and multi-catalyst systems. Through these systems, each catalyst can produce polymer with a different microstructure and properties in absence of cooperation. As a result, the obtained polymer is a blend of two or more kind of polymers [Mortazavi et al., 2015]. In this study, a binary catalytic system was used in polymerization of ethylene. A mixture of Ziegler-Natta (ZN) and a binuclear late transition metal (BLTM) catalyst was employed in presence of methylaluminoxane (MAO) and triethylaluminium (TEA) as cocatalyst and impurity scavenger, respectively. To study on binary catalytic system and polyethylene (PE) properties, the experiments were performed under similar polymerization condition for each single catalyst and mixture of them. Initial experiments showed ZN/BLTM with a moderate catalytic performance between activities of each catalysts (Figure A). Chains vinyl-end and branching, caused by β-hydrogen elimination transfer reactions to the metal center or to the monomer followed by a subsequent olefin reinsertion which called chain-walking mechanism allowing the growing center to migrate along the polymer chain, is more pronounced in polymers produced by LTM catalysts [Chen et al., 2015]. Based on aforementioned reasons, branching degree and vinyl content for ZN/BLTM catalytic system was higher than pure ZN. In comparison to pure BLTM, vinyl content increased and branching degree decreased which can be attribute to presence of the counterpart catalyst (ZN) which leads to weaker M—C bond and dissociation of the chain along with vinyl-end.

Keywords: ethylene polymerization, binary catalytic system, microstructure, Ziegler-Natta, binuclear late transition metal

Figure A

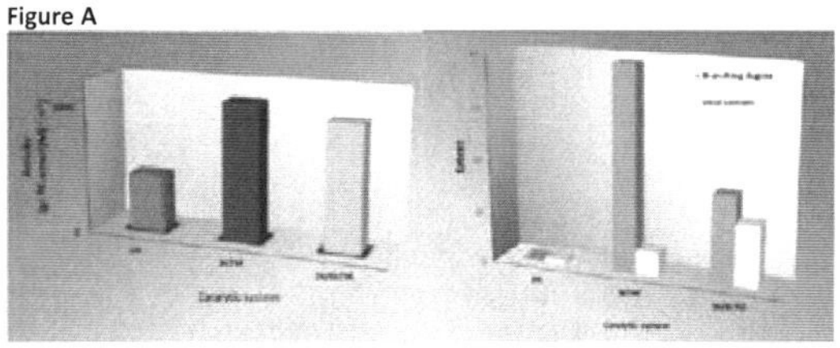

RDPS-A-P-034 [Controlled/Living Polymerization]

One-pot synthesis of well-defined polyether/polyester block copolymers and terpolymers by a highly efficient catalyst switch approach

Haleema Alamri, Nikos Hadjichristidis

King Abdullah University of Science and Technology, Physical Sciences and Engineering Division, KAUST Catalysis Center, Polymer Synthesis Lab, Thuwal 23955, Saudi Arabia

A highly efficient methodology, based on a novel catalyst switch approach with rapid crossover characteristics, was developed for the one-pot synthesis of block co/terpolymers of cyclic ethers and esters. This new approach, which takes advantage of one of the best catalysts for the epoxides1 (t-BuP4) and cyclic esters (t-BuP2) polymerization,2 opens new horizons towards the synthesis of cyclic ethers/esters complex macromolecular architectures.3 In this work we also provide conclusive evidence to support the non-toxicity of (t-BuP4) and (t-BuP2) used in this study using the cell viability measurements.

REFERENCES:

1. J. Zhao, D. Pahovnik, Y. Gnanou and N. Hadjichristidis, Macromolecules, 2014, 47, 3814-3822.
2. H. Alamri, J. Zhao, D. Pahovnik and N. Hadjichristidis, Polymer Chemistry, 2014, 5, 5471-5478.
3. H. Alamri and N. Hadjichristidis, Polymer Chemistry, 2016, 7, 3225-3228.

Keywords: polyether/polyester, block copolymers, ring opening polymerization

Scheme 1.

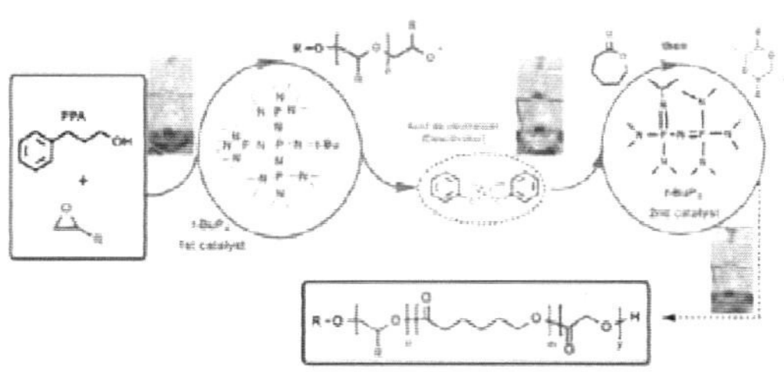

General reaction scheme of an efficient one-pot route towards the synthesis of (substituted polyether)-b-polyester di/triblock co/terpolymers via an organocatalytic switching approach with a highly efficient crossover polymerization.

RDPS-B-P-001 [Complex Macromolecular Structures]

Theoretical investigation of organic schiff base compounds on corrosion inhibition of copper

Saifi Issaadi[1], Tahar Douadi[2]

[1]*Department of Chemistry, Faculty of Science, University of Setif-1, DZ-19000, Setif-Algeria.*

[2]*Department of genie procedy, Faculty of technology, University of Setif-1, DZ-19000, Setif-Algeria.*

Quantum chemical calculations based on DFT method were performed on two schiff bases compounds, used as corrosion inhibitors for copper in HCl media to determine the relationship between the molecular structure of schiff base and inhibition efficiency. Quantum chemical parameters such as the highest occupied molecular orbital energy (EHOMO), the lowest unoccupied molecular orbital energy (ELUMO), energy gap (ΔE), dipole moment (I), electronegativity (v), electron affinity (A), global ionization potential (I), the fraction of electrons transferred (ΔN), and the total energy (TE), were calculated. The theoretically obtained results were found to be consistent with the experimental data reported.

Keywords: Optimization, DFT, Quantum chemical

RDPS-B-P-002 [Complex Macromolecular Structures]

Synthesis, Spectroscopy, X-Ray Crystallography, Electrochemestry and Theoretical Investigation of a Tetradentate Copper Schiff Base Complex Cu2L2.2H2O

Samra Rahmouni[1], Amel Djedouani[2], Salima Thabti[3], Abderrahmen Bendaas[1]

[1]*Laboratory of Electrochemistry of Molecular Materials and Complexes (LEMMC), Department of Engineering Process, Faculty of Technology, Ferhat Abbas University Setif 1, DZ-19000 Setif, Algeria.*

[2]*Laboratoire de Physicochimie Analytique et Cristallochimie des Matériaux Organométalliques et Biomoléculaires, Constantine 1.*

[3]*Faculté des sciences et technologie, département SM Université de Mohamed el Bachir el ibrahimi, El Anasser- Bordj Bou-Arréridj Algérie*

Schiff bases are widely used as ligands to coordinate various metal ions because of their good solubility [1, 2]. These compounds contain several donor atoms and have potential application in water treatment, due to their ability to readily form transition metals complexes. These complexes species are also involved in enzymatic reaction and present in proteins from conjunctive tissue. They can also be used as selective membranes or as corrosion prevent compounds [3-5]. New binuclear copper complex [Cu2L2.2H2O] has been synthesized. This complex was characterized by a variety of physicochemical techniques such as FT-IR, UV-Vis and cyclic voltammetry measurements. The molecular structure of this complex was determined by single crystal X-ray diffraction technique. This technique shows that complex belongs to the triclinic system with space group P-1(2). The redox behavior was explored by cyclic voltammetry on glassy carbon electrode in DMSO at 100 mV/s. The DFT B3LYP/6–31 G (d,p) calculations are carried out for the determination of the optimized structures. The fundamental vibrational wave numbers were calculated and a good agreement between observed and scaled calculated wave numbers was achieved.

Keywords: Copper (II) complex, Crystal structure, Cyclic voltammetric, Quantum chemical Calculations (DFT).

RDPS-B-P-003 [Complex Macromolecular Structures]

Ring-Opening Reactions of Backbone Epoxidized Polyoxanorbornene

Ufuk Saim Gunay, Erhan Demirel, Gurkan Hizal, Umit Tunca, Hakan Durmaz

Department of Chemistry, Istanbul Technical University, Istanbul, Turkey

This study is about the epoxidation of the poly (oxanorbornene imide) (PONB) main chain and its reactivity to well-known nucleophiles (amines, azide and thiols) through the amine-epoxy and thiol-epoxy ring opening reactions and azidation of epoxy ring [Reactive and Functional Polymers, 2015, 94, 35-42]. The ring opening reactions with amines yielded the amine-hydroxyl PONBs. An allylamine-hydroxyl functionalized PONB was further functionalized efficiently with 1-octanethiol by radical thiol-ene reaction. The ring-opening reaction of the main chain epoxy using thiols resulted in a lower functionalization than amines with a similar functional group. In addition, sodium azide together with NH4Cl, was employed efficiently in the ring-opening reaction of the main chain epoxy, resulting in an azide/hydroxyl-functional PONB.

Keywords: Poly(oxanorbornene imide) (PONB), Ring-Opening Metathesis Polymerization (ROMP), backbone epoxidation, amine-epoxy ring-opening reaction, thiol-epoxy ring-opening reaction

RDPS-B-P-004 [Complex Macromolecular Structures]

Post-Functionalization Of Perfluorophenyl Ester-Functional Acyclic Diene Metathesis Polymer

Ozgun Daglar, Hakan Durmaz, Gurkan Hizal, Umit Tunca

Department of Chemistry, Istanbul Technical University, Istanbul, Turkey

The acyclic diene metathesis (ADMET) polymerization was utilized for the design of the ADMET polymer (*Mn*=21200 g/mol, *Mw/Mn*=1.74) with pendant perfluorophenyl ester functionality using Grubbs 1st generation catalysis overnight in bulk at 80°C. Next, a wide variety of functional groups, like benzyl, octyl, propargyl, allyl and furfuryl was quantitatively incorporated to the ADMET polymer backbone through various amines using activated ester substitution reaction. The ADMET polymers studied in this work were characterized using ^{1}H, ^{13}C and ^{19}F NMR, GPC and DSC and displayed a monomodal distribution and a rather broad polydispersity index in the range of 1.33-1.90.

Keywords: Acyclic diene metathesis (ADMET) polymerization, activated ester substitution reaction, perfluorophenyl ester, benzyl amine, propargyl amine

Schematic representation of ADMET polymerization and activated ester substitution reactions.

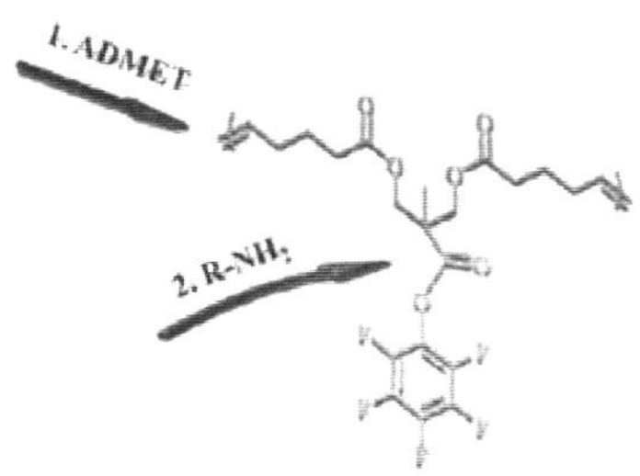

RDPS-B-P-005 [Complex Macromolecular Structures]

Main Chain Post-Functionalization of Reactive Triple Bond Containing Polyester via Diels-Alder Cycloaddition Reactions

Cansu Esen, Ozgun Daglar, Muge Cetin, Hakan Durmaz, Gurkan Hizal, Umit Tunca

Department of Chemistry, Istanbul Technical University, Istanbul, Turkey

Novel polyester obtained by an esterification reaction of acetylenedicarboxylic acid and 1,4-butanediol was prepared. Due to its electron deficient character, the alkyne functionality in the main chain was readily reacted with various electron rich compounds such as furan, anthracene or 2,3-dimethyl-1,3-butadiene via Diels-Alder (DA) [4+2] cycloaddition reactions. The suitable DA reaction temperatures to reach the quantitative functionalization of the main chain was found to be 80 °C for furan and 2,3-dimethyl-1,3-butadiene and 110 °C for anthracene in 1,4-dioxane. The progress of the reactions was easily followed by ^{1}H NMR by comparing the related signals. All studied polymers were also characterized by ^{13}C NMR, FT-IR and GPC. The thermal behaviors of the obtained polymers were revealed by DSC and TGA measurements. It was found that anthracene and 2,3-dimethyl-1,3-butadiene functionalized polymers as well as the precursor polymer were stable up to 200 °C, whereas the furan functionalized polyester demonstrated the *retro*-Diels-Alder reaction around 130 °C.

Keywords: Diels-Alder cycloaddition reactions, Condensation reaction, Acetylenedicarboxylic acid, Furan, Anthracene

Schematic Representation of Main Chain Post-Functionalization of Acetylenedicarboxylic Acid Based Polyester via Diels-Alder Cycloaddition Reactions

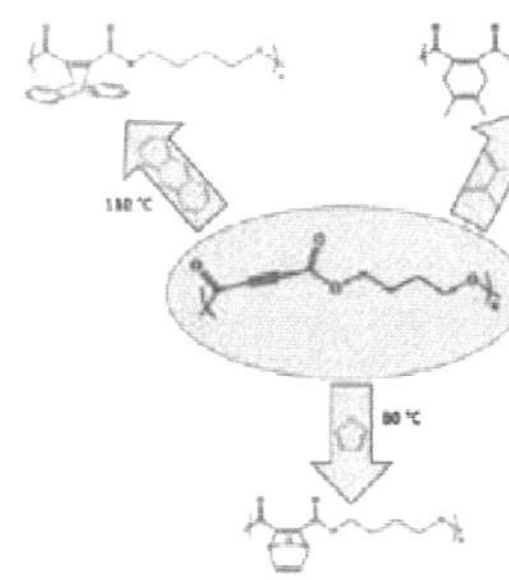

RDPS-B-P-006 [Complex Macromolecular Structures]

Active Ester Substitution And Nitrene Addition Reactions On To Multiwalled Carbon Nanotubes For Double Modification

Pinar Sinem Omurtag, Hakan Durmaz, Gürkan Hızal, Ümit Tunca

Department of Chemistry,Istanbul Technical University, Complex Macromolecular Structure Center, Maslak 34469, Istanbul, Turkey

Carbon nanotubes have speedily turned into an interdiciplinary field on account of showing their superior structural, mechanical, electrical, and electromechanical features. These features increase the usage of carbon nanotubes (CNTs) in many areas such as sensoring, nerve cell stimuli, drug delivery, cancer therapy. However, major limitation for application of CNTs is their poor solubility, dispersion, processibility and self aggregation. Hence, functionalization of CNTs is necessary. Therefore some methods have been reported such as non-covalent functionalization (π-π stacking interactons) and covalent functionalizations (Diels-Alder reaction, nitrene addition or thiol-ene click reaction)1,2,3. This study highlights the application of the active ester substitution reaction and nitrene addition reactions to perform double modification strategy to multiwalled (MWCNTs) carbon nanotubes for surface modification. For this purpose, well-defined polymers (PS, PMMA) were prepared by using ATRP and then via simple substitution reaction with NaN3 they gained azide functionality. Secondly, carboxylic groups on the surface of the MWCNTs (MWCNTs-COOH) were converted to the acyl chlorides. Reaction of the acyl chlorides with pentafluorophenol formed the active esters. Active ester modified MWCNTs (MWCNTs-COOC6F5) showed a high reactivity toward amines4. Following reaction occurred in the presence of MWCNTs-COOC6F5 and primary or secondary amines resulting in amide bonds. Finally, obtaining MWCNTs-CONHR double modified with α-azide polymer 5. The homopolymer precursors characterized by using the GPC, NMR and UV analysis. Furthermore, characterization of CNTs will be made by using TGA, FT-IR.

References::

[1] Zydziak, N.; Yameen B.; Barner-Kowollik, C. Polym. Chem. 2013, 4, 4072–4086
[2] Iskin, B.; Yilmaz, G.; Yagci, Y. Chem. Eur. J. 2012, 18, 10254 – 10257
[3] Temel, G.; Uygun, M.; Arsu, N. Polym. Bull. 2013, 70, 3563–3574.
[4] Park, H. J.; Kim, J.; Chang, J. Y.; Theato, P. Langmuir 2008, 24, 10467-10473
[5] Munirasu, S.; Albuerne, J.; Boschetti-de-Fierro, A.; Abetz, V. Macromol. Rapid Commun. 2010, 31, 574–579

Keywords: carbon nanotubes, active ester substitution, nitrene addition

RDPS-B-P-007 [Complex Macromolecular Structures]

Multiarm Star Polymer as Novel Polymeric Photoinitiator for Free Radical Polymerization

Neşe Çakır[1], Duygu S. Esen[2], Ümit Tunca[1], Nergis Arsu[2], Gürkan Hızal[1]

[1]*Department of Polymer Science and Technology, Istanbul Technical University, Istanbul, Turkey*
[2]*Department of Chemistry, Yildiz Technical University, Istanbul, Turkey*

Benzoin and derivatives have long been known to be efficient photoinitiators for free radical polymerization of vinyl monomers and have found wide application in the polymer industry.[1,2] Star polymers are branched polymers that consist of several linear polymer chains connected to a central core.[3] The interest in star-shaped polymers stems from their unique shape and possible processing advantages due to their compact structure.[4]

Here, we present a benzoin and derivative, MTB funtional PS (Benzoin-PS and MTB-PS) were prepared by ATRP of St and used as macroinitiators in a cross-linking reaction with DVB to yield multiarm star homopolymers with benzoin and MTB periphery, respectively. Obtained multiarm (Benzoin-PS)$_m$-polyDVB and (MTB-PS)$_p$-polyDVB star polymers were used as polymeric photoinitiators. Photokinetics of 1,6-hexanediol diacrylate (HDDA) with multiarm (Benzoin-PS)$_m$-polyDVB and (MTB-PS)$_p$-polyDVB star polymers were studied by Photo-DSC. These polymeric photoinitiators were used in free radical photopolymerization of methyl methacrylate (MMA) in air atmosphere.

This is the first study based on star polymers that have benzoin and MTB end-functionalities in each arm that showed high light absorption in the near-UV spectral region. It can be seen that photopolymerization using star polymer as initiator proceeds with a significantly fast polymerization rate.

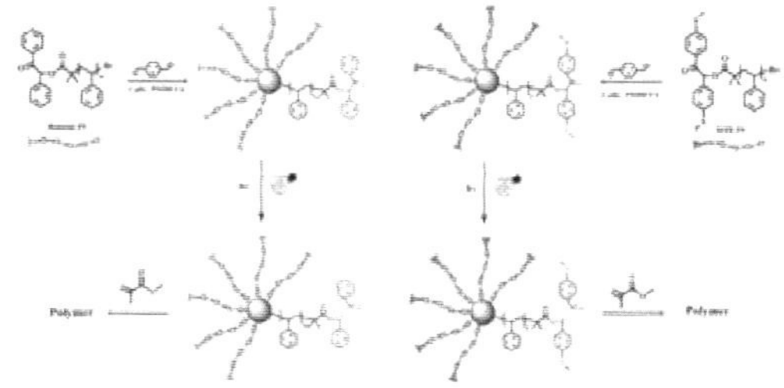

Scheme 1. Photoinitiation mechanism of multiarm (Benzoin-PS)$_m$-polyDVB and (MTB-PS)$_p$-polyDVB star polymer

References:

1) J. P. Fouassier, Photoinitiation, Photopolymerization and Photocuring; Hanser Verlag: Munich, 1995.
2) Y. Yagci, S. Jockusch, N. J. Turro, *Macromolecules* 2010, 43, 6245–6260.
3) Hadjichristidis, N. *Polym. Sci., Part A: Polym. Chem.*, 1999, 37, 857–871.
4) Xia J., Zhang X., Matyjaszewski, K., *Macromolecules*, 1999, *32* (13), 4482–4484.

Keywords: Multiarm star polymer, ATRP, polymeric photoinitiator

RDPS-B-P-008 [Complex Macromolecular Structures]

Synthesis and Self-Assembly of Fluorene-Vinylene Alternating Copolymers in "Hairy-Rod" Architecture: Side Chain – Mediated Tuning of Conformation, Microstructure and Photophysical Properties

Demet Göen Colak[1], Ioan Cianga[2], Luminita Cianga[2], Yusuf Yagci[1]

[1]*Department of Chemistry, Faculty of Science and Letters, Istanbul Technical University, Istanbul, Turkey*
[2]*"Petru Poni" Institute of Macromolecular Chemistry, Iasi, Romania*

In the present study, we report the synthesis of fluorenylene vinylene copolymers with well-defined oligomeric side chains via Wittig polycondensation by using "T"-shaped or "Cross"- shaped p-terphenyl macromonomers. The overall character of the copolymers, in "hairy-rod" architecture, was systematically investigated by attaching of hydrophilic PEG 2000, hydrophobic polar oligo-ε-caprolactone or hydrophobic and non-polar oligostyrene side chains. The results revealed that the side chain choice, as a tunable parameter, is an effective strategy to drive molecular ordering, packing motifs and overall microstructure of a conjugated polymer. Self-assembling of the copolymers by simple direct dissolution method was achieved in various solvents by modifying their selectivity in relation to the side chain or main chain. The morphology investigations demonstrated that unique nanofeatures obtained in each case (helical foldamers, vesicles, disks or helical turns) depend on the nature, number and position of the side chains which influence the photophysical properties. The"hairy-rod" topology is also responsible for the self-assembly of the materials in molten state, as thermal analysis revealed, and the propensity of the new synthesized conjugated main chain for helical folding was evidenced, as well. In conclusion, due to their distinct properties such as water dispersibility and biocompatibility, these materials could find applications in biomedical field as well as in opto and bio electronics.

Keywords: Conjugated Polymers, Poly(fluorene vinylene)s, "Hairy-rods", Self-assembly.

Schematical representation of the synthesized "Hairy-Rod" copolymers.

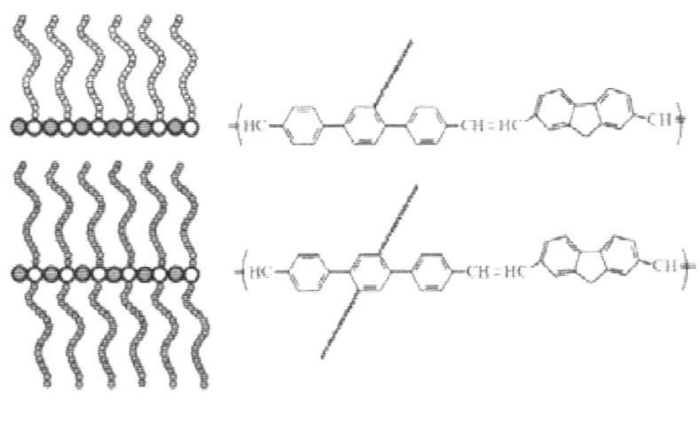

RDPS-B-P-009 [Complex Macromolecular Structures]

Synthesis of Dendronized Polymers Containing Polyphenylene Dendron via Macromonomer Approach by ROMP and Their Characterization

Ki Taek Bang, Inho Choi, Tae Lim Choi

Department of Chemistry, Seoul National University, Seoul, Korea

Dendronized polymers are unique macromolecules that consist of dendrons attached to polymer backbones. These polymers have several advantages because one can design a polymer with a precisely controlled molecular architecture and desired properties by tuning the polymer backbone and the dendritic wedge. Also dendronized polymers exhibit an extended conformation because of the steric repulsion between the repeating units (Kyung-Oh Kim. et. al., *Macromolecules* 2013, *46*, 5905). Meanwhile, polyphenylene dendrimer is highly rigid, monodisperse macromolecule that is composed of substituted benzene ring. It has several interesting characteristics of being shape-persistent, chemically stable and structural tunability. Its characteristics allow unique materials that can be utilized various kinds of applications thought bright synthetic design (Brenton A. G. Hammer et. al., *Chem. Soc. Rev.* 2015, *44*, 4072). Considering the unique features of the dendronized polymer and polyphenylene dendrimer, combining the two concepts will afford another type of very unique macromolecule. However, to the best of our knowledge, dendronized polymers containing polyphenylene dendron have never been reported. In this research, synthesis and characteristics of the dendronized polymers have been reported. The dendronized polymers were synthesized by ring-opening metathesis polymerization (ROMP) using endo-trichycle[4.2.2.0]deca-3,9-diene (TD) derivatives via macromonomer approach. The conformation of well-defined dendronized polymers in solution state has been investigated by multi-angle laser light scattering (MALLS). The film-state conformation of the polymers was revealed by Atomic force microscopy (AFM) and transmission electron microscopy (TEM). Detailed explanation about the synthesis and their unique characteristics of the dendronized polymer will be introduced.

Keywords: dendronized polymer, polyphenylene dendrimer, ring-opening metathesis polymerization(ROMP)

A structure of the dendronized polymer

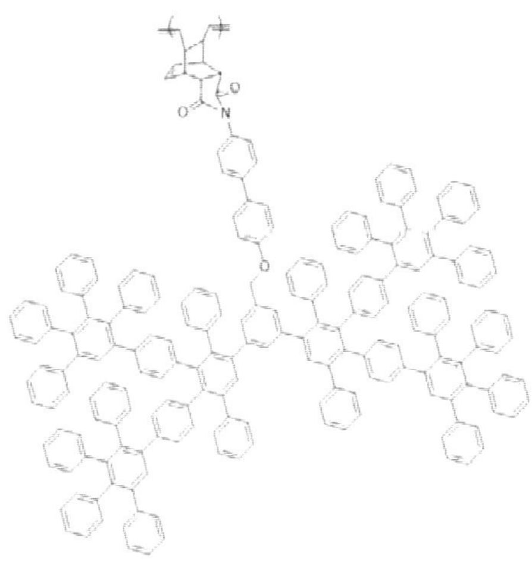

RDPS-B-P-010 [Complex Macromolecular Structures]

Synthesis of Polythioether with Clickable Side Chain Through Thiol Michael Condensation Reaciton

Serter Luleburgaz, Hande Tınas, Hakan Durmaz, Umit Tunca, Gurkan Hizal

Department of Chemistry, Istanbul Technical University, Istanbul, Turkey

The Michael addition reaction has emerged as a versatile synthetic methodology and gained much interest in recent years among polymer chemists to produce polymers with linear or non-linear architectures. The reaction can be classified under the term "click" chemistry since it requires mild reaction conditions as compared to other click reactions. [1,2] In this study, polythioether with pendant alkyne functional group, which is suitable for further click reactions, was synthesized through the Micheal addition reaction. For this purpose, 1,1,1-tris(hydroxymethyl)ethane reacted with 4-pentynoic acid and subsuquently reacted with acryloyl chloride to obtain 1. The polythioether was prepared by means of Micheal addition reaction between 1 and 1,2-ethandithiol in the presence of triethyl amine as catalyst. The side chain click modifications of polythioether was carried out using either copper catalyzed azide-alkyne cycloaddition (CuAAC) reaction employing benzyl azide and mono azide-functional polyethylene glycol (PEG-N3) or radical thiol-yne reaction employing propane thiol. The obtained polymers were characterized by NMR, GPC and DSC measurements and the results demonstrated the formation of polythioether backbone with a moderate molecular weight and quantitative post-functionalizations on the side chains.

Keywords: Michael addition reaction, polythioether, CuAAC, Thiol-yne, Click Chemistry

General Scheme

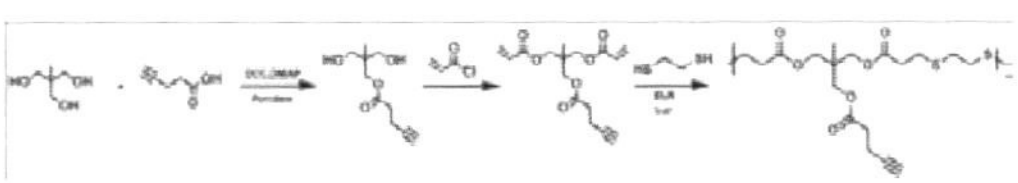

Polymer synthesis is shown in figure.

Modifications

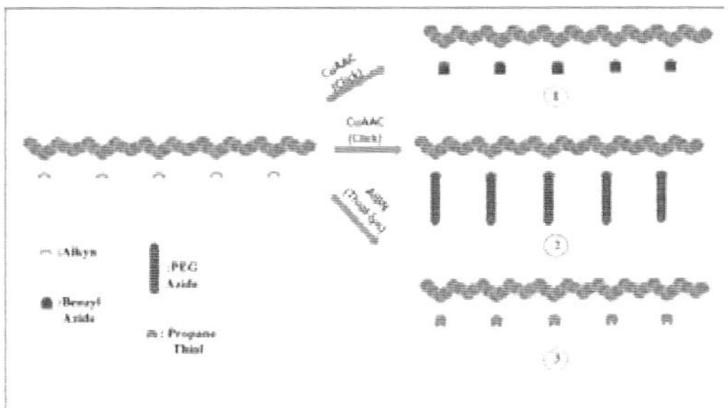

RDPS-B-P-011 [Complex Macromolecular Structures]

Synthesis and characterization of porphyrin containing commercial polymers

Aylin Bilgili[1], Fehmi Saltan[2], Hakan Akat[1]

[1]*Department of Chemistry, Ege University, İzmir, Turkey*
[2]*Department of Chemistry, Çankırı Karatekin University, Çankırı, Turkey*

In recent years, polymeric materials containing macrocyclic units with the ability to bind transition metals have attracted much interest. Potential applications for these materials are found in the field of the chemical conversion of solar energy by photocatalytic or electrochemical reactions,highly selective polymeric membrane, polymer metal complexes used as immobilized catalysts and polymer liquid crystals. Among the macrocyclic structures incorporated in the polymer chains, porphyrin and metalloporphyrin units have frequently been used especially for the construction of artificial oxygen carriers inspired by those in biological systems, the development of new catalytic systems band the synthesis of new receptors able to bind amino acids (via metal-coordination and hydrogenbonding interaction). There are two types a of polymers in which the metalcontaining units either are bound to the polymer as pendant groups or are part of the polymer main chain. In porphyrin compounds, there are methine bridges between the pyrrole rings and these compounds form the coupled system of double bonds, including 18 delocalized π electrons. Because of the porphyrin compounds have spectacular stability, unuqie photophysical and electrochemical properties, presumable and rigid structures, they show important application potentials in many fields.

In this study we will synthesize tetraphenyl porphyrin from benzaldehyde and pyrrole. And then tetraphenyl porphyrin will be respectively nitrated and amination via benzene ring. After than this step, alkyne-edged tetraphenyl porphyrin will be synthesized by propargyl bromide. Finally, PVC-N3 and ve PS-N3 in different percentages will link up tetraphenly porphyrin by click chemistry method. Structural characterizations of the synthesized polymers will be performed by 1H-NMR, 13C-NMR, FT-IR spectrums. Molecular weights will be determined by GPC. Spectrometric analysis will be performed with UV and fluorimetry devices. Thermal analysis will be performed with TG and DSC. Besides, we aim determination of glass transition temperature and the percent degradation of the synthesized polymers.

Keywords: tetra phenyl porphyrin, commercial polymer, radical polymerization, click chemistry

Figure 1. Synthesis steps of polymers

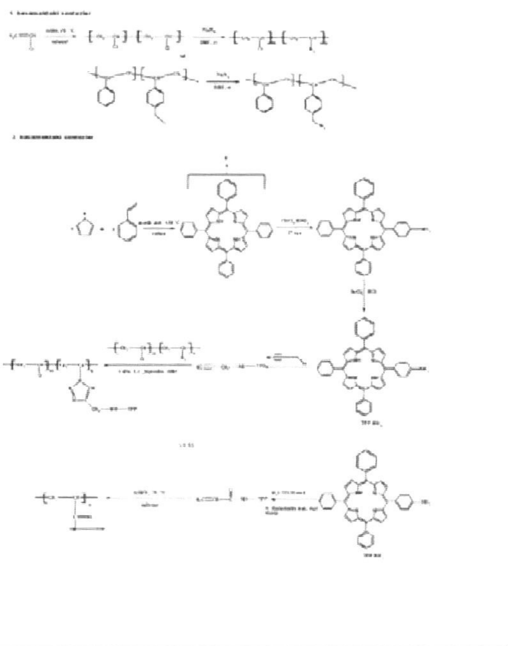

RDPS-B-P-012 [Complex Macromolecular Structures]

Coacervation between biopolymers

Ayse Basak Kayitmazer, Alaaddin Faruk Koksal, Elif Kilic Iyilik, Fatma Akcay Ogur

Department of Chemistry, Bogazici University, Istanbul, Turkey

Complex coacervation is a liquid–liquid phase separation of polyelectrolytes with oppositely charged macromolecules such as proteins, dendrimers, micelles, inorganic nanoparticles, nucleotides or other polyelectrolytes. Coacervation between two oppositely charged, weak and semiflexible biopolymers—namely, sodium hyaluronate (HA) and chitosan (CH) was experimentally studied. Phase separation is facilitated by increasing polyelectrolyte charge density; i.e., using chitosan with higher degrees of deacetylation for CH. Coacervation is favored for HA with longer chain length at constant ionic strength. On the other hand, turbidity versus ionic strength plotted at a constant pH shows a maximum critical ionic strength regardless of the total polymer concentration or pH. As pH increases, insoluble interpolyelectrolyte complexes (PEC) take different forms ranging from metastable coacervate suspensions (circular droplets) to precipitates for I = 0.15–0.75 M NaCl. Coacervate suspensions were still observed with zeta potential values much higher than zero at charge ratios less than 0.46. This result is attributed to inequivalence in charge spacing of the two biopolymers. The total polymer concentration (Cp) is directly correlated with the turbidity of the PECs; i.e., as the total polymer concentration increases, turbidity increases. Macroscopically, precipitates were observed at Cp = 1.0 mg ml-1. No precipitates, however, were observed at Cp = 0.5 mg ml-1, where phase separation led to coacervation. Higher turbidity for larger Cp would then arise from the higher number of associated PEC aggregates. Titration of the HA/CH mixture with salt allows the system to go through different phases, as evident by changes in the turbidity. Enhancement of phase separation with added salt is explained as follows: (i) charge screening reduces long-range repulsions and overcomes intracomplex disproportionation. (ii) Rearrangements in the intracomplex structure might facilitate the appearance of clusters of interpolymer complexes.

Keywords: Complex coacervation, Biopolymers, Polyelectrolytes

RDPS-C-P-002 [Light-induced Reactions in Polymer Science]

Synthesis and and evaluation of novel thioxanthone-based methacrylic acids

Tuğçe Nur Eren[1], Duygu Avci[1], Jacques Lalevée[2]

[1]*Department of Chemistry, Bogazici University, 34342 Bebek, Istanbul, Turkey*
[2]*Institut de Science des Materiaux de Mulhouse IS2M, UMR CNRS 7361, UHA, 15, Rue Jean Starcky, 68057 Mulhouse Cedex, France*

Monomeric or polymeric photoinitiators are investigated in the literature due to their advantages such as low odor, less toxicity, good solubility in the formulation and a polymer effect; compared to their low-molecular-weight nonmonomeric analogues. Among them, thioxanthone (TX) derivatives are important due to their excellent light absorption properties and high photoinitiation efficiency in the visible region. In this work, two (TX)-functionalized methacrylic acid photoinitiators were synthesized (Figure). The photopolymerization efficiencies of the synthesized photoinitiators together with isopropylthioxanthone (ITX) as a reference are studied under LED exposure at 385 and 405 nm in the polymerizations of trimethylolpropane trimethacrylate; using real-time FTIR. The photochemical mechanisms are studied by electron spin resonance, steady state photolysis, laser flash photolysis and cyclic voltammetry.

Keywords: LED, photopolymerization, thioxanthone

Figure.

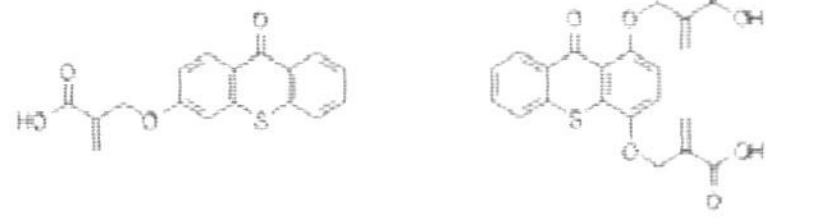

RDPS-C-P-003 [Light-induced Reactions in Polymer Science]

Extension of Linear PIBs to Higher Molecular Weight Linear and Network Products via Thiol-ene Chemistry

Elif Kurnaz[1], Nihan Nugay[1], Turgut Nugay[1], Joseph P. Kennedy[2]

[1]*Department of Chemistry, Polymer Research Center, Bogazici University, Istanbul, Turkey*
[2]*Department of Polymer Science, The University of Akron, Akron, Ohio*

This research focuses on the preparation of high molecular weight rubbery linear or crosslinked polyisobutylenes (PIBs) from low molecular weight liquid PIB precursors by thiol-ene click chemistry [1]. Thiol-ene click leads to polymer blocks by chain extension. The synthetic strategy to prepare both high molecular weight linear PIBs and PIB networks consists of three steps: (i) synthesis of the difunctional carbocationic initiator 5-tert-butyl-1,3-bis(2-chloro-2-propyl)benzene (t-Bu-m-DiCumCl), (ii) synthesis of well-defined PIBs by living carbocationic polymerization followed by end-capping with allyltrimethylsilane, and (iii) UV light induced thiol-ene click reaction in the presence of 2,2-dimethoxy-2-phenyl acetophenone (DMPA) photoinitiator. The effect of photoinitiator concentration, reaction time, nature of the thiol, and thermal initiation on the efficiency of radical thiol-ene click addition leading to extension and crosslinking were investigated.

The authors thank Bogazici University Research Foundation Projects (14B05P12) for financial support.

References :

[1] Kennedy, J.P., Nugay, T. and Nugay, N., Patent Application WO 2014/138017 A1, 2014.

Keywords: Polyisobutylene, Thiol-ene, Carbocationic Polymerization

Synthetic strategies for extension and for networks

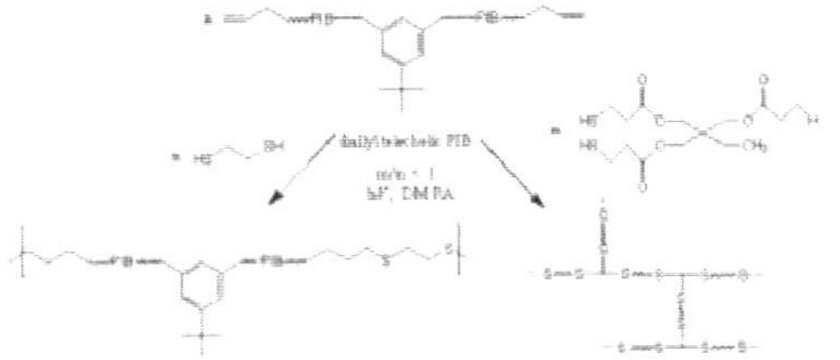

RDPS-C-P-004 [Light-induced Reactions in Polymer Science]

The use of 1-(2-anthryl)-1-phenylethylene moiety at the chain end to perform chain extension and block copolymerization reactions

Çimen Özgüç[1], Yusuf Yağcı[2], Turgut Nugay[1]

[1]*Department of Chemistry, Bogazici University, Bebek, Istanbul,Turkey*
[2]*Department of Chemistry, Istanbul Technical University, Maslak, Istanbul,Turkey*

Anthracene containing polymers have gathered significant research interest in the last decade due to the attractive features of the molecule, a major one being the ability to photodimerize by ultraviolet (UV) irradiation above 300nm via a [4π+4π] cycloaddition reaction. The resulting photodimers can be reversed either thermally at an elevated temperature or via UV irradiation at wavelengths below 300 nm. In this research anthryl and di-anthryl telechelic polymers have been synthesized with the incorporation of anthracene bearing monomer, 1-(2-anthryl)-1-phenylethylene (APE), in the initiation step or via chain end functionalization. Living anionic polymerization and quasi-living cationic polymerization techniques have been utilized to obtain well-defined polymers containing the photosensitive anthracene moieties. Reversible photocycloaddition reactions of anthracene end-functionalized polymers have been studied via performing UV irradiations of alternating wavelengths on the synthesized materials. Both the photo-induced chain extension of polymers and reverse photoscission reactions were performed successfully by the irradiation process at 365nm and 254 nm, respectively.

Keywords: anthracene, photocycloaddition, anionic polymerization, cationic polymerization

Reversible photocycloaddition reaction of anthracene end-functionalized polymers

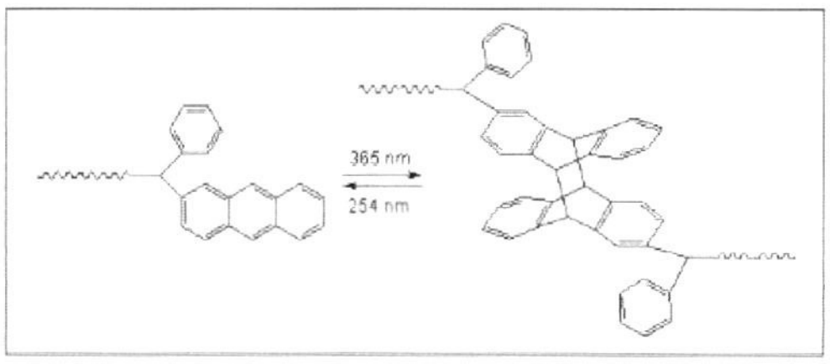

RDPS-C-P-005 [Light-induced Reactions in Polymer Science]

Synthesis of Clickable Hydrogels and Linear Polymers by Type II Photoinitiation

Mustafa Ciftci[1], Eljesa Murtezi[1], Yusuf Yagci[1], Yusuf Yagci[2]

[1]*Department of Chemistry, Faculty of Science and Letters, Istanbul Technical University, Maslak, TR-34469 Istanbul, Turkey*
[2]*Center of Excellence for Advanced Materials Research (CEAMR) and Chemistry Department, Faculty of Science, King Abdulaziz University, P.O. Box 80203, Jeddah 21589, Saudi Arabia*

A mild and highly versatile photochemical method for the preparation of clickable macromolecular structures, i.e., hydrogels and linear polymers via hydrogen abstraction type photoinitiation using aromatic carbonyl compounds such as camphorquinone (CQ) in conjunction with functional co-initiators has been developed. The strategy pertains to the photochemical generation of initiating radicals by hydrogen abstraction of excited CQ from acetylenic alcohol followed by free radical polymerization. For hydrogels, poly(ethylene glycol) methylether methacrylate (PEGMA) and poly(ethylene glycol) diacrylate (PEGDA) were used in the system as monomer and cross-linker, respectively. Fluorescent pyrene groups were anchored onto the hydrogels through copper-catalyzed Huisgen 1,3-dipolar azide/alkyne click reaction by using 2-azidomethyl pyrene (Py-N_3) as evidenced by FT-IR and fluorescence spectroscopy. The applicability of the method for macromolecular synthesis was also demonstrated. For this purpose, poly(methyl methacrylate) with alkyne terminal groups was prepared by photoinitiated free radical polymerization of methyl methacrylate (MMA) using CQ and 3-(trimethylsilyl)propargyl alcohol as sensitizer and hydrogen donor, respectively. After deprotection, the click reaction with azido polystyrene obtained by Atom Transfer Radical Polymerization (ATRP) followed by azidation resulted in the formation of polystyrene-*b*-poly(methyl methacrylate). The intermediates at various stages and final block copolymer were characterized by spectral analysis and Gel Permeation Chromatography.

References:

1. Yilmaz G, Kahveci MU, Yagci Y, Macromolecular Rapid Communications, 2011, 32(23), 1906-1909.
2. Murtezi E, Ciftci M, Yagci Y. Polymer International, 2015, 64(5), 588-594.

Keywords: photopolymerization, click chemistry, hydrogels, visible light polymerization

Synthesis of "clickable" PEG crosslinked hydrogels by photoinitiated free radical polymerization and their functionalization.

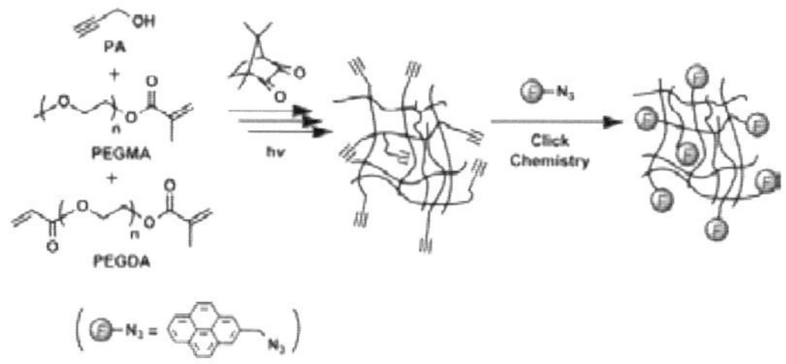

RDPS-C-P-006 [Light-induced Reactions in Polymer Science]

Dibenzoyldiethylgermane as a Visible Light Photo-reducing Agent for CuAAC Click Reactions

Mustafa Arslan[1], Gorkem Yilmaz[2], Yusuf Yagci[2]

[1]*Department of Chemistry, Kirklareli University, Kırklareli, Turkey and Department of Chemistry, Istanbul Technical University, Istanbul, Turkey*
[2]*Department of Chemistry, Istanbul Technical University, Istanbul, Turkey*

Click reactions are the most widely investigated class of organic reactions with superior features over the traditional routes such as rapidity, orthogonality, versatility and almost quantitative efficiency. Nevertheless, equimolar concentrations of the substrates are enough to benefit from such advantages and the products are easy to purify since no by-products are formed. Due to these outstanding advantages, click chemistry has a wide-range of applications in organic chemistry, polymer chemistry, supramolecular chemistry, drug discovery and material design. The most common examples of click reactions are the copper catalyzed azide–alkyne cycloaddition (CuAAC) and Diels–Alder (DA) reactions. A highly active, versatile and photoresponsive system for copper catalyzed azide–alkyne cycloaddition (CuAAC) click reaction has been developed using the dibenzoyldiethylgermane (DBDEG) photoinitiator with the copper(II) chloride (CuCl2)–N,N,N′,N″,N″-pentamethyldiethylenetriamine (PMDETA) ligand under visible light. Selected azide and alkyne compounds with various functional groups are used to conduct the click reaction and almost quantitative yields are attained. The approach pertains to the visible light generation of germyl radicals capable of reducing Cu(II) ions to Cu(I) species to catalyze the CuAAC reactions. This strategy has been applied in the achievement of various macromolecular reactions including polymer end-group functionalization, block copolymer formation and step-growth polymerization.

Keywords: Click Chemistry, Photoinitiator, Dibenzoyldiethylgermane

A highly active, versatile and visible light-responsive system for CuAAC click reaction using dibenzoyldiethylgermane photoinitiator

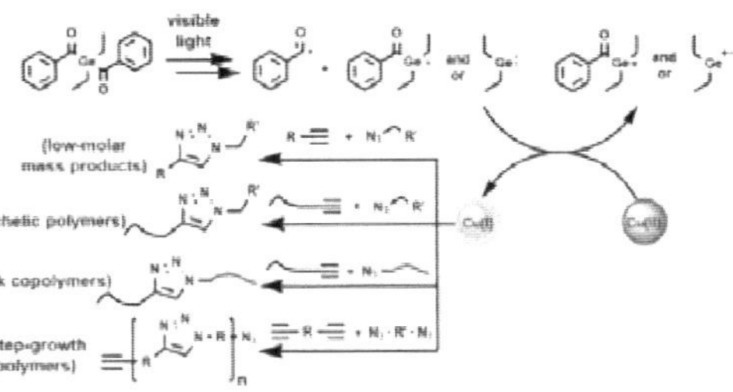

The highly active, versatile and visible light-responsive system for CuAAC click reaction using dibenzoyldiethylgermane photoinitiator with Cu(II) has been developed.

RDPS-C-P-008 [Light-induced Reactions in Polymer Science]

Visible and LED Light-Induced Metal Free ATRP Using Electron Acceptor Dyes in Conjunction with Amines

Ceren Kutahya, F. Simal Aykac, Gorkem Yilmaz, Yusuf Yagci

Department of Chemistry, Istanbul Technical University, Istanbul, Turkey

Two structurally similar dyes, namely Eosin Y and Erythrosin B were used to mediate the photoinduced ATRP of (meth)acrylates, and vinyl monomers in the absence of inorganic catalysts. The postulated mechanism considers the electron transfer from the amines to the excited state electron acceptor dyes, which in turn reduce alkyl halides to reversibly produce radicals that initiate polymerization. The polymers were characterized by NMR and GPC analyses. It was demonstrated that the method produces polymers with controllable chain end functionality and narrow molecular weight distribution. The dye/amine system was shown to be efficient under various colors of LED and industrially available visible light irradiation. The livingness nature of the polymerization and the irradiation dependency of polymerization was presented.

Keywords: photochemical processes, redox processes, living radical polymerization, atom transfer radical polymerization

General mechanism of photoinduced ATRP using electron acceptor dyes in conjunction with amines

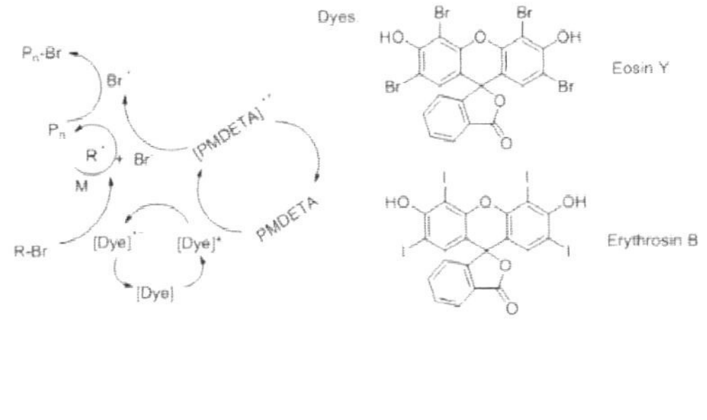

RDPS-C-P-009 [Light-induced Reactions in Polymer Science]

Benzodioxinone Photochemistry for Macromolecular Synthesis

Volkan Kumbaraci, Naciye Talinli, Yusuf Yagci

Istanbul Technical University, Department of Chemistry, 34469, Maslak, Istanbul, Turkey

Benzodioxinones were recently reported to be efficient photosensitive compounds in various organic reactions. Upon irradiation, they undergo decomposition to form α-oxoketene intermediates and ketones. The former can readily react with nucleophilic groups, such as hydroxyl or amine groups to form ester and amide, respectively (Soltani et al., 2005, Angew. Chem-Int. Ed., 44, 1696-1699). In this presentation, the use of the photochemically generated intermediates, namely ketene and ketones for macromolecular synthesis will be described. When the R1 and R2 groups in the benzodioxinone structure are phenyl groups, photochemically generated benzophenone may act as a two photon absorbing sensitizer for free radical (Tasdelen et al., 2006, Polymer, 47, 7611-7614) and cationic polymerization (Tasdelen et al., 2008, Macomolecules, 41, 295-297) and curing applications (Tasdelen et al., 2007, Macromolecules, 40, 4406-4408). Morever, the concomitantly formed keten was successfuly used as a new ligation chemistry to allowing formation of step-growth polymers block (Kumbaraci et al., 2006, Macomolecules, 39, 6031-6035), and graft copolymers (Durmaz et al., 2009, Macomolecules, 42, 3743-3749) and surface coating (Kumbaraci et al., 2007, Macromol. Rapid Commun., 28, 72-77) (Figure 1).

Keywords: Benzodioxinone, Ketene, Esterification, Photoinduced

Figure 1. Benzodioxinone Photochemistry for Macromolecular Synthesis

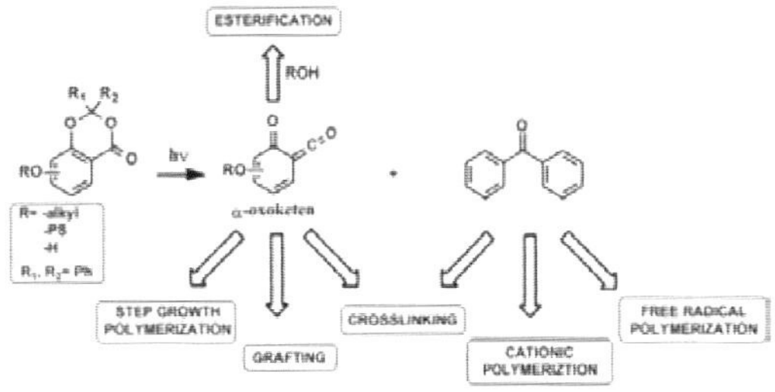

RDPS-C-P-011 [Light-induced Reactions in Polymer Science]

The investigation of photophysical and photochemical properties of thioxanthone-anthracene -9-carboxylic acid

Saliha Mutlu[1], Tolga Çeper[1], Nergis Arsu[1], Shigeru Takahara[2]

[1]Yildiz Technical University, Chemistry Department 34220, Istanbul, Turkey
[2]Chiba University, Chemistry Department, JAPAN

Thioxanthone (TX) and derivatives are among the most widely used Type II photoinitiators in various UV curing applications because of their excellent light absorption characteristics (R.S. Davidson *et al.*, Advances in Physical Chemistry, Academic Press, London, 1983). In most cases, the photoinitiating free radicals are generated by hydrogen abstraction by the triplet excited state of TX from hydrogen donors such as amines or thiols.

Amine, alkyl amino, acid or thiol linked photoinitiators are seen in the literature and many find various applications in the Radiation Curing Industry because of the great advantages of their one-component nature. Because of the one-component structure of the photoinitiator, it can serve as both a triplet photosensitizer and as a hydrogen donor during photopolymerization. Thus, these photoinitiators do not require an additional co-initiator.

An alternative approach concerns the attachment of both chromophoric and hydrogen donating groups into polymer chains (D.K. Balta, N. Arsu, J. Photochemistry and Photobiology A: Chemistry, 2013, 257, 54–59). This way the odor and toxicity problems perceived with the conventional photoinitiators and amine hydrogen donors have been overcome.

We have developed a novel thioxanthone based photoinitiator possessing an anthracene group (N. Arsu *et al.*, Macromolecules, 2007, 40, 4138-4141) and carboxylic acid namely, thioxanthone-anthracene-9-carboxylic acid. ^{1}H NMR and GC-MS were used for the characterization. Photophysical properties, fluorescence and phosphorescence studies were performed. In addition, photopolymerization reactions of mono and multifunctional acrylates were achieved.

Keywords: Photopolymerization, Photoinitiators, Thioxanthone

RDPS-C-P-012 [Light-induced Reactions in Polymer Science]

Photocoating in The Presence of Organosilica Nanoparticles from 3-Mercaptopropyltrimethoxysilane

Nurcan Karaca

Nurcan Karaca

Photopolymerization is increasingly be[illegible]d because of its advantages of fast polymerization and flexible react[illegible]ipment. This process receives continuous interest because of their wide application in UV-curable coatings, printing inks, photoresists and microelectronics. Radiation curable coatings and materials solidify by reactions initiated by radiation rather than heat initiated polymerization, solvent evaporation or oxidation. These systems can be stored indefinitely stable and safe in the absence of radiation, and curing reactions occurs on exposure to radiation at ambient temperature [1,2]. UV-Curable composites have the advantages of both curable systems and composite materials. They have been considered extensively in industries of coatings, inks, adhesives and printing. Most of the works done on radiation curing nanocomposites have been based on nanoslica. Because of the mechanical stability of composites, the polymeric materials are prepeared with slica particles and this process allow for increased toughness with negligible loss of tensile strength. These slica particles are essentially polyorganosiloxanes which have received substantial attention in the past two decades with a number of reports describing their synthesis, structural features, properties and applications. Furthermore, the thiolated organo-slicates have been special interest due to provide the advantages of functionalization of polymers [3-4]. This work presents the modification of polyorganosiloxane nanoparticles by surface initiated thiol-ene photochemistry. Polyorganosiloxane particles were prepared from (3-mercaptopropyl) trimethoxysilane (MPTMS) were modified by reaction with maleimide-styrene copolymer via 'click chemistry'. The polymer content was evaluated by thermogravimetric analysis (TGA). The resulting morphologies were characterized by electron microscopy techniques and compared to uncoated particles.

References:

[1] Fouassier, J.P., (1995). Photoinitiation, Photopolymerization and Photocuring, Munich: Hanser Publishers.
[2] Schnabel, W., (2007). Polymers and Light, Fundamentals and Technical Applications, Wiley –VCH Verlag GmbH&Co.KGAa, Weinheim.
[3] Irmukhametova, G.S., Mun, G.A. and Khuntoryanskiy, V.V., (2011). 'Thiolated Mucoadhesive and PEGylated Nonmucoadhesive Organoslica Nanoparticles from 3-Mercaptopropyltrimethoxysilane', Langmuir, 27, 9551-9556.
[4] Kurttner, C., Maier, P.C., Kunert, C., Schlaad, H., Fery, A., (2013). 'Direct Thiol-ene Photocoating of Polyorganosiloxane Microparticles', Langmuir, 29, 16119-16126.

Keywords: UV curing, nanoparticle, polyorganosiloxane, 3-Mercaptopropyltrimethoxysilane

RDPS-C-P-013 [Light-induced Reactions in Polymer Science]

Synthesis of a novel benzophenone derived UV absorber acrylate monomer

Tuğrul Cem Biçak, Ersin Acar

Department of Chemistry, Boğaziçi University, Istanbul, Turkey

UV protection of materials is of importance to avoid destructive effect of sunlight. This is achieved by adding few percent of protector material into surface coating formulations. Benzophenone derivatives are one of the most common UV protectors and they are sold under different trademarks. However, because of having a tail on 400nm range in their UV Absorption Spectrum, these compounds have slightly yellow color which disables to obtain some unique colors where they are applied. Moreover, they have miscibility problem in most organic solvents due to their hydroxyl groups which limit their application. Herein, we report synthesis of a new phenyl ester functional acrylate monomer which is colorless and highly organosoluble (Scheme.1). UV absorption is believed to occur via intra-molecular reversible hydrogen bonding. When the radiation stops, the compound returns to its original form and release the energy as heat so that the material is protected from the UV light. When polymerized, the monomer (a) was demonstrated to undergo Photo-Fries rearrangement to give hydroxyl benzophenone function (b) by UV irradiation. (Scheme 2).

In this work, both homopolymer and copolymers of the monomer with MMA were prepared and employed in UV absorbing. The results showed that the synthesized monomer can be used in coating materials without any color change. When exposed to UV light, only a thin yellow film occurs on the surface that do not change the overall color. Moreover, it can easily be polymerized in organic solvents because of being organosoluble.

Keywords: Photo-Fries Rearrangement, UV protector, benzophenone

Scheme 1

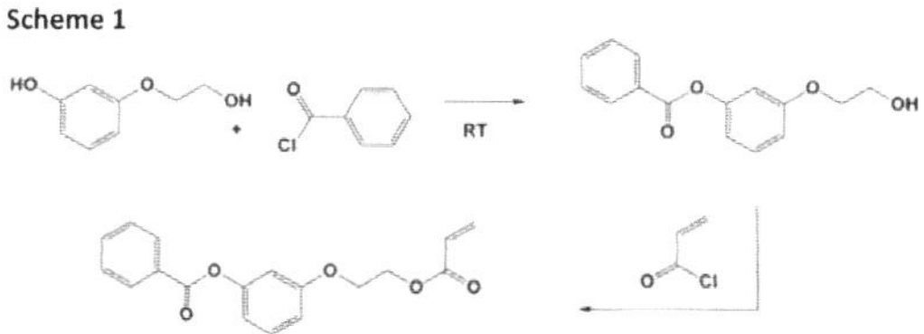

Synthesis of UV protecting monomer, 3-[2-(acryloyloxy) ethoxy] phenyl benzoate

Scheme 2

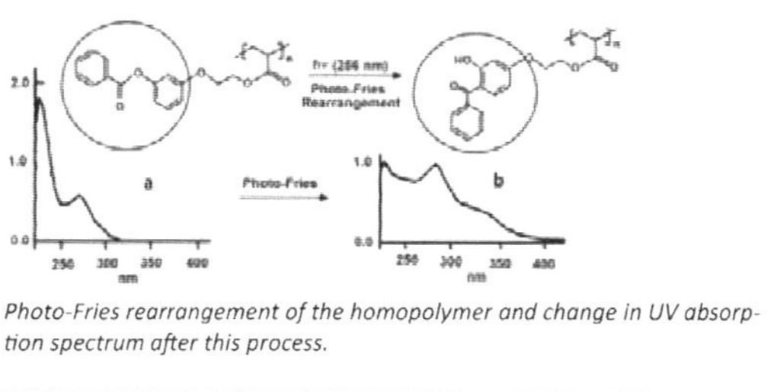

Photo-Fries rearrangement of the homopolymer and change in UV absorption spectrum after this process.

RDPS-C-P-014 [Light-induced Reactions in Polymer Science]

Photo-initiated Metal-Free Controlled Living Radical Polymerization Using Polynuclear Aromatic Compounds

Andrit Allushi[1], Steffen Jockusch[2], Gorkem Yilmaz[1], Yusuf Yagci[1], Yagci Yagci[3]

[1]Department of Chemistry,Istanbul Technical University, istanbul,Turkey
[2]Department of Chemistry, Columbia University, New York, USA
[3]Department of Chemistry,King Abdulaziz University, Saudi Arabia

Photo-initiated metal-free controlled living radical polymerization of (meth) acrylates and vinyl monomers was investigated using various polynuclear aromatic compounds namely, pyrene, anthracene and perylene. The effect of initiator type and stoichiometry on the polymer conversion and molecular weight characteristics were also examined. UV-vis and fluorescence spectral analyses along with nuclear magnetic resonance (NMR) studies were performed to determine the rates of redox processes and investigate the mechanisms of polymerizations. The obtained polymers were analyzed by NMR and gel permeation chromotography. Results show that while anthracene in conjunction with alkyl halides initiates the polymerization in an uncontrolled manner, pyrene provides the syntheses of polymers with controlled chain end functionalities and molecular weight characteristics under photo-irradiation.

Keywords: photochemical processes, redox processes, living radical polymerization, atom transfer radical polymerization

Mechanism of photoinitiated living radical polymerization using pyrene as photosensitizer

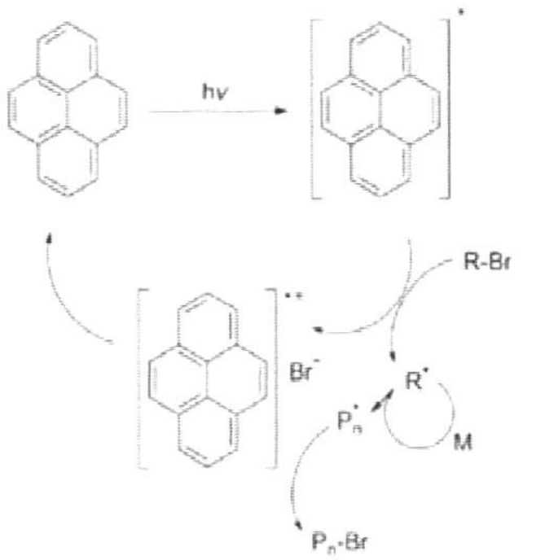

RDPS-C-P-015 [Light-induced Reactions in Polymer Science]

Novel photoinitiators based on Irgacure 2959 and methacrylates

Sesil Çınar, Duygu Avcı, Viktorya Aviyente

Department of Chemistry, Boğaziçi University, Istanbul, Turkey

Recently, development of monomeric and polymeric photoinitiators has become an important issue because of their low volatility and low migration in comparison with low-molecular weight nonmonomeric analogues. In this study, the synthesis and photochemical properties of three monomeric photoinitiators (MPI1-MPI3) based on Irgacure 2959 and two different methacrylates (tert-butyl α-bromomethacrylate and 2-isocyanatoethyl methacrylate), are reported, investigating the effect of polymerizable group structure on photoinitiation efficiency (Figure). The photopolymerization of 2-hydroxyethyl methacrylate, 1,6-hexanediol diacrylate and trimethylolpropane trimethacrylate initiated by these photoinitiators and non-polymerizable commercial low molecular weight analogs is studied through photo-DSC. Also, the electronic absorption spectra of the synthesized and commercial photoinitiators are computed with the time-dependent density functional theory at B3LYP/6-31+G(d) level using Gaussian 09 program.

Keywords: alkyl α-hydroxymethacrylates, photopolymerization, Irgacure 2959

Figure. Structures of synthesized novel photoinitiators.

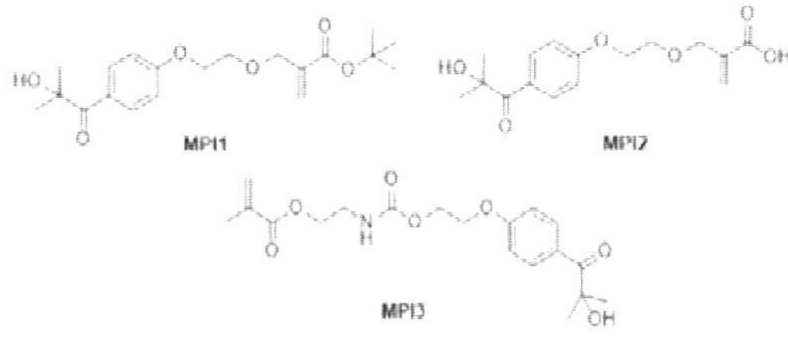

RDPS-C-P-016 [Light-induced Reactions in Polymer Science]

Synthesis of Hyperbranced Polymer Using Benzodioxine Chemistry

Cansu Aydogan, Mustafa Ciftci, Volkan Kumbaracı, Naciye Talınlı, Yusuf Yagcı

Istanbul Technical University

Hyperbranched polymers has intrinsic properties in comparison to their analogous by means of higher solubility, reduced solution viscosity and increased level of terminal functionality. These unique properties lead to their use in various industrial applications such as drug delivery systems, bioimaging agents, gene carriers, viscosity modifiers, and catalyst supports. Photosensitive benzodioxinones can be convenient alternatives for the preparation of hyperbranced polymer by photochemical pathway. The present study is about, synthesis and characterization of hyperbranched poly(methyl methacrylate) (PMMA) via self-condensing vinyl copolymerization (VL-SCVP) of methyl methacrylate (MMA) and 2-(dimethylamino)ethyl methacrylate (DMAEMA) in the presence of bisbenzodioxinone under UV light. The actual photoinitiator benzophenone released by the photolysis of bisbenzodioxinone acts as a hydrogen abstraction type photoinitiator. Depending on the concentration of DMAEMA, initiator and irradiation time, hyper-branched polymers with different branching density and cross-linked polymers were formed.

Keywords: bisbenzodioxinone, hyperbranched polymer, self-condensing vinyl copolymerization

Light Induced Synthesis of Hyperbranched Polymers Using Benzodioxinones

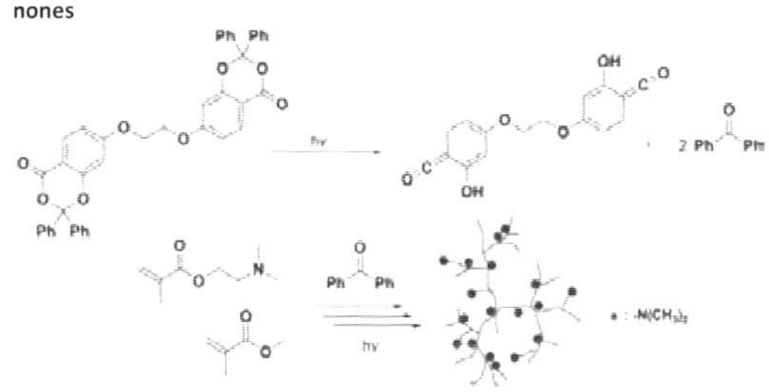

RDPS-C-P-017 [Light-induced Reactions in Polymer Science]

Photoinitiated Radical Polymerization with Iodonium Salts Bearing Polymerizable Anions

Annett Halbhuber, Bernd Strehmel

Department of Chemistry and Institut for Coatings and Surface Chemistry, Krefeld, Germany

Ion metathesis of an iodonium halide and the polymerizable anion of either a functionalized methacrylate or styrene sulfonate (compare anions a-h) results in new photoinitiating systems. These anions could be bound into the network during exposure and crosslinking resulting in a reduction of the migration tendency of the anion. Synthesis occurred according to ref. [1] with no changes of the cation; that is bis(t-butyl-phenyl) iodonium (I). The anion was selected from sulfonates (a, e), imides (b, f, h), and methides (c, g). Nevertheless, the anion i served as reference having no reactive group in the molecular skeleton. Previous works showed excellent dissolution of iodonium salts bearing the anion i. These iodonium salts were investigated with a sensitizer (Sens). Radical generation bases on electron transfer between Sens* and I. The reactive anion (a-h) helps to dissolve the salt in the monomers chosen; that is either a di- or trifunctional acrylate or methacrylate, respectively. Therefore, their solubility was quantified in the different model monomers. In addition, conductivity measurements were carried out to obtain a pattern between ion mobility and reactivity in these organic surroundings. Photopolymerization activity was studied by photo-DSC applying LEDs emitting either in the NIR (790 nm) or UV (395 nm).

Acknowledgement: We thank the BMBF for financial support within the program FHProfUnt (grant no. 03FH022PX4).

Reference:

[1] A. S. Shaplov et al, Polym. Chem., 2, 2609 (2011).

Keywords: iodonium, polymerizable anion, radical, sensitizer, NIR, solubility

Stuctures of Iodonium Salts

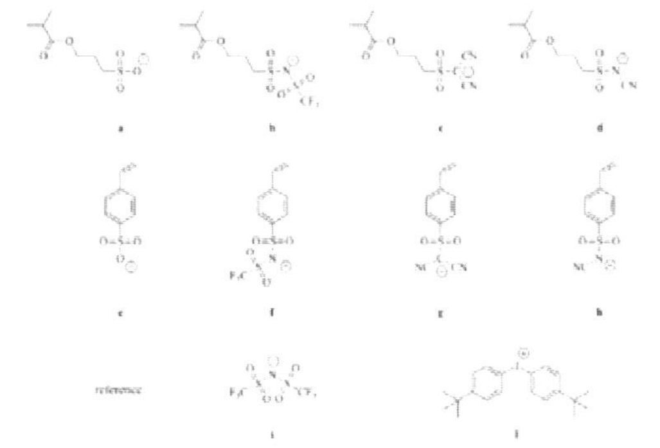

RDPS-C-P-018 [Light-induced Reactions in Polymer Science]

Synthesis of 1D tubular polymers by a supramolecular polymerization approach

Hao Yu, Robert Häner

Chemistry and Biochemistry Department, University of Bern, Bern, Swtizerland

In conventional polymerization methods, monomers are usually randomly distributed in the solution during polymerization. Therefore, the growth of the polymer chains highly depends on how efficient the active chain ends can find the monomers, which causes intrinsic drawbacks. Many factors can be detrimental to the reaction efficiency such as uneven initiator distribution, ineffective stirring, poor heat dissipation, etc. Furthermore, when the monomer gets larger in volume, the reacting sites tend to be more blocked to each other. Here, we reported the synthesis of 1D tubular polymers by a supramolecular polymerization approach. Prior to polymerization, the monomers, an amphiphilc 1,4-linked anthracene trimer (Tri-ANT), were already well pre-organized. The self-assembly of Tri-ANTs in aqueous buffer solution led to the formation of tubular 1D supramolecular structures, where Tri-ANTs were interdigitating with each other. Upon UV irradiation at 365 nm, the supramolecular structures were converted to the internally cross-linked polymers by the anthracene dimerization. Another advantage of this approach is that a "green" polymerization trigger, UV light, was employed. The polymerization was initiated evenly and thoroughly for the whole reaction mixture without adding any extra chemicals. This work provides a new method to prepare polymers in an efficient and environmentally friendly way. The pre-organization/polymerization concept may be employed to produce more complicated polymers.

Keywords: supramolecular polymerization, photo chemistry, nanotube

polymer side view

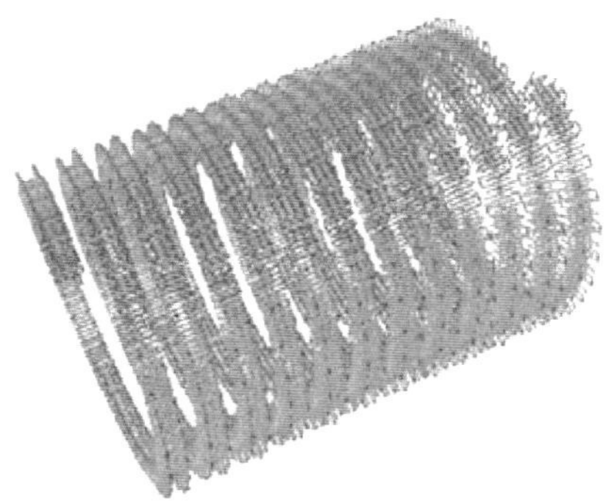

Cartoon representative of the tubular polymer from a side view.

RDPS-C-P-019 [Light-induced Reactions in Polymer Science]

Synthesis of Amide Linkage Directly from Aldehyde and Amines by Visible Light Photocatalysis

Naeem Iqbal

Health Biotechnology Division, National Institute for Biotechnology and Genetic Engineering, NIBGE Faisalabad, Pakistan

The amide bond is one of the most important bonds in biological macromolecules, and almost 25% of the drug molecules contain an amide bond. Because of the high stability, polarity, and conformational diversity of amide bonds, they have been used in synthetic intermediates to functional materials. Owing to the importance of amide bonds, intensive efforts have been made to develop methods for the construction of the amide bonds. An efficient visible-light photocatalysis-based one-pot amide synthesis method was developed; visible-light irradiation of a mixture of an aldehyde, tert-butyl hydrogen peroxide, and N-chlorosuccinimide using a Ru(bpy)3Cl2 photocatalyst afforded an acid chloride, which subsequently reacted with amine to yield the corresponding amide.

Keywords: Visible Light, Photocatalysis, Aldehyde, Amide, Oxidation

RDPS-D-P-001 [Condensation Polymerization and Thermoset]

Synthesis, Characterization and Spectroscopic Investigation of Novel Nanoporous Pyrazinoporphyrazine Network Polymer-Supported Metal (II)-Based Catalysts

Hamada H Abdel Razik[1], Basim H Asghar[2]

[1]Chemistry Department, Faculty of Science, Taif University, Taif, Saudi Arabia
[2]Chemistry Department, Faculty of Applied Sciences, Umm Al-Qura University, Makkah, Saudi Arabia

Chloranil through condensation reaction with vicinal diamine such as diaminomaleonitrile produced heterocyclic monomer, p-benzoquinonebis[2,3-b; 2`,3`-b`]pyrazine-5,6-dinitrile. The tetranitrile monomer was cyclo-tetramerised using lithium/pentanol and acetic acid affording the corresponding tetra p-benzoquinone bis[2,3-b; 2`,3`-b`]pyrazinoporphyrazine)]-based network polymer (2H-Pz). The tetranitril monomer was cyclo-tetramerised using metal salt and quinoline affording the corresponding porphyrazinato-metal II-based network polymers (M-Pz), M = Co, Ni or Cu. Elemental analytical results, IR and NMR spectral data of the prepared molecules are consistent with their assigned formulations. Molecular masses and metal contents of the synthesized polymers proved to be of high molecular masses which confirm the efficiency of tetramerization polymerization and complexation reactions. The prepared pyrazinoporphyrazines were used as efficient catalysts for the oxidation of thiophenol and benzylthiol to their disulfides in the presence of air atmosphere. The results of oxidation of thiophenol and benzylthiol show that after 15 minutes the maximum yield of the corresponding disulfides reached 95%, 91%, respectively.

Keywords: Pyrazinoporphyrazine, Tetramerization Polymerization, Polymer-Supported Metal (II)-Based Catalysts

RDPS-D-P-002 [Condensation Polymerization and Thermoset]

Epoxy Functionalized Hyperbranched polyurethane and its co-curing with benzoxazine: cure kinetics and thermal properties

Yichao Wang, Xue Jiang, Yuhong Liu

Department of Chemical Engineering, School of Chemical Engineering and Technology, Xi'an Jiaotong University

Hyperbranched polymers (HBPs) have been arousing great interest for the last two decades with the potential applications in such areas as catalysis, photoelectrics, polymer processing additives, biomedicines, pharmaceutics, and so forth. Because of their unique topologies and a number of terminal functional groups, HBP possess excellent solubility in many common organic solvents as well as the opportunity to react with thermosetting prepolymer. Novel epoxy-terminated hyperbranched polyurethanes (EHPU) with different epoxy values were prepared through functionalization of hyperbranched polyurethane (HPU) with glycidol. Fourier Transform Infrared Spectroscopy (FTIR) and Nuclear Magnetism Resonances (NMR) analysis confirmed that epoxy groups were effectively end-capped on HPU. The thermal properties of EHPU were investigated by Differential Scanning Calorimetry (DSC). Isoconversional methods (Flynn-Wall-Ozawa and Friedman methods) and FTIR were introduced to study the effect of EHPU on the thermal properties of 4'-diamono-methanebased diphenyl benzoxazine (MDA-BOZ). It was found that at low heating rates (below 1°C/min), the DSC curves of EHPU/MDA-BOZ (the mass ratio of MDA-BOZ to EHPU resin is 9:1) blend show asymmetric exothermic peak which can be separated into two peaks. The splitted two peaks were associated to the catalytic reaction between EHPU and BZ monomer and the reaction between epoxy group and BZ monomer, respectively. In addition, the differences of effective activation energy at a given conversion (Eα) between the BZ and BZ/EHPU resins can be clearly explored. Therefore, isoconversional method combined with FTIR are benefit to evaluate the effect of epoxy group on the curing reaction.

Keywords: epoxy, hyperbranched polyurethane, benzoxazine, curing kinetics, thermal properties

Figure 1 Synthesis procedure of EHPU

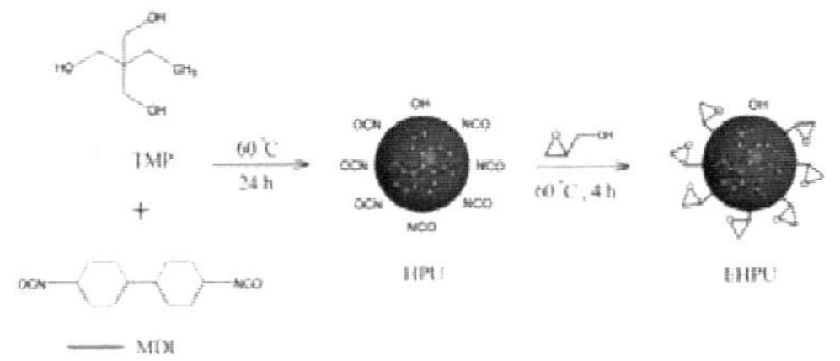

Figure 2 Comparison of curing rate between experimental data and auto-catalytic model of EHPU/MDA-BOZ

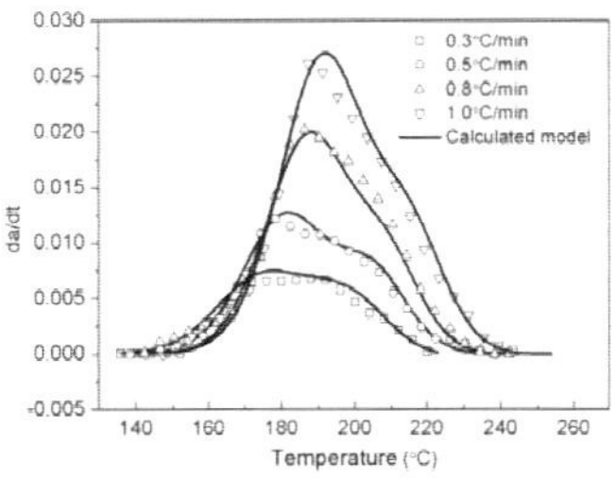

RDPS-D-P-003 [Condensation Polymerization and Thermoset]

Expression of Chirality in All-conjugated Block Copolymers without any excess of chiral centers

Marie-Paule Robert Van Den Eede, Anjan Bedi, Guy Koeckelberghs

Department of Chemistry, KU Leuven, Heverlee, Belgium

As molecular structure plays a major role in life and chemistry, the study regarding chirality remains an interesting and important research domain. However, phenomena of the expression of chirality e.g. homochirality, are still not unraveled. In order to better understand these aspects, state-of-the-art conjugated polymers are synthesized by our research group. In previous research the expression of chirality in conjugated block copolymers has already been investigated [Van den Bergh, K.; Willot, P.; Cornelis, D.; Verbiest, T.; Koeckelberghs, G. Macromolecules 2011, 44, 728][Verswyvel, M.; Monnaie, F.; Koeckelberghs, G. Macromolecules 2011, 44, 9489]. Furthermore, we were able to demonstrate that chirality can be expressed in a polymer molecule that contains no excess of chiral centers, but where chirality is evoked by a specific order of events.

A next step in this research is to approach the ideal racemic system in which we want to express chirality but this time even not evoked by a specific order of events. The polymer studied is an all-conjugated block copoly(3-alkylthiophene), in which the first block contains exactly the same amount of (*S*)-enantiomers as the second block (*R*)-enantiomers (figure 1). The only difference is the end-group. Beside the near-racemic polymer, block copolymers with different ratios of the enantiomers have been synthesized. In this way we are able to compare the results obtained for the near-racemic polymer and identify possible trends. Evidence for the formation of the desired block copolymers was provided by GPC and ^{1}H NMR analyses. Chiral expression has been studied with UV-vis and CD spectroscopy. Chiral expression was obtained for the block copolymer without any excess of chiral centers, with an even much higher chiral signal in comparison to the chiral homopolymers. This can be attributed to the influence of the initiator that installs a particular end-group at the end of the polymer chain.

Keywords: Controlled copolymerization, chain growth, KCTP, block copolymers, external initiation, chirality

Figure 1

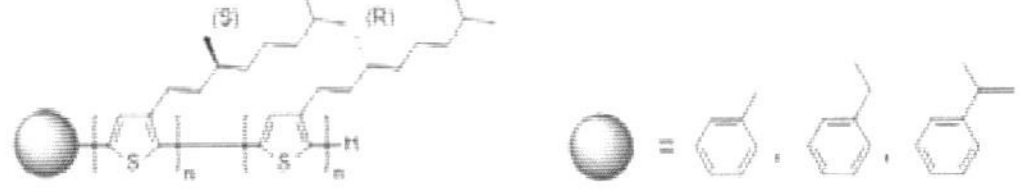

Representation of the all-conjugated block copolymer without any excess of chiral centers

RDPS-D-P-004 [Condensation Polymerization and Thermoset]

Blocked triazolinediones for easy handling, heat curable network formulations

Lucie Imbernon[1], Hannes A. Houck[1], Hannes A. Houck[2], Filip E. Du Prez[1]

[1]Polymer Chemistry Research Group, Department of Organic and Macromolecular Chemistry, Ghent University; Krijgslaan 281 S4-bis, B-9000 Ghent, Belgium
[2]Preparative Macromolecular Chemistry, Institut für Technishce Chemie und Polymerchemie, Karlsruhe Institute of Technology (KIT), Engesserstraβe 18, 76131 Karlsruhe, Germany

The reaction between unsaturated compounds and 1,2,4-triazoline-3,5-diones (TADs) recently emerged as the basis for a new click chemistry platform.[1,2] Besides all the benefits of a click reaction, this chemistry also has the advantage to be additive-free, to provide a useful colour switch and to enable ultrafast kinetics at ambient temperature. Moreover, the reaction can also be made reversible when a well-chosen partner (e.g. an indole) is selected. However, the high reactivity of TAD compounds at very low temperatures (down to -78°C in some cases) can be considered as a drawback in many applications. In search of a solution to this problem, functionalized indoles have been recently studied in our group.[3] These moieties were shown to act as efficient blocking agents for triazolinediones, and control of the unblocking temperature range could be achieved by alteration of the indole substituents. Being able to delay and/or trigger the reaction is very important when it comes to network formation for example. Indeed, if the crosslinking reaction is too fast/not triggerable, only very short timeframes are available to process the material before its gelation. In this work, blocked TADs were used as latent crosslinkers for network formation and compared to the corresponding unblocked TAD compounds. The processability of the blocked TAD based formulations was demonstrated by film formation and the obtained materials were fully characterized. Among other possibilities, this opens perspectives for the use of TADs in 1K coating formulations.

1. S. Billiet, K. De Bruycker, F. Driessen, H. Goossens, V. Van Speybroeck, J.M. Winne and F.E. Du Prez, Nature chemistry, 2014, 6, 815-821.
2. K. De Bruycker, S. Billiet, H.A. Houck, S. Chattopadhyay, J.M. Winne and F.E. Du Prez, Chemical Reviews, 2016, 116, 3919-3974.
3. H.A. Houck, K. De Bruycker, S. Billiet, C. Barner-Kowollik, J.M. Winne and F.E. Du Prez, Submitted, 2016.

Keywords: Triazolinedione, Blocking agent, Delayed crosslinking, Network

RDPS-D-P-005 [Condensation Polymerization and Thermoset]

Synthesis and characterization of amphiphilic Polyglycerol sebacate derivatives

Merve Gultekinoglu, Kezban Ulubayram
Department of Basic Pharmaceutical Sciences, Faculty of Pharmacy, Hacettepe University, Ankara 06100, Turkey

Polyglycerol sebacate (PGS) was developed in the past decades by Wang et al. and widely used in biomaterial and tissue engineering applications such as cartilage, cardiac muscle, nerve tissue engineering etc. PGS is a bioelastomer with highly tunable mechanical properties, good biocompatibility, enzymatic and hydrolytic degradability with no cytotoxicity. PGS synthesis is basically depends on polycondensation reaction and after prepolymer synthesis curing time directly affects mechanical structure and degradation time of polymer. However PGS has major drawbacks in crosslinking and water solubility. In this work, we aimed to synthesis poly ethylene glycol (PEG) and poly glycerol sebacate (PGS) derivatives as amphiphilic copolymer. This amphiphilic structure can overcome water solubility difficulties by methacrylation and/or co-polymerization with PEG. Mechanical properties, water uptake capacity, degree of functionalization were all tuned by adjusting the PEG Mw, monomer ratios and/or reaction times. All the chemical reaction steps were clarified with spectroscopic techniques such as Fourier Transform Infrared Spectroscopy (FT-IR) and Nuclear Magnetic Resonance (NMR) Spectroscopy. In addition, no cytotoxicity was observed on L929 cell line for PGS-co-PEG. We believe that the PGS-co-PEG with newly introduced functionalities will become a featured biocompatible polymer and can be used for several manufacturing techniques. Especially functionalization of PGS-co-PEG with bioconjugates looks very promising since there is a considerable big gap in the literature.

References:

1. Wang, Yadong, Yu Mi Kim, and Robert Langer. "In vivo degradation characteristics of poly (glycerol sebacate)." Journal of biomedical materials research Part A 66.1 (2003): 192-197.
2. Aydin, H. M., et al. "Microwave-assisted rapid synthesis of poly (glycerol-sebacate) elastomers." Biomaterials Science 1.5 (2013): 503-509.

Keywords: polycondensation, polyglycerolsebacate, polyethylene glycol

RDPS-D-P-006 [Condensation Polymerization and Thermoset]

Synthesis, Characterization and Investigation of Thermal Properties of POSS Containing Graphene/Polycaprolactone Composites

Fehmi Saltan[1], Aylin Bilgili[2], Hakan Akat[1]

[1]Department of Chemsitry, Ege University, İzmir, Turkey
[2]Department of Chemsitry, Çankırı Karatekin University, Çankırı, Turkey

Nowadays, developing technology that can respond to the growing needs of the pursuit of superior materials continues to increase. Hybrid materials are very important in these related studies because they carry the properties of both organic and inorganic materials. Thermal, mechanical and chemical resistance can be gained to the product by using advanced technology polymers such as polyester and polyamide. Different composites can be prepared between a hybrid molecule called as POSS (polyhedral oligomeric silsesquioxane) and commercial polymers like ε-caprolactone by using several methods (radical polymerization, click chemistry and controlled polymerization techniques). POSS is a relatively new type of organic-inorganic hybrid molecules. POSS has great potential in the synthesis of organic-inorganic composites. Also, it is known that PCL is synthetic polyester obtained by the self-condensation of the cyclic ester caprolactone. PCL has unique properties that make it attractive for biomaterials applications.

In this study, it has been demonstrated that polyhedral oligomeric silsesquioxane (POSS) based graphene-polycaprolactone nanocomposites (POSS-G-PCL) were successfully prepared by using ring opening polymerization method. For this purpose, first, graphene-POSS composite (GP) was synthesized with graphene oxide and aminopropyllsobutyl POSS (POSS-NH2). Second, POSS containing graphene-polycaprolactone composite (PCL-G-POSS) was synthesized by using ε-caprolactone, graphene-POSS, and Tin(II) octanoate as initiator in DMF at 110 oC. This reaction was repeated without of solvent (PCL-G-POSS in bulk) and by using different amount of caprolactone (PCL*-G-POSS). The synthesized composites and polymers were characterized by nuclear magnetic resonance (1H NMR), gel permeation chromatography (GPC), and fourier transform infrared spectroscopy (FTIR). Thermal behaviors of molecules were analyzed by thermogravimetry (TG). The surface morphologies of composites were analyzed by scanning electron microscope (SEM). The results of TG and GPC analyses showed that the thermal stability and molecular weight of the composites were increased with the addition of POSS.

Keywords: polyhedral oligomeric silsesquioxane (POSS), polycaprolactone, graphene, composite, thermal properties

Fig. 1 FT-IR spectra of graphite, GO, POSS-NH2, POSS-graphene, PCL-G-POSS

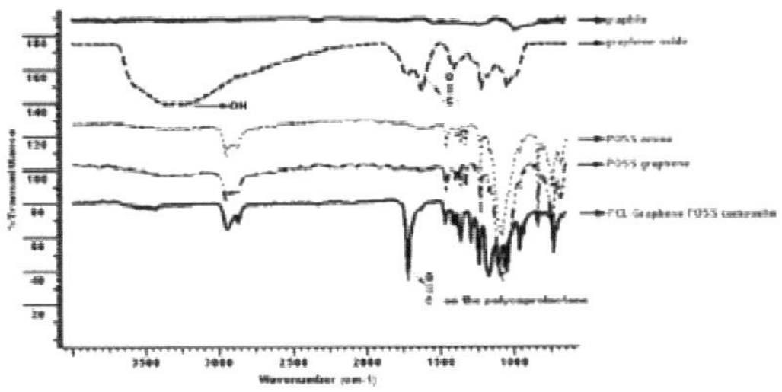

Fig. 2 1H-NMR specturms of graphene-POSS and PCL-G-POSS

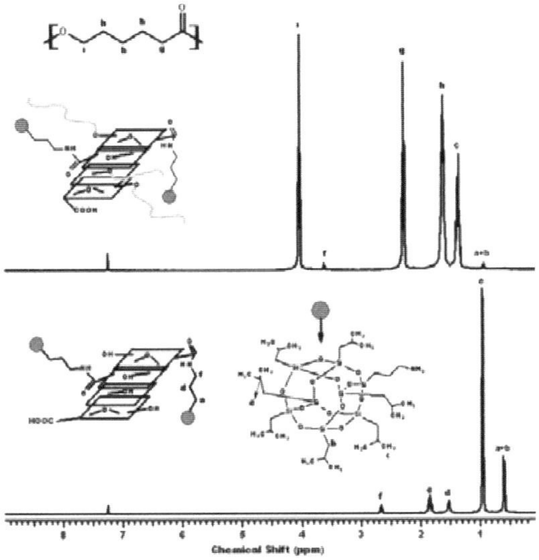

Fig. 3 TG Curves of synthesized composites..düşük

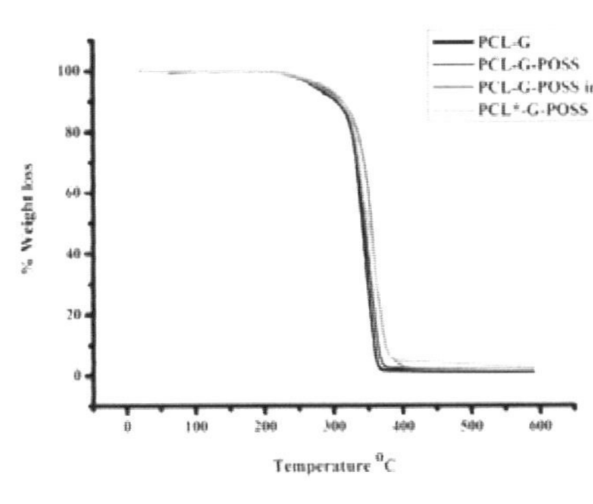

Fig. 4 TG Curves of synthesized composites (in air)

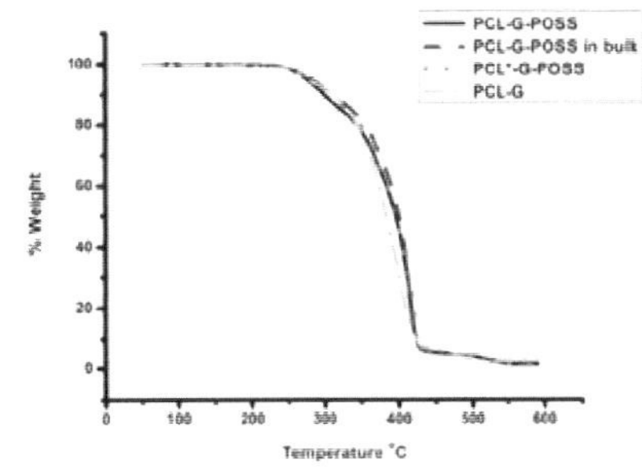

Sceheme 1. Synthesis of POSS based Graphene-Polycaprolactone (PCL-G-POSS) composite

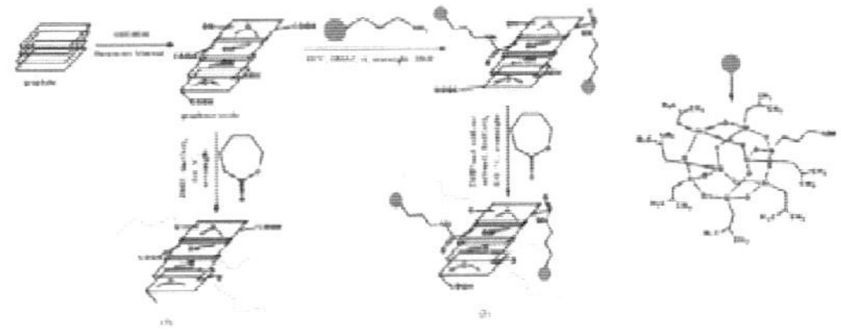

RDPS-D-P-007 [Condensation Polymerization and Thermoset]

Synthesis And Characterization of Waterborne, Branched, Functional Polyurethanes for Textile Coating Applications

Ayşe Durmuş[1], Deniz Anıl[1], Yusuf Menceloğlu[1], Serkan Ünal[2]

[1]Department of Material Science and Engineering, Faculty of Engineering and Natural Science, Sabanci University,Istanbul,Turkey
[2]Sabanci University Nanotechnology Research and Application Center,Istanbul,Turkey

Waterborne polyurethane dispersions have become significant in the market of environmentally friendly materials for advanced coating and adhesive applications due to their high performance and zero or near zero volatile organic content (VOC) in recent decades. Although the versatile chemistry of polyurethanes enables one to design high performance coating, adhesive or elastomeric materials for a wide range of applications, current synthetic routes for linear polyurethanes have limitations in the introduction of chemical functional groups on the polymer backbone. The present study focuses on the synthesis and characterization of waterborne, branched polyurethanes with a multitude of functional end-groups via the oligomeric $A_2 + B_3$ methodology. While the waterborne nature of final products makes them user and environmentally friendly, the presence of functional end-groups makes them suitable coating and adhesive materials for a variety of surfaces, specifically textiles. The influence of the degree of branching, distance between branch points, nature of the soft segment and chemical structure of end-groups on the on the dispersion properties and thermo-mechanical properties of elastomeric films obtained from these dispersions was investigated systematically and will be presented in-detail with an emphasis on textile coating applications.
This research is supported by Turkish Ministry of Science, Industry and Technology under SAN-TEZ grant number 0843.STZ.2014.

Keywords: Polyurethanes, Waterborne, Branching, Functional Polyurethanes, A2 + B3 polymerization, Textile coating

RDPS-D-P-008 [Condensation Polymerization and Thermoset]

Synthesis and Characterization of Novel Cyclomatrix-Type Polyphosphazene Microspheres Cross Linked With Octachlorocyclotetraphosphazene via Self-Assembly Approach

Simge Metinoglu, Yasemin Süzen

Department of Chemistry, Anadolu University, Eskişehir, Turkey

In the recent times, synthesis, characterization[1-3] and application studies[4-6] of cyclomatrix polyphosphazene nano/microspheres via precipitation polymerization technique has accelerated. In this study, highly crosslinked inorganic and organic hybrid polyphosphazene microspheres including amino groups on the surface have been successfully synthesized via self-assembly polycondensation reaction. In preparation of microspheres, 4,4'diaminodiphenylmethane (DADPM) and octachlorocyclotetraphosphazene (OCCP) were used for monomer and a new crosslinker, respectively. Experiments were performed in acetonitrile using triethylamine (TEA) as an acid acceptor and ultrasonic bath. Besides, effects of different concentration of OCCP and DADPM, reaction time and ultrasonic power on obtaining morphologically good spheres have been investigated. The size of acquired microspheres were approximately 1.00μm. The novel microspheres were characterized by SEM, FTIR, ZETA, TGA, UV, Fluorescense and XRD.

References :

[1] Köhler J.; Kühl S.; Keul H.; Möller M.; Pich A., J. Polym. Sci. Pol. Chem., 52, 527-536, 2014.
[2] Wang Y.; Mu J.; Li L.; Shi L.; Zhang W.; Jiang Z., High Perform. Polym., 24(3), 229-236, 2012.
[3] Zhao Z.; Ji J., Adv Eng Mater, 16, No. 8, 2014.
[4] Wei W.; Huang X.; Chen K.; Tao Y.; Tang X., RSC Adv., 2, 3765-3771, 2012.
[5] Fu J.; Wang M.; Zhang C.; Zhang P.; Xu Q., Mater. Lett., 81, 215-218, 2012.
[6] Wang D.; Hu Y.; Meng L.; Wanga X.; Lu Q., RSC Adv., 5, 92762-92768, 2015.

Keywords: polyphosphazene, octachlorocyclotetraphosphazene, 4,4-diaminodiphenylmethane

SEM image of polyphosphazene microspheres.

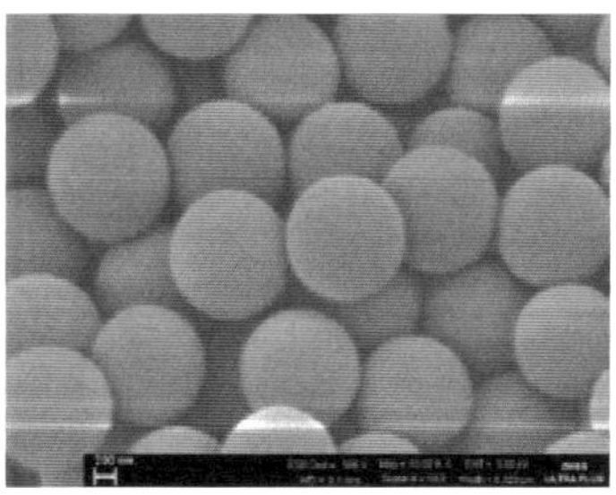

RDPS-D-P-009 [Condensation Polymerization and Thermoset]

Preparation and Characterization of Tyramine Based Inorganic-Organic Hybride Polyphosphazene Microspheres

Simge Metinoglu, Yasemin Süzen

Department of Chemistry, Anadolu University, Eskişehir, Turkey

Tyramine (4-hydroxyphenethylamine) is a naturally occurring biogenic monoamine, derived from the amino acid tyrosine. Tyramine is a neuroactive substance that is high level in the nervous system. It is reported to possess various biological properties, such as antioxidant, adrenergic, agonistic and monoamine oxidase.[1-3] Inorganic and organic hybride polyphosphazene microspheres including both an inorganic crosslinker molecule (N3P3Cl6, hexachlorocyclotriphosphazene) and an organic monomer have attracted more and more attention due to their small size, large specific surface area, high mobility, easy recovery from the dispersion and reversible dispensability, etc. [4-6] In this study, crosslinked polyphosphazene microspheres were synthesized using tyramine (TA) and hexachlorocyclotriphosphazene (HCCP) as a monomer and crosslinker, respectively by self-assembly polycondensation reaction. Microspheres have both hydroxyl and amine groups on surfaces arising from TA. The characterization of microspheres was performed by SEM, FTIR, ZETA, TGA, UV, Fluorescense and XRD.

References :

1. H. Siddiqui; M. A. Bashir; K. Javaid; A. Nizamani; H. Bano; S. Yousuf; A. Rahman; M. I. Choudhary J. Enzyme. Inhib. Med. Chem. 1475-6366, 2016.
2. S. H. Jezzini1; D. R. Colo; M. A. Sosa PLoS ONE, 9(10), e111314, 2014.
3. L. Duportets; R. B. Barrozo; F. Bozzolan; C. Gaertner; S. Anton; C. Gadenne; S. Debernard Insect Molecular Biology, 19(4), 489–499, 2010.
4. Z. Li; G. Wang; A. Zhang; L. An; Y. Tian J. Appl. Polym. Sci, 43336 (1 of 7), 2016.
5. J. Köhler; S. Kühl; H. Keul; M. Möller; A. Pich Journal of Polymer Science, Part A: Polymer Chemistry, 52, 527–536, 2014.
6. L. Zhu; Y. Zhu; Y. Pan; Y. Huang; X. Huang; X. Tang Macromol. React. Eng., 1, 45–52, 2007.

Keywords: tyramine, polyphosphazene, hexachlorocyclotriphosphazene

SEM image of polyphosphazene tyramine microspheres

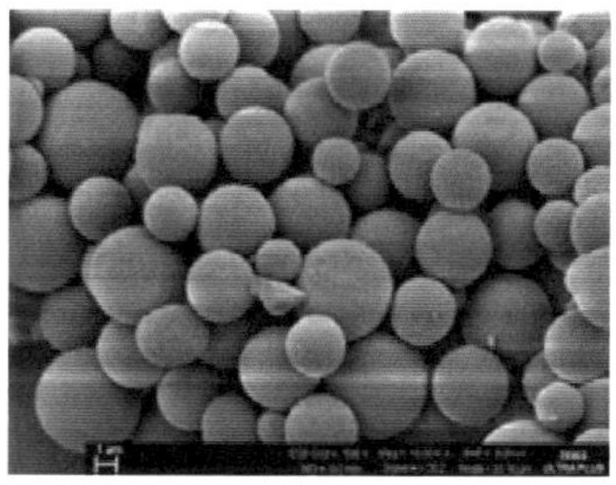

RDPS-D-P-010 [Condensation Polymerization and Thermoset]

Investigating cure advancement of a dual-cure Polyurethane-Acrylate (PUA) coatings used for metal substrates

Babak Kaffashi, Mohsen Sarrafi

Department of Polymer Engineering, School of Chemical Engineering, College of Engineering, University of Tehran, Tehran, Iran

Photo-polymerization is an applicable method to produce a polymeric network from liquid resin systems. Among several techniques, UV-curing is more common because of reduced energy consumption and less equipment costs. The important limitation for use of UV-curing systems is the coating of 3D subjects or high pigmented coatings. Here the whole area of coating is not illuminated and sufficient cure is not achieved. To overcome this, a second curing system is applied. The most common method is the use of thermal curing along with UV irradiation. The final coating properties, however, are obtained by the combination of a UV-curing process and a chemical crosslinking process. The most common coating type is the two-component urethane-acrylate, where an acrylic polyol solution (with polyester acrylate structure) is mixed with polyisocyanate just before application to the substrate. Monitoring the cure advancement of a polymeric resin gives very important information about reaction kinetics and can be useful for coating processes. It can be carried out with different methods such as real-time FTIR, photo-DSC and chemorheology techniques. In this study, a two-component polyurethane acrylate were synthesized with polyester polyol and urethane monoacrylate structure. Toluene diisocyanate (TDI) and 2-hydroxy ethyl methacrylate (HEMA) were used as received. Polyester polyol was synthesized in a way that the final coating film possesses both flexibility and surface hardness. Cure advancement of this system was examined in different conditions. Thermal curing of two components at different temperatures was characterized with the rheological measurement. Viscosity advancement was described with Boltzmann sigmoidal model to compare important parameters such as gel time and the intensity of gelation process. Photochemical curing with UV irradiation was monitored with real-time FTIR via checking the absorption peak for acrylate double bond at 810 nm-1. The effect of diluents on the conversion efficiency under irradiation was also investigated.

Keywords: Photo-polymerization, UV-curing, coating, Boltzmann sigmoidal model, viscosity, rheology

Figure 1-Curing.jpg

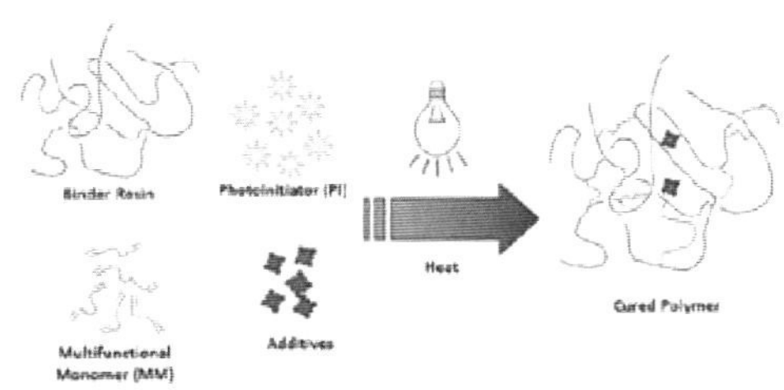

Figure 2-Graphical abstract.jpg

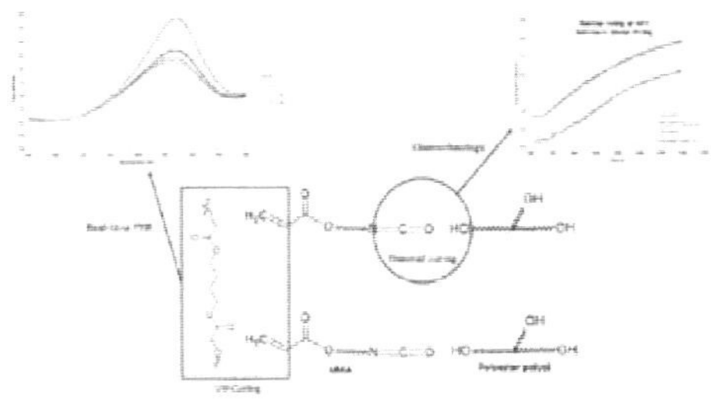

Figure 3-Thermal curing @ 80C.jpg

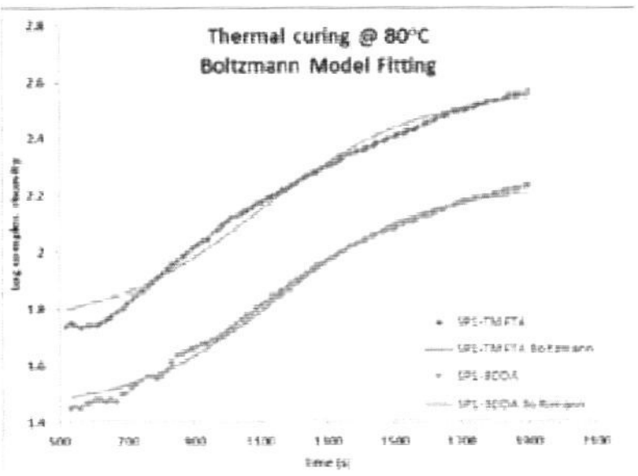

Figure 4-Thermal curing @ 100C.jpg

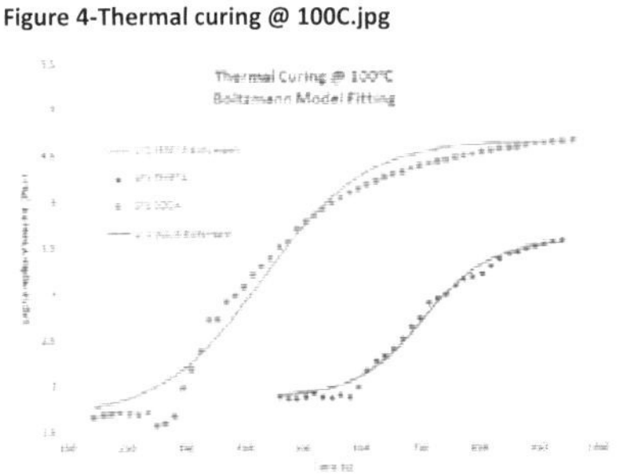

Figure 5-UV curing with BDDA @ 810cm-1.jpg

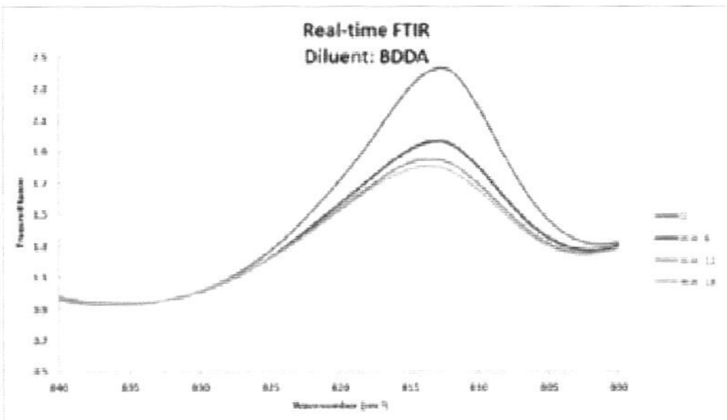

Figure 6-UV curing with TMPTA @ 810cm-1.jpg

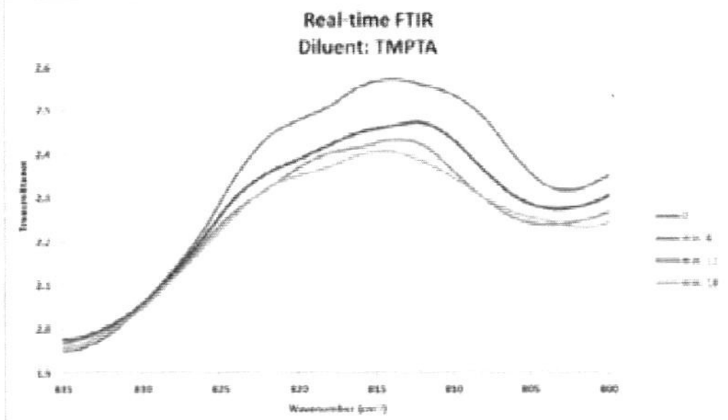

Figure 7-UV curing conversion with BDDA.jpg

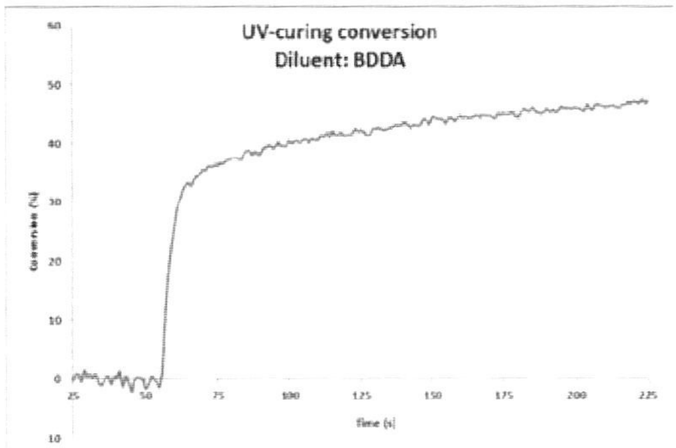

Figure 8-UV curing conversion with TMPTA.jpg

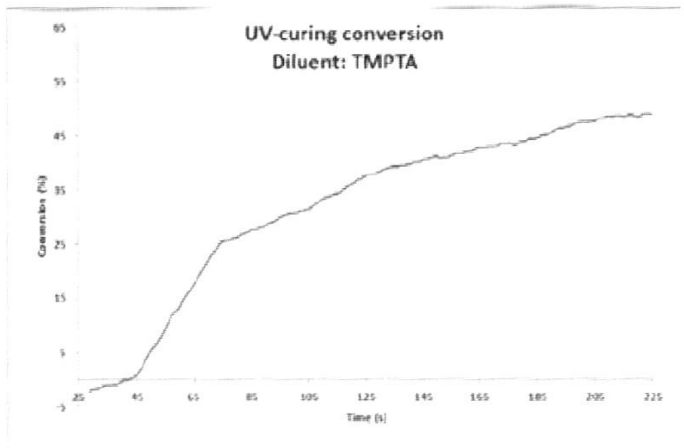

RRB-P-001 [Renewable Resources and Biopolymers]

Potassium hydroxide pulping of rice straw in biorefinery initiative

Md Sarwar Jahan, Fahmida Haris, Md Mostafizur Rahman, Purabi Rani Samaddar

Pulp and Paper Research Division, BCSIR Laboratories, Dhaka, Bangladesh

Rice straw is supposed to be one of the most important lignocellulosic raw materials for the pulp and paper industry in forest deficient countries like Bangladesh. The major problem of rice straw pulping is silica, which inhibits the recovery of black liquor in alkali pulping processes. A pulp mill cannot be economically and environmentally feasible without chemical recovery system. Therefore, potassium and ammonia based pulping have been proposed by a number of researchers to use black liquor as fertilizer. But those researches did not mention alkalinity of the black liquor. The liquor with pH above 8 cannot be used in soil. Silica is an essential component of materials which have an abundance of uses in industry and are a vital component in many things used in our everyday lives. Therefore, silica isolation can add value in the pulp mill. The present research is focused on the pulping of rice straw by potassium hydroxide (KOH) process and separation of silica and lignin from the black liquor. At the conditions of 12.7% alkali charge as NaOH, 2 h of cooking at 150oC and material to liquor ratio of 1:6 pulp yield was 44.4% with kappa number 13.4. KOH pulp bleached to 81% brightness by D0EpD1 bleaching sequences with ClO2 consumption of 20kg/ton of pulp. The papermaking properties of KOH pulp from rice straw were slightly better than those of corresponding NaOH pulp. The silica present in the rice straw was dissolved in the black liquor that was precipitated by reducing pH to 7 by adding dilute sulfuric acid. The amount of precipitated silica was 10.5% of rice straw. The membrane separation processes including ultrafiltration (UF) and nanofiltration can separate/recover lignin and hemicellulosic sugars from the silica separated black liquor. and finally potassium rich waste water can be used in irrigation.

Keywords: Potassium hydroxide pulping, Biorefinery, Hemicellulosic sugars, Lignin, Silica

Flow diagram of KOH pulping in biorefinery concept

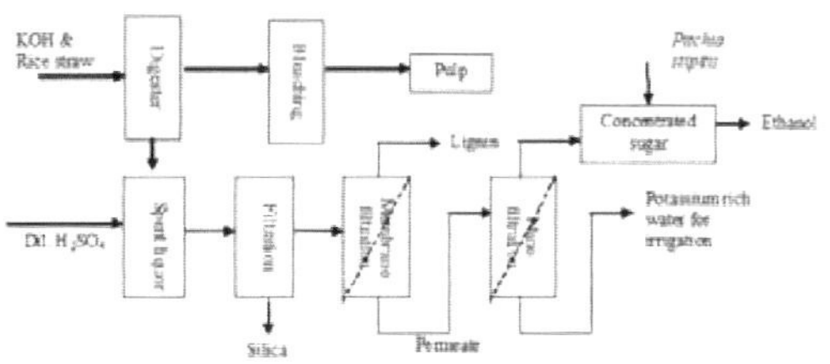

RRB-P-002 [Renewable Resources and Biopolymers]

Influence of Temperature and Itaconic Acid on Emulsion Copolymerization of Butyl Acrylate and Methyl Methacrylate in Semi-Batch Process

Leonardo Zborowski Sobrinho[1], Daniela Beirao Porto[2], Jesus Roberto Taparelli[2], Lucia Helena Innocentini Mei[2]

[1]Institute of Physical Sciences and Technology, Paulista University, Sorocaba, Brazil and Department of Materials and Bioprocess Engineering, School of Chemical Engineering, State University of Campinas, Campinas, Brazil.
[2]Department of Materials and Bioprocess Engineering, School of Chemical Engineering, State University of Campinas, Campinas, Brazil.

The need to use renewable resources to produce polymers stimulates the application of products such as itaconic acid, which is used, for instance, in the functionalization of polymers obtained via emulsion polymerization. In this report analyzes the influence of itaconic acid and temperature on semi-batch emulsion copolymerization of butyl acrylate and methyl methacrylate. The reactions were made at two different temperatures (75 and 85°C), with or without itaconic acid. The resulting lattices were evaluated in their solids content and average diameter particles. Besides, a sample of each lattice was coagulated with ethanol and the resulting solid was analyzed in its elemental and gel content, viscosity at low shear stress rate, and thermal properties. The presence of itaconic acid reduced global conversion of the co-monomers, due to the reduction in polymerization rate, and increased the gel content of the polymers, increasing the chains crooslinking. It was also observed that the presence of itaconic acid increased viscosity at low shear stress rate and slightly increased the copolymers glass transition temperatures, due to the presence of carboxylic acid groups, which increased the polymer chains polarity.

Keywords: renewable resources, itaconic acid, emulsion polymerization

Viscosity vs shear stress rate for all formulations at low shear stress rate.

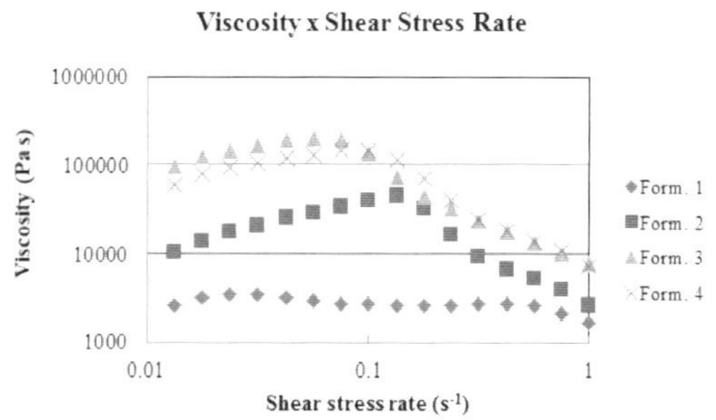

In this figure, it is presented the values of viscosity at low shear stress rate of all coagulated polymers synthesized.

Formulations utilized to obtain itaconic acid acrylate derivatives by emulsion polymerization.

	75IA4	85IA4	75IA0	85IA0
Temperature (°C)	75	85	75	85
Emulsion				
Water (g)	264.98	264.98	264.98	264.98
Sipomer COPS 1 (g)	2.13	2.13	2.13	2.13
Hostapal BVQ-9 (g)	41.83	41.83	41.83	41.83
Butyl Acrylate (g)	514.08	514.08	535.5	535.5
Methyl Methacrylate (g)	90.72	90.72	94.5	94.5
Itaconic Acid (g)	25.2	25.2	0	0
Solution 1				
Water (g)	251.07	251.07	251.07	251.07
Sodium acetate (g)	2.51	2.51	2.51	2.51
Sodium Persulphate (g)	0.18	0.18	0.18	0.18
Solution 2				
Water (g)	43.04	43.04	43.04	43.04
Ammonium Persulphate (g)	1.97	1.97	1.97	1.97
Solution 3				
Water (g)	17.93	17.93	17.93	17.93
Ammonium Persulphate (g)	0.36	0.36	0.36	0.36

This table presents the used formulations on this report.

General results from the characterization of resultant synthesized polymers

Formulation	Solids Content (%)	Average Particle Diameter (mm)	Gel content (%)	Itaconic Acid Percentage in mass	Tg (°C)
1	48.5	0.39	71.15	1.19	-33.3
2	49.1	0.35	65.14	2.33	-31.7
3	50.9	0.37	53.43	0	-34.2
4	52	0.33	47.74	0	-33.7

In this table, they are presented the results of characterization of synthesized polymers.

RRB-P-003 [Renewable Resources and Biopolymers]

Synthesis of new resins based on epoxidized sunflower oil via acrylation reaction and study of their biodegradability

Naima Belhaneche Bensemra[1], Ratiba Irinislimane[2]

[1]*Département Génie de l'Environnement, Ecole Nationale Polytechnique, Algiers, Algeria*
[2]*Faculté de Sciences, Université M'Hamed BOUGARA, Boumerdès, Algeria*

In the present work, new resins based on epoxidized sunflower oil (ESFO) were synthesised via acrylation reaction. ESFO with 6.1 % oxirane index was reacted with acrylic acid in the presence of triethylamine as catalyst. Reaction temperature and catalyst percent were optimized. The reaction conversion was evaluated by following the oxirane ring reduction which was calculated using titration of the epoxy ring and the quantitative infrared spectroscopy analysis. The acrylated epoxidized sunflower oil (AESFO) resins were characterized by Fourier transform infrared spectroscopy (FTIR) and 1H nuclear magnetic resonance (NMR). In the second step of work, the biodegradation level of sunflower oil derivatives (ESFO and AESFO) was carried out at a lab scale following a respirometry test based on determination of released carbon dioxide (CO2). The results showed that sunflower oil derivatives into soil cover a considerable level of biodegradation.

Keywords: epoxidized sunflower oil, acrylation, biodegradable polymers, respirometry

RRB-P-004 [Renewable Resources and Biopolymers]

Improved anaerobic digestion in the thermophilic phase by a thermal pretreatment

Kheiredine Bani, Derbal Kerroum, Benchikhlehocine Mosaab

Departement Of Environnement Constantine Algéria

The experimental study was conducted to evaluate the effects of thermal pretreatment of secondary sludge on anaerobic digestion using as substrate a dairy rejection consisting primarily of whey. The inoculum was subjected to heat treatment 120 ° C, 160 ° C, 180 ° C for 2 hours, while considering the slurry without a corresponding treatment température equal to 20 ° C. The trials were conducted in a series of reactors of 250ml mesophile37 stage ° C.

The effect of heat treatment on the physicochemical parameters of the sludge before incubation shows that:

The soluble matter concentration greatly increases the required minimum time of treatment to achieve the highest dissolution rate is equal to 30 min. The solubilization of the COD increases proportionally with the treatment temperature, an increase in the treatment temperature of 120 ° C results in a COD of solubilization ratio of 7% and a temperature of 180 ° C it is 25%, and solubilization degree reached 34% for a temperature of 180 ° C. The best treatments (160 ° C and 180 ° C) result in a production of biogas from 2.7 to 3.5 times that of the untreated septage to a time equal to 25 days.

Keywords: Energy, Biogas, prétraitement

RRB-P-005 [Renewable Resources and Biopolymers]

The Kinetic of Biogas Production Rate from Dairy Wastewater in Batch Mode «Effect Inoculums to Substrate Ratio in thermophilic Phase»

Kheiredine Bani, Benchikhlehoucine Mosaab, Derbal Karoum

Departement of Environnement University Constantine Algéria

In this study, the kinetics of the biogas production was studied by conducting a series of laboratory experiment using secandaire sludge of purification station as inoculums and dairy Wastewater as substrat. The effects of various reports substrate/inoculums (S/I) are considered, respectively for (0), (0.2), (0.4), (0.8) and (1.6) inthe thermophilic phase (55 °C). For the ratio tested it was determined the following kinetic parameters; the maximum rate of production of methane (NmL. gMS-1.j-1) (Rmax): the cumulative methane production at the end of the experiment (NmL.gMS-1) (Vmax); and a minimum of Time to produce biogas (λ).

The results of biogas yields show that, the best ratio compared these parameters are the ratio(S/I) =1.6 and the ratio(S/I)= 1.4.

Keywords: Modelisation, Biogas, Dairy effluents, Production Rate.

RRB-P-006 [Renewable Resources and Biopolymers]

Mechanical characterization of reversible hydration of PLA based materials

María Martín[1], Freddys Beltrán[1], María Ulagares De La Orden[4], Joaquín Martínez Urreaga[3], Vicente Lorenzo[2]

[1]*Polímeros: caracterización y aplicaciones (POLCA), Universidad Politécnica de Madrid, Spain*
[2]*Departamento de Física Aplicada e Ingeniería de Materiales, Universidad Politécnica de Madrid, Spain*
[3]*Departamento de Ingeniería Química Industrial y del Medio Ambiente, Universidad Politécnica de Madrid, Spain*
[4]*Departamento de Química Orgánica, Universidad Complutense de Madrid, Spain*

The use of PLA in food packaging, mainly in contact with wet food, is rapidly growing. Water can produce irreversible and reversible effects on PLA. On the one hand, hydrolitic degradation of PLA cannot be reversed. On the other hand, water plasticization is predominant during the first stages of the hydration. The main objective of this communication is to present some results about the influence of the water plasticization on the surface mechanical properties of PLA.

Quenched and aged (24 days at RT) amorphous PLA (L:D≈96:4) samples were immersed in deionized water at 37 ºC for a few hours. The hydrated samples were then vacuum dried at RT. Surface mechanical properties were determined by Depth Sensing Indentation, DSI. The character of the interactions PLA-water was studied by FTIR-ATR. These experimental techniques were selected because the depth of the indentations and the penetration depth of the evanescent wave are of the same order of the thickness of the surface layer where water is confined after a few hours of immersion (some micrometers).

The figure illustrates that the indentation hardness, Hit, and the indentation modulus, Eit, of PLA fall approximately 10% after 210 minutes in water and, then, both Hit and Eit reach again their initial values after 10 days at RT. This reversibility can be explained in terms of FTIR-ATR spectra that show an increment during the hydration of the 3623 and 3555 cm^{-1} bands that have been associated to free and weakly interacting water. This water is eliminated along the subsequent drying process and the mechanical properties recover their initial values.

Consequently, the results seems to indicate that DSI is an adequate technique to study the evolution of mechanical properties during reversible hydration and drying of PLA.

(The financial support of Spanish MINECO, grant MAT2013-47972-C2-2-P, is gratefully acknowledged).

Keywords: PLA, water sorption, plasticization, Depth Sensing Indentation

Fig. 1

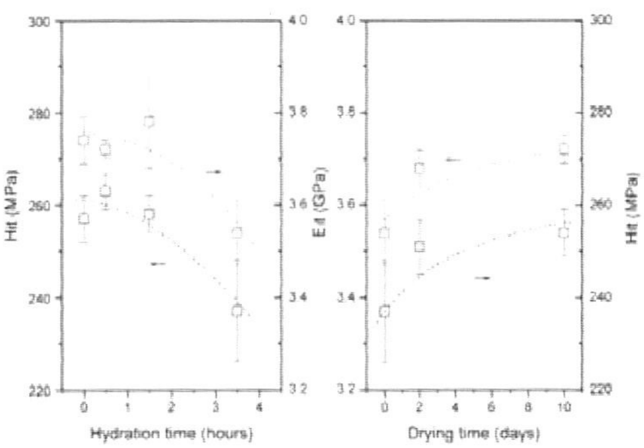

Dependence of indentation hardness, Hit; and modulus, Eit, on wetting and drying time

RRB-P-007 [Renewable Resources and Biopolymers]

Mechanical properties of biocomposites based on sunflower oil and alfa fibers

Naima Belhaneche Bensemra[1], Sihem Kadem[1], Ratiba Irinislimane[2]

[1]*Ecole Nationale Polytechnique*
[2]*Université M'Hamed BOUGARA*

Biocomposites or green composites are biodegradable materials made from bio-based materials to replace traditional petroleum-based plastics. Plant-oils (vegetable oils) can be used as resins after chemical modification. The mechanical properties of these composites can be improved by adding naturel fibers as reinforcement. In this work, sunflower oil (SO) was epoxidized and then reacted with acrylic acid to obtain acrylated epoxidized sunflower oil (AESO). The oxirane ring concentration was estimated using a titration method (gentian violet as indicator). The copolymerization was realised by blending AESO with co-monomer, initiator and catalyst to different weight percent (wt %). After the determination of the optimal concentration of resin, natural fibers were added by varying the weight percentage. Mechanical properties were characterized.

Keywords: Biocomposites, biodegradability, sunflower oil, natural fibers, mechanical properties.

RRB-P-008 [Renewable Resources and Biopolymers]

Characterization of new biodegradable blends based on ethylene-butyl acrylate copolymers with thermoplastic starch

Jesus Luis Pablos Lagartos[1], Ana Morro[2], Fernando Catalina[1], Teresa Corrales[1], Irma Marín[2], Concepción Abrusci[2]

[1]*Department of Applied Macromolecular Chemistry, Institute of Polymer Science and Technology, CSIC, Madrid, Spain*
[2]*Department of Molecular Biology, UAM, Cantoblanco, Madrid, Spain*

The protection of the environment from ever-increasing plastics waste has gained academic and industrial importance in recent years and the development of new biodegradable materials has become a great challenge. In particular, partially biodegradable materials prepared by blending natural biodegradable and non-biodegradable synthetic polymers that exhibit economic advantages and superior properties than the completely biodegradable polymers.In this work, thermoplastic starch (TPS) plasticized with glycerol has been used to prepare blends with ethylene-butyl acrylate copolymer (EBA) and ethylene-acrylic acid copolymer (EAA) at 20 % respect to EBA as compatibilizer. The blends with 10, 30 and 60 % of TPS content were characterized by X-ray diffraction, Thermal Gravimetric Analyses (TGA), ATR-Fourier Transform Infrared Spectroscopy (ATR-FTIR), Scanning Electron Microscopy (SEM), water-Contact Angle measurements (CA), Dynamic Mechanical Thermal Analysis (DMTA) and Differential Scanning Calorimetry (DSC). Initiated autoxidation of the polymer blends was studied by chemiluminescence (CL) under nitrogen and oxygen confirming that the presence of the polyolefin-TPS interphase did not substantially affect the oxidative thermostability of the materials. For biodegradation studies, firstly three bacterial species were isolated from the blend films buried in soil and identified as *B. subtilis*, *B. borstelensis* and *B. licheniformis*. A mixture of the three Bacilli (MIX) as well as *B. subtilis* were tested for evaluation of the biodegradation of the blends (28 days at 45ºC) and mineralization was determinate by carbon dioxide measurement using the indirect impedance technique [1].
Acknowledgements: The authors would like to thank the MINECO (Spain) for financial support (MAT2012-31709). One of the authors, C.A., would also like to thank the Ramón y Cajal program.

References:

[1]Abrusci, C.; Pablos, J. L.; Corrales, T.; Lopez-Marıín, J.; Marín, I.; Catalina, F. 2011. Int. Biodet. Biodeg., 65, 451.

Keywords: Thermoplastic starch, ethylene-butyl acrylate copolymer, starch blend chemiluminescence, bacterial biodegradation.

RRB-P-009 [Renewable Resources and Biopolymers]

Utilization of chemically treated waste rubber as ion exchangers for industrial waste water treatments

Yassin Abdel Latif Aggour[1], Ayed Saad Al Shihri[2]

[1]Faculty of Science, Damiatta University, New-Damiatta, Egypt
[2]College of Science, King Khalid University, Kingdom of Saudi Arabia

Contamination of environments with scraped vulcanized rubber causes significant problems due to its high resistivity to natural and biological conditions. Our previous work concerned with chemical treatments and characterization of powdered scraped tires, and converting them to an ion exchanger. These work aimed to utilize the obtained ion exchanger to remove heavy metal ions (such as Cu+2, Cd+2, Pb+2, Zn+2) and CrO4- ions from their aqueous solutions. Also the ion exchanger will used to remove organic pollutants such as phenol and naphthalene from aqueous solutions. The ion exchange capacity (IEC), were studied as function of pollutant concentrations and the mass of ion exchangers. The IEC reach to 10mg/g for removal of Cd+2, Pb+2, and Zn+2.The distribution and separation coefficient of the process were determined and discussed. The results were applied at optimum conditions to remove heavy metals, phenol and naphthalene from some industrial waste water. Also treatments of polluted agriculture soils (with sewage/heavy metals) were performed.

Keywords: waste rubber, chemical treatment, Ion exchanger, industrial waste wate

RRB-P-010 [Renewable Resources and Biopolymers]

Effect of soybean oil in crosslink density and dynamic mechanical properties of natural rubber compounds with silica

Carlos Henrique Scuracchio[1], Aline Zanchet[2], Janaina Da Silva Crespo[3]

[1]Department of Materials Engineering, Federal University of São Carlos
[2]Center of Applied Engineering, Modeling and Social Sciences, Federal University of ABC
[3]University of Caxias do Sul

Processing aids are organic substances that, when added to polymers, improve their flexibility and their processability. They are largely used to reduce the viscosity and the elasticity of the pre-vulcanized materials, so the capacity of mixing with other additives into elastomeric matrix and their fluidity during molding are enhanced. However, the growing use of these oil derivatives may generate serious environmental problems, mainly when they are disposed of in landfills and dumps. Thus, some studies have been conducted in order to develop environmentally friendly compounds, using materials from natural sources. The aim of this work is to show how the soybean oil can influence of the vulcanization (Table 1) and dynamic mechanical properties of vulcanized rubber, in comparison to the use of petroleum based oil, and its influence on the dispersion of silanized silica. Figure 1 (a) Loss modulus (E''), and (b) Storage modulus (E') as a function of temperature, which determined the glass transition temperatures of the compositions (Table 2). The results show that the lubricating effect is more effective for the naphthenic oil sample as the the sample with 5 phr of soybean oil is stiffer. The samples with soybean oils presented more relevant differences above Tgs. On the other side, the higher the oil concentration in the compound the greater the mobility of the segments of the chains present in NR, as expected.
Aknowledgements: The authors would like to thank FAPESP (process 2012/14844-3) for the financial support.

Keywords: natural rubber, renewable additivies, plastificizing, lubrication, dynamic mechanical properties

Figure 1

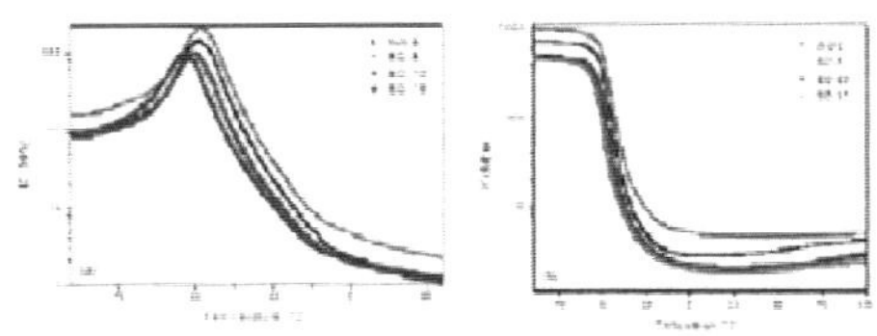

(a) Loss modulus (e''), and (b) Storage modulus (e') as a function of temperature.

Table 1

Ingredients (phr)	NO-5	SO-5	SO-10	SO-15
NR	100	100	100	100
Zinc Oxide	5	5	5	5
Stearic Acid	1	1	1	1
Silanized silica	50	50	50	50
Sulfur	2,5	2,5	2,5	2,5
MBT	2	2	2	2
TMTD	0.5	0.5	0.5	0.5
Naphtenic oil	5	-	-	-
Soybean oil	-	5	10	15

Composition of the formulations, in phr, with soybean and naphtenic oils

Table 2

Sample	Tg (°C)
nO-5	-48
SO-5	-47
SO-10	-50
SO-15	-53

Glass Transition Temperature (Tg)

RRB-P-011 [Renewable Resources and Biopolymers]

Design, synthesis and characterization of novel high bio based content dual-cure systems: high solids and water-dispersible

Ozlem Kubra AKDOGAN, Swaroop Shendre, Vijay Mannari
Coatings Research Institute,Eastern Michigan University.Michigan,USA

In the recent years there has been growing demand for "green" and sustainable materials that can effectively replace petro-based materials currently used in advanced coating systems.. Developing dual curable oligomers based on renewable resources that meet the demanding requirements of cost and performance of such coatings is important for their sustainable future growth. In the present work, using the principles of green chemistry, we have designed and synthesized novel soybean oil based acrylate and hydroxyl functional oligomers. These oligomers are characterized for their chemical structure (FT-IR), physical and chemical properties. A series of free-radical type dual curable coating compositions have been prepared using soy-based oligomers as primary components, conventional reactive diluents, type-I photo-initiators, crosslinkers and additives. The compositions have been formulated to have bio-based content of 50% by wt. of higher. Dual cured coatings of these compositions have been characterized for physical, chemical and thermo-mechanical properties to assess their suitability for useful commercial applications. The study shows soy-based acrylate and hydroxyl oligomers as promising alternative materials to petro-based ones and with suitable oligomer design and formulation of coatings, high performance dual-curable "green" coatings can be formulated.

Keywords: biobased,dual cure,high solids, water dispersible

RRB-P-012 [Renewable Resources and Biopolymers]

Biodegradable polyester synthesis from sugar beet pulp pectin

Hande Barkan, Cüneyt Hüseyin Ünlü, Oya Galioğlu Atıcı

Istanbul Technical University Science & Letters Faculty Chemistry Department Maslak 34469 Istanbul Turkey

Sugar beet pulp is an industrial plant waste arise from sugar production. It is mainly composed of polysaccharides such as cellulose, hemicellulose, and pectin. This study aims to evaluate the sugar beet pulp as a raw material for synthesis of biodegradable polyester and is composed of two parts; first extraction of pectin from industrial crop waste sugar beet pulp under different conditions then modifying pectin in order to obtain biodegradable polyester. Pectin is a polysaccharide which is composed of 1-4 linked galacturonic acid units in acidic or methyl ester form to some extent. Sugar beet pulp contains a considerable amount of pectin (25-30% of dry weight). Pectin was extracted using acidic conditions with a maximum yield of pectin was determined in terms of dry percent starting material weight as 21%. Obtained pectin compared to commercially available apple pectin. In the second phase pectin was modified using terephtalic acid with the aim of pectin-terephtalate condensation product. Conversion was found as 56% based on starting materials amounts. Spectral and chemical characterizations indicated that condensation polymer was obtained.

Keywords: Pectin, Terephtalic acid, Biodegradable polymer, Sugar beet pulp, Polyester

RRB-P-013 [Renewable Resources and Biopolymers]

A lignin-derived chemical platform for biobased polymers via ADMET

Andrea Hufendiek, Sophie Lingier, Laetitia Vlaminck, Filip E. Du Prez

Department of Organic and Macromolecular Chemistry, Polymer Chemistry Research Group, Ghent University, Ghent, Belgium

In the past years, there has been a strong development in the field of polymer synthesis from renewable monomers, to provide alternatives to or replacements for crude oil based monomers. Some polymers based on renewables are already available on an industrial level. Yet, these polymers are largely derived from carbohydrates or triglycerides as renewable resources. However, a significant share of industrially relevant polymers contains aromatic entities to ensure certain thermal and mechanical properties, and thus the quest for renewable aromatic monomers has begun. Nowadays, renewable aromatic compounds can be derived from cashew nutshell liquid, polyphenols and lignin.[1] Lignocellulosic biomass is an especially promising resource since it is abundant and does not compete with food or feed.

Key intermediates in the synthesis of renewable aromatic building blocks from lignin are vanillin and vanillin-like derivatives, which can be converted into diols. A versatile chemical platform (see scheme), consisting of dienes, was synthesized from potentially renewable lignin-derived diols containing an aromatic moiety. Subsequently, ADMET polymerization was performed on this series of dienes in order to investigate the influence of the substitution pattern in the monomers on the resulting thermal properties of the polymers. In addition, the remaining double bonds of the polymers, synthesized by ADMET polymerization, allow for further finetuning of the material properties by post-polymerization functionalization via, e.g. triazolinedinone chemistry.

[1]Caillol S.et al., Green Chemistry 2014;16(4):1987-1998.

Keywords: lignin, vanillin, ADMET

Scheme 1

RRB-P-014 [Renewable Resources and Biopolymers]

Fundamental Properties of Cellulose Derivatives Important for Thermoplastic Processing

Yu Bai, James H. Wang, Bing Zhou, Jun Kuang

Sinopec Shanghai Research Institute of Petrochemical Technology

Thermoplastic polymers have contributed to the modern convenient lifestyle of our society for more than half a century. The technology advancement in petroleum refining and polymerization enabled the availability of plastic products having excellent performance and at low cost. However, as the world is concerned about global warming, the environmental impact of thermoplastics is becoming a major concern. To deal with this challenge, there is an increasing interest to explore the utilization of natural resources for making renewable polymers, and cellulose is one of the leading natural polymers being considered for this application. However, cellulose is not a thermoplastic material and it cannot be used directly in a thermoplastic process. Through the manufacture of cellulose derivatives such as a cellulose esters and ethers, certain cellulose derivatives may have thermoplastic properties. But for adequate thermoplastic processing, a cellulose derivative must have certain thermal and rheological properties in a range required by such processes. To assist the selection of potential cellulose derivatives for thermoplastic processing, a number of cellulose esters and ethers were selected and their fundamental properties were evaluated. The melting behavior, melt rheology and thermal stability of these cellulose derivatives were determined. Based on the results, promising cellulose derivatives were subsequently subject to thermoplastic processing on a twin screw extruder. Several thermoplastic cellulose materials were identified from this work, but it was found not completely possible to predict thermoplastic processability of cellulose derivatives by judging the fundamental properties alone, melt extrusion must be used as a final validation test. This work has built a basis for further development of suitable thermoplastic materials from cellulose.

Keywords: Cellulose derivatives, thermoplastic processability, cellulose ester, cellulose ethers, extrusion processing

RRB-P-015 [Renewable Resources and Biopolymers]

Production of chitosan-based conductive hydrogel by photopolymerization technique for biomedical applications

Neslihan Alemdar, Celil Ulutürk

Department of Chemical Engineering, Marmara University, Istanbul 34722, Turkey

Chitosan (CTS) obtained from the renewable source is one of the most important of natural polymers and it is used in the polymer matrix, microcapsules, microspheres and drug delivery systems, biosensor applications. When chitosan contacts with the human body, it shows excellent properties such as bioactivity, antimicrobial activity, and enzymatically biodegradability. Owing to these properties, chitosan has been widely used in biotechnology field. Although chitosan has many advantages including mucoadhesive, pH sensitive, better absorption capacity, it has an insufficient mechanical performance limiting the applications. In order to improve the thermal and mechanical properties, chitosan is modified by various methods. The grafting of acrylic monomers onto the chitosan backbone is one of the modification methods. In the light of this information, in this work, glycidyl methacrylate (GMA), which is the acrylic monomer, was grafted on the chitosan backbone and thus CTS-g-GMA was produced. Then, CTS-g-GMA and poly(ethylene glycol) diacrylate (PEGDA) were photopolymerized under UV light to obtain chitosan-based hydrogel. Then, a conductive polymer, polyaniline was formed into the hydrogel matrix. Thus, a semi-interpenetrating network was obtained. The structure of hydrogel manufactured with this way was confirmed by FT-IR analysis. To determine water uptake capacity of this hydrogel, swelling test was carried out under the certain conditions. Conductivity of the hydrogel was determined by four-point probe method. As a result, this versatile material has broad potential for biomedical applications especially for biosensor applications because of the conductivity.

Keywords: Chitosan, Conductive hydrogel, Biosensor

RRB-P-016 [Renewable Resources and Biopolymers]

Synthesis of polyhidroxyesters derived from 10-undecenoic acid. Modification with aminoacids

Carmen Valverde

Department of organic and analytical chemitry, Suspol group, Rovira i Virgilli University, Tarragona, Spain

Aliphatic polyesters are generally considered to be well-suited for applications as polymer-based biomaterials due to their demonstrated biocompatibility and biodegradability. To widen the versatility of aliphatic polyesters, special efforts have been devoted to the functionalization of the polyester backbone. Strategies that employ a variety of routes and chemistries for the pendent functionalization of synthetic polyesters via polymerization of functionalized monomers, post-polymerization modification, or a combination of both, have been used. Synthesis of new functional biodegradable polyesters bearing hydrophilic reactive pendent groups such as hydroxyl, carboxyl, amino, and so forth become attractive as biomaterials in biomedical field. Vegetable oils constitute a convenient source of platform chemicals to prepare polyesters and other polymeric materials. Vegetable oils and their derived fatty acids not only provide great economic and environmental advantages but, conveniently designed, can afford to low toxicity, biodegradable and biocompatible materials for medical engineering. In this work, we present two routes to polyhydroxyester from 10-undecenoic acid regarding the use of 10,11-epoxyundecanoic acid as a AB type monomer and 10,11-dihydroxyundecanoic acid which has the ability to perform as AB or AB2 monomer depending on the activity and selectivity of the catalysts used, rendering linear or branched polyesters. The unsaturated 10-undecenoic acid can be easily oxidized to 10,11-epoxyundecanoic acid with H2O2 and Candida Antartica lipase B (CALB), or 10,11-dihydroxyundecanoic acid using H2O2 and renewable formic acid as reagents (Fig. 1). To test the ability of these functional polyesters to prepare protein bioconjugates three chemical modifications through covalently anchoring aminoacid units were assayed. Two of them are the direct functionalization with L-phenylalanine-Boc or a protected cysteine derivative. The other is a two-step modification, first with maleic anhydride followed by reaction with L-serine-Boc. All approaches used N-(3-dimethylaminopropyl)-N'-ethylcarbodiimidehydrochloride (EDAC) as carboxylic group activator.

The authors express their thanks to MICINN (Ministerio de Ciencia e Innovación) (MAT2014-53652-R) for financial support for this work.

REFERENCES:

H. Seyednejad, A.H. Ghassemi, C.F. van Nostrum, T. Vermonden, WE. Hennink. J. Control. Rel. 2011, 152, 168-176.
A. Gandini, T. M. Lacerda, A. J. F. Carvalho, T. Trovatti, Chem. Rev. 2016, 116, 1637–1669.
M. Kuhlmann, O. Reimann, C.P.R. Hackenberger. Macromol. Chem. Phys. 2015, 36, 472-476.
Z. You, H. Cao, J. Gao, P. H. Shin, B. W. Day, Y. Wang, Biomaterials, 2010, 31, 3129-3138.

Keywords: 10-undecenoic acid, aliphatic polyester, polyhydroxyester, chemical modification, aminoacids

Fig. 1.

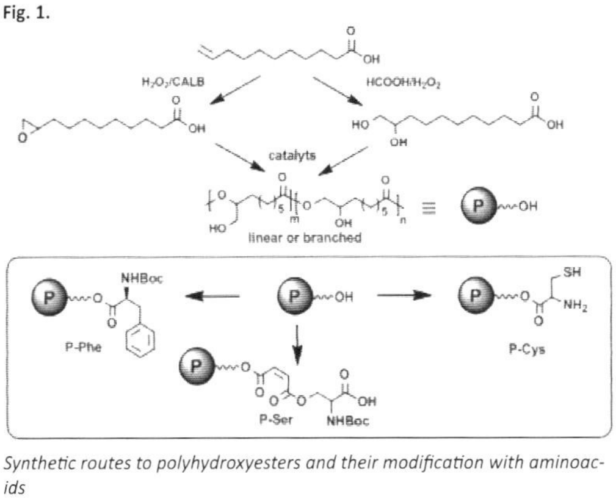

Synthetic routes to polyhydroxyesters and their modification with aminoacids

RRB-P-017 [Renewable Resources and Biopolymers]

Polyaniline/wheat stalk dust composites for effective removal of arsenic (V) from aqueous solution

Santosh Khanal, Indu Khadka, Rameshwor Adhikari

Central Department of Chemistry, Tribhuvan University, Kirtipur, Kathmandu

Conducting polymers have become one of the most attractive subjects of investigations during past decades. Among them, polyaniline (PAni) is unique due to its high electrical conductivity, good environmental stability, and ease of preparation. An attempt is made to prepare polyaniline/wheat stalk dust (PAni/WSD) composite by in situ polymerization of aniline as a novel bio-adsorbent for effective removal of arsenic from aqueous solution. The morphology and chemical structure of developed absorbent were evaluated with the aid of scanning electron microscopy (SEM) and Fourier transform (FTIR) infrared spectroscopy, respectively. For the evaluation of adsorption efficiency and the influence of various experimental parameters such as pH, adsorbent dosage, and contact time of developed adsorbents, batch adsorption experiments were carried out. The optimum conditions for Arsenic (V) removal were found to be pH 10, adsorbent dose of 1 g L-1 and equilibrium time 60 min. The equilibrium adsorption was better described by Langmuir isotherm model. The maximum adsorption capacity (qm) for monolayer adsorption was found to be 33.34 mg g-1.

Keywords: Polyaniline, Wheat stalk dust, Arsenic, Adsorption, Batch studies

RRB-P-018 [Renewable Resources and Biopolymers]

Screening of various moderately Bacillus strains for PHA production

Meliha Fetahović[1], Pınar Çağlayan[2], Ayşe Oğan[1], Özkan Danış[1], Meral Birbir[2]

[1]Department of Chemistry-Biochemistry, Faculty of Arts and Sciences, Marmara University, Istanbul, Turkey
[2]Department of Biology, Faculty of Arts and Sciences, Marmara University, Istanbul, Turkey

Polyhydroxyalkanoates (PHAs) are biodegradable polyesters that can be synthesized by various species of *Bacteria* as reserves of energy and carbon in an environment that is carbon-rich but poor in certain nutrients such as phosphorus and nitrogen. Development of eco-friendly bio-degradable bio-based plastics from cheap, renewable resources are becoming increasingly important but PHAs hold only a relatively small fraction of the biopolymer market share because of their relatively high production cost and concurrent availability of low-cost petrochemical plastics. Among the PHA producing bacterial strains, *Bacillus* are the most popular source of PHAs species being capable of growing in large scale culture that are taxonomically very diverse. Moderately halotolerant *Bacillus* strains can grow optimally in media containing 3–15% NaCl. Numerous *Bacillus* species from many different environments have been screened for PHA production and more screening studies are needed for obtaining the most efficient PHA synthesizing strains. In this study PHA production capacity of five *Bacillus* strains; *Bacillus siamensis*, *Bacillus tequilensis*, *Bacillus licheniformis*, *Bacillus safensis* and *Bacillus pumilus* which were isolated from salted sheep skins were screened. Among the strains studied here, the strain with the highest capacity for PHA biosynthesis was found to be *Bacillus safensis*. According to our results PHA content of *Bacillus safensis* was about 51% of the cell dry weight when starch was used as a carbon source. The biopolymer produced was identified and characterized. Based on our findings *Bacillus safensis* is promising candidate for producing PHAs from renewable materials.

Keywords: Biopolymers, Bacillus spp., renewable sources

RRB-P-019 [Renewable Resources and Biopolymers]

Urban Carpentry Waste Based Natural Fibers for Preparation of Reinforced Composites with Thermoplastics

Rom Nath Baral[1], Rameshwar Adhikari[2]

[1]*Research Center for Applied Science and Technology, Tribhuvan University, Kirtipur, Nepal*
[2]*Nepal Polymer Institute, Kirtipur, Nepal*

Polymer composites, derived from fibers and polymer matrix, have wide range of outdoor and indoor applications. Natural fibers such as those from woods' saw dust, pines, flax, sisal, jute etc. and those produced as wastes by carpentry may provide renewable resources in plastic industries supporting the economic growth of Nepal using locally available raw materials. In this work, we focus on preparing composite materials using biobased and biodegradable polymers as matrix for the fabrication of the composite for high loads bearing applications such as indoor furniture. The cellulosic fibers were extracted from different urban carpentries based solid wastes originated from common Nepalese timber sources, such as Sorearobusta, Pinusroxburghii and Schimawallichii, Senegalin catechu,using different thermo-mechanical and chemical treatments and incorporated into commercially available polylactic acid (PLA) via melt mixing procedure followed by compression molding. The materials were characterized by different spectroscopic and microscopic techniques as well as impact mechanical testing. The mechanical performance of the composites will be discussed in correlation with the observed morphology. The composites thus developed are expected to be useful for indoor technical applications.

Keywords: Composites, filler, matrix, thermoplastics, mechanical properties

RRB-P-020 [Renewable Resources and Biopolymers]

Synthesis and Characterization of Pectin Based Hydrogel Films: Effect of pH and Drug Loading Method on Drug Release Kinetics

Özde Zeynep Güner, Banu Arabacıoğlu Kocaağa, F. Seniha Güner

Department of Chemical Engineering, Istanbul Technical University, Istanbul, Turkey

Pectin based products are widely used in wound care and treatment products. Pectin, as a natural, inexpensive polysaccharide, creates an oxygen and moisture permeable media; also it has antimicrobial properties which increase pectin's use in medical applications (Tummalapalli et al., 2016). Despite its mucoadhesive property providing convenience for medical applications, they have low thermal stability and poor mechanical properties as a disadvantage and this can be eliminated by blending pectin with different polymers or simply cross-linking with divalent ions (such as Ca+2, Fe+2, Zn+2) to improve its thermal and mechanical stability (Reddy, Eswaramma, Rao, Lee, 2014; Patil, Chavanke, Wagh, 2012). This study aims to develop pectin based drug carrier films which are based on two different drug loading systems: swelling and conventional (mixing). All samples were prepared as egg-box model created by crosslinking pectin with calcium chloride. Theophylline was chosen as a model drug in consideration of its interactive relation with pectin. The efficiency of two methods in drug release was analyzed by UV-vis spectrophotometer. Through wound healing process, pH value decreases and pH range shows an alteration for different types of wounds. Our study aims to have an efficient product for inflammatory wounds, so pH range that covers average wound healing process which is pH 4-8 was considered. Three different buffer solutions were used: phosphate-buffered saline (pH 4.6), tris(hydroxymethyl) aminomethane (pH 6.4) and deionized water (pH 7.12). For these pH values, drug release effectiveness's were compared. The results show that as pH decreases, both percentage drug release and drug release rate decreases.

Keywords: hydrogel, pectin, biopolymer, drug release, wound healing

RRB-P-021 [Renewable Resources and Biopolymers]

Conducting Polymer Nanocomposites using Multi-walled Carbon Nanotubes in a Biodegradable Polymer Matrix

Kedar Nath Dhakal[1,2,3], Santosh Khanal[2], Ralf Lach[4], Wolfgang Grellmann[4], Gert Heinrich[1], Rameshwar Adhikari[2]

[1]*Leibniz Institute für Polymerforschung (IPF), Hohe Str. 6, D-01069 Dresden, Germany*
[2]*Central Department of Chemistry, Tribhuvan University, Kirtipur, Kathmandu, Nepal*
[3]*Nepal Polymer Institute (NPI), P. O. Box 24411, Kathmandu, Nepal*
[4]*Polymer Service GmbH Merseburg, Eberhard-Leibnitz-Str. 2, D-06217 Merseburg, Germany*

Polymer composites are usually electrical insulators but can be imparted with conducting properties by introducing some conductive fillers like carbon nanotubes, metal particles, fibres etc. In this respect, environmentally benign polymeric materials with optimum electrical and mechanical properties are desirable for several practical applications. In this work, we investigate the nanocomposites of a completely biodegradable co-polyester [poly(butylene adipate-co-terephthaate); (PBAT], commercially available as Ecoflex®, with the aim of developing biodegradable flexible electrically conducting materials. The polymer was melt compounded with multi-walled carbon nanotubes (MWCNTs) using micro-compounder and the composites were further modified into fibrous structures via melt spinning procedures. The materials were characterized by different spectroscopic and microscopic techniques as well as biodegradability and mechanical tests. The electrical and mechanical properties of the composites will be discussed in correlation to the observed morphology of the materials. The materials thus developed are expected to be useful for applications as microelectrodes and sensors.

Keywords: nanocomposites, filler, biodegradability, micro-compounder, electrical properties

RRB-P-022 [Renewable Resources and Biopolymers]

In situ Synthesis of Poly(butylene adipate)/ Moroccan Beidellite clay Bionanocomposites by Polycondensation

Raihane Mustapha[1], Ilsouk Mohamed[1], Castelvetro Valter[2], Lahcini Mohamed[1], Bronco Simona[3], Rhouta Benaissa[4], Conzatti Lucia[5], Bianchi Sabrina[3]

[1]*Laboratory of Organometallic and Macromolecular Chemistry-Composie Materials University Cadi-Ayyad; Faculty of Sciences and Techniques. Av. A. Khattabi, BP.549, 40000 Marrakech, Morocco*
[2]*Dipartimento di Chimica e Chimica Industriale - Università di Pisa Via G. Moruzzi 3, 56124 Pisa -Italy*
[3]*IPCF-CNR UOS Pisa, CNR, via Moruzzi n°1, 56124 Pisa, Italy*
[4]*Laboratory of Condensed Matter and Nanostructure (LMCN), University Cadi-Ayyad; Faculty of Sciences and Techniques. Av. A. Khattabi, BP.549, 40000 Marrakech, Morocco*
[5]*Consiglio Nazionale delle Ricerche (CNR). Istituto per lo Studio delle Macromolecole (ISMAC). Via De Marini 6, 16149, Genova (Italy)*

Currently, the bioplastics market is coming out of its infancy and capturing the plastics market a growth rate of 30 % annually. Biodegradables polyesters such poly(hydroxylalkonaote), poly(caprolactone) and poly(lactic acid) have attracted considerable attention because of their potential applications. Poly(butylene adipate) (PBA) is a typical linear synthetic biodegradable aliphatic polyester which had attracted wide attention from scientific and industry. PBA can be potentially used in biomedical and eco-friendly materials for its biodegradability. However practical applications of PBA have been limited because of its softness, slow crystallization rate, its deficiency in mechanical, and thermal and barrier properties. Scientists/engineers from academia and industries are investigating the use of nanoclays as reinforcements in order to produce a new class of bionanocomposites with improved properties with preservation of the materials biodegradability, without eco-toxicity. Natural silicates such as beidellite, appear as suitable fillers that are natural abundant resources in Morocco, being of great socio-economical relevance. Rhassoul clay (Moroccan stevensite) is a worldwide known material that being toxin free can be used as additives for food, facemask, and health-care applications. The most of bionancomposites (co) polyestes based on PBA/ clay have been prepared by meth process.

To the best of our knowelge, the present work aims to investigate, for the first time, the valuing of the Moroccan clay (Beidelite BDT) as nanochage to prepare bionanocomposites with poly(butylene adipate) by in situ polycondensation using organometallic catalyst based on titanium butoxide. In contrast to the tin octoate as standard catalyst used in situ polymerization which required inert-gas technique, titanium butoxide used in this study is more stable and toward mositure in air and it can be handked in short time without protecting gas. The effect of the various weight of organoclay on the morphology, thermal and surface properties of the resulting nanocomposites PBA/ BDT Moroccan clay will also discussed.

Financial support for cooperative research came from, the Italian MIUR Cooperlink projet No. CII118NLT and the CNRST (Morocco)-CNRi (Italy) program, are gratefully acknowledged.

Keywords: In situ polycondensation, poly(butylene adipate), Moroccan beidellite clay, bionanocomposites, characterizations and properties

RRB-P-023 [Renewable Resources and Biopolymers]

Drug-Loaded Pectin Based Wound Dressing Hydrogels: Effect of Crosslinker Ratio and Drug Amount on Release Profile

Cansu Çam, Banu Arabacıoğlu Kocaağa, Fatma Seniha Güner

Department of Chemical Engineering, Istanbul Technical University, Istanbul, Turkey

Pectin is a biocompatible and biodegradable biopolymer that has been lately increased importance in wound treatment applications due to its good gelling mechanism easily modified biochemistry, antibacterial properties and mucoadhesive structure (Munarin et al., 2012). The existence of the carboxyl groups on the pectin backbone that interact divalent cations enable to form "egg-box" model. This unique form of pectin brings opportunity to create drug-loaded system which provides controllable drug release mechanism (Espitia et al., 2013).

This work focuses on creating drug-loaded pectin based biofilms as a wound dressing of hydrogel in two different drug amounts (30 mg and 60 mg) and at three different crosslinker ratio (7, 10 and 15 %, by weight percent). In the manufacturing of the wound dressing, low methoxy pectin is used. Additionally, glycerol is used as plasticizer, and CaCl2 is crosslinker. Theophylline is chosen as a model drug molecule and drug loaded biofilms were prepared by mixing (conventional) method to form diffusion-controlled matrix-type drug release systems. In vitro drug release study was performed in tris buffer (pH=6.4) as a simulation of wound exudate. Drug release performance of the biofilms was analyzed by UV-vis spectrophotometer at λ=271 nm. It is indicated that the drug release profile is greatly affected by crosslinker variance and drug amount.

Espitia, P.J.P., Du, W-X., Avena-Bustillos, R.J., Soares, N.F.F., and McHugh, T.H., 2013. Edible Films from Pectin: Physical-Mechanical and Anti-Microbial Properties., Food Hydrocolloids, 35, 287-296.
Munarin, F., Tanzi, M.C., and Petrini, P., 2012. Advances in Biomedical Applications of Pectin Gels. International Journal of Biological Macromolecules, 51, 681-689.

Keywords: pectin, wound dressing, hydrogel, controlled drug release

RRB-P-024 [Renewable Resources and Biopolymers]

Isocyanate- and phosgene-free synthesis of polyhydroxyurethane thermosets by employing multifunctional bio-based and organic/inorganic cyclic carbonates and bio-based amine hardeners

Hannes Blattmann, Rolf Mülhaupt

Freiburg Materials Research Center and Institute for Macromolecular Chemistry, Albert-Ludwigs University Freiburg, Germany

Isocyanate-based polyurethanes (PU) represent one of the most important polymer classes in everyday life, but experience severe challenges for future regulatory procedures due to the safety issues of isocyanates. Non-isocyanate polyhydroxyurethanes (PHU), prepared by the polyaddition of environmentally benign cyclic carbonates and amine hardeners, hold great prospect to substitute PUs. Multifunctional cyclic carbonates are obtained by the catalytic fixation of CO_2 by carbonation of epoxides. Therefore, carbonated polyglycidylethers derived from bio-based polyols are among the most prominent examples. Simultaneously, the aminolysis of the same polyglycidylethers and limonene dioxide with aqueous ammonia represents a versatile one-step synthesis of multifunctional bio-based amine curing agents and does not require either organic solvents or tedious purification (H. Blattmann et al. *Green Chemistry* 2016, DOI: 10.1039/C5GC02563J). This study elucidates the influences of molecular architecture and amine content of multifunctional β-amino alcohols (AA), and the type of polyfunctional cyclic carbonates on the thermal and mechanical properties, degradation and solvent swelling of PHUs. Facile cure was achieved by blending the high-viscosity AAs with hexamethylene diamine (HMDA). As compared to HMDA, the addition of the less reactive polyfunctional AAs increases gel time and simultaneously improves PHU stiffness. In a different approach, the chemical fixation of CO_2 with glycidylether functionalized polyhedral oligomeric silsesquioxanes (POSS) yields mono- and polydisperse, multifunctional POSS cyclic carbonates (H. Blattmann et al. *Macromolecules* 2016, *49 (3)*, 742-751). On cure with diamines, POSS carbonates and their blends with polyol-based cyclic carbonates produce organic/inorganic hybrid polyhydroxyurethane thermosets. Young's modulus and tensile strength improve with increasing POSS content as a result of higher crosslink density. Sub-micron phases are formed during polymerization due to self-assembly and nanophase-separation as was verified by scanning electron microscopy (SEM). Moreover, the incorporation of POSS significantly improves scratch resistance of optically transparent and colorless coatings as examined by the surface gloss and SEM images.

Keywords: non-isocynate polyurethane, polyhydroxyurethane, carbon dioxide, POSS cyclic carbonate, polyhydroxyurethane hybrid, bio-based amine curing agent

Scheme 1

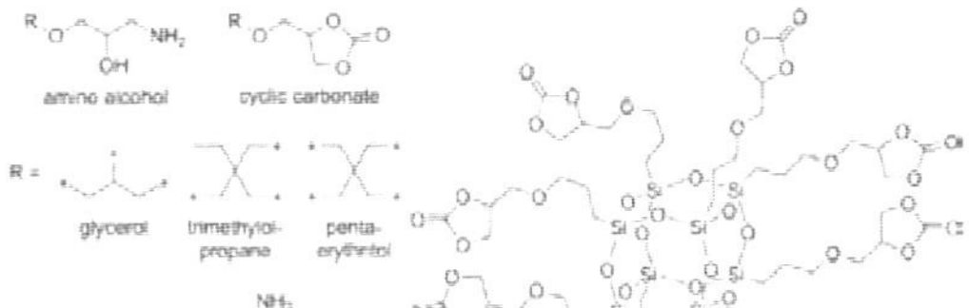

Multifunctional cyclic carbonates and amine hardeners from bio-based polyglycidylethers and limonene dioxide (left) and organic/inorganic POSS cyclic carbonate (right) for the preparation of polyhydroxyurethane thermosets.

RRB-P-025 [Renewable Resources and Biopolymers]

The Some of Physical, Chemical and Mechanical Properties of Polyurethane Foams-contained Potato Crust Wastes Polyol-I

Tülay Gürsoy[1], Aycan Gür[1], Mehmet Hakkı Alma[2], Ertuğrul Altuntaş[2], Eyyüp Karaoğul[2]

[1]*Department of Chemistry, Yüzüncü Yıl University, Van,Turkey*
[2]*Department of Industrial Forestry, Sütçü İmam University, Kahramanmaras,Turkey*

Potato crust wastes were liquefied with different kinds of organic acid catalysts (H_2SO_4, HCl and H_3PO_4) and catalyst concentration within different liquefaction periods by using PEG400/glycerin binary solvent in microwave oven soon after potato crust-based polyols were characterized. Among them, catalytic activity of H_2SO_4 was the best and the resultant liquefied potato crust-based polyol had suitable hydroxyl value, therefore this polyol was chosen for the preparation of polyurethane foam(PU). Feasible formulations for preparing liquefied potato crust-based polyurethane foams were proposed. Some mechanical and physical properties of the foam were investigated. The results indicated that the apparent density, heat conductivity value, compressive strength, elongation after pressure and elastic modulus of polyurethane foams derived from potato crust polyols were 0.0544 g/cm^3, 0.0331 W/mK, 301.4 kPa, 2.24% and 1.48 MPa, respectively. In addition, weight loss after the biodegradation of potato crust-based foam for a service test of 3 months was 17.84%. The liquefied products were also analyzed by using Attenuated Total Reflectance-Fourier Transform Infrared (ATR-FTIR) spectroscopy and X-ray Diffractometer. ATR-FTIR spectra verified the formation of bonds as urea and urethane presented in the structure of PU foams. From Wide Angle X-Ray Diffraction pattern of PU foam, it was detected that polyurethane composite was of amorphous structure, resulting in excellent physical properties. The foams prepared from liquefied potato crust-based polyols obtained by microwave-assistant liquefaction exhibited excellent physical and mechanical properties. It was found that the microwave heating was more efficient than the conventional oil bath heating for the potato crust-based liquefaction.

Keywords: Potato crust, Liquefaction, Polyurethane Foam, Physical-Mechanical Properties, Microwave-heating Method, Biodegradation

RRB-P-026 [Renewable Resources and Biopolymers]

The Some of Physical and Termal Properties of Polyurethane Foams-contained Potato Crust Wastes Polyol-II

Tülay Gürsoy[1], Aycan Gür[1], Mehmet Hakkı Alma[2], Ertuğrul Altuntaş[2], Eyyüp Karaoğul[2]

[1]*Department of Chemistry, Yüzüncü Yıl University, Van,Turkey*
[2]*Department of Industrial Forestry, Sütçü İmam University, Kahramanmaras,Turkey*

The liquefied potato crust-based polyurethane (PU) foams were prepared from the liquefied potato crust-based polyols. Potato crust wastes liquefied in the presence of polyethylene glycol-dominant reaction reagent by using three kinds of organic acid as catalyst in a microwave oven under stirring. The liquefied potato crust-based polyols were characterized for the preparation of polyurethane foams. The potato crust-based polyol which was obtained from 9% sulfuric acid catalyst within 30 min liquefaction reaction period was chosen for the preparation of PU foams. PU foams prepared according to one-shot technic. Potato crust contained-PU foams were analyzed by using Scanning Electron Microscopy (SEM) and Thermo Gravimetric Analyzer. Sound conduction and water absorption properties of PU foams were also measured. SEM images showed that optimum cell thickness was 300.4 µm and heterogeneous pore structure was in existence. In proportion to thermo gravimetric analysis, TGA curve showed only one stage of weight loss and initial degradation temperature was 388°C, degradation temperature half of mass was 423°C, and the residue remained at 800 °C was about 6.08%. Moreover, sound conduction speed of foams prepared from liquefied potato crust polyols was partially high, but close than foams prepared from fossil-based polyols. Water absorption properties were found to be higher than those of petroleum-based PU foams.

Keywords: liquefaction, potato crust-based polyol, starch, polyurethane foam, water-absorption, morphologic and thermal property

RRB-P-027 [Renewable Resources and Biopolymers]

Polysilicon Acetals via Silicon Acetal Metathesis Polymerization (SAMP)

Ersen Göktürk[1], Ertuğrul Şahmetlioglu[2], Stephen Albert Miller[3]

[1]*Department of Chemistry, Mustafa Kemal University, Hatay, Turkey*
[2]*Nanotechnology Research Center (ERNAM), Erciyes University, Kayseri - 38039, Turkey*
[3]*Department of Chemistry, University of Florida, Gainesville/Florida, United States*

The silicone (polydimethylsiloxane, PDMS) is 38% silicon and 22% oxygen and is the most commonly used silicon-containing polymer. PDMS can be used for many applications. Important and specific attributes include low glass transition temperature, high conformational flexibility, biocompatibility, high gas permeability, hydrophobicity, and good oxidative, thermal, and UV light stability. Distinct from silicones such as PDMS, polysilicon acetals are polymers containing silicon acetal units (–OSiR2O–) connected by hydrocarbon segments (–R'–), having the generalized structure [–OSiR2OR'–]. We present families of sustainable polysilicon acetals via a novel methodology, silicon acetal metathesis polymerization (Figure 1). This silicon acetal metathesis reaction is exploited for the step-growth polymerization of bis-silicon acetals (MeOSiMe2OROSiMe2OMe) with metathetical loss of Me2Si(OMe)2. Thus, a convenient and generalized SAMP method is introduced as the acid-catalyzed copolymerization of a diol (HOROH) and Me2Si(OMe)2, driven by elimination of methanol and/or Me2Si(OMe)2 to produce polysilicon acetals [1].

Keywords: polysilicon acetals, Silicon Acetal Metathesis Polymerization, polydimethylsiloxane

RRB-P-028 [Renewable Resources and Biopolymers]

Environmentally degradable biohybrids of MMT with polypyrrolidones from microbial itaconic acid

Mohammad Asif Ali, Nupur Tandon, Seiji Tateyama, Tatsuo Kaneko

Energy and Environment Area, japan advanced institute of science and technology, Ishikawa, Japan

The development of high-performance, environmentally-degradable polymers usable as engineering plastics from renewable resources are significant for the establishment of a sustainable society. Besides, the degradable resins give a way for overcoming the plastic waste problems. Conventional polyamides such as NylonTM account for a substantial percentage of engineering plastics. Then we focus on bioplastic NylonTM from widely-produced biomolecules such as itaconic acid which is derived from fermentation of glucose in the presence of Aspergillus terreus. An in-bulk polymerization of itaconic acid, which contains two carboxylic acid groups and vinylene moietiy, salts with a series of diamines by a melt poly-condensation process generated bio-derived thermally stable polymers. The structure favors two-step reactions of Michael addition/amidation with diamines to form N-substituted "cyclic amide". enhancing the polyamides performance required to be bio-based in terms of green sustainable science. Next we prepared nylon/montmorillonite (MMT) nanocomposites formed by nanocoupling of itaconate-bioderived polyamide with MMT through nylon salt method. As a result of thermomechanical evaluation by various techniques such as TGA, DSC, TEM, FT-IR, WAXD and tensile testing, it was found that the nanocomposites showed the high performance; glass transition temperature, Tg, and 10 % weight loss temperature, T10, ranged 70-80 oC and 400-430 oC respectively. The processed fibers showed tensile strength and Young's modulus ranged 2.2-5.5 GPa and 65-241 MPa, respectively. In addition, the elongation at break was largely enhanced by composite fomration of NaMMT to be 0-10 wt% and then strain energy attained to be 22 J/m3. On the other hand, the nanocomposites had less moisture absorbing capacity of 2.5-3.9 wt% than conventional polyamide such as PA6 or PA66. The resulting bio-derived composites showed degradable behaviors inside soil and under ultraviolet-irradiated in water were additionally found by photo-induced ring-opening reaction of the pyrrolidone ring to give hydrophilic side-chains of carboxylic acid.

Keywords: Nylon salts, Itaconic acid, polypyrrolidones

RRB-P-029 [Renewable Resources and Biopolymers]

Effect Of Modified Chitosan Biopolymer On Biodiesel Synthesis

Zafer Ömer Özdemir[1], Halil Mutlubas[2]

[1]Energy Technology Research and Application Center, Kirklareli University, Kirklareli,Turkey
[2]Department of Energy Systems Engineering, Kirklareli University, Kirklareli, Turkey

Chitosan is a polymer which made up natural aminopolysaccharides. It has unique structure, multi-dimensional features, advanced functions and high biochemical-industrial use in large areas. Chitosan is a derivation of Chitin, which is the most abundant natural biopolymer in which polymers bear nitrogen in the main chain.

In this study, chitosan is used to increase the efficiency in the synthesis of biodiesel which is a kind of renewable energy source.

Chitosan is used as a catalyst with amino acid modifications on it during FAME (Fatty Acid Methyl Ester) synthesis. Amino acid modification of chitosan is performed with Microwave assisted SPPS (Solid Phase Peptide Synthesis). The biodiesel product was analyzed with (UV-visible, viscosity, pH, flash test).

Keywords: Biodiesel, Biopolymer, Chitosan, Microwave

RRB-P-030 [Renewable Resources and Biopolymers]

Renewable Biocomposite Production from Agricultural Waste Biomass

Müge Sennaroğlu Bostan[1], Selim Ceylan[2]

[1]Department of Chemical Engineering, Marmara University, Istanbul, Turkey
[2]Department of Chemical Engineering, 19 Mayıs University, Samsun, Turkey

Agriculture in Turkey has kept its role as a major labor potential due to the structure of country climate, location and fertile soils. After the processing of agricultural products a large quantity of agricultural waste are produced, whereas common application of this waste is not available yet. Turkey is the world leader in hazelnut production, causing a large amount of annual hazelnut shells (HS) and hazelnut husk (HH) waste. Polycaprolactone (PCL) is a biodegredable polymer which has a low melting point, low glass transition temperature and compatibility with a range of materials. It could be used with natural fibres from agricultural wastes to have effective cost reduction and add improved mechanical properties to biocomposites. For this purpose, this study is focused on the evaluation of morphological, mechanical, thermal and degradation properties of new biocomposites based on PCL and HS and/or HH fillers at different contents. Biocomposites were prepared by melting of PCL at 135 oC and 100 rpm under magnetic stirring and introducing different amounts of HS or HH followed by mixing for 5 mins. The obtained biocomposites were molded into disc shape samples under 5mPa pressure for 1 min and further characterized by FTIR (Fourier transform infrared spectroscopy), DMA (dynamic mechanical analysis), DSC (differential scanning calorimetry) and TGA (termal gravimetric analysis). Our results confirmed that HS or HH can be considered as a biofiller in polymer composites due to its low cost as well as enriched mechanical and thermal properties. Melting and crystalization entalpies were evaluated for bio-composites and compared with neat PCL. As a result PCL/HS and PCL/HH composites are promising biomaterials. They may have an economic contribution to country for their potential to solve a great amount of agricultural waste problem as well as achieving biocompatible and environmentally friendly products.

Keywords: Hazelnut Shell, Hazelnut Husk, Polycaprolactone, Renewable Biocomposites

RRB-P-031 [Renewable Resources and Biopolymers]

Improvement of Water Sensitivity and Mechanical Properties of Xylan-Based Polymeric Materials

Merve Akkus[1], Necati Ozkan[2], Ufuk Bakir[1]

[1]Department of Chemical Engineering, Middle East Technical University, Ankara, Turkey
[2]Department of Polymer Science and Technology, Middle East Technical University, Ankara

Hemicelluloses are sustainable resources of biopolymers with a huge potential to be converted into materials substituting petroleum-based products, but their hydrophilic characteristics pose a challenge to their industrial adoption. Hydrophobic modification of hemicelluloses can be achieved by acetylation, which is a widely used technique for the modification of cellulose and starch. The main aim of this study is to decrease the water sensitivity and to increase the mechanical properties of xylan-based polymeric materials by acetylation. Xylans were isolated from corn cobs by alkaline extraction without removing potassium acetate (KAc) that is formed during the alkaline extraction. Extracted xylans were extruded into the form of strips using a twin-screw extruder and strips were acetylated in the presence of KAc using acetic anhydride as an acetylation agent, without any catalysts added. ATR-FTIR analysis verified the exchange of hydrophilic hydroxyl groups by acetyl groups upon acetylation in the presence of KAc. Moisture uptake, water solubility and water vapor transfer rate of the strips decreased significantly which is accompanied by a substantial increase in the mechanical properties. Ultimate tensile strength values of the strips at 50% relative humidity, increased from 6 MPa to 13 MPa upon acetylation. The difference in ultimate tensile strength values was more dramatic at 70% relative humidity, which increased from 1 MPa to 10 MPa upon acetylation. 10 times higher strength even at high relative humidity conditions were related with the decreased moisture uptake, and acetylation in the presence of KAc was shown to be an effective method to obtain xylans with decreased water sensitivity and improved mechanical properties.

Keywords: hemicellulose, xylan, acetylation, water sensitivity, mechanical properties

RRB-P-033 [Renewable Resources and Biopolymers]

Biodegradable Copolyester Composites Reinforced by Wheat Stalk Microcrystalline Cellulose

Jyoti Giri1, Ralf Lach2, Hai Hong Le[2], Wolfgang Grellmann[2], Hans Joachim Radush[2], Goerg H Michler[3], Jean Marc Saiter[4], Sven Henning[5], Rameshwar Adhikari[1]

[1]Central Department of Chemistry, Tribhuvan University, Kirtipur, Kathmandu, Nepal
[2]Center of Engineering, Martin Luther University Halle-Wittenberg, Merseburg/Saale, Germany
[3]Department of Physics, Martin Luther University Halle-Wittenberg, Halle/Saale, Germany
[4]Universite' de Rouen, Laboratory LECAP, Faculte des Sciences, Avenue de l'Universite, Rouen, France
[5]Fraunhofer Institute for Mechanics of Materials, Halle/Saale, Germany

In this work, agricultural waste wheat stalk was kept in a series of thermo-chemical and mechanical treatments to extract microcrystalline cellulose (MCC-JG). The MCC-JG extracted was compounded with poly(butylenesadipate-co-terephthalate), PBAT, biodegradable polymer, via melt mixing using Brabender Plasticorder followed by compression moulding. The samples were characterized by Fourier transform infrared spectroscopy (FTIR), tensile test, microindentation technique (HM), thermogravimetric analysis (TGA) and differential scanning calorimetry (DSC). The degradation behaviour of the composites was measured by microbial incubation and soil composting. The water holding capacity was measured in environmental chamber at 23 °C. Cellulosic filler were found homogeneously dispersed in the composites. The composites possessed low ductility with good thermal stability. Tensile reinforcement of the PBAT/ MCC-JG composite was studied with respect to yield stress and young's modulus with the variation of MCC-JG content, shown in Figure 1. The reinforcement was observed up to 40 phr MCC-JG content in the composites. These composites were found to be degrading under soil composting and showed biosensitivity on microbial incubation by colour change. Water holding capacity of the composites was also found to be increased with MCC-JG content.

Keywords: biodegradable polymer, microcrystalline cellulose, mechanical properties, microhardness

RRB-P-034 [Renewable Resources and Biopolymers]

Experimental and Computational Studies of Ring Opening Polymerization of α-angelica lactone with Zinc Acetate

Sıla Gümüştaş, Armağan Kınal, Mehmet Balcan

Department of Chemistry, Ege University, İzmir, Turkey

Owing to the presence of a C=C bond in α-angelica lactone, the ring opening polymerization (ROP) of five-membered cyclic lactone becomes feasible under moderate conditions and the resultant polyester may provide a potential route to produce functionalized new aliphatic polyesters [1].

Tin(II) 2-ethylhexanoate; Sn(Oct)2 has been widely used as the initiator for ROP. For biomedical applications, the cytotoxicity of Sn(Oct)2 and other tin compounds is a serious disadvantage and purification of biodegradable polyesters from tin compounds is expensive and time-consuming. Therefore, metal salts and complexes such as zinc, magnesium, calcium and iron showing a similarly good performance as a initiator or catalyst for ROP combination with a much lower toxicity are highly desirable [2,3].

In our study poly(α-angelica lactone) was successfully synthesized via ROP of α-angelica lactone with $Zn(OAc)_2 . 2H_2O$. The reaction conditions such as reaction time, reaction temperature and moomer/initiator molar ratio are determined. The polymer was characterized by FTIR, NMR and GPC. The thermal properties were investigated by DSC and TG. Furthermore, the initiation step of ROP mechanism of α-angelica lactone with Zn(OAc)2 was computationally investigated by employing the DFT-B3LYP method.

References:

1. Chen T., Qin Z., Qi Y., Deng T., Ge X., Wang J. and Hou X., Polym. Chem. 2011, 2, 1190-1194
2. Govda R. R., Chakraborty D., Journal of Momecular Catalysis A:Chemical 2010, 333, 167-172
3. Kricheldorf H.R. and Behnken G., Journal of Macromolecular Science Part A: Pure and Applied Chemistry, 2007, 44, 795-800

Keywords: alpha-angelica lactone, biodegradable polymer, ring opening polymerization

Reaction scheme of the polymerization

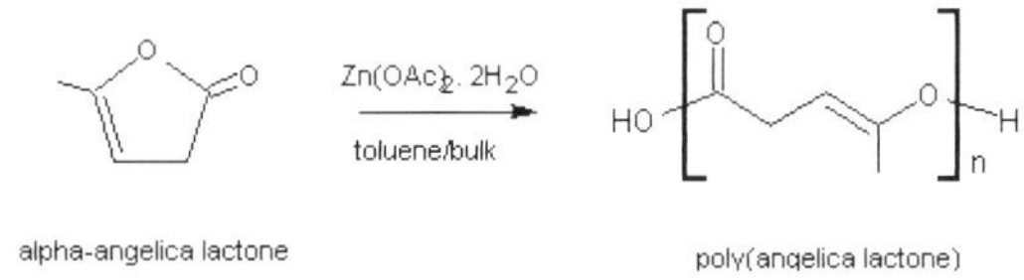

RRB-P-035 [Renewable Resources and Biopolymers]

Biopolymer based Composite Hydrogel as Food Packaging Material

Nabanita Saha[1], Kezban Kezban Akkaya[2], Ramasubba Reddy Palem[1], Ruhan Benlikaya[3], Petr Saha[1]

[1]Centre of Polymer Systems, Tomas Bata University in Zlin, Zlin, Czech Republic
[2]Department of Chemical Engineering, Faculty of Engineering, Ege University, Izmir, Turkey
[3]Department of Secondary Science and Mathematics Education, Balikesir University, Balikesir, Turkey

Materials used in food packaging play a significant role in determining the shelf life of any food product. The excellent porous structure of Polyvinylpyrrolidone-Sodium Carboxy Methyl Cellulose (PVP-CMC) hydrogel film provided the environment for keeping table grapes fresh until three weeks by allowing the pathway to diffuse the moisture from inside to outside of the package easily, while traditional packages available in the markets as a food package were unable to prevent early spoilage in the same duration [1]. In addition to these remarkable properties, only drawback of the PVP-CMC film is it's brittle in dry state [2]. Thus, this study is focused on improvement of mechanical properties of the film by adding three different fillers [Sepiolite (Sep), PVA Nanofiber (PVA) and Bacterial Cellulose (BC)] in various concentrations in order to increase the suitability of the material as a food package. The films (thickness 0.07-0.09 mm) are transparent, which were prepared via the solvent casting method [1, 2]. Structural characteristics of the modified PVP-CMC films were determined by Fourier Transform Infrared Spectroscopy (FTIR) and X-ray Diffraction (XRD). The surface morphologies of the dry films examined by Scanning Electron Microscopy (SEM) and Optical Microscopy are depicted in Fig.1.

RD and SEM analysis confirm the uniform dispersion of the fillers within the PVP-CMC films. The values of tensile modulus for the resulting materials decrease as 47.79± 13.85 MPa (PVP-CMC-BC) > 33.75±7.19 MPa (PVP-CMC-Sep) > 25.21±5.97 MPa (PVP-CMC-PVA) > 20.53±2.37 MPa (PVP-CMC). It is seen that all fillers help to improve the mechanical property of PVP-CMC film.

mong these fillers, BC, renewable polymer, performed as the best filler. Permeability and biodegradability studies of the modified films are in progress.

Acknowledgement: The work was supported by the Ministry of Education, Youth and Sports of the Czech Republic - NPU Program I (LO1504).

Keywords: Bacterial cellulose, Food packaging, PVP-CMC hydrogel, PVA nano fiber, Sepiolite

Fig.1

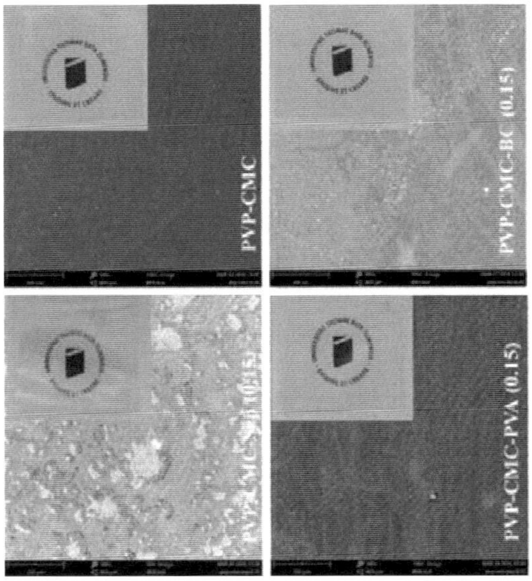

Optical and SEM images of biopolymer based composites hydrogel

Keywords: Polysaccharides, Cyanobacterium, Gels, Liquid crystals, Anisotropy, Orientation

Process of super-anisotropic hydrogels from cyanobacterial polysaccharides

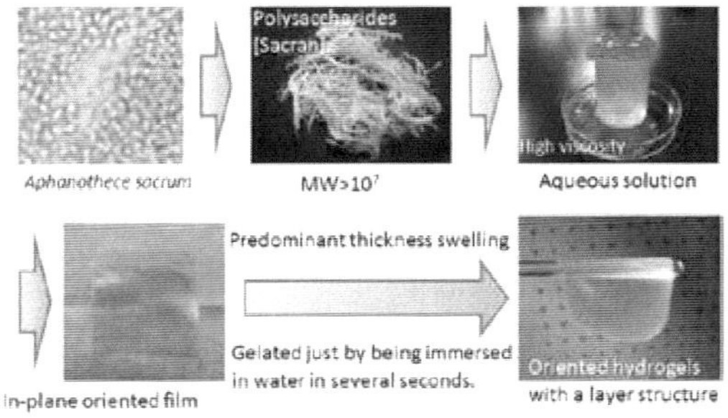

Supergiant LC polysaccharides, sacran, which was extracted Aphanothece sacrum, was cast over viscous aqueous solution to form oriented films which was soon gelated just by immersion in water.

RRB-P-036 [Renewable Resources and Biopolymers]

Super-anisotropy in cyanobacterial hydrogels of supergiant polysaccharide SACRAN

Maiko Kaneko, Kittima Amornwachirabodee, Saranyoo Sornkamnerd, Tatsuo Kaneko

Energy and Environment Area, Japan Advanced Institute of Science and Technology, Nomi, Japan

Aphanothece sacrum is a freshwater cyanobacterium indigenous to Japan, and has a huge jelly-like matrix with a cell-cluster size from several centimeters to several decimeters. Here we extracted the polysaccharides, sacran, from the matrix. From multi-angle static light scattering analyses, it was found that the sacran has very high value of absolute molecular weight over 10^7. From various spectroscopic analyses, sacran was found to have anionic groups such as carboxyl and sulfate groups. Based on ionic groups and very high molecular weight, sacran showed very high water retention capacity. Added to these properties, sacran shows the liquid crystallinity under the condition at very low solution concentration around 0.25 wt % because of the rigid molecular chains. Using the property we prepared the physical hydrogel of sacran by an in-situ gelation of the layered film made by cast-drying method. Concretely, the 0.5 wt % sacran solution was dried in oven at 60 °C for 12 h to create the thin film with a thickness of about 50 micrometer, and the film was then immersed into pure water to anisotropic hydrogels Furthermore, it was found that the swollen degree of the gels was controlled from 20 to 800 times by changing the heating treatment temperature of the film at 70, 80, 90,100 and 100 °C. The orientation of sacran film was confirmed by SEM, WAXD and crossed-polarizing microscope. The obtained anisotropic sacran hydrogel showed high elastic modulus. Additionally high-performance hydrogel was prepared by chemically cross-linking of sacran with divinylsulfone (DVS) and drying on substrate into film. The chemical hydrogel showed perfect anisotropic swelling behavior, linear swelling ratio of thickness was higher than linear swelling ratio of width for 10,000-40,000 times.

RRB-P-037 [Renewable Resources and Biopolymers]

Investigating the effect of PEG and diisocyanates on physical and viscoelastic properties of poly L-(lactic-acid) blended with gelatin

Babak Kaffashi, Nazanin Shakoury

Department of Polymer Engineering, School of Chemical Engineering, College of Engineering, University of Tehran, Tehran, Iran

The concept of biodegradable plastics is of considerable interests with respect to solid waste accumulation. Many natural biodegradable plastics are blended with synthetic polymers to produce plastics which meet the functional requirements. Polyesters play a predominant role as biodegradable plastics due to their potentially hydrolysable ester bonds. Poly (lactic-acid) is linear aliphatic polyester synthesized from raw renewable resources. But, the inherent brittleness of PLA has been a major bottleneck for its large-scale commercial applications. Other main limitations of PLA towards its wider industrial application are its poor thermal resistance, hydrophobicity and slow rate of degradation. There are several ways to fix such problems; for example plasticization, blending with other polymers, copolymerization and incorporation of fillers. In this study, PLA is melt blended with gelatin, polyethylene glycol and various diisocyanates to produce a hydrophilic composite with less brittleness which degrades more quickly. Different composition of methylene diphenyl diisocyanate, isophorone diisocyanate, hexamethylene diisocyanate, PEG400 and gelatin were prepared. In this regard, multiple tests such as rheometry, contact angle and biodegradation were done. The rheological properties data represent the effects of PEG and diisocyanate on PLLA/gelatin blends. The PEG addition of 5-25 weight percent, shift the cross-over frequency of G' and G'' to lower frequencies. In addition, composites made by IPDI and MDI have better melt behavior than HDI. Based on the rheological analysis, composites with 1.5 percent diisocyanate and 8 weight percent PEG400 show the best viscoelastic behavior. On the other hand, The contact angle measurements demonstrate that gelatin and PEG addition results in a more hydrophilic material as the contact angles were reduced. Furthermore, the presence of diisocyanate rendered the composite hydrophobic. Finally, according to this researchtests blend of 8 percent PEG400 and 1.5 MDI seems shows some superb properties.

Keywords: PLLA, Gelatin, Contact angle, Rheological measurements, diisocyanate

Figure 1(a) ***LAGxPxIx stands for the blend containing PLLA (LA), gelatin (G), PEG (P) and diisocyanate (I). In case of zero concentration of each of the above zero is shown (numbers are in percent)**

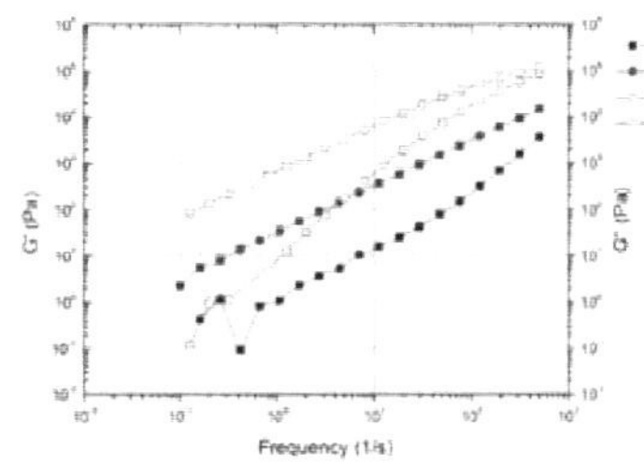

Figure 1(a)-Effect of PEG addition in constant gelatin content on dynamic rheological test (G' and G'' versus ω) of PLLA/gelatin/PEG400 samples in comparison with pure PLLA at 170°C; (a): LAG5P5I0

Figure 1(b)

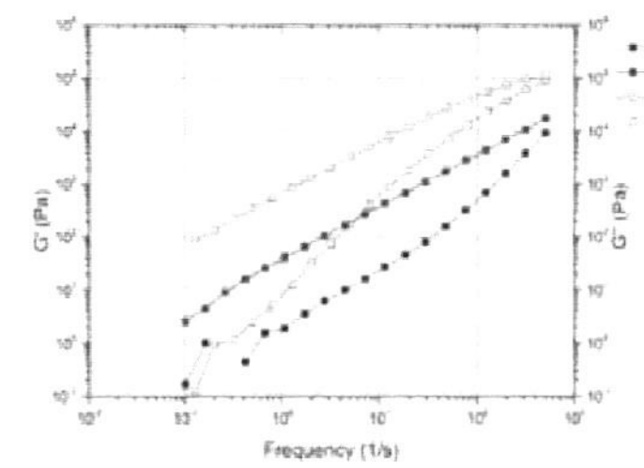

Figure 1(b)-Effect of PEG addition in constant gelatin content on dynamic rheological test (G' and G'' versus ω) of PLLA/gelatin/PEG400 samples in comparison with pure PLLA at 170°C; (b): LAG5P10I0

Figure 1(c)

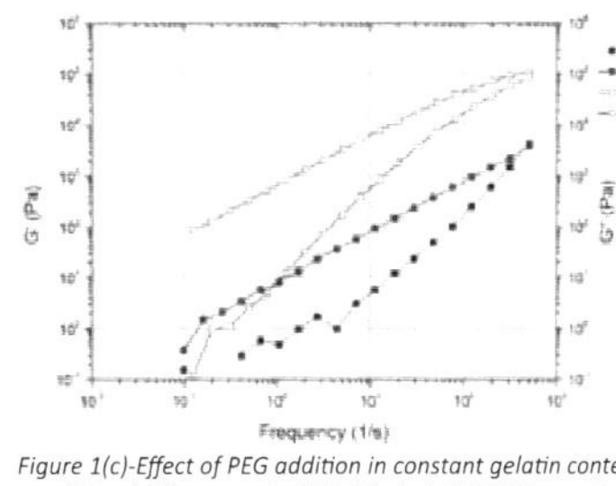

Figure 1(c)-Effect of PEG addition in constant gelatin content on dynamic rheological test (G' and G'' versus ω) of PLLA/gelatin/PEG400 samples in comparison with pure PLLA at 170°C; (c): LAG5P15I0

Figure 1(d)

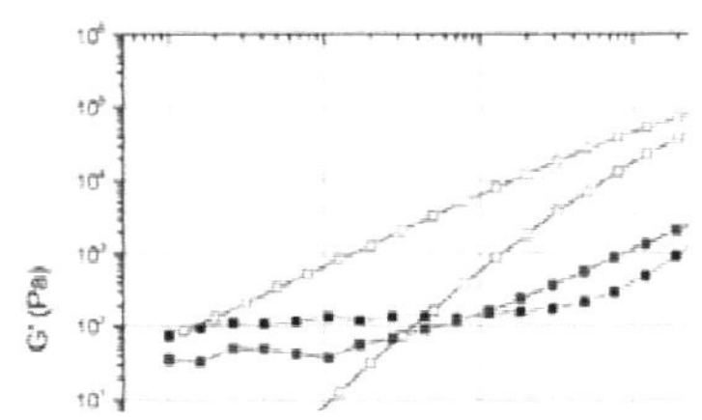

Figure 1(d)-Effect of PEG addition in constant gelatin content on dynamic rheological test (G' and G'' versus ω) of PLLA/gelatin/PEG400 samples in comparison with pure PLLA at 170°C; (d): LAG5P25I0

Figure 2(a)

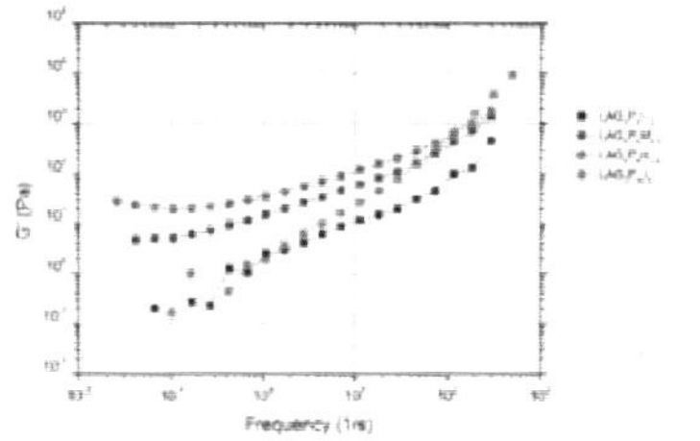

Figure 2(a)-Effect of diisocyanate addition and diisocyanate type in constant gelatin and PEG400 content on dynamic rheological test of PLLA/gelatin/PEG400 samples in comparison with pure PLLA at 170°C; (a) G' versus ω

Figure 2(b)

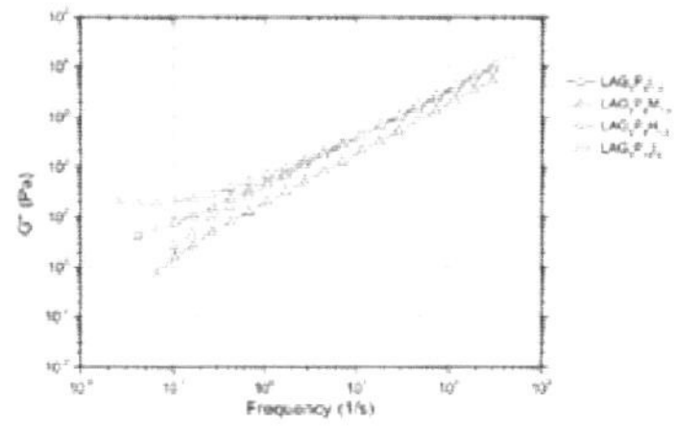

Figure 2(b)-Effect of diisocyanate addition and diisocyanate type in constant gelatin and PEG400 content on dynamic rheological test of PLLA/gelatin/PEG400 samples in comparison with pure PLLA at 170°C; (b): G'' versus ω

Figure 3(a)

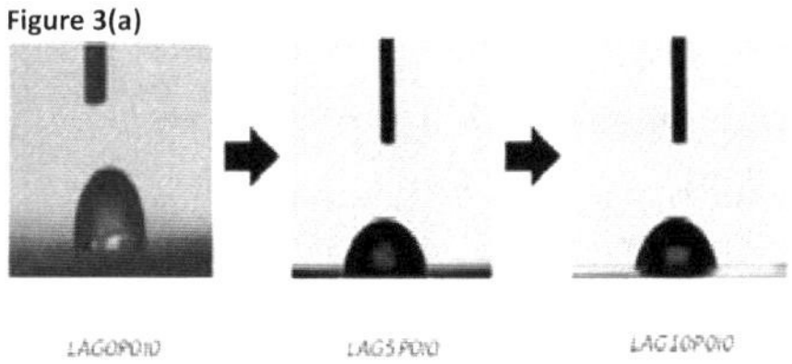

Figure 3(a)-Effect of gelatin increase in PLLA/ gelatin composite

Figure 3(b)

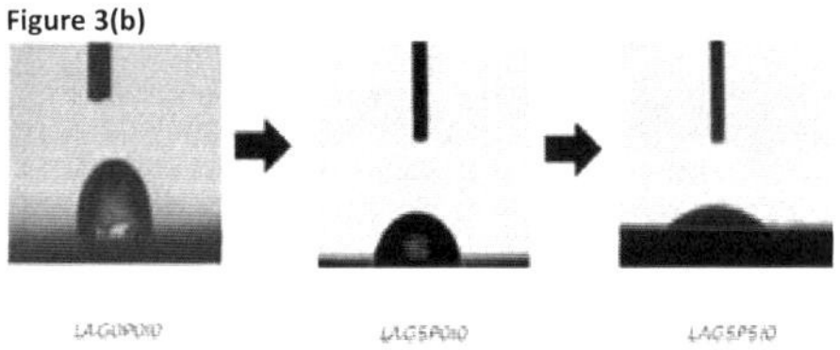

Figure 3(b)-Effect of diisocyanate on PLLA/ gelatin/ PEG400 blend

Figure 3(c)

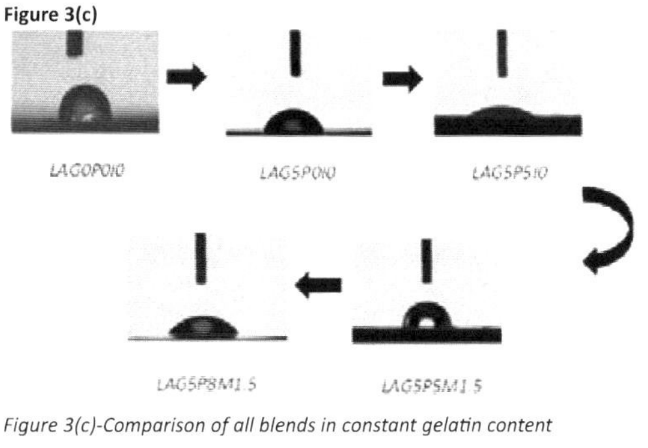

Figure 3(c)-Comparison of all blends in constant gelatin content

SFP-P-001 [Smart and Functional Polymers]

Amphiphilic polypeptides containing a hydrophobic luminescent terminal: probing critical aggregation concentration and lowest critical solution temperature by aggregation-induced emission

Jin Long Hong

Department of Materials and Optoelectronic Science

Luminogens with aggregation-induced emission (AIE) behavior are highly emissive in the aggregated solution state regardless of their non-emissive dilute solution state. When an AIE-active luminogen is connected by polymer chain, its emission is strongly correlated with the aggregated structure of the polymer chain. With this respect, an amphiphilic polypeptide TP-PPMEO (Figure 1)¬, containing a hydrophilic peptide PPMEO chain terminated with an AIE-active tetraphenylthiophene (TP) luminogen, was prepared and its AIE-related emission toward different aggregated states was characterized. Primarily, the isolated TP-PPMEO chains in dilute aqueous solution emit dimly but when polymer chains start to aggregate at concentration exceeding critical aggregation concentration (CAC), the TP-PPMEO chains emit intensely regarding the AIE-related emission triggered by aggregation. The aggregated solution was then heated to temperatures above the lowest critical solution temperature (LCST) to result in the dissociation of hydrogen bond interactions between PPMEO and water. In consequence, mass coagulation of several TP-PPMEO aggregates occurs to generate large aggregates, within which the PPMEO chains further shrunk in size in order to reduce unfavorable contacts with water. The TP luminogens inside the shrunk chain domains are therefore hampered in molecular rotation, rendering emission with intensity higher than those at lower temperatures. The AIE-related emission is therefore correlated with aggregation level induced by concentration and heating.

Keywords: Aggregation-induced emission, polypeptide, luminescent terminal, aggregation level, critical aggregation concentration, lowest critical solution temperature

SFP-P-002 [Smart and Functional Polymers]

Characterization, Synthesis of New Benzoylthiourea Polymers and Thermal Properties

Gülşah Kurt Şahan

University of Aksaray, Faculty of Arts and Sciences, Chemistry Department,68100,AKSARAY

Four new benzoyl thiourea polymers were synthesized via reaction of terephthalodiisothiocyanate which is obtained by the reaction of substituted amines, terephthaloyl dichloride and potassium thiocyanate. As the phase transfer catalyst polyethylene glycol-400 were tested. Substances; 1HNMR, SEM were characterized by elemental analysis. Molecular weights of P1 and P2 could not be detected because of their incomplete dissolution. Molecular weight of P3 and P4 was determined by GPC. Thermal properties of obtained polymers were examined by using TG/DTG, DTA, DSC analysis. Degradation reactions, mass loss, melting and decomposition points, enthalpy values were determined by analysing curves.

Acknowledgments: This study was supported financially by no:2012/06 BAP (Scientific Research Projects, University of AKSARAY, TURKEY) project and no:114Z693 TUBITAK (The Scientific and Technological Research Council of Turkey) project

Keywords: benzoylthiourea polymers, polyethylene glycol-400, TGA/DSC/DTA

FIG 1-2

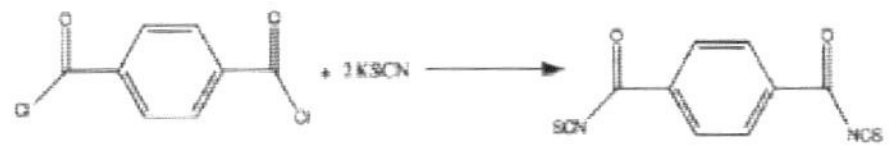

Figure 1: Synthesis of terephthaloyl diisothiocyanate

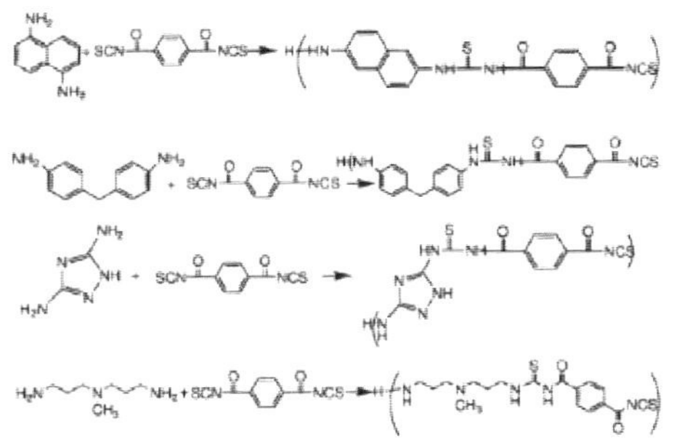

Figure 2. Synthesis of compounds p1, p2, p3 and p4

FİG 3

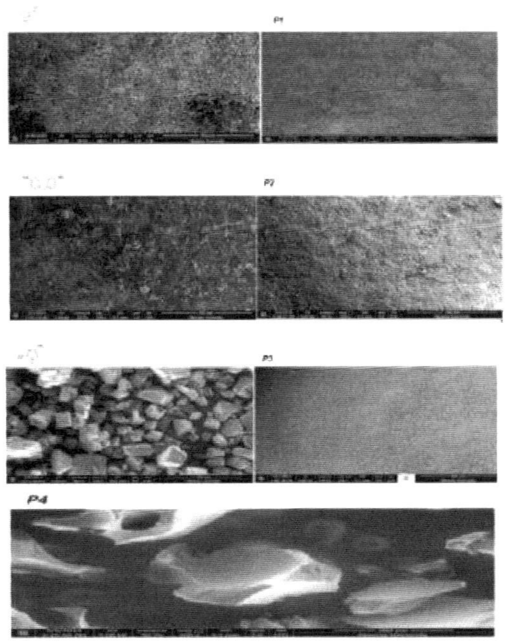

Figure 3 Comparison of SEM photographs of amines and the newly synthesized polymers

SFP-P-003 [Smart and Functional Polymers]

Elaboration in ionic liquids and characterization of polyaniline applied to nitrite ions detection

Abdoulaye Diarisso[1], Modou Fall[1], Cheikh Ahmad Elkabir Lô[1], Noureddine Raouafi[2], Ramzi Zarrougui[2]

[1]*Faculty of Sciences and Techniques, Cheikh Anta Diop University, Dakar, Sénégal*
[2]*Tunis Faculty of Sciences, E Manar University, Tunis, Tunisia*

Polyaniline films were electrosynthesized on a glassy carbon electrode and ITO plates in the presence of a ionic liquid (IL), namely butyltriéthyl ammonium trifluoroacetate (N2224-TFA). This class of compounds have the advantage of acting as solvent and supporting salt providing the dopant. The films showed a high solubility and a higher conductivity which are of great importance for a better knowledge of electronic properties of PANI generally insoluble in most common solvents, and for the development of new electronic devices. The films were characterized by electrochemical means (cyclic voltammetry and electrochemical impedance spectroscopy) and compared to PANI films obtained in 1 M HCl. They were found to be more conductive and to have a higher reversibility as regards the doping/undoping processes. The films morphology was also analyzed by MEB. PANI films synthetized in ILs (oxidized form) are blue in solution and more soluble than those synthetized in aqueous HCl. The UV measurements denoted local π-π* charge transfer transitions between quinoid and benzene cycles at 320 and 600 nm, implying the formation of the emeraldine base form during the electrosynthesis. PANI was also used as a sensor for the detection of nitrite ions using cyclic voltammetry and differential pulse voltammetry methods. It showed good stability and reversibility. A linear correlation was obtained in NO2- concentration ranges of 10-100 µM and 10-150 µM, for the first and second oxidation peaks, respectively. These results indicated a limit of detection of 0.5 mg L-1, which is the tolerable limit in drinking water in Europe (Council Directive 98/83/EC of 3 November 1998).

Keywords: Polyaniline, Ionic Liquids, Nitrite Detection

SFP-P-004 [Smart and Functional Polymers]

Molecularly imprinted HPLC column salid-phase material design for Dermatan Sulfate/ Chondroitin Sulfate Separation

Zihni Onur Uygun[1], Ferhan Sağın[1], Burcu Okutucu[3], Şükriye Hacıkara[2]

[1]*Department of Medical Biochemistry, Ege University, İzmir, Turkey*
[2]*Department of Children Nephrology, Ege University, İzmir, Turkey*
[3]*Department of Biochemistry, Ege University, İzmir, Turkey*

Proteoglycans which are among the most abundant molecules of the inter-cellular structure contain one or more glycosaminoglycan (GAG) chains that are covalently attached to the core protein. The levels of GAG excreted via urine is used as a marker to monitor different diseases (chronic renal disease, renal fibrosis, glomerular filtration abnormalities, bladder stones, breast and lung cancers, hypertension and diabetes etc.). However, GAG analysis is not simple due to the high polarity and microheterogenity of GAG structure. In this study, determination of clinically important species of GAGs, chondroitin sulfate(CS) and dermatan sulfate(DS) determination were carried out by an improved material for chromatographic column separation. Chromatographic column filling material was prepared as follows; 1mg GAG (chondroitin sulfate and dermatan sulfate separately) was dissolved in acetonitrile separately. For each tube 4mg acrylamide, 900µL trimethylolpropane trimethacrylate (TRIM) as a cross linker, 10mg Azobisisobutyronitrile (AIBN) as an initiator were added then the tubes were filled to 40 mL the total volume of acetonitrile(porogen) and treated with nitrogen gas to remove oxygen for prevent side oxidative reactions. The polymerizations were carried out at 65O degrees for 18 hours as thermo-radical polymerization reaction in dark. All fragments of GAGs were ionized in mass spectroscopy positively(ESI+). Ionization parameters were 2,7eV as capillary voltage, 350OC desolvation temperature and 650 L/h desolvation gas flow rate.. According to the MS results, six different calibration curves were prepared under 1000ng/mL (Figure1-6) and the GAG concentrations were calculated by using these calibration curves.

Keywords: glycosaminoglycan, chondroitin sulfate, dermatan sulfate, molecular imprinting, solid phase extraction

SFP-P-005 [Smart and Functional Polymers]

Electrochemical synthesis of poly (4-aminobenzenesulfonic acid) applied to lead-contaminated wastewater treatment

Matar Niang, Mamadou Fall, Mohamed Lamine Sall, Momath Lo, Abdou Karim Diagne Diaw, Diariatou Gningue Sall

Department of Chemistry, Cheikh Anta Diop University, Dakar,Sénégal

In recent years, environmental contamination and health risks (cancer, neurological disorders, birth defects) due to heavy metals have become a major preoccupation. It is therefore urgent to develop new materials easy to access, ecologically reliable and very useful in the treatment of wastewater. In this perspective we chose to study the electropolymerization of 4-aminobenzenesulfonic acid in mixed organic medium in order to develop a new highly effective polymer for the treatment of contaminated waters by lead ions (Pb2+). Thus, we have successfully attempted to electro synthesis films of poly (4-aminobenzenesulfonic acid) by cyclic voltammetry, galvanostatic and by potentiostatic modes. We also show the opportunity of the acid to play a double role as monomer and auto doping. The study of the spectroscopic characterization by IR allowed to high light the predominance of quinoid form in the film structure. While IR spectroscopy and fluorescence have shown that the phenyl group responsible for the fluorescence is not involved in the coupling process during the electropolymerization process. In the context of their use as metals traps (Pb2+), very interested results on the treatment of water containing heavy metals were obtained by fluorescence spectroscopy. In this study the observed phenomenon quenching in the presence of metals confirmed the removes of metals by formation of complexes with polymer.

Keywords: organic conductive polymer, 4-aminobenzenesulfonic acid, mixed medium (acetonitrile + soda), heavy metals

SFP-P-006 [Smart and Functional Polymers]

Functional UV Coating Composite For Steel Coating

Hye Jin Yoo

Surface Technology Research Group, POSCO Technical Research Laboratories, South Korea

Recently, considerable effort has focused on the functional coating of antibacterial properties in coated steel applications, and which seem to be quite useful in mobile devices, appliances, interior, buildings, and medical stuffs. For this, various bacteria-killing techniques have been developed, such as silver nanoparticles, layer-by-layer (LBL) nanofabrication, incorporation of various antibiotics, and so on. However, the antibacterial coating simultaneously keep up with other requirements like, anticorrosive, high gloss, suitable mechanical properties. In this work, we developed antibacterial coating formulation based on UV steel coating process by using chemically-modified urushiol lacquer. In fact, various kinds of lacquer trees (e.g., toxicodendron vernicifluum) have been used as paints and coating materials for woods and metal surfaces since 5,000 years in oriental culture. This natural lacquer exhibits superior barrier property against oxygen and water, as well as good durability, chemical resistance, and mechanical properties. However, traditional technique based on the enzymatic polymerization needs long time as well as manpower. In addition, pristine urushiol derivatives tend to inhibit free-radical polymerization due to its catechol structure of radical stabilization mechanism. Herein we present a facile method to minimize the radical inhibition for the development of antibacterial, anticorrosive, and high gloss UV coating formulation.

Keywords: Urushiol, Isocyanate, UV coating composite, Steel coating, Antibacterial property

Figure 1. FT-IR spectrum of pristine urushiol

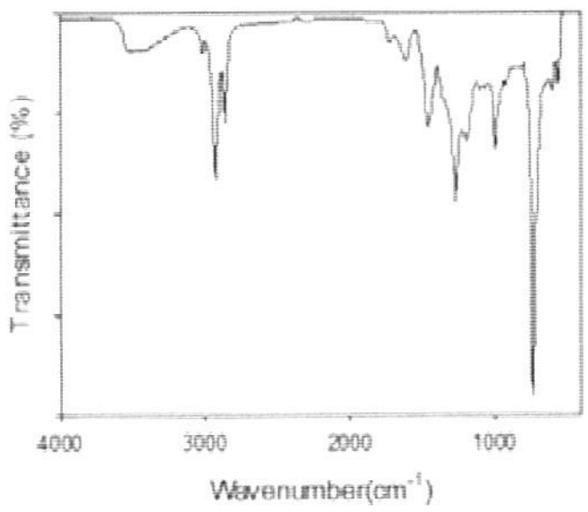

Figure 2. FT-IR spectrum of chemical modification of urushiol

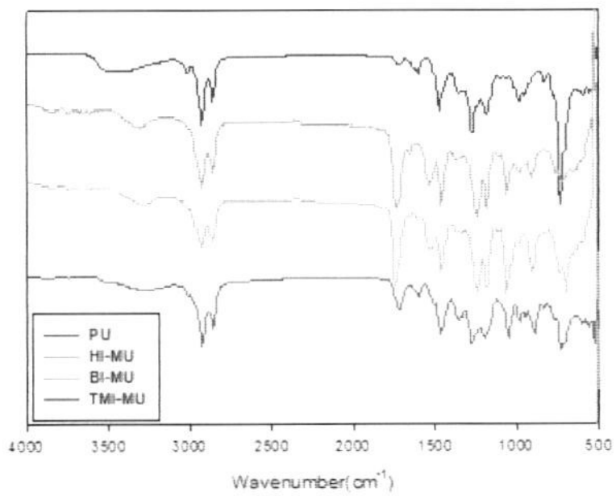

Figure 3. 1H NMR spectra of chemical modification of urushiol

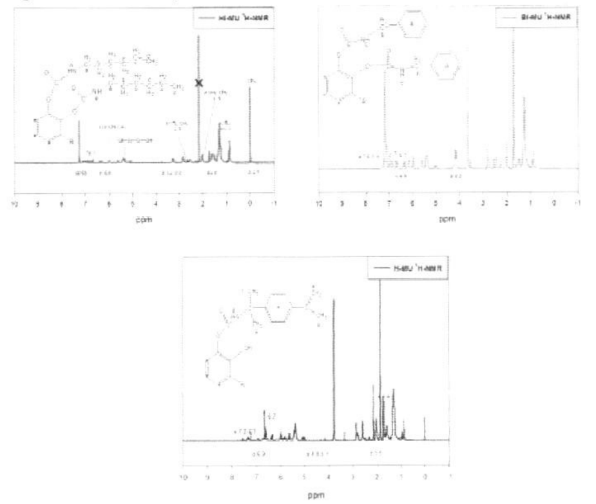

Figure 4. The effect of gel contents on the polymerization time

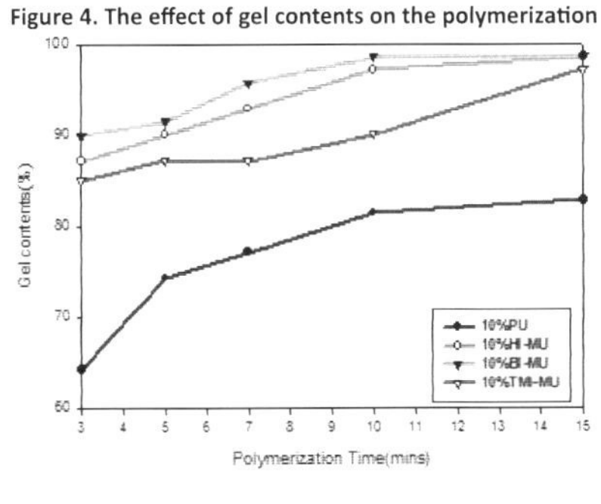

Figure 5. The antibacterial property of control, pristine urushiol and the chemical modification of urushiol films

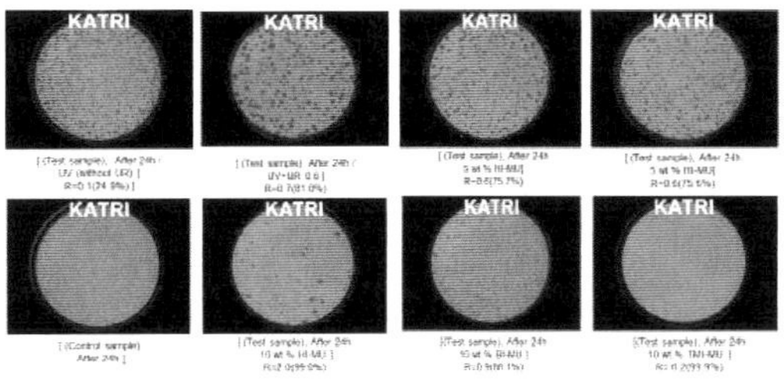

Figure 6. The Antimicrobial activity (R) of control, pristine urushiol and the chemical modification of urushiol films

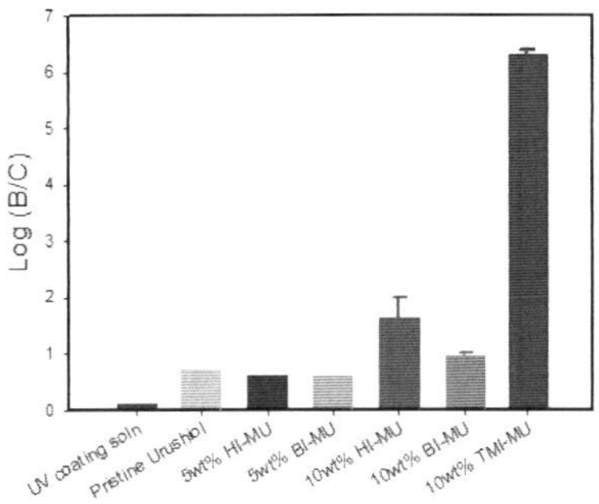

Scheme 1. Chemical modification of urushiol with mono-isocyanates

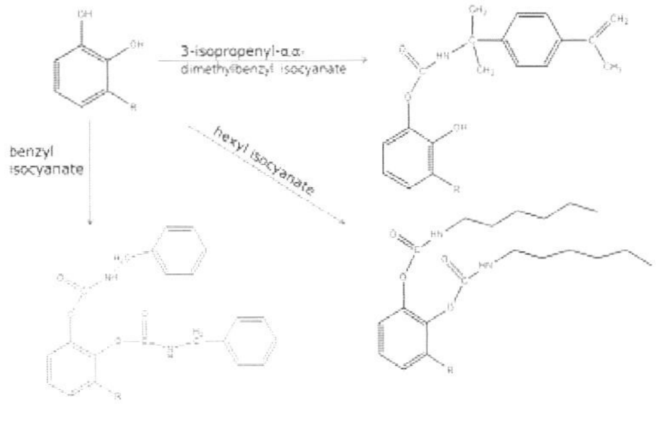

Table 1. This is the example of a table.

Test Item	Assessment of Antibacterial Activity
Test Method	JIS Z 2801(2000)
Test Result	Bacteriostatic reduction value; R
Type of sample & Size	Metal, 50×50mm
Bacteriostatic reduction value; R = log(B/C)	B : The number of bacteria on the control samples after 24h C : The number of bacteria on the test sample after 24h

SFP-P-007 [Smart and Functional Polymers]

Acid/base sensing layers on the surface of polypropylene fibers by grafting poly(acrylic acid)/poly(3,4-ethylenedioxythiophene)

Xingping Wang, Si Meng, Wujun Ma, Zhe Zhou, Bin Sun, Zongyi Qin, Mengfang Zhu

State Key Laboratory for Modification of Chemical Fibers and Polymer Materials, College of Materials Science & Engineering, Donghua University, Shanghai, China

A method for fabricating conductive Polypropylene-graft-poly(acrylic acid)/ Poly(3,4-ethylenedioxythiophene) (PP-g-PAA/PEDOT) composite fibers was developed. The total process consists of two main steps: 1) Grafting PAA onto the PP fiber surfaces by ultraviolet-induced living graft polymerization 2) In situ polymerization PEDOT onto the modified PP fiber substrate surface. The influence of experimental conditions on the conductivity of the PP-g-PAA/ PEDOT composite fibers was systematically studied. The maximum conductivities of PP-g-PAA/PEDOT composite fibers polymerized with oxidant $FeCl_3$ and ferric p-toluenesulfonic acid (FepTS) can reach 0.07 S/cm and 10 S/cm, respectively. Furthermore, Results show that the adhesion between the conductive PEDOT layer and the modified PP fiber substrate was significantly enhanced due to the electrostatic attractions between PEDOT chains and carboxylic groups. The as-prepared PP-g-PAA/PEDOT composite fibers are used to detect the HCl and NH_3 vapor. The PP-g-PAA/PEDOT composite fibers show good reversible and reproducible of resistance change upon periodic exposure to diluted analyte and dry air.

Keywords: Poly(3,4-ethylenedioxythiophene) (PEDOT), conducting polymer, chemical sensors, photografting, conductive composite fiber

The reversible and reproducible of response of PP-g-PAA/PEDOT composite fibers upon periodic exposure to chemical vapor and the possible sensing mechanism

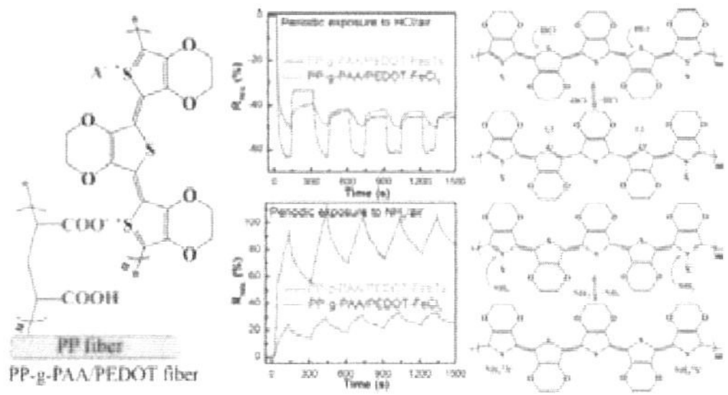

SFP-P-008 [Smart and Functional Polymers]

Synthesis, modification and controlled release of poly(isobutylene vinly ether-co-itaconic anhydride

Fatıma Zehra Dikici, Ahmet Okudan, Makbule Bilge Sağkan

Department of Chemistry, Selcuk University, Konya, Turkey

The main aim of studies in drug field for many years, has been occurred to develop a novel molecule that therapeutic various diseases. However, these researches take a long time, the economic burden and don't always give the expected result has started to search for alternatives. The goal in this the search for alternative is to improve the quality of life as well as treatment of patients. Reduce the drug dose, prolong the dosing interval, decontamination from side and detrimental effects, even sending to the target area of the drug are for this purpose and are controlled release system that best respond to these expectations. The building blocks of many of the controlled release system are polymer and drug release velecity and period are set by the polymer. According to usage path, the purpose and properties of the drug, the system is prepared using one or more polymers.[1]"" ""*In this study, copolimerizations of ısobutylene vinly ether, itaconic anhydride were obtained by free radical polymerization with various monomers in a suitable solvent and temperature medium.[2,3]Synthesized polymers were modified by chemically with ASA. Modified polymers with ASA will be examined controlled release in a certain temperature and pH.[1,4]""*

Referances:

[1]Çolak B., Acetylsalicylic Acid Modification of Maleic Anhydride Copolymer and Investigation of Controlled Release, MS Thesis, Istanbul Technical University, Institute of Science and Technology, 2009.
[2]Karasakal A., Okudan A., The effect of H-boding on Radical Copolimerization of Maleic Anhydride with N-tert-Butylacrylamide and Its Characterization, International Journal of Polymer Science, 2013.
[3]Şengöz O., Synthesis, Characterization and Modification of Copolymers Containing Maleic Anhydride, MS Thesis, Selcuk University, Institute of Science and Tecnology, 2014.
[4]Scott M. Henry, Mohamed E. H. El-Sayed, Christopher M. Pirie, Allan S. Hoffman, and Patrick S. Stayton, pH- Responsive Poly(Styrene-alt-Maleic Anhydride) Alkylamide Copolymers for Intracellular Drug Delivery, 15, 2006.

Keywords: Controlled Release, Free Radical Polimerization, Isobutylene Vinly Ether, Itaconic Anhydride

Figure 1

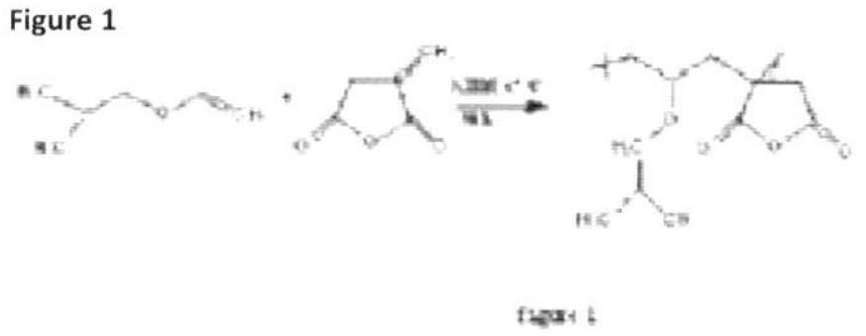

copolimerizations of ısobutylene vinly ether (IBVE), itaconic anhydride (IA)

Figure 2

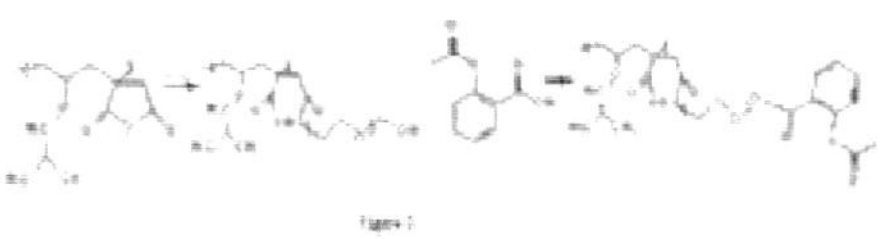

(IBVE-co-IA) modified with PEG and ASA

SFP-P-009 [Smart and Functional Polymers]

Induction of Chiral Structures to Conjugated Polymers Using Circularly Polarized Light

Tamaki Nakano, Yue Wang

Institute for Catalysis, Hokkaido University, Sapporo, Japan

Circularly polarized light (CPL) can be regarded as homochiral light and is obtained by modulation of non-polarized light. Our CPL-based methods of optically active polymer preparation are base on "enantiomer-selective excitation" of a group in polymer chain which exists as a racemic mixture in the ground state. Enantiomer-selective excitation of such a group leads to an optically active form of a polymer chain rich in the antipodes that are not selectively excited (de-racemization process) if exchange between racemates is prohibited in the ground state and is possible only in excited states. A series of work was initiated by the finding of photo racemization of a preferred-handed helical polyacrylate where side-chain biphenyl moiety in the polymer underwent "twist-coplanar transition" on photo excitation with non-polarized light.[1-3] This work was extended to CPL-driven helix formation of poly(9,9-di-n-octylfluorene-2,7-diyl) in a thin film form where 5/1-helix is formed by CPL irradiation. [4,5] In this method, interactions between chains and between chains and substrate for film formation play an important role. It has been hence difficult to induce chirality to an amorphous material which does not have significant inter-chain interactions. Recently, we found a method to overcome this point which uses achiral aid molecules to reinforce inter-chain interactions [6].

1. T. Sakamoto, Y. Fukuda, S. Satoh, T. Nakano, Angew. Chem. Int. Ed., 2009, 48, 9308-9311.
2. A. Pietropaolo, T. Nakano, J. Am. Chem. Soc.,2013, 135, 5509–5512.
3 Y. Wang, T. Sakamoto, Y. Koyama, Yuma Takanashi, Jiro Kumaki, J. Cui, X. Wan, T. Nakano, Polym. Chem., 2014, 5, 718–721.
4. Y. Wang, T. Sakamoto, T. Nakano, Chem. Comm., 2012, 48, 1871 – 1873.
5. A. Pietropaolo, Y. Wang, T. Nakano, Angew. Chem. Int. Ed., 2015, 54, 2688–2692.
6. Y. Wang, A. L. Kanibolotsky, P. J. Skabara, T. Nakano, Chem. Commun., 2016, 52, 1919 – 1922.

Keywords: circularly polarized light, helix, de-racemization

SFP-P-010 [Smart and Functional Polymers]

PDMS: a stimuli-responsive polymer for RFID sensing

Kate E. Belsey[1], Adam V. S. Parry[2], Mohamed A. Ziai[3], Christina V. Rumens[1], Stephen G. Yeates[2], John C. Batchelor[3], Simon J. Holder[1]

[1]*Functional Materials Group, School of Physical Sciences, University of Kent, Canterbury, CT2 7NH, UK*
[2]*School of Chemistry, University of Manchester, Manchester, M13 9PL, UK*
[3]*Antennas Group, School of Engineering and Digital Arts, University of Kent, Canterbury, CT2 7NT, UK*

The ability of poly(dimethylsiloxane) (PDMS) to absorb both volatile and non-volatile compounds makes it an ideal candidate for the sensing component in radio frequency identification (RFID) tag sensors. We have already reported the use of PDMS as the active component in an RFID vapour sensor, whereby the PDMS upon swelling in the vapour acted as an actuator by displacing a feed loop that modified the antenna response. The degree of swelling, and hence RF response, was directly proportional to the Hansen solubility parameters and the vapour pressure of the organic vapour [Rumens et al., *J Mat Chem C.*, 2015, 3, (39)].

In this work we present findings from the inkjet printing of colloidal silver ink solutions onto poly(dimethylsiloxane) (PDMS) substrates and which were then thermally sintered to generate conductive loops. In combination with an etched antenna and an ASIC (Application Specific Integrated Circuit) RFID silicon chip, these PDMS substrates with printed feed loops were exposed to a number of volatile organic solvent vapours. This resulted in the swelling of the substrates which led to the formation of microcracks across the printed loop. The resulting change in resistivity of the swollen conductive loops was detected by the RFID reader in the form of an increased transmitted power and a reduced read range for the tags.

Interestingly, on removal from the vapour, the substrates contracted back to their original size, returning the performance of the RFID tag to close to the original behaviour. This process could be repeated numerous times, indicating that the tags were 'multi-use' or recyclable. Siloxanes with different functionalities have been shown to respond differently to solvent vapours, meaning that incorporating these functionalities could allow for a completely selective, reusable sensor. This work paves the way to a fully ink-jet printed RFID tag.

Figure 1

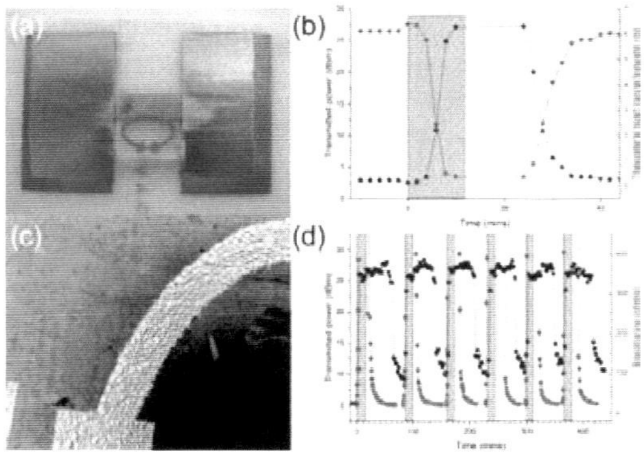

(a) prototype RFID tag with antenna, PDMS with printed feed loop and attached chip; (b) RFID response of tag to vapour exposure (grey region indicates presence of solvent); (c) microcracks in printed loop after 30 mins exposure to vapour; and (d) cycling read range and resistance of tags in vapour.

Keywords: Poly(dimethylsiloxane), stimuli-responsive, ink-jet, RFID, sensor

SFP-P-011 [Smart and Functional Polymers]

Synthesis of a Photo-Cleavable Core Prepared by Passerini Reaction and Preparation of an ABC 3-Mictoarm Star Polymer with This Core

Gözde Deveci, Muhammet Ubeydullah Kahveci

Yildiz Technical University, Faculty of Chemical Metallurgical Eng., Department of Bioengineering, 34220 Esenler, Istanbul, Turkey

Star shaped polymers consist of a core and macromolecular arms emanating from the core and these polymers are usually classified as regular and mictoarm star polymers. Star polymers are an important member of branched polymers because of their unique topological structures and physical/chemical properties, such as lower viscosity in solutions, higher encapsulation efficiency, multivalent functionality, and improved stimuli-responsive property. Therefore, star polymers have received more attention in polymer community. A new approach for synthesis of a stimuli-responsive miktoarm star polymer is proposed in this study. Briefly, we have synthesized an ABC 3-mictoarm star polymer with photo cleavable core employing a new synthetic approach. In this method, Passerini Reaction of propargyl iso-cyanoacetamide, 5-hydroxy-2-nitrobenzaldehyde and 2-bromobutanoic acid yielded an o-nitrobenzyl ester functional photo-cleavable core bearing an alkyne, alkyl bromide and hydroxyl group that were utilized in copper catalyzed Click Reaction (CuAAC), atom transfer radical polymerization (ATRP) and ring opening polymerization (ROP), respectively. Firstly, azide end functional PEG was conjugated to the core. Successively, the PEG conjugated core containing was utilized as a macro-initiator in ATRP of styrene and ROP of lactide. All polymers were characterized by GPC, NMR and FT-IR. After synthesis of photo-cleavable mictoarm terpolymer, the polymer was exposed to UV light and photo-degredation was monitored by GPC. Synthesized photo-cleavable star polymer has a big potential to apply in drug delivery, gene delivery, biotechnology, medical devices, diagnostics and sensor systems. This work was financially supported by Yildiz Technical University (Project #: 2013-07-04-KAP05).

Keywords: Passerini Reaction, star polymer, photo-cleavable polymer, ATRP, Click Reaction, ROP

SFP-P-012 [Smart and Functional Polymers]

Synthesis and Self-Assembly of High Molecular Weight Polystyrene-block-poly[2-(N-morpholino) ethyl methacrylate]: A Story about Microphase Separation, Amphiphilicity, and Stimuli-Responsivity

Steffen Eggers[1], Felix Lauterbach[1], Volker Abetz[2]

[1]*University of Hamburg, Institute of Physical Chemistry, Germany*
[2]*Helmholtz-Zentrum Geesthacht, Institute of Polymer Research, Germany*

The synthesis of polystyrene-*block*-poly[2-(*N*-morpholino)ethyl methacrylate] (PS-*b*-PMEMA) (Figure 1, top) as a new highly amphiphilic and multiple stimuli-responsive block copolymer is presented. To achieve high molecular weights far beyond 100 kDa in a highly controlled manner (dispersities < 1.1), a synthetic route via sequential combination of anionic polymerization (for the PS block) and reversible addition–fragmentation chain transfer (RAFT) polymerization (for the PMEMA block) is used. The synthesized block copolymers are investigated regarding their microphase separation in bulk, which delivers well-ordered self-assembled bulk structures as evidenced by small-angle x-ray scattering (SAXS) and atomic force microscopy (AFM). In aqueous solution, the block copolymer self-assembles into narrowly size distributed micelles with a PMEMA corona. The stimuli-triggered micelle response towards temperature (Figure 1), pH, and kosmotropic as well as chaotropic salts is shown via various dynamic light scattering (DLS) experiments. Furthermore, a dependency of aggregate size on solvent composition in polymer/tetrahydrofuran/water mixtures is described. The reported findings deliver a feasible pathway to high molecular weight block copolymers with tailored chemical properties and show the potential of PS-*b*-PMEMA as material for mechanically demanding switchable devices.

Keywords: combination of controlled polymerization methods, RAFT polymerization, self-assembly, microphase separation, micellization, stimuli-responsivity

Figure 1

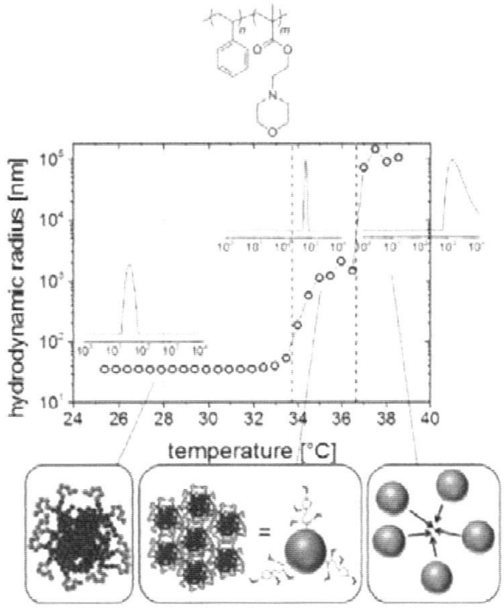

Simplified chemical structure of the investigated block copolymer polystyrene-block-poly[2-(N-morpholino)ethyl methacrylate] (top) and two-step aggregation behavior upon heating in aqueous solution.

SFP-P-013 [Smart and Functional Polymers]

PEG-Based Reactive Copolymers via Azide-Alkyne Click Cycloaddition-Mediated Step-Growth Polymerization

Mehmet Arslan[1], Özgül Gök[2], Rana Sanyal[2], Amitav Sanyal[2]

[1]*Yalova University, Engineering Faculty, Polymer Engineering Department, Yalova, Turkey*
[2]*Department of Chemistry, Bogazici University, 34342 Bebek, Istanbul, Turkey*

Design, synthesis and post-polymerization functionalization of linear polytriazoles with reactive maleimide side chain functionalities was described. Copper(I)-catalyzed Huisgen 1,3-dipolar cycloaddition reaction was utilized to prepare copolymers in a A1-A1/B1-B1/B2-B2 type combination of diazide and dialkyne functionalized triethylene glycol derivatives with diyne functionalized furan-protected maleimide containing monomer. After polymerization, side chain maleimide groups were unmasked via the retro-Diels-Alder reaction. Post-polymerization modifications of these PEG-based copolymers containing the furan-protected maleimide group are accomplished using a photoinitiated radical thiol-ene reaction, whereas polymers containing the reactive maleimide group are functionalized using the Diels–Alder and nucleophilic thiol-ene reactions. The nucleophilic thiol-ene reaction is also utilized to fabricate hydrogels by treatment of the maleimide-containing PEG polymers with dithiol-containing crosslinkers.

This work was supported by State Planning Commission of Turkey (DPT) with a grant no. 09K120520.

Keywords: Click chemistry, Diels-Alder reaction, Step growth polymerization, Thiol-maleimide conjugation

SFP-P-014 [Smart and Functional Polymers]

Smart Windows Application of Carbazole and Triazine based Star Shaped Architecture

Merve Guzel, Metin Ak

Department of Chemistry, Pamukkale University,Denizli,Turkey

A novel triazine-based, star shape and electroactive monomer, 2,4,6-tris(2-(9H-carbazol-9-yl)ethoxy)-1,3,5-triazine (TCZ) which contains 2,4,6-trichloro-1,3,5-triazine as core and 2-(9H-carbazol-9-yl)ethanol as arms, was successfully synthesized. After electrochemical polymerization of this star shape monomer, obtained polymer shows superior optoelectronic and thermal properties due to its unique three-dimensional shape and highly-branched structure in comparison with linear analogues. Electrochromic studies exhibited that PTCZ has turquoise color in oxidized state and transparent in neutral state. Due to fact that the redox color characteristics of PTCZ are indispensable for smart windows, PTCZ based electrochromic device was formed with PEDOT as complementary coloring material. A potential range of -1.5 and 1.8 V was determined to be suitable for operating the PTCZ/PEDOT device between transparent and blue colors. Characterizations of the device were performed in term of switching times, optical contrast, optical memory and redox stability.

Keywords: Star shaped polymers, Electrochromism, Spectroelectrochemistry, Electrochromic device

Properties and application of novel triazine-based, star shape polymer

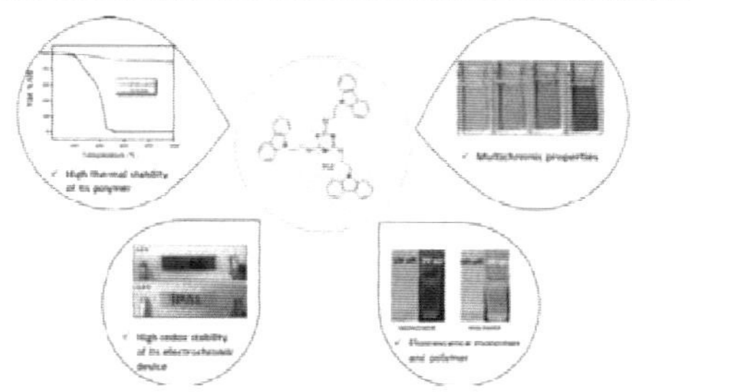

SFP-P-015 [Smart and Functional Polymers]

Shape-memory Polymers - How to Bring Hard and Soft Segments Together

Hannes Schäfer[1], Katharina Koschek[2]

[1]University of Bremen, Department of Biology / Chemistry (FB2), Leobener Strasse NW2C, 28359 Bremen, Germany
[2]Fraunhofer Institute for Manufacturing Technology and Advanced Materials (IFAM), Adhesive Bonding and Surfaces, Wiener Str. 12, 28359 Bremen, Germany

Shape-memory polymers (SMP) are a class of smart materials with a great potential for high performance applications, for example aerospace or biomedical devices.[1] They are able to regain their previous "memorized" form after being manipulated into a temporary shape by applying an external stimulus, e.g. by a change in temperature. Commonly, SMPs consists of two (or more) phases: A hard segment that is setting up a network and thereby defining the original and "memorized" shape, e.g. via chemical cross-links, and a soft segment that allows the change from a memorized to a temporary shape and its fixation. The latter segment can consist, for example, of a semicrystalline structure, able to undergo a phase transition at a distinct temperature.[2-4] The challenge here is to find a suitable way to connect hard and soft segments on the molecular level to obtain the desired properties of the SMPs. Blend polymers, only based on weak interactions, suffer from demixing of the segments, which influences the material properties. Covalent bonding between soft and hard segments would improve performance and stability of the SMPs. Therefore, we systematically modified semi-crystalline polyesters with different end groups to examine the influence of several functional groups on the polymerization behavior of the soft and hard segments. The resulting polymers were studied with respect to their mechanical and shape-memory properties which were correlated to the chemical structure. Providing a deeper understanding of the structure-property relationship is of great interest for further design strategies of SMPs.

[1] H. Meng, G. Li, Polymer 2013, 54, 199-2221.
[2] A. Lendlein, S. Kelch, Chem. Int. Ed. 2002, 41, 2034-2057.
[3] M. D. Hager, S. Bode, C. Weber, U. S. Schubert, Prog. Polym. Sci. 2015, 49–50, 3-33.
[4] H. Lützen, T. M. Gesing, B. K. Kim, A. Hartwig, Polymer 2012, 53, 6089-6095.

Keywords: Shape-memory polymers, structure-property relationship, stimuli responsive polymers

SFP-P-016 [Smart and Functional Polymers]

Reversible ring opening reaction of a model 1,3-benzoxazine with thiols

Tobias Urbaniak[1], Marc Soto[2], Katharina Koschek[2]

[1]University of Bremen, Bibliothekstrasse 1, 28359 Bremen, Germany
[2]Fraunhofer Institute for Manufacturing Technology and Advanced Materials IFAM, Wienerstrasse 12, 28359 Bremen, Germany

Crosslinked polymers with reversible covalent linkages have drawn increasing interest in the past few years because of the ability to alter their properties with external stimuli. The ring opening reaction of 1,3-benzoxazines with thiols at ambient temperature is one reversible reaction with a great potential for this application and was first described by Endo *et al* in 2013.[1-2]. Our main objective is to study and to control the reversible character of that reaction in order to gain insight into the reaction mechanism. This was done i.a. by shifting the initial concentration of the appropriate reactants. Furthermore, we have studied the effect of thiol reactivity on the reversibility of the ring opening reaction in protic and non-protic solvents as well as under solvent-free conditions. We have found a way to control the reversibility of that reaction with a suitable combination of reactants and reaction conditions.

References :

[1] I. Gorodisher, R. J. DeVoe, R. J. Webb, *Handbook of Benzoxazine Resins*, Elsevier Science, 2011, 11, 211-234.
[2] A. W. Kawaguchi, A. Sudo, T. Endo, *ACS Macro Letters*, 2013, 2, 1–4.

Keywords: 1,3-benzoxazine, thiols, reversible crosslinking, ring opening, stimuli-responsive

Reversible ring opening reaction of a model 1,3-benzoxazine with thiols

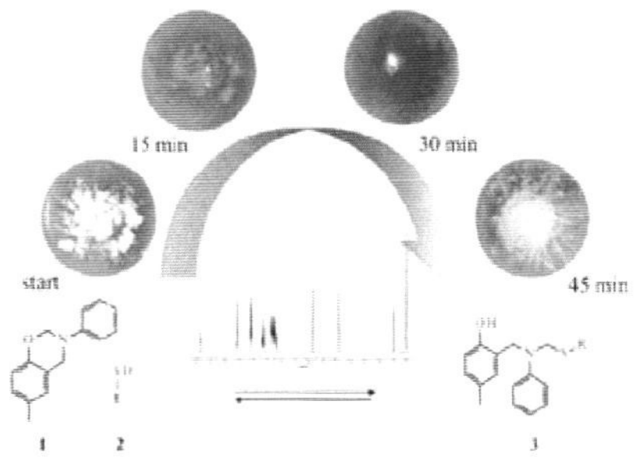

Figure 1. Reversible ring opening reaction of N-phenylbenzoxazine 1 with different thiols 2 yielding the corresponding adducts 3 at room temperature.

SFP-P-017 [Smart and Functional Polymers]

Acryl amide based thermoresponsive copolymers: total control over the lower critical solution temperature

Niklas Lucht[1], Steffen Eggers[1], Volker Abetz[2]

[1]Department of Physical Chemistry, University of Hamburg, Hamburg, Germany
[2]Helmholtz-Zentrum Geesthacht, Zentrum für Material- und Küstenforschung GmbH, Institute of Polymer Science, Geesthacht, Germany

Stimuli-responsive polymers have gained much attention in material sciences in the recent years. PNIPAM as the most extensively investigated thermoresponsive polymer has a Lower Critical Solution Temperature (LCST) in water at about 32 °C. Our work introduces a novel tailored statistical copolymer consisting of two acrylamides, namely *N*-acryloyl piperidine and *N*-acryloyl pyrrolidine, that offer a gradually tuneable LCST range between 3 °C and 50 °C. Because of a well-shaped linear dependency of the LCST with the monomer ratio in the copolymer, the desired LCST can be achieved fairly easy just by varying the monomer concentration prior to polymerisation. Figure 1 sketches the LCST as a function of the mol fraction of *N*-acryloyl piperidine. In addition, the effects of solvent mixtures on the thermoresponsivity were systematically studied. The observed cononsolvency effect, which is currently a vividly debated issue in smart-polymer science, is discussed in respect to additive structure, solvent–polymer, and solvent–solvent interactions. It is mostly postulated in literature that cononsolvency does occur because of favoured hydration of the alcohol by water, inducing a change in the water structure and therefore a lower solubility of the polymer. The polymer system introduced in this work shows that the cononsolvency-effect cannot fully be explained by just the approach of a favoured hydration of the alcohol. Therefore, a solvent–polymer-interaction is considered to explain the observations in this work.

Keywords: LCST, stimuli-responsive polymers, thermoresponsive polymers, smart materials, cononsolvency

Figure 1

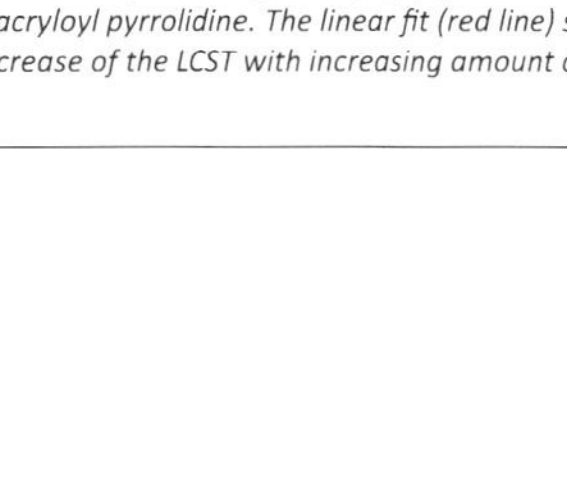

LCST vs. mol fraction of N-acryloyl piperidine in the copolymer system with N-acryloyl pyrrolidine. The linear fit (red line) shows a nearly perfect linear decrease of the LCST with increasing amount of the piperidine monomer.

SFP-P-018 [Smart and Functional Polymers]

Studies on Poly(Acrylate)S Containing Pendant Ligand With Schiff Base Azo Compound And Its Divalent Metal Complexes

Keerthiga Rajendiran

Pondicherry Engineering College

Series of monomers were prepared by reacting (meth)acryloyl chloride with azo compounds derived from various diamine Schiff base and diazonium salt of 4- amino phenol compound. The monomers were polymerized in DMF at 70^{0}C using benzoyl peroxide as free radical initiator. Polymer metal complexes were obtained in DMF solutions of polymers with aqueous solutions of metal ions. The polymers and the polymer metal complexes were characterized by elemental analysis and spectral studies. The IR spectra of these complexes suggest that the metals are coordinated through nitrogen of the azo group and the oxygen of the phenolic –OH group. The electronic spectra and magnetic moments of the polychelates shows that octahedral and square plannar structure for Ni(II) and Cu(II) complexes respectively. X-ray diffraction studies revealed that polychelates are highly crystalline. The thermal and electrical properties, catalytic activity, structure property relationships and ion uptake properties are also discussed. Further the synthesized polymer metal complexes were used for anti microbial studies.

References:

1. Tsuchida.E, and Nishide.H, ACS Symp.Ser., 1980, 121,147-163.
2. Kaliyappan.T and Kannan.P,Prog . Polym. Sci., 2000, 25,343-370.
3. Vijayalakshmi.S, Sankar.R, Subramanian.S, Rajagopan.S and Kaliyappan.T., J. Appl. Polym.Sci., 2007,104,797-802.
4. Bernabe L.Rivas, Ivan M. Peric, Sandra Villegas., Polym. Bull.,2010,65,917-928.
5. A.Z El-Sonbati, A.A.M. Belal, M.A.Diab, R.H. Mohamed., Journal of molecular structure., 2011,26-31.

Keywords: Chelating ligands, Polymer metal complexes, anti microbial studies

SFP-P-019 [Smart and Functional Polymers]

Introducing upper critical solution temperature to polymer multilayer films

Cansu Ustoglu, Eda Cagli, Irem Erel Goktepe

Department of Chemistry, Middle East Technical University, Ankara, Turkey

Preparation of nanoengineered thin films and particles via layer by layer (LbL) assembly technique for biomedical applications such as controlled release systems, targeted drug delivery have attracted attention recently. Using this technique, the amount of drug loaded within the multilayers can be easily tuned by increasing the number of layers deposited at the surface. In this study, block copolymer micellar nanoaggregates with cores exhibiting upper critical solution temperature behavior near body temperature at physiological pH were prepared. For this purpose, firstly, micellar nanoaggregates were prepared from poly [(ethylene oxide)-b-(2-vinyl pyridine)](PEO-b-P2VP) and negatively charged multivalent salts via electrostatic interactions among the positively charged P2VP and salt anions. We found that the concentration of the multivalent salt was critical on the temperature stability of the micellar aggregates. We then used such micellar nanoaggregates as building blocks to prepare layer-by-layer films. We monitored the disintegration of the micellar cores and release of functional molecules from surfaces with increasing temperature.

Keywords: Layer-by-layer films, stimuli responsive, temperature responsive, controlled release systems

SFP-P-020 [Smart and Functional Polymers]

Castor oil-derived polyurethanes engineered by dynamic crosslinks for self-healing

Marc Comí

Department of organic and analytical chemistry, suspol group, rovira i virgili University, Tarragona, Spain

Smart materials with a response to external stimulus such as light, heat or chemicals are attracting significant interest in the material science field due to their diverse potential applications.[1] Recently, ionic hydrogen bonds have been widely used to construct supramolecular polymers with versatile functions, such as facile processing/recycling, self-healing, shape memory and high responsivity to stimuli.[2,3] These bonds are usually formed by mixing acid and base, i.e. the classical acid base reaction. Engineering plant derived polyurethanes by reversible association-dissociation ionic crosslinks is a promising approach to enhance the weak mechanical strength of aliphatic plant-derived PUs as well as improve their hydration properties. In this study, we report a series of novel supramolecular PU networks (SPUNs) based on natural components which can heal itself at room/moderate temperature. As illustrated in Fig. 1, SPUNs containing ionic hydrogen bonds are easily prepared by a proton transfer reaction between natural multifunctional carboxylic acids such as sebacic (1), succinic (2), oxalic (3) and citric (4) and a castor oil-derived PU containing pendant tertiary amines. The thermoplastic PU was prepared by reacting a diol synthesized from 10,11-epoxyundecan-1-ol and diethylamine, with hexamethylene diisocyanate. The chemical structure of the multifunctional carboxylic acids greatly influences the thermal, mechanical and moisture absorption properties of the prepared SPUN materials. Due to the existence of abundant dynamic ionic hydrogen bonds in the networks, the prepared materials exhibit self-healing properties at low temperatures applying slight pressure

Keywords: Ionic crosslinks, polyurethanes networks, self-healing, castor oil

Fig1

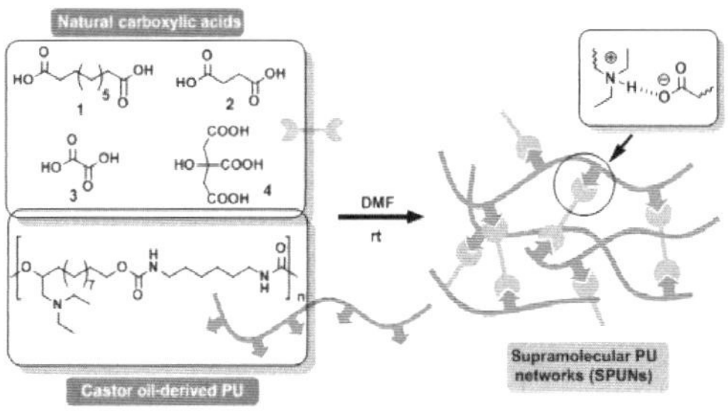

Schematic illustration of SPUNs structure

SFP-P-021 [Smart and Functional Polymers]

Synhesis,Characterization And Antibacterial Activity With 8 -Hydroxy Quinoline Of Poly(Isobutylene Vinly Ether -co-Itaconic Anhydride)

Bilge Makbule Sağkan, Ahmet Okudan, Fatıma Zehra Dikici

Department of Chemistry, Selcuk University, Konya, Turkey

Antimicrobial agents are, in general, low-molecular-weight compounds, which are in many cases compounded with polymeric matrix through physical mixing.However they leach out from the polymeric matrix containuosly and often cause toxicity to the enviroment.Morever they are usually volalite and chemically unstable.Immobilization of the low-molecular-weight antimicrobial agents to polymeric material through chemically bonding would be an ideal solution to overcome these problems.The antimicrobial agent-bound polymeric materials exhibit antimicrobial activities by slowly releasing active agents through hydrolsis,while some polmers are also antimicrobially active by themselves.[1] The use of antimicrobial polymers offers promise for enhancing the efficacy of some existing antimicrobial agents and minimizing the environmental problems accompanying conventional antimicrobial agents by reducing the residual toxicity of the agents,increasing their efficiency and selectivity, and prolonging the lifetime of the antimicrobial agents.[2]"]" "["In this study, the polymers containing anhydride group were obtained by free radical polymerization with various monomers in a suitable solvent and temperature medium.[3,4] Synthesized polymers will be prepared by radical polymerization with 8-HQ antibacterial agents.The synthesized copolymers will be tested against different microorganisms.[5]

References:

[1] Park E.S., Kim H.S., Kim M.N., Yoon J.S.,Antibacterial Activities of Polystyrene-block-poly(4-vinylpyridine) and poly(styrene-random-4-vinly pyridine),European Polymer Journal, 40, 2004.
[2] Kenawy El-Refaıe, Fouad I. Abdel-Hay, El-Raheem R. El-Shanshourt, Mohamed H. El-Newehy, Biologically Active Polymer.5. Synthesis and Antimicrobial Activity of Modified Poly(glycidlymethacrylate-co-2-hydroxyethly methacrylate) Derivatives with Quaternary Ammonium and Phosphoniım Salts, Journal of Polymer Research,40,2002.
[3] Karasakal A., Okudan A., The effect of H-boding on Radical Copolimerization of Maleic Anhydride with N-tert-Butylacrylamide and Its Characterization, International Journal of Polymer Science, 2013.
[4] Şengöz O., Synthesis, Characterization and Modification of Copolymers Containing Maleic Anhydride, MS Thesis, Selçuk University, Institute of Science and Tecnology, 2014.
[5] Patel P., Shah B., Ray A., and Patel R., Acrylic Homo- and Co-polymers Based on 2,4-dichlorophenyl Methacrylate and 8-quinolinly Methacrylate, Journal of Polymer Research, 11,2004.

Keywords: antimicrobial properties,8 hydroxyquinoline, free radical polymerization

figure1 and 2

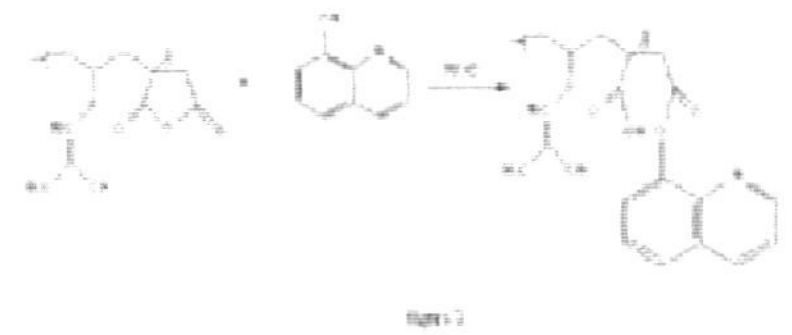

Figure1: Copolimerizations of Isobutylene Vinly Ether,Itaconic Anhydride
Fıgure2:ımmobiliized of 8-hydroxy quinoline

SFP-P-022 [Smart and Functional Polymers]

Controlled radical ring opening polymerisation for smart and degradable polyesters

Jens Gaitzsch, Wolfgang Meier

University of Basel, Department of Chemistry, Klingelbergstrasse 80, 4056 Basel

Recent years have seen considerable advances in radical and ring-opening polymerisation. One gap however, the development of smart and degradable polymers is only filling slowly.[1] This is mainly due to the fact that ring-opening polymerisation does not tolerate the unprotected amines or alcohols necessary. By using controlled radical ring opening polymerisation (CoRROP) from cyclic ketene acetals (CKAs), this issue can be overcome. CKAs can be achieved in a 2 step synthesis from the corresponding diol. In a first step the ring is closed using a transacetalisation with chloroacetaldehyd dimethyl acetal. While early reports suggest to accomplish this step using prolonged heating, a cobalt-catalysed method known in organic chemistry proved to be feasible and this step can be accomplished quickly at room temperature.[2] The second step sees the elimination of hydrogen chloride to eventually form the CKA.[3] In order to proof that functional polymers can be formed, amphiphilic block-copolymers from the CKAs will be used for self-assembly into micelles or vesicles. One of the 3 CKAs targeted is an acetal with 2 methyl units to support the radical formed later (1), another one has a hydroxyl unit in its side chain (2) and the third one contains a tertiary amine (3, all numbers relate to Figure 1). This range is to demonstrate which side-chain moieties can be tolerated by this method.

Hence, the resulting material then has the range of properties aimed at and we are certain that this chemistry can fill the gap to form degradable and responsive polyesters.

[1] J. Gaitzsch, X. Huang, B. Voit, Chemical Reviews 2016, 116, 1053-1093
[3] U. M. Battisti, C. Sorbi, S. Franchini, A. Tait, L. Brasili, Synthesis 2014, 46, 943-946.
[4] W. J. Bailey, Z. Ni, S. R. Wu, J. Polym. Sci., Part A: Polym. Chem. 1982, 20, 3021-3030.

Keywords: Polyesters, Radical Ring-Opening Polymersation, Functional Polymers, ATRP, CKAs

SFP-P-023 [Smart and Functional Polymers]

Synthesis of stimuli responsible metallosupramolecular poly-L-lysine as smart polymer structure

Imren Özcan, Süleyman Köytepe, Turgay Seçkin

İnönü University, Faculty of Arts and Science, Chemistry Department, Malatya, Turkey

Smart polymers or stimuli-responsive polymers have been widely used due to their practical interests, and several types of responsiveness, including temperature, pH, light, pressure, magnetic and electrical fields, have been reported [1]. Especially these type polymers generally use design of shape memory polymeric systems, smart hydrogels, self-healing materials. Main interactions in stimuli-responsive polymers are metal-ligand coordination, hydrogen bonding, host-guest interaction or hydrophobic association. Among them, the coordinative bond is of particular interest given the wide range of stabilities accessible depending on the metal-ligand pair [2]. In the presence of different transition metal ions, the poly-L-lysine chains have been further self-assembled into networks where macromolecular chains are held together by metal-ligand bridges. For this aim, terpyridine side-chain functionalized poly-L-lysine have been prepared from 4'-Chloro-2,2':6',2''-terpyridine and lineer poly-L-lysine structure (Figure 1). Prepared a series of stimuli-responsive metallo-supramolecular gels have been prepared terpyridine side-chain functionalized poly-L-lysine was connected to various metal ions (Ni(II), Co(II) and Cu(II) ions). Synthesized stimuli responsible metallosupramolecular poly-L-lysine gel network was characterized by UV-vis, NMR and FTIR spectroscopy. The gel morphology, gel structure, stimuli-responsive properties, particularly rheological properties, are examined as functions of metal ion composition, gel concentration, pH and salt content. The response properties of these materials to different stimulus such as different metal ions chemical and electrochemical effect were also investigated.

References:

[1] E.Sato, Y.Masuda, J.Kadota, T.Nishiyama, H.Horibe, Dual stimuli-responsive homopolymers: Thermo- and photo-responsive properties of coumarin-containing polymers in organic solvents. Eur. Polym. J. 2015, 69, 605–615.
[2] J. Brassinne, F.D. Jochum, C.-A. Fustin, J.-F. Gohy, Revealing the Supramolecular Nature of Side-Chain Terpyridine-Functionalized Polymer Networks, Int. J. Mol. Sci. 2015, 16, 990-1007

Keywords: Smart polymer, supramolecular polymer, poly-L-lysine, terpyridine

Figure 1.

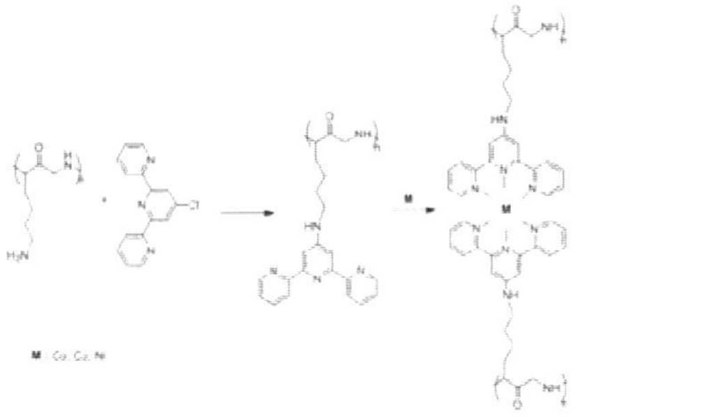

Synthesis of terpyridine side-chain functionalized poly-L-lysine.

SFP-P-024 [Smart and Functional Polymers]

Synthesis and characterization of new cationic photoinitiator

Betül Fatma Erkoç

Yildiz Technical University, Chemistry Department 34220 Istanbul Turkey

Photoinitiated cationic polymerization is an important industrial process widely used in various applications. Coatings, inks, adhesives and photolitography can be performed without the use of solvents, so a potential source of air and water pollution is eliminated. Photoinitiated cationic polymerization holds considerable promises in the future, particularly as a means of overcoming volatile emission, toxicity and molecular oxygen inhibition limitations. Photoinitiated cationic polymerization by direct irradiation of simple onium salts can be performed usually below 300 nm. To enhance the spectral sensitivity of onium salts is a substitution of the chromophores onto the aromatic groups of the salts is important. Due to limitations ocur on the potential uses of cationic polymerization at high wavelength. A number of phenacyl based compounds with pyridinium, anilinium and phosphonium salt structures were shown to be appropriate photoinitiators for cationic polymerization. Their photoactivity can further be shifted to longer wavelengths by using precursors with extended absorptivity in the synthesis. Among many free radical photoinitiators, α-aminoalkylphenone derivatives are perticularly useful for applications in pigmented systems where long wavelengthes irradiation is crucial. Synthesis and photoactivity of α-aminoalkylphenone type phenacyl onium salt possessing double photochromic groups in the structure was reported by Yagci (Yagci et.al., European Polymer Journal, 47 (2011) 792–799). To see the effect of photochromic group bromoacetonapthone was used instead of phenacylbromide and similar synthesis strategy was employed as reported earlier (Yagci et.al. Macromol. Rapid Commun. 2015). Photophysical and photochemical studies were performed.

Keywords: Photoinitiator, cationic photoinitiator, cationic polymerization

SFP-P-025 [Smart and Functional Polymers]

Synthesis and characterization of thermal stable metallo-supramolecular polymers bearing phosphazene unit

Selda SEZER, Süleyman Köytepe, Ahmet Gültek, Turgay Seçkin

İnönü University, Faculty of Arts and Science, Chemistry Department, Malatya, Turkey

Stimuli-responsive systems reacting to outer physical or chemical stimulator are pretty popular recently[1]. This kind of systems, especially as smart systems, can be potentially applied increasingly in electronic, optoelectronic and medical imaging fields [1-3]. Among these type systems, polimerical stimuli-responsive systems have a priviledged position due to their being stable and easily adaptable. In the most of stimuli-responsive polimers, there exist secondary interactions, strong hydrogen bonds and metal ligant interactions[3]. Metallo-supramolecular polimers prepared through composing coordination polimers reversibly with important transition metals have important physical properties, they possess important potentials for many technological applications. Despite their importance especially in sensor and electronic fields many metallo-supramolecular polimers have been the thermal consistency/durability problem. In this study, for overcome the this problem, thermal stable rigid phospahazen unit were connected supramolecular structure (Figure 1). Firstly, 4'-Chloro-2,2':6',2''-terpyridine as a versatile ligand was reacted with p-nitophenol. Then nitro groups was reduced to amin unit. Amine functionalized terpyridine was connected with hexachlorocyclotriphospazene. Prepared terpyridine ligand functionalized phosphazene was coordinated to various metal ions (Ni(II), Co(II) and Cu(II) ions). Chemical structure of the prepared metal containing polymers were investigated in terms of FTIR, NMR and UV-vis spectrophotometer. Obtained metallo supramolecular polymers bearing different metal atoms were also investigated situmuli-responsive properties and thermal stability. As a result, prepared metallo-supramolecular polymers bearing phosphazene unit have good thermal stability compare with general type lineer supramolecular structure.

This work was supported by TUBITAK under Grant No. KBAG115R023.

References:

[1] Ulrich S. Schubert, Harald Hofmeier and George R. Newkome, Modern Terpyridine
Chemistry, Wiley-VCH Verlag, Weinheim
[2] A. Wild, A. Winter, F. Schlütter, U.S. Schubert, Chem. Soc. Rev. 40, 1459 (2011).
[3] H. Hofmeier, S. Schmatloch, D. Wouters, U.S. Schubert, Macromol. Chem. Phys. 204, 2197 (2003).

Keywords: Smart polymer, metallo-supramolecular polymer, phosphazene, terpyridine

Figure 1.

Synthesis of metallo-supramolecular polymers bearing phosphazene unit

SFP-P-026 [Smart and Functional Polymers]

Effect of supramolecular assembly and hydrophilicity on supramolecular temperature sensor

Kanykei Ryskulova[1], Patrice Woisel[2], Richard Hoogenboom[1]

[1]*Department of Organic and Macromolecular Chemistry, Ghent University, Ghent, Belgium*
[2]*Unité des Matériaux et Transformations, UMR CNRS 8207, Equipe Ingénierie des Systèmes Polymères (ISP), Université de Lille, Villeneuve d'Ascq Cedex, France*

The combination of supramolecular chemistry and polymer science is quickly emerging to develop sophisticated smart and adaptive materials. Supramolecular interactions such as hydrogen bonding, metal-ligand and host-guest interactions can be used to impart dynamic responsive behaviour to polymer structures.Recently, we developed new type polymeric thermometers with memory function by combining a thermoresponsive polymer that undergoes a solution lower critical solution temperature transition (LCST=lower critical solution temperature) with donor-acceptor host-guest supramolecular interactions.Moreover, the supramolecular interactions induced control over metastable behavior, which could be used as a memory function). Such supramolecular polymeric thermometers were further explored to understand the effect of the polymer structure on the supramolecular assembly abilities as well as how this affects the properties and memory of the resulting polymeric thermometers. Several thermoresponsive water soluble polymers with supramolecular guest units randomly distributed along the side-chains were synthesized and supramolecular assemblies in water were studied (see Figure 1). Reversible addition fragmentation chain transfer (RAFT) controlled free radical polymerization was used to synthesize different poly(oligoethylene glycol acrylate) s (POEGA's) with side-chain naphthalene guest moieties and different cloud point temperatures. The thermoresponsive properties of these polymers and its dependence on the host-guest complexation of the naphthalene units with the cyclobis(paraquat-phenylene)(CBPQT4+) as host are reported and show more effect of polymer structure rather than hydrophilicity effect on supramolecular interaction. The temperature sensing ability of the supramolecular assemblies gives insight to the fundamental understanding of the effect of structure and supramolecular interactions on the thermoresponsive phase transition behavior.

Keywords: supramolecular chemistry, thermoresponsive polymers, polymeric thermometers, supramolecular sensors

Supramolecular complexation of the polymers

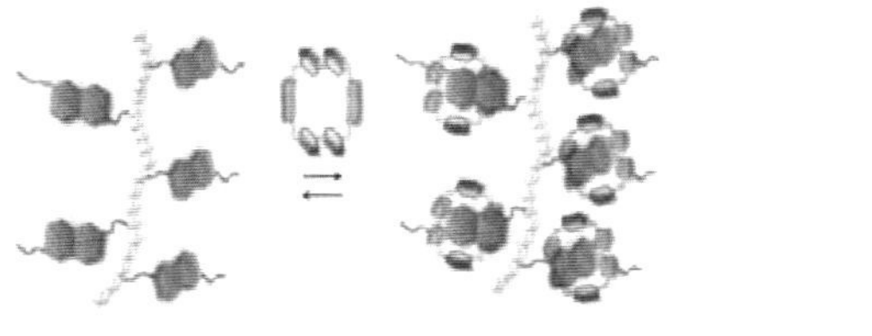

Schematic representation of the polymer structures and their temperature dependent supramolecular complexation behaviour in aqueous media. The polymeric host-guest complexes give rise to a purple solution and upon heating the polymer chains collapse yielding a white opaque solution, which can be used to develop polymeric thermometers.

SFP-P-028 [Smart and Functional Polymers]

Hybrid microspheres from zirconium butoxide and an ionic-non ionic copolymer

Bruno F Urbano[1], Leandro Toledo[1], Jorge Yañez[2], Cristian Campos[3], Bernabé L Rivas[1]

[1]*Department of Polymers, Faculty of Chemical Sciences, University of Concepción, Chile*
[2]*Department of Analytical and Inorganic Chemistry, Faculty of Chemical Sciences, University of Concepción, Chile*
[3]*Department of Physical Chemistry, Faculty of Chemical Sciences, University of Concepción, Chile*

The synthesis of hybrid materials is an increased focus of recent studies due to the potential synergism of the constituents used on the final properties. The present study describes the synthesis of hybrid microspheres hybrid of poly(4-vinylbezyltrimetil ammonium chloride-co-2-hydroxyethyl methacrylate) P(ClVBTA-co-HEMA) and zirconium oxide as inorganic moiety. The hybrid microspheres were obtained in two stages. First, the microspheres were synthesized by emulsion radical polymerization. The aqueous phase contains the monomer 4-vinylbenzyl trimethylammonium chloride (ClVBTA) and 2-hydroxyethylmethacrylate (HEMA) on diferent mole ratio, N,N-methylen-bis-acrylamide (MBAAm) as crosslinker, and the co-initiators ammonium persulfate (APS) and N,N,N',N'-tetramethylethylenediamine (TEMED). Span 80 was used as stabilizer of the emulsion and incorpored in the organic phase (n-heptane or n-decane). Subsequently, the polymeric microsphere were soaked in a solution of Zr(IV) butoxide for at least 24 h and then initiated the reaction sol-gel by adding a mixture water/ethanol 1:1. Finally, the hybrid microspheres were purified, dried and characterized by FTIR, TGA, SEM-EDS, among others. Parameters of the emulsion formulation such as the water-organic solvent volume ratio, surfactant concentration, monomers mole ratio were studied with an experimental design. The results showed that it is possible to obtain a material with well-defined morphology (spherical) using 20/80 v/v of water-organic solvent and 0.025 mol/L of Span 80. On the other hand, both the organic solvent used and the agitation speed applied during the emulsification and polymerization affect the size of the microspheres. Regarding the impregnation we found that time and the solvent's polarity affect importantly the degree of incorporation of inorganic precursor into the polymeric microsphere. This hybrid material exhibits potential applications in adsorption processes as well as in the field of catalysis. Moreover, the synthesis route allows the use of other polymers and inorganic precursor expanding the range of applications.

FONDEF IDeA ID15I10211 and CIPA-CONICYT Regional R08C1002.

Keywords: hybrid material, copolymer, emulsion polymerization, zirconium

Figure 1. SEM-EDS analysis of hybrid microspheres

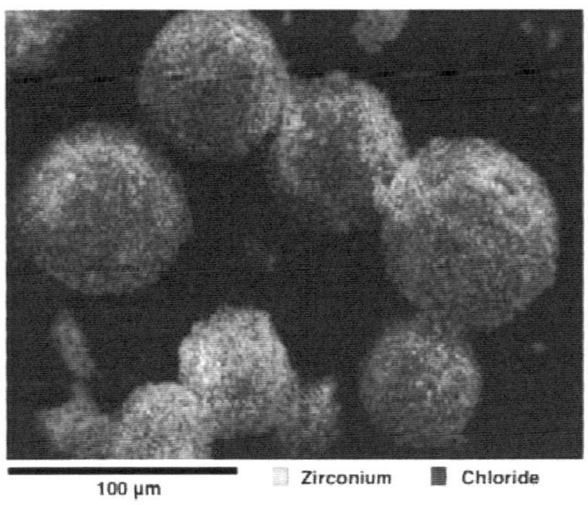

SFP-P-029 [Smart and Functional Polymers]

Synthesis and Characterization of NIPAAm-based Responsive Thin Layers

Anil Bozdogan, Esra Kasapgil, Ilke Anac

Materials Science and Engineering Department, Gebze Technical University, Kocaeli 41400, Turkey

The swelling behaviour of poly(N-isopropylacrylamide) (PNIPAAm)-based terpolymer (containing methacrylic acid and benzophenone methacrylate comonomers) films in water is controlled by an external stimuli (NaCl). PNIPAAm)-based terpolymer was synthesized by free radical polymerization and characterized by NMR and FTIR techniques. The PNIPAAm based gel films were prepared by spin coating from polymer/ethanol solution and crosslinked and attached to the modified gold substrates via UV irradiation. Layer thicknesses and polymer concentration were monitored by SPR/OWS on the same crosslinked PNIPAAm terpolymer layers anchored to gold surfaces. In addition to the expected volume phase transition at NaCl concentrations above 1 M, SPR/OWS reveals subtle thickness variations at lower salt concentrations.

[1] Junk M.J.N., Anac I., Menges B., Jonas U., Langmuir, 2010, 26, 12253-12259
[2] Gianneli M., Anac I, Rostkamp R, Menges B, Loppinet B, Jonas U, Knoll W, Fytas G. Macromolecular Chemistry and Physics 2015, 216, 277–286

Keywords: surface plasmon resonanse, poly(N-isopropylacrylamide), swelling, radical polymerization

SFP-P-030 [Smart and Functional Polymers]

Poly(Ionic Liquid)s-Derived Nitrogen-Doped Porous Carbons for Environmental Remediation and Catalysis

Jiang Gong, Huijuan Lin, Markus Antonietti, Jiayin Yuan

Department of Colloid Chemistry, Max Planck Institute of Colloids and Interfaces, Research Campus Golm, D-14476, Potsdam, Germany

Since the first conversion of poly(ionic liquid)s (PILs) into mesoporous carbons in our group in 2010, PILs have been proven to be a unique class of carbon precursors [1]. When comparing PILs over other polymers, following advantages can be identified. Firstly, PILs are thermally stable at elevated temperatures due to the ionic nature. Secondly, PILs contain incorporated/conjugated heteroatoms such as nitrogen, favorably yielding heteroatoms in the carbon products. Thirdly, PILs are surface-active polymers and can form homogeneous layers on most of surfaces.

Nitrogen-doped porous carbons have received global interest ascribed to their tunable physicochemical properties and a wide of applications in catalysis, energy and environment [2]. The nitrogen doping in the carbon framework not only adjusts the physical properties by increasing the overall electron density at the Fermi level, but also enhances the basicity to be highly active in catalysis. But many synthetic approaches are hampered by time-consuming post-treatments/complicated activation procedures.

Our recent work focuses the facile sustainable conversion of functional PILs into various nitrogen-doped porous carbons bearing high specific surface areas, including nanosheets, fibers and foams using C_3N_4 nanosheets, porous fibers and salts as templates. For example, without post-treatments or activation steps, the nitrogen-doped porous carbon nanosheets readily reach a surface area above 1100 m^2/g with hierarchical micro-/meso-/macropore structure while keeping a high nitrogen content (17.4 wt%). Their applications in the CO_2 capture and sensor, dye removal and photo-degradation, and selective aerobic oxidation of alcohols are explored. The nitrogen-doped porous carbon foams show excellent performance in the photo-degradation of Rhodamine B under visible light irradiation. The relationships of surface properties, pore structures and performances are highlighted, which would greatly advance the related research.

References:

[1] J. Yuan, C. Giordano, M. Antonietti, Chem. Mater. 2010, 22, 5003.
[2] J. Yuan, D. Mecerreyes, M. Antonietti, Prog. Polym. Sci. 2013, 38, 1009.

Keywords: Poly(Ionic Liquid)s, Nitrogen-Doped Porous Carbons, Environmental Remediation, Catalysis, Sustainable approach, CO2 Capture

SFP-P-031 [Smart and Functional Polymers]

Functionalized Chitosan by RAFT for Pharmaceutical Formulations

Catalina Natalia Yilmaz[1], Sinem Yaprak Karavana[1], Onur Yilmaz[2]

[1]Ege University, Faculty of Pharmacy, Department of Pharmaceutical Technology, Izmir, Turkey
[2]Ege University, Faculty of Engineering, Leather Engineering Department, Izmir, Turkey

Chitosan known as being a biocompatible, biodegradable natural polysaccharide and has a great interest to be used in pharmacy field have been chosen as a source material for producing polymeric matrices in the study. Chemical modification can open new ways to obtain materials based on chitosan with tailored properties. Despite of the great interest for graft modification, controlled/living free radical techniques seems more attractive nowadays and involve facile and fast procedure, high regioselectivity, quantitative yield, mild reaction conditions, without generation of by-products. Present study reports thermo-responsive chitosan derivatives obtained via a facile functionalization method. The temperature-responsive chitosan was synthesized by grafting from technique of N-isopropylacrylamide (NIPAM) using chitosan as macro-RAFT agent. Prior to grafting the chitosan was modified with phthalic anhydride and RAFT agent (i.e. 4-cyano-4-(phenylcarbonothioylthio) pentanoic acid) as illustrated within Scheme 1. The obtained hybrid polymers were further used as drug carriers for topical applications.

Acknowledgments: Authors acknowledge the financial support from "The Scientific and Technological Research Council of Turkey (TUBITAK), cofunded by Marie Curie Actions under FP7", project number 115C078.

Keywords: chitosan, RAFT polymerization, thermoresponsive polymers

Scheme 1

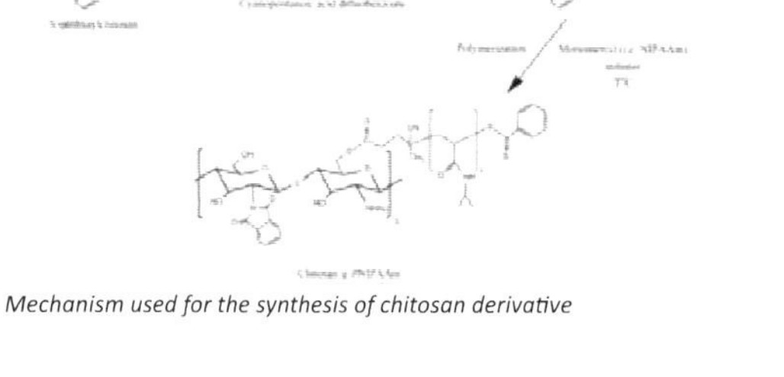

Mechanism used for the synthesis of chitosan derivative

SFP-P-033 [Smart and Functional Polymers]

Synthesis of alkoxysilane functional copolymers via RAFT polymerization technique

Onur Yılmaz[1], Ersin Önem[1], Hasan Özgünay[1], Catalina Natalia Cheaburu Yılmaz[2], Hüseyin Ata Karavana[1], Ali Yorgancıoğlu[1]

[1]*Ege University, Faculty of Engineering, Leather Engineering Department, Bornova, Izmir, Turkey*
[2]*Petru Poni Institute of Macromolecular Chemistry of the Romanian Academy, Laboratory of Physical Chemistry of Polymers, Iasi, Romania*

Polymers bearing alkoxysilane functional moieties are an important class of materials since the reactive alkoxysilane function can be used as coupling agent to design polymers for applications in optics, reactive coatings, catalysis, antifouling paints, etc. However, there are only limited number of studies reporting the synthesis of alkoxysilane bearing copolymers via controlled radical polymerization techniques leading to well define structures and molecular architectures1-3. Present study deals with the synthesis of block and random copolymers of vinyl triethoxysilane and 3-methoxysilylpropyl methacrylate with different hydrophilic and hydrophobic acrylic monomers via reversible addition fragmentation chain transfer (RAFT) polymerization technique in solvent medium. The obtained copolymers have been characterized by FTIR, H-NMR; molecular weight and weight distributions by GPC, thermal properties by DSC to confirm the success of the synthesis.

Acknowledgement: Authors acknowledge the financial support from the Scientific and Technological Research Council of Turkey (TÜBİTAK), Grant No: 115M650

References:

1. Sarıcılar, S., Knott, R., Barner-Kowollik, C., Davis, T.P. ve Heuts J.P.A. 2003. "Reversible addition fragmentation chain transfer polymerization of 3-[tris(trimetylsilyloksi)silyl] propyl methacrylate", Polymer, 44, 5169-5276.
2. Mellon, V., Rinaldi, D., Bourgeat-Lami, E. ve D'Agosto, F. 2005. "Block copolymers of γ-methacryloxypropyltrimethoxysilane and methyl methacrylate by RAFT polymerization. A new class of polymeric precursors for the sol-gel process", Macromolecules, 38, 1591-1598.
3. Nguyen, M.N., Bressy, C. ve Margaillan, A. 2005. "Controlled radical polymerization of a trialkylsilyl methacrylate by reversible addition-fragmentation chain transfer polymerization", Journal of Polymer Science:Part A: Polymer Chemistry, 43, 5680-5689

Keywords: Alkoxysilane, RAFT, functional copolymers

SFP-P-034 [Smart and Functional Polymers]

Release of Doxorubicin from Layer-by-Layer Films of Poly (2-Isopropyl-2-Oxazoline)

Meltem Haktanıyan, Süleyman Atilla, Eda Çağlı, Irem Erel Göktepe

Department of Chemistry, Middle East Technical University

Stimuli responsive polymers are attracting materials for biomedical applications, specifically for controlled delivery systems. These polymers have the unique ability to change their behavior in response to changes in pH, temperature, light, magnetic field and electric field. Due to local pH and temperature changes in the body, temperature and pH are the most extensively studied stimuli for the development of controlled delivery systems. Among temperature responsive polymers, poly (N-isopropylacrylamide) (PNIPAM) is the most well-known and studied polymer. Recently, poly (2-alky-2-oxazoline)s are getting increasing interest due to their chemical structure resembling pseudo-peptides and their lower critical solution temperature (LCST) behavior.

In this study, first, poly (2-isopropyl-2-oxazoline) (PIPOX) was synthesized via cationic ring opening polymerization. We then prepared layer-by-layer (LbL) films of poly (2-isopropyl-2-oxazoline) (PIPOX) and water soluble complexes of Tannic Acid (TA) and Doxorubicin (DOX). PIPOX is hydrogen accepting neutral polymer, whereas TA has hydrogen donor groups enabling the LbL film construction through hydrogen bonding interactions. In this way, DOX, an anticancer drug, was also loaded within the multilayers. Release of DOX from the multilayers was induced via pH- and temperature changes. Our results showed that increasing temperature resulted in an increase in the amount of DOX, released by pH-trigger.

Keywords: smart polymers, temperature responisve polymers, Poly(2-isopropyl-2-oxazoline), Layer-by-layer technique, Hydrogen bonded multilayers

SFP-P-035 [Smart and Functional Polymers]

Synthesis and characterization of some polyurethanes bearing pyridine-metal complex unit

Süleyman Köytepe, Birgül Köstek, Imren Özcan, Turgay Seçkin

İnönü University, Faculty of Arts and Science, Chemistry Department, 44280, Malatya, Turkey

Polymers bearing pyridine-metal complex have great attention because of their easy formation and rich coordination chemistry with a large variety of metal ions, that has allowed their use as catalysts, electronic material and some smart materials [1-2]. This attention is still growing; a considerable research effort is devoted to the synthesis of modified and supported reagents for materials chemistry [3]. In this study, pyridine based monomers as will be obtained using 2,6-diamino pyridine and vanillin (4-Hydroxy-3-methoxybenzaldehyde) having dihydroxyl functionality (Figure 1). Firstly, 4-hydroxy-3-methoxybenzaldehyde was reacted with diamino compounds by Schiff base reaction to produce ligand structure monomer. Then, this monomer was reacted with different diisocyanates for produce polyurethane. Prepared ligand functionalized polymer was connected to various metal ions (Ni(II), Co(II) and Cu(II) ions). Chemical structure of the prepared metal containing polymers were investigated in terms of FTIR, NMR, TGA, DTA, DSC, GPC, UV-vis spectrophotometer and amperometric techniques. Obtained polymers bearing different metal atoms were also investigated situmuli-responsive properties. As a result, prepared polyurethanes bearing pyridine-metal complex unit are good candidates for electronic, opto-electronic, and catalytic applications.

References:

[1] SR. Batten, NR. Champness, X-M Chen, J Garcia-Martinez, S Kitagawa, L Öhrström, M O'Keeffe, MP Suh, J Reedijk, Pure and Applied Chemistry(IUPAC) 85(8), 2013 1715-1724.
[2] K. Fromm, Coord. Chem. Rev. 252(8–9), 2008 856–885.
[3] NS. Venkataramanan, G Kuppuraj, S Rajagopal, Coordination Chemistry Reviews, 249(11–12), 2005 1249–1268.

Keywords: Smart polymer, pyridine, metal-complex polymer, polyurethane

Figure 1

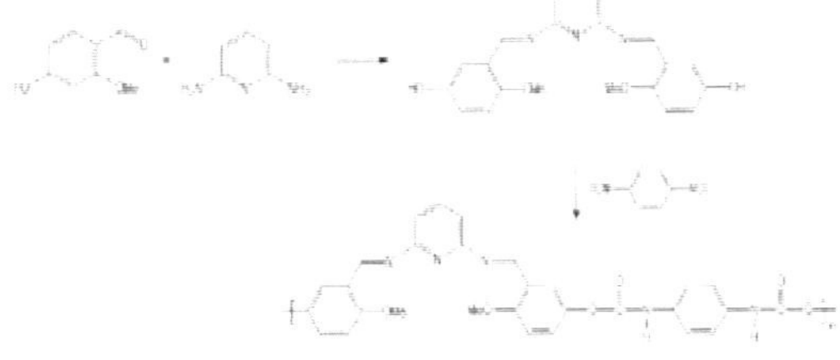

Synthesis of polyurethanes bearing pyridine unit

SFP-P-036 [Smart and Functional Polymers]

Synthesis of novel supramolecular gels containing terpyridine-metal complexes bonding units

Süleyman Köytepe, Birgül Köstek, Büşra Aksoy, Imren Özcan, Turgay Seçkin

İnönü University, Faculty of Arts and Science, Chemistry Department, 44280, Malatya, Turkey

Supramolecular chemistry has received remarkable importance in the last years [1-2]. Supramolecular polymers as important class in supramolecular chemistry prepare with units hold together by reversible secondary interaction. Non-covalent forces that hold supramolecular polymer structures together include metal-ligand coordination, π-π interactions, hydrogen bonding, donor-acceptor associations, electrostatics, hydrophilic-hydrophobic, and van der Waals forces [2-3]. One of the most important advances in the field of supramolecular chemistry, the morphology, physical properties and chain packing density of polymer molecules is rearrangement by the use of functional side groups. In this study, firstly, 5-amino-1-pentanol groups bounded polyvinyl chloride was reacted with 4-chloro-2,2':6',2''-terpyridine to produce terpyridine-side chain polymer (Figure 1). Than, prepared terpyridine side-chain functionalized polymer was connected to various metal ions (Ni(II), Co(II) and Cu(II) ions). By addition of transition metal ions, concentrated aqueous solutions of this polymer turn into metallo-supramolecular gels whose dynamic mechanical properties are investigated by rotational rheometry. Chemical structure of the prepared metallosupramolecular gels were investigated in terms of FTIR, NMR, TGA, UV-vis spectrophotometer and amperometric techniques were used for the investigation of situmuli-responsive properties of polymers. As a result, prepared terpyridine-based metallosupramolecular gels are good candidates for electronic, opto-electronic, and photovoltaic applications as a smart stimuli-responsive material.

References:

[1] Lehn, J.-M. Supramolecular Chemistry: Concepts and Perspectives, VCH, New York, 1995.
[2] S.T. Nguyen, D.L. Gin, J.T. Hupp, X. Zhang, Supramolecular chemistry: Functional structures on the mesoscale, Proc. Nat. Acad. Sci., 2001 98(21) 11849–11850.
[3] U.S. Schubert, H. Hofmeier, G.R. Newkome, Modern Terpyridine Chemistry, Wiley-VCH, New York, Weinheim 2006.

Keywords: Smart polymer, supramolecular gels, terpyridine metal complexes.

Figure 1

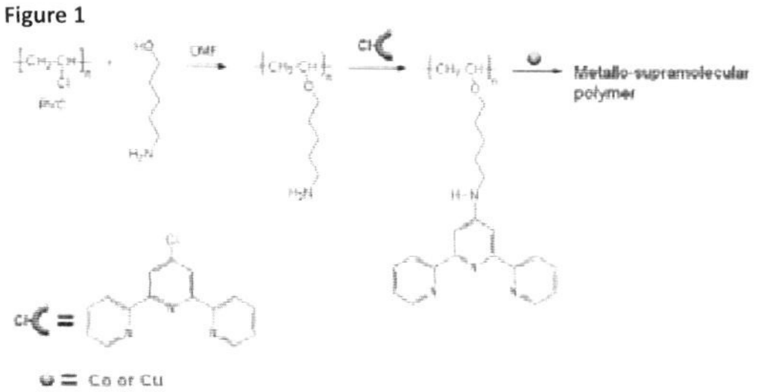

Synthesis of supramolecular gels containing terpyridine metal complexes bonding units

SFP-P-037 [Smart and Functional Polymers]

Thermoresponsive PEGMA Polymer Nanoparticles: A New Colloidal Drug Carrier

Mehmet Sahin Atas[1], Ayse Nur Ozkaya Balcı[2], Mustafa Selman Yavuz[1]

[1]Metallurgy and Materials Engineering Department, Selcuk University, Konya, Turkey
[2]Advanced Technology Research and Application Center, Selcuk University, Konya, Turkey

Over the past decade, thermoresponsive polymers have been increasingly investigated in nanotechnology and biotechnology applications. Synthetic or biological macromolecules exhibiting a lower critical solution temperature (LCST) in water are structures of prime importance in applied polymer research[1]. Recently, there has been a great interest on responsive polymeric micelles. Such structures are generally called as smart materials which have stimuli responsive nature upon heating. Indeed, these water-soluble polymers precipitate over their LCST due to shrinking phenomena [2]. In case of thermoresponsive polymers, temperature can be used as a simple external trigger to change the conformation of the polymer. Poly(N-isopropyl acrylamide) (PNIPAM) and its copolymers are one of the most frequently used thermoresponsive systems. Although some literature reports showed that only certain cell types can survive after treatment of PNIPAM containing colloids. To improve the antifouling properties of PNIPAM polymers, hydrophilic poly(ethyleneglycol) (PEG) containing methacrylate units can be incorporated into the polymerization recipes[4]. Herein, new poly (ethylene glycol methyl ether metacrylate) (PEGMA) colloids having different LCST values ranging from 32 to 90 oC are successfully synthesized and characterized by using DLS, UV, and FTIR. The thermoresponsive characteristics of all synthesized PEGMA nanoparticles are explored and compared. Controlled drug and fluorescent dye release mechanisms of these PEGMA colloids are investigated under conventional heating or laser ablation (photothermal).

This work was supported by the Scientific and Technological Research Council of Turkey (TUBITAK Grant 214M232) and Selcuk University Scientific Research Fund (BAP).

References:

[1]. Lutz, J.-F.; Hoth, A.; Schade, K., Designed Monomers & Polymers 2009,12 (4), 343-353.
[2]. Ulasan, M.; Yavuz, E.; Bagriacik, E. U.; Cengeloglu, Y.; Yavuz, M. S., Journal of biomedical materials research. Part A 2015,103 (1), 243-51.
[3]. Yavuz, M. S.; Buyukserin, F.; Zengin, Z.; Camli, S. T., Journal of Polymer Science Part a-Polymer Chemistry 2011,49 (22), 4800-4808.

Keywords: LCST, PEGMA nanoparticles, photothermal

SFP-P-038 [Smart and Functional Polymers]

New coatings based on fluoropolymer/ clay nanocomposites prepared by *in situ* polymerization: Synthesis and characterizations

Raihane Mustapha[1], Karamane Mohamed[1], Yagçi Yusuf[2], Lahcini Mohamed[1], Ilsouk Mohamed[1], Atilla Tasdelen Mehmet[3]

[1]Laboratory of Organometallic and Macromolecular Chemistry- Composites Materials, Faculty of Sciences and Technologies, Cadi Ayyad University. BP 549 Marrakech (Morocco)
[2]Istanbul Technical University, Faculty of Letters and Sciences, Department of Chemistry, Maslak, TR-34469 Istanbul (Turkey)
[3]Department of Polymer Engineering, Faculty of Engineering, Yalova University, TR-77100 Yalova, (Turkey)

Polymer–clay nanocomposites have been widely investigated in recent years due to enhancement in mechanical, thermal, gas barrier, catalytic and fire retardant properties that nanoclay imparts to polymer matrices. The improvement in properties combined with the low cost of manufacturing of polymer–clay nanocomposites has led to their use in engineering applications. The *in situ* polymerization may be the most desirable method for preparing nanocomposites because the types of nanoparticles and the nature of polymer precursors can vary in a wide range to meet the requirements.

Fluoropolymers exhibit remarkable properties, such high thermal, chemical, aging, and weather resistance, excellent inertness to solvents, to hydrocarbons, to acids, and to alkalies, low surface energy (oil and water repellency).

In frame work of developing new fluorinated caoting, nanocomposites based on poly(2,2,2-trifluoroethyl methacrylate), PMATRIF, were successfully prepared by in situ polymerization at 70°C using AIBN as radical initiator in presence of a MMT modified with triphenylvinylbenzylphosphonium chloride (Phos). Depending on the MMT-Phos loading (1, 3 or 5% by weight), nanocomposites of partially exfoliated and intercalated structures were obtained as evidenced by X-ray diffraction and SEM. These nanocomposites were also characterized by IR and NMR spectroscopy. The nanoscale dispersion of MMT-Phos was also evidenced by a moderate increase of the Tg of these nanocomposites, and an improvement of their thermal stability as evidenced by TGA. The thermal stability of PMATRIF/MMT-Phos nanocomposites also increased with MMT-Phos loading.

The Surface properties of PMATRIF/x % MMT-Phos nanocomposites were studied by contact angle measurements with water (WCA). The incorporation of fluorinated polymers (PMATRIF) into the galleries of the organomodified MMT-Phos increases the hydrophobicity of the Na-MMT clay and improves also the surface properties of obtained nanocomposites, and thus the development of new coating fluorinated nanocomposites with clay.

The authors thank the CNRST (Morocco)-TÜBITAK (Turkey) program for its financial support.

Keywords: Fluorinated polymer, In situ polymerization, nanocomposites, structural characterizations, surface properties

SFP-P-039 [Smart and Functional Polymers]

Solubility behavior of poly(oligo(ethylene glycol) methacrylate)s in water: Effect of different additives

Maryam Bozorg[1], Birgit Fischer[1], Volker Abetz[2]

[1]Department of Physical chemistry, University of Hamburg, Hamburg, Germany
[2]Institute of Polymer Research, Helmholtz-Zentrum Geesthacht, Geesthacht, Germany

In this work poly(oligo(ethylene glycol)) based copolymers with different compositions and structures are synthesized via Cu-based atom transfer radical polymerization. These copolymers are temperature responsive and show a lower critical solution temperature (LCST) in water. Their solubility behavior in water is studied thoroughly. The effect of cosolvent and cosolute addition such as different salts, surfactants and organic solvents is investigated. According to dynamic light scattering results, the lower critical solution temperature (LCST) depends on the structure of the copolymer and it is changing slightly by the additives. Good cosolvents like ethanol can increase the LCST through dissolving the collapsed copolymer chains to some extent and increase the LCST. The same effect is observed for surfactants which make the polymer solution more stable by preventing aggregation. Above a certain concentration of surfactant, the solution is stable at all temperatures and no LCST is observed.

The effect of salts on the solubility of the copolymers is following the Hofmeister series and it is relelated to the concentration of the salts. According to their affinity to the polymer, the salts can increase or decrease the LCST. The biggest change is observed for very hydrophobic salts (such as NaSCN) which increase the LCST or very hydrophilic ones (such as LiCl) which decrease the LCST.

Keywords: poly(oligo(ethylene glycol) methacrylate)s, temperature responsive polymers, LCST, Hofmeister series

SFP-P-040 [Smart and Functional Polymers]

Synthesis, Modifications and Characterization of New -Type-TER-Polymer Containing Vinly Propionate

Şerife Civil, Ahmet Okudan, Bilge Makbule Sağkan, Fatıma Zehra Dikici

Department of Chemistry, Selcuk University, Konya, Turkey

The focus of long time researches is synthesising or exploring new medication having biological effects. The important point in this issue is the usage of medication[1]. Medications using in traditional way is required frequent dosing, and to stay in thetherapeutic interval high dosage is needed. It means that they have some disadvantages. This negative effects speed up the researches about oscillation of medication by restrained way in along time. The main system to change the usage of medications is active, controlled drug delivery system which polymers are used in. In this systems, active ingredient remains as active with a single application for a long time and the system are provided to save time and cost[1]. In this experiment, by using monomers of maleic anhydride, vinyl propionate and N-isopropylacrylamide, new kinds of co- polymers and ter- polymers will be synthesized in radical way[2]. This polimers will be mixted with acetylsalicylic acid. In addition to that, polimers which have already synthesized are going to be modified by polyethylene glycol, and mixted with acetylsalicylic acid. Polymers getting in this work will be characterized. Then, their usage in pharmaceutical industry is going to be investigated[3]

References:
[1] Çetin M. krotonik asit-akrilamid-akrilik asit içeren hidrojellerin şişme davranışları ve pantoprazol-na nın kontrollü salımında kullanılabilirliğinin incelenmesi,Gazi Üniversitesi, Fen Bilimleri Enstitüsü,Ankara,2006.
[2] Karasakal A. Okudan A,Theeffect of H-boding on RadicalCopolimerization of MaleicAnhydridewith N-tert-Butylacrylamide and Its Characterization, International Journal of PolymerScience, 2013.
[3] Çolak B. AcetylsalicylicAcid Modification of MaleicAnhydride Copolymer and Investigation of Controlled Release, MS Thesis,Istanbul Technical University, Institute of Science and Technology, 2009.

Keywords: Vinyl Propionate, Maleic Anhydride,Controlled realesed

Figure 1

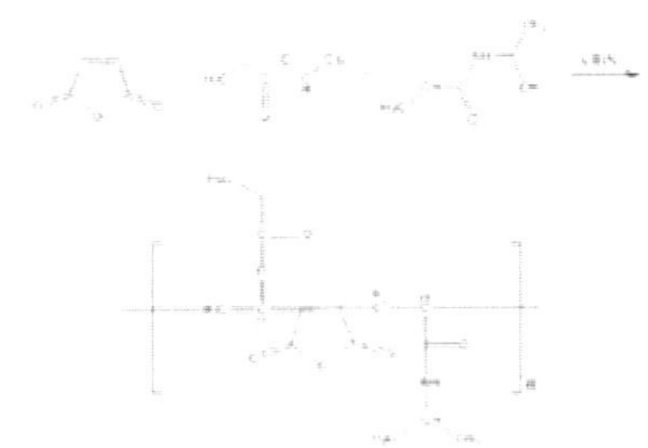

terpolimerization

Figure 2

modified

SFP-P-041 [Smart and Functional Polymers]

Polypyrrole coating on cotton textile as adsorbent for methylene blue dye

Mohamad Mohamad Ayad[1], Wael Amer[2], Sawsan Zaghlol[1], Boberc Boberc[3], Jaroslav Stejskal[3]

[1]Department of Chemistry, Faculty of Science, University of Tanta, Tanta 31527, Egypt.
[2]Egypt-Japan University of Science and Technology, Borg El Arab, Alexandria 21934, Egypt
[3]Institute of Macromolecular Chemistry, Academy of Sciences of the Czech Republic, 162 06 Prague 6, Czech Republic

The need to maintain a cleaner environment for the survival of both aquatic and terrestrial lives including human beings is very crucial and it is of increasing concern to the environmentalists. Pollution caused by agents such as dyes is amongst the list which rendered the environment unwholesome and posed serious health concern to the population. In present work, polypyrrole coating on cotton textile have been utilized as a promising adsorbent for methylene blue dye. The adsorption was assessed by using the UV-visible absorption spectroscopy. The influence of various parameters such as dye concentration and adsorbent dose has been studied. The experimental data were fitted into various kinetic models. It was observed that pseudo-second-order kinetic model described the adsorption process well and the Freundlich isotherm best fits adsorption data. The thermodynamic study revealed that the adsorption of methylene blue by polypyrrole deposited on cotton textile was feasible, spontaneous, and exothermic process.

Keywords: dye, polypyrrole, cotton textile, adsorption

SFP-P-042 [Smart and Functional Polymers]

Tailor-made Functionalized Polymers as a Basis for the Formation of Supramolecular Structures

Elisabeth Wittenberg[1], Volker Abetz[2]

[1]Department of Physical Chemistry, University of Hamburg, Hamburg, Germany
[2]Institute of Polymer Research, Helmholtz-Zentrum Geesthacht, Geesthacht, Germany

Supramolecular chemistry is an increasing field in modern chemistry and deals with the creation of highly ordered structures based on multiple non-covalent interactions. In polymer chemistry, applying the principles of supramolecular chemistry, this leads to the possibility to determine polymer properties as well as the design of new polymer architectures.[1] Our research focuses on the modification of styrene-butadiene copolymers by introduction of hydrogen-bonding motifs along the polymer chain. Due to the high reactivity of double bonds, the modification of polybutadiene can easily be done via an epoxidation step with following hydrolysis of the epoxide.[2] In a third step, we were able to modify the hydroxylated styrene-butadiene copolymer by Steglich esterification. Steglich esterification is a well-known and often used type of esterification under mild conditions, which allows the reaction of sterically demanding molecules like secondary and tertiary alcohols with carboxylic acids. A broad variety of commercially available carboxylic acids that can be used to modify styrene-butadiene copolymers arises from this approach. Here we present the modification of styrene-butadiene copolymers using *N*-Boc-β-alanine (Boc = *tert*-butyloxycarbonyl), which can form at least three different hydrogen-bonding complexes.The characterization of these hydrogen-bonding complexes can be realized by infrared spectroscopy. Investigating the C=O stretching vibration, it can be clearly distinguished between the 'complexed' and 'free' species in the equilibrium.[3] Temperature-dependent measurements show a higher population of the 'free' species absorption band at higher temperatures, while at lower temperatures the absorption band of the 'complexed' species increases (Figure 1).

[1] Rint P. Sijbesma, Felix H. Beijer, Luc Brunsveld, Brigitte J. B. Folmer, J. H. K. Ky Hirschberg, Ronald F. M. Lange, Jimmy K. L. Lowe, E. W. Meijer, Science 1997, 278, 1601-1604.
[2] Chih-Cheng Peng, V. Abetz, Macromolecules 2005, 38, 5575-5580.
[3] L. de Lucca Freitas, C. Auschra, V. Abetz, R. Stadler, Colloid Polymer Science 1991, 259, 566-575.

Keywords: Supramolecular Chemistry, Non-covalent Interactions, Functionalized Polymers, Polymer Modification

Figure 1

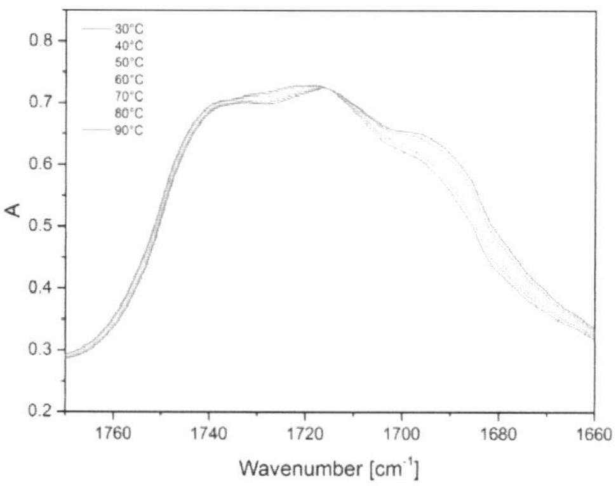

Temperature-dependent (90-30°C) infrared spectra of N-Boc-β-alanine modified styrene-butadiene copolymer.

SFP-P-043 [Smart and Functional Polymers]

A Novel Red to Transmissive Electrochromic Polymer Based on Phenanthrocarbazole

Gizem Atakan, Görkem Günbaş

Chemistry Department, Middle East Technical University, Ankara, Turkey

A novel EDOT and phenanthrocarbazole containing polymer was designed and synthesised electrochemically and electrochromic properties were investigated. To our best knowledge, this is the first example of a phenanthrocarbazole containing electrochromic material. The polymer shows red colour in its neutral state and switches to a highly transmissive state. A high optical contrast (51%) and fast switching times (0.65 s and 0.52 s) were achieved, making this polymer and important candidate for electrochromic device and simple display applications. The study also revealed that phenantrocarbazole is a promising unit towards realization of novel, high performance electrochromic materials.

Keywords: phenanthrocarbazole, electrochromic polymer, red to transmissive

SFP-P-044 [Smart and Functional Polymers]

Functionalized Polydimethylsiloxanes for specific adhesion

Robin Heedfeld, Xiaomin Zhu, Martin Möller

DWI – Leibniz Institute for Interactive Materials, RWTH Aachen University, Aachen, Germany

Polydimethylsiloxane (PDMS) is one of the most important polymers in industry. It can be find in many daily life applications due to a number of unique physic-chemical properties, such as high flexibility, low surface tension, high and low temperature resistance, electrical insulation, chemical inertness and biocompatibility [1,2]. The properties of PDMS originate from its unique chemical structure, in which a polar inorganic siloxane backbone is surrounded by organic methyl groups [3]. The interaction between the PDMS chains is dominated by the weak van der Waals force between the methyl groups. The introduction of functional groups can certainly lead to a stronger interchain interaction, which is the topic of this contribution. The goal of this study is to introduce interactions that are characteristic for proteins into PDMS in order to render and analyze new functions to this material. These compounds were prepared via a three step synthetic route, as shown in schematic 1. PDMS with SiH groups were synthesized by means of equilibrium polymerization of cyclic siloxane monomers. Afterwards, the hydrosilylation with allyl glycidyl ether results in the formation of epoxy-substituted PDMS. The final functionalization was performed via the reaction of epoxy-groups with acid or amine groups. The functional groups that we choose were inspired by nature motives. In nature, especially in proteins, strong interactions such as hydrogen bonding, ion-pair formation, acid-base interactions, π- π- stacking are present, and they give rise to the formation of secondary and tertiary structures [4]. We adapted these interactions to polysiloxanes to create new functionalities. In this study, we investigate the adhesion behavior of our new functionalized materials on organic and inorganic substrates. These new functionalized polydimethylsiloxanes are promising materials for textile and cosmetic applications, and are under current investigations.

1. James, E. Mark.; ACS Symposium Series, 2009, 729, 1.
2. Michael J. Owen,; ACS Symposium Series, 2010, 13-18, 1.
3. E. Rochow, An Introduction to the Chemistry of Silicones, John Wiley and Sons, 1946.
4. Susanne Marquse.; Protein Science, 1994, 3, 2217.

Keywords: polydimethylsiloxanes, functional polymers, specific adhesion

Synthetic strategy route for functional organopolysiloxanes

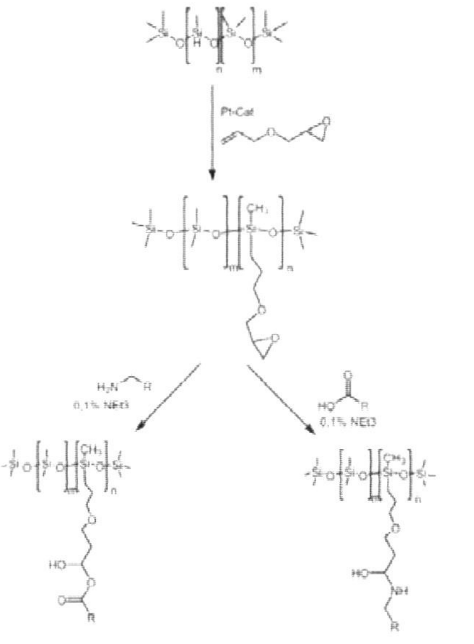

SFP-P-045 [Smart and Functional Polymers]

Introduction of Catalytical Property to the PAMAM type Dendrimers

Metin Tülü, Kevser Harmandar, Mustafa Ulvi Gürbüz

Dept.of Chem. Fac. of Arts and Sciences, Yıldız Technical University, 34210 Istanbul,TR.

Dendrimers are new class of macromolecules having good solubility, diverse functionality, controlled size and narrow polydispersity. Poly(amidoamine), PAMAM, is the most well-known dendrimer dominating with amide linkage. In general, PAMAM molecules don't possess catalytical property, however, their internal voids or their well-defined surface groups can be used to host or bind some special metals. When a transition metal is doped inside dendritic substrate, obtained nanoparticles can exhibit catalytical activity [2]. Crooks et al. has made the first Cu-Dendritic nanoparticles then continued with Au, Ag, Pt, Pd, Cu and Rh to increase the catalytic diversity[3]. A current study suggests synthesis of PAMAM type nanoparticles and paves the way for future applications as homogenous catalyst. Starting from m-xylylenediamine hydroxyl terminated PAMAM dendrimers were synthesized, and then Cu-DENs nanoparticles were prepared. Preparation of Cu-DEN and its catalytical activity was monitored by uv-vis spectroscopy.

Authors would like to thank Yıldız Technical University Scientific Research Project (2016-01-02-YL04) for financial support.

References:

[1] Donald A. Tomalia and Jean M. J. Fréchet (2002) "Discovery of dendrimers and dendritic polymers: A brief historical perspective" Journal of Polymer Science Part A: Polymer Chemistry. Volume 40, Issue 16: 2719–2893. [2] Crooks, R.M. Zhao, M. Sun, L. Chechik, V. and Yeung, L.K., (2001). "Dendrimer-encapsulated metal nanoparticles: synthesis, characterization, and applications to catalysis", Accounts of Chemical Research, 34: 181-190. [3] Balogh, L. ve Tomalia, D.A.,(1998). "Poly (amidoamine) dendrimer-templated nanocomposites. 1. Synthesis of zerovalent copper nanoclusters", Journal of the American Chemical Society, 120: 7355-7356.

Keywords: Dendrimer, PAMAM, Catalytical activity, UV-VIS

Figure 1 Cu-Dendrimer

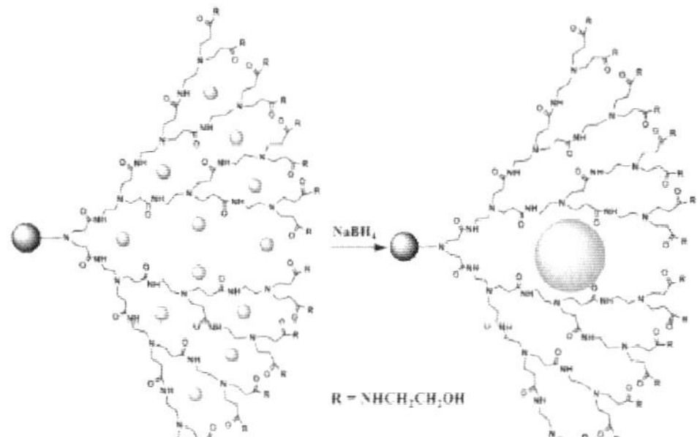

Figure 1. Cu-DENs naoparticles designed to exhibit catalytical activity

SFP-P-046 [Smart and Functional Polymers]

Dispersible Multi-Walled Carbon Nanotubes via Non-Covalent Interaction with Functional Copolymers

Busra Cengiz, Amitav Sanyal

Department of Chemistry, Bogazici University, Bebek, 34342, Istanbul, Turkey

Carbon nanotubes (CNT's) display exceptional properties like unique size distributions, hollow tube structures and high specific surface areas. All these properties enable CNT's to be used in many fields such as sensors, nanocomposites, molecular devices, nanoscale electronics, delivery of gene and medicine.1 However, the poor dispersion of CNT's in most organic and aqueous solvent is the major limitation for their processability and hindering their wide application.2 In order to overcome this problem, surface modification of CNTs can be used.3 Non-covalent (physical) or covalent (chemical) functionalization of CNT's can provide a solution for their solubility or chemical and biological compatibility problem. Here, non-covalent interaction is used since CNT's preserve their structure and original properties after such modification.

In this study, we report design and synthesis of side chain reactive copolymers and show solubilization of multi-walled CNT's in aqueous environment by attaching these polymers to CNT's via non-covalent interaction. For this purpose, PEG based copolymers which have thiol reactive units were synthesized via atom transfer radical polymerization method. These polymers were wrapped onto CNT's under aqueous conditions. These water soluble conjugates were then functionalized via nucleophilic thiol-ene reaction. Also, both CNT's with and without polymer were functionalized with thiol containing molecules and their stability was investigated.

References:

1. Pantarotto, D., Singh, R., McCarthy, D., Erhardt, M., Briand, J., Prato, M., Kostarelos, K. and Bianco, A. Angew. Chem. Int. Ed., 2004, 43, 5242-5246.
2. Koh, B., Cheng, W. Langmuir, 2014, 30 (36), 10899-10909.
3. Sinani, V., Gheith, M., Yaroslavov, A., Rakhnyanskaya, A., Sun, K., Mamedov, A., Wicksted, J. and Kotov, N. J. Am. Chem. Soc., 2005, 127 (10), 3463-3472.

Keywords: CNT's, functional polymers, thiol-ene

SFP-P-047 [Smart and Functional Polymers]

Molecularly Imprinted Microspheres for the Adsorption of Chlorpyrifos

Wei Yao[1], Yanjun Fang[2], Hong Yin[1], Zhixian Gao[2]

[1]*Center for Disease Control and Prevention, Agency for Offices Administration*
[2]*Institute of Hygiene and Environmental Medicine*

To obtain the desired specific adsorbents for chlorpyrifos to enrichment, separation and analysis of trace pesticides residue, molecularly imprinted polymer microspheres were prepared by precipitation polymerization using chlorpyrifos and methacylic acid (MAA) as template and functional monomer respectively. Computer simulation was employed to demonstrate the mechanism of the interaction between chlorpyrifos and MAA. The adsorption properties of chlorpyrifos, in acetonitrile for imprinted microspheres were evaluated by equilibrium rebinding experiments. Scatchard plot analysis revealed that there was. The specificity of the imprinted microspheres was investigated by binding analysis using chlorpyrifos and related organophosphate pesticides. The results indicated that the obtained imprinted microspheres showed a good selectivity for chlorpyrifos.

Keywords: Adsorption, Chlorpyrifos,Molecularly imprinted polymers,Precipitation polymerization

SFP-P-048 [Smart and Functional Polymers]

Synthesis and Characterization of pH-Responsive Zwitterionic ABC Triblock Copolymers and Intermediary Layer Cross-Linked Micelles

Agung Ari Wibowo, Gökhan Solmaz, Vural Bütün

Department of Polymer Science and Technology, Eskişehir Osmangazi University, Campus of Meşelik, 26480, Eskişehir, Turkey

Well-defined ABC triblock copolymers consisting of poly[(ethylene glycol) methyl ether], poly(tert-butyl methacrylate), and poly[2-(diisopropylamino) ethyl methacrylate] (PEG-b-PtBMA-b-PDPA) were synthesized via atom transfer radical polymerization. The obtained triblock copolymers were successfully converted into PEG-b-PMAA-b-PDPA by acid hydrolysis, where PMAA is poly(methacrylic acid). Depend on the solution pH, the PEG-b-PMAA-b-PDPA can self assembly to three micellar aggregates at room temperature. At pH < 4, micellar aggregates comprising of hydrogen bonding interactions between PEG and fully protonated PMAA complex cores stabilized by protonated PDPA coronas are formed [Varshney and Je, Macromolecules, 2001]. In the range of pH 5 – 7, as a result of charge compensation between partially ionized PMAA and partially protonated PDPA, micelles possessing polyion complex core are formed. Above pH 8, PDPA block gain hydrophobic character and form the core of micelles while PEG/ionized PMAA hybrid blocks form hydrated coronas. Finally, intermediary layer cross-linked (ILCL) micelles (PDPA-core in aqueous media) were also prepared from these PEG-b-PMAA-b-PDPA triblock copolymers by cross-linking the inner PMAA shell using 1,2-bis(2-iodoethoxy) ethane (BIEE) [Bütün et al. J. Am. Chem. Soc. 1999]. In intermediary layer cross-linked micelles, the outer shell (PEG) acted as steric stabilizer and prevented inter-micellar aggregation even when the cross-linking chemistry was carried out at 1% (w/v) polymer concentration. The zeta potential studies of uncross-linked and ILCL micelles also be done to confirm the occurrence of the cross linking reactions. The pH-responsive formation of three types of aggregates and ILCL micelles were characterized by dynamic light scattering (DLS), 1H NMR spectroscopy, and transmission electron microscopy (TEM).

Keywords: atom transfer radical polymerization, pH responsive polymer, micelles, intermediary layer cross-linked (ILCL)

SFP-P-049 [Smart and Functional Polymers]

Synthesis and Study of Chemical resistance of Short Bamboo fiber reinforced epoxy composites

Anu Gupta

Seth Jai Prakash Mukand Lal Institute of Engineering & Technology, Radaur, Yamunanagar - 135103, Haryana, India

Natural fiber reinforced composites have generated a lot of interest in the research community as they offer an attractive solution to the ever depleting petroleum resource. Polymers are replacing the conventional metallic materials in plenty of applications. Bamboo, a natural bio-material has emerged as a potential material for composites, owing to its unique biological structure and mechanical performance among the natural fibres. Bamboo offers significant applications due to its high strength, easy availability at low cost. In order to probe, whether a composite can find applications in chemical and storage tanks, study of its chemical resistance is necessary. Though bamboo is extensively used as a valuable forest material from times immemorial (because of its high strength-to-weight ratio), the studies on these fibers are meager. In the present paper, short bamboo fibre reinforced epoxy composite has been synthesized by hand layup technique.The composites of four different compositions (10wt%, 20wt%, 30wt% and 40wt% fiber loading) are made. Chemical resistance and Water absorption has been studied.These results indicate that bamboo can be used as a potential reinforcing material for making low load bearing thermosetting composites.

Keywords: Polymer Composites, Bamboo Fibers, Natural Fibers, Chemical Resistance

SFP-P-050 [Smart and Functional Polymers]

Microphase Separation in supramolecular poly(1-hexene) with self-complementary hydrogen bonding along the backbone

Saeed Houshmandmoayed[1], Mohammad Hossein Jandaghian[2], Mostafa Ahmadi[2], Mohammad Mahdi Mortazavi[3]

[1]*Jam Petrochemical Co*
[2]*Department of Polymer Engineering and Color Technology, Amirkabir University of Technology, Tehran, Iran*
[3]*Polymerization Engineering Department, Iran Polymer and Petrochemical Institute (IPPI), Tehran, Iran*

Polyolefinic supramolecular systems have received new interests as the complementary physical hydrogen bonding can increase their mechanical properties, while the high-temperature processing behavior can remain unchanged (Nojiri et al. Polymer 2016, 87, 308). However, deeper investigation is needed to separate the effect of microphase separation of polar hydrogen bonding moieties from those granted by the crystalline structure of these materials. We have therefore synthesized novel fully amorphous poly(1-hexene)s with different contents of maleic anhydride groups and used it to graft 2-ureido-4pyrimidinone (UPy) groups along the backbone. The transient network formed through microphase separation and self-association of UPy groups are studied by SEM, AFM and rheological analysis both in linear and nonlinear oscillatory shear experiments. SAOS experiments performed at different temperatures reveal significant thermorheological complexity as the chain dynamics and stickers' lifetime have different temperature sensitivities. The nonlinear shear experiments show significant strain-hardening behavior. These findings reveal the significant effect of physical interactions in amplification of entanglement density and network structure.

Keywords: poly(1-hexene), hydrogen bonding, supramolecule, rheology

SFP-P-051 [Smart and Functional Polymers]

Chemical Preparation and Characterization of New Biodegradable Poly(ε-caprolactone) formed from Diosgenin and Cholesterol

Sevinc Ilkar Erdagi[1], Erdinc Doganci[2], Faruk Yilmaz[3], Cavit Uyanik[1]

[1]*Department of Chemistry, Kocaeli University, Kocaeli, Turkey*
[2]*Department of Chemistry and Chemical Process Technologies, Kocaeli Vocational School, Kocaeli, Turkey*
[3]*Department of Chemistry, Gebze Technical University, Gebze, Kocaeli, Turkey*

Diosgenin (DGN) is a well-known steroidal saponin from a therapeutic herb, has been used as a starting material for production of steroidal hormones. Diosgenin plays a predominant role in the treatment of Alzheimer Disease, and it is also considered a lead compound in the discovery of novel anticancer drugs [1-2]. Cholesterol (Chol) is a very important molecule for humans as the component of cell membrane and precursor of steroid hormones and bile acids [3]. In this study, hydroxyl functional group of diosgenin was used as the initiator in the ring opening polymerization of ε-caprolactone (ε-CL) to prepare α-diosgenin-ω-hydroxy PCL polymer (DGN-PCL-OH). ω-Hydroxy functionality of DGN-PCL-OH was then successfully converted into alkyne. On the other hand, hydroxyl group of cholesterol was converted into bromide and azide (Chol-N_3). Further end-group modification of Chol-N_3 was achieved quantitatively by copper (I)-catalyzed cycloaddition of azide functional group and alkyne functional DGN-PCL, which led to the desired PCL with α-diosgenin and ω-cholesterol groups (DGN-PCL-Chol) (Fig.1). The effect of molar ratio of ε-CL to the initiator on the molecular weight of the obtained DGN-PCL-Chol was also investigated.

References:

1. C. Li, L. Dai, K. Liu, L. Deng, T. Peic, J. Lei, RSC Adv. 2015, 5, 74828–74834.
2. J. E. Chojnacki, K. Liu, J. M. Saathoff, S. Zhang, Bioorg. Med. Chem. 2015, 23, 7324–7331.
3. S. I. Erdagi, E. Doganci, C. Uyanik, F. Yilmaz, Reactive and Functional Polymers 2016, 99 49–58.

Keywords: cholesterol, click chemistry, diosgenin, poly (ε-caprolactone)

Figure 1. The structure of DGN-PCL-Chol polymers.

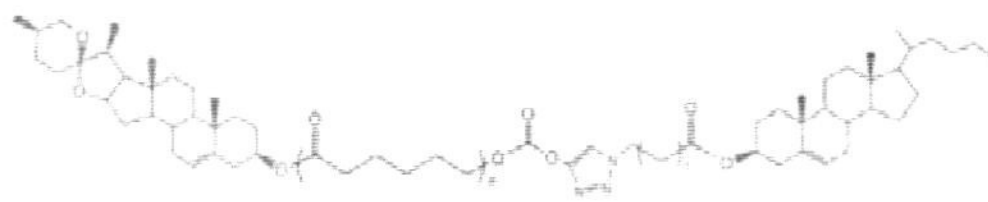

SFP-P-052 [Smart and Functional Polymers]

Specific ion effects on the friction and adhesion for zwitterionic polymer brushes

Yunlong Yu, G. Julius Vancso, Sissi De Beer

University of Twente, Materials Science and Technology of Polymers

Zwitterionic polymers, such as poly(2-(methacryloyloxy)ethylphosphorylcholine) (PMPC), can potentially be employed as efficient lubricants or antifouling surfaces, because they are strongly hydrated in pure water. For many (biomedical) applications, however, the aqueous liquids are not pure and contain different types of ions. We study the effect of specific ions on the hydration of PMPC brushes and show, using atomic force microscopy experiments, that there is a pronounced reduction in the brush swelling for high concentrations of small monovalent anions and for divalent ions in general. The weaker hydration gives rise to higher adhesion and friction upon relative motion of an AFM colloid probe and the brush.

Keywords: polymer brush, zwitterionic polymers, atomic force microscopy

SFP-P-053 [Smart and Functional Polymers]

Semi-interpenetration with Conducting Polyanilines on Poly(N-isopropylacrylamide) Hydrogels by Radiation Induced Graft Polymerization: Synthesis, Characterization and Their Application in the Controlled Drug Delivery Systems

Betül Taşdelen[1], Selçuk Poyraz[2], Aslıhan Koruyucu[2]

[1]*Department of Biomedical Engineering, Namık Kemal University, Tekirdağ, Turkey*
[2]*Department of Textile Engineering, Namık Kemal University, Tekirdağ, Turkey*

In this work, semi-IPN poly(N-isopropylacrylamide)/polyaniline composite hydrogel was successfully synthesized by two-steps gamma radiation induced polymerization in aqueous solution. In this study, we proposed a novel alternative method to incorporate a conductive linear polymer (PANI) inside a network of thermo-sensitive hydrogel under γ-irradiation. The composite hydrogel with good conductive properties also displayed unique thermo-sensitivity. The s-IPN hydrogel systems were employed for swelling and diffusion experiments in water and aqueous solutions of the Methylene Blue (MB) as a model drug in at room temperature were investigated. The s-IPN composite hydrogels showed good drug uptake and release properties.This study was carried out at physiological pH to provide useful data towards the design of drug delivery systems in biologic media.

Keywords: N-isopropylacrylamide, Polyaniline, s-IPN, gamma irradiation.

Scanning electron micrograph of the s-IPN composite hydrogel

SFP-P-054 [Smart and Functional Polymers]

Synthesis and Characterization of Self Healing Polydimethylsiloxane-graft-Poyurethane

Ali Sabri Berkem[1], Turgut Nugay[2], Ilke Anac[1]

[1]*Department of Materials Science and Engineering, Gebze Technical Univeristy, Kocaeli, Turkey*
[2]*Department of Chemistry, Bogazici University, Istanbul, Turkey*

As polymers have low mechanical durability compared to metals, the usage area of those lightweight materials are limited. In addition, upon any damage, either repair of those materials is not possible or the cost of repair is so high. These problems can be prevented by imparting a self-healing ability to a polymeric material. In this study, the self-healing polydimethylsiloxane-graft-poyurethane (PDMS-graft-PU) material was synthesized. In order to prepare PDMS-graft-PU, linear aliphatic PU having one –OH functional group (PU-OH) was first synthesized by polycondensation reaction using dibutyltinlaurate as catalyst in THF. The –OH functional group was then converted to –COOH group using citric acid. The PU-COOH was grafted onto aminopropyl functionalized PDMS by EDC/HCL coupling reaction. The chemical structure of the resulting polymers, PU-OH, PU-COOH and PDMS-graft-PU, were characterized by 1H NMR and FTIR. The self-healing ability of PDMS-graft-PU was determined by tear test. The mechanical test specimen was prepared by casting PDMS-graft-PU dissolved in chloroform on Teflon plate. Tear strength of both the virgin and healed specimens was measured and the healing of the damaged area was monitored in real-time by polarized optical microscopy.

[1] J. H. Harreld, M. S. Wong, P. K. Hansma, D. E. Morse, G. D. Stucky, "Self-healing organosiloxane materials containing reversible and energy-dispersive crosslinking domains" WO2004035644A2, 2004

Keywords: polyurethane, PDMS, EDC coupling, self-healing

SFP-P-055 [Smart and Functional Polymers]

Controlled release of Paclitaxel from multilayers of block copolymer micelles with temperature-responsive coronae and pH-responsive cores

Eda Cagli, Irem Erel Goktepe

Department of Chemistry, Middle East Technical University, Ankara, Turkey

Controlled release of drug molecules from polymer nanosystems is getting increasing attention in cancer treatment. In this study, we aim to prepare polymer multilayer films that are stable at physiological conditions with little or no drug release but induce release of an anti-cancer drug at acidic pH. First, novel stimuli responsive diblock copolymers of 2-alkyl oxazolines and 4-vinyl pyridine were synthesized. Then, block copolymer micelles (BCMs) with temperature responsive polyoxazoline-corona and polybasic core were obtained via pH-induced self-assembly in aqueous environment. Paclitaxel, a model hydrophobic anti-cancer drug will be loaded in the P4VP micellar corona prior to film assembly. Multilayers will be constructed using Paclitaxel loaded BCMs and Tannic Acid (TA) as building blocks. Release of Paclitaxel will be induced at acidic conditions due to protonation of the P4VP blocks and disintegration of the micellar cores. Considering the acidic nature of tumor tissues, these nanosystems can be promising for controlled release applications. Finally, we will investigate the effect of LCST of polyoxazoline blocks on the release of Paclitaxel from the multilayers.

The authors acknowledge the financial support of TUBITAK (Grant Number: 113Z586).

Keywords: layer-by-layer self assembly, block copolymer micelles, temperature and pH responsive systems, controlled drug release

SFP-P-056 [Smart and Functional Polymers]

Highly fluorescent porphyrin-functional styrene copolymer nanofibers

Sümeyra Bayir[1], Mesut Görür[2], Erdinç Doğancı[3], Tamer Uyar[4], Faruk Yılmaz[1]

[1]Department of Chemistry, Gebze Technical University, Kocaeli, 41400, Turkey
[2]Department of Chemistry, Istanbul Medeniyet University, Istanbul, 34700, Turkey
[3]Department of Chemistry and Chemical Processing Technologies, Kocaeli University, Kocaeli, 41380, Turkey
[4]Institute of Materials Science & Nanotechnology, Bilkent University, Ankara, 06800, Turkey

Polymer nanofibers can be used in a wide range of applications like filter materials, nano-composites, ultrafiltration, ion exchange, enzyme immobilization, optic chemical sensors, electromagnetic interference (EMI) shielding, electrochromic fibers [1], and many others. Nanofibers with fluorescence characteristics are preferred in optical chemical sensor applications due to the high surface area, high sensitivity and fast responses of fluorescence sensors [2-3]. In this study, porphyrin- functional styrene copolymer was synthesized in a three-step synthetic procedure, including nitroxy-mediated polymerization (NMP) of styrene and vinilbenzyl chloride monomers, azidification of chloride side-groups of the obtained styrene copolymer, and finally attachment of alkyne-functional porphyrin compound to the copolymer via Cu(I)-catalyzed 1,3-dipolar cycloaddition reaction. The prepared fluorescent-active porphyrin-functional styrene copolymer was employed in the production of nanofibers via electrospinning method (Figure 1).

The authors acknowledge TUBITAK for the financial support (Grant Number: 113Z577).

References:

[1] C. Shin, G.G. Chase, D.H. Reneker, Colloids and Surfaces A: Physicochemical and Engineering Aspects, 262, (2005), 211-215.
[2] Y. Yang, H. Wang, K. Su, Y. Long, Z. Peng, N. Li, F. Liu, Journal of Materials Chemistry, 21, (2011), 11895-11900.
[3] Senthamizhan A., Celebioglu A., Bayir S., Gorur M., Doganci E., Yilmaz F., Uyar T., ACS Applied Materials and Interfaces, 2015, 7(38), 21038–21046.

Keywords: nanofibers, electrospinning, dansyl, functional polymers

Figure 1. Schematic representation of the nanofiber formation using electrospinning method and SEM micrograph.

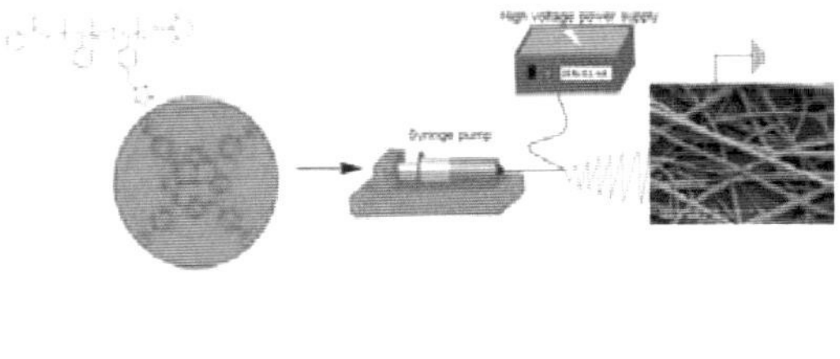

SFP-P-057 [Smart and Functional Polymers]

Synthesis, Kinetic, Equilibrium and Thermodynamic Studies on Adsorption of Cu(II) by Thiosemicarbazide Chitosan

Mudasir Ahmad

Mudasir Ahmad and Saiqa Ikram

A hybrid chitosan-thiosemicarbazide framework (TSCS) as an adsorbent for Cu (II) was synthesized from dialdehyde chitosan through condensation reaction with thiosemicarbazide, stabilized by the reduction reaction with sodium borohydride. TSCS was characterized by means of FT-IR and XPS. Surface morphologies were studied by FESEM and BET, which revealed the highly macro porous structure. The thermal analyses was done through TGA showing much stable chemical configuration at about ≥ 400oC. The influence of other parameters viz. pH, contact time, dose of adsorbent and initial metal ion concentration on adsorption were also analyzed. The experimental equilibrium data was evaluated by Langmuir, Freundlich and Dubinin-Radushkevich isotherm models. The applicability of the fit of model to experimental values was accessed on the basis of error analyses. The Langmuir adsorption model was best fitted with experimental value, suggests the existence of monolayer coverage of adsorbed molecules with a maxima of 142.85mg/g. The kinetic data was analyzed using pseudo-first-order, pseudo-second-order and intraparticle diffusion models and the pseudo-second-order kinetics were found for all the concentrations. The calculated thermodynamic parameters such as ΔH, ΔGo and ΔS were -2.33 kJmol 1, 570.40Jmol-1 and 9.75Jmol-1K-1 signifies the adsorption of Cu(II) onto TSCS is endothermic, spontaneous and a process of physisorption. The regeneration efficiency of the TSCS as an adsorbent was found to be≥90-95% using 0.1M EDTA.

Keywords: chitosan, thiosemicarbazide, Cu(II), adsorption

SFP-P-058 [Smart and Functional Polymers]

Synthesis of Core Cross-Linked Micelles from a Triblock Copolymer Containing Poly(glycidyl metharylate) Functional Block

Vural Bütün[1], Gokhan Solmaz[1], Gokhan Kocak[2]

[1]Department of Polymer Science and Technology, Eskisehir Osmangazi University, Eskisehir, Turkey
[2]Department of Chemistry, Faculty of Arts and Science, Eskisehir Osmangazi University, Eskisehir

Glycidyl methacrylate (GMA) is a reactive monomer, which is frequently used for the preparation of functional polymers due to the ability of its epoxy group to react in mild reaction conditions with nucleophiles [Strube OI, et al. Mac. Chem Phys. 2015]. It is commercial and relatively cheap monomer and commonly used to obtain different polymer architectures such as star, block, bead... Among the polymer architectures, block copolymers have been getting increasing interest because of their unique solution and associative properties as a consequence of their molecular structure. However, the practical applications of micelles are limited due to their structural instability since the micellar structure can hardly keep stable upon dilution or changes of external conditions. In the literature, various cross-linking chemistries have been reported for micelles depending on their structures and functional groups in related layer. Common types of cross-linked micelles are (i) core cross-linked (CCL), (ii) shell cross-linked and (iii) intermediary layer cross-linked micelles [Read ES, Armes SP Chem Commun 2007, van Nostrum CF. Soft Matter 2011]. An amphiphilic poly[(ethylene glycol)methyl ether]-block-poly(glycidyl methacrylate)-block-poly(methyl methacrylate) (MPEG-b-PGMA-b-PMMA) triblock copolymer was synthesized via ATRP and its core-shell micellar solutions (5 and 10 mg L-1) were prepared in acetone/water (3:7 v/v) mixture by PGMA-b-PMMA diblock forming the micelle core and MPEG block forming the micellar corona. The cores of these core-shell micelle were cross-linked by the reaction of the epoxy functional group of P(GMA) block with three different cross-linkers; ethylenediamine, hexamethylenediamine, cysteamine. The GMA/cross-linker ratios were varied to tune degrees of cross-linking. The resulting CCL structures showed different responses to solution conditions depending on their degrees of cross-linking and solvent types (such as acetone, ethanol).

Acknowledgments: This work was supported by the Commission of Scientific Research Projects of ESOGU (Grant Numbers 201119006 and 201319A205).

Keywords: core cross-linked micelles, triblock copolymer, glycidyl methacrylate, epoxy ring opening, functional polymers.

SFP-P-059 [Smart and Functional Polymers]

Polypyrrole functionalized by 4-amino-3-hydroxynaphthalene sulfonic acid for removal chromium hexavalent

Mohamed Lamine Sall, Momath Lo, Matar Niang, Abdou Karim Diagne Diaw, Diariatou Gningue Sall

Université Cheikh Anta Diop of DAKAR

In this work we synthesized chemically, a polypyrrole film, functionalized by the presence of a base salt of 4-amino-3-hydroxynaphthalene sulfonic acid. Indeed we dissolved 3.2 g of 4-amino-3-hydroxynaphthalenesulfoni c acid in 90 mL of distilled water. The acid is not soluble in water we added NaOH until it is completely soluble the resulting pH is 5.1. To this solution we add pyrrole and we let stir for 30 min. Finally we add gradually stirring 10.8 g of potassium peroxodisulfate (KPS) dissolved in 60 mL of distilled water and we leave without agitation for 12 hours. We add acetone to stop the reaction. We obtain a black powder that we wash with distilled water with acetone until the filtrate was colorless. The filtered and washed powder was dried for 6 h at 60 ° C in oven. The FT-IR study spectra of the polymer confirmed the formation of a polypyrrole film functionalized by the presence of 4-amino-3-hydroxynaphthalene-1-sulfonic acid in polymeric unit The use of these polymers for the treatment of waste water containing chromium, by the immersion method, was carried out in solutions of chromic optimized pH (pH=2).The UV-vis spectroscopy study of metallic solution showed absorbance decrease in the order of 97% in the presence of the film (solid). Removal of the metal in the solution by forming a complex with the polymer is confirmed by the changes in FTIR spectra.

Keywords: pyrrole, acide 4-amino-3-hydroxynaphtalene-1-sulfonique, Uv-visible, IR-TF chrome(VI)

SFP-P-060 [Smart and Functional Polymers]

Benzoxazole Derived Triphenylamine Based Polyimides: Synthesis, Characterization and Photophysical Properties

Humaira Masood Siddiqi[1], Asma Iqbal[1], Toheed Akhtar[2], O.ok. Park[2]

[1]Quaid-i-Azam University, Department of Chemistry, Islamabad, Pakistan
[2]2Department of Chemical and Biomolecular Engineering (BK21+ graduate program), Korea Advanced Institute of Science and Technology, 291 Daehak-ro, Yuseong-gu, Daejeon,305-701, Republic of Korea.

In recent yearspacking-disruptive triarylamine based polymeric materials have been greatly investigatedfor their use in electronic devices such as light emitting diodes, solar cells, and electrochromic devices. Among the polymeric materials aromatic polyimides are attracting the researchers due to their high thermal/mechanical stability, photoluminescent, and electroluminescent properties which are suitable for optoelectronic applications. In this contribution, we report the synthesis of a series of novel polyimides bearing benzoxazole as pendant and electron rich triphenylamine (TPA) in polymer backbone. The novelty of this work lies in the structure of polyimides where TPA linkage and benzoxazole units are fused with each other. Thus introduction of fused TPA-benzoxazole unit furnished synthesized polyimides with some significant structural advantages like (a) the para position of TPA linkage is inherently blocked avoiding the coupling reactions and enhancing charge mobility of polyimide. (b) Fused TPA-benzoxazole unit improved the mobility of charges along the polymer chain. Conventional two step condensation reaction was followed for the preparation of polyimides. In the first step poly(amic acid) was prepared which was further thermally imidized to achieve free standing polyimide films. Thermal properties were investigated by DSC and TGA, Tg were observed above 250 oC and no melting peak was observed till 300 oC which indicated the amorphous nature of theses polyimides, whereasno thermal degradation was observed prior to 400 oC. The Photophysical data of these polyimides was investigated by employing Uv-Vis and photoluminescence techniques, which exhibited blue light emission properties in solution state as shown in fig.1 as well as in spin coated thin films with good PL quantum yields. High HOMO levels with lower onset oxidation potentials were observed which make these polyimides suitable for future optoelectronic properties. Fig.1 PL emission of polyimides synthesized from benzoxazole derived triphenylamine based diamine monomer.

Keywords: triphenylamine, benzoxazole derived triphenylamine,polyimides, blue light emitting, optoelectronics.

Fig.1 PL emission of polyimides synthesized from benzoxazole derived triphenylamine based diamine monomer.

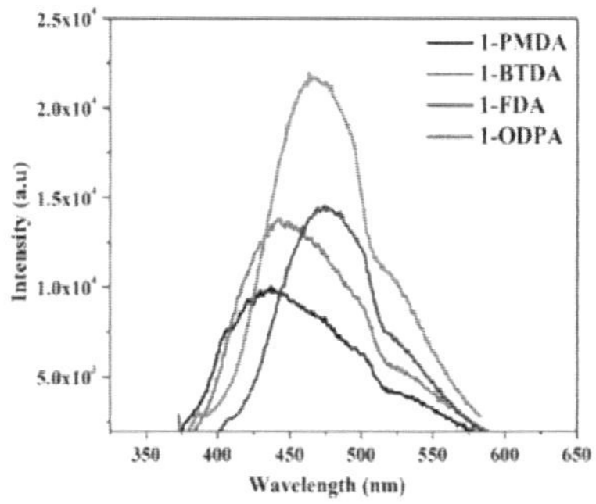

SFP-P-061 [Smart and Functional Polymers]

Polysulfones with Pendant Benzoxazine Units by Click Chemistry

Cemil Dizman[1], Çağatay Altınkök[2], Mehmet Atilla Taşdelen[2]

[1]Institute of Chemical Technology, TUBITAK Marmara Research Center, Gebze, Kocaeli, Turkey
[2]Department of Polymer Engineering, Faculty of Engineering, Yalova University, Yalova, Turkey

Aromatic polysulfones are high performance polymers with excellent chemical and thermal stability, excellent strength and flexibility, transparency, as well as a high glass transition temperature and good film forming properties. 1,2,3 The properties of the polysulfone polymers may be improved by incorporation of functional groups.

In this study, functional polysulfones with pendant benzoxazine units was synthesized via "click" chemistry methodology. First, polysulfone was modified by chloromethylation and subsequent azidation process.

,4-dihydro-3-(prop-2-ynyl)-2H-benzoxazine (P-pa) was prepared independently as a click component. Finally, azido functionalized polysulfone (PSU-N3) was coupled to P-pa with high conversion by click chemistry. According to the results, benzoxazine functionalized PSU (PSU-p-Bnz) showed improved thermal properties compared to pristine PSU.

References:

[1] Olabisi O, Handbook of Thermoplastics, Marcel Dekker, New York (1997).
[2] RaoVL, JMacromol Sci RevMacromol ChemPhys C39: 655–711 (1999).
[3] Dizman, C., Tasdelen, M. A. and Yagci, Y. (2013), Recent advances in the preparation of functionalized polysulfones. Polym. Int., 62: 991–1007.

Keywords: Benzoxazine, Click chemistry, Polysulfone

Figure 1

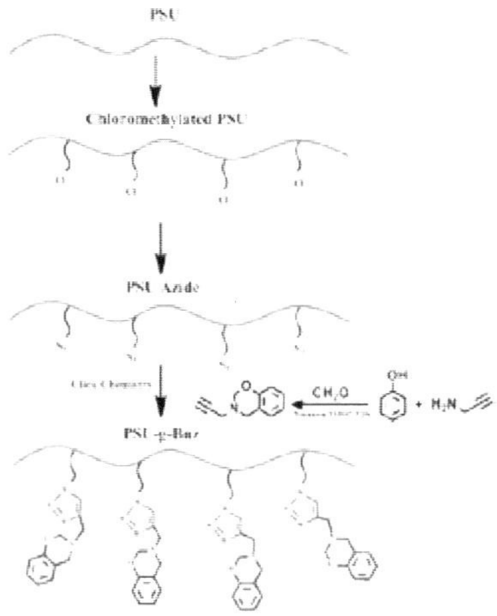

General scheme for the synthesis of benzoxazine functionalized polysulfone

SFP-P-063 [Smart and Functional Polymers]

Polyethyleneimine modified Halloysite embedding p(Acrylamide) Cryogel Composites

Sultan Butun Sengel, Selin Sagbas, Nurettin Sahiner

Department of Chemistry, Canakkale Onsekiz Mart University, Canakkale, Turkey

In this study, natural clay Halloysite nanotubes (HNT) were modified with polytethyleneimine (PEI) as PEI-HNTs and incorporated into superporous cryogel of poly(acrylamide) (p(AAm)) during cryogellation. The modification of HNTs with PEI was accomplished in alcoholic media by directly and/or indirectly by using coupling agents such as 3-chloropropyltrimethoxysilane and (3-aminopropyl)triethoxysilane, then the prepared PEI-HNTs were dispersed basic media by sonication and then combined with AAm solution containing N'N-methylenebisacrylamide (MBA) as crosslinker and then potassium persulfate (KPS) as initiator was included for cryogellation that was accomplished at -20 oC for 24h. A series of the p(AAm) based composite cryogels were prepared by varying the used amounts of PEI-HNTs. All the synthesized materials were characterized via SEM, TEM, Zeta Potential measurements (for PEI-HNTs), TGA, FT-IR spectroscopy, as well as surface area and pore size analysis using BET method. Additionally, the PEI modified HNTs and p(AAm) embed-ding PEI-HNTs cryogel composite antibacterial properties were investigated.

Keywords: Cryogel/hyrodrogel, PEI-Halloysite, cryogel-composite

SFP-P-062 [Smart and Functional Polymers]

Magnetically responsive controlled drug release of vinblastine-loaded chitosan particles

Keng Shiang Huang[1], Chih Hui Yang[2]

[1]*The School of Chinese Medicine for Post-Baccalaureate, I-Shou University, Taiwan*

[2]*Department of Biological Science and Technology, I-Shou University, Taiwan*

Vinblastine (VBL) is a major chemotherapeutic drug; however, in some cases, it may cause severe side effects in patients with cancer. Designing a novel VBL pharmaceutical formulation is a crucial and emerging concern among researchers for reducing the use of VBL. This study developed a stimuli-responsive controlled VBL drug release system from magnetically sensitive chitosan capsules. A magnetically responsive controlled drug release system was designed by embedding superparamagnetic iron oxide (SPIO) nanoparticles (NPs) in a chitosan matrix and an external magnet. In addition, droplet microfluidics, which is a novel technique for producing polymer spheres, was used for manufacturing monodispersed chitosan microparticles. The drug encapsulation rate was 67%–69%. The in vitro drug release test indicated that the VBL could be 100% released from chitosan composite particles in 80–130 min under magnetic stimulation. The pulsatile magnetically triggered tests showed individual and distinctive controlled release patterns. Thus, the timing and dose of VBL release was controllable by an external magnet. The results presume that using a magnetically responsive controlled drug release system offers a valuable opportunity for VBL drug delivery, where the delivery system is an active participant, rather than a passive vehicle, in the optimization of cancer treatment. The proposed actively-targeted magnetic drug delivery system offers many advantages over conventional drug delivery systems by improving the precision and timing of drug release, easy operation, and higher compliance for pharmaceutical applications.

Keywords: microfluidics, superparamagnetic, chitosan, vinblastine, drug delivery, controlled release

SFP-P-064 [Smart and Functional Polymers]

The Synthesis and Characterization of Self Healing Materials Based on Thioxanthone Disulfides

Umut Koyunucu, Ömer Tahir Günkara, Nergis Arsu

Yildiz Technical University Chemistry Department

Degradation, damage, and failure are natural consequences of material applications. But, nature has developed materials and systems based on the concept of damage management (healing) in order to extend survival possibilities. What we really need are artificial materials that behave like human body; sensing a failure, stopping it from getting worse and then repairing it as qucikly as possible, all by themselves. In the last 20 years, the dream of implementing this concept to engineering systems to extend service lifetime has become a reality [1]. The aim of the self-healing materials concept is to extend the service life-time of materials, structures and systems made there of, not by increasing their initial performance but by implementing the concept of autonomous or induced repair [2]. Self healing materials are a class of smart materials that have the structurally incorporated ability to repair damage caused by mechanical usage over time. Several approaches for introducing this self-healing concept in synthetic materials have been explored in recent years. To date, the most succesful examples are based on encapsulation and reversible chemistry. In this study, we synthesized a novel photoactive disulfide compound, and to achieve this thioxanthonation was made for dihydroxydiphenyl disulphide. Photophysical and photochemical properties were determined by known spectroscopic techniques. The initiating efficiency of these compound for photoinitiated polymerization of acrylates and methacrylates were performed. Acrylate/ Disulpide Composites were prepared to find the aid of this thioxanthonated disulfide on the self healing ability.

References:

1. García, S. J.; Fischer, H. R.; van der Zwaag, S.Prog. Org. Coat. 2011, 72, 211-221
2. Lafont, U.; Zeijl, H. v.; van der Zwaag, S.Microelectron. Reliab. 2012, 52,71-89.
3. J A.Syrett, C Remzi Becer, D M. Haddleton, Polym. Chem. 2010, 1, 978-987.

Keywords: Disulfide, self-healing, thioxanthone

Bridging Continents & Bridging Molecules

MACRO 2016

World Polymer Congress Istanbul 17-21 July

AUTHOR INDEX